MW00562298

CONSTANTS AND CONVERSION FACTORS
(to four significant figures)

speed of light	$c = 2.998 \times 10^8$ m/s
electron charge unit	$e = 1.602 \times 10^{-19}$ C
Coulomb force constant	$\dfrac{1}{4\pi\varepsilon_0} = 8.988 \times 10^9$ N $\cdot$ m^2/C^2
	$\dfrac{e^2}{4\pi\varepsilon_0} = 1.440$ eV $\cdot$ nm
electron mass	$m_e = 9.109 \times 10^{-31}$ kg $= 0.5110$ MeV$/c^2$
proton mass	$M_p = 1.673 \times 10^{-27}$ kg $= 938.3$ MeV$/c^2$
proton–electron mass ratio	$\dfrac{M_p}{m_e} = 1836$
Planck's constant	$h = 6.626 \times 10^{-34}$ J $\cdot$ s $= 4.136 \times 10^{-15}$ eV $\cdot$ s
	$hc = 1240$ eV $\cdot$ nm
	$\hbar = 1.055 \times 10^{-34}$ J $\cdot$ s $= 6.582 \times 10^{-16}$ eV $\cdot$ s
	$\hbar c = 197.3$ eV $\cdot$ nm
Avogadro's number	$N_A = 6.022 \times 10^{23}$ mole^{-1}
Boltzmann's constant	$k_B = 1.381 \times 10^{-23}$ J/K $= 8.617 \times 10^{-5}$ eV/K
electron Compton wavelength	$\dfrac{h}{m_e c} = 2.426 \times 10^{-12}$ m
Bohr radius	$a_0 = 5.292 \times 10^{-11}$ m
Rydberg energy unit	$E_0 = 13.61$ eV
Rydberg constant	$R_\infty = 1.097 \times 10^7$ m^{-1}
fine structure constant	$\alpha = \dfrac{1}{137.0}$
Bohr magneton	$\mu_B = 9.274 \times 10^{-24}$ A $\cdot$ m^2 $= 5.788 \times 10^{-9}$ eV/G
nuclear magneton	$\mu_N = 3.152 \times 10^{-12}$ eV/G
gravitational constant	$G = 6.673 \times 10^{-11}$ N $\cdot$ m^2/kg^2
electron volt	eV $= 1.602 \times 10^{-19}$ J
atomic mass unit	u $= 1.661 \times 10^{-27}$ kg $= 931.5$ MeV$/c^2$
cross section unit	barn $= 10^{-28}$ m^2 $= (10$ fm$)^2$
light-year	lt-y $= 9.461 \times 10^{15}$ m

INTRODUCTION
TO
THE
STRUCTURE
OF
MATTER

INTRODUCTION TO THE STRUCTURE OF MATTER

A Course in Modern Physics

John J. Brehm

and

William J. Mullin

University of Massachusetts
Amherst, Massachusetts

WILEY
John Wiley & Sons

Copyright © 1989, by John Wiley & Sons, Inc.

All rights reserved. Published simultaneously in Canada.

No part of this publication may be reproduced, stored in a retrieval system or transmitted in any form or by any means, electronic, mechanical, photocopying, recording, scanning or otherwise, except as permitted under Sections 107 or 108 of the 1976 United States Copyright Act, without either the prior written permission of the Publisher, or authorization through payment of the appropriate per-copy fee to the Copyright Clearance Center, 222 Rosewood Drive, Danvers, MA 01923, (978) 750-8400, fax (978) 750-4470. Requests to the Publisher for permission should be addressed to the Permissions Department, John Wiley & Sons, Inc., 111 River Street, Hoboken, NJ 07030, (201) 748-6011, fax (201) 748-6008, E-Mail: PERMREQ@WILEY.COM.

To order books or for customer service please, call 1(800)-CALL-WILEY (225-5945).

Library of Congress Cataloging in Publication Data:

Brehm, John J.
  Introduction to the structure of matter.

  Bibliography
  Includes index.
  1. Physics. 2. Matter—Constitution. I. Mullin,
William J. II. Title.

QC21.2.B74 1989   530   88-5768
ISBN 0-471-60531-X

Printed in the United States of America

Printed and bound by the Hamilton Printing Company.

20 19 18 17 16 15

to
Mary Ellen and Sandra
with love

PREFACE

A first course in modern physics should be exciting and rewarding for student and teacher alike. Such a course offers an opportunity to appreciate the wondrous findings of the 20th century as milestones in the growth of a whole new science. From this perspective the student can take pleasure in the discovery of new phenomena and the teacher can draw inspiration from the revolutionary new ideas.

This course should be offered in a year-long format to serious students of physics no later than their junior year. An *early* introduction to the 20th century is essential if the students are to recognize soon enough the continuing flow of developments in modern physics. The course should also be reasonably *thorough* so that undergraduates can build on it as a secure foundation. Our textbook is aimed at this body of readers with these objectives in view.

A first course in modern physics can be a daunting experience for the students, and even for the teacher. The revolutionary nature of the material invites the students to challenge all their classical learning and forces the teacher to encourage a continually questioning posture. The adoption of such an attitude toward physics is difficult to learn, and to teach, in a constructive manner. This problem is compounded by the fact that the subject has become too vast to be surveyed in its entirety in a single year-long presentation. Compromises must be faced at every turn, as students and teacher together seek a balance between the depth of the questions raised and the breadth of the topics addressed, out of a seemingly open-ended agenda.

There may have been a time, several generations ago, when a teacher could present all the great discoveries of modern physics in a single two-semester course. Those prospects are no longer realistic if the level of presentation is to be sufficiently deep for serious students. Considerable depth is desired if the students are to build a solid base for future coursework. Such students are inclined to ask many penetrating questions. Their goals are not properly served by a mere survey of factual information, with a promise of detail yet to come in some future course. These arguments in favor of depth can be accommodated through judicious planning by the teacher as to the breadth of material included in the course.

The text writer must acknowledge this conflict between depth and breadth, and make allowance for a variety of compromises. Our book takes the view that any year-long course for physics students should be thorough enough to explicate the basic questions, and yet extensive enough to cover many areas. We also believe that, within reason, the text should offer every topic desired by every teacher, even if the sum of all the topics is too much for two semesters. This diversity of offerings is necessary if a given teacher is to have access to every possible pathway through the available material.

We recommend that four foundation stages be prepared first, and one of two main routes be chosen afterward, with the adoption of this book as a text. The first four areas cover special relativity, introduction to quantum theory, quantum mechanics, and application to atoms. These topics occupy Chapters 1–9 and should consume more than one semester of lecture time. The teacher may then choose to go from atoms to larger quantum systems, such as molecules and condensed matter, or to smaller quantum systems, such as nuclei and elementary particles. The one route spans Chapters 10–13, and the other spans Chapters 14–16. Our textbook gives ready access to either of these alternatives.

Every topic in the book is developed to contain a core of subject matter for the teacher's lectures and a body of supporting material for the students' readings. We devote a substantial portion of the text to side-reading because we expect the students to turn from the lecture to the book for a properly detailed understanding of the subject. This plan assumes the usual ideal classroom setting, where the teacher introduces the students to concepts and applications, and the students refine their grasp of the material by studying on their own.

The book is aimed generally at physics majors in their junior year, although some flexibility is built into this choice of level. In fact, a large portion of the material has actually been taught at the University of Massachusetts to second-semester sophomores and first-semester juniors. We assume an adequate background to be three previous semesters of physics, à la Resnick and Halliday. We take for granted any topic found in this preparatory body of coursework, and we occasionally reiterate certain classical points if they bear directly on new ideas. The new material is always constructed in a self-contained and thorough manner, consistent with the intended level. Sometimes it is instructive to build up the material from the ground and eventually "take off" toward a rather high level. One of our main assumptions is that interested students want to "take off" on occasions, with the guidance of the teacher. These instances of excess occur here and there throughout the book and are easy to bypass if that is the teacher's wish.

The text emphasizes discoveries and ideas, and incorporates analytical formalism sufficient for the level of presentation. We regard our treatment of modern physics as a prelude to a subsequent course in quantum mechanics. Hence, the text is supposed to build a case for the quantum theory and then supply enough quantum machinery to give the students an understanding of all the different quantum systems. The text is *not* intended as a presentation of quantum formalism, however. Instead, we use only the Schrödinger equation to convey most of the quantum mechanics in the book. Thus, many of the essential "words" of quantum mechanics (Hamiltonian, commutation relation, spinor, matrix element, . . .) never actually appear, even though the concepts lie just beneath the surface in our discussions.

The mathematical demands of the book are quite consistent with the level intended for the physics. We assume that all students at this level are somewhat skilled in the use of calculus. We also take the view that exposure to new mathematics is a worthy secondary objective of the course and its accompanying text. Matrices are employed on a few occasions to elucidate certain algebraic considerations. Complex numbers are put to constant use in the solution of the Schrödinger equation and the interpretation of quantum mechanics. Differential equations are encountered throughout our description of the quantum theory. We bring forward the necessary mathematical procedures and assume no previous familiarity in this area. Special deliberation is exercised when the topics of interest are governed by partial differential equations. Our main assumption about the introduction of unfamiliar mathematical techniques is that the students are willing to move quickly into the use of differential equations in order to grasp the interpretation of the solutions. We promote the development of these capabilities for a good and healthy cause. The students' analytical powers are strengthened for the future, and their understanding of the subject is enriched as a result. Modern physics is ideal in this respect because the subject cannot be properly supported without a certain amount of mathematical scaffolding and because the new ideas can be fully brought to life with this support.

History is also included in order to present modern physics as a sort of adventure in 20th century science. We insert these narrative elements whenever we wish to illustrate the growth of ideas and put separated events in proper context. Modern physics should be viewed as an *ongoing* adventure, and historical background can be used as a suitable vehicle to convey this impression.

Every major subject area has its own chapter in the book. Some of the chapters are rather long because of the broad scope of the particular areas. This aspect of the book should be ignored by the reader since every chapter is comfortably divided into sections of manageable length. The sections themselves should be regarded as the basic logical units of the text.

At least one example appears at the end of almost every section. We put all numerical computations and some algebraic manipulations into these examples so that we can separate illustrative material from the main flow of the text. Every topical area is further illustrated by a wide selection of problems at the end of each chapter.

The book includes only one appendix containing our table of nuclear properties. We deviate from the usual practice followed in other books and choose not to relegate supporting material to a series of appendixes. Instead, we evaluate every fragment of material on its own instructional merits. In many cases these fragments are important for their mathematical content and are worthwhile for students to learn. Any item of pedagogic interest is given its own place along with other pedagogic matter in the main body of the text, and all items of less instructional value are left out of the text altogether.

It is sometimes necessary to digress and set aside a certain derivation so that the accompanying topic can be presented without interruption. We employ these digressions sparingly and call them "details." The occasional detail is inserted at the end of the relevant section as a sensible alternative to the use of an appendix at the end of the book. This practice enables the reader to locate useful information out of the main flow but still near its proper place in the text.

We have argued for a balance between depth and breadth in the teaching of modern physics. A teacher can strike such a balance with the aid of this book by exercising selectivity over all the available material. It is clear that every topic cannot be covered properly in a single year. Our hope for the students is that they find this book instructive as a presentation of the teacher's choice of topics and also enlightening as a source of further reading later in their careers.

John J. Brehm

Amherst, Massachusetts

William J. Mullin

ACKNOWLEDGMENTS

Several of our colleagues contributed bits of wisdom and advice to this project. We would like to thank Ian Aitchison, Tom Arny, Ed Chang, John Donoghue, Bob Gray, Bob Hallock, Ted Harrison, Bob Krotkov, Francis Pichanick, Kandula Sastry, Janice Shafer, and Mort Sternheim for offering us the benefits of their expertise.

We should also extend our appreciation to the reviewers who provided constructive criticisms of the manuscript. All their remarks were given serious attention, and almost all were acted on in one way or another to improve the quality of the presentation.

We were extremely fortunate to have the assistance of Nellie Bristol in the preparation of our text. Her ability to convert a disorderly handwritten draft into a typed manuscript was needed to carry out the project, and her customary good humor was always greatly appreciated.

We would express our gratitude to our families too, if we could only find the words. Their encouragement was there at the start when none of us could judge the scale of the endeavor, and continued to be there throughout as the enormity of the work became all too apparent.

J. J. B.

W. J. M.

CONTENTS

CHAPTER TEN
MOLECULES   *501*

CHAPTER ELEVEN
QUANTUM STATISTICAL PHYSICS   *538*

CHAPTER TWELVE
SOLIDS   *575*

O N E

RELATIVITY

*T*he 20th century has been a time of extraordinary achievement in physics. Every great discovery of the era has had a profound influence on our modern view of the natural world.

This age of new ideas began with an awakening anxiety over the status of the classical laws. Classical physics was regarded as well established, but only within a proven domain of applicability. In fact, classical physics had apparent limitations since the established laws were unable to account for certain puzzling phenomena. Eventually, there was enough accumulated evidence of this sort to call for a reexamination of existing principles.

The first decades of the 20th century saw the coming of the special theory of relativity and the beginning of the quantum theory. These were not mere extensions of classical physics. The new theories *revised* the foundations of the discipline and defined the framework of modern physics. A revolution has resulted, affecting the course of developments in almost every area of modern science.

Classical physics has been found to have boundaries of validity. Departures from the classical regime may be viewed as thrusts in *two* main directions across these boundaries. The first path leads us to relativity, and the second much-broader path leads us to quantum mechanics. We are confronted with new situations in either direction. The problems usually involve very large velocity in the first case, and very small size in the second. We may adopt these two extremes as preliminary criteria to identify the domains of relativistic and quantum physics.

We start with relativity because of its historical priority as the first of the two fully developed theories. It is instructive to look for historical background throughout our presentation of modern physics, and it is particularly informative to go back into the 19th century in order to set the stage for relativity.

There were no serious reasons to question the soundness of classical mechanics before 1900. J. C. Maxwell, L. E. Boltzmann, and others drew upon Newton's laws and employed classical methods to interpret the thermal behavior of systems containing large numbers of particles. Hence, it could be said that classical statistical physics grew up in an environment cultivated by the validity of Newtonian mechanics. These contributions to thermal physics were among the main advances of 19th century science.

1

The towering achievement of the period was the unification of the laws governing electricity and magnetism in Maxwell's electromagnetic theory. All the known properties of charges and currents and all the known behavior of electric and magnetic fields were described by Maxwell's equations in agreement with existing experimental evidence. The theory went further and predicted the propagation of oscillating electric and magnetic fields through empty space. It was also possible to predict the speed of wave propagation in terms of parameters appearing in Maxwell's equations. Laboratory evidence for electromagnetic waves in vacuum was subsequently demonstrated in the experiments of H. R. Hertz. The prediction for the wave speed has been confirmed many times in several different experiments during the past century and a half. The current value

$$c = 2.99792458 \times 10^8 \text{ m/s}$$

has become known as the *speed of light*, a quantity conveniently approximated as 3×10^8 m/s, to an accuracy of three significant figures. (In fact, c has recently been *defined* as exactly 299792458 meters per second, leaving the meter to be determined by experiment.)

The 19th century was a productive period, when scientists could reaffirm Newton's laws and verify Maxwell's equations as parallel bodies of doctrine. Together these classical formulations described physics successfully over an enormous range of applications. By 1900, however, a confrontation of principle had begun to develop between Newton's mechanics and Maxwell's electrodynamics. The point at issue involved the description of physical behavior in *moving* frames of reference. The conflict between the two systems of classical laws was addressed and resolved by the theory of relativity.

To appreciate the problem, we have to recall some basic notions about the use of coordinate systems for the description of particles. In particular, we wish to examine how these descriptions change when we pass to different frames of reference, especially those in motion at constant relative velocity. Newtonian mechanics tells us that an inertial frame of reference is a coordinate system in which Newton's first law holds. Any other frame in uniform relative motion is also inertial, and so the whole Newtonian scheme is valid in all such frames. We find, however, that electromagnetic effects such as the phenomenon of light appear to select a privileged frame for the validity of Maxwell's equations, and we are thereby led to a very unsatisfactory logical conclusion.

The main issue in these arguments proves to be the behavior of the classical laws under a transformation of the coordinates from one frame to another. The theory of relativity examines this question with regard to the observation of the physical quantities and the meaning of the physical laws. Principles of relativity have appeared in physics before, associated with the names of Galileo and Newton, so that this line of inquiry has a familiar previous theory. In modern physics we are concerned with the principles of special relativity put forward by A. Einstein in 1905. The propagation of electromagnetic waves and the speed of light are central concepts in the logic of this theory. When we follow the logic we find that the theory requires us to alter our naive attitudes about the treatment of space and time.

1-1 The Luminiferous Aether

The identification of light as an electromagnetic wave led scientists of the 19th century to presume that light had properties in common with *mechanical* forms of wave motion. The mathematical similarity between sound and light was apparent since both types

Albert Einstein

of wave were described by solutions of the wave equation. Of course, there were distinctions to observe as well. Sound waves were known to be longitudinal oscillations of the medium, without polarization properties, while electromagnetic waves were understood as oscillations of electric and magnetic fields, with polarization transverse to the direction of propagation. The similarities suggested that light, like mechanical waves, should require a medium for support and propagation. The medium was known as the aether, an entity that supposedly possessed some very remarkable properties. It had to fill empty space with a universal tension in order that electromagnetic waves could propagate in vacuum at the unique speed c. The medium also had to be infinitely rigid and incompressible, so that longitudinal disturbances could not exist, and yet had to offer no obstacle whatsoever to the motion of material bodies. Thus, the aether was conceived to be a mechanical medium for light with no mechanical properties in the presence of matter.

Given the advantages of hindsight, we should find it difficult to appreciate the general acceptance and appeal of such a contrivance. Physicists of the period were accustomed to the adoption of mechanical models and were not receptive to the possibility of wave motion without a mechanical medium. Indeed, Maxwell himself had advocated an "aethereal medium" in his own definitive work on electromagnetic theory. When the predicted waves were discovered after his death, the scientific community was predisposed to accept the aether as a working hypothesis and examine its further consequences.

We should realize that a philosophical position of some depth is rooted in these arguments. The existence of an aether implies mechanical wave propagation where disturbances *of* the medium travel *through* the medium. In this context, the speed of propagation refers to the speed of wave motion *relative to the medium*. Believers in the aether would therefore regard the accepted value of c as the speed of light waves with respect to their aethereal medium. This aether would then represent an *absolute frame of reference* in which all electromagnetic waves travel with the speed c. Furthermore, since the existence of the waves and the prediction of the wave speed are consequences of Maxwell's equations, it follows that the aether must constitute the unique frame of reference in which Maxwell's equations are valid. Electromagnetic theory would then have to assume an altered form in another frame in motion through the aether so that light waves would be expected to have a different velocity of propagation with respect to that moving frame.

The aether was obviously more than just a mechanical convenience for the propagation of light. It also embodied a logical attitude toward the validity of certain established physical laws. Manifestations of the aether were presumably subject to experimental investigation. Aether believers did not presuppose that the Earth was at rest in the absolute frame of reference and set out to devise an experiment capable of detecting the motion of the Earth relative to the aether. The required apparatus had to be sufficiently sensitive to measure an Earth speed presumed to be much smaller than c. The historic experiment was performed with the aid of an interferometer, which had been developed by A. A. Michelson for the measurement of lengths to extraordinary precision. Michelson was joined in the aether measurement by E. W. Morley, and the results of the famous Michelson–Morley experiment were reported in 1887.

The interferometer is sketched in Figure 1-1. Light of a definite wavelength is split into two beams by a half-silvered mirror; the two beams then travel at right angles and are reflected back by the two mirrors, 1 and 2, shown in the figure. The beam returning from mirror 1 is finally reflected to the observer where it interferes with the beam returning from mirror 2. The two beams are coherent since they originate from a single source, and the difference in their optical paths determines how they interfere at the position of the observer. If the paths are of equal length, the two beams have the same transit time, arrive in phase, and produce constructive interference. This statement suggests the following way to measure the speed of the Earth through the aether, since it is implicit in the statement that the speed of light is the same over the two paths.

In keeping with the aether hypothesis we let $\mathbf{c}$ be the velocity of light and $\mathbf{u}$ be the velocity of the Earth with respect to the aether frame. The velocity of light in the Earth or interferometer frame is therefore

$$\mathbf{c}' = \mathbf{c} - \mathbf{u}. \tag{1-1}$$

It is convenient to transfer the analysis to the Earth frame and let the aether move through the interferometer with velocity $-\mathbf{u}$. Then $\mathbf{c}'$ represents an aether-drifted light velocity, analogous to the velocity of a swimmer in moving water. We take the aether drift to be from right to left in the figure, and, for simplicity, we let the two arms of the interferometer have exactly the same length d. Equation (1-1) tells us that the light speeds to the right and to the left are $c - u$ and $c + u$, respectively. Therefore, the transit time to mirror 1 and back is

$$t_1 = \frac{d}{c-u} + \frac{d}{c+u} = \frac{2dc}{c^2 - u^2}.$$

The inset to Figure 1-1 shows how to construct $\mathbf{c}'$ for the light traveling to mirror 2

Figure 1-1

Michelson interferometer with arms of equal
length. For the Michelson–Morley
experiment, the aether is presumed to be
moving through the interferometer with speed
u. The aether hypothesis implies that the light
speeds differ over the two paths, as shown.

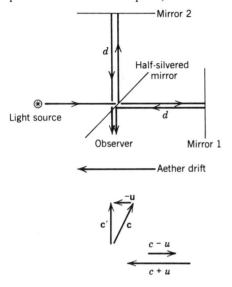

and back. The light speed c' over this path is evidently $(c^2 - u^2)^{1/2}$ each way.
Consequently, the transmit time to mirror 2 and back is

$$t_2 = \frac{2d}{\left(c^2 - u^2\right)^{1/2}}.$$

Since u/c is expected to be very small, we can use the binomial expansion to write

$$\left(1 - \frac{u^2}{c^2}\right)^{-1} = 1 + \frac{u^2}{c^2} + \cdots \quad \text{and} \quad \left(1 - \frac{u^2}{c^2}\right)^{-1/2} = 1 + \frac{1}{2}\frac{u^2}{c^2} + \cdots.$$

The two transit times are then given to order (u^2/c^2) by

$$t_1 = \frac{2d}{c}\left(1 + \frac{u^2}{c^2}\right) \quad \text{and} \quad t_2 = \frac{2d}{c}\left(1 + \frac{1}{2}\frac{u^2}{c^2}\right).$$

The difference in transit times causes the beams to arrive at the observer out of phase.
To the same order, the optical path difference is $c(t_1 - t_2)$. The number of wavelengths contained in this distance determines the number of fringes by which the
interference pattern departs from the situation for zero path difference. A rotation of
the apparatus by $90°$ introduces a change of roles between the times t_1 and t_2. The

total number of fringes shifted from the one orientation of the interferometer to the other is consequently twice the number identified above. This maneuver is particularly useful for the more realistic circumstance in which the interferometer arms are of unequal length. The final result for the fringe shift upon 90° rotation of the device is

$$\delta n = 2\frac{c(t_1 - t_2)}{\lambda} = 2\frac{d}{\lambda}\left(\frac{u}{c}\right)^2. \tag{1-2}$$

We emphasize that these assertions are predicated upon the adoption of the aether hypothesis and the use of Equation (1-1).

The experimenters were confident that a shift as small as $\frac{1}{100}$ of a fringe could be detected, whereas the expected result was around $\delta n = 0.4$. Instead, the Michelson–Morley experiment yielded a *null* result; *no* fringe shift was observed. Subsequent versions of the experiment have reproduced this same result. To the consternation of aether believers everywhere the conclusions were inescapable. The motion of the Earth through the aether frame was not detectable, and the speed of light in the interferometer was the same for the two perpendicular paths.

A number of revisionist theories came forward around 1900 with the proposition that the aether existed but nature conspired to prevent its detection. The most notable of these was a proposal offered by G. F. FitzGerald and H. A. Lorentz. According to their idea, the interferometer arm parallel to the direction of motion through the aether was supposed to contract by a factor $(1 - u^2/c^2)^{1/2}$ so that t_1 and t_2 would be identical times in the above analysis. This ad hoc suggestion turned out to have its place in future developments. However, it remained for Einstein to interpret the problem in its proper light.

Example

The Michelson–Morley experiment used light of wavelength $\lambda = 5.9 \times 10^{-7}$ m. The interferometer ultimately achieved arm lengths with $d = 11$ m by the use of multiple reflections along the two perpendicular pathways shown in Figure 1-1. The speed of the Earth around the Sun was known to be about 30 km/s, and so $u/c = 10^{-4}$ was used as an estimate for the Earth speed in the aether frame. Based on these numbers, the fringe shift in Equation (1-2) was expected to be

$$\delta n = 2\frac{11 \text{ m}}{5.9 \times 10^{-7} \text{ m}}(10^{-4})^2 = 0.4,$$

as remarked above.

1-2 Principles of Relativity

The concept of the aether as an absolute frame of reference was rendered meaningless by the Michelson–Morley experiment. Einstein was not really influenced by this result, even though he had previously entertained the desire to perform such an experiment himself. He had already become convinced that notions of absolute rest and absolute motion should have no observable consequences and that only *relative* motion could have physical meaning. This point of view had a previous history. It was accepted that the laws of mechanics should make no distinction between a state of rest and a state of motion with constant velocity. The idea that uniform motion must be relative, and hence detectable only with reference to an external point, originated with Galileo. This venerable and self-evident *relativity principle* assumed a new thrust when Einstein applied it to the propagation of light in vacuum. The question was part of the problem of reconciling electromagnetism and mechanics to the same relativistic viewpoint. Einstein investigated this problem during his period of employment at a Swiss patent office in an environment isolated from the stimulus of institutional research.

Einstein recognized a conflict between the classical theories of Newton and Maxwell. He noted that Newtonian mechanics allowed an observer to move at speed c, since an accelerating force could cause an object to reach any speed if applied long enough, and he asked how electromagnetic waves traveling in the same direction, also at speed c, might appear to the moving observer. Because the wave fronts would be accompanying the observer, the oscillations of the propagating fields would not be detectable. The electromagnetic wave motion predicted by Maxwell's equations would therefore not exist in the observer's frame of reference. The argument employed the logic of Newton's laws to deny a logical consequence of Maxwell's laws. This amounted to an inconsistency or contradiction in the classical theory. Einstein also saw that his thought experiment furnished a violation of the relativity principle regarding relative motion. He argued that a moving observer in a closed chamber could know that the chamber was in motion at light speed, without reference to any external point, by noting that light could not be observed in the chamber.

Space travel at the speed of light

Einstein's celebrated paper of 1905 demonstrates a consistent picture of mechanics and electrodynamics based on the following two principles of relativity:

The laws of electromagnetism are valid in all frames of reference in which the laws of mechanics hold.

The speed of light in vacuum is the same for all observers independent of their motion or the motion of the source.

The first postulate implies that observers in relative motion at constant velocity must agree concerning their expression of the physical laws. The second asserts that these observers must also agree that light waves propagate at speed c with respect to every frame of reference in which each particular observer is at rest. Einstein's theory sweeps the aether aside and declares it to be superfluous with this assertion.

The second postulate is alien to our rudimentary common sense. Consider the situation shown in Figure 1-2 in which a moving observer O' carries a light source and sees wave fronts of light propagating away at speed c. A stationary observer O sees those wave fronts moving also at speed c, even though O' is approaching O at a speed u, which in principle could be almost as great as c. This seems paradoxical if we react to the kind of common sense that has taught us to use Equation (1-1) as the rule for the addition of velocities. However, we have just learned that there is experimental evidence refuting this rule for the propagation of light. The correct method of adding velocities must yield a *different* rule so that the observers O and O' in the figure can agree that the light speed has the value c. Furthermore, a regime of speeds must exist for things other than light, with velocities $\mathbf{v}$ and $\mathbf{v}'$ substituted for $\mathbf{c}$ and $\mathbf{c}'$ in Equation (1-1), such that the equation is a valid approximation to the exact formula for velocity addition. The second postulate has survived every experimental test, and so our common sense has to be revised by giving cautious consideration to the space and time coordinate systems used by observers in relative motion.

Let us first define what we mean by an *observer* in more general terms. Imagine a coordinate frame in which clocks are placed at suitably spaced regular intervals. This lattice of timers can be synchronized to keep identical times by the use of light signals. A simple method for doing this is given in the example at the end of the section. Let each lattice site be equipped with a device for recording events that occur at that

---

Figure 1-2

Two observers in relative motion. O is at rest and O' moves toward O at constant speed u. O and O' agree on the speed of light coming from the source carried by O'.

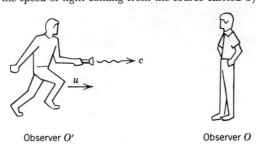

Observer O' Observer O

position. Such a lattice is equivalent to a reference frame populated by observers at all the different sites, all with synchronized timers. It is obvious that a single observer who reads the recorded data is sufficient in a lattice so equipped. The particular location of that observer in the reference frame is immaterial. In fact, even the one observer is redundant since it is the lattice of timers and recorders that makes the actual observations. From now on, when we speak of an observer we are referring to the whole lattice frame in this kind of scheme.

Next, let us return to Figure 1-2 and identify a frame S in which observer O is at rest and a frame S' in which observer O' is at rest. The frame S' moves uniformly through the frame S at speed u. These two coordinate systems are shown at two instants of time in Figure 1-3. At the first instant when S and S' coincide, a light flash is emitted from a source at rest at the origin in S', and at the later instant, the wave front from this flash reaches a detector at a location fixed in S. This wave propagates spherically from its point of origin with the same light speed c in both frames, according to the second principle of relativity. The figure indicates qualitatively that the detection of the wave front occurs at a distance Δx in S and at a *lesser* distance $\Delta x'$ in S'. In view of this the elapsed times in S and S' are different:

$$\Delta t = \frac{\Delta x}{c} \text{ in } S \quad \text{and} \quad \Delta t' = \frac{\Delta x'}{c} \text{ in } S',$$

so that $\Delta t'$ is evidently *less* than Δt. We sketch this argument here to introduce a feature of the second postulate, which we must be prepared to accept. Observers in S

Figure 1-3

Reference frames S and S' in uniform relative motion. S' moves through S with speed u, and so S moves with the same speed in the other direction through S'. A light source at the origin in S' emits a flash at the instant when S and S' coincide, and the propagating wave front is observed later at a location fixed in S. Emission and detection are shown in the frame S on the left and in the frame S' on the right.

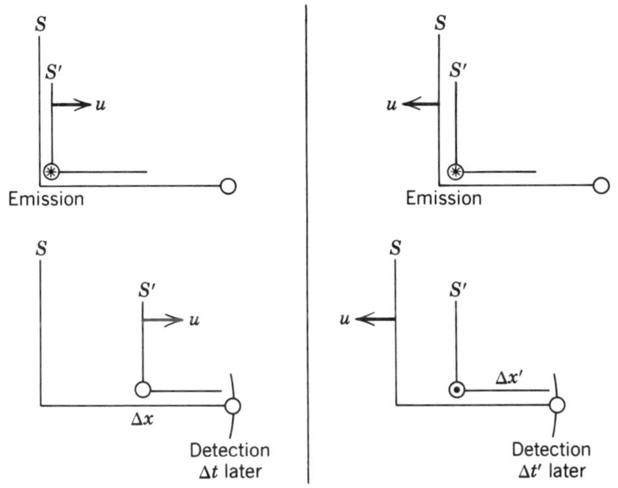

and in S' do not agree about time intervals *or* space intervals in order that they *do* agree about the light speed c. We deduce the actual relations for time and space intervals in different frames in Section 1-3.

Relative motion satisfies a reciprocal property, which we have incorporated in Figure 1-3. We note that S' moves to the right with speed u when S is the frame at rest and that S moves to the left with the *same* speed u when S' is the frame at rest. This reciprocity is necessary to prevent us from distinguishing frames and thereby identifying an absolute frame of reference.

Time can no longer be taken for granted as an absolute variable whose ticking frequency is common to all observers. Instead, time is a relative variable that differs for different observers. Time is to be regarded as an attachment intrinsic to a given frame of reference since that frame contains a lattice of synchronized clocks ticking with a particular time interval. Another frame in relative motion has its own lattice of clocks that tick with a *different* time interval to an observer in the given frame. We have to allow for this behavior of time in our revised common sense.

The relativistic meaning of time calls for a reexamination of several familiar notions. The idea of *simultaneity* is among the first to consider. In a given frame events can be observed to happen at *separate* locations at the *same* time. We have just learned, however, that the time variable is peculiar to the chosen frame, and so the same events can be observed at *different* times in another frame. Thus, *simultaneity is relative*, since events that are simultaneous in S' are not simultaneous in S if S' and S are frames in relative motion. We can illustrate this by means of the thought experiment in Figure 1-4. A railroad car is shown moving to the right with constant speed u in S, the frame of the Earth. Let S' denote the car frame in which the car is at rest. The car is equipped with a light source at its midpoint and a light-sensitive buzzer at each end of the car. The left side of the figure shows the situation in S', where the light source and the two buzzers are at rest. First, the light flashes, and later, the wave front reaches the ends of the car so that the buzzers go off

---

Figure 1-4

Propagation of a light flash from the midpoint of a moving railroad car. Light-triggered buzzers act as detectors at each end of the car. In S', the rest frame of the car, the buzzers are observed to go off simultaneously. In S, where the car is moving to the right, the left buzzer is observed to go off before the right buzzer.

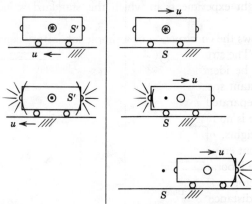

simultaneously. The right side of the figure shows what happens in S, where the light source and buzzers are moving to the right. The light flashes and the wave front propagates outward from the point of origin. It is clear from the figure that the wave front reaches the left end of the car *before* it reaches the right end so that the buzzers do not go off simultaneously in S.

We have begun to describe a revised treatment of space and time as introduced by Einstein to ensure consistency between mechanics and electromagnetism. We should be able to see that Maxwell's theory occupies the privileged position in this picture. If Maxwell's equations are valid in a particular reference frame, they are supposed to hold in any other frame in uniform relative motion. Otherwise, it would not be possible for electromagnetic theory to predict the same speed of light c for both frames. It would therefore appear that Newton's laws must be modified in order to complete Einstein's picture.

Example

Suppose that we have an array of clocks at rest with respect to each other and that we wish to synchronize these clocks to read identical times. Let us consider a linear array in which the clocks are placed along a line at locations separated by 3 m intervals. A light signal traverses a 3 m distance in 10^{-8} s, or 10 ns. Let the clocks be triggered to start ticking when a wave front of light passes, and preset the line of clocks to read sequentially $0, 10, 20, 30, \ldots$, in nanoseconds. A light flash at the location of the 0 ns clock activates the other clocks in sequence so that they all keep identical times thereafter.

1-3 Time Dilation and Length Contraction

Time is a variable whose value is recorded in the observation of an event. A pair of events spans a time interval whose duration depends on the frame of reference in which the observations are made. In our discussion of Figure 1-3 we have considered two such events defined by the emission and detection of a light signal. We have acknowledged that the elapsed time between events is relative for frames in relative motion because the speed of light is a universal quantity. We now want to show how the universality of the speed of light can be used as a standard when we compare time intervals in different frames. We proceed by devising "relativity laboratories" and conducting thought experiments in which this standard is incorporated as a basic feature.

Figure 1-5 shows the construction of a simple laboratory consisting of a floor and a mirrored ceiling. The structure is assumed to be moving uniformly so that two frames of reference can be identified. The laboratory is at rest in the frame S' and is in motion with constant speed u in the frame S. Observers in S and S' agree that floor and ceiling are separated by the *same* distance d in the two frames. (They agree on this because there is no relative motion in the vertical direction and so there is no way to distinguish heights, or vertical coordinates, from one frame to the other.) The thought experiment employs a light source and a detector located at the same position on the floor of the laboratory. The source emits a light signal, which the mirror reflects back to the detector as shown in the figure. In S', the signal travels straight up and straight down, a distance $2d$, in a time interval $\Delta t'$. In S, the signal travels out to the moving mirror and back to the moving detector, a distance $2\ell > 2d$, in a time interval

Figure 1-5

Relativity-of-time laboratory at rest in S' and moving in S with speed u. In S', the light path has length $2d$ and transit time $\Delta t'$. In S, the light path has length 2ℓ and transit time Δt. The time interval in S is greater than the time interval in S' by the factor $(1 - u^2/c^2)^{-1/2}$. In this experiment $\Delta t'$ is a proper time interval.

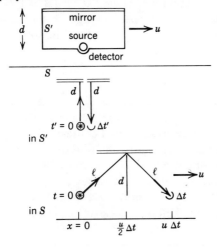

Δt. We use the universality of the speed of light to express the time intervals as

$$\Delta t' = \frac{2d}{c} \text{ in } S' \quad \text{and} \quad \Delta t = \frac{2\ell}{c} \text{ in } S.$$

It is obvious from these relations and from the figure that Δt is *greater* than $\Delta t'$. If we examine the light path in S we see that

$$\ell^2 = d^2 + \left(\frac{u}{2}\Delta t\right)^2 .$$

Elimination of ℓ and d from these equations yields the relation

$$\left(\frac{c}{2}\Delta t\right)^2 = \left(\frac{c}{2}\Delta t'\right)^2 + \left(\frac{u}{2}\Delta t\right)^2 .$$

When we solve this expression for Δt in terms of $\Delta t'$, we obtain the final formula:

$$\Delta t = \frac{\Delta t'}{\sqrt{1 - u^2/c^2}} . \tag{1-3}$$

This very important result describes the phenomenon of time dilation.

Figure 1-6

Fixed synchronized clocks showing more elapsed time than the clock in motion. The measuring rod from A to B is at rest along with the synchronized clocks in S. Its length D is a proper length determined by the spatial interval in S between points A and B.

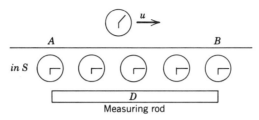

The device in the experiment is a kind of clock. The fact that the clock is at rest in S' distinguishes that frame from S, which may be any other frame where the device is moving. We have timed the interval between two events, again using the emission and detection of a light signal. The frame S' is distinguished as the only one in which these events occur at the *same* spatial location. The time interval between two such events at the same point in space is a unique quantity called the *proper time interval* between the events. According to Equation (1-3), the elapsed time Δt between the same events in any other frame is *dilated*, or enlarged, by the factor $(1 - u^2/c^2)^{-1/2}$ compared to the proper time interval $\Delta t'$.

The identification of a proper time is often the key to the resolution of many puzzling applications of time dilation. Consider clocks in relative motion as a first example. We can expect such timekeeping devices to record different times in the manner illustrated in Figure 1-6. The array of clocks between points A and B in the frame S are synchronized, and another clock in motion from A to B indicates a smaller time interval for the journey than that recorded by each of the clocks in the fixed array. In terms of Equation (1-3), the moving clock reads $\Delta t'$ while the synchronized array reads Δt for the elapsed time. The moving clock keeps proper time; the ticks of that clock are events at a fixed location in the moving frame and are being observed at a sequence of locations in the fixed frame S. The relation between Δt and $\Delta t'$ depends on the ratio u/c. Time dilation has real effects that are not readily observed unless the speed u is sufficiently close to c. The phenomenon is therefore surprising to us because the regime of such large velocities is beyond our limited experience.

Observers in relative motion must also disagree over the measured length for a spatial interval when the length is along the direction of motion. To see how their measurements differ, let us construct another kind of relativity laboratory, building again on the propagation of light signals. As before, we want a design in which light is emitted and returned to the same point in the laboratory. Figure 1-7 shows a construction consisting of two facing walls in which the one wall is mirrored while the other contains a light source and a detector at the same location. The device is in motion at constant speed u in the frame S and is at rest in the frame S'. The wall-to-wall distance depends on the frame and is denoted by L' in S' and by L in S. We can measure these lengths by timing the passage of a light signal from the source

---

Figure 1-7
Relativity-of-length laboratory at rest in S'
and moving in S with speed u. The lengths of
the laboratory in the two frames are denoted
by L' and L, respectively. The transit time of
a light signal from the source back to the
detector is observed in both frames. The
time intervals Δt and $\Delta t'$ are related by the
time-dilation formula, and the length L is
shorter than the length L' by the factor
$(1 - u^2/c^2)^{1/2}$. In this experiment L' is a
proper length.

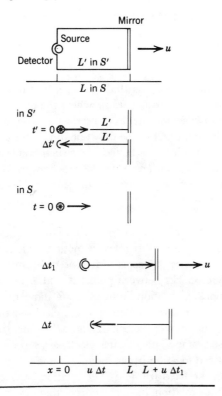

---

back to the detector after its reflection at the mirror. In S', the light flashes at $t' = 0$
and the reflected signal arrives at the detector at $t' = \Delta t'$ after traveling the distance
L' each way. The transit time is

$$\Delta t' = \frac{2L'}{c} \quad \text{in } S'.$$

Note that $\Delta t'$ is a proper time interval because the emission and detection of the
signal are events at the same location in S'. In S, the light signal propagates as shown
in the figure. The mirror moves to the right with speed u while the signal travels with
speed c and overtakes the mirror. If the flash goes off at $t = 0$ in S, then at $t = \Delta t_1$

the signal reaches the mirror, which has moved to the location $L + u \, \Delta t_1$, and at $t = \Delta t$ the reflected signal arrives at the detector, which has moved during the intervening time to the location $u \, \Delta t$. Since the light travels the distance $L + u \, \Delta t_1$ in the time Δt_1, we have

$$\frac{L + u \, \Delta t_1}{c} = \Delta t_1$$

and so

$$\Delta t_1 = \frac{L}{c - u}.$$

The reflected light travels from $L + u \, \Delta t_1$ back to $u \, \Delta t$ in the time $\Delta t - \Delta t_1$; therefore, we set

$$\frac{L + u \, \Delta t_1 - u \, \Delta t}{c} = \Delta t - \Delta t_1$$

so that

$$\Delta t - \Delta t_1 = \frac{L}{c + u}.$$

These results imply

$$\Delta t = \frac{L}{c + u} + \frac{L}{c - u} = \frac{2Lc}{c^2 - u^2} \quad \text{in } S.$$

We also know that Δt and $\Delta t'$ are related by Equation (1-3), because we have identified $\Delta t'$ as a proper time interval. The resulting equality is written as

$$\frac{2Lc}{c^2 - u^2} = \frac{\Delta t'}{\sqrt{1 - u^2/c^2}} = \frac{2L'}{\sqrt{c^2 - u^2}}.$$

The final formula then follows by solving for L to get

$$L = L' \sqrt{1 - \frac{u^2}{c^2}}. \tag{1-4}$$

This very important result describes the phenomenon of length contraction.

The laboratory in Figure 1-7 has length L' in the frame S' where the device is at rest. The rest frame is a distinguished frame and so the length L' is a unique quantity, called the *proper length*. Equation (1-4) says that the length L of the same object in any frame S in which the object is moving is *shortened* relative to the proper length by the contraction factor $(1 - u^2/c^2)^{1/2}$. Length contraction, like time dilation, is a real effect that becomes observable for sufficiently large values of u.

Some of these manipulations are reminiscent of aspects of the Michelson–Morley analysis. In fact, the conclusion in Equation (1-4) reproduces the ad hoc proposal made by FitzGerald and Lorentz to explain the null result of the Michelson–Morley experiment. The phenomenon is often called *Lorentz contraction* for this reason. Note, however, that the result is a consequence of Einstein's theory and makes no use of

Lorentz's assumption that objects contract along their direction of motion because of the mechanical properties of materials. Instead, Einstein's view holds the observable contraction to be an intrinsic property of space and time coordinate systems.

The relativity of space intervals can be deduced in another way. Let us return to Figure 1-6 and assume that an observer travels with the moving clock between the indicated points A and B. The observer measures the elapsed time $\Delta t'$ on that clock and then uses the speed u to compute

$$D' = u \, \Delta t'$$

as the distance from A to B. This result represents the length of the array of clocks in S, observed in the frame S' of the moving clock as the array passes by (traveling to the left). The corresponding distance in S from A to B is given by the length D of the measuring rod shown in the figure. An observer in S reports that the moving clock travels the distance D (to the right) in time Δt so that

$$D = u \, \Delta t.$$

We can use Equation (1-3) and immediately relate the lengths D and D':

$$D' = D\sqrt{1 - \frac{u^2}{c^2}} \, . \tag{1-5}$$

This conclusion agrees with Equation (1-4), even though the two results may not appear to be quite the same. It is important to realize that the distance D is the length of a rod in the rod's rest frame and is therefore a *proper* length. The observer on the moving clock measures this distance and obtains D', the appropriately *contracted* length. Figure 1-8 shows the situation in the frame S' where the moving clock, mounted by the primed observer, is at rest. In this frame the rod moves to the left with speed u, and so the length D' is seen by the primed observer as the contraction of the proper length D.

Some bewilderment over these formulas is inevitable. When circumstances involve two (or more) relatively moving frames, it is natural to wonder where the primes should go in the use of Equations (1-3), (1-4), and (1-5). The recommendation would be to identify wherever possible the *proper time* or *proper length* in the appropriate frame. The time or space interval in any other frame is then dilated or contracted according to the formulas.

Figure 1-8

Observation of the length of a moving rod. The proper length is D; the observer in S' measures the contracted length D'.

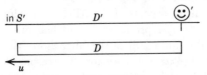

Example

An experimenter transports a clock on an eastbound transcontinential flight. The instrument has been synchronized with a clock on the West Coast and is compared with another clock on the East Coast after the journey. The West Coast and East Coast clocks remain synchronized throughout. The picture is the same as the one shown in Figure 1-6. Let the speed of the plane be $u = 300$ m/s and let the transcontinental distance be $D = 6000$ km. A ground-based observer records the flight time to be

$$\Delta t = \frac{D}{u} = \frac{6 \times 10^6 \text{ m}}{3 \times 10^2 \text{ m/s}} = 2 \times 10^4 \text{ s}.$$

The traveling experimenter records a shorter proper time interval given by $\Delta t' = \Delta t \sqrt{1 - u^2/c^2}$. The traveling clock therefore runs slower by the amount

$$\Delta t - \Delta t' = \Delta t \left[1 - \sqrt{1 - \frac{u^2}{c^2}} \right] = \Delta t \left[1 - \left(1 - \frac{1}{2} \frac{u^2}{c^2} \right) \right] = \frac{\Delta t}{2} \left(\frac{u}{c} \right)^2.$$

(We use the binomial expansion and note that the first nonvanishing term is enough since $u/c = 10^{-6}$.) The time difference between clocks at the end of the journey is

$$\Delta t - \Delta t' = \frac{2 \times 10^4 \text{ s}}{2} (10^{-6})^2 = 10^{-8} \text{ s}.$$

Time dilation has a very small effect here because jet travel is so slow compared to the speed of light. The traveler observes the transcontinental distance to be $D' = u \, \Delta t'$. This corresponds to the Lorentz-contracted version of the proper length D, where D and D' are related as in Equation (1-5). The difference between D and D' is

$$u(\Delta t - \Delta t') = u \frac{\Delta t}{2} \left(\frac{u}{c} \right)^2 = \frac{D}{2} \left(\frac{u}{c} \right)^2 = \frac{6 \times 10^6 \text{ m}}{2} (10^{-6})^2 = 3 \times 10^{-6} \text{ m},$$

a minuscule distance of only 3 micrometers.

Example

The situation is quite different in high-energy physics where relativistic effects are among the essential properties of the elementary particles. Indeed, it is here that we routinely find convincing evidence to support our belief in relativity. Let us consider the muon as an example of an unstable elementary particle. The muon has a 2 μs average lifetime and decays into lighter particles. Muons are produced in the interactions of other very-high-energy particles as they enter the Earth's atmosphere. Many fast muons are able to reach the surface of the Earth before decay, even though they must travel a considerable 10 km distance through the atmosphere to do so. Time dilation is in evidence here as the 10 km distance and the 2 μs time interval do not refer to the same space–time frame. The 2 μs lifetime represents a proper time interval in the muon's rest frame. A fast muon, traveling at 99.9% of the speed of light, would see only about 600 m of atmosphere pass by before decay:

$$D' = u \, \Delta t' = 0.999(3 \times 10^8 \text{ m/s})(2 \times 10^{-6} \text{ s}) = 6 \times 10^2 \text{ m}.$$

For $u/c = 0.999$ the time dilation factor is $(1 - u^2/c^2)^{-1/2} = 22$. Hence, an observer on Earth sees these muons live 22 times longer and travel 22 times farther. From Equation (1-5) we have

$$D = \frac{D'}{\sqrt{1 - u^2/c^2}} = 22(0.6 \text{ km}) = 13 \text{ km}.$$

Thus, the Earth observer sees the muon's lifetime dilated, and the muon in turn sees the distance through the atmosphere contracted.

Example

The relativity of simultaneity is the key to the following famous puzzle, which we present with numbers first and then with general expressions. Let us use what we have learned in order to make a 5 m pole fit into a 4 m barn. The obvious procedure would be to run with the pole at speed $u = \frac{3}{5}c$ into the barn (which we assume to be open at both ends), as described in Figure 1-9. The observer O in the figure sees that the 5 m length of the pole is contracted by the factor $\sqrt{1 - u^2/c^2} = \frac{4}{5}$ and hence can say that the pole just fits in the 4 m length of the barn. Of course, there is more to the puzzle than this. The observer O' sees the barn approaching the pole, observes that the 4 m length of the barn is contracted by the same factor $\frac{4}{5}$, and therefore says that the barn is only 3.2 m long, too short to contain the entire length of the 5 m pole. We have an obvious disagreement here between two observers and this would appear to be a contradiction in relativistic physics. In fact, the disagreement is real and is to be expected. To appreciate this we must realize that we are describing the observation of *two* events. Event B is the arrival of the right end of the pole at the right end of the barn, and event C is the arrival of the left end of the pole at the left end of the barn. Observer O sees these events as simultaneous, but observer O' cannot agree since the simultaneity of the two events is relative. Thus, the fitting of the pole in the barn is also relative, and observers in relative motion are bound to disagree over it. Let us examine their disagreement in general terms, letting S be the barn frame and S' be the pole frame. For reference, we identify event A, the arrival of the right end of the pole at the left end of the barn. Let this event occur at the origin in both frames, when $t = 0$ in S and when $t' = 0$ in S'. Call the proper length of the pole L'_0 (the length of the pole in S'), and call the proper length of the barn ℓ_0 (the length of the barn in S). The sequence of events A, B, and C is shown in the figure. In S, the length of the pole is

$$L = L'_0 \sqrt{1 - \frac{u^2}{c^2}} \, .$$

Figure 1-9

Pole-in-the-barn puzzle. The pole has proper length L_0', and the barn has proper length $\ell_0 < L_0'$. Pole and barn are in relative motion at speed u, chosen such that the Lorentz-contracted length of the pole equals the proper length of the barn. In the rest frame of the barn, events B and C are simultaneous. In the rest frame of the pole, event B precedes event C.

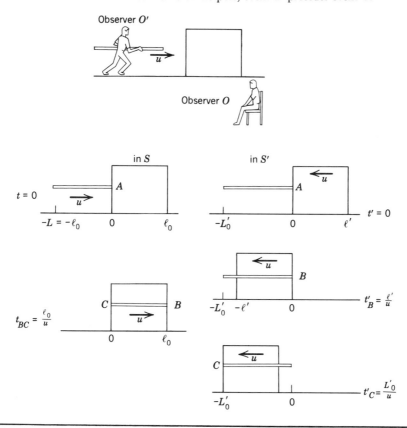

We choose u such that $L = \ell_0$, so the pole "fits" in the barn in S. Then, when $t = t_{BC} = \ell_0/u$, the events B and C are observed in S as simultaneous. In S', the length of the barn is

$$\ell' = \ell_0 \sqrt{1 - \frac{u^2}{c^2}}.$$

(Remember, we are not to be confused about the placement of primes in the equations as long as we identify the proper length and then Lorentz contract it.) In S', event B occurs when $t' = t_B' = \ell'/u$, and event C occurs later when $t' = t_C' = L_0'/u$. Note that, since $L = \ell_0$ by construction, we have

$$\ell' = L \sqrt{1 - \frac{u^2}{c^2}} = L_0'\left(1 - \frac{u^2}{c^2}\right).$$

The times in the two frames are related:

$$t'_B = \frac{\ell'}{u} = \frac{\ell_0}{u}\sqrt{1 - \frac{u^2}{c^2}} = t_{BC}\sqrt{1 - \frac{u^2}{c^2}}$$

and

$$t'_C = \frac{L'_0}{u} = \frac{\ell'}{u}\frac{1}{1 - u^2/c^2} = \frac{t_{BC}}{\sqrt{1 - u^2/c^2}}.$$

Thus, there is a time gap in S', by which the events B and C fail to be simultaneous.

1-4 Relative-Motion Symmetry

Relativity denies the concept of absolute motion and treats relative motion symmetrically. We have noted the reciprocal symmetry of frames in uniform relative motion in Figure 1-3. We have also used the symmetry implicitly in Figure 1-5 to claim that observers in such frames agree on lengths transverse to the direction of motion. This reciprocity has certain confusing aspects that we now want to consider.

Suppose that two frames S and S' are in relative motion with velocity **u**. If we can draw conclusions about S' moving at speed u in one direction through S, we must be able to draw similar conclusions about S moving at speed u in the opposite direction through S'. Let us visualize this in terms of a rocket ship traveling past a platform in deep space as in Figure 1-10. Call the rocket frame S' and the platform frame S, and assume for convenience that clocks in the two frames are synchronized when the origins of the frames coincide. We already know from Figure 1-6 that a moving rocket clock at the origin in S' runs slower than a fixed array of platform clocks in S. However, we can equally well say that the platform is moving past the rocket in the other direction since only their relative motion is physically significant. Relative-motion symmetry then requires a moving platform clock at the origin in S to run slower than a fixed array of rocket clocks in S'. A picture of the symmetrical situation is included in Figure 1-10.

It would appear that we have the possibility of a contradiction in the application of this symmetry. In fact, no difficulties can arise, but we must be careful to examine possible problems with questions that are properly posed. Note that each point of view in Figure 1-10 employs a single clock keeping proper time in a "moving" frame, compared with an array of synchronized clocks in a "fixed" frame. Note especially that the proper time of interest refers to the ticks of the single moving clock as events at the same position in that clock's frame.

The moving clocks have to be examined differently if we want to relate readings between a single rocket clock and a single platform clock. We might wish to compare a clock (call it C') at the origin in S' with a clock (call it C) at the origin in S by letting the rocket pass the platform and then return. This procedure allows C' and C to separate and then come back together for confrontation at the same location. We must realize, however, that the clock C' parts company with the frame S' as soon as it stops moving uniformly, as it must in order to reverse direction and return. Thus, C' is at rest in S'', a new frame of reference moving through S, when it comes back to meet C. Of course, the clock C remains at rest in S throughout. This method of comparing clocks evidently requires the use of three reference frames.

Figure 1-10

Rocket ship and space platform in uniform relative motion. A rocket clock, at rest in S', runs slower than an array of platform clocks in S. A platform clock, at rest in S, runs slower than an array of rocket clocks in S'.

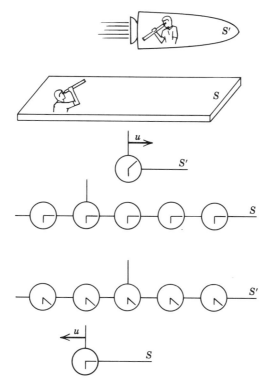

Let us apply the symmetry in a different way by letting an observer with one clock look through a telescope at another receding clock and compare readings. The clocks maintain uniform relative motion, and so the two frames S and S' suffice in this case. We recognize first of all that the act of reading the time through a telescope is accomplished by receiving the light emitted previously from the receding clock. In S, let C' be moving away and sending its light to the location of C, as described in Figure 1-11. Light that reaches the origin in S at time $t = t_1$ is light from C', emitted earlier in S at time $t = t_0$. Since the light travels the indicated distance ut_0 in the time interval $t_1 - t_0$, we set

$$\frac{ut_0}{c} = t_1 - t_0,$$

so that

$$t_0 = \frac{t_1}{1 + u/c}.$$

Figure 1-11

Reading a receding clock C' and comparing with a clock C in hand. Light received at time t_1 in S is light from C', emitted at time t_0' in S'. The reading on C' is related to the reading on C by $t_0' = rt_1$, where r determines the red shift.

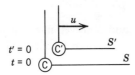

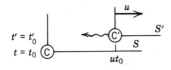

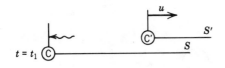

The reading on C' is a proper time, denoted by t_0'. We therefore obtain

$$t_0' = \sqrt{1 - \frac{u^2}{c^2}}\, t_0 = \sqrt{\frac{1 - u/c}{1 + u/c}}\, t_1.$$

Thus, when C' is read through a telescope and compared with C, it is found that the time recorded on C' is *behind* the time recorded on C by the factor

$$r = \sqrt{\frac{c - u}{c + u}}. \tag{1-6}$$

If we reverse the situation so that an observer in S' reads C and compares the reading with C', we reach the symmetrical conclusion in which the receding clock is observed to run behind the clock in hand by the same ratio r given in Equation (1-6).

The *Doppler effect* for light offers an interesting practical realization of these ideas. The effect for sound is familiar, and it is instructive to recall the results so that we can appreciate the distinctions.

Sound waves are observed to undergo a shift in frequency when the source of sound is in motion through the medium, either approaching or receding from the observer. A different shift is obtained if the observer is moving and the source is stationary in the medium. The shifts for sound waves are not symmetrical because the medium provides a frame of reference in which source motion and observer motion can be distinguished. Formulas for the various cases may be found in any introductory textbook. We can combine all the possibilities in a single expression by writing the observed sound frequency $\bar{f}$ as

$$\bar{f} = \frac{w + u_O}{w - u_S} f_0. \tag{1-7}$$

In this equation, f_0 is the frequency of the stationary source, w is the speed of sound in the medium, and u_S and u_O are the speeds of the source and the observer, also with respect to the medium. The signs of these last two quantities are such that u_S is positive when the source moves toward the observer, and u_O is positive when the observer moves toward the source. (As a check, let source and observer move in the same direction with the same speed so that $u_O = -u_S$. The result from Equation (1-7) is $\bar{f} = f_0$, as we might expect.)

Light waves experience a different Doppler effect because light is a unique form of wave motion. There is no medium, and so there is no distinction over source motion and observer motion; instead, source and observer are governed by relative-motion symmetry. Moreover, time intervals for the source and time intervals for the observer differ in relative motion according to the time-dilation formula. There are several ways to derive the Doppler result for light. Let us analyze the effect in the manner customarily used for sound but adapted suitably to apply to light.

We distinguish light and sound from the outset by adopting different notation. Let us call the source frequency ν_0 and the observed frequency $\bar{\nu}$, and consider the case where the source moves toward the observer with speed u, as shown in Figure 1-12. Of course, only the *sense* of the relative motion (*approaching* rather than *receding*) is physically significant. Let S' be the source frame and S be the observer frame, and consider the situation in the figure where two wave fronts are observed at instants a full cycle apart. In S', the time interval T' between the emission of these wave fronts

Figure 1-12
Doppler effect for light where the source approaches the observer with speed u. Two consecutive wavefronts of like phase are observed in order to identify the Doppler-shifted wavelength $\bar{\lambda}$. The observed frequency $\bar{\nu}$ is blue shifted relative to the source frequency ν_0 according to the formula $\bar{\nu} = \nu_0/r$.

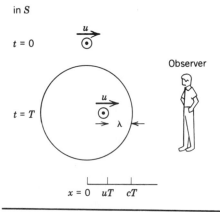

is the period

$$T' = \frac{1}{\nu_0}.$$

In S, the corresponding time interval is denoted by T. Since T' is a proper time interval, Equation (1-3) applies:

$$T = \frac{T'}{\sqrt{1 - u^2/c^2}}.$$

At $t = T$ the observer notes the current position of the wave front, previously emitted at $t = 0$, and measures its distance from the wave front of identical phase coming from the moving source at the same instant. This measurement determines the observed wavelength

$$\bar{\lambda} = cT - uT$$

as indicated in the figure. The observed frequency is therefore given by

$$\bar{\nu} = \frac{c}{\bar{\lambda}} = \frac{c}{(c - u)T}.$$

When we fold the time-dilation equation for T into this, we obtain the desired relation between $\bar{\nu}$ and ν_0:

$$\bar{\nu} = \frac{c}{(c - u)T'} \sqrt{1 - \frac{u^2}{c^2}} = \sqrt{\frac{c + u}{c - u}} \, \nu_0.$$

The result is the Doppler formula for relative motion in the approaching sense. The receding case is described simply by changing the sign of u. Note how the factor r, defined in Equation (1-6), makes its appearance in the two situations. The quantity r determines a *red shift* since, in the case of a receding source, the observer detects a lesser red-shifted frequency given by

$$\bar{\nu}_{\text{receding}} = r\nu_0. \tag{1-8}$$

In astrophysics, the red shift for a receding emitter is expressed by the parameter $z = 1/r - 1$. The formula obtained above from Figure 1-12 applies to the case of an approaching source, where the observed frequency is given by

$$\bar{\nu}_{\text{approaching}} = \frac{\nu_0}{r}. \tag{1-9}$$

In this circumstance, the observer detects a blue-shifted frequency.

The result for light in Equations (1-6) and (1-9) is obviously rather different from the result for sound in Equation (1-7). This is expected since the processes differ in several respects as noted above. In fact, the observation of red-shifted light is quite similar in concept to the act of reading the time on a receding clock and comparing with a clock in hand, the exercise that has led us to Equation (1-6) in the first place. Of course, we must be reminded again how small u/c is in practical circumstances. A sensitive measurement is necessary to verify that Equation (1-9) is correct for Doppler-shifted light. It is interesting to compare the Doppler shifts for light and sound by expanding the relevant formulas in powers of the ratio between the speed of relative motion and the wave speed. It happens that the light formula in Equation (1-9) is exactly the same as the sound formula in Equation (1-7) to first order in this ratio. The differences between the two expressions only begin to appear when the equations are examined in second order.

The Doppler shift is a very useful tool in astronomy. We can use the effect to learn how fast celestial objects such as stars or nebulas are moving toward or away from us. These bodies emit wavelengths characteristic of their constituent matter, and the observation of the Doppler-shifted wavelengths leads to a determination of their radial velocities with respect to the Earth. We can even use the effect to deduce the rotational speed of a star like the Sun. The edge of such a body emits blue-shifted light as it rotates toward us while the opposite edge emits red-shifted light as it rotates away. Characteristic wavelengths from the source are therefore observed to be broadened, by a blue shift on the one side and a red shift on the other.

Example

Let us return to the rocket ship and space platform in Figure 1-10, and suppose that "space twins" are born at the instant of coincidence between the rocket origin in S' and the platform origin in S. The "twins" are Chester, whose location is at the S' origin on the rocket, and Esther, whose position is at the S origin on the platform. Let the relative speed between frames be $u = \frac{3}{5}c$, so that the time-dilation factor is $(1 - u^2/c^2)^{-1/2} = \frac{5}{4}$. Suppose that when Esther is 5 years old, Chester's rocket turns around abruptly and comes back to the platform at the same speed $\frac{3}{5}c$. We have already noted that the frame S' carries on without Chester, who has now taken residence at the origin in a new frame S'' (not shown in the figure). This frame moves in from right to left through the frame S. Reunion takes place after the passage of 10 years in S, Esther's frame; therefore, Esther has reached "age" 10 at this time. Chester's clock, in S' on the way out and in S'' on the way back, runs slower by a factor of $\frac{4}{5}$, so that Chester has reached "age" 8 when he returns to confront his "twin." The famous *twin paradox* asks why we cannot reverse the situation symmetrically to Chester's point of view and argue instead that it is Esther who goes off on her platform and then comes back as the younger of the two. In response, we can say that relative-motion symmetry applies between S and S', and between S and S'', but does not apply over the whole excursion. Chester's entire motion is not symmetrical to Esther's, and so the age difference, whereby Esther stays at home and "ages" more, is a real asymmetrical effect. Chester knows, without reference to any point outside his rocket, that it is he who goes off and comes back because he experiences a lurching force when he reverses direction. The resulting acceleration temporarily breaks the symmetry by dividing the relative motion between its receding and approaching phases. We return to look at some more of the details of this problem later on.

1-5 The Lorentz Transformation

The dilation of proper times and the contraction of proper lengths are the first main results of relativity. Now that we know how these proper intervals are altered as we pass to other frames, it is appropriate that we generalize and ask next how arbitrary descriptions of space and time *transform* when we move from *any* one frame of reference to another. Again, we are guided by the principles of relativity and particularly by the universal property of the speed of light.

We are concerned with the space–time description of events. The word *event* has already been put in use to denote a happening associated with a location in space and an instant in time. Occurrences at the same spatial location are distinct events if they happen at different times. In more abstract language, an event defines a point in space–time. Observers in relative motion make different assignments of space–time coordinates for such a point. Thus, a given event has space–time coordinates (x, y, z, t) in S and (x', y', z', t') in S', where S and S' are frames in relative motion. Our problem is to deduce the transformation rule by which the coordinates in one frame are determined, given the set of coordinates in the other frame.

Let us restrict our attention to frames in relative motion along the direction of a common axis. Two such reference systems S and S' are shown in Figure 1-13, where S' moves through S at speed u along the x and x' axes. Each frame contains a lattice of synchronized clocks, reading t in S and t' in S'. As usual, we choose to start the clocks at $t = 0$ and $t' = 0$ when the two frames coincide. Then, at a later time t in S, the primed origin at $x' = 0$ is located at the point $x = ut$. The event shown in the figure has coordinates (x, y, z, t) in S and (x', y', z', t') in S'. For reasons that we have already discussed, the spatial coordinates transverse to the relative motion are the same in the two frames, so that $y' = y$ and $z' = z$ in this case. Hence, the relations at issue are those between the variables (x, t) and (x', t').

In pre-Einstein relativity, the time coordinates t and t' are presumed to be the same for an event, in keeping with the notion of absolute time. Let us adopt $t' = t$ temporarily and pursue the consequences before we consider the true relativistic picture. We can use the fact that $x = ut$ when $x' = 0$ to establish the equality

---

Figure 1-13

Reference systems S and S' in relative motion. An event occurs at (x, y, z, t) in S and (x', y', z', t') in S'. In this view, S' is moving through S.

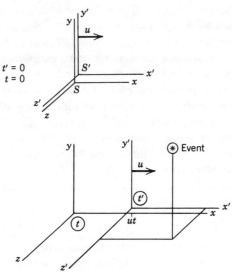

$x' = x - ut$. The collection of relations among the space–time coordinates may then be written as

$$x' = x - ut,$$
$$y' = y,$$
$$z' = z,$$
$$t' = t. \tag{1-10}$$

This particular system for relating the space–time descriptions of a given event is called the *Galilean transformation*.

The set of transformation formulas is part of the Galilean–Newtonian principle of relativity, which says that if the laws of mechanics hold in S, they hold also in S', where S and S' are related by Equations (1-10). To see this, let the event of interest denote the instantaneous position of a particle, and consider the velocity of the particle as seen in the two frames. If we introduce the vector $\mathbf{u} = u\hat{\mathbf{x}}$ as the constant velocity of S' in S and differentiate Equations (1-10) with respect to the time, we get the vector result

$$\mathbf{v}' = \frac{d\mathbf{r}'}{dt'} = \frac{d}{dt}(\mathbf{r} - \mathbf{u}t) = \mathbf{v} - \mathbf{u}. \tag{1-11}$$

Note that explicit use is made of the notion of absolute time.

A second differentiation yields $\mathbf{a}' = \mathbf{a}$; therefore, the acceleration of the particle measured in S' is the same as that measured in S. The particle accelerates when a force is applied. The components of this force are (F_x, F_y, F_z) in S and $(F_{x'}, F_{y'}, F_{z'})$ in S'. We recognize that the components are not altered by the transformation because the orientation of the axes is retained and the time variable is unchanged. We can express this equality of components by writing $\mathbf{F}' = \mathbf{F}$ and conclude that, if $\mathbf{F} = m\mathbf{a}$ governs the motion in S for a particle of mass m, then $\mathbf{F}' = m\mathbf{a}'$ holds likewise in S'. Thus, Newton's law has the same form in any two frames connected by the Galilean transformation of coordinates.

Note that Equation (1-11) immediately becomes Equation (1-1) when the event of interest is chosen to be a point on a wave front of light. We can therefore regard Equation (1-1) as a by-product of Galilean relativity. Since we have already disavowed this equation as inconsistent with the observed behavior of light, we can argue that the Galilean transformation is likewise inconsistent with the proper description of electromagnetic phenomena. In fact, it is possible to show that the transformation of coordinates defined by Equations (1-10) does not preserve the form of Maxwell's equations. Thus, the improper behavior of the speed of light is only one of the manifestations of the failure of Galilean relativity. Of course, it is Einstein's theory that provides the correct principles of relativity and produces the desired set of transformation formulas. The laws of mechanics have to be modified to accommodate those principles.

We have known all along that the absolute-time prescription $t' = t$ cannot be allowed in Equations (1-10). Instead, we expect that t', like x', should depend on x and t. It is clear that the variables (x, t) and (x', t') must be related by linear equations since the corresponding intervals are related in Equations (1-3) and (1-4) as simple proportionalities. To proceed, let us again refer to the event in Figure 1-13 and use the fact that $x = ut$ when $x' = 0$. This time we assume that x' is proportional to $x - ut$ and set

$$x' = \gamma(x - ut), \tag{1-12}$$

Figure 1-14

Another view of the frames S and S' in which S is moving through S'.

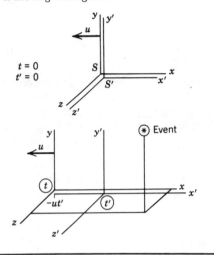

where the proportionality constant γ is to be determined. We should realize in advance that γ depends on u, the speed of S' moving through S. Next, let us reverse our point of view and take S to be moving through S' in the other direction, as in Figure 1-14. Since the unprimed origin at $x = 0$ corresponds to the point $x' = -ut'$, the coordinates of the indicated event must have the property that x is proportional to $x' + ut'$. We use the same proportionality constant as in Equation (1-12) and write

$$x = \gamma(x' + ut') \tag{1-13}$$

because the relative speed between frames is u from either point of view and because γ depends only on u.

We can now connect the viewpoints in the two figures and determine γ by considering the propagation of a light signal. Let a light flash be emitted at the coincidence of origins in the two frames when $t = 0$ and $t' = 0$, and let the given event at the later time represent a point on the propagating wave front. For simplicity, we choose $y' = y = 0$ and $z' = z = 0$ so that the event occurs at a location on the x and x' axes. The coordinates must satisfy $x = ct$ in S and $x' = ct'$ in S' because of the universality of the speed of light. When these data are inserted in Equations (1-12) and (1-13), the expressions become

$$ct' = \gamma(ct - ut) \quad \text{and} \quad ct = \gamma(ct' + ut').$$

Two relations between t and t' result:

$$t' = \gamma\left(1 - \frac{u}{c}\right)t \quad \text{and} \quad t = \gamma\left(1 + \frac{u}{c}\right)t'.$$

We can combine these to get an identity,

$$t = \gamma\left(1 + \frac{u}{c}\right)\gamma\left(1 - \frac{u}{c}\right)t,$$

and immediately solve for the unknown constant of proportionality:

$$\gamma = \frac{1}{\sqrt{1 - u^2/c^2}}. \tag{1-14}$$

The resulting quantity is recognized as the familiar time-dilation factor appearing in Equation (1-3) and in other formulas thereafter. Next, let us return to Equation (1-13) and insert Equation (1-12) to get

$$x = \gamma^2(x - ut) + \gamma ut'.$$

We can solve this equality for t' and obtain

$$t' = \gamma t + \frac{1 - \gamma^2}{\gamma u}x.$$

Equation (1-14) tells us that $(1 - \gamma^2)/\gamma u = -\gamma u/c^2$, and so the final relation takes the form

$$t' = \gamma\left(t - \frac{u}{c^2}x\right). \tag{1-15}$$

These results fulfill our objective, as Equations (1-12) and (1-15) serve to determine the transformed variables (x', t') in S' from the given variables (x, t) in S.

The rules for finding the coordinates of an event in S', given those in S, are collected as follows:

$$x' = \gamma(x - ut),$$
$$y' = y,$$
$$z' = z,$$
$$t' = \gamma\left(t - \frac{u}{c^2}x\right). \tag{1-16}$$

This famous set of formulas is known as the Lorentz transformation. Credit is given to Lorentz because of his discovery, before Einstein, of the curious mathematical property that the form of Maxwell's equations remains unchanged when the space and time variables are substituted according to Equations (1-16) and when the fields in the equations are also transformed by a suitable set of rules. Lorentz's contribution is that of a believer in the aether hypothesis and should therefore be regarded as a mathematical observation without physical foundation. Einstein's theory gives status to the contribution by providing the proper theoretical setting in which the transformation emerges logically from the principles of relativity. Einstein's famous paper of 1905 demonstrates over again Lorentz's observation that the transformation rules leave the form of Maxwell's equations unchanged. Thus, the Lorentz transformation supplies the general mathematical language to use for the description of space and time and for the formulation of dynamical principles.

The speed of light has a central role in all these proceedings. We can see that the light speed represents a speed limit inasmuch as γ in Equation (1-14) becomes infinite in the limit $u \to c$. Hence, frames of reference may have relative motion at any speed u, even u approaching c, as long as u remains smaller than c. Of course, our experience makes us more familiar with the nonrelativistic regime where u is much less than c. When we examine the Lorentz transformation in the nonrelativistic limit, we find $\gamma \to 1$ and $t' \to t$. Therefore, absolute time and the Galilean transformation become valid approximations to Equations (1-16) in this regime.

The Lorentz transformation can be inverted so that we can pass from the frame S' to the frame S. Equation (1-13) is evidently one member of such an inverse set of relations. The whole set may be obtained from Equations (1-16) simply by exchanging primed for unprimed coordinates and changing the sign of u:

$$x = \gamma(x' + ut'),$$
$$y = y',$$
$$z = z',$$
$$t = \gamma\left(t' + \frac{u}{c^2}x'\right). \tag{1-17}$$

Note that we can associate Equations (1-16) with Figure 1-13 and Equations (1-17) with Figure 1-14. Of course, relative-motion symmetry assures us that both sets of relations are Lorentz transformations with entirely symmetrical meanings. Frames of reference are called *Lorentz frames* if their space and time coordinates possess these transformation properties.

The Lorentz transformation is a general scheme of relations between any two Lorentz frames. Let us gain familiarity with some of the transformation strategies by confirming the relativistic effects that we have already investigated regarding the relativity of time and space intervals and the relativity of simultaneity.

To examine time dilation, we return to the situation shown in Figure 1-5 so that we can identify two events occurring at the same space point x_0' in S' and then transform the events to S. Denote the coordinates of the events in each frame as

$$(x_1, t_1) \text{ in } S \quad \text{and} \quad (x_0', t_1') \text{ in } S'$$

for event 1, and

$$(x_2, t_2) \text{ in } S \quad \text{and} \quad (x_0', t_2') \text{ in } S'$$

for event 2. The time interval between events is $\Delta t = t_2 - t_1$ in S and $\Delta t' = t_2' - t_1'$ in S'. When we apply Equations (1-17) to the two events, we obtain for the time coordinates in S

$$t_2 = \gamma\left(t_2' + \frac{u}{c^2}x_0'\right)$$

and

$$t_1 = \gamma\left(t_1' + \frac{u}{c^2}x_0'\right).$$

The time interval in S is found by subtracting to get

$$\Delta t = \gamma \Delta t'. \tag{1-18}$$

Since $\Delta t'$ is a proper time interval in this situation, the result describes time dilation as in Equation (1-3).

To examine length contraction, let us consider the measurement illustrated in Figure 1-8. We want to determine the length of the moving rod in S' by locating the endpoints of the rod in that frame, where the two spatial locations are to be found at the same time t_0'. Associate an event with the simultaneous arrival of each end of the rod as shown in the figure, letting the coordinates in S' be

$$\left(x_1', t_0'\right) \quad \text{and} \quad \left(x_2', t_0'\right)$$

for the arrival of the left and right ends, respectively. We can determine the corresponding spatial coordinates for the two ends in S by applying Equations (1-17) to each event:

$$x_2 = \gamma\left(x_2' + ut_0'\right)$$

and

$$x_1 = \gamma\left(x_1' + ut_0'\right).$$

When we subtract, and use $D = x_2 - x_1$ and $D' = x_2' - x_1'$ to define the lengths in the figure, we get

$$D = \gamma D', \quad \text{or} \quad D' = \frac{D}{\gamma}, \tag{1-19}$$

in agreement with Equation (1-5). As in the earlier derivation, D is the proper length of the rod and D' is its contracted length.

To examine the relativity of simultaneity, we refer back to the thought experiment in Figure 1-4, where the soundings of two buzzers at opposite ends of the moving railroad car represent two spatially separated events, occurring simultaneously in S' but not in S. Let these events be denoted in S' by the coordinates

$$\left(x_1', t_0'\right) \quad \text{and} \quad \left(x_2', t_0'\right)$$

for the buzzers on the left and right, respectively. Note that both events have the same time coordinate in S'. When we apply Equations (1-17) to each event, we find the time when each buzzer goes off in S:

$$t_2 = \gamma\left(t_0' + \frac{u}{c^2}x_2'\right)$$

and

$$t_1 = \gamma\left(t_0' + \frac{u}{c^2}x_1'\right).$$

Finally, we subtract, and let $\Delta t = t_2 - t_1$ and $L' = x_2' - x_1'$ to obtain

$$\Delta t = \gamma\frac{u}{c^2}L'. \tag{1-20}$$

This result formalizes the qualitative description given to accompany Figure 1-4. The distance L' is the length of the railroad car in S'. Since the car is at rest in S', L' is the proper length of the car. In S, where the car is moving, the left buzzer is observed to go off before the right buzzer with intervening time given by Equation (1-20).

Example

Here is a final illustration of the Lorentz transformation based on the issue of simultaneity. Two hunters are separated by a 9 m distance and fire their shotguns 10 ns apart. A passing game warden observes that the shots are simultaneous. The speed of the warden in the frame of the hunters can be deduced from this information. Regard the gunshots as two events, and introduce Lorentz frames S and S' in relative motion such that the hunters are at rest in S and the warden is at rest in S'. Let the first shot occur when $t = t' = 0$, and assume for convenience that S and S' are coincident at this instant. The two shots are events, with coordinates

$$(0,0) \text{ in } S \quad \text{and} \quad (0,0) \text{ in } S'$$

for the first shot, and

$$(x_1, t_1) \text{ in } S \quad \text{and} \quad (x_1', 0) \text{ in } S'$$

for the second shot. The given data are $x_1 = 9$ m and $t_1 = 10^{-8}$ s. Note that both events occur at $t' = 0$ in S'. Equations (1-16) may then be consulted to interpret the simultaneity hypothesis as

$$0 = \gamma\left(t_1 - \frac{u}{c^2} x_1 \right).$$

Therefore, S' must be moving through S with speed u given by

$$\frac{u}{c} = \frac{ct_1}{x_1} = \frac{(3 \times 10^8 \text{ m/s})(10^{-8} \text{ s})}{9 \text{ m}} = \frac{1}{3},$$

and so the game warden's speed in the hunter's frame is $c/3 = 10^8$ m/s. Note that the warden can observe the firing of the second shot *before* the first by traveling at a speed greater than $c/3$.

1-6 Transformation of Velocity

The motion of a particle is described by the time dependence of the particle's position vector $\mathbf{r}(t)$. The velocity vector is then obtained by differentiating:

$$\mathbf{v}(t) = \frac{d\mathbf{r}}{dt}.$$

These familiar statements pertain to the description of motion in a particular

Figure 1-15

Particles in motion with velocity **v**. The velocities refer to the frame S, while S' moves through S with speed u.

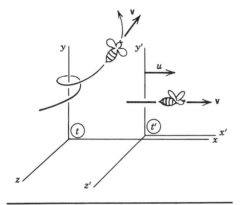

reference frame S, where the coordinates of the moving particle are (x, y, z, t). In another Lorentz frame S' in motion relative to S, the coordinates of the particle transform into (x', y', z', t'), and so the description in that frame is in terms of the transformed vectors $\mathbf{r}'(t')$ and $\mathbf{v}'(t')$. The interesting feature is the transformation of the time as we pass to the new frame. We have to take this into account when we determine how **v** transforms into **v'**.

The motion of the particle is taken to be arbitrary in the following derivations. Figure 1-15 shows some sample cases along with two frames S and S' related to each other by the Lorentz transformation developed in Section 1-5. Let us consider first the case of straight-line motion in the x direction, where the particle velocity is denoted by **v** in the frame S, as shown in the figure. In S', we have

$$v' = \frac{dx'}{dt'},$$

the limit of $\Delta x'/\Delta t'$ for small space and time intervals. Let these intervals refer to the space and time coordinates of events 1 and 2 defined by two successive locations of the particle along its path. Then, from Equations (1-16), we obtain

$$\frac{\Delta x'}{\Delta t'} = \frac{x'_2 - x'_1}{t'_2 - t'_1} = \frac{\gamma(x_2 - ut_2) - \gamma(x_1 - ut_1)}{\gamma\left(t_2 - \frac{u}{c^2}x_2\right) - \gamma\left(t_1 - \frac{u}{c^2}x_1\right)}$$

$$= \frac{\Delta x - u\,\Delta t}{\Delta t - \frac{u}{c^2}\Delta x} = \frac{\dfrac{\Delta x}{\Delta t} - u}{1 - \dfrac{u}{c^2}\dfrac{\Delta x}{\Delta t}}.$$

The result follows when the limit of vanishing time intervals is taken:

$$v' = \frac{v - u}{1 - \dfrac{uv}{c^2}}. \tag{1-21}$$

One-dimensional velocities transform their speeds from v to v' according to this rather complicated formula.

Two elementary checks may be made on the result. Suppose that the particle has constant speed and that S' is moving along with the particle; we then have $u = v$ so that $v' = 0$, as expected. Suppose instead that the particle is at rest in S; in this case $v = 0$ and so $v' = -u$, also as expected. A more critical check on Equation (1-21) is made by applying the formula to the speed of light. If we set $v = c$ we find

$$c' = \frac{c - u}{1 - uc/c^2} = c.$$

Thus, the light speed transforms into itself, in keeping with the property of universality.

Figure 1-15 also shows the more general case in which $\mathbf{v}$, the particle's velocity vector in S, has three nonzero components. The transformed components of the velocity $\mathbf{v}'$ in S' are

$$v_x' = \frac{v_x - u}{1 - \dfrac{uv_x}{c^2}},$$

$$v_y' = \frac{v_y}{\gamma\left(1 - \dfrac{uv_x}{c^2}\right)},$$

$$v_z' = \frac{v_z}{\gamma\left(1 - \dfrac{uv_x}{c^2}\right)}. \tag{1-22}$$

We can infer the first of these formulas directly from the derivation of Equation (1-21). The second and third involve identical manipulations, so let us consider only v_y':

$$\frac{\Delta y'}{\Delta t'} = \frac{y_2' - y_1'}{t_2' - t_1'} = \frac{y_2 - y_1}{\gamma\left(t_2 - \dfrac{u}{c^2}x_2\right) - \gamma\left(t_1 - \dfrac{u}{c^2}x_1\right)}$$

$$= \frac{\Delta y}{\gamma\left(\Delta t - \dfrac{u}{c^2}\Delta x\right)} = \frac{\dfrac{\Delta y}{\Delta t}}{\gamma\left(1 - \dfrac{u}{c^2}\dfrac{\Delta x}{\Delta t}\right)}.$$

The result given in Equations (1-22) is obtained when the limit defining dy'/dt' is taken. Note that v_y' and v_y are *not* the same, even though $y' = y$, because of the invalidity of absolute time.

The inverse transformation of velocity proceeds as in the treatment of coordinates, by exchanging primed for unprimed components and by making the replacement $u \rightarrow -u$. When we perform these substitutions, we get

$$v_x = \frac{v'_x + u}{1 + \dfrac{uv'_x}{c^2}},$$

$$v_y = \frac{v'_y}{\gamma\left(1 + \dfrac{uv'_x}{c^2}\right)},$$

$$v_z = \frac{v'_z}{\gamma\left(1 + \dfrac{uv'_x}{c^2}\right)}. \tag{1-23}$$

The formulas given in Equations (1-22) and (1-23) constitute the correct relativistic rules for adding velocities. These results take the place of Equation (1-11) and of course reduce to that equation in the nonrelativistic limit, when both u and v are very small compared to c.

The aberration of starlight can be discussed in these terms. We know that if we observe a "fixed" star with a terrestrial telescope we have to tip the orientation of the lenses because of the motion of the Earth relative to the star. The situation is illustrated in Figure 1-16, in which S is the star frame and S' is the Earth frame. A light ray emitted from the star at an angle θ in S is received in S' at a greater angle θ', according to the following analysis. The components of the velocity of light in S are

$$v_x = -c \sin \theta \quad \text{and} \quad v_y = -c \cos \theta.$$

The corresponding components in S' are obtained by means of Equations (1-22):

$$v'_x = -\frac{c \sin \theta + u}{1 + \dfrac{u}{c}\sin \theta} \quad \text{and} \quad v'_y = -\frac{c \cos \theta}{\gamma\left(1 + \dfrac{u}{c}\sin \theta\right)}.$$

With a little algebra we can verify that $v'^2_x + v'^2_y = c^2$. The desired direction angle θ' is given by computing the quantity

$$\tan \theta' = \frac{v'_x}{v'_y} = \gamma \frac{c \sin \theta + u}{c \cos \theta}$$

or

$$\tan \theta' = \frac{1}{\sqrt{1 - u^2/c^2}}\left(\tan \theta + \frac{u}{c \cos \theta}\right). \tag{1-24}$$

It is clear from this expression that θ' is larger than θ. Therefore, we see that we must

Figure 1-16

Abberation of light from a fixed star observed
with a moving telescope. A light ray from the
star makes an angle θ in the star frame and
is received in the Earth frame at a larger
angle θ'. The classical method of adding
velocity vectors is shown below, in the frame
of the telescope. The relativistic prediction
differs from the classical result by the factor γ.

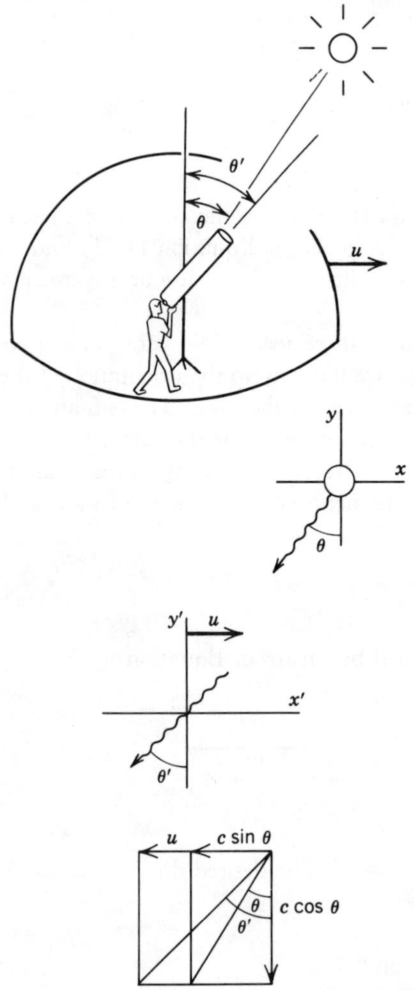

tip the telescope forward relative to the direction defined by θ, as indicated in the
figure. The classical treatment of the aberration of starlight employs nonrelativistic
velocity addition. This analysis is also sketched in the figure, in an inset taken from the
telescope's point of view. We note that the same answer results except for the presence
of the γ factor in Equation (1-24). Consequently, there is no difference between the
classical and relativistic predictions to order u/c.

Example

Relativistic velocity addition is nicely illustrated by considering the passing of two rocket ships, as shown in Figure 1-17. Let the rockets have the same 20 m proper length, and assume that in the Earth frame S the speed of each rocket is $u = \frac{4}{5}c$. The problem is to find the speed of rocket 2 relative to rocket 1 and then to find the time elapsed on rocket 1's clock while rocket 2 passes. The figure shows the motion of the rockets in two Lorentz frames; on the left the passing is seen in the Earth frame S, and on the right the passing is seen in S', the rest frame of rocket 1. The speed u' of rocket 2 in S' is obtained by inserting $v = -u$ and $v' = -u'$ in Equation (1-21):

$$u' = \frac{2u}{1 + u^2/c^2} = \frac{2\left(\frac{4}{5}c\right)}{1 + 16/25} = \frac{40}{41}c.$$

The contraction factor associated with this relative velocity is $\sqrt{1 - u'^2/c^2} = \frac{9}{41}$. Let the length of rocket 1 in S' be denoted by ℓ_0'; this specifies the rocket's proper length so $\ell_0' = 20$ m. The length ℓ' of rocket 2 in the frame S' is given by length contraction:

$$\ell' = \ell_0'\sqrt{1 - u'^2/c^2} = (20 \text{ m})\left(\frac{9}{41}\right) = \frac{180}{41} \text{ m}.$$

Figure 1-17

Two rocket ships passing each other. The proper length of each rocket is ℓ_0'. In S, the Earth frame, both rockets have speed u and contracted length ℓ. Passing takes place in S between times $t = 0$ and $t = t_1 = \ell/u$. In S', the rest frame of rocket 1, oncoming rocket 2 has speed $u' = 2u/(1 + u^2/c^2)$ and contracted length ℓ'. Passing in S' takes place from $t' = 0$ to $t' = t_1' = (\ell_0' + \ell')/u'$.

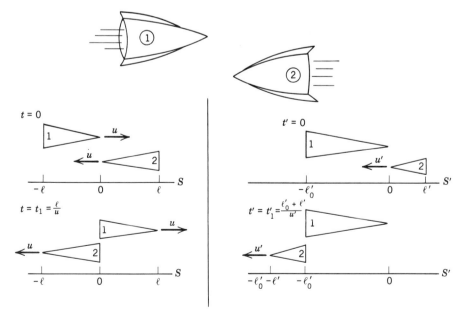

The passing time in S' is noted in the figure:

$$t_1' = \frac{\ell_0' + \ell'}{u'} = \frac{20 + \frac{180}{41}}{\frac{40}{41}} \ \mathrm{m}/c = 25 \ \mathrm{m}/c,$$

using meter/c as a convenient unit of time. These results complete the problem. Let us confirm the answer for t_1' by analyzing the motion in S, as shown on the left side of the figure. In this frame each rocket has the contracted length

$$\ell = \ell_0' \sqrt{1 - u^2/c^2} = (20 \ \mathrm{m})\left(\tfrac{3}{5}\right) = 12 \ \mathrm{m},$$

and so the passing time in S is

$$t_1 = \frac{\ell}{u} = \frac{12 \ \mathrm{m}}{\frac{4}{5}c} = 15 \ \mathrm{m}/c,$$

as indicated in the figure. This represents a proper time interval because it gives the time elapsed between two events at the same point $x = 0$ in S. Therefore, the corresponding time interval in S' is dilated relative to t_1:

$$t_1' = \frac{t_1}{\sqrt{1 - u^2/c^2}} = (15 \ \mathrm{m}/c)\left(\tfrac{5}{3}\right) = 25 \ \mathrm{m}/c,$$

in confirmation of our previous result.

1-7   Space–Time

Relativity dissolves the boundaries of definition between space and time, and mixes these variables together in the passage between Lorentz frames. We see this most clearly in the Lorentz transformation, where one observer's time for some event is related to another observer's time *and* position for the same event. We have already adopted a notion of space–time as the setting in which events happen. Now we want to extend the idea so that we can visualize several Lorentz frames at once, and let the space and time coordinates mix accordingly.

Let us first introduce some new notation by setting

$$\beta = \frac{u}{c} \quad \text{and} \quad \gamma = \frac{1}{\sqrt{1 - \beta^2}}. \tag{1-25}$$

As usual, u refers to the relative speed between two frames S and S', as in Figures 1-13 and 1-14. We then observe that the equations of the Lorentz transformation become more symmetrical when we take ct instead of t for the time variable. Equations (1-16) may be rewritten to incorporate these features:

$$
\begin{aligned}
x' &= \gamma(x - \beta ct), \\
y' &= y, \\
z' &= z, \\
ct' &= \gamma(ct - \beta x).
\end{aligned}
\tag{1-26}
$$

The resulting set of formulas employs coordinates (x, y, z, ct) that have the same dimensions of length and display an obvious interchangeability in the variables x and ct.

The appealing symmetry of the equations takes on considerably more substance when we recognize the important property of *Lorentz invariance*. For an event at (x, y, z, ct), let a "distance" s be introduced by defining the quantity

$$s^2 = x^2 + y^2 + z^2 - c^2 t^2. \tag{1-27}$$

A primed version of this is also defined by the coordinates (x', y', z', ct') for the same event in another frame. We can use Equations (1-26) to relate s' and s:

$$
\begin{aligned}
s'^2 &= x'^2 + y'^2 + z'^2 - c^2 t'^2 = \gamma^2 (x - \beta ct)^2 + y^2 + z^2 - \gamma^2 (ct - \beta x)^2 \\
&= \gamma^2 (1 - \beta^2) x^2 + y^2 + z^2 - \gamma^2 (1 - \beta^2) c^2 t^2 \\
&= x^2 + y^2 + z^2 - c^2 t^2 \\
&= s^2.
\end{aligned}
$$

The remarkable result is that s is not changed in the transformation to the new Lorentz frame. Therefore, s has the same value in all Lorentz frames and is said to be a *Lorentz-invariant quantity*.

The conclusion that $s'^2 = s^2$ is a statement of Lorentz invariance. Equation (1-27) reminds us of the expression for the square of the distance r between the origin and a spatial point (x, y, z),

$$r^2 = x^2 + y^2 + z^2.$$

This distance does not change when we perform rotations of the three-dimensional coordinate system. Similarly, we can regard s as a "distance" in four-dimensional space–time between the origin and an event (x, y, z, ct). Lorentz invariance then says that this "distance" remains unchanged when we carry out Lorentz transformations on the space–time frame. The negative sign is crucial for the time contribution in Equation (1-27) since Lorentz invariance would not follow without it. In a given frame space and time have distinct meanings, and the minus sign for the time highlights that distinction. The sign tells us that Euclidean geometry does not hold in our four-dimensional space–time. The interpretation of s as a distance must therefore be made advisedly since s^2 is not a positive quantity in all cases. We may extend the meaning of Equation (1-27) by considering two events and defining the *space–time interval* Δs between the corresponding points in space–time:

$$(\Delta s)^2 = (\Delta x)^2 + (\Delta y)^2 + (\Delta z)^2 - c^2 (\Delta t)^2. \tag{1-28}$$

This expression is Lorentz invariant and may be either positive or negative. Space–time can present a complicated picture owing to the mixing of coordinates in the passage from frame to frame. It is very useful to know that such features of the general picture as the space–time interval remain unchanged during the mixing process.

A full visualization of space–time requires a four-dimensional picture. Let us forego this complication by concentrating on problems for which there is only one spatial dimension. Consider the straight-line motion of a particle whose path, x versus t, lies in the plane of the variables (x, ct), as in Figure 1-18. The locus of points on the

Figure 1-18
World-line for a particle in straight-line motion. The motion starts from $x = 0$ when $t = 0$ and has constant velocity **v** until $t = t_1$. Three possibilities are shown thereafter, corresponding to different accelerations of the particle. The dashed line at $45°$ refers to motion at light speed from the same starting point.

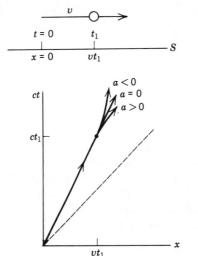

path is called the *world-line* of the particle. The figure shows the motion beginning at $x = 0$ when $t = 0$ and proceeding at constant speed v until $t = t_1$. Thereafter, the world-line may continue as a straight line, if the particle does not accelerate, or may curve downward or upward if the particle speeds up ($a > 0$) or slows down ($a < 0$). Note that the slope at any point along the world-line is

$$\frac{dct}{dx} = \frac{c}{dx/dt},$$

where dx/dt is the instantaneous velocity of the particle. This slope must always exceed unity for a moving particle and may approach unity from above as the particle motion approaches light speed.

We can use such diagrams in the (x, ct) plane to visualize various kinds of events in their space–time setting. It is especially useful to see how the interpretation of the diagrams can be extended by defining different pairs of axes corresponding to different Lorentz frames. We let the axes denoted by (x, ct) pertain to the particular choice of frame S, and we consider the transformation to another frame S' by means of Equations (1-26) so that we can introduce new axes in space–time labeled by (x', ct'). The orientation of the new primed axes in the plane of x and ct is established as follows. Equations (1-26) tell us that along the ct' axis we have

$$x' = 0 \quad \text{so that} \quad ct = \frac{x}{\beta},$$

and along the x' axis we have

$$ct' = 0 \quad \text{so that} \quad ct = \beta x.$$

The lines $ct = x/\beta$ and $ct = \beta x$ are drawn in Figure 1-19 and labeled as ct' and x' axes, respectively. Note that these axes are *not* perpendicular but skewed. Each axis is rotated inward toward the 45° line by the same angle α, where $\tan \alpha = u/c = \beta$. The axes approach the 45° line from either side in the limit $u \to c$ and $\beta \to 1$, as the speed of the frame S' approaches light speed in its uniform motion through S. A given event at (x, ct) in S and at (x', ct') in S' can be plotted with respect to either set of axes. Because the primed axes are skewed, the projection of primed coordinates parallel to these axes differs from the usual method of dropping perpendiculars onto the axes. We should not be surprised to encounter such non-Cartesian properties since we have already noted the non-Euclidean nature of space–time. The space–time interval between the origin and the event indicated in the figure is given by

$$s^2 = x^2 - c^2 t^2 = x'^2 - c^2 t'^2.$$

We may also consider a third frame S'', which, like S', moves through S at the same speed u but does so in the opposite direction. The coordinates (x'', ct'') in S'' are found by formulas identical to Equations (1-26) except for the replacement $u \to -u$.

Figure 1-19

Space–time diagram showing three pairs of axes for three different Lorentz frames. The transformed axes are skewed, inward for S' moving through S along the positive x axis, and outward for S'' moving through S along the negative x axis. In the space–time diagram, the x' axis is along the line $ct = \beta x$, and the ct' axis is along the line $ct = x/\beta$. Slopes are determined by the angle α, where $\tan \alpha = \beta$.

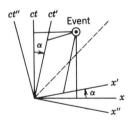

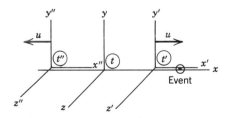

Figure 1-20

Events in a space–time diagram with axes for two Lorentz frames S and S'. Events 1 and 0 occur at the same spatial location in S'. Events 2 and 0 are separated by the propagation of a light signal. Events 3 and 0 are simultaneous in S'.

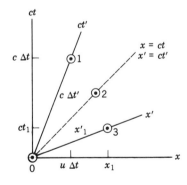

Accordingly, the axes in the space–time diagram are rotated by the same angle α in the outward direction, as shown in the figure.

Example

Let us apply these space–time perspectives to look at several kinds of events. Event 0 is introduced for reference at the space–time origin, and other events are located as in Figure 1-20. Events 1 and 0 occur at the same point in S', the spatial origin at $x' = 0$. The time $\Delta t'$ between these events in S' is therefore a proper time interval. In S, the corresponding time interval is Δt, as indicated in the figure. We can use the invariance of the space–time interval to write

$$(u\,\Delta t)^2 - (c\,\Delta t)^2 = -(c\,\Delta t')^2$$

and solve for Δt to recover Equation (1-3), the time-dilation formula. Note how the non-Euclidean geometry may tend to deceive us here; despite appearances in the figure, Δt is *larger* than $\Delta t'$. Events 2 and 0 are separated by the propagation of light. Note that event 2 satisfies $x = ct$ and $x' = ct'$, and so a light flash emitted at 0 can propagate outward and be detected later at a location on the x and x' axes as event 2. Events 3 and 0 are observed in S' to be simultaneous since both occur at $t' = 0$. Note that event 3 follows event 0 in S by the indicated time interval t_1. The relation between these events is the same as that in the example of the hunters and the game warden in Section 1-5.

Example

As a finale, let us return to the twin paradox and reconsider the motion of the space twins Chester and Esther at the end of Section 1-4. Recall that Chester has gone on a rocket trip away from Esther at speed $u = \frac{3}{5}c$ and, after 5 years on Esther's clock, abruptly turns around and comes back at the same speed. We know that the problem involves three Lorentz frames, Esther's frame S and Chester's frames S' on the way out and S'' on the way back. Figure 1-21 shows the axes for these three frames in a space–time diagram, where S is chosen as the frame at rest. We use years and light-years for the unwritten units of time and distance in the diagram and in the following discussion. (Recall that a light-year is the distance traveled by light in 1 year.) When $t = t' = t'' = 0$, the frames S and S' coincide in space, and the spatial origin in S'' is located at $x = 6$ in S. The moving frames approach each other so that S' and S'' coincide when their origins reach $x = 3$ at $t = 5$ in S. This instant corresponds to Chester's turning point, when the clocks in S' and S'' read $t' = t'' = 4$. World-lines for Chester and Esther are also shown in the figure. Esther remains at $x = 0$ for 10 years in S, while Chester stays at $x' = 0$ for 4 years out in S' and then changes his address to $x'' = 0$ for 4 years back in S''. Chester's clock reads $t'' = 8$ upon reunion with Esther. We have reached this conclusion from Esther's point of view by staying in the single Lorentz frame S and observing that Chester's time runs slower. Thus, we have recognized that her time is dilated by the factor $\frac{5}{4}$ relative to the proper time in his frame, whether in S' going out or in S'' coming back. The puzzling aspect of the twin paradox arises when we adopt Chester's point of view instead and try to confirm Esther's advanced time. Relative motion symmetry implies that Chester's time in S' is dilated relative to Esther's proper time by the same factor $\frac{5}{4}$. Hence, Chester observes that his outward transit time $\Delta t' = 4$ corresponds to only $\Delta t = 3.2$ on Esther's clock. Exactly the same relation holds between $\Delta t''$ and Δt for Chester's return. It would appear that Esther's time is coming up short in Chester's calculation. In fact, Chester knows that the passage of Esther's time gives $t = 10$, and not $t = 6.4$ (twice 3.2), because he sees that a certain aspect of the problem has been left out of the bookkeeping. We can see what the missing feature is by looking at the space–time diagram, especially at the point on

---

Figure 1-21

Lorentz frames and space–time diagram for the twin paradox. The axes are (x, ct) in S, (x', ct') in S', and (x'', ct'') in S''. Coordinates on the axes are given in light-years. Esther stays at $x = 0$ in S, while Chester goes off at $x' = 0$ in S' and comes back at $x'' = 0$ in S''. Upon reunion, Esther's clock reads $t = 10$, while Chester's inbound clock reads $t'' = 8$. Chester's time intervals, $\Delta t' = 4$ on the way out and $\Delta t'' = 4$ on the way back, are dilations of proper intervals in Esther's time of duration $\Delta t = 3.2$ at each end of the journey. The changeover from S' to S'' accounts for the gap $\Delta t = 3.6$ in Chester's view of Esther's time.

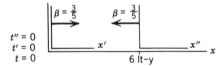

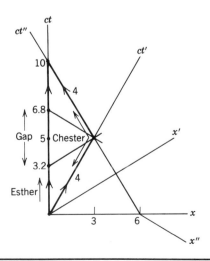

Chester's world-line where he reverses his rocket and changes over from S' to S''. Two lines are drawn through the space–time location of this event, one parallel to the x' axis and another parallel to the x'' axis. Along these two lines events occur in S' and in S'' that are simultaneous with Chester's turn-around event. On Esther's clock, the times for these events occur 3.2 units from the beginning and 3.2 units from the end of the trip, as indicated in the figure. Thus, the diagram reveals a *simultaneity gap* of duration $\Delta t = 3.6$, which Chester must include in his bookkeeping to account for all of Esther's time. Of course, the calculation is much easier in S where Chester's turning point is simultaneous with the midpoint of Esther's world-line. We have appealed to the relativity of simultaneity to resolve the paradox, and we have found the space–time diagram to be an indispensable aid in visualizing the argument.

1-8 Relativistic Momentum and Energy

The properties of space–time follow directly from Einstein's second postulate of relativity regarding the universality of the speed of light. Little need has arisen yet to invoke the other postulate, which stipulates that the physical laws maintain their form from one Lorentz frame to another. We now want to use this first postulate more prominently as a constructive influence in the development of the principles of relativistic dynamics.

We are concerned with the motion of a particle of mass m and velocity $\mathbf{v}$ in a specific frame of reference. Newtonian dynamics is valid as long as the velocity remains much smaller than the speed of light. Then we know that we can define momentum and kinetic energy for the particle as

$$\mathbf{p} = m\mathbf{v} \quad \text{and} \quad K = \tfrac{1}{2}mv^2 \quad \text{(nonrelativistic)}. \tag{1-29}$$

These quantities appear in conservation laws determined by the dynamics. Examples are the conservation of the total momentum in a free system of particles and the conservation of the total kinetic energy in an elastic collision between particles. Our objective is to extend the principles of dynamics by removing the restriction to small values of v/c and generalizing the quantities given in Equations (1-29). The newly formulated dynamical laws must continue to provide for the conservation of momentum and energy. In fact, the first postulate of relativity requires that if the conservation laws hold in one Lorentz frame they must also hold in any other Lorentz frame as well. The problem is to deduce the proper relativistic expressions for momentum and energy so that the corresponding conserved quantities are respected by the Lorentz transformation. We know that these expressions must also reduce in the nonrelativistic limit to the familiar forms in Equations (1-29).

The most direct way of presenting this problem is simply to divulge the answer immediately and then demonstrate the properties of the solution. In this spirit we assert without further ado that the relativistic momentum $\mathbf{p}$ and the relativistic energy E are given for a particle of mass m by the formulas

$$\mathbf{p} = \frac{m\mathbf{v}}{\sqrt{1 - v^2/c^2}} \tag{1-30}$$

and

$$E = \frac{mc^2}{\sqrt{1 - v^2/c^2}}.$$ (1-31)

The momentum obviously goes over into its nonrelativistic form for $v \ll c$. The energy E is a new kinematical quantity whose meaning is to be examined in due course.

The underlying strategy should be spelled out clearly so that these assertions can be grasped. We expect observations of momentum and energy to yield altered values when the observations are made in a different Lorentz frame. Our strategy focuses on the proper behavior of these kinematical quantities under transformation to any such new frame. We know that Equations (1-26) describe the passage from one space–time frame S, where the coordinates are (x, y, z, ct), to another space–time frame S', where the coordinates are (x', y', z', ct'). Let us set up a correspondence of properties between the momentum and energy variables and the space and time coordinates by first assembling the kinematical quantities into the fourfold set $(p_x, p_y, p_z, E/c)$. (We use E/c instead of E for the fourth variable so that all four quantities in the set have the same units of momentum. This convention parallels our use of ct instead of t in Equations (1-26).) We then assume that these four variables in the frame S *transform* into a new set $(p_x', p_y', p_z', E'/c)$ when we pass to the new frame S', and *require* that the transformation rules have *exactly the same form* as those used to relate (x, y, z, ct) and (x', y', z', ct'). The proposed relations are just like Equations (1-26):

$$p_x' = \gamma\left(p_x - \beta\frac{E}{c} \right),$$

$$p_y' = p_y,$$

$$p_z' = p_z,$$

$$\frac{E'}{c} = \gamma\left(\frac{E}{c} - \beta p_x \right),$$ (1-32)

where β and γ are again given by Equations (1-25). These relations specify the behavior of relativistic momentum and energy under Lorentz transformation. Note in particular that p_x and E are mixed in the passage to the new frame just as x and t are.

The argument behind Equations (1-32) should be restated for emphasis. We want to guarantee the transfer of conserved quantities from one Lorentz frame to another and ensure specifically that, if $\mathbf{p}$ and E obey conservation laws in S, then $\mathbf{p}'$ and E' automatically obey the same conservation laws in S'. We accomplish this by arguing that $\mathbf{p}$ and E should transform into $\mathbf{p}'$ and E' according to the *same* rules that govern the transformation of coordinates from S to S'. Thus, the strategy starts with the formulas for momentum and energy in Equations (1-30) and (1-31) and requires that these quantities transform like space and time. Our remaining task is to demonstrate that the strategy works.

The demonstration employs Equations (1-22) for the transformation of velocity along with the equations given above. Let us begin by recognizing the resemblance between the Lorentz transformation parameter γ and the v-dependent root factor in the expressions for **p** and E. We define

$$\gamma_v = \frac{1}{\sqrt{1 - v^2/c^2}} \qquad (1\text{-}33)$$

as a shorthand device and note that γ depends on the relative speed u between frames S and S' while γ_v depends on the velocity **v** of the particle in the frame S. Thus, γ is a constant parameter since u is constant, while γ_v is a dynamical variable that varies with time in situations where the speed of the particle is changing. Equations (1-30) and (1-31) take the form

$$\mathbf{p} = \gamma_v m \mathbf{v} \quad \text{and} \quad E = \gamma_v mc^2$$

when γ_v is introduced. In S', the particle velocity is **v'**, and so a corresponding variable $\gamma_{v'}$ is defined by analogy with Equation (1-33). The components of **v** and **v'** are related in Equations (1-22). We can use these formulas to establish a remarkable algebraic relation among the γ's:

$$\gamma_{v'} = \gamma\gamma_v \left(1 - \frac{uv_x}{c^2} \right). \qquad (1\text{-}34)$$

This result contains all the complicated algebra needed to demonstrate the transformation properties of **p** and E. We confirm the first of Equations (1-32) by making the following calculation:

$$\gamma\left(p_x - \beta\frac{E}{c} \right) = \gamma\gamma_v m(v_x - u) = \gamma_v m \frac{v_x - u}{1 - uv_x/c^2} = \gamma_{v'} m v_x' = p_x'.$$

We have used, in turn, Equations (1-30) and (1-31), (1-33), (1-34), (1-22), and finally (1-30) again in its primed form. The last of Equations (1-32) follows in similar fashion:

$$\gamma(E - \beta c p_x) = \gamma\gamma_v m(c^2 - uv_x) = \gamma_{v'} mc^2 = E'.$$

Here, we use the same series of formulas concluding with the primed version of Equation (1-31). These two derivations along with analogous ones for p_y' and p_z' furnish the desired proof that **p** and E in Equations (1-30) and (1-31) indeed transform under Lorentz transformation according to Equations (1-32).

Our discussion has focused on transformation properties, and so the presentation may seem somewhat abstract. We should realize that relativity is a treatment of frames of reference where rules for the transformation between frames are essential considerations. We have acknowledged this by introducing relativistic momentum and energy in context with the Lorentz transformation. Now that these abstract conceptual features have been presented, we can proceed to the physical interpretation of the new dynamical quantities.

It is obvious that **p** and E should depend on the velocity of the particle. We can eliminate the explicit dependence on **v** and relate **p** and E directly to each other by squaring and comparing expressions. The result is a very important equality known as the relativistic relation between energy and momentum:

$$E^2 = c^2 p^2 + m^2 c^4. \qquad (1\text{-}35)$$

We have yet to identify the meaning of the relativistic energy E. Let us note first that E has a definite nonvanishing value, even when the particle is at rest. If we set $p = 0$ in Equation (1-35), we obtain

$$E_{\text{rest}} = mc^2, \tag{1-36}$$

the minimum energy possible for a free particle. This quantity is determined by the mass of the particle and is called the *rest energy*. If we let v be nonzero but very small compared to c, we can approximate γ_v by expanding in powers of $(v/c)^2$:

$$\gamma_v = \left(1 - \frac{v^2}{c^2}\right)^{-1/2} = 1 + \frac{v^2}{2c^2} + \cdots .$$

Insertion of this expansion in Equation (1-31) results in an approximation for the relativistic energy:

$$E = mc^2 + \tfrac{1}{2}mv^2 \quad \text{(nonrelativistic)}$$

to order (v^2/c^2). We recognize the two contributions to be the rest energy and the Newtonian kinetic energy. This calculation is useful only for small v/c. The result can be generalized for any value of v by defining the *relativistic kinetic energy* as

$$K = E - mc^2 = mc^2 \left(\frac{1}{\sqrt{1 - v^2/c^2}} - 1 \right). \tag{1-37}$$

A graph of this expression is shown in Figure 1-22. The effect of the speed limit is seen clearly in the definition and in the figure. The kinetic energy of the particle is not a real-valued quantity and is therefore undefined for $v > c$. The nonrelativistic version of K is an adequate approximation to Equation (1-37) for $v \ll c$ but departs from the exact expression and eventually violates the speed limit as v grows without bound. The interpretation of the relativistic energy may be summarized by writing

$$E = mc^2 + K. \tag{1-38}$$

This formula is valid for any $v < c$ and consists of two separate pieces, the one associated with the mass of the particle and the other associated with its motion.

Equations (1-30) and (1-31) suggest the identification of a "relativistic mass" in the combination of factors $\gamma_v m$. This quantity grows as v increases and becomes infinite as $v \to c$. The factor m by itself is sometimes called the "rest mass" to distinguish it from the v-dependent relativistic mass. Figure 1-22 indicates how a practical application may be made of this nomenclature. We know that the kinetic energy of a particle can be increased by applying a force and doing work. The work done on the particle causes a gain in the particle's relativistic mass along with the expected increase in velocity. The figure shows that we cannot accelerate the particle to indefinitely large velocity since there is a speed limit at $v = c$. Close to the limit, the increase in relativistic mass becomes the dominant effect as further increase in velocity becomes more and more restricted. It is obvious from the figure that an infinite amount of work is required to accelerate the particle all the way to light speed. We have already noted that the relativistic mass approaches infinity in this limit, while the rest mass remains fixed throughout the acceleration process. Now that we have acknowledged the

Figure 1-22

Relativistic kinetic energy versus particle velocity. K is not defined beyond the speed limit where $v = c$. The nonrelativistic expression $K_{NR} = \frac{1}{2}mv^2$ violates the speed limit and holds only for $v \ll c$.

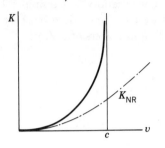

Figure 1-23

Conversion of mass to energy. The nucleus A changes into the less massive nucleus B, and γ rays carry away the energy arising from the difference in mass. The elementary particle decay $\pi^0 \to \gamma + \gamma$ illustrates the complete conversion of mass into γ-ray energy.

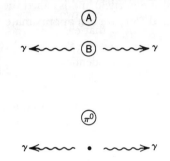

distinctive behavior of "relativistic mass," let us forego the terminology hereafter and always refer implicitly to "rest mass" whenever we speak of *mass* from now on.

The formula $E = mc^2$ appears frequently in popular science writing. In context, the usage refers generally to a nuclear process in which a small portion of the mass of an atomic nucleus is converted into a large amount of energy. Prospects for obtaining energy from mass have their origin in one of Einstein's thought experiments. His hypothetical process involves a mass at rest spontaneously changing into a lighter mass, also at rest, with the simultaneous emission in opposite directions of two high-energy electromagnetic waves, called γ rays. The described process

$$A \to B + \gamma + \gamma$$

is shown schematically in Figure 1-23, where A and B denote two species of atomic nuclei. The sum of the energies carried away by the two γ rays is evidently given by

$$\Delta E = \Delta M c^2,$$

where ΔM is the nuclear mass difference $M_A - M_B$. A more current realization of Einstein's thought experiment is found in the indicated process of π^0 decay,

$$\pi^0 \to \gamma + \gamma,$$

in which one of the elementary particles, the π^0 meson, is observed to change into γ rays and convert *all* its mass into γ-ray energy.

Mass is an intrinsic attribute of a particle. A nonvanishing value for the mass is a requisite property for any classical particle, as entities with no mass have no place as particles in classical physics. Relativity makes room for *massless particles* inasmuch as the expressions for relativistic momentum and energy make sense in the limit of zero mass. Equations (1-35) and (1-36) tell us that

$$E = cp \quad \text{and} \quad E_{\text{rest}} = 0 \qquad (1\text{-}39)$$

in this case. Equations (1-30) and (1-31) make sense as indeterminate forms in the limit $m \to 0$, provided the simultaneous limit $v \to c$ is also taken. Thus, massless particles may exist provided they travel at the speed of light. Such particles cannot be found at rest in any frame and must have speed $v = c$ in *all* Lorentz frames. The limiting behavior of energy and momentum $E \to 0$ and $p \to 0$ may be observed for a massless particle provided the small values occur together in the appropriate frame obeying Equations (1-39).

The first of these two equations may be somewhat familiar. We know from Maxwell's theory that electromagnetic waves carry energy and momentum according to the properties of the Poynting vector. The energy and momentum of the radiation satisfy a relation identical to the equation $E = cp$. The possible massless-particle aspects of the radiation remain to be explored in Chapter 2.

Example

The momentum and energy of a particle are determined by the relativistic formulas, and not by Equations (1-29), whenever the kinetic energy of the particle is comparable with its rest energy. To illustrate, let us consider beams of accelerated electrons and protons and take the approximate values 0.5 MeV and 1 GeV for the respective rest energies. (Recall that the electron volt eV is a unit of energy equal to that acquired by a single electronic charge accelerated in a potential difference of 1 volt.) It is clear that a 1 MeV beam kinetic energy is relativistic for the electron but not for the proton and that a 10 GeV beam is relativistic in both cases. We can assess any of these situations by turning to Equation (1-37) and solving for the ratio v/c in terms of the ratio K/mc^2. The result of this exercise is given in Problem 25 at the end of the chapter:

$$\frac{v}{c} = \frac{\sqrt{\kappa(\kappa + 2)}}{\kappa + 1}, \quad \text{where } \kappa = \frac{K}{mc^2}.$$

We note that the formula becomes

$$\frac{v}{c} = \sqrt{2\kappa} \quad \text{when } \kappa \ll 1,$$

in agreement with the result obtained from the nonrelativistic expression $K = \frac{1}{2}mv^2$. At 1 MeV we have

$$\kappa = 2 \quad \text{and} \quad \frac{v}{c} = \sqrt{\frac{8}{9}} \quad \text{for electrons,}$$

while

$$\kappa = \frac{1}{1000} \quad \text{and} \quad \frac{v}{c} = \sqrt{\frac{1}{500}} = 0.045 \quad \text{for protons,}$$

At 10 GeV the electron case gives

$$\kappa = 10000 \quad \text{and} \quad \frac{v}{c} = \sqrt{\frac{100020000}{100020001}} = 0.999999995,$$

while the proton case gives

$$\kappa = 10 \quad \text{and} \quad \frac{v}{c} = \sqrt{\frac{120}{121}} = 0.996.$$

Since the electron is ultrarelativistic at this energy, it is more practical to compute v/c from an expansion of the formula in powers of $1/\kappa$. The result is also to be found in Problem 25:

$$\frac{v}{c} = 1 - \frac{1}{2\kappa^2} \quad \text{to order} \quad \frac{1}{\kappa^2}.$$

The computation gives

$$1 - \frac{v}{c} = \frac{1}{2\kappa^2} = 0.5 \times 10^{-8} \quad \text{for } \kappa = 10^4,$$

in agreement with the answer given above.

1-9 Relativistic Dynamics

Classical dynamics begins with concepts of force and mass in the context of Newton's laws of motion and then proceeds to the conservation laws of energy and momentum. The empirical approach is taken to establish the equations

$$\mathbf{F} = m\mathbf{a} \quad \text{and} \quad \mathbf{F} = \frac{d\mathbf{p}}{dt} \quad \text{(classical)}$$

as principles that govern the behavior of particles. This scheme is internally consistent and amply confirmed by experiment, as long as the particles have velocity much smaller than c.

Our treatment of relativistic dynamics begins with a different premise based on the conservation of momentum as a first principle. The conservation of relativistic energy follows logically from the same starting point. We have introduced these ideas in Section 1-8 and we are now prepared to take up the concept of force, again pursuing the empirical approach to the dynamical equations. We turn to this question next, after we have made the following observation about the two conservation laws.

Let us consider a process in which an initial system of colliding particles reacts to form a final system of emerging particles. The particles in the final state of the reaction do not have to be the same as those in the initial state. We suppose that an observer in S determines the total momentum and total energy, before and after the reaction, and records the results as $(\mathbf{P}_{before}, E_{before})$ and $(\mathbf{P}_{after}, E_{after})$. Differences in momentum and energy can be defined as

$$\Delta \mathbf{P} = \mathbf{P}_{after} - \mathbf{P}_{before}$$

and

$$\Delta E = E_{after} - E_{before}.$$

The transformation rules in Equations (1-32) can then be used to deduce the corresponding differences determined by an observer in another Lorentz frame S' in motion relative to S. The relations are

$$\Delta P_x' = \gamma\left(\Delta P_x - \beta\frac{\Delta E}{c}\right),$$

$$\Delta P_y' = \Delta P_y,$$

$$\Delta P_z' = \Delta P_z,$$

$$\frac{\Delta E'}{c} = \gamma\left(\frac{\Delta E}{c} - \beta\Delta P_x\right). \tag{1-40}$$

Next, we suppose that the observer in S confirms momentum conservation by finding $\mathbf{P}_{\text{before}}$ to be the same as $\mathbf{P}_{\text{after}}$ so that $\Delta\mathbf{P} = 0$. We also assume that the observer in S' agrees, finding momentum to be conserved in the form $\Delta\mathbf{P}' = 0$. Equations (1-40) then imply that $\Delta E = 0$ in S and that $\Delta E' = 0$ in S'. Thus, both observers agree that the total relativistic energy is conserved because they have already agreed that the total momentum is conserved. This linkage of agreements is secured by the Lorentz transformation. The agreement itself is the first premise of relativistic dynamics, coming before the introduction of forces.

We express the force law for a relativistic particle by writing

$$\mathbf{F} = \frac{d\mathbf{p}}{dt} = \frac{d}{dt}\frac{m\mathbf{v}}{\sqrt{1 - v^2/c^2}}. \tag{1-41}$$

It is understood that we must establish the operational validity of such a law by turning to experiment, just as in the case of Newton's equation $\mathbf{F} = m\mathbf{a}$. The most useful candidate for investigation is the electromagnetic force on a charged particle. The dynamics of a charge e, moving in electric and magnetic fields $\mathbf{E}$ and $\mathbf{B}$, is governed by

$$\mathbf{F} = e(\mathbf{E} + \mathbf{v} \times \mathbf{B}), \tag{1-42}$$

the well-known Lorentz force from classical electrodynamics. It is an empirical fact that Equations (1-41) and (1-42) together provide a description of charged-particle motion in agreement with experiment for *any* velocity $\mathbf{v}$. Equation (1-41) is considerably more complicated than the simple expression $\mathbf{F} = m\mathbf{a}$. Often, it is possible to obviate some of the complications by relying on conservation laws wherever possible.

The concept of work links the force in Equation (1-41) with the relativistic energy. We can examine this connection by constructing the *work–energy theorem*, just as we do in Newtonian mechanics. For simplicity, let us confine our derivation to the case of motion in one dimension. The work done by the force as the particle moves along the x axis between points x_1 and x_2 is expressed in terms of the familiar integral over x. We convert this to an integral over v in the following sequence of steps:

$$W = \int_{x_1}^{x_2} F\,dx = \int \frac{dp}{dt}v\,dt = \int v\,dp = \int_{v_1}^{v_2} v\frac{dp}{dv}\,dv,$$

where v_1 and v_2 are the velocities of the particle at x_1 and x_2. We then use

$$\frac{dp}{dv} = m\frac{d}{dv}(v\gamma_v) = m\left(\gamma_v + v\frac{d\gamma_v}{dv}\right) = m\left(\gamma_v + \frac{v^2}{c^2}\gamma_v^3\right) = m\gamma_v^3,$$

in which we have inserted the relation

$$\frac{d\gamma_v}{dv} = \frac{d}{dv}\frac{1}{\sqrt{1 - v^2/c^2}} = \frac{v/c^2}{\left(1 - v^2/c^2\right)^{3/2}} = \frac{v}{c^2}\gamma_v^3.$$

These maneuvers transform the formula into an integrable expression:

$$W = \int_{v_1}^{v_2} \frac{mv}{\left(1 - v^2/c^2\right)^{3/2}}\,dv = \left.\frac{mc^2}{\sqrt{1 - v^2/c^2}}\right|_{v_1}^{v_2} = E_2 - E_1.$$

We may then use Equation (1-38) to rewrite the result as

$$W = K_2 - K_1. \tag{1-43}$$

Thus, the work done by the force on the particle is equal to the change in the particle's relativistic kinetic energy, a conclusion parallel to that found in the Newtonian version of the theorem.

Next, let us apply the dynamical equations in three-dimensional vector form to demonstrate how the relativistic momentum of a charged particle can be measured. The application is the familiar one in which the moving charge is injected into a uniform magnetic field, as sketched in Figure 1-24. The motion is governed by the force

$$\mathbf{F} = e\mathbf{v} \times \mathbf{B}. \tag{1-44}$$

Figure 1-24
Relativistic charged particle in a uniform magnetic field.

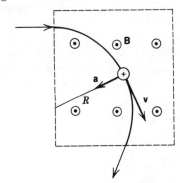

As usual, this force does no work since **F** and **v** are perpendicular vectors:

$$W = \int_1^2 \mathbf{F} \cdot d\mathbf{r} = \int_1^2 \mathbf{F} \cdot \mathbf{v}\, dt = 0.$$

It then follows from Equation (1-43) that the kinetic energy is constant, and so the speed of the particle is also constant. Therefore, γ_v is a constant parameter in Equation (1-41), and the equation of motion takes the simpler form

$$e\mathbf{v} \times \mathbf{B} = \gamma_v m \frac{d\mathbf{v}}{dt}. \qquad (1\text{-}45)$$

This expression tells us that the acceleration $\mathbf{a} = d\mathbf{v}/dt$ is perpendicular to **v** and **B**. If **v** is initially perpendicular to **B** when the particle is injected into the field, the three vectors **a**, **v**, and **B** remain mutually perpendicular throughout the motion. We recognize that a particle orbit with these properties must be planar and circular, as indicated in the figure. The familiar centripetal acceleration results, with magnitude $a = v^2/R$, where R is the radius of the circular orbit. We insert all this information into Equation (1-45) to get

$$evB = \gamma_v m \frac{v^2}{R},$$

and we rearrange to obtain the final formula:

$$eBR = \gamma_v mv = p.$$

This expression for the relativistic momentum is identical in form to the nonrelativistic result. The measurement of p then follows directly from measurements of B and R.

1-10 Collisions and Reactions

Modern physics is concerned with the structure of matter and the interactions of the constituents of matter. These basic properties of nature can be investigated by probing systems of particles in collision processes. Thus, we are able to "see" the structure of the atom and the nucleus by exposing the systems to beams of suitable particles with appropriate incident energies. Studies of matter on an even smaller scale are accomplished in collisions at relativistic momenta. The larger collision energies make possible the disintegration of the nucleus and the production of new elementary particles. These investigations at high energy are analyzed according to the principles of *relativistic kinematics*. The analytical procedures are in essence exercises in the application of momentum conservation and energy conservation. It is important to realize that we do not have to understand the specific nature of the forces between particles in order to implement these conservation laws.

To begin our survey of relativistic kinematics, let us reconsider the conversion between mass and energy and emphasize that mass does not have to be conserved. We are guided by the principle of conservation of the total energy, where we construe the energy to include all possible manifestations. Consider the collision shown in Figure 1-25 in which the two colliding particles are assumed to have the same mass m and speed v. The particles come together from opposite directions and interact in a *completely inelastic* collision to form a single mass M. We refer to the process in these

Figure 1-25

Inelastic collision in which $m + m \rightarrow M$. Conservation of momentum and energy imply that mass is not conserved.

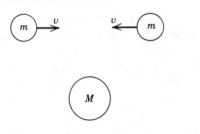

Figure 1-26

Inelastic collision in a conservative classical model. The initial kinetic energy is converted into potential energy stored in the compressed spring.

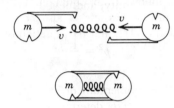

terms since all the kinetic energy of the initial system disappears in the collision. The final mass M is produced at rest because the conserved total momentum is zero. We determine this mass by invoking conservation of the total energy in either of the two forms

$$Mc^2 = 2\frac{mc^2}{\sqrt{1 - v^2/c^2}} \tag{1-46}$$

or

$$Mc^2 = 2mc^2 + 2K,$$

where K is the kinetic energy of each initial particle. It is obvious that the final mass M exceeds the total initial mass $2m$ as the initial kinetic energy is entirely converted into the increase in mass.

Let us imagine a classical model for such an inelastic collision in which the masses stick together and yet only conservative forces are involved in the balancing of all the energy. We may use the massless ratchets-and-spring device in Figure 1-26 for this purpose, and assume that the colliding particles lock onto each other with negligible friction and compress the spring between them. The spring absorbs, and stores as potential energy, the total kinetic energy brought into the collision by the initial particles. Our relativistic treatment simply identifies the final aggregate as a new system having its own mass M, whose rest energy includes the effect of the potential energy modeled by the compressed spring. If we remove the spring from the classical model, we conclude that all the initial kinetic energy is dissipated in friction and shows up as heat in the final system. Again, the relativistic treatment regards this quantity of heat as part of the final rest energy Mc^2. Thus, even thermal energy can be associated with an identifiable mass from the viewpoint of relativity.

Next, let us reverse the completely inelastic process and consider the breakup of M into two unequal masses, as shown in Figure 1-27. This phenomenon is realized physically in the spontaneous fission of a nucleus into two fragments and in the two-body decay of an unstable nucleus or elementary particle. Since M is at rest, momentum conservation demands that m_1 and m_2 have opposite momenta of the same magnitude p, as indicated. Conservation of energy then requires that

$$E_1 + E_2 = Mc^2$$

or

$$\sqrt{c^2p^2 + m_1^2c^4} + \sqrt{c^2p^2 + m_2^2c^4} = Mc^2$$

with the use of Equation (1-35). It is not difficult to solve this relation for p^2, the only unknown quantity, and obtain

$$p^2 = \frac{\left[M^2 - (m_1 + m_2)^2\right]\left[M^2 - (m_1 - m_2)^2\right]}{4M^2}c^2. \qquad (1\text{-}47)$$

Thus, the momentum of each particle in the final state of the two-body breakup process is *uniquely* determined by the values of the participating masses M, m_1, and m_2. Conservation of energy may also be written in terms of the kinetic energies of m_1 and m_2:

$$m_1c^2 + K_1 + m_2c^2 + K_2 = Mc^2.$$

The difference in rest energy between initial and final states is called the *Q-value* for the given process. In this instance, the *Q*-value has the form

$$Q = (M - m_1 - m_2)c^2 = K_1 + K_2. \qquad (1\text{-}48)$$

Note that the kinetic energy in the final state is obtained from the *excess* mass in the initial state.

Let us now introduce Lorentz frames in our discussion of the breakup process. For simplicity, we take the masses of the two final particles to be equal and consider the two-body decay $M \rightarrow m + m$ described in Figure 1-28. The upper part of the figure shows the decay of M *at rest* in the Lorentz frame S'. Note that this view of the process is the reverse of the collision in Figure 1-25, and so Equation (1-46) gives the correct

---

Figure 1-27
Two-body decay $M \rightarrow m_1 + m_2$ with M at rest.

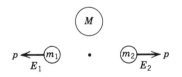

Figure 1-28
Two-body decay $M \rightarrow m + m$ for M at rest and for M in flight. The decay in flight has the simplifying feature that one of the final particles is left at rest. These decays are related by a transformation of Lorentz frames in which S' is the rest frame of M and S is the frame of the decay in flight.

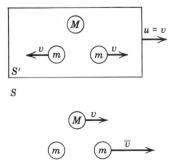

relation between the indicated speed v and the masses of the particles. The Lorentz transformation comes into the picture when we also consider the decay of M *in flight* and introduce another Lorentz frame S in which M is a moving particle. This version of the process is illustrated in the lower part of Figure 1-28 for the special situation where one of the final particles happens to be produced at rest. Our problem is to determine the velocity $\bar{v}$ of the other final particle in the frame S, given the known velocity v in the frame S'. Special circumstances characterize this problem; the initial mass is at rest in S', and one of the final masses has speed v in S' and is at rest in S. These conditions tell us that S' must be moving through S with speed $u = v$ and that M must be moving in S with the same speed, as shown in the figure.

We analyze the problem by using the Lorentz transformation to relate momenta and energies between the frames S' and S. The transformation from S' to S is performed with formulas inverse to those in Equations (1-32):

$$p_x = \gamma\left(p'_x + \beta\frac{E'}{c} \right),$$

$$\frac{E}{c} = \gamma\left(\frac{E'}{c} + \beta p'_x \right). \tag{1-49}$$

Let us apply the first of these relations to the final particle moving to the right in the figure. In S' we have

$$p'_x = \gamma_v mv \quad \text{and} \quad E' = \gamma_v mc^2,$$

and so

$$p_x = \gamma_v\left(\gamma_v mv + \frac{v}{c}\gamma_v mc \right) = 2\gamma_v^2 mv.$$

Note that we have used $u = v$ to replace γ by γ_v and β by v/c. We know, however, that the momentum in S is given in terms of the speed $\bar{v}$ by

$$p_x = \gamma_{\bar{v}} m\bar{v}.$$

These two expressions for p_x imply an equality between the factors $2\gamma_v^2 v$ and $\gamma_{\bar{v}}\bar{v}$. The desired solution for $\bar{v}$ emerges from this equality after a few algebraic steps:

$$\bar{v} = \frac{2v}{1 + v^2/c^2}. \tag{1-50}$$

We should recognize the result. Equation (1-50) expresses the relativistic addition of velocities according to the first formula in Equations (1-23).

Our point in this discussion of Figure 1-28 is to emphasize the fact that the two indicated decays, at rest and in flight, are one and the same phenomenon viewed from two different Lorentz frames. It is evidently sufficient to understand the decay at rest since the decay in flight can then be analyzed by applying the Lorentz transformation.

Many interesting phenomena are observed in relativistic collisions as a consequence of the conversion between mass and energy. The conservation laws allow processes in which particles collide and produce altogether different systems of particles in the final state. These reactions must conserve the total momentum and the total energy, so that the equations

$$\mathbf{P}_{\text{before}} = \mathbf{P}_{\text{after}} \quad \text{and} \quad E_{\text{before}} = E_{\text{after}} \tag{1-51}$$

govern the production of new particles. There also exist other conservation laws that pertain to other attributes of the interacting particles. An assessment of all possible conservation laws in a particular reaction constitutes a certain view of the interactions responsible for the process, where the observations are made at a great distance from the scene of the reaction. The subscripts *before* and *after* in Equations (1-51) are meant to convey this sense of the observation procedure.

The observations generally involve a two-body collision between an incident beam particle and a stationary target particle. An accelerator is usually the source of the incident beam. The laboratory (*lab*) frame designates the Lorentz frame in which the target particle is at rest. The Lorentz transformation can be used to describe the same process in the center of mass (*CM*) frame, defined as the Lorentz frame in which the colliding particles have *zero* total momentum. We should recognize the similarity between this special frame of reference and the rest frame of the decaying particle in Figure 1-28. It is convenient to formulate the physics of an observed process in the CM frame and then transform to the lab frame for comparison with the measured data. The transformation from CM to lab is therefore an especially important application of the Lorentz transformation.

Beams of particles can also be made to collide with each other and produce new particles in the head-on collision. The colliding beams provide the experimenter with an opportunity to observe reactions directly in the CM frame. The advantages of colliding-beam experiments are illustrated in the example below.

Let us look at the details of the lab ↔ CM transformation and adopt notation according to Figure 1-29. The lab frame is S, where the target particle M is at rest and the beam particle m has speed v. The CM frame is S', where m and M approach each other with speeds v' and u, respectively. We recognize immediately that u must also be the speed of S' moving through S. Our goal is to establish relations among the three given speeds. Let us digress for a moment and recall the following two formulas taken from the Newtonian view of the transformation between S and S':

$$v' = v - u \quad \text{and} \quad mv' = Mu \quad \text{(nonrelativistic)}.$$

Figure 1-29

Two-body collision in the lab frame and in the CM frame.

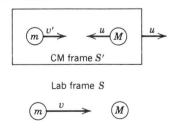

The first equation is a statement of velocity addition in Galilean relativity, and the second is a statement of nonrelativistic momentum conservation in the CM frame. We can eliminate v' between these two equations and obtain the desired relation between u and v:

$$u = \frac{m}{m + M} v \quad \text{(nonrelativistic)}. \tag{1-52}$$

To conclude the digression, let us recall that the center of mass of the Newtonian system is a well-defined point, located between m and M, whose speed remains unchanged before, during, and after the collision. The relativistic transformation between S and S' calls for more complicated equations. We express velocity addition and momentum conservation by writing

$$v' = \frac{v - u}{1 - uv/c^2} \quad \text{and} \quad \frac{mv'}{\sqrt{1 - v'^2/c^2}} = \frac{Mu}{\sqrt{1 - u^2/c^2}}.$$

It is obvious that the elimination of v' is more laborious in this case. The result is left to be derived in Problem 36 at the end of the chapter:

$$u = \frac{mv}{m + M\sqrt{1 - v^2/c^2}}. \tag{1-53}$$

Of course, the final formula goes over into Equation (1-52) for $v \ll c$.

We have only sketched this route to Equation (1-53) because we now want to show the simpler recommended way to obtain the speed u. Let us return to Figure 1-29 and identify the total momentum and total energy for the two-body system in the lab frame S:

$$P_{\text{total}} = \frac{mv}{\sqrt{1 - v^2/c^2}} \quad \text{and} \quad E_{\text{total}} = \frac{mc^2}{\sqrt{1 - v^2/c^2}} + Mc^2. \tag{1-54}$$

The Lorentz transformation generates the corresponding quantities in the CM frame S'. However, the total momentum in S' is known to be *zero* by definition of the CM frame. We use this information as a constraint on the transformed momentum:

$$P'_{\text{total}} = \gamma \left(P_{\text{total}} - \frac{\beta}{c} E_{\text{total}} \right) = 0.$$

A very general expression for the ratio u/c results:

$$\beta = c \frac{P_{\text{total}}}{E_{\text{total}}}. \tag{1-55}$$

The formula takes the form

$$\frac{u}{c} = c \frac{mv/\sqrt{1 - v^2/c^2}}{mc^2/\sqrt{1 - v^2/c^2} + Mc^2}$$

when Equations (1-54) are used, and the desired Equation (1-53) follows directly from this result. We note in passing that the relativistic relation between S and S' does *not* refer to any point called the "center of mass." Instead, S' is characterized by the momentum condition $P'_{total} = 0$. It would be more fitting to call S' the "center of momentum" frame for this reason.

Example

Formulas in relativistic kinematics become simpler when the colliding particles have equal mass. To illustrate, consider proton–proton collisions and let each particle have the same mass M in Equation (1-53):

$$u = \frac{v}{1 + \sqrt{1 - v^2/c^2}}.$$

This relation gives u, the speed of the CM frame through the lab. We can turn the equation inside out and solve for v, the speed of the beam particle in the lab:

$$v = \frac{2u}{1 + u^2/c^2}.$$

The result is easily understood by noting that the two protons must have the same speed given by u in S', the CM frame. The expression for the speed v of the beam proton then follows directly from velocity addition. We have found it instructive to refer to P'_{total} in the preceding paragraph, so let us now evaluate its partner E'_{total}, the total energy in the CM frame. The second of Equations (1-49) gives the relation between the total energies in the two frames:

$$E_{total} = \gamma(E'_{total} + \beta c P'_{total}) = \gamma E'_{total},$$

where

$$E'_{total} = 2\frac{Mc^2}{\sqrt{1 - u^2/c^2}} = 2\gamma Mc^2$$

in this equal-mass situation. We solve for the Lorentz-transformation parameter

$$\gamma = \frac{E'_{total}}{2Mc^2}$$

and combine this equation with our expression for E_{total} to find the interesting result

$$E'^2_{total} = 2Mc^2 E_{total}.$$

Let us apply our formula to an illustration involving colliding proton beams, each with 10 GeV kinetic energy. For convenience, we let the proton be assigned an approximate 1 GeV rest energy. The resulting total CM energy for the two-proton system is

$$E'_{total} = 2(10\text{ GeV} + 1\text{ GeV}) = 22\text{ GeV}.$$

The related value of the total lab energy is quite large because of the quadratic relation between the two quantities:

$$E_{\text{total}} = \frac{E'^2_{\text{total}}}{2Mc^2} = \frac{(22\ \text{GeV})^2}{2\ \text{GeV}} = 242\ \text{GeV}.$$

The beam proton must therefore have a correspondingly large kinetic energy in the lab frame, where the target proton is at rest:

$$K = E_{\text{total}} - 2Mc^2 = 240\ \text{GeV}.$$

This calculation tells us that a 10 GeV-on-10 GeV colliding-proton facility is the equivalent of a 240 GeV proton accelerator with a fixed proton target, in the sense that the total energy in the center of mass system is the same in the two laboratories. A large portion of the incident beam energy on the stationary target is evidently consumed in transporting the "center of mass" of the two-proton system through the collision process. This allocation of energy to center of mass motion makes that amount of energy unavailable for reactions in which new particles can be created. Colliding beams operate ideally in the CM frame where none of the energy is impounded in this way.

1-11 Four-Vectors‡

Relativity invites us to think in terms of four dimensions where the fourth dimension refers to the time. We already know that a geometrical conception of space and time is useful for the presentation of events and world-lines. We now want to incorporate relativistic momentum and energy and expand the presentation into a more fully developed geometrical picture. The utility and elegance of such a four-dimensional description become more apparent as the effects of the Lorentz transformation materialize in the picture. This treatment of relativity by the methods of four-dimensional geometry is the contribution of H. Minkowski. The four dimensions define a geometrical framework for relativity known as *Minkowski space*. We introduce these elegant methods at the end of the chapter so that we can enrich our outlook over the concepts of Einstein's theory.

A particular choice of Lorentz frame S corresponds to a certain set of space and time axes in the four-dimensional space. We specify the space–time coordinates of a given event with respect to these axes and assemble the coordinates in the form (x, y, z, ict). The variables define the *space–time* (or *position*) *four-vector* that locates the event in S with respect to the origin in Minkowski space. Note that the fourth dimension is taken to be an *imaginary*-valued coordinate; this feature of the definition serves to distinguish time from space in the chosen Lorentz frame.

It is convenient to represent the position four-vector for the event in S by introducing the column matrix

$$\mathcal{a} = \begin{bmatrix} x \\ y \\ z \\ ict \end{bmatrix}. \tag{1-56}$$

‡The methods and results of this section are needed again only in isolated parts of Chapters 2 and 16.

A different Lorentz frame S' corresponds to another set of space and time axes in Minkowski space. The given event is located in S' by means of a primed four-vector, with coordinates (x', y', z', ict') and column matrix

$$\mathit{s}' = \begin{bmatrix} x' \\ y' \\ z' \\ ict' \end{bmatrix}.$$ (1-57)

Thus, we use s to specify some event in one frame S, and we use s' to specify the *same* event in another frame S'.

The Lorentz transformation relates the components of s and s' according to Equations (1-26). We can rewrite these relations in more compact notation with the aid of the matrix equality

$$\begin{bmatrix} x' \\ y' \\ z' \\ ict' \end{bmatrix} = \begin{bmatrix} \gamma & 0 & 0 & i\gamma\beta \\ 0 & 1 & 0 & 0 \\ 0 & 0 & 1 & 0 \\ -i\gamma\beta & 0 & 0 & \gamma \end{bmatrix} \begin{bmatrix} x \\ y \\ z \\ ict \end{bmatrix}.$$ (1-58)

(It is important to appreciate the equivalence between Equations (1-26) and Equation (1-58). Similar techniques of matrix multiplication are put to use freely throughout this section.) We take advantage of the simplifying matrix notation to cast Equation (1-58) into its final concise form,

$$\mathit{s}' = \mathscr{L}\mathit{s},$$ (1-59)

in which $\mathscr{L}$ denotes the indicated four-by-four matrix parametrized by the constants β and γ. This square matrix depends only on the speed u of S' moving through S. It therefore contains all the information needed to relate position four-vectors for any given event as observed in the two Lorentz frames S and S'. The highly streamlined formula replaces a cumbersome set of relations and thereby facilitates our interpretation of Minkowski-space geometry.

We are immediately brought back to familiar ground when we evaluate the product of matrices $\mathit{s}^T\mathit{s}$ (where s^T denotes the transpose of the matrix s):

$$\mathit{s}^T\mathit{s} = \begin{bmatrix} x & y & z & ict \end{bmatrix} \begin{bmatrix} x \\ y \\ z \\ ict \end{bmatrix} = x^2 + y^2 + z^2 - c^2 t^2,$$ (1-60)

and, in similar fashion,

$$\mathit{s}'^T\mathit{s}' = x'^2 + y'^2 + z'^2 - c^2 t'^2.$$ (1-61)

These calculations reproduce the squared space–time intervals s^2 and s'^2 separating the origin and the event in the frames S and S'. Recall that s^2 has been defined in Equation (1-27). We have learned that this quantity does not change under Lorentz transformation; therefore, we can reexpress the Lorentz invariance of the space–time interval in terms of Minkowski-space four-vectors by writing

$$\mathit{s}'^T\mathit{s}' = \mathit{s}^T\mathit{s}.$$ (1-62)

The main reason for the imaginary number i in the time coordinate of δ is now apparent. We need the i to secure the vital minus sign in the $c^2 t^2$ contribution to $\delta^T \delta$. This sign then ensures Lorentz invariance in the form expressed by Equation (1-62).

Our strategy may be summarized and evaluated as follows. We have two space–time frames S and S' in Minkowski space, and we are given a specific event for observation. The event is located in S and in S' by the four-vectors δ and δ', respectively. Passage from S to S' is described by a *Lorentz-transformation matrix* $\mathscr{L}$ that transforms δ into δ'. The Lorentz-invariance condition in Equation (1-62) tells us that the "length" of the original four-vector is preserved in this operation. It follows that the Lorentz transformation can be interpreted as a *rotation* of the axes in Minkowski space, since rotation preserves length. The Lorentz transformation of interest refers to motion of S' through S along the x direction, and so the corresponding rotation involves the x and ict axes in the four-dimensional space. This remarkable correspondence between frames in relative motion and rotations in space–time can now be used to good advantage.

The geometrical interpretation of Lorentz invariance translates directly into a characterizing property of the transformation matrix $\mathscr{L}$. Let us return to Equation (1-59) and take the transpose of both sides:

$$\delta'^T = \delta^T \mathscr{L}^T.$$

Equation (1-62) then becomes

$$\delta^T \mathscr{L}^T \mathscr{L} \delta = \delta^T \delta$$

when Equation (1-59) is also used. The product of matrices $\mathscr{L}^T \mathscr{L}$ evidently has the same effect as the four-by-four unit matrix on *any* four-vector δ. The matrix equation

$$\mathscr{L}^T \mathscr{L} = 1 \tag{1-63}$$

follows from this observation. We can verify the equality by explicit matrix multiplication:

$$
\begin{bmatrix}
\gamma & 0 & 0 & -i\gamma\beta \\
0 & 1 & 0 & 0 \\
0 & 0 & 1 & 0 \\
i\gamma\beta & 0 & 0 & \gamma
\end{bmatrix}
\begin{bmatrix}
\gamma & 0 & 0 & i\gamma\beta \\
0 & 1 & 0 & 0 \\
0 & 0 & 1 & 0 \\
-i\gamma\beta & 0 & 0 & \gamma
\end{bmatrix}
$$

$$
=
\begin{bmatrix}
\gamma^2(1 - \beta^2) & 0 & 0 & 0 \\
0 & 1 & 0 & 0 \\
0 & 0 & 1 & 0 \\
0 & 0 & 0 & \gamma^2(1 - \beta^2)
\end{bmatrix}.
$$

Since $\gamma^2(1 - \beta^2) = 1$, the unit matrix is obtained and Equation (1-63) is confirmed.

Minkowski space is constructed on the notion of a position four-vector. Let us broaden the concept and take any collection of four physical quantities to be a four-vector, *provided* those four quantities transform just like the components of the position four-vector in the passage to another Lorentz frame. To be specific, the column matrix of quantities

$$
\ell =
\begin{bmatrix}
b_x \\
b_y \\
b_z \\
ib_t
\end{bmatrix}
$$

is defined to be a four-vector in the frame S if there is a corresponding set of quantities ℓ' in another frame S' that satisfies the transformation rule

$$\ell' = \mathscr{L}\ell, \tag{1-64}$$

where $\mathscr{L}$ is the *same* transformation matrix used in Equation (1-59) to pass from S to S'. Thus, ℓ is a four-vector if ℓ transforms like δ. Not every three-dimensional vector **b** can be promoted to four-vector status. A case in point is the velocity of a particle $\mathbf{v} = d\mathbf{r}/dt$; the transformation properties of **v** in Equations (1-22) do not conform to the necessary rule.

The best example of a four-dimensional entity that transforms like δ is the *momentum–energy* (or *momentum*) *four-vector*. If we take **p** and E from Equations (1-30) and (1-31) and assemble the column matrix

$$\not{p} = \begin{bmatrix} p_x \\ p_y \\ p_z \\ iE/c \end{bmatrix}, \tag{1-65}$$

we then know from our discussion in Section 1-8 that $\not{p}$ meets the necessary transformation criterion. The components of $\not{p}$ give the momentum and energy of a particle as observed in the frame S. These quantities transform to the frame S' according to Equations (1-32). The transformation rule has the form of the matrix equation

$$\not{p}' = \mathscr{L}\not{p}, \tag{1-66}$$

as required for a four-vector. In fact, we have anticipated this desired behavior in our discussion of the properties of relativistic momentum and energy. Note that the imaginary factor i is needed in the fourth component of $\not{p}$ to make the transformation of the momentum four-vector come out correctly.

We can now begin to appreciate the power and elegance of four-vectors. Let us start by showing that the Lorentz-invariance condition in Equation (1-62) holds for $\not{p}$ as it does for δ:

$$\not{p}'^{T}\not{p}' = \not{p}^{T}\not{p}. \tag{1-67}$$

This statement is easily proved by multiplying through Equation (1-63) with $\not{p}^{T}$ on the left and $\not{p}$ on the right, and by using Equation (1-66). The resulting equality becomes

$$p_x'^2 + p_y'^2 + p_z'^2 - \frac{E'^2}{c^2} = p_x^2 + p_y^2 + p_z^2 - \frac{E^2}{c^2},$$

or

$$p'^2 - \frac{E'^2}{c^2} = p^2 - \frac{E^2}{c^2},$$

when the components of the momentum four-vectors are inserted in the calculation. Observe again how the presence of the factor i in the fourth component of $\not{p}$ affects the identification of a Lorentz-invariant property. Equation (1-35) tells us that the invariant quantity $p^T p$ is determined by the mass of the particle:

$$p'^T p' = p^T p = -m^2 c^2. \tag{1-68}$$

Thus, Lorentz invariance holds for masses as it does for space–time intervals, requiring in this instance that all observers agree on the mass of a given particle.

Equation (1-63) can be regarded as the source of all possible Lorentz-invariance properties. If we choose ℓ to represent any four-vector, and we multiply this basic equation fore and aft by ℓ^T and ℓ and use Equation (1-64), we obtain

$$\ell'^T \ell' = \ell^T \ell \tag{1-69}$$

as a more general statement of Lorentz invariance. Let us apply the general expression to the reaction problem described in Figure 1-30. We take $\mathscr{P}$ to be the total momentum four-vector for the system of particles in the lab frame S, and we let $\mathscr{P}'$ be the analogous four-vector in the CM frame S'. The equality of four-vectors

$$\mathscr{P}_{\text{before}} = \mathscr{P}_{\text{after}}$$

expresses conservation of momentum *and* energy in S, while the analogous equality

$$\mathscr{P}'_{\text{before}} = \mathscr{P}'_{\text{after}}$$

makes a similar statement in S'. Of course, we cannot say that $\mathscr{P}'$ and $\mathscr{P}$ are equal. These four-vectors refer to different frames and must be related, component by component, via the Lorentz transformation

$$\mathscr{P}' = \mathscr{L}\mathscr{P}.$$

Figure 1-30

Collision process leading to several final particles $m + M \rightarrow m_1 + m_2 + \cdots + m_n$. The reaction is shown, before and after the collision, in the lab frame S and in the CM frame S'.

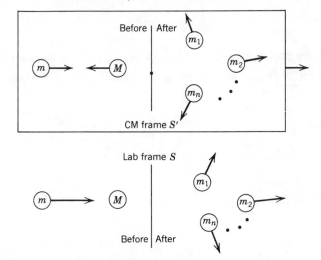

Because of Lorentz invariance, however, the statement

$$\mathscr{P}'^{T}\mathscr{P}' = \mathscr{P}^{T}\mathscr{P} \tag{1-70}$$

can be made as a valid equality. This simple formula opens the way for an endless variety of applications.

Example

Every elementary particle has its own antiparticle, where the two species have the same mass but opposite charge. The proton and antiproton are related to each other in this way. Antiprotons can be produced in reactions initiated by the collisions of a beam of protons incident on protons in a stationary target. The specific process that involves the least number of final particles is

$$p + p \rightarrow \bar{p} + p + p + p.$$

Note that the creation of the antiproton $\bar{p}$ requires the simultaneous production of an extra proton in the final state in order to conserve the total charge and the total number of nuclear particles. This reaction cannot proceed unless the beam proton has enough kinetic energy to produce the excess mass in the final system. Our problem is to determine the minimum beam kinetic energy, or *threshold*, for the process. The situation at threshold is such that, in the CM frame, the four final particles are produced at rest, and so just enough energy is available to account for the four final rest energies. In the lab frame, the four equal-mass particles travel together, in the same direction as the beam, and carry equal shares of the total momentum. If the beam energy is increased above threshold, it becomes possible for the final masses to separate from each other, as in the example shown in Figure 1-30. At threshold, the total momentum four-vector in the CM frame *after* the reaction is

$$\mathscr{P}' = \begin{bmatrix} 0 \\ 0 \\ 0 \\ i4Mc \end{bmatrix},$$

where M is the proton mass. In the lab frame, the total momentum four-vector *before* the collision is

$$\mathscr{P} = \begin{bmatrix} p \\ 0 \\ 0 \\ i(E + Mc^2)/c \end{bmatrix},$$

where p and E pertain to the beam proton. When we use the Lorentz-invariance relation in Equation (1-70), we obtain the equality

$$p^2 - \frac{(E + Mc^2)^2}{c^2} = -16M^2c^2.$$

Since p and E satisfy

$$E^2 = c^2 p^2 + M^2 c^4,$$

the equality becomes

$$2EM + 2M^2 c^2 = 16M^2 c^2 \quad \Rightarrow \quad E = 7Mc^2.$$

We want to know the beam kinetic energy, and so we introduce

$$E = K + Mc^2$$

and get

$$K = 6Mc^2$$

for the final answer. If we call the proton rest energy 1 GeV, we see that the beam kinetic energy for protons in the lab must be at least 6 GeV for the production of antiprotons.

Problems

1. Consider the Michelson–Morley experiment in Figure 1-1 for the realistic situation in which the arms of the interferometer have unequal lengths d_1 and d_2. Assume that the light speeds are aether drifted and obtain a formula for the difference in transit time for the light to travel to and from each mirror. Repeat the derivation with the apparatus rotated by 90°. Obtain an expression for the number of fringes by which the interference pattern shifts owing to this rotation.

2. A runner moves with speed $u = \frac{24}{25}c$ along a 100 m track. Determine the time on the runner's watch to run the course. What is the length of the track in the runner's frame?

3. A meter stick approaches an observer horizontally with speed $\frac{24}{25}c$. What is the length of the stick in the observer's frame? Suppose instead that the meter stick is oriented at an angle $\theta' = \cos^{-1}(\frac{5}{6})$ in the frame moving with the stick. Calculate the length of the stick and its angle of orientation in the frame of the observer.

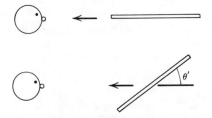

4. Barney wishes to drive his racing car the length of a straight track at high constant speed u. His serviceman Clyde has observed from the pit stop that the car consumes fuel at the rate dn/dt (in droplets of fuel injected in the carburetor per second). Clyde has filled the tank with exactly enough fuel for Barney to drive the course. What rate does Barney measure for his fuel consumption? Does he observe his car to reach exactly the end of the track, to run short, or to have fuel left over at the end?

5. Electrons from the main beam at the Stanford Linear Accelerator Center can reach speeds as large as $0.9999999997c$. Let these electrons enter a detector 1 m long, and calculate the length of the detector in the rest frame of one of the particles.

6. A rocket ship has proper length 30 m and travels at $\frac{12}{13}c$ past an installation equipped to emit and receive radar signals. The emitter sends out a signal at the instant the rear of the rocket goes by, and the signal is reflected from the nose of the rocket back to the receiver. Determine the transit time of the signal out and back, as measured at the radar station.

7. The Doppler shifts for sound and for light may be compared in terms of their power series expansions. For sound, distinguish as two separate formulas the one for source motion alone by taking $u_O = 0$, and the one for observer motion alone by taking $u_S = 0$. Expand the expressions to second order, that is, to order $(u/w)^2$ for sound and to order $(u/c)^2$ for light, to show that the differences between the two cases do not show up until the second order.

8. Unlike the case for sound waves, there is a Doppler shift for light when the source moves at speed u transverse to the orientation of the observer. Prove that the formula for the observed frequency in this situation is

$$\bar{\nu}_{\text{transverse}} = \nu_0\sqrt{1 - u^2/c^2} \ .$$

9. An observer is located between two stationary light sources emitting at the same frequency ν_0. How fast must the observer move toward one of the sources so that the observed frequencies differ by a factor of 2?

10. The radiation received from quasars contains wavelengths characteristic of the common elements, except that the wavelengths are observed to be red shifted. If the shift is entirely attributed to the Doppler effect, the amount of the shift $\Delta\lambda$ relative to the emitted wavelength λ_0 can be used to determine the recessional speed v of the quasar. Derive a formula for the ratio v/c in terms of $\Delta\lambda/\lambda_0$. Since 1986, several quasars have been found with red shifts greater than 4. Calculate the value of v/c corresponding to $\Delta\lambda/\lambda_0 = 4$.

11. The indicated frames S and S' coincide when $t = t' = 0$. In S, event A occurs at the origin when $t = 0$, and event B occurs later at $x_1 = 9$ m when $t = t_1 = 10^{-8}$ s. In S', events A and B are simultaneous, occurring at the origin and at x_1', respectively, when $t' = 0$. Determine the speed of the frame S' moving through the frame S. How far apart are the simultaneous events in S'?

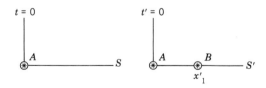

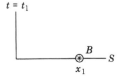

12. Two hunters are located 4 m apart. Each of them fires his shotgun, the one preceding the other by 20 ns. A passing game warden claims that the shots are fired simultaneously. Is this possible? If it is, how fast must the game warden be moving?

13. A laboratory has proper length $L' = 10$ m and moves through S with speed $u = \frac{3}{5}c$. Another Lorentz frame S' moves with the device. Three events are observed as indicated. Event A is the emission of a light signal at $(x = 0, t = 0)$ in S and $(x' = 0, t' = 0)$ in S'. Event B occurs when the signal reaches a mirror at the right end of the laboratory. Event C occurs when the signal returns to a detector at the left end of the laboratory. Determine the space and time coordinates (x, t) in S and (x', t') in S' for events B and C.

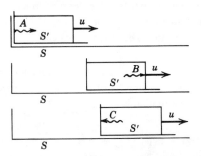

14. The runner in the figure is being filmed by a camera operator who moves with speed $v = \frac{12}{13}c$, parallel to a 100 m track. The figure shows the situation at $t = 0$ in S, the Earth frame, when runner and camera operator are 10 m apart. The runner's watch in S' reads $t' = 0$ at this instant, where S' is the runner's frame. Calculate the relative speed with which the runner overtakes the camera operator. Determine the time on the runner's watch required for overtaking. Does this occur before the runner has completed the course?

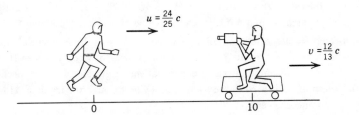

15. A skateboard has speed $u = \frac{3}{5}c$ in S, the Earth frame; its proper length is 1 m. A bug crawls in the same direction along the board with velocity $v' = \frac{4}{5}c$, relative to the board. In S', the rest frame of the board, find the time for the bug to crawl from one end of the board to the other. In S, find the speed v of the bug and the time for the bug to crawl the length of the board. The bug carries a watch; find the time on the watch for the bug to crawl the length of the board.

16. When hockey star Guy L'Einstein takes his relativistic slap shot, his body is moving along the ice at speed u_0 while his wrist is thrust forward at speed u_1' relative to his body and his hockey puck is pushed ahead at speed v'' relative to his wrist. Each of these speeds has the same value, $u_0 = u_1' = v'' = \frac{3}{5}c$. The speeds in S, the ice frame, are shown as u_0, u_1, and v in the figure. Determine the speed v of the puck in S. Let his slapshot have stroke length $d = 1$ m in S, as illustrated, and calculate the time required to get the shot off in S. Determine the corresponding time interval observed on Guy's wristwatch.

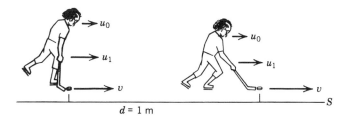

$d = 1 \text{ m}$

17. Refer to Problem 12 and identify the gunshots as events in a space–time diagram. Use the diagram to prove that the game warden's claim is impossible.

18. Refer to Figure 1-9 for the details of the pole-in-the barn puzzle, and plot all the various distances and times on a space–time diagram, using (x, ct) and (x', ct') axes. Indicate the events A, B, and C on the diagram. Calculate the Lorentz-invariant space–time intervals between events A and B to relate the times t_{BC} and t'_B and between events A and C to relate the times t_{BC} and t'_C.

19. Refer to Problem 13 and plot the events A, B, and C on a space–time diagram. Draw the world-line of the propagating light signal.

20. The elastic scattering of two equal masses is shown in two Lorentz frames S and S', before and after a collision. The total momentum is zero before and after in S', where all speeds have the same value v'. The frame S' moves through the frame S with speed u in such a way that, in S, one of the particles has no x component of velocity before and after the collision. Determine the indicated velocities u, v_1, v_{2x}, and v_{2y} in terms of v' under these conditions.

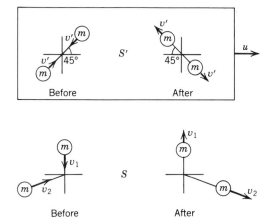

21. Continue Problem 20 and demonstrate as follows that the total relativistic momentum is conserved in S. First, show explicitly that $P_{\text{total } y} = 0$, where

$$P_{\text{total } y} = \frac{mv_{2y}}{\sqrt{1 - v_2^2/c^2}} - \frac{mv_1}{\sqrt{1 - v_1^2/c^2}}$$

before the collision. Then, argue from the symmetry of the figure that $P_{\text{total } y}$ is conserved. Finally, argue that $P_{\text{total } x}$ is also conserved, again from the symmetrical construction of the special situation shown in the figure.

22. Prove the validity of the relation

$$\gamma_{v'} = \gamma\gamma_v\left(1 - \frac{uv_x}{c^2}\right)$$

for velocities **v** and **v'** related by velocity addition, where $\gamma = (1 - u^2/c^2)^{-1/2}$, $\gamma_v = (1 - v^2/c^2)^{-1/2}$, and $\gamma_{v'} = (1 - v'^2/c^2)^{-1/2}$.

23. The Lorentz transformation equations for relativistic momentum and energy are examined in the text. The demonstration of their validity is not given for the components of the momentum transverse to the x direction. Prove that $p_y' = p_y$ holds, and argue that $p_z' = p_z$ follows by an identical derivation.

24. The angle ϕ in the figure is defined by $\sin\phi = v/c$, where v is the speed of a particle of mass m. Identify all the straight lengths in the figure in terms of the appropriate kinematical quantities.

mc^2

25. A particle has speed v and relativistic kinetic energy K. Prove that

$$\frac{v}{c} = \frac{\sqrt{\kappa(\kappa + 2)}}{\kappa + 1},$$

where $\kappa = K/mc^2$. For $\kappa \gg 1$, expand the result in powers of $1/\kappa$ and show that

$$\frac{v}{c} = 1 - \frac{1}{2\kappa^2} + \cdots$$

to order $1/\kappa^2$.

26. The value of v/c for a 1 J proton is very very close to unity. Find how close by making an explicit calculation of $1 - v/c$.

27. Draw graphs of the relativistic momentum and energy

$$p = \frac{mv}{\sqrt{1 - v^2/c^2}} \quad \text{and} \quad E = \frac{mc^2}{\sqrt{1 - v^2/c^2}}$$

as functions of the speed v. Make the substitution $m \to i\mu$ and let v exceed c to obtain a modified version of p and E. How does the relativistic relation between these quantities differ from the unmodified case? Draw graphs of p and E, in their modified form, and observe their behavior as $v \to \infty$. (These manipulations describe faster-than-light particles, called *tachyons*. Such a particle cannot be found at rest in any Lorentz frame; therefore, the mass parameter m is not interpretable in terms of a rest energy and is not required to be real valued.) Where are the world-lines of tachyons supposed to lie in a space–time diagram?

28. Prove the work–energy theorem for particle motion in three dimensions. To be specific, use the relativistic force law $\mathbf{F} = d\mathbf{p}/dt$, consider the work done by the force between locations 1 and 2 along the path of the particle

$$W = \int_1^2 \mathbf{F} \cdot d\mathbf{r},$$

and show that $W = E_2 - E_1$.

29. A particle of mass M at rest disintegrates into two fragments of equal mass m. Determine the velocity of each fragment in terms of M and m.

30. A particle of mass M at rest decays into two unequal masses m_1 and m_2. Show that the square of the momentum of each of the final particles is given by

$$p^2 = \frac{\left[M^2 - (m_1 + m_2)^2 \right]\left[M^2 - (m_1 - m_2)^2 \right]}{4M^2} c^2.$$

31. Continue Problem 30 and derive formulas for the relativistic kinetic energies of the final particles. Show that the results can be expressed as

$$K_1 = \frac{M - m_1 + m_2}{2M} Q \quad \text{and} \quad K_2 = \frac{M - m_2 + m_1}{2M} Q,$$

where Q denotes the Q-value for the decay.

32. Consider an elastic collision of equal-mass particles with one particle initially at rest. The conservation laws imply that the final momenta are perpendicular in the nonrelativistic case. Is this true for a relativistic collision? Answer the question by deriving the relativistic formula

$$c^2 \mathbf{p}_1 \cdot \mathbf{p}_2 = K_1 K_2$$

relating the final momenta and kinetic energies.

33. In S, a particle of mass m and speed $v = \frac{4}{5}c$ strikes a target particle of mass $5m$ at rest. Let the frame S' be the center of mass frame, and determine the speed u with which S' moves through S. Find the speeds v' and u' of the two masses m and $5m$ in S'.

34. The lower part of Figure 1-28 shows a particular example of the decay $M \to m + m$ in flight. Use the conservation laws to obtain $\bar{v}$ in terms of v, where the speeds are identified in the figure.

35. Consider a reaction caused by m_1 colliding with m_2, where m_2 is initially at rest. The collision is completely inelastic as the two masses stick together to form a single object of mass M. Derive a formula for M in terms of m_1, m_2, and the incident kinetic energy K_1.

36. Figure 1-29 shows a two-body collision of unequal masses m and M in the lab frame and in the CM frame. The indicated speeds v, v', and u are constrained by the velocity addition formula and by the condition on the momenta that defines the CM frame. Use these two relations to eliminate v' so that u is determined in terms of v.

37. In S, a particle of mass m has the momentum four-vector

$$\not{p} = \begin{bmatrix} p \\ 0 \\ 0 \\ iE/c \end{bmatrix}.$$

Let S' be the rest frame of the particle so that the corresponding momentum four-vector p' is obtained from p by the Lorentz transformation $p' = \mathcal{L}p$. Determine the transformation matrix $\mathcal{L}$ required to obtain the form necessary for p'. Construct $\mathcal{L}$ by deducing the parameters β and γ in terms of m, p, and E.

38.   A proton–proton collision can create a π^0 meson as an additional particle in the final state, if there is sufficient energy. The observed reaction is

$$p + p \rightarrow p + p + \pi^0.$$

Derive an expression, in terms of the masses, for the minimum kinetic energy $K_{\text{threshold}}$ for beam protons incident on protons at rest in this reaction. Calculate the numerical value of $K_{\text{threshold}}$, using 938.3 MeV and 135.0 MeV for the rest energies of the proton and the meson.

T W O

PHOTONS

*T*he beginnings of quantum physics occupy a period of history that spans a quarter of a century. Our survey of these important developments is spread across the following three chapters. The topics in this survey describe the evolution of a new treatment of radiation and matter in which certain basic notions of classical physics are questioned and discarded. A series of discoveries points the way toward a new theory and, at the end of the evolutionary period, quantum mechanics finally emerges as the governing body of principles.

The quantum theory is a revision of physics that supersedes classical mechanics and has a classical limit in the regime where Newton's laws are valid. The main inspiration for the new theory is the observed phenomenon of *quantization*, in which physical quantities are found to exist in *discrete states* rather than continuous distributions. Our plan is to introduce quantization and quantum behavior in the form of accumulated evidence. We proceed along historical lines in these three chapters as we prepare for the more formal material in the chapters to come.

We should note that quantization in nature was certainly not unknown in the 19th century. It was believed that an examination of bulk matter on a sufficiently small scale would reveal a discrete composition, supposedly made up of *molecules* and *atoms*. This microscopic particle model was invoked by Maxwell, Boltzmann, and others in the formulation of the kinetic theory of gases. Molecules and atoms were similarly anticipated in the early laws of chemistry through the findings of J. Dalton and A. Avogadro. The quantization of electric charge was also known in this same period. Faraday's experiments on the conduction of electricity in solutions gave evidence for charged ions with charges equal to multiples of a basic indivisible unit. This unit was eventually identified with the charge of the *electron*, a fundamental constituent of all atoms. Thus, the *quantization of matter* and the *quantization of charge* were recognized in context with certain discoveries and were accepted as concepts before the end of the 19th century.

These properties of nature should be set apart from the quantum revolution since they did not point directly to the need for a new fundamental theory. The real revolution began when physical motion was found to have quantized behavior in radiation and in matter. Once it became apparent that the energy of certain systems

Max Planck

could exist only in discrete states, it was clear that a serious break with classical physics was at hand.

Quantum physics came to life in the year 1900. The occasion was marked in history by a famous pronouncement put forward by M. K. E. L. Planck to explain the observed properties of the radiation emitted by incandescent objects. This commonplace phenomenon posed an unsolved problem that had lodged at the forefront of theoretical physics for several decades. Principles of thermodynamics and electromagnetism had been applied to the problem, but classical methods had failed to give a sensible explanation of the experimental results. Finally, Planck grasped in desperation for a solution based on the new idea of quantization.

The quantum hypothesis of Planck and the subsequent interpretation of the idea by Einstein gave electromagnetic radiation discrete properties somewhat similar to those of a particle. These quantized components of light became known as photons. It was surprising that such discrete characteristics should be in evidence since light was firmly established as a wave phenomenon. The quantum theory made provision for radiation to have *both* wave and particle aspects in a complementary form of coexistence. The theory was extended when matter was also found to have wave characteristics as well as particle properties. These formative notions continued to evolve, completely at variance with classical ideas, until 1925 when the formal apparatus of the quantum theory finally came into being.

The evolutionary phase of quantum mechanics is called the period of the "old quantum theory." We devote the following two chapters to the main feature of this theory, the *quantization of energy* in radiation and in atoms. The last of our three introductory chapters then concludes these developments with a preliminary picture of the wave description of matter.

2-1 Blackbody Radiation

An ordinary property of every object is its ability to emit and absorb electromagnetic radiation. The phenomenon is called *thermal radiation* because it involves an interchange between radiation energy in the electromagnetic fields around the object and thermal energy owing to the motion of particles within the object. The interchange is assumed to be an equilibrium process occurring at a certain temperature. Some of the features of this complex problem appeal to common sense. The familiar observation that an incandescent solid glows "red-hot" when heated, and "white-hot" when heated more, suggests a correlation between the temperature of the solid and the frequency of the emitted radiation. In fact, the object emits and absorbs radiation of *all* frequencies, and so a particular range of emitted frequencies tends to prevail for a particular temperature. Hence, the problem of interest for a body at a given temperature concerns the *distribution* of the radiated electromagnetic energy as a function of the emitted frequency. This distribution is called the frequency *spectrum* of the radiating object.

Let us define the various relevant quantities so that we can describe the phenomenon. The *radiant emittance* (or radiancy, or emissive power) $M(T)$ refers to the total energy radiated by the object at Kelvin temperature T per unit time per unit area. This quantity is a temperature-dependent function, given in units W/m^2. A continuous frequency spectrum is also defined since the emittance has contributions in every frequency interval $d\nu$. We express this distribution of frequencies by writing

$$M(T) = \int_0^\infty M_\nu(T)\, d\nu. \tag{2-1}$$

The integrand $M_\nu(T)$ identifies the *spectral radiant emittance*, or total energy radiated per unit time per unit area per unit frequency interval. Our notation stresses the dependence of this *spectral* quantity on the two variables, ν as well as T.

The emission of radiation is simultaneous with the incidence of radiation for an object at equilibrium. Incident energy may be either reflected or absorbed, and so two separate incident quantities are specified at frequency ν and temperature T, the spectral reflectance $\rho_\nu(T)$ and the spectral absorptance $\alpha_\nu(T)$. These fractions of the energy incident per unit time per unit area per unit frequency interval must satisfy the equality

$$\rho_\nu(T) + \alpha_\nu(T) = 1. \tag{2-2}$$

We are concerned with the equilibrium situation where the rates of emission and absorption are equal by definition. A *perfect* absorber reflects no incident radiation and therefore satisfies $\alpha_\nu(T) = 1$. This ideal radiator is a perfect emitter and is called a *blackbody*. Figure 2-1 shows a model of a blackbody constructed in the form of an evacuated cavity with walls at temperature T and with a hole in one of the walls. The hole is very small so that rays entering the cavity have essentially no chance to be reflected back out. The spherical shape shown in the figure is not a necessary feature of the model. A blackbody is a useful idealization, which we can employ as a standard by introducing the associated spectral emittance $M_\nu^b(T)$ as a standardizing function. We can use this fundamental fictitious quantity to define the spectral emissivity

$$\varepsilon_\nu(T) = \frac{M_\nu(T)}{M_\nu^b(T)}; \tag{2-3}$$

Figure 2-1

Model of a blackbody radiator. All radiation incident on the hole in the wall of the cavity is effectively absorbed. The flux rate of energy radiated at temperature T is given by the blackbody emittance $M^b(T)$. The spectrum spans all frequencies and has a universal shape for each temperature.

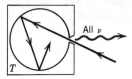

this ratio then provides a measure of the radiating efficiency for a real object whose spectral emittance is $M_\nu(T)$.

The following facts should be added as background for the blackbody problem. In 1859 G. R. Kirchhoff proved that the ratio $M_\nu(T)/\alpha_\nu(T)$ should be a *universal* function of ν and T, the same for all radiators. He used thermodynamic arguments to show that a failure of this universal property was equivalent to the existence of a perpetual motion machine that violated the second law of thermodynamics. The universal function in Kirchhoff's theorem was identified to be the spectral emittance of a blackbody $M_\nu^b(T)$. These arguments presented a challenge for the physics community to deduce the form of the unknown function. In 1879 J. Stefan conjectured, on empirical grounds, that the emittance of an object should be proportional to the fourth power of its temperature. In 1884 Boltzmann proved the conjecture theoretically, but only in the case of a blackbody. Their conclusion, the Stefan–Boltzmann law, was expressed as

$$M^b(T) = \int_0^\infty M_\nu^b(T)\,d\nu = \sigma T^4, \qquad (2\text{-}4)$$

with $\sigma = 5.67 \times 10^{-8}$ W/m$^2 \cdot$ K^4 for the value of the Stefan–Boltzmann constant.

The burden of Kirchhoff's challenge fell initially on experimenters to develop a suitable laboratory blackbody and then measure the radiation over a broad range of frequencies at fixed temperature. In time the shape of the blackbody spectrum became established. Curves like those in Figure 2-2 were obtained for various temperatures, where each curve displayed a single maximum occurring at a particular frequency ν_m. The observed variation of the spectrum with T was such that the peak of the distribution shifted to higher frequency as the temperature was increased. In 1893 W. Wien deduced from thermodynamics that ν_m and T should obey a *linear* relation. He also proposed a form for the frequency distribution $M_\nu^b(T)$ in agreement with the limited available data. Meanwhile, pioneering improvements were being made on the experimental front as O. Lummer and E. Pringsheim proceeded to broaden the range of frequencies in the measured distributions. By 1900 it was clear from these data that Wien's proposed formula was not adequate as a fit to the whole known spectrum. Thus, Kirchhoff's challenge continued to stand through the end of the century.

This brief summary sets the stage for Planck's contribution. The great importance of his idea warrants a detailed analysis of the entire blackbody problem. We

Figure 2-2

Blackbody frequency spectra for temperatures $T_1 < T_2$. The distribution of emitted frequencies has a maximum at ν_{m_1} for temperature T_1 and at ν_{m_2} for temperature T_2. According to Wien's law, ν_m is proportional to T.

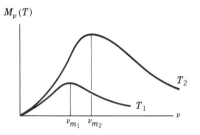

Figure 2-3

Blackbody wavelength spectra for temperatures $T_1 < T_2$. The maximum occurs at λ_m for temperature T, where the product $\lambda_m T$ is given by Wien's constant.

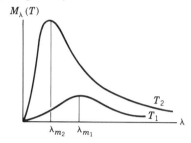

concentrate on the ideal blackbody radiator and suppress the superscript on $M_\nu^b(T)$, as we have already done in Figure 2-2. Our system is the cavity model in Figure 2-1 whose shape, we have noted, can be arbitrary. The derivation of the spectrum requires several steps, the most important of which is the use of Planck's quantum hypothesis. Any desired property of the resulting blackbody solution may also be deduced, including particularly the Stefan–Boltzmann T^4 law in Equation (2-4).

One of the deductions that we can draw from Planck's result is Wien's law for the position of the peak in the blackbody spectrum. This result is more often expressed in terms of the wavelength as the variable rather than the frequency, so let us show how these two types of distribution are related. Equation (2-1) may be consulted for this purpose and written in two ways:

$$M(T) = \int_0^\infty M_\nu(T)\,d\nu = \int_0^\infty M_\lambda(T)\,d\lambda. \tag{2-5}$$

The second integral defines $M_\lambda(T)$, the *wavelength distribution* of the emittance. We can transform the integration over ν into an integration over λ according to the familiar rule

$$\int_0^\infty M_\nu(T)\,d\nu = \int_\infty^0 M_\nu(T)\frac{d\nu}{d\lambda}\,d\lambda.$$

Note how the range of integration is reversed to accommodate the inverse relation between ν and λ. When we compare with the λ integration in Equation (2-5) we get

$$M_\lambda(T) = -M_\nu(T)\frac{d\nu}{d\lambda} = \frac{c}{\lambda^2}M_\nu(T) \quad \text{where} \quad \nu = \frac{c}{\lambda}. \tag{2-6}$$

We note that the λ dependence of $M_\lambda(T)$ is not the same as the ν dependence of $M_\nu(T)$, although both types of spectrum display peaking behavior. Figure 2-3 shows how the wavelength spectrum maximizes at λ_m for a particular temperature T, and especially how λ_m shifts to *lower* wavelength as T is increased. For wavelength distributions of blackbody radiation, Wien's law takes the form

$$\lambda_m T = 2.898 \times 10^{-3} \text{ K} \cdot \text{m} \quad \text{(Wien's constant)}. \tag{2-7}$$

It should be emphasized that there is no reason for the peak positions ν_m and λ_m in the respective distributions to be connected by the relation $c = \nu\lambda$.

Blackbody radiation is realized physically in many practical applications. The radiation from the Sun is a particularly conspicuous candidate for treatment by the blackbody model. Measurements of the solar spectrum reveal that 99% of the radiation falls in the wavelength range 270–4960 nm. The quantity of interest is the solar constant S, defined as the total solar energy received at the Earth per unit time per unit area at normal incidence, corrected for the effects of the Earth's atmosphere. A recent determination of this quantity quotes the value $S = 1351$ W/m². The wavelength distribution of the incident radiation is called the solar spectral irradiance. This function of λ peaks around $\lambda_m = 470$ nm, and so Wien's law gives $T = 6166$ K for the Sun's surface temperature. The quoted solar constant may also be used in conjunction with the Stefan–Boltzmann law to produce the slightly different result $T = 5762$ K. A blackbody distribution at this temperature can be made to fit the solar spectral irradiance and reproduce the quoted value of S for the integrated area under the curve.

Blackbody radiation also leaves its traces in more subtle areas. Experiments conducted in the 1960s by A. A. Penzias and R. W. Wilson have demonstrated the existence of isotropic background radiation that presumably permeates the universe. This electromagnetic background can be fit by a blackbody distribution with a

Solar spectral irradiance versus wavelength. The area under the curve gives the value of the solar constant $S = 1351$ W/m². Data are taken from the survey of M. P. Thekaekara et al., *Appl. Opt.* **8**:1713–1732 (1969).

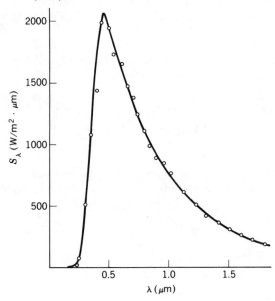

parametrizing temperature close to $T = 3$ K. The phenomenon constitutes substantial evidence for the Big Bang theory of the universe, inasmuch as the radiation energy apparently represents the residue in the current epoch of the primordial energy produced in the Big Bang.

Let us return to the analysis of the Planck law and take two preliminary steps that serve to isolate the crucial feature of the problem. We begin by shifting the focus of the investigation away from the energy flux rate out of the cavity and onto the *energy density* contained inside the cavity. The relation between the corresponding spectral quantities is

$$M_\nu(T) = \frac{c}{4} u_\nu(T), \qquad (2\text{-}8)$$

where $u_\nu(T)$ denotes the electromagnetic energy in the cavity per unit volume per unit frequency interval. We then recognize that the radiation in the cavity takes the form of *standing electromagnetic waves*. These occur in every interval $d\nu$ as discrete modes whose number per unit frequency interval depends on the frequency. We can therefore determine the spectral energy density at frequency ν if we count the number of modes and multiply by the average energy for each mode. This construction is expressed as

$$u_\nu(T) = \frac{N_\nu}{V} \langle \varepsilon \rangle, \qquad (2\text{-}9)$$

where V is the volume of the cavity, N_ν is the number of electromagnetic modes at frequency ν per unit frequency interval, and $\langle \varepsilon \rangle$ is the average energy per mode. The problem thus reduces to the determination of the two factors N_ν and $\langle \varepsilon \rangle$. Each of these quantities involves a rather complicated derivation, requiring separate sections for a careful treatment. We learn where the heart of the problem lies by separating N_ν and $\langle \varepsilon \rangle$, as we find that the quantum hypothesis makes its appearance through the T-dependent factor $\langle \varepsilon \rangle$.

Example

Consider the application of Wien's law to the following rather different blackbody situations. If we take the surface temperature of the Sun to be 5800 K and consult Equation (2-7), we find that the peak of the solar spectrum should occur at

$$\lambda_m = \frac{2.898 \times 10^{-3} \text{ K} \cdot \text{m}}{5800 \text{ K}} = 500 \text{ nm},$$

a wavelength near the center of the visible range. On the other hand, the universal 3 K background radiation gives

$$\lambda_m = \frac{2.898 \times 10^{-3} \text{ K} \cdot \text{m}}{3 \text{ K}} = 0.97 \text{ mm},$$

a wavelength in the microwave region.

Figure 2-4

Averaging over rays with velocity components to the right.

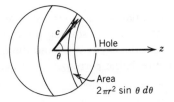

Example

The energy flux rate $M(T)$ and the energy density $u(T)$ are related for cavity radiation by a proportionality factor $c/4$, as in Equation (2-8). We can deduce this result as the product of two contributions with the aid of Figure 2-4. The relation calls for a factor of $\frac{1}{2}$ because, for all the electromagnetic radiation in the cavity, only half of the standing-wave energy corresponds to rays with velocity components to the right where the hole is located in the figure. (Recall that a standing wave is an equal-parts admixture of traveling waves having opposite directions of propagation.) These rays carry energy through the hole with an average velocity to the right given by the z component of the velocity of light, $c_z = c \cos \theta$, averaged over the right hemisphere. If we let the sphere in the figure have arbitrary radius r and average over the hemispherical area we obtain

$$\langle c_z \rangle = \frac{\int_0^{\pi/2} (c \cos \theta) 2\pi r^2 \sin \theta \, d\theta}{\int_0^{\pi/2} 2\pi r^2 \sin \theta \, d\theta} = \frac{c \int_0^1 x \, dx}{\int_0^1 dx} = \frac{c}{2},$$

in which we have used the substitution $x = \cos \theta$. The desired relation is then found by assembling factors:

$$M = \tfrac{1}{2} \langle c_z \rangle u = \frac{c}{4} u.$$

The corresponding frequency distributions are related in similar fashion, in agreement with Equation (2-8). We should note that the derivation does not require the cavity to have a spherical shape.

2-2 Standing Electromagnetic Waves

An ideal blackbody absorbs and emits equal amounts of radiation in steady equilibrium. Our cavity model of a blackbody treats the interior of the cavity as a storage volume for the incoming and outgoing radiation. We assume perfectly conducting walls for the enclosure so that the fields are completely enclosed and the energy is stored in the form of standing electromagnetic waves. We realize at once that the geometrical and physical properties of the cavity select only those particular configurations of wave fields that "fit" the enclosure. Each of these allowed configurations, or *modes*, has its own characteristic frequency of oscillation. Our first objective is to determine the number of modes $N_\nu \, d\nu$ that occur in the frequency range between ν and $\nu + d\nu$. The ν-dependent quantity N_ν has been identified in Equation (2-9). To minimize the mathematical complications, we take advantage of the assertion that the shape of the cavity does not matter, and we choose a cubical enclosure for this phase of the problem. Figure 2-5 shows the geometry of our chosen cube of dimension L by L by L.

Let us prepare ourselves for the mathematics of standing waves in a three-dimensional region by recalling the one-dimensional problem of a vibrating string with fixed ends. In this case the oscillating variable is the transverse displacement y, a function of position x along the string as well as time t. Figure 2-6 illustrates the lowest modes of vibration for a string with ends fixed at $x = 0$ and $x = L$. The nth standing wave has $n - 1$ additional points of zero vibration, called *nodes*, at equally spaced locations along the string. Each of these modes can be described by a standing-wave function of the form

$$y(x, t) = y_0 \sin \frac{2\pi x}{\lambda} \sin 2\pi ft \qquad (2\text{-}10)$$

with amplitude y_0, wavelength λ, and frequency f. The function satisfies the wave

Figure 2-5

Cubical cavity with perfectly conducting walls as an enclosure for standing electromagnetic waves.

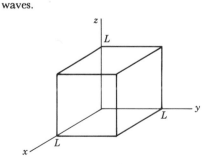

Figure 2-6

Standing waves on a vibrating string with fixed ends. The wavelength of the nth mode is $\lambda_n = 2L/n$.

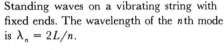

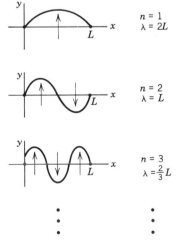

equation

$$\frac{\partial^2 y}{\partial x^2} = \frac{1}{v^2}\frac{\partial^2 y}{\partial t^2}, \tag{2-11}$$

where v is the speed of wave propagation in the string medium. It is easy to verify that the parameters obey $v = f\lambda$ by inserting Equation (2-10) into Equation (2-11). Recall that the standing-wave character of $y(x, t)$ is attributable to the multiplicative construction of the expression. This product of oscillating functions of x and t describes oscillations that remain in place and do not travel along the length of the string. The explicit form of Equation (2-10) exhibits the required node at $x = 0$ and also incorporates the other node at $x = L$ if λ satisfies the condition

$$\sin\frac{2\pi L}{\lambda} = 0.$$

The wavelengths are therefore limited to the set of values

$$\lambda_n = \frac{2L}{n}, \tag{2-12}$$

and so the allowed frequencies are correspondingly restricted:

$$f_n = \frac{v}{\lambda_n} = \frac{v}{2L}n. \tag{2-13}$$

Hence, the modes of the string are indexed by an integer and assume the form

$$y_n(x, t) = y_0\sin\frac{n\pi x}{L}\sin\frac{n\pi vt}{L} \tag{2-14}$$

when Equations (2-12) and (2-13) are inserted into Equation (2-10).

We describe standing electromagnetic waves in a cavity by similar methods, with allowances made for the three-dimensional nature of the wave configurations and for the vector character of the electromagnetic field. The oscillating quantity of interest is the electric field $\mathbf{E}(x, y, z, t)$. (The accompanying magnetic field is known in terms of $\mathbf{E}$ and does not have to be discussed explicitly.) Each component of $\mathbf{E}$ is a wave that satisfies the wave equation in three dimensions with speed of propagation c:

$$\frac{\partial^2 E_x}{\partial x^2} + \frac{\partial^2 E_x}{\partial y^2} + \frac{\partial^2 E_x}{\partial z^2} = \frac{1}{c^2}\frac{\partial^2 E_x}{\partial t^2} \quad \text{and similarly for } E_y \text{ and } E_z. \tag{2-15}$$

A boundary condition holds at each wall of the enclosure since the components of $\mathbf{E}$ tangential to the wall are required to be continuous as a consequence of Maxwell's equations. Our cubical enclosure is bounded by a perfectly conducting medium in which the $\mathbf{E}$ field *vanishes*. Therefore, the tangential components of $\mathbf{E}$ inside the cavity must vanish at each wall to meet the requirements of continuity. If we refer to Figure

2-5, we see that this implies the following specifications:

$$E_x = 0 \quad \text{at } y = 0 \text{ and } L, \text{ and at } z = 0 \text{ and } L,$$

$$E_y = 0 \quad \text{at } x = 0 \text{ and } L, \text{ and at } z = 0 \text{ and } L,$$

$$E_z = 0 \quad \text{at } x = 0 \text{ and } L, \text{ and at } y = 0 \text{ and } L. \tag{2-16}$$

For standing waves we again want multiplicative expressions like Equation (2-10), but adapted for three-dimensional configurations and three-component fields. The waves describe oscillations in place in all three spatial variables and obey boundary conditions as specified if the components of the field are

$$E_x = E_{0x}\cos\frac{n_1\pi x}{L}\sin\frac{n_2\pi y}{L}\sin\frac{n_3\pi z}{L}\sin 2\pi\nu t,$$

$$E_y = E_{0y}\sin\frac{n_1\pi x}{L}\cos\frac{n_2\pi y}{L}\sin\frac{n_3\pi z}{L}\sin 2\pi\nu t,$$

$$E_z = E_{0z}\sin\frac{n_1\pi x}{L}\sin\frac{n_2\pi y}{L}\cos\frac{n_3\pi z}{L}\sin 2\pi\nu t. \tag{2-17}$$

Note that *three* separate integers are needed here in place of the one employed in Equation (2-14). It is not difficult to demonstrate that these functions behave correctly at the walls and satisfy the wave equation in all three components. The latter point should be examined closely since it leads to a restrictive condition on the frequency ν. If the first of Equations (2-17) is inserted in the wave equation for E_x, the result is

$$-\left[\left(\frac{n_1\pi}{L}\right)^2 + \left(\frac{n_2\pi}{L}\right)^2 + \left(\frac{n_3\pi}{L}\right)^2\right]E_x = -\frac{1}{c^2}(2\pi\nu)^2 E_x.$$

This equality tells us that the allowed frequencies are given by a formula like Equation (2-13),

$$\nu = \frac{c}{2L}\sqrt{n_1^2 + n_2^2 + n_3^2}, \tag{2-18}$$

except that the role of the single integer in that result is now played by the three integers n_1, n_2, and n_3.

The modes that "fit" the cubical enclosure are not as easy to picture as the modes on a string. Equations (2-17) and (2-18) prescribe the allowed fields and frequencies according to the values assigned for the set of three independent integers (n_1, n_2, n_3). We associate each selected triplet of integers, such as $(0, 1, 1)$ or $(1, 2, 3)$ or $\ldots$, with a unique frequency, such as $(c/2L)\sqrt{2}$ or $(c/2L)\sqrt{14}$ or $\ldots$, and a unique spatial configuration of standing electromagnetic wave fields oscillating in the cavity. The most illuminating way to employ these integers in a bookkeeping scheme is to introduce a three-dimensional space whose axes are defined by the integer variables (n_1, n_2, n_3). Figure 2-7 shows how the various triplets of integers are plotted as points in a lattice so that each lattice site specifies an allowed mode and its corresponding frequency.

Figure 2-7

(n_1, n_2, n_3) space. Each lattice site refers to an oscillating configuration of fields in the cavity. The number of sites in the frequency interval $d\nu$ at frequency ν is equal to the volume of the corresponding spherical shell in the first octant. The shell has radius $2L\nu/c$ and thickness $(2L/c)\,d\nu$.

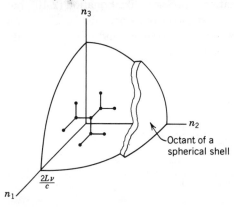

The figure can be used to count the number of modes $N_\nu\,d\nu$ in a given frequency interval $d\nu$. Let us choose a frequency ν and interval $d\nu$ and visualize the calculation with the aid of the indicated spherical shell in the three-dimensional (n_1, n_2, n_3) space. Equation (2-18) gives

$$\sqrt{n_1^2 + n_2^2 + n_3^2} = \frac{2L}{c}\nu$$

for the radius of the shell, and $(2L/c)\,d\nu$ for its thickness. A unit cube in this space contains one lattice site, and all the sites lie in the first *octant*, where each of the n's is a positive integer. Therefore, we can count the number of oscillating field configurations by calculating the *volume* of the octant shell that contains the corresponding lattice points. We take $V = L^3$ for the volume of the cavity and get

$$\frac{1}{8}4\pi\left(\frac{2L}{c}\nu\right)^2\frac{2L}{c}\,d\nu = \frac{4\pi\nu^2}{c^3}V\,d\nu$$

for the volume of the shell. The desired number of modes $N_\nu\,d\nu$ is actually twice this result, because every spatial configuration of fields has two independent polarization states. We therefore obtain

$$N_\nu = \frac{8\pi\nu^2}{c^3}V \tag{2-19}$$

as the final expression for the number of modes at frequency ν per unit frequency interval.

This construction gives the correct formula for one of the factors that make up the spectral energy density $u_\nu(T)$ in Equation (2-9). Classical methods can also be used to deduce a result for the other unknown factor $\langle \varepsilon \rangle$, the average energy per mode in the cavity. We are concerned with this second result only in passing since it leads us to a prediction for $u_\nu(T)$ that cannot be correct. Our motive in presenting the classical form of $\langle \varepsilon \rangle$ is to prepare the way for Planck's quantum hypothesis.

We want to know the average energy for a population of many radiation modes, each with its own energy. The enclosed fields are in equilibrium with the walls of the cavity at temperature T, and so the radiation exchanges energy with the many material particles bound to the walls. Statistical physics teaches us how to average over such large numbers of particles in order to deduce bulk properties for the system and average values for certain particle variables. The kinetic theory of gases is an example of this averaging procedure. Each of the three dimensions of linear particle motion in a gas at temperature T is assigned an average kinetic energy $\frac{1}{2}k_BT$, where k_B is Boltzmann's constant. We can apply this conclusion to an oscillating particle bound to the walls of the radiating cavity if we allow the oscillator to have potential energy as well as kinetic energy. We know that the averages of these two contributions to the energy are *equal* for any oscillating particle in simple harmonic motion. Therefore, we obtain k_BT as the average *total* energy for each degree of freedom of a linear oscillator in a collection of such particles at temperature T. This result is an application of the classical principle of *equipartition of energy*. We then return to our system of enclosed radiation in equilibrium with bound oscillators and simply transfer the result for the average energy from one part of the system to the other. If we equate the average energy per degree of freedom of a bound particle to the average energy per mode of radiation, we obtain

$$\langle \varepsilon \rangle = k_BT \tag{2-20}$$

as the desired factor remaining in the formula for $u_\nu(T)$.

When Equations (2-19) and (2-20) are assembled in Equation (2-9), the conclusion is

$$u_\nu(T) = \frac{8\pi\nu^2}{c^3}k_BT \tag{2-21}$$

as the *classical* prediction for the spectral energy density of a cavity radiator. Equation (2-8) then gives

$$M_\nu(T) = \frac{2\pi\nu^2}{c^2}k_BT \tag{2-22}$$

for the corresponding spectral emittance. The failure of these predictions can be ascertained from the graph of $M_\nu(T)$ in Figure 2-8. We see that the result resembles the experimental spectrum in Figure 2-2 only at low frequency. We also see that the increase at high frequency causes a catastrophic divergence of the integrated emittance as defined in Equation (2-1).

The principle of equipartition of energy was applied to radiation by J. W. Strutt Rayleigh around the turn of the century. Rayleigh's name has been attached to the unsatisfactory result in Equation (2-21) for this reason. The high-frequency catastrophe was a disastrous conclusion for the blackbody problem. This issue proved to be a critical breaking point for classical physics.

Figure 2-8

Rayleigh's form for the frequency spectrum of
the blackbody emittance.

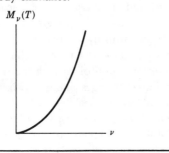

2-3 The Maxwell–Boltzmann Distribution

Planck's solution to the blackbody problem is based on an average energy $\langle \varepsilon \rangle$ that differs radically from $k_B T$. We can appreciate this feature of the problem more fully if we first understand how such average values are determined for systems containing a large number of elements in thermal equilibrium. It is instructive to digress from blackbody radiation for a while and devote an entire section to this question. The digression follows a straightforward line of *statistical* arguments to a rather general conclusion. Our purpose in the end is to apply the resulting statistical formalism to a collection of electromagnetic modes, even though the arguments are developed for a collection of particles.

Statistical physics deals with systems whose variables are far too numerous to treat individually and whose properties are therefore defined in terms of averages. This approach takes advantage of the large number of variables to analyze the unobserved behavior of the constituents of a system and extract the observed thermodynamic features of the system as a whole. The average energy of a particle in a many-body system is one such dynamical quantity of particular interest.

We are concerned with a large collection of *identical* particles, individually labeled so that each can be said to have its own well-defined energy. Of course, we recognize at once that these energies change repeatedly because of collisions among the particles. The average particle energy depends on the temperature T, while the system has a certain distribution of particle energies for a given value of T. Our main objective is to learn how a given total energy E is distributed over N identical constituents in thermodynamic equilibrium. We visualize the distribution in terms of a large collection of cells of definite energy to which the particles in the system can be assigned. We refer to these energy cells by means of a discrete integer index, and we let the increments of energy be as refined as we like. Thus, if the ith cell is reserved for particles with energy ε_i, the goal is to determine the number of particles n_i in that cell for a large system of particles in thermal equilibrium at temperature T. The population of the various energy cells gives an indication of the likelihood, or probability, for some given particle energy to occur in the system. The desired result is a formula, derived independently by Maxwell and Boltzmann around 1860, expressing n_i as a function of ε_i with T appearing as a parameter.

Our first notion to develop is the concept of a *state* describing the entire system. Classical physics allows us to introduce distinguishing labels so that we can follow each of the N identical particles and assign an energy cell for every particle. This

Figure 2-9

Three identical particles distributed over two cells. Microstates of the system are the eight distinct particle assignments to each cell noting, for example, that

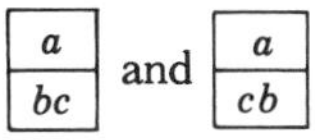

denote the same microstate. Macrostates of the system are the four different cell populations disregarding the labels of the individual particles. The number of ways to realize each macrostate is given by $N!/n_1!n_2!$, where $N = n_1 + n_2 = 3$.

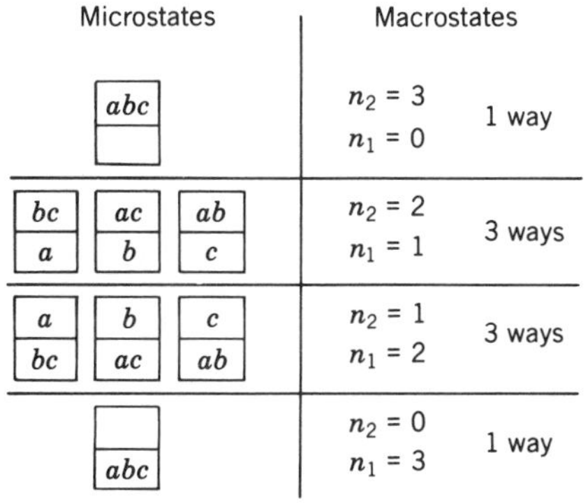

Microstates	Macrostates	
$\boxed{abc}$ $\boxed{}$	$n_2 = 3$ $n_1 = 0$	1 way
$\boxed{bc}\ \boxed{ac}\ \boxed{ab}$ $\boxed{a}\ \ \boxed{b}\ \ \boxed{c}$	$n_2 = 2$ $n_1 = 1$	3 ways
$\boxed{a}\ \ \boxed{b}\ \ \boxed{c}$ $\boxed{bc}\ \boxed{ac}\ \boxed{ab}$	$n_2 = 1$ $n_1 = 2$	3 ways
$\boxed{}$ $\boxed{abc}$	$n_2 = 0$ $n_1 = 3$	1 way

complete specification of the system is called a *microstate*. Such a detailed description is unnecessarily complete since distinct microstates should be regarded as physically equivalent if their cell population numbers are the same without regard for the labels that distinguish the particles from cell to cell. We therefore adopt another less specific scheme in which we simply list the numbers of particles that occupy each energy cell, assigning n_1 particles with energy ε_1, n_2 particles with energy ε_2, and so on. This revised method employs the original variables n_i and ε_i to generate a list of occupation numbers $(n_1, n_2, \ldots, n_r)$ for the distribution of N particles into r cells. Such a list defines a *macrostate* of the system.

We see immediately that many different microstates can often be identified with a single macrostate. This multiplicity of ways to assemble a given macrostate is a measure of the likelihood of achieving the corresponding configuration of particles. Thus, a certain macrostate has a highly probable distribution of cell occupations, and defines a correspondingly likely state of the system, if a sufficiently large number of microstates contributes to the distribution.

A very simple example is illustrated in Figure 2-9 to differentiate these two kinds of states. The microstates for $N = 3$ and $r = 2$ are shown as the eight distinct ways of assigning three particles, labeled a, b, and c, over two cells. (This illustration reproduces the exercise of flipping three coins and recording the head/tail outcome for each coin.) The macrostates are the four different pairs (n_1, n_2) of cell occupation numbers. (This list replicates the complete summary of heads/tails possibilities for the three coins.) We can assume, a priori, that each microstate has an equal chance to occur (just as head-head-head has the same likelihood as head-head-tail); however, we can see that the various macrostates have different probabilities (as three heads are less probable than two heads/one tail).

A system of particles can occupy a certain microstate at a given time and then change its microstate repeatedly thereafter because of particle collisions. We assume that all such microstates of the system are equally probable as a basic hypothesis. This assumption means that over a long time scale any one complete specification of every particle's state is expected to occur as often as any other. Of course, every system has rare configurations of particles; these are rare because there are very few ways of achieving these configurations as macrostates. Hence, macrostates are *not* equally probable because they are usually realized by differing numbers of microstates. As time passes and microstates change, the most frequently occurring macrostate, representing the most probable configuration of the system, is the one that corresponds to the greatest number of microstates. We can accomplish our objective and learn how the cell variables n_i and ε_i are related by implementing this simple idea.

The statistical problem is first of all a question of counting. We want an expression for the number of microstates corresponding to a given macrostate whose cell occupation numbers are denoted by $(n_1, \ldots, n_r)$. This number is expressed as $W_N(n_1, \ldots, n_r)$ and is called the *thermodynamic probability* for the macrostate in question. We deduce W_N by first recalling that the number of ways to assign N particles (the number of permutations of N objects) is given by the product of the N descending integer factors

$$N(N-1)(N-2) \cdots = N!.$$

Therefore, the number of distinct ways to distribute N particles so that n_1 are in cell 1, and the rest are not in cell 1, is

$$\frac{N!}{n_1!(N-n_1)!}.$$

Note that the number $N!$ alone overcounts the $n_1!$ permutations of particles within cell 1 as well as the $(N-n_1)!$ permutations of particles outside cell 1, and so these numbers are factored out of $N!$. We continue counting in this fashion by arguing next that the number of ways to distribute the remaining $N-n_1$ particles so that n_2 are in cell 2, and the rest are not in either cell 1 or cell 2, is

$$\frac{(N-n_1)!}{n_2!(N-n_1-n_2)!},$$

and so on, until all N particles are distributed over all r cells. The grand total number of such formations is given by the product of the corresponding r factors:

$$\frac{N!}{n_1!(N-n_1)!} \frac{(N-n_1)!}{n_2!(N-n_1-n_2)!} \cdots \frac{(N-n_1-\cdots-n_{r-1})!}{n_r!(N-n_1-\cdots-n_r)!}.$$

The result can be simplified in an obvious way to give the final number of microstates

$$W_N(n_1, \ldots, n_r) = \frac{N!}{n_1! \ldots n_r!}. \tag{2-23}$$

We get this answer by introducing the relation

$$n_1 + \cdots + n_r = N \tag{2-24}$$

as a condition that constrains the occupation numbers and accounts for the total number of particles. Note that the definition $0! \equiv 1$ is also used.

Now that we know the number of ways to form the macrostate $(n_1, \ldots, n_r)$, we next want to determine the most probable of all such distributions. The desired macrostate is evidently the one that maximizes W_N. We observe that the same distribution is determined by the solution that maximizes $\ln W_N$ and that the function $\ln W_N$ is more convenient to consider than W_N itself. Equation (2-23) is converted for this purpose to read

$$\ln W_N(n_1, \ldots, n_r) = \ln N! - \ln n_1! - \cdots - \ln n_r!$$

$$= \ln N! - \sum_{i=1}^{r} \ln n_i! \tag{2-25}$$

by employing some of the well-known properties of logarithms.

Let us take advantage of the large number of particles and assume that each n_i is large enough to justify a few approximations. Quantities like $\ln n!$ can be very accurately approximated with the aid of Stirling's formula:

$$\ln n! \rightarrow n \ln n - n \quad \text{for large } n. \tag{2-26}$$

This substitution enables us to rewrite Equation (2-25) as

$$\ln W_N(n_1, \ldots, n_r) = N \ln N - N - \sum_i (n_i \ln n_i - n_i)$$

$$= N \ln N - \sum_i n_i \ln n_i, \tag{2-27}$$

where we express the approximation as an equality and again use Equation (2-24).

We maximize $\ln W_N$ by taking integer increments in each n_i and seeking the largest result. Since the increments are so much smaller than the n's themselves, it is reasonable to treat each n_i as a continuous variable. We would then seek our solution by requiring

$$\frac{\partial}{\partial n_i} \ln W_N = 0$$

for every n_i, *if* the variables $(n_1, \ldots, n_r)$ could all be regarded as independent. In fact, we know that two restrictive conditions exist among the variables so that only $r - 2$ of them are actually independent. One condition fixes the total number of particles to equal N, as in Equation (2-24). Let us call this the N constraint and rewrite the condition as

$$\sum_i n_i = N. \tag{2-28}$$

The other restriction fixes the total energy of the system to equal E. Let us call this the E constraint and express the condition as

$$\sum_i n_i \varepsilon_i = E. \tag{2-29}$$

It would appear that at the last moment our problem has taken on a complication that prevents us from treating all the variables symmetrically and forces us to select two arbitrarily as the ones that are not independent.

The constraints can be handled by means of an elegant symmetrical procedure, devised by J. L. Lagrange in the 18th century. We do not directly use $\ln W_N$, a function of r variables where only $r - 2$ are independent, but instead introduce another quantity

$$F(n_1, \ldots, n_r, \alpha, \beta) = \ln W_N(n_1, \ldots, n_r) - \alpha\left(\sum_i n_i - N\right) - \beta\left(\sum_i n_i\varepsilon_i - E\right),$$

$$(2\text{-}30)$$

a function of $r + 2$ variables where *all* $r + 2$ *are* independent. The procedure then requires that we maximize F over all the variables $n_1, \ldots, n_r$, α, and β.

We find that the conditions

$$\frac{\partial F}{\partial \alpha} = 0 \quad \text{and} \quad \frac{\partial F}{\partial \beta} = 0$$

immediately reproduce the N constraint in Equation (2-28) and the E constraint in Equation (2-29). These conditions tell us that F ultimately reduces to $\ln W_N$ when F is at its maximum. We also obtain

$$\frac{\partial F}{\partial n_i} = 0 \quad \text{for all } i = 1 \text{ to } r, \tag{2-31}$$

so that the net effect is to maximize $\ln W_N$, as required, *and* simultaneously satisfy the two constraints. The advantages of Lagrange's procedure are twofold; the entire set of r occupation variables appears on the same footing, and the extra variables α and β enter as new quantities with their own physical significance.

Let us now proceed to our objective with the aid of Lagrange's method. The combination of Equations (2-27) and (2-30) produces the expression

$$F = N \ln N - \sum_i n_i \ln n_i - \alpha\left(\sum_i n_i - N\right) - \beta\left(\sum_i n_i\varepsilon_i - E\right),$$

and so the maximizing condition in Equation (2-31) becomes

$$\frac{\partial F}{\partial n_i} = -\ln n_i - 1 - \alpha - \beta\varepsilon_i = 0. \tag{2-32}$$

We can solve this equality for the ith occupation number to obtain

$$n_i = e^{-1-\alpha}e^{-\beta\varepsilon_i}. \tag{2-33}$$

This is essentially the desired result except that the quantities α and β remain to be identified. If we use the N constraint in Equation (2-28) to eliminate α we find

$$N = \sum_i n_i = e^{-1-\alpha}\sum_i e^{-\beta\varepsilon_i} = e^{-1-\alpha}Z \quad \Rightarrow \quad e^{-1-\alpha} = \frac{N}{Z},$$

where

$$Z = \sum_i e^{-\beta \varepsilon_i}. \tag{2-34}$$

Equation (2-33) can then be rewritten as

$$n_i = \frac{N}{Z} e^{-\beta \varepsilon_i}, \tag{2-35}$$

the final expression for the Maxwell–Boltzmann distribution function.

Our concluding formula specifies the number of particles in the ith energy cell for the macrostate of maximum thermodynamic probability. It can be argued that the system tends to this configuration in the approach to thermal equilibrium. We know that the maximizing state cannot represent a static final situation because of the continual rearrangement of the particles among the cells. The conclusions suggest instead that large departures from the most probable macrostate are expected to be highly improbable.

Equation (2-34) defines a new quantity Z called the *partition function*. This definition plays a recurring role in many of the applications of statistical physics. The expression for Z depends on the manner in which ε_i varies from one cell to another, and so the explicit β dependence of Z can only be expressed by examining separate cases. The variable β itself remains to be identified. A variety of arguments can be used to obtain

$$\beta = \frac{1}{k_B T}. \tag{2-36}$$

One such demonstration of this result is given in the second illustration below.

Example

The approximation in Equation (2-26) follows directly from Stirling's formula for the factorial function

$$x! = \sqrt{2\pi x}\left(\frac{x}{e}\right)^x \left[1 + O\left(\frac{1}{x}\right)\right].$$

Let us gain some feeling for the approximation by inspecting the graph in Figure 2-10. We can invoke the properties of the logarithm to write

$$\ln n! = \ln n + \ln(n-1) + \cdots + \ln 1$$

and then identify the sum of terms with the area of the rectangles shown in the figure. The area under the curve $\ln x$ is a good approximation to the area of the rectangles if n is large enough. The rest of the construction is summarized in the following steps:

$$\ln n! \rightarrow \int_1^n \ln x \, dx = (x \ln x - x)\Big|_1^n = n \ln n - n + 1 \rightarrow n \ln n - n \quad \text{for large } n.$$

It is obvious that this approximation can save enormous amounts of computational time when n has very large values.

Figure 2-10

Approximation of ln $n!$ by the integral $\int_1^n \ln x \, dx$. This construction leads to Stirling's approximation.

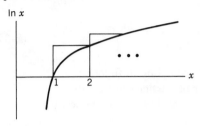

Example

The energy cells with a *discrete* index in Equation (2-35) can readily be adapted for application to the case of free particles with a *continuous* energy. We accomplish this by letting a very refined incremental cell of definite energy ε be identified with an infinitesimally thin spherical shell in the space of the three velocity components (v_x, v_y, v_z). These variables are connected by the formula for the energy of a free particle,

$$\varepsilon = \tfrac{1}{2}mv^2 = \tfrac{1}{2}m\left(v_x^2 + v_y^2 + v_z^2\right),$$

and so the radius of the shell in velocity space is given by

$$\sqrt{v_x^2 + v_y^2 + v_z^2} = \sqrt{\frac{2\varepsilon}{m}}\,.$$

The thickness of the shell dv is related to the energy interval $d\varepsilon$ by differentiating:

$$v^2 = \frac{2\varepsilon}{m} \quad \Rightarrow \quad 2v\,dv = \frac{2}{m}d\varepsilon.$$

The shell then determines an element of volume in velocity space

$$d\tau_v = 4\pi v^2\,dv = 4\pi\sqrt{\frac{2\varepsilon}{m}}\,\frac{d\varepsilon}{m}\,.$$

The number dn of particles in this energy cell is proportional to the cell volume $d\tau_v$ and is also proportional to the Maxwell–Boltzmann factor $e^{-\beta\varepsilon}$ according to Equation (2-35). We can therefore set

$$dn = Ae^{-\beta\varepsilon}\,d\tau_v,$$

where A is a proportionality constant. We have just seen that $d\tau_v$ is proportional to $\varepsilon^{1/2}\,d\varepsilon$, and so we can reexpress the number of particles in the cell as

$$dn = A'\varepsilon^{1/2}e^{-\beta\varepsilon}\,d\varepsilon,$$

where A' is another proportionality constant. Two definite integrals are needed at this point. We get the first from a table of integrals,

$$\int_0^\infty x^{1/2}e^{-\beta x}\,dx = \tfrac{1}{2}\sqrt{\pi}\,\beta^{-3/2} \quad (\text{Dwight } 860.04),$$

and we derive the second from the first,

$$\int_0^\infty x^{3/2}e^{-\beta x}\,dx = -\frac{d}{d\beta}\int_0^\infty x^{1/2}e^{-\beta x}\,dx = \tfrac{3}{4}\sqrt{\pi}\,\beta^{-5/2}.$$

These formulas enable us to calculate the average kinetic energy per particle as follows:

$$\langle\varepsilon\rangle = \frac{\int_0^\infty \varepsilon\,dn}{\int_0^\infty dn} = \frac{\int_0^\infty \varepsilon^{3/2}e^{-\beta\varepsilon}\,d\varepsilon}{\int_0^\infty \varepsilon^{1/2}e^{-\beta\varepsilon}\,d\varepsilon} = \frac{\tfrac{3}{4}\sqrt{\pi}\,\beta^{-5/2}}{\tfrac{1}{2}\sqrt{\pi}\,\beta^{-3/2}} = \frac{3}{2\beta}.$$

We know from kinetic theory that the average kinetic energy per particle is $\tfrac{3}{2}k_BT$; hence, our result leads us to the conclusion

$$\beta = \frac{1}{k_BT},$$

as asserted above in Equation (2-36).

2-4 The Quantum Hypothesis

Rayleigh's result for the spectral energy density $u_\nu(T)$ was certainly not the correct answer to Kirchhoff's problem. The fault was found to be the lack of a frequency dependence in the classical expression for the average energy $\langle\varepsilon\rangle$. It was apparent that $\langle\varepsilon\rangle$ should decrease sharply with increasing ν so that the predicted blackbody distribution could rise and then fall with frequency as observed in Figure 2-2. A drastic revision of the high-ν behavior was obviously needed to make the integral over the spectrum converge and give the desired T^4 prediction for the total emittance.

Planck discovered the proper ν dependence in his empirical studies of the blackbody problem and, on these grounds, proposed the following formula for the spectral energy density:

$$u_\nu(T) = \frac{8\pi\nu^2}{c^3}\frac{h\nu}{e^{h\nu/k_BT} - 1}. \tag{2-37}$$

The second group of factors in the formula was deduced to represent the average energy per mode at frequency ν,

$$\langle \varepsilon \rangle = \frac{h\nu}{e^{h\nu/k_B T} - 1},$$ (2-38)

replacing the classical equipartition result $k_B T$. Planck's construction of $\langle \varepsilon \rangle$ guaranteed the required asymptotic behavior of $u_\nu(T)$ and also reproduced the known low-frequency limit where the classical expression was regarded as acceptable. The construction contained two constants, now known as Planck's constant h and Boltzmann's constant k_B, and provided experimental determinations for both parameters. The values obtained from Planck's investigations of the blackbody data were not very different from the currently quoted figures

$$h = 6.6260755 \times 10^{-34} \text{ J} \cdot \text{s} \quad \text{and} \quad k_B = 1.380658 \times 10^{-23} \text{ J/K}.$$

Planck's constant made its first appearance as the basic new parameter of quantum physics on this occasion.

These findings were recognized as an inspired contribution. The proposed blackbody formula was an accurate representation of experiment and, as such, was a satisfactory response to Kirchhoff's long-standing challenge. Planck did not let the issue rest at the empirical level, however. He turned his attention to a theoretical derivation of his formula and in the process gave the quantity of energy $h\nu$ an interpretation of fundamental significance to the quantum theory.

Planck recognized that cavity radiation should be treated as an equilibrium problem involving an exchange of energy in the cavity between the radiation fields and the particles bound to the walls. He accepted the view that the bound particles could be modeled as oscillators, like masses connected to springs, with arbitrary frequencies of oscillation. The familiar classical outlook permitted these oscillating particles to have a continuous range of energies corresponding to a continuous variety of possible amplitudes of oscillation. This meant that the particles could exchange any amount of energy with the radiation fields in the cavity. Unfortunately, the classical assumptions led to the undesirable Rayleigh result, as we have seen in Equation (2-21). In desperation Planck proposed that the energy exchanged between the oscillators in the walls and the radiation in the cavity must vary in *discrete* rather than continuous amounts. The proposition implied that the radiation in the cavity at each frequency ν could be represented as a large collection of *quantized elements of energy* $\varepsilon_n(\nu)$, whose designation could be labeled by means of an integer index n. Planck wanted ε_n to depend on ν so that the average energy $\langle \varepsilon \rangle$ would have a ν dependence capable of averting the Rayleigh catastrophe at large values of ν. He argued that ε_n should correspond to n multiples of a fundamental unit of energy proportional to ν and wrote

$$\varepsilon_n(\nu) = nh\nu$$ (2-39)

to express his hypothesis. The fundamental unit of energy $h\nu$ for radiation of frequency ν was called a *quantum of energy*, in which Planck's constant h was introduced as the proportionality factor relating energy and frequency. With this assertion the *quantization* of a dynamical quantity, the energy, appeared in physics for the first time.

Our analysis of the blackbody problem implements Equation (2-39) by treating the collection of quantized energies as a statistical system. The procedure follows an interpretation of Planck's reasoning along lines developed by P. J. W. Debye in 1910. We adopt this picture because it is easier to present than Planck's thermodynamic oscillator theory and because it draws us toward a more immediate understanding of the nature of the radiation.

Our model of cavity radiation is based on a collection of standing electromagnetic modes in an enclosure bounded by perfectly conducting walls. The effect of Planck's hypothesis is to restrict these modes to a discrete set of allowed energies in which the energy of a particular mode at frequency ν occurs as an integral multiple of $h\nu$ in accord with Equation (2-39). This restriction is obviously a radical departure from the classical point of view. The energy of an oscillating mode depends on the squares of the components of the amplitude E_0 appearing in Equations (2-17); these amplitudes are regarded classically as continuously varying quantities, implying continuous values of the energy.

We should also make clear the thrust of the quantum hypothesis as a device for making the blackbody distribution fall with increasing ν. Let us fix the temperature T and consider a certain total amount of energy to be distributed over all modes at all frequencies. The Maxwell–Boltzmann distribution function determines the probabilities for the assignments of energy to the various modes at frequency ν. Let us denote the governing factor in the distribution function as

$$f(\varepsilon) = e^{-\varepsilon/k_B T}. \tag{2-40}$$

This expression gives the likelihood, in Debye's interpretation, for the occurrence of a given energy ε in the distribution of energies in the cavity. Figure 2-11 shows how the larger energies are exponentially suppressed by the behavior of $f(\varepsilon)$. The modes at large ν are similarly suppressed because of the Planck relation between energy and

Figure 2-11

Maxwell–Boltzmann distribution factor $f(\varepsilon) = e^{-\varepsilon/k_B T}$ giving the relative likelihood for an energy ε in a distribution of energies at temperature T. Quantized energies $nh\nu$ are sampled and weighted by the distribution factor for three different values of $h\nu$.

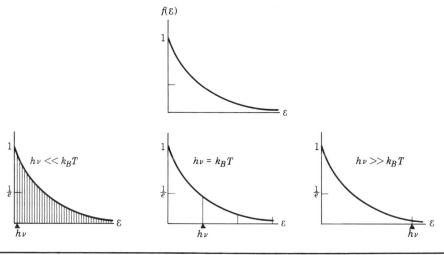

frequency. The figure indicates the details of this suppression for three different choices of frequency. For $h\nu \ll k_B T$, a large number of multiples of $h\nu$ are sampled by $f(\varepsilon)$ before the value of the function drops to $1/e$. For $h\nu \gg k_B T$, the reverse is true and even the first allowed quantum of energy is greatly suppressed. The case $h\nu = k_B T$ is also included, showing the weight of the first quantum to be $f(\varepsilon) = 1/e$. Thus, the direct connection between energy and frequency in Planck's hypothesis accomplishes the suppression of high frequencies in the blackbody spectrum through the influence of the thermal distribution of energies.

Let us now proceed with the derivation of Planck's formula for $\langle \varepsilon \rangle$. At frequency ν, the average energy per mode is computed as the sum of the various weighted energies, with weights given by the likelihood factors, divided by the sum of the weights:

$$\langle \varepsilon \rangle = \frac{\sum\limits_{n} \varepsilon_n f(\varepsilon_n)}{\sum\limits_{n} f(\varepsilon_n)}. \tag{2-41}$$

Since we have discrete energies to average, the averaging process involves discrete summations in which the sums range from $n = 0$ to $n = \infty$. When Equation (2-39) is incorporated, the expression for the average energy becomes

$$\langle \varepsilon \rangle = \frac{\sum\limits_{n} nh\nu e^{-nh\nu/k_B T}}{\sum\limits_{n} e^{-nh\nu/k_B T}} = k_B T \frac{\sum\limits_{n} nxe^{-nx}}{\sum\limits_{n} e^{-nx}}, \tag{2-42}$$

where

$$x = \frac{h\nu}{k_B T}. \tag{2-43}$$

We establish the functional form of $\langle \varepsilon \rangle$ by starting with the denominator and writing

$$Z(x) = \sum_{n=0}^{\infty} e^{-nx} = 1 + e^{-x} + e^{-2x} + \cdots.$$

We recognize this summation to be a geometric series in which the ratio of successive terms is e^{-x}. The sum of the series is

$$Z(x) = \frac{1}{1 - e^{-x}}.$$

Next, we note that the numerator in Equation (2-42) is found from $Z(x)$ by performing the following derivative operation:

$$-x\frac{d}{dx}Z(x) = -x\frac{d}{dx}\sum_{n} e^{-nx} = x\sum_{n} ne^{-nx}.$$

Finally, we assemble the results of these observations and carry out one last series of maneuvers:

$$\langle \varepsilon \rangle = -\frac{k_B Tx}{Z(x)} \frac{d}{dx} Z(x) = -k_B Tx \frac{d}{dx} \ln Z(x)$$

$$= k_B Tx \frac{d}{dx} \ln(1 - e^{-x}) = k_B Tx \frac{e^{-x}}{1 - e^{-x}} = \frac{k_B Tx}{e^x - 1}.$$

Planck's formula for $\langle \varepsilon \rangle$ is then obtained when Equation (2-43) is reinstated in this result.

The remarkable formula for the spectral energy density in Equation (2-37) encompasses all that we might wish to know about blackbody radiation. The following examples illustrate this with derivations of Wien's law and the Stefan–Boltzmann law.

Example

We analyze Wien's law in terms of the wavelength spectrum of the blackbody emittance, as shown in Figure 2-3. Equations (2-6), (2-8), and (2-37) may be combined to give

$$M_\lambda(T) = \frac{c}{\lambda^2} \frac{c}{4} \frac{8\pi \nu^2}{c^3} \frac{h\nu}{e^{h\nu/k_B T} - 1} = \frac{2\pi hc^2}{\lambda^5} \frac{1}{e^{hc/\lambda k_B T} - 1},$$

using $\nu = c/\lambda$. This expression can be made more compact by recalling the dimensionless variable x from Equation (2-43) and rewriting x in terms of λ:

$$x = \frac{hc}{\lambda k_B T}.$$

The wavelength distribution then becomes

$$M_\lambda(T) = \frac{2\pi(k_B T)^5}{h^4 c^3} g(x),$$

where

$$g(x) = \frac{x^5}{e^x - 1}.$$

The function $g(x)$ describes the *universal shape* of the wavelength spectrum for a blackbody at any temperature. A spectral curve is defined with a single peak occurring at the special value of x given by

$$\hat{x} = 4.965.$$

To prove this assertion, let the derivative of $g(x)$ vanish and solve for x:

$$\frac{dg}{dx} = \frac{x^4}{e^x - 1}\left(5 - \frac{x}{1 - e^{-x}}\right) = 0 \quad \text{when} \quad 1 - e^{-x} = \frac{x}{5}.$$

We can use a calculator to get $x = 5$ as an approximate solution and then obtain $x = 4.965$ from there by trial and error. Wien's law pertains to the peak of $M_\lambda(T)$ at $\lambda = \lambda_m$. Since $dM_\lambda(T)/d\lambda = 0$ when $dg/dx = 0$, the parameters λ_m and $\hat{x}$ are immediately related by the equality

$$\hat{x} = \frac{hc}{\lambda_m k_B T}.$$

A final calculation gives

$$\lambda_m T = \frac{hc}{\hat{x} k_B} = \frac{(6.626 \times 10^{-34} \text{ J} \cdot \text{s})(2.998 \times 10^8 \text{ m/s})}{4.965 \left(1.381 \times 10^{-23} \text{ J/K}\right)}$$

$$= 2.897 \times 10^{-3} \text{ K} \cdot \text{m},$$

in excellent agreement with Wien's constant in Equation (2-7).

Example

The derivation of the Stefan–Boltzmann law begins with the formula for the frequency spectrum of the emittance. Equations (2-8) and (2-37) are consulted to get

$$M_\nu(T) = \frac{2\pi h}{c^2} \frac{\nu^3}{e^{h\nu/k_B T} - 1}.$$

The total emittance is then found by applying Equation (2-1):

$$M(T) = \int_0^\infty \frac{2\pi h}{c^2} \frac{\nu^3 \, d\nu}{e^{h\nu/k_B T} - 1} = \frac{2\pi h}{c^2} \left(\frac{k_B T}{h}\right)^4 \int_0^\infty \frac{x^3 \, dx}{e^x - 1},$$

in which Equation (2-43) is again used, this time to change the variable of integration. The dimensionless integral in the computation has a known value, tabulated as

$$\int_0^\infty \frac{x^3 \, dx}{e^x - 1} = \frac{\pi^4}{15} \quad \text{(Dwight 860.33)}.$$

Therefore, the T dependence of the integrated emittance has the form

$$M(T) = \sigma T^4,$$

where the constant σ represents the accumulation of factors

$$\sigma = \frac{2\pi^5 k_B^4}{15 h^3 c^2} = \frac{2\pi^5}{15} \frac{\left(1.381 \times 10^{-23} \text{ J/K}\right)^4}{\left(6.626 \times 10^{-34} \text{ J} \cdot \text{s}\right)^3 \left(2.998 \times 10^8 \text{ m/s}\right)^2}$$

$$= 5.676 \times 10^{-8} \text{ W/m}^2 \cdot \text{K}^4,$$

in excellent agreement with the value of the Stefan–Boltzmann constant in Equation (2-4).

Example

Suppose that the radiation from a blackbody cavity at 5000 K is being examined through a filter that passes a 2 nm wavelength band centered at the peak of the blackbody wavelength spectrum. Our problem is to compute the radiated power transmitted by the filter, given a 1 cm radius for the circular aperture of the device. We proceed by locating the wavelength peak from Wien's law,

$$\lambda_m = \frac{2.90 \times 10^{-3} \text{ K} \cdot \text{m}}{5000 \text{ K}} = 580 \text{ nm},$$

and noting that the filter passes wavelengths in the interval 579–581 nm. The transmitted portion of the blackbody emittance would usually be found by integrating the distribution $M_\lambda(T)$ over the wavelength range of the filter. In this case, the range is narrow enough to permit an approximation in which $M_\lambda(T)$ is evaluated at $\lambda = \lambda_m$ and multiplied by the wavelength interval $\Delta\lambda = 2$ nm. We then obtain the transmitted power when we multiply this result by the area of the filter aperture. Thus, the final formula for the power reads

$$P = \pi r^2 M_{\lambda_m}(T) \, \Delta\lambda = \pi r^2 \frac{2\pi hc^2}{\lambda_m^5} \frac{1}{e^{\hat{x}} - 1} \Delta\lambda.$$

Note that we have used the relation between λ_m and $\hat{x}$ from our first example. The final calculation gives

$$P = 2\pi^2 \frac{\left(10^{-2} \text{ m}\right)^2 \left(6.63 \times 10^{-34} \text{ J} \cdot \text{s}\right)\left(3 \times 10^8 \text{ m/s}\right)^2 \left(2 \times 10^{-9} \text{ m}\right)}{\left(5.80 \times 10^{-7} \text{ m}\right)^5 \left(e^{4.96} - 1\right)}$$

$$= 25.3 \text{ W}.$$

We have already remarked that the use of a broad-band filter would necessitate an integration over the relevant wavelength interval. This task would probably have to be done numerically.

2-5   The Photoelectric Effect

Planck's quantum hypothesis was a radical concept whose effects were discovered in a very complicated setting. Real inspiration was necessary to recognize the influence of quantization beneath the surface of the blackbody problem, and further inspiration was needed to ascertain the meaning of quantization as a new physical principle. Planck's contribution was not immediately acknowledged as a major turning point in theoretical physics. Planck himself was among those who hesitated over the question of interpretation, while Einstein was the one who came forward with the next definitive idea.

Planck's assumption allowed oscillating particles and radiation fields at frequency ν to exchange energy only in integral multiples of the quantum of energy $h\nu$. Einstein's proposal interpreted the radiation directly as an intrinsically discrete system composed of these quanta of energy. (Our analysis of the blackbody problem has already been

Figure 2-12

Schematic representation of the photoelectric process.

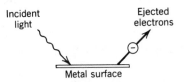

Figure 2-13

Investigation of the photoelectric current as a function of the applied voltage.

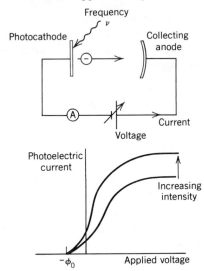

adapted to this interpretation in Section 2-4). The notion that light should have a discrete composition was greeted with skepticism. Everyone was inclined to resist such a reckless proposition because it seemed to contradict the established picture of wave behavior. Einstein looked for indications from experiment that supported his conjecture and found the desired evidence in 1905 in the photoelectric effect. (We should recall that he also put forward his special theory of relativity in the same remarkable year.) In time his proposed quanta of light came to be known as photons.

The photoelectric effect is one of the many phenomena that involve the interaction of electromagnetic radiation with matter. Figure 2-12 shows a sketch of the process, in which the incidence of light on a metal surface causes electrons to be ejected from the body of the metal. A collecting plate at a higher potential can be added to the picture, as in Figure 2-13, so that the photoelectrons can be detected and a photoelectric current can be measured in an external circuit. This simple mechanism is the basis for several familiar electronic devices found in everyday use. The process also occupies a prominent position in the development of 20th century physics.

The first observation of the photoelectric effect was made by Hertz in 1887, very soon after his demonstration of the electromagnetic nature of light. Like so many early discoveries, the effect was noticed by accident. The light-ejected negative charges were proved to be electrons in experiments conducted by J. J. Thomson in 1899. The puzzling aspect of the phenomenon was the fact that the energy of the emitted electrons did not vary with the intensity of the incident light. This was quite surprising since it was thought that electrons with greater energy should be seen if the metal surface was exposed to a greater flux of electromagnetic energy. The photoelectron energy did show an increase, however, when the frequency of the incident light was increased.

These results bewildered everyone but Einstein, who recognized the process as a concrete application of his light-quantum hypothesis. In his view the effect occurred when a single incident quantum of light was absorbed in the metal and all its energy was given up to a single electron. The electron could then be ejected if it acquired enough energy from the absorbed photon to separate the particle from the metal and release the excess energy as free-electron kinetic energy. An increase in the intensity of the light would flood the metal surface with a greater number of quanta and cause more electrons to be ejected. These photoelectrons would produce a greater photoelectric current but would carry no more energy individually unless more was delivered to them by the incident photons. An increase in the *frequency* of the incident light would result in a greater photon energy. The photons would then have more energy to give up to the electrons, and so each ejected electron could leave the surface with greater kinetic energy. Thus, Einstein was able to explain every aspect of the puzzle by assuming an interaction between light and matter, represented by the absorption of discrete quanta of light. The photon concept eventually gained credibility in the decade that followed, as Einstein's predictions were confirmed in a series of photoelectric experiments undertaken by R. A. Millikan.

Let us examine Einstein's predictions in the context of a typical photoelectric experiment, following the procedure illustrated in Figure 2-13. We take light of fixed frequency and variable intensity to be incident on a photocathode, and we let the liberated electrons be accelerated to a collecting anode by applying a difference in potential between the two elements. An ammeter records the observation that the photoelectric current increases, and eventually levels off, as the applied voltage is increased for a given light intensity. This *saturation* of the current indicates a limiting regime where the ejected photoelectrons are being collected at their maximal rate. An increase in the intensity of the light causes the incidence of photons and the liberation of electrons to proceed at a greater rate so that the photoelectric current rises to a higher saturation level. If we reduce the applied voltage to negative values we observe that the negative potential difference retards the collection of photoelectrons at the anode until the current finally vanishes at a particular applied voltage $-\phi_0$. The quantity ϕ_0 is called the *stopping potential*. Its value provides a direct measure of the maximum kinetic energy for an ejected electron,

$$K_{max} = e\phi_0, \qquad (2\text{-}44)$$

because the applied voltage $-\phi_0$ must be such as to retard the fastest electron leaving the cathode. (Note that the electron charge is expressed as $-e$ to obtain this equation.) We observe that the graph of current versus voltage follows a different curve for each intensity of the light and that the various curves have varying levels for the saturation current but the *same* values for the stopping potential. We therefore conclude that ϕ_0 does not vary with the flux of the incident radiation as long as the frequency of the light remains fixed. This striking feature of the experiment is at odds with classical expectations and bears out Einstein's proposition.

The light-quantum hypothesis describes the incident light at frequency ν as a stream of photons, each with energy

$$\varepsilon = h\nu. \qquad (2\text{-}45)$$

A photoelectron is ejected from the metal surface of the cathode whenever it absorbs a photon, provided enough energy is acquired to release the electron from confinement inside the metal. The basic mechanism is expressed as

$$\gamma + e_{bound} \rightarrow e_{free},$$

Figure 2-14
Stopping potential versus frequency as
predicted by Einstein's equation. The slope of
the straight line gives the value of the ratio
h/e.

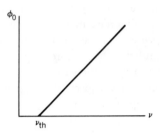

where the symbol γ refers to the absorbed photon. A portion of the absorbed energy ε
is spent to remove the electron from the metal, and the remainder appears as
free-electron kinetic energy. The minimum energy required to free an electron by this
process is called the photoelectric *work function* W_0. This quantity is characteristic of
the specific material in the surface so that W_0 varies from one type of metal to
another. Every electron leaving a given metal with *maximum* kinetic energy must pay a
cost in energy equal to W_0 in order to obtain its freedom. We can account for the
energy absorbed in this case by writing

$$h\nu = W_0 + K_{\max} \quad \text{or} \quad K_{\max} = h\nu - W_0. \tag{2-46}$$

The second formula is Einstein's equation for the photoelectric effect.

We can combine Equations (2-44) and (2-46) to obtain a prediction for the
stopping potential:

$$\phi_0 = \frac{h}{e}\nu - \frac{W_0}{e}. \tag{2-47}$$

This result tells us that the experimental value observed for ϕ_0 should increase linearly
with an increasing light frequency ν. Figure 2-14 shows a representation of the
straight-line relation between ϕ_0 and ν. The graph is typical of those found for any
given metal and is representative of the results obtained by Millikan in confirmation
of Einstein's hypothesis. Note that the slope of the straight line is predicted to be h/e
in Equation (2-47). A photoelectric experiment can therefore provide another determination of Planck's constant, by a method completely different from Planck's
blackbody approach. Note also that the figure indicates a minimum light frequency
below which the photoelectric effect cannot occur. This *threshold* frequency ν_{th} is
obtained by setting $\phi_0 = 0$ in Equation (2-47) to get

$$h\nu_{\text{th}} = W_0. \tag{2-48}$$

Light with this frequency contains photons with just enough energy to release
electrons from the metal with zero speed of ejection. Light of lesser frequency cannot
produce photoelectrons no matter how intense the illumination might be.

Example

Lithium is one of the metals studied by Millikan in his tests of Einstein's equation. The Li photoelectric work function has the tabulated value $W_0 = 2.42$ eV. Equation (2-48) gives the threshold frequency for the emission of photoelectrons as

$$\nu_{th} = \frac{W_0}{h} = \frac{(2.42 \text{ eV})(1.60 \times 10^{-19} \text{ J/eV})}{6.63 \times 10^{-34} \text{ J} \cdot \text{s}} = 5.84 \times 10^{14} \text{ Hz},$$

in good agreement with Millikan's original results. The corresponding maximum wavelength for the photoelectric effect in Li is then given by

$$\lambda_{th} = \frac{c}{\nu_{th}} = \frac{3 \times 10^8 \text{ m/s}}{5.84 \times 10^{14} \text{ Hz}} = 514 \text{ nm},$$

a value in the visible range. Suppose that a 1 W monochromatic source of light with this wavelength illuminates a Li surface at a distance of 1 m. The number of photons emitted by the source per second is

$$\frac{1 \text{ J/s}}{(2.42 \text{ eV/photon})(1.60 \times 10^{-19} \text{ J/eV})} = 2.58 \times 10^{18} \text{ photons/s}.$$

The corresponding number striking unit area of the Li surface per second is

$$\frac{2.58 \times 10^{18} \text{ photons/s}}{4\pi(1 \text{ m})^2} = 2.06 \times 10^{17} \text{ photons/s} \cdot \text{m}^2,$$

assuming a uniform distribution of photons over a sphere of 1 m radius.

2-6 X Rays

The quantum hypotheses of Planck and Einstein pertain to frequencies and wavelengths of radiation across the entire electromagnetic spectrum. Quanta of energy in the electron-volt regime are associated with wavelengths of light in the visible range, where the photoelectric effect is observed as an important process. Longer wavelengths in the infrared and microwave regions of the spectrum correspond to quanta whose energies fall below the photoelectric thresholds of matter. These forms of radiation make their appearance in other kinds of photon interactions. An altogether different domain of interesting radiative processes is found at the shorter wavelengths where the corresponding photon energies are of kilo-electron-volt order. The radiation known as x rays occurs in this part of the electromagnetic spectrum.

X rays were detected for the first time in 1895 by W. K. Roentgen. The discovery was a by-product of experiments on the behavior of electron currents, or cathode rays, in the space between the terminals of a rarefied gas tube. The radiation was observed to come from the high-potential end of the tube and was found to have a remarkable ability to penetrate material and cause ionization in matter. The nature of the radiation remained obscure until 1912, when x rays were conclusively demonstrated to

Figure 2-14

Electromagnetic spectrum. Frequencies and
wavelengths are plotted on logarithmic scales
and satisfy the relation $\nu\lambda = c = 3 \times 10^8$ m/s.

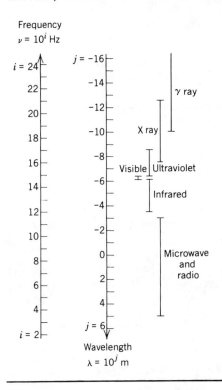

have wave properties as a result of experiments devised by M. T. F. von Laue. These
studies involved the diffraction of beams of x rays using crystals whose lattice spacings
were small enough to accommodate the nanometer wavelengths of the radiation. The
wavelengths were observed to become shorter as the applied voltage on the x-ray tube
was increased through kilovolt values. The radiation was also shown to have propagation and polarization characteristics in common with other kinds of electromagnetic
waves.

Let us examine the principles of x-ray diffraction at this point so that we can have
the results on hand for later reference. We use classical wave optics and present the

Figure 2-15

Bragg reflection of x rays from neighboring
crystal planes.

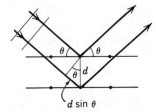

arguments with the aid of Figure 2-15. The diffraction of x rays by a crystal lattice is described in terms of the interference of waves reflected from parallel crystal planes. A crystal plane is defined by a particular two-dimensional array of atoms at their lattice sites inside the crystal. Incident waves are scattered by all the atoms in the crystal, and of course each atom scatters radiation in all directions. The role of the crystal plane is to single out a certain direction for the scattered waves such that the angle of scattering equals the angle of incidence, as in the law of reflection familiar from geometrical optics. Atoms in the crystal produce this collective effect because waves scattered from individual atoms are in phase and interfere constructively if their direction corresponds to reflection from a crystal plane. The resulting wave geometry is illustrated in the figure.

The figure goes on to show two equal-angle reflections from neighboring parallel planes. Waves reflected from planes with lattice spacing d undergo a second stage of constructive interference at reflection angle θ if the optical path difference for the two reflected waves equals a whole number of wavelengths. The figure tells us that the difference between paths is given by $2d \sin \theta$, and so we expect to observe constructive interference whenever the angle obeys the relation

$$\frac{2d}{\lambda} \sin \theta = \text{an integer.} \tag{2-49}$$

This formula can be employed to determine x-ray wavelengths using crystals of known lattice spacing and, conversely, to investigate crystal structures using x rays of known wavelength. The formula in Equation (2-49) is called the *Bragg condition*, and the reflection angle in the formula is called the Bragg angle. These methods of x-ray analysis are named after W. H. and W. L. Bragg, pioneers (as father and son) of the science of x-ray crystallography.

Figure 2-16 shows a sketch of a certain type of x-ray tube in which the x rays are produced in the collisions of energetic electrons with a metal target. The electrons boil away from a hot filament and are accelerated through a large potential difference to the target, where they collide with the heavy atoms in the metal and undergo an

Figure 2-16

Elements of an x-ray tube and features of an x-ray spectrum. The spectrum consists of continuous and characteristic components.

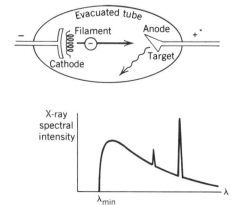

Figure 2-17

Bremsstrahlung process for an electron entering the field of an atomic nucleus.

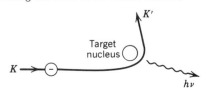

abrupt deceleration. Much of the electrons' loss of energy goes into heating the target, while the rest is emitted directly in the form of radiation. The observed x rays exhibit all wavelengths above a certain minimum value, as illustrated by the distribution of x-ray intensities in the figure. This spectrum shows a superposition of *two* distinct kinds of x-ray distributions. The radiation has a continuous component containing all wavelengths λ above the minimum value λ_{min}; the shape of this distribution is essentially the same for any kind of metal target in the x-ray tube. A discrete collection of sharp wavelengths is also observed in the spectrum; these x-ray lines are characteristic of the target and differ from one metal to another.

The continuous distribution of x rays is called a *bremsstrahlung* spectrum, from the German word for "braking radiation." This component is associated with the classical radiation that results from the acceleration of charged particles, as the electrons are brought to a sudden stop in the metal target. The minimum wavelength is explained by including the quantum hypothesis in the description. Let us refer to Figure 2-17 and visualize the process in terms of a single electron entering the target, encountering an atom, and emitting a single x-ray photon. (Actually, the nucleus of the atom is responsible for the indicated Coulomb attraction and acceleration of the radiating electron.)

The basic process of energy loss may be described as

$$e \rightarrow e + \gamma \quad \text{near a nucleus,}$$

a sort of inverse to the photoelectric effect. We identify the energy of the emitted photon as the difference of kinetic energies shown in the figure:

$$h\nu = K - K'.$$

Note that the recoil energy of the atom can be neglected because the nucleus of the atom is very massive. The energy of the photon is a maximum for those collisions in which the electron is brought to rest. In this situation we have

$$K' = 0 \Rightarrow h\nu_{max} = K.$$

The kinetic energy of the incoming electron is determined by the voltage ϕ across the terminals of the x-ray tube:

$$K = e\phi.$$

We know that a maximum frequency corresponds to a minimum wavelength because of the relation $\nu = c/\lambda$. We can therefore obtain the quantity λ_{min} by assembling the relations that define ν_{max}:

$$\frac{hc}{\lambda_{min}} = h\nu_{max} = e\phi$$

and so

$$\lambda_{min} = \frac{hc}{e\phi}. \tag{2-50}$$

It is obvious that K' is free to assume any value from zero on up to the value of K. Therefore, ν must vary continuously between zero and ν_{max}, while λ ranges continuously between λ_{min} and infinity.

It was noted long ago that a measurement of the threshold wavelength in a bremsstrahlung spectrum could be used in conjunction with Equation (2-50) to furnish another accurate determination of Planck's constant. Studies of the short-wavelength limit were conducted for this purpose in experiments performed by W. Duane and F. L. Hunt in 1915.

The discrete lines in the x-ray spectrum are peculiar to the specific metal in the target of the x-ray tube. These *characteristic x rays* are significant because they offer direct evidence for the quantum properties of matter.

X rays have enjoyed a wide variety of practical applications as a result of their remarkable penetrating power. X-ray sources have been put to use as diagnostic and therapeutic devices in medicine and in dentistry. The radiation has also been employed in industry to examine structures for possible hidden defects.

X rays have recently been observed from sources in the universe beyond the solar system. Galactic sources of the radiation are believed to exist in binary stars, where bremsstrahlung x rays are supposedly produced as ionized matter streams from one star to the other. The exciting new field of x-ray astronomy has developed out of these discoveries.

Example

The combination of constants hc appears often enough in atomic physics to warrant a separate calculation:

$$hc = \frac{(6.626 \times 10^{-34} \text{ J} \cdot \text{s})(2.998 \times 10^{8} \text{ m/s})(10^{9} \text{ nm/m})}{1.602 \times 10^{-19} \text{ J/eV}}$$

$$= 1240 \text{ eV} \cdot \text{nm} = 1.240 \text{ keV} \cdot \text{nm}.$$

We can apply the result immediately to Equation (2-50) and express the short-wavelength cutoff for the continuous x-ray spectrum as

$$\lambda_{min} = \frac{1.240 \text{ kV} \cdot \text{nm}}{\phi}.$$

Thus, a 40 kV x-ray tube has a cutoff at

$$\lambda_{min} = \frac{1.24}{40} \text{ nm} = 0.031 \text{ nm}$$

and produces a continuous spectrum of x rays containing all longer wavelengths.

2-7   The Compton Effect

Einstein's original quantum theory of light proposed that the photon be regarded only as a quantum of *energy*. His conception of the photon continued to evolve until 1917, as he came to believe that a notion of *directedness* should also be incorporated in his idea. He made provision for this by letting the light quantum have characteristics of *momentum* as well as energy. It might seem odd that the father of special relativity should hesitate for more than a decade to bring together these complementary

Figure 2-18

Compton scattering of radiation. The scattered wavelength λ' is longer than the incident wavelength λ. In the classical picture, the incident plane wave and the scattered spherical wave have the same frequency and wavelength.

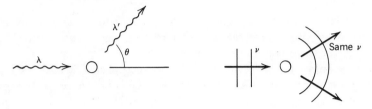

relativistic properties, but such was Einstein's provisional attitude toward the developing quantum theory. He looked for support from experiment to establish momentum and energy as joint aspects of the same hypothesis and was not able to find immediate evidence. A test of his revised photon concept was finally proposed by Debye and, independently, by A. H. Compton. The decisive experiment was then performed by Compton in 1923.

The Compton effect pertains to the scattering of monochromatic x rays by atomic targets and refers to the observation that the wavelength of the scattered x rays is *greater* than that of the incident radiation. Figure 2-18 illustrates the process and identifies the Compton wavelength shift in terms of the wavelength difference $\lambda' - \lambda$. This quantity is observed to vary as a function of the scattering angle θ shown in the figure. The experiment is performed with x rays because the short wavelengths are needed to have an observable effect. A pronounced x-ray wavelength shift is associated with a scattering of the x ray by an *electron* in an atom rather than by the atom as a whole. We demonstrate this experimentally by finding that the shift does not depend on the identity of the atomic scatterer, and so we attribute the effect to the electron as the common constituent of all target atoms. Our discussion of the Compton effect is given in terms of the scattering of an x ray by a *free* electron, since an x-ray quantum carries enough energy to make the distinction between a bound electron and a free electron essentially irrelevant.

The figure goes on to show how the phenomenon presents another unexplainable problem for classical physics. The classical mechanism for the scattering of an electromagnetic wave is explained in a radiation theory developed by (and named after) Thomson. His model assumes that the incident wave comes in at frequency ν and causes the target electron to oscillate and radiate the outgoing wave. The oscillation frequency is also given by ν since the electron is driven at this frequency by the incoming wave. The outgoing classical radiation oscillates at the frequency of the source and therefore has to have the same frequency ν. Thus, the classical picture cannot account for the observed wavelength shift.

The quantum theory of radiation treats the x-ray beam as a stream of photons. For x rays of wavelength λ the photon energy is given by

$$\varepsilon = h\nu = \frac{hc}{\lambda}. \tag{2-51}$$

The photon is also assigned a momentum according to Einstein's revision of the

Figure 2-19

Kinematics of Compton scattering.

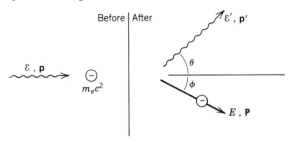

photon concept:

$$p = \frac{\varepsilon}{c} = \frac{h\nu}{c} = \frac{h}{\lambda}. \tag{2-52}$$

Note that this assignment of energy and momentum reproduces the relativistic relation given in Equation (1-39). Recall that the equation holds for a particle of zero mass whose speed is equal to c in all Lorentz frames. Thus, the revised light-quantum prescription begins to attach some of the properties of a *massless particle* to the behavior of a photon.

The Compton process is then described in terms of a relativistic collision involving the elastic scattering of a photon by an electron,

$$\gamma + e \rightarrow \gamma + e.$$

We proceed by imposing the familiar conservation laws for the total relativistic momentum and energy, where the various kinematic quantities are identified in Figure 2-19. We take the electron to be at rest initially and to have energy and momentum E and $\mathbf{P}$ after the collision. These variables are related by the relativistic formula given in Equation (1-35):

$$E^2 = P^2 c^2 + m_e^2 c^4. \tag{2-53}$$

We may write conservation of momentum as

$$\mathbf{p} - \mathbf{p}' = \mathbf{P}$$

and square the equality to get

$$c^2 p^2 - 2c^2 \mathbf{p} \cdot \mathbf{p}' + c^2 p'^2 = c^2 P^2.$$

Since $cp = \varepsilon$ and $cp' = \varepsilon'$, the result can be rewritten as

$$\varepsilon^2 - 2\varepsilon\varepsilon'\cos\theta + \varepsilon'^2 = E^2 - m_e^2 c^4,$$

where Equation (2-53) is used to eliminate P^2. We also employ conservation of relativistic energy by writing

$$\varepsilon + m_e c^2 = \varepsilon' + E.$$

Rearranging and squaring produce the following equality:

$$\varepsilon^2 - 2\varepsilon\varepsilon' + \varepsilon'^2 = E^2 - 2Em_e c^2 + m_e^2 c^4.$$

The two quadratic equations represent information obtained from independent conservation laws. We subtract equalities to get

$$2\varepsilon\varepsilon'(1 - \cos\theta) = 2m_e c^2 (E - m_e c^2)$$
$$= 2m_e c^2 (\varepsilon - \varepsilon'),$$

and we divide by $2\varepsilon\varepsilon'$ to obtain

$$1 - \cos\theta = m_e c^2 \left(\frac{1}{\varepsilon'} - \frac{1}{\varepsilon} \right)$$
$$= m_e c^2 \left(\frac{\lambda'}{hc} - \frac{\lambda}{hc} \right),$$

using Equation (2-51) at the last step. Finally, we solve for the wavelength difference and find

$$\Delta\lambda = \lambda' - \lambda = \frac{h}{m_e c}(1 - \cos\theta) \qquad (2\text{-}54)$$

as the desired result for the Compton wavelength shift.

It is interesting that the shift in wavelength depends only on θ and does not vary with the wavelength of the incident radiation. This feature of the result is seen clearly if we rewrite Equation (2-54) in the form

$$\Delta\lambda = \lambda_C (1 - \cos\theta).$$

The parameter λ_C is known as the *Compton wavelength* of the electron, a constant defined in terms of fundamental constants as

$$\lambda_C = \frac{h}{m_e c}. \qquad (2\text{-}55)$$

This quantity is numerically equal to 0.00243 nm, a length that sets the scale for the wavelength shift in the Compton effect. It is clear that the incident wavelength must be comparable to λ_C if λ' is to be noticeably different from λ. For this reason the effect is not detectable for visible light and only begins to be measurable for x rays.

Compton's experiment was performed with incident radiation at one of the characteristic wavelengths of molybdenum, taken from a Mo target in an x-ray tube. These x rays were scattered from graphite, and the scattered radiation was observed in a detector set at 90° with respect to the direction of incidence, as shown in Figure 2-20. Compton's data confirmed his prediction of a wavelength shift and thereby verified the formula in Equation (2-54) as a valid consequence of the premises

Figure 2-20

X-ray scattering at 90°. The observed distribution of scattered wavelengths shows Compton scattering by electrons at $\lambda' = \lambda + \lambda_C$ and Thomson scattering by atoms at $\lambda' = \lambda$.

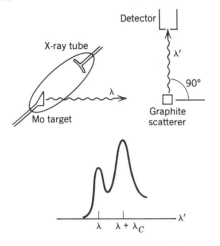

underlying Equations (2-51) and (2-52). Further studies at several values of λ supported the conclusion that the shift was independent of wavelength. These results firmly established the quantum nature of electromagnetic radiation with respect to both energy and momentum aspects of Einstein's proposition.

Let us draw a final connection between these conclusions and the classical treatment of the scattering of radiation. We have alluded to the latter viewpoint, and to Thomson's classical theory, in our discussion of Figure 2-18. The classical picture requires $\lambda' = \lambda$, while the quantum picture predicts $\Delta\lambda \ll \lambda$ and $\lambda' \approx \lambda$ for $\lambda \gg \lambda_C$. This observation tells us that Compton scattering becomes Thomson scattering in the limit where the wavelength becomes much larger than the Compton wavelength. We can see the Thomson limit in another guise by referring to the lower part of Figure 2-20. The observed wavelength distribution for x rays scattered at 90° exhibits a peak at $\lambda' = \lambda$, in addition to the result expected at $\lambda' = \lambda + \lambda_C$. The second peak agrees with the Compton prediction in Equation (2-54) for $\theta = 90°$. The first peak represents no shift and evidently corresponds to photon scattering from the atom as a whole. In this circumstance, the factor $h/m_e c$ in Equation (2-54) is replaced by h/Mc, where M is the mass of the atom. The Compton wavelength of the atom is much smaller than λ_C since M is much larger than m_e. Hence, even x-ray wavelengths experience no observable shift in scattering from the whole atom, and the corresponding feature of the data exhibits the classical Thomson prediction.

The photon is often called a *particle* because it occurs in radiation discretely and because it has energy and momentum properties appropriate for a relativistic particle of zero mass. We refrain from adopting this usage in all its implications, however. We especially avoid contemplating any sort of localization of the photon, as we might imagine in the case of ordinary types of particles. None of our applications of the light-quantum concept includes any notion of electromagnetic energy and momentum at some localized position in space. We learn that the photon is absorbed by an electron in the photoelectric effect and is scattered by an electron in the Compton effect. The electron can be expected to have certain localization properties, but the interaction of the electron with the photon does not require the photon to have any localization of its own. Thus, when we speak of radiation as a system of photons with particle attributes, we realize that we are beginning to entertain a new *quantum–particle* idea in which discreteness and localization are entirely separate considerations.

Example

Let us begin with Equation (2-55) and compute the value for the Compton wavelength of the electron. We use the electron rest energy and the convenient constant hc to get

$$\lambda_C = \frac{hc}{m_e c^2} = \frac{1240 \text{ eV} \cdot \text{nm}}{0.5110 \times 10^6 \text{ eV}} = 0.002427 \text{ nm},$$

as quoted above. If we then let the incident x ray have wavelength $\lambda = 0.0711$ nm as in Compton's 1923 experiment, we find from Equation (2-54) that the wavelength of the x ray scattered at $\theta = 90°$ is

$$\lambda' = \lambda + \lambda_C = (0.0711 + 0.0024) \text{ nm} = 0.0735 \text{ nm}.$$

Figure 2-20 shows how this feature would appear alongside the classical Thomson component at $\lambda' = 0.0711$ nm. (The broadening of the observed wavelengths may be attributed to the fact that the target particle is not necessarily at rest, as assumed in the analysis.) For 90° x-ray scattering, the recoil of the electron has momentum components, parallel to the incident photon and antiparallel to the scattered photon, given by

$$P_x = p = \frac{h}{\lambda} = \frac{6.63 \times 10^{-34} \text{ J} \cdot \text{s}}{0.711 \times 10^{-10} \text{ m}} = 9.32 \times 10^{-24} \text{ kg} \cdot \text{m/s}$$

and

$$P_y = p' = \frac{h}{\lambda'} = \frac{6.63 \times 10^{-34} \text{ J} \cdot \text{s}}{0.735 \times 10^{-10} \text{ m}} = 9.02 \times 10^{-24} \text{ kg} \cdot \text{m/s}.$$

We can compute the electron recoil angle ϕ from these quantities:

$$\tan \phi = \frac{P_y}{P_x} = \frac{9.02}{9.32} = 0.968 \quad \Rightarrow \quad \phi = 44°.$$

This calculation provides further illustration of the Compton effect as an exercise in relativistic elastic collisions.

Figure 2-21

Doppler effect for a photon from a moving source.

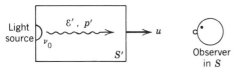

Example

Let us take the photon's energy and momentum from Equations (2-51) and (2-52) and introduce a momentum four-vector for the photon as

$$\not{p} = \begin{bmatrix} \mathbf{p} \\ i\varepsilon/c \end{bmatrix}, \quad \text{where} \quad p = \frac{\varepsilon}{c} = \frac{h\nu}{c}.$$

We know how the four-vector $\not{p}$ transforms under the Lorentz transformation to produce the photon's energy and momentum in another Lorentz frame. We can use this approach as an elegant alternative to our earlier derivation of the Doppler shift for the frequency of light, since both ε and p are directly related to the frequency ν. Figure 2-21 shows the notation for the case of a light source of frequency ν_0 at rest in S', where S' moves with speed u toward an observer at rest in S. We let the photon be directed at the observer so that we need only treat two components of the photon's momentum four-vector in the two Lorentz frames. Since S' is the rest frame of the source, the primed four-vector is

$$\not{p}' = \begin{bmatrix} p' \\ i\varepsilon'/c \end{bmatrix} = \frac{h\nu_0}{c}\begin{bmatrix} 1 \\ i \end{bmatrix}.$$

The corresponding components in S are found by applying the inverse of the Lorentz transformation defined in Equation (1-66):

$$\not{p} = \mathscr{L}^{-1}\not{p}' = \begin{bmatrix} \gamma & -i\gamma\beta \\ i\gamma\beta & \gamma \end{bmatrix}\begin{bmatrix} p' \\ i\varepsilon'/c \end{bmatrix},$$

where

$$\gamma = \frac{1}{\sqrt{1-\beta^2}} \quad \text{and} \quad \beta = \frac{u}{c}.$$

We let the observed photon frequency be called $\bar{\nu}$ in S and perform the transformation as follows:

$$\frac{h\bar{\nu}}{c}\begin{bmatrix} 1 \\ i \end{bmatrix} = \gamma\frac{h\nu_0}{c}\begin{bmatrix} 1 & -i\beta \\ i\beta & 1 \end{bmatrix}\begin{bmatrix} 1 \\ i \end{bmatrix}$$

$$= \frac{h\nu_0}{c}\gamma\begin{bmatrix} 1+\beta \\ i(1+\beta) \end{bmatrix} = \frac{h\nu_0}{c}\gamma(1+\beta)\begin{bmatrix} 1 \\ i \end{bmatrix}.$$

The calculation tells us that $\bar{\nu}$ and ν_0 are related by the formula

$$\bar{\nu} = \nu_0\gamma(1 + \beta) = \nu_0\frac{1 + \beta}{\sqrt{1 - \beta^2}} = \nu_0\sqrt{\frac{1 + \beta}{1 - \beta}} \, ,$$

in agreement with our previous result in Equation (1-9).

2-8   γ Rays and Electron–Positron Pairs

Photons with extremely large energy are found in the γ-ray regime, beyond the x-ray region of frequencies in the electromagnetic spectrum. In practice, any wavelength of nanometer order or less refers to γ radiation, so that the x-ray and γ-ray regions of the spectrum actually overlap. High-energy γ rays appear abundantly in nature in certain very energetic physical processes. Some of the nuclear reactions that take place at very high temperatures in stars produce γ radiation. An appreciable γ-ray component is also present in the secondary cosmic radiation that results from high-energy particle interactions in the Earth's atmosphere. Nuclear γ rays are also observed in the laboratory when accelerated particles are used to excite nuclei and stimulate radiation.

Antimatter can be formed in the interactions of γ rays if the radiation has sufficiently large energy. The production process occurs when an incident γ-ray photon is absorbed in the vicinity of an atomic nucleus, and a matter–antimatter pair of particles is created from the absorbed photon's energy. The least massive of these pair systems consists of the electron and its antiparticle, the positron. Typical particle–antiparticle properties are observed for the electron and positron; the two species have equal mass m_e and opposite charge $-e$ and $+e$. The electron and positron are denoted as e^- and e^+, and the pair-production process is expressed as

$$\gamma \rightarrow e^- + e^+ \quad \text{near a nucleus.}$$

Figure 2-22 shows a sketch of the reaction in which a high-energy incoming photon disappears and materializes into massive particles of opposite charge. Pair production takes place in the Coulomb field of the nucleus, where the photon can be absorbed

Figure 2-22
Electron–positron pair production. A high-energy photon is absorbed in the field of an atomic nucleus, and an $e^- e^+$ pair is created.

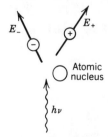

and where the nucleus can act as a massive body to ensure the conservation of momentum and energy. The nucleus is an essential participant as a *spectator* to the process. It is clear that a spectator is needed because, if the photon could spontaneously convert into an e^-e^+ pair in empty space, a Lorentz frame could then be found in which e^- and e^+ would have equal and opposite momenta, and γ would be at rest. Such a conclusion is untenable since γ rays must have speed c in all frames.

To analyze pair production we consider a large photon energy that is still small compared to the very large rest energy of the nucleus. We are then allowed to neglect the recoil kinetic energy of the spectator and account for the energies shown in the figure by writing

$$hν = E_- + E_+$$
$$= K_- + K_+ + 2m_e c^2, \tag{2-56}$$

where K_- and K_+ are the kinetic energies of e^- and e^+. The photon energy must exceed a minimum value in order for the γ ray to initiate pair production. We obtain

$$hν_{min} = 2m_e c^2 \tag{2-57}$$

from Equation (2-56) when we set $K_- = 0$ and $K_+ = 0$. This formula determines the threshold for pair production in the limit of infinite mass for the spectator nucleus. The numerical value of this threshold γ-ray energy is 1.02 MeV. The corresponding maximum γ-ray wavelength is given by

$$λ_{max} = \frac{c}{ν_{min}} = \frac{h}{2m_e c},$$

a result equal to half the Compton wavelength of the electron. Radiation of shorter wavelength contains photons whose energies exceed the threshold and produce electrons and positrons with nonzero kinetic energy.

Pair creation has a reverse process known as pair annihilation. This reaction occurs at any energy when a positron encounters an electron and the two particles disappear, converting their total relativistic energy to γ radiation. Positrons suffer this fate inevitably whenever they come into proximity with matter. The e^-e^+ system cannot annihilate into a single γ ray because the photon would then be found at rest in the CM frame of the e^-e^+ pair. Pair annihilation occurs most rapidly in the two-photon mode

$$e^- + e^+ \rightarrow 2γ,$$

although annihilation into three γ rays is also possible for certain configurations of the electron–positron system.

An interesting e^-e^+ system actually exists in the form of *positronium*, in which e^- and e^+ are bound together by the force of Coulomb attraction. This quasiatomic structure has been synthesized in the laboratory and has been analyzed thoroughly in experiment and in theory. Positronium has a very short lifetime because of the pair-annihilation mechanism; the system lives only 10^{-10} s on average before decaying by its 2γ annihilation mode. Figure 2-23 shows an illustration of the decay in the positronium rest frame, where the indicated photon energies are determined by the relation

$$2hν = 2m_e c^2.$$

Figure 2-23
Positronium annihilation into two photons.

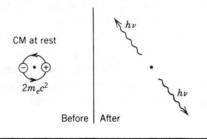

It follows that each γ-ray wavelength is equal in value to the electron Compton wavelength h/m_ec in this particular frame.

We have discussed certain photon phenomena in this chapter as episodes in the growth of the early quantum theory. The photoelectric effect, Thomson and Compton scattering, and pair production are especially important examples of photon interactions because these processes determine how radiation energy is dissipated in its passage through matter. To illustrate, let us consider a beam of radiation of given intensity impinging on a material medium, and assume radiation of a definite wavelength with a corresponding photon energy $h\nu$. The different photon interactions cause different kinds of *attenuation* in the beam intensity as the radiation penetrates the medium. The scattering processes deflect photons out of the beam direction, and the photoelectric and pair-production reactions remove photons from the beam. These three possibilities are drawn schematically in Figure 2-24. Each process has a range of values in the variable $h\nu$, where the particular interaction acts as the dominant cause of beam attenuation. We have just learned that pair production has a γ-ray threshold near 1 MeV. This process is the primary contributor to attenuation of the beam when $h\nu$ exceeds 10 MeV. Of course, the other two processes are the only ones possible when $h\nu$ is less than 1 MeV. The photoelectric effect dominates below 100 keV, while scattering takes over in the intermediate range above and below 1 MeV. We can make these qualitative assertions more specific after we have learned how to measure and compare the various processes.

Figure 2-24
Attenuation of a beam of photons in matter.

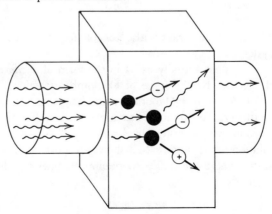

Example

The threshold formula for pair production in Equation (2-57) holds in the limiting case of an infinite-mass spectator. The nucleus of an atom approximates this limiting condition quite satisfactorily. Let us instead consider pair production in the neighborhood of an electron, so that we are compelled to account for the kinematics of all participants in the reaction

$$\gamma + e^- \rightarrow e^- + e^- + e^+.$$

The threshold formula turns out to be quite different in this case. We express the situation at threshold by assigning the incident photon an energy $\varepsilon = h\nu_{\min}$ and by letting the three final particles have the same speed u in the same direction. Momentum conservation requires

$$\frac{\varepsilon}{c} = \frac{3m_e u}{\sqrt{1 - u^2/c^2}}, \qquad\qquad (*)$$

and energy conservation requires

$$\varepsilon + m_e c^2 = \frac{3m_e c^2}{\sqrt{1 - u^2/c^2}}. \qquad\qquad (**)$$

Dividing $(*)$ by $(**)$ results in a determination of the speed u:

$$\frac{\varepsilon}{\varepsilon + m_e c^2} = \frac{u}{c}$$

and so

$$1 - \frac{u^2}{c^2} = 1 - \frac{\varepsilon^2}{\left(\varepsilon + m_e c^2\right)^2} = \frac{2\varepsilon m_e c^2 + m_e^2 c^4}{\left(\varepsilon + m_e c^2\right)^2}.$$

We insert these results in $(*)$ and get

$$\varepsilon = \frac{3m_e u c}{\sqrt{1 - u^2/c^2}} = 3m_e c^2 \frac{\varepsilon}{\varepsilon + m_e c^2} \frac{\varepsilon + m_e c^2}{\sqrt{m_e c^2 \left(2\varepsilon + m_e c^2\right)}},$$

or

$$\sqrt{2\varepsilon + m_e c^2} = 3\sqrt{m_e c^2}.$$

Squaring then leads to the final result:

$$\varepsilon = h\nu_{\min} = \frac{9m_e c^2 - m_e c^2}{2} = 4m_e c^2.$$

The methods of Section 1-11 can also be used to obtain the same conclusion.

Problems

1. Show that the universal function in Kirchhoff's theorem is identical with the spectral emittance of a blackbody. Prove that, for an arbitrary radiating object, the theorem can be stated as

$$\varepsilon_\nu(T) = \alpha_\nu(T),$$

an equality between the spectral emissivity and the spectral absorptance.

2. Calculate the power received from the Sun at the Earth, the total power radiated by the Sun, and the temperature of the Sun. The following data are provided:

$$S = 1350 \text{ W/m}^2 \quad \text{solar constant,}$$
$$r_E = 6.37 \times 10^6 \text{ m} \quad \text{radius of the Earth,}$$
$$R = 1.50 \times 10^{11} \text{ m} \quad \text{radius of the Earth's orbit,}$$
$$r_S = 6.96 \times 10^8 \text{ m} \quad \text{radius of the Sun.}$$

At what wavelength should the peak in the solar spectrum occur?

3. As a hypothetical situation (call it "radiation in flatland"), consider standing electromagnetic waves in a *two*-dimensional "cavity," letting the enclosure be square as shown. Assume that **E** has only two components, E_x and E_y, and that **B** must be perpendicular to the plane of the enclosure. Obtain expressions for the components of the electric field, subject to the condition $E_{tan} = 0$ at the boundaries. Determine the allowed frequencies of the standing-wave modes.

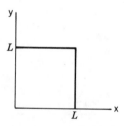

4. Continue with the hypotheses of Problem 3, and show that the number of modes at frequency ν per unit frequency interval is

$$N_\nu = \frac{2\pi\nu}{c^2} A,$$

where A is the area of the enclosure.

5. Still continuing, let "radiation" be emitted from a small opening in the enclosure, noting that the opening is *one*-dimensional so that the radiation streams across an arclength. Let M_ν be the energy emitted per unit time per unit arclength per unit frequency interval, and let u_ν be the energy stored in the enclosure per unit area per unit frequency interval. Show that these quantities are related by

$$M_\nu = \frac{c}{\pi} u_\nu.$$

6. Itemize all the microstates for the assignment of four particles into three cells. Organize these according to macrostates of the system and verify that the number of microstates

corresponding to each macrostate is in agreement with the general formula for the thermodynamic probability.

7. Consider the application of the Maxwell–Boltzmann distribution to a system of classical oscillators representing the radiation modes in a cavity. Classically, a continuous range of the energy ε is allowed for each oscillator. Assume that the number of modes dn per unit energy interval $d\varepsilon$ is proportional to the factor $e^{-\beta\varepsilon}$ (and not $\sqrt{\varepsilon}\,e^{-\beta\varepsilon}$, as in the particle case). Show that the average energy per mode is

$$\langle\varepsilon\rangle = \frac{1}{\beta} = k_B T$$

by making the appropriate averaging calculation.

8. Show that the Planck result for $\langle\varepsilon\rangle$, the average energy per mode, agrees with the classical equipartition result $k_B T$ in the limit of small frequency.

9. A 50 g mass hangs from a spring whose spring constant is 80 N/m. The oscillations of the mass have amplitude 10 cm. Can it be argued that this macroscopic classical oscillator is a Planck oscillator by assigning the system to a quantized energy state whose energy is a multiple of $h\nu$? How accurately would the energy of the oscillator need to be known in order to insist on this interpretation?

10. The wavelength distribution of a 5000 K blackbody radiator is studied for a range of wavelengths extending to either side of λ_m. The endpoints of the range $(\lambda_{\min}, \lambda_{\max})$ correspond to the values of λ where the spectral emittance is half the peak value. Determine the values of $\lambda_{\min}$ and $\lambda_{\max}$.

11. For a blackbody, the frequency distribution peaks at ν_m and the wavelength distribution peaks at λ_m. Consider the derivations of the dependence of ν_m and λ_m on the temperature to prove that $\nu_m\lambda_m \neq c$. Obtain the actual expression for the product $\nu_m\lambda_m$.

12. The peak value M_{λ_m} at the wavelength λ_m in the distribution of blackbody radiation increases with temperature as indicated in Figure 2-3. Show that M_{λ_m} depends on T as

$$M_{\lambda_m} = CT^p,$$

determining the exponent p and the constant C.

13. Continue with "flatland radiation" following Problem 5, and deduce a formula for the frequency dependence of M_ν. (Should the Planck formula for $\langle\varepsilon\rangle$ be different from the case for a three-dimensional cavity?) Integrate M_ν over all ν to obtain the total emittance M. Cast the integration in a form that contains the dimensionless integral

$$\int_0^\infty \frac{x^2\,dx}{e^x - 1}$$

as an explicit factor. Establish the form of the Stefan–Boltzmann law for "flatland," and determine the analogue of the Stefan–Boltzmann constant in terms of the dimensionless integral and other constant factors.

14. Determine the range of photon energies that corresponds to visible light, with wavelengths lying between 400 and 700 nm. Repeat the calculation for radiation in the ultraviolet range from 5 to 400 nm, and in the infrared range from 700 to 500,000 nm.

15. Radio station WTTT (Amherst) broadcasts at 1430 kHz with 5 kW power. Calculate the energy of a WTTT photon and the number of photons broadcast per second.

16. Estimate the number of photons emitted per second by a 100 W light bulb, assuming the average wavelength for the light to be 500 nm. At what distance from the source is the photon flux equal to 100 photons per second per square centimeter?

17. Consider the photoelectric process for a free electron,

$$\gamma + e_{\text{free}} \rightarrow e_{\text{free}},$$

and prove that the process is not possible.

18. A 40 W ultraviolet light source of wavelength 248 nm illuminates a magnesium surface 2 m away. Determine the number of photons emitted from the source per second and the number incident on unit area of the Mg surface per second. The photoelectric work function for Mg is 3.68 eV. Calculate the kinetic energy of the fastest electrons ejected from the surface. Determine the maximum wavelength for which the photoelectric effect can be observed with a Mg surface.

19. The radiation from a 500 K blackbody strikes a metal surface whose work function is 0.214 eV. Determine the wavelength for which the peak of the blackbody spectrum occurs, and determine the longest wavelength in the spectrum capable of ejecting photoelectrons from the surface. What portion of the blackbody's total emittance $M(T)$ is effective in producing photoelectrons from the metal surface? Express the result in terms of a dimensionless integral over the Planck distribution.

20. A γ ray with wavelength 0.005 nm is incident on an electron at rest and is scattered straight backward. Calculate the wavelength of the scattered γ ray and the kinetic energy (in keV) of the recoiling electron.

21. A γ ray of wavelength 0.0062 nm is incident on an electron initially at rest. The electron is observed to recoil with kinetic energy 60 keV. Calculate the energy of the scattered γ ray (in keV), and determine the direction in which it is scattered.

22. Obtain a formula for the fractional loss of energy $(\varepsilon - \varepsilon')/\varepsilon$ for a Compton-scattered photon in terms of the incident wavelength λ and the Compton shift $\Delta\lambda$. Calculate values of this quantity for 90° photon scattering as λ varies from 0.1 nm down to 0.001 nm.

23. Consider Compton scattering of photons with wavelength λ, and show that the scattered photon direction θ and the recoil electron direction ϕ are related by the expression

$$\cot\frac{\theta}{2} = \left(1 + \frac{\lambda_C}{\lambda}\right)\tan\phi,$$

where λ_C is the electron Compton wavelength.

24. Bremsstrahlung x rays of the shortest wavelength from the target in a given x-ray tube have just enough energy to create an electron–positron pair in a dense medium. Deduce the value of the voltage on the terminals of the x-ray tube.

25. In pair production, a photon of energy $h\nu$ is absorbed in the neighborhood of a particle of mass M to make an $e^- e^+$ pair, each with mass m_e. The threshold photon energy for this process is known to be

$$h\nu_{\text{min}} = 2m_e c^2 \quad \text{for } M \rightarrow \infty$$

and

$$h\nu_{\text{min}} = 4m_e c^2 \quad \text{for } M = m_e.$$

Derive a formula for $h\nu_{\min}$ that is valid for *any* M, and show that the result gives agreement in these two special cases.

26. Consider the Compton scattering of photons from a *beam* of electrons, as shown in the figure. Let a photon and an electron collide head-on with relativistic energies ε and E, respectively, and derive an expression for the energy ε' of the final photon, in the special case of *backward* photon scattering. Show that, for $E \gg m_e c^2$, the result reduces to

$$\varepsilon' = \frac{E}{1 + m_e^2 c^4 / 4\varepsilon E}.$$

Calculate the value of ε' for the collision of a 700 nm photon with a 20 GeV electron.

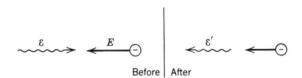

Before | After

T H R E E

INTRODUCTION
TO
THE
ATOM

*Q*uantum physics began to receive enthusiastic attention in 1911 on the occasion of the First Solvay Congress. The most prominent names in physics convened and gave the new theory a forum, where all in attendance could respond to the radical ideas of Planck and Einstein. It was clear to adherents and skeptics alike that the quantum theory was in need of a proper formulation and that the theory was likely to have applications to matter as well as radiation. It was generally recognized that the structure of the atom presented the next urgent problem to be solved.

The modern view of the atom began to develop during the same year. E. Rutherford was the one responsible for devising the basic model and for conducting the decisive experiment. Rutherford's conception of the atom drew upon classical principles alone and contained a serious flaw that posed another insoluble problem for classical physics. The remedy was found by appealing to quantum concepts to complement the classical picture. This next major contribution to the quantum theory was introduced in 1913 by N. H. D. Bohr. The proposal had the effect of broadening the new theory into a combined quantum treatment of matter and radiation. The Rutherford atom and the Bohr atom were the most important developments in physics during this period.

3-1 The Reality of Molecules and Atoms

The molecular and atomic nature of matter attracted speculative interest long before the real existence of molecules and atoms could be proved. These constituent particles were believed to exist as identical units in a given substance and were supposed to make up the composition of any sample of that substance. This belief began to gain ground on two fronts during the 19th century. The concept of submicroscopic units was given consideration in chemistry in a scheme to organize the regularities of the elements and was directly employed in the kinetic theory in a model to describe the behavior of gases. Skeptics could still argue, however, that even the most successful

Ernest Rutherford

Niels Bohr

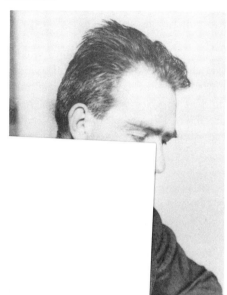

scheme or model could not e   particles. This
skepticism had to be overcome   y of the whole
molecular picture.

The periodic table of the ele   . The idea was
originally put forward in 1869   arranging the
known chemical elements in o   atomic number
supplanted the atomic mass   , although the
ultimate interpretation of the   ntil much later.
Mendeleev's table listed the a   where each row
contained a series of elements   ern from left to
right, and where each column   the same from
top to bottom. This organiz   d order out of
disarray, as elements were ar   properties and
vacancies were left available   stration of these
regularities supported the beli   ticles were to be
associated with the various sp

Avogadro's hypothesis was   of molecules and
atoms. The principle stated   d pressure and
temperature must contain eq   number was not
immediately determined by e   al determination
of this quantity proved to be   of the molecular
theory.

A discovery early in the   rown played an
important part in these devel   examine grains of
pollen suspended in a fluid and observed that the microscopic particles executed a
distinctive random motion. The explanation for this effect came decades later, after
the formulation of the kinetic theory. The movement of pollen grains was seen as
evidence for the thermal motion of particles in the suspending fluid, as the collisions of

fluid particles with suspended particles resulted in the observed Brownian motion. Einstein analyzed this behavior in 1905 (again, his phenomenal year) and showed that the conclusions could be used experimentally to deduce a value for Avogadro's number.

We can reproduce Einstein's treatment of Brownian motion in a few steps by following P. Langevin's more elementary analysis of the derivation. Let us consider a large number of identical spherical particles with radius a and mass m, suspended in a fluid of viscosity η, and examine the forces that act in one direction on one of the particles. The force of viscous damping depends on the velocity of the particle according to the law of G. G. Stokes:

$$F_{vis} = -\beta \frac{dx}{dt}, \tag{3-1}$$

where

$$\beta = 6\pi\eta a. \tag{3-2}$$

The suspended particle also experiences a random force F_{col} whenever it collides with particles in the suspending fluid. Newton's law for the mass m takes the form

$$m\frac{d^2x}{dt^2} = F_{col} - \beta\frac{dx}{dt}, \tag{3-3}$$

since there are no other forces acting on m. (We neglect gravity and buoyancy because these effects tend to cancel each other.)

We are not concerned with a detailed solution for $x(t)$, the instantaneous position of the suspended particle. Such a function would be rather difficult to predict because of the complex t dependence of the random external force F_{col}. Instead, we are interested in *averaged* quantities associated with x, where the averaging operation $\langle\ \rangle$ is computed over the large number of suspended particles. The average of $x(t)$ can be expected to vanish since the random motion causes positive and negative values of x to occur with equal likelihood over the collection of particles. Hence, the *fluctuation* of suspended particles about zero displacement becomes the main quantity of interest. We define this notion by means of the averaged variable $\langle x^2 \rangle$. The fluctuation corresponding to motion in one dimension is then given by

$$x_{rms} = \sqrt{\langle x^2 \rangle}, \tag{3-4}$$

the root-mean-square (rms) displacement in the x direction.

Provisions are made for this analysis by casting Equation (3-3) in terms of the variable x^2. We use the relations

$$\frac{d}{dt}x^2 = 2x\frac{dx}{dt} \quad \text{and} \quad \frac{d^2}{dt^2}x^2 = 2x\frac{d^2x}{dt^2} + 2\left(\frac{dx}{dt}\right)^2$$

and then multiply Equation (3-3) by x to obtain the equality

$$m\left[\frac{1}{2}\frac{d^2}{dt^2}x^2 - \left(\frac{dx}{dt}\right)^2\right] = xF_{col} - \frac{\beta}{2}\frac{d}{dt}x^2. \tag{3-5}$$

The next step is to average this result over all the suspended particles. We note first that the term xF_{col} must average to zero because of the random nature of the collisions throughout the suspending fluid. The force F_{col} drops out of the equality at this point; however, the mechanism for Brownian motion is still present in one of the other remaining terms. Note that the second contribution on the left is twice the kinetic energy of the particle for one degree of freedom. We recall that the average of this energy is given by $\frac{1}{2}k_BT$, where T is the temperature of the suspending fluid. Thus, when we introduce

$$\left\langle m\left(\frac{dx}{dt}\right)^2\right\rangle = k_BT$$

for the average of the term in question, we find that thermal-equilibrium aspects of the problem are still securely in place. The averaging procedure is finally performed as follows:

$$\left\langle \frac{m}{2}\frac{d^2}{dt^2}x^2 + \frac{\beta}{2}\frac{d}{dt}x^2\right\rangle = \left\langle m\left(\frac{dx}{dt}\right)^2\right\rangle \quad \Rightarrow \quad \frac{m}{2}\frac{dg}{dt} + \frac{\beta}{2}g = k_BT, \quad (3\text{-}6)$$

where the final equation is written with the shorthand notation

$$g(t) = \frac{d}{dt}\langle x^2\rangle.$$

Note that m and β are common to all particles and therefore act as constants when the averages are taken.

We draw our main conclusion from this differential equation for $g(t)$ by passing immediately to a certain limiting case. The phenomenon of Brownian motion becomes more and more pronounced for diminishing values of the mass of the suspended particle, so that Equation (3-6) describes the situation of interest in the $m \to 0$ limit. The dg/dt term then drops out of the equation, and so the solution for $g(t)$ assumes the simple limiting form

$$\frac{d}{dt}\langle x^2\rangle = \frac{2k_BT}{\beta}. \tag{3-7}$$

We take $\langle x^2\rangle = 0$ at $t = 0$ and obtain the t dependence of the Brownian fluctuation by integrating Equation (3-7):

$$\langle x^2\rangle = \frac{2k_BT}{\beta}t = \frac{k_BT}{3\pi\eta a}t, \tag{3-8}$$

using Equation (3-2) at the last step. We obtain the final formula

$$\langle x^2\rangle = \frac{RT}{3\pi\eta aN_A}t \tag{3-9}$$

when we introduce the relation

$$k_B = \frac{R}{N_A}$$

to eliminate Boltzmann's constant in terms of the universal gas constant R and Avogadro's number.

Einstein saw that his result for $\langle x^2\rangle$ could be applied in a suitably designed study of Brownian motion to make an accurate determination of Avogadro's number N_A.

This remarkable deduction was the basis for a series of experiments undertaken by J. B. Perrin in 1908. Perrin's measurements of N_A were reasonably close to the current value

$$N_A = 6.0221367 \times 10^{23} \text{ molecules/mole.}$$

It soon became clear that all sorts of different determinations of N_A were in essential agreement. The fact that a consensus prevailed among many types of measurement constituted a strong case for the existence of molecules and atoms as real constituents of matter.

Example

Brownian motion can be investigated with the aid of a microscope, a stopwatch, and a fluid containing very small spheres of the same known size. An estimate of the expected fluctuation serves to illustrate the ideas. Let the suspending fluid be water at room temperature ($17°C$), and suppose that spheres of 1 μm radius are observed for 1 min. We take $\eta = 10^{-2}$ poise $= 10^{-3}$ N · s/m^2 for the viscosity of water and consult Equation (3-8) to obtain

$$\langle x^2 \rangle = \frac{k_B T t}{3\pi \eta a} = \frac{(1.38 \times 10^{-23} \text{ J/K})(300 \text{ K})(60 \text{ s})}{3\pi(10^{-3} \text{ N · s/m}^2)(10^{-6} \text{ m})} = 26.4 \times 10^{-12} \text{ m}^2.$$

Equation (3-4) then gives the rms displacement as

$$x_{\text{rms}} = 5.13 \times 10^{-6} \text{ m},$$

or approximately 5 μm after 1 min.

3-2   The Electron

The atom was believed to have an internal charged structure even before the real existence of atoms could be firmly ascertained. Several pieces of evidence supported this belief. Faraday's electrolysis experiments detected the presence of charged atomic particles, or ions, in solutions. Radiation from atoms suggested the influence of some kind of oscillating charge inside the atomic system. Radioactivity demonstrated the ability of some atoms to change certain aspects of their internal composition. These indications of structure were already in evidence before 1900. At the end of the 19th century the electron was finally identified as a universal charged constituent that appeared in the construction of all atoms.

Electrons were discovered in electrical discharges in gases at low pressure. These phenomena were observed in the application of a potential difference to gaseous systems containing electrons, ions, and neutral atoms. The electrons in a discharge tube could be accelerated by an applied voltage to produce a beam of *cathode rays*, so called because the negative charges moved in the applied field toward the anode and appeared to originate at the cathode. The beam particles were known to have negative charge because the deflection of the beam by transverse magnetic and electric fields could be detected and correlated with the sign of the charge. The identity of the particles was established by Thomson in 1897, in an experiment designed to measure the charge-to-mass ratio e/m of the cathode rays.

Figure 3-1

Thomson's cathode ray tube. Electrons are accelerated from the cathode to the anode and are collimated by the slit to produce a beam. The electrons are deflected by the transverse field applied to the beam on its way toward the phosphorescent screen.

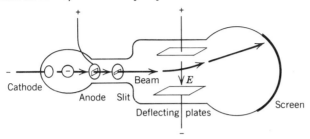

A sketch of Thomson's cathode ray tube is shown in Figure 3-1. Let us recall the familiar details of the classic e/m experiment by referring to Figure 3-2, where we describe the path of a negative charge $-e$ in a transverse electric field. We regard the strength E of the deflecting field as known, given a plate separation d and an applied voltage ϕ:

$$E = \frac{\phi}{d}.$$

The velocity in the x direction is equal to the speed v of the particle as it enters the field, while the acceleration in the y direction is equal to eE/m since eE is the magnitude of the force on the mass m. If we let the particle traverse the field in time t, we get

$$x = vt$$

for the distance across the field and

$$y = \frac{1}{2}\frac{eE}{m}t^2 = \frac{E}{2}\frac{e}{m}\left(\frac{x}{v}\right)^2 \tag{3-10}$$

for the vertical deflection of the beam as it leaves the field. The speed v is determined by a separate procedure at another stage of the experiment, where a magnetic field of strength B is also applied in a direction perpendicular to both $\mathbf{v}$ and $\mathbf{E}$. The desired configuration of $\mathbf{v}$, $\mathbf{E}$, and $\mathbf{B}$ is such that the electric and magnetic forces cancel, so

Figure 3-2

Electron deflection in an electric field.

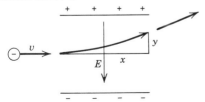

that the particle moves through the fields undeflected. We tune the B field to ensure the equality of forces,

$$eE = evB,$$

and use this condition to obtain the velocity

$$v = \frac{E}{B}. \tag{3-11}$$

Equations (3-10) and (3-11) are then combined to produce a formula for the ratio e/m in terms of measurable quantities.

Thomson's determination of e/m for the electron gave a value that was rather different from the current figure

$$\frac{e}{m_e} = 1.7588196 \times 10^{11} \text{ C/kg}.$$

His result was significant nevertheless, because the order of magnitude was too large to be interpretable in terms of ions (given what was known at the time about ionic charges and masses) and because the value did not vary appreciably when different gases and cathodes were used in the tube. It was clear that he had discovered a unique particle of small mass whose occurrence was common to atoms of all species.

Since the cathode rays in Thomson's discharge tube were electrons with negative charge, it would follow that ions with positive charge should also be detected drifting through the gas in the opposite direction. These ions were in fact observed in a similar kind of tube in which the cathode was perforated to allow the reverse flow of particles with positive charge. Thomson called the particles *positive rays*. Rutherford proposed that the most elementary ions should be those obtained from hydrogen atoms, and he gave these the name *protons*. An e/m experiment of the Thomson type would find all such atomic ions to have charge-to-mass ratios thousands of times smaller than the value of e/m_e.

Another experiment was needed in order to determine the quantities e and m_e separately. The magnitude of the electron charge was measured next by Millikan in a series of studies beginning in 1906.

The famous Millikan oil-drop experiment is like its partner, the Thomson e/m experiment, to the extent that both employ concepts from classical physics and both produce information for the modern era. A sketch of Millikan's device in Figure 3-3 shows an oil drop of mass M suspended in air in an electric field E between a pair of charged metal plates. We let the selected drop have a net positive charge q, and we express the charge as n multiples of the electron charge unit e. A microscope is used to observe the rise and fall of the drop through a fixed fiducial region as indicated in the figure. The forces on the drop are the weight Mg, the Stokes-law force βv due to the viscosity of air, and the electric force neE. We assume that the motion in the fiducial region occurs at *terminal velocity* for all observations of rise and fall. The forces acting on the drop are therefore in equilibrium so that the drop moves with *constant* speed v.

The procedural aspects of the experiment are described in the figure. First, we examine the fall of the drop with the electric field turned off. The figure tells us that

Figure 3-3

Schematic plan of Millikan's oil-drop experiment. A droplet of mass M and charge q is allowed to rise and fall in air between a pair of charged plates. The forces are in equilibrium, so the motion is at terminal velocity, in the fiducial region labeled by the vertical distance y. The drop falls with E turned off in time t_0 and rises with E turned on in time t. The damping force due to the viscosity of air acts upward on the falling drop and downward on the rising drop.

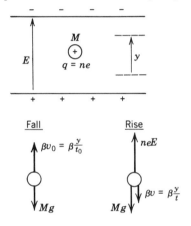

the equilibrium of forces is expressed as

$$\beta \frac{y}{t_0} = Mg, \tag{3-12}$$

where t_0 is the measured fall time. Next, we measure the rise time t with the electric field turned on. The figure shows the equilibrium condition to be

$$neE = Mg + \beta \frac{y}{t}$$

$$= \beta y \left(\frac{1}{t_0} + \frac{1}{t} \right),$$

where Equation (3-12) is used to get the final expression. We then alter the charge on the drop by exposing the air gap to a burst of x rays. The ionizing radiation changes the charge multiple from n to n' and introduces a new measured rise time t'. The previous equation changes accordingly into the new form

$$n'eE = \beta y \left(\frac{1}{t_0} + \frac{1}{t'} \right).$$

The desired formula for the analysis of the experiment is obtained by subtracting the last two results and solving for the difference in charge:

$$(n' - n)e = \frac{\beta y}{E}\left(\frac{1}{t'} - \frac{1}{t}\right). \tag{3-13}$$

We insert data into this formula for a series of repeated trials, corresponding to a variety of integer differences $n' - n$, and thus deduce the value of e from the resulting survey.

The results of a Millikan oil-drop experiment and a Thomson e/m experiment can be combined to determine the mass of the electron. More current experimental methods are also available in which greater accuracy is achieved in the measurement of e and m_e. The values quoted for these basic parameters of the first known elementary particle are

$$e = 1.60217733 \times 10^{-19}\,\text{C} \quad \text{and} \quad m_e = 9.1093897 \times 10^{-31}\,\text{kg}.$$

The fundamental constant e is especially significant in view of the quantization of charge since the parameter gives the basic charge unit for all known observable particles.

Example

Let us get a feeling for the numbers involved in a Thomson e/m experiment by choosing the following reasonable values for the measurements. Take the deflection plates in Figure 3-1 to be at 200 volts with a 2 cm separation so that the deflecting electric field is

$$E = \frac{\phi}{d} = \frac{200\,\text{V}}{0.02\,\text{m}} = 10^4\,\text{V/m}.$$

Let the length of the plates be 4 cm and let the vertical deflection of the electron beam be 0.5 cm so that the speed of the electrons is found from Equation (3-10) to be

$$v = \sqrt{\frac{Eex^2}{2my}} = \left[\frac{(10^4\,\text{V/m})(1.60 \times 10^{-19}\,\text{C})(0.04\,\text{m})^2}{2(9.11 \times 10^{-31}\,\text{kg})(0.005\,\text{m})}\right]^{1/2}$$

$$= 1.68 \times 10^7\,\text{m/s}.$$

The strength required for the B field may then be computed from Equation (3-11):

$$B = \frac{E}{v} = \frac{10^4\,\text{V/m}}{1.68 \times 10^7\,\text{m/s}} = 5.95 \times 10^{-4}\,\text{T},$$

corresponding to an approximate 6 G field.

Example

We need to know β in order to use Equation (3-13) in a Millikan oil-drop experiment, and Equation (3-2) reminds us that β depends on the droplet radius a as well as the viscosity η for air. Let us eliminate a in terms of other known or measurable quantities, since the radius of the drop is not measured directly. We assume a spherical drop and express the mass as

$$M = \tfrac{4}{3}\pi a^3 \rho,$$

where ρ is the density of oil. We then recall the formula for β from Equation (3-2) and rewrite Equation (3-12) in the form

$$6\pi\eta a \frac{y}{t_0} = \frac{4}{3}\pi a^3 \rho g.$$

This equality allows us to calculate a from other more accessible parameters.

3-3 The Nuclear Model of the Atom

By 1900 it was generally acknowledged that Thomson's electrons were present somewhere within the structure of every atom. The distribution of the internal charge of the atom was the next important issue to be decided. The decade leading up to 1910 was a time for atomic models to be put forward so that specific predictions could be tested by experiment. Classical principles were the only tools available for the construction of these models. Quantum physics was still an infant science and would have nothing to contribute until the classical picture of the atom had had a chance to develop.

It was accepted that an atom was electrically neutral in its normal state and that the removal of electrons from an atom resulted in the formation of a positively charged ion. Every atom of a particular element was supposed to contain a certain number of electrons to form a full complement. The chemical significance of this number, the *atomic number Z*, did not emerge until somewhat later. It was clear, however, that a given atom with Z electrons could be neutral only if the atom also contained positively charged matter with a total charge equal to $+Ze$. Furthermore, it was known that the electron was thousands of times less massive than any atom, and so it was obvious that almost all the mass of the atom had to reside in its positive component.

Thomson visualized the distribution of charge and mass in the atom by means of a spherical model, shown in Figure 3-4, in which the whole atomic volume was supposed to be filled by the massive positive component, except for isolated locations throughout the volume where the Z electrons were to be found. He assumed that the electrons were subject to the Coulomb force and were able to move about positions of

Figure 3-4

Thomson's model of the atom.

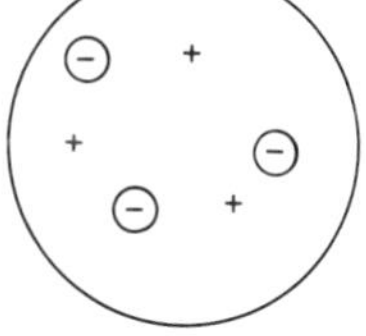

Figure 3-5

Rutherford's model of the atom.

equilibrium in their charged environment. The emission of light by atoms was believed to be attributable to the motion of the atomic electrons. Thomson hoped that the observed frequencies of the light could be explained in terms of the periodic motion of his embedded electrons. This enterprise was doomed to fail, as the Thomson atom was soon proved by experiment to be incorrect.

Direct tests of Thomson's model were carried out by means of scattering experiments. The observed process involved the elastic collision of charged particles, in which the Coulomb force between the incident and target particles could act as a probe for the distribution of charge in the structure of the bombarded particle. The scattering of a beam of particles was analyzed to determine the dependence on the scattering angle and the beam energy. This information provided a test for specific predictions of the atomic model as well as a method for exploring unknown features of the scattering process. The first investigations of the structure of the atom by these techniques were Rutherford's α-particle scattering experiments.

Rutherford made his monumental contribution to the understanding of the atom as an outgrowth of his studies of the radioactive elements. These earlier experiments enabled him to explain radioactivity as a characteristic change of properties in certain atoms, accompanied by the observed emission of very energetic particles. Greater achievements followed, as Rutherford undertook the exploration of the atom using α particles to probe the unknown structure. He had discovered α particles in the radioactive disintegrations of certain heavy elements and identified the radiated particles to be fully ionized helium atoms stripped of their two electrons. Rutherford realized that these energetic positive particles could be used for probing matter on an atomic scale. In 1909 he and his associate H. W. Geiger began their program of α-particle scattering experiments. These studies refuted Thomson's model and established an entirely different picture for the structure of the atom.

Rutherford proposed the nuclear model of the atom in 1911. He assumed that the massive positive component of the atom was contained within a very small region of the atomic volume, called the *atomic nucleus*. A scale of size could be determined for the nucleus by α-particle scattering. The atomic radius was known to be of order 10^{-10} m, and the nuclear radius was found from experiment to be of order 10^{-14} m. The atomic electrons were supposed to move in stable orbits around the nucleus, as in Figure 3-5. Thus, the Rutherford atom resembled the solar system, with the force of gravity replaced by the force of Coulomb attraction and with the system reduced to atomic scale.

The famous Rutherford scattering experiment was laid out according to the plan described in Figure 3-6. A radioactive source provided α particles with energies of

Figure 3-6

Design of the Rutherford α-particle scattering experiment.

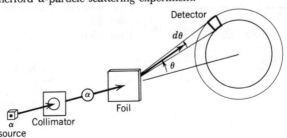

several MeV, and a collimated beam of these particles was directed at a very thin foil target. Scattered particles were detected on a scintillating screen that responded to the arrival of a charged particle by emitting a burst of visible light. The distribution of the scattered α particles was studied at fixed energy as a function of the *scattering angle* θ shown in the figure. Different beam energies were obtained by using various α sources, and different atoms were probed by substituting foils of various metals. The purposes of the experiment were to determine the angular behavior of the scattering and to compare observations with predictions in order to assess the validity of the atomic model. Two qualitative conclusions were immediately drawn from the first experiments. Almost all the incident particles were transmitted through the foil with very little deflection, and a very few incident particles (approximately 1 in 10,000) were scattered *backward* into detectors located on the beam side of the foil. These amazing observations were explainable only in terms of a nuclear model of the atom.

Let us see how this evidence bears on the choice of atomic models. We note first that it is safe to ignore atomic electrons when α particles are incident on atoms because the electrons are too light to cause appreciable scattering of the much heavier α particles. The observed scattering is therefore attributable to the interaction between the α particle and the massive positive component of the atom. This interaction is expected to have very different effects in the two models of Thomson and Rutherford. The Thomson atom has a very *diffuse* distribution of positive charge, and so the encounters between beam particles and target particles are likely to result in only small random deflections as the α particles pass through the target atom. Rutherford's model of the atom adopts a very *concentrated* distribution of positive charge so that most of the atom presents itself as empty space to an incident particle. Therefore, most encounters with the Rutherford atom also result in small deflections, *except* for the rare occasion when the encounter between the α particle and the target nucleus is nearly *head-on*. These collisions produce the rare backward scattering events that distinguish the Rutherford atom from the Thomson atom. Thus, the qualitative predictions of the nuclear model are borne out in the experimental results, while the diffuse model is decisively refuted. The details of the Rutherford scattering problem are analyzed quantitatively in Section 3-4.

Example

We discover other failures of Thomson's model when we look at the Thomson version of the hydrogen atom. Let the single electron of hydrogen be embedded in a spherical medium having radius R, constant charge density ρ, and total charge e. These quantities are connected by the formula for the charge

$$e = \tfrac{4}{3}\pi R^3 \rho.$$

The force on the embedded electron depends on the charge $\tfrac{4}{3}\pi r^3 \rho$, the portion of the total positive charge that lies within the electron's instantaneous location at radius r. Gauss' law determines this force to be

$$F = -\frac{1}{4\pi\varepsilon_0}\left(\tfrac{4}{3}\pi r^3 \rho\right)\frac{e}{r^2},$$

where the sign denotes an attractive force directed toward the center of the atom. We note that F is proportional to the variable r as in the case of the restoring force of a spring:

$$F = -kr$$

with effective "spring constant"

$$k = \frac{1}{4\pi\varepsilon_0} \frac{4}{3}\pi\rho e = \frac{1}{4\pi\varepsilon_0} \frac{e^2}{R^3}.$$

We therefore expect the electron to execute simple harmonic motion with frequency

$$f = \frac{1}{2\pi} \sqrt{\frac{k}{m_e}},$$

and we also expect the oscillating electron to emit radiation of the same frequency. These considerations have nothing to do with reality, as the spectrum of hydrogen has infinitely many frequencies and indicates a very different physical picture.

3-4 Rutherford Scattering

We treat the scattering of α particles by atomic nuclei as a two-stage problem in classical physics. The first part of the analysis pertains to the orbit of a *single* charged particle under the influence of a force of Coulomb repulsion. The second part deals with the behavior of a *beam* of particles whose individual orbits are known from the first stage of the problem. We cast the treatment in terms of a particle beam because our ultimate concern is the experimental question of detecting *many* scattered particles as the beam impinges on *many* target nuclei. This approach leads us to the important concept of the scattering cross section.

Nonrelativistic principles are more than adequate here since the kinetic energy of the α particles is much smaller than their rest energy. We begin by studying the scattering of a single beam particle by a single target nucleus, with charges denoted by ze and Ze, respectively. (Since α particles are ionized He atoms, the corresponding charge index is actually given by $z = 2$.) Both charges are positive and so the scattering is caused by a repulsive Coulomb interaction. We let M refer to the mass of the incident particle, and we take the mass of the target particle to be much larger than M. We do this for simplicity so that we can assume a picture in which the target charge Ze remains fixed at the origin and does not recoil during the collision. (The no-recoil assumption does not reduce the generality of our treatment. We comment briefly on this point again in Section 3-6.)

Let us look first at the special case of a head-on collision. The repulsive nature of the force entails a *distance of closest approach* that varies with the kinetic energy of the incident beam. We call this useful parameter D and define the distance with the aid of Figure 3-7. The total energy of the system is equal to the beam kinetic energy K when the particle is at the extreme position $r = \infty$ and is equal to the Coulomb potential energy when the particle is at the turning position $r = D$. We use conservation of energy to equate the two quantities,

$$K = \frac{1}{4\pi\varepsilon_0} \frac{Zze^2}{D},$$

Figure 3-7

Distance of closest approach in a head-on collision.

and then solve for the distance to the turning point:

$$D = \frac{1}{4\pi\varepsilon_0}\frac{Zze^2}{K} = \frac{1}{4\pi\varepsilon_0}\frac{2Zze^2}{Mv^2}. \tag{3-14}$$

The speed of the particle at $r = \infty$ appears in the final step through the nonrelativistic expression for the beam kinetic energy, $K = \frac{1}{2}Mv^2$.

We turn next to the general case of a non-head-on collision and consider the orbit of a scattered particle. Figure 3-8 shows how the incident particle is initially directed along an incoming asymptote that passes the target at a distance b, known as the *impact parameter*. This distance is an important property of the orbit in any scattering problem. The conservation laws of energy and angular momentum require the outgoing asymptote to have the *same* impact parameter. It then follows that the orbit is symmetrical about a bisector that divides the motion into incoming and outgoing parts. These features of the orbit are indicated in the figure. We note that the orbit in Figure 3-8 can be identified with the path of the scattered particle in Figure 3-6 and that the same *scattering angle* θ appears in both figures. Our main objective in this part of the analysis is to establish the relation between the impact parameter and the scattering angle.

Angular momentum plays a major role in the solution for the orbit. A conservation law holds for the angular momentum vector $\mathbf{L} = \mathbf{r} \times \mathbf{p}$ because the Coulomb force is a *central* force that imposes no torque on the particle. Therefore, the direction of $\mathbf{L}$ is fixed (so that the orbit lies in a plane containing the origin), and the magnitude of $\mathbf{L}$

Figure 3-8

Orbit of a charged particle in a non-head-on collision. The orbit has incoming and outgoing asymptotes parametrized by the impact parameter b and the scattering angle θ. The polar coordinates r and φ denote the instantaneous orbit variables for the scattered particle.

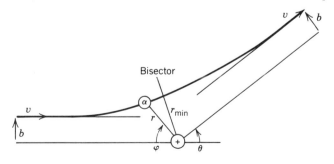

is a constant of the motion given by

$$L = Mvb. \qquad (3\text{-}15)$$

We find this result by examining the orbit at either of its two extremes. Let us also specify the conserved quantity L at an *arbitrary* point along the orbit, where the orbit variables are denoted by the polar coordinates r and φ as in Figure 3-8. The component of velocity perpendicular to the position vector $\mathbf{r}$ is given by $r\,d\varphi/dt$, and so the constant of the motion takes the form

$$L = Mr^2 \frac{d\varphi}{dt}. \qquad (3\text{-}16)$$

We can use this formula immediately to eliminate the unknown variable φ from the dynamics.

The desired solution for the orbit is supposed to be an expression for r as a function of φ. Newton's law leads us instead to a description in terms of the joint variables $r(t)$ and $\varphi(t)$, where the time serves as a variable parameter. We recall that the radial acceleration contains the expected contribution from the t dependence of r as well as the familiar centripetal term:

$$a_r = \frac{d^2 r}{dt^2} - r \left(\frac{d\varphi}{dt} \right)^2 .$$

We then introduce the Coulomb force on the mass M and write the radial equation of motion as

$$M \left[\frac{d^2 r}{dt^2} - r \left(\frac{d\varphi}{dt} \right)^2 \right] = \frac{1}{4\pi\varepsilon_0} \frac{Zze^2}{r^2}. \qquad (3\text{-}17)$$

This expression can be manipulated into a more recognizable differential equation by making the change of variable

$$r = \frac{1}{u} \qquad (3\text{-}18)$$

and by using the relation

$$\frac{d\varphi}{dt} = \frac{L}{Mr^2} = \frac{L}{M} u^2$$

from Equation (3-16). The following maneuvers are executed to change the variables from $r(t)$ and $\varphi(t)$ to the single unknown function $u(\varphi)$:

$$\frac{dr}{dt} = \frac{dr}{du} \frac{du}{d\varphi} \frac{d\varphi}{dt} = -\frac{1}{u^2} \frac{du}{d\varphi} \frac{L}{M} u^2 = -\frac{L}{M} \frac{du}{d\varphi},$$

and

$$\frac{d^2 r}{dt^2} = \frac{d}{d\varphi} \left(\frac{dr}{dt} \right) \frac{d\varphi}{dt} = -\frac{L}{M} \frac{d^2 u}{d\varphi^2} \frac{L}{M} u^2 .$$

We insert these results in Equation (3-17) and find

$$M \left[- \frac{L^2}{M^2} u^2 \frac{d^2 u}{d\varphi^2} - \frac{1}{u} \frac{L^2}{M^2} u^4 \right] = \frac{Zze^2}{4\pi\varepsilon_0} u^2.$$

We then divide by the factor $-L^2 u^2 / M$ and obtain a differential equation for $u(\varphi)$:

$$\frac{d^2 u}{d\varphi^2} + u = - \frac{Zze^2}{4\pi\varepsilon_0} \frac{M}{L^2}. \tag{3-19}$$

The constant on the right can be rewritten in a more compact form by consulting Equations (3-14) and (3-15) and by using $K = \frac{1}{2}Mv^2$:

$$\frac{Zze^2}{4\pi\varepsilon_0} \frac{M}{L^2} = DK \frac{M}{M^2 v^2 b^2} = \frac{D}{2b^2}.$$

These steps lead us to the final version of the equation for $u(\varphi)$:

$$\frac{d^2 u}{d\varphi^2} + u = - \frac{D}{2b^2}. \tag{3-20}$$

Our equation of motion is now in a form amenable to immediate solution.

The second-order differential equation for $u(\varphi)$ is solved by noting that $-D/2b^2$ is a particular solution of the given equation and by observing that $\sin \varphi$ and $\cos \varphi$ are independent solutions of the corresponding homogeneous equation. We obtain the general solution of Equation (3-20) by combining these special solutions as

$$u(\varphi) = A \sin \varphi + B \cos \varphi - \frac{D}{2b^2}. \tag{3-21}$$

This result is subject to two conditions, both imposed at $r = \infty$ on the *incoming* portion of the orbit in Figure 3-8. We specify the first requirement,

$$u = 0 \quad \text{when } \varphi = 0, \tag{3-22}$$

to designate the initial position on the orbit. We also observe that the incoming velocity is

$$v = - \frac{dr}{dt} = + \frac{L}{M} \frac{du}{d\varphi},$$

and so we specify the second requirement,

$$\frac{du}{d\varphi} = \frac{Mv}{L} = \frac{1}{b} \quad \text{when } \varphi = 0, \tag{3-23}$$

to designate the initial derivative on the orbit. The resulting *orbit equation* for $u(\varphi)$

must therefore be given by the solution

$$\frac{1}{r} = \frac{1}{b}\sin\varphi - \frac{D}{2b^2}(1 - \cos\varphi), \tag{3-24}$$

because this expression has the structure of Equation (3-21) and obeys the conditions in Equations (3-22) and (3-23).

Our final result contains more information than we need, since the orbit equation describes every point on the orbit of the particle. In a typical experiment, a beam of particles is directed onto a target and the scattered particles are counted in a detector far from the target, so that only the *asymptotes* of the orbit are under actual observation. We recall from Figure 3-8 that these asymptotes are described by the orbit parameters b and θ, and we note that the orbit equation tells us how b and θ are related. When the orbit reaches the outgoing asymptote we have

$$r = \infty \quad \text{and} \quad \varphi = \pi - \theta.$$

Equation (3-24) becomes

$$\begin{aligned}
0 &= \frac{1}{b}\left(\sin\theta - \frac{D}{2b}(1 + \cos\theta)\right) \\
&= \frac{1}{b}\left(2\sin\frac{\theta}{2}\cos\frac{\theta}{2} - \frac{D}{2b}2\cos^2\frac{\theta}{2}\right)
\end{aligned}$$

for these values of the orbit variables. The desired relation between the scattering angle and the impact parameter follows as a direct result:

$$\sin\frac{\theta}{2} - \frac{D}{2b}\cos\frac{\theta}{2} = 0 \quad \Rightarrow \quad \cot\frac{\theta}{2} = \frac{2b}{D}. \tag{3-25}$$

We need only this much information from the orbit equation to continue our analysis of Rutherford scattering.

The first stage of the problem has been devoted to the collision of a single beam particle with a single target nucleus. Let us now turn to the second stage and consider the observational aspects of the scattering process. An experiment usually involves a beam carrying many incident particles with the same kinetic energy and a target sample containing many nuclei of the same species. We identify two factors pertaining to these considerations—the number of nuclear scatterers exposed to the beam and the rate of incidence of particles onto the target. The factors are deduced from properties of the beam and the target as follows. First, introduce ρ and m as the density and atomic mass per mole of target material, and assemble these quantities along with Avogadro's number to form the number of nuclei per unit volume

$$\frac{\rho N_A}{m}.$$

Then, let δ be the thickness of the sample and suppose that a surface area A_0 of target is exposed to the beam. The number n of nuclear scatterers is evidently equal to the product of these accumulated factors:

$$n = \frac{\rho N_A}{m}A_0\delta. \tag{3-26}$$

The *beam intensity* I_0 is defined as the number of incident particles crossing unit area per unit time, and so the number N_0 of beam particles incident on the target per unit time is written as

$$N_0 = I_0 A_0. \tag{3-27}$$

The parameters introduced in the last two equations are controlled in the design of the experiment.

The scattered particles are counted by a detector system that has an axis of symmetry along the direction of the incident beam. Figure 3-6 shows a small detector at the scattering angle θ, where scattered particles are collected at a rate that does not vary with the placement of the detector around the beam axis. All particles scattered at angle θ are therefore detected uniformly over a thin *ring* of detectors around this axis. Let the ring lie on a sphere of radius R centered at the target, and let the small width of the ring be specified by means of the angular increment $d\theta$ as indicated in the figure. The element of area swept out by this detector is given by

$$dS = (2\pi R \sin\theta)(R\,d\theta).$$

The distance R from the target to the detector has no significance in the interpretation of the scattering distribution. Only the direction of the scattered particles is of interest, and so the factor R^2 is removed from the element of area to define

$$d\Omega = \frac{dS}{R^2} = 2\pi \sin\theta\,d\theta \tag{3-28}$$

as the more effective measure of the scope of the detector. This definition introduces the element of *solid angle* as a mathematical quantity whose span takes in all directions that emanate from the target and pass through the detector at the angle θ. The solid angle $d\Omega$ is dimensionless and is conventionally given in *steradians*, by analogy with the angle $d\theta$ given in radians.

We let dN denote the number of particles scattered per unit time into a given solid angle $d\Omega$. This measurable quantity is proportional to $d\Omega$ so that the ratio $dN/d\Omega$ is independent of the size of the subtended solid angle. The resulting number of counts per unit time per unit solid angle varies with the scattering angle θ and the beam energy K. Hence, the ratio $dN/d\Omega$ gives the angular distribution of scattered particles, the main object of immediate experimental interest.

Particles are scattered into $d\Omega$ by every target nucleus exposed to the beam. Figure 3-9 shows how a single orbit is influenced by *one* nucleus and how the behavior of all such orbits is incorporated in the treatment of a beam containing many particles. According to the figure, every incident particle whose impact parameter lies between b and $b + db$ follows an orbit whose scattering angle lies between θ and $\theta + d\theta$. Axial symmetry applies to this statement, so that all particles entering the axially symmetric element of beam area $2\pi b\,db$ are scattered by the one nucleus into the indicated element of solid angle $d\Omega$. We implement this essential observation with the aid of quantities introduced in the preceding paragraphs. The number of particles incident per unit time is given by the product of factors

$$2\pi b\,db\,I_0,$$

and the number scattered into $d\Omega$ per unit time by a single nucleus is given by the

Figure 3-9

Ingredients in the definition of the scattering cross section. All particles passing through the beam area $2\pi b\, db$ are scattered into the solid angle $d\Omega$. The relation between the scattering angle θ and the impact parameter b is such that $d\theta$ and db have opposite sign.

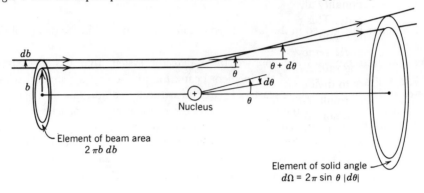

ratio

$$\frac{dN}{n}.$$

These two expressions are set equal, and the beam intensity I_0 is divided out, in order to define the *scattering cross section*

$$d\sigma = \frac{dN}{nI_0} = 2\pi b\, db. \qquad (3\text{-}29)$$

The identification of this new quantity is the final goal in our analysis of the scattering problem.

Let us examine the significance of the cross section by recalling Equation (3-27) and rewriting Equation (3-29) as

$$d\sigma = \frac{dN}{N_0} \bigg/ \frac{n}{A_0}.$$

We can interpret this expression for $d\sigma$ in the following words:

$$\frac{\#\text{ particles scattered into } d\Omega/\text{time}}{\#\text{ particles incident}/\text{time}} \bigg/ \frac{\#\text{ target nuclei encountered}}{\text{area of beam}}$$

or

$$\#\text{ scattered into } d\Omega \text{ by one nucleus}/\#\text{ incident per unit area of beam}.$$

The interpretation tells us that the cross section has units of *area*; this simple observation leads us to the basic meaning of the new concept. The cross section $d\sigma$ represents the effective target area presented to a single beam particle by a single target nucleus for scattering into the solid angle $d\Omega$. Cross sections are given in *barns*, where the unit of area is defined as

$$1 \text{ barn} = 10^{-28} \text{ m}^2.$$

A direct proportionality always holds between $d\sigma$ and $d\Omega$ so that the ratio $d\sigma/d\Omega$ is a more basic quantity. This ratio is called the *differential scattering cross section*, a function of scattering angle and beam energy, given in units of barns/steradian (b/st). The definition of $d\sigma/d\Omega$ removes the last experimentally controlled ingredient from the measured counting rate dN. Removal of the factors n, I_0, and $d\Omega$ begins in Equation (3-29) and serves to divide out those elements of dN that are peculiar to a particular experiment. The result $d\sigma/d\Omega$ is left behind as a measure of the basic two-body process that is common to all such experiments.

The definition of the cross section holds generally for any kind of elastic scattering. We apply the concept directly to the Rutherford scattering problem by recalling the relation between b and θ for Coulomb scattering and by implementing Equation (3-29). We relate db and $d\theta$ by differentiating Equation (3-25) to find

$$-\csc^2\frac{\theta}{2}\frac{d\theta}{2} = \frac{2}{D}\,db. \qquad (3\text{-}30)$$

We note that db and $d\theta$ are of *opposite* sign. This property of the orbits can be seen in Figure 3-9, where the repulsive nature of the Coulomb force causes particles incident with impact parameters larger than b to be scattered through angles smaller than θ. We insert Equations (3-25) and (3-30) into Equation (3-29) and obtain the formula for the cross section in the following series of steps:

$$\begin{aligned}
d\sigma &= 2\pi\frac{D}{2}\cot\frac{\theta}{2}\left(-\frac{D}{2}\csc^2\frac{\theta}{2}\frac{d\theta}{2}\right) \\
&= \frac{\pi}{4}D^2\frac{\sin(\theta/2)\cos(\theta/2)}{\sin^4(\theta/2)}(-d\theta) \\
&= \frac{\pi}{8}D^2\frac{\sin\theta|d\theta|}{\sin^4(\theta/2)} \\
&= \frac{D^2}{16}\frac{d\Omega}{\sin^4(\theta/2)}. \qquad (3\text{-}31)
\end{aligned}$$

The solid angle expression in Equation (3-28) is used at the final step. We can then recall Equation (3-14) and immediately write the differential cross section as

$$\frac{d\sigma}{d\Omega} = \frac{D^2}{16\sin^4(\theta/2)} = \left(\frac{1}{4\pi\varepsilon_0}\frac{Zze^2}{4K}\right)^2\csc^4\frac{\theta}{2}. \qquad (3\text{-}32)$$

These two famous formulas describe the angle and energy dependence of the *Rutherford cross section*.

Geiger and his student E. Marsden completed a very painstaking series of α-particle scattering experiments in 1913. They made observations over an angular range from 5° to 150° and counted more than 10^5 scattered particles on a zinc sulfide scintillating screen. Their results verified the predictions of the Rutherford formula as to the $\csc^4(\theta/2)$ behavior in the scattering angle and the $1/K^2$ dependence in the α-particle energy. They also established that the scattering was proportional to the foil thickness δ and the square of the atomic mass of the foil material. This last observation was taken as an indicator of the expected dependence on the nuclear charge factor Z, since the atomic mass was believed to be proportional to Z. These findings confirmed Rutherford's theory in all predictable respects.

The nuclear model of the atom treats the nucleus as a core of small but *finite* size, while the analysis of Rutherford scattering assumes a point-like nucleus. The derivation is valid nevertheless, provided the α particle does not have enough energy to penetrate the volume of the nucleus. The two spherical particles interact through the Coulomb force between point charges until the beam energy becomes sufficiently large to cause penetration, whereupon the strong nuclear force takes over and becomes the dominant effect. These considerations suggest a way of using Rutherford scattering to deduce the radius R of a particular nucleus. We employ the energy-dependent distance D for reference and compare results from experiment for the differential cross section with predictions from Equation (3-32) for $d\sigma/d\Omega$. Agreement implies the validity of the Coulomb scattering assumption, and this in turn implies that D exceeds R. We then reduce D by increasing K until we observe the onset of a departure from the Coulomb prediction. The corresponding critical value of D provides a determination of the nuclear radius R.

Let us finally be reminded that the derivation of the Rutherford formula is based entirely on classical principles. It is reasonable to wonder whether Newtonian mechanics should apply on a scale as small as the atomic nucleus, particularly because we anticipate the influence of quantum principles in atomic and subatomic systems. It happens that a proper quantum mechanical treatment of nonrelativistic Coulomb scattering leads to exactly the same conclusion as the classical Rutherford derivation.

Example

Let us generalize the notion of solid angle by identifying an element of area dS on a sphere of radius R according to the construction in Figure 3-10. In this case we define dS with *two* variables, a polar angle θ and an azimuthal angle ϕ, unlike the ring of surface area used to derive Equation (3-28). The figure gives the area as

$$dS = R^2 \sin\theta \, d\theta \, d\phi,$$

Figure 3-10

Element of surface area dS on a sphere of radius R. The element of solid angle is defined by $d\Omega = dS/R^2$.

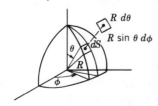

and so the definition of the solid angle becomes

$$d\Omega = \frac{dS}{R^2} = \sin\theta \, d\theta \, d\phi.$$

We obtain the total solid angle by integrating $d\Omega$ over the range of the angular coordinates, $0 \leq \theta \leq \pi$ and $0 \leq \phi \leq 2\pi$. Let us first integrate over ϕ to get

$$\int_{\text{all } \Omega} d\Omega = \int_0^\pi \sin\theta \, d\theta \int_0^{2\pi} d\phi = \int_0^\pi 2\pi \sin\theta \, d\theta,$$

a result to be compared with the expression for the ring in Equation (3-28). The final integration over θ gives

$$\int_{\text{all } \Omega} d\Omega = 2\pi \int_0^\pi \sin\theta \, d\theta = -2\pi \cos\theta \Big|_0^\pi = 4\pi,$$

so that the total solid angle is equal to 4π steradians.

Example

Rutherford scattering is illustrated nicely by choosing calculations based on typical α-particle scattering experiments. Let us consider a beam of 5.30 MeV α particles incident on a gold foil as a good example of such an experiment. We begin by computing the following useful combination of constants that recurs continually in atomic physics:

$$\frac{e^2}{4\pi\varepsilon_0} = \frac{(8.988 \times 10^9 \text{ N} \cdot \text{m}^2/\text{C}^2)(1.602 \times 10^{-19} \text{ C})^2}{1.602 \times 10^{-19} \text{ J/eV}}$$

$$= 1.440 \times 10^{-9} \text{ eV} \cdot \text{m}.$$

We then take $z = 2$ and $Z = 79$ and consult Equation (3-14) to find the distance of closest approach:

$$D = \frac{e^2}{4\pi\varepsilon_0} \frac{Zz}{K} = \frac{(1.44 \times 10^{-9} \text{ eV} \cdot \text{m})(79)(2)}{5.30 \times 10^6 \text{ eV}} = 4.29 \times 10^{-14} \text{ m}.$$

The density and molar mass of gold are

$$\rho = 19.3 \times 10^3 \text{ kg/m}^3 \quad \text{and} \quad m = 197 \text{ g/mole};$$

therefore, the number of Au nuclei per unit volume is given by

$$\frac{\rho N_A}{m} = \frac{(19.3 \times 10^3 \text{ kg/m}^3)(6.02 \times 10^{23} \text{ nuclei/mole})}{0.197 \text{ kg/mole}}$$

$$= 5.90 \times 10^{28} \text{ nuclei/m}^3.$$

Let the foil thickness be 2.10×10^{-7} m and suppose that 10^4 α particles strike the foil per second. Equation (3-26) determines the number of target nuclei encountered per unit beam area as

$$\frac{n}{A_0} = \frac{\rho N_A}{m} \delta = \left(5.90 \times 10^{28} \text{ nuclei/m}^3\right)\left(2.10 \times 10^{-7} \text{ m}\right)$$

$$= 1.24 \times 10^{22} \text{ nuclei/m}^2.$$

Our main objective is to predict the total number ΔN of α particles scattered *backward* by the foil per second. The number dN scattered per second into solid angle $d\Omega$ is found by using the known Rutherford cross section in conjunction with Equations (3-27) and (3-29):

$$dN = nI_0 \, d\sigma = \frac{n}{A_0} N_0 \, d\sigma.$$

The desired quantity ΔN is obtained by integrating this equality over the backward solid angle, where the scattering angle ranges over $\pi/2 \leq \theta \leq \pi$. We express the total number of backscattered α particles relative to the number incident by recalling Equations (3-28) and (3-31) to get

$$\frac{\Delta N}{N_0} = \int_{\substack{\text{backward} \\ \Omega}} \frac{n}{A_0} \frac{d\sigma}{d\Omega} \, d\Omega = \int_{\pi/2}^{\pi} \frac{n}{A_0} \frac{D^2}{16} \frac{2\pi \sin\theta \, d\theta}{\sin^4(\theta/2)}.$$

The integration proceeds as follows:

$$2\pi \int_{\pi/2}^{\pi} \frac{\sin\theta \, d\theta}{\sin^4(\theta/2)} = 4\pi \int_{\pi/2}^{\pi} \frac{\cos(\theta/2) \, d\theta}{\sin^3(\theta/2)} = 4\pi \int_{\sqrt{1/2}}^{1} \frac{2 \, du}{u^3} = \left. \frac{8\pi}{-2u^2} \right|_{\sqrt{1/2}}^{1} = 4\pi,$$

using the change of variable $u = \sin(\theta/2)$. Our final result for the fraction of incident α particles scattered backward is

$$\frac{\Delta N}{N_0} = 4\pi \frac{n}{A_0} \frac{D^2}{16} = \frac{\pi}{4}\left(1.24 \times 10^{22} \text{ m}^{-2}\right)\left(4.29 \times 10^{-14} \text{ m}\right)^2 = 1.79 \times 10^{-5}.$$

We then find

$$\Delta N = \left(10^4 \text{ } \alpha \text{ particles/s}\right)\left(1.79 \times 10^{-5}\right) = 0.179 \text{ } \alpha \text{ particles/s}$$

for the predicted number of counts per second. This value is comparable with the observations of Geiger and Marsden.

3-5 The Quantum Picture of the Atom

Rutherford's planetary model assigned the nucleus and the electrons to separate inner and outer regions of the atom. The success of the model inspired Bohr to imagine a corresponding separation of physical domains of influence, in which the electrons in the atom accounted for the chemical properties of the element while the nucleus was responsible for any radioactive behavior. This picture began to reveal the correlation between the number of electrons in the atom and the location of the element in the periodic table. The meaning of the atomic number finally emerged after the construction of Bohr's model of the atom.

The Rutherford atom evolved into the Bohr atom in 1913. The evolution of models resulted from the introduction by Bohr of certain quantum concepts in a new treatment of matter. This revolutionary contribution to the understanding of the atom initiated the next phase in the development of the quantum theory.

Indications of quantum behavior in matter had been accumulating long before the time of Rutherford and Bohr. The best source of evidence was to be found in the spectra of light emitted by atoms. Investigations of atomic spectra began early in the 19th century, after J. von Fraunhofer discovered the existence of dark lines in the spectrum of light from the Sun. These studies gathered impetus when it was learned that heated samples of the elements emitted light whose wavelengths appeared in discrete patterns. Kirchhoff showed that each observed spectrum was *characteristic* of the chemical species contained in the source of the light. Atomic spectroscopy came into being, as investigators proceeded to measure and tabulate the wavelengths of light in the spectra of the various elements. Cesium and rubidium were among the new elements identified with the aid of these spectroscopic methods. Atomic spectra gave precise experimental information in the form of a unique signature for each of the atoms. These measurements offered access to the internal structure of the atoms, once it was learned how the different spectra were to be decoded. The appropriate decoding principles were eventually established in the quantum theory of atoms.

Most of our knowledge of the properties of matter has come to us through some branch of spectroscopy. The oldest spectroscopic techniques employ optical types of apparatus. We can describe how these instruments are used to observe *emission* and *absorption* spectra by referring to the schematic illustrations in Figure 3-11. An emission spectrum is the result of the spectral analysis of light emitted by atoms in a gas discharge tube. The atoms in the tube gain energy in collisions with electrons and lose energy by emitting radiation. The figure shows how the light from the tube is directed through a slit to a prism spectrometer and is dispersed by the prism into different component wavelengths. We see the resulting emission spectrum as a series of discrete *spectral lines*, corresponding to different colored images of the slit. The positions of the lines in the image plane of the spectrometer are unique to the atoms in the discharge tube. An absorption spectrum is formed when atoms are used to *remove* certain wavelengths from the familiar continuous spectrum of white light. The figure shows how the light from an incandescent source is directed through a slit and is then allowed to traverse a gas-filled cell before entering the prism spectrometer. The dispersion of the prism produces a continuous spectrum of colors due to the thermal radiation from the source, accompanied by a superimposed set of discrete dark lines associated with the intervening gas. These dark images of the slit are just like the dark lines seen by Fraunhofer in his observations of the solar spectrum. The absence of lines is attributed to the absorption of specific wavelengths of the incident white light by the atoms in the gas. Thus, the absent wavelengths are unique characteristics of the atoms in the gas-filled cell. We find that the bright lines in the emission spectrum occur in the same positions as the dark lines in the absorption spectrum if the gas in the discharge tube is the same as the gas in the absorption cell.

Optical line spectra were collected, without further analysis or interpretation, through much of the 19th century. The first successful empirical understanding of the

Figure 3-11

Spectrometers for the study of emission spectra and absorption spectra. Focusing lenses are also needed as additional elements in the two optical systems.

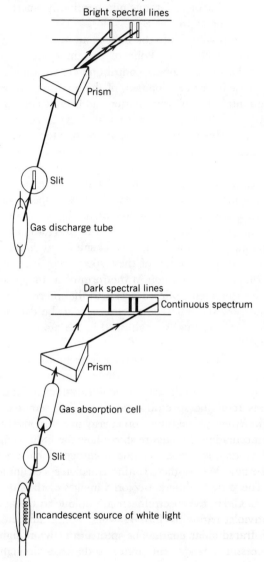

observed lines was achieved in 1885 in the case of hydrogen, the simplest atom. J. J. Balmer showed that the visible spectrum of wavelengths in hydrogen could be fitted by the formula

$$\lambda = \left(3645.6 \times 10^{-10} \text{ m}\right) \frac{n^2}{n^2 - 4} \qquad n = 3, 4, \ldots. \qquad (3\text{-}33)$$

His fit applied to the so-called Balmer series of lines in which the integers $n = 3, 4, \ldots$ corresponded to the spectral labels $\alpha, \beta, \ldots$ shown in Figure 3-12. The measured

Figure 3-12

Balmer series of spectral lines in hydrogen. The observed wavelengths are in good agreement with the indicated predictions from Bohr's model for the hydrogen atom.

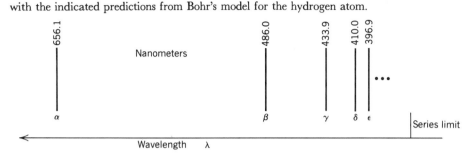

wavelengths diminished in separation according to the given dependence on n and converged to a series limit for the limiting value of the index $n \to \infty$. Immediate steps were taken to extend the fitting procedure to other series of lines in other elements. The hydrogen result was generalized in 1890 by J. R. Rydberg, who rewrote Balmer's prescription in the form

$$\frac{1}{\lambda} = R_{\mathrm{H}}\left(\frac{1}{2^2} - \frac{1}{n^2} \right)$$

and then interpreted his expression as a special case of the following more general formula for hydrogen:

$$\frac{1}{\lambda} = R_{\mathrm{H}}\left(\frac{1}{n'^2} - \frac{1}{n^2} \right). \tag{3-34}$$

Rydberg's analysis reproduced the Balmer series for $n' = 2$, with a value for the constant given by $R_{\mathrm{H}} = 10972160 \text{ m}^{-1}$. This parameter has since been determined with very great accuracy and has been named the Rydberg constant. Rydberg's formula defined a double sequence in the two integers n and n', and so the introduction of the new index n' raised the possibility of other as-yet unknown series of hydrogen lines. The predicted $n' = 3$ series was discovered in the infrared region of the spectrum by L. C. H. F. Paschen in 1908, and the fundamental $n' = 1$ series was found in the ultraviolet region by T. Lyman in 1914. Other series were also revealed later on, in the infrared spectrum and beyond. Thus, the empirical results of Balmer and Rydberg successfully broke the spectral code for hydrogen and remained to be properly explained in some more comprehensive theoretical framework.

It had become apparent by 1913 that the planetary model of the atom was in urgent need of reevaluation. The atomic electrons were supposed to be moving in orbits around the nucleus under the influence of the attractive Coulomb force. Unfortunately, these orbiting charged particles were expected to emit classical radiation as a result of their accelerated motion. The laws of classical electrodynamics implied that the atomic system should undergo a radiative loss of energy, which would cause the atom to be intrinsically *unstable*. Hence, the electron orbits would steadily shrink in size and rapidly collapse onto the nucleus, while the radiation was being emitted with continuously increasing frequency. This fatal flaw in the classical picture threatened to undermine Rutherford's model of the atom. It turned out that the picture was fundamentally flawed by the adoption of the classical point of view.

Bohr solved the problem of radiative instability by making a break with the principles of classical physics. His proposed solution contributed another new idea to the developing quantum theory, in the same spirit as the radical proposals put forward by Planck and Einstein in the previous decade. Bohr believed in the nuclear model of the atom but questioned whether classical electrodynamics should always apply to the behavior of electrons. He began to entertain the possibility of a quantum approach, thinking that Planck's constant h should have a natural place in the description of the atom. He noted that the classical picture could not explain why all atoms of a given species should have similar electron orbits, as suggested by the evidence obtained from atomic spectra. He also noted that Rutherford's model gave no indication of the size of the atom, and he regarded this failure as a basic weakness of the classical approach. It was clear that a natural measure of length could not be constructed from the physical constants of the problem unless h was included along with the classical parameters. It was observed that the product of factors $(4\pi\varepsilon_0/e^2)(h^2/m_e)$ gave an appropriate unit of length and an approximate order of magnitude to use as a practical scale of length for the atom.

Bohr speculated that electron motion in the atom should have quantized properties in which only certain states of orbital motion were allowed. He proposed a new quantum condition by requiring discrete values of the *energy* for these allowed states. He could then show that his requirement was consistent with the observation of discrete wavelengths in the spectra of atoms. His idea was partially inspired by the form of the Rydberg formula, Equation (3-34), in which the wavelengths in hydrogen were expressed as differences of pairs of discrete-valued quantities.

The Bohr theory of the atom is based on the following two postulates. First, the electrons in the atom may exist in a discrete set of *stationary states* of definite energy, defined so that radiation is not emitted by an atom in such a state. Second, the atom may undergo a nonclassical *transition* from one of these allowed states to another and thereby emit or absorb a single quantum of electromagnetic radiation. The new concept of *quantization of energy* in matter originates in these two propositions.

We can interpret Bohr's postulates with the aid of the illustrations in Figure 3-13. The proposed set of stationary states is represented by a sequence of discrete levels of increasing energy in an *energy level diagram*. We can then exhibit upward or downward transitions on such a diagram as quantum jumps between pairs of states with quantized energies E_n and $E_{n'}$. An upward transition occurs with the absorption of a

Figure 3-13

Energy level diagram for the stationary states of an atom. A transition to a state of higher or lower energy is accompanied by the absorption or emission of a photon.

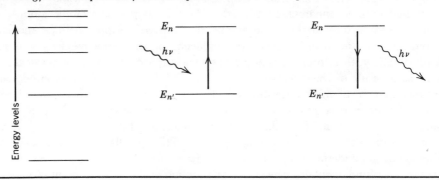

photon, and a downward transition occurs with the emission of a photon. In either case we obtain Bohr's formula for the energy of the photon by applying conservation of energy in the overall system of atom plus radiation to get

$$hv = \frac{hc}{\lambda} = E_n - E_{n'}. \qquad (3\text{-}35)$$

Only quantized energies appear in this expression, and so only certain allowed wavelengths are predicted for the emission and absorption of radiation.

Bohr's postulated energy levels gave a natural explanation for the discrete characteristics of atomic line spectra. His theory also offered a solution-by-decree for the problem of atomic instability. The existence of stationary states and quantized energy levels was simply declared, while the applicability of classical radiation theory was simply denied, and no theoretical arguments were put forward to support these new ideas. This treatment of the atom associated the phenomenon of atomic radiation with the possibility of transitions between pairs of states, and *not* with the acceleration of electrons in orbits. Consequently, radiation frequencies were not to be identified with frequencies of orbital electron motion in the atom. We should note that the theory did not say what the atom was doing while the radiative transition was in progress. This classical question was at variance with the new quantum point of view and was not supposed to have an immediate answer.

Example

We have suggested the following natural scale of length in the context of Bohr's quantum treatment of the atom:

$$\frac{4\pi\varepsilon_0}{e^2} \frac{h^2}{m_e} = \frac{4\pi\varepsilon_0}{e^2} \frac{(hc)^2}{m_e c^2} = \frac{(1240 \text{ eV} \cdot \text{nm})^2}{(1.440 \text{ eV} \cdot \text{nm})(0.5110 \times 10^6 \text{ eV})} = 2.090 \text{ nm}.$$

(Note that the electron mass is given in MeV/c^2 units in this calculation.) We learn in Section 3-6 how a scale of length proportional to the suggested quantity arises naturally in Bohr's model of the hydrogen atom.

3-6   The Bohr Model of the One-Electron Atom

Bohr's theory of the atom proposed the existence of stationary states; however, the procedures needed to determine these states were not discovered until 1925. In the meantime it was possible to make progress in quantum physics by judiciously blending new quantum ideas with certain concepts carried over from classical mechanics. Bohr took such an approach in 1913 to develop his model of the hydrogen atom.

Our treatment of Bohr's model generalizes the case of hydrogen to cover any system of two bodies bound together by Coulomb attraction. We let the charges and masses be denoted as $-e$ and m for the one particle, and $+Ze$ and M for the other. Our intention is to construct a model of an arbitrary bound system containing one electron, and so we intend to set $m = m_e$ and let M range over a variety of choices. The special case of the hydrogen atom is obtained if we take $Z = 1$ and choose M to be the proton mass M_p. Arbitrary one-electron ions can also be described by choosing

Figure 3-14

Circular motion of two charged particles with center of mass at rest.

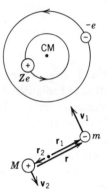

other integer values of Z and other larger values of M. We restrict the discussion to particles in *circular* orbits, and we take account of the motion of *both* particles. Of course, we expect that the mass M should remain at rest in the limit $M/m \rightarrow \infty$, and we know that the hydrogen atom is close to this limiting situation since the proton-to-electron mass ratio is quite large:

$$\frac{M_p}{m_e} = 1836.$$

The small effect of a finite nuclear mass M is detectable in atoms and should therefore be retained as a small correction. Allowance should be made for finite M also as a matter of principle, because the properties of two masses in motion are of some interest and because the effects can be incorporated without difficulty.

We take the center of mass of the two-body system to be at rest at the origin and define position vectors for the two particles as in Figure 3-14. Our first objective is to reduce the two-body problem to an equivalent one-body description in terms of a single vector variable given by the *relative coordinate*

$$\mathbf{r} = \mathbf{r}_1 - \mathbf{r}_2. \tag{3-36}$$

The origin is located at the center of mass so that $\mathbf{r}_1$ and $\mathbf{r}_2$ must satisfy the relation

$$m\mathbf{r}_1 + M\mathbf{r}_2 = 0. \tag{3-37}$$

We can solve this pair of equations to obtain the particle position vectors

$$\mathbf{r}_1 = \frac{M}{M+m}\mathbf{r} \quad \text{and} \quad \mathbf{r}_2 = -\frac{m}{M+m}\mathbf{r} \tag{3-38}$$

and then differentiate with respect to t to find the corresponding velocities

$$\mathbf{v}_1 = \frac{M}{M+m}\mathbf{v} \quad \text{and} \quad \mathbf{v}_2 = -\frac{m}{M+m}\mathbf{v}. \tag{3-39}$$

The *relative velocity* is defined as

$$\mathbf{v} = \frac{d\mathbf{r}}{dt} \tag{3-40}$$

in passing from Equations (3-38) to (3-39).

We achieve our objective by casting the various two-body dynamical quantities in terms of the relative coordinate and velocity. Equations (3-39) are used to reexpress the total kinetic energy:

$$K = \frac{m}{2}v_1^2 + \frac{M}{2}v_2^2 = \frac{m}{2}\left(\frac{M}{M+m}\right)^2 v^2 + \frac{M}{2}\left(\frac{m}{M+m}\right)^2 v^2 = \frac{1}{2}\frac{mM}{M+m}v^2.$$

Thus, we obtain the desired one-body expression for K in the form

$$K = \tfrac{1}{2}\mu v^2 \tag{3-41}$$

by defining the *reduced mass* of the two-body system as

$$\mu = \frac{mM}{M+m}. \tag{3-42}$$

This mass parameter also appears along with the relative dynamical variables in the formula for the angular momentum vector $\mathbf{L}$. The special problem at hand involves two particles in circular orbits. We recall that the total angular momentum is a constant of the motion because the force of Coulomb attraction is a *central* force. The magnitude of the conserved total angular momentum is found by using Equations (3-38) and (3-39) to get

$$L = mv_1 r_1 + Mv_2 r_2 = m\left(\frac{M}{M+m}\right)^2 vr + M\left(\frac{m}{M+m}\right)^2 vr,$$

or

$$L = \mu vr \tag{3-43}$$

as the final result. We can put the equations to further use and express the centripetal force on each particle as

$$F = \frac{mv_1^2}{r_1} = \frac{Mv_2^2}{r_2} = \frac{\mu v^2}{r}. \tag{3-44}$$

The potential energy for the system of two charges is given by the Coulomb formula

$$V = -\frac{1}{4\pi\varepsilon_0}\frac{Ze^2}{r}. \tag{3-45}$$

We note that this final dynamical quantity is expressed directly in terms of the relative variable r.

These results illustrate the consequences of a general theorem in mechanics. Every two-body central-force problem is reducible to an equivalent one-body problem in which the single particle is described by reduced mass μ, position vector $\mathbf{r}$, and potential energy $V(r)$. In this instance the function $V(r)$ is provided by the attractive Coulomb potential energy given in Equation (3-45).

We observe in passing that the theorem has immediate implications in the analysis of Rutherford scattering. The theorem allows us to replace the problem of scattering from a target of finite mass by the problem of reduced-mass scattering from a fixed repulsive center of force. Recall that the original treatment has been given in Section 3-4 in terms of scattering from a fixed target of infinite mass. We can apply the theorem and adapt our fixed-target results simply by substituting the reduced mass for the mass of the scattered particle. Our conclusions are then interpreted to hold in a reference frame where the center of mass is at rest.

Let us return to the bound system in Figure 3-14 and evaluate the total energy of the two charged particles. We need not include the rest energies of the particles in these nonrelativistic considerations. The constituents of systems in atomic physics always maintain their identities, and so the corresponding rest energies act as constant quantities in all such problems. With this understanding we express the total energy as

$$E = K + V$$

and consult Equations (3-41) and (3-45) to find

$$E = \frac{1}{2}\mu v^2 - \frac{1}{4\pi\varepsilon_0}\frac{Ze^2}{r}.$$

The centripetal force in Equation (3-44) is identified with the attractive Coulomb force as

$$\frac{\mu v^2}{r} = \frac{1}{4\pi\varepsilon_0}\frac{Ze^2}{r^2}, \tag{3-46}$$

and so the total energy takes the form

$$E = \left(\frac{1}{2} - 1\right)\frac{Ze^2}{4\pi\varepsilon_0 r} = -\frac{1}{2}\frac{Ze^2}{4\pi\varepsilon_0 r}. \tag{3-47}$$

Note that E is a *negative* quantity because V is negative and because the potential energy is larger in magnitude than the kinetic energy whenever the energies pertain to a *bound* system.

The construction of the model has adhered to classical principles up to this point. Bohr followed these steps to the same point and then introduced a new quantum hypothesis in order to produce the desired stationary states. He argued that the orbits in the atom should be *discrete* and that an allowed classical orbit was one whose angular momentum L was an integral multiple of the quantum unit $h/2\pi$. This basic quantity was recognized to have dimensions of angular momentum and was found to recur sufficiently often in quantum physics to warrant a special symbol, given by the current value

$$\hbar = \frac{h}{2\pi} = 1.05457266 \times 10^{-34}\ \text{J} \cdot \text{s}.$$

Bohr's quantum condition restricted the angular momentum of the atomic system to quantized units of $\hbar$ according to the formula

$$L = \mu v r = n\hbar. \tag{3-48}$$

Thus, an integer n was introduced as a *quantum number* through the new concept of *quantization of angular momentum.* Justification for this ad hoc rule came later in another of the early developments in the old quantum theory.

We adopt Bohr's hypothesis and combine the classical expression in Equation (3-46) with the quantum statement in Equation (3-48) to find

$$v^2 r^2 = \frac{Ze^2}{4\pi\varepsilon_0} \frac{r}{\mu} = \frac{n^2 \hbar^2}{\mu^2}.$$

This remarkable result implies that the orbit radius r is a *discrete* quantity whose allowed values are given in terms of fundamental constants as

$$r_n = \frac{n^2}{Z} \frac{4\pi\varepsilon_0 \hbar^2}{e^2 \mu}. \tag{3-49}$$

We rewrite this formula in the more compact form

$$r_n = \frac{n^2}{Z} \frac{m_e}{\mu} a_0 \tag{3-50}$$

by defining

$$a_0 = \frac{4\pi\varepsilon_0 \hbar^2}{e^2 m_e}. \tag{3-51}$$

The important parameter a_0 is approximately equal to 0.05 nm and is called the *Bohr radius.* This quantity appears in Equation (3-50) as a unit of length and serves as a useful scale of length for all problems in atomic physics.

The energies in the various orbits must be quantized since the orbit radii are discrete. The prediction of these energies is the main conclusion of the Bohr model. We obtain discrete values for E by inserting Equation (3-49) into Equation (3-47):

$$E_n = -\frac{1}{2} \frac{Ze^2}{4\pi\varepsilon_0 r_n} = -\frac{1}{2} \frac{Ze^2}{4\pi\varepsilon_0} \frac{Z}{n^2} \frac{e^2 \mu}{4\pi\varepsilon_0 \hbar^2} = -\frac{Z^2}{n^2} \left(\frac{e^2}{4\pi\varepsilon_0} \right)^2 \frac{\mu}{2\hbar^2}.$$

The result can be written compactly as

$$E_n = -\frac{Z^2}{n^2} \frac{\mu}{m_e} E_0 \tag{3-52}$$

by introducing the *Rydberg energy unit*

$$E_0 = \left(\frac{e^2}{4\pi\varepsilon_0} \right)^2 \frac{m_e}{2\hbar^2}. \tag{3-53}$$

The parameter E_0 is significant because it sets the scale for the energy levels in the hydrogen atom. We note that E_0 can also be expressed as

$$E_0 = \frac{1}{2}\left(\frac{e^2}{4\pi\varepsilon_0\hbar c}\right)^2 m_e c^2 = \frac{\alpha^2}{2}m_e c^2, \tag{3-54}$$

where the final form involves the electron rest energy along with a new constant given by

$$\alpha = \frac{e^2}{4\pi\varepsilon_0\hbar c}. \tag{3-55}$$

This last combination of fundamental physical parameters is a *dimensionless* quantity known as the *fine structure constant*. Both E_0 and α are known very accurately from experiment:

$$E_0 = 13.6056981 \text{ eV}$$

and

$$\frac{1}{\alpha} = 137.0359895.$$

We note that α also makes its appearance in Equation (3-51):

$$a_0 = \frac{4\pi\varepsilon_0\hbar c}{e^2}\frac{\hbar}{m_e c} = \frac{1}{\alpha}\frac{\hbar}{m_e c}. \tag{3-56}$$

This observation tells us that the Bohr radius and the electron's Compton wavelength $h/m_e c$ are related by a rather interesting numerical quantity.

The reduced mass μ appears explicitly in the formulas for r_n and E_n via the ratio μ/m_e. This reduced-mass correction factor is very close to unity in almost all applications. We recall that m stands for m_e in every case, and so Equation (3-42) gives the factor as

$$\frac{\mu}{m_e} = \frac{M}{M + m_e}.$$

Thus, the ratio is equal to $\frac{1836}{1837}$ when M is set equal to M_p in the case of hydrogen and is even closer to unity when M refers to other nuclear masses. The correction is interesting nonetheless, because it is possible to observe the resulting small differences in energy levels for atoms whose nuclei have the same value of Z but different values of M. These nuclear species are called *isotopes*, and the correction arising from the factor μ/m_e is called the *isotope effect*. The ratio μ/m_e becomes rather different from unity in the special case of positronium. We describe this electron–positron bound system in the context of the Bohr model by taking $Z = 1$ and setting $M = m_e$ to get $\mu/m_e = \frac{1}{2}$. Obviously, the reduced-mass correction has an appreciable effect on the energy levels in this instance.

Bohr's quantization condition for angular momentum has led us to discrete orbits and quantized energies. The corresponding allowed states of motion in the Bohr atom are labeled by means of the single quantum number n, whose origin is traced directly to Equation (3-48). Quantization of energy presents the main results of the model in terms of a system of energy levels ordered according to the values of n. Let us choose $Z = 1$ in Equation (3-52) so that we can visualize these results for the special case of the hydrogen atom. The energy level diagram is shown in Figure 3-15 to consist of an infinite sequence of values of E_n as n ranges over all positive integers. The system has a lowest level called the *ground state* in which

$$n = 1 \quad \text{and} \quad E_1 = -\frac{\mu}{m_e} E_0 = -13.60 \text{ eV}.$$

Equation (3-50) gives the radius for the first (and smallest) Bohr orbit as

$$r_1 = \frac{m_e}{\mu} a_0.$$

Larger values of n refer to *excited states* with higher energies and greater orbit radii. The limit $n \to \infty$ defines a special limiting state of the system in which

$$r_\infty = \infty \quad \text{and} \quad E_\infty = 0.$$

The two bound particles have infinite separation and no motion in this state. Therefore, the limiting value of n corresponds to the configuration of the atom at the threshold for *ionization*. The system is allowed to have all values of the energy above this threshold; however, the associated states pertain to unbound particles in relative motion and are not interpreted as states of the atom. The *ionization energy* is determined by the difference in energy between the ground state and the ionization threshold:

$$E_\infty - E_1 = \frac{\mu}{m_e} E_0 = 13.60 \text{ eV}.$$

This quantity represents the minimum energy that the atom must absorb in its natural state in order to release the bound electron. The absorbed energy can be supplied in the form of an incident photon, and so the ionization energy is equal to the work function for the photoelectric effect in hydrogen.

Bohr's model is not based on a consistent set of quantum principles and does not qualify as a genuine quantum theory. The model is remarkable nevertheless, because the results provide *correct* predictions for the energy levels of the one-electron atom. We can verify these predictions by examining the wavelengths in the emission spectrum of the atom. Let us consider a radiative transition from an excited initial state i to a less energetic final state f, and let us label the states by the quantum numbers n_i and n_f as in Figure 3-16. We adapt Bohr's basic formula in Equation (3-35) to the notation in the figure by writing

$$h\nu = \frac{hc}{\lambda} = E_i - E_f. \tag{3-57}$$

The energies are obtained from Equation (3-52), and so the wavelengths are given by

Figure 3-15

Energy levels in Bohr's model of the hydrogen atom. The energies are given in terms of the quantum number n by the formula $E_n = -(E_0/n^2)(\mu/m_e)$, where E_0 is the Rydberg energy unit and μ/m_e is the reduced-mass correction factor.

Figure 3-16

Emission of a photon in the transition $i \rightarrow f$.

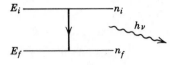

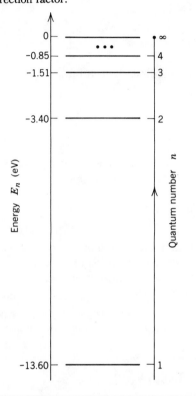

the expression

$$\frac{1}{\lambda} = \frac{E_i - E_f}{hc} = -Z^2 \frac{\mu}{m_e} \frac{E_0}{hc}\left(\frac{1}{n_i^2} - \frac{1}{n_f^2}\right).$$

We can rewrite this result as

$$\frac{1}{\lambda} = Z^2 \frac{\mu}{m_e} R_\infty \left(\frac{1}{n_f^2} - \frac{1}{n_i^2}\right) \tag{3-58}$$

if we define

$$R_\infty = \frac{E_0}{hc} = \frac{\alpha^2}{2}\frac{m_e c}{h}, \tag{3-59}$$

using Equation (3-54). The result in Equation (3-58) is essentially the same as Rydberg's empirical formula. We can make the identification with Equation (3-34) if we take $Z = 1$ and set

$$R_{\mathrm{H}} = \frac{\mu}{m_e} R_{\infty}.$$

Thus, it is evident that the energy levels are correctly given by Bohr's model since the predicted transitions are in agreement with the observed spectral lines. Both R_{H} and R_{∞} are known with very great accuracy from experiment; the latter constant has the current quoted value

$$R_{\infty} = 10973731.534 \text{ m}^{-1}.$$

(The notation can be understood with the aid of Equation (3-42); R_{∞} denotes the Rydberg constant for an atom with a nucleus of *infinite* mass M.)

Equation (3-58) describes families of spectral lines according to the various series predicted by Rydberg. These series are classified by systematically choosing various fixed values for the final quantum number n_f. The choice $n_f = 1$ designates the Lyman series of transitions to the ground state from all initial levels in the range $n_i \geq 2$. Likewise, transitions to the $n_f = 2$ state from $n_i \geq 3$ give the Balmer series, transitions to the $n_f = 3$ state from $n_i \geq 4$ give the Paschen series, and so on. The scheme of transitions is shown for hydrogen in Figure 3-17. It is interesting to note

Figure 3-17

Transitions in the emission spectrum of hydrogen. The spectral lines are grouped into series according to values of the final quantum number. The final state of the atom has $n = 1$ in the Lyman series, $n = 2$ in the Balmer series, $n = 3$ in the Paschen series, and so on.

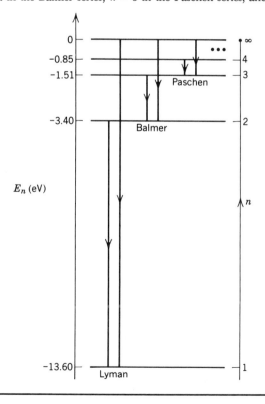

Figure 3-18
Effect of atomic recoil in the emission of a photon.

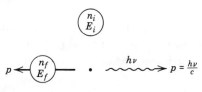

that Bohr's theoretical model predates the discovery of all the known series in hydrogen except for the Balmer lines in the visible spectrum and the Paschen lines in the infrared.

Let us now go back and look more critically at Figure 3-16 and Equation (3-57). If we visualize the radiative transition according to the before-and-after picture in Figure 3-18, we see that the recoil motion of the atom has been left out of the analysis. Let us investigate this oversight and show that no serious error has been made in our prediction of the photon energy $h\nu$. If the atom is assumed to be initially at rest, as in the figure, then conservation of momentum requires the photon and the final atom to have opposite momenta of the same magnitude p. We denote the energy difference between the two states of the atom as

$$\Delta E = E_i - E_f.$$

This quantity is the same as the difference in rest energies for the initial and final atom. The kinetic energy of the recoiling atom is given by $p^2/2M$, where M is the mass of the atom. Conservation of energy provides the determining relation among all these quantities:

$$E_i = h\nu + E_f + \frac{p^2}{2M}. \tag{3-60}$$

Since the momentum of the photon is

$$p = \frac{h\nu}{c},$$

Equation (3-60) becomes a quadratic equation in the unknown energy $h\nu$:

$$E_i = h\nu + E_f + \frac{(h\nu)^2}{2Mc^2}$$

or

$$(h\nu)^2 + 2Mc^2(h\nu) - 2Mc^2\Delta E = 0.$$

The desired solution for the photon energy is given by the root

$$h\nu = -Mc^2 + \sqrt{M^2c^4 + 2Mc^2\Delta E} = Mc^2\left[\left(1 + \frac{2\Delta E}{Mc^2}\right)^{1/2} - 1\right].$$

The ratio $2\,\Delta E/Mc^2$ is certainly very small if the energy difference ΔE is in the eV-to-keV range. Therefore, an expansion can be used to get a very good approximation:

$$h\nu = Mc^2\left[1 + \frac{1}{2}\frac{2\,\Delta E}{Mc^2} - \frac{1}{8}\left(\frac{2\,\Delta E}{Mc^2}\right)^2 - 1\right] = \Delta E\left(1 - \frac{\Delta E}{2\,Mc^2}\right). \quad (3\text{-}61)$$

This result tells us that Equation (3-57) should be modified by the correction factor $(1 - \Delta E/2\,Mc^2)$. We conclude that the effect of recoil is negligible whenever the transition energy ΔE is much smaller than the rest energy of the radiating system.

The Bohr model was a success for its time even though its limitations were immediately apparent. The treatment could be generalized to noncircular orbits, but the model could not be extended successfully to atoms with more than one electron. The wavelengths in the spectrum of the hydrogen-like atoms could be calculated, but the intensities of the spectral lines could not be predicted. All these deficiencies were to be remedied in a more comprehensive quantum theory of matter and radiation.

Bohr regarded his model as a tentative blend of classical and quantum ideas. He recognized that classical physics must have its domain of applicability, and he believed that results from the quantum theory should merge into a correspondence with classical predictions. Bohr formulated a *correspondence principle* in which he argued that nature should be described by a quantum theory and that the theory should have a limiting regime in which classical physics would become valid. He treated Planck's constant as a small unique unit of quantum behavior and proposed that quantum mechanics should somehow reduce to classical mechanics in the limit $h \to 0$. Operational evidence for such a limit was supposed to appear as a convergence of energy levels for large values of the relevant quantum number.

Bohr's model of the one-electron atom provides an excellent illustration of the correspondence principle in operation. Let us return to Equation (3-57) and consider the frequency of the radiation emitted in the transition $i \to f$. Equations (3-52) and (3-54) are used to obtain

$$\nu_{i\to f} = \frac{E_i - E_f}{h} = Z^2\frac{\mu}{m_e}\frac{E_0}{h}\left(\frac{1}{n_f^2} - \frac{1}{n_i^2}\right) = Z^2\frac{\mu}{m_e}\frac{\alpha^2 m_e c^2}{2h}\left(\frac{1}{n_f^2} - \frac{1}{n_i^2}\right).$$

We are interested in the specially defined photon frequency ν_n that corresponds to transitions between the *adjacent* levels $n_i = n$ and $n_f = n - 1$, and we are particularly concerned with the behavior of ν_n for large n:

$$\nu_n = Z^2\frac{\mu}{m_e}\frac{\alpha^2 m_e c^2}{2h}\left(\frac{1}{(n-1)^2} - \frac{1}{n^2}\right) \to \frac{Z^2}{n^3}\frac{\mu}{m_e}\frac{\alpha^2 m_e c^2}{h}.$$

Let us also consider the electron orbit frequency, given by

$$f_n = \frac{v_n}{2\pi r_n} = \frac{\dfrac{Z}{n}\alpha c}{2\pi\dfrac{n^2}{Z}\dfrac{m_e}{\mu}a_0} = \frac{Z^2}{n^3}\frac{\mu}{m_e}\frac{\alpha c}{2\pi}\alpha\frac{m_e c}{\hbar} = \frac{Z^2}{n^3}\frac{\mu}{m_e}\frac{\alpha^2 m_e c^2}{h}.$$

To obtain this expression we recall Equations (3-50) and (3-56) for the quantized radius r_n, and we also employ a result for the quantized orbit speed v_n as quoted in Problem 14 at the end of the chapter. We know that the frequency of the emitted radiation is not to be identified directly with the frequency of orbital motion. However, these last two calculations demonstrate the equality of v_n and f_n for large values of n, as required by Bohr's correspondence principle.

Example

Let us begin by computing all the basic parameters that arise in the Bohr model. Equation (3-55) gives the fine structure constant:

$$\alpha = \frac{e^2}{4\pi\varepsilon_0} \frac{2\pi}{hc} = \frac{1.440 \text{ eV} \cdot \text{nm}}{1240 \text{ eV} \cdot \text{nm}} 2\pi = \frac{1}{137.0}.$$

Equation (3-56) gives the Bohr radius in terms of the electron's Compton wavelength:

$$a_0 = \frac{1}{2\pi\alpha} \frac{h}{m_e c} = \frac{137.0}{2\pi} (0.002427 \text{ nm}) = 0.05292 \text{ nm}.$$

Equation (3-54) gives the Rydberg energy unit:

$$E_0 = \frac{\alpha^2}{2} m_e c^2 = \frac{0.5110 \times 10^6 \text{ eV}}{2(137.0)^2} = 13.61 \text{ eV}.$$

Equation (3-59) gives the Rydberg constant in the limit of an infinite-mass nucleus:

$$R_\infty = \frac{E_0}{hc} = \frac{13.61 \text{ eV}}{1240 \text{ eV} \cdot \text{nm}} = 0.01098 \text{ nm}^{-1}.$$

Next, let us choose $Z = 1$ and calculate the energy levels of hydrogen using Equation (3-52). The reduced-mass correction for hydrogen can almost be ignored since

$$\frac{\mu}{m_e} = \frac{M_p}{M_p + m_e} = \frac{1836}{1837}.$$

The four lowest levels have energies

$$-\frac{1836}{1837}(13.61 \text{ eV}) \begin{bmatrix} 1 \\ 1/4 \\ 1/9 \\ 1/16 \end{bmatrix} = -\begin{bmatrix} 13.60 \\ 3.40 \\ 1.51 \\ 0.85 \end{bmatrix} \text{eV},$$

as indicated in the energy level diagrams in Figures 3-15 and 3-17. Finally, let us look at some of the hydrogen wavelengths that follow from Equation (3-58).

The formula for λ becomes

$$\lambda = \left(\frac{\mu}{m_e}R_\infty\right)^{-1}\left(\frac{1}{n_f^2} - \frac{1}{n_i^2}\right)^{-1}$$

$$= \frac{1837}{1836(0.01098 \text{ nm}^{-1})}\left(\frac{1}{n_f^2} - \frac{1}{n_i^2}\right)^{-1} = \frac{91.12 \text{ nm}}{1/n_f^2 - 1/n_i^2}.$$

In the Lyman series we have $n_f = 1$, and so we find the longest wavelength for $n_i = 2$:

$$\lambda = \frac{91.12 \text{ nm}}{1 - 1/4} = \frac{4}{3}(91.12 \text{ nm}) = 121.5 \text{ nm},$$

an ultraviolet spectral line. In the Balmer series we have $n_f = 2$, and so we obtain the following values for λ:

$$\begin{bmatrix} 36/5 \\ 16/3 \\ 100/21 \\ 9/2 \\ 196/45 \end{bmatrix}(91.12 \text{ nm}) = \begin{bmatrix} 656.1 \\ 486.0 \\ 433.9 \\ 410.0 \\ 396.9 \end{bmatrix} \text{nm} \quad \text{for} \quad n_i = \begin{bmatrix} 3 \\ 4 \\ 5 \\ 6 \\ 7 \end{bmatrix}.$$

These wavelengths are included in Figure 3-12; all the values are within 0.1 nm of Balmer's original results. Each series has a shortest wavelength that occurs in the series limit as $n_i \to \infty$. The limiting wavelengths are found from the formula

$$\lambda_\infty = \left(\frac{\mu}{m_e}R_\infty\right)^{-1}n_f^2.$$

In the Lyman series

$$n_f = 1 \quad \text{gives} \quad \lambda_\infty = 91.1 \text{ nm},$$

and in the Balmer series

$$n_f = 2 \quad \text{gives} \quad \lambda_\infty = 364.5 \text{ nm}.$$

All higher series have their limits in the infrared and beyond.

3-7 Characteristic X Rays

The spectrum of radiation from an x-ray tube is generally composed of continuous and discrete wavelengths. A sketch of these superimposed distributions has been furnished in Figure 2-16, and an explanation for the continuous component has been given in Section 2-6. The discrete lines in the spectrum are peculiar to the specific metal in the target of the x-ray tube. These characteristic x rays represent a signature whose interpretation is similar to that of the optical spectrum for the given element.

Figure 3-19

Orbital model of atomic excitation and x-ray emission. The collision of a beam electron with a target atom causes an atomic electron to be ejected from an inner orbit. Another atomic electron fills the vacancy and emits an x-ray photon. The deexcitation of the atom occurs between Bohr orbits labeled by the quantum numbers n_i and n_f.

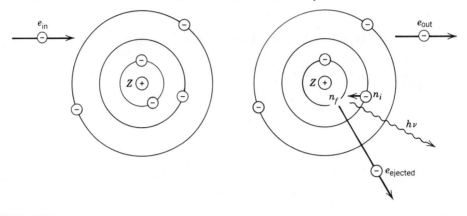

The discrete lines are attributed in either case to transitions between the quantized energy levels of the corresponding atom. X-ray spectral lines have short wavelengths of nanometer order, and so x-ray transition energies of the atom are in the keV range. The large energy required to excite the atom is supplied by the electron beam in the high-voltage tube.

X-ray spectra are associated with complex atoms containing many electrons. This apparent complication is not a serious analytical problem in the x-ray regime because the excitation energies are large enough to remove a tightly bound electron from an *inner* orbit near the nucleus of the atom. In this circumstance the emitted x rays are assumed to result from the transitions of a *single* electron while the other electrons are regarded as spectators. Bohr's simple one-electron model is quite useful in such a situation.

Let us visualize the x-ray excitation and emission processes in the many-electron atom in terms of a system of independent electrons occupying a collection of quantized Bohr orbits. We label the orbits allowed for an individual electron by the familiar quantum number n, in the manner of Equations (3-50) and (3-52). The electrons of interest for the excitation process are those in the innermost orbits of the atom, the so-called K and L orbits, where the quantum numbers are given by $n = 1$ and $n = 2$. Figure 3-19 shows a model of such a system in which the atomic electrons are distributed over several different Bohr orbits. The figure also shows how the atom becomes excited when the system is struck by an energetic incident electron, as in the bombardment of the target in an x-ray tube. The collision causes the ejection of an atomic electron from an inner orbit, so that the resulting singly ionized atom is left in a highly excited state. The ion deexcites when one of the remaining electrons makes a quantum jump from an outer orbit and fills the vacancy left by the ejected electron. This transition to a state of lower energy is accompanied by the emission of an x-ray photon. The so-called K and L series of x rays result from all such electron transitions to the $n = 1$ and $n = 2$ inner orbits of the ion.

The first comprehensive study of characteristic x rays was conducted in 1913 by H. G. J. Moseley. He investigated the K and L spectra of many of the elements in the periodic table and based his analysis on the notion of inner-electron behavior in the context of Bohr's model. His survey of the elements revealed regularities that finally established the identity between the nuclear charge index Z and the atomic number in the periodic table.

Let us show how the Bohr model is applied in Moseley's analysis. We refer to Figure 3-19 and consider the transition of orbits for a single electron from n_i to n_f, ignoring all the other electrons in the atom. The wavelength of the emitted x ray is given in terms of these quantum numbers by Equation (3-58). Let us rewrite the formula as an expression for the x-ray frequency:

$$\nu = Z_{\text{eff}}^2 cR \left(\frac{1}{n_f^2} - \frac{1}{n_i^2} \right). \tag{3-62}$$

This equation contains an all-purpose Rydberg constant R representing the quantity $\mu R_\infty / m_e$ and an effective charge parameter Z_{eff} replacing the usual nuclear charge index Z. We use Z_{eff} instead of Z because the electron in transition sees a nuclear charge *smaller* than Ze as the other atomic electrons tend to screen the nucleus from view. Even an inner electron in transition experiences this screening effect to some degree. We can build the effect into the model if we introduce an empirical screening constant z_f and set

$$Z_{\text{eff}} = Z - z_f. \tag{3-63}$$

It is expected that z_f should *increase* with the final quantum number n_f, because the final orbit radius increases with n_f and because the screening is greater for transitions to larger final orbits. Equation (3-63) may be inserted into Equation (3-62), and the result can be rearranged to give the following expression for Z_{eff}:

$$Z - z_f = \frac{\sqrt{\nu}}{\sqrt{cR\left(1/n_f^2 - 1/n_i^2\right)}}.$$

This formula becomes

$$Z = z_K + \frac{\sqrt{\nu}}{\sqrt{cR\left(1 - 1/n_i^2\right)}} \quad \text{with } n_i = 2, 3, \ldots, \tag{3-64}$$

when we choose $n_f = 1$ and consider the K series, and becomes

$$Z = z_L + \frac{\sqrt{\nu}}{\sqrt{cR\left(1/4 - 1/n_i^2\right)}} \quad \text{with } n_i = 3, 4, 5, \ldots, \tag{3-65}$$

when we choose $n_f = 2$ and consider the L series. The equations predict *linear* relations between Z and $\sqrt{\nu}$, with different intercepts and different sets of slopes in each series. The K series includes K_α and K_β spectral lines for $n_i = 2$ and 3, with corresponding slopes $\sqrt{4/3cR}$ and $\sqrt{9/8cR}$. The L series includes L_α, L_β, and L_γ lines for $n_i = 3$, 4, and 5, with slopes $\sqrt{36/5cR}$, $\sqrt{16/3cR}$, and $\sqrt{100/21cR}$. Figure 3-20 shows sets of Z versus $\sqrt{\nu}$ graphs whose linear behavior reflects these predictions of the one-electron model.

Figure 3-20

Graphs of the atomic number versus the square root of the x-ray frequency in Moseley's analysis.

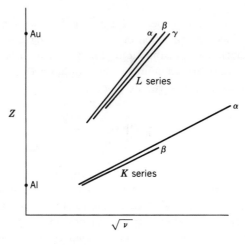

Moseley was able to change targets in his x-ray tube and observe the frequencies of x rays for more than 40 of the elements between aluminum and gold in the periodic table. He treated Z as the atomic number for each of the elements in the table and plotted his observations in the manner of Figure 3-20. The linear relations between Z and $\sqrt{\nu}$ were confirmed, and the slopes were found to agree with expectations from the Bohr model for the K_α and L_α lines. Approximate values were also deduced for the intercepts of the straight lines in the two series:

$$z_K = 1 \quad \text{and} \quad z_L = 7.4.$$

Moseley's procedures resolved certain questionable assignments of elements in the periodic table and eventually superseded existing chemical methods for identifying new elements. The most important result of his work was the conclusive demonstration that the atomic number, the nuclear charge index, and the number of electrons in the neutral atom were all given by the same quantity Z.

Figure 3-19 can be used to describe two other related phenomena. Both of these possible processes involve the ejection of *two* atomic electrons instead of the one shown in the figure. The primary types of x-ray lines are produced in the transitions of singly ionized atoms, as discussed above. Doubly ionized atoms can also be formed when electrons collide with atoms in the target of an x-ray tube. The orbit structure in the resulting ion is somewhat different from the singly ionized case, and so the radiative transitions in this system produce x rays with correspondingly different frequencies. These x rays appear in the spectrum as secondary satellites to the primary spectral lines. The ejection of a second electron can also occur in an entirely different *radiationless* process, first observed by P. Auger in 1925. The phenomenon, known as the *Auger effect*, is initiated by the excitation of the atom to a singly ionized configuration containing the usual sort of inner-electron vacancy. The ion does not deexcite by photon emission, however; instead, the system spontaneously ejects another electron and becomes a doubly ionized atom. In such a process the second electron carries away kinetic energy

$$K = E_i^{(1)} - E_f^{(2)},$$

where the notation for the initial and final energies refers to the singly and doubly ionized configurations. Thus, the system in the figure releases energy in the transition $n_i \rightarrow n_f$, and the transition energy is immediately absorbed through an *internal conversion* mechanism in which the second electron is detached and *no* radiation is produced. The emission of an Auger electron is also called *autoionization*; this effect is generally observed in competition with the emission of an x ray.

Example

Moseley's x-ray spectrum of zinc includes the wavelength $\lambda = 0.1445$ nm. We can analyze this particular line by choosing $n_i = 2$ in Equation (3-64) to get

$$Z - z_K = \sqrt{\frac{\nu}{(3/4)cR}} = \sqrt{\frac{4}{3\lambda R}}$$

$$= \sqrt{\frac{4}{3(1.445 \times 10^{-10}\ \text{m})(1.097 \times 10^7\ \text{m}^{-1})}} = 29.00.$$

The intercept for the K series is given empirically as $z_K = 1$. Hence, our calculation tells us that Z should be equal to 30, in excellent agreement with the known atomic number of Zn.

3-8 Atomic Processes and the Excitation of Atoms

Every atomic system is represented in Bohr's theory by a set of stationary quantized energy states. The theory requires all atoms of a given element to have the same ground state and the same excited states, and so the basic observable properties of the element must somehow reflect the energy level structure of the corresponding atom. All sorts of atomic processes can be examined experimentally to deduce various aspects of this structure.

The atom has been presented as a system of electron orbits in Figure 3-19. We should regard the orbital picture as a primitive quasiclassical device and turn instead to Bohr's discrete energy states for a proper quantum view of the atom. Let us visualize the atomic processes in terms of this new language and look at some of the different forms of interaction between the atom and radiation. Figure 3-21 illustrates several of these phenomena, where each is initiated by an incident photon of energy $h\nu$. We represent the atom by the associated set of energy levels, and we assume the effect of atomic recoil to be negligible in each case.

Elastic photon scattering appears in part (a) of the figure. The atom remains in its ground state as shown, so that the incoming and outgoing photons have the same energy $h\nu$. No other form of interaction can occur if the wavelength of the incident radiation is large compared to the size of the atom. The process is given the name *Rayleigh scattering* in this circumstance because the results are adequately described by Rayleigh's classical scattering theory for electromagnetic waves.

Figure 3-21
Atomic processes induced by incident
radiation.

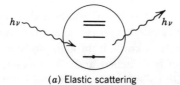

(*a*) Elastic scattering

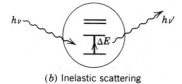

(*b*) Inelastic scattering

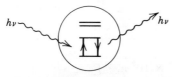

(*c*) Resonance radiation

(*d*) Fluorescence

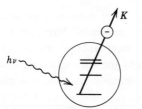

(*e*) Photoelectric effect

(*f*) Compton scattering

The incident photon may have an energy large enough to cause a transition of the atom to an excited state. This possibility is illustrated in part (b) of the figure. The process is an example of inelastic scattering since the radiation loses energy to excite the atom. The energy of the scattered photon is given by

$$h\nu' = h\nu - \Delta E, \tag{3-66}$$

where ΔE is the indicated *excitation energy* between the excited state and the ground state. Each excited state of the system corresponds uniquely to an observable shift in the frequency of the scattered radiation. Hence, a measurement of the shifts in frequency constitutes a determination of the energy levels of the given scatterer. This procedure is used extensively in molecular spectroscopy. The process is called *Raman scattering* after C. V. Raman, discoverer of the effect in 1928.

The energy of the incident photon may happen to be equal to one of the excitation energies of the atom. The fundamental absorption and emission processes of Figure 3-13 can occur in tandem at this energy, as indicated in part (c) of the figure. Elastic scattering is the result since the initial and final photons have the same energy $h\nu$. The phenomenon is called *resonance radiation* because the effect is similar to the resonant behavior of a driven harmonic oscillator. The photon energy matches the energy required to excite the atom, and so the cross section for elastic photon scattering is *enhanced*, indicating a large response of the atom to the incident radiation.

The figure shows a situation in part (d) where enough energy is absorbed to excite the atom into one of the higher-energy states. Deexcitation may then proceed in a sequence of downward transitions, accompanied by a cascade of emitted photons. This phenomenon is known as *fluorescence*. A common example occurs when an atom absorbs ultraviolet light and deexcites by emitting several wavelengths of visible light. The figure illustrates the effect by showing two fluorescent photons $h\nu'_1$ and $h\nu'_2$, along with the inelastically scattered photon $h\nu'$.

Part (e) presents the photoelectric effect as the absorption of an incident photon accompanied by the ejection of a bound electron. Note that the excitation produces an ionized system with energy K above the ionization level. Note also that the work function of the atom is equal to the difference in energy between the ionization state and the ground state.

Part (f) shows the Compton effect for the atom in its ground state. The energy of the incident photon is large enough to eject a bound electron and produce outgoing radiation of longer wavelength. The energies in the figure obey the familiar Compton relation $h\nu - h\nu' = K$.

The excitation of an atom does not always have to be a radiation-induced process. Figure 3-22 shows how an atom may gain energy and become excited in a collision with another particle. We have already spoken of excitations by collision in our remarks about the gas atoms in a discharge tube and the target atoms in an x-ray tube. The colliding particle is an electron in these applications and in the two situations shown in the figure. We let the incident electron have kinetic energy K as indicated, and we consider two possibilities for the behavior of the atom. The recoil of the massive atom is again assumed to have a negligible effect in each case. The upper part of the figure shows an elastic collision, where the atom remains in its ground state and the electron scatters without change in kinetic energy. Inelastic scattering can also take place if the initial kinetic energy is large enough to raise the atom into an excited state, as in the lower part of the figure. In this case the scattered electron has less

Figure 3-22

Electron–atom collisions. The scattered electron loses kinetic energy when the atom is excited to a higher energy state.

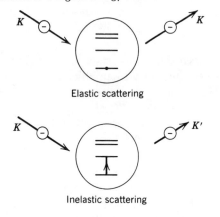

Elastic scattering

Inelastic scattering

kinetic energy, by an amount equal to the excitation energy of the atom:

$$K' = K - \Delta E. \tag{3-67}$$

We note that Equations (3-66) and (3-67) refer to parallel processes of excitation whose net effects on the atom are exactly the same. The excited atom may give up energy in either case by the emission of radiation.

The first measurements of collisional excitation in atoms were made in 1914 by J. Franck and G. L. Hertz. Their experiment studied a current of electrons in a tube containing mercury vapor and revealed an abrupt change in the current at a certain critical value of the applied voltage. They were able to interpret this observation as evidence of a threshold for inelastic scattering in the collisions of electrons with mercury atoms. The behavior of the current was an indication that electrons could lose a discrete amount of energy and excite mercury atoms in their passage through mercury vapor, provided the electrons were given enough energy by the accelerating voltage. They backed up their conclusion by detecting the emission of ultraviolet light caused by the deexcitation of the mercury excited state. The Franck–Hertz observations constituted direct and decisive confirmation of the existence of quantized energy levels in atoms.

Figure 3-23 shows a picture of a typical Franck–Hertz experiment. The apparatus consists of a mercury-vapor tube equipped with cathode, grid, and anode terminals. A variable accelerating voltage is applied between cathode and grid, while a smaller retarding voltage is applied between anode and grid, as indicated in the figure. Electrons can acquire full kinetic energy K from the accelerating voltage and reach the anode to produce a current, provided they suffer no loss of energy owing to inelastic collisions with mercury atoms. Inelastic scattering results in a lesser electron kinetic energy K', as described in Figure 3-22 and Equation (3-67). The retarding voltage is large enough to turn back any inelastically scattered electrons but is not so large as to prevent the other electrons from reaching the anode. The two possible types of electron behavior are denoted in the figure by the labels K and K'. The

Figure 3-23

Schematic apparatus and results for a Franck–Hertz experiment.

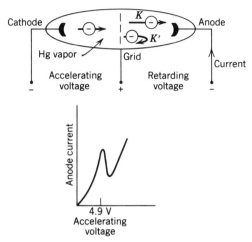

anode current is observed to grow with the accelerating voltage until the corresponding value of K becomes equal to the mercury excitation energy ΔE. An abrupt
reduction in current occurs, signaling the onset of inelastic scattering, when the
voltage reaches this critical value. The sudden drop in the observed current is
illustrated in the lower part of Figure 3-23.

Example

The critical voltage for the Franck–Hertz experiment in mercury is given as 4.9
V. The wavelength of ultraviolet light from mercury atoms in the Franck–Hertz
tube is quoted as $\lambda = 253.67$ nm. Let us verify the consistency of these two
measurements. The energy of the emitted ultraviolet photon is the same as the
excitation energy ΔE, so that λ and ΔE are related by

$$\Delta E = \frac{hc}{\lambda}$$

according to Bohr's basic formula in Equation (3-35). Inelastically scattered
electrons have kinetic energy $K' = 0$ at the inelastic threshold. Hence, Equation
(3-67) relates the critical voltage to the excitation energy through the equality

$$K = \Delta E.$$

Thus, the corresponding threshold value of the electron kinetic energy is

$$K = \frac{hc}{\lambda} = \frac{1240 \text{ eV} \cdot \text{nm}}{253.7 \text{ nm}} = 4.888 \text{ eV}.$$

The critical voltage should therefore be 4.888 V, in agreement with the
measured value 4.9 V.

Figure 3-24

Three fundamental photon–atom processes in which $h\nu = \Delta E$.

Spontaneous emission

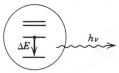

Parameter A

Absorption

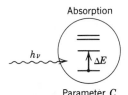

Parameter C

Stimulated emission

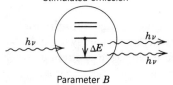

Parameter B

Figure 3-25

Growth and decay of populations of atoms at two energy levels. The atoms are in a radiation field of frequency $\nu = (E_2 - E_1)/h$.

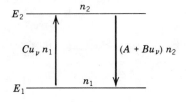

3-9   The Laser

Let us reconsider radiative transitions of the kind described in Figure 3-21 so that we can bring together three basic types of photon–atom interaction. The processes of interest are illustrated diagrammatically in Figure 3-24. In each case we are concerned with a photon energy $h\nu$ that exactly matches the difference in energy ΔE between two particular atomic energy levels. We continue our practice of the previous section and assume that the photon momentum is not large enough to cause appreciable recoil of the atom.

The first two diagrams in the figure show the related phenomena of *spontaneous emission* and *absorption*. We let A^* denote an excited state of atom A and express the two processes as

$$A^* \to A + \gamma \quad \text{and} \quad \gamma + A \to A^*.$$

We should recognize these familiar excitation properties of the atom by recalling Figure 3-13. The third diagram introduces the new process of *stimulated emission*,

$$\gamma + A^* \to A + \gamma + \gamma,$$

in which an incident photon induces the deexcitation of the atom from a higher to a lower energy state. The transition mechanism is such that the triggering photon and the emitted photon occur together, necessarily with the same frequency ν, in the final state of the system. This simultaneous emergence of two photons from the deexcited atom corresponds to the emission of coherent radiation in which two electromagnetic waves are generated in phase. The emitted photons can proceed to interact in like manner throughout a medium containing other excited atoms of the same species A. The resulting cascade of photon-doubling transitions over the whole medium produces monochromatic waves with the property of coherence on a large scale. This amplification effect is the basis for the operation of the laser. (The word *laser* is an acronym for *light amplification* by the *stimulated emission* of *radiation*.)

To generate the cascade of photons it is necessary that atoms in the medium be found in the appropriate excited state. Since the atoms are in the presence of radiation with frequency ν, the population of the two relevant energy levels is governed by the dynamical interplay of all three processes in Figure 3-24. The original analysis of these interconnected phenomena can be traced to investigations by Einstein dating back to 1917. We can discuss the problem with the aid of Figure 3-25 by considering two energy levels E_1 and E_2 with time-dependent populations of atoms given by n_1 and n_2. The population of the upper level is fed by the absorption process, so that the indicated growth rate $Cu_\nu n_1$ is proportional to the number of atoms in the lower level as well as the spectral energy density of radiation at frequency ν. The population of the lower level grows by spontaneous emission at a rate An_2 and also grows by stimulated emission at a rate $Bu_\nu n_2$. Both of these growth rates are proportional to the number of atoms in the upper level. Notice, however, that the rate of spontaneous emission does not contain the spectral energy density as a factor because this effect does not involve the presence of the stimulating radiation field. The three constants A, B, and C are called *Einstein coefficients*. These quantities have been assigned in Figure 3-24 to parameterize the quantum behavior of the three basic processes.

The growth and decay of the upper and lower populations are evidently described by the rate equations

$$\frac{dn_2}{dt} = Cu_\nu n_1 - (A + Bu_\nu)n_2 \tag{3-68a}$$

and

$$\frac{dn_1}{dt} = (A + Bu_\nu)n_2 - Cu_\nu n_1. \tag{3-68b}$$

The numbers n_1 and n_2 must approach constant values after a sufficient length of time. Hence, the equations imply a relation between the final populations:

$$(A + Bu_\nu)n_2 = Cu_\nu n_1. \tag{3-69}$$

Let us combine this result with other information about n_1 and n_2 to learn some properties of the Einstein coefficients. We can assume that the atoms are distributed according to the Maxwell–Boltzmann distribution for a system in thermal equilibrium at temperature T. It then follows from Equations (2-35) and (2-36) that n_1 and n_2 obey a second relation of the form

$$\frac{n_1}{n_2} = \frac{e^{-E_1/k_B T}}{e^{-E_2/k_B T}} = e^{h\nu/k_B T}, \tag{3-70}$$

where $h\nu = E_2 - E_1$. An expression for the spectral energy density results from these last two observations:

$$A + Bu_\nu = Cu_\nu e^{h\nu/k_B T} \quad \Rightarrow \quad u_\nu = \frac{A}{Ce^{h\nu/k_B T} - B}.$$

Let us now assume that the atoms are in the presence of radiation characteristic of a blackbody field and recall the Planck formula from Equation (2-37):

$$u_\nu = \frac{8\pi\nu^2}{c^3} \frac{h\nu}{e^{h\nu/k_B T} - 1}.$$

We can then deduce the following connections between the parameters A, B, and C by comparing the two expressions for u_ν:

$$C = B \quad \text{and} \quad \frac{A}{B} = \frac{8\pi h\nu^3}{c^3}. \tag{3-71}$$

(These arguments can be assembled somewhat differently to redirect the inferences. For example, it is possible to *derive* Planck's formula if it is learned independently that the parameters satisfy Equations (3-71).) We have pursued this question out of curiosity. Let us turn next to applications and examine the practical consequences of Equation (3-70).

The laser operates through an active medium of atoms whose energy states can be populated selectively by radiative means. Equation (3-70) tells us that the lower state in any pair of levels is preferentially occupied at thermal equilibrium. Consequently, the basic laser process of stimulated emission cannot be operative unless some mechanism is included in the system for *pumping* atoms to the higher energy. This added feature must be introduced to maintain a *population inversion* that overturns the normal thermal distribution. Figure 3-26 shows how a three-level laser incorporates such a pumping scheme. An electrical discharge is used to raise atoms from the ground state to the energy level E_3, and then a rapid spontaneous decay brings the excited atoms down to the level E_2. This state is described as *metastable* since its properties are such as to inhibit spontaneous decay back to the ground state. An incident photon of energy $h\nu = E_2 - E_1$ can then stimulate the desired laser transition to the level E_1.

The ruby laser is an example of such a three-level system. Green light from a flash lamp pumps chromium ions in the ruby crystal to an excited level, and nonradiative deexcitation takes the ions promptly to a long-lived state at slightly lower energy. Stimulated emission then follows, generating a coherent beam of red (694 nm) output light.

Laser start-up cannot proceed in the three-level system until the ground-state population has been reduced by more than half. Large input pumping power is needed to achieve this condition. Further inefficiency results from the fact that ground-state atoms in the medium are likely to absorb the radiation emitted by other atoms. These problems do not arise in the four-level laser because the device is

Figure 3-26

Three-level laser with a pumping mechanism for populating level E_2 at the expense of level E_1.

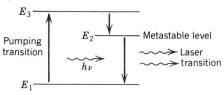

Figure 3-27

Four-level laser with a population inversion between levels E_3 and E_2.

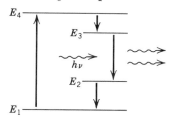

designed so that the ground state cannot participate in the laser transition. Figure 3-27 shows a possible level scheme with such a property. Pumping raises atoms from the ground state to level E_4, and the rapid population of level E_3 follows by spontaneous emission. Laser transitions are stimulated from E_3 to E_2 by photons of energy $h\nu = E_3 - E_2$. The population inversion between these two levels is easily maintained as atoms at E_2 undergo rapid spontaneous decay to the ground state. It should also be clear from the figure that atoms in the ground state are not able to absorb laser light at the frequency ν.

The He–Ne gas laser is a familiar laboratory instrument with the performance properties of the four-level scheme shown in Figure 3-28. The active medium for this device is typically a 10 : 1 mixture of helium and neon gases at low pressure (of order 1 mm Hg). An electrical discharge pumps helium atoms to an excited state, which happens to have approximately the same energy above the ground state as one of the upper levels in neon. A resonant transfer of excitation energy occurs between helium

Figure 3-28

Highly schematic plan of the He–Ne gas laser.

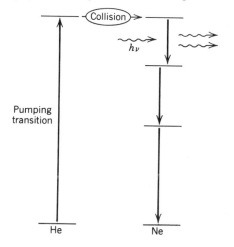

Figure 3-29

Basic components of a gas laser.

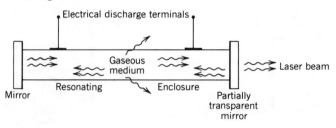

and neon via the collision process

$$He^* + Ne \rightarrow He + Ne^*.$$

The Ne* level then deexcites through a photon-induced laser transition, as indicated qualitatively in the figure. Helium is used for the pumping phase because the lowest excited states of the helium atom are uniquely situated for the collisional excitation of specific neon levels. A discharge in pure neon would pump atoms less selectively and populate too many different neon states. The He–Ne laser generates a red (633 nm) output beam as one of several monochromatic possibilities. The beam can deliver quite large amounts of power per unit area in a very narrow wavelength interval.

A resonating cavity of suitable length is an essential feature of the typical laser. Figure 3-29 shows how the resonant characteristics of a gas laser are achieved by means of parallel plane mirrors at the ends of the gas enclosure. The separation of the mirrors is designed to store radiation whose frequency corresponds to the desired laser transition. The reflection system can accomplish a well-defined separation of frequencies even though the length of the resonating cavity accommodates many closely spaced optical modes. One of the two mirrors is partially transparent and transmits a portion of the repeatedly reflected light to give the output of the laser. The storage system also tends to eliminate photons whose directions are not parallel to the axis of the resonator. As a result the long dimension of the cavity determines a precise directionality for the output beam.

All the properties of monochromaticity, coherence, intensity, and directionality are attainable at extraordinary levels in the laser. Many unique applications have been discovered for the use of laser beams. These include the modulation and signal-carrying capability of noise-free light in laser communication systems, the accurate determination of very large distances (e.g., from Earth to Moon) by the reflection of laser light, the investigation of multiphoton interactions with atoms using the methods of laser spectroscopy, and the proposed fusion of nuclei in samples compressed by laser radiation.

We have already noted that the theory of the laser began with Einstein's analysis of stimulated emission. These basic ideas were eventually put into use almost four decades later. The first laser was actually a *maser*, built in 1954 by C. H. Townes to generate coherent *microwave* radiation from laser transitions in ammonia. The first instrument to operate at optical frequencies was the ruby laser, constructed in 1960 by T. H. Maiman. The invention of the optical gas laser followed soon thereafter.

Example

It is apparent that stimulated emission competes with spontaneous emission to depopulate the higher energy level in Figure 3-25. Let us examine this competition by constructing the ratio of the two rates from Equations (2-37) and (3-71):

$$\frac{Bu_\nu}{A} = \frac{1}{e^{h\nu/k_BT} - 1}.$$

The result may be recognized as the average number $\langle n \rangle$ of photons with frequency ν in blackbody radiation. To see this recall Equation (2-38) for the average energy $\langle \varepsilon \rangle$ per mode at frequency ν, and write

$$\langle n \rangle = \frac{\langle \varepsilon \rangle}{h\nu} = \frac{1}{e^{h\nu/k_BT} - 1}.$$

We can readily assess the effects of frequency and temperature by evaluating this ratio of rates for different operating conditions. Thus, for optical transitions at room temperature we might consider $h\nu = 2$ eV and $k_BT = \frac{1}{40}$ eV and find

$$\frac{h\nu}{k_BT} = 80 \quad \Rightarrow \quad \frac{Bu_\nu}{A} = \frac{1}{e^{80} - 1}.$$

It is clear that these conditions favor spontaneous emission over stimulated emission by an enormous factor. The numbers are quite different for microwave transitions at the same temperature. If we change ΔE and choose $h\nu = 10^{-4}$ eV, we obtain

$$\frac{h\nu}{k_BT} = 4 \times 10^{-3} \quad \Rightarrow \quad \frac{Bu_\nu}{A} \to \frac{1}{h\nu/k_BT} = 250.$$

The conditions favor stimulated emission in this case.

3-10   The Quantum of Action

Bohr's quantum picture of the atom did not develop at once into a fundamental quantum theory. Quantization of energy was soon accepted as a demonstrable property of the states of matter, and yet the underlying quantum principles remained unknown for more than a decade. Some tentative insights were provided by Bohr's quasiclassical model of the one-electron atom. Similar methods were also adopted by A. J. W. Sommerfeld in his attempt to formulate the basic quantum principles. Sommerfeld's approach to the mystery of quantization was based on a generalizing concept known as the quantum of action.

The new ideas were focused on periodic behavior since periodicity was a common feature of the earliest quantum systems. The frequencies of Planck's oscillators and the orbits in Bohr's atom were instances where aspects of quantization were linked directly to periodic properties. This connection was the key to Sommerfeld's proposal of a generalized quantization principle. His procedure employed *phase space* as a scheme for unifying all forms of periodic motion in quantum systems.

Let us introduce the idea of phase space by considering the evolution in time of the classical motion of a particle. The dynamical equations of motion determine a complete specification of position and velocity at any instant, if the particle's position and velocity are prescribed at some initial value of the time. We may use position and momentum as an alternative pair of variables for an equivalent description of particle motion.

Phase space is spanned by a system of axes whose definition is given in terms of these position and momentum variables. The axes are constructed in *pairs* by taking a position axis and a momentum axis for *each* of the particle's degrees of freedom. Hence, a point in phase space denotes all the physical conditions of the system at a specific instant since the point is located by the instantaneous values of all the components of position and momentum. The point moves with time so that the evolution of the system is described by a directed *path* in phase space. This locus of connected points forms a closed path that repeats itself in cycles for the special case of continuous periodic motion.

Sommerfeld's proposal expresses a general quantization principle for any kind of periodic motion. The principle constrains the motion for each pair of *conjugate* phase-space variables (q, p_q), where the notation refers to the coordinate and momentum for a particular degree of freedom of the particle. Every (q, p_q) plane in phase space exhibits periodic motion as a cyclic path surrounding a fixed area. Sommerfeld's rule employs Planck's constant as a unit of area and defines an allowed quantum state of the system to be a locus of points in phase space whose enclosed area satisfies the formula

$$\int_{\text{cycle}} p_q \, dq = n_q h. \tag{3-72}$$

This area integral over a complete cycle in phase space is called the *action* integral. The units of action are evidently the same as the units of h, and so Planck's constant appears in the formula as a primitive quantum unit of action. Thus, the rule associates a quantum number n_q with the coordinate q and specifies an allowed path in phase space to be one of a *discrete* set of paths whose separation in area is given by the small quantum of action h.

Let us illustrate Sommerfeld's idea by examining the one-dimensional motion of a mass on a spring. Phase space is the (x, p) plane, and the oscillating time dependence of the coordinate is given for amplitude A and phase angle δ as

$$x = A \cos(\omega t - \delta).$$

The angular frequency is related to the mass and spring constant by the familiar expression

$$\omega = \sqrt{\frac{k}{m}}.$$

The other phase space variable is the momentum of the particle,

$$p = m\frac{dx}{dt} = -m\omega A \sin(\omega t - \delta).$$

These oscillating variables are connected through the definition of the conserved total

Figure 3-30

Loci in phase space for a harmonic oscillator. The quantization condition separates consecutive allowed paths by an annular area equal to Planck's constant, the quantum of action.

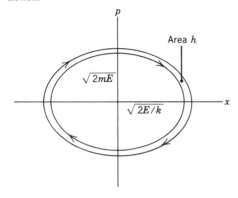

energy

$$E = \frac{p^2}{2m} + \frac{kx^2}{2}, \tag{3-73}$$

where the conserved quantity is given by

$$E = \frac{m\omega^2}{2} A^2 \sin^2(\omega t - \delta) + \frac{k}{2} A^2 \cos^2(\omega t - \delta) = \frac{k}{2} A^2.$$

We can rewrite Equation (3-73) as

$$\frac{p^2}{2mE} + \frac{x^2}{2E/k} = 1 \tag{3-74}$$

and thus identify the closed path in phase space to be an *ellipse* parameterized by the constant energy E. Figure 3-30 shows how two such ellipses are constrained by the quantization condition in Equation (3-72). We avoid computation of the action integral by recalling the formula for the area of an ellipse, and we write the condition in terms of the indicated semimajor and semiminor axes as

$$\pi\sqrt{2mE}\sqrt{\frac{2E}{k}} = nh.$$

Hence, the allowed values of the energy are quantized according to the formula

$$E = n\frac{h}{2\pi}\sqrt{\frac{k}{m}} = n\hbar\omega. \tag{3-75}$$

We recognize the condition for Planck's quantized oscillators in this result.

Let us turn next to Bohr's circular orbits and demonstrate another application of the quantization principle. This special case of two-dimensional motion has only one degree of freedom defined by the angle θ. The conjugate momentum variable p_θ is identified as the *angular* momentum L. We write Equation (3-72) as

$$\int_{\text{cycle}} L \, d\theta = n_\theta h$$

and immediately use the fact that L is a constant of the motion to obtain

$$L \int_{\text{cycle}} d\theta = L \cdot 2\pi = n_\theta h.$$

We note in passing that this simple calculation is valid even when the orbit is not circular. Our result takes the form

$$L = n_\theta \hbar \qquad (3\text{-}76)$$

and thus reproduces Bohr's expression for the quantization of angular momentum as written in Equation (3-48).

Sommerfeld's quantization conditions were of some interest in the days of the old quantum theory. The principle found extensive use in the generalization of Bohr's model to the case of noncircular orbits. Degrees of freedom were associated with both θ and r in this case, so that phase space was enlarged to include the two pairs of conjugate variables (θ, p_θ) and (r, p_r). The angular momentum quantum number n_θ retained its validity, while a second quantum number n_r was also introduced by applying the quantization principle to the radial variables as well:

$$\int_{\text{cycle}} p_r \, dr = n_r h.$$

The two quantum numbers were employed by Sommerfeld and others to account for a broader array of quantized energy states in the analysis of the hydrogen atom. Of course, all these quasiclassical considerations passed rapidly into history with the coming of quantum mechanics.

Problems

1. Suppose that, in an observation of Brownian motion in water at 300 K, spherical particles of 1 μm diameter experience an rms displacement of 12 μm in 3 min. Calculate the value of Avogadro's number from these data. (Recall that the universal gas constant has the value $R = 8.314$ J/mole $\cdot$ K.)

2. Obtain the general solution to the differential equation for Brownian motion,

    $$\frac{dg}{dt} + \frac{\beta}{m} g = \frac{2k_B T}{m},$$

    and compare the result with the particular solution

    $$g(t) = \frac{2k_B T}{\beta}$$

    for the limiting situation where $m \to 0$.

3. Consider Thomson's e/m experiment, and let θ denote the angular deflection of the electron beam as shown in the drawing. Deduce a formula to show how θ depends on the ratio e/m and on other parameters of the experiment such as the field strengths E and B.

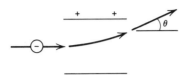

4. The figure shows what appears to be an e/m apparatus similar to Figure 3-1. This conventional cathode ray tube generates electrons from a hot filament and accelerates the particles through a potential difference ϕ_0. The electron beam passes between plates of length x and separation d, where a voltage ϕ is applied across the plates. Obtain an expression for the vertical deflection of the beam, showing that this measurable quantity has no dependence on either e or m_e.

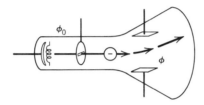

5. Here are some data from a Millikan oil-drop experiment:

    plate voltage = 5080 V   rise time (field on) = 22.1 s

    plate separation = 16.0 mm   air viscosity = 182 μpoise

    rise–fall distance = 10.2 mm   oil density = 920 kg/m^3

    fall time (field off) = 11.8 s

    Determine the radius and the mass of the oil drop, and use the known value of e to deduce the number of electron charge units on the drop.

6. A Thomson model of the hydrogen atom predicts a unique frequency of oscillation for the electron. Assume that radiation of the same frequency is emitted by the atom, and compute the wavelength of the radiation. Take the radius of the atom to be $R = 0.05$ nm.

7. Prove that the α-particle orbit in Figure 3-8 has the same incoming and outgoing impact parameters.

8. Refer to Figure 3-8 and show that the distance of closest approach r_{min} for a non-head-on collision in Rutherford scattering is related to the scattering angle θ by

$$r_{min} = \frac{D}{2}\left(1 + \csc\frac{\theta}{2}\right),$$

    where D is the head-on distance of closest approach. Check the validity of the formula by considering the special case of a head-on collision.

9. Grains of sand are elastically scattered by a bowling ball as shown. Obtain a formula for the differential scattering cross section. Let the total cross section be defined as

$$\sigma = \int_{\text{all }\Omega} \frac{d\sigma}{d\Omega}\, d\Omega,$$

and show that σ is equal to the cross-sectional area of the ball.

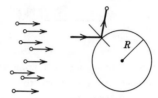

10. A 3 MeV beam of α particles strikes an aluminum target. Determine the distance of closest approach between α particles and aluminum nuclei at this energy, and calculate the number of aluminum nuclei per unit volume in the target. (Aluminum has atomic number 13, atomic mass 27, and density 2.70 g/cm^3.)

11. Suppose that the α-particle beam in Problem 10 carries 10^5 α particles/s to the aluminum target, and let the target thickness be 10^{-4} cm. Calculate the number of α particles scattered per second into the backward hemisphere.

12. A beam of 6 MeV protons is scattered by a gold foil of 10^{-5} cm thickness. The proton detector is ring shaped as in Figure 3-6. The ring collects protons scattered at 60° and subtends a small element of angle equal to 1°. Calculate the distance of closest approach between protons and gold nuclei at this energy, the number of gold nuclei per unit volume in the target, the size of the solid angle subtended by the detector, and the fraction of incident protons scattered into the detector. (Gold has atomic number 79, atomic mass 197, and density 19.3 g/cm^3.)

13. Consider Rutherford scattering at beam energy K, and derive a formula for the fraction of incident α particles scattered through angles greater than a given angle θ_0.

14. Show that the Bohr orbits have quantized speeds given by

$$v_n = \frac{Z}{n}\alpha c,$$

in which α is the fine structure constant. Use this result to assess the validity of the nonrelativistic treatment of the Bohr atom.

15. In the hydrogen spectrum, the Brackett series and the Pfund series are sets of spectral lines corresponding to transitions to the $n = 4$ and $n = 5$ levels, respectively. For these two series calculate the maximum and minimum values of the emitted radiation energy, frequency, and wavelength.

16. The photon of largest energy in the hydrogen spectrum occurs at the Lyman series limit. Calculate the momentum of the photon, and determine the velocity of the recoiling atom. Take the mass of the atom to be 1.67×10^{-27} kg.

17. Obtain values (in eV) for the energies in the energy level diagram for singly ionized helium He$^+$. Identify all transitions in He$^+$ for which the emitted wavelengths are in the visible range, 350–700 nm.

18. The $Z = 1$ neutral atom has three isotopic species: hydrogen, deuterium, and tritium. The nuclear masses for these atoms are (approximately) M, $2M$, and $3M$, with $M = 1.67 \times 10^{-27}$ kg. Obtain a formula and calculate a value for the difference of the Lyman α wavelengths for hydrogen and tritium.

19. What are the Bohr-model formulas for the energy levels and the orbit radii in positronium? What are the orbit speeds for the electron and positron? Obtain a formula for the wavelengths in the positronium Balmer series, and calculate the corresponding maximum and minimum wavelengths.

20. The doubly ionized lithium ion Li^{++} is a one-electron atom having $Z = 3$. Identify all the transitions in the Li^{++} emission spectrum that lie in the far-ultraviolet region of the spectrum and beyond, where the wavelengths are less than 50 nm.

21. A muonic atom is formed when a negative muon replaces an electron in a normal atom. Determine the radius of the first Bohr orbit and the ground-state energy in a muonic hydrogen atom. The muon mass is 105.7 MeV/c^2.

22. The following K_α wavelengths are found in Moseley's data:

$$0.8364, \quad 0.2111, \quad \text{and} \quad 0.1798 \text{ nm.}$$

Identify the elements responsible for each of these x rays.

23. The following L_α, L_β, and L_γ wavelengths are measured for one of the elements in Moseley's survey:

$$0.4385, \quad 0.4168, \quad \text{and} \quad 0.3928 \text{ nm.}$$

Use the L_α line to identify the element, and compare the L_β and L_γ wavelengths with the values predicted from the Bohr model.

24. A Franck–Hertz experiment in hydrogen shows dips in the anode current at 10.20 and 12.09 V. What wavelengths should be observed in the radiation emitted from the Franck–Hertz tube?

25. Consider the motion of a bead of mass m on a wire of length a. The bead is free to slide back and forth with constant speed between $x = -a/2$ and $x = +a/2$. What is the locus of the bead's motion in phase space? Impose Sommerfeld's quantization condition to determine the energies for the allowed states of motion.

F O U R

MATTER
WAVES

*T*he goals of the quantum revolution neared fulfillment in the years just before 1925. One of the decisive contributions of this period came forward in 1924 in the doctoral thesis of L. V. P. R. de Broglie. The dissertation proposed that matter should possess *both particle and wave* characteristics, in parallel with the quantum properties of radiation. It was very bold to suppose that discrete matter should have wave behavior since there were no known indications from experiment to support the proposal. Fortunately, the skeptical response to de Broglie's hypothesis was not shared by Einstein, who found the idea promising and urged that it be given serious consideration. The matter wave continued to be controversial even after experimental evidence for it was found. It was obvious that original insight was needed to interpret this new *nonclassical* type of wave.

An altogether different approach was adopted in W. K. Heisenberg's treatment of the quantum theory. His view of the situation was more like that of Bohr, inasmuch as classical concepts were kept in sight while waves were not directly employed. Deviation from classical physics was a central theme in Heisenberg's thinking. He reexamined the meaning of the basic observable quantities in particle mechanics and found that *uncertainties* were inherent in the measurement of such variables. He then argued that the principles of dynamics should be given by a formalism that made allowance for the act of measurement. The familiar classical variables were to appear prominently in this theory, and the intrinsic observational uncertainties were also supposed to be built into the definition of all the quantities. Heisenberg formulated these ideas, and thus arrived at his version of quantum mechanics, in 1925.

Our main concerns in this chapter are de Broglie's matter wave and Heisenberg's uncertainty principle. We anticipate some of the arguments of quantum mechanics to interpret the matter-wave concept, and we look critically at certain notions from classical physics when we take up the question of uncertainty.

4-1 De Broglie's Hypothesis

The matter-wave conjecture was motivated in part by the similarity between quantized states and standing waves. The quantization of energies was associated with the existence of discretely allowed configurations in a quantum system. This phenomenon was comparable to the occurrence of harmonic frequencies for discrete modes in a vibrating medium, such as a length of string or a column of air. De Broglie's proposition was also an appeal to symmetry. He noted that the quantum theory of radiation was based on a coexistence of particle and wave attributes, and he argued that the quantum theory of matter should adopt a similar kind of dual picture.

The skeptics needed to be reminded of the circumstances that attended the proof of the wave nature of light. It was originally thought that light traveled in straight lines, or rays, much like the motion of free particles. The wave properties of light were demonstrated by detecting the effects of diffraction over intervals of space whose extent was of the same order as the wavelength of the light. This observation encouraged de Broglie to believe that a wavelength might be associated with a moving particle and that the resulting diffraction effects were such as to have gone unnoticed in the more conspicuous forms of matter.

The matter-wave hypothesis assigns a wavelength λ to a particle with momentum of magnitude p. These quantities are related by the formula

$$\lambda = \frac{h}{p}, \tag{4-1}$$

in obvious analogy to the relation for a photon given in Equation (2-52). Thus, a free particle with mass m and speed v has a *de Broglie wavelength*

$$\lambda = \frac{h}{mv} \tag{4-2}$$

for the special case of nonrelativistic motion. Equation (4-1) is understood to hold more generally in situations where p denotes the relativistic momentum.

We can appreciate Bohr's condition for the quantization of angular momentum by appealing immediately to de Broglie's relation. Let us suppose that an allowed orbit in Bohr's model is such that the de Broglie wave for the bound particle is a *standing* wave that just fits the circumference of the orbit. We illustrate this intuitive notion in Figure 4-1 and write the condition for the standing wave in terms of Equation (4-1) as

$$2\pi r = n\lambda = n\frac{h}{p}.$$

Figure 4-1

Standing de Broglie wave on an allowed circular Bohr orbit.

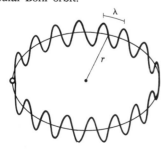

The result can be rearranged to produce the desired expression for the angular momentum:

$$L = pr = n\frac{h}{2\pi} = n\hbar,$$

as stated without argument in Equation (3-48). Of course, we should be aware that this construction is still only an ad hoc quantum rule applied to a classical system.

De Broglie did not explain the physical nature of the oscillations in his proposed wave. It was obvious that the matter wave (or phase wave, in his language) could *not* be identified with the vibration of a material medium because it would not be mechanically possible to reconcile the integrity of a physical particle with the diffusing property of a physical wave. An altogether new kind of wave was being introduced, and the interpretation was being deferred for later reflection.

Example

Let us show how the de Broglie wavelength is used to distinguish regimes of different physical behavior. An electron with kinetic energy 100 eV has momentum

$$p = \sqrt{2m_e K} = \sqrt{2(9.11 \times 10^{-31} \text{ kg})(100 \text{ eV})(1.60 \times 10^{-19} \text{ J/eV})}$$

$$= 5.40 \times 10^{-24} \text{ kg} \cdot \text{m/s},$$

and de Broglie wavelength

$$\lambda = \frac{h}{p} = \frac{6.63 \times 10^{-34} \text{ J} \cdot \text{s}}{5.40 \times 10^{-24} \text{ kg} \cdot \text{m/s}} = 1.23 \times 10^{-10} \text{ m} = 0.123 \text{ nm}.$$

This result is comparable with the wavelength of an x ray, so that a suitable diffraction experiment should be sensitive to the proposed wave behavior of such a particle. In contrast, a 1 g pellet with speed 100 m/s has de Broglie wavelength

$$\lambda = \frac{h}{Mv} = \frac{6.63 \times 10^{-34} \text{ J} \cdot \text{s}}{(10^{-3} \text{ kg})(100 \text{ m/s})} = 6.63 \times 10^{-33} \text{ m}.$$

The value of λ for this macroscopic object is many orders of magnitude smaller than the actual size of the object, and so no experiment can be devised to demonstrate its alleged wave properties.

Example

It might appear from Equation (4-2) that the de Broglie wavelength is always larger than the Compton wavelength h/mc. This conclusion is false because Equation (4-2) is applicable only for nonrelativistic motion. Let us adopt the more general form in Equation (4-1) and observe that the de Broglie and Compton wavelengths can actually be equated as

$$\frac{h}{p} = \frac{h}{mc} \quad \text{when} \quad p = mc.$$

This condition determines the corresponding relativistic energy of the particle:

$$\sqrt{E^2 - m^2c^4} = mc^2 \quad \Rightarrow \quad E = \sqrt{2}\,mc^2.$$

The kinetic energy of the relativistic particle is then given by

$$K = (\sqrt{2} - 1)mc^2.$$

We have used Equations (1-35) and (1-37) to obtain these results.

4-2   Electron Diffraction

Evidence for matter waves was found in 1927 in two separate laboratories. The investigators were C. J. Davisson and L. H. Germer in one of the experiments, and G. P. Thomson in the other. Both experiments demonstrated the effects of diffraction for beams of electrons and obtained results in agreement with de Broglie's wave predictions. (G. P. Thomson was the son of J. J. Thomson. It was noteworthy that the one should discover the wave nature of the electron 30 years after the other had discovered the electron as a particle.)

The apparatus of the Davisson–Germer experiment was laid out as in Figure 4-2. Electrons from a hot filament were accelerated through various potential differences to produce beams of different energy. The electron beam was directed onto a surface cut in a single crystal of nickel, and the scattered electrons were observed at different angles by varying the angular position of a detector with respect to the direction of the incident beam. The distribution was plotted as a function of the indicated scattering

Figure 4-2

Design of the Davisson–Germer electron diffraction experiment. The electron beam experiences wave-like Bragg reflection from parallel planes in the scattering crystal.

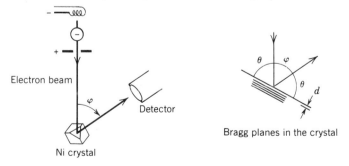

Figure 4-3

Intensity of scattered electrons versus scattering angle φ in the Davisson–Germer experiment. Distributions are shown for beam energies of 40, 54, and 68 eV.

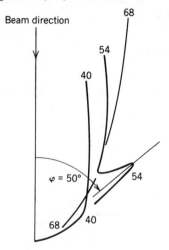

angle φ for each beam energy in the manner shown in Figure 4-3. These observations indicated that the scattering of electrons from the crystal target was just like the diffraction of x rays by the same crystal. The intensity of scattered electrons showed a large background of backscattered electrons and revealed a pronounced diffraction peak in a certain direction for a particular value of the beam energy. This peak was detected in the original experiment at an angle $\varphi = 50°$ for a beam energy of 54 eV, as indicated in the figure.

We can explain these observations if we apply the matter-wave hypothesis to the electrons and recognize the effects of *Bragg reflection* in the resulting wave behavior. The de Broglie wavelength varies with the beam energy, so that diffraction is observed when this wavelength is comparable in size with the spacing between parallel crystal planes in the nickel crystal. Thus, the matter wave exhibits the same diffraction phenomenon as that found in the Bragg reflection of x rays. If we recall the Bragg condition from Equation (2-49) and apply the formula to matter waves, we find that a diffraction maximum is expected in a direction φ given by the prediction

$$\frac{\lambda}{2d} = \sin \theta = \cos \frac{\varphi}{2}. \tag{4-3}$$

Note that Figure 4-2 has been consulted to obtain

$$\theta = \frac{\pi}{2} - \frac{\varphi}{2}$$

as the relation between the angle φ and the usual Bragg angle θ. Note also that the spacing d between crystal planes can be regarded as known, since this quantity can be measured separately by diffracting x rays of known wavelength from the given crystal. Finally, observe that the Bragg angle is fixed in Figure 4-2 and that the diffraction peak appears at this angle for a unique value of the beam energy.

Figure 4-4

Electron diffraction pattern from one of Thomson's experiments.

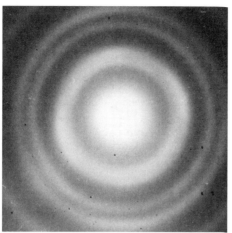

Thomson's experiment was a study of electron diffraction for electron beams of greater energy and shorter wavelength. The first scattering targets were thin films made of organic materials; these were later replaced by metals such as aluminum and gold. The metallic films had no large-scale crystalline structure, so that the scattering of electrons was caused by minute crystals distributed at random in each of the targets. Electrons transmitted through such a film formed exactly the same diffraction pattern as that observed for an x-ray beam of the same wavelength. A typical pattern was similar to the one shown in Figure 4-4. Concentric sharp rings of maxima and minima appeared symmetrically around the beam axis, just like the diffraction resulting from a small aperture or a small obstacle.

The demonstration of wave properties by beams of particles was an indication of quantum behavior by *free* electrons. This discovery came at a time when most of the investigations of quantum properties were focused on electrons bound in atoms. It should also be remarked that the charge of the particle was not a relevant consideration and that beams of *neutral* particles such as neutrons were later found to undergo the same diffraction effects.

Example

Let us ask why diffraction is seen in the Davisson–Germer experiment at 54 eV, but not at 40 eV, as Figure 4-3 suggests. We need the de Broglie wavelength at each energy in order to answer this question. The momentum at 54 eV is

$$p = \sqrt{2m_e K} = \sqrt{2(9.11 \times 10^{-31} \text{ kg}) (54 \text{ eV}) (1.60 \times 10^{-19} \text{ J/eV})}$$

$$= 3.97 \times 10^{-24} \text{ kg} \cdot \text{m/s},$$

and so the corresponding de Broglie wavelength is

$$\lambda = \frac{h}{p} = \frac{6.63 \times 10^{-34} \text{ J} \cdot \text{s}}{3.97 \times 10^{-24} \text{ kg} \cdot \text{m/s}} = 1.67 \times 10^{-10} \text{ m} = 0.167 \text{ nm}.$$

We find λ at 40 eV by using the inverse proportionality between λ and $\sqrt{K}$:

$$\lambda = \sqrt{\frac{54}{40}}\,(0.167\text{ nm}) = 0.194\text{ nm}.$$

The Bragg condition compares these wavelengths with the quantity $2d$. The nickel crystal in the experiment is known (from an x-ray determination) to have interplanar spacing $d = 0.091$ nm, and so $2d = 0.182$ nm. Equation (4-3) has no solution when $K = 40$ eV, since λ exceeds $2d$ in this case; therefore, no diffraction peak is expected at this energy. The calculation at $K = 54$ eV gives

$$\cos\frac{\varphi}{2} = \frac{\lambda}{2d} = \frac{0.167\text{ nm}}{0.182\text{ nm}} = 0.918 \quad \Rightarrow \quad \varphi = 46.9°,$$

in rather good agreement with the Davisson–Germer measurement of 50°.

4-3 Particle–Wave Duality

Electron diffraction is a display of wave behavior by a beam of particles in a *wave-type* experiment. The experiment is sensitive to the wave nature of the electron because the de Broglie wavelength of the particle is of the same order of magnitude as the interplanar spacing in the diffracting crystal. No diffraction is observed, however, when an electron beam of the same energy passes between the plates of a cathode-ray tube, because the wavelength and the plate separation do not compare favorably for an observation of wave behavior. Thus, the same electron has matter-wave properties in the one experiment and classical-particle properties in the other.

A similar dichotomy has already made its appearance in our discussion of radiation. We know that the wave nature of light has been fully verified by interference and diffraction experiments, and we have recently learned that discrete properties are also in evidence in such quantum phenomena as the photoelectric effect and the Compton effect.

It is clear that the discrete nature of matter and the wave nature of radiation do not provide a complete description, as classical physics has led us to believe. We have to allow for complementary behavior that is wave-like in matter and particle-like in radiation, because we are confronted with these properties when we perform the appropriate experiments. Quantum physics thus requires the *dual* aspects of particle *and* wave to be present simultaneously in matter and in radiation. This concept of particle–wave duality also embraces a *principle of complementarity*. The dual characteristics are complementary since matter and radiation are completely described by adopting both particle and wave points of view. These two properties are not found to be in contradiction because it is not possible to devise a single experiment that tests both particle and wave aspects at once.

Particle–wave duality introduces the notion of a *quantum particle*. The extraordinary behavior of this new entity is quite unlike any of the usual phenomena encountered in our previous experience with classical systems. To illustrate, let us consider the electron as a prime example of such a particle and explore some of the further consequences of its dual qualities.

We expect the wave nature of the electron to carry over from electron diffraction to *any* wave-type experiment. Hence, the remarkable properties of the matter wave may also be seen if a double-slit experiment is performed with a beam of electrons, provided the de Broglie wavelength is of the same order as the spacing between the two slits. Let us proceed by detecting many individual electrons at specific locations on a screen, after the particles have passed through the pair of slits. We confirm our expectation of wave behavior by obtaining a distribution of detected electrons that conforms exactly to the familiar *interference pattern* produced by light waves in a comparable Young's double-slit apparatus. If we reduce the incident intensity and collect only a few electrons on the screen, we find that the sparse distribution of collected electrons still resembles the original pattern to the extent that the interference minima are observed at the same locations as before.

The observations are totally different in a *macroscopic* double-slit experiment where, for instance, a much larger pair of slits is used to transmit a beam of pellets. We recall from a previous example that the de Broglie wavelength of such a particle is too small to be detectable, and so we expect to find no interference effects. If we let the pellets be embedded in a screen after passing through the slits, we see that the embedding occurs in two regions of the screen directly downstream from the two slits. This distribution tells us that each of the pellets has passed through one slit or the other, as expected for a beam of classical particles.

The double-slit experiment for electrons cannot be interpreted to say whether a given electron goes through a given slit. Instead, the two slits transmit electrons in the manner of a wave passing through *both* slits at once. Let us suppose that the experiment is altered by mounting a separate current loop around each slit. An induced current can then be observed in one loop or the other to detect the passage of an electron through the corresponding slit. This induced current affects the passing electron, however, so that the particle is likely to be found at a different location on the screen owing to the modification of the slits. Thus, we destroy the wave behavior of the original two-slit interference experiment if we attempt to get classical information about the particles by tracking the electrons through the separate slits. In fact, we succeed in changing one wave-type experiment into another whenever we monitor the two slits. The two-slit interference pattern appears when the monitoring device is off, while two overlapping single-slit diffraction patterns appear when the device is on.

The real meaning of de Broglie's matter wave remains to be explained. We know, of course, that the wave is not to be associated with the oscillations of a material medium, but we are not yet in position to identify the actual oscillating entity. Let us approach this ultimate question of interpretation in the quantum treatment of matter by looking first at the implications of particle–wave duality in the more familiar case of radiation.

The radiation field is a continuous system of propagating electric and magnetic oscillations. We follow convention and choose the electric field $\mathbf{E}(\mathbf{r}, t)$ to specify the form of the propagating wave. We also assume a monochromatic wave so that the oscillations have a single frequency and wavelength given by ν and λ. The flow of radiated energy is expressed by constructing the electromagnetic Poynting vector in terms of the total field $\mathbf{E}$ at a given point P in space. Thus, we find that the flow of energy per unit time across unit area at P is given by the expression

$$c\varepsilon_0 \mathbf{E}^2.$$

Figure 4-5 provides a specific example in which the monochromatic radiation field is produced by the illumination of a double slit and the point P is identified by the position of a small photoelectric cell. The total field $\mathbf{E}$ at P is the sum of coherent waves $\mathbf{E}_1$ and $\mathbf{E}_2$ arriving at the given cell from the two slits. An interference pattern is observed in the flow of energy as P varies over an array of such cells. Constructive interference occurs at a given cell if the two waves arrive in phase, as in the situation shown in the figure. The indicated slit separation d, wavelength λ, and direction θ

Figure 4-5

Interference of electromagnetic waves in a double-slit experiment. The same interference pattern is obtained with a beam of electrons.

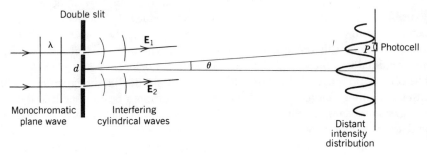

then satisfy the condition

$$\frac{d}{\lambda}\sin\theta = \text{an integer,} \tag{4-4}$$

provided the interference is observed at a large distance from the slits.

The general expression $c\varepsilon_0 \mathbf{E}^2$ oscillates with time, and so a more useful observable quantity is found by averaging with respect to time over a complete cycle. We therefore define $\mathbf{E}_P^2$ as the time average of $\mathbf{E}^2$ at the point P and introduce the *radiation intensity*

$$c\varepsilon_0 \mathbf{E}_P^2$$

to obtain a quantity whose value depends only on the position P. The figure shows how this intensity varies with the location of the photocell in the case of double-slit interference.

The implications of particle–wave duality emerge when these familiar results are cast in terms of photons. We know that the photon carries radiation energy $h\nu$, and so we may express the radiation intensity at P by means of the alternative expression

$$h\nu I_P,$$

where I_P denotes the average number of photons entering unit area at P per unit time. The conclusion is a simple equality between the two ways of writing the radiation intensity:

$$h\nu I_P = c\varepsilon_0 \mathbf{E}_P^2. \tag{4-5}$$

This formula is a statement of particle–wave duality since the left side is written in the language of discrete quanta while the right side is written in the language of wave fields. Note that our example in the figure adapts at once to the introduction of the quantity I_P because the indicated photoelectric cell is sensitive to the arrival of individual photons.

Double-slit diffraction for increasing levels of incident intensity. Increasing numbers of photons are detected per second at random locations on a distant screen. The probability distribution of detected photons conforms to the wave diffraction pattern. The intensity of this pattern is given at large distance by the angle-dependent function

$$4I_0\cos^2\frac{\phi_d}{2}\frac{\sin^2(\phi_a/2)}{(\phi_a/2)^2},$$

where $\phi_d = (2\pi d/\lambda)\sin\theta$ and $\phi_a = (2\pi a/\lambda)\sin\theta$.

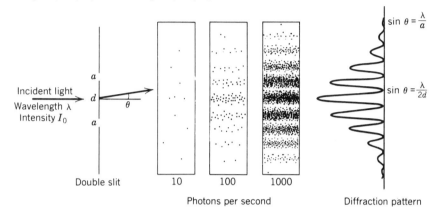

The photon description on the left side of Equation (4-5) makes no reference to any concept of path for the arrival of photons at position P. We entertain no such notion, even when we imagine a very low level of intensity and consider only a single photon in transit. Instead, the detection of a photon at a particular location is understood to be a *random* event that represents the one and only operational specification of the position of the photon. Our double-slit example illustrates this act of locating a photon as a single random occurrence in which the photon is absorbed once and for all by an atom in a photoelectric cell.

The wave description on the right side of the equation tells us where the random events of photon detection are likely to occur. To draw this conclusion we let the relation between $\mathbf{E}_P^2$ and I_P establish a correspondence between an *intensity distribution* for the behavior of waves and a *probability distribution* for the detection of photons. The quantity I_P is interpreted as a measure of the probability of observing a photon in a small region located at P. The equation connects this probability to the square of the field at that location. The familiar wave field $\mathbf{E}(\mathbf{r}, t)$ is thus interpreted as a mathematical *wave function*, or guiding field in Einstein's words, that furnishes statistical information regarding the likelihood for the random arrival of photons. Our double-slit experiment illustrates this interpretation even at very low levels of intensity. The resulting photon distributions are very sparsely populated and show no isolated events in anyregion where destructive interference occurs for the corresponding wave. A more marked resemblance then develops between the photon population and the interference pattern as the incident intensity increases.

Let us add two final remarks about the wave function $\mathbf{E}(\mathbf{r}, t)$. We recall that the electric field for waves in free space satisfies the familiar wave equation in the form given in Equation (2-15). We also know that **E** is an observable physical quantity since we can determine the field by measuring the electrical force on a small test charge.

The implications of particle–wave duality can be transferred by analogy from radiation to matter. Let us explore the possibilities of such an analogy and gain a preliminary understanding of de Broglie's matter wave. We anticipate the definition of a mathematical wave function called $\Psi(\mathbf{r}, t)$ for matter, in place of $\mathbf{E}(\mathbf{r}, t)$ for radiation. We assume that Ψ is found by solving a certain wave equation, of a form yet to be discovered. Observable properties of Ψ are proposed according to arguments analogous to the probability interpretation of radiation, as just deduced on the basis of Equation (4-5). We first suppose that a quantity symbolized as Ψ_P^2 is derivable from $\Psi(\mathbf{r}, t)$, as $\mathbf{E}_P^2$ is derivable from $\mathbf{E}(\mathbf{r}, t)$. We then interpret Ψ_P^2 as a measure of the probability of finding a quantum particle in a small region located at position P. The notation Ψ_P^2 is introduced temporarily to convey the analogy with Equation (4-5), so that the probability of finding the particle at P has the form of an intensity determined by the square of the wave. The proposed wave function $\Psi(\mathbf{r}, t)$ is the basic element in a formal theory of a quantum particle. Since the formalism is based on a treatment of waves, it is understood that the addition, or superposition, of waves is supposed to be incorporated as a main feature. Interference and diffraction effects for quantum particles are then immediately explainable in terms of a probability interpretation of the observed wave behavior.

These anticipatory remarks are offered as a statement of our objectives. We know that evidence for de Broglie's matter wave is to be found in electron diffraction or any other wave-type electron experiment. Our arguments imply that such observations of wave behavior are to be interpreted probabilistically, in the manner of the probability interpretation of radiation. To illustrate, let us return once more to our double-slit experiment for light in Figure 4-5 and recall our discussion earlier in the section regarding an analogous experiment for a beam of electrons. Particle–wave duality predicts a complete parallel between radiation and matter, as an interference pattern is mapped out by an array of photoelectric cells in the one case and by a similar array of electron detectors in the other. The common probabilistic interpretation tells us how the observed pattern describes a distribution of discrete observations at each point P in the array. A photoelectric cell at P measures the probability of detecting a photon while an electron detector at P measures the probability of detecting an electron, as random detection events in both instances.

Example

Both Thomsons, father and son, performed experiments with beams of electrons. We should be able to understand why the father saw no evidence for matter waves by computing the relevant de Broglie wavelength. A typical cathode-ray experiment has been discussed in the first example of Section 3-2. The electrons were found to have a speed of 1.68×10^7 m/s and were allowed to pass between a pair of plates with 2 cm separation. The corresponding de Broglie wavelength would have the value

$$\lambda = \frac{h}{m_e v} = \frac{6.63 \times 10^{-34}\ \text{J} \cdot \text{s}}{(9.11 \times 10^{-31}\ \text{kg})(1.68 \times 10^7\ \text{m/s})} = 0.0433\ \text{nm}.$$

This wavelength would be far too small for an observation of diffraction effects due to the 2 cm spacing of the pair of plates.

4-4 Determinism and Randomness

Newtonian mechanics describes the motion of a classical particle by means of a well-defined time-dependent coordinate vector $\mathbf{r}(t)$. The goal of classical dynamics is to predict the time dependence of $\mathbf{r}(t)$ for a given applied force through the use of Newton's law

$$m\frac{d^2\mathbf{r}}{dt^2} = \mathbf{F}.$$

Two pieces of initial data must be given if this second-order differential equation is to have a unique solution. Such initial information may be expressed in terms of position $\mathbf{r}$ and velocity $\mathbf{v}$, or position $\mathbf{r}$ and momentum $\mathbf{p}$, at time $t = 0$. The resulting dynamical scheme is said to be *deterministic*, or causal, since precisely specified initial conditions control the prediction of a precisely determined particle trajectory.

These familiar principles and objectives are at variance with the considerations that pertain to a quantum particle. Recall that this new conception of a particle is adopted whenever the corresponding de Broglie wavelength becomes a detectable property. Our illustration of double-slit interference for a beam of electrons is one example where the observations in the experiment are not addressed by the predictions of classical determinism. Single-slit diffraction for a beam of electrons is another instance in which a wave-type phenomenon is controlled by the properties of the electron's de Broglie wave. In either case the matter wave is responsible for an interference or diffraction pattern, and the distribution of intensity is interpreted as a distribution of probability. Thus, each electron in the beam has a likelihood to be collected in *any* of the given detectors that map out the observed pattern. No concept of electron path is involved in these observations. Instead, the detection of a particular electron is a *random* event whose occurrence represents the one and only opportunity to specify the location of the particle.

It is clear that there is *no necessity* for particle dynamics to be based on the analysis of detailed motion if the experimental observations are not able to confront the analytical predictions. We may take the random detection of electrons in double-slit interference or single-slit diffraction to be representative of such an experimental situation. The results of this type of experiment can be understood if we set deterministic principles aside and turn instead to the question of predicting probabilities. There is a *clear necessity* for the latter kind of particle theory whenever randomness characterizes the observable behavior of the particle.

We have anticipated the formulation of this theory in the previous section by referring to the expected properties of the wave function $\Psi(\mathbf{r}, t)$ in our interpretation of the matter wave. The determination of $\Psi(\mathbf{r}, t)$ for the quantum particle becomes the objective in quantum physics, replacing the problem of solving for $\mathbf{r}(t)$ in the case of the classical particle. We incorporate the random detectability of the quantum particle in the probability interpretation of Ψ and make no reference at all to the deterministic classical trajectory defined by $\mathbf{r}(t)$. Note that, while the detection of a particle at a given location is an entirely random event, the probability of random

Figure 4-6

Examples of classical particle motion.

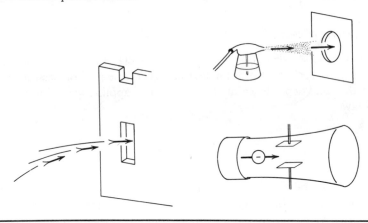

detection at any location is supposed to be predictable from the determination of the matter wave.

Of course, Newtonian mechanics retains its applicability for macroscopic nonrelativistic phenomena. In fact, we need not hesitate to follow the deterministic approach even in certain cases of submicroscopic particle motion, as noted in the example of Section 4-3. We can decide on the proper treatment of the particle by comparing the de Broglie wavelength with the size of potentially diffracting components in the given physical system. Figure 4-6 shows several situations in which diffraction effects are regarded as unobservable on the basis of this comparison.

Let us suggest another way to assess the applicability of the classical theory by noting that every theory is described in terms of measurable quantities. Classical physics takes the approach that predictions can be confirmed by means of measuring devices whose intrusive influence on the system of interest can be made arbitrarily small. The predictions become abstractions divorced from reality, however, when the act of measurement is so disturbing as to prevent the verification of the theory. Let us consider these criteria in the context of the deterministic trajectory $\mathbf{r}(t)$, the precise time-dependent position of a particle. We can measure this quantity by illuminating the particle and observing the scattered radiation. We know, of course, that the scattering of a photon can impart momentum to the particle in any random direction. We should therefore ask whether this act of observation constitutes an intolerable disturbance on the motion of the particle.

Figure 4-7 shows an example of a macroscopic object whose behavior can be watched with the aid of visible light. The corresponding photon momentum is much too small to have a detectable effect in this case. The figure also shows how we might attempt to observe an orbiting atomic electron. X rays are used for the purpose since short wavelengths are needed to achieve the desired high resolution. A very severe disturbance results because the x-ray photon carries a large momentum and can transfer enough energy to the atom to cause the removal of the electron from its atomic orbit. Thus, we can easily verify a prediction of classical motion in the upper part of the figure, but we cannot even contemplate such a verification in the lower part.

Figure 4-7

Observations of particle motion by means of scattered illumination. When the incident wavelength is reduced to accommodate the size of the particle, the momentum transferred by the photon becomes large enough to disturb the observed motion.

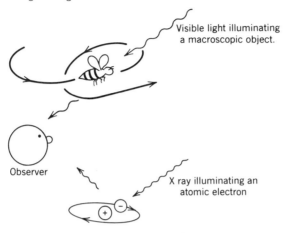

Visible light illuminating a macroscopic object.

Observer

X ray illuminating an atomic electron

These considerations of the measurement process are fundamental to the understanding of quantum physics. We examine the specific consequences of the problems of measurement in the next section.

4-5 The Uncertainty Principle

Heisenberg's formulation of quantum particle dynamics was introduced in 1925. His picture of quantum mechanics contained the usual time-dependent dynamical variables, such as the coordinate $\mathbf{r}(t)$ and the momentum $\mathbf{p}(t)$, but not in the form of ordinary functions of the time. Instead, the observable quantities of the theory were represented by means of matrices whose mathematical properties were tailored to allow for the problem of measurement. This theory of matrix mechanics was soon joined by an alternative theory of wave mechanics based on the concept of de Broglie's matter wave. The famous Heisenberg uncertainty principle was deduced in 1927 as a common feature in the interpretation of the two equivalent formalisms.

We are not going to discuss Heisenberg's picture of quantum mechanics any further because the alternative wave approach is easier to understand in a first exposure to the theory. Since the uncertainty principle can be established in both matrix and wave pictures, we can regard the resulting idea as a deduction fundamental to *any* version of the quantum theory.

The principle delimits our ability to make accurate measurements of the observable properties of a quantum particle. It is proved in the theory that corresponding components of the conjugate variables $\mathbf{r}$ and $\mathbf{p}$ cannot both be known with absolute precision at the same time and that uncertainties must exist in their simultaneous determination. It is specifically shown that the conjugate quantities (x, p_x) have uncertainties $(\Delta x, \Delta p_x)$, which satisfy the inequality

$$\Delta x \, \Delta p_x \geq \frac{\hbar}{2}, \tag{4-6}$$

Werner Heisenberg

and that identical statements hold for (y, p_y) and (z, p_z). The conclusion identifies a *minimum* product of uncertainties, which constrains the allowed accuracy of specification for one of the conjugate observables at a certain time, given the accuracy of specification for the other at the same time. It is therefore possible to know either coordinate or momentum with certainty only at the expense of total ignorance regarding the other conjugate quantity. Note that the principle applies to conjugate pairs of components such as x and p_x and does not relate uncertainties for nonconjugate variables such as x and p_y.

The minimum product of uncertainties in Equation (4-6) is proportional to the small quantum of action. Figure 4-8 shows how this small lower bound defines a tight boundary between prohibited and allowed regions of uncertainty. It is clear that the uncertainty in the coordinate can be made quite small unless the uncertainty in the momentum is nearly vanishing, and vice versa. The figure indicates a very thin "neighborhood of certainty" in which values of x and p_x are simultaneously assigned with more accuracy than the principle permits for a quantum particle. Again we see the role played by the smallness of Planck's constant as we consider physical systems of vastly different scale. The limitations imposed by the uncertainty principle can be ignored when measurements are performed on macroscopic objects, because experimental errors are always much larger for these systems. On the other hand, the inequality can be expected to have a decisive influence in the treatment of quantum particles such as electrons bound in atoms. These observations are quite consistent with our remarks about the measurement problem in the context of Figure 4-7.

We should note that we have not yet given a mathematical definition for the uncertainties Δx and Δp_x. In fact, we have not even defined the concepts of position and momentum for a quantum particle and cannot do so until we have made some progress into Chapter 5. This shortcoming is not an immediate concern since it is perfectly correct to think of the observables in terms of their measured values. The

Figure 4-8

Domain of uncertainties allowed by Heisenberg's uncertainty principle. The origin represents certainty in both variables. A nearby point $(\delta x, \delta p)$ denotes a pair of small uncertainties on the boundary of the allowed region.

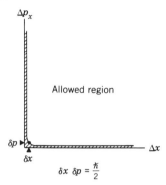

lack of a proof of Equation (4-6) should not concern us either, provided we recognize Planck's constant as the main factor in the minimum product of uncertainties.

Thought experiments can be performed to illustrate the logic of the uncertainty principle and explain the appearance of Planck's constant in the result. A simple device for the purpose is the elementary thin-lens apparatus illustrated in Figure 4-9. Let us use the device to form a real image of a particle and consider the uncertainties involved in the determination of the particle's position. We orient the axis along the vertical y direction and let the particle be illuminated vertically from below, as shown in the figure. Formation of an image means that the particle must scatter at least one photon into the aperture of the lens. The transverse x component of momentum of a scattered photon may then have any value in the range

$$(-p \sin \theta, \, p \sin \theta),$$

where θ is half the angle subtended by the aperture and p is the scattered photon momentum. Planck's constant makes its appearance when we use $p = h/\lambda$ to express the momentum in terms of the indicated wavelength. The act of measurement causes the particle to recoil with an opposite transverse momentum. We therefore obtain

$$\Delta p_x = 2p \sin \theta = 2\frac{h}{\lambda} \sin \theta \qquad (4\text{-}7)$$

as the uncertainty in the x component of momentum of the observed particle.

The apparatus cannot fix the exact transverse position of the particle because diffraction by the aperture of the lens causes the image to spread out. The lens sees a particle as a circular blur formed by the central peak in the diffraction pattern of the aperture, and it sees two particles as overlapping patterns like those shown in the figure. The images are said to be just resolved if they satisfy *Rayleigh's criterion*, whereby the central maximum of the one image pattern coincides with the first

Figure 4-9

Real image formation to determine the location of a particle. The position uncertainty Δx can be reduced by enlarging the lens aperture, but the momentum uncertainty Δp_x is made larger as a result.

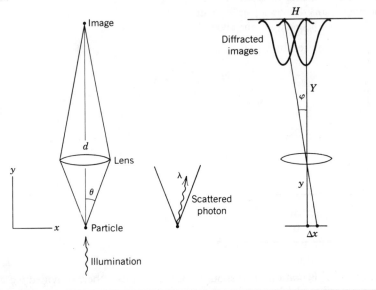

minimum of the other. This condition is illustrated in the figure as the means of defining the uncertainty Δx in the measurement of a particle's transverse coordinate. We can use the geometry in the figure to construct an expression for Δx once we know the indicated diffraction angle φ. The angle for the first diffraction minimum is found from the wavelength and the lens diameter according to the formula

$$\sin \varphi = 1.22 \frac{\lambda}{d}.$$

(We recognize the analogous formula $d \sin \varphi = \lambda$ as the condition for the first diffraction minimum due to a slit. The additional numerical factor 1.22 is needed in the case of a circular aperture.) Two geometrical relations connect the various distances and angles:

$$\frac{\Delta x}{y} = \frac{H}{Y} = \tan \varphi$$

and

$$\frac{d/2}{y} = \tan \theta.$$

The related quantities are the object and image distances y and Y and the image separation H. We eliminate y and d as follows to get a result for Δx in terms of the angles:

$$\Delta x = y \tan \varphi = \frac{d}{2 \tan \theta} \tan \varphi = \frac{1.22\lambda}{\sin \varphi} \frac{\tan \varphi}{2 \tan \theta}.$$

The angle φ is intended to be quite small, so that the position uncertainty becomes

$$\Delta x = \frac{0.61\lambda}{\tan \theta}. \tag{4-8}$$

The other angle is constrained only by the condition $\theta < 90°$ for a finite aperture. Let us write our results in Equations (4-7) and (4-8) side by side as

$$\Delta x = \frac{0.61\lambda}{\tan \theta} \quad \text{and} \quad \Delta p_x = \frac{2h \sin \theta}{\lambda} \tag{4-9}$$

and multiply the two uncertainties to get

$$\Delta x \, \Delta p_x = (1.22 \cos \theta)h.$$

The result is an estimated product of uncertainties equal to Planck's constant times a fixed nonzero numerical factor. This quantity cannot be reduced to zero, in agreement with the uncertainty principle, because θ is always bounded away from $90°$ for any finite aperture. We note that we can improve our image resolution by enlarging the aperture and introducing a larger value of θ. Equations (4-9) tell us that we then succeed in reducing Δx, but at the expense of increasing Δp_x. The wavelength can also be decreased to gain better resolution, with exactly the same result.

It is possible to examine the meaning of the uncertainty principle by another approach based on the wave nature of matter. Let us suppose that a beam of particles is incident on a slit and that all the particles have the same momentum p with no uncertainty. The associated de Broglie wavelength $\lambda = h/p$ refers to an incident monochromatic matter wave, as illustrated in Figure 4-10. The infinite extent of this plane wave corresponds to a uniform probability for finding a beam particle any-where in the region to the left of the slit and hence represents complete uncertainty in position for any particle in the incident beam. The wave is diffracted by the slit and produces a distant intensity distribution just like the pattern obtained for light. The figure describes a distribution with a large central diffraction peak whose first minimum is given in terms of the indicated angle and slit width by the formula

$$\sin \theta = \frac{\lambda}{a}.$$

Figure 4-10

Diffraction of a beam of particles by a single slit.

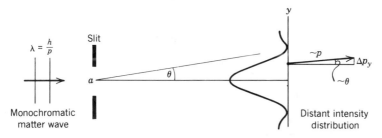

We cannot predict the point of detection for any particle in the region to the right of the slit. Instead, we use the wave intensity distribution to determine the probability for detection of a particle at any given location. The slit localizes the probability of finding the particle in the transverse y direction, as demonstrated by the intensity in the central peak. We may estimate the uncertainty in the transverse coordinate for any such particle by setting

$$\Delta y = a.$$

The diffraction peak is actually broader than the slit, and so there must be a transverse momentum uncertainty Δp_y to account for the divergence of the beam. We may estimate this quantity for a particle in the central peak by referring to the sketch shown in the figure and writing

$$\Delta p_y = p \sin \theta.$$

The product of these estimated uncertainties is

$$\Delta y \Delta p_y = ap \sin \theta = a \frac{h}{\lambda} \frac{\lambda}{a} = h,$$

again in agreement with the uncertainty principle. We note that a narrower slit corresponds to a reduced uncertainty in the transverse coordinate. We also note that a broadening of the diffraction peak results, indicating an increased uncertainty in the transverse momentum.

The uncertainty principle imposes a conceptual limitation on our ability to determine the properties of a particle. The foregoing illustrations refer to inaccuracies of measurement that cannot be reduced arbitrarily by any amount of improvement or ingenuity in experimental technique. Instead, the observed coordinate and momentum are involved in a conspiracy that prevents us from knowing both conjugate variables with arbitrary accuracy at the same time. We therefore regard these uncertainties as operational qualities of the variables themselves. The basic principles of the quantum theory must somehow embody this inherent feature of the observable quantities, since the formalism is supposed to contain the uncertainty principle as a fundamental deduction.

Example

Let us use Heisenberg's uncertainty principle to make quantitative statements about the two situations shown in Figure 4-7. Suppose that a 10 g object is visually observed with 1% accuracy to have a speed of 10 m/s. The momentum of the particle is

$$p = mv = (0.01 \text{ kg})(10 \text{ m/s}) = 0.1 \text{ kg m/s},$$

and the experimental uncertainty in the momentum is

$$\Delta p = (0.01)p = 0.001 \text{ kg m/s}.$$

The minimum position uncertainty must be

$$\Delta x = \frac{\hbar}{2 \Delta p} = \frac{1.05 \times 10^{-34} \text{ J} \cdot \text{s}}{2(10^{-3} \text{ kg m/s})} = 0.53 \times 10^{-31} \text{ m}.$$

This result of the uncertainty principle is far below any realistic measuring error and has no practical significance for such a measurement. On the other hand, imagine that an electron is to be observed in its orbit in a hydrogen atom. We assume that we can make a position measurement with sufficient refinement to distinguish between two adjacent orbits. Let us take Equations (3-50) and (3-56) from the Bohr model, ignore the reduced-mass effect, and express the nth Bohr radius as

$$r_n = n^2 a_0 = n^2 \frac{\hbar}{\alpha m_e c}.$$

Our stipulated maximum uncertainty in the measurement of an orbit radius is

$$\Delta r = r_n - r_{n-1} = \frac{\hbar}{\alpha m_e c} \left[n^2 - (n-1)^2 \right] = \frac{\hbar}{\alpha m_e c} (2n - 1).$$

The corresponding momentum uncertainty must then be at least as large as

$$\Delta p_r = \frac{\hbar}{2\Delta r} = \frac{\alpha m_e c}{2(2n-1)}.$$

We interpret this quantity to be the radial momentum imparted to the electron when the radial position measurement is made. The magnitude of the electron momentum in the nth orbit is found to be

$$m_e v_n = \frac{\alpha m_e c}{n}$$

from Problem 14 at the end of Chapter 3. We note that Δp_r and $m_e v_n$ are of the same order, and we conclude that the act of measurement is disruptive enough to affect the orbit of the observed particle.

Example

We may specify the precise location of a particle if we are prepared to give up all knowledge of the particle's momentum. Equal likelihood may then be assigned to any momentum, no matter how large the value. We can achieve a certain degree of boundedness in the momentum if we can tolerate a lesser degree of localization. This argument implies that an increase in localization is accomplished at the expense of an increase in energy, or simply that localization costs energy. To illustrate, let us assume that an electron is to be located somewhere within a region of atomic size. The position uncertainty is of order $\Delta x = 10^{-10}$ m, and so the minimum momentum uncertainty is of order

$$\Delta p = \frac{\hbar}{\Delta x} = \frac{1.05 \times 10^{-34} \text{ J} \cdot \text{s}}{10^{-10} \text{ m}} = 1.05 \times 10^{-24} \text{ kg} \cdot \text{m/s}.$$

The actual electron momentum may be at least this large; therefore, the kinetic energy of the electron may be as large as

$$K = \frac{p^2}{2m_e} = \frac{\left(1.05 \times 10^{-24} \text{ kg} \cdot \text{m/s}\right)^2}{2\left(9.11 \times 10^{-31} \text{ kg}\right)\left(1.60 \times 10^{-19} \text{ J/eV}\right)} = 3.78 \text{ eV}.$$

This result is quite reasonable since the order of magnitude compares favorably with values of the total energy in the Bohr model of the atom. If we attempt to localize an electron in a region of nuclear size, we find a considerable increase in our estimate of the energy. A position uncertainty of order $\Delta x = 10^{-14}$ m implies a much larger momentum uncertainty Δp. The electron is now likely to have a large relativistic kinetic energy, of a scale appreciably larger than that found for nuclear particles. We are then able to argue that electrons should not occur as constituents of the nucleus. The numerical details of this argument are left to Problems 7 and 8 at the end of the chapter.

Example

The uncertainty principle can be used to estimate a lower bound for the energy of a particle. Consider an oscillating mass m on a spring with force constant k. The energy is a constant of the motion, given in terms of the variables x and p by the formula

$$E = \frac{p^2}{2m} + \frac{kx^2}{2}.$$

Classical physics allows a minimum energy equal to zero for the trivial case of an oscillator at rest at the origin, where $p = 0$ and $x = 0$. This simple configuration violates the uncertainty principle, as the oscillator must have some minimal motion due to the necessary uncertainties in x and p. We can represent the constant value of E by means of averages of the kinetic and potential energies over a cycle of the motion by writing

$$E = \frac{\langle p^2 \rangle}{2m} + \frac{k}{2}\langle x^2 \rangle.$$

The average values of x and p should vanish for an oscillating particle, and so the average values of x^2 and p^2 can be identified with the squares of the corresponding uncertainties:

$$\langle x^2 \rangle = \left(\Delta x\right)^2 \quad \text{and} \quad \langle p^2 \rangle = \left(\Delta p\right)^2 = \left(\frac{\hbar}{2\,\Delta x}\right)^2.$$

We set $\Delta x = \delta$ to abbreviate the notation and rewrite the energy as

$$E = \frac{\hbar^2}{8m\delta^2} + \frac{k}{2}\delta^2.$$

Figure 4-11 shows that this expression has a minimum for a certain value of δ.

Figure 4-11

Estimation of the zero-point energy for a linear oscillator.

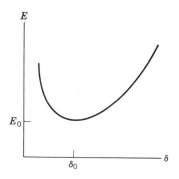

We determine the minimizing uncertainty δ_0 by requiring

$$\frac{dE}{d\delta} = 0 = -\frac{\hbar^2}{4m\delta^3} + k\delta$$

to get

$$\delta_0^2 = \sqrt{\frac{\hbar^2}{4mk}} \, .$$

The minimum energy is found from this result:

$$E_0 = \frac{\hbar^2}{8m}\sqrt{\frac{4mk}{\hbar^2}} + \frac{k}{2}\sqrt{\frac{\hbar^2}{4mk}} = \frac{\hbar}{2}\sqrt{\frac{k}{m}} = \frac{\hbar}{2}\omega_0,$$

where ω_0 appears in the final answer as the familiar angular frequency of the classical oscillator. Our estimation of the minimum energy from the uncertainty principle happens to agree with the exact formula obtained from quantum mechanics. The result is called the *quantum zero-point energy* for the one-dimensional oscillator.

Example

Consider the image cast on a screen by a beam of particles projected through a slit. The width of the image grows with the width of the slit, and also grows when the slit is made very narrow owing to the diffraction of the beam. Hence, there must exist an intermediate width for the slit that produces a minimum width for the image. We can use the uncertainty principle to study this beam-divergence effect since we expect some degree of uncertainty in the components of coordinate and momentum transverse to the direction of the beam. We suppose that a particle passing through the slit is localized within a distance y_0 of the beam axis, and we argue that the corresponding transverse

Figure 4-12

Uncertain initial conditions for a particle passing through a slit.

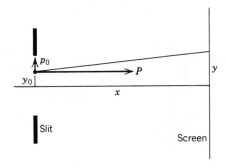

momentum can be as large as

$$p_0 = \frac{\hbar}{2y_0}.$$

These uncertainties appear in Figure 4-12 as uncertain initial conditions for a classical treatment of the subsequent motion of the particle. The indicated off-axis displacement is given by

$$y = y_0 + \frac{p_0}{m}t = y_0 + \frac{\hbar}{2my_0}t,$$

where t is the time when the particle reaches the screen. The distance to the screen is determined by the beam momentum P as

$$x = \frac{P}{m}t,$$

and so the off-axis displacement becomes

$$y = y_0 + \frac{\hbar}{2my_0}\frac{mx}{P} = y_0 + \frac{\hbar x}{2Py_0}.$$

This expression has a minimum for a certain choice of y_0, as obtained from the condition

$$\frac{dy}{dy_0} = 0 = 1 - \frac{\hbar x}{2Py_0^2}.$$

The minimizing value of y_0 and the corresponding minimum value of y are

$$y_0 = \sqrt{\frac{\hbar x}{2P}} \quad \text{and} \quad y = y_0 + \frac{y_0^2}{y_0} = 2y_0.$$

We conclude that the image cast by the particle beam has its smallest width when the half-width of the slit is taken to be $\sqrt{\hbar x/2P}$. Note that the result for y_0 can be rewritten as

$$\sqrt{\frac{\hbar x}{2P}} = \sqrt{\frac{\lambda x}{4\pi}},$$

where λ is the de Broglie wavelength of the particle.

4-6   Waves and Wave Packets

We begin our formal treatment of the quantum theory in Chapter 5. Our purpose in this concluding section is to prepare a foundation for the wave picture of quantum mechanics by reviewing the properties of waves. We are especially concerned with the use of a wave to convey the idea of *localization*, since we know that a wave distribution is supposed to represent a distribution of probability for locating a quantum particle.

This review deals only with waves described by a single spatial variable. We begin with the properties of a monochromatic traveling wave and illustrate by means of the wave function

$$\Psi = A \cos 2\pi \left(\frac{x}{\lambda} - \nu t \right) = A \cos(kx - \omega t). \tag{4-10}$$

Note that factors of 2π are eliminated by introducing the wave number k and the angular frequency ω as

$$k = \frac{2\pi}{\lambda} \quad \text{and} \quad \omega = 2\pi\nu. \tag{4-11}$$

The argument of Ψ is called the *phase* of the simple harmonic wave. This quantity determines the cyclic behavior of the wave as the function Ψ varies with the independent variables x and t. Figure 4-13 shows how two configurations of a traveling wave on a string are seen as snapshots of Ψ versus x, taken at two different times during a cycle.

Equation (4-10) is not necessarily restricted to waves in a one-dimensional medium, even though the expression has only one spatial variable. We can also let Ψ represent

Figure 4-13

Simple harmonic wave at times $t = 0$ and $t = 1/4\nu$. The wave propagates with phase velocity $v_\phi = \nu\lambda$.

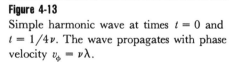

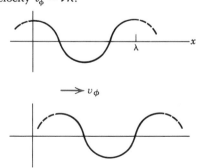

a *plane wave* in three dimensions and visualize the wave as the propagation of a plane surface of constant phase. This wave front is defined by setting the argument of Ψ equal to a constant:

$$kx - \omega t = \phi_0.$$

The resulting plane surface has a varying position along the x axis, given by

$$x = \frac{\phi_0 + \omega t}{k} \quad \text{at time } t.$$

Hence, the surface of constant phase propagates in the *positive* x direction with *phase velocity*

$$v_\phi = \frac{\omega}{k} = \nu\lambda. \tag{4-12}$$

A similar conclusion is drawn for the traveling wave in Figure 4-13. The alternative wave function

$$\Psi = A \sin(kx - \omega t)$$

Figure 4-14

Analogue model of phase propagation. Clocks are arrayed at 3 m intervals along an axis, and each clock is set in sequence to run 3 h out of synchronism with its predecessor. The hour hands simulate wave motion, with wavelength $\lambda = 12$ m and frequency $\nu = (12 \text{ h})^{-1}$. The system of clocks is read at times $t = 0$, $t = 3$ h, $t = 6$ h, ... In each instance special notice is taken of the indicated 12-o'clock reading as a particular choice of phase. The selected orientation of the hour hand is observed to propagate with phase velocity $v_\phi = 1$ m/h. Thus, the phase travels along the axis while the clocks remain in place.

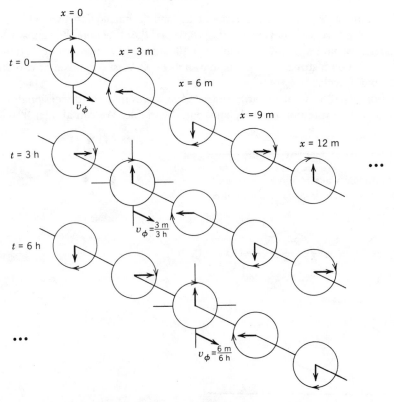

describes a wave with the same properties as the original Ψ in Equation (4-10), apart from the obvious quarter-cycle difference in phase. Another entirely different example of phase propagation is described in Figure 4-14.

A complex-number representation is often employed as a convenient mathematical device for the handling of phase in an oscillating system. A brief summary of the properties of complex numbers is given in Table 4-1. The main result for our purposes

---

Table 4-1   Properties of Complex Numbers

Imaginary number $i = \sqrt{-1}$

Complex number $z = x + iy$ (Cartesian form)
Real part $x = \text{Re } z$
Imaginary part $y = \text{Im } z$
Complex conjugate $z^* = x - iy$

Modulus $|z|$ where $|z|^2 = zz^* = x^2 + y^2 = r^2$

Phase θ where $\tan \theta = \dfrac{y}{x}$

Complex plane

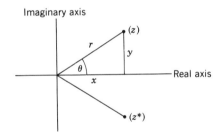

Power series

$$x = r\cos\theta = r(1 - \frac{\theta^2}{2!} + \cdots)$$

$$y = r\sin\theta = r(\theta - \frac{\theta^3}{3!} + \cdots)$$

$$z = r(\cos\theta + i\sin\theta)$$

$$= r[1 + i\theta + \frac{(i\theta)^2}{2!} + \frac{(i\theta)^3}{3!} + \cdots]$$

Complex number $z = re^{i\theta}$ (polar form)

Phase factors $e^{i\theta} = \cos\theta + i\sin\theta$ and $e^{-i\theta} = \cos\theta - i\sin\theta$
Real part $\text{Re } e^{i\theta} = \cos\theta = (e^{i\theta} + e^{-i\theta})/2$
Imaginary part $\text{Im } e^{i\theta} = \sin\theta = (e^{i\theta} - e^{-i\theta})/2i$

---

is the formula for the complex exponential function

$$e^{i\phi} = \cos\phi + i\sin\phi,$$

in which the cosine and sine functions appear as the real and imaginary parts. This formula can be applied advantageously to represent a wave in the form

$$\Psi = Ae^{i(kx-\omega t)} = A[\cos(kx - \omega t) + i\sin(kx - \omega t)]. \tag{4-13}$$

Notice that the expression describes both cosine and sine monochromatic traveling waves at once. The complex form can be used to make calculations of wave behavior in a physical medium, and then the real or imaginary part can be taken, as appropriate, at the end of the analysis.

The foregoing illustrations and formulas pertain to waves traveling in the positive x direction. The mathematical expressions are immediately adapted for propagation in the opposite direction by changing the phase variable from $(kx - \omega t)$ to $(-kx - \omega t)$. In either case the various forms of the wave function are easily seen to satisfy the one-dimensional wave equation, written (with speed of propagation v) in Equation (2-11). Of course, we are not yet able to write the wave equation for use in the case of matter waves.

Let us consider the possible adoption of Equation (4-13) as a matter wave for a quantum particle. We see at once that the parametrization defines a unique de Broglie wavelength $\lambda = 2\pi/k$ and a correspondingly unique momentum for the particle:

$$p = \frac{h}{\lambda} = \hbar k.$$

This assignment of momentum has no uncertainty, and so the uncertainty principle implies no localization in the position of the particle. The given matter wave evidently describes a particle whose likelihood of detection is uniform over all locations in space.

We begin to achieve a degree of localization when we *superpose* waves with slightly different wavelength and frequency. Let us illustrate this point by adding two such wave functions, using the real-valued form in Equation (4-10):

$$\Psi = A\cos[(k + \delta k)x - (\omega + \delta\omega)t] + A\cos[(k - \delta k)x - (\omega - \delta\omega)t].$$

The complex exponential representation provides an efficient way to perform the addition:

$$\Psi = A\,\mathrm{Re}\{e^{i[(k+\delta k)x-(\omega+\delta\omega)t]} + e^{i[(k-\delta k)x-(\omega-\delta\omega)t]}\}$$

$$= A\,\mathrm{Re}\{[e^{i(\delta kx-\delta\omega t)} + e^{-i(\delta kx-\delta\omega t)}]e^{i(kx-\omega t)}\}$$

$$= 2A\cos(\delta kx - \delta\omega t)\mathrm{Re}[e^{i(kx-\omega t)}]$$

$$= 2A\cos(\delta kx - \delta\omega t)\cos(kx - \omega t). \tag{4-14}$$

Figure 4-15

Composition of two monochromatic waves with different wave numbers and frequencies $(k + \delta k, \omega + \delta \omega)$ and $(k - \delta k, \omega - \delta \omega)$. The composite system propagates with group velocity $v_g = \delta \omega / \delta k$.

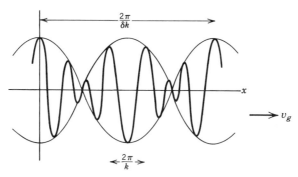

This superposition of waves exhibits the familiar phenomenon of *beats*. Figure 4-15 shows a snapshot of Ψ versus x in which the oscillations of the wave are contained in an *envelope* defined by the first of the two wave factors in Equation (4-14). This factor by itself describes a monochromatic wave with wavelength $2\pi/\delta k$ and frequency $\delta \omega / 2\pi$. The whole wave pattern in the figure travels in the x direction with a speed determined by the propagation of the envelope. The speed of such a composite wave system is known as the *group velocity*, given in this instance by the formula

$$v_g = \frac{\delta \omega}{2\pi} \frac{2\pi}{\delta k} = \frac{\delta \omega}{\delta k}. \tag{4-15}$$

Note that the group velocity is quite different from the phase velocities

$$\frac{\omega + \delta \omega}{k + \delta k} \quad \text{and} \quad \frac{\omega - \delta \omega}{k - \delta k}$$

for the two monochromatic waves in the composition of Ψ.

A semblance of localization is apparent in Figure 4-15 since the resultant wave clusters into a sequence of groups traveling together with a well-defined speed. This formation of localized groups becomes a more marked effect with the superposition of many different monochromatic waves. The resulting localization is still not satisfactory as a representation of a quantum particle because the addition of *finite* numbers of different monochromatic waves always produces an *infinite* sequence of repeating groups. This undesirable repetitive feature can be eliminated by summing an infinite number of component waves according to the following prescription.

Figure 4-16 shows schematically how a *wave packet* displays the desired property of localization. The composition produces oscillations confined to a *single* region of space and thus provides an idealized picture of a localized matter wave. This structure may be employed to derive a correspondingly localized probability distribution for a

Figure 4-16

Schematic wave packet representing the localization of a particle. The packet travels with group velocity v_g.

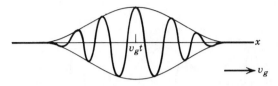

quantum particle. We construct such a wave packet by superposing complex mono-chromatic waves of the sort given in Equation (4-13), where the superposition runs over a *continuous* range of wave numbers according to the integration formula

$$\Psi = \int_{-\infty}^{\infty} A(k) e^{i(kx - \omega t)} \, dk. \tag{4-16}$$

An amplitude function $A(k)$ is supplied to specify the contribution of each component wave in the composition of Ψ. Let us suppose that the distribution of wave numbers assigns a maximum weight to a particular wave number $k = \bar{k}$, as illustrated in Figure 4-17. The construction of the wave packet has the property that the angular frequency ω appears as a function of k in the integration of Equation (4-16). Hence, the component waves are summed for infinitesimal increments dk in wave number and $d\omega$ in angular frequency and are controlled by $A(k)$ so that destructive inter-ference of waves tends to occur outside the localized portion of the wave packet. The resulting wave system propagates with a group velocity similar to the ratio given in Equation (4-15). The formula reads

$$v_g = \frac{d\omega}{dk}, \tag{4-17}$$

Figure 4-17

Distribution of wave numbers in the composition of a wave packet. The amplitudes of the component waves are heavily weighted around wave number $k = \bar{k}$ and become vanishingly small outside the interval $(\bar{k} - \bar{\kappa}, \bar{k} + \bar{\kappa})$.

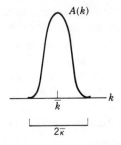

where the infinitesimal increments are taken at the value of the heavily weighted wave number $k = \bar{k}$ in the figure.

We should note the important distinction between the group velocity v_g in Equation (4-17) and the phase velocity v_ϕ in Equation (4-12). A value of v_ϕ is defined for every wave number k in the integration of Equation (4-16), while a single value of v_g is associated with the entire wave packet and the heavily weighted wave number $k = \bar{k}$. The two quantities can be connected by the relation

$$v_g = \left.\frac{d\omega}{dk}\right|_{\bar{k}} = \left.\frac{d}{dk}(kv_\phi)\right|_{\bar{k}} = \left.\left(v_\phi + k\frac{dv_\phi}{dk}\right)\right|_{\bar{k}}. \tag{4-18}$$

Thus, the group velocity and the phase velocity at $\bar{k}$ differ whenever v_ϕ depends on k.

Phase velocity is independent of wave number for the special case of light waves in vacuum. We see this from the simple formula

$$v_\phi = \nu\lambda = c,$$

and so we obtain the expected result $v_g = c$ for the group velocity of any wave packet (or pulse) of light in vacuum. A different result is found for the propagation of light in a medium because of the effect of the index of refraction as a factor in the phase velocity

$$v_\phi = \frac{c}{n}.$$

The refractive index n has a dependence on k, and so the determination of the group velocity for a light pulse in a medium calls for an identification of the wave number $\bar{k}$ and a proper evaluation of Equation (4-17). This k dependence (or λ dependence) of the index of refraction causes the phenomenon of *dispersion*.

The dispersive property of a wave can also be described as the result of a nonlinear relation between the wave parameters ω and k. We see such an effect when we compute the group velocity of a matter-wave packet for a localized free particle. Let us introduce energy and momentum for this purpose and adapt the parameters in Equations (4-11) accordingly, using the Planck–Einstein and de Broglie relations

$$\omega = 2\pi\frac{E}{h} = \frac{E}{\hbar} \quad \text{and} \quad k = 2\pi\frac{p}{h} = \frac{p}{\hbar}. \tag{4-19}$$

We can then regard the matter-wave packet as a superposition of matter waves with definite momentum, summed over a continuous range of momenta and energies. In this parametrization the formula for the group velocity of the wave packet becomes

$$v_g = \frac{dE}{dp}, \tag{4-20}$$

where the heavily weighted momentum in the superposition is given as

$$\bar{p} = \hbar\bar{k}. \tag{4-21}$$

The effect of dispersion arises because the parameters satisfy the nonlinear relation

$$E = \frac{p^2}{2m}$$

for a free particle. The group velocity of the corresponding wave packet is then found to be

$$v_g = \frac{\bar{p}}{m} \tag{4-22}$$

from Equations (4-20) and (4-21). We identify this result immediately as the velocity of the localized free particle.

The mathematical properties of the wave packet in Equation (4-16) are not difficult to analyze. Let us examine these details in the context of Figure 4-17 specifically so that we can visualize the idea of localization. We note that the equation defines the wave packet as a sum over all wave numbers and that the range of summation becomes finite when the amplitude $A(k)$ has the shape indicated in the figure. The formula for Ψ therefore assumes the simpler form

$$\Psi = \int_{\bar{k}-\bar{\kappa}}^{\bar{k}+\bar{\kappa}} A(k) e^{i(kx-\omega t)} \, dk \;\rightarrow\; A(\bar{k}) \int_{\bar{k}-\bar{\kappa}}^{\bar{k}+\bar{\kappa}} e^{ikx} e^{-i\omega t} \, dk.$$

The last step approximates the integral by taking the amplitude function to be evaluated at $k = \bar{k}$, where the distribution of wave numbers has its peak.

The remaining integration is then performed via the following series of maneuvers. We first define a new integration variable κ by setting

$$k = \bar{k} + \kappa,$$

and we note the κ dependence of the angular frequency:

$$\omega = \frac{E}{\hbar} = \frac{p^2}{2m\hbar} = \frac{\hbar}{2m} k^2 = \frac{\hbar}{2m}\left(\bar{k}^2 + 2\bar{k}\kappa + \kappa^2\right).$$

The resulting expression for the wave packet becomes

$$\Psi = A(\bar{k}) e^{i\bar{k}x} \int_{-\bar{\kappa}}^{\bar{\kappa}} e^{i\kappa x} e^{-i(\hbar t/2m)(\bar{k}^2 + 2\bar{k}\kappa + \kappa^2)} \, d\kappa$$

$$= A(\bar{k}) e^{i\bar{k}x} e^{-i\bar{\omega}t} \int_{-\bar{\kappa}}^{\bar{\kappa}} e^{i\kappa x} e^{-i(\hbar\bar{k}/m)t\kappa} e^{-i(\hbar/2m)t\kappa^2} \, d\kappa, \tag{4-23}$$

in which the value of ω at $k = \bar{k}$ is introduced as

$$\bar{\omega} = \frac{\hbar\bar{k}^2}{2m}.$$

We then make a further approximation and replace the third exponential factor in the integrand by unity. Our treatment of this function of κ^2 is justified since κ remains small when the integration is restricted to such a small k interval as that shown in Figure 4-17. These steps lead us to the result

$$\Psi = A(\bar{k}) e^{i(\bar{k}x - \bar{\omega}t)} \int_{-\bar{\kappa}}^{\bar{\kappa}} e^{i(x - v_g t)\kappa} \, d\kappa, \quad \text{where} \quad v_g = \frac{\hbar\bar{k}}{m},$$

in which Equations (4-21) and (4-22) are recalled to identify the group velocity v_g.

The elementary integral yields the desired formula for the wave packet:

$$\Psi = A(\bar{k})e^{i(\bar{k}x - \bar{\omega}t)}\frac{e^{i(x - v_g t)\bar{k}} - e^{-i(x - v_g t)\bar{k}}}{i(x - v_g t)}$$

$$= 2A(\bar{k})e^{i(\bar{k}x - \bar{\omega}t)}\frac{\sin(x - v_g t)\bar{k}}{x - v_g t}. \tag{4-24}$$

The significant part of the final expression is the rightmost ratio of factors containing the quantity $(x - v_g t)$. The shape of this function of x at a given time t is just like the behavior of the wave packet in Figure 4-16. We note that the oscillations are localized within a region centered at $x = v_g t$ and conclude that the wave system must be traveling at group velocity v_g in the positive x direction.

Let us conclude our discussion of wave packets by offering a few additional mathematical comments without proof. These remarks should appeal to our intuition and are meant to apply quite generally for waves of any kind. We take a proper wave packet Ψ to be a system of waves whose structure is localized within a definable spatial interval Δx. We construct such a system by choosing an amplitude function $A(k)$ whose distribution is confined to a wave-number range of width Δk. It can then be proved that Δx and Δk satisfy a relation of the form

$$\Delta x \, \Delta k \sim 1 \tag{4-25}$$

for any consistent definition of the two quantities. This result turns into a statement of the uncertainty principle when the second of Equations (4-19) is employed to relate wave number k and momentum p. The range of k values in the construction of the wave packet corresponds to an interval $\Delta\omega$ in the angular frequency. This localized system of waves travels past a given location in a time interval Δt. It can be proved that the intervals $\Delta\omega$ and Δt satisfy the additional relation

$$\Delta\omega \, \Delta t \sim 1. \tag{4-26}$$

If we again consult Equations (4-19) to relate angular frequency and energy, we find that this second conclusion becomes an uncertainty principle involving energy and time:

$$\Delta E \, \Delta t \sim \hbar. \tag{4-27}$$

The generality of Equations (4-25) and (4-26) should be emphasized as both are applicable to any type of localized wave. We should also acknowledge the primitive status of the uncertainty principle

$$\Delta x \, \Delta p \geq \frac{\hbar}{2}$$

since the inequality can be proved from the basic quantum mechanical properties of the observables x and p. All the proofs cited in this paragraph fall outside the scope of our mathematical presentation.

Figure 4-18

Diffraction of waves by a single slit. Each element of slit width dy emits a wave element $d\Psi$. The emitted waves are in phase at the plane of the slit and travel parallel paths to a distant point of observation.

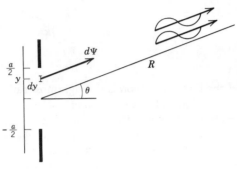

Example

The analysis of single-slit diffraction is an instructive exercise in the addition of waves and the use of complex numbers. We refer to Figure 4-18 and consider the superposition of waves arriving at a point located by the indicated distance R and off-axis angle θ. Huygens' construction tells us that each element of slit width dy in a slit of width a acts as a source of waves assumed to emanate in phase from the plane of the slit. We let R be large enough so that we can assume parallel paths for all waves out to the point of observation. A wave emitted at coordinate y in the slit travels a distance $(R - y \sin \theta)$ to the remote point. We express the contribution of the emitted wave element in complex form as

$$d\Psi = Ae^{i[k(R - y \sin \theta) - \omega t]} \frac{dy}{a}$$

and evaluate the sum of such waves as follows:

$$\Psi = \int_{\text{slit}} d\Psi = \int_{-a/2}^{a/2} Ae^{i(kR - \omega t)} e^{-iky \sin \theta} \frac{dy}{a}$$

$$= \frac{A}{a} e^{i(kR - \omega t)} \int_{-a/2}^{a/2} \cos(ky \sin \theta)\, dy = \frac{2A}{a} e^{i(kR - \omega t)} \int_{0}^{a/2} \cos(ky \sin \theta)\, dy$$

$$= Ae^{i(kR - \omega t)} \frac{\sin\left(\dfrac{ka}{2} \sin \theta\right)}{\dfrac{ka}{2} \sin \theta}.$$

Oddness and evenness properties are employed in the integrand along the way to the final result. The rightmost ratio of factors is the main part of the expression since the square of this ratio describes the θ dependence of the distribution of intensity for single-slit diffraction.

Figure 4-19

Gaussian distributions for the amplitude function $A(k)$ and the wave packet Ψ. The widths Δx and Δk obey a reciprocal relation in which the product $\Delta x \, \Delta k$ is equal to a pure number.

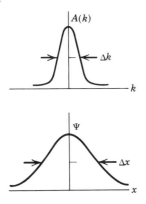

Example

The following construction of a wave packet illustrates the uncertainty relation in Equation (4-25). We consider a superposition of complex waves at $t = 0$, given by

$$\Psi = \int_{-\infty}^{\infty} A(k) e^{ikx} \, dk,$$

and let the distribution of wave numbers be specified by the amplitude function

$$A(k) = e^{-(k/\bar{\kappa})^2}.$$

The shape of this *gaussian distribution* is shown in Figure 4-19. We note that the peak of the function occurs at $k = 0$, and so we can assume that we are describing a wave packet Ψ for a particle at rest. The integration becomes

$$\Psi = \int_{-\infty}^{\infty} e^{-(k/\bar{\kappa})^2}(\cos kx + i \sin kx) \, dk = \int_{-\infty}^{\infty} e^{-(k/\bar{\kappa})^2} \cos kx \, dk$$

since the sine portion of the integrand is odd and integrates to zero. A table of definite integrals yields the result for the wave packet at $t = 0$:

$$\Psi = \sqrt{\pi} \, \bar{\kappa} e^{-(\bar{\kappa}x/2)^2} \quad \text{(Dwight 861.20)}.$$

This function is also shown in the figure to have the form of another gaussian distribution. Each of the two shapes has a readily identifiable width, defined as the indicated spread in the distribution evaluated at half the maximum value of the function. These definitions of Δx and Δk can be shown to satisfy Equation (4-25) for any choice of $\bar{\kappa}$; the demonstration of this property is left as Problem 17 at the end of the chapter. Note the reciprocal sense in which the parameter $\bar{\kappa}$

controls the widths of the two distributions. We see that $A(k)$ is sharp and Ψ is broad if $\bar{\kappa}$ is small, and we find that the reverse is true if $\bar{\kappa}$ is large. This striking feature makes the gaussian wave packet an especially appealing prototype to illustrate the mathematical concept of localization.

Problems

1. Let the kinetic energy of a particle at room temperature be expressed as $\frac{3}{2}k_BT = \frac{1}{40}$ eV. Calculate the de Broglie wavelength for electrons and for neutrons at room temperature. (Take the neutron mass to be 1.67×10^{-27} kg.)

2. Consider the de Broglie relation for a relativistic particle and show that wavelength and kinetic energy are related by

$$\lambda = \frac{hc}{\sqrt{K(K + 2mc^2)}}.$$

This formula implies

$$\lambda \approx \frac{hc}{K} \quad \text{for } K \gg mc^2,$$

and

$$\lambda \approx \frac{h}{\sqrt{2mK}} \quad \text{for } K \ll mc^2.$$

Obtain the leading correction to each of these approximations.

3. A potential difference of 1000 V accelerates a beam of electrons in an electron diffraction experiment. The beam strikes a crystal whose interplanar spacing is 0.05 nm. Determine the directions of the observed diffraction maxima in terms of the angle φ indicated in Figure 4-2.

4. A beam of 1 eV neutrons strikes a crystal whose crystal planes are spaced by 0.025 nm. Determine the angle φ for which the first diffraction maximum is observed. Are there higher orders in the diffraction pattern?

5. Plane waves of 500 nm light are incident on a photocell whose receptor is 1 cm square. Suppose that the cell records a counting rate of one photon per second with 100% efficiency. Calculate the intensity of the light (in W/m^2), and determine the amplitude of oscillation of the propagating electric field.

6. An object is dropped from rest so that its subsequent vertical position is given by

$$y(t) = \frac{g}{2}t^2.$$

This deterministic result becomes blurred if the initial position and velocity are not so accurately known. Let $y(0)$ and $v(0)$ lie somewhere in the intervals $(-\delta y, \delta y)$ and $(-\delta v, \delta v)$, respectively, and deduce the resulting behavior of the variable $y(t)$. Draw a graph of the deterministic result and show how the graph is modified by these considerations.

7. A proton bound in a nucleus experiences the strong force of nuclear attraction when the particle is within about 10^{-15} m range of another nuclear particle. Estimate the kinetic

energy of the proton for this sort of localization. What can then be said regarding the potential energy?

8. Reevaluate the calculations of the previous problem for an electron localized in a region of nuclear size of order 10^{-14} m.

9. A hydrogen atom has total energy

$$E = \frac{p^2}{2m_e} - \frac{e^2}{4\pi\varepsilon_0 r},$$

neglecting the reduced-mass effect. Let r denote the radial distance within which the bound electron is localized, and estimate the corresponding momentum p. Continue the estimation procedure to obtain expressions for the minimum value of E and the minimizing value of r. Compare the results with those found in the Bohr model.

10. A particle of mass m is confined to a one-dimensional region of length a. Use the uncertainty principle to obtain an expression for the minimum energy of the particle. Calculate the value of this energy for a 1 g bead on a 10 cm wire and for an electron in a region of 0.1 nm length.

11. Let pellets be dropped onto a small spot on the floor. The uncertainty principle implies that the pellets cannot be assumed to fall straight down from rest. Allow for uncertain initial conditions, especially those transverse to the vertical, and deduce an expression for the minimum range R on the floor at a distance H below the location of release. Calculate the range for 1 g pellets dropped from a 2 m height.

12. Show that any function of the form $f(kx \mp \omega t)$ is a solution of the one-dimensional wave equation. Identify the speed of wave propagation in terms of the parameters in f.

13. Carry out the addition of the two waves

$$\Psi = A\sin\left[(k + \delta k)x - (\omega + \delta\omega)t\right] + A\cos\left[(k - \delta k)x - (\omega - \delta\omega)t\right]$$

by means of the complex-number representation, and interpret the result.

14. Consider the interference of two waves Ψ_1 and Ψ_2, emitted in phase from two very narrow parallel slits. The waves have the same amplitude, wavelength, and frequency and are observed as shown, at an angle θ and at a large distance R. Construct the superposition $\Psi_1 + \Psi_2$ using the complex wave format, and deduce the θ dependence of the resulting interference pattern.

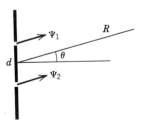

15. Consider the addition of waves emitted in phase from two parallel slits with width a and separation d. The resultant wave is observed as shown, at an angle θ and at a large distance R. Let the wave element emitted at the indicated location y be written as

$$d\Psi = Ae^{i[k(R - y\sin\theta) - \omega t]}\frac{dy}{a}.$$

Determine the resultant wave and deduce the θ dependence of the observed diffraction pattern. Evaluate the limit of these results as $a \to 0$.

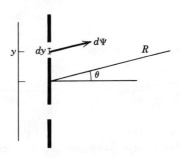

16. The width of a wave packet and the width of the associated wave-number distribution are connected by the "equality" $\Delta x \, \Delta k \sim 1$. Use this result to construct a simple proof of the analogous "equality" $\Delta \omega \, \Delta t \sim 1$ involving the related intervals of angular frequency and time.

17. Refer to Figure 4-19 and consider the gaussian wave packet discussed in the second example of Section 4-6. Express the widths Δx and Δk in terms of $\bar{\kappa}$, the parameter appearing in the amplitude function

$$A(k) = e^{-(k/\bar{\kappa})^2},$$

and show that the product $\Delta x \, \Delta k$ is equal to a pure number independent of $\bar{\kappa}$.

18. Determine the form of the complex wave packet Ψ at $t = 0$, given the amplitude distribution

$$A(k) = \begin{cases} A_0 & \text{for } k \text{ in the interval } (\bar{k} - \bar{\kappa}, \bar{k} + \bar{\kappa}) \\ 0 & \text{otherwise} \end{cases}.$$

Obtain the squared modulus $|\Psi|^2$ and sketch a graph of this function.

19. Let the amplitude function for a complex wave packet Ψ be given as

$$A(k) = A_0 \frac{\sin ka}{ka},$$

and determine the form of Ψ at $t = 0$.

F I V E

QUANTUM
MECHANICS

*T*he formalism of the quantum theory was substantially in place by 1926. Heisenberg's nonclassical treatment of observable physical quantities constituted one of the two original versions of the theory. The other alternative presentation of quantum mechanics was put forward around the same time by E. Schrödinger. These formulations were offered from two different mathematical points of view and were soon proved to be equivalent theories.

Schrödinger drew some of his inspiration from studies of the theory of light undertaken by W. R. Hamilton in the 19th century. Hamilton had investigated the connection between the equations of geometrical optics and wave optics, and had demonstrated mathematically how the one description could be regarded as an approximation to the other. Schrödinger argued that classical particle dynamics should bear a similar relation to a proposed wave theory for a quantum particle. De Broglie's matter wave was in immediate harmony with this notion, and so Schrödinger adopted the hypothesis as a main ingredient in his theory. The matter-wave approach to quantum physics made use of solutions of a wave equation, while Heisenberg's picture gave a representation of measurable quantities by means of matrices. Schrödinger's wave mechanics had more intuitive appeal than Heisenberg's matrix mechanics because the wave picture was easier to visualize. The mathematics of waves was also well understood, while the matrices in Heisenberg's formalism had infinite dimension and were therefore rather unfamiliar.

A comprehensive interpretation was essential to the mathematical consistency and physical content of this new theory. The solutions of Schrödinger's wave equation were correctly interpreted by M. Born (over Schrödinger's objections) in 1926. The probabilistic aspects of the theory were formalized through this important contribution.

Quantum physics became a quantitative science as a consequence of these discoveries. The new theory was based on a coherent system of principles and furnished a predictive scheme by which the formalism could be tested. These goals had been awaiting fulfillment for more than a decade and were at last achieved.

This chapter is intended as an introduction to the basic concepts of quantum mechanics. We include all that we need, here and in Chapter 6, to conduct a

Erwin Schrödinger

satisfactory survey of modern physics. Schrödinger's wave equation is the keystone of the entire presentation. The solutions of this differential equation enable us to describe such phenomena as the occurrence of stationary states and the quantization of energy. The probability interpretation of the wave function is also introduced in this chapter. The probabilistic treatment then tells us how to define the physical observables for a quantum particle. All these new ideas and methods are needed as we depa.. from classical physics and proceed toward the quantum theory of the structure of matter.

5-1 The Schrödinger Equation

Most of our discussion in this chapter is devoted to the special case of a particle in *one-dimensional motion*. Our intention is to introduce the new mathematical procedures in certain familiar situations where the behavior of the particle is described by a single linear degree of freedom. Figure 5-1 shows an oscillating mass on a spring and a freely sliding bead on a wire as two such examples. Each situation involves *constrained motion* in which the quantum particle is highly localized with respect to the two coordinates transverse to the particle's one degree of freedom. This constraint allows us to ignore two out of three coordinates in all stages of the treatment. We also assume that the motion of the particle is governed by a conservative force so that we can define a potential energy as a function of the remaining spatial variable.

Let us assume that the particle is able to move in the x direction and that the motion is nonrelativistic. We let the potential energy be expressed by means of a function $V(x)$ so that the force on the particle is given as $F = -dV/dx$. Classical mechanics would have us determine $x(t)$, the position of the particle at time t. This sort of result is not the objective of interest in quantum mechanics. Instead, we want to find the wave function $\Psi(x, t)$ for the matter wave and then use Ψ to make predictions about the probability of random detection of the particle. These goals

Figure 5-1

Particle motion constrained to one dimension.

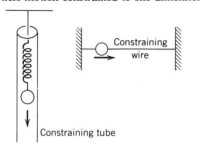

cannot be addressed until the wave equation for Ψ is specified. It is obvious that we are looking for a partial differential equation in the independent variables x and t. The equation is evidently supposed to contain the given potential energy $V(x)$ in one of its terms.

Let us turn for guidance to the problem of wave propagation in a one-dimensional medium. We recall the appropriate wave equation as

$$\frac{\partial^2 u}{\partial x^2} = \frac{1}{v^2} \frac{\partial^2 u}{\partial t^2},$$ (5-1)

where v is the speed of propagation and $u(x, t)$ is the displacement of the vibrating medium. Equation (5-1) is *linear* in u; therefore, the existence of independent solutions $u_1(x, t)$ and $u_2(x, t)$ implies the existence of another solution given by the sum of waves

$$u_1(x, t) + u_2(x, t).$$

In general, the linearity of the equation allows us to take any set of solutions $u_k(x, t)$ and construct arbitrary linear combinations in the form

$$\sum_k a_k u_k(x, t)$$

to obtain other solutions. This property is known as the *principle of superposition*. The concept is fundamental to the description of interference and the construction of wave packets, since both of these procedures involve the summing of waves. We want the desired equation for the matter wave to be linear in the wave function $\Psi(x, t)$ so that matter waves can also enjoy these important consequences of superposition.

We observe that Equation (5-1) is a differential equation of *second order* in the time. It follows that *two* pieces of initial data must be given to determine the subsequent time dependence of the solution. Thus, we may take an initial specification of displacement and velocity for all points in the medium as

$$u(x, 0) \quad \text{and} \quad \left(\frac{\partial u}{\partial t} \right)_{t=0}$$

and then predict a unique solution $u(x, t)$ for any later value of t. The matter wave

$\Psi(x, t)$ is subject to rather different considerations. We learn as follows that the Schrödinger wave equation is of *first order* in the time so that the wave function is uniquely determined from a *single* initial condition given by the behavior of $\Psi(x, t)$ at $t = 0$.

Let us examine this feature of the quantum theory by considering the case of a *free* particle. We choose the arbitrary reference level in the definition of the potential energy $V(x)$ so that the constant value of V is equal to zero for a zero force. A localized free particle is described by means of a wave packet of the form

$$\Psi(x, t) = \int_{-\infty}^{\infty} A(k) e^{i(kx - \omega t)} \, dk,$$

where Ψ has properties as discussed in Section 4-6. The wave parameters satisfy the relation

$$\omega = \frac{\hbar k^2}{2m} \tag{5-2}$$

for a particle of mass m, and so the formula for the free-particle wave packet becomes

$$\Psi(x, t) = \int_{-\infty}^{\infty} A(k) e^{i[kx - (\hbar/2m)k^2 t]} \, dk. \tag{5-3}$$

We note that the k dependence of the complex exponential function establishes a correspondence between second-order differentiation with respect to x and first-order differentiation with respect to t, whereby

$$\frac{\partial^2 \Psi}{\partial x^2} = \int_{-\infty}^{\infty} A(k)(ik)^2 e^{i[kx - (\hbar/2m)k^2 t]} \, dk$$

and

$$\frac{\partial \Psi}{\partial t} = \int_{-\infty}^{\infty} A(k) \left(-\frac{i\hbar}{2m} k^2 \right) e^{i[kx - (\hbar/2m)k^2 t]} \, dk.$$

These two results tell us that the quantities $(i\hbar/2m)\partial^2 \Psi/\partial x^2$ and $\partial \Psi/\partial t$ are identical. We express this equality as

$$-\frac{\hbar^2}{2m} \frac{\partial^2}{\partial x^2} \Psi = i\hbar \frac{\partial}{\partial t} \Psi \tag{5-4}$$

and thus obtain the Schrödinger equation for the wave function of a free particle.

Note that the partial derivatives in Equation (5-4) are of different order because the wave parameters ω and k appear with different powers in Equation (5-2). We are again reminded of the Planck–Einstein and de Broglie formulas

$$E = \hbar\omega \quad \text{and} \quad p = \hbar k, \tag{5-5}$$

and we recall that the relation between ω and k expresses the nonrelativistic *energy relation* for a free particle of mass m:

$$\frac{p^2}{2m} = E. \tag{5-6}$$

This connection holds for each component wave in the matter-wave packet Ψ. Our observations suggest a simple procedure that produces the Schrödinger equation directly from the energy relation. We find that Equation (5-4) is obtained immediately if the two sides of Equation (5-6) are interpreted as *differential operators* according to the substitutions

$$p^2 \rightarrow -\hbar^2 \frac{\partial^2}{\partial x^2} \quad \text{and} \quad E \rightarrow i\hbar \frac{\partial}{\partial t}, \tag{5-7}$$

where the operations are allowed to act on the wave function $\Psi(x, t)$. This proposition introduces a pair of *representation rules* for the new quantum treatment of the physical quantities p^2 and E.

It has already been noted in Section 4-6 that the monochromatic wave functions $\cos(kx - \omega t)$, $\sin(kx - \omega t)$, and $e^{i(kx - \omega t)}$ are possible solutions of the ordinary wave equation. We now observe that the cosine and sine wave functions cannot possibly occur as solutions of Equation (5-4) because of the appearance of the first-order time derivative. In fact, the presence of the imaginary number i in the time-derivative term leads us to expect a complex-valued time dependence in $\Psi(x, t)$.

Let us not be tempted to regard our deduction of the Schrödinger equation as a derivation. We should realize that Equation (5-3) is an *assumed* free-particle wave packet whose chosen form is desirable as a matter wave for a localized particle. Equation (5-4) is then shown to follow from this assumption. Such a heuristic approach is quite proper since the Schrödinger equation is not to be considered as a derivable result. Instead, the equation represents a hypothesis whose validity is to be ascertained by examining the predicted consequences. Schrödinger's theory is intended to describe a certain body of conceivable measurements and, like Newton's theory, is supposed to work within a certain regime of applicability. Equation (5-4) is somewhat akin to Newton's first law, since the principles refer to a free particle in both situations. Our next concern is the analogue of Newton's second law, as we proceed to Schrödinger's wave equation for the case of a nonfree particle.

As preparation, let us return to the free-particle case and include the potential energy as a *nonvanishing* constant in the analysis. This tactic is not as contrived as it may seem since the results are going to lead us to something new. A constant potential energy V is incorporated in the energy relation for a particle of mass m by writing

$$\frac{p^2}{2m} + V = E, \tag{5-8}$$

where both E and p are constant quantities for a free particle. We may use the simple monochromatic wave function

$$\Psi(x, t) = A e^{i(kx - \omega t)} \tag{5-9}$$

and thereby assign a precise momentum $\hbar k$, since we are not concerned with the question of localization in this part of the discussion. The wave number k and the angular frequency ω satisfy the equality

$$\frac{\hbar^2 k^2}{2m} + V = \hbar\omega \tag{5-10}$$

by virtue of Equations (5-5) and (5-8). This relation between ω and k tells us that the wave function in Equation (5-9) must obey the differential equation

$$-\frac{\hbar^2}{2m}\frac{\partial^2}{\partial x^2}\Psi + V\Psi = i\hbar\frac{\partial}{\partial t}\Psi. \tag{5-11}$$

Thus, we obtain another version of the Schrödinger equation for a free particle, and we conclude that no real physical distinction exists between Equation (5-4) and Equation (5-11).

The same comments hold for the alternative wave function

$$\Psi(x, t) = Ae^{i(-kx-\omega t)}. \tag{5-12}$$

We observe that Equations (5-9) and (5-12) refer to waves traveling in opposite directions, since the expressions differ only in the sign of k, and that Equation (5-10) controls the wave parameters without regard for this sign. It is therefore obvious that both wave functions are equally valid solutions of the free-particle Schrödinger equation.

We see the purpose of using a nonzero constant potential energy for a free particle when we turn to the situation of interest, where the particle is *not* free and the corresponding potential energy is not constant. A typical varying potential energy might resemble the function shown in Figure 5-2. The energy relation maintains the form of Equation (5-8), although the momentum p can no longer be treated as a constant. It follows that a constant wave number k cannot be employed to parametrize the wave function as in Equations (5-9) and (5-12). We can still regard E as a constant in Equation (5-8) and retain a constant angular frequency ω. (We should also say that our discussion assumes a choice of constant E that exceeds $V(x)$ for all values of x. This stipulation is already built into Figure 5-2.) It would seem that Equation (5-10) has no further use now that V enters as a function of x. Let us suppose, however, that we approximate the potential energy function $V(x)$ in the manner illustrated in the figure, so that the x axis is divided into pieces and average values of the potential energy are substituted for $V(x)$ in all the intervals. The figure shows a new potential energy function that approximates $V(x)$ by a stepwise construction of *constant* potential energy segments. The particle acts like a *free* particle in every interval, and so the conclusions of the preceding paragraphs are still applicable on this basis. We therefore identify a different *constant* wave number k interval-by-interval and again rely on Equation (5-10) with a correspondingly different constant V in each of the intervals. Hence, Equations (5-9) and (5-12) become valid forms for $\Psi(x, t)$, as Equation (5-10) holds in every interval for either of these wave functions or for any combination. This piecewise construction of Ψ satisfies the Schrödinger equation as given in Equation (5-11) for the case of a stepwise potential energy. We can approximate the original potential energy function $V(x)$ to any desired accuracy by this procedure. Therefore, we argue that the corresponding exact wave function $\Psi(x, t)$ can be found as an exact solution of Equation (5-11) by allowing the equation to contain a potential energy that varies with x.

Figure 5-2

Approximation to a potential energy function $V(x)$ by a sequence of constant potential energy segments $V_0, V_1, V_2, \ldots$.

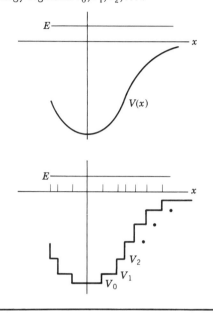

We should look for plausibility instead of rigor in these arguments. The result asserts that quantum mechanics in one dimension is expressed in terms of a wave function $\Psi(x, t)$, where Ψ satisfies the Schrödinger equation as written in Equation (5-11) for *any* given potential energy $V(x)$. We are supposed to accept this assertion and become confident of its validity by using the procedures and examining the consequences.

Example

The functions $e^{i(kx - \omega t)}$ and $e^{i(-kx - \omega t)}$ are solutions of the Schrödinger equation for a free particle, so any combination of the two functions is also a solution. Let us gain some practice with the differential equation by considering the particular combination

$$\Psi(x, t) = A \cos kx\, e^{-i\omega t} = \frac{A}{2}\left[e^{i(kx - \omega t)} + e^{i(-kx - \omega t)} \right].$$

We confirm directly that we have a solution by performing the following differential operations:

$$\left(-\frac{\hbar^2}{2m}\frac{\partial^2}{\partial x^2} + V \right)\Psi = \left(+\frac{\hbar^2}{2m}k^2 + V \right) A \cos kx\, e^{-i\omega t}$$

and

$$i\hbar\frac{\partial}{\partial t}\Psi = \hbar\omega A \cos kx\, e^{-i\omega t}.$$

> It is clear that the Schrödinger equation is satisfied by Ψ if the wave parameters ω and k obey Equation (5-10). This free-particle wave function is a superposition of two component waves traveling in opposite directions.

5-2 Probability Interpretation

Schrödinger's wave equation is only one of the ingredients in the Schrödinger theory. Attention must also be given immediately to the proper meaning of the solutions of the new equation. Our remarks in Chapter 4 suggest that we adopt a probability interpretation for these matter-wave solutions. We now wish to pursue this suggestion so that we can learn how the matter wave is used to determine probabilities for the random detection of a quantum particle. We continue with the one-dimensional approach in order to introduce these important considerations while the theory is in its simplest mathematical stage.

Schrödinger originally regarded the wave function Ψ as a visualizable physical wave. Born took Ψ to be a *mathematical* wave function instead and identified $|\Psi|^2$ as the basic observable quality of the matter wave. Born's interpretation prevailed, despite the fact that his proposal did not meet with Schrödinger's immediate approval.

An acceptable solution of the Schrödinger equation is supposed to provide a *complete* description of a *physical state* for a given system. This statement means that all observed properties of the system are obtainable from the solution, once the wave function Ψ has been found. A given physical system is defined for a particle in one dimension by a particular specification of the potential energy $V(x)$. The resulting differential equation for Ψ is of first order in the time, and so a unique wave function is determined if the initial behavior of the state is prescribed. This aspect of the wave function is associated with the differential operator $i\hbar\,\partial/\partial t$ in the equation. The

Max Born

presence of the imaginary factor implies that the time dependence of Ψ must be given by a *complex-valued* solution. This general property of Ψ suggests that the wave function itself cannot be a measurable physical quantity, as in the case of solutions of the ordinary wave equation. Instead, the wave function must be a purely mathematical device to furnish probabilistic information about the state of the system. The Schrödinger theory is then supposed to include further rules that determine the various observable properties of the particle when the system is in such a state.

The fundamental observable entity in this interpretation is the probability of finding a particle at a certain location. Such a quantity must be real and positive, mathematical properties not possessed by the complex function Ψ. Physical properties of matter waves have been proposed in Chapter 4, where distributions of probability have been likened to distributions of wave intensity. The definition of the desired probability is expressed accordingly in terms of the square of the modulus of Ψ,

$$|\Psi|^2 = \Psi^*\Psi,$$

because this (and only this) operation on the wave function conveys the meaning of an intensity for a complex-valued wave.

Let us postpone the actual definition of the probability for a moment and call immediate attention to an especially desirable feature of the quantity $\Psi^*\Psi$. We suppose that we have a system in which particles are transmitted by a double slit, and we assume that the wave function for a single particle is written as the sum of two complex waves

$$\Psi = \Psi_1 + \Psi_2.$$

The separate parts of Ψ are identified as single-slit wave functions with modulus and phase such that

$$\Psi_1 = |\Psi_1|e^{i\phi_1} \quad \text{and} \quad \Psi_2 = |\Psi_2|e^{i\phi_2}.$$

We are allowed to superpose these waves because of the *linear* property of the Schrödinger equation. The double-slit probability involves the square of the wave modulus:

$$\begin{aligned}
|\Psi|^2 &= \left(\Psi_1^* + \Psi_2^*\right)\left(\Psi_1 + \Psi_2\right) \\
&= |\Psi_1|^2 + \Psi_2^*\Psi_1 + \Psi_1^*\Psi_2 + |\Psi_2|^2 \\
&= |\Psi_1|^2 + |\Psi_1||\Psi_2|\left[e^{i(\phi_1 - \phi_2)} + e^{-i(\phi_1 - \phi_2)}\right] + |\Psi_2|^2.
\end{aligned}$$

The result contains the two single-slit quantities $|\Psi_1|^2$ and $|\Psi_2|^2$, as well as the additional *interference term*

$$2|\Psi_1||\Psi_2|\cos\left(\phi_1 - \phi_2\right).$$

We observe the effect of this phase-dependent contribution in the phenomenon of double-slit interference. The phases ϕ_1 and ϕ_2 vary for different points of observation and produce constructive and destructive interference in a pattern just like that found for light in Figure 4-5. We note that the interference pattern results from the superposed form of the wave function Ψ for a single particle. The result determines the distribution of probability for detection of any such particle in a beam of particles

Figure 5-3

Probability density for a localized particle in one dimension.

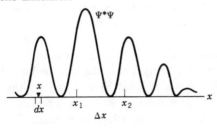

incident on the double slit. We now clarify these observations by expressing the probabilistic properties of Ψ in terms of a new defined quantity.

Born's interpretation identifies $\Psi*\Psi$ as the *probability density* for a system in a state with wave function Ψ. This quantity depends on the variables that specify the degrees of freedom of the system. Thus, a single particle in one-dimensional motion has probability density

$$P(x, t) = |\Psi(x, t)|^2, \tag{5-13}$$

in which $\Psi*\Psi$ may exhibit a variation with time. Figure 5-3 shows a sketch of a possible shape for the x dependence of P at a particular time t. If a certain value of the coordinate x is selected and an infinitesimal interval dx is taken, as in the figure, then the probability of finding the particle in the given interval at time t is defined by the expression

$$P(x, t) \, dx = |\Psi(x, t)|^2 \, dx.$$

The probability of finding the particle in the finite interval Δx between the indicated points x_1 and x_2 is determined by computing the integral

$$\int_{x_1}^{x_2} |\Psi(x, t)|^2 \, dx.$$

The total probability of finding the particle somewhere in the entire one-dimensional space is then found by extending the range of integration to get

$$\int_{-\infty}^{\infty} |\Psi(x, t)|^2 \, dx.$$

Note that $\Psi(x, t)$ must be a suitably localized wave function to ensure the convergence of this integral. Note also that the numerical magnitude of Ψ is not fixed by solving the Schrödinger equation, since any solution Ψ may always be multiplied by any constant and still serve as a solution. The probability integral is employed to remove this arbitrariness and set the overall scale of Ψ. The wave function is said to be *normalized* if the solution Ψ satisfies the additional restriction

$$\int_{-\infty}^{\infty} |\Psi(x, t)|^2 \, dx = 1. \tag{5-14}$$

This *normalization condition* means that the particle has unit probability to be found somewhere in all of one-dimensional space.

The probability density makes allowance for a possible time dependence in the local behavior of the probability. This feature of the interpretation of Ψ suggests the introduction of another local quantity to represent the *flow* of probability. An appropriate definition emerges from consideration of the free-particle equations obeyed by $\Psi(x, t)$ and by $\Psi^*(x, t)$:

$$-\frac{\hbar^2}{2m}\frac{\partial^2}{\partial x^2}\Psi = i\hbar\frac{\partial}{\partial t}\Psi \quad \text{and} \quad -\frac{\hbar^2}{2m}\frac{\partial^2}{\partial x^2}\Psi^* = -i\hbar\frac{\partial}{\partial t}\Psi^*.$$

Note that the two equations are related by complex conjugation. These formulas may be used to analyze the time dependence of the probability density for a free particle:

$$\frac{\partial}{\partial t}(\Psi^*\Psi) = \Psi^*\frac{\partial \Psi}{\partial t} + \frac{\partial \Psi^*}{\partial t}\Psi$$

$$= \Psi^*\left(-\frac{\hbar}{2im}\frac{\partial^2 \Psi}{\partial x^2}\right) + \left(\frac{\hbar}{2im}\frac{\partial^2 \Psi^*}{\partial x^2}\right)\Psi$$

$$= -\frac{\hbar}{2im}\left(\Psi^*\frac{\partial^2 \Psi}{\partial x^2} - \frac{\partial^2 \Psi^*}{\partial x^2}\Psi\right)$$

$$= -\frac{\partial}{\partial x}\left[\frac{\hbar}{2im}\left(\Psi^*\frac{\partial \Psi}{\partial x} - \frac{\partial \Psi^*}{\partial x}\Psi\right)\right].$$

The equality can be rewritten to read

$$\frac{\partial}{\partial t}P(x, t) + \frac{\partial}{\partial x}j(x, t) = 0. \tag{5-15}$$

The result serves to introduce the *probability current density*

$$j(x, t) = \frac{\hbar}{2im}\left(\Psi^*\frac{\partial \Psi}{\partial x} - \frac{\partial \Psi^*}{\partial x}\Psi\right). \tag{5-16}$$

This new quantity appears in the defining equation as a *flux* of probability. Thus, Equation (5-15) has the form of a *conservation law* in which a time variation of the local probability is compensated by a space variation of the flux across the local region.

It is possible to cast the probability conservation law in integral rather than differential form by considering the probability of finding the particle in some finite interval $\Delta x = x_2 - x_1$. The rate of change of this quantity is evaluated as follows:

$$\frac{d}{dt}\int_{x_1}^{x_2}P(x, t)\, dx = \int_{x_1}^{x_2}\frac{\partial}{\partial t}P(x, t)\, dx$$

$$= -\int_{x_1}^{x_2}\frac{\partial}{\partial x}j(x, t)\, dx = j(x_1, t) - j(x_2, t). \tag{5-17}$$

The result represents the net flow of probability into the interval Δx. Equation (5-15) carries over in the case of a nonfree particle, with the same definition for the probability current density $j(x, t)$. The details are left to be examined in Problem 2 at the end of the chapter.

The probability interpretation of the wave function has an immediate extension from one to three dimensions. Let us take a brief excursion into three-dimensional space and identify some of the more self-evident facts for later reference. We let the wave function Ψ represent the state of a three-dimensional system, and we use the notation $\Psi(\mathbf{r}, t)$ to indicate the dependence on the three coordinates (x, y, z) as well as the time. The probability of finding the particle in an infinitesimal volume element $d\tau$ at time t is then written as

$$|\Psi(\mathbf{r}, t)|^2 \, d\tau,$$

so that the corresponding probability in a finite volume $\Delta\tau$ is given by the volume integral

$$\int_{\Delta\tau} |\Psi(\mathbf{r}, t)|^2 \, d\tau.$$

The overall scale of the wave function is fixed by requiring Ψ to obey the restriction

$$\int_{\text{all space}} |\Psi(\mathbf{r}, t)|^2 \, d\tau = 1. \tag{5-18}$$

This normalization condition generalizes the formula in Equation (5-14) for application in three dimensions.

Section 4-6 has provided us with an intuitive picture of a matter wave for a localized quantum particle. We have just learned that the property of localization is described by the distribution in space of the real positive probability density $\Psi^*\Psi$. Let us hasten to clarify our understanding of this quantity, since the notion certainly does *not* mean that the particle itself is distributed over all space. Instead, the probability of *locating* the particle is understood to have this interpretation. To illustrate, let us appeal to the measurement process in a single-slit experiment and recall that the location of the particle is not determined with certainty after passing through the slit. We let the infinitesimal element $d\tau$ represent the volume of a certain cell in a particle detector, and we take the value of $\Psi^*\Psi \, d\tau$ to be the probability of detecting the particle in the given cell. Note that the *entire* particle is either found or not found in the cell at a particular instant. Hence, the process of measurement for a single particle results in a random count somewhere in the detector system. If we make the same observations on a large number of similar particles in separate single-slit experiments, we obtain a distribution of random counts that conforms to the distribution in space predicted by the probability density $\Psi^*\Psi$. Thus, we find that the wave function Ψ for a single particle determines the single-slit diffraction pattern for a *beam* of particles according to this probabilistic interpretation.

Example

We have glossed over a subtle point in the normalization of the wave function in Equation (5-14). It is obvious that the element of probability $\Psi^*\Psi \, dx$ may depend on t because of the t dependence of Ψ. This dependence on t remains after the integration over x, and yet the total probability is set equal to unity with no account taken of the apparent t dependence on the left side of the equation. We can justify the result, specifically for the case of a free particle, if we consult Equation (5-17) and rewrite the formula for the time derivative of the probability as

$$\frac{d}{dt}\int_{x_1}^{x_2}\Psi^*\Psi\,dx = -\frac{\hbar}{2im}\left(\Psi^*\frac{\partial\Psi}{\partial x} - \frac{\partial\Psi^*}{\partial x}\Psi\right)\Bigg|_{x_1}^{x_2}.$$

It is clear that the probability of finding the particle in the *finite* interval $\Delta x = x_2 - x_1$ may vary with t, since the right side of this equality is not necessarily required to vanish. Equation (5-14) pertains to the *infinite* interval, however, and so the question concerns the behavior of the equality in the limits $x_1 \to -\infty$ and $x_2 \to +\infty$. A properly localized wave function must vanish rapidly at infinity so that

$$\left(\Psi^*\frac{\partial\Psi}{\partial x} - \frac{\partial\Psi^*}{\partial x}\Psi\right)\Bigg|_{x_1}^{x_2} \to 0 \quad \text{as } x_1 \to -\infty \text{ and } x_2 \to \infty.$$

It follows that the normalization integral $\int_{-\infty}^{\infty}|\Psi(x,t)|^2\,dx$ has no t dependence, and so Equation (5-14) can be introduced to set the resulting *constant* value of the integral equal to unity.

5-3 Stationary States

The Schrödinger theory is based on a *partial* differential equation whose solution determines the space dependence and time dependence of the wave function. The relevant mathematics goes beyond that employed in Newtonian mechanics, where the dynamics entails the solution of an *ordinary* differential equation. Fortunately, the mathematical treatment of partial differential equations of the Schrödinger type is well developed and straightforward. The most useful method of solution is provided by the technique of *separation of variables*. This procedure leads us directly to one of the central ideas in quantum physics—the concept of stationary states.

The wave function has independent variables x and t in the one-dimensional version of quantum mechanics. We can separate the x dependence from the t dependence if we seek solutions in which the wave function is written as a *product* of functions of a single variable:

$$\Psi(x,t) = \psi(x)f(t). \tag{5-19}$$

This expression is supposed to satisfy the Schrödinger equation, and so the product ψf must obey a modified form of Equation (5-11):

$$-\frac{\hbar^2}{2m}\frac{d^2\psi}{dx^2}f + V\psi f = i\hbar\psi\frac{df}{dt}.$$

Note that the partial derivatives acting on the wave function Ψ turn into ordinary derivatives acting on the factors ψ and f. If we divide both sides of the equation by the product of factors ψf, we obtain the equality

$$\frac{-\dfrac{\hbar^2}{2m}\dfrac{d^2}{dx^2}\psi(x) + V(x)\psi(x)}{\psi(x)} = \frac{i\hbar\dfrac{d}{dt}f(t)}{f(t)}. \tag{5-20}$$

This result has an interesting structure, inasmuch as the left side of the equation depends only on x while the right side depends only on t. It is possible to have an equality between a function of x and a function of t only if each function is a constant. We therefore set each side of Equation (5-20) equal to a *separation constant* λ and obtain separate equations for ψ and f in which λ appears as a common ingredient:

$$-\frac{\hbar^2}{2m}\frac{d^2\psi}{dx^2} + V\psi = \lambda\psi \tag{5-21}$$

and

$$i\hbar\frac{df}{dt} = \lambda f. \tag{5-22}$$

Note that V must occur as a function of x *alone*; the method of separation of variables would not apply otherwise. We conclude from this procedure that the Schrödinger equation has solutions in the form expressed in Equation (5-19), provided the two ordinary differential equations have solutions for the factors ψ and f.

Let us look first at the solution of Equation (5-22). The first-order differential equation can be solved at once to yield

$$f(t) = e^{-i\lambda t/\hbar}, \tag{5-23}$$

apart from an arbitrary multiplicative constant. This complex exponential solution describes an oscillating function of time:

$$f(t) = \cos\frac{\lambda}{\hbar}t - i\sin\frac{\lambda}{\hbar}t.$$

The wave function constructed in Equation (5-19) is therefore characterized by a unique angular frequency ω, connected to the separation constant λ through the relation

$$\omega = \frac{\lambda}{\hbar}.$$

A unique energy may also be associated with this type of wave function by virtue of the Planck–Einstein formula:

$$E = \hbar\omega = \lambda. \tag{5-24}$$

Thus, Equation (5-19) describes a system in which the quantum particle is found in a state of precisely specified energy.

Equation (5-24) tells us that the separation constant is the same as the energy parameter of the state. The actual value of the energy remains to be determined from

Equation (5-21), the other of the two ordinary differential equations in the construction of the wave function. We substitute E for λ and rewrite the equation as

$$-\frac{\hbar^2}{2m}\frac{d^2}{dx^2}\psi(x) + V(x)\psi(x) = E\psi(x). \tag{5-25}$$

The resulting differential equation for $\psi(x)$ with energy parameter E is called the *time-independent Schrödinger equation*. It is obvious that we must be given the potential energy $V(x)$ before we can proceed to solve for the unknown function $\psi(x)$. We note that we are able to establish the form of the t dependence of Ψ in Equation (5-19) without any knowledge of the form of $V(x)$. Equation (5-25) then plays the main role in the remaining construction of Ψ. The product expression in Equation (5-19) is indeed a valid solution of the Schrödinger equation, *provided* the value of E corresponds to a physically allowable solution $\psi(x)$ from the time-independent equation. An allowed value of E is called an *energy eigenvalue*. The associated physically allowed solution $\psi(x)$ is called the corresponding *eigenfunction*. We must learn how to solve Equation (5-25) in order to find these last two ingredients and assemble the wave function.

The method of separation of variables produces solutions of the Schrödinger equation in the special form

$$\Psi(x, t) = \psi(x)e^{-iEt/\hbar}. \tag{5-26}$$

There is certainly no reason to suppose that every wave function is necessarily of this sort. Indeed, we know that any number of these allowed solutions can be superposed to generate a valid wave function. We also know that the determination of $\Psi(x, t)$ in a particular problem depends on given information regarding the initial wave function $\Psi(x, 0)$. The point to emphasize in this section is that solutions of the type expressed in Equation (5-26) are special wave functions whose mathematical properties describe the *stationary states* of the system. To appreciate the terminology let us examine the probability density associated with the state in Equation (5-26). The entire time dependence of Ψ is contained in a single complex exponential factor of unit modulus, and so the square of the modulus of Ψ has *no t* dependence in such a state:

$$\left|\Psi(x, t)\right|^2 = \left|\psi(x)\right|^2. \tag{5-27}$$

Hence, the state is said to be *stationary* because the probabilistic aspects of the corresponding wave function do not vary with time. We have already attached additional significance to this wave function by noting above that the energy of the state is a precisely defined quantity.

We expect to find a *set* of allowed E's and ψ's as solutions of Equation (5-25) for any given potential energy $V(x)$. These solutions represent the energy levels and stationary state eigenfunctions for the particular dynamical system. We elaborate on this observation by turning to special cases of quantum systems in the next two sections.

Example

Let us accept the fact that the Schrödinger theory always yields a set of energy eigenvalues $E_1, E_2, \ldots$, corresponding to the set of stationary state wave functions

$$\psi_1(x)e^{-iE_1t/\hbar}, \quad \psi_2(x)e^{-iE_2t/\hbar}, \quad \ldots .$$

A particle in one of these states has a stationary probability density, as noted in Equation (5-27). Suppose, however, that the state is a superposition of two stationary states described by the wave function

$$\Psi(x, t) = a_1\psi_1(x)e^{-iE_1t/\hbar} + a_2\psi_2(x)e^{-iE_2t/\hbar}.$$

We obtain this function as the unique solution of the Schrödinger equation if we are given

$$\Psi(x, 0) = a_1\psi_1(x) + a_2\psi_2(x)$$

as the initial condition for $\Psi(x, t)$. We then find that $\Psi * \Psi$ is *not t* independent:

$$|\Psi(x, t)|^2 = \left(a_1^*\psi_1^*e^{iE_1t/\hbar} + a_2^*\psi_2^*e^{iE_2t/\hbar}\right)\left(a_1\psi_1e^{-iE_1t/\hbar} + a_2\psi_2e^{-iE_2t/\hbar}\right)$$

$$= |a_1|^2|\psi_1(x)|^2 + a_1^*a_2\psi_1^*(x)\psi_2(x)e^{i(E_1-E_2)t/\hbar}$$

$$+ a_2^*a_1\psi_2^*(x)\psi_1(x)e^{-i(E_1-E_2)t/\hbar} + |a_2|^2|\psi_2(x)|^2.$$

The expression contains two *oscillating* terms with angular frequency

$$\omega = \frac{|E_1 - E_2|}{\hbar}.$$

The state represented by the wave function $\Psi(x, t)$ is not stationary, and so all its probabilistic features can be expected to oscillate with the same angular frequency ω.

5-4 The One-Dimensional Box

Several first principles of quantum mechanics have already made their appearance in this chapter. It is appropriate that we pause and reflect on the new ideas by considering models like the ones shown in Figure 5-1. We devote this section to the simpler of the two indicated systems.

The illustrated example of a sliding bead on a wire is known in quantum mechanics as the problem of a particle in a one-dimensional box. The classical particle travels freely between the two ends of its course and abruptly reverses direction at either end with no loss of energy. We describe this model by means of the potential energy shown in Figure 5-4. The classical motion is confined to an interval of length a, and the origin is taken to be at the center of the interval. The potential energy function is then defined as

$$V(x) = \begin{cases} 0 & \text{for } -a/2 < x < a/2 \\ \infty & \text{for } |x| > a/2, \end{cases} \tag{5-28}$$

so that a nonvanishing force acts only at $x = \pm a/2$. An energy E is assigned to the

Figure 5-4

Potential energy function $V(x)$ for a particle in a one-dimensional box.

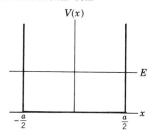

system as in the figure; hence, the kinetic energy of the particle is equal to E in the region of vanishing potential energy. Classical motion is forbidden outside this region because the infinite value of V exceeds any possible choice of E. Thus, the given potential energy function provides barriers to confine the particle in the interval $[-a/2, a/2]$ for any assigned energy.

Let us now insert the function $V(x)$ in the Schrödinger equation and find the stationary-state wave functions for the quantum particle. We learn as follows that only certain *discrete* values of the energy E are obtained from Equation (5-25), the time-independent equation for the eigenfunction $\psi(x)$. The equation takes the form

$$-\frac{\hbar^2}{2m}\frac{d^2\psi}{dx^2} = E\psi$$

when x is in the interval $(-a/2, a/2)$ where $V = 0$. We rewrite this differential equation as

$$\frac{d^2\psi}{dx^2} + k^2\psi = 0,$$

where

$$k = \sqrt{\frac{2mE}{\hbar^2}}, \tag{5-29}$$

and recognize the immediate solutions

$$\cos kx \quad \text{and} \quad \sin kx.$$

The potential energy becomes infinite when x is not in the range $[-a/2, a/2]$, and so Equation (5-25) makes sense only if $\psi(x)$ *vanishes* everywhere in this outside region. We conclude that there is *zero* probability of finding the quantum particle anywhere beyond the interval $[-a/2, a/2]$.

The probability density is physically observable and should therefore be continuous in x for all problems. It follows that the wave function must always be a continuous function of x. This requirement is supposed to hold for our particle-in-a-box eigenfunction $\psi(x)$. The piecewise behavior of the function can be described by the general solution

$$\psi(x) = \begin{cases} A\cos kx + B\sin kx & \text{for } -a/2 < x < a/2 \\ 0 & \text{for } |x| > a/2. \end{cases}$$

Continuity at $x = \pm a/2$ implies two conditions on the parameters:

$$A\cos\frac{ka}{2} + B\sin\frac{ka}{2} = 0$$

and

$$A\cos\frac{ka}{2} - B\sin\frac{ka}{2} = 0.$$

(We use the evenness and oddness of the cosine and sine to write the second result.) The two equalities can be combined to read

$$A\cos\frac{ka}{2} = 0 \quad\text{and}\quad B\sin\frac{ka}{2} = 0.$$

We cannot allow both A and B to be zero, and we know that $\cos(ka/2)$ and $\sin(ka/2)$ cannot both vanish for the same value of k. Only two other possibilities remain:

$$B = 0 \quad\text{and}\quad \cos\frac{ka}{2} = 0 \quad\text{so that}\quad \frac{ka}{2} = \frac{\pi}{2}, \frac{3\pi}{2}, \ldots,$$

or

$$A = 0 \quad\text{and}\quad \sin\frac{ka}{2} = 0 \quad\text{so that}\quad \frac{ka}{2} = \pi, 2\pi, \ldots.$$

Both options are permitted, and both make equivalent predictions for the parameter k. The allowed values of this quantity are evidently restricted to the *discrete* set

$$k_n = \frac{\pi}{a}n, \tag{5-30}$$

where n is any integer, odd or even.

We observe that there are two classes of eigenfunctions in this model. Cosine solutions are found for $B = 0$ and *odd n*:

$$\psi_n(x) = A\cos\frac{n\pi x}{a} \quad\text{for } x \text{ in the interval } [-a/2, a/2].$$

Sine solutions are found for $A = 0$ and *even n*:

$$\psi_n(x) = B\sin\frac{n\pi x}{a} \quad\text{for } x \text{ in the interval } [-a/2, a/2].$$

The multiplicative constants A and B remain to be determined in these formulas. Note that the integer n serves as a *quantum number* to index the allowed eigenfunctions and the allowed values of k. The energy eigenvalues are obtained from Equations

(5-29) and (5-30) and are labeled in similar *quantized* fashion:

$$E_n = \frac{\hbar^2}{2m}k_n^2 = \frac{\hbar^2}{2m}\frac{\pi^2}{a^2}n^2. \tag{5-31}$$

We note that the possibility of an $n = 0$ state is ruled out because the corresponding eigenfunction vanishes. The lowest allowed value of E_n occurs for $n = 1$ and gives the energy of the *ground state* as

$$E_1 = \frac{\hbar^2\pi^2}{2ma^2}. \tag{5-32}$$

The nonzero result is interpreted as quantum zero-point energy in the ground state of the system. We have encountered this notion before in our comments on localization in the context of the uncertainty principle.

Let us summarize our conclusions and express the stationary-state wave functions in *normalized* form. We write the results as follows:

$$\Psi_n(x, t) = \begin{cases} \sqrt{\dfrac{2}{a}}\cos\dfrac{n\pi x}{a}\, e^{-iE_n t/\hbar} & (\text{odd } n) \\[2mm] \sqrt{\dfrac{2}{a}}\sin\dfrac{n\pi x}{a}\, e^{-iE_n t/\hbar} & (\text{even } n) \end{cases} \quad \text{for } -\frac{a}{2} \leq x \leq \frac{a}{2} \tag{5-33a}$$

and

$$\Psi_n(x, t) = 0 \quad \text{for } |x| \geq a/2. \tag{5-33b}$$

The factor $\sqrt{2/a}$ ensures that each Ψ_n satisfies the normalization condition

$$1 = \int_{-\infty}^{\infty} |\Psi_n(x, t)|^2 \, dx = \int_{-a/2}^{a/2} |\Psi_n(x, t)|^2 \, dx.$$

We verify this property of the wave functions in the first example below. The energy eigenvalues E_n appear in the t dependence of Equations (5-33) as multiples of the ground-state energy:

$$E_n = n^2 E_1. \tag{5-34}$$

It is convenient to organize this information about the stationary states according to the scheme of energy levels and normalized eigenfunctions shown in Figure 5-5.

Note that the zeroes, or *nodes*, of the eigenfunctions increase in number as the energy E_n and the quantum number n increase. This property is a general characteristic of the stationary-state eigenfunctions for any problem. It is apparent from the figure that the eigenfunction $\psi_n(x)$ has $n + 1$ nodes in the interval $[-a/2, a/2]$. In fact, a comparison between Figure 5-5 and Figure 2-6 reveals how closely the various functions resemble the standing waves on a vibrating string.

The two classes of wave functions in Equations (5-33) are distinguished by their *even* and *odd* behavior under a change of sign in the variable $x \to -x$. We see from Figure 5-5 that the states alternate in this respect as the energies increase. The

Figure 5-5

Energy levels and normalized eigenfunctions for a particle in a one-dimensional box.

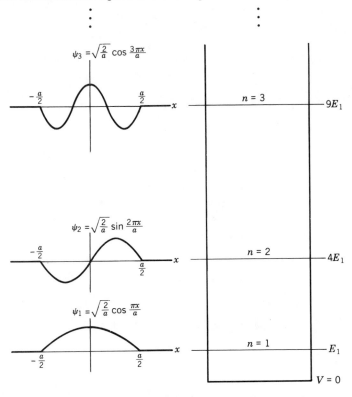

evenness or oddness of a given particle-in-a-box eigenfunction is expressed by writing

$$\psi_n(-x) = +\psi_n(x) \quad \text{for odd } n \tag{5-35a}$$

and

$$\psi_n(-x) = -\psi_n(x) \quad \text{for even } n. \tag{5-35b}$$

This symmetry or antisymmetry of a stationary-state wave function under the substitution $x \rightarrow -x$ is called the *parity* of the state. Equations (5-35) may be restated to say that $\Psi_{\text{odd } n}$ has even parity and $\Psi_{\text{even } n}$ has odd parity. The parity is always found to be a manifest property of the states whenever the potential energy of the particle is an *even*, or *symmetric*, function of x. We should expect the observable probability density in a stationary state to be symmetric under the replacement $x \rightarrow -x$ if the dynamics of the particle is governed by a symmetric potential energy $V(x)$. We then conclude from the even behavior of $\Psi_n^*\Psi_n$ that the wave function Ψ_n must be either an even or an odd function of x. The particle in a box is the simplest such example in which the system has a symmetric potential energy and displays stationary states of definite parity.

The eigenfunctions $\psi_n(x)$ illustrate another mathematical concept known as *orthogonality*. The property of interest refers to integrals of pairs of eigenfunctions written as

$$\int_{-a/2}^{a/2} \cos\frac{n\pi x}{a}\cos\frac{n'\pi x}{a}\,dx, \quad \int_{-a/2}^{a/2} \sin\frac{n\pi x}{a}\cos\frac{n'\pi x}{a}\,dx,$$

and

$$\int_{-a/2}^{a/2} \sin\frac{n\pi x}{a} \sin\frac{n'\pi x}{a} \, dx.$$

These expressions occur frequently in applications of the model. The integrals are found to *vanish*, and the functions are said to be *orthogonal*, whenever the quantum numbers n and n' have *different* values. Orthogonality is a general result for families of eigenfunctions appearing in all kinds of problems. It is obvious in the case at hand that the second of the three integrations is equal to zero, because the integrand is odd and the range of integration is symmetric about the origin. The other two integrals are left to be examined in Problem 20 at the end of the chapter.

Example

The problem of the particle in a box has many illustrations and applications. Let us look first at the normalization of the wave functions in Equations (5-33). We note immediately that the range of the normalization integral shrinks from $\int_{-\infty}^{\infty}$ to $\int_{-a/2}^{a/2}$ since the wave functions vanish outside the interval $[-a/2, a/2]$. The cosine eigenfunctions produce the following integration:

$$\int_{-\infty}^{\infty} |\Psi_n|^2 \, dx = \int_{-a/2}^{a/2} \frac{2}{a}\cos^2\frac{n\pi x}{a} \, dx = \frac{2}{a}\int_{-a/2}^{a/2} \frac{1}{2}\left(1 + \cos\frac{2n\pi x}{a}\right) dx$$

$$= \frac{1}{a}\left(x + \frac{a}{2n\pi}\sin\frac{2n\pi x}{a}\right)\bigg|_{-a/2}^{a/2}.$$

The sine eigenfunctions produce the same integration except for the sign of the second term in the end result. This contribution vanishes at the two limits, and so both calculations reduce to the same first term and give the value unity, as announced.

Example

Let us turn next to the formula for the energy of the ground state in Equation (5-32) and apply the expression to a neutron confined in a linear region of nuclear size. We take $a = 10^{-14}$ m for the length of the box and get

$$E_1 = \frac{\hbar^2\pi^2}{2ma^2} = \frac{\left[(1.05 \times 10^{-34} \text{ J}\cdot\text{s})\pi/10^{-14} \text{ m}\right]^2}{2(1.67 \times 10^{-27} \text{ kg})(1.60 \times 10^{-13} \text{ J/MeV})} = 2.04 \text{ MeV}.$$

This amount of zero-point energy is comparable with the scale of energies for the constituents of nuclei.

Example

The quantization of states is the main nonclassical result of this model. Figure 5-5 reveals a separation between the discrete energy levels which increases with the quantum number n. Let us interpret this observation in the light of Bohr's correspondence principle and examine the limit of large quantum numbers. The separation of adjacent levels is obtained from Equation (5-34):

$$\Delta E_n = E_{n+1} - E_n = \left[(n+1)^2 - n^2 \right] E_1 = (2n+1)E_1.$$

We note that the energy difference ΔE_n grows linearly with n while the energy E_n grows quadratically. Hence, the ratio of these quantities tends to zero for large n:

$$\frac{\Delta E_n}{E_n} = \frac{(2n+1)E_1}{n^2 E_1} \longrightarrow \frac{2}{n}.$$

We recognize this as a sort of classical limit in which the discreteness of the levels becomes increasingly difficult to establish, even as the levels grow farther apart.

Example

Let us investigate the classical limit from another angle by applying the quantum formulas to a *macroscopic* particle, such as a 5 g bead on a 40 cm wire. We learn immediately that the nonzero ground-state energy of the particle is extraordinarily small:

$$E_1 = \frac{\hbar^2 \pi^2}{2ma^2} = \frac{\left[(1.05 \times 10^{-34}\,\text{J}\cdot\text{s})\pi/0.4\,\text{m} \right]^2}{2(5 \times 10^{-3}\,\text{kg})} = 6.80 \times 10^{-65}\,\text{J}.$$

It would be difficult to distinguish this quantum ground state from the classical state of rest. Let us suppose instead that the bead is moving with speed 2 m/s. We can try to associate a quantum number with this state of motion by consulting Equation (5-31) and identifying the kinetic energy

$$\frac{m}{2}v^2 = n^2 \frac{\hbar^2 \pi^2}{2ma^2}.$$

The resulting value of the quantum number is enormous:

$$n = \frac{mva}{\hbar\pi} = \frac{2mva}{h} = \frac{2(5 \times 10^{-3}\,\text{kg})(2\,\text{m/s})(0.4\,\text{m})}{6.63 \times 10^{-34}\,\text{J}\cdot\text{s}} = 1.21 \times 10^{31}.$$

It would be difficult to insist that such a particle is in a discrete quantum state since the energy would have to be known with incredible accuracy to distinguish the state from its immediate neighbors. Discreteness of energy can scarcely be regarded as a detectable property of this macroscopic particle.

5-5 The Harmonic Oscillator

The other of the two illustrations in Figure 5-1 shows the example of an oscillating mass on a spring. The classical oscillator is an important topic in physics, and so the quantum oscillator is a natural model to study in our introduction to the Schrödinger theory. We begin by recalling the following classical results as background for this investigation.

The classical particle is subject to a restoring force proportional to the displacement from equilibrium. The oscillator has potential energy

$$V(x) = \frac{k}{2}x^2, \tag{5-36}$$

so that the particle experiences a force

$$F = -\frac{dV}{dx} = -kx,$$

where k is the force constant of the spring. Newton's law then provides the differential equation for the displacement of the particle:

$$m\frac{d^2x}{dt^2} = -kx.$$

The solution determines the displacement and velocity as

$$x(t) = A\cos(\omega_0 t + \phi)$$

and

$$v(t) = \frac{dx}{dt} = -\omega_0 A\sin(\omega_0 t + \phi)$$

and thus parametrizes the motion in terms of amplitude A, phase angle ϕ, and angular frequency

$$\omega_0 = \sqrt{\frac{k}{m}}.$$

These expressions for x and v appear in the formula for the total energy and produce a constant of the motion:

$$E = K + V = \frac{m}{2}v^2 + \frac{k}{2}x^2$$

$$= \frac{m}{2}\omega_0^2 A^2 \sin^2(\omega_0 t + \phi) + \frac{k}{2}A^2\cos^2(\omega_0 t + \phi)$$

$$= \frac{k}{2}A^2.$$

This result can be rearranged to give v as a function of x:

$$v = \sqrt{\frac{2}{m}\left(E - \frac{k}{2}x^2\right)} = \sqrt{\frac{k}{m}(A^2 - x^2)}.$$

Figure 5-6

Potential energy $V(x)$ for a harmonic oscillator. Classical motion with total energy E is not allowed in the region $|x| > A$, where V exceeds E.

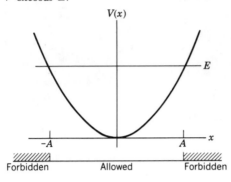

Figure 5-7

Classical probability density for a harmonic oscillator.

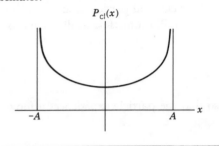

It is obvious that x^2 must not exceed A^2 if v is to be a physical velocity for a classical particle.

The classical picture can be visualized with the aid of Figure 5-6, which shows the potential energy $V(x)$ along with a chosen value for the constant total energy E. We note that any choice of energy is permitted in the classical treatment of the problem. The figure indicates that

$$V(x) = E \quad \text{at } x = \pm A = \pm \sqrt{\frac{2E}{k}}$$

and that classical motion is forbidden if $|x| > A$. If x is in the allowed region $[-A, A]$, a classical probability density can be defined to express the likelihood of finding the classical particle in a specified interval dx. The corresponding probability $P_{cl}(x)\, dx$ must be proportional to $\omega_0\, dt$, the portion of the cycle of oscillation that the particle spends in the given interval. We find that the probability density is given by the formula

$$P_{cl}(x) = \frac{1}{\pi\sqrt{A^2 - x^2}}. \tag{5-37}$$

The proof of this result is left to Problem 11 at the end of the chapter. Figure 5-7 shows how the function $P_{cl}(x)$ describes an increasing probability for intervals near the turning points $x = \pm A$, where the oscillating particle reverses its motion.

We can use the harmonic-oscillator potential energy in a variety of physical applications that do not actually contain a mass-and-spring system. The interaction of the two atoms in a diatomic molecule is a case in point. Figure 5-8 shows a graph in which the potential energy V varies with the separation r between the two atoms to simulate the main dynamical properties of the molecule. The function V has an equilibrium position at separation r_0, where dV/dr is equal to zero. We can place the origin of the x axis at this point by defining $x = r - r_0$ and expand V in a power

Figure 5-8

Model of the potential energy of a diatomic molecule. The harmonic-oscillator potential energy $V(x)$ serves as an adequate approximation if the separation of the two atoms stays close to the equilibrium value r_0.

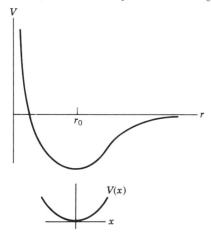

series about the equilibrium position:

$$V = V(r_0) + \frac{1}{2}\left(\frac{d^2V}{dr^2}\right)_{r_0} x^2 + \cdots .$$

Note that no term linear in x is present since the coefficient $(dV/dr)_{r_0}$ vanishes. We may neglect the higher powers of x if we confine our attention to bounded motion in the neighborhood of the point $r = r_0$. The choice of reference level is immaterial and can be shifted to eliminate the constant $V(r_0)$ and leave only the term quadratic in x. The resulting potential energy is just like the harmonic-oscillator function in Equation (5-36), as suggested in the figure. Any diatomic system is adequately described in terms of harmonic-oscillator behavior as long as the energy stays near the minimum value of V. The same conclusion holds whenever the physical problem of interest involves a similar stable-equilibrium configuration.

The quantum description of the harmonic oscillator is based on the eigenfunction solutions of the time-independent Schrödinger equation. These solutions determine the stationary states of the oscillating quantum particle. We define this problem in terms of Equation (5-25) by taking $V(x)$ to be the potential energy given in Equation (5-36). We then proceed to find a solution in the form of an eigenfunction $\psi(x)$ with energy eigenvalue E. The differential equation for ψ contains several bothersome constant factors. Let us rearrange these constants so that the equation reads

$$\frac{d^2\psi}{dx^2} = \frac{2m}{\hbar^2}\left(\frac{k}{2}x^2 - E\right)\psi = \frac{mk}{\hbar^2}\left(x^2 - \frac{2E}{k}\right)\psi$$

and then manipulate the result to obtain the awkward-looking equality

$$\frac{\hbar}{\sqrt{mk}}\frac{d^2\psi}{dx^2} = \left(\frac{\sqrt{mk}}{\hbar}x^2 - \frac{2E}{\hbar}\sqrt{\frac{m}{k}}\right)\psi .$$

Our purpose in these steps is to construct a streamlined version of the differential equation in which all the cluttering constants are out of the way. We accomplish this in one more step by introducing a dimensionless variable ξ and a dimensionless parameter λ according to the definitions

$$\xi^2 = \frac{\sqrt{mk}}{\hbar} x^2 \tag{5-38}$$

and

$$\lambda = \frac{2E}{\hbar} \sqrt{\frac{m}{k}} = \frac{2E}{\hbar\omega_0}. \tag{5-39}$$

These maneuvers result in the compact equation

$$\frac{d^2\psi}{d\xi^2} = \left(\xi^2 - \lambda\right)\psi \tag{5-40}$$

in which the eigenfunction ψ is regarded as a function of the new variable ξ. Note that λ assumes the role of an eigenvalue analogous to E in the original equation for ψ. The quantum oscillator presents a challenging mathematical problem since Equation (5-40) is a rather unfamiliar second-order differential equation.

Let us make some classical observations to illuminate the structure of this equation. We recall that the classical particle is confined to the region $|x| \leq A$, and we note from Equation (5-38) that the corresponding maximum value of $|\xi|$ is given by

$$\xi^2_{\max} = \frac{\sqrt{mk}}{\hbar} A^2.$$

This expression simplifies when we use the classical connection between the amplitude A and the energy E:

$$\xi^2_{\max} = \frac{\sqrt{mk}}{\hbar} \frac{2E}{k} = \lambda \quad \text{(classical)}. \tag{5-41}$$

Our final result tells us how the dimensionless quantities ξ and λ are related in classical physics. The relation has a special significance in the quantum problem because of the presence of the factor $(\xi^2 - \lambda)$ in Equation (5-40). We observe from the differential equation that the factors $d^2\psi/d\xi^2$ and ψ have *unlike* signs for $\xi^2 < \lambda$ and *like* signs for $\xi^2 > \lambda$. We also realize that $\psi(\xi)$ must have smooth behavior at the matching points $\xi = \pm\sqrt{\lambda}$. (Recall that ψ must always be continuous, and note that $d\psi/d\xi$ must also be continuous since $d^2\psi/d\xi^2$ exists at every point.) The two domains $\xi^2 < \lambda$ and $\xi^2 > \lambda$ correspond to the allowed and forbidden regions of Figure 5-6. Thus, it becomes apparent that the quantum problem has a nonvanishing solution in the forbidden region, in sharp contrast to the classical situation. It follows that there are nonzero probabilities of finding the quantum particle in locations where the classical particle is *never* found.

We expect small probabilities for intervals in the region $\xi^2 > \lambda$. This expectation suggests the necessity for a suppression of the eigenfunction ψ at large values of ξ^2. It is easy to show that a possible ψ is provided by the gaussian function

$$\psi(\xi) = e^{-\xi^2/2}.$$

The derivatives of ψ are

$$\frac{d\psi}{d\xi} = -\xi e^{-\xi^2/2}$$

and

$$\frac{d^2\psi}{d\xi^2} = \xi^2 e^{-\xi^2/2} - e^{-\xi^2/2} = \left(\xi^2 - 1\right)e^{-\xi^2/2}.$$

Thus, we see that the given function satisfies the differential equation *provided* λ has the special value

$$\lambda_0 = 1.$$

Equation (5-39) gives the energy eigenvalue for this solution as

$$E_0 = \frac{\hbar\omega_0}{2}\lambda_0 = \frac{\hbar\omega_0}{2}. \tag{5-42}$$

The eigenfunction $e^{-\xi^2/2}$ corresponds to a normalized stationary-state wave function of the form

$$\Psi_0(x, t) = \left(\frac{mk}{\pi^2\hbar^2}\right)^{1/8} e^{-\sqrt{mk}\,x^2/2\hbar} e^{-iE_0t/\hbar}. \tag{5-43}$$

This simplest solution of Equation (5-40) produces the smallest possible result for the parameter λ and the energy eigenvalue E. Hence, the wave function Ψ_0 and the energy E_0 refer to the *ground state* of the oscillator, and the quantity $\hbar\omega_0/2$ represents the quantum zero-point energy of the system. The normalization and other features of the ground-state wave function are discussed in the example below.

If we continue the analysis we learn that all other possible solutions of Equation (5-40) are suppressed in the forbidden region by the same gaussian factor $e^{-\xi^2/2}$. The solutions have the form

$$\psi(\xi) = H(\xi)e^{-\xi^2/2}, \tag{5-44}$$

where the factor $H(\xi)$ denotes a *polynomial*. This function is obtained first by deducing the differential equation satisfied by H and then by solving the equation. We elaborate on these mathematical details at the end of the section.

The main conclusion of the analysis is the fact that the parameter λ is allowed to have only the following discrete values:

$$\lambda_n = 2n + 1. \tag{5-45}$$

We designate these special values by an integer n and thereby introduce a *quantum number* to label the stationary states of the system. If we insert the expression for λ_n in the defining relation, Equation (5-39), we obtain the formula for the nth energy

Figure 5-9

Energy levels and eigenfunctions for the first four stationary states of the harmonic oscillator. Shaded areas represent the penetration of the wave function into regions where classical motion is forbidden.

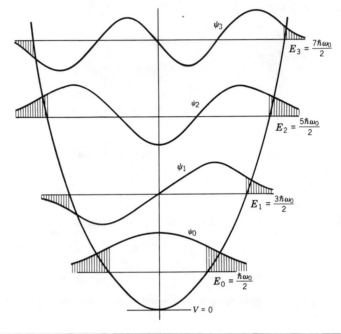

eigenvalue:

$$E_n = \frac{\hbar\omega_0}{2}\lambda_n = \hbar\omega_0\left(n + \tfrac{1}{2}\right). \tag{5-46}$$

The corresponding eigenfunction ψ_n is sketched along with its energy level for each of the lowest values of n in Figure 5-9. We observe, from Equation (5-46) and from the figure, that the discrete energies obey the *equal-spacing rule*

$$E_{n+1} - E_n = \hbar\omega_0.$$

The spacing is equal to the quantum oscillator energy hypothesized by Planck in his treatment of the radiating blackbody cavity. The figure also shows how the eigenfunctions "leak" into the classically forbidden region and how the nodes of ψ_n increase in number with the energy E_n. This behavior is a general feature of any solution of Schrödinger's wave equation.

Detail

The solutions of Equation (5-40) are found by a standard procedure in the treatment of differential equations. We begin by assuming an eigenfunction ψ of the form given by Equation (5-44), and we proceed to deduce the differential equation obeyed by H. The derivatives of ψ are

$$\frac{d\psi}{d\xi} = \frac{dH}{d\xi}e^{-\xi^2/2} - H\xi e^{-\xi^2/2}$$

and

$$\frac{d^2\psi}{d\xi^2} = \frac{d^2H}{d\xi^2}e^{-\xi^2/2} - 2\frac{dH}{d\xi}\xi e^{-\xi^2/2} - He^{-\xi^2/2} + H\xi^2 e^{-\xi^2/2}.$$

We insert these results in Equation (5-40) and find

$$\left[\frac{d^2H}{d\xi^2} - 2\xi\frac{dH}{d\xi} + (\xi^2 - 1)H\right]e^{-\xi^2/2} = (\xi^2 - \lambda)He^{-\xi^2/2}.$$

Therefore, our construction of ψ produces a valid solution of Equation (5-40) if the unknown function $H(\xi)$ satisfies the equation

$$\frac{d^2H}{d\xi^2} - 2\xi\frac{dH}{d\xi} + (\lambda - 1)H = 0. \tag{5-47}$$

The analysis of the quantum oscillator thus reduces to the problem of solving this differential equation for every possible function H and parameter λ. We see that there is an immediate solution of the form

$$H = \text{any constant} \quad \text{with } \lambda = 1,$$

corresponding to the ground-state wave function in Equation (5-43). The physically acceptable solutions of Equation (5-47) are restricted to *polynomials* for $H(\xi)$, since any other type of function fails to produce the desired suppression of ψ for large values of ξ^2. Let us accept this fact without proof and introduce the integer n to denote the order of these polynomials. We observe that the potential energy $V(x)$ is a *symmetric* function of x, and so we expect to find stationary-state eigenfunctions of *definite parity*. Since the factor $e^{-\xi^2/2}$ in Equation (5-44) is an even function of x, it follows that each polynomial $H(\xi)$ must be either even or odd in its dependence on ξ and must therefore consist of either all-even or all-odd powers of the variable. We express the acceptable even and odd solutions of Equation (5-47) as

$$H_n(\xi) = \begin{cases} \displaystyle\sum_{\text{even } k=0}^{n} a_k\xi^k & \text{for even } n \\[2em] \displaystyle\sum_{\text{odd } k=1}^{n} a_k\xi^k & \text{for odd } n. \end{cases} \tag{5-48}$$

The differential equation is then solved by finding the coefficients a_k. Our main goal is to secure the important result in Equation (5-45) for the allowed values of

λ. It is sufficient to examine the case of the even-order polynomial

$$H_n = a_0 + a_2\xi^2 + a_4\xi^4 + \cdots + a_n\xi^n \quad \text{where } a_n \neq 0.$$

The derivatives of H_n are

$$H_n' = 2a_2\xi + 4a_4\xi^3 + \cdots + na_n\xi^{n-1}$$

and

$$H_n'' = 2a_2 + 12a_4\xi^2 + \cdots + n(n-1)a_n\xi^{n-2}.$$

We insert these expressions in Equation (5-47) and obtain

$$\left[2a_2 + 12a_4\xi^2 + \cdots + n(n-1)a_n\xi^{n-2}\right]$$
$$-2\xi\left[2a_2\xi + \cdots + na_n\xi^{n-1}\right]$$
$$+(\lambda - 1)\left[a_0 + a_2\xi^2 + \cdots + a_n\xi^n\right] = 0.$$

The sum of the coefficients of each power of ξ must vanish, power by power, if this equality is to hold for all values of ξ. The coefficient of ξ^n produces the condition

$$-2na_n + (\lambda - 1)a_n = 0.$$

We note that a_n cannot be zero, and so we arrive at our main conclusion:

$$\lambda = 2n + 1.$$

At the other end of the polynomial we find

$$2a_2 + (\lambda - 1)a_0 = 0 \qquad \text{from the coefficient of } \xi^0,$$
$$12a_4 - 4a_2 + (\lambda - 1)a_2 = 0 \quad \text{from the coefficient of } \xi^2,$$
$$\cdots$$

These conditions determine each coefficient a_{k+2} in terms of its predecessor a_k:

$$a_2 = \frac{1 - \lambda}{2}a_0 = -na_0,$$

$$a_4 = \frac{5 - \lambda}{12}a_2 = \frac{2 - n}{6}a_2,$$

$$\cdots$$

The results can be generalized to read

$$a_{k+2} = \frac{2(k - n)}{(k + 1)(k + 2)}a_k.$$

The derivation of this general *recursion relation* between successive coefficients in the polynomial is left to be proved in Problem 17 at the end of the chapter.

Equation (5-45) can be combined with Equation (5-47) to give the Hermite differential equation:

$$\frac{d^2 H_n}{d\xi^2} - 2\xi \frac{dH_n}{d\xi} + 2nH_n = 0. \qquad (5\text{-}49)$$

The first few Hermite polynomials may be listed as follows:

$$H_0(\xi) = 1,$$
$$H_1(\xi) = 2\xi,$$
$$H_2(\xi) = 4\xi^2 - 2,$$
$$H_3(\xi) = 8\xi^3 - 12\xi,$$
$$\cdots$$

Both the equation and its polynomial solutions are named after C. Hermite, a French mathematician of the 19th century.

Example

Much can be said about the ground-state wave function Ψ_0 in Equation (5-43). Let us begin by verifying the normalization. The probability density is given by the time-independent quantity

$$P(x) = \Psi_0^* \Psi_0 = \left(\frac{mk}{\pi^2 \hbar^2} \right)^{1/4} e^{-\xi^2}, \quad \text{where } \xi = \left(\frac{mk}{\hbar^2} \right)^{1/4} x,$$

and so the normalization integral $\int_{-\infty}^{\infty} P(x)\, dx$ becomes

$$\int_{-\infty}^{\infty} \left(\frac{mk}{\pi^2 \hbar^2} \right)^{1/4} e^{-\xi^2}\, dx = \frac{1}{\sqrt{\pi}} \int_{-\infty}^{\infty} e^{-\xi^2}\, d\xi = \frac{2}{\sqrt{\pi}} \int_0^{\infty} e^{-\xi^2}\, d\xi.$$

The last step can be taken because the integrand is an even function. The final result is equal to the value at ∞ of a tabulated function known as the error integral

$$\text{erf}(u) = \frac{2}{\sqrt{\pi}} \int_0^u e^{-\xi^2}\, d\xi.$$

The normalization of Ψ_0 is confirmed if we can prove that $\text{erf}(\infty) = 1$. To demonstrate this we define the numerical quantity

$$J = \int_0^{\infty} e^{-x^2}\, dx$$

and evaluate the square of J as an area integral:

$$J^2 = \int_0^{\infty} e^{-x^2}\, dx \cdot \int_0^{\infty} e^{-y^2}\, dy = \iint_{\text{1st quadrant}} dx\, dy\, e^{-(x^2 + y^2)}.$$

We then transform the integration variables to polar coordinates and obtain

$$J^2 = \int_0^{\pi/2} d\theta \int_0^\infty r \, dr \, e^{-r^2} = \frac{\pi}{2} \left(\frac{e^{-r^2}}{-2} \right) \Bigg|_0^\infty = \frac{\pi}{4}.$$

The result is $J = \sqrt{\pi}/2$, and so the calculation finally gives the desired value

$$\text{erf}(\infty) = \frac{2}{\sqrt{\pi}} J = 1.$$

The probability density $\Psi_0^* \Psi_0$ is shown as a function of the convenient variable ξ in Figure 5-10. The indicated points $\xi = \pm 1$ are significant for this state because these values of the variable correspond to the extreme displacements of the classical particle at energy $E_0 = \hbar \omega_0 / 2$, as implied by Equation (5-41). It is apparent from the figure that there is a nonnegligible probability of finding the particle in the classically forbidden region. The probability in the allowed region is found in two steps, by computing

$$\int_{-A}^{A} P(x) \, dx = \frac{2}{\sqrt{\pi}} \int_0^1 e^{-\xi^2} \, d\xi = \text{erf}(1)$$

and by consulting a table of the error integral to get $\text{erf}(1) = 0.84$. Hence, we see that in the ground state the particle has 16% likelihood to be found *outside* the range of classical motion. The $n = 0$ state (or any other low-lying quantum state) is not expected to exhibit classical behavior. We recall that the classical probability density is given by the quantity $P_{cl}(x)$ in Figure 5-7, and we note that this distribution looks nothing like the behavior of $\Psi_0^* \Psi_0$. Figure 5-11 shows a graph of $\Psi_{10}^* \Psi_{10}$ to illustrate the fact that the distributions begin to compare with P_{cl} when the quantum numbers become sufficiently large. Ob-

Figure 5-10

Ground-state probability density as a function of the variable $\xi = (mk/\hbar^2)^{1/4} x$. Classical motion with energy $\hbar \omega_0 / 2$ is forbidden outside the interval $-1 \leq \xi \leq 1$.

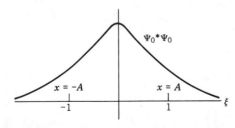

Figure 5-11

Probability density for the $n = 10$ state of the harmonic oscillator. The average of the distribution agrees with the classical probability density in Figure 5-7.

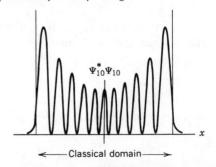

serve that all the wiggles in this $n = 10$ probability density occur in the classically allowed region, and note that the average value of the distribution approximates the shape of the classical function P_{cl}.

5-6 Eigenfunctions and Eigenvalues

The models in the last two sections provide insight into the workings of the Schrödinger equation. These studies illustrate the properties of quantization, since the stationary states of the two systems occur in discrete sets where only certain values of the energy are allowed. We are now in position to examine the generality of this all-important aspect of quantum physics.

Let us continue to focus on stationary states and wave functions with the elementary time dependence

$$\Psi(x, t) = \psi(x)e^{-iEt/\hbar}.$$

The eigenfunction $\psi(x)$ and the energy eigenvalue E are obtained together by solving the time-independent Schrödinger equation

$$-\frac{\hbar^2}{2m}\frac{d^2\psi}{dx^2} + V(x)\psi = E\psi.$$

Our goal is to explain the general circumstances behind the *discrete* occurrence of ψ and E. We take the potential energy to be some given function $V(x)$ and base our claim of generality on the arbitrariness of this function. A representative potential energy is provided in Figure 5-12 for use throughout the following discussion. The graph also includes a chosen value for the total energy E, to be considered as a candidate for an allowed energy eigenvalue in the Schrödinger equation. Note that the selected energy satisfies the equality

$$V(x) = E \quad \text{at } x = x_1 \text{ and } x_2.$$

These values of the coordinate play a significant role in the dynamics of the classical and the quantum particle.

We begin with some *classical* observations regarding the kinetic energy

$$K = E - V(x) = \frac{m}{2}v^2.$$

Rearrangement yields a formula for the velocity as a function of x:

$$v^2 = \frac{2}{m}(E - V(x)). \tag{5-50}$$

A real value of v is obtained from this formula only if the value of x is such that $E \geq V(x)$. We visualize this condition in Figure 5-12 by noting that one-dimensional classical motion is *bounded* between the indicated points x_1 and x_2 for the given energy E. These special values of x are called the classical *turning points*. The figure shows that the two points approach or recede from each other as E decreases or increases. We observe that no turning points exist, and so no configuration of the system is possible,

Figure 5-12

Potential energy function $V(x)$ and total energy E for classical motion with two turning points. The classical particle is not allowed in the regions to the left of x_1 and to the right of x_2. The eigenfunction $\psi(x)$ is constructed so that the curvature-to-value ratio is positive in the forbidden region and negative in the allowed region.

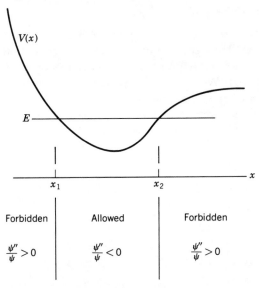

if E falls below the minimum of $V(x)$. We also note that the left turning point x_1 remains finite while the right turning point x_2 recedes to infinity when E reaches a certain critical value. The motion is unbounded in the positive x direction for any larger choice of the energy. Our two models in the preceding sections can be seen from Figures 5-4 and 5-6 to have the property that bounded motion persists, no matter how large E becomes. The main thrust of our discussion in this section pertains to bounded motion in the system described by Figure 5-12.

The conclusions drawn from Equation (5-50) are classical and do not apply to the quantum particle. It is clear that the equation itself is in conflict with the uncertainty principle since the formula defines an exact value for the momentum of a particle, given an exact value for the coordinate. The classical turning points x_1 and x_2 are nevertheless important in the determination of quantum behavior. We can appreciate their importance immediately if we rewrite the time-independent Schrödinger equation as

$$\frac{d^2\psi}{dx^2} = \frac{2m}{\hbar^2}(V(x) - E)\psi, \tag{5-51}$$

or as

$$\frac{\psi''(x)}{\psi(x)} = \frac{2m}{\hbar^2}(V(x) - E). \tag{5-52}$$

The right side of the second equation represents *given* information with regard to the variable x and indicates a change of sign at the turning points like that observed for the quantity v^2 in Equation (5-50). The quantum problem deviates from its classical

Figure 5-13

Eigenfunction behavior on either side of a classical turning point.

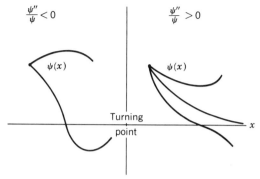

counterpart, however, and offers a solution for *all* x, in the form of the eigenfunction $\psi(x)$, with nonzero probability density in the *forbidden* as well as the allowed regions of classical motion.

Equation (5-52) prescribes a change of sign at a turning point as a property of the *ratio* of unknowns ψ''/ψ. The equation expresses the ratio of the curvature of ψ to the value of ψ at each point x and hence determines the *shape* of the eigenfunction on a point-by-point basis. It is evident that the curvature of ψ vanishes, so that ψ has a point of inflection, at each of the turning points. Thus, there is a one-to-one correspondence between regions in which the curvature-to-value ratio is positive or negative, and regions in which the classical motion is forbidden or allowed. This observation is summarized in the lower part of Figure 5-12. Figure 5-13 goes on to show the qualitative shapes of possible eigenfunctions deduced from Equation (5-52). Note that ψ may change its sign away from a turning point as long as ψ'' has a coincident change of sign. The eigenfunction then has a node and a point of inflection at the same point. The eigenfunction may even oscillate and produce a succession of nodes, provided the region is one in which ψ''/ψ is negative.

Equation (5-52) is not the only ingredient in the determination of ψ. Appeal must also be made to additional physical considerations beyond those expressed by the differential equation. We recall that the probability density is a measurable quantity, and so we require that $\Psi^*\Psi$ be a continuous function of x with a unique and finite value at every point, including the limits $x \to \pm\infty$. The behavior at $\pm\infty$ is especially important, since $\Psi^*\Psi$ is supposed to be integrable over $-\infty < x < \infty$ if Ψ is to be a normalizable wave function. The properties of continuity and finiteness must hold for the eigenfunction $\psi(x)$ if they hold for $\Psi^*\Psi$. These requirements on the behavior of ψ usually include one more condition. Equation (5-51) implies that $d^2\psi/dx^2$ is unique and finite wherever the given potential energy $V(x)$ is unique and finite. It then follows that $d\psi/dx$ must also be a continuous function of x. (An exceptional case is the problem of the eigenfunctions in a one-dimensional box. Continuity of $d\psi/dx$ is not imposed in that problem because the value of V becomes infinite at the ends of the box.) The additional conditions on $\psi(x)$ are to be incorporated in the procedure for selecting a physically allowable solution. Some of the shapes drawn in Figure 5-13 can be eliminated from consideration on these grounds.

We can now assemble our new mathematical ideas and draw our main conclusions about quantization. First, let us realize that we are concerned with an ordinary

Figure 5-14

Construction of an allowable eigenfunction. The slope of ψ at x_0 can be adjusted to make $\psi \to 0$ as $x \to \infty$. For this selection it is possible to have $\psi \to 0$ as $x \to -\infty$ only if a suitable choice has been made for the eigenvalue E.

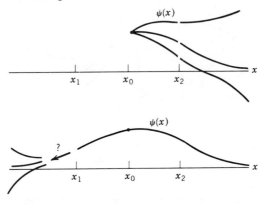

differential equation of second order, whose solution $\psi(x)$ is uniquely determined if we specify values for $\psi(x_0)$ and $\psi'(x_0)$, where x_0 is any chosen point. Let us take x_0 to be in the classically allowed region between the points x_1 and x_2 in Figure 5-12. We assign $\psi(x_0)$ an arbitrary positive value and then examine different choices for $\psi'(x_0)$ to see how the shape of ψ develops for $x > x_0$. The upper part of Figure 5-14 illustrates some of the possibilities. Note that each possible ψ turns downward to the left of x_2 and fits smoothly onto another piece of ψ that turns upward to the right of x_2. Only one of the three indicated curves remains finite everywhere on the right; the prospective solution has the property that $\psi \to 0$ as $x \to \infty$. We can tune $\psi'(x_0)$ so that $\psi(x)$ assumes this particular shape for the given value of E. If we then follow the selected $\psi(x)$ to the left of x_0, as in the lower part of the figure, we are likely to find that ψ has unsatisfactory behavior as $x \to -\infty$. We cannot expect ψ to tend asymptotically to zero on the left as well as the right unless we also tune a *second* parameter in addition to $\psi'(x_0)$. The only remaining adjustable quantity is the energy E. A properly behaved eigenfunction results when a suitable choice of E is made. If we let E depart even slightly from its determined value we damage the asymptotic behavior of the solution by forcing ψ to diverge either above or below the x axis. This argument rules out the occurrence of nodes in any classically forbidden region extending to infinity, since ψ must diverge once it crosses the axis in such a region.

Let us recall that the entire discussion pertains to values of E for which the corresponding classical motion is *bounded* between turning points, as in Figure 5-12. Our arguments imply that the energy E is *quantized* because only certain *discrete* choices of E are found to have allowable solutions for $\psi(x)$. These discretely determined stationary states are called *bound states* since the probability of finding the quantum particle vanishes at asymptotically large distance.

It is also possible to have an eigenfunction for a value of E such that only a *single* turning point exists. For instance, Figure 5-12 shows that x_1 becomes a lone turning point when E is chosen large enough. The classically allowed region then has infinite extent to the right, and so vanishing asymptotic behavior cannot be required as $x \to \infty$ since ψ is supposed to oscillate instead. We see that *any* such value of E is allowed above a certain threshold, once this constraint on ψ is removed. The resulting

Figure 5-15

Two eigenfunctions with different numbers of nodes. At x_0, ψ_0 and ψ_1 have equal values, but ψ_1 has more negative curvature than ψ_0. Equation (5-51) then implies that E_1 must be larger than E_0.

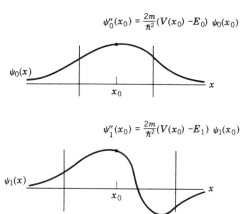

$$\psi_0''(x_0) = \frac{2m}{\hbar^2}(V(x_0) - E_0)\ \psi_0(x_0)$$

$\psi_0(x)$

x_0

x

$$\psi_1''(x_0) = \frac{2m}{\hbar^2}(V(x_0) - E_1)\ \psi_1(x_0)$$

$\psi_1(x)$

x_0

x

stationary states are called *continuum states*. The associated wave functions have the property that the probability of finding the quantum particle at large distance does not approach zero.

We introduce a quantum number n to enumerate the discrete energy eigenvalues E_n and eigenfunctions $\psi_n(x)$. The wave functions have the form

$$\Psi_n(x, t) = \psi_n(x)e^{-iE_n t/\hbar}$$

for the corresponding stationary states. We have devoted Sections 5-4 and 5-5 to circumstances in which only this discrete category of stationary states can arise. These models exemplify the general situation, where an eigenfunction $\psi_n(x)$ has certain nodes, all occurring at points in the classically allowed region, and where the quantum number n denotes an ordering of the number of nodes with increasing energy E_n. The fact that the nodes of ψ_n increase in number as the energy E_n increases can be established by means of a rather technical proof. Let us summarize the content of the proof qualitatively by noting that the chain of argument takes the following steps:

$$\text{more nodes} \longrightarrow \text{more oscillation} \longrightarrow \text{more curvature} \longrightarrow \text{larger energy}.$$

The link between the curvature and the energy eigenvalue is supplied by Equation (5-51). Figure 5-15 illustrates this part of the reasoning with the aid of a simple construction.

The potential energies in Figures 5-4 and 5-6 have the special property of symmetry under the parity replacement $x \to -x$. The resulting stationary states are therefore found to have definite even or odd parity. Our arbitrarily chosen function $V(x)$ in Figure 5-12 lacks this special symmetry feature. Hence, the eigenfunctions in the general discussion do not necessarily have any predictable behavior relating positive and negative values of x.

It is clear that the time-independent Schrödinger equation can have a collection of distinct eigenfunctions $\psi_n(x)$ with energy eigenvalues E_n. The members of this set of

Figure 5-16

Qualitative forms of the lowest eigenfunctions for a V-shaped potential energy.

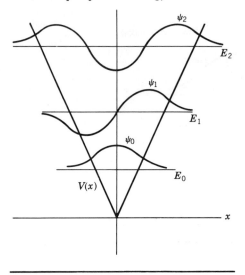

solutions obey the *orthogonality property*

$$\int_{-\infty}^{\infty} \psi_{n'}^*(x)\psi_n(x)\,dx = 0 \qquad (5\text{-}53)$$

for any two different quantum numbers n and n'. We have already seen orthogonality in the context of the particle-in-a-box problem at the end of Section 5-4. The proof of the general formula in Equation (5-53) is demonstrated in the last example below.

Our description of the eigenvalue problem is taken from an area of mathematics known as Sturm–Liouville theory. This general analysis of the solutions of second-order ordinary differential equations is named after the 19th century mathematicians C.-F. Sturm and J. Liouville.

Example

Consider the V-shaped potential energy shown in Figure 5-16. The associated differential equation has unfamiliar analytical solutions; however, there is no mathematical obstacle to deter us from making the following qualitative remarks. Each choice of energy corresponds to bound classical motion with two turning points, and so the sign of the curvature of ψ varies according to Equation (5-51) in three regions of the x axis. The shape of $V(x)$ tells us at once that *all* the stationary states are discrete and that the eigenfunctions ψ_n have definite parity. We expect the nodes of ψ_n to increase in number with E_n, so that a given eigenfunction must have one more node than its predecessor. It follows that the parities alternate with increasing n and ascending energy. These observations on the curvature, nodes, and parity of the eigenfunctions are displayed qualitatively in the figure.

Figure 5-17

Square-well potential energy and ground-state energy E. The eigenfunction for the ground state must be symmetric with no finite nodes.

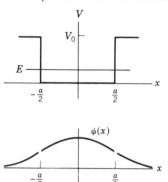

Example

Next, we examine an eigenvalue problem that we can solve in more detail. Let the potential energy be given by the square-well function in Figure 5-17, where

$$V(x) = \begin{cases} 0 & \text{for } -a/2 < x < a/2 \\ V_0 & \text{for } |x| > a/2, \end{cases}$$

and consider the determination of the energy and the eigenfunction for the ground state. We note that V is a symmetric function of x and conclude that $\psi(x)$ must be an even function with no finite nodes. (The eigenfunction has to be either even or odd. An odd function would have at least the one node at $x = 0$; however, the even function with no nodes would have the least energy.) The definite parity of $\psi(x)$ implies that the solution can be found by concentrating on the behavior of ψ for positive values of x. In the region $0 \leq x < a/2$ where $V = 0$, Equation (5-51) has the form

$$\frac{d^2\psi}{dx^2} = -k_1^2\psi \quad \text{with } k_1^2 = \frac{2mE}{\hbar^2}.$$

The general solution in this interval is

$$\psi(x) = A \cos k_1 x$$

since the other possible solution $\sin k_1 x$ is an odd function. In the region $x > a/2$ where $V = V_0$, the differential equation becomes

$$\frac{d^2\psi}{dx^2} = k_2^2\psi \quad \text{with} \quad k_2^2 = \frac{2m(V_0 - E)}{\hbar^2}.$$

The general solution in this interval is

$$\psi(x) = Be^{-k_2 x}$$

since the other possible solution $e^{k_2 x}$ diverges as $x \to \infty$. The two expressions for ψ are required to match smoothly at $x = a/2$. Continuity of ψ implies

$$A \cos \frac{k_1 a}{2} = B e^{-k_2 a/2},$$

and continuity of $d\psi/dx$ implies

$$-k_1 A \sin \frac{k_1 a}{2} = -k_2 B e^{-k_2 a/2}.$$

We divide the second of these equations by the first and make several cancellations to find

$$k_1 \tan \frac{k_1 a}{2} = k_2.$$

The defining relations for k_1 and k_2 are then used to convert this result into a formula for E in terms of the parameters of the model:

$$\tan \sqrt{\frac{ma^2}{2\hbar^2} E} = \sqrt{\frac{V_0 - E}{E}}.$$

The formula has an interesting limit as $V_0 \to \infty$. We see that the argument of

---

Figure 5-18

Determination of the ground-state energy in a square well. The left and right curves refer to the two sides of the equality

$$\tan \frac{\pi}{2} \sqrt{\frac{E}{E_1}} = \sqrt{\frac{V_0 - E}{E}},$$

and the intersection of the two curves gives the solution. The limit $V_0 \to \infty$ results in the solution $E = E_1$, corresponding to the ground state of a particle in a one-dimensional box.

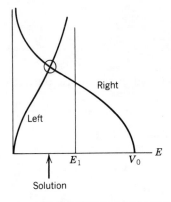

the tangent approaches $\pi/2$, and we find that the ground-state energy becomes equal to E_1, the energy obtained in Equation (5-32) for the problem of the one-dimensional box:

$$\sqrt{\frac{ma^2}{2\hbar^2}E} = \frac{\pi}{2} \quad \Rightarrow \quad E_1 = \frac{\hbar^2\pi^2}{2ma^2}.$$

Note also that $k_2 \rightarrow \infty$ as $V_0 \rightarrow \infty$, and so $\psi(x)$ no longer penetrates the classically forbidden region in this limit. The original formula for E assumes a tidier form when we introduce E_1 as a parameter:

$$\tan\frac{\pi}{2}\sqrt{\frac{E}{E_1}} = \sqrt{\frac{V_0 - E}{E}}.$$

The most useful way to solve this transcendental equation for the ground-state energy is to plot both sides of the relation versus E and look for an intersection of the two graphs. The procedure is employed in Figure 5-18 to provide a solution below the limiting energy E_1.

Example

Finally, we undertake a proof of the orthogonality relation in Equation (5-53). We let ψ_n and $\psi_{n'}$ be different eigenfunctions with different discrete energy eigenvalues E_n and $E_{n'}$. Thus, $\psi_n(x)$ satisfies the equation

$$-\frac{\hbar^2}{2m}\frac{d^2\psi_n}{dx^2} + V(x)\psi_n = E_n\psi_n,$$

while $\psi_{n'}^*(x)$ satisfies the conjugated equation

$$-\frac{\hbar^2}{2m}\frac{d^2\psi_{n'}^*}{dx^2} + V(x)\psi_{n'}^* = E_{n'}\psi_{n'}^*.$$

(We assume that $V(x)$ is real, and we use the fact that $E_{n'}$ is also real.) If we multiply the first equation by $\psi_{n'}^*$ and the second by ψ_n, and then subtract the second equation from the first, we obtain

$$-\frac{\hbar^2}{2m}\left(\psi_{n'}^*\frac{d^2\psi_n}{dx^2} - \frac{d^2\psi_{n'}^*}{dx^2}\psi_n\right) = (E_n - E_{n'})\psi_{n'}^*\psi_n.$$

The left side of this equality can be rewritten and then integrated as follows:

$$-\frac{\hbar^2}{2m}\int_{-\infty}^{\infty}\frac{d}{dx}\left(\psi_{n'}^*\frac{d\psi_n}{dx} - \frac{d\psi_{n'}^*}{dx}\psi_n\right)dx = -\frac{\hbar^2}{2m}\left(\psi_{n'}^*\frac{d\psi_n}{dx} - \frac{d\psi_{n'}^*}{dx}\psi_n\right)\Bigg|_{-\infty}^{\infty}.$$

The result of the integration is *zero*, since properly behaved discrete eigenfunctions tend to zero at $\pm\infty$. The right side of the equality must therefore vanish upon integration:

$$(E_n - E_{n'})\int_{-\infty}^{\infty}\psi_{n'}^*\psi_n\,dx = 0.$$

Our hypotheses rule out the possibility that E_n might be equal to $E_{n'}$. It then follows that the integral factor must be equal to zero as in Equation (5-53).

5-7   Expectation Values

The wave function and the probability density are fundamental to the analysis of any quantum system. We use the wave function Ψ to specify a physical state, and we then suppose that the probability interpretation of Ψ determines all observable aspects of the system in a manner consistent with the constraints of the uncertainty principle. The probability density is only the first of a list of measurable quantities to be associated with the given state. The list goes on to include position, momentum, energy, and all the other observables of a quantum particle. We may regard the values of these physical variables as information encoded in the wave function. Our next objective is to learn how this information is extracted for any specified state of the system.

Classical mechanics treats the position of the particle at time t as the primitive observable quantity. The momentum $p(t)$ is then found from the coordinate $x(t)$ by computing

$$p(t) = m \frac{dx}{dt}.$$

Quantum mechanics takes a different approach beginning with the description of the state at time t. The normalized wave function $\Psi(x, t)$ is then employed to construct the probability density

$$P(x, t) = \Psi^*(x, t)\Psi(x, t).$$

We know that the concept of localization is embodied in $P(x, t)$; hence, we expect the probabilistic definition for the position of a quantum particle to reside in the interpretation of this quantity. Let us visualize our localized particle in terms of Figure 5-3 and recall that such a distribution at time t assigns a *likelihood* $P(x, t)dx$ for finding the particle in an interval dx at location x. We cannot determine the position of the particle with certainty, and so we turn to a statistical treatment of the observed coordinate instead. A number of experiments performed on similarly prepared systems would result in a variety of values for the measured position of the particle. The average of sufficiently many such measurements would provide the logical operative specification of the *value expected* for that observable. We therefore define the position of a quantum particle by means of an *average*, which reflects the statistical information contained in the state. Every other observable is likewise defined in terms of an average, or expectation value, whose determination depends on the wave function. Uncertainties in these observables are also to be defined as averages over the particular state.

Let us digress briefly to recollect the procedure for averaging over a *discrete* sample of objects. We assume that there are N objects of several different types and that the ith type of object occurs n_i times in the sample. We wish to evaluate a quantity A whose value is equal to A_i for every object of type i. The fraction n_i/N denotes the likelihood that a given object is of this type, and so the average value of A is computed by weighting the value A_i with the ith likelihood factor and by summing over all the different types:

$$\langle A \rangle = \sum_i A_i \frac{n_i}{N} = \frac{\sum_i A_i n_i}{\sum_j n_j}. \tag{5-54}$$

Note the use of the relation $\Sigma_j\, n_j = N$ in the last step. We have calculated averages based on this familiar method in Section 2-4.

The wave function $\Psi(x, t)$ describes an analogous sample of probable positions for a particle at time t. Figure 5-3 shows an example in which the coordinate x serves as a *continuous* index and the probability $P(x, t)\, dx$ acts as a likelihood factor for each value of x. Any x-dependent quantity A has an average value defined as in Equation (5-54), except that the summation over the discrete index is replaced by an integration over the continuous variable:

$$\langle A \rangle = \int_{-\infty}^{\infty} A(x)P(x, t)\, dx. \qquad (5\text{-}55)$$

This formula may be recast to exhibit the explicit wave function in the form

$$\langle A \rangle = \int_{-\infty}^{\infty} \Psi^*(x, t)A(x)\Psi(x, t)\, dx. \qquad (5\text{-}56)$$

Note that the quantity of interest $A(x)$ enters multiplicatively and may therefore appear anywhere in the integrand. We discover the motive for inserting $A(x)$ *between* Ψ^* and Ψ as we proceed with further related developments.

The average value of the coordinate has a special significance in this formalism. We write

$$\langle x \rangle = \int_{-\infty}^{\infty} \Psi^*(x, t)x\Psi(x, t)\, dx, \qquad (5\text{-}57)$$

and refer to $\langle x \rangle$ as the expectation value of the position of the particle at time t. The terminology is linked to our earlier remarks regarding the value expected for the measured quantity. It is clear from Equation (5-57) that $\langle x \rangle$ may depend on t. Thus, the role of $x(t)$ in classical physics is transferred to the expectation value $\langle x \rangle$ in quantum mechanics. Consequently, we anticipate a definition for $\langle p \rangle$, the expectation value of the momentum at time t, which satisfies the relation

$$\langle p \rangle = m\frac{d}{dt}\langle x \rangle \qquad (5\text{-}58)$$

for a nonrelativistic particle. In fact, this formula tells us how to define $\langle p \rangle$ when we return to that question below.

A good example of an x-dependent observable is provided by the potential energy function $V(x)$. Equation (5-56) expresses the corresponding expectation value as

$$\langle V \rangle = \int_{-\infty}^{\infty} \Psi^*V(x)\Psi\, dx. \qquad (5\text{-}59)$$

Another useful expectation value of a function of x is given by the expression

$$\langle x^2 \rangle = \int_{-\infty}^{\infty} \Psi^*x^2\Psi\, dx.$$

We note that $\langle x^2 \rangle$ and $\langle x \rangle^2$ are determined by evaluating *different* integrals, and we therefore realize that there is no necessary reason to expect an equality between the two quantities. It is apparent that a special type of state Ψ must be involved if $\langle x^2 \rangle$ and $\langle x \rangle^2$ are to have equal values.

The definition of the uncertainty Δx assumes that the average value $\langle x \rangle$ has already been computed, as in Equation (5-57), for a particle in the state Ψ. We then imagine that we conduct a large number of experiments on particles in the same state and observe the departure of the position of each particle from the expected value, as measured by the behavior of the variable $x - \langle x \rangle$. The root-mean-square of this deviation is defined according to Equation (3-4) as

$$(x - \langle x \rangle)_{\text{rms}} = \sqrt{\langle (x - \langle x \rangle)^2 \rangle} ,$$

to give the desired formula for the uncertainty in the position of the particle. We can simplify the resulting expression for Δx by the following manipulations:

$$\begin{aligned} (\Delta x)^2 &= \left\langle (x - \langle x \rangle)^2 \right\rangle = \langle x^2 - 2x\langle x \rangle + \langle x \rangle^2 \rangle \\ &= \langle x^2 \rangle - 2\langle x \rangle \langle x \rangle + \langle x \rangle^2 \\ &= \langle x^2 \rangle - \langle x \rangle^2. \end{aligned} \tag{5-60}$$

We know that the integration over x, denoted by the symbol $\langle\ \rangle$, always results in a quantity independent of x. We use this observation above to factor $\langle x \rangle$ out of the integral expression $\langle x \langle x \rangle \rangle$ in the second of the three terms obtained from $\langle (x - \langle x \rangle)^2 \rangle$. We also note that $\langle x - \langle x \rangle \rangle$ is always equal to zero and therefore cannot be used to represent the uncertainty in x. The final formula in Equation (5-60) tells us that $\langle x^2 \rangle$ and $\langle x \rangle^2$ are equal only for special states in which the variable x has *no* uncertainty.

We turn next to the momentum of a quantum particle and consider our first instance of a dynamical variable that is not expressible as a function of x. Equation (5-58) provides the requisite guidance as we subject the quantity $m\,d\langle x \rangle/dt$ to the following series of maneuvers:

$$\begin{aligned} m\frac{d}{dt}\langle x \rangle &= m\frac{d}{dt}\int_{-\infty}^{\infty} \Psi^* x \Psi\, dx \\ &= m\int_{-\infty}^{\infty} \frac{\partial}{\partial t}(\Psi^* x \Psi)\, dx = m\int_{-\infty}^{\infty} \left(\frac{\partial \Psi^*}{\partial t} x \Psi + \Psi^* x \frac{\partial \Psi}{\partial t} \right) dx. \end{aligned}$$

The potential energy $V(x)$ is taken to be real valued, and the Schrödinger equations for Ψ and for Ψ^* are consulted, to obtain the relations

$$\frac{\partial \Psi}{\partial t} = \frac{i\hbar}{2m}\frac{\partial^2 \Psi}{\partial x^2} + \frac{V}{i\hbar}\Psi \quad \text{and} \quad \frac{\partial \Psi^*}{\partial t} = -\frac{i\hbar}{2m}\frac{\partial^2 \Psi^*}{\partial x^2} - \frac{V}{i\hbar}\Psi^*.$$

We insert these formulas in the expression for $m\,d\langle x \rangle/dt$ and continue to manipulate the resulting integral:

$$\begin{aligned} m\int_{-\infty}^{\infty} &\left[\left(-\frac{i\hbar}{2m}\frac{\partial^2 \Psi^*}{\partial x^2} - \frac{V}{i\hbar}\Psi^* \right) x\Psi + \Psi^* x \left(\frac{i\hbar}{2m}\frac{\partial^2 \Psi}{\partial x^2} + \frac{V}{i\hbar}\Psi \right) \right] dx \\ &= \frac{i\hbar}{2}\int_{-\infty}^{\infty} \left(\Psi^* x \frac{\partial^2 \Psi}{\partial x^2} - \frac{\partial^2 \Psi^*}{\partial x^2} x\Psi \right) dx \\ &= \frac{i\hbar}{2}\int_{-\infty}^{\infty} \left[\frac{\partial}{\partial x}\left(\Psi^* x \frac{\partial \Psi}{\partial x} - \frac{\partial \Psi^*}{\partial x} x\Psi + \Psi^* \Psi \right) - 2\Psi^* \frac{\partial \Psi}{\partial x} \right] dx \\ &= \frac{i\hbar}{2}\left(\Psi^* x \frac{\partial \Psi}{\partial x} - \frac{\partial \Psi^*}{\partial x} x\Psi + \Psi^* \Psi \right)\Bigg|_{-\infty}^{\infty} - i\hbar\int_{-\infty}^{\infty} \Psi^* \frac{\partial \Psi}{\partial x}\, dx. \end{aligned}$$

The endpoint contributions vanish because Ψ approaches zero rapidly at $\pm\infty$, and the integral term remains as the desired result for $m\,d\langle x\rangle/dt$. We return to Equation (5-58) and write this important conclusion in the form

$$\langle p\rangle = \int_{-\infty}^{\infty}\Psi^*(x,t)\frac{\hbar}{i}\frac{\partial}{\partial x}\Psi(x,t)\,dx. \tag{5-61}$$

Note that the differential operation $(\hbar/i)\partial/\partial x$ *must* appear between Ψ^* and Ψ in this formula.

An important procedural result emerges when the expression for $\langle p\rangle$ is compared with our previous formulas for $\langle x\rangle$ and $\langle A\rangle$. Whenever a wave function $\Psi(x,t)$ is used to describe the state of the system, it is correct to represent the momentum variable by means of the differential operator $(\hbar/i)\partial/\partial x$, where the operator acts on the x dependence of the wave function. We write this representation rule symbolically as

$$p \to \frac{\hbar}{i}\frac{\partial}{\partial x}, \tag{5-62}$$

and we then represent the square of the momentum by setting

$$p^2 \to \left(\frac{\hbar}{i}\frac{\partial}{\partial x}\right)^2 = -\hbar^2\frac{\partial^2}{\partial x^2} \tag{5-63}$$

as another operator acting on $\Psi(x,t)$. These rules can be invoked to compute the momentum uncertainty Δp according to the formula

$$(\Delta p)^2 = \langle p^2\rangle - \langle p\rangle^2, \tag{5-64}$$

since the development of Equation (5-60) for Δx is equally valid for Δp or any other uncertainty. With these definitions of the uncertainties in x and p it is possible to prove the uncertainty principle as stated in Equation (4-6).

The energy of a quantum particle is another primary observable to be identified by these means. Let us proceed by regarding the classical nonrelativistic energy relation as an equality of expectation values:

$$\langle E\rangle = \left\langle\frac{p^2}{2m}\right\rangle + \langle V\rangle.$$

We use Equations (5-59) and (5-63) to write the equality in terms of an integral:

$$\begin{aligned}
\langle E\rangle &= \int_{-\infty}^{\infty}\Psi^*\left[-\frac{\hbar^2}{2m}\frac{\partial^2}{\partial x^2}\Psi + V(x)\Psi\right]dx\\
&= \int_{-\infty}^{\infty}\Psi^*i\hbar\frac{\partial}{\partial t}\Psi\,dx. \tag{5-65}
\end{aligned}$$

The final formula holds because the wave function Ψ obeys the Schrödinger equation.

The evaluation of $\langle E\rangle$ employs a time derivative inside an integral, just as the evaluation of $\langle p\rangle$ employs a space derivative. The above result for $\langle E\rangle$ is the basis for a second representation rule

$$E \to i\hbar\frac{\partial}{\partial t}, \tag{5-66}$$

whereby the energy variable is represented by letting the differential operator $i\hbar\, \partial/\partial t$ act on the wave function $\Psi(x, t)$. We have entertained the possibility that p and E are expressed by means of derivatives in Section 5-1, and now we see the underlying reasons for the proposition. The operator assignments for p and E are general and apply beyond the context of our one-dimensional treatment. We should be prepared to use these representation rules, and others yet to come, whenever we wish to pass from classical physics to Schrödinger's quantum theory.

It is obvious from Equation (5-57) that $\langle x \rangle$ is a real-valued quantity, as must be the case since the value of $\langle x \rangle$ is measurable. The same must be true of $\langle p \rangle$ and $\langle E \rangle$, despite appearances in Equations (5-61) and (5-65). The real-valued property of these quantities is left to be examined in Problem 26 at the end of the chapter.

We can give a further interpretation to the stationary states of a system if we look at the expectation value of the energy in such a state. We let the system be described by a wave function with energy eigenvalue E_n,

$$\Psi_n(x, t) = \psi_n(x)e^{-iE_n t/\hbar},$$

and obtain

$$i\hbar\frac{\partial}{\partial t}\Psi_n = \psi_n(x)E_n e^{-iE_n t/\hbar} = E_n\Psi_n \tag{5-67}$$

under application of the energy operator. The expectation value of the energy then becomes

$$\langle E \rangle = \int_{-\infty}^{\infty} \Psi_n^* i\hbar\frac{\partial}{\partial t}\Psi_n\, dx = E_n\int_{-\infty}^{\infty}\Psi_n^*\Psi_n\, dx = E_n$$

since Ψ_n is a normalized wave function. In similar fashion we also find

$$\left(i\hbar\frac{\partial}{\partial t}\right)^2\Psi_n = E_n^2\Psi_n$$

and

$$\langle E^2 \rangle = \int_{-\infty}^{\infty}\Psi_n^*\left(i\hbar\frac{\partial}{\partial t}\right)^2\Psi_n\, dx = E_n^2.$$

The fact that $\langle E^2 \rangle$ and $\langle E \rangle^2$ have the same value tells us that the energy of any stationary state is equal to the energy eigenvalue for that state with *zero uncertainty*.

We recall from Equation (5-27) that the probability density is independent of time in a stationary state:

$$\left|\Psi_n(x, t)\right|^2 = \left|\psi_n(x)\right|^2.$$

Time independence implies a sort of nonlocalization of the particle with respect to the time, since the probability of finding the particle has no time variation. We know that localization in time and uncertainty in energy obey an uncertainty relation, and we associate this observation directly with the vanishing of the energy uncertainty ΔE in the given state. Equation (5-67) formalizes the observation by stating that the energy operator $i\hbar\, \partial/\partial t$ acts on a stationary state to produce the *same* state multiplied by its energy eigenvalue. This kind of eigenvalue equation does not hold necessarily for a more general type of wave function. We stress the importance of the special property conveyed by Equation (5-67) by letting the stationary states be known as eigenfunctions of the energy operator, or *energy eigenfunctions*.

Example

We learn more about the special nature of the stationary states when we consider a superposition of two such wave functions:

$$\Psi = c\Psi_n + c'\Psi_{n'}, \quad \text{where} \quad E_n \neq E_{n'}.$$

Application of the energy operator yields the result

$$i\hbar \frac{\partial}{\partial t}\Psi = cE_n\Psi_n + c'E_{n'}\Psi_{n'}$$

by virtue of the eigenvalue property in Equation (5-67). The expression on the right does not contain the state Ψ as a specific factor; hence, the given wave function does not satisfy the eigenvalue equation, even though Ψ is certainly a solution of the Schrödinger equation. To interpret the state we begin with the normalization condition:

$$1 = \int_{-\infty}^{\infty} \Psi^*\Psi\, dx = \int_{-\infty}^{\infty} \left(c^*\Psi_n^* + c'^*\Psi_{n'}^*\right)\left(c\Psi_n + c'\Psi_{n'}\right) dx = |c|^2 + |c'|^2.$$

We recall that Ψ_n and $\Psi_{n'}$ are normalized and orthogonal, as in Equations (5-14) and (5-53), and we use these properties to obtain this simple conclusion. Note that no cross-term contributions arise from the two stationary states. The expectation value of the energy also follows at once with the aid of the same properties:

$$\langle E \rangle = \int_{-\infty}^{\infty} \Psi^* i\hbar \frac{\partial}{\partial t}\Psi\, dx$$

$$= \int_{-\infty}^{\infty} \left(c^*\Psi_n^* + c'^*\Psi_{n'}^*\right)\left(cE_n\Psi_n + c'E_{n'}\Psi_{n'}\right) dx = |c|^2 E_n + |c'|^2 E_{n'}.$$

These two results imply that the energy $\langle E \rangle$ lies *between* the eigenvalues E_n and $E_{n'}$. We interpret the quantities $|c|^2$ and $|c'|^2$ as probabilities of finding the system in the stationary states with energies E_n and $E_{n'}$. It is obvious that the state Ψ is not stationary and that the corresponding energy uncertainty is not zero.

Example

There is no simpler illustration of the ideas in this section than the problem of a particle in a box. Let us take the particle to be in its ground state so that, for x in the interval $[-a/2, a/2]$, the wave function is

$$\Psi = \sqrt{\frac{2}{a}} \cos\frac{\pi x}{a} e^{-iE_1 t/\hbar} \quad \text{with } E_1 = \frac{\hbar^2 \pi^2}{2ma^2}.$$

The expectation value of x is given by the integral

$$\langle x \rangle = \int_{-\infty}^{\infty} \Psi^* x \Psi \, dx = \int_{-a/2}^{a/2} \frac{2}{a} x \cos^2 \frac{\pi x}{a} \, dx.$$

This integral vanishes on inspection, because the integrand is odd and the range of integration is symmetric. The expectation value of p has the form

$$\langle p \rangle = \int_{-\infty}^{\infty} \Psi^* \frac{\hbar}{i} \frac{\partial}{\partial x} \Psi \, dx = \int_{-a/2}^{a/2} \frac{2}{a} \frac{\hbar}{i} \left(-\frac{\pi}{a} \right) \cos \frac{\pi x}{a} \sin \frac{\pi x}{a} \, dx$$

and vanishes for the same reason. The evaluation of $\langle x^2 \rangle$ produces a nonvanishing result, with the aid of a table of integrals:

$$\langle x^2 \rangle = \int_{-\infty}^{\infty} \Psi^* x^2 \Psi \, dx = \int_{-a/2}^{a/2} \frac{2}{a} x^2 \cos^2 \frac{\pi x}{a} \, dx = \frac{2}{a} \frac{a^3}{\pi^3} \left(\frac{\pi^3}{24} - \frac{\pi}{4} \right)$$

$$= \frac{a^2}{12\pi^2} (\pi^2 - 6).$$

The calculation of $\langle p^2 \rangle$ is easier to perform because the Schrödinger equation can be used:

$$\langle p^2 \rangle = \int_{-\infty}^{\infty} \Psi^* \left(-\hbar^2 \frac{\partial^2}{\partial x^2} \right) \Psi \, dx$$

$$= \int_{-a/2}^{a/2} \Psi^* \left(2mi\hbar \frac{\partial}{\partial t} \right) \Psi \, dx = 2mE_1 \int_{-a/2}^{a/2} \Psi^* \Psi \, dx = 2mE_1.$$

To obtain the uncertainties in x and p we compute

$$(\Delta x)^2 = \langle x^2 \rangle - \langle x \rangle^2 = \frac{a^2}{\pi^2} \frac{\pi^2 - 6}{12}$$

and

$$(\Delta p)^2 = \langle p^2 \rangle - \langle p \rangle^2 = 2mE_1 = \frac{\hbar^2 \pi^2}{a^2}.$$

The product of the uncertainties is then given by

$$\Delta x \, \Delta p = \sqrt{\frac{\pi^2 - 6}{12}} \, \hbar,$$

a result slightly larger than Heisenberg's lower bound $\hbar/2$.

Example

We now combine the main features of the previous two examples and assume that our particle in a box is described by the wave function

$$\Psi(x, t) = \frac{1}{\sqrt{a}} \left(\cos \frac{\pi x}{a} e^{-iE_1 t/\hbar} + \sin \frac{2\pi x}{a} e^{-4iE_1 t/\hbar} \right).$$

We recognize the state to be a superposition of the two lowest-energy eigenfunctions,

$$\Psi = \frac{1}{\sqrt{2}}(\Psi_1 + \Psi_2),$$

in a normalized equal-parts admixture. The expectation value of x is given by a rather lengthy expression:

$$\langle x \rangle = \int_{-\infty}^{\infty} \Psi^* x \Psi \, dx$$

$$= \int_{-a/2}^{a/2} \frac{x}{a} \left[\cos^2 \frac{\pi x}{a} + \cos \frac{\pi x}{a} \sin \frac{2\pi x}{a} \left(e^{3iE_1 t/\hbar} + e^{-3iE_1 t/\hbar} \right) + \sin^2 \frac{2\pi x}{a} \right] dx.$$

The first and last terms in the integrand are odd and do not survive the integration. Only the middle contribution remains to be computed with the help of a table of integrals:

$$\langle x \rangle = \frac{1}{a} \left(e^{3iE_1 t/\hbar} + e^{-3iE_1 t/\hbar} \right) \int_{-a/2}^{a/2} x \cos \frac{\pi x}{a} \sin \frac{2\pi x}{a} \, dx$$

$$= \frac{2}{a} \cos \frac{3E_1 t}{\hbar} \cdot \frac{8a^2}{9\pi^2} = \frac{16a}{9\pi^2} \cos \frac{3E_1 t}{\hbar}.$$

This result describes an interesting time dependence for the average position of the particle whereby $\langle x \rangle$ oscillates with amplitude $16a/9\pi^2$ and angular frequency $3E_1/\hbar$. The behavior of $\langle x \rangle$ is sketched in Figure 5-19, along with the shape of the initial wave function $\Psi(x, 0)$. The expectation value of p calls for a

Figure 5-19

Oscillation of the expectation value $\langle x \rangle$ for a superposition of particle-in-a-box states. The initial wave function is

$$\Psi(x, 0) = \frac{1}{\sqrt{a}} \left(\cos \frac{\pi x}{a} + \sin \frac{2\pi x}{a} \right),$$

and the angular frequency of oscillation of $\langle x \rangle$ is $3E_1/\hbar$, where E_1 is the ground-state energy.

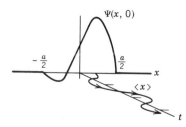

more laborious calculation:

$$\langle p \rangle = \int_{-\infty}^{\infty} \Psi^* \frac{\hbar}{i} \frac{\partial}{\partial x} \Psi \, dx$$

$$= \int_{-a/2}^{a/2} \frac{\hbar}{ia} \left(\cos \frac{\pi x}{a} e^{iE_1 t/\hbar} + \sin \frac{2\pi x}{a} e^{4iE_1 t/\hbar} \right)$$

$$\times \left(-\frac{\pi}{a} \sin \frac{\pi x}{a} e^{-iE_1 t/\hbar} + \frac{2\pi}{a} \cos \frac{2\pi x}{a} e^{-4iE_1 t/\hbar} \right) dx.$$

Nonvanishing contributions arise solely from the even terms in the integrand:

$$\langle p \rangle = -\frac{\hbar \pi}{ia^2} \left(e^{3iE_1 t/\hbar} \int_{-a/2}^{a/2} \sin \frac{2\pi x}{a} \sin \frac{\pi x}{a} \, dx \right.$$

$$\left. - 2e^{-3iE_1 t/\hbar} \int_{-a/2}^{a/2} \cos \frac{\pi x}{a} \cos \frac{2\pi x}{a} \, dx \right)$$

$$= -\frac{\hbar \pi}{ia^2} \left(e^{3iE_1 t/\hbar} \cdot \frac{4a}{3\pi} - 2e^{-3iE_1 t/\hbar} \cdot \frac{2a}{3\pi} \right)$$

$$= -\frac{\hbar \pi}{ia^2} \frac{4a}{3\pi} 2i \sin \frac{3E_1 t}{\hbar} = -\frac{8\hbar}{3a} \sin \frac{3E_1 t}{\hbar}.$$

Let us finally make a test of the correctness of the t dependence in $\langle x \rangle$ and $\langle p \rangle$:

$$m \frac{d}{dt} \langle x \rangle = m \frac{16a}{9\pi^2} \left(-\frac{3E_1}{\hbar} \right) \sin \frac{3E_1 t}{\hbar}$$

$$= -\frac{16ma}{3\pi^2 \hbar} \frac{\hbar^2 \pi^2}{2ma^2} \sin \frac{3E_1 t}{\hbar} = -\frac{8\hbar}{3a} \sin \frac{3E_1 t}{\hbar}.$$

The result is equal to $\langle p \rangle$ as required by Equation (5-58).

5-8 Radiative Transitions

The framework of quantum physics supports the theory of quantized energy levels as originally proposed by Bohr. Features of Bohr's picture can already be seen in the assembled formalism, even though the presentation has been limited to one dimension. It is actually quite instructive to look at the picture in this context because the one-dimensional models are useful sources of insight for the more realistic situations that lie ahead.

Bohr's quantum theory postulates the existence of stationary states and the occurrence of radiative transitions. The quantization of states has found its place in the Schrödinger theory, but the mechanism for transitions has been left unmentioned. Thus, the states of the quantum particle are presented in a kind of isolation, as no provision has yet been made for the particle to change its energy, and so no means has been provided for the states to reveal their existence.

We are concerned with radiative transitions, brought on by the presence of electromagnetic fields in interaction with the quantum particle. Fortunately, we are able to adopt an approximation procedure for the treatment of this complex problem. We continue to use the familiar quantized states that arise from the internal dynamics of the quantum particle system. We then let the resulting levels experience the small effect of a *perturbing* interaction between the system and the electromagnetic fields. The most natural way to admit the added influence of electromagnetism is to assign a charge to the quantum particle and let the accelerated motion of the particle produce classical radiation. This hybrid semiclassical approximation takes the energy levels from the Schrödinger theory and includes the excitation and deexcitation of the levels by appealing to classical radiation theory.

The procedure immediately accommodates Bohr's hypotheses. In fact, the desired stationary states and radiative transitions are easily recognized, just by inspecting the time dependence of the probability density $|\Psi(x, t)|^2$. We let e denote the assumed charge of the particle so that we can use the expression for the expectation value of the coordinate x to introduce the *electric dipole moment*

$$e\langle x \rangle = e \int_{-\infty}^{\infty} x |\Psi(x, t)|^2 \, dx. \tag{5-68}$$

We are especially interested in circumstances where this quantity *oscillates* with time, because we know that an oscillating dipole emits classical electromagnetic radiation.

No time dependence can appear in the dipole moment, and so no radiation is observed, if Ψ represents a stationary state. We are led directly to this conclusion when we combine Equations (5-27) and (5-68) to find

$$e\langle x \rangle = e \int_{-\infty}^{\infty} x |\psi(x)|^2 \, dx.$$

Thus, Bohr's hypothesis of nonradiating energy levels is automatically realized in the behavior of the stationary states.

A radiating system corresponds to a situation where Ψ is given as a *superposition* of stationary states. Any two energy eigenfunctions may be added together to illustrate this property, as long as the two energy eigenvalues are not equal. Let us choose quantum numbers n and n' to construct the combination

$$\Psi(x, t) = c\psi_n(x)e^{-iE_n t/\hbar} + c'\psi_{n'}(x)e^{-iE_{n'}t/\hbar},$$

and let us assume for definiteness that E_n exceeds $E_{n'}$. We call this sort of superposition a *transition state* because the wave function describes simultaneous probabilities for finding the system in states of different energy. The probability density in Equation (5-68) consists of time-independent and time-dependent pieces, and the electric dipole moment takes the form

$$e\langle x \rangle = e \int_{-\infty}^{\infty} x \Big\{ |c\psi_n|^2 + c^*c'\psi_n^*\psi_{n'}e^{i(E_n - E_{n'})t/\hbar}$$

$$+ c'^*c\psi_{n'}^*\psi_n e^{-i(E_n - E_{n'})t/\hbar} + |c'\psi_{n'}|^2 \Big\} \, dx. \tag{5-69}$$

The integrals containing the two stationary terms are equal to zero for eigenfunctions of definite parity and are not associated with radiation in any case. The other two contributions have an oscillating time dependence, as desired, with frequency

$$\nu = \frac{E_n - E_{n'}}{h}.$$

This one-dimensional oscillation of a charge-bearing particle is just like the behavior of a radiating linear antenna, where charge is driven at a certain frequency and is able to emit electromagnetic radiation of the same frequency. The associated quantum system has a determinable probability of transition from a higher to a lower energy level, resulting in the emission of a discrete spectral line. Thus, the transition states automatically offer the means by which Bohr's hypothesized radiative transitions are to be understood.

The wave function is supposed to contain the answer to every possible question pertaining to a given quantum system. Transition states are equipped especially with the information needed to predict intensities for the observed spectral lines. Information of this sort appears specifically in the complex-valued coefficients, or amplitudes, which accompany the oscillating time-dependent terms in Equation (5-69). We know from classical physics that the radiation intensity is proportional to the square of the amplitude of oscillation of the electric dipole moment, because the electric and magnetic fields are directly proportional to that quantity. The complex amplitudes in Equation (5-69) are called quantum mechanical transition amplitudes since the corresponding squared moduli determine the probabilities for the transitions $n \rightarrow n'$ and $n' \rightarrow n$. We introduce the *dipole transition amplitude* for the transition $n \rightarrow n'$ by choosing the third term in the equation and selecting the particular integral factor

$$x_{n'n} = \int_{-\infty}^{\infty} \psi_{n'}^{*}(x) x \psi_n(x)\, dx. \tag{5-70}$$

This integral specifies uniquely the mutual contribution of the two states involved in the transition. Indeed, the strength of the transition is largely determined by the magnitude of $x_{n'n}$, as the predicted intensity of the spectral line depends on the square of the modulus of this amplitude. Note that Equation (5-69) also contains a similar quantity, with n and n' interchanged, pertaining to the other transition $n' \rightarrow n$. This second integral is not independent of the first, since the two amplitudes are clearly connected by the relation $x_{nn'}^{*} = x_{n'n}$.

It may happen that the integral in Equation (5-70) vanishes for some pair of quantum numbers n and n'. The transition $n \rightarrow n'$ is then not allowed to occur with the emission of radiation characteristic of an oscillating electric dipole. We say that the electric dipole transition is forbidden in this circumstance. The conditions of a given problem produce a rule, called an *electric dipole selection rule*, to tell how the quantum numbers must be related in order that the transition from n to n' is allowed. An illustration of such a selection rule in one-dimensional quantum mechanics is given as follows.

Example

Let the quantum system be the harmonic oscillator of Section 5-5, and consider the possibility of electric dipole transitions between the energy levels. We wish to show that the amplitude $x_{n'n}$ is equal to zero unless the quantum numbers n and n' are linked in a very selective way. Recall that the eigenfunctions $\psi_n(x)$ for this model are written in terms of Hermite polynomials as

$$H_n(\xi)e^{-\xi^2/2}, \quad \text{where} \quad \xi = \left(\frac{mk}{\hbar^2}\right)^{1/4} x.$$

Hence, the amplitude defined by Equation (5-70) is proportional to the integral

$$\int_{-\infty}^{\infty} H_{n'}(\xi)\xi H_n(\xi)e^{-\xi^2}\,d\xi.$$

We consult Problem 16 at the end of the chapter to find the recursion relation

$$H_{n+1} - 2\xi H_n + 2nH_{n-1} = 0,$$

and we substitute for ξH_n in the integral to obtain

$$\tfrac{1}{2}\int_{-\infty}^{\infty} H_{n'}(\xi)\big[H_{n+1}(\xi) + 2nH_{n-1}(\xi)\big]e^{-\xi^2}\,d\xi.$$

The resulting two integrals can be rewritten as

$$\int_{-\infty}^{\infty} \psi_{n'}(x)\psi_{n+1}(x)\,dx + 2n\int_{-\infty}^{\infty} \psi_{n'}(x)\psi_{n-1}(x)\,dx,$$

apart from an overall multiplicative constant. Our last step is to invoke Equation (5-53), the general orthogonality property of eigenfunctions. (Note that complex conjugation is immaterial here since these eigenfunctions are real.) We conclude that the integrals must vanish whenever the labels on the ψ's do not match. This conclusion applies to the first term when $n' \neq n + 1$, and it applies to the second term when $n' \neq n - 1$. Thus, we find that dipole transitions are allowed between states n and n' if and only if the quantum numbers are related by the selection rule

$$n' = n \pm 1.$$

It follows that the excitation and deexcitation of oscillator states by this mechanism can only take place between *adjacent* levels. Since the energy levels of the oscillator are equally spaced, the energy of the photon absorbed or emitted must be equal to the spacing $\hbar\omega_0$ in *all* electric dipole transitions. We recognize the substance of Planck's quantum hypothesis in this result.

5-9 Barrier Penetration

We have had little opportunity to discuss *continuum* states since almost all our attention has been devoted to systems with quantized energy. Recall that stationary states may exist in the continuum, where the energy eigenvalues vary continuously in excess of some threshold energy. Figure 5-20 indicates how such a state might occur and shows the general behavior of a typical eigenfunction. We observe that any energy in the continuum range corresponds to classical motion that is *not* bounded by a pair of turning points. Consequently, the associated eigenfunction is able to oscillate indefinitely at large distance, like the example shown in the figure. We wish to study this type of wave function in terms of an elementary model whose analytical solution is immediately recognizable. The quantum phenomenon of barrier penetration can then be treated in detail via this simplified approach.

Let us first review the properties of the continuum states of a free particle. This simplest of all quantum systems has been employed in the beginning of the chapter,

Figure 5-20

Eigenfunction for an energy level in the continuum. There is only one turning point on the left for any energy E above the threshold. The eigenfunction oscillates in the region of classical motion, which extends to infinity on the right.

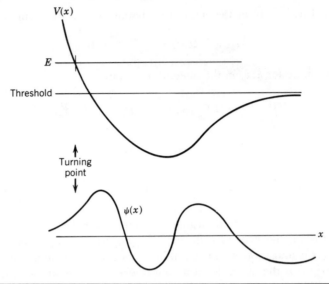

where the monochromatic traveling waves

$$e^{i(kx-\omega t)} \quad \text{and} \quad e^{i(-kx-\omega t)}$$

are introduced as elementary wave functions. We can obtain standing waves instead by constructing the combinations

$$e^{i(kx-\omega t)} + e^{i(-kx-\omega t)} = 2\cos kx\, e^{-i\omega t}$$

and

$$e^{i(kx-\omega t)} - e^{i(-kx-\omega t)} = 2i\sin kx\, e^{-i\omega t}.$$

All these functions are stationary-state solutions of the free-particle Schrödinger equation, with the *same* unrestricted energy eigenvalue

$$E = \hbar\omega = \frac{\hbar^2 k^2}{2m}.$$

We note that the two traveling forms are distinguished by their respective behavior as *eigenfunctions of the momentum operator*:

$$\frac{\hbar}{i}\frac{\partial}{\partial x}e^{i(\pm kx-\omega t)} = \pm\hbar k\, e^{i(\pm kx-\omega t)}.$$

The momentum eigenvalues $\pm\hbar k$ refer to an unlocalized particle moving in the positive and negative x directions. The standing waves do not have this unique-momentum property because the expressions are formed by superposing the two

Figure 5-21

Reflection and transmission of a classical particle by a rectangular potential energy barrier.

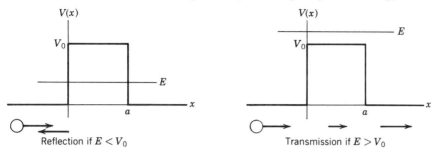

momentum eigenfunctions. We note instead that the two standing waves have their own distinctive attributes of *definite positive and negative parity*. We are going to use the relations among these free-particle eigenfunctions as we proceed with our main topic.

Barrier penetration is a process that enables a quantum particle to leak through a classically forbidden region where $V(x)$ is greater than E. We simplify the mathematics of this problem by assuming that $V(x)$ consists of piecewise-constant segments in the shape of a rectangular obstacle to the motion of the particle. Figure 5-21 illustrates the encounter of a classical particle with a potential energy barrier of height V_0. Both classical and quantum versions of the problem require that we specify whether or not the energy of the particle exceeds the height of the barrier. The figure shows that the classical particle experiences either reflection or transmission in the two instances. These cleanly separated alternatives do not apply to the quantum particle, as some transmission occurs for $E < V_0$ and some reflection occurs for $E > V_0$. We are especially concerned with the first situation since the case $E < V_0$ pertains to barrier penetration.

As preparation, let us solve for the wave function in the presence of a *single-step* potential energy. We employ the function

$$V(x) = \begin{cases} 0 & x < 0 \\ V_0 & x > 0 \end{cases}$$

and consider a stationary state with $E < V_0$, as indicated in Figure 5-22. It is clear that the point $x = 0$ divides the axis into two regions such that classical motion is

Figure 5-22

Eigenfunction describing reflection by a single-step potential energy.

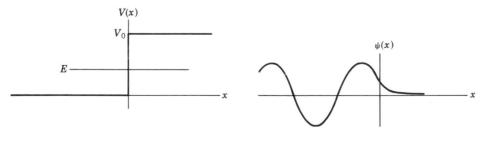

allowed on the left and forbidden on the right. We wish to find the eigenfunction solutions in each region and then match the functions smoothly at the turning point. For $x < 0$, the eigenfunction satisfies

$$\frac{d^2\psi}{dx^2} = -k_1^2\psi \quad \text{with} \quad k_1^2 = \frac{2m}{\hbar^2}E. \tag{5-71}$$

Both $\cos k_1 x$ and $\sin k_1 x$ are needed as solutions, so that the general form of the eigenfunction may be written as

$$\psi = A\cos k_1 x + B\sin k_1 x \quad \text{in the region } x < 0.$$

(The complex exponential functions $e^{ik_1 x}$ and $e^{-ik_1 x}$ are alternative solutions to be employed below.) For $x > 0$, the differential equation becomes

$$\frac{d^2\psi}{dx^2} = k_2^2\psi \quad \text{with} \quad k_2^2 = \frac{2m}{\hbar^2}(V_0 - E). \tag{5-72}$$

Both $e^{k_2 x}$ and $e^{-k_2 x}$ are possible solutions, but only the latter choice remains finite as $x \to \infty$. We therefore write the eigenfunction as

$$\psi = Ce^{-k_2 x} \quad \text{in the region } x > 0.$$

Continuity requirements are imposed on ψ and $d\psi/dx$ at $x = 0$ to yield the following two conditions on the three unknown coefficients:

$$\text{continuity of } \psi(0) \quad \Rightarrow \quad A = C$$
$$\text{continuity of } \psi'(0) \quad \Rightarrow \quad k_1 B = -k_2 C.$$

We choose to write A and B in terms of C, and leave C undetermined in the final expression for the eigenfunction:

$$\psi(x) = \begin{cases} C\left(\cos k_1 x - \dfrac{k_2}{k_1}\sin k_1 x\right) & x < 0 \\ Ce^{-k_2 x} & x > 0. \end{cases}$$

A sketch of this result in Figure 5-22 shows a standing-wave configuration, with nodes at fixed locations along the negative x axis.

It is instructive to recast our solution in terms of the complex exponential representation. The eigenfunction for negative x takes the form

$$\psi = C\left[\frac{1}{2}\left(e^{ik_1 x} + e^{-ik_1 x}\right) - \frac{k_2}{2ik_1}\left(e^{ik_1 x} - e^{-ik_1 x}\right)\right]$$
$$= \frac{C}{2}\left[\left(1 + i\frac{k_2}{k_1}\right)e^{ik_1 x} + \left(1 - i\frac{k_2}{k_1}\right)e^{-ik_1 x}\right],$$

and so the stationary-state wave function becomes

$$\Psi = \frac{C}{2}\left(1 + i\frac{k_2}{k_1}\right)e^{i(k_1 x - \omega t)} + \frac{C}{2}\left(1 - i\frac{k_2}{k_1}\right)e^{i(-k_1 x - \omega t)} \quad \text{in the region } x < 0.$$

This expression is a superposition of right- and left-propagating momentum eigen-functions with momentum eigenvalues $\pm \hbar k_1$. The complex coefficients have equal moduli because of the equality

$$\left| 1 + i \frac{k_2}{k_1} \right|^2 = \left| 1 - i \frac{k_2}{k_1} \right|^2 .$$

Consequently, the incident and reflected traveling waves are able to combine and produce the standing wave shown in the figure. The wave function becomes

$$\Psi = C e^{-k_2 x} e^{-i\omega t} \quad \text{in the region } x > 0.$$

This expression exhibits penetration and attenuation into the forbidden region in the manner discussed in Section 5-6.

Our stationary-state wave function describes the reflection of a quantum particle with *definite* wave number k_1. The momentum uncertainty is zero and so the description is not that of a localized particle. To achieve localization we should make a wave packet by superposing a continuous range of k_1 values as in the construction of Equation (5-3).

The comparison between real- and complex-valued solutions completes our pre-paration for the problem of barrier penetration. We parametrize the barrier prob-lem according to the model on the left in Figure 5-21. The barrier height V_0 exceeds E and is defined by the potential energy

$$V(x) = \begin{cases} 0 & x < 0 \\ V_0 & 0 < x < a \\ 0 & x > a. \end{cases}$$

The solution is more involved in this case because there are three regions and hence three pieces to the eigenfunction. For $x < 0$ and for $x > a$, the differential equation for $\psi(x)$ is

$$\frac{d^2\psi}{dx^2} = -k_1^2 \psi$$

as in Equation (5-71). The complex exponential solutions $e^{ik_1 x}$ and $e^{-ik_1 x}$ are chosen this time, for reasons that become clear below. If x is in the interval $(0, a)$, the eigenfunction satisfies

$$\frac{d^2\psi}{dx^2} = k_2^2 \psi$$

as in Equation (5-72), and the real exponential solutions $e^{k_2 x}$ and $e^{-k_2 x}$ are again obtained. The variable is bounded in this case, so that neither of the exponentials can be discarded because of unacceptable behavior at large distance. It would therefore appear that our eigenfunction should contain all possible terms in the general form

$$\psi(x) = \begin{cases} A e^{ik_1 x} + B e^{-ik_1 x} & x < 0 \\ C e^{-k_2 x} + D e^{k_2 x} & 0 < x < a \\ \hat{A} e^{ik_1 x} + \hat{B} e^{-ik_1 x} & x > a. \end{cases} \qquad (5\text{-}73)$$

In fact, we can remove a term from one of the pieces of ψ if we appeal as follows to the physical conditions of the problem.

Let us refer again to the classical picture, and let us choose to consider a quantum particle incident on the barrier from the *left* as in Figure 5-21. This choice eliminates the term $\hat{B}e^{-ik_1x}$ from the eigenfunction in the region $x > a$ because the corresponding wave function $\hat{B}e^{i(-k_1x-\omega t)}$ describes incidence from the right. We build our choice of incident direction into Equations (5-73) by setting $\hat{B} = 0$. The five remaining unknown coefficients satisfy the following four conditions of continuity on the functions ψ and $d\psi/dx$ at the points $x = 0$ and $x = a$:

$$A + B = C + D \quad \text{from } \psi(0),$$

$$ik_1(A - B) = -k_2(C - D) \quad \text{from } \psi'(0),$$

$$Ce^{-k_2a} + De^{k_2a} = \hat{A}e^{ik_1a} \quad \text{from } \psi(a),$$

$$-k_2(Ce^{-k_2a} - De^{k_2a}) = ik_1\hat{A}e^{ik_1a} \quad \text{from } \psi'(a). \tag{5-74}$$

A solution can be extracted from these equations by finding four of the unknowns in terms of a fifth arbitrarily chosen coefficient. The algebra is rather lengthy and is relegated to the second example at the end of the section. We quote the main results as ratios of squared moduli for two of the coefficients:

$$\left|\frac{A}{\hat{A}}\right|^2 = 1 + \frac{1}{4}\left(\frac{k_2}{k_1} + \frac{k_1}{k_2}\right)^2 \sinh^2 k_2 a \tag{5-75}$$

and

$$\left|\frac{B}{\hat{A}}\right|^2 = \frac{1}{4}\left(\frac{k_2}{k_1} + \frac{k_1}{k_2}\right)^2 \sinh^2 k_2 a. \tag{5-76}$$

The unknown $\hat{A}$ is left undetermined in these formulas. (We choose $\hat{A}$ instead of A, a more natural physical choice, in order to simplify the algebra.)

Our findings are interpreted by examining the nature of the wave function on either side of the barrier:

$$\Psi(x, t) = \begin{cases} Ae^{i(k_1x-\omega t)} + Be^{i(-k_1x-\omega t)} & x < 0 \\ \hat{A}e^{i(k_1x-\omega t)} & x > a. \end{cases} \tag{5-77}$$

We observe that Ψ consists of *incident*, *reflected*, and *transmitted* waves by noting the directions of propagation in the two regions:

$$\Psi = \begin{cases} \Psi_{\text{inc}} + \Psi_{\text{refl}} & x < 0 \\ \Psi_{\text{trans}} & x > a. \end{cases}$$

Equations (5-75) and (5-76) imply that $|A|^2$ exceeds $|B|^2$; hence, the incident and reflected waves cannot combine to produce nodes as in the case of the single-step standing wave in Figure 5-22. It is a straightforward matter to calculate the probability density and prove that $\Psi^*\Psi$ has no zeros in any of the three regions. The qualitative behavior of the probability density is sketched in Figure 5-23.

We note that $|\Psi|^2 = |\hat{A}|^2$ in the region $x > a$. The fact that there is a nonzero probability of finding the particle in the region to the right tells us that there is a

Figure 5-23

Probability density in a model of barrier penetration.

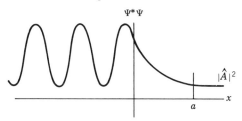

nonzero probability for the particle to *tunnel* through the barrier. This remarkable quantum result is attributable to the wave nature of the particle. Such wave behavior is readily demonstrated in the case of light, using optical devices like the ones illustrated in Figure 5-24. We should expect the effect of quantum tunneling to be undetectable for macroscopic particles. On the other hand, we find numerous manifestations of the effect in such atomic and subatomic phenomena as the behavior of electrons in metals and the emission of α particles from nuclei. We reserve these examples of tunneling for discussion in their proper context later on.

It should be obvious from Figure 5-23 that our wave function is not normalizable. This should come as no surprise, as we have already acknowledged the need for a wave packet to describe a localized particle. The wave packet would have the form of an integral over the wave number k_1, in which the coefficients A to D would be expressed in terms of $\hat{A}$ for each value of k_1, as in Equations (5-75) and (5-76). Such a construction would involve an excessive amount of mathematics. Let us turn instead to another method of extracting measurable information from the unnormalizable state. The technique makes use of the probability current density

$$j(x, t) = \frac{\hbar}{2im}\left(\Psi^* \frac{\partial \Psi}{\partial x} - \frac{\partial \Psi^*}{\partial x} \Psi \right).$$

Recall that we have introduced this quantity in Equation (5-16) to define a *flow of probability* in the x direction. A current of probability provides a natural way to compare the incident, reflected, and transmitted components of the wave function for barrier penetration.

Figure 5-24

Transmission of light through an optical barrier. Total internal reflection takes place at a prism-to-air interface if the angle of incidence exceeds the critical angle. Partial transmission through a two-prism combination occurs when the intervening air gap is sufficiently small.

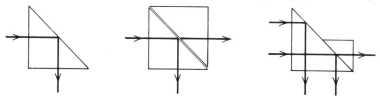

The piecewise form of the wave function $\Psi(x, t)$ generates a similar structure in the current density $j(x, t)$. The pieces of interest pertain to the two regions on either side of the barrier, where Ψ has terms as given in Equations (5-77). The calculation of j appears to have several contributions in the region $x < 0$:

$$j = \frac{\hbar}{2im}\left\{\left(A^*e^{-ik_1x} + B^*e^{ik_1x}\right)(ik_1)\left(Ae^{ik_1x} - Be^{-ik_1x}\right)\right.$$
$$\left. -(-ik_1)\left(A^*e^{-ik_1x} - B^*e^{ik_1x}\right)\left(Ae^{ik_1x} + Be^{-ik_1x}\right)\right\}.$$

All the cross-terms cancel, however, so that only the incident and reflected pieces remain:

$$j = \frac{\hbar}{2im}ik_1\left(2|A|^2 - 2|B|^2\right)$$
$$= \frac{\hbar k_1}{m}|A|^2 - \frac{\hbar k_1}{m}|B|^2 = j_{\text{inc}} + j_{\text{refl}}. \tag{5-78}$$

Note that a one-to-one correspondence holds between j_{inc} and Ψ_{inc} and between j_{refl} and Ψ_{refl}, because of the disappearance of the cross-terms. The calculation of j yields a transmitted contribution alone in the region $x > a$:

$$j = \frac{\hbar k_1}{m}|\hat{A}|^2 = j_{\text{trans}}. \tag{5-79}$$

Each of the current densities j_{inc}, j_{refl}, and j_{trans} represents a flux of probability in the direction of the associated traveling wave. We recognize at once that the factor $\hbar k_1/m$ denotes the *speed* of the particle. Let us take special note of the fact that this speed is common to the results in Equations (5-78) and (5-79) because the height of the barrier is the *same* from either side. It should be apparent that a different step height on the right side of the barrier in Figure 5-21 would imply a different speed for $x > a$ and a correspondingly different factor in the transmitted flux.

The main conclusions of the barrier problem are found in the form of ratios representing the transmission and reflection of probability relative to the incident flux. The relevant defined quantities are the *transmission* and *reflection* coefficients

$$T = \left|\frac{j_{\text{trans}}}{j_{\text{inc}}}\right| \quad \text{and} \quad R = \left|\frac{j_{\text{refl}}}{j_{\text{inc}}}\right|. \tag{5-80}$$

We use Equations (5-75), (5-76), (5-78), and (5-79) to obtain

$$T = \left|\frac{\hat{A}}{A}\right|^2 = \frac{1}{1 + \dfrac{1}{4}\left(\dfrac{k_2}{k_1} + \dfrac{k_1}{k_2}\right)^2 \sinh^2 k_2 a} \tag{5-81}$$

and

$$R = \left|\frac{B}{A}\right|^2 = \left|\frac{B}{\hat{A}}\right|^2\left|\frac{\hat{A}}{A}\right|^2 = \frac{\dfrac{1}{4}\left(\dfrac{k_2}{k_1} + \dfrac{k_1}{k_2}\right)^2 \sinh^2 k_2 a}{1 + \dfrac{1}{4}\left(\dfrac{k_2}{k_1} + \dfrac{k_1}{k_2}\right)^2 \sinh^2 k_2 a}. \tag{5-82}$$

Note that the speeds always cancel in the calculation of R and that the speeds also cancel in the calculation of T because of the symmetrical height of the barrier. We see immediately that Equations (5-81) and (5-82) obey a sort of conservation law:

$$T + R = 1.$$

Thus, the coefficients T and R are able to give well-defined physical information about probabilities for particle transmission and reflection, even though the results are deduced from an unnormalizable wave function.

The transmission coefficient is especially interesting because it measures the probability for a truly nonclassical phenomenon. We may recall the defining relations for k_1 and k_2 in Equations (5-71) and (5-72) and derive an approximation to this quantity for use whenever a is substantially larger than $1/k_2$:

$$T = 16 \frac{E}{V_0} \left(1 - \frac{E}{V_0} \right) e^{-2k_2 a}. \tag{5-83}$$

The proof of this formula is offered as Problem 37 at the end of the chapter. It should be clear that the exponential in Equation (5-83) is the dominating factor since $e^{-2k_2 a}$ becomes a very small number for $k_2 a \gg 1$.

Example

Let us apply Equations (5-81) and (5-83) to a numerical illustration in which a 2 eV electron encounters a 10 eV potential energy barrier of 0.1 nm thickness. We wish to try both formulas for T in order to test the approximation, and we also want to assess the smallness of the quantum tunneling effect. The wave number for the incident electron is

$$k_1 = \frac{\sqrt{2mE}}{\hbar} = \frac{\sqrt{2(9.11 \times 10^{-31} \text{ kg})(2 \text{ eV})(1.60 \times 10^{-19} \text{ J/eV})}}{1.05 \times 10^{-34} \text{ J} \cdot \text{s}}$$

$$= 7.27 \times 10^9 \text{ m}^{-1} = 7.27 \text{ nm}^{-1}.$$

The similarly defined quantity inside the barrier is given by

$$k_2 = \frac{\sqrt{2m(V_0 - E)}}{\hbar} = \sqrt{\frac{V_0 - E}{E}} \, k_1.$$

We have $V_0 = 10$ eV, and so we find $(V_0 - E)/E = 4$ and $k_2 = 2k_1$. It follows that $k_2 a$ is larger than unity, but not by much:

$$k_2 a = 2(7.27 \text{ nm}^{-1})(0.1 \text{ nm}) = 1.45.$$

We use this number to compute

$$e^{k_2 a} = 4.26, \quad e^{-k_2 a} = 0.235, \quad \text{and} \quad \sinh k_2 a = \tfrac{1}{2}(e^{k_2 a} - e^{-k_2 a}) = 2.01.$$

The transmission coefficient can now be calculated according to each formula.

Equation (5-81) gives

$$T_{\text{exact}} = \cfrac{1}{1 + \cfrac{1}{4}\left(\cfrac{k_2}{k_1} + \cfrac{k_1}{k_2}\right)^2 \sinh^2 k_2 a} = \frac{1}{1 + \frac{1}{4}(2.5)^2(2.01)^2} = 0.137,$$

while Equation (5-83) gives

$$T_{\text{approx}} = 16\frac{E}{V_0}\left(1 - \frac{E}{V_0}\right)e^{-2k_2 a} = 16\left(\frac{2}{10}\right)\left(\frac{8}{10}\right)(0.235)^2 = 0.141.$$

The agreement indicates that $k_2 a$ is large enough after all for the approximate formula to be reliable with this choice of parameters. The answer itself tells us that the electron has a remarkable 14% chance to tunnel through the potential energy barrier.

Example

The continuity conditions for the rectangular barrier are straightforward to apply, although the algebra is somewhat tiresome. Let us outline an efficient algebraic strategy that determines four of the unknowns in terms of a fifth undetermined coefficient. The algebra simplifies if we define

$$\alpha = \frac{k_2}{k_1} \quad \text{and} \quad \Gamma = \hat{A}e^{ik_1 a}$$

as shorthand notation, and if we choose $\hat{A}$ to be the undetermined quantity. The last two of Equations (5-74) are

$$Ce^{-k_2 a} + De^{k_2 a} = \Gamma \quad \text{and} \quad i\alpha\left(Ce^{-k_2 a} - De^{k_2 a}\right) = \Gamma.$$

We can solve these easily to get

$$Ce^{-k_2 a} = \Gamma\frac{1 + i\alpha}{2i\alpha} \quad \text{and} \quad De^{k_2 a} = -\Gamma\frac{1 - i\alpha}{2i\alpha}.$$

The first two of Equations (5-74) then become

$$A + B = C + D = \frac{\Gamma}{2i\alpha}\left[e^{k_2 a}(1 + i\alpha) - e^{-k_2 a}(1 - i\alpha)\right]$$

$$= \Gamma\left(\cosh k_2 a - \frac{i}{\alpha}\sinh k_2 a\right)$$

and

$$A - B = i\alpha(C - D) = \frac{\Gamma}{2}\left[e^{k_2 a}(1 + i\alpha) + e^{-k_2 a}(1 - i\alpha)\right]$$

$$= \Gamma(\cosh k_2 a + i\alpha \sinh k_2 a).$$

We can also solve these easily to obtain

$$A = \Gamma\left[\cosh k_2 a + \frac{i}{2}\left(\alpha - \frac{1}{\alpha}\right)\sinh k_2 a\right]$$

and

$$B = -\frac{i}{2}\Gamma\left(\alpha + \frac{1}{\alpha}\right)\sinh k_2 a.$$

The moduli of these complex-valued quantities are needed in the interpretation of the wave function. We note that $|\Gamma|^2 = |\hat{A}|^2$, and so we find

$$\left|\frac{A}{\hat{A}}\right|^2 = \cosh^2 k_2 a + \frac{1}{4}\left(\alpha - \frac{1}{\alpha}\right)^2\sinh^2 k_2 a = 1 + \frac{1}{4}\left(\alpha + \frac{1}{\alpha}\right)^2\sinh^2 k_2 a$$

and

$$\left|\frac{B}{\hat{A}}\right|^2 = \frac{1}{4}\left(\alpha + \frac{1}{\alpha}\right)^2\sinh^2 k_2 a.$$

These results are cited above as Equations (5-75) and (5-76).

Example

The investigation of barriers and tunneling is also applicable to bound states. Consider the double-well potential energy for a particle in Figure 5-25 as a case in point. The symmetry of $V(x)$ implies the existence of eigenfunctions with definite parity. We can use these symmetry properties along with the curvature arguments of Section 5-6 to deduce shapes for some of the solutions. The figure shows the bound-state eigenfunctions ψ_1 and ψ_2 for the two lowest energy levels E_1 and E_2. Let us suppose that the wave function of the particle is given by some combination of these states, such as

$$\Psi = \frac{1}{\sqrt{2}}\left[\psi_1(x)e^{-iE_1 t/\hbar} + \psi_2(x)e^{-iE_2 t/\hbar}\right].$$

We know that this superposition of stationary states produces an oscillating expectation value for the position of the particle, with frequency of oscillation

$$\nu = \frac{E_2 - E_1}{h}.$$

The qualitative shape of Ψ at $t = 0$ is shown in the figure, along with the oscillations of $\langle x \rangle$ as a function of time. We can think of the oscillatory motion as a periodic tunneling with frequency ν, in which the particle starts in the well on the right and tunnels to a symmetrical location in the well on the left in a time $1/2\nu$. The particle has two symmetrical positions of stable equilibrium for the given potential energy $V(x)$. These locations are separated classically by a barrier; however, the tunneling process enables the quantum particle to pass

Figure 5-25

Double-well potential energy and two bound-state eigenfunctions. A superposition of the two states describes a tunneling oscillation of the particle from one well to the other. This behavior is analogous to the periodic inversion of the NH_3 molecule.

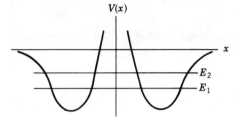

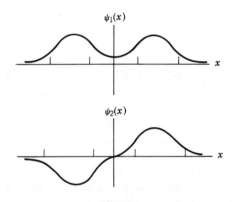

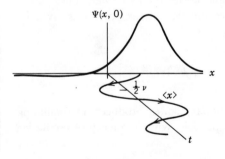

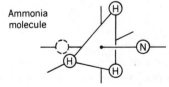

from a neighborhood of one stable point to a corresponding neighborhood of the other. This type of oscillation provides a simple model for the inversion behavior of the ammonia molecule NH_3. The molecule has a tetrahedral structure in which the three hydrogen atoms define a plane at the base of a pyramid, while the nitrogen atom has a position of stable equilibrium at one of two symmetrical locations on either side of the plane. The nitrogen atom acts like the particle in the double-well model, as the periodic inversion of the molecule corresponds to an oscillation between the two stable configurations. The NH_3 inversion frequency $\nu = 23870$ MHz is accurately known and has been employed as a precise standard of time in one of the earliest atomic clocks.

5-10 Two-Particle Systems in One Dimension

Many of the important problems in quantum physics are concerned with assemblies of several particles. The two-body system is already rich enough to generate a number of new ideas for consideration in the Schrödinger theory. Our attention turns now to the description of such systems, while the treatment continues to focus on behavior in one dimension. This investigation leads us directly to a profound new concept when we consider the symmetry of two identical particles in quantum mechanics.

Let us begin with some general remarks about the analysis of two quantum particles in one dimension. We assume that the particles can be distinguished, at least by their masses, for the first part of our discussion. There are two coordinates, x_1 and x_2, so that the state of the system is described by a wave function $\Psi(x_1, x_2, t)$, which depends on these two independent variables. Note that we do not introduce separate wave functions for each particle because, in general, we do not want the constituents of the system to be isolated from each other.

The classical energy relation contains the kinetic energy of each particle and the potential energy of the system:

$$\frac{p_1^2}{2m_1} + \frac{p_2^2}{2m_2} + V(x_1, x_2) = E.$$

We return to the representation rules in Equations (5-62) and (5-66) and write

$$p_1 \rightarrow \frac{\hbar}{i}\frac{\partial}{\partial x_1}, \quad p_2 \rightarrow \frac{\hbar}{i}\frac{\partial}{\partial x_2}, \quad \text{and} \quad E \rightarrow i\hbar\frac{\partial}{\partial t}$$

as differential operators acting on $\Psi(x_1, x_2, t)$. We then use these operators to convert the energy relation into the two-particle Schrödinger equation:

$$-\frac{\hbar^2}{2m_1}\frac{\partial^2}{\partial x_1^2}\Psi - \frac{\hbar^2}{2m_2}\frac{\partial^2}{\partial x_2^2}\Psi + V(x_1, x_2)\Psi = i\hbar\frac{\partial}{\partial t}\Psi. \qquad (5\text{-}84)$$

The probability interpretation of the wave function makes a *joint* statement regarding the detection of the two particles. We define the probability density in terms of the normalized wave function as

$$P(x_1, x_2, t) = \Psi^*(x_1, x_2, t)\Psi(x_1, x_2, t),$$

and then we specify

$$P(x_1, x_2, t)\, dx_1\, dx_2 = \left|\Psi(x_1, x_2, t)\right|^2 dx_1\, dx_2 \qquad (5\text{-}85)$$

as the probability at time t for finding particle 1 in the interval dx_1 at location x_1 and particle 2 in the interval dx_2 at location x_2. The wave function is normalized to give unit probability of finding both particles somewhere, according to the condition

$$\int_{-\infty}^{\infty} dx_1 \int_{-\infty}^{\infty} dx_2 |\Psi|^2 = 1.$$

We assume that V has no dependence on the time so that we can find stationary-state solutions of the Schrödinger equation. Each of these states has an energy eigenvalue E appearing as a parameter in the t dependence of the wave function

$$\Psi(x_1, x_2, t) = \psi(x_1, x_2)e^{-iEt/\hbar}.$$

The spatial eigenfunction $\psi(x_1, x_2)$ and the energy E are found by solving an eigenvalue problem defined by the time-independent partial differential equation

$$-\frac{\hbar^2}{2m_1}\frac{\partial^2}{\partial x_1^2}\psi - \frac{\hbar^2}{2m_2}\frac{\partial^2}{\partial x_2^2}\psi + V(x_1, x_2)\psi = E\psi. \qquad (5\text{-}86)$$

Our general discussion concludes at this point. We can proceed further if we are told how V depends on its two variables. Only one special type of dependence on x_1 and x_2 is of interest to us in the remainder of this section.

We suppose that the potential energy has an additive structure in the two coordinates, of the form

$$V = V_1(x_1) + V_2(x_2). \qquad (5\text{-}87)$$

Hence, the two particles may be subject to some external force but are not influenced by each other since V includes no contribution linking the coordinates x_1 and x_2. The two particles are dynamically isolated from one another in this special situation, and the solution for $\psi(x_1, x_2)$ simplifies as a consequence.

We construct ψ by identifying separate eigenfunctions for each particle. The eigenfunctions are defined as $\hat{\psi}(x_1)$ and $\tilde{\psi}(x_2)$, with energy eigenvalues $\hat{E}$ and $\tilde{E}$, according to the *one-particle* equations

$$-\frac{\hbar^2}{2m_1}\frac{d^2\hat{\psi}}{dx_1^2} + V_1(x_1)\hat{\psi} = \hat{E}\hat{\psi}$$

and

$$-\frac{\hbar^2}{2m_2}\frac{d^2\tilde{\psi}}{dx_2^2} + V_2(x_2)\tilde{\psi} = \tilde{E}\tilde{\psi}.$$

We then find that, if we multiply the first of these equations by $\tilde{\psi}(x_2)$ and the second by $\hat{\psi}(x_1)$, and add the two results, we produce the equality

$$-\frac{\hbar^2}{2m_1}\frac{d^2\hat{\psi}}{dx_1^2}\tilde{\psi} - \frac{\hbar^2}{2m_2}\hat{\psi}\frac{d^2\tilde{\psi}}{dx_2^2} + V\hat{\psi}\tilde{\psi} = (\hat{E} + \tilde{E})\hat{\psi}\tilde{\psi},$$

using Equation (5-87) to write the third term on the left side. The result may be compared with Equation (5-86) to reveal that a solution for $\psi(x_1, x_2)$ has been found. The two-particle eigenfunction is evidently of *product* form,

$$\psi(x_1, x_2) = \hat{\psi}(x_1)\tilde{\psi}(x_2), \tag{5-88}$$

while the energy eigenvalue is given as a *sum* of energies:

$$E = \hat{E} + \tilde{E}. \tag{5-89}$$

The particles share the total energy E as independent entities because the dynamics of the system is based on a potential energy of additive form in the two coordinates. Thus, the stationary state of the two *independent particles* is represented by a *multiplicative* energy eigenfunction with an *additive* energy eigenvalue. It is apparent that this construction would not hold in the absence of the hypothesis in Equation (5-87).

Our problem takes an entirely new and surprising turn when we consider particles that cannot be distinguished. The constituents of such a system must have the same mass and also the same value for every other identifying characteristic. We find that the probability interpretation forces the probability density to have a certain unique property in any state of such a system. The issue becomes clear if we return to Equation (5-85) and realize that we cannot always say which particle is in dx_1 and which is in dx_2 if the two particles are indistinguishable. We do not encounter this problem in classical physics because we are able to follow the trajectories of classical particles precisely and are therefore able to tag identical particles visually in order to tell them apart. The uncertainty principle precludes these possibilities for quantum particles. Figure 5-26 shows schematically two distinct pairs of classical orbits whose outgoing particles are detected indistinguishably in the quantum mechanical treatment. These which-is-which considerations imply the existence, and indeed the necessity, of a new type of symmetry property in a system of identical particles.

If measurements cannot determine which of the two particles is which, it follows that predictions of observable quantities must be unaffected by the replacement of one particle by the other. This *identical-particle symmetry* is built into the probability density by requiring that quantity to be symmetric under the interchange of particle coordi-

Figure 5-26

Distinguishable classical orbits in a collision of two identical particles.

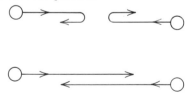

nates in the wave function. Thus, the probability density in Equation (5-85) is supposed to obey the condition

$$|\Psi(x_1, x_2, t)|^2 = |\Psi(x_2, x_1, t)|^2 \tag{5-90}$$

in a system of two identical particles. The wave function can realize this symmetry in one of two ways. It must be the case that Ψ is a *symmetric* wave function satisfying

$$\Psi(x_1, x_2, t) = +\Psi(x_2, x_1, t), \tag{5-91}$$

or it must be the case that Ψ is an *antisymmetric* wave function satisfying

$$\Psi(x_1, x_2, t) = -\Psi(x_2, x_1, t). \tag{5-92}$$

We observe that the symmetry of $|\Psi|^2$ is dictated by familiar quantum arguments and that no new principle is involved in Equation (5-90). On the other hand, it is a deeper question to ask whether Ψ itself should choose symmetry or antisymmetry. The actual choice between Equations (5-91) and (5-92) turns out to be ordained as an *intrinsic* property of the particles in question.

The independent-particle eigenfunction in Equation (5-88) is not correct for identical particles because the expression does not satisfy identical-particle symmetry. We can easily adapt the solution, however, and generate either a symmetric or an antisymmetric eigenfunction. Equation (5-88) is *symmetrized* by defining

$$\psi^S(x_1, x_2) = \frac{1}{\sqrt{2}} \left(\hat{\psi}(x_1)\tilde{\psi}(x_2) + \hat{\psi}(x_2)\tilde{\psi}(x_1) \right),$$

or *antisymmetrized* by defining

$$\psi^A(x_1, x_2) = \frac{1}{\sqrt{2}} \left(\hat{\psi}(x_1)\tilde{\psi}(x_2) - \hat{\psi}(x_2)\tilde{\psi}(x_1) \right).$$

We observe that both $\hat{\psi}(x_1)\tilde{\psi}(x_2)$ and $\tilde{\psi}(x_1)\hat{\psi}(x_2)$ are solutions of Equation (5-86) with energy $E = \hat{E} + \tilde{E}$. We then insist that symmetry or antisymmetry is also required for a proper eigenfunction, and we construct ψ^S and ψ^A by superposing equal parts of the two solutions with the appropriate signs. The factor $1/\sqrt{2}$ is included to ensure the normalization of the wave functions. The resulting expressions assign each particle an equal share in the occupation of the given single-particle states described by $\hat{\psi}$ and $\tilde{\psi}$. Hence, the symmetry has the remarkable effect that each of the two particles is "aware" of the configuration of the other, even in the absence of an interaction between particles.

We should note that it is possible to use the solution $\hat{\psi}(x_1)\tilde{\psi}(x_2)$ as a symmetric eigenfunction in the special circumstance where $\hat{\psi}$ and $\tilde{\psi}$ are identical functions. Let us also note that the antisymmetric eigenfunction becomes

$$\psi^A = 0 \quad \text{when } \hat{\psi} \text{ and } \tilde{\psi} \text{ are identical functions.}$$

This interesting observation means that no antisymmetric state can exist for particles whose single-particle quantum states are labeled by the same quantum number.

Identical-particle symmetry has far-reaching implications in the structure of atoms and nuclei, and in the theory of other systems of identical constituents. We take up the investigation of these matters in due course.

Example

A simple nontrivial illustration of an independent-particle system is provided by two noninteracting particles in a one-dimensional box. This rather hypothetical problem is defined by adapting the potential energy in Section 5-4 to the two-body expression

$$V(x_1, x_2) = \begin{cases} 0 & \text{when both } x_1 \text{ and } x_2 \text{ are in } (-a/2, a/2) \\ \infty & \text{otherwise.} \end{cases}$$

Equation (5-86) tells us that the eigenfunction $\psi(x_1, x_2)$ satisfies

$$-\frac{\hbar^2}{2m_1}\frac{\partial^2}{\partial x_1^2}\psi - \frac{\hbar^2}{2m_2}\frac{\partial^2}{\partial x_2^2}\psi = E\psi$$

if both x_1 and x_2 are in the interval $(-a/2, a/2)$ and that the eigenfunction vanishes if either coordinate is outside this region. We appeal to our independent-particle result in Equation (5-88) and write the solutions for $\psi(x_1, x_2)$ as products of particle-in-a-box eigenfunctions taken from Section 5-4. Each of these single-particle states has a single quantum number, and so every two-particle eigenfunction has *two* such quantum numbers, one for each factor in the product:

$$\psi_{n_1 n_2}(x_1, x_2) = \psi_{n_1}(x_1)\psi_{n_2}(x_2).$$

The corresponding energy eigenvalues are labeled in similar fashion. Equation (5-89) gives these energies as sums of single-particle eigenvalues taken from Equation (5-31):

$$E_{n_1 n_2} = \frac{\hbar^2\pi^2}{2m_1 a^2}n_1^2 + \frac{\hbar^2\pi^2}{2m_2 a^2}n_2^2 = \frac{1}{2}\left(\frac{\hbar\pi}{a}\right)^2\left(\frac{n_1^2}{m_1} + \frac{n_2^2}{m_2}\right).$$

The stationary-state wave functions have the form

$$\Psi_{n_1 n_2} = \psi_{n_1}(x_1)\psi_{n_2}(x_2)e^{-iE_{n_1 n_2}t/\hbar},$$

and so the two-particle probability in Equation (5-85) becomes a simple product of separate probabilities for each independent particle:

$$P(x_1, x_2, t)\,dx_1\,dx_2 = \left|\psi_{n_1}(x_1)\right|^2 dx_1 \cdot \left|\psi_{n_2}(x_2)\right|^2 dx_2.$$

It is instructive to list a few of the eigenfunctions and energies for some of the lowest values of the quantum numbers n_1 and n_2:

$$\psi_{11} = \frac{2}{a}\cos\frac{\pi x_1}{a}\cos\frac{\pi x_2}{a} \qquad E_{11} = \frac{1}{2}\left(\frac{\hbar\pi}{a}\right)^2\left(\frac{1}{m_1} + \frac{1}{m_2}\right)$$

$$\psi_{12} = \frac{2}{a}\cos\frac{\pi x_1}{a}\sin\frac{2\pi x_2}{a} \qquad E_{12} = \frac{1}{2}\left(\frac{\hbar\pi}{a}\right)^2\left(\frac{1}{m_1} + \frac{4}{m_2}\right)$$

$$\psi_{21} = \frac{2}{a}\sin\frac{2\pi x_1}{a}\cos\frac{\pi x_2}{a} \qquad E_{21} = \frac{1}{2}\left(\frac{\hbar\pi}{a}\right)^2\left(\frac{4}{m_1} + \frac{1}{m_2}\right)$$

$$\cdots$$

We can apply these results to the case of identical particles if we set $m_1 = m_2 = m$ and either symmetrize or antisymmetrize the eigenfunctions. The stationary-state wave functions become

$$\left.\begin{array}{c}\psi^S_{n_1 n_2}\\[1ex]\psi^A_{n_1 n_2}\end{array}\right\} = \frac{1}{\sqrt{2}}\Big(\psi_{n_1}(x_1)\psi_{n_2}(x_2) \pm \psi_{n_1}(x_2)\psi_{n_2}(x_1)\Big)e^{-iE_{n_1 n_2}t/\hbar},$$

where

$$E_{n_1 n_2} = \frac{1}{2m}\left(\frac{\hbar\pi}{a}\right)^2\left(n_1^2 + n_2^2\right).$$

A list of the eigenfunctions and energies includes the following entries:

$$\psi^S_{11} = \frac{2}{a}\cos\frac{\pi x_1}{a}\cos\frac{\pi x_2}{a} \qquad E_{11} = \frac{1}{2m}\left(\frac{\hbar\pi}{a}\right)^2 \cdot 2,$$

$$\psi^S_{12} = \psi^S_{21} = \frac{1}{\sqrt{2}}\frac{2}{a}\left(\cos\frac{\pi x_1}{a}\sin\frac{2\pi x_2}{a} + \sin\frac{2\pi x_1}{a}\cos\frac{\pi x_2}{a}\right)$$

$$E_{12} = \frac{1}{2m}\left(\frac{\hbar\pi}{a}\right)^2 \cdot 5.$$

$$\psi^A_{12} = -\psi^A_{21} = \frac{1}{\sqrt{2}}\frac{2}{a}\left(\cos\frac{\pi x_1}{a}\sin\frac{2\pi x_2}{a} - \sin\frac{2\pi x_1}{a}\cos\frac{\pi x_2}{a}\right)$$

Observe that $\psi_{n_1 n_2}$ is automatically symmetric, while $\psi^A_{n_1 n_2}$ is nonexistent, whenever n_1 and n_2 are equal.

5-11 The Three-Dimensional Box

The final topic in this chapter is intended as an introduction to quantum mechanics in three dimensions. Our purpose is to let the new three-dimensional formalism begin in a limited context where the familiar mathematics of one dimension can still be used. We return to the investigation of a *single* particle in this section and continue the problem into the next three chapters.

We describe a quantum particle in three-dimensional Cartesian coordinates by a straightforward extension of the one-dimensional Schrödinger theory. A state of the particle has a wave function $\Psi(x, y, z, t)$. The dependence on the three coordinates is determined by solving a three-dimensional version of the Schrödinger equation. We deduce the form of the equation by referring to the classical energy relation

$$\frac{p^2}{2m} + V(x, y, z) = E$$

and noting that the momentum $\mathbf{p}$ is a vector with three components. The kinetic energy therefore consists of three terms:

$$\frac{p^2}{2m} = \frac{p_x^2 + p_y^2 + p_z^2}{2m}.$$

The customary rule for representing the momentum by a differential operator applies to each Cartesian component of the vector $\mathbf{p}$. We employ this technique to introduce the set of differential operators

$$p_x \to \frac{\hbar}{i}\frac{\partial}{\partial x}, \quad p_y \to \frac{\hbar}{i}\frac{\partial}{\partial y}, \quad p_z \to \frac{\hbar}{i}\frac{\partial}{\partial z}, \quad \text{and} \quad E \to i\hbar\frac{\partial}{\partial t},$$

all acting on the wave function $\Psi(x, y, z, t)$. The energy relation is thereby converted directly into the equality

$$-\frac{\hbar^2}{2m}\left(\frac{\partial^2}{\partial x^2}\Psi + \frac{\partial^2}{\partial y^2}\Psi + \frac{\partial^2}{\partial z^2}\Psi\right) + V(x, y, z)\Psi = i\hbar\frac{\partial}{\partial t}\Psi.$$

We abbreviate the combination of spatial derivatives by adopting the convenient symbol

$$\nabla^2 = \frac{\partial^2}{\partial x^2} + \frac{\partial^2}{\partial y^2} + \frac{\partial^2}{\partial z^2},$$

and we then rewrite the equality as

$$-\frac{\hbar^2}{2m}\nabla^2\Psi + V(x, y, z)\Psi = i\hbar\frac{\partial}{\partial t}\Psi. \tag{5-93}$$

The result expresses the Cartesian form of the Schrödinger equation in three dimensions.

The probability interpretation of Ψ has already been described in the discussion leading to Equation (5-18). We identify an infinitesimal neighborhood of the point (x, y, z) by means of the volume element

$$d\tau = dx\,dy\,dz$$

and define the expression

$$|\Psi(x, y, z, t)|^2\,dx\,dy\,dz$$

as the probability of finding the particle in $d\tau$ at time t. The wave function is supposed to satisfy the normalization condition

$$\int_{-\infty}^{\infty} dx \int_{-\infty}^{\infty} dy \int_{-\infty}^{\infty} dz \, |\Psi(x, y, z, t)|^2 = 1,$$

so that the particle has unit probability to be found somewhere in all space.

A stationary state of the particle is an energy eigenfunction with energy eigenvalue E. The corresponding wave function is written as

$$\Psi(x, y, z, t) = \psi(x, y, z)e^{-iEt/\hbar}.$$

The spatial eigenfunction $\psi(x, y, z)$ and energy E are found by solving the eigenvalue problem conveyed by the time-independent partial differential equation

$$-\frac{\hbar^2}{2m}\nabla^2\psi + V(x, y, z)\psi = E\psi. \tag{5-94}$$

The dependence of V on its three coordinates has to be specified before any more progress can be made.

A simple but instructive illustration is provided by the free motion of a particle in a three-dimensional box. We confine the particle by means of an infinite exterior potential energy:

$$V(x, y, z) = \begin{cases} 0 & x \text{ in } [-a/2, a/2], \ y \text{ in } [-b/2, b/2], \ z \text{ in } [-c/2, c/2] \\ \infty & \text{otherwise.} \end{cases}$$

Thus, the region of motion is bounded by a rectangular parallelepiped with dimensions (a, b, c), and the center of the region is chosen to be the origin as in Figure 5-27.

Figure 5-27

Rectangular boxes for the confinement of a particle. The energy levels are indicated below by the quantum numbers $n_1 n_2 n_3$. The states pass through different stages of degeneracy as the box assumes higher degrees of symmetry.

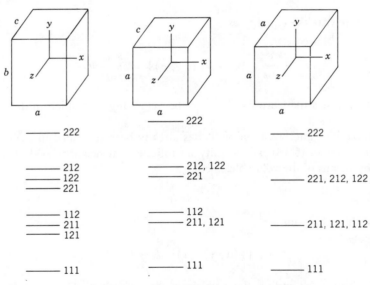

The allowed solutions of Equation (5-94) must vanish outside the box and satisfy continuity at the six walls

$$x = \pm \frac{a}{2}, \quad y = \pm \frac{b}{2}, \quad \text{and} \quad z = \pm \frac{c}{2}.$$

We can assemble such functions of (x, y, z) by using our solutions $\psi_n(x)$ from Section 5-4 in each of the three variables. We denote the one-dimensional particle-in-a-box eigenfunctions as

$$\psi_{n_1}(x), \quad \psi_{n_2}(y), \quad \text{and} \quad \psi_{n_3}(z)$$

in the regions defined by

$$x \text{ in } \left[-\frac{a}{2}, \frac{a}{2}\right], \quad y \text{ in } \left[-\frac{b}{2}, \frac{b}{2}\right], \quad \text{and} \quad z \text{ in } \left[-\frac{c}{2}, \frac{c}{2}\right].$$

We then take products of these solutions to form the desired eigenfunctions in three dimensions:

$$\psi_{n_1 n_2 n_3}(x, y, z) = \psi_{n_1}(x)\psi_{n_2}(y)\psi_{n_3}(z). \tag{5-95}$$

(A similar argument has led us to the expressions for the standing electromagnetic modes in Equations (2-17).) Each factor on the right side of Equation (5-95) is independently indexed by its own quantum number, and so a set of *three* quantum numbers is needed to label all the requisite functions of the three Cartesian coordinates. We can write an explicit formula for these eigenfunctions inside the box as

$$\psi_{n_1 n_2 n_3} = \sqrt{\frac{2}{a}} \left\{ \begin{matrix} \cos \\ \sin \end{matrix} \right\} \left(\frac{n_1 \pi x}{a}\right) \cdot \sqrt{\frac{2}{b}} \left\{ \begin{matrix} \cos \\ \sin \end{matrix} \right\} \left(\frac{n_2 \pi y}{b}\right) \cdot \sqrt{\frac{2}{c}} \left\{ \begin{matrix} \cos \\ \sin \end{matrix} \right\} \left(\frac{n_3 \pi z}{c}\right), \tag{5-96}$$

where the notation { } ( ) tells us to choose the cosine or sine function of the indicated variable when the relevant quantum number is odd or even.

The energy eigenvalues are labeled by the same three quantum numbers. We determine the allowed energies by direct use of Equation (5-94) inside the box:

$$-\frac{\hbar^2}{2m}\nabla^2\psi = E\psi.$$

If we insert Equation (5-96) in this differential equation, we obtain the following result:

$$-\frac{\hbar^2}{2m}\left(\frac{\partial^2}{\partial x^2} + \frac{\partial^2}{\partial y^2} + \frac{\partial^2}{\partial z^2}\right)\psi_{n_1 n_2 n_3}$$

$$= -\frac{\hbar^2}{2m}\left[-\left(\frac{n_1 \pi}{a}\right)^2 \psi_{n_1 n_2 n_3} - \left(\frac{n_2 \pi}{b}\right)^2 \psi_{n_1 n_2 n_3} - \left(\frac{n_3 \pi}{c}\right)^2 \psi_{n_1 n_2 n_3}\right]$$

$$= \frac{\hbar^2 \pi^2}{2m}\left[\left(\frac{n_1}{a}\right)^2 + \left(\frac{n_2}{b}\right)^2 + \left(\frac{n_3}{c}\right)^2\right]\psi_{n_1 n_2 n_3}.$$

It follows that the energy eigenvalues are given in terms of the three quantum numbers by the formula

$$E_{n_1 n_2 n_3} = \frac{\hbar^2 \pi^2}{2m} \left[\left(\frac{n_1}{a} \right)^2 + \left(\frac{n_2}{b} \right)^2 + \left(\frac{n_3}{c} \right)^2 \right]. \tag{5-97}$$

The corresponding normalized stationary-state wave functions are expressed in terms of the eigenfunctions in Equation (5-96) as

$$\Psi_{n_1 n_2 n_3} = \psi_{n_1 n_2 n_3}(x, y, z) e^{-i E_{n_1 n_2 n_3} t / \hbar}.$$

These few steps produce a complete set of solutions to the problem and do so without the need for any new mathematical machinery.

An interesting new feature appears in our results when the box has two or more dimensions of equal length. Figure 5-27 includes the special case of a second box with lengths (a, a, c), in which the energies are found from Equation (5-97) to be

$$E_{n_1 n_2 n_3} = \frac{\hbar^2 \pi^2}{2m} \left(\frac{n_1^2 + n_2^2}{a^2} + \frac{n_3^2}{c^2} \right). \tag{5-98}$$

The significance of this case resides in the fact that the formula for the energy levels is not affected by an interchange of the two quantum numbers n_1 and n_2. We therefore have a situation in which the wave functions $\Psi_{n_1 n_2 n_3}$ and $\Psi_{n_2 n_1 n_3}$ represent different states whose energies $E_{n_1 n_2 n_3}$ and $E_{n_2 n_1 n_3}$ are the same. These states are indeed distinct because the two wave functions $\Psi_{n_1 n_2 n_3}(x, y, z, t)$ and $\Psi_{n_2 n_1 n_3}(x, y, z, t)$ differ in their dependence on x and y. This set of circumstances exemplifies a general quantum phenomenon of far-reaching importance. The situation is called *degeneracy*, and the states are said to be *degenerate*, whenever a quantum system has two or more distinct energy eigenfunctions with the *same* energy eigenvalue. In the case at hand, we associate the degeneracy of the states $\Psi_{n_1 n_2 n_3}$ and $\Psi_{n_2 n_1 n_3}$ with the *symmetry* of the box under an interchange of the coordinates x and y. The figure shows that this symmetry distinguishes the second box from the first, where the assignment of unequal lengths (a, b, c) precludes such an interchange.

The problem takes on a higher degree of symmetry in the further special case of a cube with dimensions (a, a, a). This third box is also shown in the figure, along with the energy levels obtained from the formula:

$$E_{n_1 n_2 n_3} = \frac{1}{2m} \left(\frac{\hbar \pi}{a} \right)^2 (n_1^2 + n_2^2 + n_3^2). \tag{5-99}$$

The complete symmetry of the box causes the expression for the energy to be completely symmetric under any interchange of the three quantum numbers. The resulting degeneracy of the states is maximal since *all* the different states $\Psi_{n_1 n_2 n_3}, \Psi_{n_3 n_2 n_1}, \ldots$ have the same energy, regardless of the order in which the quantum numbers occur in the eigenfunctions. A numerical example is given below to illustrate how successive degrees of symmetry of the box result in sequential stages of degeneracy.

We have encountered degeneracy before, and not taken note, in our discussions of a free particle. It is clear that the monochromatic wave functions

$$e^{i(kx - \omega t)} \quad \text{and} \quad e^{i(-kx - \omega t)}$$

represent distinct degenerate states. These right- and left-propagating waves have different momentum eigenvalues $\pm \hbar k$, but the same energy $\hbar \omega$.

A degeneracy always reflects the existence of some symmetry in a given problem. Degeneracy occurs more frequently in the chapters that lie ahead. The interpretation in terms of symmetry always provides a deeper understanding of the particular physical phenomenon.

Example

The following calculations demonstrate the evolution of degeneracy for the three systems indicated in Figure 5-27. First, we let the dimensions of the box have different, nearly equal lengths $(a, \frac{10}{9}a, \frac{10}{11}a)$. Equation (5-97) determines the energies as

$$E_{n_1 n_2 n_3} = \frac{\hbar^2 \pi^2}{2ma^2}\left(n_1^2 + \frac{81}{100}n_2^2 + \frac{121}{100}n_3^2 \right) = E_1 \frac{100n_1^2 + 81n_2^2 + 121n_3^2}{100},$$

in terms of the energy parameter E_1 defined in Equation (5-32). Next, we let two of the lengths be equal by choosing dimensions $(a, a, \frac{10}{11}a)$, and we use Equation (5-98) to find

$$E_{n_1 n_2 n_3} = E_1 \frac{100n_1^2 + 100n_2^2 + 121n_3^2}{100}.$$

Finally, we let the box be a cube with lengths (a, a, a) and use Equation (5-99) to obtain

$$E_{n_1 n_2 n_3} = E_1\left(n_1^2 + n_2^2 + n_3^2 \right).$$

We observe a developing simplicity in the energy formula as we pass toward maximal degeneracy. The lowest energies are tabulated in multiples of $E_1/100$ to display this behavior as follows:

n_1	n_2	n_3	$\dfrac{100}{E_1}E_{n_1 n_2 n_3}$	$\sqrt{\dfrac{a^3}{8}}\,\psi_{n_1 n_2 n_3}$
1	1	1	$302 \rightarrow 321 \rightarrow 300$	$\cos\dfrac{\pi x}{a}\cos\dfrac{\pi y}{a}\cos\dfrac{\pi z}{a}$
1	1	2	$665 \quad 684 \quad 600$	$\cos\dfrac{\pi x}{a}\cos\dfrac{\pi y}{a}\sin\dfrac{2\pi z}{a}$
1	2	1	$545 \rightarrow 621 \rightarrow 600$	$\cos\dfrac{\pi x}{a}\sin\dfrac{2\pi y}{a}\cos\dfrac{\pi z}{a}$
2	1	1	$602 \quad 621 \quad 600$	$\sin\dfrac{2\pi x}{a}\cos\dfrac{\pi y}{a}\cos\dfrac{\pi z}{a}$
1	2	2	$908 \quad 984 \quad 900$	$\cos\dfrac{\pi x}{a}\sin\dfrac{2\pi y}{a}\sin\dfrac{2\pi z}{a}$
2	1	2	$965 \rightarrow 984 \rightarrow 900$	$\sin\dfrac{2\pi x}{a}\cos\dfrac{\pi y}{a}\sin\dfrac{2\pi z}{a}$
2	2	1	$845 \quad 921 \quad 900$	$\sin\dfrac{2\pi x}{a}\sin\dfrac{2\pi y}{a}\cos\dfrac{\pi z}{a}$
2	2	2	$1208 \rightarrow 1284 \rightarrow 1200$	$\sin\dfrac{2\pi x}{a}\sin\dfrac{2\pi y}{a}\sin\dfrac{2\pi z}{a}$

· · ·

(We include in the rightmost column a listing of the eigenfunctions from Equation (5-96) for the case where the particle is confined to a cube.) The passage to maximal degeneracy is clearly seen in the energy levels, as listed in the table and as drawn in the lower portion of the figure.

Problems

1. Suppose that the functions $\Psi_1(x, t), \Psi_2(x, t), \Psi_3(x, t), \ldots$ are solutions of the Schrödinger equation in one dimension. Let these functions be superposed by constructing

    $$\Psi = a_1\Psi_1 + a_2\Psi_2 + a_3\Psi_3 + \cdots,$$

    in which the a's are constants, and show that Ψ is also a solution. What mathematical step needs to be justified when the construction of Ψ involves an infinite series of functions?

2. The probability current density is defined as

    $$j(x, t) = \frac{\hbar}{2im}\left(\Psi^* \frac{\partial \Psi}{\partial x} - \frac{\partial \Psi^*}{\partial x} \Psi \right).$$

    Use the Schrödinger equation to deduce the probability conservation law

    $$\frac{\partial}{\partial x}j(x, t) + \frac{\partial}{\partial t}P(x, t) = 0.$$

    What condition must the potential energy $V(x)$ satisfy in the derivation of this result?

3. Suppose that the state of a particle is described by the monochromatic wave function

    $$\Psi = Ae^{i(kx - \omega t)}.$$

    Calculate the probability current density for this state and interpret the result. Replace Ψ by the wave function in the example of Section 5-1 and repeat the calculation.

4. Show that the normalization integral $\int_{-\infty}^{\infty}|\Psi(x, t)|^2 \, dx$ is time independent for the case of a nonfree particle. What condition must $V(x)$ satisfy to ensure this result?

5. Let $\psi_1(x)e^{-iE_1t/\hbar}$ and $\psi_2(x)e^{-iE_2t/\hbar}$ be two stationary-state wave functions with real-valued eigenfunctions ψ_1 and ψ_2. Suppose that the state of a particle is described by the wave function

    $$\Psi(x, t) = a_1\psi_1(x)e^{-iE_1t/\hbar} + a_2\psi_2(x)e^{-iE_2t/\hbar}.$$

    Show that the probability density is given by

    $$|\Psi(x, t)|^2 = |a_1|^2(\psi_1(x))^2 + |a_2|^2(\psi_2(x))^2$$
    $$+ 2|a_1||a_2|\psi_1(x)\psi_2(x)\cos\left(\frac{E_1 - E_2}{\hbar}t - \phi_1 + \phi_2 \right),$$

    where ϕ_1 and ϕ_2 are the phases of the complex constants a_1 and a_2.

6. A particle in a one-dimensional box is confined to the interval $[-a/2, a/2]$ and is in its first excited state. Calculate the probability of finding the particle in the subinterval $(a/8, 3a/8)$.

7. Show that the particle-in-a-box wave functions can be expressed as combinations of traveling waves that propagate in opposite directions inside the box. Obtain the phase velocity of these traveling waves.

8. Prove the orthogonality properties

$$\int_{-a/2}^{a/2} \cos\frac{n\pi x}{a} \cos\frac{n'\pi x}{a} \, dx = 0 \quad \text{and} \quad \int_{-a/2}^{a/2} \sin\frac{n\pi x}{a} \sin\frac{n'\pi x}{a} \, dx = 0$$

for unequal quantum numbers n and n'.

9. Suppose that the wave function for a particle in a one-dimensional box is given by the superposition

$$\Psi = c\Psi_n + c'\Psi_{n'},$$

where Ψ_n and $\Psi_{n'}$ represent any two of the normalized stationary states of the particle. What condition must the complex constants c and c' satisfy in order for Ψ to be a normalized wave function? Interpret this result.

10. Suppose that the wave function at $t = 0$ for a particle in a one-dimensional box is given by

$$\Psi(x,0) = \frac{1}{\sqrt{a}}\left(\sin\frac{2\pi x}{a} + \cos\frac{3\pi x}{a}\right).$$

What is the subsequent form of the wave function $\Psi(x, t)$? Use this form to compute the probability density, and interpret the time dependence of the result.

11. Show that the probability of finding a classical oscillator in an interval dx between $-A$ and $+A$ is given by

$$P_{\text{cl}}(x) \, dx = \frac{dx}{\pi\sqrt{A^2 - x^2}},$$

where A is the amplitude of oscillation. What is the probability of finding the oscillating particle somewhere in the finite interval $[0, A/2]$?

12. Apply direct differentiation to the ground-state wave function for the harmonic oscillator,

$$\Psi = e^{-\sqrt{mk}\,x^2/2\hbar}\,e^{-i\omega_0 t/2} \quad \text{(unnormalized)},$$

and show that Ψ has points of inflection at the extreme positions of the particle's classical motion.

13. For an oscillator in its ground state, calculate the probability of finding the particle between $x = 0$ and $x = A/2$, where A is the classical amplitude of oscillation. Obtain a numerical result and compare with the answer to Problem 11.

14. Assume that an atom in a metallic crystal behaves like a mass on a spring. Let the spring constant for a copper atom correspond to angular frequency $\omega_0 = 10^{13}$ rad/s, and calculate the atom's amplitude of zero-point motion. Take the Cu mass to be equal to 63 H masses.

15. Prove that the Hermite polynomials $H_n(\xi)$ satisfy

$$H_{n+1} = 2\xi H_n - \frac{dH_n}{d\xi}$$

by showing that $(2\xi H_n - H_n')$ obeys the differential equation for H_{n+1}.

16. Prove that the Hermite polynomials $H_n(\xi)$ satisfy

$$H_{n+1} - 2\xi H_n + 2n H_{n-1} = 0$$

by showing that $(2\xi H_n - 2n H_{n-1})$ obeys the differential equation for H_{n+1}. (The result from Problem 15 might be helpful.)

17. Construct a general proof of the recursion relation

$$a_{k+2} = \frac{2(k-n)}{(k+1)(k+2)} a_k$$

for the coefficients in the nth-order Hermite polynomial $H_n(\xi)$.

18. Let the potential energy of a particle be given by the function

$$V = V_0 \left[\left(\frac{x_0}{x} \right)^2 - \frac{x_0}{x} \right],$$

where V_0 and x_0 are positive constants. Sketch a graph of the function and determine the equilibrium position of the particle. Deduce the form of the harmonic-oscillator approximation for this potential energy.

19. Let E be the total energy of the particle in the previous problem, and obtain a formula for the turning points of the classical motion. Consider the results for $E < 0$ and for $E > 0$ as separate cases.

20. Consult a table of integrals to demonstrate the orthogonality relation

$$\int_{-\infty}^{\infty} \psi_2^*(x) \psi_0(x) \, dx = 0,$$

where ψ_0 and ψ_2 are harmonic-oscillator eigenfunctions for $n = 0$ and 2.

21. A W-shaped potential energy function $V(x)$ produces a set of bound states, the three lowest of which are arranged as shown in the figure. Sketch the qualitative behavior of the three corresponding eigenfunctions.

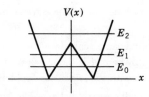

22. An eigenfunction in a square-well penetrates the classically forbidden region as indicated in the drawing. Define the penetration depth d to be the distance into the region over

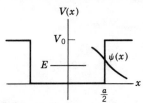

which the probability density falls by the factor $1/e$. Deduce the formula for d and calculate the penetration depth for an electron with $V_0 - E = 3$ eV.

23.   Assume that the indicated square-well potential energy is capable of producing at least a ground state and a first excited state, as shown. Sketch the behavior of the eigenfunction for the first excited state.

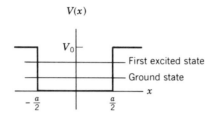

24.   Show analytically how to determine the energy eigenvalue for the first excited state in Problem 23.

25.   Write out the proof of the formula for the square of the uncertainty in x,

$$(\Delta x)^2 = \langle x^2 \rangle - \langle x \rangle^2,$$

using explicit integral expressions for all the relevant expectation values.

26.   Prove that $\langle p \rangle$ and $\langle E \rangle$ are real by showing that $\langle p \rangle^* = \langle p \rangle$ and $\langle E \rangle^* = \langle E \rangle$.

27.   Let the wave function $\Psi(x, t)$ be expressed as the superposition of two stationary states having different energy eigenvalues,

$$\Psi = c\Psi_n + c'\Psi_{n'},$$

and obtain a formula for the energy uncertainty ΔE in this state. As a special case let Ψ be an equal-parts admixture of Ψ_n and $\Psi_{n'}$, and interpret the corresponding result for ΔE.

28.   Assume that a particle in a one-dimensional box is in its first excited state, and calculate the expectation values $\langle x \rangle$, $\langle x^2 \rangle$, $\langle p \rangle$, and $\langle p^2 \rangle$. Evaluate the uncertainties Δx and Δp, and compute the product of these quantities.

29.   Use the wave function $\Psi(x, t)$ for the superposition of states in Problem 10, and evaluate the expectation values $\langle x \rangle$ and $\langle p \rangle$. Verify that these quantities satisfy the relation $\langle p \rangle = m d\langle x \rangle / dt$.

30.   Write down a suitable transition state for a particle in a box describing transitions from the $n = 2$ level to the $n = 1$ level. Evaluate the corresponding dipole transition amplitude.

31.   Repeat the steps of Problem 30 for the case of particle-in-a-box transitions from $n = 3$ to $n = 2$.

32.   Generalize the considerations of Problems 30 and 31 to allow for transitions between arbitrary pairs of levels, and deduce the appropriate selection rule for electric dipole transitions. Obtain a general formula for the nonvanishing dipole transition amplitude, and compare with the results of Problems 30 and 31.

33.   Refer to Figure 5-22 and solve for the nodes of the single-step eigenfunction

$$\psi(x) = C\left(\cos k_1 x - \frac{k_2}{k_1} \sin k_1 x \right) \quad \text{in the region } x < 0,$$

where $k_1 = \sqrt{2mE}/\hbar$ and $k_2 = \sqrt{2m(V_0 - E)}/\hbar$. What happens to this solution in the limit $E \to V_0$?

34.   A particle with energy E is incident on a single-step potential energy barrier with step height V_0. Incidence is from the left and E exceeds V_0, as indicated. Determine the form of the eigenfunction that describes this situation, and interpret the various contributions to the wave function.

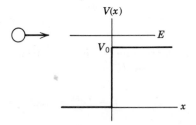

35.   Use the results of Problem 34 to derive expressions for the transmission and reflection coefficients T and R, and show that $T + R = 1$.

36.   A particle with energy E is incident on a rectangular potential energy barrier with height V_0. Assume $E < V_0$ as indicated, and let the particle be incident from the left. The transmitted wave function has the form

$$\Psi_{\text{trans}} = \hat{A}e^{i(k_1 x - \omega t)} \quad \text{in the region } x > a.$$

Show that the probability density in the region of the barrier $0 < x < a$ is given by

$$\Psi^*\Psi = |\hat{A}|^2 \left[\cosh^2 k_2(a - x) + \left(\frac{k_1}{k_2} \right)^2 \sinh^2 k_2(a - x) \right],$$

where k_1 and k_2 are defined in Problem 33.

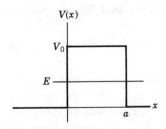

37. The transmission coefficient for the particle in Problem 36 is given by

$$T = \left[1 + \frac{1}{4} \left(\frac{k_2}{k_1} + \frac{k_1}{k_2} \right)^2 \sinh^2 k_2 a \right]^{-1}.$$

Derive the approximate version of this formula,

$$T = 16 \frac{E}{V_0} \left(1 - \frac{E}{V_0} \right) e^{-2k_2 a},$$

which holds when a is substantially larger than $1/k_2$.

38. A 50 g particle slides with speed 20 cm/s toward a bump 1 cm high and 2 cm wide. Adopt a rectangular-barrier model to estimate the probability of finding the particle on the other side of the bump.

39. Let the rectangular-barrier problem be defined by the symmetric potential energy in the figure, and assume the condition $E < V_0$. Use the symmetry to construct eigenfunctions of definite parity.

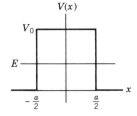

40. A particle is free to move inside a pizza box with dimensions $(a, a, a/10)$. Determine the six lowest energy levels, and identify the quantum numbers of the degenerate states at each level.

41. A particle moves freely inside a cookie box with dimensions $(a/5, a/5, a)$. Determine the ten lowest energy levels, and tabulate the results according to the quantum numbers of the corresponding states. What degeneracies occur among these states?

S I X

QUANTIZATION
OF
ANGULAR
MOMENTUM

Many important features of quantum me-
chanics are revealed when the Schrödinger theory is extended to three dimensions. We
find again, as we have found already in one dimension, that the Schrödinger equation
and the probability interpretation of the wave function lead directly to stationary
states, quantized energy levels, and expectation values. We also discover a powerful
new quantum concept that does not arise in the restrictive framework of one
dimension. The entire chapter is given over to the development of this fundamental
topic.

A limited version of three-dimensional quantum mechanics has already been
introduced at the end of Chapter 5. The Cartesian coordinate system of Section 5-11
must be set aside, however, because such a choice of coordinates would be very ill
advised for the situation of interest here. We are concerned in this chapter with the
behavior of a system under the influence of a *central force*. We know from our
experience with classical physics that the general central-force problem exhibits
conservation of angular momentum and that the solution of the problem is expedited when
we build this property into the analysis. Our objective is to learn how the concept of
angular momentum appears in quantum mechanics and how the conservation law
takes effect as a quantum principle.

It would seem that the central force embraces a rather restricted class of dynamical
problems. In fact, the ideas associated with angular momentum prove to be surprisingly general and make their appearance on many different occasions throughout
modern physics. The topic is a vital ingredient in the theory of atoms, since the
Coulomb interaction between an electron and the nucleus of an atom is a prime
example of a central force. We treat the problem of central forces in generality so that
our results can be regarded as comprehensive. We can then apply our conclusions
specifically to the atom, or to any other appropriate quantum problem.

 Our treatment is based on the Schrödinger equation in three-dimensional *polar*
coordinates. We are faced with certain complications when we adopt this coordinate

system and proceed to solve the resulting differential equations. We give these equations serious consideration so that we can appreciate the interpretation of the solutions. Our efforts are rewarded in the end as several very important principles emerge from this investigation.

6-1   Central Forces

Let us begin by recalling the classical treatment of the two-body central force. We suppose that the two masses m and M are separated by a variable distance r and that the center of mass is at rest, as in Figure 6-1. The force on each particle is assumed to be oriented along the line defined by the separation vector $\mathbf{r}$. We have learned in Section 3-6 that this two-body problem is completely equivalent to a *one-body problem* in which a reduced mass μ is attracted or repelled by a fixed center of force located at the origin. We recall that the reduced mass is given by

$$\mu = \frac{mM}{M + m}$$

and that the separation vector $\mathbf{r}$ is the coordinate vector for the mass μ.

It is also assumed that the force on μ is conservative and hence derivable from a potential energy V. The central nature of the force implies that the potential energy depends on the magnitude, but *not* the direction, of the vector $\mathbf{r}$. We therefore introduce a unit vector $\mathbf{r}$ and write the central force as

$$\mathbf{F} = -\hat{\mathbf{r}}\frac{dV(r)}{dr}. \tag{6-1}$$

This expression tells us that $\mathbf{F}$ points away from the origin if dV/dr is negative and toward the origin if dV/dr is positive. We take the function $V(r)$ to be completely *arbitrary* to ensure the generality of our conclusions.

The force law in Equation (6-1) allows us to identify a constant of the motion. We know from classical mechanics that insights are always gained whenever a conserved quantity is revealed, and we expect the same experience to carry over in quantum physics. It is evident that the angular momentum $\mathbf{L}$ is conserved since the force $\mathbf{F}$

Figure 6-1

Two interacting particles and the corresponding reduced-mass particle.

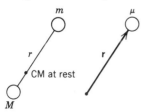

Figure 6-2

Classical orbit in the plane perpendicular to the angular momentum $\mathbf{L}$. The momentum $\mathbf{p}$ has polar components p_r and $p_\perp$.

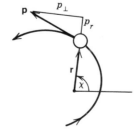

produces no torque on the particle:

$$\mathbf{p} = \mu \frac{d\mathbf{r}}{dt} \quad \text{and} \quad \mathbf{L} = \mathbf{r} \times \mathbf{p}$$

$$\Rightarrow \quad \frac{d\mathbf{L}}{dt} = \frac{d\mathbf{r}}{dt} \times \mathbf{p} + \mathbf{r} \times \frac{d\mathbf{p}}{dt} = \mathbf{r} \times \mathbf{F} = 0.$$

Since $\mathbf{L}$ is a constant vector, it follows that the classical orbit of the particle lies in a fixed plane through the origin, perpendicular to the direction of $\mathbf{L}$. In that plane the momentum $\mathbf{p}$ has components along $\mathbf{r}$ and normal to $\mathbf{r}$, as shown in Figure 6-2. These components are expressed in terms of the radial distance r and angle χ as

$$p_r = \mu \frac{dr}{dt} \quad \text{and} \quad p_\perp = \mu r \frac{d\chi}{dt},$$

so that the constant magnitude of the angular momentum is given by

$$L = r p_\perp .$$

We use this result to rewrite the expression for the kinetic energy,

$$\frac{p^2}{2\mu} = \frac{p_r^2 + p_\perp^2}{2\mu} = \frac{p_r^2}{2\mu} + \frac{L^2}{2\mu r^2},$$

and we find that the classical energy relation becomes

$$\frac{p_r^2}{2\mu} + \frac{L^2}{2\mu r^2} + V(r) = E. \tag{6-2}$$

This equation is our starting point for an examination of the Schrödinger equation in a coordinate system appropriate to the central-force problem.

Equation (6-2) is an interesting formula in classical mechanics. The equality is a statement about radial variables alone, in which L and E are constants parametrizing the classical orbit. It is instructive to call the second term on the left a *centrifugal* potential energy and identify an *effective* potential energy function as

$$V_{\text{eff}}(r) = V(r) + \frac{L^2}{2\mu r^2}. \tag{6-3}$$

The equality becomes

$$\frac{p_r^2}{2\mu} + V_{\text{eff}}(r) = E \tag{6-4}$$

and thus assumes a form just like the energy relation for one-dimensional motion. We can use this result to find the radial turning points for the orbit of a particle, as in the following illustration.

Figure 6-3

Effective potential energy for an attractive Coulomb force. The two shaded regions indicate allowed ranges of r for classical motion with total energies E_1 and E_2. A bound orbit results for energy E_1, and an open orbit results for energy E_2.

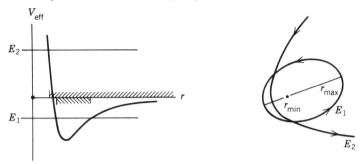

Example

Let V be an attractive potential energy of Coulomb form:

$$V(r) = -\frac{b}{r} \quad (b > 0) \qquad \text{so} \qquad V_{\text{eff}}(r) = -\frac{b}{r} + \frac{L^2}{2\mu r^2}.$$

Figure 6-3 shows a sketch of the effective potential energy for some chosen value of the constant L, along with two selected values of the constant E. Observe that the centrifugal part of V_{eff} is always a repulsive contribution. The figure tells us that this term dominates $V(r)$ at small r and prevents the particle from approaching the point $r = 0$. The quantity $E - V_{\text{eff}}$ cannot be negative because of Equation (6-4), and so the equality $E = V_{\text{eff}}(r)$ defines as turning points the minimum and maximum radial distances for the given energy E. The shaded domains in the figure indicate the allowed physical regions for classical motion in the variable r. For $E_1 < 0$ we have a bound orbit with two turning points, and for $E_2 > 0$ we have an open orbit with one turning point. Pictures of these two orbits are also shown in the figure.

6-2 The Schrödinger Equation in Spherical Coordinates

We have made the transition from classical to quantum mechanics on several previous occasions. This time the appropriate form of Schrödinger's wave equation is to be found from the classical energy relation in Equation (6-2). It is expected that the terms in this equation should turn into derivative and multiplicative operations acting on the wave function Ψ. We know in advance that E becomes the differential operator $i\hbar\, \partial/\partial t$ and that the Schrödinger equation has the general three-dimensional structure

$$-\frac{\hbar^2}{2\mu}\nabla^2\Psi + V\Psi = i\hbar\frac{\partial}{\partial t}\Psi, \tag{6-5}$$

where $\nabla^2 = \partial^2/\partial x^2 + \partial^2/\partial y^2 + \partial^2/\partial z^2$. The central nature of the potential energy function V is our main consideration.

Figure 6-4

Cartesian and spherical coordinate systems.

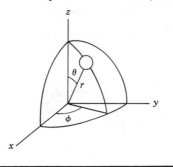

We are concerned with a general situation where V depends *only* on r. The radial variable is a rather complicated function of the Cartesian coordinates:

$$r = \sqrt{x^2 + y^2 + z^2}.$$ (6-6)

Hence, it is obvious that the Cartesian variables (x, y, z) are not suitable and that spherical polar coordinates (r, θ, ϕ) should be adopted instead. Figure 6-4 shows how the two coordinate systems are related through the formulas

$$x = r \sin \theta \cos \phi,$$
$$y = r \sin \theta \sin \phi,$$
$$z = r \cos \theta.$$ (6-7)

We want the wave function to appear as $\Psi(r, \theta, \phi, t)$, and so we must reexpress the differential operator ∇^2 in terms of the transformed coordinates (r, θ, ϕ). It would be a simple matter to find a reference book and look up the desired formula for the transformed operator. We gain considerable insight if, instead, we return to our starting point and look separately at each of the contributions to Equation (6-2).

We know that momentum components are supposed to be written as differential operators, as in the representation rules $p_x \rightarrow (\hbar/i)\partial/\partial x$ and $p^2 \rightarrow (\hbar/i)^2 \nabla^2$. It is tempting to suppose that p_r^2 becomes the square of the radial operator $(\hbar/i)\partial/\partial r$; however, this supposition is not correct. On the other hand, it is not wrong to presume instead that Equation (6-2) can be implemented by substituting purely radial and purely angular differential operators for p_r^2 and L^2. We appeal to this equation for guidance because the terms indicate a clear separation of radial and angular components according to the classical construction in Figure 6-2.

We determine the differential form of p_r^2 by considering the special case of a wave function $\Psi(r, t)$, which has *no* dependence on θ and ϕ. The derivation begins with the derivatives

$$\frac{\partial \Psi}{\partial x} = \frac{\partial \Psi}{\partial r} \frac{\partial r}{\partial x} = \frac{x}{r} \frac{\partial \Psi}{\partial r}$$

and

$$\frac{\partial^2 \Psi}{\partial x^2} = \frac{1}{r} \frac{\partial \Psi}{\partial r} - \frac{x^2}{r^3} \frac{\partial \Psi}{\partial r} + \frac{x^2}{r^2} \frac{\partial^2 \Psi}{\partial r^2}.$$

Substitution of y for x and z for x then yields the other two derivatives:

$$\frac{\partial^2 \Psi}{\partial y^2} = \frac{1}{r}\frac{\partial \Psi}{\partial r} - \frac{y^2}{r^3}\frac{\partial \Psi}{\partial r} + \frac{y^2}{r^2}\frac{\partial^2 \Psi}{\partial r^2}$$

and

$$\frac{\partial^2 \Psi}{\partial z^2} = \frac{1}{r}\frac{\partial \Psi}{\partial r} - \frac{z^2}{r^3}\frac{\partial \Psi}{\partial r} + \frac{z^2}{r^2}\frac{\partial^2 \Psi}{\partial r^2}.$$

Thus, the special form of $\nabla^2\Psi$ is found by adding the three results and using Equation (6-6):

$$\nabla^2\Psi(r, t) = \frac{2}{r}\frac{\partial \Psi}{\partial r} + \frac{\partial^2 \Psi}{\partial r^2}.$$

We may express this formula in two equally useful alternative ways:

$$\nabla^2\Psi(r, t) = \frac{1}{r^2}\frac{\partial}{\partial r}\left(r^2\frac{\partial \Psi}{\partial r}\right)$$

$$= \frac{1}{r}\frac{\partial^2}{\partial r^2}(r\Psi). \tag{6-8}$$

The rule $p^2 \rightarrow (\hbar/i)^2\nabla^2$ can now be used along with the first of these expressions to write

$$p_r^2 \rightarrow \left(\frac{\hbar}{i}\right)^2 \frac{1}{r^2}\frac{\partial}{\partial r}r^2\frac{\partial}{\partial r} \tag{6-9}$$

as the radial differential operator that acts on the wave function $\Psi(r, t)$ in the Schrödinger equation. We then argue that, if it is correct to express L^2 with no radial derivatives and p_r^2 with no angular derivatives, it must therefore be correct to express p_r^2 by the purely radial differential operator in Equation (6-9), acting on *any* wave function $\Psi(r, \theta, \phi, t)$.

We expect the formula for L^2 to contain no radial derivatives because the classical construction in Equation (6-2) and Figure 6-2 has the effect of separating changes in angle from changes in r. Let us clarify this point, and secure an important formula for later use, by examining a specific component of the angular momentum. We choose the z component

$$L_z = xp_y - yp_x$$

and consult Equations (6-7) to manipulate the corresponding differential operator:

$$L_z \rightarrow \frac{\hbar}{i}\left(x\frac{\partial}{\partial y} - y\frac{\partial}{\partial x}\right) = \frac{\hbar}{i}\left(r\sin\theta\cos\phi\frac{\partial}{\partial y} - r\sin\theta\sin\phi\frac{\partial}{\partial x}\right).$$

We also invoke the chain rule to find

$$\frac{\partial}{\partial \phi} = \frac{\partial x}{\partial \phi}\frac{\partial}{\partial x} + \frac{\partial y}{\partial \phi}\frac{\partial}{\partial y} + \frac{\partial z}{\partial \phi}\frac{\partial}{\partial z} = -r\sin\theta\sin\phi\frac{\partial}{\partial x} + r\sin\theta\cos\phi\frac{\partial}{\partial y},$$

and we then compare the two differential expressions. The result is an important new representation rule:

$$L_z \rightarrow \frac{\hbar}{i} \frac{\partial}{\partial \phi}. \qquad (6\text{-}10)$$

The z component of $\mathbf{L}$ is distinguished because the z direction has been selected as the polar axis in Figure 6-4. The other two components of $\mathbf{L}$ also become purely angular differential operators. We quote the relevant formulas as

$$L_x \rightarrow \frac{\hbar}{i} \left(-\sin\phi \frac{\partial}{\partial \theta} - \cot\theta \cos\phi \frac{\partial}{\partial \phi} \right)$$

and

$$L_y \rightarrow \frac{\hbar}{i} \left(\cos\phi \frac{\partial}{\partial \theta} - \cot\theta \sin\phi \frac{\partial}{\partial \phi} \right) \qquad (6\text{-}11)$$

and relegate the details to the end of the section.

Equations (6-10) and (6-11) can be used to construct the three terms in the square of the magnitude of $\mathbf{L}$:

$$L^2 = L_x^2 + L_y^2 + L_z^2.$$

The derivation is a straightforward task, which we leave to Problem 3 at the end of the chapter. The result is another important new representation rule:

$$L^2 \rightarrow \left(\frac{\hbar}{i} \right)^2 \Lambda^2 = \left(\frac{\hbar}{i} \right)^2 \left(\frac{1}{\sin\theta} \frac{\partial}{\partial \theta} \sin\theta \frac{\partial}{\partial \theta} + \frac{1}{\sin^2\theta} \frac{\partial^2}{\partial \phi^2} \right). \qquad (6\text{-}12)$$

Note that a special symbol Λ^2 is introduced in this formula to denote the differential portion of the operator. We explore the detailed structure of Λ^2 later on, when we examine the form of the wave function.

The two rules in Equations (6-9) and (6-12) can now by employed to convert Equation (6-2). The operators act on the wave function $\Psi(r, \theta, \phi, t)$ to generate the Schrödinger equation in spherical coordinates:

$$-\frac{\hbar^2}{2\mu r^2} \left(\frac{\partial}{\partial r} r^2 \frac{\partial}{\partial r} \Psi + \Lambda^2 \Psi \right) + V(r)\Psi = i\hbar \frac{\partial}{\partial t} \Psi. \qquad (6\text{-}13)$$

It is interesting that the kinetic energy term in Equation (6-5) has turned into such a complicated differential expression. The complications arise because the choice of coordinate system has been tailored to the behavior of the potential energy term. We pay this price willingly in order to take advantage of the fact that V depends only on r. The mathematical strategy pays enormous dividends when we turn to the problem of solving Equation (6-13).

Our next step is to seek solutions for wave functions of the form

$$\Psi = \psi(r, \theta, \phi) e^{-iEt/\hbar}, \qquad (6\text{-}14)$$

where the time dependence is parametrized by an energy eigenvalue E. Such a wave function satisfies the eigenvalue equation

$$i\hbar \frac{\partial}{\partial t}\Psi = E\Psi \tag{6-15}$$

and therefore has the properties of an energy eigenfunction. Thus, the proposed state Ψ is a stationary state, whose wave function gives a time-independent probability density $|\Psi|^2 = |\psi|^2$, and whose energy occurs with zero uncertainty. The spatial eigenfunction $\psi(r, \theta, \phi)$ and the energy eigenvalue E are determined jointly by solving an eigenvalue problem based on the time-independent partial differential equation

$$-\frac{\hbar^2}{2\mu r^2}\left(\frac{\partial}{\partial r}r^2\frac{\partial}{\partial r}\psi + \Lambda^2\psi\right) + V(r)\psi = E\psi. \tag{6-16}$$

Our previous experience in one dimension has taught us to expect a set of allowed solutions for the eigenvalues and eigenfunctions.

It should be noted that the passage from Equation (6-5) to Equation (6-13) by way of the classical energy relation has led to a derivation of ∇^2 in polar form, and a great deal more. Our goal is to understand the concept of angular momentum in quantum mechanics, and our approach has kept that quantity in view throughout. The main conclusions are embodied in Equations (6-10) and (6-12), along with Equation (6-13). We should regard the operator formulas for L_z and L^2 as essential adjuncts to the Schrödinger equation. The eventual interpretation of the angular momentum stems from this association.

Detail

Let us sketch the algebra involved in the derivation of the operator expressions for L_x and L_y. We begin by using Equations (6-7) to derive relations between the sets of operators $(\partial/\partial x, \partial/\partial y, \partial/\partial z)$ and $(\partial/\partial r, \partial/\partial \theta, \partial/\partial \phi)$. The chain rule has already been used to compute $\partial/\partial \phi$, and may be used again to find

$$\frac{\partial}{\partial r} = \frac{\partial x}{\partial r}\frac{\partial}{\partial x} + \frac{\partial y}{\partial r}\frac{\partial}{\partial y} + \frac{\partial z}{\partial r}\frac{\partial}{\partial z} = \sin\theta\cos\phi\frac{\partial}{\partial x} + \sin\theta\sin\phi\frac{\partial}{\partial y} + \cos\theta\frac{\partial}{\partial z}$$

and

$$\frac{\partial}{\partial \theta} = \frac{\partial x}{\partial \theta}\frac{\partial}{\partial x} + \frac{\partial y}{\partial \theta}\frac{\partial}{\partial y} + \frac{\partial z}{\partial \theta}\frac{\partial}{\partial z}$$

$$= r\cos\theta\cos\phi\frac{\partial}{\partial x} + r\cos\theta\sin\phi\frac{\partial}{\partial y} - r\sin\theta\frac{\partial}{\partial z}.$$

The results for the operators $(\partial/\partial r, \partial/\partial \theta, \partial/\partial \phi)$ can be assembled in matrix form as

$$
\begin{bmatrix} \dfrac{\partial}{\partial r} \\[2ex] \dfrac{\partial}{\partial \theta} \\[2ex] \dfrac{\partial}{\partial \phi} \end{bmatrix} = \begin{bmatrix} \sin\theta\cos\phi & \sin\theta\sin\phi & \cos\theta \\[2ex] r\cos\theta\cos\phi & r\cos\theta\sin\phi & -r\sin\theta \\[2ex] -r\sin\theta\sin\phi & r\sin\theta\cos\phi & 0 \end{bmatrix} \begin{bmatrix} \dfrac{\partial}{\partial x} \\[2ex] \dfrac{\partial}{\partial y} \\[2ex] \dfrac{\partial}{\partial z} \end{bmatrix}.
$$

The square matrix can then be inverted to obtain expressions for the operators $(\partial/\partial x, \partial/\partial y, \partial/\partial z)$:

$$
\begin{bmatrix} \dfrac{\partial}{\partial x} \\[2ex] \dfrac{\partial}{\partial y} \\[2ex] \dfrac{\partial}{\partial z} \end{bmatrix} = \begin{bmatrix} \sin\theta\cos\phi & \dfrac{\cos\theta\cos\phi}{r} & -\dfrac{\sin\phi}{r\sin\theta} \\[3ex] \sin\theta\sin\phi & \dfrac{\cos\theta\sin\phi}{r} & \dfrac{\cos\phi}{r\sin\theta} \\[3ex] \cos\theta & -\dfrac{\sin\theta}{r} & 0 \end{bmatrix} \begin{bmatrix} \dfrac{\partial}{\partial r} \\[2ex] \dfrac{\partial}{\partial \theta} \\[2ex] \dfrac{\partial}{\partial \phi} \end{bmatrix}.
$$

(It is a straightforward matter to multiply the two square matrices and get the unit matrix as the expected result.) We proceed next to the calculation of the desired components of **L**:

$$
L_x = yp_z - zp_y \rightarrow \frac{\hbar}{i}\left(y\frac{\partial}{\partial z} - z\frac{\partial}{\partial y} \right)
$$

$$
= \frac{\hbar}{i}\left[r\sin\theta\sin\phi\left(\cos\theta\frac{\partial}{\partial r} - \frac{\sin\theta}{r}\frac{\partial}{\partial \theta} \right) \right.
$$

$$
\left. -r\cos\theta\left(\sin\theta\sin\phi\frac{\partial}{\partial r} + \frac{\cos\theta\sin\phi}{r}\frac{\partial}{\partial \theta} + \frac{\cos\phi}{r\sin\theta}\frac{\partial}{\partial \phi} \right) \right]
$$

and

$$
L_y = zp_x - xp_z \rightarrow \frac{\hbar}{i}\left(z\frac{\partial}{\partial x} - x\frac{\partial}{\partial z} \right)
$$

$$
= \frac{\hbar}{i}\left[r\cos\theta\left(\sin\theta\cos\phi\frac{\partial}{\partial r} + \frac{\cos\theta\cos\phi}{r}\frac{\partial}{\partial \theta} - \frac{\sin\phi}{r\sin\theta}\frac{\partial}{\partial \phi} \right) \right.
$$

$$
\left. -r\sin\theta\cos\phi\left(\cos\theta\frac{\partial}{\partial r} - \frac{\sin\theta}{r}\frac{\partial}{\partial \theta} \right) \right].
$$

These expressions reduce in one more step to the formulas quoted in Equations (6-11).

6-3 Rotational Motion

Angular momentum takes on several curious properties in the passage from classical to quantum mechanics. The peculiarities are fundamental to the new quantum interpretation of this important physical quantity. We encounter some of these

Figure 6-5

Particle motion constrained to a circle.

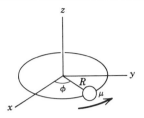

features immediately when we look at simple quantum systems in purely rotational forms of motion.

The simplest rotational problem concerns the free motion of a particle of mass μ constrained to a circle of radius R. We can visualize the particle as a bead sliding freely on a circular wire and compare the system to a particle in a one-dimensional box. Figure 6-5 shows that the problem has only one coordinate variable, the azimuthal angle ϕ. The particle is free, apart from the constraint, and so the kinetic energy is the sole contributor to the classical energy relation:

$$\frac{p_\perp^2}{2\mu} = \frac{L_z^2}{2\mu R^2} = E. \tag{6-17}$$

We have introduced the angular momentum, noting that L_z is the only nonzero component of **L**.

The quantum particle has a wave function $\Psi(\phi, t)$ and a probability density $P(\phi, t) = |\Psi(\phi, t)|^2$. We obtain the Schrödinger equation from Equation (6-17) by letting the operator substitutions $E \rightarrow i\hbar\, \partial/\partial t$ and $L_z \rightarrow (\hbar/i)\partial/\partial\phi$ act on Ψ:

$$-\frac{\hbar^2}{2\mu R^2}\frac{\partial^2}{\partial\phi^2}\Psi = i\hbar\frac{\partial}{\partial t}\Psi.$$

A stationary state with energy E is described by the wave function

$$\Psi = \psi(\phi)e^{-iEt/\hbar},$$

where ψ obeys the equation

$$\frac{d^2\psi}{d\phi^2} = -\frac{2\mu R^2}{\hbar^2}E\psi.$$

This familiar differential equation has complex exponential solutions of the form

$$\psi = Ae^{\pm i\lambda\phi} \quad \text{with} \quad \lambda^2 = \frac{2\mu R^2}{\hbar^2}E.$$

We choose these functions instead of $\cos\lambda\phi$ and $\sin\lambda\phi$ to prepare the way for a further property of the stationary states.

At this point we bring forward the essential distinction between the circular system and the particle in a box. The coordinate ϕ in the problem at hand is a *cyclic* variable that repeats itself periodically after the basic interval $[0, 2\pi]$. Therefore, if we want the stationary-state wave function Ψ to be *single valued*, we must impose a *periodicity condition* on the eigenfunction:

$$\psi(\phi + 2\pi) = \psi(\phi). \tag{6-18}$$

This requirement applies to our complex exponential solutions with the following implication:

$$Ae^{\pm i\lambda(\phi + 2\pi)} = Ae^{\pm i\lambda\phi} \Rightarrow \quad e^{\pm 2\pi i\lambda} = 1.$$

Thus, the requirement tells us that the parameter λ must satisfy

$$\cos 2\pi\lambda \pm i \sin 2\pi\lambda = 1,$$

so that the values of λ are restricted by the condition

$$\lambda = m, \quad \text{an integer}.$$

The energy eigenvalue is directly related to λ:

$$E = \frac{\hbar^2}{2\mu R^2}\lambda^2 \quad \text{so} \quad E_m = \frac{\hbar^2}{2\mu R^2}m^2. \tag{6-19}$$

The resulting *quantized* energy E_m defines a set of discrete energy levels in which the integer m appears as an *azimuthal quantum number*.

These remarkable stationary states have an additional property with regard to the angular momentum L_z. We note that the complex exponential solutions are *eigenfunctions of the angular momentum operator*:

$$\frac{\hbar}{i}\frac{\partial}{\partial\phi}Ae^{\pm im\phi} = \pm\hbar m Ae^{\pm im\phi}. \tag{6-20}$$

Consequently, the angular momentum L_z is also quantized in these states, with eigenvalues given by integer multiples of $\hbar$. The signs of the eigenvalues $\pm\hbar m$ correspond directly to counterclockwise and clockwise directions of motion. We recognize the Bohr quantization condition of Figure 4-1 in these observations. We also observe that the alternative solutions $\cos m\phi$ and $\sin m\phi$ do not possess this angular momentum interpretation and have therefore not been employed to describe the stationary states of the particle. Finally, note that the allowed values of E_m in Equation (6-19) are insensitive to the sign of m, so that every energy above the $m = 0$ level has a two-fold degeneracy. In other words, clockwise and counterclockwise motions have the same energy.

The multiplicative constant in the solution is determined by normalizing Ψ to unit probability. The probability of finding the particle in the azimuthal interval $d\phi$ is given by $|\Psi|^2\, d\phi$, and so the normalization condition becomes

$$1 = \int_0^{2\pi}|\Psi|^2\, d\phi = \int_0^{2\pi}|A|^2\, d\phi = 2\pi|A|^2.$$

Therefore, we take $A = 1/\sqrt{2\pi}$ and write the normalized wave function as

$$\Psi_m = \frac{e^{im\phi}}{\sqrt{2\pi}} e^{-iE_m t/\hbar} \quad \text{with } m = 0, \pm 1, \pm 2, \cdots. \quad (6\text{-}21)$$

Each member of this set is *simultaneously* an energy eigenfunction and an angular momentum eigenfunction, with energy eigenvalue E_m and angular momentum eigenvalue $\hbar m$. Consequently, both of the observables E and L_z are known with *zero* uncertainty in any state Ψ_m.

Equation (6-18) has played the key role in these arguments. We have deduced quantization of energy and quantization of angular momentum from the single-valued property of the wave function. Let us reconsider this condition and acknowledge that the property is not as self-evident as it may seem. We realize that the wave function Ψ is not observable and that single-valuedness should be imposed on the probability density $|\Psi|^2$. For a stationary state we have

$$\Psi = Ae^{i\lambda\phi}e^{-iEt/\hbar} \quad \text{so} \quad |\Psi|^2 = |A|^2, \quad \text{a constant.}$$

Thus, we see that $|\Psi|^2$ is already single valued and continuous, whether or not λ is equal to an integer. Of course, this assertion is restricted to situations where Ψ is a stationary state. Let us reapply the requirement of single-valuedness in the context of a *localized* particle and write the wave function as a superposition of stationary states. To make our point we consider a simple combination of two states,

$$\Psi = Ae^{i\lambda\phi}e^{-iEt/\hbar} + A'e^{i\lambda'\phi}e^{-iE't/\hbar},$$

and examine the ϕ dependence of the probability density:

$$|\Psi|^2 = |A|^2 + A^*A'e^{-i(\lambda-\lambda')\phi}e^{i(E-E')t/\hbar}$$
$$+ A'^*Ae^{i(\lambda-\lambda')\phi}e^{-i(E-E')t/\hbar} + |A'|^2.$$

We replace ϕ by $\phi + 2\pi$ and find that $|\Psi|^2$ becomes

$$|A|^2 + A^*A'e^{-i(\lambda-\lambda')\phi}e^{-2\pi i(\lambda-\lambda')}e^{i(E-E')t/\hbar}$$
$$+ A'^*Ae^{i(\lambda-\lambda')\phi}e^{2\pi i(\lambda-\lambda')}e^{-i(E-E')t/\hbar} + |A'|^2.$$

Thus, single-valuedness of the probability density implies

$$e^{2\pi i(\lambda-\lambda')} = 1 \quad \text{or} \quad \lambda - \lambda' = \text{an integer.}$$

We can implement this conclusion by taking every λ to be integer valued and thereby recover the results of the preceding paragraphs.

This entire illustration of rotational motion is flawed from the outset by the presence of the constraint. Our assumption of a precisely circular path for the particle violates the uncertainty principle and causes the fixed direction of $\mathbf{L}$ to be a violation of basic quantum principles. These difficulties are handled properly in the full three-dimensional quantum treatment of rotational motion.

Example

Let us introduce rotation in three dimensions by considering the rotating rigid body in Figure 6-6. The classical rotator consists of two point masses connected by a massless rigid rod. We assume free rotation about an axis perpendicular to

Figure 6-6

Rigid body in free rotational motion about a
perpendicular axis through the center of mass.
The classical angular momentum vector **L** has
a fixed orientation. The direction of **L** is
uncertain in quantum physics.

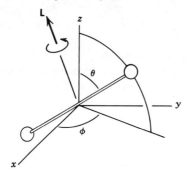

the rod and write the classical energy relation in terms of the kinetic energy
alone:

$$\frac{L^2}{2I} = E.$$

The angular momentum **L** and the moment of inertia I appear in this formula.
The quantum mechanical rigid body is described by a wave function $\Psi(\theta, \phi, t)$,
where θ and ϕ refer to the orientation of the rod as shown in the figure. We take
the operator rule for L^2 from Equation (6-12) and apply the familiar operator
rule for E to obtain the Schrödinger equation for the system:

$$-\frac{\hbar^2}{2I}\Lambda^2\Psi = i\hbar\frac{\partial}{\partial t}\Psi.$$

The wave function for a stationary state is written as

$$\Psi = \psi(\theta, \phi)e^{-iEt/\hbar},$$

where ψ and E are found by solving

$$-\frac{\hbar^2}{2I}\Lambda^2\psi = E\psi,$$

a partial differential equation in the variables (θ, ϕ). We must learn to analyze
this kind of equation in order to finish the problem. When we are able to return
to the solution we find that quantization of energy and quantization of angular
momentum are again obtained as joint results. In this case, the behavior of the
angular momentum is found to be consistent with the uncertainty principle.

6-4 Separation of Variables

The Schrödinger equation determines a unique wave function Ψ if a description of Ψ is given at $t = 0$. Our procedure in one dimension has been to find the eigenfunctions of the time-independent equation first, thereby establishing the spatial behavior of the stationary states. This complete family of energy eigenfunctions can then be used to assemble a unique solution Ψ for any given description of the state at $t = 0$. We take the same approach in three dimensions and look first at the time-independent equation for the allowed spatial eigenfunctions $\psi(r, \theta, \phi)$. Equation (6-16) is the starting point for this investigation.

We solve the partial differential equation for ψ by appealing again to the method of separation of variables. The straightforward procedure runs into complications this time, because of the intricate nature of the differential operations in spherical coordinates. The first step is to propose a product form of solution, where the radial dependence and the angular dependence occur in ψ as separate factors:

$$\psi(r, \theta, \phi) = R(r)Y(\theta, \phi). \tag{6-22}$$

The motive for this construction can be seen in the following rearranged version of Equation (6-16):

$$\frac{\partial}{\partial r} r^2 \frac{\partial}{\partial r} \psi + \frac{2\mu r^2}{\hbar^2} (E - V(r))\psi = -\Lambda^2 \psi.$$

We note that a purely radial operation appears on the left side of the equality, while a purely angular operation appears on the right. The validity and usefulness of this observation hinge on the fact that the central potential energy V depends only on r. Hence, the adoption of the product form $\psi = RY$ causes the radial and angular operations in the equation to act separately on the radial and angular factors in the eigenfunction:

$$\frac{d}{dr} r^2 \frac{dR}{dr} Y + \frac{2\mu r^2}{\hbar^2} (E - V(r))RY = -R\Lambda^2 Y.$$

We divide through by RY to find

$$\frac{1}{R} \left[\frac{d}{dr} r^2 \frac{dR}{dr} + \frac{2\mu r^2}{\hbar^2} (E - V(r))R \right] = -\frac{\Lambda^2 Y}{Y},$$

and we observe that the left side of the result depends only r, while the right side depends only on θ and ϕ. Such an equality can hold between functions of different variables only if each side of the equality is a constant. We therefore set the two sides equal to a separation constant λ and obtain the following pair of equations for Y and R:

$$-\Lambda^2 Y = \lambda Y \tag{6-23}$$

and

$$\frac{d}{dr} r^2 \frac{dR}{dr} + \frac{2\mu r^2}{\hbar^2} (E - V(r))R = \lambda R. \tag{6-24}$$

Note that the two differential equations are linked together because the separation constant appears in each equation. Thus, we conclude our first step by learning that eigenfunctions exist in the product form $\psi = RY$, provided properly behaved solutions of Equations (6-23) and (6-24) can be found.

Our main goal in this section is to determine the *angular* dependence of ψ. We set aside the equation for $R(r)$ so that we can concentrate on Equation (6-23), the partial differential equation for $Y(\theta, \phi)$. For this purpose we need to recall the detailed structure of the differential operator Λ^2, as found in Equation (6-12). We apply this operator in Equation (6-23), multiply through by $\sin^2\theta$, and rearrange terms to generate the equality

$$-\frac{\partial^2 Y}{\partial \phi^2} = \sin\theta \frac{\partial}{\partial \theta} \sin\theta \frac{\partial Y}{\partial \theta} + \lambda \sin^2\theta \, Y. \tag{6-25}$$

This equation is set up for our second step in the separation of variables, since the equality involves a pure ϕ operation on the left and a pure θ operation on the right. The next move is to look for solutions in which the function Y depends on these angles through two separate factors:

$$Y(\theta, \phi) = \Theta(\theta)\Phi(\phi). \tag{6-26}$$

We insert this construction into Equation (6-25) and divide through by $\Theta\Phi$ to find

$$-\frac{1}{\Phi}\frac{d^2\Phi}{d\phi^2} = \frac{1}{\Theta}\left(\sin\theta \frac{d}{d\theta}\sin\theta \frac{d\Theta}{d\theta} + \lambda \sin^2\theta \, \Theta\right)$$

$$= m^2. \tag{6-27}$$

The result is an equality between a function of ϕ on the left and a function of θ on the right. We argue as before that each side must be equal to a constant, and we build this property into Equation (6-27) by introducing the indicated parameter m^2 as a second separation constant. We then conclude that solutions of Equation (6-23) exist in the product form $Y = \Theta\Phi$, provided properly behaved solutions can be found for the two ordinary differential equations

$$\frac{d^2\Phi}{d\phi^2} + m^2\Phi = 0 \tag{6-28}$$

and

$$\frac{1}{\sin\theta}\frac{d}{d\theta}\sin\theta \frac{d\Theta}{d\theta} + \left(\lambda - \frac{m^2}{\sin^2\theta}\right)\Theta = 0. \tag{6-29}$$

Equation (6-27) has been divided through by $\sin^2\theta$ in the second of these results.

We have seen differential equations like Equation (6-28) many times and are well aware that the solutions are oscillating functions. In fact, the ϕ-dependent part of this analysis can be identified immediately with the purely azimuthal problem of Section 6-3. Accordingly, we choose the desired solutions of Equation (6-28) to have the complex exponential form

$$\Phi(\phi) = e^{im\phi} \quad \text{with } m \text{ positive and negative,} \tag{6-30}$$

and we reject the alternative choices provided by $\cos m\phi$ and $\sin m\phi$. Section 6-3 has taught us that these solutions have proper azimuthal behavior only if the functions are *periodic* in ϕ with period 2π. This requirement implies that m must be integer valued in Equation (6-30), as it is in the case of the rotational wave functions in Equation (6-21). Thus, m appears again as an azimuthal quantum number.

The θ dependence of the eigenfunction is found by solving Equation (6-29). This part of the problem presents a type of differential equation that has not come up in any of our previous studies of quantum systems. Let us make the following brief remarks about the solutions of the equation, without proof, and reserve the mathematical details for later discussion. We note that the domain of the polar angle θ spans the interval $[0, \pi]$ and that the differential equation contains infinities at the endpoints owing to the zeros of the factor $\sin \theta$ at $\theta = 0$ and π. The only solutions for $\Theta(\theta)$ that are finite and single valued over the whole interval are those for which the separation constant λ has the special values

$$\lambda_\ell = \ell(\ell + 1) \quad \text{with } \ell = 0, 1, 2, \ldots . \tag{6-31}$$

Thus, the integer m is joined by *another* integer ℓ in the determination of the allowed solutions of Equation (6-29). These two integers are linked by a further condition, as each given nonnegative value of ℓ is assigned a set of allowed values of m in the range

$$m = -\ell, -\ell + 1, \ldots, \ell - 1, \ell. \tag{6-32}$$

The resulting properly behaved solutions are labeled by the two integers as $\Theta_{\ell m}(\theta)$. The labels are attached to reflect the fact that the functions vary with ℓ and m through the explicit appearance of the two parameters in the differential equation. Again, we emphasize that the unfamiliar assertions in this paragraph are presented without proof and are not supposed to be self-evident. The particular statements in Equations (6-31) and (6-32) are very important results; their content is essential to our understanding of angular momentum. We provide some of the mathematics behind these conclusions at the end of the section.

Our main interest in the quoted results lies in the physical interpretation of the stationary states. Let us set the stage for this discussion by combining the acceptable solutions for $\Theta(\theta)$ and $\Phi(\phi)$ and forming a family of functions of the two angular variables:

$$Y_{\ell m}(\theta, \phi) = \Theta_{\ell m}(\theta)e^{im\phi}. \tag{6-33}$$

We call these functions *spherical harmonics*, and we characterize their properties by means of the following *eigenvalue formulas*. The combination of Equations (6-23) and (6-31) implies that $Y_{\ell m}$ must satisfy

$$-\Lambda^2 Y_{\ell m} = \ell(\ell + 1)Y_{\ell m}. \tag{6-34}$$

The elementary ϕ dependence in Equation (6-33) tells us that $Y_{\ell m}$ must also satisfy

$$\frac{1}{i}\frac{\partial}{\partial\phi}Y_{\ell m} = mY_{\ell m}. \tag{6-35}$$

The members of this set of finite single-valued functions are indexed according to the scheme indicated in Equations (6-31) and (6-32). We arrange the indices represented by m in steps of increasing ℓ as follows:

$$\ell = 0 \quad m = 0$$
$$\ell = 0 \quad m = -1, 0, 1$$
$$\ell = 2 \quad m = -2, -1, 0, 1, 2$$
$$\cdots$$

The labels ℓ and m are known as the *angular momentum quantum numbers*. The physical meaning of these symbols is the primary topic of interest in Section 6-5.

We do not intend to write down specific expressions for the functions $\Theta_{\ell m}(\theta)$. Instead, we follow established practice and present the θ dependence along with the ϕ dependence in the combined form of the spherical harmonics $Y_{\ell m}(\theta, \phi)$. Table 6-1 furnishes a partial list of these functions for the ℓ values 0, 1, and 2. The tabulated expressions are subject to a normalization condition and include certain multiplicative constants as a result of this property. According to universal convention, all the spherical harmonics are normalized to unity by the requirement

$$1 = \int_{\text{all } \Omega} |Y_{\ell m}(\theta, \phi)|^2 \, d\Omega, \tag{6-36}$$

Table 6-1   Spherical Harmonics

$\ell = 0$	$m = 0$	$Y_{00} = \sqrt{\dfrac{1}{4\pi}}$
$\ell = 1$	$m = 1$	$Y_{11} = -\sqrt{\dfrac{3}{8\pi}} \sin\theta \, e^{i\phi}$
	$m = 0$	$Y_{10} = \sqrt{\dfrac{3}{4\pi}} \cos\theta$
	$m = -1$	$Y_{1-1} = \sqrt{\dfrac{3}{8\pi}} \sin\theta \, e^{-i\phi}$
$\ell = 2$	$m = 2$	$Y_{22} = \sqrt{\dfrac{15}{32\pi}} \sin^2\theta \, e^{2i\phi}$
	$m = 1$	$Y_{21} = -\sqrt{\dfrac{15}{8\pi}} \sin\theta \cos\theta \, e^{i\phi}$
	$m = 0$	$Y_{20} = \sqrt{\dfrac{5}{16\pi}} (3\cos^2\theta - 1)$
	$m = -1$	$Y_{2-1} = \sqrt{\dfrac{15}{8\pi}} \sin\theta \cos\theta \, e^{-i\phi}$
	$m = -2$	$Y_{2-2} = \sqrt{\dfrac{15}{32\pi}} \sin^2\theta \, e^{-2i\phi}$

$$\cdots$$

where the integration ranges over $0 \leq \theta \leq \pi$ and $0 \leq \phi \leq 2\pi$, and where the element of solid angle is given by

$$d\Omega = \sin \theta \, d\theta \, d\phi.$$

It is straightforward to verify that any given $Y_{\ell m}$, in Table 6-1 or in any other tabulation, has the properties expressed by Equations (6-34), (6-35), and (6-36).

One very significant point should finally be emphasized. We observe that the potential energy $V(r)$ has no influence whatsoever in the solution of Equation (6-23), and we conclude that the determination of the angular behavior is independent of the specific nature of the central force. This conclusion means that the dependence of the wave function on θ and ϕ is *always* given by the spherical harmonics for *every* central-force problem. We also note that $V(r)$ appears in Equation (6-24), the equation for the radial function $R(r)$, and that the energy eigenvalue E makes its appearance in the same radial equation. Thus, the energy and the radial behavior of the eigenfunction are controlled by the choice of central force, but the angular behavior is controlled by the mere fact that the force is central.

Detail

Equation (6-31) is sufficiently important to warrant the following mathematical discussion. Let us proceed by transforming the structure of Equation (6-29). We define a new variable by setting

$$\zeta = \cos \theta \quad \text{so that} \quad d\zeta = -\sin \theta \, d\theta,$$

and we introduce a new notation for the solution by writing

$$\Theta(\theta) = P(\zeta).$$

The result is a differential equation for the function $P(\zeta)$ of the form

$$\frac{d}{d\zeta}(1 - \zeta^2)\frac{dP}{d\zeta} + \left(\lambda - \frac{m^2}{1 - \zeta^2}\right)P = 0,$$

in which the domains of the variables are related by the transformation

$$\theta \text{ in the interval } [0, \pi] \quad \Rightarrow \quad \zeta \text{ in the interval } [-1, 1].$$

We know in advance that m is integer valued, and we want to show that λ must satisfy Equation (6-31). For this purpose we consider first the special $m = 0$ case and examine the solution of the corresponding differential equation:

$$\frac{d}{d\zeta}(1 - \zeta^2)\frac{dP}{d\zeta} + \lambda P = 0.$$

Let us assert, without proof, that the only finite single-valued solutions of this equation are polynomial functions. We *define* the integer ℓ to denote the *order* of each polynomial and write

$$P(\zeta) = \cdots + a_\ell \zeta^\ell \quad \text{with } a_\ell \neq 0.$$

This highly abbreviated expression suffices because only the highest power of ζ is needed in the next few steps. We return to the differential equation and compute the derivatives

$$\frac{dP}{d\zeta} = \cdots + \ell a_\ell \zeta^{\ell-1}$$

and

$$\frac{d}{d\zeta}(1 - \zeta^2)\frac{dP}{d\zeta} = \frac{d}{d\zeta}\left[\cdots + \ell a_\ell(\zeta^{\ell-1} - \zeta^{\ell+1})\right]$$

$$= \cdots + \ell(\ell - 1)a_\ell \zeta^{\ell-2} - \ell(\ell + 1)a_\ell \zeta^\ell.$$

We then compare this result with the quantity

$$-\lambda P = -\lambda\left(\cdots + a_\ell \zeta^\ell\right)$$

and find that the coefficients of ζ^ℓ agree if λ obeys the formula

$$\lambda = \ell(\ell + 1),$$

as quoted in Equation (6-31). The ℓth order solutions in the $m = 0$ case are denoted as $P_\ell(\zeta)$ and are called Legendre polynomials, after the 18th century mathematician A.-M. Legendre.

Detail

Equation (6-32) is also important enough to justify a few mathematical comments. We observe that the original differential equation for $P(\zeta)$ is not affected by the sign of m and that only *positive* values of the integer need to be considered in the search for solutions. Let us write the equations for $m > 0$ and for $m = 0$ together,

$$\frac{d}{d\zeta}(1 - \zeta^2)\frac{dP}{d\zeta} + \left[\ell(\ell + 1) - \frac{m^2}{1 - \zeta^2}\right]P = 0$$

and

$$\frac{d}{d\zeta}(1 - \zeta^2)\frac{dP_\ell}{d\zeta} + \ell(\ell + 1)P_\ell = 0,$$

and make a series of swift remarks without further demonstration. The equation for P_ℓ can be differentiated m times to generate a new equation of the form

$$(1 - \zeta^2)\frac{d^2}{d\zeta^2}\frac{d^m}{d\zeta^m}P_\ell - 2\zeta(m + 1)\frac{d}{d\zeta}\frac{d^m}{d\zeta^m}P_\ell + (\ell - m)(\ell + m + 1)\frac{d^m}{d\zeta^m}P_\ell = 0.$$

A new function can also be identified by defining the expression $(1 - \zeta^2)^{m/2} d^m P_\ell / d\zeta^m$. The new equation can then be used to show that the new function satisfies the *original* differential equation for P. These arguments tell us that a whole family of solutions is obtained for each ℓ and for every positive m by the construction

$$P_{\ell m}(\zeta) = \left(1 - \zeta^2\right)^{m/2} \frac{d^m}{d\zeta^m} P_\ell(\zeta).$$

Since the highest power in P_ℓ is ζ^ℓ, the mth-order derivative produces a vanishing result for $P_{\ell m}$ if m exceeds ℓ. It follows that we must have $m \leq \ell$, as we have stated in Equation (6-32).

Example

The (θ, ϕ)-dependent solutions are easy enough to confirm as specific cases. Let us turn to Table 6-1 for the spherical harmonic with $\ell = m = 1$,

$$Y_{11} = -\sqrt{\frac{3}{8\pi}} \sin \theta e^{i\phi},$$

and verify that this function satisfies Equation (6-34). We recall Equation (6-12) to perform the differentiation:

$$-\Lambda^2 Y_{11} = \sqrt{\frac{3}{8\pi}} \left(\frac{1}{\sin \theta} \frac{\partial}{\partial \theta} \sin \theta \frac{\partial}{\partial \theta} + \frac{1}{\sin^2 \theta} \frac{\partial^2}{\partial \phi^2} \right) \sin \theta e^{i\phi}$$

$$= \sqrt{\frac{3}{8\pi}} \left[\frac{1}{\sin \theta} \frac{\partial}{\partial \theta} \sin \theta \cos \theta e^{i\phi} + \frac{1}{\sin \theta} (-e^{i\phi}) \right]$$

$$= \sqrt{\frac{3}{8\pi}} \frac{e^{i\phi}}{\sin \theta} (\cos^2 \theta - \sin^2 \theta - 1) = -2\sqrt{\frac{3}{8\pi}} \sin \theta e^{i\phi} = 2 Y_{11}.$$

The answer agrees with Equation (6-34) since $\ell(\ell + 1) = 2$ for $\ell = 1$. Equation (6-35) is easily checked and gives the result

$$\frac{1}{i} \frac{\partial}{\partial \phi} Y_{11} = -\sqrt{\frac{3}{8\pi}} \sin \theta e^{i\phi} = Y_{11},$$

as appropriate for $m = 1$. The normalizing constant in Y_{11} is also in order:

$$\int_{\text{all } \Omega} |Y_{11}|^2 \, d\Omega = \int_0^{2\pi} d\phi \int_0^\pi \sin \theta \, d\theta \left(\frac{3}{8\pi} \sin^2 \theta \right)$$

$$= \frac{3}{8\pi} \cdot 2\pi \int_{-1}^1 d\zeta (1 - \zeta^2) = \frac{3}{4} \left(\zeta - \frac{\zeta^3}{3} \right) \Bigg|_{-1}^1 = \frac{3}{4} \cdot \frac{4}{3} = 1,$$

using $\zeta = \cos\theta$ to transform the θ integration. Note that the minus sign in Y_{11} plays no part in these exercises. This sign is due to a phase convention and does not need to concern us.

6-5 Angular Momentum Quantum Numbers

Bohr's model of the atom introduces quantization of angular momentum as a supplementary quantum hypothesis. Consequently, an angular momentum quantum number appears in the model as an *ad hoc* property of the classical angular momentum. These primitive notions can now be replaced by turning to the Schrödinger formalism, where the actual behavior of the angular momentum is found to be a consequence of the theory and not a matter of hypothesis. The way is clear for this deduction of quantum properties, now that the mathematical results of Section 6-4 are properly in place.

Our approach to angular momentum in quantum mechanics is by way of the spherical harmonics in Equation (6-33). The characteristics of this family of functions are embodied in the two eigenvalue formulas found in Equations (6-34) and (6-35). We can interpret these results at once if we recall the angular momentum representation rules

$$L^2 \rightarrow -\hbar^2\Lambda^2 \quad \text{and} \quad L_z \rightarrow \frac{\hbar}{i}\frac{\partial}{\partial\phi},$$

and if we rewrite the eigenvalue equations as

$$-\hbar^2\Lambda^2 Y_{\ell m} = \hbar^2\ell(\ell+1)Y_{\ell m} \quad \text{and} \quad \frac{\hbar}{i}\frac{\partial}{\partial\phi}Y_{\ell m} = \hbar m Y_{\ell m}. \tag{6-37}$$

The two equations tell us that the spherical harmonics are simultaneous eigenfunctions of the L^2 operator and the L_z operator. Equation (6-22) transfers these properties directly to the stationary states, since the spherical harmonics provide the (θ, ϕ) dependence of the various stationary-state eigenfunctions.

Let us summarize our observations in the following two statements. The stationary states are eigenfunctions of the square of the angular momentum such that

the eigenvalues of L^2 are equal to $\hbar^2\ell(\ell+1)$,

where ℓ is any nonnegative integer.

The stationary states are also eigenfunctions of the z component of the angular momentum such that

the eigenvalues of L_z are equal to $\hbar m$,

where the integer m lies between $-\ell$ and ℓ

for each given value of the integer ℓ.

These statements express the first principles of angular momentum quantization. The occurrence of integers means that the allowed values of L^2 and L_z are *discrete*. The eigenvalue properties also imply that the allowed values occur in the stationary states with *zero uncertainty*. These quantized results for the observables L^2 and L_z are determined by the angular momentum quantum numbers ℓ and m, the integers introduced as indices for the spherical harmonics. The stationary states are labeled by the same quantum numbers as an indication that any such state is assigned the precise value $\hbar^2\ell(\ell+1)$ for L^2 and the precise value $\hbar m$ for L_z. This quantum mechanical picture of angular momentum quantization is more complex than the rudimentary concept proposed by Bohr.

The quantization scheme describes a peculiar treatment of the vector $\mathbf{L}$ in which quantized values are assigned to L_z, but no statements are made about L_x and L_y. This property of the angular momentum stems from fundamental quantum principles and deserves a closer examination. Let us identify the three Cartesian components of $\mathbf{L}$ according to the rule

$$\mathbf{L} \to \frac{\hbar}{i}\Lambda$$

by writing

$$\Lambda_x = y\frac{\partial}{\partial z} - z\frac{\partial}{\partial y}, \quad \Lambda_y = z\frac{\partial}{\partial x} - x\frac{\partial}{\partial z}, \quad \text{and} \quad \Lambda_z = x\frac{\partial}{\partial y} - y\frac{\partial}{\partial x}. \quad (6\text{-}38)$$

It is not difficult to prove that the three differential operators obey the following multiplication formulas:

$$\Lambda_x\Lambda_y - \Lambda_y\Lambda_x = -\Lambda_z,$$
$$\Lambda_y\Lambda_z - \Lambda_z\Lambda_y = -\Lambda_x,$$
$$\Lambda_z\Lambda_x - \Lambda_x\Lambda_z = -\Lambda_y. \quad (6\text{-}39)$$

The proof of these remarkable relations is left to Problem 9 at the end of the chapter. We wish to show that the uncertainty principle is at work in Equations (6-39) by demonstrating that exact nonzero values cannot be assigned to all three components of $\mathbf{L}$. We proceed by assuming the existence of a state Ψ in which $\mathbf{L}$ has the precisely specified vector value

$$\hbar\left(\ell_x\hat{\mathbf{x}} + \ell_y\hat{\mathbf{y}} + \ell_z\hat{\mathbf{z}}\right).$$

The factor $\hbar$ is included for convenience, and the ℓ's are introduced as dimensionless parameters. This hypothesis about the state Ψ can be formulated in terms of eigenvalue equations:

$$\frac{\hbar}{i}\Lambda_x\Psi = \hbar\ell_x\Psi,$$
$$\frac{\hbar}{i}\Lambda_y\Psi = \hbar\ell_y\Psi,$$
$$\frac{\hbar}{i}\Lambda_z\Psi = \hbar\ell_z\Psi. \quad (6\text{-}40)$$

Let us select the last of these statements and use the first of Equations (6-39) to obtain

$$\ell_z\Psi = -i\Lambda_z\Psi = i\left(\Lambda_x\Lambda_y - \Lambda_y\Lambda_x\right)\Psi.$$

Figure 6-7

Quantized angular momentum vectors for ℓ = 1. The allowed z components of **L** are $\pm\hbar$ and 0.

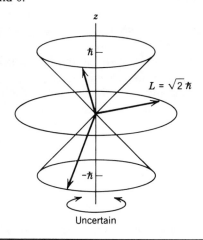

We then employ the first two of Equations (6-40) to find

$$\ell_z\Psi = i\left(i\ell_x i\ell_y - i\ell_y i\ell_x\right)\Psi.$$

We conclude that the parameter ℓ_z vanishes in the given state. Similar procedures can also be used to obtain vanishing results for ℓ_x and ℓ_y in the same state. It follows that the vector **L** is allowed to have a precise *vector* value only if the state has angular momentum equal to zero. Only one of the observables L_x, L_y, and L_z can be specified with zero uncertainty in more general circumstances, and L_z has been taken as the chosen component.

Our quantization statement concerning L_z tells us that each value of the quantum number ℓ is assigned $2\ell + 1$ values of the quantum number m between $-\ell$ and ℓ. We can interpret this property of the angular momentum by visualizing a collection of possible **L** vectors with the same allowed length $\hbar\sqrt{\ell(\ell+ 1)}$ and with $2\ell + 1$ allowed projections onto the z axis. Every member of this collection has a definite value of L_z, given by $\hbar m$ with zero uncertainty, and completely uncertain values of L_x and L_y. A precise direction for **L** is not defined, and so the vector may lie anywhere on a cone whose axis points in the z direction and whose apex angle is determined by the quantum numbers ℓ and m. There are $2\ell + 1$ such cones corresponding to the different m values for the given ℓ value. Figure 6-7 illustrates this state of affairs in the $\ell = 1$ case. There are three uncertain orientations for **L**, with three definite values of L_z, and with random values of L_x and L_y, subject to the length constraint $L_x^2 + L_y^2 + L_z^2 = 2\hbar^2$. This representation of **L** vectors should not be confused with the classical precession of **L** around a fixed axis. Precessional motion is a dynamical process caused by an external torque, whereas Figure 6-7 is a schematic picture of quantum kinematical behavior for a free **L** vector.

A quantized **L** vector is never permitted to point along the z axis because L_x and L_y would then have the value zero with no uncertainty. We have just learned that the largest eigenvalue of L_z is $\hbar\ell$ and that the length of **L** is $\hbar\sqrt{\ell(\ell+ 1)}$. Therefore, the smallest angle between **L** and the z axis is equal to $\cos^{-1}\sqrt{\ell/(\ell+1)}$, an angle that approaches zero only for very large values of the quantum number ℓ. In this limit the states having $m = \pm\ell$ are as close as possible to the case of a classical **L** vector with a fixed direction. Figure 6-7 shows a situation far from this limit, where $\ell = 1$ gives a minimum angle of $45°$ between **L** and the z axis.

The central-force problem is said to have *rotational symmetry* since the dynamics is governed by a potential energy that is independent of angles. The analysis of the problem appears to give the z component of **L** a privileged role associated with the choice of direction for the polar axis. Since the central force does not single out a preferred direction in space, it follows that the polar axis can be chosen in *any* arbitrary direction as far as the dynamics is concerned. Our conclusions about the quantization of angular momentum may be stated in more general terms as a consequence of this symmetry. The arbitrariness in the choice of z axis allows us to refer to the component of **L** along *any* direction and assign quantized values defined by the eigenvalue $\hbar m$ accordingly.

The quantization of angular momentum should seem rather unnatural from our macroscopic point of view. We have looked at quantum ideas in this light before, and so we should be prepared to see $\hbar$ play its part again as a very small quantity. We should also recall that $\hbar$ carries the units of angular momentum. Let us return to Figure 6-5 and consider a macroscopic object orbiting a fixed point at constant speed. If we let the angular momentum be of order $1 \text{ J} \cdot \text{s}$, we find that the smallness of $\hbar$ causes the corresponding angular momentum quantum number to be of order 10^{34}. It would be unrealistic to insist (or deny) that the angular momentum must have discrete values, and to claim that the direction of **L** must be uncertain, for such a classical object. However, we can expect to find firm evidence for these notions in submicroscopic systems, where angular momentum quantization becomes a vital consideration.

Example

Let us recollect the rigid-body problem from the example at the end of Section 6-3 and show that a connection develops between the quantization of angular momentum and the quantization of energy. This direct link comes about because the problem is defined by angular variables alone, as in Figure 6-6. The stationary-state eigenfunctions are known to satisfy the partial differential equation

$$-\Lambda^2\psi = \frac{2I}{\hbar^2}E\psi.$$

If we compare this with Equation (6-34), we can immediately identify the spatial eigenfunctions

$$\psi_{\ell m} = Y_{\ell m}(\theta,\phi)$$

and the energy eigenvalues

$$E_\ell = \frac{\hbar^2}{2I}\ell(\ell+1).$$

Hence, the desired stationary-state wave functions are uniquely determined by the angular momentum quantum numbers:

$$\Psi_{\ell m} = Y_{\ell m}(\theta, \phi)e^{-iE_\ell t/\hbar}.$$

Note that the energy depends only on ℓ and that each ℓ has $2\ell + 1$ different wave functions corresponding to the different values of m. Thus, each of the quantized energy levels E_ℓ has $2\ell + 1$ *degenerate* states. Equation (6-36) tells us that the wave functions are automatically normalized to unity over the total solid angle:

$$1 = \int_{\text{all } \Omega} |\Psi_{\ell m}|^2 \, d\Omega.$$

Equations (6-37) imply that the stationary states are eigenfunctions of the L^2 operator and the L_z operator:

$$-\hbar^2 \Lambda^2 \Psi_{\ell m} = \hbar^2 \ell(\ell + 1)\Psi_{\ell m} \quad \text{and} \quad \frac{\hbar}{i}\frac{\partial}{\partial \phi}\Psi_{\ell m} = \hbar m \Psi_{\ell m}.$$

Because of the elementary ϕ dependence of the spherical harmonics, the probability density is independent of ϕ in each of these states:

$$|\Psi_{\ell m}|^2 = |Y_{\ell m}|^2 = |\Theta_{\ell m}(\theta)|^2.$$

This simple observation has deeper roots in the uncertainty principle. We note that the L_z operator $(\hbar/i)\partial/\partial\phi$ is related to ϕ as the momentum operator $(\hbar/i)\partial/\partial x$ is related to x. Therefore, we might suppose that an uncertainty principle for ϕ and L_z should hold by analogy:

$$\Delta x \, \Delta p_x \geq \frac{\hbar}{2} \quad \Rightarrow \quad \Delta\phi \, \Delta L_z \geq \frac{\hbar}{2}.$$

Since $\Psi_{\ell m}$ is an eigenfunction of L_z, it follows that the uncertainty in L_z is zero and that the uncertainty in the azimuthal variable is maximal. The probability distribution is uniform in ϕ and displays no trace of localization in the azimuthal angle as a reflection of this property. It is instructive to sample some of the angular probability densities by examining the $\ell = 1$ case. We consult Table 6-1 to find

$$|\Psi_{11}|^2 = |\Psi_{1-1}|^2 = \frac{3}{8\pi}\sin^2\theta \quad \text{and} \quad |\Psi_{10}|^2 = \frac{3}{4\pi}\cos^2\theta.$$

The corresponding θ-dependent distributions are sketched in Figure 6-8. The absence of a ϕ dependence allows us to visualize the probability distributions in space by rotating these drawings around the z axis, as suggested in the figure. We let the distance from the origin to the resulting surface represent the value of $|\Psi_{\ell m}|^2$ in that direction for the given value of ℓ and m. Therefore, the quantity $|\Psi_{\ell m}(\theta, \phi, t)|^2 \, d\Omega$ expresses the probability of finding the rigid body in an element of solid angle $d\Omega$ oriented in a direction defined by the angles (θ, ϕ). Let us suppose that observations of this orientation are made for a large number

Figure 6-8

Angular probability densities for $\ell = 1$.

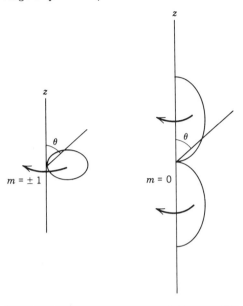

of systems with energy E_ℓ. Since E_ℓ has a $(2\ell + 1)$-fold degeneracy, the results of the measurement must conform to the *average* of the $2\ell + 1$ probability densities for the set of degenerate states with the given energy. This quantity always turns out to be a constant, independent of angles for any ℓ. To illustrate, let us consider our results for $\ell = 1$ and compute the corresponding average:

$$\frac{1}{3}\left\{|\Psi_{11}|^2 + |\Psi_{10}|^2 + |\Psi_{1-1}|^2\right\} = \frac{1}{3}\left\{\frac{3}{4\pi}(\sin^2\theta + \cos^2\theta)\right\} = \frac{1}{4\pi}.$$

The rotational symmetry of the problem is the basis for this conclusion. The result cannot depend on angles because the dynamics does not identify a spatial direction to use as a natural choice of orientation for the z axis. The $(2\ell + 1)$-fold degeneracy of the energy levels is the manifestation of this symmetry in wave-function language.

6-6 Parity

Rotational symmetry guides the central-force problem, and angular momentum provides the guiding principles. The symmetry is reflected in the angular momentum interpretation of the stationary states and in the generality of the angle dependence furnished by the spherical harmonics. These angular eigenfunctions exhibit another symmetry associated with the concept of parity. This additional property of the stationary states can be developed now, even though our solution of the quantum central-force problem is still incomplete, since the angular part of the solution is the only necessary ingredient.

Parity pertains to the reflection properties of the wave function. The concept is unique to quantum physics and has no natural place in classical mechanics. We have

Figure 6-9

Space inversion of the coordinate axes. The coordinates transform from (x, y, z) to $(-x, -y, -z)$.

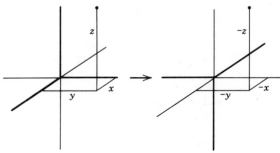

seen the idea of parity in one dimension. In particular, we have learned that a symmetric potential energy $V(x)$ leads to eigenfunctions $\psi(x)$ with either even or odd behavior under the reflection $x \rightarrow -x$. The parity operation in three dimensions is defined by reflecting all three axes according to the coordinate transformation $(x, y, z) \rightarrow (-x, -y, -z)$ indicated in Figure 6-9. The operation is also called *space inversion* as the transformation changes a right-handed system of coordinates into a left-handed system.

Parity complements the rotational symmetry of the quantum central-force problem. It is clear from Equation (6-6) that the transformation $(x, y, z) \rightarrow (-x, -y, -z)$ has no effect on r and that $V(r)$ is likewise indifferent to the inversion of the axes. This additional symmetry of V means that parity offers other information about the states in addition to the angular momentum interpretation. Figure 6-10 shows how the angular coordinates are affected by space inversion. Since the radial aspects of the solution have nothing to do with this operation, and since the general spherical-harmonic form of the angular solution is already known, it follows that the parity properties are available for immediate inspection just by looking at the behavior of the spherical harmonics.

Figure 6-10

Location of a point in spherical coordinates and in parity-inverted spherical coordinates. When the Cartesian axes are reversed, the polar angle is measured from the new z direction as $\pi - \theta$ and the azimuthal angle is measured from the new x direction as $\pi + \phi$. The radial coordinate r is the same in the two coordinate systems.

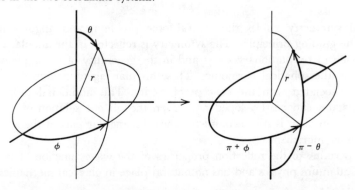

Figure 6-10 shows that the spherical coordinates (r, θ, ϕ) are transformed in the inverted system into the new spherical coordinates $(r, \pi - \theta, \pi + \phi)$. Let us apply this transformation to a stationary state with angular momentum quantum numbers ℓ and m, and with angle dependence given by $Y_{\ell m}(\theta, \phi)$. The parity operation causes the spherical harmonic to undergo the replacement of variables

$$Y_{\ell m}(\theta, \phi) \rightarrow Y_{\ell m}(\pi - \theta, \pi + \phi).$$

The new function turns out to be unchanged in value except for a possible change in sign. We refer back to Equation (6-33) and obtain a sign-changing factor through the relation

$$Y_{\ell m}(\pi - \theta, \pi + \phi) = \Theta_{\ell m}(\pi - \theta)e^{im(\pi + \phi)} = (-1)^m \Theta_{\ell m}(\pi - \theta)e^{im\phi}.$$

The connection between $\Theta_{\ell m}(\pi - \theta)$ and $\Theta_{\ell m}(\theta)$ acts as another source of minus signs. We investigate this detail below and learn that the final expression for the transformed spherical harmonic is

$$Y_{\ell m}(\pi - \theta, \pi + \phi) = (-1)^{\ell} Y_{\ell m}(\theta, \phi). \tag{6-41}$$

This formula tells us that the stationary states have *definite parity* given by the factor $(-1)^{\ell}$, in addition to the angular momentum properties associated with the quantum numbers ℓ and m. The parity is determined solely by ℓ such that states with even ℓ have even parity, and states with odd ℓ have odd parity.

Our deductions about parity are going to seem rather sterile until some physical use for the concept arises. Parity makes its appearance here as part of a classification scheme for certain types of wave function. Let us accept this notion on kinematical grounds and wait for dynamical applications to come up in due course.

Detail

Let us establish the relation between $\Theta_{\ell m}(\pi - \theta)$ and $\Theta_{\ell m}(\theta)$, and supply the missing link leading to Equation (6-41). We return to Section 6-4 and recall that our solutions for $\Theta(\theta)$ have the form

$$P_{\ell m}(\zeta) = (1 - \zeta^2)^{m/2} \frac{d^m}{d\zeta^m} P_{\ell}(\zeta), \qquad \zeta = \cos \theta.$$

We note that the parity operation produces the sign change

$$\zeta \rightarrow -\zeta \quad \text{since} \quad \cos(\pi - \theta) = -\cos \theta.$$

It is a property of the Legendre polynomials that the functions $P_{\ell}(\zeta)$ are composed of even powers of ζ for even ℓ and odd powers of ζ for odd ℓ. (The differential equation for $P_{\ell}(\zeta)$ is not affected by the sign change $\zeta \rightarrow -\zeta$, and so the solutions can be classified as even or odd.) The resulting parity of the polynomials is described by writing

$$P_{\ell}(-\zeta) = (-1)^{\ell} P_{\ell}(\zeta).$$

We use this formula to establish the corresponding behavior of the $P_{\ell m}$'s:

$$P_{\ell m}(\zeta) \rightarrow P_{\ell m}(-\zeta) = \left(1 - \zeta^2\right)^{m/2} \frac{d^m}{d(-\zeta)^m} P_\ell(-\zeta)$$

$$= \left(1 - \zeta^2\right)^{m/2}(-1)^m \frac{d^m}{d\zeta^m}(-1)^\ell P_\ell(\zeta) = (-1)^{\ell+m} P_{\ell m}(\zeta).$$

The desired result for the θ dependence is expressed as

$$\Theta_{\ell m}(\pi - \theta) = (-1)^{\ell+m}\Theta_{\ell m}(\theta).$$

Hence, the operation of space inversion causes the spherical harmonics to become

$$Y_{\ell m}(\pi - \theta, \pi + \phi) = \Theta_{\ell m}(\pi - \theta)e^{im(\pi+\phi)}$$

$$= (-1)^{\ell+m}\Theta_{\ell m}(\theta)(-1)^m e^{im\phi} = (-1)^\ell Y_{\ell m}(\theta, \phi),$$

as quoted in Equation (6-41).

Example

Let us test the parity formula by checking the behavior of the spherical harmonic Y_{11}:

$$Y_{11}(\pi - \theta, \pi + \phi) = -\sqrt{\frac{3}{8\pi}} \sin(\pi - \theta)e^{i(\pi+\phi)}$$

$$= -\sqrt{\frac{3}{8\pi}} \sin\theta(-e^{i\phi}) = -Y_{11}(\theta, \phi).$$

The result agrees with Equation (6-41), since the parity of Y_{11} is supposed to be odd.

6-7 Quantization of Energy

The central-force problem is solved in two stages after the separation of variables is introduced in Equation (6-22). All aspects of the angle dependence are established first on grounds of rotational symmetry, without any regard for the actual nature of the central force. The specific dynamics enters the problem through the choice of potential energy $V(r)$, only when consideration turns to the solution for the radial dependence. This next phase of the procedure also includes the determination of the allowed energies of the system. Thus, the two-stage approach demonstrates quantization of angular momentum and then quantization of energy, as the investigation proceeds from angular behavior in Equation (6-23) to radial behavior in Equation (6-24).

Our attention shifts to the differential equation for the radial factor $R(r)$ in the stationary-state eigenfunction. When we take Equation (6-31) into account and rearrange terms, we find that Equation (6-24) becomes

$$-\frac{\hbar^2}{2\mu r^2}\frac{d}{dr}r^2\frac{dR}{dr} + \left[V(r) + \frac{\hbar^2}{2\mu r^2}\ell(\ell+1)\right]R = ER. \tag{6-42}$$

An alternative radial equation can also be obtained by using the second version of the formula for ∇^2 given in Equation (6-8). If we substitute this expression for the radial derivative in Equation (6-42) and multiply through by r, we generate the *equivalent* radial differential equation

$$-\frac{\hbar^2}{2\mu}\frac{d^2}{dr^2}(rR) + \left[V(r) + \frac{\hbar^2}{2\mu r^2}\ell(\ell+1)\right](rR) = E(rR). \tag{6-43}$$

Our interest is drawn immediately to the close resemblance between the second form of the equation and the classical energy relation in Equation (6-2). We see that the radial differential equation follows directly from the classical formula when the two replacements

$$p_r^2 \to \left(\frac{\hbar}{i}\frac{d}{dr}\right)^2 \quad \text{and} \quad L^2 \to \hbar^2\ell(\ell+1) \quad \text{(one dimension)}$$

are introduced as operations acting on the radial function $rR(r)$. Note that the substitutions are specifically labeled as one-dimensional rules, intended only for application to the *radial* dependence of the stationary states. These replacements should not be confused with the three-dimensional operator formulas for p_r^2 and L^2 found in Equations (6-9) and (6-12).

We collect the terms in brackets in Equations (6-42) and (6-43) and define the effective potential energy for the quantum central-force problem as

$$V_{\text{eff}}(r) = V(r) + \frac{\hbar^2}{2\mu r^2}\ell(\ell+1), \tag{6-44}$$

by analogy with the classical expression in Equation (6-3). This ℓ-dependent quantity provides the means by which the given potential energy $V(r)$ makes its entrance in the problem. The energy eigenvalue E also makes its appearance along with $V(r)$ in each of the two versions of the radial differential equation. The structure of Equation (6-43) is especially interesting because this differential equation for $rR(r)$ is identical in form to Equation (5-25), the differential equation for the eigenfunction $\psi(x)$ in one dimension. The parallel between these equations suggests that our insights about the solution $\psi(x)$ for a given potential energy $V(x)$ can be transferred virtually intact to the solution $rR(r)$ for a given effective potential energy $V_{\text{eff}}(r)$. Of course, we must allow for a certain distinction between the variables, since x varies over $(-\infty, \infty)$ in one dimension while r varies over $[0, \infty)$ in three dimensions. This observation tells us that the question of evenness or oddness does not apply to the r dependence, and it reminds us that the parity of the three-dimensional eigenfunction is a property of the angular behavior alone.

Our experience with Equation (5-25) leads us to suppose that the radial differential equation should have a *family* of properly behaved solutions, corresponding to a set of allowed energies. A particular value of E is acceptable as an energy eigenvalue if the accompanying solution $R(r)$ is continuous and finite everywhere, including the point $r = 0$ and the limit $r \to \infty$. The ℓ dependence of V_{eff} is an important feature of this problem, since the explicit appearance of the quantum number ℓ in the radial equation causes the form of the differential equation to vary with the choice of ℓ. It follows that every ℓ has its own sequence of allowed results for E and $R(r)$ and that a *pair* of integer indices must be introduced to label all the radial solutions. We see that ℓ is needed as one of the labels to define the differential equation through the ℓ dependence of V_{eff} and that another *new quantum number*, denoted by n, is also required to list the quantized solutions for a given choice of ℓ. We identify the family of radial functions and energy eigenvalues by the notation

$$R_{n\ell}(r) \quad \text{and} \quad E_{n\ell},$$

and we summarize the arguments by rewriting Equation (6-42) as an *eigenvalue equation*:

$$\left\{ -\frac{\hbar^2}{2\mu r^2} \frac{d}{dr} r^2 \frac{d}{dr} + V(r) + \frac{\hbar^2}{2\mu r^2} \ell(\ell+1) \right\} R_{n\ell} = E_{n\ell} R_{n\ell}. \qquad (6\text{-}45)$$

This formula indicates a different operation on the left side of the equality for each integer ℓ and produces a correspondingly different set of allowed solutions as a result. We use ℓ to label the various sets, and we use n to index the members of each set.

We have proceeded to the solution of the central-force problem by following a series of steps beginning with Equations (6-14) and (6-22). The result of these procedures is a collection of stationary-state wave functions with the general structure

$$\Psi_{n\ell m}(r, \theta, \phi, t) = R_{n\ell}(r) Y_{\ell m}(\theta, \phi) e^{-iE_{n\ell}t/\hbar}. \qquad (6\text{-}46)$$

We note that the states of this three-dimensional problem are specified by the assignment of *three* quantum numbers. The angular momentum quantum numbers ℓ and m have already been interpreted in Section 6-5, and especially in Equations (6-37). Let us return to these two eigenvalue formulas and observe that the eigenfunction behavior can be transferred directly to the stationary states:

$$-\hbar^2 \Lambda^2 \Psi_{n\ell m} = \hbar^2 \ell(\ell+1) \Psi_{n\ell m} \qquad (6\text{-}47)$$

and

$$\frac{\hbar}{i} \frac{\partial}{\partial \phi} \Psi_{n\ell m} = \hbar m \Psi_{n\ell m}. \qquad (6\text{-}48)$$

The equations tell us that the wave function $\Psi_{n\ell m}$ is a simultaneous eigenfunction of the L^2 operator and the L_z operator, with L^2 eigenvalue $\hbar^2 \ell(\ell+1)$ and L_z eigenvalue $\hbar m$. Equation (6-41) then tells us that $\Psi_{n\ell m}$ has definite parity, since the state has the space-inversion property

$$\Psi_{n\ell m} \to (-1)^{\ell} \Psi_{n\ell m} \quad \text{under the transformation } (x, y, z) \to (-x, -y, -z). \quad (6\text{-}49)$$

The stationary states and energy eigenvalues are also designated by the third quantum number n, whose meaning has been described in general terms in the preceding paragraph. Those observations can be formulated by the further property

$$ i\hbar \frac{\partial}{\partial t} \Psi_{n\ell m} = E_{n\ell} \Psi_{n\ell m}, \tag{6-50} $$

whereby the wave function $\Psi_{n\ell m}$ is identified as an energy eigenfunction with energy eigenvalue $E_{n\ell}$.

We may assume that n labels the energy levels for a given ℓ such that the energies increase with n, as is the case in one dimension where n is the only quantum number. The one-dimensional problem has taught us that the larger energy eigenvalues belong to the eigenfunctions $\psi(x)$ with the greater numbers of nodes. The same must be true for the function $rR(r)$ in three dimensions, since $\psi(x)$ and $rR(r)$ satisfy similar differential equations. This correspondence suggests that we call n the *radial node quantum number* and identify the integer with the number of nodes of the radial function $R_{n\ell}$. We illustrate these remarks in the example below.

The main conclusion of the central-force problem is the fact that wave functions exist as *simultaneous* eigenfunctions of the energy, the square of the angular momentum, and the z component of the angular momentum. We have just observed that the stationary states $\Psi_{n\ell m}$ are endowed with these remarkable properties. Such findings are noteworthy indeed, because the situation enables us to determine all three independent physical quantities at the same time with *zero* uncertainty. Again, the rotational symmetry of the central force is the essential ingredient behind this coincidence. Let us recall that angular momentum is a conserved quantity in the classical version of the central-force problem. We now find that the conservation law for angular momentum emerges in the quantum problem through the occurrence of states in which E, L^2, and L_z are observable without uncertainty. The labeling scheme provided by the quantum numbers n, ℓ, and m is supposed to be interpreted in this context.

It should be noted that the quantum number m is not among the indices for the quantized energy in Equation (6-50). This integer does not appear anywhere in the radial differential equation, and so the absence of m from the allowed solutions for E and $R(r)$ is an expected result. The existence of a *degeneracy* is signaled by the fact that $E_{n\ell}$ does not depend on m. Degenerate states occur for a given choice of n and ℓ because the corresponding energy eigenvalue $E_{n\ell}$ has a collection of $2\ell + 1$ distinct wave functions, one for each possible value of m. The different wave functions with the same energy vary with m according to the expression in Equation (6-46) and differ among themselves according to the various z components of the quantized angular momentum. This degeneracy is another consequence of the rotational symmetry of the central-force problem. The force does not provide a natural direction for the choice of polar axis, and so observable quantities like the energy cannot depend on m, the quantum number associated with this choice. We also note that E contains an L^2 term from the outset in Equation (6-2), and we therefore expect the energy eigenvalue to vary with the L^2 quantum number ℓ.

An important general observation can be made by examining Equation (6-45) at small r. We see that two of the three terms on the left side of the equality contain the factor $\hbar^2/2\mu r^2$ and hence become infinite as $r \to 0$. Let us assume that the potential energy is less divergent than $1/r^2$ (as is the case for the $1/r$ Coulomb potential energy), so that the $V(r)$ term in the equation can be ignored for small r. We can then show that the radial solution near the origin must obey the general power law

$$ R(r) \to Ar^{\ell} \quad \text{as } r \to 0, \tag{6-51} $$

Figure 6-11

Radial behavior of the eigenfunctions near $r = 0$ for various values of ℓ.

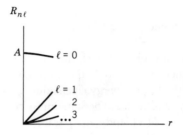

where A is an unspecified constant. We verify this behavior in Equation (6-45) by noting that the first and third terms on the left behave as $r^{\ell-2}$ and dominate at small r if Equation (6-51) holds, and by demonstrating that the two dominant contributions satisfy the differential equation. The proposed power law results in the equality

$$\frac{d}{dr} r^2 \frac{dR}{dr} = \ell(\ell+1) A r^\ell = \ell(\ell+1) R \quad \text{near } r = 0,$$

and so the first and third terms achieve the desired solution for small values of r. (Note that a $1/r^{\ell+1}$ power law also satisfies the equality but introduces an unacceptable divergence at the origin.) Figure 6-11 illustrates the power behavior of $R_{n\ell}(r)$ for various increasing values of ℓ. Our assumption about $V(r)$ means that the given potential energy becomes important only away from the origin and that the ℓ-dependent centrifugal part of $V_{\text{eff}}(r)$ is the decisive term near $r = 0$. We interpret our result by noting that the centrifugal contribution is always *repulsive* and tends to push the wave function away from the origin, as the name of the term suggests and as the figure indicates.

We know from Chapter 5 that the Schrödinger equation has familiar analytical solutions for only a few special choices of potential energy. We also know that progress can always be made from qualitative arguments pertaining to the solutions. The following illustration falls in the latter category.

Example

Let the potential energy be a linear function of r and consider the $\ell = 0$ case, so that the effective potential energy is written as

$$V_{\text{eff}} = V = cr.$$

We are concerned with radial solutions of the form $rR(r)$, and so we turn to Equation (6-43) and rearrange terms to get

$$\frac{d^2}{dr^2}(rR) = -\frac{2\mu}{\hbar^2}[E - V(r)](rR).$$

The analytical solutions of this equation are rather unfamiliar for a linear

Figure 6-12

Energy levels and radial functions $rR(r)$ for a linear potential energy with $\ell = 0$.

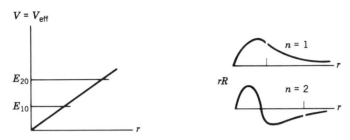

potential energy. Let us proceed graphically instead and deduce the shape of $rR(r)$ for the two lowest $\ell = 0$ energy levels. The differential equation describes the curvature of the solution relative to the value of the function at each point r. The function $rR(r)$ is subject to the condition that the solution must vanish at the endpoint $r = 0$, because of Equation (6-51), and in the limit $r \rightarrow \infty$, because of the requirement of normalizability. As the quantized energies increase, the corresponding solutions exhibit more curvature over larger intervals and display more nodes in the region where nodes are allowed. These arguments tell us that the first solution should have only the one node at $r = 0$ and that the second solution should have a second node. Figure 6-12 shows V_{eff} along with the two lowest energies $E_{n\ell}$ for $n = 1$ and $n = 2$ in the $\ell = 0$ case. A qualitative sketch of the two $\ell = 0$ solutions is also included to illustrate our remarks about curvature and nodes. Note that each energy gives a single classical turning point where the curvature of $rR(r)$ changes sign and that each solution decays without nodes for values of r beyond the turning point. The $\ell = 0$ case is special because the centrifugal contribution does not appear in the effective potential energy. We leave the $\ell \neq 0$ situation to Problem 14 at the end of the chapter.

6-8   Observables in Spherical Coordinates

The main goals of this chapter are fulfilled in Equation (6-46), the expression for the stationary states. These wave functions exhibit quantization of angular momentum and quantization of energy as compatible properties of the central-force system. The methods can now be used to evaluate and interpret certain physical quantities in terms of the angular momentum framework.

We begin with the probability properties of a general wave function Ψ. The corresponding probability density $|\Psi(r, \theta, \phi, t)|^2$ represents a probability per unit volume in spherical polar coordinates. Figure 6-13 shows that the element of volume in these coordinates is given by the infinitesimal quantity

$$d\tau = r^2\,dr\,\sin\theta\,d\theta\,d\phi = r^2\,dr\,d\Omega,$$

where $d\Omega$ denotes the familiar element of solid angle. The wave function is normal-

Figure 6-13

Volume element $d\tau$ in spherical polar coordinates.

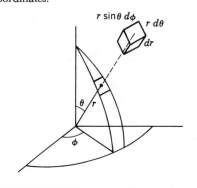

ized to unit probability over all space if Ψ satisfies the condition

$$1 = \int_{\text{all space}} |\Psi|^2 \, d\tau.$$

Thus, the element of probability $|\Psi(r, \theta, \phi, t)|^2 \, d\tau$ expresses the likelihood of finding the particle in the volume $d\tau$ at the location (r, θ, ϕ) at time t.

Let us examine the normalization condition for the special case of a stationary state of the kind described in Equation (6-46). The multiplicative dependence on the variables enables us to split the three-dimensional integration into separate radial and angular factors:

$$1 = \int_{\text{all space}} |\Psi_{n\ell m}|^2 \, d\tau$$

$$= \int_0^\infty r^2 \, dr \int_{\text{all }\Omega} d\Omega |R_{n\ell}(r)|^2 |Y_{\ell m}(\theta, \phi)|^2.$$

If we then recall the normalization of the spherical harmonics from Equation (6-36), we find that the normalization condition for a stationary state reduces at once to a single integration over r:

$$1 = \int_0^\infty |R_{n\ell}(r)|^2 r^2 \, dr. \tag{6-52}$$

This result invites us to introduce a *radial probability density*

$$P_{n\ell}(r) = r^2 |R_{n\ell}(r)|^2, \tag{6-53}$$

subject to the condition in Equation (6-52):

$$1 = \int_0^\infty P_{n\ell}(r) \, dr.$$

We interpret the differential element $P_{n\ell}(r) \, dr$ as the probability of finding the particle anywhere in a spherical shell of thickness dr at a distance r from the origin. We note that Equation (6-53) contains *no* dependence on the time, and we emphasize that $P_{n\ell}(r)$ is defined *only* for a stationary state (hence the indices n and ℓ). Attention should also be drawn to the explicit appearance of r^2 in Equation (6-53), since this factor causes the radial probability density to vanish at the origin for every $\Psi_{n\ell m}$. Equation (6-51) tells us that $P_{n\ell}(r)$ behaves as $r^{2\ell+2}$, while $|\Psi_{n\ell m}|^2$ behaves as $r^{2\ell}$, near $r = 0$. This distinction in behavior at the origin arises because the radial probability density is defined with respect to the thickness of a spherical shell, whose surface area supplies the extra factor of r^2.

The probability density has an interesting spatial structure in the case of a stationary state. Equations (6-33) and (6-46) tell us that the ϕ dependence and t dependence of $\Psi_{n\ell m}$ are given by the simple phase factors $e^{im\phi}$ and $e^{-iE_{n\ell}t/\hbar}$. These functions have unit modulus, and so the probability density is independent of ϕ and t in such a state:

$$|\Psi_{n\ell m}|^2 = |R_{n\ell}(r)|^2|\Theta_{\ell m}(\theta)|^2. \tag{6-54}$$

The resulting distribution in the variables r and θ forms a "cloud of probability" whose configuration in space is stationary and symmetric about the z axis. The cloud has vanishing density on certain nodal surfaces defined by the zeros of the two factors in Equation (6-54). Nodes occur in $R_{n\ell}(r)$ at definite radii for given values of n and ℓ and produce *spherical* surfaces of zero density. Nodes also occur in $\Theta_{\ell m}(\theta)$ at definite polar angles for given values of ℓ and m and produce *conical* surfaces of zero density, oriented symmetrically around the z axis. These nodal surfaces intersect at right angles within the distribution of probability and divide the cloud into regions that become more and more numerous as the quantum numbers become larger and larger. We should emphasize that the cloud does *not* represent a "spatial smearing of the particle," where $|\Psi|^2\,d\tau$ defines a "fraction of the particle" found in the cell $d\tau$. Instead, a spatial smearing of the *probability* is intended, where $|\Psi|^2\,d\tau$ gives the likelihood of finding the *entire* particle at the location of interest. We should also recall the implications of the uncertainty principle in connection with the ϕ and t independence of Equation (6-54). The expression for $|\Psi_{n\ell m}|^2$ indicates maximal uncertainty in the azimuthal angle and the time, since the distribution describes complete nonlocalization of the particle with respect to these variables. This observation reflects the fact that the z component of the angular momentum and the energy are known with zero uncertainty in the state $\Psi_{n\ell m}$. Let us finally remark, once again, that the properties described are those of a *stationary* state.

The evaluation of any physical quantity depends on the state of the system and involves the determination of an expectation value. We evaluate a given observable by performing measurements of the quantity for a large number of separate systems in the same state Ψ. The average of these measurements corresponds to the expectation value of the observable in the given state. We follow the rules established in Chapter 5 and express this expectation value in terms of an integral

$$\int \Psi^*Q\Psi\,d\tau,$$

where Q is inserted as a multiplicative or differential operator to represent the particular observable. The central-force problem calls for the evaluation of two main types of physical quantity. We find that the angular part of the spatial integration reduces to a simple exercise in each case, because of the angular momentum properties of the states.

We look first at the expectation value of r in the stationary state $\Psi_{n\ell m}$:

$$\langle r \rangle = \int_{\text{all space}} \Psi_{n\ell m}^* r \Psi_{n\ell m} \, d\tau. \tag{6-55}$$

The calculation simplifies immediately when Equation (6-46) is inserted and Equation (6-36) is employed:

$$\langle r \rangle = \int_0^\infty r^2 \, dr \int_{\text{all } \Omega} d\Omega \, r |R_{n\ell}(r)|^2 |Y_{\ell m}(\theta, \phi)|^2$$

$$= \int_0^\infty r^3 |R_{n\ell}(r)|^2 \, dr = \int_0^\infty r P_{n\ell}(r) \, dr. \tag{6-56}$$

Note that the final result can be given in terms of the radial probability density and that the expectation value of any function of r can be similarly expressed:

$$\langle F(r) \rangle = \int_0^\infty F(r) P_{n\ell}(r) \, dr. \tag{6-57}$$

The same expectation value has the more general form

$$\langle F(r) \rangle = \int \Psi^* F(r) \Psi \, d\tau = \int_0^\infty r^2 \, dr \int d\Omega \, F(r) |\Psi(r, \theta, \phi, t)|^2 \tag{6-58}$$

when Ψ is not a stationary state. We note that Equation (6-58) may depend on t, while Equation (6-57) must be time independent and stationary, as expected for a stationary state.

Let us turn next to the computation of expectation values for some of the important angular operators. We start with the evaluation of L^2 in the state $\Psi_{n\ell m}$ and recall the operator formula from Equation (6-12) to find

$$\langle L^2 \rangle = \int \Psi_{n\ell m}^* (-\hbar^2 \Lambda^2) \Psi_{n\ell m} \, d\tau. \tag{6-59}$$

No differentiation is necessary here, despite the appearance of the operator in the integrand. Equation (6-47) is put to use, along with the normalization of the wave function, to generate a simple calculation instead:

$$\langle L^2 \rangle = \hbar^2 \ell(\ell + 1) \int \Psi_{n\ell m}^* \Psi_{n\ell m} \, d\tau$$

$$= \hbar^2 \ell(\ell + 1). \tag{6-60}$$

We proceed in similar fashion to the expectation value of L_z in the same state and use the operator formula in Equation (6-10) to write

$$\langle L_z \rangle = \int \Psi_{n\ell m}^* \frac{\hbar}{i} \frac{\partial}{\partial \phi} \Psi_{n\ell m} \, d\tau. \tag{6-61}$$

Again, the indicated differentiation is bypassed, as the eigenvalue property in Equation (6-48) and the normalization condition lead directly to the desired conclusion:

$$\langle L_z \rangle = \hbar m \int \Psi_{n\ell m}^* \Psi_{n\ell m} \, d\tau$$

$$= \hbar m. \tag{6-62}$$

Equations (6-60) and (6-62) tell us what the measured values of L^2 and L_z should be, as *averages* over a large number of systems all in the same state.

We can make a stronger statement if we consider the uncertainty properties of the state in the light of Equations (6-47) and (6-48). To this end we consider the following quantities and apply the eigenvalue relations twice to obtain

$$\langle (L^2)^2 \rangle = \int \Psi_{n\ell m}^* (-\hbar^2 \Lambda^2)(-\hbar^2 \Lambda^2) \Psi_{n\ell m} \, d\tau = \left[\hbar^2 \ell (\ell + 1) \right]^2$$

and

$$\langle L_z^2 \rangle = \int \Psi_{n\ell m}^* \left(\frac{\hbar}{i} \frac{\partial}{\partial \phi} \right) \left(\frac{\hbar}{i} \frac{\partial}{\partial \phi} \right) \Psi_{n\ell m} \, d\tau = (\hbar m)^2.$$

These results imply the properties

$$\langle (L^2)^2 \rangle = \langle L^2 \rangle^2 \quad \text{and} \quad \langle L_z^2 \rangle = \langle L_z \rangle^2$$

for a stationary state, and hence confirm the claim that the uncertainties ΔL^2 and ΔL_z are equal to zero in any such state. We conclude that the determination of L^2 and L_z *always* gives the answers expressed in Equations (6-60) and (6-62) for every measurement of every system described by $\Psi_{n\ell m}$. The examples at the end of the section illustrate these arguments further in situations where Ψ is a different type of state.

The last item on our agenda pertains to the orthogonality of the stationary states. The proof of this property involves the use of vector calculus and is left as Problem 16 at the end of the chapter. Orthogonality has already appeared as a mathematical concept in one dimension and is formulated as before in terms of the spatial eigenfunctions in the stationary states

$$\Psi_{n\ell m} = \psi_{n\ell m}(r, \theta, \phi) e^{-iE_{n\ell} t/\hbar}.$$

We let $\psi_{n_1 \ell_1 m_1}$ and $\psi_{n_2 \ell_2 m_2}$ be any two different members of the family of eigenfunctions and write the *orthogonality condition* as

$$\int_{\text{all space}} \psi_{n_1 \ell_1 m_1}^* \psi_{n_2 \ell_2 m_2} \, d\tau = 0. \tag{6-63}$$

This integral vanishes whenever the two sets of quantum numbers $(n_1 \ell_1 m_1)$ and $(n_2 \ell_2 m_2)$ are not the same in all three integers. The spherical harmonics have a similar property whereby

$$\int_{\text{all } \Omega} Y^*_{\ell_1 m_1} Y_{\ell_2 m_2} \, d\Omega = 0 \tag{6-64}$$

whenever $(\ell_1 m_1)$ and $(\ell_2 m_2)$ are different pairs of quantum numbers. These integral formulas are used frequently as calculational tools to supplement the normalization properties of the various functions. Their usefulness is illustrated in the following exercises.

Example

We have concentrated on the properties of a pure state $\Psi_{n\ell m}$ and should now consider a more general kind of wave function. We know that a superposition of these states is a solution of the Schrödinger equation, so let us construct our wave function by combining two pure states with different L_z quantum numbers:

$$\Psi = \frac{1}{\sqrt{2}} (\Psi_{111} + \Psi_{11-1}) = \frac{1}{\sqrt{2}} R_{11}(Y_{11} + Y_{1-1}) e^{-iE_{11}t/\hbar}.$$

We note that Ψ_{111} and Ψ_{11-1} are degenerate, so that the t dependence of Ψ has a unique frequency. It follows that a measurement of the energy in this state must yield the result

$$\langle E \rangle = E_{11} \quad \text{with zero uncertainty.}$$

The factor $1/\sqrt{2}$ is included to ensure that Ψ is normalized. Let us verify this point by computing the normalization integral:

$$\int |\Psi|^2 \, d\tau = \frac{1}{2} \int |\Psi_{111}|^2 \, d\tau + \frac{1}{2} \int |\Psi_{11-1}|^2 \, d\tau = \frac{1}{2} + \frac{1}{2}.$$

We observe that the cross-terms drop out of the calculation because of the orthogonality property, and we recall that every $\Psi_{n\ell m}$ is normalized by the condition

$$1 = \int_{\text{all space}} |\Psi_{n\ell m}|^2 \, d\tau.$$

Each piece of Ψ is an eigenfunction of L^2 with the same $\ell = 1$ eigenvalue $2\hbar^2$. Therefore, we can write

$$-\hbar^2\Lambda^2\Psi = 2\hbar^2\Psi \quad \text{and} \quad (-\hbar^2\Lambda^2)(-\hbar^2\Lambda^2)\Psi = (2\hbar^2)^2\Psi$$

and apply these eigenvalue relations along with the normalization condition to obtain the expectation values

$$\langle L^2 \rangle = \int \Psi^*(-\hbar^2\Lambda^2)\Psi \, d\tau = 2\hbar^2$$

and

$$\langle (L^2)^2 \rangle = \int \Psi^*(-\hbar^2\Lambda^2)(-\hbar^2\Lambda^2)\Psi \, d\tau = (2\hbar^2)^2.$$

The equality between $\langle (L^2)^2 \rangle$ and $\langle L^2 \rangle^2$ implies that a measurement of L^2 must yield the result

$$\langle L^2 \rangle = 2\hbar^2 \quad \text{with zero uncertainty.}$$

The uncertainty in L_z is not zero for a particle in the state Ψ. To establish this property we first consult Equation (6-48) to find

$$\frac{\hbar}{i}\frac{\partial}{\partial\phi}(\Psi_{111} + \Psi_{11-1}) = \hbar\Psi_{111} - \hbar\Psi_{11-1}$$

and

$$\left(\frac{\hbar}{i}\frac{\partial}{\partial\phi}\right)\left(\frac{\hbar}{i}\frac{\partial}{\partial\phi}\right)(\Psi_{111} + \Psi_{11-1}) = \hbar^2\Psi_{111} + \hbar^2\Psi_{11-1}.$$

We then compute the expectation values

$$\langle L_z \rangle = \int \Psi^* \frac{\hbar}{i}\frac{\partial}{\partial\phi}\Psi\, d\tau = \frac{1}{2}\int (\Psi_{111}^* + \Psi_{11-1}^*)\hbar(\Psi_{111} - \Psi_{11-1})\, d\tau = 0$$

and

$$\langle L_z^2 \rangle = \int \Psi^* \left(\frac{\hbar}{i}\frac{\partial}{\partial\phi}\right)\left(\frac{\hbar}{i}\frac{\partial}{\partial\phi}\right)\Psi\, d\tau$$

$$= \frac{1}{2}\int (\Psi_{111}^* + \Psi_{11-1}^*)\hbar^2(\Psi_{111} + \Psi_{11-1})\, d\tau = \hbar^2.$$

Again, the orthogonality and normalization of the individual stationary states are put to use. The uncertainty in L_z is obtained from these results:

$$(\Delta L_z)^2 = \langle L_z^2 \rangle - \langle L_z \rangle^2 = \hbar^2 \quad \text{so} \quad \Delta L_z = \hbar.$$

Hence, measurements of L_z for a large number of systems in the state Ψ must yield an average value

$$\langle L_z \rangle = 0 \quad \text{with fluctuation } \pm\hbar \text{ about the average.}$$

We consult Table 6-1 and find that the angle dependence of the wave function is given by

$$Y_{11} + Y_{1-1} = -\sqrt{\frac{3}{8\pi}}\sin\theta(e^{i\phi} - e^{-i\phi})$$

$$= -i\sqrt{\frac{3}{2\pi}}\sin\theta\sin\phi = -i\sqrt{\frac{3}{2\pi}}\frac{y}{r}.$$

Consequently, the angular distribution of $|\Psi|^2$ is described by the expression

$$|Y_{11} + Y_{1-1}|^2 = \frac{3}{2\pi}\sin^2\theta\sin^2\phi = \frac{3}{2\pi}\left(\frac{y}{r}\right)^2.$$

This probability density is not symmetric about the z axis, unlike the situation for a pure eigenfunction of L_z. Instead, the distribution has lobes of maximum density aligned along the $\pm y$ axes, so that the probability of finding the particle is somewhat localized in the two directions $\phi = \pi/2$ and $\phi = 3\pi/2$. Some localization in ϕ is expected since the uncertainty in L_z is not zero. The fact that $|\Psi|^2$ behaves as $(y/r)^2$ tells us that the probability distribution is symmetric about the y axis and suggests that our state is an eigenfunction of L_y. It is therefore apparent that a choice of polar axis in the y direction would yield a simpler description of the state. We should recognize from the outset that we are considering a rather contrived wave function. The state Ψ is stationary since $|\Psi|^2$ is not time dependent, even though Ψ is not a pure state of the form $\Psi_{n\ell m}$. In fact, *any* superposition

$$a\Psi_{n\ell m} + a'\Psi_{n\ell m'} + a''\Psi_{n\ell m''} + \cdots$$

is perfectly acceptable as a stationary state, since each of the contributing members has the same energy $E_{n\ell}$. This degeneracy with respect to the L_z quantum number has already been interpreted as a consequence of rotational symmetry. We can appeal to the symmetry and choose a new direction for the polar axis so that any such superposition of degenerate states transforms into a simpler single-term wave function.

Example

Let us construct a nonstationary state by forming an equal-parts combination of two nondegenerate states with different values of ℓ:

$$\Psi = \frac{1}{\sqrt{2}}(\Psi_{100} + \Psi_{110}) = \frac{1}{\sqrt{2}}\left(R_{10}Y_{00}e^{-iE_{10}t/\hbar} + R_{11}Y_{10}e^{-iE_{11}t/\hbar}\right).$$

It is easy to show that the probability density oscillates with frequency $(E_{11} - E_{10})/h$ in such a state. The following calculations demonstrate that the energy of the particle has nonzero uncertainty as a further related property. We apply the energy operator to Ψ and find

$$i\hbar\frac{\partial}{\partial t}(\Psi_{100} + \Psi_{110}) = E_{10}\Psi_{100} + E_{11}\Psi_{110}$$

and

$$\left(i\hbar\frac{\partial}{\partial t}\right)\left(i\hbar\frac{\partial}{\partial t}\right)(\Psi_{100} + \Psi_{110}) = E_{10}^2\Psi_{100} + E_{11}^2\Psi_{110}.$$

The necessary expectation values are then given by the expressions

$$\langle E \rangle = \int \Psi^* i\hbar\frac{\partial}{\partial t}\Psi\, d\tau$$

$$= \frac{1}{2}\int(\Psi_{100}^* + \Psi_{110}^*)(E_{10}\Psi_{100} + E_{11}\Psi_{110})\, d\tau = \frac{1}{2}(E_{10} + E_{11})$$

and

$$\langle E^2 \rangle = \int \Psi^* \left(i\hbar \frac{\partial}{\partial t} \right) \left(i\hbar \frac{\partial}{\partial t} \right) \Psi \, d\tau$$

$$= \frac{1}{2} \int \left(\Psi_{100}^* + \Psi_{110}^* \right) \left(E_{10}^2 \Psi_{100} + E_{11}^2 \Psi_{110} \right) d\tau = \frac{1}{2} \left(E_{10}^2 + E_{11}^2 \right).$$

The uncertainty in E follows from these two results:

$$(\Delta E)^2 = \langle E^2 \rangle - \langle E \rangle^2 = \tfrac{1}{2} \left(E_{10}^2 + E_{11}^2 \right) - \tfrac{1}{4} \left(E_{10} + E_{11} \right)^2 = \tfrac{1}{4} \left(E_{11} - E_{10} \right)^2,$$

and so

$$\Delta E = \tfrac{1}{2} \left(E_{11} - E_{10} \right).$$

Our conclusions should seem reasonable for an equal-parts wave function, since the average energy $\langle E \rangle$ lies midway between the two levels E_{10} and E_{11}, and the energy uncertainty ΔE gives the deviation in energy on either side of the average. Let us apply the same logic to the calculation of $\langle L^2 \rangle$. We have an equal-parts admixture of $\ell = 1$ and $\ell = 0$ eigenfunctions, and so we expect to find

$$\langle L^2 \rangle = \tfrac{1}{2} \left(2\hbar^2 + 0\hbar^2 \right) = \hbar^2 \quad \text{and} \quad \Delta L^2 = \tfrac{1}{2} \left(2\hbar^2 - 0\hbar^2 \right) = \hbar^2.$$

Direct confirmation of these results is left to Problem 18 at the end of the chapter. It is clear that Ψ is an eigenfunction of L_z with eigenvalue zero, since both Ψ_{100} and Ψ_{110} are $m = 0$ states. We must therefore have

$$\langle L_z \rangle = 0 \quad \text{with zero uncertainty}.$$

Hence, measurements of L_z must always yield the value zero for any system in the state Ψ.

Problems

1. Determine an expression for $r_{\min}$, the distance of closest approach on the orbit of a classical particle, for the case of a repulsive Coulomb potential energy

$$V(r) = \frac{b}{r}, \qquad b > 0.$$

   Assume that the particle has nonzero angular momentum.

2. Consider the orbit of a classical particle with an attractive Coulomb potential energy

$$V(r) = -\frac{b}{r}, \qquad b > 0,$$

   and determine expressions for the radial turning points when E is negative and when E is positive. Note that, for $E < 0$, there is a minimum possible E for each nonzero value of the angular momentum L.

3. Use the operator results for L_x, L_y, and L_z, as found in Section 6-2, to derive the operator formula for the quantity $L^2 = L_x^2 + L_y^2 + L_z^2$.

4. The particle in Figure 6-5 has a wave function $\Psi(\phi, t)$ and a probability density $P(\phi, t) = |\Psi(\phi, t)|^2$. The wave function satisfies the equation

$$-\frac{\hbar^2}{2\mu R^2}\frac{\partial^2}{\partial \phi^2}\Psi + V(\phi)\Psi = i\hbar\frac{\partial}{\partial t}\Psi,$$

in which $V(\phi)$ is included as a real-valued potential energy. A probability current density $j(\phi, t)$ can be defined such that j and P obey the probability conservation law

$$\frac{\partial}{\partial \phi}j(\phi, t) + \frac{\partial}{\partial t}P(\phi, t) = 0.$$

Deduce the appropriate formula for j.

5. Set $V = 0$ in the previous problem, and let the wave function have the form

$$\Psi(\phi, t) = Ae^{i(m\phi - \omega t)},$$

where m is an integer. Use the result of the problem to compute $j(\phi, t)$ for this state, and interpret the answer.

6. Consider a bead constrained to move on a circular wire, as in Figure 6-5, and assume the existence of a barrier at $\phi = 0$ that the bead cannot cross. (Imagine an inaccessible region $-\phi_0 < \phi < \phi_0$ in the limit $\phi_0 \to 0$.) What condition on the wave function $\Psi(\phi, t)$ does this stipulation imply? Determine the allowed energies and the corresponding eigenfunctions. Are there any degeneracies among the states?

7. A particle of mass μ moves freely on the surface of a sphere of radius R. Express the classical energy relation in terms of the angular momentum $\mathbf{L}$, and use the operator formulas to deduce the Schrödinger equation for the system. Let the wave function Ψ be an energy eigenfunction with energy eigenvalue E and spatial eigenfunction ψ. What is the form of the differential equation for ψ?

8. Refer to Table 6-1 for the spherical harmonic Y_{22}, and verify that this function satisfies the eigenvalue equation $-\Lambda^2 Y_{\ell m} = \ell(\ell + 1)Y_{\ell m}$ and the normalization condition $\int |Y_{\ell m}|^2\, d\Omega = 1$.

9. Let the angular momentum $\mathbf{L}$ be represented in terms of the differential operators Λ_x, Λ_y, and Λ_z introduced in Section 6-5, and derive the relation

$$\Lambda_x \Lambda_y - \Lambda_y \Lambda_x = -\Lambda_z.$$

The other multiplication formulas are obtained in like manner.

10. A hydrogen molecule in purely rotational motion may be visualized as a rigid dumbbell like the example in Figure 6-6. Draw an energy level diagram for the stationary states of the system, and indicate the quantum numbers and degeneracy of each level. Let the separation between the hydrogen atoms in the molecule be 0.075 nm, and calculate the numerical value (in eV) for the energy scale unit $\hbar^2/I$, where I denotes the moment of inertia. Determine the allowed transition energies if the transitions obey the selection rule $\Delta\ell = \pm 2$. Calculate the emitted wavelength corresponding to the transition $\ell = 2 \to \ell = 0$. (Symmetry principles prevent transitions in which ℓ changes by one unit.)

11. Refer to the rigid-body example in Section 6-5, and consider the state obtained by superposing two degenerate normalized $\ell = 1$ wave functions:

$$\Psi = \frac{1}{\sqrt{2}}(\Psi_{11} + \Psi_{1-1}).$$

Write out the explicit expression for Ψ and show that Ψ is normalized. Calculate the expectation value

$$\langle L_z \rangle = \int \Psi^* \frac{\hbar}{i} \frac{\partial}{\partial \phi} \Psi \, d\Omega,$$

and determine the uncertainty ΔL_z in this state. Show that the probability density depends on ϕ, and interpret this observation in terms of the uncertainty principle.

12. Refer to the rigid-body problem at the end of Section 6-5, and calculate the average of the five probability densities for $\ell = 2$:

$$\frac{1}{5} \sum_{m=-2}^{2} |\Psi_{2m}|^2.$$

Interpret the result.

13. Determine the parity of the spherical harmonic Y_{21} by examining the explicit behavior of the function under the angle-inversion operation $(\theta, \phi) \rightarrow (\pi - \theta, \pi + \phi)$.

14. Consider a linear potential energy $V(r) = cr$, and assume a nonzero value for the angular momentum quantum number ℓ. Make a drawing of the effective potential energy and identify the regions in which classical motion is not allowed. Consider the two lowest energy levels, $E_{1\ell}$ and $E_{2\ell}$, and sketch the behavior of the solution $rR(r)$ for each case. Make a qualitative argument to explain why $E_{n\ell}$ exceeds E_{n0} for $\ell \neq 0$.

15. Consult Table 6-1, and determine the values of θ for which the probability density vanishes in each of the five $\ell = 2$ stationary states. Sketch the resulting conical surfaces of constant θ.

16. The eigenfunctions $\psi_{n\ell m}(r, \theta, \phi)$ satisfy the orthogonality property

$$\int_{\text{all space}} \psi^*_{n_1 \ell_1 m_1} \psi_{n_2 \ell_2 m_2} \, d\tau = 0.$$

Prove this result for the case of two nondegenerate eigenfunctions by proceeding as follows. Write out the time-independent Schrödinger equations satisfied by $\psi_{n_2 \ell_2 m_2}$ and $\psi^*_{n_1 \ell_1 m_1}$. Multiply the first equation by $\psi^*_{n_1 \ell_1 m_1}$ and the second by $\psi_{n_2 \ell_2 m_2}$. Take the difference between the two expressions and integrate the resulting equality over all space. Assume a real-valued potential energy, and use an identity from vector calculus to complete the proof.

17. Verify the orthogonality properties of the pairs of spherical harmonics Y_{21} and Y_{20}, and Y_{21} and Y_{11}, by explicit calculation of the integrals $\int Y_{21}^* Y_{20} \, d\Omega$ and $\int Y_{21}^* Y_{11} \, d\Omega$.

18. Consider the nonstationary state

$$\Psi = \frac{1}{\sqrt{2}} (\Psi_{100} + \Psi_{110}),$$

in which an equal-parts mixture is constructed using two of the normalized states $\Psi_{n\ell m}$. Calculate the expectation value of L^2 and the uncertainty in L^2 for a particle in this state.

19. The wave function $\Psi = a_1 \Psi_{n_1 \ell_1 m_1} + a_2 \Psi_{n_2 \ell_2 m_2}$ is a combination of the normalized stationary-state wave functions $\Psi_{n\ell m}$ and is therefore a solution of the Schrödinger equation. Show that, for Ψ to be normalized, the complex coefficients a_1 and a_2 must satisfy the condition $|a_1|^2 + |a_2|^2 = 1$. Calculate the expectation values $\langle L^2 \rangle$ and $\langle L_z \rangle$, and the uncertainties ΔL^2 and ΔL_z, for a particle in this state.

S E V E N

THE
ONE-ELECTRON
ATOM

Schrödinger discovered his version of the quantum theory in 1925 when he found that de Broglie's matter wave could be applied to the problem of atomic structure. His goal was to explain the atom as a wave system whose discrete frequencies were related to Bohr's quantized energy levels. At first Schrödinger tried a relativistic treatment of the atomic electron, but his attempt to blend relativity with quantum theory was prematurely conceived and had disappointing results. The nonrelativistic formalism followed with more success and furnished a correct picture for the main properties of the atom. In 1926 he published a series of articles on the principles of quantization, beginning with his theory of the hydrogen atom. The Schrödinger equation made its first appearance in the literature in the course of this project.

The solution of the Schrödinger equation for hydrogen constitutes the fundamental problem in the theory of atoms. Hydrogen offers a unique opportunity to test the Schrödinger theory since the one-electron atom is the only atomic system that admits an exact solution to the Schrödinger equation. The importance of the hydrogen wave functions extends beyond considerations of the simplest atom, because solutions much like those for hydrogen can be used as the basis of an approximate formulation for the more complex atoms.

We have already learned about the prominent status of the hydrogen atom in the early development of the quantum theory. We know that the Balmer–Rydberg formula for the spectral lines and the Bohr model for the energy levels are consistent with the spectroscopic data. Results from the Schrödinger equation should agree with these observations if the equation is indeed a valid foundation for quantum mechanics. In fact, we might expect that such a fundamental approach should predict all observable nonrelativistic properties of the hydrogen system. We examine the predictions for a system governed by Coulomb attraction in this chapter, and we leave certain important refinements to be discussed later in Chapter 8.

7-1 The Radial Differential Equation

Let us generalize the hydrogen problem immediately to include two features that have already appeared in the Bohr model. We assume that a single electron is bound to a nucleus of charge Ze, and we treat the atom as a two-body system in motion with center of mass at rest. The groundwork for the quantum construction of this problem has been laid in Chapter 6, beginning with our discussion of Figure 6-1. For the electron–nucleus system with masses m_e and M, we define a reduced mass

$$\mu = m_e \frac{M}{M + m_e}$$

and then cast the analysis in terms of an equivalent one-body problem to describe the behavior of a particle of mass μ. The correction factor $M/(M + m_e)$ is very near unity; nevertheless, this generalization is kept in the problem (at no extra cost) because the reduced-mass effect is subject to experimental test. The binding force is due to the Coulomb attraction between the charges Ze and $-e$. Therefore, the potential energy of the system is given by

$$V(r) = -\frac{Ze^2}{4\pi\varepsilon_0 r}, \tag{7-1}$$

a rotationally symmetric function. This electrostatic interaction provides the dynamics behind all the conclusions discussed in the rest of the chapter.

Our main objective is to find the stationary states of the atom. We have a central force determined by an r-dependent potential energy, and so we turn right away to the general procedures of Chapter 6 and use the spherical harmonics $Y_{\ell m}(\theta, \phi)$ to express the angle dependence of the stationary-state wave functions

$$\Psi_{n\ell m}(r, \theta, \phi, t) = R_{n\ell}(r) Y_{\ell m}(\theta, \phi) e^{-iE_{n\ell}t/\hbar}. \tag{7-2}$$

The angular momentum quantum numbers ℓ and m convey the angular momentum properties of these states, as discussed in Section 6-5, while the third quantum number n remains to be clearly identified. We can determine the r dependence of a wave function with angular momentum quantum number ℓ by solving the differential equation found in Equation (6-42):

$$-\frac{\hbar^2}{2\mu r^2} \frac{d}{dr} r^2 \frac{dR}{dr} + \left[V(r) + \frac{\hbar^2}{2\mu r^2} \ell(\ell+1) \right] R = ER, \tag{7-3}$$

where $R(r)$ and E denote the radial function and energy eigenvalue appearing in Equation (7-2). The Coulomb potential energy in Equation (7-1) is employed for $V(r)$ to give the explicit form for this radial equation. The result is an unfamiliar second-order ordinary differential equation whose solution must be investigated in some detail.

We gain considerable insight into the radial behavior if we first restrict our attention to wave functions $\Psi_{n\ell m}$ with no dependence on angle. Necessarily, these *spherically symmetric* solutions have angular momentum quantum numbers

$$\ell = 0 \quad \text{and} \quad m = 0,$$

since $Y_{00}(\theta, \phi)$ is the only possibility for a constant angular function. We insert this information along with Equation (7-1) into Equation (7-3) and find that $R(r)$ satisfies the differential equation

$$-\frac{\hbar^2}{2\mu}\left(R'' + \frac{2}{r}R'\right) - \frac{Ze^2}{4\pi\varepsilon_0 r}R = ER. \tag{7-4}$$

Since the solution is expected to approach zero as $r \to \infty$, it is reasonable to try an *exponentially damped* radial dependence:

$$R(r) = Ae^{-r/a},$$

where the parameter a has units of length and sets the scale of distance to describe the damping. This trial function provides a solution to Equation (7-4) for all values of r. To prove our point we compute

$$R' = -\frac{A}{a}e^{-r/a} = -\frac{R}{a}$$

and

$$R'' = \frac{A}{a^2}e^{-r/a} = \frac{R}{a^2}.$$

Then, when we put these results into Equation (7-4) and divide by $R(r)$, we obtain

$$-\frac{\hbar^2}{2\mu}\left(\frac{1}{a^2} - \frac{2}{ar}\right) - \frac{Ze^2}{4\pi\varepsilon_0 r} = E.$$

The equality holds for all r, provided the coefficient of $1/r$ vanishes:

$$\frac{\hbar^2}{\mu a} - \frac{Ze^2}{4\pi\varepsilon_0} = 0.$$

Thus, our first conclusion is a formula for the length parameter:

$$a = \frac{4\pi\varepsilon_0 \hbar^2}{Ze^2\mu}.$$

We consult Equation (3-51) so that we can rewrite the formula in terms of the Bohr radius a_0:

$$a = \frac{m_e}{\mu}\frac{a_0}{Z}, \quad \text{where} \quad a_0 = \frac{4\pi\varepsilon_0 \hbar^2}{e^2 m_e}. \tag{7-5}$$

The energy eigenvalue is also obtained by returning to the equality and comparing the two constant terms:

$$E = -\frac{\hbar^2}{2\mu a^2}$$

$$= -\frac{\hbar^2}{2\mu}\left(\frac{\mu}{m_e}\right)^2 Z^2\left(\frac{e^2}{4\pi\varepsilon_0}\right)^2\frac{m_e^2}{\hbar^4} = -\frac{\mu}{m_e}Z^2\left(\frac{e^2}{4\pi\varepsilon_0}\right)^2\frac{m_e}{2\hbar^2}.$$

We recognize this result as the Bohr formula for the energy of the *ground state*. Let us introduce the Rydberg energy E_0 into the expression by writing

$$E = -\frac{\mu}{m_e}Z^2 E_0, \quad \text{where} \quad E_0 = \left(\frac{e^2}{4\pi\varepsilon_0}\right)^2\frac{m_e}{2\hbar^2}, \tag{7-6}$$

as in Equation (3-53).

These conclusions are reassuring since our attempt at a simple solution has produced an $\ell = 0$ wave function with the correct ground-state energy for the atom. The full wave function for this stationary state is written in the form

$$\Psi_{100} = \frac{e^{-r/a}}{\sqrt{\pi a^3}}e^{-iE_1^{\text{Bohr}}t/\hbar}, \tag{7-7}$$

in which the multiplicative constant remains to be explained. We take $n = 1$ for the value of the third quantum number, and we call the energy eigenvalue E_1^{Bohr} as a reminder of the Bohr result found in Section 3-6. We recall that the spherical harmonic for $\ell = m = 0$ is given by the function

$$Y_{00} = \frac{1}{\sqrt{4\pi}},$$

and we observe that our result can be expressed according to Equation (7-2) by identifying the radial function and energy eigenvalue as

$$R_{10} = \sqrt{\frac{4}{a^3}}\,e^{-r/a}$$

and

$$E_{10} = E_1^{\text{Bohr}} = -\frac{\mu}{m_e}Z^2 E_0.$$

Other $\ell = m = 0$ wave functions with $n > 1$ also exist as solutions of Equation (7-4) for other discrete values of E. We proceed more systematically to find *all* the radial solutions, for $\ell = 0$ and for $\ell \neq 0$, in Section 7-2.

Let us conclude these opening remarks by quoting a useful integral formula:

$$\int_0^\infty r^n e^{-r/r_0}\,dr = n!\,r_0^{n+1}. \tag{7-8}$$

This expression has numerous applications throughout the chapter.

Example

The multiplying factor in Equation (7-7) is confirmed by applying the normalization condition to the wave function Ψ_{100}:

$$\int |\Psi_{100}|^2 \, d\tau = \int d\Omega \int_0^\infty r^2 \, dr \frac{e^{-2r/a}}{\pi a^3}$$

$$= \frac{4\pi}{\pi a^3} \int_0^\infty r^2 e^{-2r/a} \, dr = \frac{4}{a^3} 2! \left(\frac{a}{2} \right)^3 = 1.$$

We have made immediate use of Equation (7-8) to complete this calculation.

7-2   Solutions of the Radial Equation

We are concerned with the discrete bound states produced by the force of Coulomb attraction in the one-electron atom. The corresponding energy levels are found by solving the radial differential equation with $V(r)$ given by the Coulomb potential energy. This mathematical problem has a remarkable solution with a very interesting physical interpretation.

We begin by adopting an equivalent alternative to Equation (7-3). Let us return to Equation (6-43) and insert the Coulomb potential energy to obtain

$$-\frac{\hbar^2}{2\mu r} \frac{d^2}{dr^2} rR - \frac{Ze^2}{4\pi\varepsilon_0 r} R + \frac{\hbar^2}{2\mu r^2} \ell(\ell+1)R = ER. \tag{7-9}$$

We have learned in the Bohr model, and again in Section 7-1, that the Bohr radius a_0 and the Rydberg energy E_0 are natural parameters to use as scales of length and energy in the atom. It is fruitful to introduce these scale parameters in Equation (7-9), since the result is a greatly simplified expression for the differential equation. To this end we refer to Equations (7-5) and (7-6) and define a dimensionless variable ρ and a dimensionless eigenvalue η by writing

$$r = a\rho = \frac{4\pi\varepsilon_0 \hbar^2}{Ze^2\mu} \rho \tag{7-10}$$

and

$$E = -\frac{\mu}{m_e} Z^2 E_0 \eta = -\frac{\hbar^2}{2\mu a^2} \eta. \tag{7-11}$$

Equation (7-9) then becomes

$$-\frac{\hbar^2}{2\mu a^3 \rho} \frac{d^2}{d\rho^2} a\rho R - \frac{\hbar^2}{\mu a^2 \rho} R + \frac{\hbar^2}{2\mu a^2 \rho^2} \ell(\ell+1)R = -\frac{\hbar^2}{2\mu a^2} \eta R,$$

and cancellation of the factor $-\hbar^2/2\mu a^2 \rho$ produces the final form

$$\frac{d^2}{d\rho^2} \rho R + 2R - \frac{\ell(\ell+1)}{\rho} R = \eta\rho R.$$

Thus, the use of the quantities ρ and η enables us to remove all the excess constants from the radial differential equation.

We take R to be a function of ρ, and we let this dimensionless variable range over the interval $[0, \infty)$. The solution is required to vanish at infinity, and exponential damping is expected, in view of the result obtained for the ground state in the previous section. We can incorporate this behavior, and transform the differential equation into a more useful form, if we introduce a new function $F(\rho)$ by defining

$$R = e^{-\sqrt{\eta}\,\rho} \frac{F(\rho)}{\rho}. \tag{7-12}$$

The equation for R contains the term

$$\frac{d^2}{d\rho^2} \rho R = \frac{d^2}{d\rho^2} e^{-\sqrt{\eta}\,\rho} F = \frac{d}{d\rho} \left(e^{-\sqrt{\eta}\,\rho} F' - \sqrt{\eta}\, e^{-\sqrt{\eta}\,\rho} F \right)$$

$$= e^{-\sqrt{\eta}\,\rho} \left(F'' - 2\sqrt{\eta}\, F' + \eta F \right),$$

and so the differential equation can be stripped of the factor $e^{-\sqrt{\eta}\,\rho}$ to read:

$$F'' - 2\sqrt{\eta}\, F' + \eta F + \frac{2}{\rho} F - \frac{\ell(\ell + 1)}{\rho^2} F = \eta F.$$

These maneuvers lead us to the final version of the differential equation for $F(\rho)$:

$$F'' - 2\sqrt{\eta}\, F' + \left[\frac{2}{\rho} - \frac{\ell(\ell + 1)}{\rho^2} \right] F = 0, \tag{7-13}$$

where F must satisfy the further condition

$$F(0) = 0. \tag{7-14}$$

We impose this requirement so that the radial function R in Equation (7-12) remains finite at the origin.

It is not obvious that any real progress is made by this procedure, since the unknown function F obeys a differential equation that is just as unfamiliar as the original equation for the desired function R. Let us look again at the solution for the ground state to relieve some of this skepticism. We know from Section 7-1 that the radial function has the form

$$R = Ae^{-\rho} \quad \text{for the } \ell = 0 \text{ ground state.}$$

Equation (7-12) tells us that we must have

$$F(\rho) = A\rho \quad \text{and} \quad \sqrt{\eta} = 1$$

for this state, in clear agreement with Equations (7-13) and (7-14). Thus, we see that F is a *simpler* function than R in the case of the ground state, and we might suppose that this property of F relative to R persists for the other states of the atom.

The quantum number ℓ can have any integer value in Equation (7-13). It is instructive to isolate the $\ell = 0$ case and begin the investigation with the differential equation

$$F'' - 2\sqrt{\eta}\, F' + \frac{2}{\rho}F = 0. \tag{7-15}$$

We may not recognize this as one of our familiar differential equations, but we can still solve the equation by means of standard methods. The results of these procedures include a determination of the dimensionless parameter η in terms of an *integer n*:

$$\sqrt{\eta} = \frac{1}{n}. \tag{7-16}$$

This integer arises from the fact that F must be a *polynomial* in ρ, where n *defines* the *order* of the polynomial. Thus, the integer n serves as a quantum number to label a whole family of $\ell = 0$ solutions. We examine the mathematics behind these conclusions in more detail at the end of the section.

Equation (7-16) is our main result for the $\ell = 0$ case. The result tells us, through Equation (7-11), that the energy eigenvalue is given by

$$E_{n0} = -\frac{\mu}{m_e}Z^2\frac{E_0}{n^2} \tag{7-17}$$

in the nth $\ell = 0$ state. We follow the notation of Equation (7-2) and label the values of E by the integers n and ℓ, taking $\ell = 0$. This conclusion is very striking because the formula reproduces the one given in Equation (3-52) for the nth energy level of the Bohr model. Let us draw attention to this point by setting

$$E_{n0} = E_n^{\text{Bohr}} \tag{7-18}$$

and writing

$$\Psi_{n00} = R_{n0}(r)Y_{00}(\theta, \phi)e^{-iE_n^{\text{Bohr}}t/\hbar}.$$

Taken together, these expressions give the eigenvalue and eigenfunction of the energy in the nth $\ell = 0$ state. The radial function must have the form

$$R = e^{-\rho/n}\frac{F(\rho)}{\rho},$$

according to Equations (7-12) and (7-16). The details at the end of the section tell us why F must be a polynomial and how n defines the order of the polynomial. We see that the ground-state wave function of Section 7-1 is found among these results when the $n = 1$ level is chosen.

The solution of Equation (7-13) proceeds in similar fashion for the $\ell \neq 0$ states. Again, we find that F is a polynomial in ρ of order n. Our purpose in separating the cases for $\ell = 0$ and $\ell \neq 0$ is to call special attention to the outcome of this investigation. We learn in the mathematical details below that Equation (7-16) holds, not only for $\ell = 0$, but also for *any* angular momentum state. Equation (7-11) then leads us to a most surprising conclusion:

The energies allowed for each ℓ are independent of the value of ℓ.

We express the general result for the energy levels of the one-electron atom in the manner of Equations (7-17) and (7-18) by writing

$$E_{n\ell} = -\frac{\mu}{m_e} Z^2 \frac{E_0}{n^2} = E_n^{\text{Bohr}}. \qquad (7\text{-}19)$$

Thus, only the *one* quantum number n is needed to determine the energies, as in the Bohr model, even though two quantum numbers are normally required to label the energy eigenvalues in a central-force problem. Of course, the radial functions vary with ℓ as well as n, since the differential equation for $F(\rho)$ contains an explicit ℓ-dependent term.

We have introduced the integer n in the stationary state $\Psi_{n\ell m}$ of Equation (7-2) by defining n as the order of the radial polynomial F. This definition of a third quantum number sets the hydrogen problem apart from the general central-force problem of Chapter 6. Recall that in the general treatment the third index is introduced as a radial node quantum number. Hydrogen is a special case, and so n is given a different meaning and a special name. In the case of hydrogen we refer to n as the *principal* quantum number, and we note that n alone determines the energy. Equations (7-12) and (7-16) define a collection of radial functions

$$R_{n\ell} = e^{-\rho/n} \frac{F_{n\ell}(\rho)}{\rho} \quad \text{with } \rho = \frac{r}{a}. \qquad (7\text{-}20)$$

Note that $F(\rho)$ bears the labels n and ℓ to display the fact that both of these quantum numbers control the composition of the polynomial. If we examine $R_{n\ell}(r)$ for various n and ℓ, we see that the principal quantum number n does *not* count

---

Table 7-1   Radial Functions[a] for the One-Electron Atom

$n = 1$	$\ell = 0$	$R_{10} = \dfrac{2}{\sqrt{a^3}} e^{-\rho}$
$n = 2$	$\ell = 0$	$R_{20} = \dfrac{1}{\sqrt{2a^3}} \left(1 - \dfrac{\rho}{2}\right) e^{-\rho/2}$
	$\ell = 1$	$R_{21} = \dfrac{1}{2\sqrt{6a^3}} \rho e^{-\rho/2}$
$n = 3$	$\ell = 0$	$R_{30} = \dfrac{2}{3\sqrt{3a^3}} \left(1 - \dfrac{2}{3}\rho + \dfrac{2}{27}\rho^2\right) e^{-\rho/3}$
	$\ell = 1$	$R_{31} = \dfrac{8}{27\sqrt{6a^3}} \rho \left(1 - \dfrac{\rho}{6}\right) e^{-\rho/3}$
	$\ell = 2$	$R_{32} = \dfrac{4}{81\sqrt{30a^3}} \rho^2 e^{-\rho/3}$

$\cdots$

---

[a]The variable is denoted by $\rho = r/a$ and each of the functions obeys the normalization condition $\int_0^\infty (R_{n\ell})^2 r^2 \, dr = 1$.

radial nodes. Several of the radial functions are listed in Table 7-1, and graphs of the functions are shown in Figure 7-1. It is easy to infer from either the table or the figure that each $R_{n\ell}$ has $n - \ell - 1$ nodes in the interval $(0, \infty)$. Hence, the combination of integers $n - \ell - 1$ plays the role of a radial node quantum number for the one-electron atom. We emphasize once again that n has a new meaning in this special problem and that the energy depends only on n. We also observe, from the ρ^{ℓ} behavior of the functions in Table 7-1, that each $R_{n\ell}$ has an ℓ th-order zero at $r = 0$, as required by Equation (6-51).

Figure 7-1

Radial functions $\sqrt{a^3}\, R_{n\ell}$ taken from Table 7-1. The variable is the dimensionless quantity $\rho = r/a$. Each function has $n - \ell - 1$ nodes in the interval $(0, \infty)$.

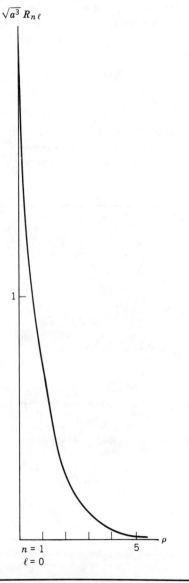

Figure 7-1
Continued.

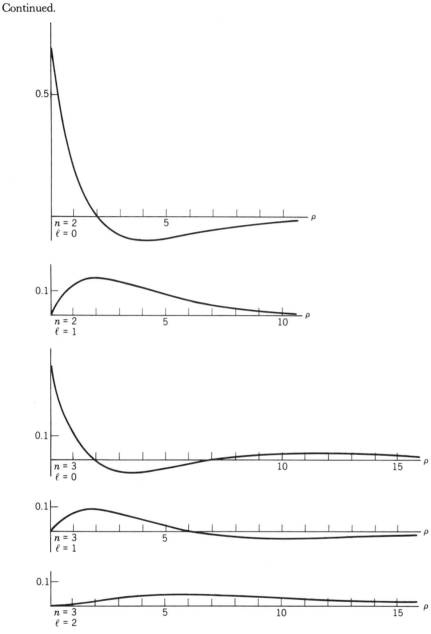

The entries in the table contain certain multiplicative factors to ensure the normalization of the wave function $\Psi_{n\ell m}$. We know from Equation (6-36) that the spherical harmonics $Y_{\ell m}$ are normalized separately over solid angle, and we recall from Equation (6-52) that the radial functions $R_{n\ell}$ also satisfy their own normalization over r:

$$1 = \int_0^\infty (R_{n\ell})^2 r^2 \, dr. \tag{7-21}$$

Finally, we observe that the radial functions obey the orthogonality property

$$\int_0^\infty R_{n'\ell} R_{n\ell} r^2 \, dr = 0 \quad \text{when } n \neq n',$$

as a consequence of Equation (6-63).

Detail

Equation (7-13) can have only polynomial solutions for $F(\rho)$ because, otherwise, the associated radial function R would not remain finite in the limit $\rho \to \infty$. Let us accept this statement without proof so that we can determine these polynomials as our next step. The differential equation for the $\ell = 0$ case is considered first, so that F is supposed to satisfy Equation (7-15). We immediately assume a polynomial of the form

$$F(\rho) = A\left(\rho + a_2\rho^2 + a_3\rho^3 + \cdots + a_n\rho^n\right) \quad \text{with } a_n \neq 0, \qquad (7\text{-}22)$$

where $a_2, a_3, \ldots, a_n$ are coefficients to be determined in the analysis. Note that the polynomial has order n and satisfies Equation (7-14). The first two derivatives of F are

$$F' = A\left(1 + 2a_2\rho + 3a_3\rho^2 + \cdots + na_n\rho^{n-1}\right)$$

and

$$F'' = A\left(2a_2 + 6a_3\rho + \cdots + n(n-1)a_n\rho^{n-2}\right).$$

If we insert these expressions into Equation (7-15), we obtain the equality

$$\left(2a_2 + 6a_3\rho + \cdots + n(n-1)a_n\rho^{n-2}\right) - 2\sqrt{\eta}\left(1 + 2a_2\rho + \cdots + na_n\rho^{n-1}\right)$$
$$+ \frac{2}{\rho}\left(\rho + a_2\rho^2 + \cdots + a_n\rho^n\right) = 0.$$

This string of terms in powers of ρ can vanish for all values of the variable only if the assembled coefficients are equal to zero for each power. We examine the highest power and find

$$-2\sqrt{\eta}\, na_n + 2a_n = 0 \quad \text{from the } \rho^{n-1} \text{ term.}$$

The stipulation in Equation (7-22) rules out the possibility that a_n vanishes. The important result is the determination of η quoted in Equation (7-16):

$$\sqrt{\eta} = \frac{1}{n}.$$

The lower powers of ρ in the equality provide the following further information:

$$2a_2 - 2\sqrt{\eta} + 2 = 0 \qquad \text{from the } \rho^0 \text{ term,}$$
$$6a_3 - 4\sqrt{\eta}\, a_2 + 2a_2 = 0 \quad \text{from the } \rho^1 \text{ term,}$$

$$\cdots$$

We can solve these equations for the unknown coefficients in ascending order, using Equation (7-16) in the process, and obtain

$$a_2 = \sqrt{\eta} - 1 = \frac{1}{n} - 1,$$

$$a_3 = \frac{2\sqrt{\eta} - 1}{3} a_2 = \frac{1}{3}\left(\frac{2}{n} - 1\right)\left(\frac{1}{n} - 1\right),$$

$$\cdots$$

The coefficients are related successively according to the general formula

$$a_{k+1} = \frac{2(k/n - 1)}{k(k + 1)} a_k. \qquad (7\text{-}23)$$

The implied sequence of coefficients must start with the assignment $a_1 = 1$ to conform to Equation (7-22). This formula concludes our summary of properties of the $\ell = 0$ solutions.

Detail

The general case for $\ell \neq 0$ is considered next, so that F is now supposed to satisfy Equation (7-13). We generalize the nth-order $\ell = 0$ polynomial in Equation (7-22), and thereby accommodate the additional ℓ-dependent term in the differential equation, by adopting an F function of the form

$$F(\rho) = A\left(\rho^{\ell+1} + a_{\ell+2}\rho^{\ell+2} + a_{\ell+3}\rho^{\ell+3} + \cdots + a_n\rho^n\right) \quad \text{with } a_n \neq 0. \quad (7\text{-}24)$$

We note that the leading term in F must behave as $\rho^{\ell+1}$ if the radial function in Equation (7-12) is to have r^ℓ behavior near the origin, as required by the general rule in Equation (6-51). The coefficients $a_{\ell+2}, a_{\ell+3}, \ldots, a_n$ are determined as before, by inserting the polynomial into the ℓ-dependent differential equation for F. The required derivatives are

$$F' = A\left[(\ell + 1)\rho^\ell + (\ell + 2)a_{\ell+2}\rho^{\ell+1} + (\ell + 3)a_{\ell+3}\rho^{\ell+2} + \cdots + na_n\rho^{n-1}\right]$$

and

$$F'' = A\left[\ell(\ell + 1)\rho^{\ell-1} + (\ell + 1)(\ell + 2)a_{\ell+2}\rho^\ell + (\ell + 2)(\ell + 3)a_{\ell+3}\rho^{\ell+1} + \right.$$
$$\left. \cdots + n(n - 1)a_n\rho^{n-2}\right].$$

Equation (7-13) then provides the following lengthy constraint among the various powers of ρ:

$$\left[\ell(\ell + 1)\rho^{\ell-1} + (\ell + 1)(\ell + 2)a_{\ell+2}\rho^\ell + (\ell + 2)(\ell + 3)a_{\ell+3}\rho^{\ell+1} + \right.$$
$$\left. \cdots + n(n - 1)a_n\rho^{n-2}\right]$$
$$- 2\sqrt{\eta}\left[(\ell + 1)\rho^\ell + (\ell + 2)a_{\ell+2}\rho^{\ell+1} + \cdots + na_n\rho^{n-1}\right]$$
$$+ \frac{2}{\rho}\left[\rho^{\ell+1} + a_{\ell+2}\rho^{\ell+2} + \cdots + a_n\rho^n\right]$$
$$- \frac{\ell(\ell + 1)}{\rho^2}\left[\rho^{\ell+1} + a_{\ell+2}\rho^{\ell+2} + a_{\ell+3}\rho^{\ell+3} + \cdots + a_n\rho^n\right] = 0.$$

Note that the highest and lowest powers are ρ^{n-1} and ρ^{ℓ}, and that the terms in $\rho^{\ell-1}$ *cancel* because of the choice of leading term in Equation (7-24). Again we assemble the coefficients of each power of ρ and obtain

$$-2\sqrt{\eta}\, na_n + 2a_n = 0 \quad \text{from the } \rho^{n-1} \text{ term,}$$

as well as

$$(\ell + 1)(\ell + 2)a_{\ell+2} - 2\sqrt{\eta}\,(\ell + 1) + 2 - \ell(\ell + 1)a_{\ell+2} = 0$$

$$\text{from the } \rho^{\ell} \text{ term,}$$

$$(\ell + 2)(\ell + 3)a_{\ell+3} - 2\sqrt{\eta}\,(\ell + 2)a_{\ell+2} + 2a_{\ell+2} - \ell(\ell + 1)a_{\ell+3} = 0$$

$$\text{from the } \rho^{\ell+1} \text{ term,}$$

$$\cdots$$

The first of these equalities is the same as in the $\ell = 0$ case. This most remarkable result tells us that Equation (7-16) holds for *any* value of ℓ. The remaining equations can be solved as before to yield the coefficients of the polynomial in ascending order:

$$a_{\ell+2} = \frac{\sqrt{\eta}\,(\ell + 1) - 1}{\ell + 1} = \frac{1}{n} - \frac{1}{\ell + 1},$$

$$a_{\ell+3} = \frac{\sqrt{\eta}\,(\ell + 2) - 1}{2\ell + 3}a_{l+2} = \frac{1}{2\ell + 3}\left(\frac{\ell + 2}{n} - 1\right)\left(\frac{1}{n} - \frac{1}{\ell + 1}\right),$$

$$\cdots$$

The successive coefficients are connected by a general *recursion relation* of the form

$$a_{k+1} = \frac{2(k/n - 1)}{k(k + 1) - \ell(\ell + 1)}a_k. \tag{7-25}$$

This formula generalizes the $\ell = 0$ expression in Equation (7-23). We leave the derivation to Problem 3 at the end of the chapter and illustrate the use of the formula as follows.

Example

Let us pick quantum numbers $n = 2$ and $\ell = 1$ and construct the wave function Ψ_{21m} with the aid of Equation (7-25). We must begin with the definition $a_{\ell+1} = 1$ to adhere to Equation (7-24), and so we set $a_2 = 1$ in this application. Equation (7-25) gives the next coefficient as

$$a_3 = \frac{2(2/2 - 1)}{6 - 2}(1) = 0.$$

The succession of contributions to the polynomial terminates at once so that

Equation (7-24) has only a single term:

$$F_{21}(\rho) = A\rho^2.$$

From Equations (7-19) and (7-20) we have

$$E_{21} = E_2^{\text{Bohr}} \quad \text{and} \quad R_{21} = A\rho e^{-\rho/2}.$$

The desired wave function can therefore be written as

$$\Psi_{21m} = \frac{Ar}{a} e^{-r/2a} Y_{1m}(\theta, \phi) e^{-iE_2^{\text{Bohr}} t/\hbar} \quad \text{with } m = 1, 0, -1.$$

The normalization condition in Equation (7-21) determines the remaining constant A:

$$1 = \int_0^\infty (R_{21})^2 r^2 \, dr = \int_0^\infty \frac{A^2 r^2}{a^2} e^{-r/a} r^2 \, dr$$

$$= \frac{A^2}{a^2} \int_0^\infty r^4 e^{-r/a} \, dr = \frac{A^2}{a^2} 4! \, a^5 = 24 a^3 A^2,$$

using the integral formula in Equation (7-8). Thus, we obtain $A = 1/\sqrt{24a^3}$ for the normalizing factor, and we observe that the resulting expression for R_{21} agrees with the entry in Table 7-1.

7-3 Degeneracy

It is remarkable that the main features of the hydrogen atom can be rigorously understood with so little difficulty. We have been able to solve the Schrödinger equation exactly and obtain convincing results from a detailed study of the radial solutions. Let us move past this important mathematical phase of the problem and show that the physical interpretation is even more interesting.

We have found that the energy levels in the Schrödinger theory agree with the Bohr formula and vary only with the principal quantum number n. This unique feature of the one-electron atom is expressed in Equation (7-19). We know from the general central-force problem that every stationary state of the atom is assigned a set of three quantum numbers $(n\ell m)$. It is expected that the energy eigenvalue should be independent of m on grounds of rotational symmetry, but it is surprising to find that the energy does not vary with ℓ for a given value of n. This circumstance implies an unusual degeneracy among the states, since a choice of n fixes the same energy for several possible states with different values of ℓ.

We can establish the nature of the degeneracy if we examine the structure of the radial portion of the wave function

$$\Psi_{n\ell m} = e^{-\rho/n} \frac{F_{n\ell}(\rho)}{\rho} Y_{\ell m}(\theta, \phi) e^{-iE_n t/\hbar} \quad \text{with } \rho = \frac{r}{a}. \tag{7-26}$$

This rewritten version of Equation (7-2) incorporates Equations (7-19) and (7-20) and introduces E_n as shorthand notation for the Bohr energy E_n^{Bohr}. A glance at Equation (7-24) reminds us that the polynomial $F_{n\ell}(\rho)$ has its lowest and highest powers in the terms $\rho^{\ell+1}$ and ρ^n. It is obvious from the construction that n cannot be smaller than $\ell + 1$. Therefore, a given choice of the angular momentum quantum number ℓ allows the principal quantum number to range over the values

$$n = \ell + 1, \ell + 2, \ldots .$$

A sequence of levels of increasing energy arises with this list of n values for the given ℓ. The degeneracy pattern is easier to see if the bookkeeping scheme is turned around so that a particular energy level, and hence a particular n, is chosen. The ℓ values with the same energy must be those for which

$$n \geq \ell + 1,$$

namely

$$\ell = 0, 1, 2, \ldots, n - 1. \tag{7-27}$$

This degeneracy with respect to ℓ is compounded further by the familiar degeneracy with respect to m, whereby each ℓ admits $2\ell + 1$ possibilities for the azimuthal quantum number

$$m = -\ell, -\ell + 1, \ldots, \ell - 1, \ell.$$

The combined array of allowed values of ℓ and m embraces a collection of states $\Psi_{n\ell m}$ of the form expressed in Equation (7-26), all with the *same* Bohr energy E_n. We summarize these conclusions by listing the wave functions in ascending order of n in Table 7-2 and by drawing the energy levels in Figure 7-2. Note that Table 7-1 has already been organized to give the functions $R_{n\ell}$ in the same fashion. Note also from the figure that the degeneracy of the nth energy level is evidently given by

$$d_n = n^2, \tag{7-28}$$

an expression for the number of distinct states with the same energy E_n.

Table 7-2   Degenerate States of the One-Electron Atom

$n = 1$	$\ell = 0$	$m = 0$	$\Psi_{100} = R_{10}Y_{00}e^{-iE_1t/\hbar}$	nondegenerate
$n = 2$	$\ell = 0$	$m = 0$	$\Psi_{200} = R_{20}Y_{00}e^{-iE_2t/\hbar}$	four
	$\ell = 1$	$m = 1, 0, -1$	$\Psi_{21m} = R_{21}Y_{1m}e^{-iE_2t/\hbar}$	degenerate states
$n = 3$	$\ell = 0$	$m = 0$	$\Psi_{300} = R_{30}Y_{00}e^{-iE_3t/\hbar}$	
	$\ell = 1$	$m = 1, 0, -1$	$\Psi_{31m} = R_{31}Y_{1m}e^{-iE_3t/\hbar}$	nine
	$\ell = 2$	$m = 2, 1, 0, -1, -2$	$\Psi_{32m} = R_{32}Y_{2m}e^{-iE_3t/\hbar}$	degenerate states

$\cdots$

Figure 7-2

Energy levels and degeneracies of the one-electron atom. There are n^2 distinct states $\Psi_{n\ell m}$ with the same energy E_n. Spectroscopic notation is used to label the various shells and orbitals. The nth shell contains n orbitals, and the $n\ell$ orbital consists of $2\ell + 1$ states.

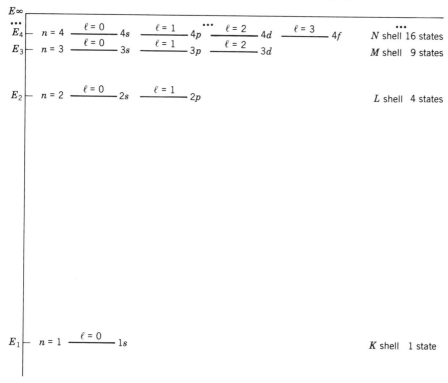

An ancient form of *spectroscopic notation* is still in common use as a means of designating these single-electron atomic states. We recall from Section 6-8, and especially from Figure 6-8, that the different assignments of the angular momentum quantum numbers ℓ and m have their own characteristic angular distributions of probability. We let ℓ be known as the *orbital quantum number*, and we call the distribution of probability for each ℓ an *electronic orbital*. Spectroscopic notation identifies these configurations by the letter code

$$s\,p\,d\,f\,g\,h\,\cdots,$$

in one-to-one correspondence with the ℓ values

$$0\;1\;2\;3\;4\;5\cdots.$$

The single-electron energy levels are called *shells*, with labels

$$K\,L\,M\,N\,\cdots,$$

corresponding to the n values

$$1\;2\;3\;4\cdots$$

as in the old Bohr model. These notational schemes are included in Figure 7-2. We can become acquainted with their use through the following illustration.

Example

The $n = 2$ level of the one-electron atom is known as the L shell. The figure indicates the existence of degenerate s and p orbitals in this shell, comprising one $2s$ state and three $2p$ states. The four different normalized wave functions have the same energy E_2 and appear in Table 7-2 as follows:

$$\Psi_{2\ell m} = \psi_{2\ell m} e^{-iE_2 t/\hbar},$$

where

$$\psi_{200} = R_{20}Y_{00} = \left[\frac{1}{\sqrt{2a^3}} \left(1 - \frac{\rho}{2} \right) e^{-\rho/2} \right] \left[\frac{1}{\sqrt{4\pi}} \right],$$

for the $2s$ state, and

$$\psi_{21\pm1} = R_{21}Y_{1\pm1} = \left[\frac{1}{2\sqrt{6a^3}} \rho e^{-\rho/2} \right] \left[\mp \sqrt{\frac{3}{8\pi}} \sin\theta e^{\pm i\phi} \right]$$

$$\psi_{210} = R_{21}Y_{10} = \left[\frac{1}{2\sqrt{6a^3}} \rho e^{-\rho/2} \right] \left[\sqrt{\frac{3}{4\pi}} \cos\theta \right]$$

for the $2p$ states. Entries from Tables 6-1 and 7-1 have been used to assemble the (r, θ, ϕ) dependence of these energy eigenfunctions.

7-4 Probability Distributions

The Bohr model of the atom is based on the classical notion of planar electron orbits. We know from the quantum treatment of angular momentum that this picture has only qualitative validity, and then only in the classical limit of large angular momentum quantum numbers. The proper way to visualize the atom in any of its states is provided by the relevant distribution of probability.

We deduce the physical properties of the atom in a stationary state by applying the probability interpretation to the appropriate wave function $\Psi_{n\ell m}$. The probability of locating the electron relative to the nucleus, in a volume element $d\tau = r^2 dr\, d\Omega$, is given by the familiar expression

$$|\Psi_{n\ell m}|^2 d\tau = \left[(R_{n\ell})^2 r^2 dr \right] \left[|Y_{\ell m}|^2 d\Omega \right]. \tag{7-29}$$

We have considered the spatial configuration of the probability density $|\Psi_{n\ell m}|^2$ in general terms in Section 6-8, and we have found that the distribution in angle is determined by the universal angular factor $|Y_{\ell m}|^2$. This ϕ-independent quantity varies with θ according to some graphical representation of the sort described in Figure 6-8. Figure 7-3 shows a larger assortment of these axially symmetric angular distributions, corresponding to the angular momentum states listed in Table 6-1. We have also found that the probability in Equation (7-29) can be integrated over solid angle to

Figure 7-3

Polar-angle distributions $|Y_{\ell m}|^2$ in various angular momentum states. Axially symmetric angular distributions of the probability density are generated by rotating these graphs around the z axis. The distributions satisfy the normalization condition $2\pi\int_0^\pi |Y_{\ell m}|^2 \sin\theta \, d\theta = 1$.

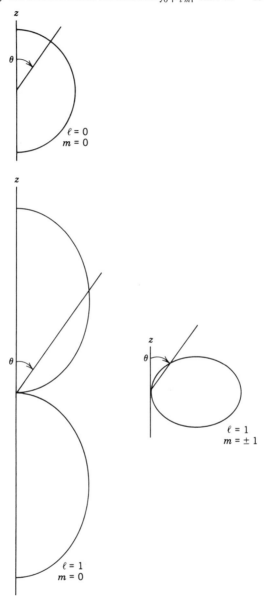

generate the radial probability density

$$P_{n\ell}(r) = r^2(R_{n\ell})^2. \tag{7-30}$$

This quantity is introduced only for stationary states and has a different structure for each central-force problem. We interpret $P_{n\ell}(r)\,dr$ in the case of the atom as the

Figure 7-3
Continued.

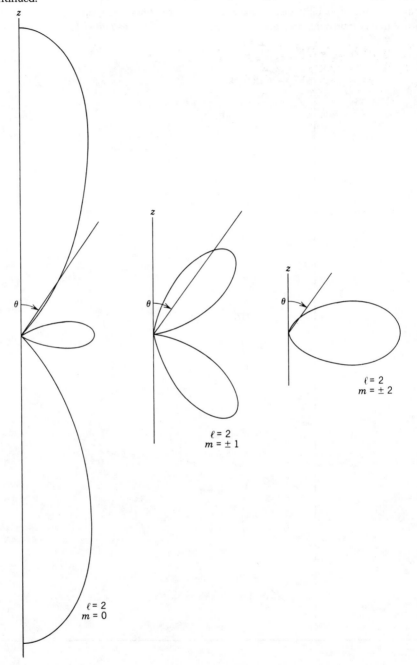

probability of finding the electron at a distance between r and $r + dr$ from the nucleus.

The ground state of the atom is described by the wave function in Equation (7-7). Spherical symmetry is an obvious property of the corresponding angle-independent probability density

$$|\Psi_{100}|^2 = \frac{e^{-2r/a}}{\pi a^3},$$

in marked contrast with the plane circular orbit of the simpler Bohr model. We can integrate $|\Psi_{100}|^2$ over the surface of a sphere of radius r to find the radial probability density

$$P_{10}(r) = \frac{4}{a^3} r^2 e^{-2r/a}$$

(or we can take R_{10} from Table 7-1 and apply Equation (7-30) to obtain the same answer). This radial distribution is shown in Figure 7-4 to have a sharp peak at $\rho = 1$, where $r = a$. Thus, we make contact with the Bohr model to the extent that the most probable distance of the electron from the nucleus is equal to the radius of the first Bohr orbit.

The figure also describes the radial probability distributions for the other states listed in Table 7-1. We note that the existence of nodes in $R_{n\ell}$ causes the occurrence of peaks in $P_{n\ell}$ and that the distributions with a single peak are those for which $\ell = n - 1$. These states of maximum ℓ correspond to the circular Bohr orbits, since a single most-probable radial distance is defined in each case.

The average distance between the electron and the nucleus is given by the expectation value of r. In a stationary state we find

$$\langle r \rangle = \int \Psi_{n\ell m}^* r \Psi_{n\ell m} \, d\tau = \int_0^\infty r^3 (R_{n\ell})^2 \, dr = \int_0^\infty r P_{n\ell}(r) \, dr, \qquad (7\text{-}31)$$

as in Equation (6-56). General formulas exist for $R_{n\ell}$, with arbitrary values of n and ℓ, and so general expressions also exist for $\langle r \rangle$ and for other expectation values involving functions of r. We are not concerned with the derivation of these quantities, so let us just quote the following results:

$$\langle r \rangle = an^2 \left\{ 1 + \frac{1}{2} \left[1 - \frac{\ell(\ell+1)}{n^2} \right] \right\}, \qquad (7\text{-}32)$$

$$\left\langle \frac{1}{r} \right\rangle = \frac{1}{an^2}, \qquad (7\text{-}33)$$

$$\left\langle \frac{1}{r^2} \right\rangle = \frac{2}{a^2 n^3 (2\ell+1)}, \qquad (7\text{-}34)$$

$$\left\langle \frac{1}{r^3} \right\rangle = \frac{2}{a^3 n^3 \ell(\ell+1)(2\ell+1)}. \qquad (7\text{-}35)$$

Note that the appropriate power of the length scale a appears in each expression. We can apply Equation (7-32) right away and calculate $\langle r \rangle$ for $n = 1$ and $\ell = 0$ to find the average distance $\langle r \rangle = \frac{3}{2}a$ for the ground state. The point $\langle \rho \rangle = \langle r \rangle / a$ is

Figure 7-4

Radial probability distributions for the radial functions in Figure 7-1. The graphs describe the dimensionless quantity $aP_{n\ell} = a^3\rho^2(R_{n\ell})^2$ versus the dimensionless variable $\rho = r/a$. The expectation value $\langle\rho\rangle = \langle r\rangle/a$ is shown for each of the states. The distributions satisfy the normalization condition $\int_0^\infty aP_{n\ell}\,d\rho = 1$.

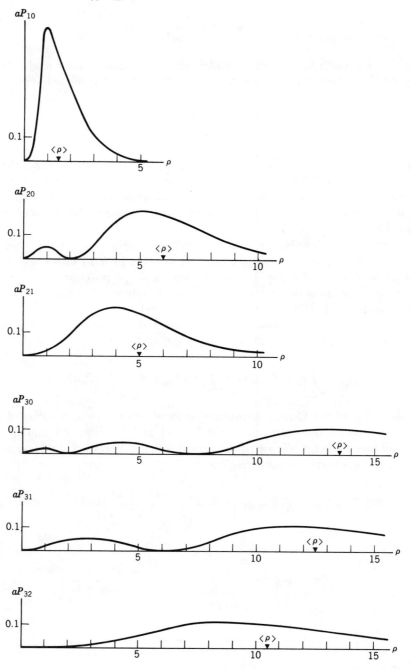

Distributions of probability in various states of the one-electron atom

indicated for this case, and for each of the other radial probability distributions, in Figure 7-4. Equation (7-33) can also be used immediately to determine the expectation value of the Coulomb potential energy:

$$\langle V \rangle = -\frac{Ze^2}{4\pi\varepsilon_0}\left\langle \frac{1}{r} \right\rangle = -\frac{Ze^2}{4\pi\varepsilon_0 an^2}.$$

All of these formulas have applications in atomic physics.

Example

Let us confirm our observations about the ground state. We can obtain the probability density by computing

$$P_{10} = 4\pi r^2 |\Psi_{100}|^2 = \frac{4}{a^3}r^2 e^{-2r/a}.$$

The derivative of this function is

$$\frac{dP_{10}}{dr} = \frac{4}{a^3}\left(2r - \frac{2}{a}r^2\right)e^{-2r/a} = \frac{8}{a^3}r\left(1 - \frac{r}{a}\right)e^{-2r/a},$$

and so P_{10} maximizes at $r = a$, as claimed above. The expectation value of r is given by Equation (7-31):

$$\langle r \rangle = \int_0^\infty rP_{10}\,dr = \frac{4}{a^3}\int_0^\infty r^3 e^{-2r/a}\,dr = \frac{4}{a^3}3!\left(\frac{a}{2}\right)^4 = \frac{3}{2}a,$$

in agreement with the result from Equation (7-32) for this state. Note that we have again found use for the integral formula in Equation (7-8).

Example

We have seen how $\langle V \rangle$ follows in one step from Equation (7-33) for any stationary state of the atom. Let us pursue this result and extract some additional physical information about the system. We first use Equations (7-5) to find

$$\langle V \rangle = -\frac{Ze^2}{4\pi\varepsilon_0 n^2 a} = -\frac{Ze^2}{4\pi\varepsilon_0 n^2}\frac{\mu}{m_e}Z\frac{e^2 m_e}{4\pi\varepsilon_0 \hbar^2} = -Z^2\left(\frac{e^2}{4\pi\varepsilon_0}\right)^2\frac{\mu}{\hbar^2 n^2}.$$

We then note that $\langle E \rangle$ is the same as E_n in the state $\Psi_{n\ell m}$, and we employ Equations (7-6) and (7-19) to write

$$\langle E \rangle = -\frac{\mu}{m_e}\frac{Z^2}{n^2}E_0 = -\frac{\mu}{m_e}\frac{Z^2}{n^2}\left(\frac{e^2}{4\pi\varepsilon_0}\right)^2\frac{m_e}{2\hbar^2} = -\frac{Z^2}{2}\left(\frac{e^2}{4\pi\varepsilon_0}\right)^2\frac{\mu}{\hbar^2 n^2}.$$

The two results for $\langle V \rangle$ and $\langle E \rangle$ satisfy the equality

$$\langle V \rangle = 2\langle E \rangle,$$

and so the expectation value of the kinetic energy becomes

$$\langle K \rangle = \langle E \rangle - \langle V \rangle = -\langle E \rangle = \frac{Z^2}{2}\left(\frac{e^2}{4\pi\varepsilon_0}\right)^2\frac{\mu}{\hbar^2 n^2}.$$

These relations among the average energies $\langle V \rangle$, $\langle E \rangle$, and $\langle K \rangle$ hold also in classical mechanics, where the result is a special case of the *virial theorem*.

7-5 Electric Dipole Selection Rules

The states of the atom reveal their existence by the radiation emitted in transitions between pairs of energy levels. These radiative processes cause the instability of the excited states and generate the emission spectrum of the atom. Figure 7-5 reminds us

Figure 7-5

Energy shells and radiative transitions in hydrogen.

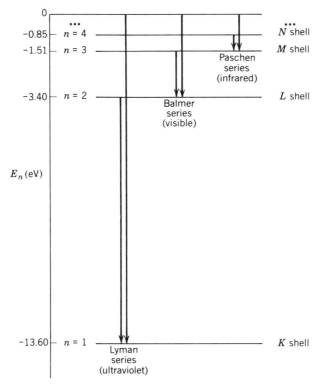

of the organization scheme for the familiar spectral lines of hydrogen, in which each series of lines corresponds to a complete set of transitions downward to a particular energy shell. We have learned in Section 5-8 how quantum mechanics predicts the probabilities for the different transitions and determines the intensities for the various emitted wavelengths. Selection rules emerge from these predictions to tell us that the transition mechanism allows only certain changes in the quantum numbers between the initial and final states of the atom.

We convert our one-dimensional treatment of transitions in Section 5-8 to the case of the one-electron atom by introducing transition states for the three-dimensional system. Let us construct such a state by superposing two stationary states of the atom:

$$\Psi = c\Psi_{n\ell m} + c'\Psi_{n'\ell'm'}$$

$$= c\psi_{n\ell m}e^{-iE_n t/\hbar} + c'\psi_{n'\ell'm'}e^{-iE_{n'}t/\hbar} \quad \text{with } E_n > E_{n'}. \qquad (7\text{-}36)$$

This wave function represents a situation in which the atom is found in both stationary states at once, with probabilities $|c|^2$ and $|c'|^2$, and thus describes an atom in transition between the given energy levels E_n and $E_{n'}$. The transition itself is caused by the interaction of the atom with the electromagnetic radiation field. The perturbing effect of the radiation is time dependent, and so the probabilities $|c|^2$ and $|c'|^2$ vary with time to indicate the course of the transition. The oscillatory time dependence of the probability density is the main feature of this construction. We know from Section 5-8 that $|\Psi|^2$ oscillates with the Bohr frequency

$$\nu = \frac{E_n - E_{n'}}{h},\tag{7-37}$$

and we conclude that the expectation value of the electric dipole moment of the atom oscillates with the same frequency. Our semiclassical picture of the interaction of the atom with the electromagnetic field identifies this system of oscillating charge as a source of electromagnetic waves emitted in a characteristic electric dipole pattern. Hence, the construction describes the emission of electric dipole radiation at the Bohr frequency ν.

We express the electric dipole moment of the atom as $-e\langle \mathbf{r} \rangle$, where $\mathbf{r}$ denotes the coordinate vector of the electron relative to the nucleus. (Actually, the electric dipole expression is given exactly by $-e\mathbf{r}$ only in the case of the hydrogen atom or in the limit of infinite nuclear mass. This subtle point is addressed in Problem 11 at the end of the chapter.) We then use Equation (7-36), and set $\omega = 2\pi\nu$, to obtain

$$-e\langle \mathbf{r} \rangle = -e \int \Psi^* \mathbf{r} \Psi \, d\tau$$

$$= -e \int \mathbf{r} \big\{ |c\psi_{n\ell m}|^2 + c^*c'\psi_{n\ell m}^*\psi_{n'\ell'm'}e^{i\omega t}$$

$$+ c'^*c\psi_{n'\ell'm'}^*\psi_{n\ell m}e^{-i\omega t} + |c'\psi_{n'\ell'm'}|^2 \big\} \, d\tau,\tag{7-38}$$

in analogy with Equation (5-69). The coefficient of the time dependence $e^{-i\omega t}$ may be taken to represent the amplitude of the oscillating dipole. The squared modulus of this complex quantity determines the probability for the transition between the initial and final atomic states, labeled respectively by the quantum numbers $(n\ell m)$ and $(n'\ell'm')$. The amplitude coefficient contains the integral factor

$$\int \psi_{n'\ell'm'}^* \mathbf{r} \psi_{n\ell m} \, d\tau\tag{7-39}$$

as its principal contributor. This important quantity is called the *dipole transition amplitude* for the transition $(n\ell m) \to (n'\ell'm')$.

We observe that this integral carries all the essential quantum mechanical information about the two atomic states involved in the electric dipole transition. The integral defines a vector quantity in which $\mathbf{r}$ has the Cartesian components

$$r \sin\theta \cos\phi, \quad r \sin\theta \sin\phi, \quad \text{and} \quad r \cos\theta.$$

Let us recall that the spatial eigenfunctions have the form

$$\psi_{n\ell m}(r, \theta, \phi) = R_{n\ell}(r) Y_{\ell m}(\theta, \phi)$$

so that we can write the three components of the transition amplitude as

$$\int_0^\infty R_{n'\ell'} r R_{n\ell} r^2 \, dr \cdot \begin{cases} \int Y_{\ell'm'}^* \sin\theta \cos\phi \, Y_{\ell m} \, d\Omega \\[2mm] \int Y_{\ell'm'}^* \sin\theta \sin\phi \, Y_{\ell m} \, d\Omega \\[2mm] \int Y_{\ell'm'}^* \cos\theta \, Y_{\ell m} \, d\Omega. \end{cases}\tag{7-40}$$

Note that only the angular part of the integration differs from one component to another. The selection rules for the radiative process follow upon inspection of these integrations, since the integrals over θ and ϕ are nonvanishing only if the quantum numbers (ℓm) and $(\ell' m')$ are related in a special way.

The ϕ integration presents the more straightforward part of this analysis because of the elementary ϕ dependence of the spherical harmonics:

$$Y_{\ell m}(\theta, \phi) = \Theta_{\ell m}(\theta) e^{im\phi}.$$

We transform $\cos \phi$ and $\sin \phi$ to complex-exponential form and find that the x, y, and z components of the transition amplitude contain the azimuthal integrals

$$\int_0^{2\pi} e^{-im'\phi} \frac{e^{i\phi} + e^{-i\phi}}{2} e^{im\phi}\, d\phi, \quad \int_0^{2\pi} e^{-im'\phi} \frac{e^{i\phi} - e^{-i\phi}}{2i} e^{im\phi}\, d\phi, \quad \text{and} \quad \int_0^{2\pi} e^{-im'\phi} e^{im\phi}\, d\phi.$$

These expressions are easily integrated to yield the following conclusions:

$$\int_0^{2\pi} e^{-im'\phi} e^{\pm i\phi} e^{im\phi}\, d\phi = 0 \quad \text{unless } m - m' \pm 1 = 0,$$

and

$$\int_0^{2\pi} e^{-im'\phi} e^{im\phi}\, d\phi = 0 \quad \text{unless } m - m' = 0.$$

The first result implies a restriction on m and m' for the x and y components of the dipole amplitude, while the second result applies only to the z component. We summarize these restrictions by writing

$$\Delta m = 0 \text{ or } \pm 1. \tag{7-41}$$

This selection rule expresses the allowed change $m' - m$ for the azimuthal quantum number of the atom in a transition accompanied by the emission of electric dipole radiation.

The θ integration is not so easy to evaluate in general terms. The three components of the transition amplitude contain the following two integrals in the polar angle:

$$\int_0^\pi \Theta_{\ell' m'} \sin \theta\, \Theta_{\ell m} \sin \theta\, d\theta \quad \text{and} \quad \int_0^\pi \Theta_{\ell' m'} \cos \theta\, \Theta_{\ell m} \sin \theta\, d\theta.$$

We may assume that the quantum numbers m and m' are already constrained as in Equation (7-41) when we consider these expressions. Evaluation of the integrals for arbitrary quantum numbers involves the recursion relations and orthogonality properties of the Legendre functions $P_{\ell m}(\cos \theta)$. We have looked briefly at these functions in Sections 6-4 and 6-6, but we have not developed the mathematics far enough to perform the desired integrations. The general result is easy enough to state, however:

The θ integrations vanish unless ℓ and ℓ' differ by *one* unit.

We express this selection rule for the orbital quantum number by writing

$$\Delta \ell = \pm 1, \tag{7-42}$$

and we illustrate the result by evaluating certain specific integrals in the example below.

Figure 7-6

Lyman transitions allowed by the $\Delta \ell = \pm 1$ selection rule.

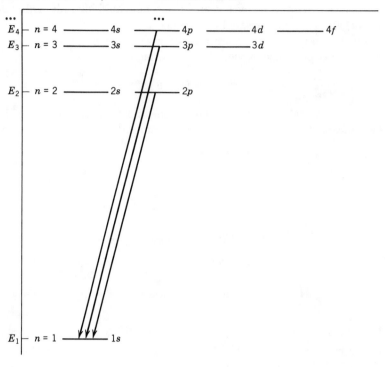

It should be emphasized that radiative transitions are not strictly forbidden if the quantum numbers of the initial and final states fail to obey Equations (7-41) and (7-42). Such transitions are actually allowed to occur, but not with the emission of radiation characteristic of an oscillating electric dipole. These other kinds of radiative processes are highly suppressed whenever the wavelength of the radiation is very large compared to the size of the radiating system.

The hydrogen transitions in Figure 7-5 can be displayed in more detail if the degeneracy with respect to ℓ is included and if the $\Delta \ell$ selection rule is taken into account. Figure 7-6 provides such a display in which all the wavelengths of the Lyman series are seen to arise from transitions of the type $np \rightarrow 1s$. Figure 7-7 goes on to show that all the wavelengths of the Balmer series are due to transitions connecting

Figure 7-7

Balmer transitions allowed by the $\Delta \ell = \pm 1$ selection rule.

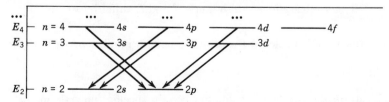

$\ell = 0$ to $\ell' = 1$, or $\ell = 1$ to $\ell' = 0$, or $\ell = 2$ to $\ell' = 1$. We can look for experimental evidence of this detailed structure and verify the $\Delta\ell$ selection rule, provided we have a suitable experiment to distinguish levels with different values of ℓ for the same n. The Δm selection rule can also be confirmed if the experiment is able to split the degeneracy of the states with different values of m for the same n and ℓ. The means to these ends are found in Chapter 8.

Let us return to the transition amplitude and inspect the parity properties of the integral $\int \psi^*_{n'\ell'm'} \mathbf{r} \psi_{n\ell m} \, d\tau$. We observe that a nonvanishing result is obtained only from integrands that are *even* under the inversion $\mathbf{r} \to -\mathbf{r}$. (Odd integrands contribute with opposite signs for volume elements in opposite directions. These contributions to the integral cancel pairwise in the integration over all space.) We then recall from Section 6-6 that each factor in the integrand has a well-defined behavior under the parity operation. The vector $\mathbf{r}$ is odd by definition, while the eigenfunctions $\psi_{n'\ell'm'}$ and $\psi_{n\ell m}$ have parities given by $(-1)^{\ell'}$ and $(-1)^{\ell}$, respectively. Therefore, inversion produces an overall-even integrand only if the two atomic states have *opposite* parity. We can state this conclusion as a selection rule:

> The parity of the state must change in an electric dipole transition.

It follows that the quantum numbers ℓ and ℓ' must differ by an odd integer, in agreement with the more restrictive result in Equation (7-42).

It is instructive to put these mathematical selection rules in terms of *conservation laws* involving parity and angular momentum. Let us describe the radiation process as

$$(n\ell m) \to (n'\ell'm') + \gamma,$$

in which γ denotes an *odd* parity to γ and use Equation (7-42) to find that the overall parity of the atom–radiation system is conserved according to the *multiplicative* law

$$(-1)^{\ell} = (-1)^{\ell'}(-1).$$

The overall angular momentum of the system must also be conserved. In fact, the rules in Equations (7-41) and (7-42) tell us that the electric dipole photon carries away a *single* $\hbar$ unit of angular momentum in the radiative process. We express the vector conservation law for the angular momentum of the system in the schematic form

$$\ell = \ell' + 1,$$

where the three vectors have lengths $\sqrt{\ell(\ell+1)}$, $\sqrt{\ell'(\ell'+1)}$, and $\sqrt{2}$, with ℓ and ℓ' constrained to differ by one unit. This interpretation of the selection rules by means of conservation laws is used often in atomic and nuclear physics.

Example

The Lyman transition $2p \to 1s$ is allowed by the $\Delta\ell$ selection rule. Let us confirm this fact by explicit evaluation of the three components of the dipole transition amplitude $\int \psi^*_{n'\ell'm'} \mathbf{r} \psi_{n\ell m} \, d\tau$. The relevant quantum numbers in the transition $(n\ell m) \rightarrow (n'\ell'm')$ are

$$n = 2 \quad \ell = 1 \quad m = 1, 0, -1 \quad \text{and} \quad n' = 1 \quad \ell' = 0 \quad m' = 0.$$

The radial integral is the same for each component in the calculation. We consult Table 7-1 and carry out the following integration over r, again with the aid of Equation (7-8):

$$\int_0^\infty R_{10} R_{21} r^3 \, dr = \int_0^\infty \frac{2}{\sqrt{a^3}} e^{-\rho} \frac{1}{2\sqrt{6a^3}} \rho e^{-\rho/2} a^4 \rho^3 \, d\rho$$

$$= \frac{a}{\sqrt{6}} \int_0^\infty \rho^4 e^{-3\rho/2} \, d\rho = \frac{a}{\sqrt{6}} 4! \left(\frac{2}{3}\right)^5 = \frac{128}{243} \sqrt{6} \, a.$$

The Δm selection rule implies that the angular integration of the z component is nonvanishing only for the $m = 0$ case. We use Table 6-1 and Equation (6-36) to perform the relevant angular integration:

$$\int Y_{00}^* \cos \theta \, Y_{10} \, d\Omega = \int \frac{1}{\sqrt{4\pi}} \sqrt{\frac{4\pi}{3}} Y_{10}^* Y_{10} \, d\Omega = \frac{1}{\sqrt{3}}.$$

We find the calculations of the x and y components easier to perform if we consider the combinations $x + iy$ and $x - iy$ first and then extract the desired results for x and y. This procedure takes advantage of the complex-exponential ϕ dependence in the expression

$$x \pm iy = r \sin \theta (\cos \phi \pm i \sin \phi) = r \sin \theta e^{\pm i\phi}$$

and blends the result with the similar ϕ dependence of the spherical harmonics. Thus, the angular integrations for the $(x \pm iy)$ components assume the general form

$$\int Y_{\ell'm'}^* \sin \theta \, e^{\pm i\phi} Y_{\ell m} \, d\Omega.$$

We again refer to the Δm selection rule and observe that a nonzero result follows for the case at hand only if we take $m = -1$ for the $(x + iy)$ component and $m = 1$ for the $(x - iy)$ component. The angular integrals in the two instances are evaluated with the aid of Table 6-1 and Equation (6-36):

$$\int Y_{00}^* \sin \theta \, e^{i\phi} Y_{1-1} \, d\Omega = \int \frac{1}{\sqrt{4\pi}} \sqrt{\frac{8\pi}{3}} Y_{1-1}^* Y_{1-1} \, d\Omega = \sqrt{\frac{2}{3}} \qquad \text{for } x + iy$$

and

$$\int Y_{00}^* \sin \theta \, e^{-i\phi} Y_{11} \, d\Omega = \int \frac{1}{\sqrt{4\pi}} \left(-\sqrt{\frac{8\pi}{3}} Y_{11}^*\right) Y_{11} \, d\Omega = -\sqrt{\frac{2}{3}} \qquad \text{for } x - iy.$$

We combine the radial and angular portions of the three integrations as follows:

$$\frac{128}{243}\sqrt{6}\,a\cdot\begin{cases}\sqrt{\dfrac{2}{3}}=\dfrac{256}{243}a & \text{for the }(x+iy)\text{ component if }m=-1\\[3mm]-\sqrt{\dfrac{2}{3}}=-\dfrac{256}{243}a & \text{for the }(x-iy)\text{ component if }m=1\\[3mm]\dfrac{1}{\sqrt{3}}=\dfrac{128}{243}\sqrt{2}\,a & \text{for the }z\text{ component if }m=0.\end{cases}$$

Vanishing results are found if the three components of the dipole transition amplitude are evaluated for any of the other possible values of the quantum number m.

Problems

1. Show that the function

    $$R(r)=A\left(1-\frac{r}{2a}\right)e^{-r/2a}$$

    is a solution of the radial differential equation for the one-electron atom in the $\ell=0$ case. What is the corresponding energy eigenvalue?

2. Determine the normalizing constant A for the wave function in Problem 1.

3. For any ℓ, the radial solutions for the one-electron atom have the form

    $$R=e^{-\rho/n}\,\frac{F(\rho)}{\rho}\quad\text{with }\rho=\frac{r}{a},$$

    where F is the nth-order polynomial

    $$F(\rho)=A\sum_{k=\ell+1}^{n}a_{k}\rho^{k}\quad\text{with }a_{\ell+1}=1.$$

    Derive the recursion formula,

    $$a_{k+1}=2\frac{k/n-1}{k(k+1)-\ell(\ell+1)}a_{k},$$

    relating the successive coefficients in the polynomial.

4. Use the recursion relation in Problem 3 to construct the radial polynomial in the case where $n=2$ and $\ell=0$.

5. Use the recursion relation in Problem 3 and apply the normalization condition to construct the wave function Ψ_{31m}.

6. List the degenerate orbital states that belong to the M shell, giving the explicit (r,θ,ϕ) dependence for each of the wave functions.

7. Let the one-electron atom be in a state with quantum numbers $n=2$ and $\ell=1$, and determine the most probable distance between the electron and the nucleus. Calculate the

expectation values $\langle r \rangle$ and $\langle V \rangle$ by explicit integration, and compare the results with the general formulas given in the text.

8. Repeat the calculations of Problem 7 for a state with quantum numbers $n = 3$ and $\ell = 1$.

9. Let the one-electron atom be in its ground state, and calculate the probability of finding the electron beyond the radius of the first Bohr orbit.

10. Consider a one-electron atom in a stationary state, and derive a formula for the rms speed of the electron relative to the nucleus. Compare the answer with the result obtained in the Bohr model.

11. Derive a formula for the electric dipole moment of the one-electron atom shown in the figure, taking the origin to be at the center of mass. The result should be proportional to $-e\mathbf{r}$ and should reduce to $-e\mathbf{r}$ if $Z = 1$ or if $M \to \infty$.

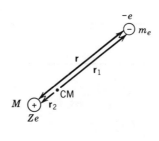

12. Derive the results

$$\int_0^{2\pi} e^{-im'\phi} e^{\pm i\phi} e^{im\phi}\, d\phi = 0 \quad \text{unless } m' - m = \pm 1,$$

and

$$\int_0^{2\pi} e^{-im'\phi} e^{im\phi}\, d\phi = 0 \quad \text{unless } m' - m = 0.$$

These calculations appear in the demonstration of the selection rules for electric dipole transitions.

13. Show by direct calculation that all three components of the dipole transition amplitude vanish for the forbidden transition $(n = 2,\ \ell = 0) \to (n' = 1,\ \ell' = 0)$.

14. Evaluate the z component of the dipole transition amplitude for the Balmer transition $3p \to 2s$.

15. Evaluate the z component of the dipole transition amplitude for the Balmer transition $3d \to 2p$.

16. Stationary states of the one-electron atom can exist in the form of superpositions of degenerate states. An example is given by the wave function

$$\Psi = \frac{1}{\sqrt{2}} \left(\psi_{200} + \psi_{210} \right) e^{-iE_2 t / \hbar}.$$

Comment on the parity properties of Ψ, and compute the components of the electric dipole moment vector $-e\mathbf{r}$ for an atom in such a state.

E I G H T

SPIN
AND
MAGNETIC
INTERACTIONS

*T*he Schrödinger theory of the atom is based on a nonrelativistic treatment of electron motion. The effects of relativity are appropriately ignored in first approximation, but the resulting picture of the atom is incomplete until such behavior is taken into account. One of our goals in this chapter is to introduce relativistic corrections as secondary contributions to the leading approximation. We are interested in these corrections because we can detect their influence in experiments of sufficient accuracy.

The Schrödinger theory has also been restricted to a treatment of the electron in which the spatial coordinates are the only variables. Our main goal in this chapter is to show that the behavior of the electron cannot be fully understood unless the particle is endowed with an additional *intrinsic* variable known as *electron spin*. This new observable is quantized like angular momentum and is detectable in experiments where the electron system interacts with an applied magnetic field. Several of the key experiments, and even a few of the important concepts, predate quantum mechanics. In fact, the earliest experimental evidence for spin stems from roots in the 19th century.

The first clues to spin came from studies of the influence of magnetic fields on the emission of light by atoms. An experiment by P. Zeeman in 1896 showed that two of the prominent lines in the visible spectrum of sodium were broadened when a sodium flame was placed between the poles of a strong electromagnet. Zeeman concluded that the frequency of the emitted light was altered by the application of the magnetic field. Lorentz interpreted this effect in terms of his classical electron theory and predicted that the magnetic field should split each spectral line into a fixed number of separate components with different frequencies and definite polarizations. Line splitting was later observed when the sodium lines were examined with improved resolution; however, the number of split components did not agree with Lorentz's predictions. This puzzling result became known as the *anomalous* Zeeman effect because the

Normal Lorentz triplet in zinc and anomalous Zeeman patterns in sodium and in zinc

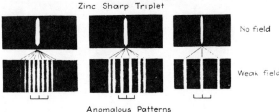

observations differed inexplicably from the *normal* effect associated with the classical theory. Normal splitting could be detected only for certain types of lines in certain elements. The early quantum ideas were also applied to this phenomenon, but the application shed light only on the normal effect. Thus, the observed anomaly in the Zeeman effect remained an unsolved problem through the period of the old quantum theory.

Improvements in spectroscopic resolving power also revealed the splitting of lines into *multiplets* of closely spaced components *without* the influence of an applied magnetic field. This behavior in *free* atoms was given the name *fine structure*. It was discovered that the split multiplets were the ones that developed an anomalous Zeeman pattern when an external magnetic field was applied. Hence, a link was believed to exist between the phenomenon of fine structure and the anomalous Zeeman effect, indicating the possibility of a common solution for the two problems. It was also found that the spectra of the elements in a given column of the periodic table displayed multiplets with common features. This connection between the structure of multiplets and the periodic table was the crucial piece of evidence that inspired W. Pauli to make the first contribution toward the eventual discovery of spin. Pauli was able to explain the occurrence of multiplets by letting the electron have a *nonclassical two-valued* property, which appeared as a *fourth quantum number* in the description of every atomic electron. Others later identified this peculiar quantized variable as the electron spin. All these remarkable developments preceded the coming of the Schrödinger equation and were carried over without difficulty into the new formulation of quantum mechanics.

We confine our discussion in this chapter to the role of electron spin in the one-electron atom, and we extend the treatment to atoms with many electrons in Chapter 9.

8-1 Atoms and Light in a Magnetic Field

Faraday believed that magnetism should influence the light emitted by atoms, but he could not develop apparatus sensitive enough to demonstrate an observable effect. His expectations were not realized until Zeeman's experiment was performed on the

Figure 8-1

Classical Lorentz-triplet predictions for an atomic source in a magnetic field. The source emits a spectral line of frequency ν_0, and the application of the field splits the line into three frequencies $\nu_0 - \delta\nu$, ν_0, and $\nu_0 + \delta\nu$, where the shift $\delta\nu$ depends on the field strength B. Light observed along the direction of **B** has only the two shifted components with circular polarizations in opposite directions. Note that the up-shifted frequency $\nu_0 + \delta\nu$ corresponds to circular polarization in the same sense as the current in the electromagnet. Light observed transverse to **B** has the two shifted components with linear polarizations perpendicular to **B**, and the one unshifted component with linear polarization parallel to **B**.

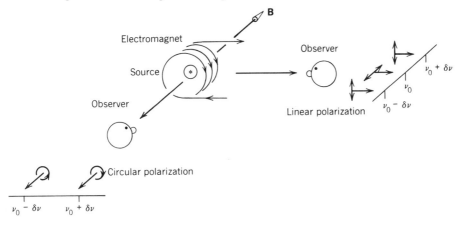

spectrum of sodium in 1896. Lorentz's interpretation of the Zeeman effect attributed the emitted light to the oscillation of charge within the atom and showed that the oscillation would be modified by the application of a magnetic field. His theory selected a given spectral line, with frequency ν_0 in the absence of the applied field, and predicted a splitting of the line into *three* different frequencies owing to the interaction between the field and the atom. He determined a frequency shift $\delta\nu$ such that this *triplet* of frequencies would occur as $\nu_0 - \delta\nu$, ν_0, and $\nu_0 + \delta\nu$. The members of the triplet were expected to have certain polarization properties depending on the direction of observation of the light relative to the direction of the applied field, as described in Figure 8-1.

The classical theory of the Zeeman effect is interesting because the analysis identifies the basic parameters of the interacting system. A similar parameterization of the interaction between the atom and the applied field appears in the quantum version of the problem. Lorentz's model associates the effect with the oscillation of the electrons in the atom and treats such an oscillation at frequency ν_0 as a source emitting classical radiation of the same frequency.

The application of a constant magnetic field **B** fixes a direction in space to which the electron motion can be referred. Let us look first at the oscillation of a single electron in a plane perpendicular to this direction. Any such linear oscillation at frequency ν_0 (with no field) can be resolved into clockwise and counterclockwise circular motions in phase, each with angular velocity $2\pi\nu_0$, as shown in Figure 8-2. We can attribute the oscillation of the electron to a simple-harmonic binding force of the form $-kr$ and define the force constant by means of the frequency formula

$$\nu_0 = \frac{1}{2\pi}\sqrt{\frac{k}{m_e}}. \tag{8-1}$$

Figure 8-2

Resolution of linear oscillation into two counter-rotating motions in phase.

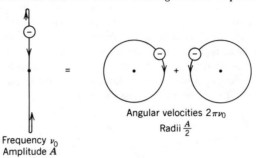

Angular velocities $2\pi\nu_0$

Radii $\frac{A}{2}$

Frequency ν_0
Amplitude A

The magnetic field subjects the electron to the additional Lorentz force $-e\mathbf{v} \times \mathbf{B}$, which acts in *opposite* radial directions for the two counter-rotating motions of the oscillating charge. This effect produces two different values for the new angular velocity $2\pi\nu$ and for the new orbit speed

$$v = 2\pi r\nu.$$

We use this expression to write the centripetal force

$$\frac{m_e v^2}{r} = 4\pi^2\nu^2 r m_e,$$

and we assume a small Lorentz force so that we can take the radius of the orbit to be unaffected by the field as a first approximation. Figure 8-3 describes the centripetal

Figure 8-3

Effect of the Lorentz force $-e\mathbf{v} \times \mathbf{B}$ for clockwise and counterclockwise rotations. The magnetic field is directed away from the reader as in Figure 8-1.

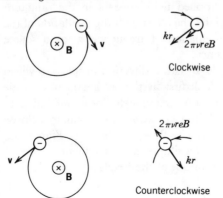

kr
$2\pi\nu reB$

Clockwise

$2\pi\nu reB$

kr

Counterclockwise

force as

$$4\pi^2\nu^2 rm_e = 2\pi\nu reB + kr$$

for clockwise motion, and

$$4\pi^2\nu^2 rm_e = -2\pi\nu reB + kr$$

for counterclockwise motion. We employ Equation (8-1) to eliminate k and rewrite the results as quadratic equations for the unknown frequency ν:

$$\nu^2 - \frac{eB}{2\pi m_e}\nu - \nu_0^2 = 0 \quad \text{(clockwise)}$$

and

$$\nu^2 + \frac{eB}{2\pi m_e}\nu - \nu_0^2 = 0 \quad \text{(counterclockwise)}.$$

The linear terms in ν must be small compared to ν_0^2 because of our assumption of a small Lorentz force. Consequently, when we solve the quadratic equations for the frequencies we find

$$\nu = \nu_0 + \frac{eB}{4\pi m_e} \quad \text{for the clockwise case} \tag{8-2a}$$

and

$$\nu = \nu_0 - \frac{eB}{4\pi m_e} \quad \text{for the counterclockwise case.} \tag{8-2b}$$

The two electron motions are therefore supposed to generate classical radiation at these two shifted frequencies. Note that the frequency shift has the form

$$\delta\nu = \frac{eB}{4\pi m_e} \tag{8-3}$$

in each case.

If we view the source along the direction of $\mathbf{B}$, as suggested in Figure 8-3, we expect the observed light to have circular polarization in the clockwise sense for $\nu = \nu_0 + \delta\nu$, and in the counterclockwise sense for $\nu = \nu_0 - \delta\nu$. This conclusion agrees with Zeeman's observations of the light from his broadened spectral line. Note that the result also confirms the fact that the sign of the oscillating charge is *negative*. We have predicted circularly polarized shifted components of light for electrons oscillating in a plane perpendicular to $\mathbf{B}$. When we view the source in a direction transverse to the applied field, we see these shifted components as light with linear polarization in the plane of the oscillating charge. The electrons in the atom may also oscillate along $\mathbf{B}$ to produce light with polarization in that direction. No shift in frequency is predicted in this case, since $\mathbf{v} \times \mathbf{B}$ vanishes. Radiation of this kind is seen unshifted if the source is viewed transverse to the direction of the applied field. All these conclusions from Lorentz's theory are illustrated in Figure 8-1.

We should note the presence of the factor e/m_e in Equation (8-3). This aspect of the analysis was appreciated at the time of Thomson's cathode-ray experiments, because it was possible to measure the frequency shift in a magnetic field of known strength and thereby determine the charge-to-mass ratio for the radiating charge in the atom. The resulting determination agreed with the value obtained for Thomson's cathode rays and thus supported the idea of the electron as a universal constituent of the atom.

Example

The frequency shift $\delta\nu$ is quite small compared to the line frequency ν_0, even for rather large values of the magnetic field. We illustrate by choosing $B = 1$ T in Equation (8-3):

$$\delta\nu = \frac{eB}{4\pi m_e} = \frac{(1.76 \times 10^{11} \text{ C/kg})(1 \text{ T})}{4\pi} = 1.40 \times 10^{10} \text{ Hz.}$$

(The figure for e/m_e is quoted from Section 3-2.) This result is to be compared with values of order 10^{15} Hz for the frequency of visible light. The units in this calculation should also be mentioned. We recall Faraday's law relating the rate of change of magnetic flux to the induced emf and pass from teslas to volts by means of the conversion

$$V = \frac{T \cdot m^2}{s}.$$

We then obtain the units

$$\frac{C}{kg} T = \frac{C}{kg} \frac{V \cdot s}{m^2} = \frac{J \cdot s}{kg \cdot m^2} = s^{-1}$$

for the computation of $\delta\nu$.

8-2 Orbital Magnetic Moments

The quantum version of the Zeeman effect involves familiar observable quantities and makes no use of the oscillatory motion described in the classical model. Quantum mechanics teaches us to recognize a shift in the frequency of a spectral line as evidence for a change in the transition energy of the emitted photon. This change in turn implies a shift in the quantized energy levels of the atom owing to the application of the magnetic field. Hence, the observation of Zeeman splittings in spectral lines translates directly into the deduction of Zeeman energy splittings among the states of the atom.

We are concerned with the *additional* energy of the atom associated with the presence of the applied magnetic field. We treat the atom as a system of moving charges and identify a *magnetic dipole moment* μ, so that we can introduce a *magnetic interaction energy*

$$V_M = -\mu \cdot \mathbf{B} \tag{8-4}$$

as an added contribution to the potential energy of the system. The magnetic moment vector parametrizes the magnetic structure of the atom, and the applied magnetic

Figure 8-4

Magnetic moment μ for a circular Bohr orbit. The direction of the angular momentum **L** is opposite to the direction of μ.

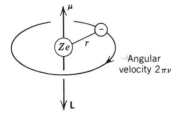

field acts as an external probe of this structure. We take **B** to be constant in space and time and observe from Equation (8-4) that the magnetic configurational energy is minimized when μ is aligned along **B**. The dynamical behavior of μ is determined by angular momentum considerations, since the magnetic moment of the atom is directly related to the angular momentum. This section is devoted to the relation between these quantities for the case of the one-electron atom. We are dealing with small energy shifts, and so we can ignore the small correction for nuclear motion and simply replace the reduced mass by the electron mass throughout. (Thus, the symbol μ for reduced mass does not appear in this chapter and cannot be confused with the new symbol μ for magnetic moment.)

We consider a circular Bohr orbit for an electron around a fixed nucleus, and we take the angular velocity of the electron to be $2\pi\nu$. The current in the electron orbit has magnitude $e\nu$, and the orbit has area πr^2, so that the product of these quantities gives the magnitude of the magnetic moment vector

$$\mu = \pi r^2 e \nu.$$

The direction of μ is shown along with the various parameters of the orbit in Figure 8-4. We recall that the magnetic field of such a current loop is similar to that of a bar magnet, at distances far from the respective dipoles. The electron orbit also represents mechanical motion with angular momentum of magnitude

$$L = m_e v r = 2\pi r^2 m_e \nu.$$

We note that the quantities μ and L contain the kinematical factor $r^2\nu$ and that their ratio depends only on fundamental constants:

$$\frac{\mu}{L} = \frac{e}{2m_e}.$$

The figure shows that the vectors μ and **L** have opposite directions for an orbiting particle of negative charge. We take this observation into account and relate μ to **L** by the vector formula

$$\mu = -\frac{e}{2m_e}\mathbf{L}. \tag{8-5}$$

The final result holds for elliptical as well as circular electron orbits.

The proportionality between $\boldsymbol{\mu}$ and $\mathbf{L}$ is a general property of a rotating charge distribution. The magnetic moment and angular momentum of an arbitrary rotating body with mass M and charge Q always satisfy a relation of the form

$$\boldsymbol{\mu} = g\frac{Q}{2M}\mathbf{L},$$

where the details of the rotating distribution determine the value of the g-factor denoted by the symbol g. We are using $Q = -e$ for an electron orbit in Equation (8-5), and so we identify the corresponding *orbital g-factor* as

$$g = 1.$$

A special symbol is introduced to represent this dimensionless quantity, in preparation for other g-factors and other kinds of magnetic moments to be encountered later on. Let us also adopt standard practice and give the angular moment in $\hbar$ units so that we can rewrite Equation (8-5) as

$$\boldsymbol{\mu} = -g\mu_B\frac{\mathbf{L}}{\hbar} \quad \text{with } g = 1, \tag{8-6}$$

where

$$\mu_B = \frac{e\hbar}{2m_e}. \tag{8-7}$$

This combination of physical constants is known as the *Bohr magneton*. The dimensions of μ_B are the same as those of $\boldsymbol{\mu}$, so that μ_B serves as a natural unit for the magnetic moment of the atom. We express the numerical value of the Bohr magneton in two ways,

$$\mu_B = 9.274 \times 10^{-24} \, \text{A} \cdot \text{m}^2 = 5.788 \times 10^{-9} \, \text{eV/G},$$

and observe that the second figure has the especially useful units of energy per magnetic field strength.

We can use the relation between $\boldsymbol{\mu}$ and $\mathbf{L}$ to determine the classical behavior of $\boldsymbol{\mu}$ in an applied $\mathbf{B}$ field. The magnetic moment experiences a torque equal to $\boldsymbol{\mu} \times \mathbf{B}$, and so the angular momentum vector varies with time as

$$\frac{d\mathbf{L}}{dt} = \boldsymbol{\mu} \times \mathbf{B}.$$

We employ Equation (8-5) to obtain the dynamical equation

$$\frac{d\mathbf{L}}{dt} = -\frac{e}{2m_e}\mathbf{L} \times \mathbf{B}$$

$$= \boldsymbol{\omega} \times \mathbf{L}, \tag{8-8}$$

introducing the constant vector

$$\boldsymbol{\omega} = \frac{e}{2m_e}\mathbf{B}. \tag{8-9}$$

Figure 8-5

Behavior of μ and $\mathbf{L}$ according to the dynamical equation $d\mathbf{L}/dt = \omega \times \mathbf{L}$. The dynamics causes a Larmor precession of the magnetic moment about the direction of the applied field.

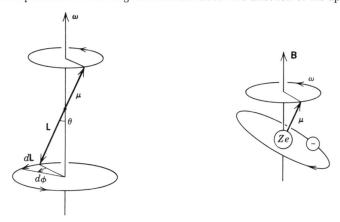

The dynamics described by Equation (8-8) is exactly the same as that for a spinning top in precessional motion about a fixed axis defined by the direction of ω. Figure 8-5 summarizes the physical properties of the oppositely directed vectors μ and $\mathbf{L}$. We observe that the magnitude of $\mathbf{L}$ is constant:

$$\frac{d}{dt}L^2 = \frac{d}{dt}\mathbf{L} \cdot \mathbf{L} = 2\mathbf{L} \cdot \frac{d\mathbf{L}}{dt} = 0,$$

using Equation (8-8) in the last step. We then use conservation of the energy $\mu \cdot \mathbf{B}$ to argue that the polar angle θ between $\mathbf{L}$ and ω also remains constant. It follows that the angular momentum varies only with respect to the azimuthal orientation of $\mathbf{L}$ about the ω axis. We consider a time interval dt and identify an increment in the azimuthal angle

$$d\phi = \frac{dL}{L \sin \theta}$$

according to the figure. We can then express the precessional angular velocity of $\mathbf{L}$ and μ by writing

$$\frac{d\phi}{dt} = \frac{1}{L \sin \theta} \frac{dL}{dt} = \frac{\omega L \sin \theta}{L \sin \theta} = \omega,$$

again using Equation (8-8). This behavior of the magnetic moment vector is an example of *Larmor precession*, a phenomenon in mechanics treated by a general theorem due to J. Larmor. Equation (8-9) gives the frequency of precession as

$$\frac{\omega}{2\pi} = \frac{eB}{4\pi m_e},$$

an expression identical to the formula for the frequency shift $\delta\nu$ in Equation (8-3). Thus, the classical dynamics of the magnetic moment presents a picture in which the applied field affects the orbital motion in the atom by superimposing a precession at the Larmor frequency about the direction of **B**. A schematic rendition of this effect is shown in the figure.

The main result of this section is the relation between μ and **L** in Equations (8-6) and (8-7). We are aware that μ and **L** have been handled classically, and we should question whether the resulting formula can be carried over to quantum mechanics. A proper quantum approach would show how electrodynamics is incorporated in the Schrödinger theory and would prove the validity of the relation between μ and **L** as quantum observables. We leave the demonstration of these steps to a more thorough treatment of quantum mechanics. It should also be noted that precessional motion is a classical phenomenon involving the behavior of a well-defined **L** vector. We must reinterpret the word if we wish to speak of precession as a concept in quantum physics.

Example

Let us prove that Equation (8-5) holds for any *elliptical* Bohr–Sommerfeld orbit. We consider an element of angle $d\theta$ swept out by the electron in a time interval dt and identify the area of the resulting sector as

$$dA = \frac{r^2}{2} \, d\theta.$$

This element of area makes a contribution to the magnetic moment given by

$$d\mu = dA \, \frac{dq}{dt} = \frac{r^2}{2} \frac{d\theta}{dt} \, dq.$$

The force on the electron is a central force, and so the quantity $r^2 \, d\theta/dt$ is a constant proportional to the angular momentum:

$$L = m_e r^2 \frac{d\theta}{dt} = \text{constant}.$$

We use this observation to obtain the formula for the magnetic moment:

$$\mu = \int_{\text{orbit}} d\mu = \frac{L}{2m_e} \int dq = -\frac{e}{2m_e} L,$$

as in the case of a circular Bohr orbit.

Example

Let us also verify the value quoted for the Bohr magneton:

$$\mu_B = \frac{e\hbar}{2m_e} = \frac{\left(1.602 \times 10^{-19}\,\text{C}\right)\left(1.055 \times 10^{-34}\,\text{J} \cdot \text{s}\right)}{2\left(9.110 \times 10^{-31}\,\text{kg}\right)}$$

$$= 9.274 \times 10^{-24}\,\text{A} \cdot \text{m}^2$$

$$= \frac{9.274 \times 10^{-24}\,\text{J/T}}{\left(1.602 \times 10^{-19}\,\text{J/eV}\right)\left(10^4\,\text{G/T}\right)} = 5.788 \times 10^{-9}\,\text{eV/G},$$

using the unit conversion from the example in Section 8-1.

8-3 The Normal Zeeman Effect

The angular momentum **L** is quantized according to the rules developed in Chapter 6. These quantization properties are passed along to the magnetic moment **μ**, since **μ** is directly related to **L**. The magnetic moment determines the behavior of an atom in an applied magnetic field via the **μ** · **B** interaction, and so the quantization of **μ** accounts for the Zeeman splitting in the energy levels of the atom. Thus, the Zeeman effect provides evidence for the quantization of **L** as a result of the quantization of **μ**.

Let us suppose that the one-electron atom is in the stationary state $\Psi_{n\ell m}$ in the absence of an applied field. The quantity L^2 has the value $\hbar^2\ell(\ell+1)$, and so the magnetic moment of the atom has magnitude

$$\sqrt{\langle \mu^2 \rangle} = \frac{g\mu_B}{\hbar}\sqrt{\langle L^2 \rangle} = g\mu_B\sqrt{\ell(\ell+1)}$$

by virtue of Equation (8-6). The z component of **L** is given by

$$\langle L_z \rangle = m\hbar$$

in this state, and so the z component of **μ** has the expectation value

$$\langle \mu_z \rangle = -\frac{g\mu_B}{\hbar}\langle L_z \rangle = -g\mu_B m. \tag{8-10}$$

These remarks suggest that we employ the interaction of the magnetic moment to gain access to the angular momentum properties of the state through the dependence of $\langle \mu_z \rangle$ on the quantum number m.

The applied **B** field is a constant vector whose fixed orientation in space offers a natural direction to choose as the axis of quantization. We align the z axis along **B** for the definition of L_z, and we specify the state $\Psi_{n\ell m}$ with azimuthal quantum number m according to this choice. The interaction energy in Equation (8-4) then assumes the simpler form

$$V_M = -\mu_z B, \tag{8-11}$$

so that the expectation value becomes

$$\langle V_M \rangle = -\langle \mu_z \rangle B = +g\mu_B B m \tag{8-12}$$

with the aid of Equation (8-10).

The quantity $\langle V_M \rangle$ represents the additional energy acquired by the atom in the state $\Psi_{n\ell m}$ when the magnetic field is applied. We note that the added energy depends only on the integer m in a field of given strength. Hence, the effect of the field is to produce a *discrete energy shift* for each of the $2\ell+1$ values of m in any $n\ell$-orbital configuration. We illustrate this phenomenon by adapting the energy level diagram in Figure 7-2 to the new situation displayed in Figure 8-6. The previous diagram reminds us that the nth energy level of the atom has energy E_n and that the states $\Psi_{n\ell m}$ are degenerate with respect to the quantum numbers ℓ and m in the absence of a field. The new diagram shows how these states split apart so that the various m values yield energies different from E_n by the amount predicted in Equation (8-12). We note that the states $\Psi_{n\ell m}$ are still degenerate in their dependence on ℓ and that the states with successive m values have the same energy spacing

$$\delta E_M = g\mu_B B \tag{8-13}$$

Figure 8-6

Normal Zeeman splittings in the energy levels of the one-electron atom. The magnetic field breaks the degeneracy of the states with respect to the quantum number m. States $\Psi_{n\ell m}$ with adjacent values of m are spaced in energy by an amount $\delta E_M = g\mu_B B$, for any choice of n and ℓ. The scale of the Zeeman splitting is grossly exaggerated in the diagram.

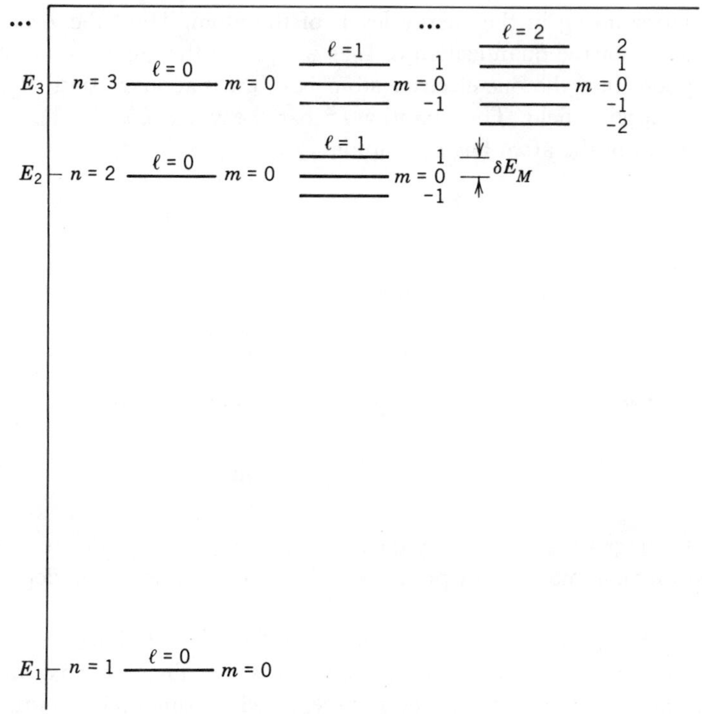

in *every* $n\ell$-orbital system. The sign of the energy shift $\langle V_M \rangle$ is the same as the sign of m, and only the $m = 0$ states are unaffected by the field. We let each of the split levels shown in Figure 8-6 refer to a distinct *precessional state* of the atom, whose energy maintains the fixed value

$$E_n + g\mu_B Bm$$

in the presence of the constant magnetic field. The azimuthal quantum number m is also called the *magnetic quantum number* because of this property.

The most obvious feature that distinguishes Figure 8-6 from Figure 7-2 is the removal of the degeneracy with respect to m. We interpret the m dependence of the energy levels as an indication of a breakdown of rotational symmetry. The applied **B** field introduces a preferred direction in space for our selection of z axis, and so the energies of the states vary with the L_z quantum number as a result.

The Zeeman splitting of the energy levels causes a frequency shift in the radiation from the atom. We determine the shift by identifying the allowed radiative transitions according to the selection rules discussed in Section 7-5. The rule $\Delta\ell = \pm 1$ has already been used to pick out the electric dipole transitions for the Lyman and Balmer series in Figures 7-6 and 7-7. The other rule requires

$$\Delta m = 0 \text{ or } \pm 1$$

Figure 8-7

Lyman α transition and normal Zeeman splittings.

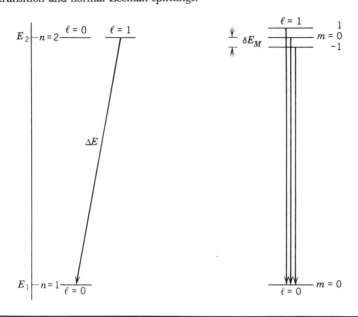

and gives more pertinent information about the Zeeman effect, as illustrated in Figures 8-7 and 8-8. We observe that all the indicated transitions involve only *three* distinct emitted-photon energies,

$$\Delta E - \delta E_M, \quad \Delta E, \quad \text{and} \quad \Delta E + \delta E_M,$$

where ΔE denotes the usual Bohr transition energy without the applied field. A *triplet* of spectral lines is the predicted result, implying a shift in frequency to either side of the usual line by an amount

$$\delta \nu = \frac{\delta E_M}{h} = \frac{g\mu_B B}{2\pi\hbar} = \frac{eB}{4\pi m_e},$$

in agreement with the classical formula in Equation (8-3). The illustrations refer specifically to the Lyman α line (where $n = 2 \to n = 1$) and the Balmer α line (where

Figure 8-8

Balmer α transitions and normal Zeeman splittings.

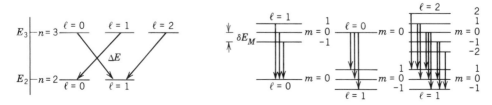

$n = 3 \rightarrow n = 2$). It is clear, however, that the triplet prediction is quite general, since the normal Zeeman splitting δE_M is common to *all* $n\ell$ orbitals. The figures tell us that the photon energies $\Delta E - \delta E_M$ and $\Delta E + \delta E_M$ are associated with the $\Delta m = +1$ and $\Delta m = -1$ transitions, and that the unshifted photon energy ΔE is due to the $\Delta m = 0$ transitions. It follows that the $\Delta m = \pm 1$ linesw correspond to the shifted frequencies in Figure 8-1. We know that these lines exhibit circular polarization when viewed along the direction of the field. We also know from Section 7-5 that the single unit in $\Delta \ell$ and Δm represents a single $\hbar$ unit of angular momentum carried away by the radiated electric dipole photon.

The predictions of the normal Zeeman effect are not borne out in observations of the spectrum of hydrogen, as anomalous splittings are seen instead of the expected triplets. The classical results have not been altered by taking a quantum approach to the problem. We have included the flawed classical treatment as background in order to point out the fact that the preferred quantum solution also fails to resolve the anomaly. We are forced to concede that our picture of the magnetic structure of the atom is incomplete.

Example

The Balmer α processes in Figure 8-8 have transition energy $\Delta E = 1.89$ eV and emitted wavelength

$$\lambda = \frac{hc}{\Delta E} = 656.1 \text{ nm}$$

with no applied field, as in the calculations at the end of Section 3-6. Let us consider the abundant collection of normal Zeeman transitions for the illustrated case

$$(n = 3, \ell = 2) \rightarrow (n = 2, \ell = 1).$$

The Δm selection rule allows nine different transitions, as shown in the figure, with three distinct photon frequencies:

$$\frac{\Delta E - \delta E_M}{h}, \quad \frac{\Delta E}{h}, \quad \text{and} \quad \frac{\Delta E + \delta E_M}{h}.$$

We define the shift in wavelength by the relation

$$\lambda + \delta \lambda = \frac{hc}{\Delta E - \delta E_M},$$

and we use the condition $\Delta E \gg \delta E_M$ to make the following approximation:

$$\lambda + \delta \lambda = \frac{hc}{\Delta E}\left(1 + \frac{\delta E_M}{\Delta E}\right) = \lambda\left(1 + \frac{\delta E_M}{\Delta E}\right) \quad \Rightarrow \quad \delta \lambda = \lambda \frac{\delta E_M}{\Delta E}.$$

For $B = 1$ T we find

$$\delta E_M = g\mu_B B = (1)(5.788 \times 10^{-9} \text{ eV/G})(10^4 \text{ G}) = 5.788 \times 10^{-5} \text{ eV}$$

and

$$\delta\lambda = (656.1 \text{ nm}) \frac{5.788 \times 10^{-5}}{1.89} = 0.0201 \text{ nm},$$

a very small departure from the zero-field value for the Balmer α line.

<hr/>

Example

Let us prove that the stationary-state wave function of the atom is unaffected by the application of the constant magnetic field, while the energy of the state undergoes the normal Zeeman energy shift. Without the field, the stationary-state eigenfunction $\psi_{n\ell m}$ and the energy eigenvalue E_n obey the equation

$$\left[-\frac{\hbar^2}{2\mu}\nabla^2 + V(r) \right]\psi_{n\ell m} = E_n\psi_{n\ell m},$$

where $V(r)$ denotes the Coulomb potential energy. Application of the field introduces another differential operator on the left, arising from Equation (8-11). We employ Equation (8-6) and recall Equation (6-10) to assemble this additional term:

$$-\mu_z B = +\frac{g\mu_B B}{\hbar}L_z \rightarrow \frac{g\mu_B B}{\hbar}\frac{\hbar}{i}\frac{\partial}{\partial\phi}.$$

The newly defined problem calls for a new stationary state,

$$\Psi = \psi e^{-iEt/\hbar},$$

whose eigenfunction ψ and eigenvalue E satisfy the equation

$$\left[-\frac{\hbar^2}{2\mu}\nabla^2 + V(r) + \frac{g\mu_B B}{i}\frac{\partial}{\partial\phi} \right]\psi = E\psi.$$

We note that $\psi_{n\ell m}$ is an L_z eigenfunction since

$$\frac{1}{i}\frac{\partial}{\partial\phi}\psi_{n\ell m} = m\psi_{n\ell m},$$

and we recognize at once that $\psi_{n\ell m}$ is a valid solution of the new differential equation:

$$\left[-\frac{\hbar^2}{2\mu}\nabla^2 + V(r) + \frac{g\mu_B B}{i}\frac{\partial}{\partial\phi} \right]\psi_{n\ell m} = E_n\psi_{n\ell m} + g\mu_B Bm\psi_{n\ell m}.$$

We may therefore take $\psi = \psi_{n\ell m}$ and claim the new energy eigenvalue to be

$$E = E_n + g\mu_B Bm,$$

in agreement with Equation (8-12).

8-4   The Stern–Gerlach Experiment

It was known by 1920 that spectral lines were split into multiplets without the influence of an applied magnetic field. The phenomenon was originally attributed to a force within the atom, where the outer electrons interacted with a proposed "magnetic core" consisting of the inner electrons and the nucleus. This interacting system of inner and outer magnetic moments was supposed to explain the structure of multiplets in the absence of a field, as well as the anomaly in the Zeeman effect in the presence of a field. These ideas were tested in 1922 by O. Stern and W. Gerlach. Their classic experiment employed a magnetic field of special design to study the behavior of a beam of atoms. Their observations demonstrated the quantization of the magnetic moment and also appeared to confirm the magnetic-core hypothesis. Soon, however, it became necessary to reject this idea and reinterpret the outcome of the experiment as unequivocal evidence for the spin of the electron.

The Stern–Gerlach experiment examines the dynamics of a magnetic dipole in a *nonuniform* magnetic field. Let us refer to Figure 8-9*a* and recall that a uniform field produces a torque but exerts no net force on such a dipole. Figure 8-9*b* goes on to show that the forces on the two poles of a magnet yield a net force, as well as a torque, if the field is not uniform. Let us assume that the nonuniform field varies appreciably in the z direction and consider only the z component of the net force according to the details shown in the figure. We express the z component of the field at the locations of the N and S poles of the magnet by the expansions

$$B_z(\mathrm{N}) = B_{z0} + z_0 \frac{\partial B_z}{\partial z} \quad \text{and} \quad B_z(\mathrm{S}) = B_{z0} - z_0 \frac{\partial B_z}{\partial z},$$

to first order in the indicated distance z_0. The coefficients B_{z0} and $\partial B_z/\partial z$ denote the field strength and the field gradient, evaluated at the center of the dipole. We let the magnet have hypothetical pole strengths $\pm g_0$, and we write the net force as

$$F_z = g_0 B_z(\mathrm{N}) - g_0 B_z(\mathrm{S}) = 2g_0 z_0 \frac{\partial B_z}{\partial z}$$

$$= \mu_z \frac{\partial B_z}{\partial z}. \tag{8-14}$$

Note that the z component of the magnetic dipole moment appears in the last step. The force causes a beam of magnets entering the field along the y axis to experience a vertical deflection, up or down depending on the sign of μ_z, as illustrated in Figure 8-10. The classical quantity μ_z is continuous in value because of the arbitrariness in the observable orientation of the dipole. The beam therefore produces a continuously distributed image in a detector beyond the region of the field.

Very different results are found in such an experiment if the beam of magnets is replaced by a beam of atoms. Each atom has a certain probability to be in some quantized state of the angular momentum component L_z. The corresponding magnetic moment μ_z is then observed with the same probability to have one of several *discrete* values. If we consider a beam of hydrogen atoms as an example, we expect to find magnetic moments given by

$$\langle \mu_z \rangle = -g\mu_B m \quad \text{with } m = -\ell, \dots, \ell$$

Figure 8-9

Forces on the poles of a magnet in a uniform field and in a nonuniform field. The field gradient exerts a net force on the magnet in the second case.

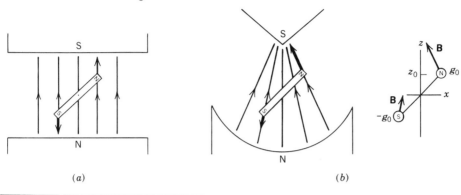

(a)   (b)

for atoms in definite L_z states. The force F_z becomes a discrete-valued quantity in this experiment, so that a fixed number of vertical deflections of the beam is detected in a given nonuniform field. Thus, an arbitrary hydrogen atom in the beam may have L_z equal to $\hbar m$ and μ_z equal to $-g\mu_B m$ with probability $|c_m|^2$, and it may then be deflected by F_z along a certain discrete path with the same probability. We call this phenomenon *space quantization* because we imagine the effect to occur via discrete trajectories in space. The result is actually observed as a discrete distribution of the beam image in a detector beyond the region of the field, as sketched in Figure 8-11. Hydrogen atoms with orbital quantum number ℓ are expected to display a $(2\ell + 1)$-fold splitting of the beam, and no deflection is supposed to be seen for a beam of ground-state $\ell = 0$ atoms. Complex atoms are also expected to undergo an *odd*

Figure 8-10

Deflection of a beam of classical magnets in a nonuniform magnetic field.

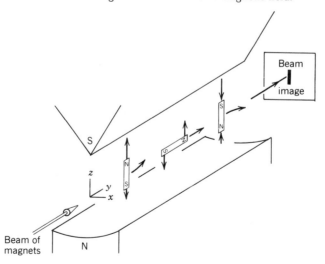

Figure 8-11

Space quantization for a beam of hydrogen atoms. An odd number of deflections is expected for a purely orbital magnetic moment, but a twofold splitting of the beam is observed in a Stern–Gerlach experiment.

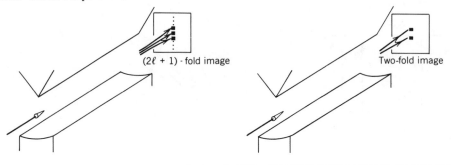

number of discrete deflections if the magnetic moments are due entirely to orbital electron motion.

The original Stern–Gerlach experiment employed a beam of neutral atoms obtained by evaporating silver in a heated oven. The silver atoms were passed through a strong transverse field gradient and were deposited on a glass plate where their deflections could be measured. This image of the beam was found to be discrete rather than continuous, in striking agreement with the notion of space quantization. In fact, the beam was observed to form a *two-fold* image, the result predicted on the basis of the prevailing magnetic-core theory of the silver atom. The core interpretation was discredited, however, when a subsequent experiment of the same design was performed with a beam of hydrogen atoms. It was expected that hydrogen should show no deflection for atoms in the ground state and should give an odd number of beam splittings for excited atoms if the magnetic moment of the atom were attributed solely to the orbital motion of the electron. Instead, the beam of hydrogen atoms again displayed a two-fold splitting, so that expectations and observations conflicted in the manner described in Figure 8-11. The "magnetic core" could not account for this result, since the core consisted of the nucleus alone and the nuclear magnetic moment was too small to explain the observed effect. (It was recognized that the natural unit for nuclear moments should be the nuclear magneton $e\hbar/2M$ instead of the Bohr magneton $e\hbar/2m_e$. The substitution of the proton mass M for the electron mass m_e made the scale smaller for nuclear moments than for atomic moments by three orders of magnitude.)

The hydrogen experiment demonstrated that $\langle \mu_z \rangle$ had a nonvanishing two-valued quality in the ground state of the atom. The two-fold splitting indicated a quantization of the magnetic moment in which two, and only two, quantized states contributed probabilities to the discrete deflections of the beam. It was apparent that this aspect of the magnetic moment of the atom was attributable to the electron but was independent of the orbital motion of the electron. Thus, a new *intrinsic* property of the electron seemed to be in evidence as a quantized *two-valued* variable. The bifurcation of the beam of silver atoms in the original Stern–Gerlach experiment also appeared to be due to this same two-valued magnetic moment for the outermost electron in the silver atom.

Pauli's contribution to the understanding of this problem came in 1924. He argued that the structure of multiplets and the anomaly in the Zeeman effect could be explained if a new formal two-valued degree of freedom was assigned to the electron. The research students S. A. Goudsmit and G. E. Uhlenbeck took up Pauli's idea in the following year and expressed this proposed two-valued quality in terms of a quantized variable with the properties of an *angular momentum*. They imagined an additional spinning motion for the orbiting electron in the atom, somewhat like the axial rotation of a planet in its orbital revolution around the Sun. The spin of the electron was supposed to produce an intrinsic magnetic moment whose magnitude was to be found from experiment. Goudsmit and Uhlenbeck put forward these properties of the electron as a *quantum* hypothesis, acknowledging that the classical picture had serious flaws on relativistic grounds. Pauli did not accept this representation of his own fundamental idea until he was convinced that spin was essential to a correct theory of multiplet structure. He finally sanctioned the concept in 1927 by showing how the new principles should be incorporated in the nonrelativistic Schrödinger theory. By then it was clear that spin was needed in the quantum theory to explain the Stern–Gerlach experiment, the fine structure of spectral lines, and the anomalous Zeeman effect.

Example

Equation (8-4) offers an alternative route to the expression for the force on a magnetic dipole. We write out the interaction energy as

$$V_M = -\mu_x B_x - \mu_y B_y - \mu_z B_z$$

and treat V_M as a function of the coordinates (x, y, z) of the dipole. The z component of the force is then found from the formula

$$F_z = -\frac{\partial}{\partial z} V_M.$$

If we assume that B_z makes the dominant z-dependent contribution to V_M, we find

$$F_z = +\mu_z \frac{\partial B_z}{\partial z}$$

as in Equation (8-14). The deflection of the beam can be predicted on the basis of familiar classical arguments. We use the temperature of the sample to obtain the beam velocity according to the relation

$$\frac{M}{2} v^2 = \frac{3}{2} k_B T,$$

where M is the mass of the atom and k_B is Boltzmann's constant. We then let y denote the beam distance across the field as in Figure 8-10 and identify the time of traversal as $t = y/v$. The vertical acceleration F_z/M is a constant, and so the final result for the vertical deflection is easily derived:

$$z = \frac{1}{2}\frac{F_z}{M}t^2 = \frac{1}{2M}\mu_z\frac{\partial B_z}{\partial z}\frac{y^2}{v^2} = \mu_z\frac{\partial B_z}{\partial z}\frac{y^2}{6k_BT}.$$

The predicted deflections follow for a given field gradient and a known magnetic dipole moment. We take the latter quantity to be quantized when we apply the formula to a beam of atoms.

8-5 The Properties of Electron Spin

We have taken the Stern–Gerlach experiment and the Goudsmit–Uhlenbeck hypothesis as grounds for the concept of electron spin. The hypothesis introduces a spin vector **S** as a new quantum mechanical variable for the electron and endows **S** with quantization rules patterned after the behavior of the orbital angular momentum **L**. We continue to follow this inferential approach, based on experiment, as an alternative to a more formal treatment of spin.

Let us choose an arbitrary direction for the z axis and consider states of the electron in which the z component of **S** is known with zero uncertainty. These determinations of S_z are expressed in terms of a new *spin quantum number* m_s by eigenvalues of the form

$$S_z = \hbar m_s. \tag{8-15}$$

We identify this variable with Pauli's formal two-valued degree of freedom, and we attribute the results of the Stern–Gerlach experiment to its discrete properties. The vector **S** is to be treated like the quantized angular momentum vector **L**. We know that L_z has $2\ell + 1$ discrete eigenvalues for some choice of the quantum number ℓ, and so we argue that there must exist a number s such that m_s has the analogous set of $2s + 1$ allowed values

$$m_s = -s, \ldots, s.$$

The fact that m_s must be two-valued then implies

$$s = \tfrac{1}{2} \quad \text{and} \quad m_s = -\tfrac{1}{2} \text{ or } +\tfrac{1}{2}. \tag{8-16}$$

The quantization rules also require **S** to have a fixed magnitude determined by the eigenvalue

$$S^2 = \hbar^2 s(s + 1) = \tfrac{3}{4}\hbar^2. \tag{8-17}$$

These properties are illustrated schematically in Figure 8-12. The z direction has been selected arbitrarily, and so the quantized behavior of **S** holds for *any* chosen direction in space along which the component of the vector may have either of the two definite values $\pm\hbar/2$. The figure reminds us that only one component of **S** can be specified with certainty, as must be the case for any angular momentum vector. The two illustrated orientations of **S** describe the *spin-up* and *spin-down* states of the electron. We call the electron a spin-$\tfrac{1}{2}$ particle to summarize the quantization properties contained in this description.

Figure 8-12

Quantized spin angular momentum vectors. The z component of $\mathbf{S}$ has the values $\pm \hbar/2$ with zero uncertainty.

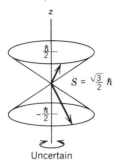

The quantum number m_s becomes part of a revised picture for the states of the atom. Now that we have adopted the symbol m_s to define the discrete values of S_z, we must go back and refine our previous notation for the magnetic quantum number m associated with L_z. Hereafter we denote the eigenvalues of L_z as

$$\hbar m_\ell, \quad \text{where } m_\ell = -\ell, \dots, \ell,$$

and employ the *two* independent indices m_ℓ and m_s to specify states according to the z components of $\mathbf{L}$ *and* $\mathbf{S}$. A stationary state of the one-electron atom is then assigned a set of *four* quantum numbers $(n\ell m_\ell m_s)$ to determine a complete configuration of the quantum system. We express the corresponding wave function $\Psi_{n\ell m_\ell m_s}$ for the given energy state of the atom, with definite orbital description in space and definite up or down orientation of electron spin, by writing

$$\Psi_{n\ell m_\ell m_s} = R_{n\ell}(r) Y_{\ell m_\ell}(\theta, \phi) e^{-iE_n t/\hbar} (\uparrow \text{ or } \downarrow). \tag{8-18}$$

We use the rightmost symbol to convey the meaning of the new electron quantum number; spin up ($\uparrow$) denotes $m_s = \frac{1}{2}$, and spin down ($\downarrow$) denotes $m_s = -\frac{1}{2}$. This new two-valued property *doubles* the number of states that appear in the energy level diagram of Figure 7-2. Thus, Equation (8-18) is the basis for the revised scheme shown in Figure 8-13, in which the energy levels depend only on the principal quantum number n, and the degeneracy of the states at energy E_n is equal to $2n^2$, twice the degeneracy found in Figure 7-2.

The properties of S_z and S^2 are rather unlike those of L_z and L^2 from the viewpoint of the correspondence principle. We recall that a macroscopic orbital angular momentum $\mathbf{L}$ is obtained in the classical limit $\hbar \to 0$, provided the limit of large ℓ and m_ℓ is also taken. No such regime of large quantum numbers exists in the case of electron spin, since Equations (8-15) and (8-17) define a vector $\mathbf{S}$ of *fixed* magnitude with only *two* quantized orientations. It follows that $\mathbf{S}$ disappears in the limit $\hbar \to 0$, as befits the purely quantum nature of the concept of spin. We should not look for much insight from the classical planetary picture of a spinning orbital electron because of this observation. In fact, we should expect spin to behave quite differently from any classical angular momentum.

Figure 8-13

Energy levels E_n and degenerate states $\Psi_{n\ell m_\ell m_s}$ for the one-electron atom. The dynamics of the atom is governed only by the Coulomb force, as in Figure 7-2. The spin of the electron may be either up ($\uparrow$) or down ($\downarrow$) for each assignment of the set of quantum numbers ($n\ell m_\ell$).

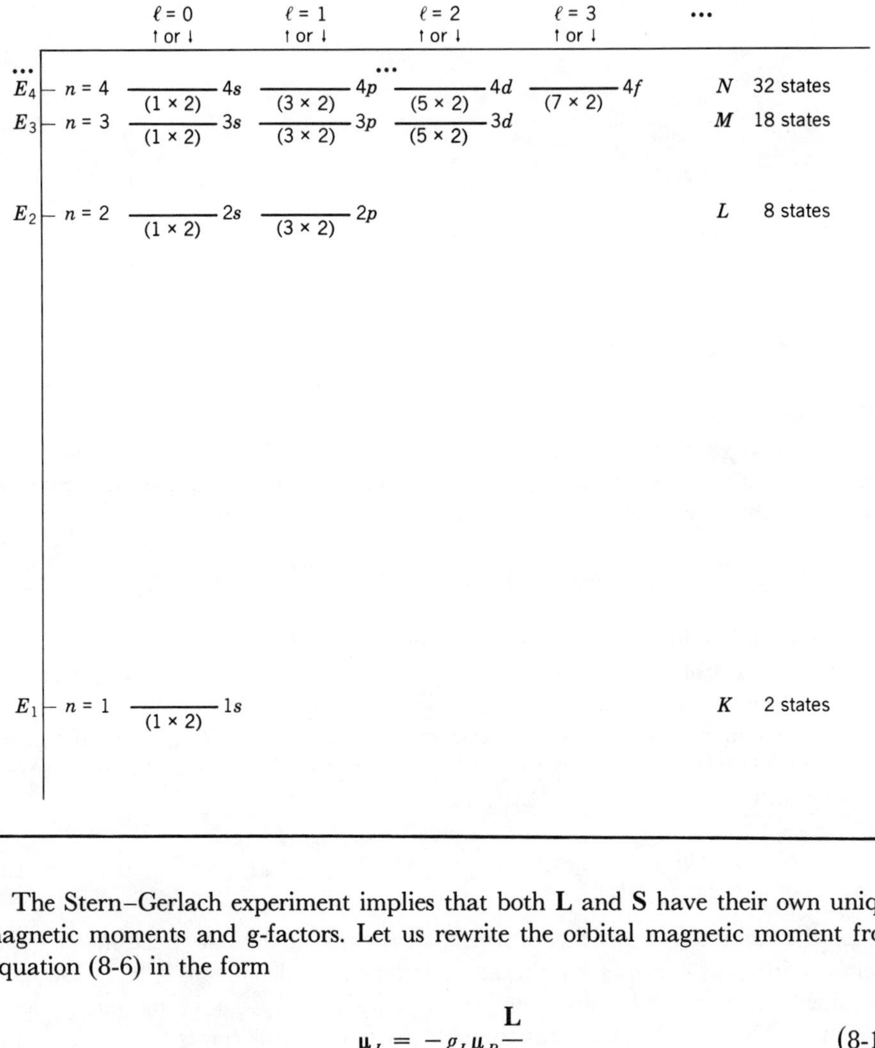

	$\ell = 0$ $\uparrow$ or $\downarrow$	$\ell = 1$ $\uparrow$ or $\downarrow$	$\ell = 2$ $\uparrow$ or $\downarrow$	$\ell = 3$ $\uparrow$ or $\downarrow$	$\cdots$		
$\cdots$			$\cdots$				
$E_4 \vdash n = 4$	(1×2) $4s$	(3×2) $4p$	(5×2) $4d$	(7×2) $4f$		N	32 states
$E_3 \vdash n = 3$	(1×2) $3s$	(3×2) $3p$	(5×2) $3d$			M	18 states
$E_2 \vdash n = 2$	(1×2) $2s$	(3×2) $2p$				L	8 states
$E_1 \vdash n = 1$	(1×2) $1s$					K	2 states

The Stern–Gerlach experiment implies that both **L** and **S** have their own unique magnetic moments and g-factors. Let us rewrite the orbital magnetic moment from Equation (8-6) in the form

$$\boldsymbol{\mu}_L = -g_L \mu_B \frac{\mathbf{L}}{\hbar} \tag{8-19}$$

and take the orbital g-factor to be

$$g_L = 1$$

as before. We then introduce the new *spin magnetic moment* of the electron by the analogous relation

$$\boldsymbol{\mu}_S = -g_S \mu_B \frac{\mathbf{S}}{\hbar}. \tag{8-20}$$

We regard this formula as the defining equation for the *spin g-factor* g_S, a parameter to be determined by experiment. Classical mechanics is unable to provide any guidance about the value of this quantity. In the relativistic quantum theory of the electron, however, it is possible to show that the spin g-factor should be given by

$$g_S = 2.$$

This number is interesting because the predicted value is exactly twice that of g_L and, furthermore, is in good agreement with experiment. It is customary to express measurements of g_S in terms of the deviation from unity for the ratio $g_S/2$. The current value for this deviation is quoted to astounding accuracy as

$$\frac{g_S - 2}{2} = 0.001159652193.$$

A technique for measuring magnetic moments to very high accuracy is discussed in Section 8-6.

Example

Figure 8-12 provides a way to visualize the two electron spin states in which S_z has the values $\hbar/2$ and $-\hbar/2$ with zero uncertainty. The spin vector **S** makes a fixed angle with the z axis, given by

$$\cos^{-1}\frac{S_z}{\sqrt{S^2}} = \cos^{-1}\frac{\hbar/2}{\sqrt{3}\,\hbar/2} = 54.7° \quad \text{in the spin-up state.}$$

Notice that **S** has a completely random azimuthal orientation in each case. The spin magnetic moment $\boldsymbol{\mu}_S$ is aligned antiparallel to **S**, according to Equation (8-20). The magnitude of this vector is given in Bohr magnetons by the value

$$\sqrt{\mu_S^2} = \frac{g_S \mu_B}{\hbar}\sqrt{S^2} = \frac{2\mu_B}{\hbar}\frac{\sqrt{3}\,\hbar}{2} = \sqrt{3}\,\mu_B.$$

This fixed result should be contrasted with the magnitude of the orbital magnetic moment for a one-electron atom,

$$\sqrt{\mu_L^2} = g_L \mu_B \sqrt{\ell(\ell+1)}\,.$$

The latter expression gives the sequence of values $0, \sqrt{2}\,\mu_B, \sqrt{6}\,\mu_B, \ldots$ for orbital quantum numbers $\ell = 0, 1, 2, \ldots$.

8-6 Magnetic Resonance

The Stern–Gerlach apparatus is useful as a device for separating magnetic states, or spin states, in a beam of atoms. We can employ the device to observe the phenomenon of magnetic resonance in atoms and use the resonance effect to make precise measurements of magnetic properties. This type of resonance is analogous to the large response obtained for a driven oscillator when the frequency of the driving force matches the natural frequency of oscillation. It is possible to subject a beam of atoms to certain contrived magnetic fields and produce the same behavior.

A constant magnetic field causes a precession of the magnetic moment of an atom. We have described the classical problem of precessional motion in Section 8-2 and Figure 8-5, and we have introduced the quantum concept of precessional states in Section 8-3 and Figure 8-6. The Larmor frequency appears in each situation as the natural frequency for the behavior of the dipole in the applied field. The energy of such a system remains fixed if no mechanism is included to bring about a change in energy. In this circumstance, the classical dipole maintains a constant angle with the applied field in Figure 8-5, and the quantum dipole makes no transitions between the split energy levels in Figure 8-6. We must impose some kind of additional driving influence on the system if we wish to change the energy and demonstrate magnetic resonance.

Let us look first at the classical problem as displayed in Figure 8-5. We let $\mathbf{B}_0$ denote the applied constant field, and we write the Larmor angular frequency and magnetic interaction energy as

$$\omega_0 = \frac{eB_0}{2m_e} \quad \text{and} \quad V_{M_0} = -\boldsymbol{\mu} \cdot \mathbf{B}_0.$$

The energy V_{M_0} is altered, and the angle between $\boldsymbol{\mu}$ and $\mathbf{B}_0$ is changed, if a *second* weaker field is imposed so that its direction varies to accommodate the precessional motion of $\boldsymbol{\mu}$. Figure 8-14 illustrates the effect of a second field $\mathbf{B}_\omega$, which oscillates with frequency $\omega/2\pi$ along the indicated y axis. The additional torque $\boldsymbol{\mu} \times \mathbf{B}_\omega$ acts in the left part of the figure to give an increase in the polar angle θ. If ω is chosen to equal ω_0, the direction of $\mathbf{B}_\omega$ reverses after half of a Larmor cycle and acts in the right part of the figure to give another increase in θ. Hence, a resonance occurs at $\omega = \omega_0$, as periodic increases appear in the configurational energy V_{M_0}. We note that the precessing system absorbs these increments in energy from the applied oscillating field.

The quantum version of the magnetic-resonance problem treats the precessing system as a collection of magnetic substates like the ones shown in Figure 8-6. The magnetic splitting of these energy levels is given according to Equation (8-12) by an expression of the form

$$\langle V_{M_0} \rangle = g\mu_B B_0 m.$$

Figure 8-14

Influence of a transverse oscillating $\mathbf{B}$ field on the classical precession of a magnetic dipole. Resonance occurs when the frequency of oscillation matches the Larmor frequency.

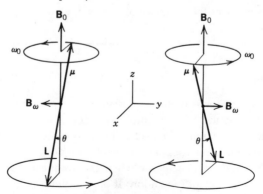

Figure 8-15

Microwave absorption causing a transition between adjacent magnetic substates.

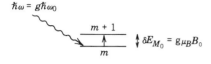

(We refrain temporarily from specifying g and m so that we can discuss magnetic resonance as a generic process.) Figure 8-15 shows two such levels for successive values of the magnetic quantum number, with energy spacing

$$\delta E_{M_0} = g\mu_B B_0$$

as found in Equation (8-13). An applied oscillating field $\mathbf{B}_\omega$ provides oscillating electromagnetic energy and acts as a source of photons of energy $\hbar\omega$. A photon is absorbed by the precessing system when ω satisfies the resonance condition

$$\hbar\omega = \delta E_{M_0} = g\frac{e\hbar}{2m_e}B_0 = g\hbar\omega_0, \tag{8-21}$$

so that the result is the upward transition indicated in the figure. (Note that ℓ may remain unchanged since the electric dipole selection rules do not apply in this instance.) We know from our calculations in Section 8-3 that the magnetic splitting is only of order 10^{-4} eV in a 10^4 G field. These numbers tell us that the oscillating field must operate at gigahertz (microwave) frequencies if the atom is to exhibit magnetic resonance in a constant magnetic field of tesla strength.

The resonance condition in Equation (8-21) can be used to make precise g-factor determinations from precise measurements of the resonance frequency. Of course, a prior calibration of the field B_0 is necessary in such an experiment. The resonance technique can be employed in a situation where the transition occurs between the spin-down and spin-up states of an atomic electron. In this case the relevant g-factor is identified as g_S, and the process is called *electron spin resonance*. The quantity g_S is very well known, and so a measurement of the frequency at resonance provides a determination of the field strength at the site of the electron. This procedure has been used in chemistry to probe the local magnetic fields inside molecules.

Figure 8-16 describes an ingenious device for making high-precision measurements of magnetic moments and g-factors. The apparatus employs nonuniform magnetic fields, of the kind used in the Stern–Gerlach experiment, to guide beams of neutral atoms and molecules. The magnetic-resonance principle is implemented by two other magnetic fields, one constant and one oscillating, located at the center of the apparatus. A typical experiment is sketched in the figure in terms of the behavior of a beam of hydrogen atoms. We assume that the atoms emerge from the source in the ground state so that we can attribute the atomic magnetic moment solely to the spin of the electron. The discrete vertical force on the magnetic dipoles splits the paths of the atoms when they enter the first nonuniform field, in the manner of the original Stern–Gerlach experiment. If we apply Equations (8-14), (8-20), and (8-15) in that

Figure 8-16

Schematic layout of a molecular-beam magnetic-resonance apparatus.

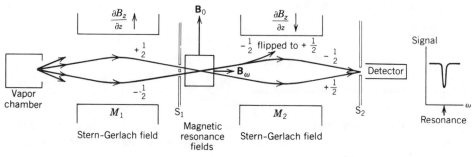

order, we obtain a force of the form

$$\langle \mu_z \rangle \frac{\partial B_z}{\partial z} = -g_S \mu_B m_s \frac{\partial B_z}{\partial z} \tag{8-22}$$

for an atom in a definite m_s state. We let the field strength increase vertically as indicated and find that an $m_s = +\frac{1}{2}$ atom experiences a downward force and assumes a downward-curving trajectory, while an $m_s = -\frac{1}{2}$ atom follows an opposite course. The figure shows how the slit S_1 selects two such orbits, one curving down and one curving up in M_1, the first Stern–Gerlach field. These paths cross and continue into a second nonuniform field M_2 where the field gradient is reversed relative to M_1. The split beams with $m_s = +\frac{1}{2}$ and $-\frac{1}{2}$ are steered through M_2 as shown and are finally recombined at the slit S_2 where the atoms are detected. Intervening between M_1 and M_2 is a magnetic-resonance element consisting of a vertical constant field $\mathbf{B}_0$ and a horizontal oscillating field $\mathbf{B}_\omega$. This part of the apparatus does not introduce transitions and does not affect the paths of the atoms unless the frequency ω is at resonance. The magnetic-resonance effect flips the electrons from spin-down to spin-up, so that atomic transitions from $m_s = -\frac{1}{2}$ to $m_s = +\frac{1}{2}$ occur where the beams cross. Those atoms that flip from $-\frac{1}{2}$ to $+\frac{1}{2}$ are deflected up rather than down in the field M_2 and fail to take the proper path for detection at S_2. Hence, fewer atoms are collected at resonance, and a sharp dip in signal is observed in the detector. A high-resolution measurement of the resonance frequency and an accurate calibration of the constant field B_0 can then be used to give a precise determination of the g-factor (g_S in this case), with the aid of Equation (8-21).

The first studies of directed beams of neutral molecules were carried out in 1911 by L. Dunoyer, and the earliest molecular-beam techniques were developed subsequently by Stern. The magnetic-resonance method was built into atomic- and molecular-beam experiments by I. I. Rabi in 1938. This innovation enabled Rabi and others to make precise measurements of various kinds of magnetic moments, first for nuclei, later for molecules, and finally for atoms. P. Kusch developed another resonance technique and used it to measure the spin magnetic moment of the electron.

8-7 Spin-$\frac{1}{2}$ Thought Experiments

We can illuminate the unfamiliar concept of spin by "conducting" hypothetical experiments on the behavior of beams of particles. The experiments are aimed specifically at the properties of spin-$\frac{1}{2}$ particles and are "performed" with the aid of

Figure 8-17

Splitting of a beam of spin-$\frac{1}{2}$ particles in a Stern–Gerlach field.

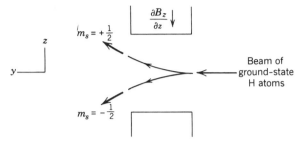

Stern–Gerlach fields. These studies are interesting because they illustrate some of the basic language and notation of quantum mechanics, without the intrusion of differential equations.

We suppose that the beam particles are hydrogen atoms in the ground state. The atoms have orbital angular moment $\ell = 0$ and exist in two quantum states labeled by the quantum number m_s. Let us first consider a beam directed along the horizontal axis of a Stern–Gerlach magnet, where the field gradient points in the negative z direction. We find from Equation (8-22) that atoms in the $m_s = +\frac{1}{2}$ state curve upward and atoms in the $m_s = -\frac{1}{2}$ state curve downward, as indicated in Figure 8-17. Note that the orientation of the magnet defines the z axis for the specification of the S_z eigenvalue $\hbar m_s$.

We can guide the two split components of the beam back together by introducing other Stern–Gerlach magnets in the manner of Figure 8-18. The three indicated field sections are aligned in tandem with their z axes parallel to the same direction and are assigned field gradients of equal magnitude and alternating sign. The central section is twice as long as each end section so that the curvature of the trajectories causes the two paths to rejoin at the left end of the apparatus. (We let the beam travel from right to left to accommodate the quantum mechanical notation introduced below.)

It is clear that this system of fields has no net effect on the composition of the beam. Atoms enter together on the right and follow discrete orbits up or down with random probabilities. The same atoms rejoin and leave together on the left, as long as the device contains nothing in between to affect one or the other of the two m_s pathways. We can convert the instrument into a polarizing *filter* if we insert an absorbing obstacle in one of the paths. Figure 8-18 indicates how such a barrier removes atoms in the $m_s = -\frac{1}{2}$ state and transmits a *polarized* beam in which every outgoing atom has spin up with respect to the z axis defined by the direction of the fields. The figure also shows a simplified version of the apparatus, with and without the barrier, for use in the following thought experiments. We emphasize that the indicated signs of the quantum number m_s refer throughout to a particular choice of z direction. We denote this direction by the symbol Z in the figures as a reminder of this important point.

Let us now examine what happens when we send the beam of atoms through two such devices in series. The polarizing filter of Figure 8-18 is used to prepare a polarized beam for transmission into a second apparatus with the same coaxial alignment. We assume that the fields are designed with equal strengths and equal gradients. The fields are oriented in this part of the experiment so that the z axis defined in the second device has the *same* direction as the z axis defined in the first.

Figure 8-18

Stern–Gerlach fields with alternating gradients. The nonuniform fields divide and rejoin a spin-$\frac{1}{2}$ beam in the first apparatus. The second apparatus acts as a polarizing filter to transmit only $m_s = +\frac{1}{2}$ atoms. Schematic pictures of these transmitting and polarizing devices are shown at the bottom of the figure.

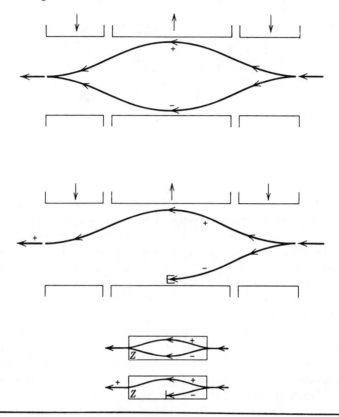

Figure 8-19 shows the results of three simple thought experiments. The transmitting apparatus in (a) presents no barrier to upward- or downward-curving trajectories and simply transmits any polarized beam without modification. The filter at the second stage in (b) allows only $m_s = +\frac{1}{2}$ atoms to pass and has no effect on the given polarized beam. The second filter in (c) terminates the beam because the apparatus passes only $m_s = -\frac{1}{2}$ atoms. Note that the quantum number m_s retains its meaning from the first filter to the second, since the z axis has the same direction in each device.

It is instructive to associate quantum mechanical probabilities with experiments of this kind. We begin by defining the complex-valued quantity

$$\chi(d, p)$$

as the quantum mechanical *amplitude* for an atom, prepared in the state p by the first apparatus, to be detected in the state d by the second apparatus. We then express the

Figure 8-19

Stern–Gerlach apparatuses in series. The z axes have the same direction in each case, as indicated by the common label Z.

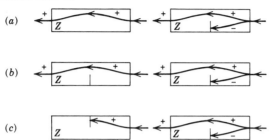

corresponding probability of detection by the squared modulus

$$|\chi(d, p)|^2.$$

(Note that the preparation and detection states read from right to left in these expressions, while the atoms proceed from right to left in the accompanying figures. The final $\leftarrow$ initial notation for a quantum process follows a commonly adopted convention. We have already used this notation in the transition amplitude of Equation (5-70).)

The experiments in Figure 8-19 are described in terms of states specified by the sign of m_s, where the quantum number is defined according to a certain direction for the axis of quantization. We are using the label Z to indicate this direction in the figure, so let us represent the experiments in parts (b) and (c) with the aid of the same notation. We introduce the amplitudes

$$\chi(+Z, +Z) \quad \text{and} \quad \chi(-Z, +Z)$$

for the detection of an atom with spin up and down along Z, given that the atom is prepared with spin up along Z. These two particularly simple situations have rather obvious probabilities:

$$|\chi(+Z, +Z)|^2 = 1 \quad \text{and} \quad |\chi(-Z, +Z)|^2 = 0$$

in case (b) and in case (c). Another simple thought experiment can also be performed with a beam of atoms prepared in the $m_s = -\frac{1}{2}$ state. The resulting probabilities are

$$|\chi(+Z, -Z)|^2 = 0 \quad \text{and} \quad |\chi(-Z, -Z)|^2 = 1$$

for the two alternatives at the detection stage of the experiment.

We now want to make the analysis more interesting by allowing a rotation of the second filter. Let us confine our attention to rotations that maintain the coaxial alignment of the two Stern–Gerlach magnet systems. The scheme is illustrated in Figure 8-20 by means of two apparatuses designated as the polarizer and the analyzer. We observe that the second set of Stern–Gerlach fields defines a new z' axis of quantization, oriented at an angle β with the z axis defined by the first set of fields. Consequently, the spin quantum numbers denote *different* kinds of states in the two

Figure 8-20

Coaxial Stern–Gerlach devices. The quantization axis in the analyzer makes an angle β with the quantization axis in the polarizer.

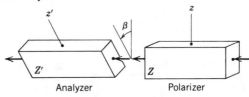

devices, since the notation $m_s = \pm \frac{1}{2}$ refers to spin up or down with respect to the z axis in the polarizer and the z' axis in the analyzer. We use the label Z' in the analyzer to distinguish its spin-up and spin-down states from those with the label Z in the polarizer.

We are concerned with the detection probabilities for all the different polarization states. The probability that an atom prepared with spin up along Z is detected with spin either up or down along Z' depends on the angle β between the two axes. We therefore expect new probabilities such that

$$|\chi(+Z', +Z)|^2 \neq 1 \quad \text{and} \quad |\chi(-Z', +Z)|^2 \neq 0$$

when we deviate from the simple $\beta = 0$ situation. The two alternatives are shown schematically in Figure 8-21. Again, note that the indicated signs specify values of m_s peculiar to the particular quantization axis. Only these *two* detection alternatives exist, and so the sum of the two expressions must give unit probability:

$$|\chi(+Z', +Z)|^2 + |\chi(-Z', +Z)|^2 = 1. \tag{8-23}$$

If we let the polarizer prepare a beam of atoms with spin down along Z and conduct another thought experiment, we find another pair of probabilities whose sum must obey a similar relation:

$$|\chi(+Z', -Z)|^2 + |\chi(-Z', -Z)|^2 = 1. \tag{8-24}$$

These equations are constraints on the four amplitudes parametrized by the angle β. It is reasonable to suppose that the amplitudes are periodic functions of β. We already know the probabilities for the special case $\beta = 0$, and we can also deduce the

Figure 8-21

Thought experiment with nonzero angle of rotation between analyzer and polarizer. Atoms prepared with spin up along Z are detected with spin up and with spin down along Z'.

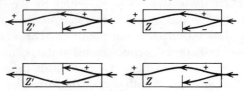

probabilities for the special case of a rotation with angle $\beta = \pi$. It is possible to use these logical arguments, and no other mathematics, to obtain the following general results:

$$|\chi(+Z', +Z)|^2 = |\chi(-Z', -Z)|^2 = \cos^2\frac{\beta}{2} \tag{8-25}$$

and

$$|\chi(-Z', +Z)|^2 = |\chi(+Z', -Z)|^2 = \sin^2\frac{\beta}{2}. \tag{8-26}$$

We leave the proof to Problem 10 at the end of the chapter.

These important formulas provide the answer to an interesting question. We suppose that a given state refers to a definite value for the component of spin along a given axis, and we ask for the probability of finding the particle with a certain component of spin along *another* direction. It is enlightening to find that the question can be formulated operationally and that the answer can be found in terms of a conceivable experiment.

Example

Let us suppose that the directions Z and Z' make a 180° angle and that the polarizer prepares atoms with spin down along Z. Equations (8-25) and (8-26) give the probabilities

$$|\chi(-Z', -Z)|^2 = \cos^2\frac{\pi}{2} = 0 \quad \text{and} \quad |\chi(+Z', -Z)|^2 = \sin^2\frac{\pi}{2} = 1$$

for finding the atoms with spin down and up along Z'. We might have anticipated this result on logical grounds and put the argument to use in the deduction of the general equations. If we take $\beta = \pi/2$ as another case, we find the probabilities

$$|\chi(+Z', +Z)|^2 = \cos^2\frac{\pi}{4} = \frac{1}{2} \quad \text{and} \quad |\chi(-Z', +Z)|^2 = \sin^2\frac{\pi}{4} = \frac{1}{2}.$$

Hence, we conclude that there is a 50-50 chance for an atom with spin up along some direction Z to be found with spin up or down along another direction Z' at right angles to Z.

8-8 Addition of Orbital and Spin Angular Momenta

The angular momentum of the one-electron atom includes both orbital motion and electron spin. Each contribution has its own magnetic moment, and hence its own interaction in an applied magnetic field. We have been able to concentrate on the spin of the electron by selecting states in which the orbital angular momentum is equal to zero. The quantized vectors **L** and **S** must be added together in situations where more general kinds of states are considered.

We define the *total angular momentum* of the atom by the addition of vectors

$$\mathbf{J} = \mathbf{L} + \mathbf{S}. \tag{8-27}$$

This sum of terms is analogous to the combined angular momentum of an orbiting spinning body, except for the fact that in the classical case the vectors $\mathbf{L}$ and $\mathbf{S}$ may have any magnitude and direction and may add up to any magnitude in the range $|\,|\mathbf{L}| - |\mathbf{S}|\,|$ to $|\mathbf{L}| + |\mathbf{S}|$.

Equation (8-27) is intended to hold for orbital and spin angular momenta with *quantized* magnitudes and orientations. The resulting vector $\mathbf{J}$ is itself a quantum mechanical angular momentum, and so the quantities J^2 and J_z are supposed to obey quantization rules akin to the quantum properties of $\mathbf{L}$ and $\mathbf{S}$. Hence, there must exist a discrete nonnegative number j such that the quantized values of J^2 occur as eigenvalues of the form

$$\hbar^2 j(j + 1).$$

The numerical variable j is called the *total angular momentum quantum number*. Each allowed value of j has an associated set of quantum numbers m_j and a corresponding set of J_z eigenvalues

$$\hbar m_j \quad \text{with } m_j = -j, -j + 1, \ldots, j - 1, j.$$

These rules employ the quantum numbers j and m_j to express the quantized magnitudes and orientations of the vector $\mathbf{J}$, just as the quantum numbers ℓ and m_ℓ determine the quantized behavior of the vector $\mathbf{L}$.

The quantum number m_j takes on $2j + 1$ possible values in integer steps between $m_j = -j$ and $m_j = +j$. It follows that j must be either an integer or a half-integer. Equation (8-27) implies a relation among the various z-component quantum numbers:

$$m_j = m_\ell + m_s \quad \text{since } J_z = L_z + S_z. \tag{8-28}$$

Hence, we conclude that m_j must be half-integral, as m_ℓ is integral and m_s is half-integral, so that j itself must range over the positive *half-integers* $\frac{1}{2}, \frac{3}{2}, \frac{5}{2}, \ldots$. The allowed values of j vary with the orbital quantum number ℓ. We note that $\mathbf{L}$ makes no contribution to $\mathbf{J}$ in the special $\ell = 0$ case, and so we find

$$j = \tfrac{1}{2} \quad \text{for any } \ell = 0 \text{ state.} \tag{8-29}$$

The determination of j for nonzero ℓ requires the addition of two quantized angular momenta $\mathbf{L}$ and $\mathbf{S}$ to yield a third quantized angular momentum $\mathbf{J}$. Only *two* possible values of j result from this procedure:

$$j = \ell + \tfrac{1}{2} \text{ or } \ell - \tfrac{1}{2} \quad \text{for any } \ell \neq 0 \text{ state.} \tag{8-30}$$

We look more closely at the addition of these vectors in the example below.

The vector-addition problem is described schematically in Figure 8-22. We choose for illustration the $\ell = 1$ states of the vector $\mathbf{L}$ and combine these configurations with the spin-up and spin-down states of the vector $\mathbf{S}$. This combination of angular

Figure 8-22

Vector addition of orbital and spin angular momenta. The factors of $\hbar$ are suppressed, and the $\ell = 1$ case is chosen for **L**. The sum $1 + \frac{1}{2}$ produces the results $\frac{3}{2}$ and $\frac{1}{2}$ for the vector **J**.

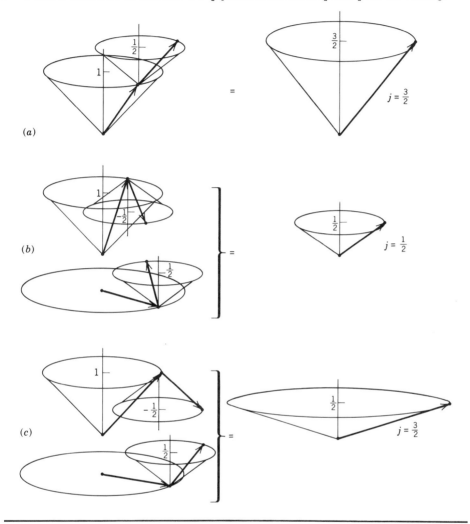

momenta can be represented (without the $\hbar$ factors) according to the addition formula

$$1 + \tfrac{1}{2} = \tfrac{3}{2} \text{ or } \tfrac{1}{2}$$

as stipulated in Equation (8-30). It is understood that the lengths of the vectors in the figure are $\sqrt{2}$ for $\mathbf{L}/\hbar$, $\sqrt{3}/2$ for $\mathbf{S}/\hbar$, and either $\sqrt{15}/2$ or $\sqrt{3}/2$ for $\mathbf{J}/\hbar$. In part (a) we add $m_\ell = 1$ and $m_s = \frac{1}{2}$ to get $m_j = \frac{3}{2}$ and obtain $j = \frac{3}{2}$ as the only possible result. In part (b) we show that the combination of $m_\ell = 1$ with $m_s = -\frac{1}{2}$ and the combination of $m_\ell = 0$ with $m_s = \frac{1}{2}$ contribute together to $m_j = \frac{1}{2}$. Consequently, both of these vector sums can be arranged to produce a $j = \frac{1}{2}$ configuration. In part (c) we find that the same orbital and spin states can also be combined in another arrangement with $m_j = \frac{1}{2}$ to form a $j = \frac{3}{2}$ state. Note that the pictures incorporate the feature of azimuthal uncertainty, as required for any angular momentum with definite z component. This aspect of the vector-addition problem is rather difficult to

Figure 8-23

Vector model for the angular momentum sum
J = L + S. The drawings symbolize the two
results $j = \ell + \frac{1}{2}$ and $j = \ell - \frac{1}{2}$.

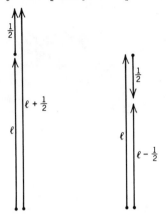

visualize and represent in a figure. A more intuitive view of Equation (8-30) is
provided for convenience in Figure 8-23.

Let us now return to Figure 8-13 and reconsider the energy level diagram for the
one-electron atom. We recall that the states are organized by the quantum numbers
$(n\ell m_\ell m_s)$ and that the wave function $\Psi_{n\ell m_\ell m_s}$ is expressed according to Equation
(8-18). Our purpose is to introduce an *alternative* method of bookkeeping that employs
the total angular momentum quantum numbers j and m_j, in place of m_ℓ and m_s. The
new scheme is based on the construction of wave functions of the form $\Psi_{n\ell jm_j}$, where
states assigned to specific values of j and m_j are assembled as combinations of states
with definite assignments of m_ℓ and m_s. Figure 8-22 suggests how such constructions
might be represented. The states $(m_\ell = 1, m_s = -\frac{1}{2})$ and $(m_\ell = 0, m_s = \frac{1}{2})$
are shown together as a $(j = \frac{1}{2}, m_j = \frac{1}{2})$ state in configuration (b) and are again
shown together as a $(j = \frac{3}{2}, m_j = \frac{1}{2})$ state in configuration (c).

A transformation of the wave functions takes the original set of quantum numbers
$(n\ell m_\ell m_s)$ into the alternative set $(n\ell jm_j)$. We observe that a given energy level E_n
has orbital quantum numbers

$$\ell = 0, 1, 2, \ldots, n - 1$$

as before, where each ℓ value (except $\ell = 0$) allows two choices for j,

$$j = \ell + \tfrac{1}{2} \text{ or } \ell - \tfrac{1}{2},$$

and each of these j values permits $2j + 1$ assignments of m_j. This transformation of
quantum numbers reorganizes the array of states in the manner described in Figure
8-24. We note that the number of degenerate states at energy E_n is now given in terms
of the total number of m_j values and is still equal to $2n^2$, as expected. The figure also
introduces a new spectroscopic notation in which the states appear with the designation nL_j, labeled by a subscript j and a new capital letter L. We use the capital letter
to denote the orbital quantum number according to the letter code

$$S \text{ for } \ell = 0, \quad P \text{ for } \ell = 1, \quad D \text{ for } \ell = 2, \quad F \text{ for } \ell = 3, \ldots.$$

Figure 8-24

Energy levels E_n and degenerate states $\Psi_{n\ell jm_j}$ for the one-electron atom. Each assignment of quantum numbers $(n\ell j)$ implies $2j + 1$ possible values of m_j, as indicated in parentheses at each level. The spectroscopic notation nL_j is used to designate the states. This scheme is an alternative to the one described in Figure 8-13. The Coulomb force provides the only interaction in each of the two figures.

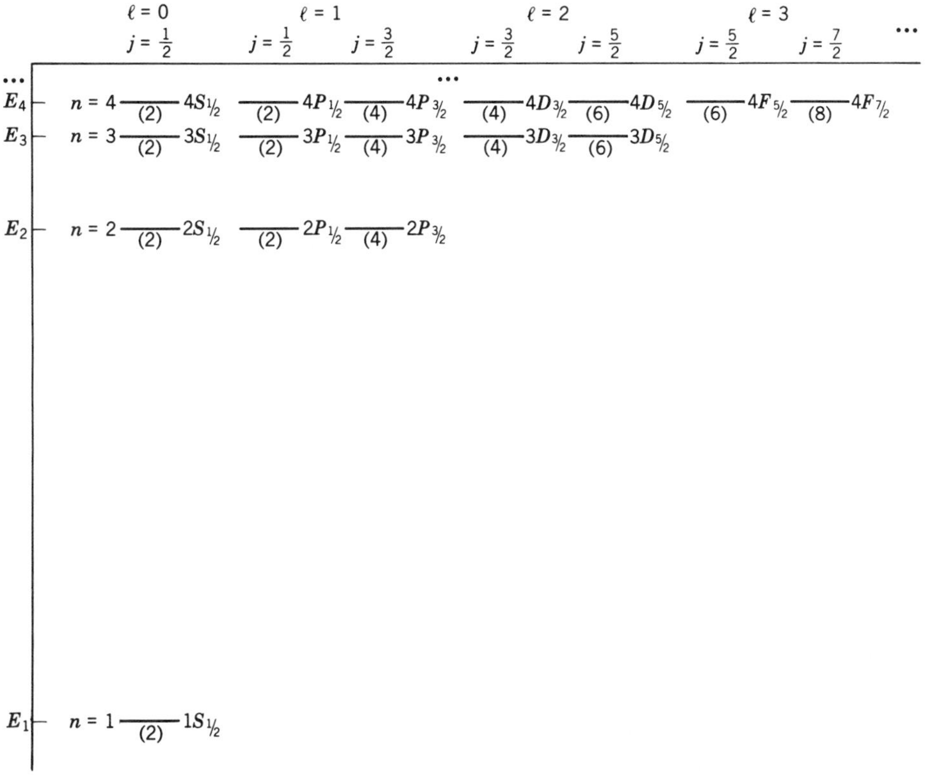

This part of the designation is equivalent to the use of the symbols $s, p, d, f, \ldots$ for single-electron states.

It should be emphasized that the energy levels in the figure are still those determined by the effects of the Coulomb force alone. We have yet to be given a physical reason for setting aside the quantum numbers $(n\ell m_\ell m_s)$ of Figure 8-13 in favor of the quantum numbers $(n\ell jm_j)$ of Figure 8-24. We may continue to use m_ℓ and m_s as long as L_z and S_z maintain their status as separately conserved angular momenta. The necessity for the alternative scheme based on j and m_j emerges in Section 8-9.

Example

Let us examine the reasoning that leads to the possible values of j in Equation (8-30). We consult Equations (8-28) and argue that j cannot exceed $\ell + \frac{1}{2}$ because

$$\max m_j = j, \quad \max m_\ell = \ell, \quad \text{and} \quad \max m_s = \tfrac{1}{2}.$$

This conclusion is supported by the inequality

$$|\mathbf{J}| \leq |\mathbf{L}| + |\mathbf{S}|.$$

The vectors must also satisfy another inequality

$$|\mathbf{J}| \geq |\mathbf{L}| - |\mathbf{S}|,$$

which turns into the following constraint on the quantum numbers:

$$j(j+1) \geq \left(\sqrt{\ell(\ell+1)} - \sqrt{3}/2\right)^2$$

or

$$j(j+1) - \ell(\ell+1) - \tfrac{3}{4} \geq -\sqrt{3\ell(\ell+1)}.$$

We may test this constraint by inserting any value of j in the quantized range of possibilities

$$j = \ell + \tfrac{1}{2}, \ell - \tfrac{1}{2}, \ell - \tfrac{3}{2}, \ell - \tfrac{5}{2}, \ldots .$$

The proof that j cannot be smaller than $\ell - \tfrac{1}{2}$ is left as Problem 13 at the end of the chapter.

8-9   The Spin–Orbit Interaction

The energy levels of the one-electron atom deviate from the results shown in Figure 8-24 because of two additional dynamical effects. Both of these further contributions to the energy of the atom are relativistic in origin and secondary in influence compared to the Coulomb potential energy. One of the effects is associated with the spin of the electron and is known as the spin–orbit interaction. This source of interaction energy introduces a *fine structure* among the degenerate states in the figure and splits the levels into multiplets of states with slightly different energies. The multiplet structure of the energy levels is observed in the emission spectrum of the atom and is interpreted as direct evidence for electron spin.

A coupling occurs between the spin of the electron and the orbital angular momentum, owing to the existence of a magnetic field *internal* to the atom. We employ Figure 8-25 to deduce the nature of this interaction. The figure shows an electron in orbit around a nucleus at rest and also shows the motion with respect to a reference frame moving with the instantaneous velocity $\mathbf{v}$ of the electron. The nuclear charge Ze has velocity $-\mathbf{v}$ in this frame, and the resulting current produces a magnetic field at the instantaneous location of the electron. We can use the Biot–Savart law to write the field in terms of the variables indicated in the figure:

$$\mathbf{B} = \frac{\mu_0}{4\pi} \frac{Ze(-\mathbf{v}) \times \mathbf{r}}{r^3}.$$

Let us express the Coulomb field of the nucleus as

$$\mathbf{E} = \frac{Ze}{4\pi\varepsilon_0} \frac{\mathbf{r}}{r^3}$$

Figure 8-25

Atomic orbital motion in two frames. The nucleus is at rest in the original frame, and the electron is instantaneously at rest in the transformed frame.

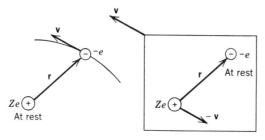

and recall the relation

$$\mu_0 \varepsilon_0 = \frac{1}{c^2}$$

in order to rewrite the magnetic field in the form

$$\mathbf{B} = -\frac{\mathbf{v} \times \mathbf{E}}{c^2}.$$

This expression describes a relativistic effect in which a *motional magnetic field* is experienced by an electron moving at small velocity $\mathbf{v}$ through an electric field $\mathbf{E}$. Let us return to the original formula for the internal magnetic field and write

$$\mathbf{B}_{internal} = \frac{Ze}{4\pi\varepsilon_0} \frac{\mathbf{r} \times \mathbf{v}}{c^2 r^3} = \frac{Ze}{4\pi\varepsilon_0} \frac{\mathbf{L}}{m_e c^2 r^3}.$$

We note that the final result contains the orbital angular momentum $\mathbf{L} = m_e \mathbf{r} \times \mathbf{v}$.

The electron spin magnetic moment interacts with the internal $\mathbf{B}$ field at the site of the electron, with the same interaction energy as in the case of Larmor precession. We make a provisional definition of this potential energy by setting

$$V_{SL} = -\boldsymbol{\mu}_S \cdot \mathbf{B}_{internal},$$

and we consult Equation (8-20) to obtain

$$V_{SL} = \left(+g_S \frac{\mu_B}{\hbar} \mathbf{S} \right) \cdot \left(\frac{Ze}{4\pi\varepsilon_0 m_e c^2 r^3} \mathbf{L} \right).$$

Equation (8-7) is recalled and the value $g_S = 2$ is inserted in the following steps:

$$V_{SL} = \left(\frac{e}{m_e} \mathbf{S} \right) \cdot \left(\frac{Ze}{4\pi\varepsilon_0 m_e c^2 r^3} \mathbf{L} \right)$$

$$= \frac{Ze^2}{4\pi\varepsilon_0} \frac{\mathbf{S} \cdot \mathbf{L}}{m_e^2 c^2 r^3}. \qquad (8\text{-}31)$$

We emphasize that this derivation is based on a classical picture of Larmor precession, taken in the instantaneous rest frame of the electron.

Our formula for the interaction energy must be transferred to the original frame where the nucleus is at rest. An additional relativistic term is needed along with Equation (8-31) to allow for the fact that the orbiting electron is in accelerated motion. This extra kinematic ingredient is known as the Thomas precession term (after L. H. Thomas, the first person to analyze the entire problem correctly). It is possible to show that Larmor precession and Thomas precession give similar contributions of opposite sign and that the overall effect is a modification of Equation (8-31) by a factor of $\frac{1}{2}$, the so-called Thomas factor. Hence, the spin–orbit interaction energy assumes the final form

$$V_{SL} = \frac{Ze^2}{4\pi\varepsilon_0} \frac{\mathbf{S}\cdot\mathbf{L}}{2m_e^2c^2r^3}. \tag{8-32}$$

The Thomas precession term can be deduced as a technical exercise in relativistic kinematics. Let us not pursue this point but remark instead that relativistic quantum mechanics generates Equation (8-32) in straightforward fashion. The latter approach is the more satisfactory one to follow in the first place, and so the final result is simply quoted on that authority.

The spin–orbit interaction undermines the usefulness of the states $\Psi_{n\ell m_\ell m_s}$ in Equation (8-18) and Figure 8-13. We can use m_ℓ and m_s as *good quantum numbers* to determine the stationary states as long as we are able to specify eigenvalues independently for the observables L_z and S_z. The circumstances require that L_z and S_z obey separate angular momentum conservation laws. These two quantities are separately conserved whenever there exist states of definite energy in which L_z and S_z also have definite values. We illustrate this situation in Figure 8-26 by showing mutually unrelated angular momentum vectors $\mathbf{L}$ and $\mathbf{S}$ whose orientations about the z axis are completely random with respect to each other. The spin–orbit interaction changes this picture because the factor $\mathbf{S}\cdot\mathbf{L}$ in Equation (8-32) implies a dependence of the energy on the *relative* orientation of $\mathbf{L}$ and $\mathbf{S}$. The two vectors are *coupled* together as a result of this new dynamical variation of the energy. We can see the coupling in the figure if we fix the energy by fixing the angle between $\mathbf{L}$ and $\mathbf{S}$ and attempt to maintain the z components of the two vectors. Such states of definite energy generally

Figure 8-26

Independently oriented $\mathbf{L}$ and $\mathbf{S}$ vectors representing a state with good quantum numbers m_ℓ and m_s. In this case $\mathbf{L}$ refers to the $\ell = 1$ state with $m_\ell = 1$.

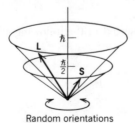

Random orientations

Figure 8-27

Coupling of spin and orbital angular momenta owing to the spin–orbit interaction. The effect is represented as a precession of **L** and **S** about **J**. The indicated vector additions correspond to the configurations displayed in parts (*b*) and (*c*) of Figure 8-22.

require combinations of **L** and **S** in which L_z and S_z cannot be assigned definite values. A state of definite energy can still have a definite value of J_z, however. The total angular momentum of the system is conserved as long as the system consists of the isolated atom. We therefore turn to j and m_j as good quantum numbers and use these parameters instead of m_ℓ and m_s to designate the stationary states. The wave functions $\Psi_{n\ell j m_j}$ have been introduced and Figure 8-24 has been drawn in anticipation of this relabeling maneuver.

We can illustrate the strategy by recalling parts (*b*) and (*c*) of Figure 8-22. The indicated configurations are composed of two distinct $(m_\ell m_s)$ states with the same value of m_j, where the combinations of **L** and **S** are arranged to form states with $j = \ell - \frac{1}{2}$ and $j = \ell + \frac{1}{2}$. Let us picture the synthesis of the two states with the aid of Figure 8-27. The influence of the spin–orbit interaction is indicated in the figure as a *precession* of the coupled vectors **L** and **S** around the vector **J**, while **J** assumes an uncertain azimuthal orientation about the axis of quantization. Precession is used in this instance as a schematic device to represent the behavior of the two $(m_\ell m_s)$ states in the construction of the two possible values of j. We are about to learn that these constructions of the quantized total angular momentum correspond to states of the atom with different values of the energy. This mechanism causes the degenerate states $\Psi_{n\ell j m_j}$ in Figure 8-24 to split apart so that atoms with the same n and ℓ acquire slightly different energies for different choices of j. Let us also observe in passing that the properties of L_z and S_z in Figure 8-26 become interesting again in cases where the spin–orbit coupling has a negligible effect. This situation prevails when the atom is in an *external* magnetic field whose strength greatly exceeds that of the internal magnetic field leading to Equation (8-31).

The spin–orbit interaction is one of two relativistic effects that contribute to the fine structure of the atom. We recall from Section 3-6 that a useful dimensionless parameter is furnished by the fine structure constant

$$\alpha = \frac{e^2}{4\pi\varepsilon_0 \hbar c},$$

and now we are able to appreciate the meaning of this parameter. We incorporate α into Equation (8-32) to get

$$V_{SL} = Z\alpha \frac{\hbar}{2m_e^2 c} \frac{\mathbf{S} \cdot \mathbf{L}}{r^3},$$

and we use the relation

$$J^2 = (\mathbf{L} + \mathbf{S}) \cdot (\mathbf{L} + \mathbf{S}) = L^2 + 2\mathbf{L} \cdot \mathbf{S} + S^2$$

to reexpress the spin–orbit interaction energy:

$$V_{SL} = Z\alpha \frac{\hbar}{4m_e^2 c} \frac{J^2 - L^2 - S^2}{r^3}. \tag{8-33}$$

When the expectation value of this quantity is evaluated in the state $\Psi_{n\ell j m_j}$, the expression takes the form

$$\langle V_{SL} \rangle = Z\alpha \frac{\hbar}{4m_e^2 c} \hbar^2 \left[j(j+1) - \ell(\ell+1) - \frac{3}{4} \right] \left\langle \frac{1}{r^3} \right\rangle. \tag{8-34}$$

The quantum numbers appear in this result as a consequence of the relabeling scheme for the wave functions. Each wave function is denoted by the quantum numbers $(n\ell j m_j)$ so that the observables J^2, L^2, and S^2 have the definite values $\hbar^2 j(j+1)$, $\hbar^2\ell(\ell+1)$, and $\frac{3}{4}\hbar^2$ in the given state. The expectation value $\langle V_{SL} \rangle$ is interpreted in terms of Figure 8-24 as the amount by which the spin–orbit interaction *shifts* the energy levels for the various assignments of these quantum numbers. Note that the square-bracketed quantity in Equation (8-34) vanishes for $\ell = 0$ states, where j can only be equal to $\frac{1}{2}$, and is double-valued for $\ell \neq 0$ states, where j can be either $\ell + \frac{1}{2}$ or $\ell - \frac{1}{2}$.

The computation in Equation (8-34) is completed by inserting Equation (7-35) from Section 7-4:

$$\left\langle \frac{1}{r^3} \right\rangle = \frac{2}{a^3 n^3 \ell(\ell+1)(2\ell+1)}.$$

(We observe that the expectation value of any function of r in the state $\Psi_{n\ell j m_j}$ can only depend on n and ℓ, the quantum numbers for orbital motion. The calculation of $\langle 1/r^3 \rangle$ therefore reduces immediately to the integral

$$\int_0^\infty \frac{1}{r^3} P_{n\ell}(r)\, dr$$

as in Equation (7-31), where $P_{n\ell}(r)$ is the radial probability density for the given state.) We also recall Equation (7-5) and again use the fine structure constant to get

$$a = \frac{4\pi\varepsilon_0 \hbar^2}{Ze^2 m_e} = \frac{\hbar}{Z\alpha m_e c}.$$

(The reduced-mass correction m_e/μ is replaced by unity for this purpose.) The desired expectation value is then written as

$$\left\langle \frac{1}{r^3} \right\rangle = \left(\frac{Z\alpha m_e c}{\hbar n} \right)^3 \frac{2}{l(l+1)(2l+1)}$$

for $\ell \neq 0$ states. We insert this result in Equation (8-34) and obtain the final formula for the energy level shift due to the spin–orbit interaction:

$$\langle V_{SL} \rangle = Z\alpha \frac{\hbar^3}{4m_e^2c} \left[j(j+1) - \ell(\ell+1) - \frac{3}{4} \right] \left(\frac{Z\alpha m_e c}{\hbar n} \right)^3 \frac{2}{\ell(\ell+1)(2\ell+1)}$$

$$= \frac{Z^4\alpha^4}{2n^3} m_e c^2 \frac{j(j+1) - \ell(\ell+1) - \frac{3}{4}}{\ell(\ell+1)(2\ell+1)}. \qquad (8\text{-}35)$$

This version of the formula gives the energy shift in terms of $m_e c^2$, the rest energy of the electron. We can refer back to Equation (3-54) and substitute the Rydberg energy unit instead. We use the equality

$$E_0 = \frac{\alpha^2}{2} m_e c^2$$

and find

$$\langle V_{SL} \rangle = \frac{Z^4\alpha^2}{n^3} E_0 \frac{j(j+1) - \ell(\ell+1) - \frac{3}{4}}{\ell(\ell+1)(2\ell+1)}. \qquad (8\text{-}36)$$

This version of the formula gives the shift in terms of E_0, the natural scale of energy in the atom. The spin–orbit energy is proportional to $\alpha^2 E_0$ and is therefore a rather small correction to any given energy level.

These results can be applied to the states with nonzero ℓ in Figure 8-24. We see that the values of the quantum numbers $(n\ell j)$ are indicated at each of the energy levels in the diagram, and we note that j is equal to either $\ell + \frac{1}{2}$ or $\ell - \frac{1}{2}$ for each selection of n and ℓ. Equation (8-36) determines an energy shift at every $\ell \neq 0$ level. Thus, a state with $j = \ell + \frac{1}{2}$ has

$$j(j+1) - \ell(\ell+1) - \frac{3}{4} = \ell \quad \text{and} \quad \langle V_{SL} \rangle = \frac{Z^4\alpha^2 E_0}{n^3(\ell+1)(2\ell+1)},$$

while a state with $j = \ell - \frac{1}{2}$ has

$$j(j+1) - \ell(\ell+1) - \frac{3}{4} = -\ell - 1 \quad \text{and} \quad \langle V_{SL} \rangle = -\frac{Z^4\alpha^2 E_0}{n^3\ell(2\ell+1)}.$$

The *spin–orbit splitting* is found by taking the difference of these two expressions:

$$\delta E_{SL} = \frac{Z^4\alpha^2 E_0}{n^3(2\ell+1)} \left(\frac{1}{\ell+1} + \frac{1}{\ell} \right) = \frac{Z^4\alpha^2 E_0}{n^3\ell(\ell+1)}. \qquad (8\text{-}37)$$

Hence, the coupling of **L** and **S** causes the state $nL_{\ell+1/2}$ to lie higher than the state $nL_{\ell-1/2}$ and results in a *doublet* of states with the same n and ℓ as shown in Figure 8-28. The remaining contribution to the fine structure of these levels arises from another relativistic effect, to be discussed in Section 8-10.

Figure 8-28

Spin–orbit splitting of the state nL_j in the one-electron atom.

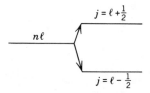

Example

Let us apply Equation (8-36) to the $2P$ states of the atom and compute the magnitude of the splitting illustrated in Figure 8-28. The energy shifts for $n = 2$ and $\ell = 1$ are

$$\langle V_{SL} \rangle = \frac{Z^4 \alpha^2}{8} E_0 \frac{\frac{15}{4} - 2 - \frac{3}{4}}{1 \cdot 2 \cdot 3} = \frac{Z^4 \alpha^2 E_0}{48} \quad \text{for} \quad j = \frac{3}{2}$$

and

$$\langle V_{SL} \rangle = \frac{Z^4 \alpha^2}{8} E_0 \frac{\frac{3}{4} - 2 - \frac{3}{4}}{1 \cdot 2 \cdot 3} = -\frac{Z^4 \alpha^2 E_0}{24} \quad \text{for} \quad j = \frac{1}{2}.$$

We take $Z = 1$ in the case of the hydrogen atom, and we find the difference of the two shifts to be

$$\delta E_{SL} = \alpha^2 E_0 \left(\frac{1}{48} + \frac{1}{24} \right) = \frac{\alpha^2 E_0}{16} = \frac{13.6 \text{ eV}}{16(137)^2} = 4.53 \times 10^{-5} \text{ eV}.$$

The smallness of this splitting is due to the smallness of the factor α^2. The spin–orbit interaction has an even smaller influence on the states of the one-electron atom for larger values of n and ℓ.

8-10 Results from Relativistic Quantum Mechanics

The Bohr energy levels reveal a doublet substructure when the analysis of the atom is refined to include relativistic contributions. We have just found spin–orbit splitting to be one such effect. We now want to identify the other contributor to the fine structure of the levels so that we can combine the two corrections. Both of these features of the atom are properties that emerge naturally from the relativistic quantum theory of the atomic electron.

We can assess the relativistic nature of the spin–orbit interaction by establishing the manner in which $\langle V_{SL} \rangle$ depends on the velocity ratio v/c. Let us recall that the Bohr energy is

$$E_n = -\frac{Z^2}{n^2} E_0 \quad \left(\text{ignoring the factor } \frac{\mu}{m_e} \right)$$

and observe from Equation (8-36) that the ratio of $\langle V_{SL} \rangle$ to E_n is of order $Z^2\alpha^2$. We know (from Problem 14 in Chapter 3 and Problem 10 in Chapter 7, for example) that v/c is proportional to $Z\alpha$ for orbital motion in the atom. It follows that the spin–orbit energy shift is a relativistic correction of order $(v/c)^2$ relative to the Bohr energy. The other fine structure effect is of the same order in v/c and should therefore appear to the same order in powers of $Z\alpha$.

This second correction involves the relativistic treatment of the kinetic energy of the electron. We recall that the Schrödinger equation for a particle of mass m is associated with the classical energy relation

$$\frac{p^2}{2m} + V = E,$$

in which the energy E excludes the rest energy mc^2 and the kinetic energy varies with p in nonrelativistic fashion. The relativistic formulas in Equations (1-35) and (1-37) give the desired version of the kinetic energy as

$$\sqrt{c^2p^2 + m^2c^4} - mc^2.$$

The expansion of this expression in powers of $(p/mc)^2$ yields the following result:

$$mc^2\left(1 + \frac{p^2}{m^2c^2}\right)^{1/2} - mc^2 = mc^2\left[\left(1 + \frac{1}{2}\frac{p^2}{m^2c^2} - \frac{1}{8}\frac{p^4}{m^4c^4} + \cdots\right) - 1\right]$$

$$= \frac{p^2}{2m} - \frac{p^4}{8m^3c^2} + \cdots .$$

We recognize the leading term as the familiar nonrelativistic kinetic energy, and we regard the next-to-leading term as the first relativistic correction. The leading contribution turns into the usual differential operator $-(\hbar^2/2m)\nabla^2$ acting on the wave function in the Schrödinger equation. We define the correction term for application to the atomic electron by the formula

$$K_{\text{rel}} = -\frac{p^4}{8m_e^3c^2}. \tag{8-38}$$

It is easy to verify that this correction is of order $(v/c)^2$ relative to the leading nonrelativistic term.

The extra kinetic energy K_{rel} is treated as another small interaction to be added to the energy of the atom along with V_{SL}. We compute these corrections by obtaining the expectation value of K_{rel} in the state

$$\Psi_{n\ell jm_j} = \psi_{n\ell jm_j} e^{-iE_n t/\hbar}$$

and by combining the result with the already-derived evaluation of $\langle V_{SL} \rangle$ in the same state. The formula for $\langle K_{\text{rel}} \rangle$ is found in the derivation at the end of the section:

$$\langle K_{\text{rel}} \rangle = -\frac{Z^4\alpha^4}{n^3}m_e c^2\left(\frac{1}{2\ell + 1} - \frac{3}{8n}\right). \tag{8-39}$$

This expression contains the same power of $Z\alpha$ as occurs in Equation (8-35) for $\langle V_{SL}\rangle$. We then determine the *total fine structure shift* in energy for any $\ell \neq 0$ state by adding the two corrections:

$$\langle V_{SL}\rangle + \langle K_{\text{rel}}\rangle = \frac{Z^4\alpha^4}{2n^3}m_e c^2 \left\{ \frac{j(j+1) - \ell(\ell+1) - \frac{3}{4}}{\ell(\ell+1)(2\ell+1)} - \frac{2}{2\ell+1} + \frac{3}{4n} \right\}.$$

This combination of terms simplifies considerably to become

$$\langle V_{SL}\rangle + \langle K_{\text{rel}}\rangle = -\frac{Z^4\alpha^4}{2n^3}m_e c^2 \left(\frac{2}{2j+1} - \frac{3}{4n} \right). \tag{8-40}$$

We leave the intervening algebra to Problem 17 at the end of the chapter.

The $\ell = 0$ states require special handling. Let us accept the fact that the relativistic corrections to these states happen to give the *same* final conclusion as in Equation (8-40). We can then rewrite our results in terms of the Rydberg energy E_0 and add the Bohr energy E_n to obtain the final formula for the energy in *any* state of the one-electron atom:

$$E_{nj} = E_n - \frac{Z^4\alpha^2}{n^3}E_0 \left(\frac{2}{2j+1} - \frac{3}{4n} \right). \tag{8-41}$$

The corresponding energy level diagram is shown in Figure 8-29. We magnify the small effect of the fine structure so that we can see the emergence of a remarkable pattern of degeneracies. We observe that the energies of the states nL_j depend on the values of n and j, but *not* on the value of ℓ. The resulting degeneracies appear throughout the figure in the pairs of states $(2S_{1/2}, 2P_{1/2}), (3S_{1/2}, 3P_{1/2}), (3P_{3/2}, 3D_{3/2}), \ldots$, where each pair has a certain energy for the given n and j, without regard for ℓ.

The fine structure of the energy levels affects the wavelengths and splits the spectral lines in the emission spectrum of the atom. The electric dipole selection rules determine which initial and final states of the atom are linked in the various radiative transitions. We have studied these rules in Section 7-5, prior to our introduction of electron spin. The properties of the electric dipole vector are spatial and do not pertain to spin, so that the spin quantum number plays no part in these selection rules. We know that atomic states can participate in a given transition if a change in parity and a change in the quantum number ℓ occur such that

$$\Delta\ell = \pm 1. \tag{8-42}$$

We also know that the emitted electric dipole photon carries away one $\hbar$ unit of angular momentum. Therefore, the initial and final quantum numbers j and j' in the process

$$\left(n\ell j m_j \right) \rightarrow \left(n'l'j'm_j' \right) + \gamma$$

must obey the vector-addition formula

$$\mathbf{j} = \mathbf{j}' + \mathbf{1}$$

in order for the total angular momentum to be conserved. It follows that changes in j

Figure 8-29

Energy levels E_{nj} for the one-electron atom. Each choice of n and j gives a pair of degenerate states nL_j with two different values of ℓ. The fine structure effects shift the levels away from their positions in Figure 8-24. The greatly exaggerated splittings indicate how the shifts diminish with increasing values of the quantum numbers.

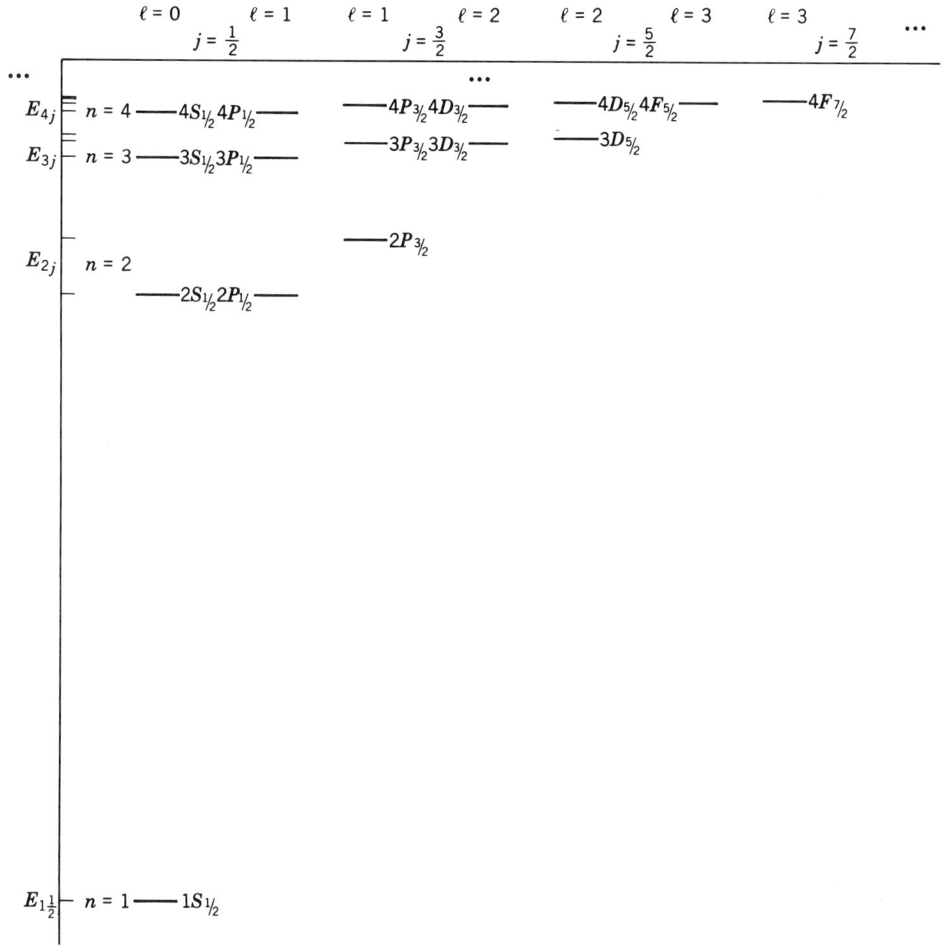

must satisfy the selection rule

$$\Delta j = 0 \text{ or } \pm 1 \quad (\text{but } 0 \nrightarrow 0). \tag{8-43}$$

The provision in parentheses notes that transitions from $j = 0$ to $j' = 0$ violate conservation of angular momentum and are not allowed. Our previous statement about Δm_ℓ in Equation (7-41) converts directly into the new selection rule

$$\Delta m_j = 0 \text{ or } \pm 1. \tag{8-44}$$

This conclusion holds because $m_j = m_\ell + m_s$ and because m_s is not affected in an electric dipole process.

Figure 8-30

Lyman α transitions allowing for the fine structure of the energy levels. The splitting is greatly exaggerated, as in Figure 8-29.

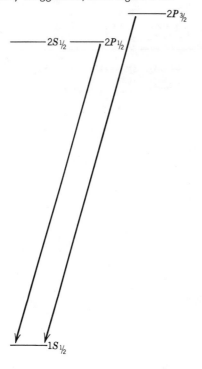

Figure 8-30 shows the Lyman α transition in hydrogen as an illustration of these selection rules. It is clear that the Lyman α process $2p \rightarrow 1s$ actually consists of the *two* atomic transitions

$$2P_{3/2} \rightarrow 1S_{1/2} \quad \text{and} \quad 2P_{1/2} \rightarrow 1S_{1/2},$$

resulting in two distinct wavelengths for the two different transition energies. The difference is small enough to require high-resolution interferometry as a means of distinguishing the spectral lines.

A consistent fine structure theory was developed by Pauli in 1927 as part of his method for incorporating spin in the Schrödinger equation. This treatment of relativistic corrections belonged to the framework of nonrelativistic quantum mechanics and offered no theoretical explanation for the origin of electron spin. Another theory of the relativistic electron was put forward with more spectacular consequence by P. A. M. Dirac in 1928. Dirac's relativistic quantum theory was constructed as a single-particle formalism based jointly on the principles of relativity and the principles of quantum mechanics. His differential equation for the wave function employed the relativistic relation between energy and momentum and assumed a form entirely different from Schrödinger's wave equation. This new relativistic theory *automatically* allowed for the spin of the electron and reduced to Pauli's theory in the nonrelativistic limit.

Paul Dirac

Let us recount the successes of Dirac's theory and leave the elegant formulation to a more advanced treatment of quantum mechanics. The theory starts with the free electron and imposes the classical relativistic energy relation appropriate for a particle of mass m:

$$c^2 p^2 + m^2 c^4 = E^2.$$

Since the relativistic momentum p and energy E appear with the same exponent, the corresponding differential operators $\partial/\partial x$, $\partial/\partial y$, $\partial/\partial z$, and $\partial/\partial t$ occur to the *same* order in the differential equation for the wave function $\Psi(x, y, z, t)$. A first-order time derivative is assumed so that the wave function evolves in time from a specification of the state at $t = 0$, as in the Schrödinger theory. The Dirac equation is therefore of *first* order in the spatial derivatives of Ψ. It follows that the differential equation for Ψ can be properly constrained by the relativistic relation between p and E only if Ψ has a certain *multivalued* structure. The wave function has a two-valuedness corresponding to the spin-up and spin-down properties of a spin-$\frac{1}{2}$ particle, as described by the Pauli quantum number m_s. The wave function also has another two-valuedness associated with the characteristics of *particle* and *antiparticle*. Thus, the free-particle Dirac equation predicts the existence and describes the behavior of spin-$\frac{1}{2}$ particles and antiparticles (such as electron and positron) with spin up and down.

An external magnetic field **B** can be introduced in Dirac's theory of the electron. The result is an interaction with **B** via a spin magnetic moment whose g-factor is *exactly* given by $g_S = 2$. The Dirac equation can also be applied to an atomic electron in the Coulomb field of a nucleus. When the interactions of the electron are examined in powers of v/c, the relativistic corrections describe K_{rel} as given in Equation (8-38) and also V_{SL} as written in Equation (8-32). The latter result automatically expresses the spin–orbit interaction in correct form, complete with the Thomas factor of $\frac{1}{2}$. The Dirac equation can also be solved exactly to determine the energy eigenvalues for the hydrogen atom. These solutions contain the effects of relativistic motion to all orders in v/c and all orders in the fine structure constant α. The resulting exact energy levels depend only on the quantum numbers n and j and reduce to the energies E_{nj} in Equation (8-41) when the solution is expanded through terms of order $\alpha^2 E_0$.

Dirac's theory was verified decisively in every relevant experiment. The prediction of antiparticles was especially significant as a bold new innovation in theoretical physics. The proposed existence of the positron was finally confirmed by the discovery of the antiparticle in 1932. The theory p rovided the general basis for a complete understanding of the relativistic quantum behavior of any spin-$\frac{1}{2}$ particle and was hailed as one of the great accomplishments in 20th century physics.

Figure 8-31

Highly schematic plan of the Lamb-shift experiment. A beam of ground-state hydrogen atoms is passed through an electron-bombardment region where the atoms can be raised to excited states by electron collision. Atoms excited by the $1S_{1/2} \rightarrow 2S_{1/2}$ transition are left in a metastable state where the electric dipole selection rules prevent radiative transitions back to the ground state. The excited atoms proceed through the waveguide to a detector designed for the collection of metastable atoms. A radio-frequency field is applied to the beam in the waveguide and induces transitions from $2S_{1/2}$ to $2P_{1/2}$ when the frequency corresponds to the difference in energy between the two nondegenerate states. The radiative transition $2P_{1/2} \rightarrow 1S_{1/2}$ is allowed so that excited atoms in the $2P_{1/2}$ state radiate to the ground state before they reach the detector. This depopulation of the metastable $2S_{1/2}$ state is observed as a decrease in signal at the detector for an rf frequency around 1060 MHz.

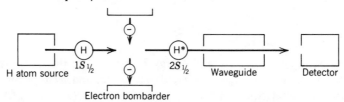

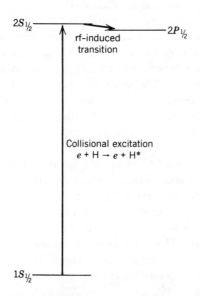

The multiplet structure of hydrogen has been presented in Equation (8-41) and Figure 8-29. This system of levels was investigated experimentally with high-resolution radio-frequency techniques by W. E. Lamb and co-workers during the late 1940s. In studies like the one sketched in Figure 8-31, it was found that the predicted degeneracy of the $(n = 2, j = \frac{1}{2})$ states was actually broken and that the $2S_{1/2}$ state had a slightly higher energy than its partner the $2P_{1/2}$ state. This measurement became known as the Lamb shift. The observation was important because it indicated the influence of phenomena that were not yet included in the relativistic quantum theory. The slight departure of the spin g-factor from the value $g_S = 2$ was another property of the electron that had no explanation in Dirac's theory. These discoveries took place in the next period of exciting developments in quantum physics.

Detail

We derive the formula for the extra kinetic energy in Equation (8-39) by first relating the relativistic correction K_{rel} to the nonrelativistic kinetic energy:

$$K_{rel} = -\frac{1}{2m_e c^2}\left(\frac{p^2}{2m_e}\right)^2 = -\frac{1}{2m_e c^2}(E - V)^2.$$

The expectation value of this quantity consists of three terms,

$$\langle K_{rel}\rangle = -\frac{1}{2m_e c^2}(\langle E^2\rangle - 2\langle VE\rangle + \langle V^2\rangle),$$

to be evaluated in the stationary state $\Psi = \psi_{n\ell j m_j}e^{-iE_n t/\hbar}$. We are using a state with energy eigenvalue E_n, and so we find

$$\langle E^2\rangle = E_n^2$$

and

$$\langle VE\rangle = \int \Psi^* V i\hbar \frac{\partial}{\partial t}\Psi\, d\tau = E_n \int \Psi^* V\Psi\, d\tau = E_n\langle V\rangle.$$

The Coulomb potential energy appears in the second and third terms of $\langle K_{rel}\rangle$. We employ Equations (7-33) and (7-34) to obtain the expectation values

$$\langle V\rangle = -\frac{Ze^2}{4\pi\varepsilon_0}\left\langle\frac{1}{r}\right\rangle = -\frac{Ze^2}{4\pi\varepsilon_0 an^2}$$

and

$$\langle V^2\rangle = \left(\frac{Ze^2}{4\pi\varepsilon_0}\right)^2\left\langle\frac{1}{r^2}\right\rangle = \left(\frac{Ze^2}{4\pi\varepsilon_0}\right)^2\frac{2}{a^2 n^3(2\ell + 1)}.$$

We then introduce α wherever possible, using the relation

$$\frac{e^2}{4\pi\varepsilon_0} = \alpha\hbar c,$$

and we ignore the reduced-mass factor μ/m_e to rewrite the Bohr formulas:

$$E_n = -\frac{Z^2}{n^2}E_0 = -\frac{Z^2}{n^2}\alpha^2\frac{m_ec^2}{2} \quad \text{and} \quad a = \frac{\hbar}{Z\alpha m_ec}.$$

The expectation values of V and V^2 are thereby rewritten as

$$\langle V \rangle = -\frac{Z\alpha\hbar c}{n^2}\frac{Z\alpha m_ec}{\hbar} = -\frac{Z^2\alpha^2}{n^2}m_ec^2$$

and

$$\langle V^2 \rangle = (Z\alpha\hbar c)^2\left(\frac{Z\alpha m_ec}{\hbar}\right)^2\frac{2}{n^3(2\ell+1)} = 2\frac{Z^4\alpha^4 m_e^2 c^4}{n^3(2\ell+1)}.$$

All three contributions to the expectation value of K_{rel} can now be assembled:

$$\langle K_{\text{rel}} \rangle = -\frac{1}{2m_ec^2}\left\{ \frac{Z^4\alpha^4 m_e^2 c^4}{4n^4} - 2\left(-\frac{Z^2\alpha^2 m_ec^2}{2n^2}\right)\left(-\frac{Z^2\alpha^2 m_ec^2}{n^2}\right)\right.$$

$$\left. +2\frac{Z^4\alpha^4 m_e^2 c^4}{n^3(2\ell+1)}\right\}$$

$$= -\frac{Z^4\alpha^4}{n^3}m_ec^2\left(\frac{1}{8n} - \frac{1}{2n} + \frac{1}{2\ell+1}\right).$$

The formula quoted in Equation (8-39) follows immediately from this result.

Example

Let us investigate the small shifts in wavelength for the Lyman α transitions shown in Figure 8-30. We set $Z = 1$ in Equation (8-41) and obtain the following energy levels:

$$E(2P_{3/2}) = E_2 - \frac{\alpha^2}{8}E_0\left(\frac{1}{2} - \frac{3}{8}\right) = E_2 - \frac{\alpha^2}{64}E_0,$$

$$E(2P_{1/2}) = E_2 - \frac{\alpha^2}{8}E_0\left(1 - \frac{3}{8}\right) = E_2 - \frac{5\alpha^2}{64}E_0,$$

$$E(1S_{1/2}) = E_1 - \alpha^2 E_0\left(1 - \frac{3}{4}\right) = E_1 - \frac{\alpha^2}{4}E_0.$$

The two transition energies are

$$E(2P_{3/2}) - E(1S_{1/2}) = E_2 - E_1 - \alpha^2 E_0\left(\frac{1}{64} - \frac{1}{4}\right) = \Delta E + \frac{15\alpha^2}{64}E_0$$

and

$$E(2P_{1/2}) - E(1S_{1/2}) = E_2 - E_1 - \alpha^2 E_0\left(\frac{5}{64} - \frac{1}{4}\right) = \Delta E + \frac{11\alpha^2}{64}E_0,$$

where ΔE denotes the difference in Bohr energies:

$$\Delta E = E_2 - E_1 = 10.2 \text{ eV}.$$

The corresponding wavelengths are

$$\lambda\left(2P_{3/2} \to 1S_{1/2}\right) = \frac{hc}{\Delta E\left(1 + 15\alpha^2 E_0/64\,\Delta E\right)} = \frac{hc}{\Delta E}\left(1 - \frac{15\alpha^2 E_0}{64\,\Delta E}\right)$$

and

$$\lambda\left(2P_{1/2} \to 1S_{1/2}\right) = \frac{hc}{\Delta E\left(1 + 11\alpha^2 E_0/64\,\Delta E\right)} = \frac{hc}{\Delta E}\left(1 - \frac{11\alpha^2 E_0}{64\,\Delta E}\right),$$

to order α^2 in the respective correction terms. We recognize the leading factor $hc/\Delta E$ to be the familiar Lyman α wavelength from the Bohr model:

$$\frac{hc}{\Delta E} = \frac{1240 \text{ eV} \cdot \text{nm}}{10.2 \text{ eV}} = 122 \text{ nm}.$$

The wavelength shifts are given by the residual quantities

$$\frac{15\alpha^2 E_0}{64\,\Delta E}\frac{hc}{\Delta E} = \frac{15}{64}\left(\frac{1}{137}\right)^2\frac{13.6}{10.2}(122 \text{ nm}) = 2.02 \times 10^{-3} \text{ nm}$$

and

$$\frac{11\alpha^2 E_0}{64\,\Delta E}\frac{hc}{\Delta E} = \frac{11}{64}\left(\frac{1}{137}\right)^2\frac{13.6}{10.2}(122 \text{ nm}) = 1.48 \times 10^{-3} \text{ nm}.$$

We conclude that the two wavelengths must be measured with at least six-figure accuracy to be resolved as separate lines.

8-11 The Zeeman Effect

The Stern–Gerlach phenomenon and the multiplet structure of the atom demonstrate the influence of electron spin. Further evidence for spin is found when the spectral lines of the atom are observed in an applied magnetic field. We have discussed these observations of the Zeeman effect early in the chapter in Sections 8-1 and 8-3. We are now prepared to learn why the prediction of triplet line splitting is not confirmed, and why the so-called anomalous pattern is seen instead. Once the Zeeman splitting is fully understood, there is no longer any reason to regard the effect as an anomaly.

An external constant **B** field interacts with an atom through the magnetic-moment interaction given in Equation (8-4):

$$V_M = -\boldsymbol{\mu} \cdot \mathbf{B}.$$

Both orbital and spin magnetic moments contribute to $\boldsymbol{\mu}$, but only the former contribution appears in our earlier treatment of the Zeeman effect. We obtain the total magnetic moment of the one-electron atom by recalling Equations (8-19) and (8-20) and inserting the g-factors $g_L = 1$ and $g_S = 2$:

$$\mu = \mu_L + \mu_S = -\frac{\mu_B}{\hbar}(\mathbf{L} + 2\mathbf{S}). \tag{8-45}$$

We then determine the magnetic-interaction shift for a particular energy level of the atom by calculating the expectation value

$$\langle V_M \rangle = -\langle \mu_z \rangle B.$$

This formula employs the usual choice of z axis along the direction of $\mathbf{B}$ and calls for the evaluation of $\langle \mu_z \rangle$ in a state that describes the atom without the application of the field.

Let us first assume conditions in which the applied $\mathbf{B}$ field is much stronger than the internal magnetic field responsible for spin–orbit coupling. This situation has already been mentioned in our discussion of Figure 8-26. We neglect the spin–orbit interaction and let the vectors $\mathbf{L}$ and $\mathbf{S}$ be *decoupled* so that the orbital and spin magnetic moments perform independent Larmor precessions in the strong applied field. The calculation of the magnetic interaction is then based on states of the atom defined by m_ℓ and m_s as good quantum numbers. We use the wave function $\Psi_{n\ell m_\ell m_s}$, with properties given by the expectation values

$$\langle L_z \rangle = \hbar m_\ell \quad \text{and} \quad \langle S_z \rangle = \hbar m_s,$$

and take the z component of Equation (8-45) to find

$$\langle V_M \rangle = +\frac{\mu_B}{\hbar}\langle L_z + 2S_z \rangle B = \mu_B B(m_\ell + 2m_s). \tag{8-46}$$

A given orbital configuration has quantum numbers n and ℓ and contains $2(2\ell + 1)$ different magnetic substates. These magnetic energy levels are shifted by the energy $\langle V_M \rangle$ according to the values of m_ℓ and m_s. We emphasize that we cannot label the states by these quantum numbers unless the spin–orbit interaction V_{SL} is dominated by the magnetic interaction V_M. This strong-field phenomenon is called the Paschen–Back effect, a special case of the Zeeman effect, named after the spectroscopists Paschen and E. Back.

The more interesting case arises when *both* V_{SL} and V_M come into play, particularly in situations where the latter is the weaker of the two interactions. Our solution of this problem is based on a synthesis of techniques in which the quantum approach is supplemented by the classical picture of precessional motion. We already know that we can account for V_{SL} by computing the expectation value in a state nL_j, where j and m_j are good quantum numbers. Our plan is to use the wave function $\Psi_{n\ell j m_j}$ for $\langle V_M \rangle$ as well.

The problem is difficult because a mismatch exists between the interaction and the state. The interaction pertains to the total magnetic moment $\mu = -(\mu_B/\hbar)(\mathbf{L} + 2\mathbf{S})$, while the state refers to the total angular momentum $\mathbf{J} = \mathbf{L} + \mathbf{S}$. Figure 8-32 shows that this feature of the problem prevents the vectors μ and $-\mathbf{J}$ from pointing in the

Figure 8-32
Vector diagram for the constructions
$\mathbf{J} = \mathbf{L} + \mathbf{S}$ and $\boldsymbol{\mu} = \boldsymbol{\mu}_L + \boldsymbol{\mu}_S$. The g-factors
g_L and g_S are not equal, and so the vectors $\boldsymbol{\mu}$
and $\mathbf{J}$ are not collinear.

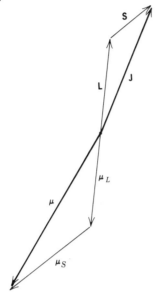

same direction. The component of $\boldsymbol{\mu}$ along $\mathbf{J}$ is given by

$$\mu_J = \frac{\boldsymbol{\mu} \cdot \mathbf{J}}{J} = -\frac{\mu_B}{\hbar J}(\mathbf{L} + 2\mathbf{S}) \cdot (\mathbf{L} + \mathbf{S})$$

$$= -\frac{\mu_B}{\hbar J}(L^2 + 2S^2 + 3\mathbf{S} \cdot \mathbf{L}).$$

We relate $\mathbf{S} \cdot \mathbf{L}$ to J^2, L^2, and S^2 as before and find

$$\mu_J = -\frac{\mu_B}{\hbar J}\left[L^2 + 2S^2 + \frac{3}{2}(J^2 - L^2 - S^2)\right]$$

$$= -\frac{\mu_B}{2\hbar J}(3J^2 + S^2 - L^2). \tag{8-47}$$

We realize, however, that the desired quantity is not μ_J but μ_z, the component of $\boldsymbol{\mu}$
along the direction of $\mathbf{B}$.

Figure 8-33 indicates how μ_J and μ_z are distinguished and how the one can be
found from the other. We introduce the three vectors $\boldsymbol{\mu}_J$, $\boldsymbol{\mu}_{\perp J}$, and $\boldsymbol{\mu}_{\perp B}$ according to
the figure and write

$$\mu_z = |\boldsymbol{\mu}_J + \boldsymbol{\mu}_{\perp J} + \boldsymbol{\mu}_{\perp B}|\cos\theta, \tag{8-48}$$

where θ is defined such that

$$J_z = J\cos\theta.$$

Figure 8-33

Construction of μ_z from μ_J.

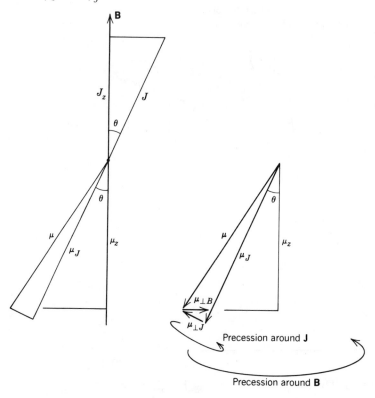

Our construction of μ_z identifies $\mu_{\perp J}$ and $\mu_{\perp B}$ as vectors perpendicular to $\mathbf{J}$ and $\mathbf{B}$, respectively. Our objective is to determine the expectation value of μ_z by the following argument based on the dynamics of precessional motion. The effect of spin–orbit coupling is represented by a precession of $\mathbf{L}$ and $\mathbf{S}$ about $\mathbf{J}$, as introduced in Figure 8-27, while the effect of the magnetic interaction is represented by another simultaneous precession of $\mathbf{J}$ about $\mathbf{B}$. Figure 8-33 shows how $\mu_{\perp J}$ rotates around the direction of $\mathbf{J}$ while $\mu_{\perp B}$ rotates around the direction of $\mathbf{B}$ during these two precessions. When we take the expectation value in the given state, we perform an *average* over the two periodic motions and obtain *vanishing* results for both vectors $\mu_{\perp J}$ and $\mu_{\perp B}$. This argument implies that Equation (8-48) can be replaced by the simpler expression

$$\mu_z \rightarrow \mu_J \cos \theta = J_z \frac{\mu_J}{J}, \qquad (8\text{-}49)$$

with the understanding that the expectation value must be taken if the replacement is to have the status of an equality.

Let us now combine Equations (8-47) and (8-49) and complete the determination of $\langle \mu_z \rangle$. We proceed by calculating the expectation value of $\mu_z J^2$ instead:

$$\langle \mu_z J^2 \rangle = \langle J_z J \mu_J \rangle$$
$$= -\frac{\mu_B}{2\hbar} \langle J_z (3J^2 + S^2 - L^2) \rangle.$$

The two sides of this equality are to be evaluated for a level nL_j, where eigenvalues are specified for the observables J^2, L^2, and S^2. We use these eigenvalue properties to substitute

$$\langle \mu_z J^2 \rangle = \hbar^2 j(j+1)\langle \mu_z \rangle$$

on the left side of the equation and

$$\left\langle J_z(3J^2 + S^2 - L^2) \right\rangle = \hbar^2[3j(j+1) + s(s+1) - \ell(\ell+1)]\langle J_z \rangle$$

on the right side. The resulting equality can then be rearranged to read

$$\langle \mu_z \rangle = -\frac{\mu_B}{2\hbar}\frac{3j(j+1) + s(s+1) - \ell(\ell+1)}{j(j+1)}\langle J_z \rangle$$

$$= -g\mu_B\frac{\langle J_z \rangle}{\hbar}, \tag{8-50}$$

where

$$g = 1 + \frac{j(j+1) + s(s+1) - \ell(\ell+1)}{2j(j+1)}. \tag{8-51}$$

This last coefficient in the final expression for $\langle \mu_z \rangle$ is called the *Lande g-factor* after A. Lande, a pioneer in the period of the old quantum theory. We take note of the fact that the spin g-factor has been explicitly written as $g_S = 2$ from the beginning of the calculation.

Equation (8-50) takes the place of Equation (8-10) in all applications of the magnetic moment. The revised formula is more involved since the Lande g-factor is not just a constant like g_L or g_S. We get $g = 1$ from Equation (8-51) and thus reproduce the orbital g-factor g_L, if we arbitrarily set $s = 0$ and let $\mathbf{J}$ and $\mathbf{L}$ become identical. Of course, we must take $s = \frac{1}{2}$ and write

$$g = 1 + \frac{j(j+1) - \ell(\ell+1) + \frac{3}{4}}{2j(j+1)} \tag{8-52}$$

in the case of the one-electron atom. This expression reproduces the spin g-factor $g_S = 2$ if the given state has quantum numbers $\ell = 0$ and $j = \frac{1}{2}$. Equation (8-50) can be recast in vector form as

$$\langle \boldsymbol{\mu} \rangle = -\frac{g\mu_B}{\hbar}\langle \mathbf{J} \rangle, \tag{8-53}$$

since only the z component can have a nonvanishing expectation value in the state $\Psi_{n\ell jm_j}$. This result tells us that the observable magnetic dipole moment vector is directly proportional to the total angular momentum, the only vector available to characterize the state of the atom.

The Zeeman energy shift is deduced directly from Equation (8-50). We evaluate the expression

$$\langle V_M \rangle = -\langle \mu_z \rangle B = +g\mu_B B\frac{\langle J_z \rangle}{\hbar}$$

in the state $\Psi_{n\ell j m_j}$ and use the fact that the wave function is an eigenfunction of J_z to write

$$\langle J_z \rangle = \hbar m_j.$$

The conclusion follows at once:

$$\langle V_M \rangle = g\mu_B B m_j, \tag{8-54}$$

a result to be compared with Equation (8-12) for the fictitious normal Zeeman energy shift. The actual situation pertains to a typical level nL_j whose $2j + 1$ magnetic substates have the same energy E_{nj} in the absence of an external magnetic field. Equation (8-54) represents an energy shift due to the application of the field, causing the magnetic substates to split apart and acquire different energies for each value of the quantum number m_j. These Zeeman splittings differ from one nL_j to another because of the variation of the Lande g-factor with ℓ and j. We therefore predict *more than three* different transition energies for the various transitions allowed by the electric dipole selection rules. The anomalous pattern of Zeeman spectral lines is observed as a result, instead of the normal Lorentz triplet. The following illustration demonstrates this effect.

Example

Let us look at the Lyman α transitions of Figure 8-30 in the presence of an applied **B** field. The transitions of interest are

$$2P_{3/2} \rightarrow 1S_{1/2} \quad \text{and} \quad 2P_{1/2} \rightarrow 1S_{1/2},$$

where each level nL_j is split into its $2j + 1$ magnetic sublevels. The Lande g-factors and Zeeman energy shifts are given by Equations (8-52) and (8-54) according to the values of m_j as follows:

in the $2P_{3/2}$ state $\quad g = 1 + \dfrac{\frac{15}{4} - 2 + \frac{3}{4}}{\frac{15}{2}} = 1 + \frac{1}{3} = \frac{4}{3}$

$$\text{and} \quad \langle V_M \rangle = \tfrac{4}{3}\mu_B B \begin{bmatrix} 3/2 \\ 1/2 \\ -1/2 \\ -3/2 \end{bmatrix},$$

in the $2P_{1/2}$ state $\quad g = 1 + \dfrac{\frac{3}{4} - 2 + \frac{3}{4}}{\frac{3}{2}} = 1 - \frac{1}{3} = \frac{2}{3}$

$$\text{and} \quad \langle V_M \rangle = \tfrac{2}{3}\mu_B B \begin{bmatrix} 1/2 \\ -1/2 \end{bmatrix},$$

in the $1S_{1/2}$ state $\quad g = 1 + \dfrac{\frac{3}{4} - 0 + \frac{3}{4}}{\frac{3}{2}} = 1 + 1 = 2$

$$\text{and} \quad \langle V_M \rangle = 2\mu_B B \begin{bmatrix} 1/2 \\ -1/2 \end{bmatrix}.$$

The resulting sets of split levels are shown in Figure 8-34. Note that the Zeeman

Figure 8-34

Zeeman splitting in the ground state and first excited states of hydrogen. An applied magnetic field splits the levels for the various values of m_j. The Lyman α fine structure transitions $2P_{3/2} \rightarrow 1S_{1/2}$ and $2P_{1/2} \rightarrow 1S_{1/2}$ result in six and four distinct spectral lines.

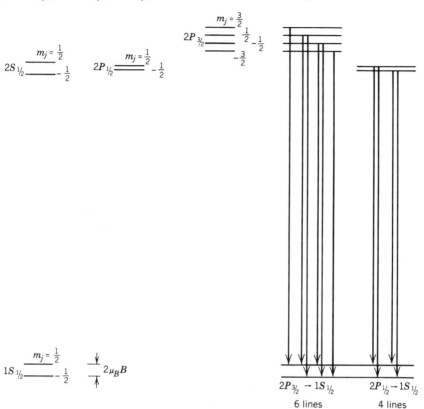

splitting for the $2S_{1/2}$ states is identical to that for the $1S_{1/2}$ states. Electric dipole transitions occur from $2P_{3/2}$ and $2P_{1/2}$ to $1S_{1/2}$, provided the initial and final m_j values obey the selection rule

$$\Delta m_j = 0 \text{ or } \pm 1$$

as shown in the figure. It is apparent that *all* the transition energies are different and that the fine structure doublet of Figure 8-30 turns into six-plus-four distinct Zeeman lines.

8-12 Hyperfine Structure

The proton is a spin-$\frac{1}{2}$ particle, like the electron, with its own magnetic moment due to spin. In fact, every atomic nucleus with spin is likewise endowed with intrinsic magnetization determined by the distribution of constituents in the nuclear system. The implications for the one-electron atom are that the nuclear magnetic moment sets up a permanent magnetic field inside the atom, and the electron magnetic moment experiences a weak Zeeman interaction with the internal field. This additional magnetic effect within the atom is known as the hyperfine interaction. Its influence is detectable in measurements of energy-level splittings that are orders of magnitude smaller than the splittings due to spin–orbit coupling. We focus our discussion on the $\ell = 0$ states of the one-electron atom and finally concentrate on the ground state of hydrogen. Our first objective is to describe the hyperfine interaction in terms of a *spin–spin coupling* between the nucleus and the electron.

The nucleus has a *nuclear spin* **I** with quantum behavior like that of an angular momentum vector. Every nucleus is assigned a *nuclear spin quantum number i* from the list of allowed integral or half-integral values

$$0, \tfrac{1}{2}, 1, \tfrac{3}{2}, \ldots,$$

such that the magnitude of the nuclear spin is given by the eigenvalue property

$$I^2 = \hbar^2 i(i + 1). \tag{8-55}$$

The assignment of the quantum number i depends on the particular quantum state of the given nuclear species. A magnetic moment can be defined for any nucleus by analogy with Equation (8-20), the formula for the magnetic moment due to electron spin. First, we reverse the sign of the charge and alter the mass scale of the Bohr magneton in the equation. These steps replace the factor $-\mu_B$ by the *nuclear magneton*

$$\mu_N = \frac{e\hbar}{2M_p} = \frac{m_e}{M_p}\mu_B. \tag{8-56}$$

The magnetic moment and the nuclear spin are then related by the vector equality

$$\boldsymbol{\mu}_I = +g_I\mu_N\frac{\mathbf{I}}{\hbar}. \tag{8-57}$$

This equation introduces the *nuclear g-factor* g_I as an experimentally measurable numerical quantity. We assume a point-like nucleus described solely by the parameters Z, i, and g_I, and we use the values $Z = 1$ and $i = \tfrac{1}{2}$ in the special case of the proton. We also use the notation

$$\frac{g_p}{2} = 2.792847386$$

to express the measured value of the proton g-factor.

The hyperfine spin–spin interaction involves a coupling between the spin magnetic moments of the electron and the nucleus. We sketch some of the details of this interaction at the end of the section and obtain the following result for an atom in an $\ell = 0$ state:

$$V_{SS} = \frac{1}{4\pi\varepsilon_0 c^2}\frac{2}{3}\boldsymbol{\mu}_S \cdot \boldsymbol{\mu}_I \nabla^2\frac{1}{r}. \tag{8-58}$$

The two spin vectors appear in the formula when Equations (8-20) and (8-57) are inserted:

$$V_{SS} = \frac{1}{4\pi\varepsilon_0 c^2}\frac{2}{3}\left(-\frac{g_S\mu_B}{\hbar}\right)\left(\frac{g_I\mu_N}{\hbar}\right)\mathbf{S} \cdot \mathbf{I} \nabla^2\frac{1}{r}$$

$$= -\frac{1}{4\pi\varepsilon_0 c^2}\frac{2}{3}g_S g_I\left(\frac{e}{2m_e}\right)^2\frac{m_e}{M_p}\mathbf{S} \cdot \mathbf{I} \nabla^2\frac{1}{r}.$$

Note that Equations (8-7) and (8-56) are also employed to eliminate the magneton factors in favor of fundamental constants.

The treatment of $\mathbf{S} \cdot \mathbf{I}$ in the expression for V_{SS} parallels the technique applied to $\mathbf{S} \cdot \mathbf{L}$ in the problem of spin–orbit coupling. A *grand total* atomic angular momentum vector $\mathbf{F}$ is defined to include the nuclear spin $\mathbf{I}$ along with the usual contributions $\mathbf{L}$ and $\mathbf{S}$. This definition reduces to the two terms

$$\mathbf{F} = \mathbf{S} + \mathbf{I} \tag{8-59}$$

in the $\ell = 0$ case. We then use the relation

$$F^2 = S^2 + I^2 + 2\mathbf{S} \cdot \mathbf{I}$$

to rewrite the formula for the hyperfine interaction:

$$V_{SS} = -\frac{1}{4\pi\varepsilon_0 c^2}\frac{g_s g_I}{3}\left(\frac{e}{2m_e}\right)^2 \frac{m_e}{M_p}(F^2 - S^2 - I^2)\nabla^2\frac{1}{r}. \tag{8-60}$$

The hyperfine energy shift in any $\ell = 0$ state is determined by computing the expectation value of this expression. The state in question has a grand total quantum number f associated with the angular momentum behavior of the vector $\mathbf{F}$, such that quantized values of F^2 are given in terms of f by the eigenvalue property

$$F^2 = \hbar^2 f(f+1). \tag{8-61}$$

We also specify the state by the familiar principal quantum number n, orbital quantum number $\ell = 0$, and total angular momentum quantum number $j = s = \frac{1}{2}$. The addition of spin vectors in Equation (8-59) takes the form

$$\tfrac{1}{2} + \mathbf{i}$$

and yields two allowed values for the grand total quantum number:

$$f = i + \tfrac{1}{2} \text{ or } i - \tfrac{1}{2}. \tag{8-62}$$

These two f values refer to two different states of the atom with two different energies, split by the small effect of the hyperfine interaction. This splitting is present in every $nS_{1/2}$ energy level and is of greatest interest in the $1S_{1/2}$ ground level of the hydrogen atom.

We calculate $\langle V_{SS} \rangle$ in an $\ell = 0$ state with quantum numbers n and f by first identifying definite values for the angular momentum factor in Equation (8-60):

$$F^2 - S^2 - I^2 = \hbar^2\left[F(f+1) - \tfrac{3}{4} - i(i+1)\right]. \tag{8-63}$$

The expectation value of V_{SS} also requires an evaluation of the remaining factor $\nabla^2(1/r)$. We examine this last detail at the end of the section and deduce a prescription in terms of the radial portion of the wave function, evaluated at the origin:

$$\left\langle \nabla^2\frac{1}{r} \right\rangle = -\left| R_{n0}(0) \right|^2. \tag{8-64}$$

These two contributions are combined to produce the general formula for the $\ell = 0$

hyperfine energy shift:

$$\langle V_{SS} \rangle = \frac{e^2}{4\pi\varepsilon_0} \frac{g_S g_I}{3} \left(\frac{\hbar}{2m_e c} \right)^2 \frac{m_e}{M_p} \left[f(f+1) - \frac{3}{4} - i(i+1) \right] |R_{n0}(0)|^2. \quad (8\text{-}65)$$

Let us specialize to the $n = 1$ case and consider the hyperfine splitting of the ground level. We consult Table 7-1 to find $R_{10}(0)$, and we also express the Bohr radius in terms of the fine structure constant α to obtain

$$|R_{10}(0)|^2 = \frac{4}{a^3} = 4 \left(\frac{Z\alpha m_e c}{\hbar} \right)^3.$$

Equation (8-65) can then be rewritten in powers of α:

$$\langle V_{SS} \rangle = \alpha \hbar c \frac{g_S g_I}{3} \left(\frac{\hbar}{2m_e c} \right)^2 \frac{m_e}{M_P} \left[f(f+1) - \frac{3}{4} - i(i+1) \right] \left[4 \left(\frac{Z\alpha m_e c}{\hbar} \right)^3 \right]$$

$$= Z^3 \alpha^4 \frac{g_S g_I}{3} \frac{m_e}{M_p} m_e c^2 \left[f(f+1) - \frac{3}{4} - i(i+1) \right]. \quad (8\text{-}66)$$

This final result is valid in the $\ell = 0$ ground level of the one-electron atom for any given nucleus.

The hyperfine energy shift depends on the value of the grand total quantum number f. Two possibilities appear in the bracketed factor in Equation (8-66):

$$f(f+1) - \tfrac{3}{4} - i(i+1) = \begin{cases} i & \text{for } f = i + \tfrac{1}{2} \\ -i - 1 & \text{for } f = i - \tfrac{1}{2}. \end{cases}$$

We see that the state with the larger energy is the one with the larger value of f. The difference in energy between the two states determines the hyperfine splitting

$$\delta E_{\text{hf}} = Z^3 \alpha^4 \frac{g_S g_I}{3} \frac{m_e}{M_p} m_e c^2 (2i + 1). \quad (8\text{-}67)$$

The combination of factors $\alpha^4 (m_e/M_p) m_e c^2$ should be compared with the analogous combination $\alpha^4 m_e c^2$ in Equation (8-40) for the fine structure splitting. It is clear that the electron–proton mass ratio suppresses spin–spin coupling in the atom relative to spin–orbit coupling by some three orders of magnitude.

The $1S_{1/2}$ ground level of hydrogen has $f = 1$ and $f = 0$ states. We find the splitting of these states by taking $Z = 1$, $i = \tfrac{1}{2}$, and $g_I = g_p$ in Equation (8-67):

$$\delta E_{\text{hf}} = \frac{2}{3} \alpha^4 g_S g_p \frac{m_e}{M_p} m_e c^2. \quad (8\text{-}68)$$

Transitions between the $1S_{1/2}$ hyperfine levels of hydrogen have been observed in the laboratory in the radio range of photon frequencies. The measured frequency is perhaps the most accurately determined quantity in physics:

$$\nu_{\rm hf} = \frac{\delta E_{\rm hf}}{h} = 1.420405751800 \text{ GHz},$$

corresponding to approximately 21 cm in wavelength. We can visualize the transition as a spin-flip phenomenon, where the electron and proton have parallel spins in the initial $f = 1$ state and antiparallel spins in the final $f = 0$ state. This effect does not change the orbital quantum number ℓ and does not have the behavior of a radiating electric dipole. The probability for the transition is severely suppressed because of these considerations.

The famous 21 cm line has played an important role in radio astronomy ever since its discovery in 1951. This radiation is detectable in the radio spectrum of galactic hydrogen clouds, despite the extreme suppression of the transition probability. Detection is made possible by the appreciable abundance of hydrogen atoms in the galaxy and by the inappreciable attenuation of radio waves by interstellar dust. Observations of the intensity of these 21 cm emissions are employed to map the distributions of hydrogen in the galaxy, and measurements of the Doppler-shifted wavelengths are used to determine the radial motion of the emitting sources. Such techniques are in current use in the investigation of galactic structures.

Detail

Let us outline the steps leading to the formula for the hyperfine spin–spin interaction. We turn to Problems 2 and 3 at the end of the chapter and use the relation $\mu_0 \varepsilon_0 c^2 = 1$ to find the magnetic field due to the nuclear magnetic moment $\boldsymbol{\mu}_I$:

$$\mathbf{B} = -\frac{1}{4\pi\varepsilon_0 c^2} \nabla \times \left(\boldsymbol{\mu}_I \times \nabla \frac{1}{r} \right)$$

$$= -\frac{1}{4\pi\varepsilon_0 c^2} \left[\boldsymbol{\mu}_I \nabla^2 \frac{1}{r} - \nabla \left(\boldsymbol{\mu}_I \cdot \nabla \frac{1}{r} \right) \right].$$

This internal **B** field interacts with the spin of the electron in an $\ell = 0$ state through the spin magnetic-moment interaction

$$V_{SS} = -\boldsymbol{\mu}_S \cdot \mathbf{B} = \frac{1}{4\pi\varepsilon_0 c^2} \left\{ \boldsymbol{\mu}_S \cdot \boldsymbol{\mu}_I \nabla^2 \frac{1}{r} - (\boldsymbol{\mu}_S \cdot \nabla) \left(\boldsymbol{\mu}_I \cdot \nabla \frac{1}{r} \right) \right\}.$$

We reexpress the interaction in the form

$$V_{SS} = \frac{1}{4\pi\varepsilon_0 c^2} \left\{ \frac{2}{3} \boldsymbol{\mu}_S \cdot \boldsymbol{\mu}_I \nabla^2 \frac{1}{r} + \left[\frac{1}{3} \boldsymbol{\mu}_S \cdot \boldsymbol{\mu}_I \nabla^2 \frac{1}{r} - (\boldsymbol{\mu}_S \cdot \nabla) \left(\boldsymbol{\mu}_I \cdot \nabla \frac{1}{r} \right) \right] \right\}$$

and remark without proof that the expectation value of the terms in square brackets vanishes in any $\ell = 0$ state. The remaining portion of V_{SS} is quoted in the text as Equation (8-58).

Detail

The calculation of $\langle V_{SS} \rangle$ calls for the expectation value of $\nabla^2(1/r)$ in a state whose $\ell = 0$ spatial eigenfunction is given as

$$\psi = \frac{R_{n0}(r)}{\sqrt{4\pi}}.$$

We differentiate $1/r$ away from the point $r = 0$ and find

$$\nabla^2 \frac{1}{r} = \frac{1}{r^2}\frac{d}{dr}r^2\frac{d}{dr}\frac{1}{r} = 0.$$

We also note that the expectation value of $\nabla^2(1/r)$ picks up a nonzero contribution in an infinitesimal spherical neighborhood of the origin:

$$\left\langle \nabla^2 \frac{1}{r} \right\rangle = \int \Psi^* \nabla^2 \frac{1}{r} \Psi\, d\tau = \int_{r=0} \psi^* \left(\frac{1}{r^2}\frac{d}{dr}r^2\frac{d}{dr}\frac{1}{r} \right) \psi\, 4\pi r^2\, dr$$

$$= 4\pi |\psi(0)|^2 \int_{r=0} \left(\frac{d}{dr}r^2\frac{d}{dr}\frac{1}{r} \right) dr$$

$$= |R_{n0}(0)|^2 \left(r^2 \frac{d}{dr}\frac{1}{r} \right) \Bigg|_{r=0} = -|R_{n0}(0)|^2.$$

This result appears in the text as Equation (8-64).

Example

The following numerical exercises illustrate the orders of magnitude involved in the analysis of hyperfine structure. First, we note that the nuclear magneton is three orders smaller than the Bohr magneton:

$$\mu_N = \frac{m_e}{M_p}\mu_B = \frac{5.788 \times 10^{-9} \text{ eV/G}}{1836} = 3.152 \times 10^{-12} \text{ eV/G}.$$

We can compare the hyperfine splitting in hydrogen to the Bohr energy by inserting the relation $E_0 = (\alpha^2/2)m_e c^2$ in Equation (8-68):

$$h\nu_{\text{hf}} = \frac{4}{3}\alpha^2 g_s g_p \frac{m_e}{M_p}E_0.$$

We find the transition frequency to be

$$\nu_{\text{hf}} = \frac{4}{3}\frac{(2)(2)(2.79)}{(137)^2}\frac{13.6 \text{ eV}}{1836}\frac{1.60 \times 10^{-19} \text{ J/eV}}{6.63 \times 10^{-34} \text{ J} \cdot \text{s}} = 1.42 \times 10^9 \text{ Hz},$$

and so we obtain

$$\lambda_{hf} = \frac{c}{\nu_{hf}} = \frac{3 \times 10^8 \text{ m/s}}{1.42 \times 10^9 \text{ Hz}} = 0.211 \text{ m}$$

for the wavelength of the 21 cm spectral line.

Problems

1. Obtain the formula for the magnetic field due to a circular current loop and for the electric field due to a pair of opposite charges, in each case at a location on the axis a distance z from the origin. Show that for large z the two results are of identical dipole form.

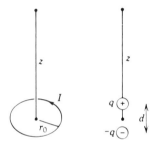

2. The general formula for the magnetic field due to a magnetic dipole moment $\boldsymbol{\mu}$ is

$$\mathbf{B} = -\frac{\mu_0}{4\pi} \nabla \times \left(\boldsymbol{\mu} \times \nabla \frac{1}{r} \right).$$

Let $\boldsymbol{\mu} = \mu\hat{\mathbf{z}}$ and show that at the axial point $(0, 0, z)$ this expression reproduces the large-z result obtained in Problem 1.

3. Prove the vector-calculus identity

$$\nabla \times (\mathbf{C} \times \nabla f) = \mathbf{C}(\nabla \cdot \nabla f) - \nabla(\mathbf{C} \cdot \nabla f),$$

where $\mathbf{C}$ is a constant vector and f is a scalar function. Apply this identity to the magnetic dipole formula in Problem 2, and reduce the form of the result.

4. A uniform solid cylinder of mass M and total charge Q rotates about its axis. Determine the g-factors relating the magnetic moment and the angular momentum, assuming that the charge is uniformly distributed in two ways: (a) over the cylindrical surface and (b) through the cylindrical volume.

5. A rectangular current loop is oriented with respect to a uniform applied $\mathbf{B}$ field so that the magnetic moment $\boldsymbol{\mu}$ makes an angle θ with $\mathbf{B}$. Show that the loop experiences a torque given by $\boldsymbol{\mu} \times \mathbf{B}$, and use this result to determine the work required in order to increase the angle between $\boldsymbol{\mu}$ and $\mathbf{B}$ by an amount $d\theta$.

6. The Paschen α line in the hydrogen spectrum is due to the transition $n = 4 \rightarrow n = 3$. Sketch the normal Zeeman splitting for the $4p$ and $3d$ energy levels of the hydrogen atom, and compute the magnitude of the splitting in a 2 T applied magnetic field.

Identify the allowed $4p \to 3d$ transitions and determine the shift of the Paschen α wavelength for each case.

7. Consider a classical model of a spinning electron in terms of a uniformly dense rotating sphere of mass m_e and radius r_e. Estimate r_e by identifying the rest energy with the electrostatic energy $e^2/4\pi\varepsilon_0 r_e$. Obtain an expression for the classical angular momentum and equate this quantity to $\hbar/2$. Use the result to solve for the speed v of an equatorial point on the surface of the rotating sphere, and show that v is considerably larger than the speed of light.

8. In a Stern–Gerlach experiment, a hydrogen beam from a 500 K oven traverses a 20 T/m field gradient along a half-meter of beam length. Calculate the separation between the deflected beams upon emergence from the magnetic field. Why is it valid to assume that the hydrogen atoms are in the ground state?

9. A magnetic-resonance experiment is conducted using a beam of hydrogen atoms in the ground state. A constant field B_0 splits the magnetic energy levels of the atoms, and an oscillating field B_ω is tuned to the frequency corresponding to transitions between these levels. Calculate the value of the resonance frequency for a field $B_0 = 2000$ G.

10. A beam of hydrogen atoms in the ground state passes through two coaxial Stern–Gerlach filters in series. The analyzing filter makes an angle β with the polarizing filter, as shown in Figure 8-20. Let the polarizer transmit atoms in the spin-up state with respect to the z axis defined by this first filter. Deduce the probability that any such atom is then detected in the spin-up state with respect to the z' axis defined by the analyzer.

11. An electron is known to have spin down along a certain direction Z. Calculate the probability of finding the electron to have spin up along another direction Z' at a $60°$ angle with Z. What is the probability of finding spin down along Z'? Repeat the calculations for the case of a $30°$ angle between Z' and Z.

12. Consider two of the $\ell = 2$ states of the one-electron atom in which the L_z and S_z quantum numbers are given as $(m_\ell = 2, m_s = -\frac{1}{2})$ and $(m_\ell = 1, m_s = \frac{1}{2})$. What two values of j are possible for these $m_j = \frac{3}{2}$ states? Draw sketches to show how $\mathbf{L}$ and $\mathbf{S}$ may be added in the two given states in order to produce the allowed values of j.

13. The vector-addition formula $\mathbf{J} = \mathbf{L} + \mathbf{S}$ implies that the inequality

$$|\mathbf{J}| \geq |\mathbf{L}| - |\mathbf{S}|$$

must hold for $\ell = 1, 2, 3, \ldots$. Prove (using a suitable graph) that, out of the range of possibilities

$$j = \ell + \tfrac{1}{2},\, \ell - \tfrac{1}{2},\, \ell - \tfrac{3}{2}, \ldots,$$

only the $\ell + \frac{1}{2}$ and $\ell - \frac{1}{2}$ cases satisfy the inequality for every integer $\ell > 0$.

14. Choose a $3D$ state of the one-electron atom and carry out the explicit integration required to obtain the expectation value of $1/r^3$. Compare the answer with the general formula $\langle 1/r^3 \rangle = 2/a^3 n^3 \ell(\ell + 1)(2\ell + 1)$.

15. The $n = 3$ level of the one-electron atom comprises the states $3S_{1/2}, 3P_{1/2}, 3P_{3/2}, 3D_{3/2}$, and $3D_{5/2}$. These states are degenerate if relativistic effects are ignored. Evaluate the spin–orbit energy shifts for the $3P$ and $3D$ states, and calculate the spin–orbit splittings of the $3P$ and $3D$ levels in hydrogen.

16. How many $4F$ states of hydrogen have the same Bohr energy E_4 in the absence of fine structure effects? Evaluate the spin–orbit energy shifts for the two possible values of j at the $4F$ level. Identify the degeneracy that remains among the $4F$ states after this splitting is taken into account.

17. Prove that the energy shifts owing to the spin–orbit interaction and the relativistic correction to the electron kinetic energy combine to give

$$\langle V_{SL} \rangle + \langle K_{rel} \rangle = -\frac{Z^4 \alpha^4}{2n^3} m_e c^2 \left(\frac{2}{2j+1} - \frac{3}{4n} \right).$$

The formulas in the text are derived for $\ell \neq 0$ states; however, the final result is valid for any nL_j state.

18. Identify all the electric dipole transitions that contribute to the Balmer α line in the hydrogen spectrum. Determine the fine structure shifts for each of the $n = 2$ and $n = 3$ states, and calculate all the different Balmer α transition energies.

19. Use the value of the Lamb-shift frequency, quoted in Figure 8-31, to compute the difference in energy between the $2S_{1/2}$ and $2P_{1/2}$ levels in hydrogen. Compare this result with the fine structure splitting between the $2P_{1/2}$ and $2P_{3/2}$ states.

20. The $3D$ level of the one-electron atom comprises ten states whose energies are equal if relativistic effects are ignored and if no magnetic field is applied. Let an external **B** field be introduced, and consider the case in which the strength of **B** greatly exceeds the internal magnetic field responsible for spin–orbit coupling. Determine the energy shift for each magnetic substate at the $3D$ level, and draw an energy level diagram in which the split levels are labeled by the quantum numbers of the states. Is the tenfold degeneracy of these $3D$ states completely broken by this effect?

21. Reconsider the atom of Problem 20 for the situation in which the applied **B** field is weaker than the internal magnetic field. Determine the energy shift of the $3D$ states resulting from the magnetic interaction. Draw an energy level diagram indicating the quantum numbers of the states and the magnitude of the splittings. Is the tenfold degeneracy at the $3D$ level completely removed in this case?

22. Let the atom of Problem 20 have the fictitious g-factor $g_S = 1$ (instead of $g_S = 2$) for the spin magnetic moment, and consider the effect of a weak applied **B** field. Derive an expression for the Zeeman energy shifts at the $3D$ level in terms of the quantum numbers of the $3D$ states. Draw an energy level diagram labeled by these quantum numbers. Is the tenfold degeneracy at the $3D$ level completely removed under these circumstances?

23. Calculate the fine structure splitting at the $4F$ energy level of the hydrogen atom. Determine the magnetic splitting of the $4F$ states when a weak **B** field is applied to the atom, and calculate the value of the magnetic energy shifts in a 5 G field. Draw a diagram of the final $4F$ array of energy levels, identifying the quantum numbers of each state.

N I N E

COMPLEX
ATOMS

*T*he evolution of quantum physics has been guided from the beginning by discoveries in atomic spectroscopy. These observations of atomic structure furnish the background for the Bohr theory of the atom, the concept of electron spin, and the Schrödinger theory of quantum mechanics. The primitive atom with one electron has provided our first opportunity to test the predictions of the quantum theory. The complex atom with many electrons is our next system to analyze for further evidence of quantum structure. We find that the analysis is complicated by the large number of degrees of freedom and that an exact solution is out of the question for any atom with more than one electron. We must approximate the many-particle dynamical problem, and so we obtain a less-quantitative body of predictions as a result. Our main goal is to understand the complex atom in terms of a set of quantized energy states with a suitable assignment of quantum numbers. The spectroscopic consequences are examined for guidance throughout the investigation.

We encounter a new principle when we consider the atom with several electrons as a system of identical quantum particles. The principle identifies a notion of identical-particle symmetry and associates a corresponding quantum property with the half-integral spin of the electron. These ideas constitute the decisive contribution of Pauli to the theory of atoms. We devote a large portion of the chapter to the remarkable implications of this principle.

The complex atoms are organized by the periodic table of the elements. Only a minimum of new quantum physics is needed for an understanding of this organizational scheme. We construct these arguments in the first few sections of the chapter. We then turn to a more detailed picture of the complex atom based on further considerations of quantum mechanics.

9-1 The Central-Field Model

We concentrate on the *neutral* atom and use the atomic number Z to locate the species in the periodic table. The corresponding system consists of Z electrons bound to a nucleus of charge Ze, assumed to be at rest. We must approximate a certain essential

Wolfgang Pauli

aspect of this problem, and so we regard any allowance for nuclear motion as an unwarranted correction.

It is instructive to return to the hydrogen atom for purposes of reference. The Coulomb potential energy and the stationary-state wave functions are given by the well-known expressions

$$V = -\frac{e^2}{4\pi\varepsilon_0 r} \quad \text{and} \quad \Psi_{n\ell m_\ell m_s} = \psi_{n\ell m_\ell m_s}(\mathbf{r})e^{-iE_n t/\hbar}.$$

We recall that the angle independence of V makes possible the separation of radial and angular variables in the eigenfunction $\psi_{n\ell m_\ell m_s}(\mathbf{r})$. The results for the energy eigenvalue E_n are found in Figure 7-2 and are transferred to Figure 9-1. Note that the scale of energy is shifted in the new figure so that the excited levels are displayed relative to the position of the ground state. The figure shows the familiar $n\ell$-orbital labels for the single-electron configurations as well as the spectroscopic notation for the states. A left superscript is added to these designations to indicate the doublet property of the states nL_j with j values equal to either $\ell + \frac{1}{2}$ or $\ell - \frac{1}{2}$. (The s states of the one-electron system are labeled as ^2S doublets, even though $j = \frac{1}{2}$ is the only possible choice. This usage anticipates a development yet to come.) The figure also shows transitions allowed by the electric dipole selection rule $\Delta\ell = \pm 1$ and includes a scale of energy in inverse nanometers along with the usual electron volts. We relate these units with the aid of the Rydberg definition in Equation (3-59),

$$R_\infty = \frac{E_0}{hc},$$

Figure 9-1

Grotrian diagram for the hydrogen atom. The fine structure of the doublet levels is too small to be seen on this scale. Electric dipole transitions are labeled by the emitted wavelengths in nanometers.

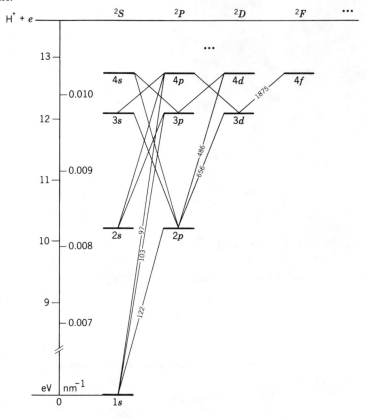

and note that 13.61 eV is equivalent to 0.01097 nm^{-1}. Figures containing this sort of information for the various atoms are called Grotrian diagrams, after the German astronomer W. R. W. Grotrian.

The exact Schrödinger treatment of the atom becomes much more complicated when there are two or more electrons to consider. We can appreciate the problem if we examine the helium atom and allow for all the interactions, as indicated in Figure 9-2. The Coulomb potential energy for the $Z = 2$ system contains two terms corresponding to electron–nucleus attraction and a third term describing electron–electron repulsion:

$$V = -\frac{2e^2}{4\pi\varepsilon_0 r_1} - \frac{2e^2}{4\pi\varepsilon_0 r_2} + \frac{e^2}{4\pi\varepsilon_0|\mathbf{r}_1 - \mathbf{r}_2|}. \tag{9-1}$$

Figure 9-2

Coulomb interactions in the helium atom. The attractive force between electron and nucleus is central, while the repulsive force between electrons is noncentral.

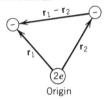

The two-electron eigenfunction $\psi(\mathbf{r}_1, \mathbf{r}_2)$ for energy E obeys the time-independent equation

$$-\frac{\hbar^2}{2m_e}\left(\nabla_1^2 + \nabla_2^2\right)\psi + V\psi = E\psi, \qquad (9\text{-}2)$$

in which the differential operators ∇_1^2 and ∇_2^2 refer to the two sets of coordinate variables (r_1, θ_1, ϕ_1) and (r_2, θ_2, ϕ_2). We observe that the electron–nucleus contributions to V have the familiar central form and do not complicate the solution for ψ. Unfortunately, the electron–electron piece of the potential energy is *noncentral*, and so the method of separation of variables is not applicable to an exact solution. In fact, it is clear at the outset that the third term in Equation (9-1) prevents the separation of the one coordinate from the other.

We are faced with a worsening situation when we consider atoms with larger values of Z. The general Z-electron problem involves an eigenfunction $\psi(\mathbf{r}_1, \ldots, \mathbf{r}_Z)$ depending on Z independent coordinate vectors. The eigenfunction for energy E satisfies the differential equation

$$-\frac{\hbar^2}{2m_e}\left(\nabla_1^2 + \cdots + \nabla_Z^2\right)\psi + V\psi = E\psi, \qquad (9\text{-}3)$$

where the Coulomb potential energy contains central-attractive and noncentral-repulsive terms for all the electrons:

$$
\begin{aligned}
V &= -\frac{Ze^2}{4\pi\varepsilon_0}\left(\frac{1}{r_1} + \cdots + \frac{1}{r_Z}\right) \\
&\quad + \frac{e^2}{4\pi\varepsilon_0}\left(\frac{1}{|\mathbf{r}_1 - \mathbf{r}_2|} + \cdots + \frac{1}{|\mathbf{r}_1 - \mathbf{r}_Z|} + \cdots + \frac{1}{|\mathbf{r}_{Z-1} - \mathbf{r}_Z|}\right) \\
&= -\frac{Ze^2}{4\pi\varepsilon_0}\sum_{i=1}^{Z}\frac{1}{r_i} + \frac{e^2}{4\pi\varepsilon_0}\sum_{i<j=1}^{Z}\frac{1}{|\mathbf{r}_i - \mathbf{r}_j|}.
\end{aligned}
\qquad (9\text{-}4)
$$

The complicating noncentral part of V consists of $Z(Z-1)/2$ distinct contributions from all pairwise combinations of independent coordinate vectors. It is obvious that we must adopt a simplifying model of the atom if we are to make any progress toward a solution of this problem.

Let us suppose for the moment that electron–electron repulsion can be regarded as a secondary correction so that the eigenfunction ψ and the eigenvalue E can be obtained in first approximation from electron–nucleus attraction alone. This procedure decouples the electrons from one another and treats each electron independently, with its own central Coulomb potential energy for every independent coordinate vector $\mathbf{r}_i$. The approach allows the application of separation of variables and produces a solution in which the eigenfunctions for the one-electron atom are used to describe each of the Z electrons. Of course, the approach is valid only if the noncentral Coulomb forces are small enough to be ignored. It is clear that the attractive potential energy between the nucleus and the electrons is the main effect since the binding of the electrons is the net result. Equation (9-4) represents a competitive situation where a given electron makes a central-attractive contribution with an enhancing factor of Z and also makes $Z - 1$ noncentral-repulsive contributions to offset the binding effect. The repulsive terms add up to an appreciable correction whose order of magnitude is not much smaller than that of the overall nuclear attraction.

It is possible to improve the approximation and still retain the independent-particle approach. We observe that each atomic electron is shielded from the nucleus by the other $Z - 1$ electrons, and we argue that every independent electron experiences a *screened nuclear attraction* attributed to an effective nuclear charge smaller than Ze. This screening of the nucleus is expected to vary with the radial distance to the given electron and is supposed to be representable on the average by a spherically symmetric potential energy function. Thus, our improved first approximation to the eigenfunction for the atom is based on an *independent-electron model* in which every electron interacts with a *central field* and has a potential energy $V_c(r)$ with the limiting behavior

$$
V_c(r) \rightarrow
\begin{cases}
- \dfrac{Ze^2}{4\pi\varepsilon_0 r} & \text{as } r \rightarrow 0 \\[3mm]
- \dfrac{e^2}{4\pi\varepsilon_0 r} & \text{as } r \rightarrow \infty.
\end{cases}
$$

We introduce an effective screened nuclear charge as a function of r by writing

$$
V_c(r) = - \frac{e^2}{4\pi\varepsilon_0 r} Z_{\text{eff}}(r), \tag{9-5}
$$

and we observe that Z_{eff} approaches Z near the nucleus and tends to unity at large distance.

A coordinate vector $\mathbf{r}_i$ is defined with spherical coordinates (r_i, θ_i, ϕ_i) for each independent electron. The dynamics of the electron is governed by the central potential energy $V_c(r_i)$, and so an independent state of the particle is determined by solving the differential equation

$$
- \frac{\hbar^2}{2m_e} \nabla_i^2 \psi_i + V_c(r_i)\psi_i = E_i\psi_i \tag{9-6}
$$

to obtain an eigenfunction $\psi_i(\mathbf{r}_i)$ with energy E_i. We let the index i run over all Z

electrons and express the total potential energy of the system by the sum of terms

$$V = \sum_{i=1}^{Z} V_c(r_i). \tag{9-7}$$

We can then prove that Equation (9-3) is satisfied by an atomic eigenfunction of *product* form,

$$\psi(\mathbf{r}_1, \ldots, \mathbf{r}_Z) = \psi_1(\mathbf{r}_1) \cdots \psi_Z(\mathbf{r}_Z), \tag{9-8}$$

whose total energy eigenvalue is given by the sum of electron energies

$$E = \sum_{i=1}^{Z} E_i.$$

The product construction for the eigenfunction of the atom in Equation (9-8) is the basic feature of the independent-particle model.

The potential energy V_c is central, and so the single-particle eigenfunction $\psi_i(\mathbf{r}_i)$ is separable in its radial and angular variables. The solution of Equation (9-6) can therefore be patterned after the wave function found in Equation (8-18). We introduce a single Greek index α as a convenient shorthand to denote the complete set of single-particle quantum numbers $(n \ell m_\ell m_s)$ and represent the desired eigenfunctions for each particle by expressions of the form

$$\psi_\alpha(\mathbf{r}) = R_{n\ell}(r) Y_{\ell m_\ell}(\theta, \phi)(\uparrow \text{ or } \downarrow). \tag{9-9}$$

This solution of Equation (9-6) for a single electron is called a *spin orbital*. The eigenfunction for the entire atom is then constructed as a product of Z such factors in the manner of Equation (9-8). The familiar spherical harmonics appear in ψ_α with indices ℓ and m_ℓ bearing the usual angular momentum interpretation. The assignment of a unique angular momentum to each individual electron is a consequence of the central-field property of the independent-electron model. Equation (9-7) conveys this dynamical property by excluding any explicit coupling between the electrons. The radial function $R_{n\ell}(r)$ obeys a differential equation like the one for the hydrogen atom, except that the purely Coulombic potential energy is replaced by the central-field function $V_c(r)$:

$$-\frac{\hbar^2}{2m_e r} \frac{d^2}{dr^2} r R_{n\ell} + \left[V_c(r) + \frac{\hbar^2 \ell(\ell+1)}{2m_e r^2} \right] R_{n\ell} = E_{n\ell} R_{n\ell}. \tag{9-10}$$

Thus, the eigenfunctions of the independent-electron model exhibit the familiar correspondence between radial functions $R_{n\ell}(r)$ and energy eigenvalues $E_{n\ell}$. These quantities cannot depend on the quantum number m_ℓ according to the usual spherical-symmetry arguments, and cannot depend on the spin quantum number m_s either since spin effects are not included in the potential energy of the atom. This procedure is expected to result in single-electron functions and energies *different* from the analogous solutions for the hydrogen atom because of the non-Coulombic nature of V_c.

We are concerned with a determination of the eigenfunction ψ based on a model for the single-electron potential energy $V_c(r)$. A successful approach to the solution of this problem exists in the form of a method, originally proposed by D. R. Hartree and eventually improved by V. Fock and J. C. Slater. Their solution treats the average behavior of each atomic electron and also incorporates the identical-particle symmetry of the multielectron system. The whole procedure is supposed to be self-consistent, since the central potential energy $V_c(r)$ for any given electron depends on the eigenfunctions determined from $V_c(r)$ for every other electron. Many useful predictions can be made by this approach to the structure of complex atoms.

Example

The ground state of the helium atom has energy -79.0 eV (on an energy level diagram where the zero level refers to a He^{++} ion plus two electrons, all at rest with infinite separation). Let us make a primitive attempt to understand this number. The crudest approximation to the He eigenfunction is found from Equation (9-2) by retaining only the first two terms from the potential energy in Equation (9-1). The resulting ground state is described by the product of two hydrogen-like $n = 1$ eigenfunctions,

$$\psi(\mathbf{r}_1, \mathbf{r}_2) = \psi_{100}(\mathbf{r}_1)\psi_{100}(\mathbf{r}_2) = \frac{e^{-(r_1 + r_2)/a}}{\pi a^3},$$

in which the spins of the two electrons are left unspecified and the radius parameter is given by $a = a_0/2$ for $Z = 2$. The ground-state energy in first approximation is the sum of the Bohr energies of the electrons:

$$E = 2E_1 = -2Z^2 E_0 = -8(13.6 \text{ eV}) = -108.8 \text{ eV}.$$

We should not expect this result to be close to the correct answer since we have ignored the repulsion of the two electrons. Let us estimate the repulsive contribution by making a gross classical calculation. If we replace the electron separation $|\mathbf{r}_1 - \mathbf{r}_2|$ in the neglected term of Equation (9-1) by the constant value $2a = a_0$, we find

$$\frac{e^2}{4\pi\varepsilon_0 a_0} = \alpha\hbar c \frac{\alpha m_e c}{\hbar} = \alpha^2 m_e c^2 = 2E_0 = 27.2 \text{ eV}.$$

The sum of the two numbers -108.8 and 27.2 yields -81.6 eV as an estimate for the ground-state energy. This computation is inadequate for a quantum system and serves only to suggest the relative sizes of the two competing effects.

9-2   The Exclusion Principle

Electron spin has proved to be essential for the understanding of multiplet structure and Zeeman splitting in the one-electron atom. Complex atoms also reveal the influence of the spin quantum number, particularly in the periodic table of the elements. This revelation of a special role for electron spin in many-electron atoms is an indication of a new kind of symmetry in systems of identical particles. The deduction of the spin degree of freedom originates in these observations, and the exclusion principle provides the underlying deductive framework.

The exclusion principle is a statement, due to Pauli, about the assignment of electron quantum numbers in the complex atom:

> No more than one electron is allowed to occupy a given quantum state specified by the complete set of single-particle quantum numbers $(n \ell m_\ell m_s)$.

This prohibitive rule acts as a constraint to exclude the existence of certain states in the many-electron system. The constraint is an additional quantum condition imposed on solutions of the Schrödinger equation for the many-particle problem, in the special circumstance where the particles are identical. When the principle is applied to the construction of the ground states of atoms, the result is the organization of the periodic table.

Regularities among the elements are presented in the periodic table according to values of the atomic number Z. The periodicity of such characteristics as chemical valence and binding energy is evidence for a periodicity in the structure of the atoms. The cycling of properties with increasing Z is explained by distributing the electrons in a succession of "energy shells" and requiring a given shell to be fully occupied for a certain number of electrons. These closed shells occur in the ground states of the noble-gas atoms helium, neon,... at $Z = 2, 10, \ldots$. The occupation of succeeding shells then continues with the ground states of the alkali atoms lithium, sodium,... at $Z = 3, 11, \ldots$.

The theory of atomic shell structure is supported by observations of the optical spectra of the elements. These spectral lines in the visible range of wavelengths are generally associated with the excitations of loosely bound electrons in the outer shells. The alkali elements are especially interesting because their optical spectra contain doublets of lines like those found in the spectrum of hydrogen. This behavior is explained by attributing the radiative transitions to a single optical electron, whose initial and final energy shells lie outside an inert core of closed shells.

Optical spectra are not the only sources of information about the shell structure of atoms. The excitations of tightly bound electrons in the inner shells are also detectable as higher-energy x-ray transitions. Characteristic x-ray lines in the various series $K, L, \ldots$ are interpreted in terms of electron transitions to inner atomic shells denoted by the same letters $K, L, \ldots$. Optical and x-ray emission processes are represented schematically as complementary forms of outer- and inner-electron activity in Figure 9-3.

A link exists between spectra and shell structure on the one hand, and the spin of the electron and the exclusion principle on the other. The following brief history describes the discovery of this connection.

A shell picture of the atom has been in force since the time of the old quantum theory. In fact, the exclusion principle emerged along with spin to explain the picture *before* the coming of quantum mechanics. The energy shells were originally defined to be the quantized electron orbits of Bohr's model, adapted to include three empirical quantum numbers and extended to apply to atoms with many electrons. The electrons were assumed to occupy the various energy shells on the basis of assignments of the three quantum numbers. Moseley used this orbit model to analyze the characteristic x rays of many elements and demonstrated the connection between the atomic number and the nuclear charge. No theoretical argument was put forward, however, to account for the idea of shell closure. The model failed especially to say why all the

Figure 9-3

Transitions of outer- and inner-shell electrons in optical and x-ray spectra. The excitation of the atom is by electron collision in each case.

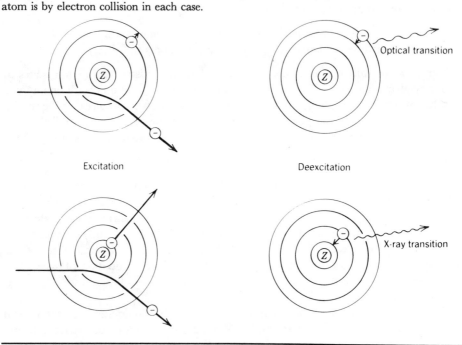

electrons in the ground state of a given atom were not simply assigned to the innermost shell of minimum energy.

Pauli was an ardent advocate of the notion of closed shells. He argued that these distributions of electrons did not participate in the optical properties of the atom because the closed shells did not contribute to the total angular momentum and magnetic moment. His argument drew only on the Bohr model and therefore did not address the basic question of shell closure.

Another problem with Bohr's model was its failure to predict all the lines observed in the x-ray spectra of the atoms. This defect was remedied by E. C. Stoner, who suggested on empirical grounds that a fully occupied shell should contain *twice* the number of electrons deduced on the basis of Bohr's three quantum numbers. The suggestion influenced Pauli to propose that the doublet structure of the alkali spectra and the anomalous Zeeman effect could be explained if a nonclassical two-valued degree of freedom was associated with the electron. The new variable turned out to be the spin quantum number and the new idea proved to be the inspiration for electron spin.

Pauli went on to observe that *four* quantum numbers were needed to specify a state for each electron in the atom. He then showed that the shell structure of the atom followed as a natural consequence if *no more than one* electron occupied any given single-particle state in the many-electron system. His proposition of the exclusion principle resolved the problem of shell closure and explained the organization of the periodic table. The original statement of the principle was put forward in 1925. Pauli's idea was later recast in more general quantum mechanical language and became one of the cornerstones of quantum mechanics.

9-3 The Ground States of Atoms and the Periodic Table

The ground state of an atom with atomic number Z is the minimum-energy configuration for the bound system of Z electrons. We proceed to construct such a state in the central-field model by letting each electron have a set of independent states denoted by the four quantum numbers $(n\ell m_\ell m_s)$. We can obtain these single-particle states from Equation (9-10) by solving for the set of energy eigenvalues $E_{n\ell}$. The solution provides an ordered sequence of energy levels for any electron in the Z-electron system.

The central field in the complex atom is not a pure Coulomb field. The single-particle energies should therefore depend on *both* quantum numbers n and ℓ, unlike the situation for the hydrogen atom. These energies must not depend on m_ℓ and m_s because the potential energy of each electron is spherically symmetric and spin independent. Hence, there are $2(2\ell + 1)$ degenerate states with the same energy $E_{n\ell}$, corresponding to the two possible values of m_s and the $2\ell + 1$ possible values of m_ℓ. This group of $2(2\ell + 1)$ spin-orbital states constitutes an atomic *subshell* at energy $E_{n\ell}$ for each pair of quantum numbers n and ℓ.

We distribute electrons into subshells and define an *electron configuration* in terms of the resulting inventory of $n\ell$ assignments for all Z electrons. We must observe the exclusion principle and assign electrons to spin-orbital states so that no two electrons have the same values of the four quantum numbers $(n\ell m_\ell m_s)$. This stipulation limits a given subshell to a maximum occupancy of $2(2\ell + 1)$ electrons with the same energy $E_{n\ell}$. The minimum-energy configuration of the atom is then found by allocating the Z electrons to the subshells of least energy, subject to this constraint.

We have noted that the energy levels of a single electron depend on ℓ as well as n and that the degeneracy of states for fixed n and different ℓ is a special property of the Coulomb potential energy. A principal quantum number is identified for the electron in the central-field model so that a close parallel is maintained with the solutions in the case of hydrogen. We select a value of ℓ and consider solutions of Equation (9-10) with potential energy $V_\ell(r)$. The solutions comprise a sequence of radial functions whose nodes increase in number with increasing values of the energy level. We then *define* the principal quantum number n as a label for $R_{n\ell}(r)$ and $E_{n\ell}$ by taking $n - \ell - 1$ as the number of nodes of $R_{n\ell}(r)$ in the interval $0 < r < \infty$. This definition of n agrees with the radial-node properties of the hydrogen solutions found in Section 7-2.

Each value of n determines an electron *shell* for a given atom. A shell consists of n subshells labeled by n and ℓ, as ℓ ranges from 0 to $n - 1$, and every $n\ell$ subshell contains $2(2\ell + 1)$ spin-orbital states as observed above. Thus, the single-electron energies in the central field depend on ℓ and n according to a scheme in which the principal quantum number retains the same meaning for all atoms. The radial functions and probability distributions for a single electron can therefore be given the same interpretation from one atom to another, in the manner of Figures 7-1 and 7-4.

We wish to know the ordering of the energies $E_{n\ell}$ so that we can assign electrons to subshells of increasing energy. The determination of these single-electron energies is a technical problem that must be solved anew for every choice of Z. A representative sample of the energy levels for a particular Z might resemble the diagram shown in Figure 9-4. If we compare this picture with the analogous diagram for hydrogen in Figures 7-2 and 8-13, we can observe certain points of similar and dissimilar behavior. The energy $E_{n\ell}$ increases (becomes less negative) with n for fixed ℓ, reflecting the usual relation between the energy eigenvalue and the number of nodes of the radial function. The energy also increases with ℓ for n fixed in a given shell. We can

Figure 9-4

Energy levels and subshells for a single electron in a non-Coulombic central field. The number of spin-orbital states is given in parentheses for each of the $n\ell$ subshells.

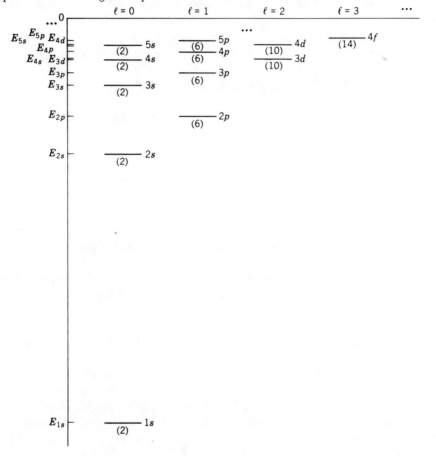

understand this new feature by considering the following arguments. The smaller values of ℓ give larger probabilities for finding the electron near $r = 0$, as a consequence of Equation (6-51). In these states the electron is less effectively screened from the nucleus and therefore experiences a greater share of the nuclear attraction. The larger values of ℓ refer to states in which there are larger probabilities for finding the electron farther from the nucleus where the nuclear charge is more fully screened. In these states the effective nuclear charge is more like that seen by the electron in the hydrogen atom. Thus, we observe a distribution in energy for subshells in a given shell where the subshell of maximum ℓ lies near the analogous level in hydrogen while the subshell of minimum ℓ occurs at a substantially lower level.

We have remarked that the central-field model must be reconstructed for every atomic number Z. The self-consistency of the model complicates the procedure by which $V_c(r)$, $R_{n\ell}(r)$, and $E_{n\ell}$ are determined. We are supposed to make a starting hypothesis regarding $Z_{\mathrm{eff}}(r)$ in Equation (9-5) and use this assumption for $V_c(r)$ to find eigensolutions $R_{n\ell}(r)$ and $E_{n\ell}$ from Equation (9-10). We must then assemble the state of the given atom by invoking the exclusion principle and assigning the collection of Z electrons to the single-particle energy levels. The resulting probability distribution for the system of particles determines an electron charge density from which a classical electrostatic potential can be computed. The potential determines a potential energy for each electron, and the average of this result over the state of the atom generates a predicted central potential energy $V_c(r)$, to be compared with the hypothesized quantity at the beginning of the cycle of calculations. Self-consistency is achieved when satisfactory agreement is obtained between the input and output versions of $V_c(r)$. This procedure summarizes the technique developed by Hartree. The subsequent modification of the method by Fock and Slater incorporates the properties of identical-particle symmetry to produce the best possible eigenfunctions for the atom.

Figure 9-5 contains a few graphs of the single-electron energy $E_{n\ell}$ versus Z, taken from calculations based on the self-consistent method. We can use this sort of information to *order* the subshell energies and construct an energy level diagram like the one in Figure 9-4. The exclusion principle allows no more than $2(2\ell + 1)$ electrons in each subshell, as noted by the numbers in parentheses at each of the indicated energy levels. The ordering of subshells is difficult to ascertain whenever the energies at two or more levels are nearly equal. Figures 9-4 and 9-5 suggest, for example, that the $3d$ and $4s$ subshells are sufficiently close together to compete in the allocation of electrons to the lowest unoccupied states. The lesser degree of screening in the case of the $4s$ electron implies a greater degree of attraction in that state, and so the competition favors $4s$ over $3d$ for values of Z within a certain range. A similar conclusion holds for the $5s$ and $4d$ levels in Figure 9-4, and for other groups of competing subshells at higher energy. Let us summarize the solution of the ordering problem by listing the subshells with increasing energy as follows, noting by parentheses the subshells of nearly equal energy:

$$1s \quad 2s \quad 2p \quad 3s \quad 3p \quad (4s \; 3d) \quad 4p \quad (5s \; 4d) \quad 5p \quad (6s \; 4f \; 5d) \quad 6p \; \ldots$$

We can use this list to build up the ground-state configuration for any atom in the periodic table.

The lowest $1s$ subshell is occupied by as many as two electrons to account for the elements hydrogen and helium ($Z = 1$ and 2). We describe the electron configurations of these atoms by the following notation:

$$\begin{array}{cc} \text{H} & \text{He} \\ 1s & 1s^2 \end{array}$$

A right superscript is introduced to denote the occupation number whenever a particular subshell is occupied by more than one electron. The $1s^2$ configuration of the He atom defines a fully occupied subshell and closes the K shell. It is convenient to visualize this system in terms of two $1s$ electrons with spins up and down.

The $2s$ subshell is next to be filled by as many as two more electrons. The corresponding two elements in the periodic table are lithium and beryllium ($Z = 3$ and 4), with the following configurations:

$$\begin{array}{ccccc} \text{Li} & \text{Be} & & \text{Li} & \text{Be} \\ 1s^2 2s^2 & 1s^2 2s^2 & \text{or} & [\text{He}]\, 2s & \ldots 2s^2 \end{array}$$

Figure 9-5

One-electron energy $E_{n\ell}$ versus atomic number Z for the lower subshells. The energies are expressed on a logarithmic scale in terms of the hydrogen-atom ground-state energy. The data are taken from tables compiled by F. Herman and S. Skillman.

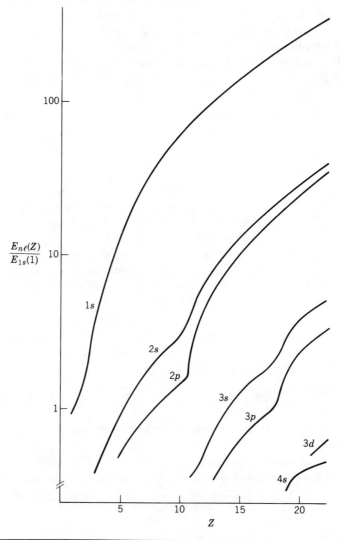

Note that the symbol [He] refers to an inner helium-like closed shell in the structure of each atom. It should also be noted that the Be configuration constitutes a filled subshell but not a closed shell. The $2p$ subshell follows with a maximum occupancy of six electrons. We fill these states by adding one electron at a time to generate the configurations of the next six atoms from boron to neon ($Z = 5$ to 10):

B	C	N	O	F	Ne
[He]$2s^2 2p$	$\ldots 2s^2 2p^2$	$\ldots 2s^2 2p^3$	$\ldots 2s^2 2p^4$	$\ldots 2s^2 2p^5$	$\ldots 2s^2 2p^6$

Each of these structures contains an inner closed [He] shell and an inner filled $2s^2$ subshell. The Ne configuration at $Z = 10$ closes the L shell, and the eight L-shell atoms from $Z = 3$ to 10 make up the second row in the periodic table.

The $3s$ and $3p$ subshells are filled in similar fashion by two and six additional electrons. This construction produces the ground states of the eight atoms from sodium to argon ($Z = 11$ to 18), corresponding to the eight elements in the third row of the periodic table. The configurations of these atoms are listed in Table 9-1, along with other pertinent information for all the ground states through atomic number 54. Note that the second and third fully occupied configurations at $Z = 10$ and 18 are abbreviated in the table by the symbols [Ne] and [Ar]. We use these abbreviations when we fill the succeeding subshells beginning at $Z = 11$ and 19.

The Ar atom closes the third family of subshells, even though the $3d$ subshell remains unoccupied. This type of closure is akin to the closing of the K and L shells by the He and Ne atoms. An explanation is suggested in Figure 9-5 by the existence of a substantial energy gap between the $3p$ level and the $(4s\ 3d)$ pair of levels. This gap is just like the two previous gaps at the $1s$ and $2p$ levels where the K- and L-shell closures occur. The $4s$ subshell is actually the next to be filled after $3p$ because the screening effect of the [Ar] core tends to favor $4s$ over $3d$ for atoms in this region of the periodic table. We see from Table 9-1 that the fourth family of subshells opens with the beginning of the $4s$ subshell by the K atom at $Z = 19$ and closes with the filling of the $4p$ subshell by the Kr atom at $Z = 36$. The $3d$ subshell is occupied in ten steps along the way, at *interior* values of Z in the fourth row of the periodic table. Notice that this resolution of the $(4s\ 3d)$ ambiguity undergoes a reversal at $Z = 24$ and again at $Z = 29$. In these special cases the Cr and Cu ground states prefer to keep only one $4s$ electron and add two more $3d$ electrons instead.

The same pattern of subshell formation is reenacted in the fifth row of the periodic table. The $5s$ subshell is opened by Rb at $Z = 37$, and the $5p$ subshell is closed by Xe at $Z = 54$. Occupation of the $4d$ subshell occurs at interior values of Z, and several reversals in the resolution of the $(5s\ 4d)$ ambiguity take place along the way.

The periodic table unifies these observations about the filling of subshells. Figure 9-6 shows how the shell structure of the ground-state atoms is organized by rows in the table and how the various rows are subdivided according to the occupation of the successive subshells. Closures occur at the ends of rows, at the noble-gas locations

$$Z = 2, 10, 18, 36, 54, \ldots .$$

These special atoms are characterized by the closing of a p subshell (except in the case of He), followed invariably by the opening of an s subshell in a shell of higher n. Each successive row in the table refers to states of electrons assigned to lie *outside closed subshells*. These assignments of n and ℓ determine the *outer* electrons of each atom because the corresponding spin orbitals give weight to the *larger* distances between electron and nucleus. Columns of the periodic table contain different atomic species with the same numbers of electrons outside closed subshells. These atoms are expected to have similar physical properties since their electron configurations have similar patterns of quantum numbers. The alkali elements in the first column are prime examples. The common properties of these atoms are clearly demonstrated in their hydrogen-like optical spectra.

Table 9-1 includes a list of experimental values of the ionization energy for each neutral atom Z. This quantity is defined as the negative of the ground-state energy on an energy level diagram in which the zero level denotes an ionized system consisting of

Table 9-1   Ground-State Properties of Atoms (Z = 1 to 54)

Z Atom	Element	Configuration	Ionization Energy (eV)	Z Atom	Element	Configuration	Ionization Energy (eV)
1 H	Hydrogen	$1s$	13.60	28 Ni	Nickel	$\ldots 4s^2 3d^8$	7.64
2 He	Helium	$1s^2$	24.59	29 Cu	Copper	$\ldots 4s\, 3d^{10}$	7.73
3 Li	Lithium	$[\text{He}]2s$	5.39	30 Zn	Zinc	$\ldots 4s^2 3d^{10}$	9.39
4 Be	Beryllium	$\ldots 2s^2$	9.32	31 Ga	Gallium	$\ldots 4s^2 3d^{10} 4p$	6.00
5 B	Boron	$\ldots 2s^2 2p$	8.30	32 Ge	Germanium	$\ldots 4s^2 3d^{10} 4p^2$	7.90
6 C	Carbon	$\ldots 2s^2 2p^2$	11.26	33 As	Arsenic	$\ldots 4s^2 3d^{10} 4p^3$	9.81
7 N	Nitrogen	$\ldots 2s^2 2p^3$	14.53	34 Se	Selenium	$\ldots 4s^2 3d^{10} 4p^4$	9.75
8 O	Oxygen	$\ldots 2s^2 2p^4$	13.62	35 Br	Bromine	$\ldots 4s^2 3d^{10} 4p^5$	11.81
9 F	Fluorine	$\ldots 2s^2 2p^5$	17.42	36 Kr	Krypton	$\ldots 4s^2 3d^{10} 4p^6$	14.00
10 Ne	Neon	$\ldots 2s^2 2p^6$	21.56	37 Rb	Rubidium	$[\text{Kr}]5s$	4.18
11 Na	Sodium	$[\text{Ne}]3s$	5.14	38 Sr	Strontium	$\ldots 5s^2$	5.70
12 Mg	Magnesium	$\ldots 3s^2$	7.65	39 Y	Yttrium	$\ldots 5s^2 4d$	6.38
13 Al	Aluminum	$\ldots 3s^2 3p$	5.99	40 Zr	Zirconium	$\ldots 5s^2 4d^2$	6.84
14 Si	Silicon	$\ldots 3s^2 3p^2$	8.15	41 Nb	Niobium	$\ldots 5s\, 4d^4$	6.88
15 P	Phosphorus	$\ldots 3s^2 3p^3$	10.49	42 Mo	Molybdenum	$\ldots 5s\, 4d^5$	7.10
16 S	Sulfur	$\ldots 3s^2 3p^4$	10.36	43 Tc	Technetium	$\ldots 5s^2 4d^5$	7.28
17 Cl	Chlorine	$\ldots 3s^2 3p^5$	12.97	44 Ru	Ruthenium	$\ldots 5s\, 4d^7$	7.37
18 Ar	Argon	$\ldots 3s^2 3p^6$	15.76	45 Rh	Rhodium	$\ldots 5s\, 4d^8$	7.46
19 K	Potassium	$[\text{Ar}]4s$	4.34	46 Pd	Palladium	$\ldots 4d^{10}$	8.34
20 Ca	Calcium	$\ldots 4s^2$	6.11	47 Ag	Silver	$\ldots 5s\, 4d^{10}$	7.58
21 Sc	Scandium	$\ldots 4s^2 3d$	6.54	48 Cd	Cadmium	$\ldots 5s^2 4d^{10}$	8.99
22 Ti	Titanium	$\ldots 4s^2 3d^2$	6.82	49 In	Indium	$\ldots 5s^2 4d^{10} 5p$	5.79
23 V	Vanadium	$\ldots 4s^2 3d^3$	6.74	50 Sn	Tin	$\ldots 5s^2 4d^{10} 5p^2$	7.34
24 Cr	Chromium	$\ldots 4s\, 3d^5$	6.77	51 Sb	Antimony	$\ldots 5s^2 4d^{10} 5p^3$	8.64
25 Mn	Manganese	$\ldots 4s^2 3d^5$	7.44	52 Te	Tellurium	$\ldots 5s^2 4d^{10} 5p^4$	9.01
26 Fe	Iron	$\ldots 4s^2 3d^6$	7.87	53 I	Iodine	$\ldots 5s^2 4d^{10} 5p^5$	10.45
27 Co	Cobalt	$\ldots 4s^2 3d^7$	7.86	54 Xe	Xenon	$\ldots 5s^2 4d^{10} 5p^6$	12.13
				$\cdots$			

Figure 9-6

Periodic table of the elements from $Z = 1$ to 54. Atoms in the same column have similar physical and chemical properties. The rows are divided into blocks of atoms whose outer electrons occupy the indicated subshells.

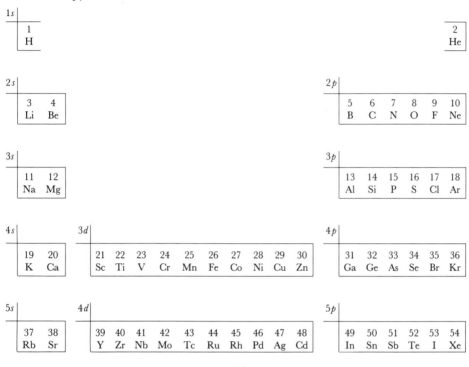

an electron at rest far from a ground-state Z^+ ion also at rest. In practical terms, the ionization energy represents the minimum photon energy that the atom must absorb in order to free a bound electron. We plot the tabulated energies against the atomic number in Figure 9-7 so that we can display the evidence of periodic shell structure. The largest values of the ionization energy are observed for the noble gases, while the smallest values are found for the alkalis at the opening of each succeeding shell. We can understand this behavior if we recall how the screening effect operates. Electrons in the same subshell tend not to be screened by each other since their spatial distributions are equivalent. Consequently, their binding is governed by an effective nuclear charge that grows with Z to a maximum at the noble-gas closure. The next electron must be added with a larger n assignment, implying a larger average distance from the nucleus. The result is an abrupt rise in screening accompanied by an abrupt fall in nuclear attraction for the added electron, as seen in the figure. Hence, we identify a closure to be the most tightly bound structure, and we recognize a single electron beyond a closure to be the least tightly bound constituent.

Electrons in unfilled outer subshells are called *valence* electrons. These weakly bound particles control the chemical properties of the atom and determine the interaction of one atom with another. The noble gases have no unfilled subshells in the ground state, and so these elements tend to be chemically inert. On the other hand, the alkalis are chemically active because their ground-state configurations contain a

Figure 9-7

Ionization energy versus atomic number up to $Z = 54$. Extreme values of the ionization energy occur as indicated for the noble gases and the alkali elements.

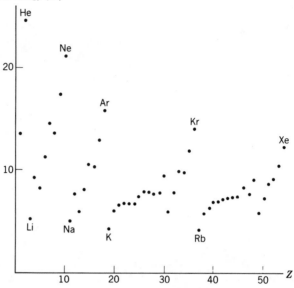

single weakly bound valence electron. These atoms are easily ionized and readily give up their lone valence electron in the formation of molecules. The chemical characteristics of the elements recur across the periodic table in a columnwise pattern of regularities, as discovered originally in the design of Mendeleev's table. This pattern is attributable to the recurrence of quantum properties in the assignment of electrons to the succession of atomic subshells. The comprehension of the whole scheme of regularities is only one of the remarkable consequences of Pauli's exclusion principle.

Example

It would be grossly inaccurate to represent the single outer electron of an alkali atom by an ordinary hydrogen eigenfunction, even though the one-electron behavior of the atom is observed to be hydrogen-like. To illustrate, let us assume maximal screening and set $Z_{\text{eff}} = 1$ for the outer electron in each alkali atom. The ionization energy is then given by E_0/n^2, the Bohr expression for hydrogen, with principal quantum numbers $n = 2$ for Li, $n = 3$ for Na, $n = 4$ for K, and so on. The resulting energies are

$$\frac{13.6 \text{ eV}}{n^2} = 3.40 \text{ eV}, 1.51 \text{ eV}, \text{ and } 0.85 \text{ eV} \quad \text{for } n = 2, 3, \text{ and } 4.$$

These predictions are not close to the actual values 5.39, 5.14, and 4.34 eV listed for Li, Na, and K in Table 9-1. It is obvious that a better model is needed to include the screening effect.

Example

Oxygen, sulfur, and selenium occur in the same column of the periodic table at $Z = 8$, 16, and 34. The ground-state configurations describe an unfilled p subshell with two electron vacancies in each case:

$$
\begin{array}{ccc}
\text{O} & \text{S} & \text{Se} \\
[\text{He}]2s^2 2p^4 & [\text{Ne}]3s^2 3p^4 & [\text{Ar}]4s^2 3d^{10}4p^4
\end{array}
$$

These elements can fill their vacancies by accepting electrons from two hydrogen atoms to form the molecules H_2O, H_2S, and H_2Se. The two vacant sites in O, S, and Se have angular probability distributions characteristic of an $\ell = 1$ subshell. We can interpret the three m_ℓ assignments for $\ell = 1$ in terms of large distributions of probability along three perpendicular directions. Therefore, we expect to find right-angle shapes when we attach two H atoms to O, S, and Se to make the three triatomic molecules. This expectation is borne out since the hydrogen bonds are at angles of $105°$, $93°$, and $90°$ in H_2O, H_2S, and H_2Se. (The angle tends to exceed $90°$ because of Coulomb repulsion between the two hydrogens and approaches $90°$ in the largest molecule where the repulsion is weakest.)

9-4 X-Ray Spectra

The periodic table organizes the states of electrons outside closed subshells as the basis for the low-energy regularities of atoms. The underlying shell theory also governs the *inner* shells of the atom where the electrons engage in higher-energy processes such as the emission of x rays. This inner-shell behavior can be examined in a given atom by analyzing an x-ray spectrum like the one shown in Figure 9-8. We have discussed the production of x rays in Section 2-6, and we have taken a preliminary look at characteristic x rays as a source of inner-electron information in Section 3-7. We now return to the study of x-ray spectra with a proper quantum theory at our disposal and find that the independent-electron model provides an especially suitable approach.

Let us reconsider our previous picture of x-ray emission in Figure 9-3 and visualize the excitation and deexcitation of the system from the viewpoint presented in Figure 9-9. The revised picture describes these processes in terms of the single-electron energy levels in the central-field model. We represent the collisional excitation of the atom by the creation of a vacancy, or *hole*, in one of the fully occupied inner subshells. The

Figure 9-8

X-ray lines in the K and L series of tungsten. The heights of the lines are indicative of the observed intensities. The $K\alpha_1$ line at 0.0209100 nm is used to define an x-ray wavelength standard.

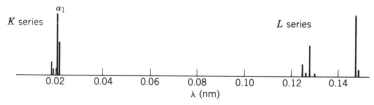

Figure 9-9

Excitation and deexcitation processes associated with x-ray emission.

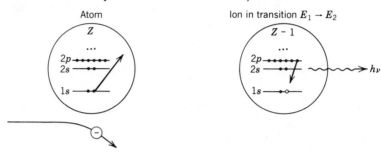

result is the formation of a highly excited ionic state with initial energy E_1. We then represent the radiative deexcitation of the system by letting an electron from a higher subshell fill the hole and emit an x-ray photon. The emitted photon has energy

$$hv = E_1 - E_2, \tag{9-11}$$

where E_2 is the energy of the excited final state in which the vacancy appears in the higher subshell. Observe from the figure that the x-ray transition occurs in the *ion* containing $Z - 1$ electrons. Note especially that a vacancy in a higher-energy subshell corresponds to a *lower* energy state of the ion, so that a hole stands for an *absence* of energy, as well as charge, in the overall electron system.

Let us also consider the related quantum phenomenon of x-ray absorption. This process is an example of the familiar photoelectric effect, where the absorption of an x-ray photon excites the atom above its ionization level and ejects a bound electron. A quantum mechanical probability can be introduced to describe the photon–atom interaction, and an absorption cross section can be defined to account for the behavior of a beam of x rays incident on the atoms in a sample of matter. We measure absorption in the laboratory by observing the attenuation of an x-ray beam in its passage through a thickness of material. The fractional decrease in intensity $-dI/I$ is related to the element of thickness dx by the proportionality

$$-\frac{dI}{I} = \mu_x \, dx,$$

where the constant μ_x defines the absorption coefficient of the material. This expression is easily integrated to give the intensity as a function of distance x through the sample, starting with incident intensity I_0:

$$\int_{I_0}^{I} \frac{dI}{I} = -\int_{0}^{x} \mu_x \, dx \quad \Rightarrow \quad \ln\frac{I}{I_0} = -\mu_x x \quad \Rightarrow \quad I = I_0 e^{-\mu_x x}.$$

The absorption coefficient varies with the material and depends on the wavelength of the x rays. We can use measurements of the attenuation to determine this dependence, and we can then infer the related behavior of the absorption cross section for the given element.

Figure 9-10

K and L absorption edges of lead. The wavelength thresholds occur where the x-ray photon energy becomes insufficient to eject a K- or L-shell electron. Emission lines of lead in the K and L series are also shown.

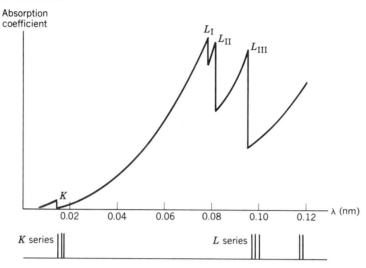

Figure 9-10 shows a typical graph of μ_x as a function of the wavelength λ. We observe zero absorption in the limit $\lambda \to 0$. This observation tells us that the absorbing medium is transparent to x rays when the beam energy is very large. We then observe a steady growth in absorption as the photon energy decreases from large values and as λ increases from zero, until we reach a sharp value of λ where the medium suddenly becomes transparent again. This feature of the graph is called an *absorption edge*, the first of several to appear with increasing λ in the figure. The indicated K absorption edge occurs at wavelength λ_K, where the photon energy is the minimum needed to ionize the atom and leave a vacancy in the K shell. When λ becomes larger than λ_K, the x-ray photon energy becomes too small to free a K-shell electron but remains large enough to eject an electron from an L (or higher) shell. We again observe a steady growth in absorption as the wavelength continues to increase until we reach one of the indicated L absorption edges. The various absorption thresholds are tabulated along with the characteristic x-ray emission lines. Both features provide a signature of the particular atom, and both give an indication of the energy levels of the system. We include the emission lines of the K and L series in the figure so that we can note the positions of these spectral lines relative to the absorption edges.

X-ray absorption can be interpreted with the aid of our representation of the atom as a collection of occupied subshells. Figure 9-11 describes the process in these terms at two different wavelength thresholds. The wavelengths λ_1 and λ_2 correspond to photons whose energies just suffice to eject an electron with no kinetic energy from the particular subshells shown in the figure. We leave the excited ion in states of energy E_1 and E_2 in the two cases, and we note the relation

$$E_1 > E_2 \quad \text{for} \quad \lambda_1 < \lambda_2.$$

Figure 9-11

X-ray absorption processes at two different absorption edges.

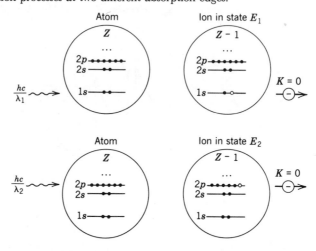

The figure tells us that the wavelengths and energies obey the formulas

$$\frac{hc}{\lambda_1} = E_1 - E_{\text{atom}} \quad \text{and} \quad \frac{hc}{\lambda_2} = E_2 - E_{\text{atom}}, \tag{9-12}$$

where E_{atom} denotes the energy of the initial atom. If we compare Figures 9-9 and 9-11, we see that the two ionic energies are the same as the energies E_1 and E_2 in Equation (9-11). If we combine Equations (9-11) and (9-12), we establish the following connection between x-ray absorption and x-ray emission:

$$h\nu = \frac{hc}{\lambda_1} - \frac{hc}{\lambda_2}. \tag{9-13}$$

This formula relates the wavelengths for two absorption edges to the frequency for a certain emission line in the spectrum of a given element. The equality implies an inequality of the form

$$\frac{hc}{\lambda} < \frac{hc}{\lambda_1} \quad \text{or} \quad \lambda > \lambda_1$$

for an emitted x ray of wavelength $\lambda = c/\nu$. This observation explains why the lines in a given series have wavelengths *above* the corresponding absorption edge, as indicated in Figure 9-10.

The absorption and emission of x rays reveal a new problem that requires an amendment to the central-field model. Figure 9-10 shows the existence of *three L* absorption edges instead of the two expected for the $2s$ and $2p$ subshells in the L shell. This behavior is a clear indication that the $n\ell$ assignments of the model are not adequate to describe the energy levels of the independent electrons. The effect is reminiscent of the multiplet structure of the one-electron atom and is attributable to a

Figure 9-12

Single-electron subshells including the effect of spin–orbit coupling.

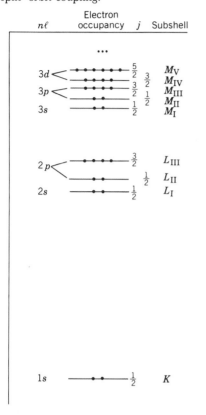

Figure 9-13

X-ray emission associated with the transition $K \rightarrow L_{\mathrm{III}}$. An electron in the L_{III} subshell fills a vacancy in the K shell while the hole makes a transition from K to L_{III}. The standard $K\alpha_1$ line of tungsten corresponds to this transition.

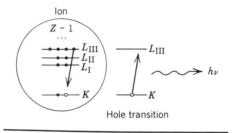

spin–orbit coupling experienced by each electron in addition to its interaction with the central field. We discuss some of the details of this coupling at the end of the section.

Figure 9-12 shows how these considerations alter the independent-electron picture of the atom by the introduction of the quantum number j. We let each energy level with subshell quantum numbers n and ℓ be *split* for $\ell \neq 0$ in the manner of Figure 8-28, and we recall the familiar quantum numbers $(n\ell jm_j)$ to specify the new single-electron states. We also label the energy levels by the *x-ray spectroscopic notation*

$$K \quad L_{\mathrm{I}}\,L_{\mathrm{II}}L_{\mathrm{III}} \quad M_{\mathrm{I}}\,M_{\mathrm{II}}\,M_{\mathrm{III}}M_{\mathrm{IV}}M_{\mathrm{V}} \quad \cdots$$

according to the tabulation given in the figure. The exclusion principle is built into this new scheme of states by stipulating that no two electrons are allowed to have the same four quantum numbers $(n\ell jm_j)$. Hence, each of the split $n\ell j$ subshells has a maximum occupancy equal to $2j + 1$, corresponding to the number of degenerate m_j states for the given value of j. (We note in passing that the figure orders the energies of the inner subshells in ascending fashion as

$$1s \quad 2s\,2p \quad 3s\,2p\,3d \quad \cdots.$$

This ordering of values of $E_{n\ell}$ holds for the *inner* electrons of a large-Z atom.) We note especially that the occurrence of the three split L subshells in Figure 9-12 is directly correlated with the existence of the three L absorption edges in Figure 9-10.

Let us now return to our picture of x-ray emission in Figure 9-9 and revise the diagrams to account for the splitting of the subshells. We illustrate the result in Figure 9-13 in terms of a particular x-ray transition. The indicated process shows the filling of an inner-shell hole by an electron from a higher subshell, along with the corresponding transition of the hole in the *opposite* direction. Since the initial and final states of the ion are associated with the location of the hole, it is conventional to represent the process as a *hole transition*. The convention employs an energy level diagram in which the highest energy state refers to a hole in the K shell, and in which the zero level refers to the ionic ground state where the hole occurs just beyond all the electron subshells. This presentation of the energy states of the ion *inverts* our view of the levels found in Figure 9-12 and produces an *x-ray level diagram* of the kind shown in Figure 9-14. The figure lists the values of the hole quantum numbers $(n\ell j)$ and indicates the allowed hole transitions. These radiative processes are governed by the electric dipole selection rules, so that the single-particle quantum numbers are

Figure 9-14

X-ray levels of lead and transitions allowed by the electric dipole selection rules. The energies are plotted on a logarithmic scale.

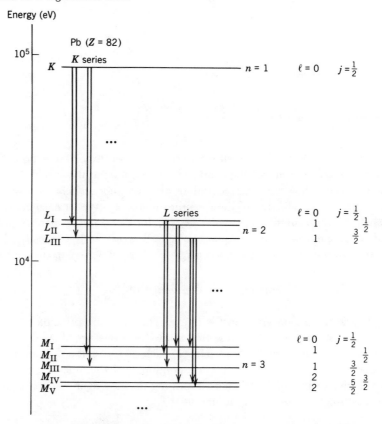

supposed to change according to the conditions

$$\Delta\ell = \pm 1 \quad \text{and} \quad \Delta j = 0 \text{ or } \pm 1 \quad (\text{but } 0 \not\rightarrow 0),$$

as in Equations (8-42) and (8-43). The various transitions are organized in the diagram into the different series of emission lines. We can employ this same general scheme whenever we wish to analyze the x-ray spectrum of any element.

Detail

We can formulate the spin–orbit interaction for an electron in a complex atom by constructing a parallel with the expression for the one-electron atom. Let us return to Equation (8-32) and rewrite the expression in terms of a derivative of the Coulomb potential energy:

$$V_{SL} = \frac{\mathbf{S} \cdot \mathbf{L}}{2m_e^2 c^2} \frac{1}{r} \frac{dV}{dr},$$

where

$$V = -\frac{Ze^2}{4\pi\varepsilon_0 r} \quad \text{and} \quad \frac{1}{r} \frac{dV}{dr} = \frac{Ze^2}{4\pi\varepsilon_0 r^3}.$$

An electron in the central-field model has central potential energy $V_c(r)$, and so the spin–orbit interaction of the electron in the complex atom must have the form

$$V_{SL} = \mathbf{S} \cdot \mathbf{L}\xi_c(r)$$

in which

$$\xi_c(r) = \frac{1}{2m_e^2 c^2} \frac{1}{r} \frac{dV_c}{dr}.$$

The function $V_c(r)$ contains the nuclear screening factor $Z_{\text{eff}}(r)$. This quantity varies between the value Z at small r and the value unity at large r, so that spin–orbit coupling becomes stronger with increasing Z, especially in states where r has a small average value. We know from Section 8-9 how the interaction V_{SL} perturbs the energy states in a central field. Let us consider an independent electron in an $n\ell$ subshell and describe the electron states by means of the quantum numbers $(n\ell j m_j)$ instead of the spin–orbital set $(n\ell m_\ell m_s)$. We use the eigenfunction $\psi_{n\ell j m_j}$ as before to calculate the expectation value of V_{SL}, and we obtain the following results:

$$\langle V_{SL} \rangle = \left\langle \frac{J^2 - L^2 - S^2}{2} \xi_c(r) \right\rangle$$

$$= \frac{\hbar^2}{2} \left[j(j+1) - \ell(\ell+1) - \frac{3}{4} \right] \langle \xi_c(r) \rangle$$

$$= \frac{\hbar^2}{2} \langle \xi_c(r) \rangle \cdot \begin{cases} \ell & \text{for } j = \ell + \frac{1}{2} \\ -\ell - 1 & \text{for } j = -\ell - \frac{1}{2}. \end{cases}$$

This energy shift affects the energy level of an electron in an $n\ell$ subshell according to the value of j and causes a splitting in every subshell with $\ell \neq 0$.

Example

The K and L_{III} absorption edges of lead in Figure 9-10 are observed at 0.01408 and 0.09511 nm. Let us consult Equation (9-13) and use these numbers to compute the wavelength emitted in the $K \to L_{\text{III}}$ transition:

$$\frac{1}{\lambda(KL_{\text{III}})} = \frac{1}{\lambda_K} - \frac{1}{\lambda_{L_{\text{III}}}} = \frac{1}{0.01408 \text{ nm}} - \frac{1}{0.09511 \text{ nm}} = \frac{1}{0.01653 \text{ nm}}.$$

This emission line can also be predicted from the energies of the K and L_{III} x-ray levels of lead. We use the tabulated values $E_K = 88.00$ keV and $E_{L_{\text{III}}} = 13.03$ keV, and we find

$$\lambda(KL_{\text{III}}) = \frac{hc}{E_K - E_{L_{\text{III}}}} = \frac{1.240 \text{ keV} \cdot \text{nm}}{74.97 \text{ keV}} = 0.01654 \text{ nm}.$$

Both calculations agree with the wavelength listed in the tables for the KL_{III} line of lead. The line and the transition are included among the information given in Figures 9-10 and 9-14.

9-5 Electron Antisymmetry

We have been able to implement the exclusion principle and explain the periodic table with only minimal use of quantum mechanics. The exclusion principle itself has been expressed in a limited context, based entirely on the assignment of electron quantum numbers in the central-field model. We now want to examine the fundamental quantum nature of Pauli's principle from the viewpoint of *electron indistinguishability* and develop the broader concept of *identical-particle symmetry*. We recall that we have already introduced this notion in the framework of one-dimensional quantum mechanics in Section 5-10.

Let us begin with the two-electron system and identify the degrees of freedom to be the pair of spatial coordinates and spin orientations

$$(\mathbf{r}_1, S_{1z}) \quad \text{and} \quad (\mathbf{r}_2, S_{2z}).$$

It is convenient to symbolize these variables in the wave function by adopting the abbreviated notation

$$\Psi(\mathbf{r}_1, S_{1z}, \mathbf{r}_2, S_{2z}, t) = \Psi(1, 2, t).$$

We then determine the probability of finding the two particles in two volume elements $d\tau_1$ and $d\tau_2$ at time t by evaluating the expression

$$|\Psi(1, 2, t)|^2 \, d\tau_1 d\tau_2.$$

The electrons in the system have identical physical attributes of charge $-e$, mass m_e, and spin $s = \frac{1}{2}$. It is not meaningful to refer to these particles as electron 1 and electron 2 because the notation implies that the two electrons can always be distinguished in any state. The intrinsic indistinguishability of identical quantum par-

Figure 9-15

Distinguishable orbits of two identical classical particles.

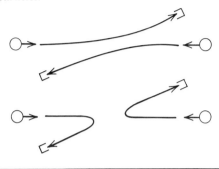

ticles has been discussed previously in Section 5-10. We are reminded in Figure 9-15 that any pair of classical particles can always be analyzed in terms of distinguishable particle orbits, but we recognize that such a treatment violates the uncertainty principle in the case of identical quantum particles. Since the two particles in the figure are identical, their indistinguishability allows us to prepare a state of separately identified colliding particles but prevents us from ascertaining the identity of the detected particles after the collision. We build this property of electrons into the probability density by imposing the requirement of identical-particle symmetry:

$$|\Psi(1,2,t)|^2 = |\Psi(2,1,t)|^2. \tag{9-14}$$

This statement follows Equation (5-90) and says that $|\Psi|^2$ is not altered by the exchange of the two sets of space and spin variables in the wave function.

The exchange symmetry of $|\Psi|^2$ can be realized in two independent ways. These alternatives correspond to mutually exclusive symmetry properties of the wave function itself. The two-particle state may satisfy either the *exchange-symmetric* condition

$$\Psi(1,2,t) = +\Psi(2,1,t) \tag{9-15}$$

or the *exchange-antisymmetric* condition

$$\Psi(1,2,t) = -\Psi(2,1,t). \tag{9-16}$$

A system containing more than two identical particles obeys a similar relation with regard to the exchange of *any* two sets of degrees of freedom in the many-variable wave function. We emphasize that the symmetry or antisymmetry of Ψ is an additional quantum property to be applied to solutions of the Schrödinger equation and that the choice of sign ($+$ for symmetry or $-$ for antisymmetry) is not meant to be taken arbitrarily. Either symmetry or antisymmetry is *imposed* on the constituents of a system as a law of nature. Two identical particles are called *bosons* if Equation (9-15) holds, or *fermions* if Equation (9-16) holds, and all particles observed in nature are classified as either bosons or fermions. (The classification is named after S. N. Bose and E. Fermi for their part in the development of statistical laws governing thermal distributions of the two types of particle.)

The exclusion principle can now be reexpressed in terms of exchange behavior in the space and spin variables of the wave function. This more general version of Pauli's rule is known as the *Pauli principle*. The application of the rule to electrons classifies these particles as fermions by the following assertion:

> The wave function for a system of electrons must be *antisymmetric* with respect to the exchange of any two sets of space and spin variables.

Pauli's original exclusion principle follows directly from this statement by arguments to be demonstrated below. The Pauli principle is more comprehensive because its implementation is not tied to any specific treatment of the electron system, and especially because its implications are immediately applicable to any particles certified as fermions. Thus, the shell structure of atoms is only one indication of the influence of this new quantum property.

Let us see how the exclusion principle is contained in the statement of antisymmetry by looking first at the case of the two-electron atom. We consider a stationary-state wave function of the form

$$\Psi(1,2,t) = \psi(1,2)e^{-iEt/\hbar}$$

and note that the Pauli principle requires the eigenfunction $\psi(1,2)$ to be antisymmetric under the exchange of variables $1 \leftrightarrow 2$. We determine ψ from Equation (9-2) by applying the central-field model to the pair of independent electrons. We know from Section 9-1 that solutions for ψ exist as products of two single-particle eigenfunctions, each having the spin-orbital form

$$\psi_\alpha(\mathbf{r}) = R_{n\ell}(r)Y_{\ell m_\ell}(\theta,\phi)(\uparrow \text{ or } \downarrow).$$

(Recall that we are using a single Greek index to represent a complete set of four spin-orbital quantum numbers for a single particle.) Thus, if α and β denote any two spin-orbital states, it is clear that

$$\psi_\alpha(1)\psi_\beta(2) \quad \text{and} \quad \psi_\beta(1)\psi_\alpha(2)$$

are degenerate solutions of Equation (9-2) since both product expressions have the same energy eigenvalue $E_\alpha + E_\beta$. Any linear combination of the two expressions is also a solution with the same energy, and so the requirement of antisymmetry can be met by constructing the particular combination

$$\psi(1,2) = \frac{1}{\sqrt{2}}\big(\psi_\alpha(1)\psi_\beta(2) - \psi_\beta(1)\psi_\alpha(2)\big). \tag{9-17}$$

This two-electron eigenfunction has the desired antisymmetric behavior under the exchange of variables $1 \leftrightarrow 2$:

$$\psi(2,1) = \frac{1}{\sqrt{2}}\big(\psi_\alpha(2)\psi_\beta(1) - \psi_\beta(2)\psi_\alpha(1)\big) = -\psi(1,2).$$

(We include the factor $1/\sqrt{2}$ to secure the normalization of ψ. This point is investigated in the first example below.) It is an immediate consequence of the antisymmetry of Equation (9-17) that ψ *vanishes* whenever α and β refer to the *same* set of single-particle quantum numbers. We conclude that no such two-electron state exists, and we thus affirm the outcome of the exclusion principle.

These arguments are readily extended beyond the case of the $Z = 2$ atom. We observe that Equation (9-17) has the structure of a determinant, and we write

$$\psi(1,2) = \frac{1}{\sqrt{2!}} \begin{vmatrix} \psi_\alpha(1) & \psi_\beta(1) \\ \psi_\alpha(2) & \psi_\beta(2) \end{vmatrix}.$$

An obvious generalization gives

$$\psi(1,2,3) = \frac{1}{\sqrt{3!}} \begin{vmatrix} \psi_\alpha(1) & \psi_\beta(1) & \psi_\gamma(1) \\ \psi_\alpha(2) & \psi_\beta(2) & \psi_\gamma(2) \\ \psi_\alpha(3) & \psi_\beta(3) & \psi_\gamma(3) \end{vmatrix} \tag{9-18}$$

for the three-electron atom, and ultimately gives

$$\psi(1,2,\ldots,Z) = \frac{1}{\sqrt{Z!}} \begin{vmatrix} \psi_\alpha(1) & \psi_\beta(1) & \cdots & \psi_\zeta(1) \\ \psi_\alpha(2) & \psi_\beta(2) & \cdots & \psi_\zeta(2) \\ \vdots & \vdots & & \vdots \\ \psi_\alpha(Z) & \psi_\beta(Z) & \cdots & \psi_\zeta(Z) \end{vmatrix} \tag{9-19}$$

for the Z-electron atom. These expressions for the eigenfunctions are known as Slater determinants. The general algebraic properties of determinants imply that the expressions change sign when any two variables are exchanged and vanish when any two spin-orbital indices denote the same state. Consequently, Slater determinants automatically satisfy the Pauli principle and furnish the basic solutions for use in the Hartree–Fock self-consistent procedure.

We have discovered the joint concepts of spin and fermion antisymmetry as new quantum mechanical properties associated with the behavior of electrons. It is intriguing to suppose that these two intrinsic characteristics of particles might be somehow interrelated, a possibility originally explored by Pauli. The possibility has indeed been realized in nature since the following observations have been established by experiment:

Particles with half-integral spin behave as fermions,
and
particles with integral spin behave as bosons.

The interrelations apply to all known particles and suggest a profound underlying idea worthy of purely theoretical investigation. The connection between spin and identical-particle symmetry has in fact been proved under certain general hypotheses in the so-called spin-statistics theorem.

Example

We can verify the normalization of the stationary-state eigenfunction in Equation (9-17) by the following computation:

$$\int d\tau_1 \int d\tau_2 |\psi(1,2)|^2$$

$$= \frac{1}{2} \int d\tau_1 \int d\tau_2 |\psi_\alpha(1)\psi_\beta(2) - \psi_\beta(1)\psi_\alpha(2)|^2$$

$$= \frac{1}{2} \int d\tau_1 \int d\tau_2 \Big\{ |\psi_\alpha(1)|^2 |\psi_\beta(2)|^2 - \psi_\alpha^*(1)\psi_\beta(1)\psi_\beta^*(2)\psi_\alpha(2)$$

$$\qquad\qquad - \psi_\beta^*(1)\psi_\alpha(1)\psi_\alpha^*(2)\psi_\beta(2) + |\psi_\beta(1)|^2 |\psi_\alpha(2)|^2 \Big\}$$

$$= \tfrac{1}{2}\{1 - 0 - 0 + 1\}.$$

The single-particle eigenfunctions are normalized to satisfy

$$\int |\psi_\alpha(1)|^2 \, d\tau_1 = 1 \quad \text{and similarly for } \psi_\beta.$$

The functions also obey the orthogonality condition in Equation (6-63):

$$\int \psi_\alpha^*(1)\psi_\beta(1) \, d\tau_1 = 0 \quad \text{if } \alpha \neq \beta.$$

Both formulas have been used to complete the above calculation.

Example

Antisymmetrization is required, but may not always be essential, for a description of electrons. To illustrate, let us suppose that independent-electron states are defined for two particles $\hat{e}$ and $\tilde{e}$ in terms of the single-particle eigenfunctions $\hat{\psi}$ and $\tilde{\psi}$. A non-antisymmetrized eigenfunction for the system has the form

$$\psi^A(1,2) = \hat{\psi}(1)\tilde{\psi}(2)$$

with probability density

$$P^A(1,2) = |\hat{\psi}(1)|^2 |\tilde{\psi}(2)|^2.$$

Equation (9-17) expresses the properly constructed eigenfunction as

$$\psi(1,2) = \frac{1}{\sqrt{2}}\big(\hat{\psi}(1)\tilde{\psi}(2) - \tilde{\psi}(1)\hat{\psi}(2)\big).$$

The corresponding probability density is

$$P(1,2) = \tfrac{1}{2}\Big\{ |\hat{\psi}(1)|^2 |\tilde{\psi}(2)|^2 - \hat{\psi}^*(1)\tilde{\psi}(1)\tilde{\psi}^*(2)\hat{\psi}(2)$$

$$\qquad\qquad + |\tilde{\psi}(1)|^2 |\hat{\psi}(2)|^2 - \tilde{\psi}^*(1)\hat{\psi}(1)\hat{\psi}^*(2)\tilde{\psi}(2) \Big\},$$

as in the previous example. The probability of finding $\hat{e}$ (an electron in state $\hat{\psi}$) in a volume $\Delta\tau$ is given by the expression

$$\hat{P} = \frac{1}{2}\int_{\Delta\tau}\left|\hat{\psi}(1)\right|^2 d\tau_1\int\left|\tilde{\psi}(2)\right|^2 d\tau_2 - \frac{1}{2}\int_{\Delta\tau}\hat{\psi}*(1)\tilde{\psi}(1)\,d\tau_1\int_{\Delta\tau}\tilde{\psi}*(2)\hat{\psi}(2)\,d\tau_2$$

$$+ \frac{1}{2}\int_{\Delta\tau}\left|\hat{\psi}(2)\right|^2 d\tau_2\int\left|\tilde{\psi}(1)\right|^2 d\tau_1 - \frac{1}{2}\int_{\Delta\tau}\hat{\psi}*(2)\tilde{\psi}(2)\,d\tau_2\int_{\Delta\tau}\tilde{\psi}*(1)\hat{\psi}(1)\,d\tau_1.$$

Note that we restrict the region of integration to $\Delta\tau$ wherever $\hat{\psi}$ is involved, and we integrate over all space in the variables of $\tilde{\psi}$ when there is no such restriction. The integration coordinates are dummy variables, and so the probability can be rewritten as

$$\hat{P} = \int_{\Delta\tau}\left|\hat{\psi}(1)\right|^2 d\tau_1 - \int_{\Delta\tau}\hat{\psi}*(1)\tilde{\psi}(1)\,d\tau_1\int_{\Delta\tau}\tilde{\psi}*(2)\hat{\psi}(2)\,d\tau_2$$

$$= \int_{\Delta\tau}\left|\hat{\psi}(1)\right|^2 d\tau_1 - \left|\int_{\Delta\tau}\hat{\psi}*(1)\tilde{\psi}(1)\,d\tau_1\right|^2.$$

We have used the fact that $\tilde{\psi}$ is a normalized single-particle eigenfunction to get the first line of this result. Note that the orthogonality of $\hat{\psi}$ and $\tilde{\psi}$ does not cause the integral of the product of these functions to vanish unless $\Delta\tau$ refers to all space. If we compute the same quantity from the non-antisymmetrized probability density $P^{A}(1,2)$, we obtain the simpler result

$$\hat{P}^{A} = \int_{\Delta\tau}\left|\hat{\psi}(1)\right|^2 d\tau_1\int\left|\tilde{\psi}(2)\right|^2 d\tau_2 = \int_{\Delta\tau}\left|\hat{\psi}(1)\right|^2 d\tau_1.$$

Thus, we see that $\hat{P}^{A}$ is an excellent approximation to $\hat{P}$ whenever the overlap integral

$$\int_{\Delta\tau}\hat{\psi}*(1)\tilde{\psi}(1)\,d\tau_1$$

for the states $\hat{\psi}$ and $\tilde{\psi}$ is negligible. Such a circumstance arises when $\hat{e}$ and $\tilde{e}$ are electrons in separate fields of force, an appreciable distance apart. The overlap is small in this case because the eigenfunctions $\hat{\psi}$ and $\tilde{\psi}$ do not have large magnitudes in the same regions of space.

9-6   The Helium Atom

The Grotrian diagram for the He atom is shown in Figure 9-16. The striking feature of these energy levels and transitions is the apparent existence of two distinct He varieties, denoted in the figure as parahelium and orthohelium. The two species present themselves in separate families of spectral lines since the transitions occur within, but not between, the para and ortho systems of energy levels. This behavior indicates the dynamical influence of a certain quantum number that distinguishes the two forms of helium. One of our main objectives is to show how this new quantum number arises from considerations of the Pauli principle.

Figure 9-16

Energy levels and electric dipole transitions of helium. Emitted wavelengths are given to the nearest nanometer, as in Figure 9-1. All states below the ionization level have $1s\,n\ell$ configurations denoted by the indicated $n\ell$ assignments. The levels are organized by the notation at the top of the diagram according to total orbital and total spin quantum numbers.

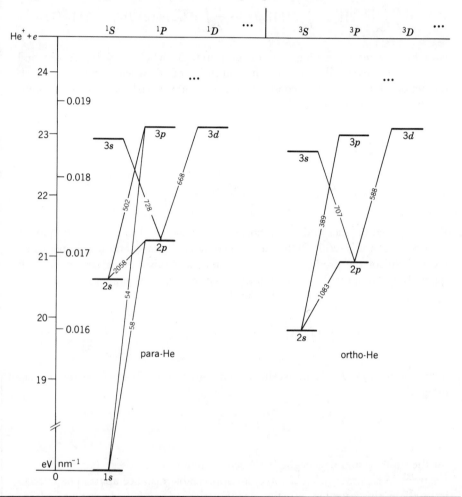

We know from Section 9-3 that the He ground state is a closed-shell system of two $1s$ electrons. The higher-energy states must therefore involve configurations in which at least one of the electron orbitals is beyond the $n = 1$ shell. In fact, *only* one particle can be excited beyond $n = 1$ since the energy required to excite both particles exceeds the 24.6 eV ionization threshold for the ionized system $\text{He}^+(1s) + e$. We can therefore classify all the discrete energy states of the bound He atom as *singly excited $1s\,n\ell$* configurations. Accordingly, the levels in Figure 9-16 are labeled by the $n\ell$-orbital designation of the excited electron, and the energy values are given by the sum of energies $E_{10} + E_{n\ell}$ in the central-field model.

It is instructive to combine the orbital angular momenta of the two electrons and define the *total orbital angular momentum* of the system as

$$\mathbf{L} = \mathbf{L}_1 + \mathbf{L}_2. \tag{9-20}$$

A *total orbital quantum number* ℓ can then be associated with **L** by the usual quantization rules. Since the one electron is never excited, it is clear that the value of ℓ in the He atom is always equal to the ℓ value of the lone excited electron. We indicate this assignment of the total orbital quantum number by the familiar spectroscopic label $S, P, D, \ldots$ at the top of each column of energy levels shown in the figure.

We get to the heart of the He problem when we consider the imposition of the Pauli principle on the two-electron system. Let us start with the ground state and rewrite Equation (9-17) in the form of a Slater determinant for the $1s^2$ configuration:

$$\psi(1,2) = \frac{1}{\sqrt{2}} \begin{vmatrix} R_{10}(r_1)Y_{00}(\theta_1,\phi_1)\uparrow_1 & R_{10}(r_1)Y_{00}(\theta_1,\phi_1)\downarrow_1 \\ R_{10}(r_2)Y_{00}(\theta_2,\phi_2)\uparrow_2 & R_{10}(r_2)Y_{00}(\theta_2,\phi_2)\downarrow_2 \end{vmatrix}.$$

The expanded result exhibits a separation of the space and spin dependence into a spatial eigenfunction multiplied by a spin eigenfunction:

$$\psi(1,2) = \left\{ \left(R_{10}(r_1)Y_{00}(\theta_1,\phi_1)\right)\left(R_{10}(r_2)Y_{00}(\theta_2,\phi_2)\right) \right\} \left\{ \frac{1}{\sqrt{2}}(\uparrow_1\downarrow_2 - \downarrow_1\uparrow_2) \right\}.$$

$$(9\text{-}21)$$

We observe that the required antisymmetrization of ψ is accomplished in a *factorized* fashion. The first bracketed quantity is obviously symmetric under the exchange of the two spatial coordinates $\mathbf{r}_1 \leftrightarrow \mathbf{r}_2$, while the second is antisymmetric under the exchange of the two spin orientations $1 \leftrightarrow 2$.

A similar factorization can be employed to obtain overall antisymmetry in any He eigenfunction. The procedure can be followed even when the construction of the state calls for a superposition of Slater determinants. We antisymmetrize the two-electron eigenfunction under the exchange $1 \leftrightarrow 2$ by assembling a *product* of symmetric and antisymmetric space and spin eigenfunctions in either of the following two ways:

$$\psi(1,2) = \begin{cases} \psi^S(\mathbf{r}_1,\mathbf{r}_2)\chi^A(1,2) \\ \quad\quad\text{or} \\ \psi^A(\mathbf{r}_1,\mathbf{r}_2)\chi^S(1,2). \end{cases} \quad\quad (9\text{-}22)$$

The antisymmetric and symmetric *spatial eigenfunctions* are obtained by constructing

$$\psi^A(\mathbf{r}_1,\mathbf{r}_2) = \frac{1}{\sqrt{2}}\left[\psi_a(\mathbf{r}_1)\psi_b(\mathbf{r}_2) - \psi_b(\mathbf{r}_1)\psi_a(\mathbf{r}_2) \right] \quad\quad (9\text{-}23)$$

and

$$\psi^S(\mathbf{r}_1,\mathbf{r}_2) = \frac{1}{\sqrt{2}}\left[\psi_a(\mathbf{r}_1)\psi_b(\mathbf{r}_2) + \psi_b(\mathbf{r}_1)\psi_a(\mathbf{r}_2) \right], \quad\quad (9\text{-}24)$$

letting each subscript a and b denote a set of *three* spatial single-particle quantum numbers $(n\ell m_\ell)$. The antisymmetric and symmetric *spin eigenfunctions* are formed in like manner by defining

$$\chi^A(1,2) = \frac{1}{\sqrt{2}}(\uparrow_1\downarrow_2 - \downarrow_1\uparrow_2) \quad\quad (9\text{-}25)$$

and

$$\chi_1^S(1,2) = \uparrow_1 \uparrow_2,$$

$$\chi_0^S(1,2) = \frac{1}{\sqrt{2}}(\uparrow_1 \downarrow_2 + \downarrow_1 \uparrow_2),$$

$$\chi_{-1}^S(1,2) = \downarrow_1 \downarrow_2. \tag{9-26}$$

As usual, the factor $1/\sqrt{2}$ is included in the formulas to maintain the normalization. We see that the exchange antisymmetry and exchange symmetry of these expressions are in correspondence with the superscripts A and S introduced in Equations (9-22).

The four spin eigenfunctions in Equations (9-25) and (9-26) exhaust all combined up and down orientations for the two electron spins. We have found it possible to construct *one* antisymmetric and *three* symmetric arrangements of these orientations. The states are designated accordingly as *singlet* and *triplet* spin eigenfunctions. We note that the lone antisymmetric spin eigenfunction χ^A already appears as a factor in Equation (9-21) for the ground state. The accompanying spatial factor is a special version of Equation (9-24) in which the indices a and b refer to the *same* single-particle orbital state. Other special cases of the symmetric spatial eigenfunction may also occur, taking the form of a single product

$$\psi^S(\mathbf{r}_1, \mathbf{r}_2) = \psi_a(\mathbf{r}_1)\psi_a(\mathbf{r}_2)$$

without the usual normalizing factor.

The lowest excited levels of helium belong to the $1s2s$ electron configuration. Let us organize the four different states in this system by the scheme described in Equations (9-22). We represent the electron spins by the singlet eigenfunction χ^A in one of the states and by the three triplet eigenfunctions χ^S in the other three states. Equation (9-22) then tells us that the spatial eigenfunction must be symmetric to accompany the singlet state and must be antisymmetric to accompany the triplet states. We use this strategy to write the $1s2s$ eigenfunctions as

$$\psi(1,2) = \left\{ \frac{1}{\sqrt{2}}\left[(R_{10}Y_{00})_1(R_{20}Y_{00})_2 + (R_{20}Y_{00})_1(R_{10}Y_{00})_2\right]\right\}\left\{\frac{1}{\sqrt{2}}(\uparrow_1 \downarrow_2 - \downarrow_1 \uparrow_2)\right\} \tag{9-27}$$

for the singlet state and

$$\psi(1,2) = \left\{ \frac{1}{\sqrt{2}}\left[(R_{10}Y_{00})_1(R_{20}Y_{00})_2 - (R_{20}Y_{00})_1(R_{10}Y_{00})_2\right]\right\}$$

$$\times \left\{ \begin{array}{c} \uparrow_1 \uparrow_2 \\ \dfrac{1}{\sqrt{2}}(\uparrow_1 \downarrow_2 + \downarrow_1 \uparrow_2) \\ \downarrow_1 \downarrow_2 \end{array} \right\} \tag{9-28}$$

for the triplet states. (We adopt the abbreviating subscripts 1 and 2 to denote the spatial variables $\mathbf{r}_1$ and $\mathbf{r}_2$.) The total orbital quantum number has the value $\ell = 0$ for each of these eigenfunctions, and so the resulting singlet and triplet states appear in

Figure 9-16 with the assignments ^1S and ^3S at the top of the energy level diagram. Singlets and triplets are also indicated in the figure for other configurations and other values of ℓ.

The singlet and triplet states are distinguished by their exchange properties under the abstract concept of fermion antisymmetry. Let us attach a more concrete meaning to the two kinds of spin eigenfunctions by recalling our interpretation of electron spin as an angular momentum vector. The two-electron system has spins $\mathbf{S}_1$ and $\mathbf{S}_2$, and so a *total spin angular momentum* can be defined by the vector sum

$$\mathbf{S} = \mathbf{S}_1 + \mathbf{S}_2. \tag{9-29}$$

The individual spin vectors have the eigenvalue properties of separate spin-$\frac{1}{2}$ particles, as discussed in Section 8-5:

$$S_1^2 = \hbar^2 s_1(s_1 + 1) = \frac{3}{4}\hbar^2 \qquad S_{1z} = \hbar m_{s_1} = \pm\frac{\hbar}{2}$$

and

$$S_2^2 = \hbar^2 s_2(s_2 + 1) = \frac{3}{4}\hbar^2 \qquad S_{2z} = \hbar m_{s_2} = \pm\frac{\hbar}{2}.$$

The total spin $\mathbf{S}$ is supposed to obey the same general quantization rules in terms of the quantities S^2 and S_z. Hence, there must exist a nonnegative numerical quantity s, called the *total spin quantum number*, such that quantized values are given for the square of $\mathbf{S}$ by the eigenvalues

$$S^2 = \hbar^2 s(s + 1). \tag{9-30}$$

Each possible s then implies a set of $2s + 1$ quantized values for the z component of $\mathbf{S}$, given by the eigenvalues

$$S_z = \hbar m_s \quad \text{with } m_s = -s, \ldots, s. \tag{9-31}$$

These equations describe the familiar behavior of an angular momentum in which the $\mathbf{S}$ vectors have definite magnitudes determined by s and discrete z-axis projections determined by m_s.

Equation (9-29) implies a relation among the z-component quantum numbers:

$$m_s = m_{s_1} + m_{s_2}.$$

It follows that m_s must be integer-valued, since m_{s_1} and m_{s_2} are half-integral, and that the possible values of s must be integers. Only two possibilities exist,

$$s = 0 \quad \text{and} \quad s = 1,$$

since any larger s would permit m_s to exceed unity, the maximum allowed for the sum of m_{s_1} and m_{s_2}. The respective z-component quantum numbers are

$$m_s = 0 \quad \text{for } s = 0 \qquad \text{and} \qquad m_s = -1, 0, 1 \quad \text{for } s = 1.$$

These four assignments of quantum numbers (sm_s) refer precisely to the four spin

Figure 9-17

Vector addition of two spin-$\frac{1}{2}$ angular momenta. The sum $\mathbf{S} = \mathbf{S}_1 + \mathbf{S}_2$ is constructed, without factors of $\hbar$, for the case $m_{s_1} = m_{s_2} = \frac{1}{2}$. The resulting spin state has $s = 1$ and $m_s = 1$, as described by the symmetric spin eigenfunction χ_1^S.

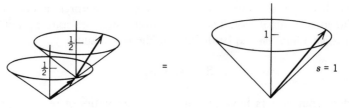

eigenfunctions identified in Equations (9-25) and (9-26). The total spin quantum number and the spin-exchange property are correlated such that

$$s = 0 \text{ denotes the antisymmetric singlet } \chi^A$$

and

$$s = 1 \text{ denotes the symmetric triplet } \chi_{m_s}^S.$$

Note that the χ^S states in Equations (9-26) are already labeled by their m_s values in anticipation of this development.

Figure 9-17 shows how the addition of two quantized spin-$\frac{1}{2}$ vectors $\mathbf{S}_1$ and $\mathbf{S}_2$ produces a quantized result for $\mathbf{S}$. We illustrate the selected spin state χ_1^S by the sum of two spin-up vectors, and we argue that χ_{-1}^S is represented by an inverted version of the same picture. Any combination of spin-up and spin-down states can be taken to form an $m_s = 0$ state. The particular construction χ^A corresponds to a total $\mathbf{S}$ vector of vanishing magnitude, while the χ_0^S state describes a combination of spin up with spin down in which the total $\mathbf{S}$ vector has magnitude $\sqrt{2}\,\hbar$.

The new quantum number s has dynamical as well as notational significance. We recognize the total spin $\mathbf{S}$ to be a conserved angular momentum for two electrons subject to the Coulomb interactions described in Equation (9-1). We also make a similar claim regarding the total $\mathbf{L}$ vector defined in Equation (9-20). These statements mean that we can regard the associated total quantum numbers (sm_s) and (ℓm_ℓ) as good quantum numbers, provided we ignore spin–orbit coupling. This interaction is of some interest since it causes a small splitting of the helium levels with nonzero ℓ in Figure 9-16. If we take account of the effect we find that L_z and S_z are no longer conserved, as in the one-electron problem, so that m_ℓ and m_s do not remain good. We then introduce the conserved total angular momentum

$$\mathbf{J} = \mathbf{L} + \mathbf{S},$$

and we use the corresponding quantum number j along with ℓ and s to label the split energy levels. The complete spectroscopic designation for a He level has the form

$$n^{\,2s+1}L_j.$$

This notation includes a value of n for the excited electron and assigns a superscript

$2s + 1$ giving the *spin multiplicity* of the level as singlet or triplet for $s = 0$ or $s = 1$. Thus, to cite again our previous examples, we denote the $1s^2$ ground state in Equation (9-21) by

$$1\,^1S_0 \quad \text{with } n = 1,\, s = 0,\, \ell = 0,\, j = 0,$$

and we denote the $1s2s$ excited states in Equations (9-27) and (9-28) by

$$2\,^1S_0 \quad \text{with } n = 2,\, s = 0,\, \ell = 0,\, j = 0$$

and

$$2\,^3S_1 \quad \text{with } n = 2,\, s = 1,\, \ell = 0,\, j = 1.$$

Only the partial notation ^{2s+1}L is employed to classify the helium levels in Figure 9-16. Notice that the same scheme has already been adopted for the hydrogen levels in Figure 9-1 by setting $s = \frac{1}{2}$ and $2s + 1 = 2$ for every state in the one-electron system.

We observe from Figure 9-16 that the singlet and triplet helium levels separate into the two varieties parahelium and orthohelium. We also note that the triplet always lies *below* the singlet in a given electron configuration, except for the special case of the $1s^2$ configuration where the triplet possibility does not exist. To choose a typical example, consider again the $1s2s$ system and note that the level $2\,^3S_1$ lies lower than its partner $2\,^1S_0$. We know that the solutions of the central-field model describe configurations of *degenerate* spin-orbital states. We also know that the central field takes in only a part of the noncentral interaction between electrons. The obvious conclusion is that the separation of singlet and triplet He energies must be due to some aspect of the interaction *between* the electrons, *beyond* the effects included in the central-field potential energy.

The repulsive electron–electron interaction favors the triplet over the singlet as a consequence of the Pauli principle. The two electrons in a triplet state must have a spatial eigenfunction of antisymmetric form, as given by Equation (9-23). Such a state vanishes when $\mathbf{r}_1 = \mathbf{r}_2$, so that there is small probability for the particles to be found near each other. Consequently, the two electrons tend to remain at a distance where the repulsive interaction has a reduced effect. On the other hand, a symmetric spatial eigenfunction governs the singlet state and permits the particles to be found closer together where they undergo a greater Coulomb repulsion. Thus, electron–electron repulsion is more effective for the singlet state in a given configuration, and so the triplet states experience more net attraction and have lower-lying energy levels. These arguments imply that the choice of spin state influences the spatial distribution of the particles and *simulates* a sort of interaction. The effect is called an *exchange force* keeping triplet-state particles apart and bringing singlet-state particles together. The coupling of space and spin variables operates through Equation (9-22) by virtue of the Pauli principle and generates this important nonclassical contribution to the behavior of indistinguishable particles.

Figure 9-16 also shows that no radiative transitions occur between para and ortho states. We understand this situation at once when we recognize that the observed spectral lines are due solely to electric dipole transitions. These processes are associated with the oscillations of a spatial vector, the electric dipole moment of the atom, and do not involve the spin vectors of the electrons. The total spin quantum number s cannot change in an electric dipole transition, and so the singlet and triplet energy levels can only participate in transitions among themselves. The electric dipole selection rules apply to the orbital quantum number of the excited electron and, therefore, to the total orbital quantum number ℓ. We summarize the constraints on the total spin and total orbital quantum numbers by writing

$$\Delta s = 0 \quad \text{and} \quad \Delta\ell = \pm 1 \tag{9-32}$$

and by noting that all the transitions in the figure obey these conditions.

Example

We have identified all the bound helium levels above the ground state with singly excited two-electron configurations. Let us examine this claim by describing the atom in first approximation without the electron–electron interaction. The energy of a state with independent-particle quantum numbers $(n_1 \; n_2)$ is then given by the Bohr formula with $Z = 2$:

$$E(n_1, n_2) = -\frac{Z^2}{n_1^2}E_0 - \frac{Z^2}{n_2^2}E_0 = -\left(\frac{1}{n_1^2} + \frac{1}{n_2^2}\right)(54.4 \text{ eV}).$$

We compile the results in the following brief table:

$n_1 \; n_2$	1 1	1 2	1 3	$\cdots$	1 ∞	2 2	$\cdots$
$E(n_1, n_2)$ in eV	-108.8	-68.0	-60.4		-54.4	-27.2	
$E(n_1, n_2) - E(1, 1)$ in eV	0	40.8	48.4		54.4	81.6	

Note that the $(1 \; \infty)$ level refers to the ionized system $He^+(1s) + e$, and observe in the third line of the table that the doubly excited energies lie above $E(1, \infty)$. Of course, the numbers in the line cannot accurately approximate the actual values plotted in Figure 9-16. The point of this simple calculation is that the neglected electron–electron repulsion contributes a positive energy shift that elevates all values of E to higher levels. The energy of the $(1 \; \infty)$ level is scarcely affected by the shift since the electrons stay far apart in this state. Appreciable upward shifts occur for levels with small n values, however, particularly in the $(1 \; 1)$ ground state and even in the $(2 \; 2)$ doubly excited system. We conclude that all doubly excited states remain higher than the ionization level after the repulsive shift is taken into account. These systems are known as autoionizing states since they transform spontaneously into the ionized atom $He^+ + e$.

9-7 Alkali Atoms

The optical spectrum of an atom spans the visible range of wavelengths and reflects the activity of the outer-shell electrons. We have illustrated the excitation and deexcitation of these optically active electrons in Figure 9-3. This picture of optical phenomena is clearly realized by the alkali atoms corresponding to the elements directly below hydrogen in the periodic table. Hydrogen-like qualities appear in the spectra of the alkali atoms and give evidence that the transitions are attributable to a single valence electron. The one electron occupies an orbital state outside a *core* of closed subshells and thus provides a simple structure to study as a testing ground for the central-field model.

We learn from the model that the lone valence electron of an alkali atom has an ns-orbital assignment in the ground state. In each case (lithium, sodium, potassium, ...) the quantum number n opens a new shell after all available orbitals of lesser energy are filled by the core electrons. We consult Table 9-1 and find that the ground states of Li, Na, K, ... have core configurations

$$1s^2, \quad 1s^2 2s^2 2p^6, \quad 1s^2 2s^2 2p^6 3s^2 3p^6, \quad \ldots$$

and valence-electron assignments

$$2s, \quad 3s, \quad 4s, \quad \ldots .$$

Each core electron has its own individual orbital and spin angular momenta $\mathbf{L}_i$ and $\mathbf{S}_i$. Both quantities add up to *zero* when the sums are taken over all electrons in the core. To understand this point we examine the quantum numbers of the core electrons and recognize that the individual z-component quantum numbers have vanishing sums

$$\sum_{\text{core}} m_{\ell_i} = 0 \quad \text{and} \quad \sum_{\text{core}} m_{s_i} = 0, \tag{9-33}$$

since every available assignment of the quantum numbers is taken to assemble the closed subshells. It follows that the overall core angular momenta

$$\sum_{\text{core}} \mathbf{L}_i \quad \text{and} \quad \sum_{\text{core}} \mathbf{S}_i$$

must have vanishing magnitudes, otherwise core states with nonzero z-component quantum numbers would also exist, in conflict with Equations (9-33).

We summarize this situation by describing a core of closed subshells as a ^1S system with total orbital and total spin quantum numbers equal to zero. The result implies that the single valence electron sees the core as a spherically symmetric distribution of charge. We should expect the central-field potential energy $V_c(r)$ to be especially good as an approximation for such a quasi-one-electron atom.

The energy levels and radiative transitions of lithium and sodium are plotted in Figures 9-18 and 9-19. We appeal to the one-electron properties of these atoms immediately and use the $n\ell$-orbital quantum numbers of the valence electron to identify the various levels. The excited states lie above the $2s$ ground state in lithium and above the $3s$ ground state in sodium, as noted in our previous remarks. The figures also show the energies, relative to the ionization level, for comparable excited states in hydrogen. A close correspondence can be seen between alkali levels and hydrogen levels, provided allowance is made for the non-Coulombic features of the alkali model. We know that the central-field potential energy $V_c(r)$ approaches its hydrogen-atom form at large r, and so we consider those orbital states in which the valence electron has large $\langle r \rangle$ and look for good agreement between the alkali and hydrogen energies. We meet this criterion in states of maximum ℓ for given n. The

Figure 9-18

Grotrian diagram for the lithium atom showing wavelengths to the nearest nanometer. Hydrogen levels for $n \geq 2$ are taken from Figure 9-1 for comparison on the right side of the diagram.

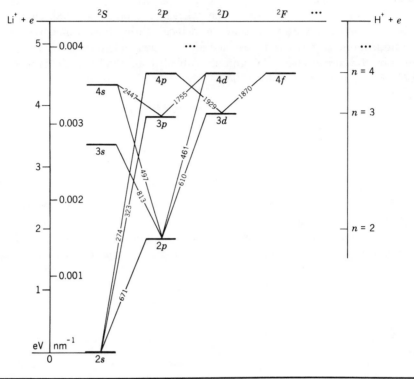

figures tell us that these particular alkali states are in fact quite close in energy to their hydrogen counterparts. These observations of the lithium and sodium levels are in keeping with our general comments in Section 9-3 about the n dependence and ℓ dependence of the energy eigenvalues for a non-Coulombic potential energy.

The angular momentum of an alkali atom is easy to analyze in view of the simple properties associated with the closed-subshell core. We define the total orbital and total spin angular momenta of the atom by the expressions

$$\mathbf{L} = \sum_{core} \mathbf{L}_i + \mathbf{L}_e \qquad (9\text{-}34)$$

and

$$\mathbf{S} = \sum_{core} \mathbf{S}_i + \mathbf{S}_e, \qquad (9\text{-}35)$$

where $\mathbf{L}_e$ and $\mathbf{S}_e$ refer to the valence electron. The core makes no contribution to $\mathbf{L}$ and $\mathbf{S}$, and so the whole atom assumes the angular momentum properties of the one electron outside the core. Thus, each atomic state has total spin quantum number $s = \frac{1}{2}$ and total orbital quantum number ℓ given by the particular $n\ell$ orbital state of the valence electron. The resulting spectroscopic notation for the alkali levels has the same

Figure 9-19

Grotrian diagram for the sodium atom showing spectral-line splittings of at least 1 nm. Levels of the hydrogen atom for $n \geq 3$ are included as in the previous figure.

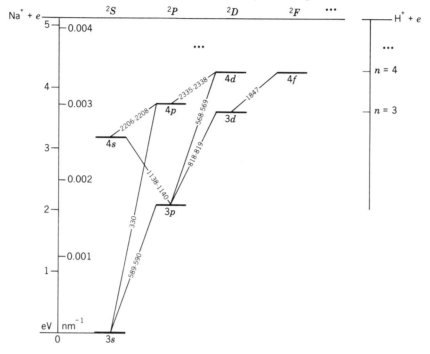

doublet form as in the case of hydrogen. Therefore, the same ^2L designations appear at the top of the energy level diagrams in Figures 9-18 and 9-19 as in Figure 9-1.

Every alkali doublet level with nonzero ℓ has a fine structure like that observed in hydrogen. (The effect is not large enough to be seen on the scale of the two figures for lithium and sodium.) This splitting of the energy levels is due to the spin–orbit interaction of the valence electron. Spin–orbit coupling has a small effect on the levels of hydrogen and becomes more appreciable in complex atoms with increasing atomic number. The interaction has already been applied to the behavior of an inner-shell electron in our discussion of x-ray spectra. We reconsider its influence in the case of alkali spectra and again restrict the interaction to a single particle, the valence electron.

We express the spin–orbit interaction for a single particle in the central-field model by the formula

$$V_{SL} = \mathbf{S} \cdot \mathbf{L}\xi_c(r) \quad \text{where} \quad \xi_c(r) = \frac{1}{2m_e^2c^2r}\frac{dV_c}{dr}. \tag{9-36}$$

These equations are carried over from our treatment of x-ray levels at the end of Section 9-4. We then introduce $\mathbf{J}$, the total angular momentum of the atom, as the sum of the quantities $\mathbf{L}$ and $\mathbf{S}$ defined in Equations (9-34) and (9-35). Again the core makes no contribution, so that the quantization properties of $\mathbf{J}$ are the same as those

Figure 9-20

Spin–orbit splitting of the $3p$ level of sodium. The $3p \to 3s$ transitions produce the two D lines in the sodium optical spectrum.

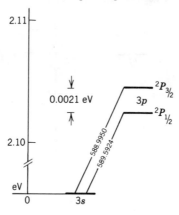

of the single valence electron. As usual, the eigenvalues of J^2 are equal to $\hbar^2 j(j+1)$, where the total angular momentum quantum number j takes on either of the two possible values $\ell + \frac{1}{2}$ or $\ell - \frac{1}{2}$ for every nonzero ℓ. The spin–orbit interaction in Equation (9-36) causes the energy of a given $n\ell j$-orbital level to shift by an amount that depends on the values of the three quantum numbers. The formula for the shift is the same as that found in our detailed remarks at the end of Section 9-4:

$$\langle V_{SL} \rangle = \frac{\hbar^2}{2} \langle \xi_c(r) \rangle \cdot \begin{cases} \ell & \text{for } j = \ell + \frac{1}{2} \\ -\ell - 1 & \text{for } j = \ell - \frac{1}{2}. \end{cases}$$

This result implies a fine structure splitting in the form of a *doublet* for every $\ell \neq 0$ level in Figures 9-18 and 9-19. The final outcome is a splitting of spectral lines that bears a certain resemblance to the line splitting in the spectrum of hydrogen.

Example

We find conspicuous evidence of spin–orbit coupling in the alkali atoms when we look at the radiative transitions of the lowest ^2P level of sodium. Figure 9-20 shows the relevant $3p$ and $3s$ states on an enlarged portion of the Na Grotrian diagram. The energies of the $^2P_{3/2}$ and $^2P_{1/2}$ states at the $3p$ level appear in the figure above the energy of the unsplit $^2S_{1/2}$ ground state at the $3s$ level. The spectroscopic notation

$$n\ell \, ^2L_j$$

summarizes all the quantum numbers $(n\ell sj)$ needed to identify any of these states. The figure also quotes the wavelengths of the radiation emitted in the two transitions

$$3p \, ^2P_{3/2} \to 3s \, ^2S_{1/2} \quad \text{and} \quad 3p \, ^2P_{1/2} \to 3s \, ^2S_{1/2}.$$

These radiative processes obey the electric dipole selection rules appropriate to the transitions of a single electron:

$$\Delta\ell = \pm 1 \quad \text{and} \quad \Delta j = 0 \text{ or } \pm 1.$$

We note that the familiar single-electron conditions are transferred directly to the quantum numbers of the whole atom. In general, the selection rules allow *three* different transitions between the pairs of doublet levels. Note, however, that Figure 9-20 describes an exceptional case since an unsplit $\ell = 0$ state is involved. We indicate the resulting ranges in wavelength for the various spectral lines of sodium in Figure 9-19.

9-8 Angular Momentum Coupling

The central-field model describes the complex atom in *leading* approximation, and the Pauli principle defines the electron configurations of the atom within that context. Two additional dynamical effects also influence the structure of the atom. One of these corrections arises from the fact that the central field includes only part of the noncentral Coulomb repulsion between the electrons. The spin–orbit coupling of the electrons provides the other correction. These *secondary* considerations have already been encountered in our discussions of certain atoms.

We wish to organize the corrections to the central field on a systematic basis and apply the resulting scheme to all electrons outside filled subshells. We continue to be concerned with optical excitations, and we recognize that every outer electron participates equally in the optical activity of the atom. The configurations of the atom are known in terms of the orbital and spin angular momenta of each electron. Our intention is to sort out the secondary corrections in stages with the aid of conservation laws among these quantities.

Let us examine the construction of the central field more closely so that we can specify the first correction to the model. The Coulomb potential energy for the Z-electron system consists of central and noncentral contributions, as written in Equation (9-4):

$$V = -\frac{Ze^2}{4\pi\varepsilon_0}\sum_{i=1}^{Z}\frac{1}{r_i} + \frac{e^2}{4\pi\varepsilon_0}\sum_{i<j=1}^{Z}\frac{1}{|\mathbf{r}_i - \mathbf{r}_j|}.$$

We know that the noncentral repulsion between electrons is not small enough to be treated as a secondary effect. We proceed instead to extract a *central approximation* from this part of the potential energy by rewriting V in the following manner:

$$V = \sum_i V_c(r_i) + V_{ee}, \tag{9-37}$$

where

$$\sum_i V_c(r_i) = -\frac{Ze^2}{4\pi\varepsilon_0}\sum_i\frac{1}{r_i} + \frac{e^2}{4\pi\varepsilon_0}\left\langle\sum_{i<j}\frac{1}{|\mathbf{r}_i - \mathbf{r}_j|}\right\rangle \tag{9-38}$$

and

$$V_{ee} = \frac{e^2}{4\pi\varepsilon_0}\sum_{i<j}\frac{1}{|\mathbf{r}_i - \mathbf{r}_j|} - \frac{e^2}{4\pi\varepsilon_0}\left\langle\sum_{i<j}\frac{1}{|\mathbf{r}_i - \mathbf{r}_j|}\right\rangle. \tag{9-39}$$

Figure 9-21

Coordinate vectors in the definition of a central
field for the ith electron in an atom.

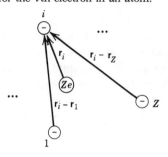

Note that a new quantity is added and subtracted to make this construction. We
define the new expression by an averaging procedure in which an ith particle is
singled out, as in Figure 9-21, and an average is taken over all other particle
coordinates and over all directions of the vector $\mathbf{r}_i$. The result for the ith particle is the
desired spherically symmetric potential energy V_c with only one remaining variable,
the radial distance r_i. Equation (9-39) then defines the *residual electrostatic interaction* V_{ee}
as the leftover part of the repulsion between electrons, after the removal of the central
approximation. Our treatment of the independent-electron model in Equations
(9-6)–(9-10) is based on the leading central contribution in Equation (9-37). We have
given the main role to this part of V, and so we regard the residual term V_{ee} as a small
perturbing effect.

The spin–orbit couplings of the electron account for another secondary interaction.
We employ the central-field function $\xi_c(r)$ in Equations (9-36) and express the sum of
these couplings by the formula

$$V_{SL} = \sum_i \xi_c(r_i)\mathbf{S}_i \cdot \mathbf{L}_i. \tag{9-40}$$

The two corrections V_{ee} and V_{SL} perturb the central field with different relative
strengths depending on the position of the atom in the periodic table. We have learned
that the spin–orbit effect is quite weak for small Z and grows much stronger with
increasing Z. We therefore identify two broad regimes of atomic number, and we
argue that V_{ee} dominates over V_{SL} for low-Z atoms, while V_{SL} dominates over V_{ee} for
high-Z atoms. Different conserved angular momenta are associated with the two
effects, and so different procedures are needed to analyze the corrections in the two
regimes.

We devote most of our attention to atoms at low Z and apply the so-called
Russell–Saunders coupling scheme. (The method is named after the astronomers H.
N. Russell and F. A. Saunders for their pioneering interpretations of complex atomic
spectra.) We temporarily ignore the spin–orbit interaction V_{SL}, and we use the
Coulomb potential energy in Equation (9-37) to find the states of the atom, in first *and*
second approximation. Figure 9-21 tells us that the noncentral Coulomb interactions
of the ith electron cause the orbital angular momentum $\mathbf{L}_i$ to be a nonconserved
quantity. The forces produce torques, but the torques sum to zero over the whole

atom, so that a conservation law holds for the total orbital angular momentum

$$\mathbf{L} = \sum_i \mathbf{L}_i. \tag{9-41}$$

The interactions in Equation (9-37) do not affect the spins of the particles, and so another conservation law also holds for the total spin

$$\mathbf{S} = \sum_i \mathbf{S}_i. \tag{9-42}$$

The separate conservation of $\mathbf{L}$ and $\mathbf{S}$ implies that the states of the atom can be specified by the associated *total orbital and spin quantum numbers* (ℓm_ℓ) and $(s m_s)$. We define these parameters in terms of the quantities L^2, L_z, S^2, and S_z by the following eigenvalue assignments:

$$L^2 \rightarrow \hbar^2 \ell(\ell + 1) \quad \text{and} \quad L_z \rightarrow \hbar m_\ell,$$
$$S^2 \rightarrow \hbar^2 s(s + 1) \quad \text{and} \quad S_z \rightarrow \hbar m_s.$$

The total z-component quantum numbers are given as sums of the individual z-component quantum numbers,

$$m_\ell = \sum_i m_{\ell_i} \quad \text{and} \quad m_s = \sum_i m_{s_i}, \tag{9-43}$$

in keeping with Equations (9-41) and (9-42). The allowed values of the quantum numbers ℓ and s are determined by the rules of vector addition and angular momentum quantization. This method of combining angular momenta produces a set of good quantum numbers $(\ell m_\ell s m_s)$ for each atomic state. Since the procedure is based on the construction of $\mathbf{L}$ and $\mathbf{S}$, the more suggestive name for the method is the *LS coupling scheme*.

We interpret *LS* coupling according to the roles played by the two parts of Equation (9-37). The central contribution $\sum_i V_c(r_i)$ gives the familiar electron configurations in which the good quantum numbers refer to the individual angular momenta $\mathbf{L}_i$ and $\mathbf{S}_i$. The noncentral interaction V_{ee} violates the conservation of $\mathbf{L}_i$ and $\mathbf{S}_i$ but maintains the conservation of $\mathbf{L}$ and $\mathbf{S}$. This violation affects the configurations in the following way. A typical configuration may contain many assignments of the individual quantum numbers $(n_i \ell_i m_{\ell_i} m_{s_i})$. The corresponding states can be associated with definite values of m_ℓ and m_s, by virtue of Equations (9-43), and can be superposed in linear combinations to form states with definite values of ℓ and s. Electrons in such a configuration experience the noncentral V_{ee} effect and assume different energies for different orientations of their angular momenta. Thus, the correction V_{ee} produces a *splitting* of degenerate states in the given configuration so that separate energy levels emerge for different pairs of total quantum numbers ℓ and s. We have seen a simple example of this splitting in the case of the $1s2s$ configuration of the He atom. Equations (9-27) and (9-28) represent product combinations of $(n_1 \ell_1 m_{\ell_1} m_{s_1})$ and $(n_2 \ell_2 m_{\ell_2} m_{s_2})$, with $m_\ell = 0$, in which the $m_s = 0$ eigenfunctions separate into an $(\ell = 0, s = 0)$ state and an $(\ell = 0, s = 1)$ state. The energies of the states are different because of the V_{ee} effect. We have given these He energy levels the names $2s\,^1S$ and $2s\,^3S$ in Figure 9-16.

The conservation of $\mathbf{L}$ and $\mathbf{S}$ implies a rotational symmetry and a corresponding degeneracy among the states of the LS coupling scheme. We find that the energies of the states may depend on ℓ and s but cannot depend on m_ℓ and m_s as the latter two quantum numbers range in integer steps from $-\ell$ to ℓ and $-s$ to s. Therefore, a given energy level for a particular ℓ and s comprises $(2\ell + 1)$ times $(2s + 1)$ degenerate states labeled by the quantum numbers $(\ell m_\ell s m_s)$. We can also perform a final addition of angular momenta and introduce the conserved total angular momentum of the atom

$$\mathbf{J} = \mathbf{L} + \mathbf{S}.$$

The *total angular momentum quantum numbers* j and m_j are then defined by the usual eigenvalues

$$\hbar^2 j(j + 1) \text{ for } J^2 \quad \text{and} \quad \hbar m_j \text{ for } J_z.$$

It is possible to rearrange the $(2\ell + 1)(2s + 1)$ degenerate states with quantum numbers $(\ell m_\ell s m_s)$ and construct as many alternative states with quantum numbers $(\ell s j m_j)$. This rearrangement from one set of good quantum numbers to another exactly parallels the transformation from $(n\ell m_\ell m_s)$ to $(n\ell j m_j)$ for the one-electron atom, as described in Section 8-8 and Figure 8-22. The two sets of quantum numbers $(\ell m_\ell s m_s)$ and $(\ell s j m_j)$ afford equally good representations of all the states with the given energy, at this stage of the LS coupling procedure. Such a collection of states with definite ℓ and s is denoted spectroscopically as

$$^{2s+1}L$$

and is called a *term*. The $(2\ell + 1)(2s + 1)$ degenerate states of a given term are designated by the final spectroscopic notation

$$^{2s+1}L_j$$

when the states are sorted according to the quantum number j. We have already employed this notation to describe the $^1S_0$ ground state and the $^1S_0$ and $^3S_1$ excited states of helium. We emphasize that it is meaningful to assign total orbital and spin quantum numbers ℓ and s to the levels of an atom only in circumstances where it is legitimate to follow the LS coupling scheme.

The next stage of the procedure is the incorporation of the spin–orbit interaction V_{SL} from Equation (9-40). We find that this weak coupling between the orbital and spin angular momenta of each electron introduces a weak breakdown of the separate conservation laws for $\mathbf{L}$ and $\mathbf{S}$ but maintains the conservation of $\mathbf{J}$ since the atom remains an isolated system. Hence, the total angular momentum quantum numbers j and m_j are still good for labeling the states. The spin–orbit interaction produces a small energy shift among the previously degenerate states of a given ℓs term and splits the term energy level into distinct *fine structure components* for all the possible values of j. These final energies cannot vary with m_j, and so each component has a remaining $(2j + 1)$-fold degeneracy with respect to this quantum number for each assignment of j.

Figure 9-22 summarizes the LS coupling treatment of the two corrections to the central field. We begin with the central potential energy $\Sigma_i V_c(r_i)$ and specify configurations by sets of independent-electron orbital quantum numbers $(n_i \ell_i)$. We then

Figure 9-22

Development of atomic energy levels $^{2s+1}L_j$ in the LS coupling scheme. The corrections to the central field break the degeneracy of a given electron configuration in two stages.

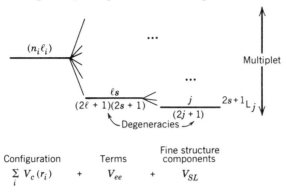

add the residual electrostatic interaction V_{ee} and divide each highly degenerate configuration into term levels with quantum numbers ℓ and s. We finally include the spin–orbit interaction and split each term into its j-dependent fine structure components. The final collection of all levels $^{2s+1}L_j$ constitutes a *multiplet* originating from a single configuration. Note that ℓ and s continue to be used through the final stage of the procedure, even though these quantum numbers are only approximately good in the presence of weak spin–orbit coupling. It is clear that the utility of ℓ and s, as well as the validity of the entire LS scheme, must eventually break down as the spin–orbit interaction grows in strength with increasing Z.

We turn to another method known as *jj coupling* when V_{SL} becomes dominant over V_{ee}. The two corrections have this property for atoms in the high-Z part of the periodic table. The states of such an atom are determined in first and second approximation by the central-field and spin–orbit interactions

$$\sum_i V_c(r_i) + V_{SL},$$

while the weaker electrostatic correction V_{ee} is set aside until the final stage. We see at once that the conservation laws for the various angular momenta take a different route under these conditions. The leading potential energy $\sum_i V_c(r_i)$ conserves $\mathbf{L}_i$ and $\mathbf{S}_i$ for each electron. The next-to-leading contribution V_{SL} does not conserve these angular momenta individually but does conserve the sum

$$\mathbf{J}_i = \mathbf{L}_i + \mathbf{S}_i.$$

We can therefore employ the eigenvalue assignments

$$\hbar^2 j_i(j_i + 1) \text{ for } J_i^2 \quad \text{and} \quad \hbar m_{j_i} \text{ for } J_{iz}$$

to introduce j_i and m_{j_i} as good quantum numbers, and we can label the states of the electrons as $(j_1 m_{j_1} j_2 m_{j_2} \cdots)$ in the absence of the residual interaction V_{ee}. The total

---

Figure 9-23

Comparison of LS and jj coupling schemes. Orbital and spin angular momenta are combined horizontally-then-vertically in LS coupling and vertically-then-horizontally in jj coupling.

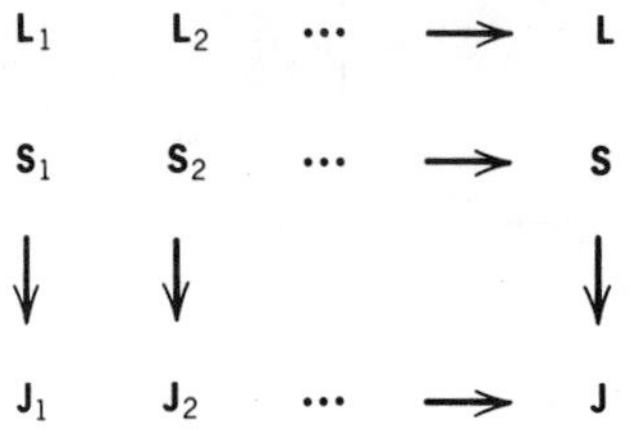

$$\mathbf{L}_1 \quad \mathbf{L}_2 \quad \cdots \quad \longrightarrow \quad \mathbf{L}$$

$$\mathbf{S}_1 \quad \mathbf{S}_2 \quad \cdots \quad \longrightarrow \quad \mathbf{S}$$

$$\downarrow \qquad \downarrow \qquad\qquad\qquad \downarrow$$

$$\mathbf{J}_1 \quad \mathbf{J}_2 \quad \cdots \quad \longrightarrow \quad \mathbf{J}$$

---

angular momentum of the atom is given by the sum of the individual $\mathbf{J}$ vectors

$$\mathbf{J} = \sum_i \mathbf{J}_i$$

and is conserved whether V_{ee} is included or not. Consequently, we are able to use the already-defined total angular momentum quantum numbers j and m_j, along with the individual quantum numbers j_i for each electron, as an alternative set of good quantum numbers $(j_1 j_2 \cdots j m_j)$. We are then in position to use this specification of the states when we finally include the correction V_{ee}.

These brief remarks summarize all that we intend to say about the method of jj coupling. Figure 9-23 shows a schematic plan of the LS and jj coupling procedures. It is understood that a specific pattern of conservation laws is selected according to the relative strengths of the two corrections, so that a corresponding pathway is taken across the figure in the two different schemes. We use the horizontal-then-vertical route and concentrate on the LS coupling scheme in the applications that follow.

We have yet to deduce the quantum numbers, ℓ, s, and j associated with the quantities

$$\mathbf{L} = \sum_i \mathbf{L}_i, \quad \mathbf{S} = \sum_i \mathbf{S}_i, \quad \text{and} \quad \mathbf{J} = \mathbf{L} + \mathbf{S}.$$

These exercises in angular momentum addition are quite similar to the vector-addition problems encountered in Section 8-8 and Figure 8-22. We begin with the observation that electrons in filled subshells form a ^1S core and make no contribution to $\mathbf{L}$ and $\mathbf{S}$. It follows that the vector sums $\sum_i \mathbf{L}_i$ and $\sum_i \mathbf{S}_i$ pertain only to the optically active electrons in the atom. Let us be content to examine the case of *two* electrons outside filled subshells. We recall that $\mathbf{S}$ has already been constructed for two electrons in Section 9-6 and that the results $s = 0$ and $s = 1$ are obtained when we add the two spin-$\frac{1}{2}$ vectors. The remaining constructions

$$\mathbf{L} = \mathbf{L}_1 + \mathbf{L}_2 \quad \text{and} \quad \mathbf{J} = \mathbf{L} + \mathbf{S}$$

then reduce to a single problem involving the addition of two integer-valued angular momenta.

The quantized vectors $\mathbf{L}_1$ and $\mathbf{L}_2$ have magnitudes $\hbar\sqrt{\ell_1(\ell_1 + 1)}$ and $\hbar\sqrt{\ell_2(\ell_2 + 1)}$, and z components $\hbar m_{\ell_1}$ and $\hbar m_{\ell_2}$. The L_z quantum number satisfies the relations

$$m_\ell = m_{\ell_1} + m_{\ell_2} \quad \text{and} \quad -\ell \le m_\ell \le \ell.$$

This parameter has its largest value when $m_{\ell_1} = \ell_1$ and $m_{\ell_2} = \ell_2$, and so the total orbital quantum number ℓ cannot exceed $\ell_1 + \ell_2$. Every lesser integer down to $|\ell_1 - \ell_2|$ is also allowed in the addition of the two quantized vectors. We therefore obtain the following list of possibilities for the total orbital quantum number:

$$\ell = |\ell_1 - \ell_2|, \ |\ell_1 - \ell_2| + 1, \ \ldots, \ \ell_1 + \ell_2 - 1, \ \ell_1 + \ell_2. \tag{9-44}$$

As usual, m_ℓ ranges from $-\ell$ to ℓ in integer steps for each of these values of ℓ. Thus, for the case $\ell_1 = \ell_2 = 1$, we have

$$\ell = 0 \quad \text{with } m_\ell = 0,$$
$$\ell = 1 \quad \text{with } m_\ell = -1, 0, 1,$$
$$\ell = 2 \quad \text{with } m_\ell = -2, -1, 0, 1, 2.$$

These nine different states describe all nine possible orientations of the selected pair of vectors $\mathbf{L}_1$ and $\mathbf{L}_2$.

The vector addition of $\mathbf{L}$ and $\mathbf{S}$ proceeds along exactly the same lines to a similar conclusion regarding the total angular momentum quantum number j. For given values of ℓ and s we find the following list of possible results:

$$j = |\ell - s|, \ |\ell - s| + 1, \ \ldots, \ \ell + s - 1, \ \ell + s. \tag{9-45}$$

Each choice of j admits a range of values for the z-component quantum number m_j from $-j$ to j in $2j + 1$ steps. It should be clear that Equation (9-45) holds equally well for integral or half-integral values of the total spin quantum number s. We add in closing that these same rules for combining angular momenta are readily adaptable to the jj coupling scheme.

Example

Let us apply LS coupling to an atom with two optically active electrons, taking the configuration to consist of one p electron and one d electron. The spin-orbital assignments for the two particles are $(\ell_1 = 1, s_1 = \frac{1}{2})$ and $(\ell_2 = 2, s_2 = \frac{1}{2})$. We combine the spins in the usual way to form total spin states with $s = 0$ and 1, and we use Equation (9-44) to obtain the values $\ell = 1$, 2, and 3 for the total orbital quantum number. The terms in this configuration are classified as singlets $(2s + 1 = 1$ for $s = 0)$ and triplets $(2s + 1 = 3$ for $s = 1)$. The three choices of ℓ generate the singlet terms ^1P, ^1D, and ^1F and the triplet terms ^3P, ^3D, and ^3F. Each term level has fine structure components with values of j as given in Equation (9-45). The $s = 0$ terms must have $j = \ell$, and so the singlet components are

$$^1P_1, \quad ^1D_2, \quad \text{and} \quad ^1F_3.$$

The $s = 1$ terms can have

$$j = 0, 1, \text{ and } 2 \quad \text{for } \ell = 1,$$
$$j = 1, 2, \text{ and } 3 \quad \text{for } \ell = 2,$$
$$j = 2, 3, \text{ and } 4 \quad \text{for } \ell = 3.$$

The corresponding triplet components are

$$^3P_{0,1,2}, \quad ^3D_{1,2,3}, \quad \text{and} \quad ^3F_{2,3,4}.$$

Thus, the entire multiplet consists of 12 different energy levels in all.

9-9 Spectroscopic Aspects of *LS* Coupling

The dynamics of *LS* coupling produces a multiplet from a configuration via the two steps shown in Figure 9-22. This procedure has clear implications for theory and experiment, since the breakdown of configurations into multiplets is a predictable phenomenon with observable spectroscopic consequences.

We are concerned with the evolution of multiplets in electron configurations of two types. A configuration is said to describe *equivalent* electrons if the same orbital quantum numbers are assigned to all the electrons outside filled subshells. Examples are the $1s^2$ configuration of helium and the $1s^2 2s^2 2p^2$ configuration of carbon. The electrons are called *inequivalent* if their $(n_i \ell_i)$ assignments are different, as in the He configuration $1s2s$ and the C configurations $1s^2 2s^2 2p3s$ and $1s^2 2s^2 2p3p$.

We find that particular attention must be given to the Pauli principle whenever we analyze a system of electrons in the equivalent category. The analysis is assisted by three observations, put forward originally by F. Hund during the early period of the quantum theory. These empirical rules refer specifically to the application of *LS* coupling in ground-state configurations containing equivalent electrons. The first two of Hund's rules pertain to the V_{ee} effect and the splitting of a configuration into ℓs-term levels:

> The term of least energy corresponds to the largest possible s while
> other term energies increase with decreasing values of s,
> and
> for maximum s the lowest-energy term occurs for the largest possible ℓ.

These guidelines can be used to address the problem of finding the lowest-energy level in a given configuration. The statements do not imply that every term level of maximum s must lie below all those of the next largest s and do not imply that the term of maximum ℓ must have the lowest energy when s is not maximum. We illustrate the limited applicability of the rules later in this section.

The first Hund rule is an extension of our arguments about the energies of the singlet and triplet states in the two-electron atom. We know from Section 9-6 that there is less repulsion between electrons in the spin-symmetric triplet states of helium because the accompanying antisymmetric spatial eigenfunctions give smaller probabilities for finding the two electrons close together. An atom with several electrons is in a state of maximum s when the system has parallel spins. Such a state is spin symmetric, and so again the accompanying spatial eigenfunction is antisymmetric in space and admits minimal repulsion between electrons.

The second Hund rule associates minimal electron–electron repulsion with the observation that electrons stay farthest apart in states of maximum ℓ. To see the plausibility of this argument we first note that the largest choice of ℓ allows the largest possibility for $|m_\ell|$. We then refer back to Figure 7-3 and recall that these states have highly equatorial polar probability distributions in which the electrons are able to have maximal separation.

We deduce the terms in a particular configuration by applying the addition of angular momenta and imposing the Pauli principle. Our concern is still limited to electrons outside filled subshells. Our attention is further restricted to the case of two electrons so that the two pairs of quantum numbers $(n_1\ell_1 \; n_2\ell_2)$ suffice to identify any configuration. The two spin-$\frac{1}{2}$ particles can form singlet and triplet term states with total spin quantum numbers $s = 0$ and $s = 1$. To specify the terms completely we must also determine the allowed values of the total orbital quantum number ℓ. This straightforward exercise in angular momentum addition is already in harmony with the Pauli principle as long as the electrons are inequivalent, since the individual subshell quantum numbers $(n_i\ell_i)$ are different and automatically satisfy the exclusion principle. The exercise becomes more involved for equivalent electrons because the list of proposed terms must be examined for possible violations of electron antisymmetry. We illuminate these remarks and illustrate Hund's rules in the following paragraphs.

The inequivalent-electron configuration $np \; n'p$ contains singlets and triplets with ℓ values 0, 1, and 2, as noted in our comments following Equation (9-44). The singlet terms are ^1S, ^1P, and ^1D, and the triplet terms are ^3S, ^3P, and ^3D. Hund's first two rules tell us that the ^3D term should have the lowest energy.

The simplest equivalent-electron situation arises in the configuration ns^2. Our discussion of the He atom in Section 9-6 has already covered the application of the Pauli principle to this problem. We see that both ^1S and ^3S terms are possible under the rules for angular momentum addition, but we find that only the ^1S term is allowed by the exchange antisymmetry of the two electrons.

The np^2 configuration provides our first opportunity to sample the real difficulties posed by equivalent electrons. To analyze this case we could list all the possible values for the electron quantum numbers $(m_{\ell_1} m_{s_1} m_{\ell_2} m_{s_2})$ and then strike those entries with the same assignments for $(m_{\ell_1} m_{s_1})$ and $(m_{\ell_2} m_{s_2})$. Instead, we adopt a more efficient way of treating two electrons antisymmetrically by following the approach introduced in Equations (9-22). These constructions sort out the choices for m_{s_1} and m_{s_2} into singlet ($s = 0$) and triplet ($s = 1$) combinations and associate antisymmetric and symmetric behavior under exchange of the two electron spins. The spin states are accompanied by spatial eigenfunctions whose exchange properties are appropriate for the overall antisymmetric description of the two electrons. We can take advantage of this procedure to reduce and organize our list of possible np^2 states. The detailed arguments are discussed at the end of the section.

The main result of our analysis of the np^2 configuration is the list of values of ℓ and s in Table 9-2. We find that $\ell = 0$ and $\ell = 2$ occur only for $s = 0$ and that $\ell = 1$ occurs only for $s = 1$. Hence, the only singlet terms are ^1S and ^1D, and the only triplet term is ^3P. The six $s = 0$ entries in the table can be organized so that one state belongs to ^1S and five states belong to ^1D. The ^3P term contains nine states since there are three m_ℓ values for $\ell = 1$ to be coupled with three m_s values for $s = 1$. Thus, the ^1S, ^1D, and ^3P terms of the np^2 configuration represent 15 different states in all, appreciably less than the number found in the $np \; n'p$ configuration.

Table 9-2   Orbital Quantum Numbers in the Configuration np^2

Total Spin $s=0$				Total Spin $s=1$		
$m_{\ell_1}\ \ m_{\ell_2}$	$m_\ell = m_{\ell_1}+m_{\ell_2}$	ℓ		$m_{\ell_1}\ \ m_{\ell_2}$	$m_\ell = m_{\ell_1}+m_{\ell_2}$	ℓ
$1\ \ \ 1$	2	2				
$\begin{bmatrix}1&0\\ \&\\ 0&1\end{bmatrix}^S$	1	2		$\begin{bmatrix}1&0\\ \&\\ 0&1\end{bmatrix}^A$	1	1
$\begin{bmatrix}1&-1\\ \&\\ -1&1\end{bmatrix}^S$	0	$2,0$		$\begin{bmatrix}1&-1\\ \&\\ -1&1\end{bmatrix}^A$	0	1
$0\ \ \ 0$	0	$2,0$				
$\begin{bmatrix}0&-1\\ \&\\ -1&0\end{bmatrix}^S$	-1	2		$\begin{bmatrix}0&-1\\ \&\\ -1&0\end{bmatrix}^A$	-1	1
$-1\ \ -1$	-2	2				

A third Hund rule is also available as another guideline to use in the final stage of multiplet formation. We recall from Figure 9-22 that the spin–orbit correction V_{SL} acts as the last step to split the various ℓs-term levels into their j-dependent fine structure components. The empirical rule identifies an incomplete subshell of an atom as either less than or more than half-filled and arranges the fine structure in either normal or inverted order:

> The lowest energy level in the lowest-energy term corresponds to the smallest j value $j = |\ell - s|$ in the normal case, or the largest j value $j = \ell + s$ in the inverted case.

We offer the following remarks to give some theoretical justification for this rule.

Equation (9-40) describes the spin–orbit interaction V_{SL} in terms of the coupled orbital and spin angular momenta of the *individual* atomic electrons. It is possible to prove that electrons in filled subshells make no contribution to this expression. It is also possible to reexpress the V_{SL} effect in terms of the *total* orbital and spin angular momenta $\mathbf{L}$ and $\mathbf{S}$. In the end we find that the expectation value of Equation (9-40) turns into a simple formula for the spin–orbit energy shift:

$$\langle V_{SL}\rangle = \left\langle \sum_i \xi_c(r_i)\mathbf{S}_i \cdot \mathbf{L}_i \right\rangle$$

$$= A\langle \mathbf{S}\cdot\mathbf{L}\rangle. \tag{9-46}$$

The expectation value is taken in a term state with quantum numbers $(\ell s j m_j)$ and produces a proportionality constant A whose value is independent of j. We use the familiar inequality

$$\langle \mathbf{S}\cdot\mathbf{L}\rangle = \frac{1}{2}\langle J^2 - L^2 - S^2\rangle = \frac{\hbar^2}{2}\left[j(j+1) - \ell(\ell+1) - s(s+1)\right],$$

and we immediately reduce the formula to a final expression involving the quantum numbers ℓ, s, and j:

$$\langle V_{SL} \rangle = \frac{A\hbar^2}{2} \left[j(j+1) - \ell(\ell+1) - s(s+1) \right]. \tag{9-47}$$

This result predicts a splitting of fine structure components in a given ℓs term and orders the energies monotonically with j for the given choice of ℓ and s. It is easy to derive two conclusions from Equation (9-47):

$$\langle V_{SL} \rangle_{j_{max}} = A\ell s\hbar^2 \quad \text{for } j = j_{max} = \ell + s$$

and

$$\langle V_{SL} \rangle_{j_{min}} = \begin{cases} -As(\ell+1)\hbar^2 \\ -A\ell(s+1)\hbar^2 \end{cases} \quad \text{for } j = j_{min} = \begin{cases} \ell - s \\ s - \ell. \end{cases}$$

It follows that the shift to the lowest energy occurs either in the state with $j = j_{min}$ if A is positive, or in the state with $j = j_{max}$ if A is negative, as indicated in Hund's third rule.

The difference in the energy of adjacent components is an interesting observable quantity. We can predict this energy interval directly from Equation (9-47):

$$\delta E_{SL} = \langle V_{SL} \rangle_j - \langle V_{SL} \rangle_{j-1}$$

$$= \frac{A\hbar^2}{2} \left[j(j+1) - (j-1)j \right] = Aj\hbar^2. \tag{9-48}$$

The result confirms Lande's interval rule, another relic from the early era of atomic spectroscopy:

> The energy separation of successive components in a given term is proportional to the larger of the two j values.

Experimental support for this rule is well documented wherever *LS* coupling is applicable.

Inverted fine structure is seen instead of normal fine structure when the incomplete subshell is more than, instead of less than, half-filled. These situations are connected by the fact that a fully occupied subshell has vanishing angular momenta and vanishing spin–orbit interactions. We can relate occupancy and vacancy in a partially filled subshell if we interpret summations of the type found in Equations (9-40) to (9-42) as

$$\sum_{\substack{n \\ \text{electrons}}} = \sum_{\substack{N \\ \text{electrons}}} - \sum_{\substack{N-n \\ \text{holes}}} = - \sum_{\substack{N-n \\ \text{holes}}},$$

where N denotes the number of electrons needed for full occupation of the subshell. Thus, n electrons and $N - n$ holes make equivalent contributions to quantities like **L**, **S**, and $\langle V_{SL} \rangle$, except for the change in sign. Fine structure inversion follows from this observation when the more-than-half-filled case $n > N/2$ replaces the less-than-half-

Figure 9-24

Schematic diagram of terms and fine structure components for an $np\,n'p$ configuration and for an np^2 configuration. Fine structure splittings are greatly exaggerated relative to the energy differences between terms.

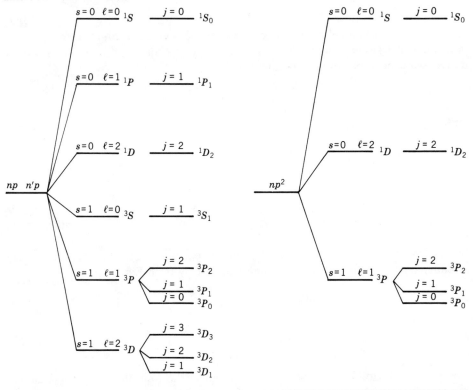

filled case $n < N/2$. The argument also implies that no fine structure is expected when the subshell is exactly half-filled.

We can use Hund's rules to deduce the state of *lowest* energy in a normal multiplet by selecting the largest s, the largest ℓ, and the smallest j, in that order. Let us also imagine a conjectured extension of this procedure in which we order *all* the levels of a multiplet with increasing values of the energy first by decreasing s, next by decreasing ℓ, and last by increasing j. This strategy is applied in Figure 9-24 to the terms ^1S, ^1P, ^1D, ^3S, ^3P, and ^3D in the $np\,n'p$ configuration. Note that all $s = 1$ terms are predicted to lie below all $s = 0$ terms and that fine structure is expected only in the terms ^3P and ^3D. The entire multiplet is supposed to consist of the ascending array of energy levels

$$^3D_{1,2,3} \quad ^3P_{0,1,2} \quad ^3S_1 \quad ^1D_2 \quad ^1P_1 \quad ^1S_0.$$

The figure goes on to show the altered prediction of levels for the equivalent-electron configuration np^2. We have just learned that the Pauli principle excludes half of the above terms so that the resulting multiplet contains only the levels

$$^3P_{0,1,2} \quad ^1D_2 \quad ^1S_0.$$

It is instructive to count the possible values of m_j for each assignment of j in the np^2 case:

$$^3P_{0,1,2} \text{ has } m_j = 0, \ m_j = -1 \text{ to } 1, \text{ and } m_j = -2 \text{ to } 2,$$

$$\text{while } {}^1D_2 \text{ has } m_j = -2 \text{ to } 2, \text{ and } {}^1S_0 \text{ has } m_j = 0.$$

The exercise demonstrates again the existence of 15 different states in this equivalent-electron system.

The proposed ordering of energy levels in Figure 9-24 is actually realized by many atoms, at least in equivalent-electron situations of the type np^2. Figure 9-25 shows the extent to which the carbon atom fulfills the predictions. We know that the ground state of carbon has the configuration $1s^2 2s^2 2p^2$, and we see in the figure that the lowest configurations of the two optical electrons are $2p^2$, $2p3s$, and $2p3p$. It is clear that the actual levels of carbon agree with the predictions for $2p^2$ but disagree considerably with the ordering proposed for $2p3p$.

The *LS* coupling scheme is based on certain good angular momentum quantum numbers in the system of interacting electrons. Parity furnishes another good quantum number to associate with the levels of the complex atom. This property is expected because the dynamics of the atom is governed by the interactions

$$\sum_i V_c(r_i) + V_{ee} + V_{SL}$$

and because these quantities are not affected by the parity operation $\mathbf{r} \rightarrow -\mathbf{r}$. We use the orbital quantum number ℓ_i to obtain $(-1)^{\ell_i}$ as the parity of the ith electron. We then use all the independent-particle quantum numbers $(\ell_1 \ell_2 \cdots \ell_Z)$ to establish the parity for an *entire multiplet* of states. Thus, Figure 9-22 tells us that the space-inversion property of every term and component is decided at the configuration level and remains good even after the conservation of each $\mathbf{L}_i$ is broken by the secondary corrections. A fully occupied subshell contains an even number of identical ℓ_i assignments and contributes even parity, so that the optically active electrons determine the overall parity of the multiplet system. As examples, both of the multiplets for $np\, n'p$ and np^2 have two odd-parity electrons in their starting configurations, and so an assignment of *even* overall parity follows for every level sketched in Figure 9-24. Note that the evenness or oddness of the total orbital quantum number ℓ has no bearing on this conclusion.

The energy levels in Figures 9-16, 9-18, 9-19, and 9-25 show many differences in energy of order 1–10 eV. Radiative transitions among these levels generate spectral lines in the visible range of wavelengths. We can attribute such optical transitions to oscillations of the electric dipole moment of the atom, and we can assume the validity of the electric dipole selection rules. We have argued that these rules act as conservation laws for the system of atom-plus-radiation, in which an electric dipole photon carries away one $\hbar$ unit of vector angular momentum. Changes in the total angular momentum quantum numbers of the atom must therefore obey the conditions

$$\Delta j = 0 \text{ or } \pm 1 \quad (\text{but } 0 \nrightarrow 0) \quad \text{and} \quad \Delta m_j = 0 \text{ or } \pm 1,$$

as in Equations (8-43) and (8-44). The parity of the atomic state must undergo a change, either odd $\rightarrow$ even or even $\rightarrow$ odd, because the emitted photon also carries away odd parity. We can go beyond the domain of strict conservation laws and

Figure 9-25

Energy levels of the carbon atom. Singlet and triplet terms are shown for the configurations $2p^2$, $2p3s$, and $2p3p$.

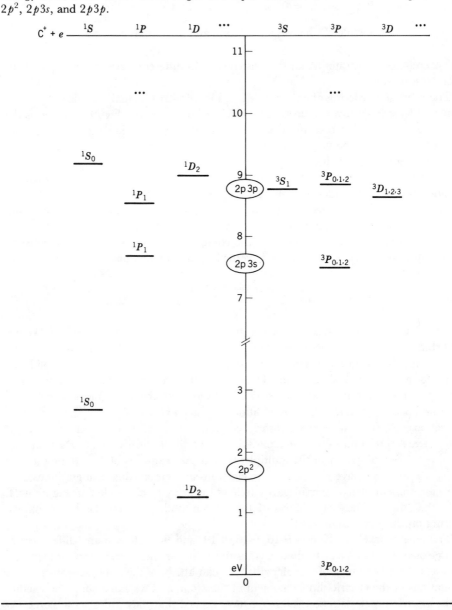

include the quantum numbers of the LS coupling scheme in these selection rules. The properties of the electric dipole moment are independent of spin, and so the total spin quantum number is not affected in the transition. This observation implies that the condition on j becomes an immediate condition on the total orbital quantum number and results in the additional pair of selection rules

$$\Delta s = 0 \quad \text{and} \quad \Delta \ell = 0 \text{ or } \pm 1 \quad (\text{but } 0 \nrightarrow 0).$$

These constraints on ℓ and s hold as long as ℓ and s are sufficiently good quantum numbers for the description of the atom. The independent-electron quantum numbers $(\ell_1\ell_2 \cdots \ell_Z)$ have a similar status in the radiation process. If the radiation can be attributed to a single electron, then the electric dipole moment refers to that electron and the orbital selection rule takes the form

$$\Delta\ell_i = \pm 1.$$

(Note that the possibility $\Delta\ell_i = 0$ is ruled out because the parity of the atom must change.) We observe that the $\Delta s = 0$ rule forbids singlet–triplet transitions in atoms with two optical electrons, as in the case of the para and ortho systems in helium. Our diagram for the carbon atom in Figure 9-25 is patterned after the helium levels in Figure 9-16 so that attention can be drawn to this analogy.

We have been assuming that the atom is isolated from any applied external field. We therefore have no preferred direction in space to select as an axis of quantization. This existence of a symmetry with respect to the choice of z direction implies that the energy of the atom cannot depend on the z-component quantum number m_j, so that a $(2j + 1)$-fold degeneracy holds for every energy level with total angular momentum quantum number j. We can introduce a natural choice of axis and break this degeneracy by applying a constant magnetic field $\mathbf{B}$. The additional magnetic interaction between the atom and the field produces an m_j-dependent energy level shift. This shift of levels is observed in the emission spectrum of the atom as the familiar Zeeman effect.

We take the z axis in the direction of $\mathbf{B}$ and express the magnetic interaction by the usual formula

$$V_M = -\boldsymbol{\mu} \cdot \mathbf{B} = -\mu_z B.$$

The magnetic moment of the atom is constructed from the orbital and spin contributions of each electron in the manner of Equation (8-45):

$$\boldsymbol{\mu} = \sum_{i=1}^{Z} \boldsymbol{\mu}_i = -\frac{\mu_B}{\hbar}\sum_i (\mathbf{L}_i + 2\mathbf{S}_i) = -\frac{\mu_B}{\hbar}(\mathbf{L} + 2\mathbf{S}).$$

Observe that $g_s = 2$ is again used for the electron spin g-factor, and note especially that the result is ideally suited to the strategy of LS coupling. Let us consider an energy level for a fine structure component $^{2s+1}L_j$ with quantum numbers $(\ell s j m_j)$. We write the formula for the energy shift due to the magnetic interaction in terms of the expectation value of V_M in this state:

$$\langle V_M \rangle = -B\langle \mu_z \rangle = \frac{\mu_B B}{\hbar}\langle L_z + 2S_z \rangle.$$

We then recall the derivation of $\langle \mu_z \rangle$ in Equation (8-50) and realize that the same techniques can be carried over by substituting the quantum numbers of the many-electron atom. The final formula for the energy shift can therefore be quoted directly from Equation (8-54):

$$\langle V_M \rangle = g\mu_B B m_j,$$

where

$$g = 1 + \frac{j(j + 1) + s(s + 1) - \ell(\ell + 1)}{2j(j + 1)}.$$

Note that the Lande g-factor varies with ℓ, s, and j as in Equation (8-51).

These results admit the real possibility for an $s = 0$ atom to exhibit a *normal* Zeeman effect. An $s = 0$ atom is in a singlet state with $j = \ell$ and $g = 1$, just like the hypothetical hydrogen atom without spin in Section 8-3. The selection rules allow transitions only to other singlet states, and so the application of a magnetic field causes the emission lines of the singlet-state atoms to trifurcate in the classical manner predicted by Lorentz. This possibility is left for further exploration in Problem 24 at the end of the chapter.

Detail

Table 9-2 lists the quantum numbers m_{ℓ_1} and m_{ℓ_2} for two equivalent electrons in an np^2 configuration. The compilation is separated into columns for $s = 0$ and $s = 1$ and is assembled in symmetric and antisymmetric combinations for the three ℓ values 0, 1, and 2. Spatial states are symmetric for $s = 0$ and antisymmetric for $s = 1$, and are tabulated as pairs of $(m_{\ell_1} m_{\ell_2})$ assignments by the notation [ ]^S and [ ]^A. We convey the meaning of these symbols by letting entries in the table correspond to the following spatial eigenfunctions for two electrons with quantum numbers $n_1 = n_2 = n$ and $\ell_1 = \ell_2 = 1$:

$$\psi_{22}^S = R_{n1}(r_1)R_{n1}(r_2)Y_{11}(\theta_1, \phi_1)Y_{11}(\theta_2, \phi_2)$$

$$\text{for } \left[m_{\ell_1} = 1 \quad m_{\ell_2} = 1\right],$$

$$\psi_{21}^S = R_{n1}(r_1)R_{n1}(r_2)\frac{1}{\sqrt{2}}\left[Y_{11}(\theta_1, \phi_1)Y_{10}(\theta_2, \phi_2) + Y_{10}(\theta_1, \phi_1)Y_{11}(\theta_2, \phi_2)\right]$$

$$\text{for } \begin{bmatrix} m_{\ell_1} = 1 & m_{\ell_2} = 0 \\ m_{\ell_1} = 0 & m_{\ell_2} = 1 \end{bmatrix}^S,$$

$$\cdots$$

$$\psi_{11}^A = R_{n1}(r_1)R_{n1}(r_2)\frac{1}{\sqrt{2}}\left[Y_{11}(\theta_1, \phi_1)Y_{10}(\theta_2, \phi_2) - Y_{10}(\theta_1, \phi_1)Y_{11}(\theta_2, \phi_2)\right]$$

$$\text{for } \begin{bmatrix} m_{\ell_1} = 1 & m_{\ell_2} = 0 \\ m_{\ell_1} = 0 & m_{\ell_2} = 1 \end{bmatrix}^A,$$

$$\cdots$$

Note that the angle-dependent factors control the exchange behavior of these expressions in a manner consistent with the definition of the functions ψ^S and ψ^A in Equations (9-22). It is clear that the symmetric function ψ_{22}^S has $m_\ell = 2$ and $\ell = 2$, given the assigned values $m_{\ell_1} = m_{\ell_2} = 1$. Both functions ψ_{21}^S and ψ_{11}^A have $m_\ell = 1$, in view of their $(m_{\ell_1} \ m_{\ell_2})$ assignments, and can belong to either $\ell = 2$ or $\ell = 1$. We note that ψ_{21}^S is symmetric, like the $\ell = 2$ function ψ_{22}^S, and must also have $\ell = 2$, so that the antisymmetric partner ψ_{11}^A must have $\ell = 1$. We have anticipated the demonstration of these total orbital quantum numbers by labeling the functions as $\psi_{\ell m_\ell}^{S,A}$. Other entries and ℓ values in the table can be understood in similar fashion. (We note that our use of the eigenfunctions $\psi_{\ell m_\ell}^{S,A}$ is a device intended to impose the Pauli principle on the two outer electrons alone. Thus, our procedure is not a rigorous application of the Pauli principle to the entire Z-electron system and is meant to be applied only for the determination of the term quantum numbers ℓ and s.)

Example

The ground-state configurations at $Z = 12, 13,$ and 14 are given in Table 9-1 as

$$[\text{Ne}]3s^2 \text{ for Mg}, \quad [\text{Ne}]3s^23p \text{ for Al}, \quad \text{and} \quad [\text{Ne}]3s^23p^2 \text{ for Si}.$$

The Mg configuration is composed of filled subshells and is expected to have a $^1S_0$ ground state. The Al atom has one optical p electron, implying a doublet term with $s = \frac{1}{2}$ and $\ell = 1$. The possible j values are $\frac{1}{2}$ and $\frac{3}{2}$ so that the ^2P term splits into $^2P_{1/2}$ and $^2P_{3/2}$ components. The $^2P_{1/2}$ state is predicted to have the lower energy. The Si atom has two equivalent p electrons outside filled subshells. The lowest levels should be just like those sketched on the right side of Figure 9-24, where the $^3P_0$ state is found to have the lowest energy.

Example

The ground state of the Fe atom at $Z = 26$ is a $1s^22s^22p^63s^23p^64s^23d^6$ configuration. The occupation of the outer subshell by six d electrons is equivalent to a subshell with four d holes. The largest possible value for the total spin quantum number corresponds to the case of four parallel spins, where $s = 2$ and $2s + 1 = 5$. It happens that the ^5D term has the lowest energy. The combination of quantum numbers $s = 2$ and $\ell = 2$ admits the possible j values 0, 1, 2, 3, and 4, and so the fine structure components $^5D_{0,1,2,3,4}$ are expected to form an inverted system in which the $^5D_4$ component occurs at the level of least energy.

Example

Let us conclude the section and the chapter with some remarks about the He–Ne laser to add to our previous description of the device in Section 3-9. We illustrate the relevant transitions of the two atoms in Figure 9-26. Operation of the laser begins with a pumping process that raises helium atoms from the ground state to the singlet and triplet $1s2s$ excited states. Note that the electric dipole selection rules do not affect these transitions since the excitation of atoms in the gaseous medium of the laser is collisional rather than radiational. The excited He states are metastable because the selection rules inhibit spontaneous transitions back to the ground state. Helium and neon atoms transfer their energies of excitation via the collision process

$$\text{He*} + \text{Ne} \rightarrow \text{He} + \text{Ne*}.$$

This mechanism populates the higher $2p^55s$ and $2p^54s$ levels of neon so that population inversions exist with respect to the $2p^53p$ levels at lower energy. Laser transitions can then be induced between these states, as indicated in the figure. The particular stimulated emission of 1.96 eV photons in transitions from $2p^55s$ to $2p^53p$ results in the generation of a 633 nm laser beam. The subsequent spontaneous decays $2p^53p \rightarrow 2p^53s \rightarrow 2p^6$ are allowed by the electric dipole selection rules.

Figure 9-26

Energy levels and transitions of He–Ne laser. Two four-level laser systems operate in neon through population inversions in the $2p^55s$ and $2p^54s$ states with laser transitions to the $2p^53p$ levels.

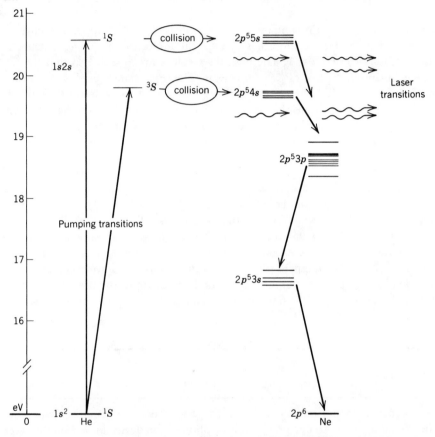

Problems

1. Show that the expression $\sum_{i<j=1}^{Z}(1/|\mathbf{r}_i - \mathbf{r}_j|)$ consists of $Z(Z-1)/2$ terms. Show also that the double summation can be rewritten as

$$\frac{1}{2}\sum_{\substack{i,\,j=1 \\ j\neq i}}^{Z}\frac{1}{|\mathbf{r}_i - \mathbf{r}_j|}.$$

2. The ground state of the helium atom lies 24.6 eV below the ionization level for the singly ionized system $\mathrm{He}^+ + e$. Deduce the position of the helium ground state relative to the double-ionization level for $\mathrm{He}^{++} + 2e$.

3. Suppose that, in first approximation, the helium eigenfunction $\psi(\mathbf{r}_1,\mathbf{r}_2)$ is assumed to obey the equation

$$-\frac{\hbar^2}{2m_e}\left(\nabla_1^2 + \nabla_2^2\right)\psi - \frac{e^2}{4\pi\varepsilon_0}\left(\frac{Z-\frac{5}{16}}{r_1} + \frac{Z-\frac{5}{16}}{r_2}\right)\psi = E\psi$$

with $Z = 2$. What would the ground-state eigenfunction and energy be in this case?

4. The sixth row of the periodic table lists the elements from $Z = 55$ to 86. What electron subshells are filled in assembling the ground states of these 32 atoms? Identify the ground-state configurations for $Z = 55$ and 56 and for $Z = 81$ to 86.

5. Consider an alkali atom in its ground state, and assume that the single electron outside closed shells experiences constant screening with $Z_{eff} > 1$. Use the tabulated values of the ionization energy to estimate Z_{eff} for lithium, sodium, and potassium.

6. The MN x-ray lines in the M series of any element correspond to $M \to N$ hole transitions in which an M-shell hole is filled by an electron from an N subshell. Sketch an x-ray level diagram showing all the M and N subshells labeled by their single-particle quantum numbers, and identify all the allowed MN hole transitions.

7. The x-ray lines of tungsten include the following wavelengths:

KL_{II}	KL_{III}	KM_{III}	$L_I M_{III}$
0.02138	0.02090	0.01844	0.12627 nm

The K absorption edge occurs at 0.01784 nm and the K-shell energy is given as 69.52 keV. Use these data to calculate the wavelength for each of the L absorption edges and the energy for each of the L subshells. Compare the computed values with the results given for tungsten in tables of x-ray information.

8. Let the eigenfunction for the three-electron atom be written as a Slater determinant:

$$\psi(1,2,3) = \frac{1}{\sqrt{3!}} \begin{vmatrix} \psi_\alpha(1) & \psi_\beta(1) & \psi_\gamma(1) \\ \psi_\alpha(2) & \psi_\beta(2) & \psi_\gamma(2) \\ \psi_\alpha(3) & \psi_\beta(3) & \psi_\gamma(3) \end{vmatrix}.$$

Demonstrate the exchange antisymmetry of this construction, and show that $\psi(1,2,3)$ satisfies the exclusion principle. Verify also that $\psi(1,2,3)$ is properly normalized.

9. Show that a definite parity can be identified with any Slater determinant for the Z-electron atom.

10. Express the singlet and triplet He eigenfunctions in the $1s2s$ configuration in terms of Slater determinants.

11. Let there be a spin–spin interaction for two electrons given by the formula

$$V_{SS} = \zeta \mathbf{S}_1 \cdot \mathbf{S}_2$$

in which ζ is a constant. Evaluate the interaction energy in the singlet and triplet states of the two electrons.

12. Refer to the wavelengths given in Figure 9-18 for the spectral lines of lithium, and calculate the energies above the ground state for the lithium levels shown in the same figure.

13. The spectrum of sodium includes lines with wavelengths (in nanometers)

$$818.3256, \quad 819.4790, \quad 819.4824$$

corresponding to the $3d \to 3p$ transitions indicated in Figure 9-19. Use the spectroscopic notation $n\ell\,^2 L_j$ to specify the states involved in each of these transitions. The deduction should allow for the fact that the sodium ^2D states form inverted doublets in which $j = \frac{3}{2}$ lies above $j = \frac{5}{2}$. Use the wavelength data to calculate the splittings of the initial and final doublet levels.

14. The ^2F levels of sodium consist of pairs of states whose closely spaced energies are difficult to resolve. Assume a normal fine structure ordering of the $j = \frac{7}{2}$ and $j = \frac{5}{2}$ states, and use a hydrogenic model of the valence electron to predict the spin–orbit splitting in the lowest ^2F doublet.

15. Consider two optically active electrons in orbital states with quantum numbers ($\ell_1 = 1$, $m_{\ell_1} = 0$) and ($\ell_2 = 1$, $m_{\ell_2} = 0$). Use graphical constructions like those in Figures 8-22 and 9-17 to represent the possible results for the total orbital quantum number ℓ.

16. Let an atom have one p electron and one d electron outside filled subshells, and use jj coupling to determine the possible values of the total angular momentum quantum number j. Show that the occurrence of j values is the same as in the LS coupling scheme.

17. The lowest excited states of the Be atom correspond to a $2s2p$ configuration of optically active electrons. What $^{2s+1}L_j$ components originate in this configuration? Assume that Hund's rules hold for the multiplet and deduce the ordering of the energy levels. Refer to a table of beryllium levels to see if the ordering occurs as predicted.

18. The spin–orbit energy shift takes the form

$$\langle V_{SL} \rangle = A\langle \mathbf{S} \cdot \mathbf{L} \rangle \quad (A \text{ independent of } j)$$

in the LS coupling scheme. Obtain expressions for the shift in states of maximum and minimum j for given values of ℓ and s.

19. Examine the terms and fine structure components in the configuration $np\, n'd$, and use Hund's rules to identify the term and component of lowest energy. Extend the scope of Hund's rules to predict an energy level diagram for the whole multiplet, showing the ordering of all the various terms and components.

20. The atoms potassium, calcium, scandium, and titanium have atomic numbers $Z = 19, 20, 21$, and 22. Identify the ground-state configuration of each atom, and organize the energy levels of each of these configurations according to the total quantum numbers s, ℓ, and j. Deduce the quantum numbers of the lowest-energy state of each atom, and identify the state by the spectroscopic notation $^{2s+1}L_j$.

21. Predict an energy level diagram showing the terms and fine structure components in the $2p3s$ configuration of the carbon atom. Compare the prediction with the known results shown in Figure 9-25.

22. Refer to the energy level diagram for the carbon atom in Figure 9-25 and indicate all the transitions allowed by the electric dipole selection rules.

23. The fine structure components for the lowest term of the iron atom have the following tabulated energies (in cm^{-1} units):

$^5D_4$	$^5D_3$	$^5D_2$	$^5D_1$	$^5D_0$
0	415.932	704.004	888.129	978.072

Use these numbers to check the validity of the Lande interval rule.

24. The excited singlet states of carbon $2p3s\,^1P_1$ and $2p^2\,^1D_2$ are shown on the energy level diagram in Figure 9-25. The two excitation energies are quoted in tables as 61982 and 10194 cm^{-1}. Let a 1 T magnetic field be applied to the atom and identify all the allowed transitions between the two split levels. Show that the normal Zeeman effect should be observed, and calculate the corresponding wavelength shifts.

T E N

MOLECULES

*I*n previous chapters we have studied the properties of the atom, that is, a single nucleus surrounded by a cloud of electrons. In our everyday dealings with matter, including that making up our own bodies, it would seem that such isolated atoms are rather rare and are encountered considerably less frequently than are *molecules*, which are the aggregates of two or more atoms. Indeed, all of our bodily processes, including breathing air, digesting food, and the formation of DNA in the cell, include chemical reactions that involve taking apart one molecule and recombining its constituent atoms to form other molecules. Our very existence is as inextricably linked to the nature of molecules as it is to that of atoms or nuclei.

Chemistry is the science of molecules. The various forces that hold molecules together include the covalent, ionic, and van der Waals interactions as manifestations of the *chemical bond*. All these forces are basically electromagnetic in nature but also involve, as we see in this chapter, some rather subtle quantum mechanical effects.

The simplest molecular form is the diatomic hydrogen molecular ion H_2^+, consisting of two protons and a single electron. The next in order of increasing complexity is H_2, with one more electron than H_2^+. These two molecules illustrate some of the basic principles involved in molecular structure, including in the case of H_2 the *covalent bond* in which the two nuclei "share" the two electrons equally. Other molecules, such as NaCl in which one nucleus steals an electron from the other, are held together by a somewhat different force known as the *ionic bond*. Atoms with closed shells such as argon attract each other with only a much weaker force known as the *van der Waals interaction*.

As with atoms, the most fruitful way to probe the structure of molecules is by spectroscopy. Not surprisingly, molecular spectra are often more complicated than those of atoms. This additional complexity arises because, in addition to electronic transitions, there are energy changes involving the relative motions of the nuclei that make up the molecule. Such motions are classified as *rotational* and *vibrational*. Because the rotational transitions involve relatively small energy changes, the spectra occur in characteristic groupings of very closely spaced lines called *bands*.

While the most mundane features of our lives depend on molecular processes, molecules are by no means restricted to our local environment. Astronomical observations at radio-wave frequencies have shown in recent years that molecular clouds are spread throughout the universe. Molecules that do not occur naturally on Earth have been discovered in space. Molecular clouds are found to be closely associated with regions giving birth to stars in galaxies. It would seem that molecules may be equally fundamental to our understanding of nature on every scale from the cosmological to the biological.

10-1   Binding by Quantum Tunneling: H_2^+

A molecule is a collection of two or more nuclei and their associated electrons with the whole complex bound together by electromagnetic forces. Solving the problem of a complex atom is difficult enough; here we have an even more complicated multicenter problem so that the electrons can be associated with one or more of several nuclei. To simplify the difficulties inherent in studying molecules we look primarily at the diatomic (two-atom) system. Furthermore, we begin our discussion with the simplest of these, H_2^+, which consists of two protons and just a single electron.

Electrons are considerably less massive than protons so that electron rearrangements occur much more rapidly than those of the more ponderous nuclei. This fact results in a useful technique, introduced by Born and J. R. Oppenheimer and known as the *Born–Oppenheimer approximation*. The nuclei are assumed to be at fixed points and the electron energies and wave functions are then found as functions of the fixed nuclear positions. The total electronic energy is then used, together with the direct internuclear forces, to form an effective potential energy for the nuclear motion.

Figure 10-1 shows the arrangement of the two protons in H_2^+; both are on the z axis, one at $-\mathbf{R}/2$ and one at $\mathbf{R}/2$, so the distance between them is R. The electron is at arbitrary position $\mathbf{r}$. The nuclear positions are taken as fixed so we need write the

Figure 10-1

Molecular coordinates for H_2^+. The protons are at $-\mathbf{R}/2$ and $\mathbf{R}/2$; the electron is at $\mathbf{r}$. The z axis is taken along the internuclear line.

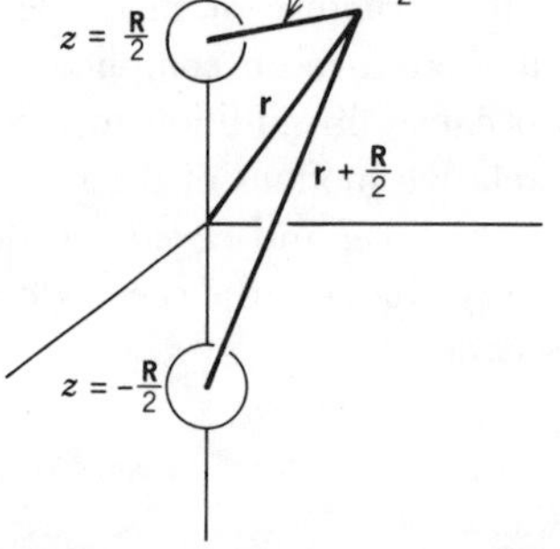

Figure 10-2

Potential energy seen by the electron in H_2^+ as it moves along the z axis. The dotted line gives that part of the potential arising just from the proton at $z = -R/2$. The dashed line shows the same for the proton at $z = R/2$.

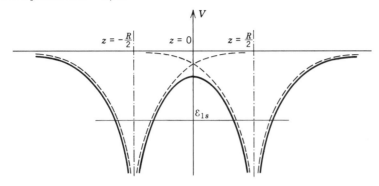

Schrödinger equation for only the *electronic* eigenfunction $\psi_e(\mathbf{r})$:

$$\left\{ -\frac{\hbar^2}{2m_e}\nabla_r^2 - \frac{ke^2}{|\mathbf{r} + \mathbf{R}/2|} - \frac{ke^2}{|\mathbf{r} - \mathbf{R}/2|} + \frac{ke^2}{R} \right\} \psi_e(\mathbf{r}) = \varepsilon_e \psi_e(\mathbf{r}), \quad (10\text{-}1)$$

where m_e is the electron mass, $k = 1/4\pi\varepsilon_0$, and ε_e is an energy eigenvalue of the electron. The first term in curly brackets is the electron kinetic energy operator. The second and third terms are the attractive potential energies caused by the protons at $-\mathbf{R}/2$ and $\mathbf{R}/2$. We have also included the last term, the proton–proton repulsion, as a part of the electronic potential energy for convenience.

Figure 10-2 shows a cut through the potential energy function as one travels along the z axis. The potential energy diverges at $z = -R/2$ and $z = R/2$ as it must for the Coulomb force. Near $z = 0$ the function is a bit lower than it would be if there were only one atomic potential energy curve. The potential energy is symmetric about this point. This double-welled potential energy function is analogous to that considered in Chapter 5 (Figure 5-25) and the wave functions have properties similar to those found there.

Suppose that only one of the potential wells is present (e.g., the one around $z = R/2$ as indicated by the dashed curve). Then the problem is quite familiar and the ground-state wave function is just $\phi_{1s}(\mathbf{r} - \mathbf{R}/2)$, the hydrogenic $1s$ state centered at $\mathbf{r} = \mathbf{R}/2$. Similarly, a particle in the potential well centered at $\mathbf{r} = -\mathbf{R}/2$ has wave function $\phi_{1s}(\mathbf{r} + \mathbf{R}/2)$. Because both potential wells are the same, just shifted in space, the energy levels in each are identical and equal to ε_{1s}.

Now consider the situation in the combined problem represented by the solid line in Figure 10-2. While the electron is in the neighborhood of $\mathbf{r} = \mathbf{R}/2$ we expect the ground-state wave function to be approximately equal to $\phi_{1s}(\mathbf{r} - \mathbf{R}/2)$ because in much of that region the dashed line and the solid line coincide. While the electron is in the neighborhood of $\mathbf{r} = -\mathbf{R}/2$, the wave function is approximately $\phi_{1s}(\mathbf{r} + \mathbf{R}/2)$. A further necessary property of the wave function arises from the symmetry of the potential energy about the point $\mathbf{r} = 0$. We expect the probability $P(\mathbf{r})$ of a particle being around $\mathbf{r} = -\mathbf{R}/2$ to be equal to that of it being around $\mathbf{r} = \mathbf{R}/2$. More precisely, we require parity invariance $P(-\mathbf{r}) = P(\mathbf{r})$. In terms of the eigenfunction this is equivalent to

$$\psi(-\mathbf{r}) = \pm\psi(\mathbf{r}). \tag{10-2}$$

Molecular wave functions that are even under a parity change (the upper sign) are said to be *gerade*; those that are odd are *ungerade*. (These are the German words for "even" and "odd.")

Eigenfunctions that satisfy both of the above sets of requirements are given by

$$\psi_+(\mathbf{r}) = \frac{1}{\sqrt{2}}\left(\phi_{1s}(\mathbf{r} - \mathbf{R}/2) + \phi_{1s}(\mathbf{r} + \mathbf{R}/2)\right) \tag{10-3a}$$

and

$$\psi_-(\mathbf{r}) = \frac{1}{\sqrt{2}}\left(\phi_{1s}(\mathbf{r} - \mathbf{R}/2) - \phi_{1s}(\mathbf{r} + \mathbf{R}/2)\right). \tag{10-3b}$$

When $\mathbf{r}$ is near $\mathbf{R}/2$, $\phi_{1s}(\mathbf{r} + \mathbf{R}/2) \approx \phi_{1s}(\mathbf{R})$ is small and $\psi_\pm(\mathbf{r}) \approx (1/\sqrt{2})\phi_{1s}(\mathbf{r} - \mathbf{R}/2)$ as required. (The $1/\sqrt{2}$ is just an approximate normalization factor.) To see the parity property of $\psi_\pm(\mathbf{r})$ we note that

$$\psi_\pm(-\mathbf{r}) = \frac{1}{\sqrt{2}}\left(\phi_{1s}(-\mathbf{r} - \mathbf{R}/2) \pm \phi_{1s}(-\mathbf{r} + \mathbf{R}/2)\right).$$

From Chapter 7 we know that the $1s$ state for hydrogen is also even under parity inversion, that is,

$$\phi_{1s}(-\mathbf{r}) = \phi_{1s}(\mathbf{r}),$$

so that Equation (10-2) is satisfied as well. A graph of the functions $\psi_\pm$ is shown in Figure 10-3.

Note that the $\psi_\pm$ of Equations (10-3) are not exact solutions of Equation (10-1), even though they behave reasonably. They are *approximations* to the two lowest states

Figure 10-3
Graph along the z axis of the approximate eigenfunctions $\psi_\pm(\mathbf{r})$ (Equations (10-3)) for H_2^+.

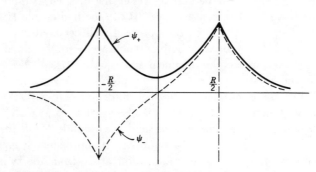

of H_2^+. An exact solution of Equation (10-1) is possible but the complications involved in finding it do not offer as much physical insight into the problem.

The energies associated with these wave functions are found by computing the expectation value

$$\langle \varepsilon \rangle \equiv \varepsilon_\pm = \int \psi_\pm^* \left[-\frac{\hbar^2}{2m_e} \nabla_r^2 + V(\mathbf{r}) \right] \psi_\pm \, d\mathbf{r}. \tag{10-4}$$

This quantity is a simple generalization to three dimensions of the one-dimensional energy expectation value given by Equation (5-65) in Section 5-7. The quantity $-(\hbar^2/2m)\nabla_r^2$ is the three-dimensional kinetic energy operator that replaces the one-dimensional form $-(\hbar^2/2m)\,\partial^2/\partial x^2$, and $V(\mathbf{r})$ is the sum of the three potential energy terms in Equation (10-1).

To simplify the calculation of the energies we establish the following shorthand notation:

$$T = -\frac{\hbar^2}{2m_e} \nabla_r^2;$$

$$V_a = -\frac{ke^2}{|\mathbf{r} - \mathbf{R}/2|}; \quad V_b = -\frac{ke^2}{|\mathbf{r} + \mathbf{R}/2|}; \quad V_R = \frac{ke^2}{R^2};$$

$$\phi_a = \phi_{1s}(\mathbf{r} - \mathbf{R}/2); \quad \phi_b = \phi_{1s}(\mathbf{r} + \mathbf{R}/2).$$

The total potential energy is then

$$V(\mathbf{r}) = V_a + V_b + V_R.$$

Note that the $\psi_\pm$ are real because ϕ_a and ϕ_b are real. Putting the $\psi_\pm$ of Equations (10-3) into Equation (10-4) gives for the energy estimate

$$\varepsilon_\pm = \frac{1}{2} \int (\phi_a \pm \phi_b)[T + V_a + V_b + V_R](\phi_a \pm \phi_b) \, d\mathbf{r}$$

$$= \frac{1}{2} \int (\phi_a \pm \phi_b)[(T + V_a)\phi_a \pm (T + V_b)\phi_b] \, d\mathbf{r}$$

$$+ \frac{1}{2} \int \phi_a^2 (V_b + V_R) \, d\mathbf{r} + \frac{1}{2} \int \phi_b^2 (V_a + V_R) \, d\mathbf{r}$$

$$\pm \frac{1}{2} \int \phi_a (V_a + V_R)\phi_b \, d\mathbf{r} \pm \frac{1}{2} \int \phi_b (V_b + V_R)\phi_a \, d\mathbf{r}.$$

The fact that ϕ_a and ϕ_b are hydrogenic functions allows us to write

$$(T + V_a)\phi_a = \varepsilon_{1s}\phi_a \quad \text{and} \quad (T + V_b)\phi_b = \varepsilon_{1s}\phi_b.$$

We also note that

$$G \equiv \int \phi_a^2 (V_b + V_R) \, d\mathbf{r} = \int \phi_b^2 (V_a + V_R) \, d\mathbf{r} \tag{10-5}$$

corresponds to the average value of the Coulomb interaction between the electron and nucleus on site a (at $\mathbf{R}/2$) and the nucleus on b (at $-\mathbf{R}/2$) or vice versa.

A *quantum interference* or *overlap* term also occurs in the energy. This is

$$S = \int \phi_a (V_a + V_R) \phi_b \, d\mathbf{r} = \int \phi_b (V_b + V_R) \phi_a \, d\mathbf{r}. \qquad (10\text{-}6)$$

This quantity arises as a cross-term in $\varepsilon_{\pm}$ because the electron has probability amplitude for being on either site. The overlap term is extremely important and is the source of the binding of the atoms to form a molecule. Note that the term involves the product $\phi_a \phi_b$. The function ϕ_a is localized about $\mathbf{R}/2$ and falls off exponentially as $\mathbf{r}$ moves away from $\mathbf{R}/2$, and similarly for ϕ_b. The product is appreciable only in the vicinity of $\mathbf{r} = 0$ where the two functions overlap. For this reason S is usually a good deal smaller than the *direct energy* terms, such as ε_{1s} or G, which depend on ϕ_a^2 or ϕ_b^2. Classically, a particle is trapped in a potential well (e.g., the electron near $\mathbf{r} = \mathbf{R}/2$ can not get into the well at $\mathbf{r} \approx -\mathbf{R}/2$ if its energy is less than the central barrier height as indicated in Figure 10-2). But because of the quantum mechanical barrier penetration effect, the wave function does enter this classically forbidden region, resulting in a nonzero overlap integral. Thus, we obtain

$$\varepsilon_{\pm} = \varepsilon_{1s} + G \pm S \qquad (10\text{-}7)$$

as the final result.

In the limit that the nuclear separation R becomes large, the overlap of ϕ_a and ϕ_b, and hence S, becomes exponentially small. It can be shown that G also becomes exponentially small. Then we have $\varepsilon_{\pm} \to \varepsilon_{1s}$. However, for smaller separations the two energies ε_+ and ε_- are not equal but are split by $2S$. The exchange integral S turns out to be negative so that ε_+ (corresponding to ψ_+, the eigenfunction with no nodes) is lower than ε_- and is the ground-state energy. More importantly, even though G is positive, ε_+ is lower than ε_{1s}, which can be considered to be the energy of the dissociated molecule. Thus, the molecule is bound.

The other state that we have formed, ψ_-, has an energy greater than ε_{1s}. This state turns out not to be bound.

Figure 10-4 shows the two energies ε_+ and ε_- as functions of internuclear separation R. These electronic energies can be interpreted as effective potential energy functions of the two protons. They either cause the protons to be drawn toward each other or repelled from one another. We write

$$V_{\pm} = \varepsilon_{\pm} - \varepsilon_{1s}$$

as the nuclear potential energy functions. Written in this way, the potential energy goes to zero as $R \to \infty$. V_- is always repulsive but V_+ is attractive until R gets so small that the Coulomb repulsion between the protons dominates. There is a minimum in V_+, defining the equilibrium separation of the two protons in the molecule.

The binding of the H_2^+ molecule arises from the exchange integral S in Equation (10-7). The physical origin of this term can be seen from the behavior of $\psi_{\pm}$ in Equations (10-3). From the plot of these two functions in Figure 10-3 we see that ψ_+ is larger than either $\phi_a = \phi_{1s}(\mathbf{r} - \mathbf{R}/2)$ or $\phi_b = \phi_{1s}(\mathbf{r} + \mathbf{R}/2)$ in the region between the two protons. The probability density associated with the two functions is

$$|\psi_{\pm}|^2 = \tfrac{1}{2}\big[|\phi_a|^2 + |\phi_b|^2 \pm 2\phi_a \phi_b \big]. \qquad (10\text{-}8)$$

The last term is the contribution from the overlap region. In $|\psi_+|^2$ this extra electron

Figure 10-4

Effective internuclear potential energies for H_2^+ as functions of proton–proton separation R. These functions are related to the electronic energies by $V_\pm = \varepsilon_\pm - \varepsilon_{1s}$. The minimum in V_+ gives R_0, the equilibrium internuclear distance.

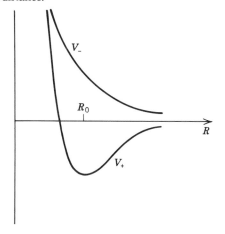

density implies a region of extra negative charge, which attracts each proton to it and partially shields one proton from the repulsion of the other. This is the effect causing molecular binding in H_2^+; hence ψ_+ is said to be a *bonding molecular orbital*. On the other hand, $|\psi_-|^2$ has less electron density than it would have without the overlap term (note ψ_- vanishes at the midpoint between the protons), and ψ_- is said to be *antibonding*.

The explicit calculation of G and S in Equation (10-7) leads to an equilibrium distance of 0.132 nm and a depth of the potential V_+ of 1.77 eV compared to the results obtained from the exact wave function of 0.106 nm and 2.79 eV. The approximate function ψ_+ of Equation (10-3) tends to be inaccurate at small R; the wave functions for the individual atomic wells become quite distorted from $1s$ states at such separations. Indeed, for $R = 0$, ψ_+ ought to transform into the ground state of He^+ but does not in our approximation (see Problem 3 at the end of the chapter).

It is possible to construct other molecular orbitals analogous to those of Equations (10-3) by substituting other atomic functions for the $1s$ states. These orbitals represent excited molecular states, some of which are bonding and some antibonding.

We next examine the angular momentum classification of the molecular orbitals of H_2^+. To do this most easily we shift the molecular coordinate system so that the origin coincides with one of the protons as shown in Figure 10-5. The atomic wave function for the electron centered on the proton at the origin is now $\phi_{1s}(\mathbf{r})$. This function is clearly an eigenstate of electron angular momentum $\ell = 0$. To examine the angular momentum properties of the molecular orbital ψ_+ of Equation (10-3a), we must also look at the atomic function for the electron centered at $\mathbf{r} = \mathbf{R}$, that is, $\phi_{1s}(\mathbf{r} - \mathbf{R})$. Because this latter state is displaced from the origin, it is *not* an eigenfunction of the electron's squared angular momentum L^2. Adding a second center of force at $\mathbf{R}$ has

Figure 10-5

Molecular coordinates for H_2^+ with the origin taken at one of the protons.

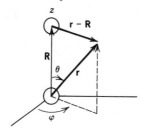

ruined the spherical symmetry of the electron's potential energy. The problem is no longer one having a simple central force. Thus, molecular orbitals cannot be classified according to ℓ values.

Molecular wave functions for diatomic molecules like H_2^+ do have *cylindrical symmetry* and therefore turn out to be eigenstates of the z component of angular momentum L_z. In molecular physics the eigenvalues of L_z for a single electron are denoted by $\lambda \hbar$, where $\lambda = 0, \pm 1, \pm 2, \ldots$. Just as hydrogenic atomic orbitals have the letter code $s, p, d, \ldots$ representing $\ell = 0, 1, 2, \ldots$, molecular orbitals corresponding to $|\lambda| = 0, 1, 2, \ldots$ are designated by the Greek letters $\sigma, \pi, \delta, \ldots$, respectively. As is shown below, $\psi_\pm$ of Equations (10-3) are both σ states.

Detail

We can easily show that ψ_+ is an eigenfunction of L_z. Using the coordinate system of Figure 10-5, write

$$|\mathbf{r} - \mathbf{R}| = \left(r^2 + R^2 - 2rR \cos \theta \right)^{1/2}.$$

This arrangement of ψ_+, and hence ψ_+ itself, depends on r and $\cos \theta$ but not on the azimuthal angle ϕ shown in Figure 10-5. From Chapter 6,

$$L_z \to \frac{\hbar}{i} \frac{\partial}{\partial \phi}$$

so that

$$L_z \psi_+ = 0,$$

and ψ_+ therefore corresponds to $\lambda = 0$.

10-2 Covalent Bonding: H_2

The H_2^+ molecule discussed in Section 10-1 is useful in showing the elements of molecular binding in a reasonably straightforward way. It is the "hydrogen atom" of molecules. However, the simplest *neutral* molecule is H_2, which carries two electrons. We now proceed to treat this fundamental problem.

Figure 10-1 continues to describe the proton configuration for H_2. However, there are now two electrons at positions $\mathbf{r}_1$ and $\mathbf{r}_2$. We construct an approximate electronic wave function first introduced by W. Heitler and F. London in 1927. Suppose for a moment that the two protons are at the same point $\mathbf{r} = 0$. We then have the charge equivalent of a helium nucleus. The electronic ground state for this system is a $1s^2$ configuration with eigenfunction

$$\psi(1,2) = \phi_{1s}(\mathbf{r}_1)\phi_{1s}(\mathbf{r}_2)\chi^A(1,2), \qquad (10\text{-}9)$$

where $\chi^A(1,2)$ is the antisymmetric or singlet spin function given by (see Equation (9-25))

$$\chi^A(1,2) = \frac{1}{\sqrt{2}}(\uparrow_1\downarrow_2 - \downarrow_1\uparrow_2). \qquad (10\text{-}10)$$

The spin function must have one electron with spin up and the other with spin down because both electrons are in the same atomic orbital. In this way eigenfunction $\psi(1,2)$ satisfies the Pauli principle; that is, it is antisymmetric and $\psi(2,1) = -\psi(1,2)$.

Next, let the two protons be drawn apart along the z axis, one to $\mathbf{R}/2$ and the other to $-\mathbf{R}/2$. We might expect one electron to be dragged along with each proton. Then Equation (10-9) would be transformed into

$$\psi(1,2) = \phi_{1s}(\mathbf{r}_1 - \mathbf{R}/2)\phi_{1s}(\mathbf{r}_2 + \mathbf{R}/2)\chi^A(1,2). \qquad (10\text{-}11)$$

This might seem a good initial guess at a molecular eigenfunction except for two important deficiencies. First, Equation (10-11) has not incorporated the physical features found to be necessary for molecular binding in the H_2^+ analysis. Each electron should have a probability of being at *each* proton, thereby spending extra time in the intermediate region between the protons. This extra electron density yields the net attraction that binds the molecule. A second deficiency arises from the fact that the proposed function does not obey the Pauli principle; it is not properly antisymmetric as it must be for fermions.

Fortunately, we can easily solve both problems at once by simply adding on to Equation (10-11) the interchanged state with electron 1 at $-\mathbf{R}/2$ and electron 2 at $+\mathbf{R}/2$. The result is

$$\psi_S^{\text{HL}}(1,2) = \frac{1}{\sqrt{2}}\left[\phi_{1s}(\mathbf{r}_1 - \mathbf{R}/2)\phi_{1s}(\mathbf{r}_2 + \mathbf{R}/2) + \phi_{1s}(\mathbf{r}_2 - \mathbf{R}/2)\phi_{1s}(\mathbf{r}_1 + \mathbf{R}/2)\right]$$
$$\times \chi^A(1,2), \qquad (10\text{-}12)$$

where the superscript HL stands for Heitler–London and the subscript S stands for singlet spin state. The factor $1/\sqrt{2}$ is put in to make ψ^{HL} approximately normalized. Because $\chi^A(1,2)$ is antisymmetric under electron interchange, the entire wave function is now antisymmetric. Also, each electron now has a probability of being at each nucleus. Note, however, that the electrons move in a rather coherent way, so that if electron 1 moves from the nucleus at $-\mathbf{R}/2$ to that at $+\mathbf{R}/2$, electron 2 flips its position in just the opposite way. In this way the wave function never allows both electrons to appear simultaneously on the same nucleus. We might think that this a positive feature because it would tend to minimize electron–electron repulsion; however, we show later that mixing in a little of this double-occupancy state actually improves binding.

There is an alternative antisymmetric wave function that we can form from the $1s$ atomic orbitals. First, we construct triplet spin states from the spin eigenfunctions as done in Chapter 9, Equation (9-26). These are

$$\chi_{m_s}^{S} = \begin{cases} \chi_1^{S} = \uparrow_1 \uparrow_2 & (10\text{-}13) \\[2mm] \chi_0^{S} = \dfrac{1}{\sqrt{2}}(\uparrow_1 \downarrow_2 + \downarrow_1 \uparrow_2) & (10\text{-}14) \\[2mm] \chi_{-1}^{S} = \downarrow_1 \downarrow_2. & (10\text{-}15) \end{cases}$$

Since these states are symmetric under electron exchange, the spatial part of the wave function must be antisymmetric. By analogy with Equation (10-12) we have

$$\psi_T^{\text{HL}} = \frac{1}{\sqrt{2}} \left[\phi_{1s}(\mathbf{r}_1 - \mathbf{R}/2)\phi_{1s}(\mathbf{r}_2 + \mathbf{R}/2) - \phi_{1s}(\mathbf{r}_2 - \mathbf{R}/2)\phi_{1s}(\mathbf{r}_1 + \mathbf{R}/2) \right] \chi_{m_s}^{S}.$$

$$(10\text{-}16)$$

The T subscript stands for triplet spin state.

As with H_2^{+}, one of the two states formed from $1s$ atomic functions is bonding and the other antibonding. To see which is which, consider the squares of the two spatial portions of the wave functions. The spatial probability densities are

$$P_{S,T}^{(12)} = \tfrac{1}{2}\left[\phi_{1s}^2(\mathbf{r}_1 - \mathbf{R}/2)\phi_{1s}^2(\mathbf{r}_2 + \mathbf{R}/2) + \phi_{1s}^2(\mathbf{r}_2 - \mathbf{R}/2)\phi_{1s}^2(\mathbf{r}_1 + \mathbf{R}/2) \right.$$

$$\left. \pm 2\big(\phi_{1s}(\mathbf{r}_1 - \mathbf{R}/2)\phi_{1s}(\mathbf{r}_1 + \mathbf{R}/2)\big)\big(\phi_{1s}(\mathbf{r}_2 - \mathbf{R}/2)\phi_{1s}(\mathbf{r}_2 + \mathbf{R}/2)\big) \right]. \quad (10\text{-}17)$$

The interference term (the last one, like that of Equation (10-8)) depends on the overlap of electronic wave functions in the region halfway between the nuclei, and implies that there is extra electronic charge there in the singlet case (plus sign) and diminished charge for the triplet (minus sign). While the added charge density of the singlet wave function does enhance electron–electron repulsion, this effect is more than offset by the added attraction between the extra charge and the two protons. It is this effect that provides the *covalent bonding* in the singlet case. The triplet wave function, on the other hand, is antibonding.

In a molecule more complicated than H_2 there is often a closed-shell core of electrons, quite tightly bound to their originating nucleus, which participate weakly or not at all in this exchange phenomenon. The outer electrons that do exchange are known as *valence* electrons. Since these electrons are paired in the sense that they exchange cooperatively from one atom to the other, the bond is called *covalent*.

The energies associated with the two eigenfunctions ψ_S^{HL} and ψ_T^{HL} may be calculated. When this is done, forms analogous to those of Equations (10-7) are found:

$$\varepsilon_{S,T} = 2\varepsilon_{1s} + G' \pm S', \quad (10\text{-}18)$$

where the upper sign refers to the singlet case and the lower to the triplet. The exchange integral S' depends on *two*-particle overlap integrals and contains terms like

$$\int\int \phi_{1s}(\mathbf{r}_1 - \mathbf{R}/2)\phi_{1s}(\mathbf{r}_2 + \mathbf{R}/2)V(\mathbf{r}_1,\mathbf{r}_2)\phi_{1s}(\mathbf{r}_1 + \mathbf{R}/2)\phi_{1s}(\mathbf{r}_2 - \mathbf{R}/2)\,d\mathbf{r}_1\,d\mathbf{r}_2,$$

where $V(\mathbf{r}_1, \mathbf{r}_2)$ can be any one of the various terms occurring in the energy operator. G' is also a two-particle quantity somewhat analogous to the G of Equation (10-5). S' turns out to be negative so that the singlet energy is lower than the triplet. The singlet energy has a minimum as a function of R and exhibits binding while the triplet has no bound state.

As mentioned briefly above, the approximate Heitler–London eigenfunction ψ_S^{HL} can be improved somewhat by including double-occupancy states:

$$\psi_S(1,2) = \psi_S^{HL}(1,2) + \gamma \frac{1}{\sqrt{2}} \left[\phi_{1s}(\mathbf{r}_1 - \mathbf{R}/2)\phi_{1s}(\mathbf{r}_2 - \mathbf{R}/2) \right.$$
$$\left. + \phi_{1s}(\mathbf{r}_1 + \mathbf{R}/2)\phi_{1s}(\mathbf{r}_2 + \mathbf{R}/2) \right] \chi^A(1,2), \quad (10\text{-}19)$$

where γ is an adjustable coefficient. The first term of the new addition has both electrons near the same nucleus at $\mathbf{R}/2$; the second has both near that at $-\mathbf{R}/2$. While each of these terms increases the average value of the electron–electron repulsion, it also gives the system an *ionic structure* in which there is some probability of the formation of the ion combinations (H^+, H^-) or (H^-, H^+). The ions attract one another and add to the molecular binding. At the equilibrium separation of the molecule (the minimum of the electronic energy), it is found that $\gamma \approx 0.2$, implying that only a small percentage of the binding results from this process. Note that the configurations (H^-, H^+), with both electrons on proton 1, and (H^+, H^-), with both on proton 2, are equally likely so that H_2 does not have a permanent electric dipole moment. In Section 10-3 we study systems that owe their bonding to the existence of a permanent separation of charge and the resulting electric dipole configuration.

It is possible to construct bonding or antibonding molecular states from excited atomic orbitals as well as from $1s$ states. For example, simply replace the $1s$ orbitals in Equations (10-12) and (10-16) with $2s$ functions. Excited-state molecular energies vary with internuclear distance R in a manner similar to the ground-state energy as shown schematically in Figure 10-6. The $R \to \infty$ limit of such energies need not be zero or all equal. As $R \to \infty$, G' and S' of Equation (10-18) go to zero and the energy $\varepsilon_S \to 2\varepsilon_{1s}$, which is the electronic energy associated with the two isolated hydrogen atoms. The bonding state constructed of two $2s$ states would have the limit $2\varepsilon_{2s}$ as $R \to \infty$. A bonding excited molecular state such as ε_4 in Figure 10-6 has a minimum at some R value and does not necessarily dissociate (i.e., separate into its atomic constituents) even though its $R \to \infty$ limit is greater than that of the ground state. On the other hand, excitation of the molecule from the ground state to an excited state such as ε_2 or ε_3 in Figure 10-6 would result in dissociation.

As a last topic of this section we consider why a closed shell (noble gas) atom does not interact covalently with any other atom. Consider the case of hydrogen and helium. The bonding molecular wave function has been seen to have a singlet spin configuration with one spin up and one spin down. Consider a covalent bond involving the hydrogen $1s$ electron with spin down, for example, and one of the helium $1s$ electrons with spin up. The wave function involves an exchange term that brings the down electron over to the He, which then has two spins down and violates the Pauli principle. This is a somewhat simplified view; a rigorous argument would construct three-electron wave functions for the two nuclear centers. These wave functions can be shown to vanish because of the Pauli principle for all but antibonding molecular orbitals. Thus, the molecule H—He does not form.

While similar arguments can be made to show that two noble gas atoms, such as He—He or He—Ar, cannot interact covalently, molecules such as Ar_2 do exist.

Figure 10-6

Electronic states for a diatomic molecule as functions of internuclear separation R. Bonding states have minima as in ε_1 or ε_4. Antibonding states do not support bound states.

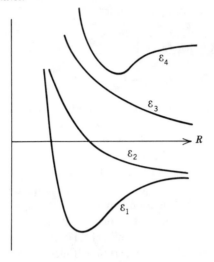

These occur because there is another attractive force mechanism known as the van der Waals interaction, which is discussed in Section 10-4.

10-3   Ionic Bonding: LiF

Hydrogen has a single electron, and the alkalis (Li, Na, Rb, Cs) have a single electron outside a closed shell. When any one of these react with any one of the halogens (F, Cl, Br, I), which are one electron short of completing a shell, they form *ionic bonds*. Basically, what happens is that the single s electron of the alkali element (or of hydrogen) is taken by the halogen to fill its shell; the resulting positive and negative ions attract one another and the molecule becomes bound.

Consider the molecule lithium fluoride, LiF. The primary question to be answered is why an electron would be pulled off the lithium atom to form Li^+ and move over to fluorine to form F^-. The case rests mainly in the fact that all the electrons in a single shell have roughly the same average atomic radius. The electronic structure of fluorine is $1s^2 2s^2 2p^5$. The $n = 1$ shell is complete and the $n = 2$ shell is missing one electron. All the $n = 2$ electrons reside at approximately the same average radius, which, of course, is larger than that corresponding to the $n = 1$ electrons. Each of the seven $n = 2$ electrons is thus screened efficiently from the nucleus by only the two $1s$ electrons. The resulting Z value seen by each of the $n = 2$ electrons is not much less than 7. An extra tenth electron borrowed from the lithium atom sees roughly the same effective Z value and is bound. The binding energy of this electron to form a negative ion is known as the *electron affinity*. This electron affinity, however, is usually smaller than the ionization potential associated with forming the positive ion. In the case of lithium (whose structure is $1s^2 2s$) it costs 5.4 eV to pull off the $2s$ electron (the ionization potential), and in forming the negative fluorine ion 3.4 eV is released (the electron affinity)—a net cost of energy of 2.0 eV. However, the energy released by the binding of the resulting ions (about 8 eV) more than compensates for this energy expense. This ionization process goes only in the direction indicated, as is easily seen by considering the energy required to form (Li^-F^+). The electron affinity for lithium is 0.6 eV; the ionization potential of fluorine is 17.4 eV. The energy requirement is 16.8 eV, while the net gain through binding of the two ions is the same as for (Li^+F^-).

Note that homonuclear diatomic molecules (those made up of like nuclei, such as H_2) have constituents with equal attractions for the electrons and so do not form permanent ions within the molecule, although they may have ionic structure in their wave functions as discussed in Section 10-2. Only *heteronuclear* molecules (those made up of unlike nuclei, as HF) can have the permanent charge separation described above.

There is a simple test by which we can judge the accuracy of the above idea of the ionic bond in LiF or in any diatomic molecule. If the lithium's electron spends all its time on the fluorine atom, then the electric dipole moment of the molecule, from Figure 10-7, is

$$p = -e\left(-\frac{R}{2}\right) + e\frac{R}{2} = eR.$$

The interatomic separation in LiF is found to be $R = 0.156$ nm so we expect $p = (1.6 \times 10^{-19}\text{ C})(1.56 \times 10^{-10}\text{ m}) = 2.5 \times 10^{-29}$ C · m. The experimental value of the electric dipole moment is $p_{\text{exp}} = 2.11 \times 10^{-29}$ C · m, 85% of the above model value. The lithium's electron does maintain some probability of being on its home-base atom in addition to spending time at the fluorine; this reduces charge separation from the model value. There are also other small effects that tend to reduce the polarization.

The above result indicates the bonding of LiF is only partly ionic; it is also partly covalent. The lithium's $2s$ electron pairs with one of the fluorine's $2p$ electrons to form this covalent bond. To see how this takes place consider the possible angular parts of the $2p$ wave function. There are three degenerate states with angular components

$$Y_{1\pm 1} = \mp\sqrt{\frac{3}{8\pi}}\sin\theta e^{\pm i\phi} \tag{10-20a}$$

and

$$Y_{10} = \sqrt{\frac{3}{4\pi}}\cos\theta. \tag{10-20b}$$

(Refer back to Table 6-1.) Figure 6-8 shows how the angular lobes of the probability density corresponding to these states are positioned. Combinations of the $Y_{1\pm 1}$ functions have lobes in the xy plane. The Y_{10} function shown in Figure 10-8 has lobes along the z axis. The upper lobe is positive and the lower one negative. Of the three functions, the one expected to have the largest overlap with a $2s$ function centered at a distance R away along the z axis is Y_{10}. We pair this so-called $2p_z$ function of fluorine with the lithium's $2s$ function for a covalent type of wave function.

Figure 10-7

Model of an ionically bound diatomic molecule. Atom B has the greater electron affinity and takes an electron from atom A. The two atoms form an electric dipole with charges $+e$ and $-e$ separated by distance R.

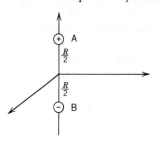

Figure 10-8

Angular function Y_{10} – a factor in the $2p_z$ wave function of fluorine.

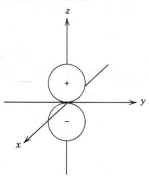

The eigenfunction describing the bonding orbitals in this molecule is then a generalization of Equation (10-19). We place the fluorine at $z = -R/2$ and the lithium at $z = +R/2$ so

$$\psi(1,2) = \left\{ A\frac{1}{\sqrt{2}}\left[\phi_{2s}(\mathbf{r}_1 - \mathbf{R}/2)\phi_{2p_z}(\mathbf{r}_2 + \mathbf{R}/2) + \phi_{2s}(\mathbf{r}_2 - \mathbf{R}/2)\phi_{2p_z}(\mathbf{r}_1 + \mathbf{R}/2)\right] \right.$$
$$+ B\phi_{2s}(\mathbf{r}_1 - \mathbf{R}/2)\phi_{2s}(\mathbf{r}_2 - \mathbf{R}/2)$$
$$\left. + C\phi_{2p_z}(\mathbf{r}_1 + \mathbf{R}/2)\phi_{2p_z}(\mathbf{r}_2 + \mathbf{R}/2)\right\}\chi^A(1,2), \tag{10-21}$$

where ψ is normalized by taking $\sqrt{A^2 + B^2 + C^2} = 1$ (see Problem 9 at the end of the chapter). The term with coefficient A is the covalent part of the wave function and those in B and C involve the ionic structures. Unlike the case of H_2, we have no reason to take $B = C$ and indeed the ionic bonding mechanism we have described requires $B \neq C$. Because we know the electrons may double up on fluorine but are very unlikely to do so on lithium, it is reasonable to set $B = 0$ as an approximation.

It can be shown that the electric dipole moment of the molecule may be related to the parameters in Equation (10-21). The result, for the case of arbitrary A, B, and C, is

$$p = eR(C^2 - B^2). \tag{10-22}$$

For H_2, whose eigenfunction is given by Equation (10-19), we have the equivalent of $B = C$ in Equation (10-21) and $p = 0$ as expected. In the present case we have

$$p = eRC^2. \tag{10-23}$$

The experimental ratio of $p/eR = 0.85$ for LiF gives us $C = \sqrt{0.85} = 0.9$, $A = \sqrt{1 - C^2} = 0.4$, the latter equality being the normalization condition. Thus, we see that the percentage of ionic bonding in the total bond of LiF is quite high.

Given the wave function of Equation (10-21) we can determine the average energy as a function of internuclear separation. This result tells us many of the properties of

Figure 10-9
Electronic energy for an ionically bound
molecule as a function of internuclear distance.
The energy is adjusted to $V(\infty) = 0$.

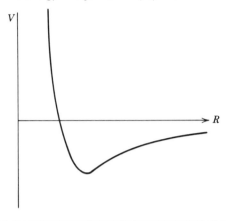

the molecule. However, a more direct approach is one in which we guess the form of
the electronic energy on the basis of general physical principles. Such a "phenomenological" form contains coefficients that vary from molecule to molecule and are
determined by fitting the predictions derived from the potential energy function to
experiment.

The potential energy function usually chosen for a general ionically bound alkali
halide molecule has the form

$$V(R) = \alpha e^{-aR} - \frac{e^2}{4\pi\varepsilon_0 R}. \tag{10-24}$$

The second term obviously represents the Coulomb attraction, at large separation R,
between the positive and negative ions. When the ions are so close to each other that
their closed-shell electronic clouds overlap, the electronic energy becomes repulsive
due to the Pauli principle as represented by the first term. The constants α and a,
which control the strength and slope of the repulsive part, are determined from
comparison with experimental data as shown in the example at the end of this section.
A qualitative plot of $V(R)$ is shown in Figure 10-9. While $V(R)$ looks much like V_+ of
Figure 10-4 or ε of Figure 10-6, it falls off at large R as $1/R$, which makes it much
longer ranged than either of those energies, since each drops off exponentially.

The equilibrium separation of the nuclei in the molecule is very near the position
R_0 of the minimum in the potential energy. (Zero-point motion changes this slightly.)
This position satisfies

$$\frac{dV}{dR}\bigg|_{R=R_0} = -a\alpha e^{-aR_0} + \frac{e^2}{4\pi\varepsilon_0 R_0^2} = 0$$

or

$$\alpha = \frac{e^2}{4\pi\varepsilon_0}\frac{e^{aR_0}}{aR_0^2}. \tag{10-25}$$

The curvature of the potential energy at $R = R_0$ is

$$K = \frac{d^2V}{dR^2}\bigg|_{R=R_0} = a^2\alpha e^{-aR_0} - \frac{2e^2}{4\pi\varepsilon_0 R_0^3}$$

from which it is found that

$$a = \frac{1}{R_0}\left(2 + 4\pi\varepsilon_0\frac{KR_0^3}{e^2}\right). \qquad (10\text{-}26)$$

Spectral data give R_0 and K directly, as we see in Section 10-8. Putting these values in Equations (10-25) and (10-26) yields α and a.

The dissociation energy is the work needed to pull the two atoms completely apart into isolated *neutral* atoms. If we were to separate the *ions* to infinity the work needed would be $-V(R_0)$, the depth of the potential energy. The removal of the electron from the negative ion and its return to the positive ion net a positive energy

$$Q = I_0 - A_0, \qquad (10\text{-}27)$$

where I_0 is the ionization potential and A_0 is the electron affinity. The dissociation energy then is related to the potential energy by

$$D = -V(R_0) - Q \qquad (10\text{-}28)$$

if we neglect the zero-point energy.

Covalent and ionic forces are the primary molecular bonding mechanisms. There are, however, other physical principles that contribute to interatomic interactions. We consider one of these, the van der Waals attraction, in Section 10-4.

Example

For the molecule lithium fluoride spectral data yield the values of the interatomic separation $R_0 = 0.156$ nm and the potential energy function curvature $K = 248$ J/m^2. From these values Equation (10-26) yields

$$a = \frac{1}{0.156 \text{ nm}}\left(2 + \frac{(248 \text{ J/m}^2)(1.56 \times 10^{-10} \text{ m})^3}{(8.99 \times 10^9 \text{ N} \cdot \text{m}^2/\text{C}^2)(1.6 \times 10^{-19} \text{ C})^2}\right) = \frac{6.13}{0.156 \text{ nm}}$$

$$= 39.2 \text{ nm}^{-1}.$$

Then the parameter α is found from Equation (10-25) to be

$$\alpha = \frac{(8.99 \times 10^9 \text{ N} \cdot \text{m}^2/\text{C}^2)(1.6 \times 10^{-19} \text{ C})^2}{6.13(1.56 \times 10^{-10} \text{ m})}e^{6.13} = 1.1 \times 10^{-16} \text{ J}$$

$$= 688 \text{ eV}.$$

Thus, the potential energy in Equation (10-24) for LiF becomes

$$V(R) = \left(688e^{-39.2R} - \frac{1.44}{R}\right) \text{ eV},$$

where R is given in nanometers.

To get the dissociation energy from $V(R)$ we need the value of Q in Equation (10-27). The ionization potential of lithium is 5.4 eV, and the electron affinity of fluorine is 3.4 eV, so

$$Q = 5.4 - 3.4 = 2.0 \text{ eV}.$$

The potential depth at R_0 is

$$V(R_0) = 688e^{-(39.2)(0.156)} - \frac{1.44}{0.156}$$
$$= 1.5 - 9.2 = -7.7 \text{ eV}.$$

From Equation (10-28)

$$D = 7.7 - 2.0 = 5.7 \text{ eV}.$$

The experimental value is 6.0 eV, so our prediction from the phenomenological potential energy function is only about 5% in error. This is quite good considering the fact that there are several small corrections to the potential energy to be taken into account. These refinements include covalent, induced dipole (see Problem 11 at the end of the chapter), and van der Waals (Section 10-4) forces.

10-4   Van der Waals Interaction

Atoms of the noble gases, having closed electronic shells, do not interact with one another via covalent or ionic forces. However, there is a weaker attraction between such atoms, called the *van der Waals force*. This force is also present in molecules that do interact covalently or ionically but it is much weaker and of lesser importance. Despite its relative weakness the heavier noble gas atoms—neon, argon, and so on—are able to form molecules via the van der Waals interaction. Helium liquifies at low temperatures only because of the presence of this force.

The Dutch physicist J. D. van der Waals proposed that all molecules have attractive forces that, at sufficiently low temperatures, can cause liquification. The force we discuss here has thus been named after him. However, it was London who in 1930 first explained the physical origin of this interaction, which is occasionally called the *London force*.

The attraction arises from an induced dipole–dipole effect. We consider here a classical model of the force that is not totally adequate but does have many of the elements of the quantum mechanical picture. Suppose two neutral H atoms approach one another, and assume no covalent force. At any instant in time an electron and the nucleus that it is orbiting form an electric dipole. This dipole creates an electric field, which is felt by the second atom whose positive charge is pulled one way and its negative another. It becomes polarized; that is, it develops an *induced* dipole moment as shown in Figure 10-10.

We can easily find the behavior of the electric field as a function of distance R from the first atom. Consider for simplicity a point A on the perpendicular to the dipole as shown in Figure 10-11. At this position only the vertical components $E_i\sin\theta$ of each field do not cancel. From the figure we see that $\sin\theta$ satisfies $\sin\theta = \frac{1}{2}a/\sqrt{R^2 + (a/2)^2}$. When the field point is far away from the dipole so $R \gg a$, we

Figure 10-10

Electric-dipole behavior of atoms. An atom with an instantaneous dipole moment p_1 creates an electric field that induces a dipole moment p_2 in a second atom.

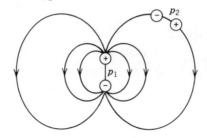

have $\sin \theta \approx a/2R$. The net electric field is then given by

$$E_d = \frac{2Q}{4\pi\varepsilon_0} \frac{\sin \theta}{R^2 + (a/2)^2} \approx \frac{2Qa}{4\pi\varepsilon_0 R^2} \frac{a}{2R} = \frac{p_1}{4\pi\varepsilon_0 R^3}, \qquad (10\text{-}29)$$

where $p_1 = Qa$ is the dipole moment of the first atom. From this we see that E_d drops off as $1/R^3$ as we move away from the dipole, a result that holds even when the point A is not on the perpendicular.

The second atom in the field $\mathbf{E}_d$ acquires an induced dipole moment $\mathbf{p}_2$ having size proportional to the electric field that it feels. So

$$\mathbf{p}_2 = \alpha\mathbf{E}_d, \qquad (10\text{-}30)$$

where α is called the *polarizability* of the molecule. This is a kind of Hooke's law approximation in which the separation of the charges is proportional to the applied force. (See Problem 11 at the end of the chapter.)

Figure 10-11

Electric field $\mathbf{E}_d$ due to a dipole $\mathbf{p}_1$. The field is the sum of the electric fields $\mathbf{E}_+$ and $\mathbf{E}_-$ of the individual charges $\pm Q$. Position A in the xy plane at a distance R from the dipole is considered. The charges $\pm Q$ are separated by distance a. The distance from either charge to A is $\sqrt{R^2 + \frac{1}{4}a^2}$.

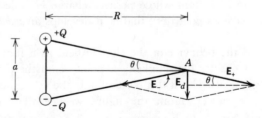

The energy of a dipole $\mathbf{p}_2$ aligned with an electric field $\mathbf{E}_d$ is

$$U = -p_2 E_d. \tag{10-31}$$

To derive this note that if a field is along the z axis, then while a dipole rotates from the xy plane into alignment with the field, the field does work $QEa/2$ on the charge $+Q$ and work $-QE(-a/2)$ on the charge $-Q$. Thus, the total dipole energy is $-2Q(a/2)E = -pE$.

If we combine Equations (10-30) and (10-31) and then use Equation (10-29), we find

$$U = -\alpha E_d^2 = -\alpha \frac{p_1^2}{(4\pi\varepsilon_0)^2} \frac{1}{R^6} = -\frac{\alpha Q^2 a^2}{(4\pi\varepsilon_0)^2} \frac{1}{R^6}.$$

We can do some dimensional analysis on α to estimate its size. Equation (10-30) gives $\alpha = p/E$, which *dimensionally* can be written as

$$\alpha = \frac{q\ell}{\dfrac{1}{4\pi\varepsilon_0}\dfrac{q}{\ell^2}} = 4\pi\varepsilon_0 \ell^3,$$

where ℓ is some length in the problem. For a dipole the appropriate length is the charge separation a, which in this case is approximately the atomic radius. So we substitute

$$\alpha = 4\pi\varepsilon_0 a^3. \tag{10-32}$$

This result agrees with that of Problem 11. We get

$$U = -\left(\frac{Q^2}{4\pi\varepsilon_0 a}\right)\left(\frac{a}{R}\right)^6. \tag{10-33}$$

The leading factor is of the same order as I_0, the ionization potential of the atom. Thus the van der Waals potential energy function has the form

$$V_{\text{vdW}}(R) \sim -I_0\left(\frac{a}{R}\right)^6. \tag{10-34}$$

An accurate calculation for two hydrogen atoms gives

$$V_{\text{vdW}}(R) = -13.0 I_0\left(\frac{a}{R}\right)^6 \tag{10-35}$$

instead of Equation (10-34).

This interaction is longer ranged than the exponentially diminishing covalent bond. Atoms such as hydrogen, which do interact covalently, also have the van der Waals attraction acting effectively at distances beyond the range of the covalent force.

Atoms of the noble gases are attracted to one another by only the van der Waals force. When they approach so closely that their electron shells overlap, the Pauli principle causes a repulsive force that increases rapidly with decreasing internuclear

---

Table 10-1   Parameters of the Lennard-Jones 12-6 Potential

	σ (nm)	ε (eV)
He	0.256	8.79×10^{-4}
Ne	0.275	3.08×10^{-3}
Ar	0.340	1.05×10^{-2}
Kr	0.368	1.44×10^{-2}
Xe	0.407	1.94×10^{-2}

Source: *Data from E. R. Dobbs and G. O. Jones*, Rep. Prog. Phys. **20**: 516 (1957).

separation. In 1925 J. E. Lennard-Jones suggested a phenomenological potential energy function that included both attractive and repulsive effects in a convenient mathematical form. This energy function is

$$V_{LJ}(R) = 4\varepsilon \left[\left(\frac{\sigma}{R} \right)^n - \left(\frac{\sigma}{R} \right)^6 \right]. \tag{10-36}$$

The constants ε, σ, and n are parameters to be determined to fit gas, liquid, and solid data. (Even though the energy function is sufficiently weak that the so-called noble gases are indeed gases at room pressure and temperature, they do liquify and solidify at sufficiently low temperature.) The most frequently used value of n is 12, in which case we have the *Lennard-Jones 12-6 potential.* It is easy to verify that the potential depth is ε and that σ is the value of R for which $V_{LJ}(R) = 0$ (see Problem 13 at the end of the chapter).

Some values of ε and σ for the 12-6 potential for various noble gases are given in Table 10-1. Note that as one progresses to heavier atoms the polarizability, which depends on atomic volume, increases and the strength of the interaction also increases.

Molecules with one member an alkali atom (which has a high polarizability) and the other a noble gas atom can be bound by the van der Waals interaction and can be observed spectroscopically. Molecules made up completely of noble gas atoms are difficult to observe but have been detected by use of mass spectrometers.

Example

The noble gases neon, argon, krypton, and xenon are able to form molecules because of the van der Waals interaction. However, they are weakly bound by comparison with molecules held together by other forces. For the ionically bound molecule LiF we know that the dissociation energy is about 6 eV. For the heavy noble gases the dissociation energy is approximately equal to the well depth. The Lennard-Jones 12-6 potential has a depth equal to the parameter ε given in Table 10-1. For Xe, $\varepsilon = 0.0194$ eV. A thermal energy equivalent to ε occurs for a temperature approximately equal to $T_{Xe} = (\varepsilon_{Xe}/k_B)$, where k_B is the Boltzmann constant 1.38×10^{-23} J/K or 8.62×10^{-5} eV/K. For xenon this is

$$T_{Xe} = \frac{0.0194 \text{ eV}}{8.62 \times 10^{-5} \text{ eV/K}} = 225 \text{ K.}$$

At this temperature, which is a little below room temperature, most of the molecules become dissociated by thermal agitation. A corresponding dissociation temperature for LiF is

$$T_{\text{LiF}} = \frac{6 \text{ eV}}{8.62 \times 10^{-5} \text{ eV/K}} = 7 \times 10^4 \text{ K}.$$

Obviously the van der Waals interaction is quite weak by comparison. For a pair of helium atoms the well depth is only 8.8×10^{-4} eV. The equivalent thermal energy is only 10.2 K. However, below this temperature we do not find bound He_2 molecules. Quantum mechanical zero-point motion is large enough to prevent these light atoms from forming any bound state at all. (See Problem 19 at the end of the chapter.)

10-5   Polyatomic Molecules: H₂O and CH₄

The electronic configuration of oxygen is $1s^2 2s^2 2p^4$. As discussed in connection with Figures 10-8 and 6-8 the p functions have threefold degeneracy; one (call it $2p_x$) has lobes along the x axis, another ($2p_y$) along y, and the third ($2p_z$) along z. If the four $2p$ electrons occupy only two of these three states, with two electrons each, then each of those two states is full. Then, according to the discussion at the end of Section 10-2, none of the electrons is available to form a covalent bond. However, if the four occupied states are, say, $2p_z^2 2p_x 2p_y$, then two of the states each have only one electron, which is said to be *unpaired*. Such a situation allows the possibility of the formation of *two* covalent bonds, one with each of the unpaired electrons. In water this results in the bonding with two hydrogen atoms. A first-approximation model of this molecule involves simply the one hydrogen s function overlapping with an oxygen $2p_x$ and the other with $2p_y$ as shown in Figure 10-12. This model has the bonds at a $90°$ angle.

Figure 10-12
Simple model of the water molecule H_2O. The electron of one hydrogen has its $1s$ function involved in a covalent bond with an oxygen $2p_x$ function. Another is bonded with the oxygen $2p_y$ function. The angle between the bonds is $90°$ in this model instead of the experimental value of $104.5°$.

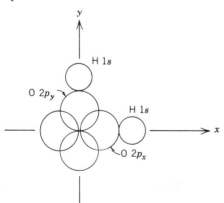

Figure 10-13
Methane molecule CH_4.

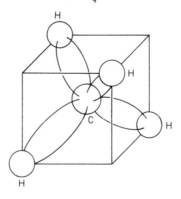

The repulsion between the two hydrogen ions causes the angle in the actual molecule to increase to 104.5°. We have seen this example briefly at the end of Section 9-3.

The bonds in H_2O have a fixed angle. Further examples of such *directed bonds* occur in the many compounds of carbon. One such molecule is methane, CH_4. Carbon's electronic state is $1s^2 2s^2 2p^2$. For a $2p^2$ configuration of $2p_x 2p_y$ it seems that carbon should form just two directed bonds like oxygen. However, carbon can bond to four hydrogens by a kind of trick. One of the $2s$ electrons goes into an excited p state so that the configuration becomes $1s^2 2s 2p_x 2p_y 2p_z$. There are then four unpaired electrons that can form bonds.

These bonds are not formed by use of the bare states $\phi_{2s}, \phi_{2p_x}, \phi_{2p_y}, \phi_{2p_z}$. Four linear combinations of these four functions can be formed as follows:

$$\psi_1 = \tfrac{1}{2}\left(\phi_{2s} + \phi_{2p_x} + \phi_{2p_y} + \phi_{2p_z}\right), \tag{10-37a}$$

$$\psi_2 = \tfrac{1}{2}\left(\phi_{2s} - \phi_{2p_x} + \phi_{2p_y} - \phi_{2p_z}\right), \tag{10-37b}$$

$$\psi_3 = \tfrac{1}{2}\left(\phi_{2s} + \phi_{2p_x} - \phi_{2p_y} - \phi_{2p_z}\right), \tag{10-37c}$$

$$\psi_4 = \tfrac{1}{2}\left(\phi_{2s} - \phi_{2p_x} - \phi_{2p_y} + \phi_{2p_z}\right). \tag{10-37d}$$

It can be shown that these states are directed toward the vertices of a tetrahedron with the carbon at the center as shown in Figure 10-13. These four states are completely equivalent to one another in contrast to the original four functions. The ψ_i of Equations (10-37) are known as *hybrid orbitals*. The energy cost of having a $2s$ electron move to an excited $2p$ state is more than offset by the binding energy of the four hydrogens whose $1s$ functions each pair up with one of the states of Equations (10-37). It costs 8.3 eV to excite the $2s$ electron but the binding of the four hydrogens yields an energy of 25.3 eV.

10-6   Rotation

In the previous sections of this chapter we have assumed the nuclei are fixed while solving for the electronic states. Starting with this section we consider how the nuclei move if we allow the electronic energy found in the previous sections to act as an interaction potential energy function between the nuclei. In a stable state the electronic energy has a minimum at some internuclear separation R_0. This distance is usually an approximate equilibrium position for the molecule. As a first approximation we can consider the nuclei to be rigidly fixed at that equilibrium separation but allow the molecule to rotate freely in space. In Section 10-7 the other possible nuclear motion, vibration of the molecule about R_0 along the internuclear line, is considered.

A rotational problem analogous to the one taken up here is treated in the example at the end of Section 6-5 and the system in question is shown in Figure 6-6. The rotational energy levels we seek are those given in that example. However, because it is important to see precisely what approximations are involved we start here from scratch.

The time-independent Schrödinger equation for the two nuclei of a homonuclear diatomic molecule is

$$-\frac{\hbar^2}{2m}\left(\nabla_1^2 + \nabla_2^2\right)\psi(\mathbf{R}_1, \mathbf{R}_2) + V(R_{12})\psi(\mathbf{R}_1, \mathbf{R}_2) = E\psi(\mathbf{R}_1, \mathbf{R}_2), \tag{10-38}$$

where $\mathbf{R}_1, \mathbf{R}_2$ are the positions of the two nuclei and $R_{12} = |\mathbf{R}_1 - \mathbf{R}_2|$. If we separate variables into center-of-mass and relative coordinates, we find, as with the hydrogen atom, that the center-of-mass motion corresponds to that of a free particle, while the Schrödinger equation for the relative motion is (compare with Equation (6-16))

$$\left[-\frac{\hbar^2}{2\mu} \left\{ \frac{1}{R^2} \frac{\partial}{\partial R} R^2 \frac{\partial}{\partial R} + \frac{\Lambda^2}{R^2} \right\} + V(R) \right] \psi_{\text{rel}}(R, \Theta, \Phi) = E_{\text{rel}} \psi_{\text{rel}}(R, \Theta, \Phi),$$

(10-39)

where R, Θ, Φ are the spherical coordinates of the vector $\mathbf{R}_{12}$, μ is the reduced mass $m_1 m_2/(m_1 + m_2)$ equal in this case to $m/2$, and Λ^2 is the angular operator (Equation (6-12))

$$\Lambda^2 = \frac{1}{\sin \Theta} \frac{\partial}{\partial \Theta} \sin \Theta \frac{\partial}{\partial \Theta} + \frac{1}{\sin^2 \Theta} \frac{\partial^2}{\partial \Phi^2}.$$

(10-40)

From Chapter 6 it is known that ψ can be written in the separable form

$$\psi_{\text{rel}}(R, \Theta, \Phi) = F(R) Y_{\ell m}(\Theta, \Phi),$$

(10-41)

where $Y_{\ell m}(\Theta, \Phi)$ is a spherical harmonic. The operator Λ^2 acting on $Y_{\ell m}$ yields $-\ell(\ell + 1)$, where $\ell = 0, 1, 2, \ldots$ with ℓ the *nuclear* orbital angular momentum quantum number. That is, ℓ is a measure of the angular momentum of the pair of nuclei rotating about an axis perpendicular to the line between the nuclei.

In general, the total angular momentum of the molecule is the sum of the nuclear plus electronic angular momenta. In the case where the electrons are in a state of zero angular momentum, the nuclear quantum number ℓ may be replaced by the total quantum number j. In many situations in which there is nonzero electronic angular momentum we can replace $L^2 \rightarrow -\hbar^2 \Lambda^2$ by the operator $J^2 = (\text{total angular momentum})^2$ with the difference being absorbed into the electronic energy $V(R)$. In either case we end up with the relative motion equation

$$\left\{ -\frac{\hbar^2}{2\mu} \left[\frac{1}{R^2} \frac{d}{dR} R^2 \frac{d}{dR} - \frac{j(j+1)}{R^2} \right] + V(R) \right\} F(R) = E_{\text{rel}} F(R). \quad (10\text{-}42)$$

Given a functional form for $V(R)$ we could now solve this equation for $F(R)$ and then find the average value of the rotational energy $(\hbar^2/2\mu) j(j+1)\langle 1/R^2 \rangle$. However, a simpler approach is possible. $V(R)$ has a minimum at R_0 so that the equilibrium separation of the nuclei, that is, the mean value of R, is approximately R_0. The kinetic and rotational energies are usually small enough that they do not significantly alter this estimate of the mean value. So the molecular rotation energy is very nearly

$$E_{\text{rot}} = \frac{\hbar^2}{2\mu} \frac{j(j+1)}{R_0^2}, \qquad j = 0, 1, 2, \ldots .$$

(10-43)

The quantity μR_0^2 in Equation (10-43) is the moment of inertia of the molecule. E_{rot} is the energy associated with the rotation of a rigid dumbbell as obtained in Section 6-5. The energy levels represented by Equation (10-43) lead to spectral lines whose separation is so small that they are often unresolved and appear as *bands*, as discussed in Section 10-8.

Occasionally, it is necessary to go beyond the approximation made in replacing $\langle R^{-2} \rangle$ by R_0^{-2} to arrive at Equation (10-43). Some spectra involving high rotational quantum numbers show the effect of "stretching" of the dumbbell, and the resulting increase of the moment of inertia becomes evident.

Example

A numerical estimate of the rotational energy is easily obtained. Consider the hydrogen molecule H_2 for which 2μ is the proton mass and $R_0 \approx 0.07$ nm. We have from Equation (10-43)

$$E_{\text{rot}} \sim \frac{\left(1 \times 10^{-34} \text{ J} \cdot \text{s}\right)^2}{\left(1.7 \times 10^{-27} \text{ kg}\right)\left(0.7 \times 10^{-10} \text{ m}\right)^2} = 1.2 \times 10^{-21} \text{ J}$$

or

$$E_{\text{rot}} \sim \frac{1.2 \times 10^{-21} \text{ J}}{1.6 \times 10^{-19} \text{ J/eV}} = 7 \times 10^{-3} \text{ eV}.$$

This energy is to be compared to a typical size of a molecular electronic energy, which can be shown (see Problem 16 at the end of the chapter) to be of order

$$E_{\text{el}} \sim \frac{\hbar^2}{m_e R_0^2} = \frac{M_p}{m_e} \frac{\hbar^2}{M_p R_0^2} = \left(1.8 \times 10^3\right)\left(7 \times 10^{-3} \text{ eV}\right) = 13 \text{ eV},$$

where m_e and M_p are the electron and proton masses, respectively.

10-7 Vibration

The problem of the relative motion of two nuclei in a diatomic molecule (as described by Equation (10-42)) is that of a particle in a potential well, $V(R)$, having a minimum. If the well is deep and narrow, it is reasonable to assume that the nuclear rotation is the most easily excited type of motion. However, we know that a particle in any well—for example, in a parabolic well—vibrates about the position of the minimum, even in the ground state. The harmonic oscillator treated in Section 5-5 is our prime model in this regard. We now consider molecular vibrations and show this problem can usually be reduced to the case of a one-dimensional harmonic oscillator.

Returning to Equation (10-42), we write

$$E_{\text{rel}} = E_{\text{vib}} + E_{\text{rot}},$$

where E_{vib} is the vibrational energy that we want to determine and E_{rot} is the rotational energy of Equation (10-43). We then have

$$-\frac{\hbar^2}{2\mu} \frac{1}{R^2} \frac{d}{dR}\left(R^2 \frac{dF}{dR}\right) + V(R)F(R) = E_{\text{vib}}F(R).$$

In analogy with the hydrogen-atom problem we let $U(R) = RF(R)$. This results in

Figure 10-14

Morse potential energy $V(R)$ with depth $-A$ at R_0. The dashed line is a parabolic potential $\tilde{V}$ that fits the Morse potential at the minimum.

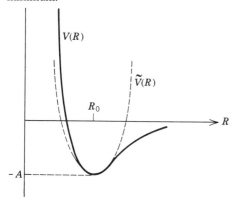

the equation

$$-\frac{\hbar^2}{2\mu}\frac{d^2U(R)}{dR^2} + V(R)U(R) = E_{\text{vib}}U(R), \qquad (10\text{-}44)$$

so the problem has been reduced to that of a one-dimensional Schrödinger equation.

For many covalently bonded molecules a suitable form for $V(R)$ is the *Morse potential energy*. This expression, deduced by P. M. Morse, is given by

$$V(R) = A\left(e^{-2a(R-R_0)} - 2e^{-a(R-R_0)}\right), \qquad (10\text{-}45)$$

where A, a, and R_0 are parameters varying from molecule to molecule. The form of this function is basically just an intelligent guess, as in the cases of Equations (10-24) and (10-36), but it does do an adequate job of correlating many molecular properties. For only a very few molecules, notably H_2, can accurate theoretical potential energy functions be derived from first principles. Most molecules are much too complicated for one to do much better than to fit Equation (10-45) or an equivalent form to several pieces of molecular data. Figure 10-14 indicates that the depth of the Morse minimum is $-A$, which occurs at $R = R_0$. The parameter a determines how fast the potential energy falls off with distance and curvature of the well.

For small oscillations about equilibrium such that the nuclear separation does not deviate very far from R_0, it is reasonable to approximate $V(R)$ by a parabola centered on R_0. Such a parabolic potential energy is shown as the dashed curve in Figure 10-14 and is denoted as $\tilde{V}(R)$. We can write

$$\tilde{V}(R) = V(R_0) + \tfrac{1}{2}K(R - R_0)^2, \qquad (10\text{-}46)$$

where the constant K is to be determined so that $V(R)$ and $\tilde{V}(R)$ have the same curvature at R_0. This means the second derivatives of $V(R)$ and $\tilde{V}(R)$ have to be the same:

$$\tilde{V}''(R_0) = K = V''(R_0) = 2a^2A. \tag{10-47}$$

Equation (10-46), with K given as $2a^2A$, is the result obtained with a Taylor expansion of $V(R)$ about $R = R_0$ if only up to quadratic terms in $R - R_0$ are kept.

With the substitution of $\tilde{V}$ for V, the Schrödinger equation becomes

$$-\frac{\hbar^2}{2\mu}\frac{d^2U(R)}{dR^2} + \frac{1}{2}K(R - R_0)^2 U(R) = \left[E_{\text{vib}} - V(R_0)\right]U(R).$$

If we let $y = R - R_0$, $\varepsilon = E_{\text{vib}} - V(R_0)$, and $u(y) = U(y + R_0)$, this becomes

$$-\frac{\hbar^2}{2\mu}\frac{d^2u(y)}{dy^2} + \frac{1}{2}Ky^2u(y) = \varepsilon u(y). \tag{10-48}$$

This equation is almost the Schrödinger equation for simple harmonic motion, as treated in Section 5-5. The one small difference is that the radial coordinate R is restricted to $R > 0$ so that $y > -R_0$. However, this is really an unimportant restriction because if the particle motion had such an amplitude that it got to $y = -R_0$ it would certainly be in a region where the parabolic approximation is invalid anyway. The gaussian wave functions for the low states of the harmonic oscillator restrict y to avoid this forbidden region and we can simply forget the restriction. On the other hand, the highly excited states in the oscillator well have much wider amplitudes than the ground state and they should not be expected to be reliable approximations to the true states of $V(R)$.

The radial Schrödinger equation with the exact Morse function in Equation (10-45) can actually be solved analytically. However, the mathematics involved is sufficiently complicated that we lose sight of the simple physics of the vibratory motion if we attempt to review the exact solution here.

The energy eigenvalues derived from Equation (10-48) are just those of the harmonic oscillator (see Equation (5-46)) with quantum number v:

$$\varepsilon_v = \hbar\omega_0\left(v + \tfrac{1}{2}\right), \qquad v = 0, 1, 2, \ldots$$

where

$$\omega_0 = \sqrt{\frac{K}{\mu}}. \tag{10-49}$$

We therefore obtain

$$E_{\text{vib}} = -A + \hbar\omega_0\left(v + \tfrac{1}{2}\right), \qquad v = 0, 1, 2, \ldots. \tag{10-50}$$

These are the energy levels corresponding to oscillations in the length of the diatomic molecule, in the approximation in which we think of the two nuclei as being connected by a stiff spring. For highly excited states this approximation fails and the Morse or other appropriate potential energy must be used to find E_{vib}.

Polyatomic molecules can often be thought of as several atoms connected by a number of springs. Such a system has a number of oscillatory patterns known as *normal modes*. The techniques of classical mechanics may be used to find the frequencies of these modes from which the vibration spectrum may then be found.

Example

The values of the Morse potential energy parameters for the O_2 molecule are

$$A = 5.2 \text{ eV}, \qquad R_0 = 0.12 \text{ nm}, \qquad a = 27 \text{ nm}^{-1}.$$

From these values we find

$$K = 2a^2A = 2(27 \text{ nm}^{-1})^2(5.2 \text{ eV}) = 7600 \text{ eV/nm}^2 = 1.2 \times 10^3 \text{ J/m}^2$$

so that a characteristic vibrational energy is

$$\hbar\omega_0 = \hbar\sqrt{\frac{K}{\mu}} = \frac{1.05 \times 10^{-34} \text{ J} \cdot \text{s}}{1.6 \times 10^{-19} \text{ J/eV}}\sqrt{\frac{1.2 \times 10^3 \text{ J/m}^2}{8(1.7 \times 10^{-27} \text{ kg})}} = 0.2 \text{ eV}.$$

Note that the reduced mass used is one-half the oxygen atomic mass. This vibrational energy should be compared with typical electronic and rotational energies of about 10 eV and 10^{-3} eV, respectively, as computed in Section 10-6.

Example

The dissociation energy D, the energy to pull the molecule apart, is approximately A, the depth of the potential energy well. This is not exactly right because the molecule has zero-point vibrational energy $\frac{1}{2}\hbar\omega_0$. Thus, we have

$$D = A - \tfrac{1}{2}\hbar\omega_0.$$

For O_2 we obtain

$$D = 5.2 - 0.1 = 5.1 \text{ eV},$$

in agreement with the experimental value.

10-8 Spectra

Much of what is known about molecules comes from examining spectra. Because a molecule is inherently more complex than an atom, molecular spectra are often more complicated than atomic spectra. From the spectral lines the various effects caused by electronic, vibrational, and rotational transitions must be disentangled. The energy scales of these three types of transition are quite different, as we have seen above, with typical values being of order 10, 10^{-1}, and 10^{-3} eV, respectively. Because the rotational energy is so small, the spectra often appear as *bands* of very closely spaced lines, the most noticeable characteristic of molecular spectra.

We examine three main spectral types: pure rotational, rotation–vibrational, and electronic. To keep the discussion simple only those transitions between molecular states having the angular momentum quantum number $\lambda = 0$ are considered. Our analysis is therefore not quite as general as possible but does illustrate most of the essential elements.

Figure 10-15

Molecular absorption transitions. Transition I is pure rotational with $\Delta j = 1$ and no change in vibrational or electronic state. Transition II involves a vibrational change $\Delta v = 1$; the rotational change is $\Delta j = \pm 1$. Transition III involves an electronic change of state with possible vibrational and rotational changes as well.

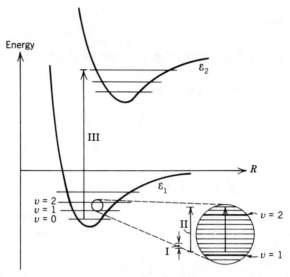

A molecular energy level within the approximations discussed in previous sections is, by Equations (10-43) and (10-50),

$$\varepsilon = \varepsilon_{el} + \hbar\omega_0\left(v + \tfrac{1}{2}\right) + Bj(j + 1), \tag{10-51}$$

where

$$B = \frac{\hbar^2}{2\mu R_0^2}. \tag{10-52}$$

Figure 10-15 presents a schematic illustration of transitions involving all three terms in Equation (10-51).

The first transition to be considered is a pure rotational change of state. Such a change is shown in Figure 10-15 as I. Because of symmetry considerations, a pure rotational transition, that is, one in which j changes but the electronic and vibrational quantum numbers remain the same, cannot occur for a homonuclear diatomic molecule, such as H_2. It can occur, however, in a heteronuclear molecule having a permanent electric dipole moment, such as LiF.

There is a selection rule for a pure rotational transition that forbids transitions other than

$$\Delta j = \pm 1.$$

For absorption spectra $j \rightarrow j + 1$ and from Equation (10-51) the frequencies are given by

$$h\nu = B[(j + 1)(j + 2) - j(j + 1)] = 2B(j + 1). \tag{10-53a}$$

Figure 10-16
Spectral form for a pure rotational transition as I in Figure 10-15. The lines are equally spaced with energy interval $2B$.

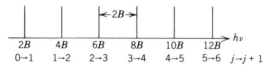

For emission spectra $j \rightarrow j - 1$ so

$$h\nu = B[j(j+1) - (j-1)j] = 2Bj. \tag{10-53b}$$

In both cases the frequencies vary linearly with j, which means the lines are evenly spaced with frequency separation

$$\Delta\nu = \frac{2B}{\hbar} \tag{10-54}$$

as shown in Figure 10-16. These transitions are in the far-infrared or microwave range.

From pure rotational spectra, and in particular from Equation (10-54), we can find B. The equilibrium interatomic spacing R_0 is then found from Equation (10-52).

Vibrational transitions cannot take place in a homonuclear molecule without an accompanying electronic change but can in a molecule with a permanent dipole moment. In such a case the selection rules are

$$\Delta v = \pm 1 \quad \text{and} \quad \Delta j = \pm 1.$$

A typical vibrational transition is shown as II in Figure 10-15. Absorption always implies $\Delta v = +1$ since the vibrational energy is so much larger than the rotational. In the usual situation the initial vibrational state is $v = 0$ since the higher vibrational states are not thermally excited at room temperature. Then the energy change is either that with $j \rightarrow j + 1$ given by

$$\begin{aligned} h\nu_R &= \hbar\omega_0\left(1 + \tfrac{1}{2}\right) + B(j+1)(j+2) - \tfrac{1}{2}\hbar\omega_0 - Bj(j+1) \\ &= \hbar\omega_0 + 2B(j+1), \qquad j = 0, 1, 2, \ldots, \end{aligned} \tag{10-55}$$

or that with $j \rightarrow j - 1$ given by

$$\begin{aligned} h\nu_P &= \hbar\omega_0\left(1 + \tfrac{1}{2}\right) + B(j-1)j - \tfrac{1}{2}\hbar\omega_0 - Bj(j+1) \\ &= \hbar\omega_0 - 2Bj, \qquad j = 1, 2, \ldots. \end{aligned} \tag{10-56}$$

As indicated by the subscripts on the ν's, these spectral lines are known as the R and P *branches*. Each of these branches gives a group of lines much like that due to the pure rotational transitions except that the frequencies are in the range about $\hbar\omega_0$ so that they fall in the infrared.

Figure 10-17

Spectral form for a vibrational–rotational transition. There are two branches with a missing line at $\hbar\omega_0$. The corresponding transition in Figure 10-15 is II.

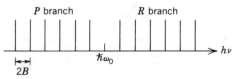

Figure 10-17 shows how the vibrational–rotational spectrum splits into the two branches in accord with Equations (10-55) and (10-56). Note that there is a gap between the two bands at the vibrational frequency $\omega_0/2\pi$. This exists because $\Delta j = 0$ is not allowed and because in Equation (10-56) $j = 0$ does not occur. (That would correspond to the transition $j = 0$ to $j = -1$.)

Transitions in which there is an electronic change of energy are represented by III in Figure 10-15. The transition frequencies can again be deduced from Equation (10-51). However, now note that ω_0 and B are not generally the same for the initial and final electronic states. B depends on the molecular moment of inertia and hence on the interatomic separation R_0, which varies with the electronic state. In Figure 10-15, for example, the minima in ε_1 and ε_2 occur at different values of R. The vibrational frequency ω_0 depends on the curvature of the electronic energy in the neighborhood of its minimum. (See Equations (10-47) and (10-49).) The curvature is also expected to be different for different states. For absorption the change in electronic and vibrational energy is

$$hv_{ev} = \varepsilon_{s's} + \hbar\omega_0'\left(v' + \tfrac{1}{2}\right) - \hbar\omega_0\left(v + \tfrac{1}{2}\right), \qquad (10\text{-}57)$$

where ω_0 and ω_0' are the vibrational frequencies for initial and final states, s and v are the initial state electronic and vibrational quantum numbers, and s' and v' are those of the final state. In absorption the initial electronic and vibrational states are usually ground states so $s = 0$ and $v = 0$.

Equation (10-57) sets the scale of the transitions, which run from the near infrared to the ultraviolet. The fine structure of the spectrum is determined by examining the rotational transitions that occur simultaneously with the above change hv_{ev}. What the rotational structure does is to spread the electronic–vibrational line into a *band* of many very closely spaced lines. The selection rule is again $\Delta j = \pm 1$ so that there is an R branch corresponding to $j \to j + 1$ with energy changes

$$hv_R = hv_{ev} + B'(j + 1)(j + 2) - Bj(j + 1), \qquad j = 0, 1, 2, \ldots, \quad (10\text{-}58a)$$

and a P branch for $j \to j - 1$ given by

$$hv_P = hv_{ev} + B'j(j - 1) - Bj(j + 1), \qquad j = 1, 2, \ldots. \quad (10\text{-}58b)$$

Because $B \neq B'$ the quadratic terms in j no longer cancel out and the spectra are not made up of evenly spaced lines as in the previous situations. We can write

$$hv_R = hv_{ev} + (B' - B)j^2 + (3B' - B)j + 2B' \qquad (10\text{-}59a)$$

Figure 10-18

Spectral form for an electronic transition. The electronic–vibrational transition energy is $h\nu_{ev}$. The band due to rotational transitions has a band head at the left (red) and degrades to the right (blue).

Figure 10-19

Pure rotational absorption spectrum of HCl gas. Note the even line spacing as indicated schematically in Figure 10-16. The frequency axis is $\tilde{\nu} = \nu/c$ in units of cm^{-1}.

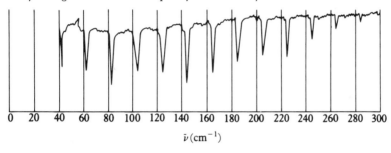

Figure 10-20

Vibrational–rotational absorption spectrum of a diatomic molecule like that shown schematically in Figure 10-17. The absent line is denoted by $\tilde{\nu}_v$.

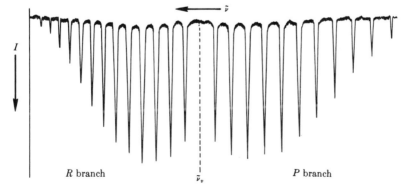

Figure 10-21

Electronic spectrum of the molecule AlO. See Figure 10-18 for a similar schematic version. Note the bands caused by the small spacing of the rotational states.

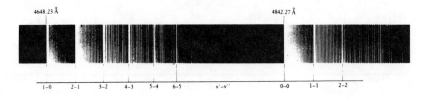

and

$$h\nu_P = h\nu_{ev} + (B' - B)j^2 - (B' + B)j. \tag{10-59b}$$

If $B' > B$ these ν_R and ν_P results increase quadratically toward higher frequency. The function $\nu_P(j)$ has a minimum at a frequency somewhat lower than ν_{ev}. In this region the lines are also more closely spaced than at high j values; at high j the spacing becomes greater because of the increasing influence of the j^2 term. A schematic drawing of this spectrum is shown in Figure 10-18. The band is said to be *degraded* to the blue (i.e., the high-frequency end) and to have a *band head* (the minimum) at the red (low-frequency) end. If $B' < B$, the band head is at the blue end and the degradation is toward the red. The details of just how the spectrum comes to look like that of Figure 10-18 are given in the example below. Such a band appears for each possible $v \rightarrow v'$ transition. The spectra appear as a series of adjacent bands.

Examples of real spectra of the types discussed in this section are shown in Figures 10-19, 10-20, and 10-21.

Example

To see how the frequencies in Equations (10-59) produce a spectrum like that shown in Figure 10-18, rewrite the equations in the form

$$y_R(j) = 10^2\left(\frac{\nu_R}{\nu_{ev}} - 1\right) = (b' - b)j^2 + (3b' - b)j + 2b' \tag{10-60a}$$

and

$$y_P(j) = 10^2\left(\frac{\nu_P}{\nu_{ev}} - 1\right) = (b' - b)j^2 - (b' + b)j, \tag{10-60b}$$

where

$$b = \frac{10^2 B}{h\nu_{ev}} \quad \text{and} \quad b' = \frac{10^2 B'}{h\nu_{ev}}.$$

Then let b and b' take on the typical values $b = 3$, $b' = 4$, so that

$$y_R(j) = j^2 + 9j + 8 \quad \text{and} \quad y_P(j) = j^2 - 7j.$$

Figure 10-22

Frequency functions $y_R(j)$ and $y_P(j)$ of Equations (10-60) for $b = 3$ and $b' = 4$. The y values corresponding to integers are the allowed frequencies and are indicated by arrows on the y axis. The band head is at the left end corresponding to the minimum in y_P.

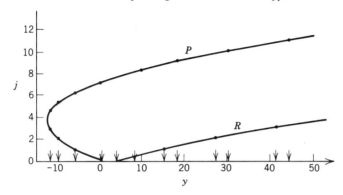

If we consider y_R or y_P to be a function of j as a continuous variable, we can plot each as shown in Figure 10-22. The plot is laid on its side so that the frequency variable y runs horizontally. Integral values of j are shown as dots along the curves; their projections on the y axis give the frequencies of the allowed transitions. It is easy to see that y_P has a minimum:

$$\frac{dy_P}{dj} = 2j - 7 = 0 \quad \text{at } j = \tfrac{7}{2} = 3.5.$$

The value of y_P at the minimum is $(\tfrac{7}{2})^2 - 7(\tfrac{7}{2}) = -12.25$. The minimum frequency occurs at the band head. The y_P parabola opens to the right so that the allowed y values become more separated in moving to larger y values. This is band degradation to the blue. The R branch does not have a minimum for positive j but also degrades to the blue.

Problems

1. A particle of mass m moves in a "double-oscillator" potential energy given by $V(x) = \tfrac{1}{2}k(|x| - a)^2$, as plotted in the figure. When the separation $2a$ between wells is sufficiently large, good approximations to the eigenfunctions for the two lowest states are

$$\psi_{\pm} = \frac{1}{\sqrt{2}}(\phi_a \pm \phi_{-a}),$$

where

$$\phi_{\pm a}(x) = \left(\frac{2\alpha}{\pi}\right)^{1/4} e^{-\alpha(x \mp a)^2}$$

is the single oscillator ground-state eigenfunction centered on $x = \pm a$, with $\alpha = \tfrac{1}{2}m\omega_0/\hbar$ and $\omega_0 = \sqrt{k/m}$.

(a)   Given the above approximate eigenfunction, show that the energies $\varepsilon_\pm$ of the two lowest states can be written in the form of Equation (10-7), where

$$G = \int \phi_{-a}^2 \bar{V}(x)\, dx \quad \text{and} \quad S = \int \phi_a \bar{V}(x)\phi_{-a}\, dx$$

with $\bar{V} = -k(x + |x|)a$, and where ε_{1s} is replaced by $\varepsilon_0 = \frac{1}{2}\hbar\omega_0$, the ground state of a single oscillator.

(b)   Evaluate S and show that the separation between the two lowest states is proportional to $e^{-2\alpha a^2}$.

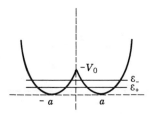

2.   The ammonia molecule NH_3 has a tetrahedral structure as shown in Figure 5-25. The nitrogen atom has two equivalent positions either on the right of the plane of the 3 hydrogens or in the "inverted" position on the left. It can tunnel between these two positions and so has a wave function analogous to that of the H_2^+ electron or to the particle in Problem 1. The splitting between the two lowest energy levels is $\varepsilon_- - \varepsilon_+ = 9.8 \times 10^{-5}$ eV.

(a)   Find the time it takes the nitrogen atom to invert.

(b)   Use the potential energy of Problem 1 as a model for that seen by the nitrogen atom. Take $a = 0.038$ nm. What is the height V_0 of the potential barrier separating the two wells?

3.   What is the electronic ground-state eigenfunction of He^+? Show that ψ_+ of Equation (10-3a) does not reduce to it for $R \to 0$.

4.   (a)   By what constant should the eigenfunction of Equation (10-19) be multiplied in order to be normalized properly? Neglect all overlap integrals.

(b)   For H_2, γ is about 0.2. What is the probability of both electrons residing in a single nucleus?

5.   An alternative to the Heitler–London approach in approximating the H_2 electronic wave function is the *molecular orbital* method. *Two* electrons, of opposite spins, are considered to be in the same H_2^+-type state ψ_+ (Equation (10-3a)) just as two electrons are put into the $1s$ orbital to describe atomic helium. The eigenfunction then is $\psi_+(r_1)\psi_+(r_2)\chi^A(1, 2)$.

(a)   Show this function is properly antisymmetric.

(b)   Express this eigenfunction in terms of the $1s$ atomic orbitals and compare with the form of Equation (10-19). In particular, determine the value of γ that governs the amount of ionic structure implied by this function.

(c)   What is the probability of both electrons residing on a single nucleus in this approximation?

6.   In the binding of H_2^+ a portion of the electronic charge is shifted from the regions immediately around the nuclei to a position in between. A simple model of this is shown in

the figure. The total charge in the system is $+e$. What portion of electronic charge δe must be placed in the intermediate position for binding to result? [Hint: Note that binding implies a negative total energy of the system.]

$$\tfrac{1}{2}(e + \delta e) \qquad -\delta e \qquad \tfrac{1}{2}(e + \delta e)$$
$$\underset{\bullet}{} \qquad \underset{x}{} \qquad \underset{\bullet}{}$$
$$\left|\!\leftarrow\!\frac{R}{2}\!\rightarrow\!\right|\!\leftarrow\!\frac{R}{2}\!\rightarrow\!\left|\right.$$

7. Other molecular orbital combinations, besides that in Problem 5, may be formed out of products like $\psi_+(r_1)\psi_-(r_2)$ and $\psi_-(r_1)\psi_-(r_2)$ times singlet or triplet spin functions. Five additional functions, all representing excited states, may be formed. What are these functions? Be sure they are all antisymmetric in electron coordinates. By writing these out in atomic functions and examining the form of the covalent part of each function, determine which are bonding and which are antibonding.

8. (a) The ionization potential for hydrogen is 13.6 eV. The electron affinity for fluorine is 0.8 eV. Evaluate the Q value of Equation (10-27) for HF. Compare this with the value for LiF as worked out in Section 10-3. From this result do you expect HF to have a greater or lesser ratio of ionic to covalent bonding than LiF?

   (b) The equilibrium separation of the two nuclei in HF is $R = 0.092$ nm. The experimental value of the electric dipole moment of this molecule is $p_{exp} = 0.6 \times 10^{-29}$ C $\cdot$ m. Calculate p/eR. Does the result here confirm your answer to the question in part (a)?

9. (a) Verify that the eigenfunction of Equation (10-21) is normalized if $A^2 + B^2 + C^2 = 1$.

   (b) Construct an eigenfunction analogous to that of Equation (10-21) for HF.

   (c) Determine the coefficients A and C (assume $B = 0$) for this eigenfunction by use of the value of p/eR_0 found in Problem 8(b).

10. (a) The internuclear separation for NaCl is $R_0 = 0.236$ nm. Spectral data show that the curvature K of the potential energy is 109 J/m^2. The ionization potential of sodium is 5.1 eV and the electron affinity of chlorine is 3.65 eV. From these values determine the parameters α and a in Equation (10-24), and find the dissociation energy D. How does your result compare with the experimental value of 4.22 eV?

    (b) Find the zero-point energy correction to the calculation of D.

11. As a classical model of atomic polarizability consider a positive point charge $+e$ (the nucleus) surrounded by a uniform charged sphere of radius a and charge density $\rho = -e/(\tfrac{4}{3}\pi a^3)$ (the electron "cloud").

    (a) Show that if the nucleus is displaced by a distance z from the center of the sphere it feels a restoring force $F = -(e\rho/3\varepsilon_0)z$.

    (b) Show that if this system is put into an electric field E it develops a dipole moment given by $p = \alpha E$ with $\alpha = 4\pi\varepsilon_0 a^3$.

12. Show that a hydrogen atom placed a distance R from H$^+$ has a potential energy U of the form $U \sim -I_0(a/R)^4$, where I_0 and a are the ionization potential and the atomic radius of the hydrogen atom, respectively. How should the phenomenological potential energy, Equation (10-24), for ionically bound molecules be modified to take account of this effect?

13. Show that the Lennard-Jones 12-6 potential of Equation (10-36) has a minimum of $-\varepsilon$ at an interparticle separation $R_0 = 2^{1/6}\sigma$.

14. Above approximately what temperature is a diatomic argon molecule expected to be dissociated?

15. The Lennard-Jones interaction between a carbon atom and a helium atom is given by Equation (10-36) with $n = 12$, $\varepsilon = 1.5 \times 10^{-3}$ eV, and $\sigma = 0.3$ nm. Suppose a helium atom is at position $\mathbf{r}_0 = (0, 0, h)$ above the flat surface of a piece of solid carbon. The origin $(0, 0, 0)$ is on the surface and the carbon solid extends from $-\infty$ to ∞ in x and y, and $-\infty$ to 0 in z. The carbon density is n_0.

    (a) Give a brief argument that shows that the helium atom has potential energy $U(\mathbf{r}_0) = n_0 \int V_{LJ}(|\mathbf{r}_0 - \mathbf{r}|)\, d\mathbf{r}$, where the integral extends over all the carbon, which is taken as continuous.

    (b) Carry out the above integral and show that U has the form

    $$U = \frac{a}{h^9} - \frac{b}{h^3}.$$

    What are a and b?

    (c) Find the value of the helium position h for which U is a minimum when the carbon density is $n_0 = 113$ particles/nm³. What is the potential energy function depth there?

    (d) To what value must the temperature be lowered for a layer of helium atoms to begin forming on the surface?

16. In the example at the end of Section 10-6 the expression $E_{el} = \hbar^2/m_e R_0^2$ is used to estimate typical electronic molecular energy. Justify this formula by use of (a) the energy levels of a particle in a box and (b) the uncertainty principle.

17. The Born–Oppenheimer approximation, the basis of studies of molecules, claims the nuclear motions are so much slower than electronic that the electronic energies can be computed for fixed nuclear positions. Compute the orders of magnitude of the periods of rotational and vibrational nuclear motion for H_2 and compare them with the period of an electron in a Bohr orbit. The Morse potential parameters for the H–H interaction are $A = 4.75$ eV, $R_0 = 0.074$ nm, and $a = 19.5$ nm⁻¹.

18. Give arguments that show that $E_{rot} \sim (m/M)E_{el}$ and $E_{vib} \sim (m/M)^{1/2}E_{el}$, where E_{el}, E_{rot}, and E_{vib} are electronic, rotation, and vibrational energies, respectively, m is the electron mass, and M is a nuclear mass. Evaluate these relations numerically (with $E_{el} \sim 10$ eV) and compare with the estimates given in the examples at the ends of Sections 10-6 and 10-7.

19. (a) By determining the curvature K at the minimum of the Lennard-Jones 12-6 potential, evaluate the harmonic oscillator approximation $\frac{1}{2}\hbar\omega_0$ to the zero-point energy of the argon molecule Ar_2. Compare this value with the depth of the potential and estimate the binding energy of Ar_2. See Table 10-1.

    (b) Repeat the above for the He_2 molecule. What does your result suggest about the possibility of ever forming a He_2 molecule?

20. The Morse potential for two interacting hydrogen atoms has the parameters given in Problem 17. From this compute the vibrational frequency ω_0 and the dissociation energy D for H_2. Compare the results with experimental values.

21. The internuclear separation for the CO molecule is $R_0 = 0.113$ nm; the dissociation energy is 9.60 eV; the vibrational frequency is $\nu_0 = \omega_0/2\pi = 6.51 \times 10^{13}$ Hz. Use these data to find the parameters of the Morse potential for the interaction between carbon and oxygen.

22. The separation between spectral lines in the pure rotational spectra of HCl is 6.35×10^{11} Hz. Use this information to find the internuclear spacing R_0.

23. Draw a diagram analogous to that of Figure 10-22 for the case $b = 2.5$, $b' = 2$. Is the band head to the red or blue? In which direction does the degradation take place?

24. Suppose you are a molecular spectroscopist and you have collected the data shown in Figure 10-18. The values of the first few y's (see Equations (10-60)) are -4.4, -4.2, -4.0, -3.4, -3.0, -2.0, -1.4, 0.8, 2.6, and 3.6. From these data find the values of the molecular rotation constants b and b'. [Hint: What value of the initial state quantum number j corresponds to the band head at $y = -4.4$?] Give the values of the j's corresponding to each of the quoted transitions and tell to which branch each belongs.

E L E V E N

QUANTUM
STATISTICAL
PHYSICS

*I*n Sections 2-3 and 2-4 we have seen how statistical reasoning could be used in conjunction with the quantum hypothesis to derive the Planck law of blackbody radiation. The procedures applied by Planck involved the kinetic theory techniques that had been developed in the 19th century by Maxwell and Boltzmann. This approach developed into the discipline of statistical mechanics with the work of J. W. Gibbs, an American physicist who was a contemporary of Planck.

The powerful techniques of statistical mechanics exploit the fact that macroscopic systems are made up of a large number of particles. It is clearly impossible to follow everything going on in so large a system. Usually only certain average quantities, such as pressure, volume, and temperature, are measured in experiments or calculated by theory. These variables are the quantities that have their origins in the joint behavior of all the particles. For example, the pressure of a gas is the result of an enormous number of individual molecular collisions with the walls of the container. The position or any other detailed dynamical property of an individual gas molecule is a variable that cannot be observed.

The seeming disadvantage of working with a small number of averaged quantities is offset by the fact that fluctuations in these averages are extremely small in a large system. The readings on a pressure gauge and on a thermometer are usually quite steady. The mechanism of probability theory at work here is the same as in coin tossing. If a coin is tossed ten times, a result, for example, of three heads and seven tails, quite far from a 50-50 split, is not rare. However, if the coin has been tossed a billion times any result not very close to an equal number of heads and tails (within $\sim 10^{-3}\%$) is quite surprising.

The use of a billion coins to determine the probability of tossing a head or tail is an example of the ensemble method of determining the averages. This technique is introduced and used in the present chapter to study the thermal properties of collections of particles having three different kinds of probabilistic behavior: those obeying Maxwell–Boltzmann statistics, Bose–Einstein statistics, or Fermi–Dirac statistics.

Before quantum mechanics, all particles had been thought of as completely distinguishable. When two billiard balls collide it is easy to tell which one is which after the collision. However, for microscopic particles, we now realize that the uncertainty principle makes it impossible to figure out which particle was incident and which particle was target. This feature of particle indistinguishability implies that all elementary, and many composite, particles obey either Bose–Einstein or Fermi–Dirac statistics. Microscopic particles of the same species that are distinguishable cannot actually exist in nature. However, they may be treated theoretically and are said to obey Maxwell–Boltzmann or classical statistics.

The macroscopic thermal properties of a system depend crucially on the statistics obeyed by the constituent particles. Electrons in a metal (Fermi–Dirac particles) behave very differently from the atoms in liquid helium (Bose–Einstein particles). However, at high temperatures all three kinds of particles have the same general type of thermal behavior.

Radiation may also be treated by the use of a particular form of quantum statistics. Planck had originally treated the properties of blackbody radiation by considering the effect of quantizing the charged oscillators in the solid walls of the cavity that emitted the radiation. Einstein pointed out that Planck's argument was flawed: Planck had calculated the energy density in a particular frequency mode of the radiation in terms of the energy of the oscillators by assuming the oscillators could emit radiation continuously. He then quantized the oscillator energy. The work of Bose on radiation and Einstein's subsequent generalization of it to massive particles led to a slightly different interpretation that we follow in this chapter. Light is made up of discrete indistinguishable massless entities called photons and the oscillator state $nh\nu$ is interpreted as describing the energy of n photons each of energy $h\nu$. This interpretation is consistent with Einstein's treatment of the photoelectric effect. Photons are thus found to obey Bose–Einstein statistics.

Blackbody radiation certainly did not provide the only evidence for the need to modify classical mechanics. The classical theory of solids predicted the heat capacity C to be independent of temperature. While this result held for high temperatures, at low temperatures C was found to drop toward zero. Einstein explained this phenomenon by modeling the solid as an independent set of quantized oscillators. This model was improved upon in 1912 by Debye who considered the solid as a set of particles coupled harmonically to one another. This theory, which gave the correct low-temperature behavior, has turned out to be remarkably similar to that of blackbody radiation. Indeed, the concept of massless Bose–Einstein particles called phonons traveling through the crystal lattice has proved to be very analogous to the photon idea.

Fermi and Dirac are responsible for the development of the statistics obeyed by electrons, protons, neutrons, and a host of other particles. The statistics forms the basis of the Pauli principle according to which one Fermi–Dirac particle is allowed per quantum state. The thermal properties of electrons in atoms and metals, protons and neutrons in nuclei, the atoms in liquid ^{3}He, and even the nature of white dwarf stars can be understood with Fermi–Dirac statistics, as described in this chapter.

11-1   Particle Indistinguishability

If two identical particles A and B pass within a de Broglie wavelength of each other, then after the encounter we are unable to tell which of the two particles is A and which is B. To do so requires knowledge of their positions to such an accuracy that, by the uncertainty principle, the encounter is essentially changed.

In Sections 5-10 and 9-5 we have seen how this property of microscopic encounters can be built into the wave functions of the particle pair in terms of their symmetry or antisymmetry. In the case that the potential energy function V is a sum of individual one-particle functions,

$$V(1,2) = V(1) + V(2),$$

the solutions to the Schrödinger equation have the form

$$\psi_{\alpha\beta}(1,2) = \begin{cases} \psi_\alpha(1)\psi_\beta(2) & \text{(11-1a)} \\ \text{or} \\ \psi_\beta(1)\psi_\alpha(2) & \text{(11-1b)} \end{cases}$$

with the $\psi_\alpha(i)$ satisfying

$$\left[-\frac{\hbar^2}{2m}\nabla_i^2 + V(i) \right]\psi_\alpha(i) = E_\alpha\psi_\alpha(i),$$

where E_α is an energy eigenvalue and ψ_α is a one-particle eigenfunction. The discussion of Section 9-5 leads us to two possible forms for the two-particle eigenfunction. One of these, the antisymmetric state

$$\psi_{\alpha\beta}^A(1,2) = \frac{1}{\sqrt{2}}\left(\psi_\alpha(1)\psi_\beta(2) - \psi_\beta(1)\psi_\alpha(2) \right), \tag{11-2}$$

is discussed at length in that section. It satisfies a relation like Equation (9-16),

$$\psi_{\alpha\beta}^A(2,1) = -\psi_{\alpha\beta}^A(1,2). \tag{11-3}$$

The other form is constructed in an analogous way to be

$$\psi_{\alpha\beta}^S(1,2) = \frac{1}{\sqrt{2}}\left(\psi_\alpha(1)\psi_\beta(2) + \psi_\beta(1)\psi_\alpha(2) \right), \tag{11-4}$$

and is symmetric under particle interchange:

$$\psi_{\alpha\beta}^S(2,1) = +\psi_{\alpha\beta}^S(1,2) \tag{11-5}$$

as in Equation (9-15).

We know from Chapter 9 that particles with half-integral spin values $(\tfrac{1}{2}\hbar, \tfrac{3}{2}\hbar, \ldots)$ have antisymmetric eigenfunctions and are known as *fermions*. These include electrons, protons, and neutrons, as well as composites of odd numbers of particles having half-integral spin such as ^{3}He atoms, and others. All these particles are said to obey *Fermi–Dirac statistics*.

Particles whose eigenfunctions behave like Equation (11-5) are said to obey *Bose–Einstein statistics* and are known as *bosons*. All such particles have integral spin $(0\hbar, 1\hbar, 2\hbar, \ldots)$. These include photons, certain elementary particles called mesons, and certain so-called collective modes in many-body systems (phonons), as well as composites of even numbers of fermions such as ^{4}He atoms, and others.

Figure 11-1

Intuitive description of the terms in the probability density Equation (11-6). The coefficient g has the value $g = 1$ for bosons, $g = -1$ for fermions, and $g = 0$ for Maxwell–Boltzmann particles.

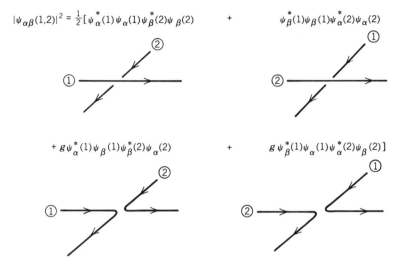

Let us now examine more closely the properties of these two-particle eigenfunctions. The measurable quantity, the probability density, is, for both cases simultaneously,

$$\left|\psi_{\alpha\beta}(1,2)\right|^2 = \tfrac{1}{2}\left[\psi_\alpha^*(1)\psi_\alpha(1)\psi_\beta^*(2)\psi_\beta(2) + \psi_\beta^*(1)\psi_\beta(1)\psi_\alpha^*(2)\psi_\alpha(2)\right.$$
$$\left. + g\psi_\alpha^*(1)\psi_\beta(1)\psi_\beta^*(2)\psi_\alpha(2) + g\psi_\beta^*(1)\psi_\alpha(1)\psi_\alpha^*(2)\psi_\beta(2)\right], \quad (11\text{-}6)$$

where $g = +1$ for bosons or -1 for fermions. A possible intuitive interpretation of the meaning of the various terms in Equation (11-6) is illustrated in Figure 11-1 and goes as follows: The first term on the right represents an encounter in which particle 1 has state α and particle 2 has state β both before and after the meeting; the second term puts 1 in β and 2 in α before and after. If only these terms exist (i.e., if $g = 0$), the final measurable results are equivalent to the situation in which the particles are distinguishable with, for example, particle 1 painted blue and particle 2 painted red. The new effects due to symmetry are contained in the third and fourth terms, the quantum interference or "exchange" terms. The third term has particle 1 starting out in β and particle 2 in α. However, after the encounter the two switch places so that 1 is in α and 2 is in β. The last term describes a similar process with 1 and 2 interchanged in roles. We have already seen how such interference terms can affect the energy of a system in our treatment of the H_2 molecule in Chapter 10. In Problem 2 at the end of the chapter we consider some further aspects of this phenomenon.

Particles of the same species, yet somehow distinguishable, are described by either of the Equations (11-1a) or (11-1b). We can tell the difference between the two forms; for example, the blue particle is in α and the red is in β, or vice versa. Such particles do not actually occur in nature but are useful conceptually. They are said to obey

Maxwell–Boltzmann statistics. Later in this chapter we show that the behavior of fermions and bosons becomes identical to that of Maxwell–Boltzmann particles at sufficiently high temperature. We can produce the effects of the statistics in Equation (11-6) by keeping only the first two terms, that is, by taking $g = 0$. (We can tell the difference between the two forms in Equations (11-1a) and (11-1b), and still have the Maxwell–Boltzmann case in Equation (11-6) while keeping both with equal amplitudes, as long as we do not let the two forms interfere.)

We can see when quantum effects become important by considering the pair wave function in the case that $\psi_\alpha(i)$ and $\psi_\beta(i)$ are localized, or peaked up at some position in space. We assume each of $\psi_\alpha(1)$, $\psi_\beta(1)$, $\psi_\alpha(2)$, and $\psi_\beta(2)$ has the same spatial width d. It is fairly obvious that the appropriate de Broglie wavelength with which to characterize such a localized state is $\lambda \sim d$. (This can be proved but we are not going to do it here.) Suppose $\psi_\alpha(1)$ and $\psi_\beta(1)$ differ mainly in being centered at different positions separated by a distance a. (An example of this situation occurs in Chapter 10 in which the H_2 molecular eigenfunction is written in terms of $1s$ atomic orbitals, each localized on one of the two protons.) If $\lambda \ll a$, the interference terms in Equation (11-6) vanish; that is, where $\psi_\alpha(1)$ is large, $\psi_\beta(1)$ is zero, and vice versa. Maxwell–Boltzmann statistics is then valid. On the other hand, if $\lambda > a$ the single-particle eigenfunctions overlap, the interference term is nonzero, and quantum effects are present. In Section 11-3 we introduce the thermal de Broglie wavelength λ_{th}, which is a measure of the average value of λ as a function of temperature. The ratio λ_{th}/a determines the importance of quantum effects in the thermal properties of a system.

We have seen previously that the fermion function in Equation (11-2) vanishes, in accord with the Pauli principle, when both particles are in the same state:

$$\psi_{\alpha\alpha}^A(1,2) = 0.$$

Fermions are antisocial. Bosons, on the other hand, are gregarious. They actually show a preference for being in the same state or place when compared to distinguishable particles. To see this take both particles at the same position, that is, $\mathbf{r}_1 = \mathbf{r}_2 = \mathbf{r}$, in Equation (11-6). Then it is easy to see that we get

$$\left|\psi_{\alpha\beta}(\mathbf{r},\mathbf{r})\right|^2 = \left|\psi_\alpha(\mathbf{r})\right|^2\left|\psi_\beta(\mathbf{r})\right|^2(1 + g).$$

For fermions $g = -1$ and this expression vanishes as one expects. For the Maxwell–Boltzmann case $g = 0$ and $(1 + g) = 1$, while for bosons $(1 + g) = 2$. This preference that bosons have for being in the same place (or state) is important in understanding several physical phenomena including the superfluidity of liquid ^{4}He.

As illustrated by the Slater-determinant eigenfunctions of Section 9-5, it is possible to generalize our symmetry principles to functions for more than two particles. To get the boson version of the Slater determinant, expand it into a series of terms and change all the minus signs to plus signs. The result is a function symmetric under the interchange of any pair of the variables.

Example

Suppose we have three particles in three states α, β, γ. The fermion version of the eigenfunction for this case is written in Problem 8 of Chapter 9. The boson version is found, from the prescription given above, to be

$$\psi^S_{\alpha\beta\gamma}(1,2,3) = \frac{1}{\sqrt{3!}} \big\{ \psi_\alpha(1)\psi_\beta(2)\psi_\gamma(3) + \psi_\beta(1)\psi_\alpha(2)\psi_\gamma(3)$$

$$+ \psi_\gamma(1)\psi_\beta(2)\psi_\alpha(3) + \psi_\alpha(1)\psi_\gamma(2)\psi_\beta(3)$$

$$+ \psi_\beta(1)\psi_\gamma(2)\psi_\alpha(3) + \psi_\gamma(1)\psi_\alpha(2)\psi_\beta(3) \big\}.$$

(In the fermion version the second, third, and fourth terms are negative.) The indices α, β, γ on the right appear in all possible permutations of three quantities so it is not surprising that, when two variables are exchanged, or when any possible permutation of the three variables $1,2,3$ occurs, the eigenfunction is unchanged.

11-2 Thermal Distribution Functions

A *distribution function* n_i gives the average number of particles to be found in an energy state ε_i at temperature T. If a system is heated (T raised) particles are redistributed to higher energy states and n_i is altered. We also expect n_i to depend on the nature of the particles. For fermions n_i must be zero or one because of the Pauli principle; for bosons n_i behaves in a manner consistent with the particles' tendency to be in the same state.

Section 2-3 contains a presentation of the Maxwell–Boltzmann distribution function. The derivation given there yields correct results but is seen to suffer from a slight flaw when examined closely. The average number of particles in a state where the energy is very large drops to less than unity and yet Stirling's approximation, which is valid only for numbers much greater than unity, has been used for *all* states. We might argue that such sparsely occupied states do not carry much weight and that the error is therefore unimportant. However, what is an unimportant difficulty for Maxwell–Boltzmann particles becomes an intolerable one for fermions for which $n_i \leq 1$ for *all* energy states ε_i. Consequently, we must refine the techniques of Chapter 2 somewhat before we can apply them to quantum particles. The need to do this provides us with the opportunity to look more closely at the ensemble concept that is at the heart of all statistical reasoning.

In the introduction to this chapter, we mention the flipping of a coin a very large number of times to determine the probability of finding a particular head or tail outcome. An alternative procedure is to solve the classical equations of motion of the coin given the prevailing conditions of thumb speed, height above the table, distribution of metal in the coin, and so on. Such a direct approach is nearly impossible.

Another such situation involves an insurance company interested in the likelihood that an individual might die before the company has accumulated sufficient premiums to cover the awards to all the beneficiaries. Someday, perhaps, medical science might be able to predict from the results of a medical exam precisely how long a person can expect to live. Since this prediction is not possible today, an insurance organization uses the medical exam to place the individual in a group or *ensemble* of people with similar medical histories. The ensemble varies depending on whether the individual is, for example, a smoker or has had a heart attack. The ensemble consists of people of known medical history all of whom have died so that an average age of death may be computed. Whether our individual dies much earlier or much later than the average is not terribly important to the finances of the company. On the *average* they are going to collect sufficient premiums to cover both possibilities.

The last sentence tells us that a statistical analysis gives information not only about average values but about fluctuations about those values. An instructor may construct a bell-shaped curve by using the grades on an exam; the curve is described by a width as well as a mean or most probable value. An interesting aspect of physical systems containing very many particles is that fluctuations of most variables about their averages are so small that we need concern ourselves only with computing the mean values.

We are interested in the behavior of a system composed of N particles. To measure any quantity, such as pressure, we put a gauge on the system and follow its readings over a sufficiently long period. A theoretical calculation of the same quantity might in principle be carried out by solving the time-dependent many-body Schrödinger equation and following the evolution of the wave function. This is extraordinarily difficult; what is done instead is to examine the average behavior of a great number of similarly constructed systems at one single time. There are reasonable assumptions built into the latter approach which make it much the simpler procedure.

Suppose our system is a box containing a gas of N particles. We assume there are energy levels $\varepsilon_1, \varepsilon_2, \varepsilon_3, \ldots$ for each *particle* in the box. This assumption is usually possible only if the particles are *noninteracting*. The energy levels are, at any one time, occupied by $n_1, n_2, n_3, \ldots$ particles just as in the notation of Section 2-3. To find the average values of the n_i a large number M of apparently identical boxes, each with N particles, is examined. In accord with the fact that it is impossible to follow individual particle motions, no attempt is made to select M systems all with identical sets of particle trajectories. On the contrary, to infer accurate average behavior of a typical system, we choose random initial conditions and random subsequent behavior for the particles in the various boxes. This assembly constitutes our ensemble. Figure 11-2 illustrates the idea. All the systems have the same set of single-particle energy levels; any energy level ε_i of system γ contains $n_i^{(\gamma)}$ particles. The numbers $n_i^{(\gamma)}$ may

Figure 11-2

Ensemble of M identical systems. Each system contains an average of N particles and has single-particle energy states $\varepsilon_1, \varepsilon_2, \varepsilon_3, \ldots$. Any state ε_i of system γ contains $n_i^{(\gamma)}$ particles. The total number of particles at level ε_i for all the members of the ensemble is $\sum_\gamma n_i^{(\gamma)} = N_i$.

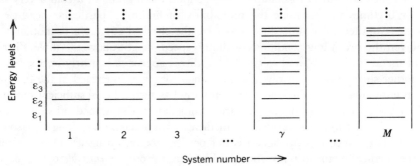

differ from system to system because of the random variation of the states of particle motion from system to system.

If each system has a total of exactly N particles the sum over energy levels (a *vertical* sum in Figure 11-2) yields

$$\sum_{i=1}^{\infty} n_i^{(\gamma)} = N.$$

However, in the following derivation we are going to relax this condition somewhat. The particular procedure we use results in an *average* of N particles per system, with fluctuations about the average that are quite negligible. With this in mind we replace the last equation with

$$\sum_{i=1}^{\infty} n_i^{(\gamma)} = N^{(\gamma)}, \tag{11-7}$$

where $N^{(\gamma)}$ is the actual or instantaneous number of particles in system γ. Since the average value of $N^{(\gamma)}$ is N in each of the M systems, the following relation holds:

$$\sum_{\gamma=1}^{M} N^{(\gamma)} = MN. \tag{11-8}$$

If we consider a given energy level ε_i and sum *horizontally* in Figure 11-2, the result is the total number of particles N_i in level i for *all* members of the ensemble. That is, we have

$$\sum_{\gamma=1}^{M} n_i^{(\gamma)} = N_i. \tag{11-9}$$

The quantity of interest is the mean of the $n_i^{(\gamma)}$ averaged over all the systems of the ensemble. This average is denoted simply by n_i without the superscript γ and is

$$n_i = \frac{N_i}{M}. \tag{11-10}$$

Another important attribute of each system is its total energy. Just as in the case of particle number we relax the restriction that each system have exactly energy E and assume only that it has that value on average. System γ has actual energy $E^{(\gamma)}$ so that

$$\sum_i n_i^{(\gamma)} \varepsilon_i = E^{(\gamma)}, \tag{11-11}$$

which is another vertical sum in the scheme of Figure 11-2. Since the average value of $E^{(\gamma)}$ is E in each of the M systems, we have

$$\sum_{\gamma=1}^{N} E^{(\gamma)} = ME. \tag{11-12}$$

To determine the average value of n_i we imagine repeatedly throwing particles at random into the ensemble's systems and levels and counting how often any particular N_i occurs. The most probable N_i value is the one to be used in Equation (11-10).

(i) Maxwell–Boltzmann Statistics   Let us begin by reproducing the Maxwell–Boltzmann results of Section 2-3 in order to illustrate the method. M systems each with an average of N particles imply a total of NM particles. We distribute them all throughout the entire ensemble without worrying about getting exactly N of them into each system. The number of particles in all energy levels ε_1 throughout the ensemble is N_1; there are N_2 particles in all levels ε_2, and so on. We want to count the number of ways to distribute N_i particles horizontally in level ε_i across the ensemble. The first of the N_i particles can be put in any one of M systems so that there are M ways of placing it. There are also M ways for the second particle. For each of the ways of placing the second particle there remain the original M ways for the first so that there are M^2 ways of positioning the two particles in the M sites. One of the ways involves, for example, two particles in level ε_i of system 1. The method counts this situation only once; an interchange of those two particles does not produce a new configuration even when they are distinguishable. For the third particle placed in the ensemble on a level ε_i there are again M positions so that there are M^3 total ways to place three particles, and so on. The total number of ways of distributing the N_i particles *horizontally* on level ε_i among all the systems of the ensemble is

$$W_h^{(i)} = M^{N_i}$$

with the subscript h referring to a horizontal distribution.

If the above procedure is repeated for a different energy level ε_j, it is obvious that the number of ways of distributing N_j particles among all the ensemble's systems on level ε_j is

$$W_h^{(j)} = M^{N_j}.$$

The total number of arrangements of particles on *both* levels ε_i and ε_j is the product $W_h^{(i)}W_h^{(j)}$. If this process is repeated for all energy levels the number of all possible horizontal arrangements is found to be

$$W_h = W_h^{(1)}W_h^{(2)}W_h^{(3)} \cdots .$$

Since the particles are distinguishable in classical statistics, interchanges of two particles vertically between two different energy levels in the same system or diagonally between two different energy levels of different systems produce a new configuration or *microstate*, in the terminology introduced in Section 2-3. The total number of ways of interchanging MN particles is $(MN)!$ However, this includes all horizontal interchanges on each of the levels ε_i; these have already been considered in the calculation of W_h. To remove them from the total number we must divide by the number of ways of interchanging N_1 particles on level ε_1, given by $N_1!$, and the number of ways on level ε_2, given by $N_2!$, and so on. It follows that the total number of vertical or diagonal interchanges that can be made to obtain new microstates from one of our original configurations is

$$W_{vd} = \frac{(MN)!}{N_1! \, N_2! \, N_3! \cdots},$$

where the subscripts v and d refer to vertical and diagonal interchanges.

Figure 11-3

Ensemble with three Maxwell–Boltzmann particles in two systems. Each system has two energy levels with $N_1 = 2$ and $N_2 = 1$. (In this example each system does not contain the same number of particles.) The particles are distinguishable and so can be labeled 1, 2, and 3; microstates 2 and 3 are then distinct. There are 24 possible configurations of which four are shown. We can generate other configurations in two ways by interchanging particle 1 with 2 or with 3 for $3 \times 4 = 12$ microstates. The other 12 microstates are found by putting any single particle in energy level ε_2 of system 2.

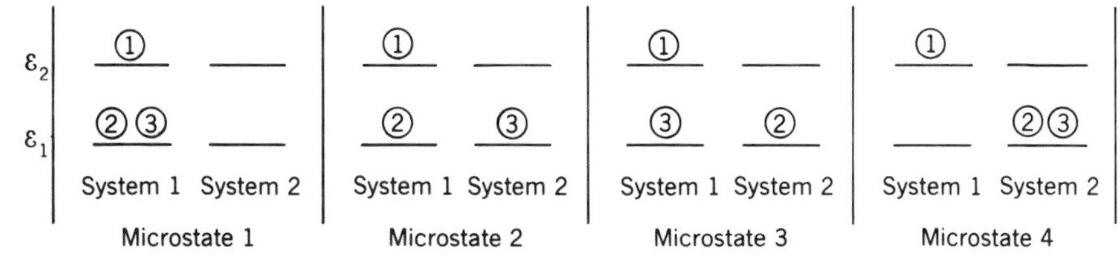

System 1 System 2	System 1 System 2	System 1 System 2	System 1 System 2
Microstate 1	Microstate 2	Microstate 3	Microstate 4

Since there are W_{vd} new configurations for each original one previously considered, the total number of microstates that can occur for a fixed set of the numbers $N_1, N_2, N_3, \ldots$ is the product of W_h and W_{vd}:

$$W = W_{vd}W_h = W_{vd}W_h^{(1)}W_h^{(2)} \cdots = (NM)! \frac{M^{N_1}}{N_1!} \frac{M^{N_2}}{N_2!} \cdots .$$

This can be written symbolically as

$$W = (NM)! \prod_i \frac{M^{N_i}}{N_i!} \quad \text{(Maxwell–Boltzmann)}, \tag{11-13}$$

where the product is over all energy levels ε_i.

As a special case consider two systems ($M = 2$) with a total of three particles ($MN = 3$). Take $N_1 = 2$ and $N_2 = 1$. Note that we do not have an equal number N of particles in each system; that condition has been relaxed as mentioned above. Equation (11-13) tells us that the total number of configurations is

$$3! \frac{2^2}{2!} \frac{2^1}{1!} = 24.$$

Figure 11-3 demonstrates how these 24 microstates are generated.

The method of Section 2-3 may be adapted here to search for the most probable set of N_i's. Since $W(N_1, N_2, \ldots)$ gives the number of microstates for a given set of N_i's, the most likely set is the one occurring most often, that is, the one causing W to be largest. We therefore maximize $W(N_1, N_2, \ldots)$ with respect to the values of $N_1, N_2, \ldots$. As in Section 2-3 we find it easier to maximize $\ln W$ rather than W itself. This process gives the same result because the logarithm of any function has its maxima and minima in the same locations as those of the function itself.

Unfortunately, we cannot maximize $\ln W$ directly because not all the N_i's are independent variables; there are two relations involving the N_i's. One such constraint is that the sum of all particles on all energy levels is the total number of particles in the ensemble:

$$\sum_{i=1}^{\infty} N_i = \sum_{i=1}^{\infty} \sum_{\gamma=1}^{M} n_i^{(\gamma)} = \sum_{\gamma=1}^{M} \sum_{i=1}^{\infty} n_i^{(\gamma)} = \sum_{\gamma=1}^{M} N^{(\gamma)} = MN, \qquad (11\text{-}14\text{a})$$

which follows from Equations (11-7), (11-8), and (11-9). Similarly, the total energy of the ensemble is given as

$$\sum_{i=1}^{\infty} \varepsilon_i N_i = \sum_{i=1}^{\infty} \varepsilon_i \sum_{\gamma=1}^{M} n_i^{(\gamma)} = \sum_{\gamma=1}^{M} \sum_{i=1}^{\infty} \varepsilon_i n_i^{(\gamma)} = \sum_{\gamma=1}^{M} E^{(\gamma)} = ME, \qquad (11\text{-}14\text{b})$$

which makes use of Equations (11-9), (11-11), and (11-12). By using Equation (11-10) we can put these constraints into a slightly different form for later use:

$$N = \frac{1}{M} \sum_i N_i = \sum_i n_i \qquad (11\text{-}15\text{a})$$

and

$$E = \frac{1}{M} \sum_i N_i \varepsilon_i = \sum_i n_i \varepsilon_i. \qquad (11\text{-}15\text{b})$$

In order to incorporate the constraints into the maximization procedure, we follow the method of Lagrange introduced in Section 2-3. Recall that, instead of using just $\ln W$ we form the auxiliary function

$$F(N_1, N_2, \ldots, \alpha, \beta) = \ln W - \alpha \left(\sum_i N_i - NM \right) - \beta \left(\sum_i \varepsilon_i N_i - EM \right),$$

where α and β, often called *Lagrange multipliers*, are as yet undetermined. We then apply the conditions

$$\frac{\partial F}{\partial N_i} = 0 \qquad (11\text{-}16)$$

for all i, treating *all* N_i as independent. The multipliers α and β are found, and can be eliminated, by use of the constraint equations. The net result is that $\ln W$ is maximized and the two constraints are also satisfied simultaneously.

Before applying Equation (11-16) it is helpful to simplify $\ln W$ by use of Stirling's approximation, Equation (2-26). For the W of Maxwell–Boltzmann statistics, Equation (11-13), we find

$$\ln W = \ln(NM!) + \ln M^{N_1} + \ln M^{N_2} + \cdots - \ln N_1! - \ln N_2! - \cdots$$
$$= \ln(NM!) + \sum_i \{ N_i \ln M - N_i \ln N_i + N_i \}.$$

The function F becomes

$$F = \ln(NM!) + \sum_i \{ N_i \ln M - N_i \ln N_i + N_i - \alpha N_i - \beta N_i \varepsilon_i \} + \alpha NM + \beta EM.$$

It is now easy to carry out the maximization:

$$0 = \frac{\partial F}{\partial N_i} = \ln M - \ln N_i - \alpha - \beta\varepsilon_i.$$

Solving this for N_i yields

$$N_i = Me^{-\alpha}e^{-\beta\varepsilon_i}. \tag{11-17}$$

The Lagrange multiplier α may be eliminated by using the constraint Equation (11-15a):

$$N = \frac{1}{M}\sum_i N_i = e^{-\alpha}\sum_i e^{-\beta\varepsilon_i}.$$

If, as in Chapter 2, the *partition function* Z is defined by

$$Z = \sum_i e^{-\beta\varepsilon_i}, \tag{11-18}$$

then

$$e^{-\alpha} = \frac{N}{Z}. \tag{11-19}$$

Putting this back into Equation (11-17) gives the result for the most probable N_i value,

$$N_i = \frac{MN}{Z}e^{-\beta\varepsilon_i}.$$

By Equation (11-10), the Maxwell–Boltzmann distribution function n_i becomes

$$n_i = \frac{N}{Z}e^{-\beta\varepsilon_i} \quad \text{(Maxwell–Boltzmann)}, \tag{11-20}$$

in agreement, of course, with the result of Section 2-3, Equation (2-35).

The second Lagrange multiplier β can be eliminated by the other constraint relation, Equation (11-15b). However, as shown in Chapter 2, β is $1/k_BT$, so we choose to keep it as an independent variable.

(ii) Fermi–Dirac Statistics We again suppose our MN particles have been spread over the levels so that N_1 are in all the ε_1 levels, N_2 in all the ε_2 levels, and so on. We ask how many ways a horizontal dispersal of N_i particles can be made. Since these particles are fermions, no more than one can be put in a single energy level of one system. Obviously, we must have $M \geq N_i$. In any configuration we have N_i filled levels (particles) and $M - N_i$ empty ones ("holes") as shown in Figure 11-4. To develop new microstates interchange particles and holes. The number of ways all M objects can be interchanged is $M!$; however, this involves interchanges of particles with particles and holes with holes which do not generate distinct microstates for indistinguishable particles. Consequently, we must divide by the number of interchanges of each of these quantities, given by, respectively, $N_i!$ and $(M - N_i)!$. We find then that

Figure 11-4

Possible configuration of N_i Fermi–Dirac particles in M systems on energy level ε_i. A new configuration can be generated by interchanging a particle and a hole (empty level).

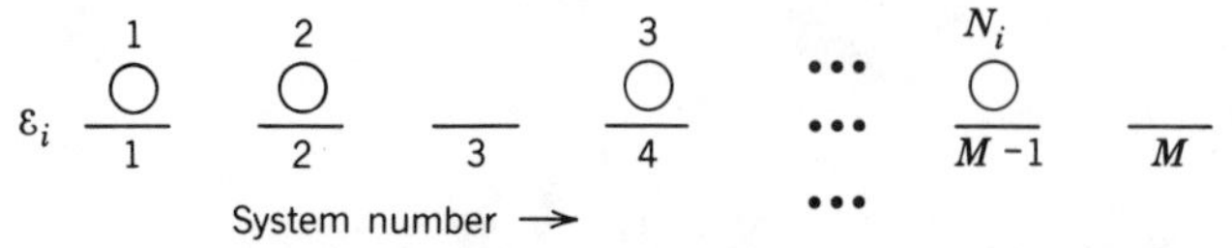

the number of horizontal configurations is

$$W_h^{(i)} = \frac{M!}{N_i!(M - N_i)!},$$

an expression that the reader may recognize as the coefficient of the general term in the formula for the binomial expansion.

In the Maxwell–Boltzmann case the next step is to consider vertical or diagonal rearrangements. However, here the particles are indistinguishable so that an interchange of a particle with energy ε_i with one of energy ε_j does *not* produce a new state. Accordingly, we have finished counting and the total number of configurations is given by the product of the $W_h^{(i)}$ for all the various levels ε_i:

$$W = \frac{M!}{N_1!(M - N_1)!} \frac{M!}{N_2!(M - N_2)!} \cdots.$$

In terms of the product symbol this is written as

$$W = \prod_i \frac{M!}{N_i!(M - N_i)!} \quad \text{(Fermi–Dirac)}. \tag{11-21}$$

Figure 11-5 illustrates this counting scheme for the example of $M = 2$, $N_1 = 2$, $N_2 = 1$. Equation (11-21) gives

$$\frac{2!}{0!\,2!} \frac{2!}{1!\,1!} = 2.$$

In this case there are only two ways of interchanging particles to generate distinct microstates.

Next we maximize $\ln W$ subject to the constraints. From Equation (11-21) and the Stirling approximation we find

$$F = \sum_i \left\{ \ln M! - (M - N_i)\ln(M - N_i) + (M - N_i) - N_i\ln N_i \right.$$

$$\left. + N_i - \alpha N_i - \beta\varepsilon_i N_i \right\} + \alpha NM + \beta EM.$$

The differentiation proceeds as before:

$$0 = \frac{\partial F}{\partial N_i} = \ln(M - N_i) - \ln N_i - \alpha - \beta\varepsilon_i$$

Figure 11-5

Ensemble with three Fermi–Dirac particles in
two systems. Each system has two energy levels
with $N_1 = 2$ and $N_2 = 1$. Because the particles
are indistinguishable they are not numbered
as in the Maxwell–Boltzmann case. Since the
particles are fermions an energy level can be
occupied by no more than one particle. Thus,
there are only two possible configurations,
both of which are shown.

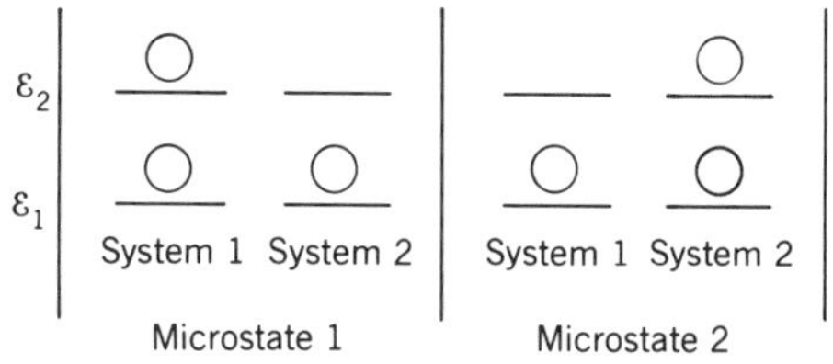

so that

$$\frac{M - N_i}{N_i} = e^{\alpha} e^{\beta \varepsilon_i}.$$

The solution for the most probable value of N_i then gives N_i/M as

$$n_i = \frac{1}{e^{\alpha} e^{\beta \varepsilon_i} + 1} \quad \text{(Fermi–Dirac)}. \tag{11-22}$$

The quantity $e^{\alpha} e^{\beta \varepsilon_i}$ is nonnegative for all values of the parameters. This property
leads to a result expected for fermions, namely, that for all energy levels ε_i,

$$n_i \leq 1.$$

Instantaneously, no state can hold more than one particle; on the average, there is
usually less than one.

In the above discussions we have not considered the spin of the particles. Since the
fermions do have spin, this effect can be included simply by supposing that each spin
orientation corresponds to a distinct state ε_i, even though each of these states may
have the same value of the energy. In this way, for fermions, we still always have
$n_i \leq 1$.

Now that we have gone through the procedure of deriving the fermion distribution
function it is possible to see why the simpler method used in deriving the
Maxwell–Boltzmann distribution in Chapter 2 could not have been used here. That
derivation assumes that $n_i!$ can be represented by the Stirling approximation, which
requires $n_i \gg 1$ for its validity. This condition does not hold for fermions. The
derivation presented here requires only that $N_i \gg 1$, which can be satisfied for
particles of any statistics by making M large enough.

The constant e^{α} in Equation (11-22) is not as easy to eliminate as in the
Maxwell–Boltzmann case. For this system α has important physical significance, as
we see in Section 11-6. Nevertheless, it is still related to the particle-number condition

Figure 11-6

Possible configuration of N_i Bose–Einstein particles in M systems on energy level ε_i. New configurations of indistinguishable particles can be generated by interchanges of the N_i particles and $M - 1$ partitions.

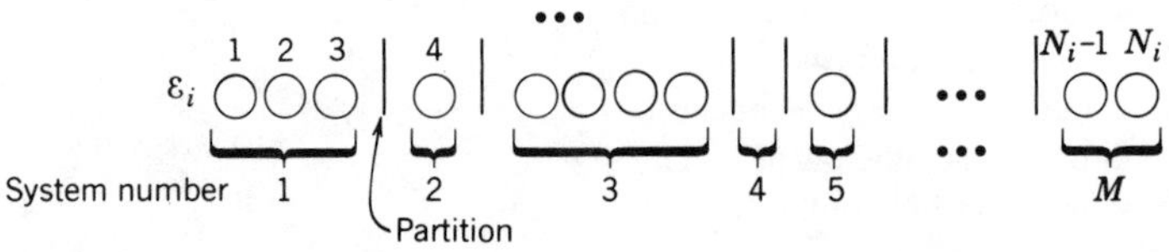

in Equation (11-15a) so that we can write

$$N = \sum_i n_i = \sum_i \frac{1}{e^{\alpha}e^{\beta \varepsilon_i} + 1} \quad \text{(Fermi–Dirac).} \tag{11-23}$$

In certain limiting cases this equation can be solved analytically and α can be expressed in terms of N, as we show in Sections 11-3 and 11-6.

(iii) Bose – Einstein Statistics   In this case we are concerned with indistinguishable particles but are not limited to assigning just zero or one particle per energy level. Horizontal counting is a bit tougher here, but there is a trick that can facilitate the task. A possible configuration of N_i particles among the M systems is shown in Figure 11-6. The systems are separated from one another by partitions. We generate all configurations by considering all possible interchanges of the partitions and particles shown in the figure. For example, if we interchange the second particle and the first partition we get a new microstate having one particle in system 1 and three in system 2. The number of ways of making interchanges of all N_i particles and all $M - 1$ partitions is $(N_i + M - 1)!$; however, this includes the $N_i!$ interchanges of particles among themselves and the $(M - 1)!$ interchanges of partitions among themselves. To get the correct count we must divide out these interchanges. It follows that the number of horizontal configurations on the ith energy level is

$$W_h^{(i)} = \frac{(N_i + M - 1)!}{N_i!(M - 1)!}.$$

Again we need not consider interchanges of particles on different energy levels for a given $N_1, N_2, \ldots$ because of particle identity. So the total number of states is found by taking the product of all the $W_h^{(i)}$ to get

$$W = \frac{(N_1 + M - 1)!}{N_1!(M - 1)!} \frac{(N_2 + M - 1)!}{N_2!(M - 1)!} \cdots$$

or

$$W = \prod_i \frac{(N_i + M - 1)!}{N_i!(M - 1)!} \quad \text{(Bose–Einstein).} \tag{11-24}$$

Figure 11-7

Ensemble with three Bose–Einstein particles in two systems. Each system has two energy levels with $N_1 = 2$ and $N_2 = 1$. The particles are indistinguishable but more than one can be in a state. Three of the possible configurations are shown. The other three are obtained by changing the single particle in ε_2 from system 1 to system 2. Note that an interchange of the two ε_1 particles in the middle configuration does not generate a new state, unlike the Maxwell–Boltzmann case.

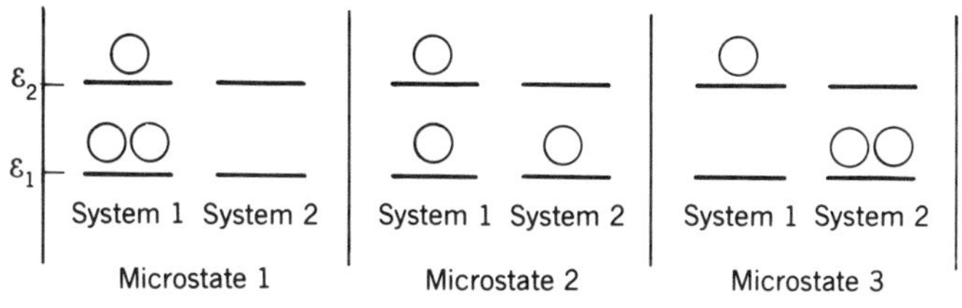

As an example of the formula we again consider the possible microstates of an ensemble with $M = 2$, $N_1 = 2$, $N_2 = 1$, but with bosons now populating the states. Equation (11-24) tells us that W should be

$$\frac{3!}{2!\,1!}\frac{2!}{1!\,1!} = 6.$$

A total of 6 states agrees with the counting shown in Figure 11-7.

If we now repeat the maximization procedure with use of Equation (11-24), we find

$$n_i = \frac{1}{e^{\alpha}e^{\beta\varepsilon_i} - 1} \quad \text{(Bose–Einstein)}. \tag{11-25}$$

(The proof is left to Problem 8 at the end of the chapter.) The Lagrange multiplier α is again determined by the condition in Equation (11-15a) so that the average number of particles in each system is given as

$$N = \sum_i \frac{1}{e^{\alpha}e^{\beta\varepsilon_i} - 1} \quad \text{(Bose–Einstein)}. \tag{11-26}$$

To see the effect of statistics on the distribution function in this case let us consider the lowest energy level, which we can take to be $\varepsilon_1 = 0$ by adjusting the arbitrary zero of energy. Equation (11-25) gives

$$n_1 = \frac{1}{e^{\alpha} - 1}. \tag{11-27}$$

Since n_1 is a particle number it must be positive; Equation (11-27) then tells us that the parameter α must be positive for a Bose system. It turns out that at a sufficiently low temperature α becomes quite small and n_1 then increases dramatically. This sudden congregating of particles into the lowest state is called the *Bose–Einstein condensation* (discovered by Einstein alone). This phenomenon is a vivid example of our claim that bosons prefer to be in the same state, and is used in Chapter 13 to illustrate some aspects of the superfluid behavior of liquid ^{4}He at low temperatures.

We end this section by stressing the limitations of our results. The derived distribution functions are useful in treating independent, or noninteracting, entities (e.g., elementary particles, atoms, molecules, phonons) that have well-defined single-particle energy states. Many (perhaps most) systems are made up of particles that interact with one another so that single-particle energy eigenvalues are not identifiable. There are procedures for dealing with the thermal properties of these systems; however, we do not treat them here.

11-3 High-Temperature, Low-Density Limit

There are circumstances in which quantum indistinguishability effects are not particularly important. Our discussion of eigenfunctions in Section 11-1 shows that this occurs when exchange terms (as in Equation (11-6)) become small. The special case given there, of localized single-particle functions each with de Broglie wavelength small compared to the interparticle separation, illustrates one way this could happen. Such a situation actually arises in a tightly bound solid where strong interatomic forces hold the particles close to their lattice positions and the eigenfunction for any given particle is highly peaked.

For a system of independent particles like those described by our thermal distribution functions, we can define an average or *thermal de Broglie wavelength* λ_{th}. This quantity depends inversely on the *average* particle momentum. If the temperature is sufficiently high or the density low enough that λ_{th} is much smaller than the interparticle spacing then a Fermi or Bose gas behaves classically and Maxwell–Boltzmann statistics does a good job in describing the gas.

To see how this happens refer back to Equation (11-19) where the Lagrange multiplier α in the Maxwell–Boltzmann case is given by

$$e^{\alpha} = \frac{Z}{N}.$$

By evaluating the partition function Z, defined in Equation (11-18), we can show that the conditions of low density and high temperature make e^{α} very large. This gives us a clue about the appropriate limit to take to find classical behavior in the Fermi and Bose distributions. In the example at the end of this section we show that Z is given by

$$Z = \frac{V}{\lambda_{\text{th}}^3}, \tag{11-28}$$

where λ_{th} is given by

$$\lambda_{\text{th}} = \left(\frac{h^2}{2\pi m k_B T} \right)^{1/2}. \tag{11-29}$$

In Problem 6 at the end of the chapter this quantity is shown to be proportional to $h/\langle p^2 \rangle^{1/2}$, which qualifies it as an average de Broglie wavelength. We find from these results that

$$e^{\alpha} = \frac{V}{N} \frac{1}{\lambda_{\text{th}}^3}, \tag{11-30}$$

which we see obeys $e^{\alpha} \gg 1$ if we have small density N/V or high temperature.

If, in the Fermi or Bose distributions (Equations (11-22) or (11-25)), we take $e^\alpha e^{\beta \varepsilon_i}$ very large compared to unity we immediately get

$$n_i = e^{-\alpha} e^{-\beta \varepsilon_i};$$

this indeed reproduces the Maxwell–Boltzmann distribution of Equation (11-20).

Next, let us calculate the pressure in an ideal gas at high temperatures and low densities. To perform this computation we need to show how a formula for pressure may be derived.

Consider a many-particle quantum system in a single energy state E_n. If some external variable, on which E_n depends, is very slowly changed, the system remains in that state while E_n slowly changes. For example, if a simple pendulum is in its ground state and the string length is slowly shortened, the pendulum rides up in energy with the ground state. However, if a change is made rapidly an oscillator might change its energy state as well as have the levels themselves change.

Suppose we compress a gas by a *slow* volume decrease dV. An energy level ε_i increases to $\varepsilon_i + d\varepsilon_i$ and particles in this energy level gain an energy $d\varepsilon_i$. The total gain in energy by the entire system is

$$dE = \sum_i n_i d\varepsilon_i. \tag{11-31}$$

This increase in energy comes from the work done on the system by the external agent compressing the system. If the gas is in a box having edge of length L, the length is diminished by dL and the work done by force F is

$$dW = -F \, dL.$$

(The minus sign occurs because positive work is done *on* the system when L decreases.) The force F is opposed by the pressure of the gas on the walls of the container, each of which has area A. Since pressure is force per unit area we have

$$F = PA$$

or

$$dW = -PA \, dL = -P \, dV. \tag{11-32}$$

This work dW must be equal to the change in energy in Equation (11-31) so that the pressure becomes

$$P = -\sum_i n_i \frac{d\varepsilon_i}{dV}. \tag{11-33}$$

For an ideal three-dimensional gas in a cubical box of length L the energy levels are given by Equation (5-99),

$$\varepsilon_{\ell_1 \ell_2 \ell_3} = \left(\ell_1^2 + \ell_2^2 + \ell_3^2\right) \frac{\pi^2 \hbar^2}{2mL^2} \tag{11-34}$$

with ℓ_1, ℓ_2, ℓ_3 positive integers. Since L is $V^{1/3}$ we can write $\varepsilon_i = C_i V^{-2/3}$, where C_i

is a quantity independent of volume. It follows that

$$\frac{d\varepsilon_i}{dV} = -\frac{2}{3}C_i V^{-5/3} = -\frac{2}{3}C_i \frac{V^{-2/3}}{V} = -\frac{2}{3}\frac{\varepsilon_i}{V}.$$

Thus, for any ideal gas, whether classical, Fermi, or Bose, we have

$$P = \frac{2}{3}\frac{1}{V}\sum_i n_i \varepsilon_i = \frac{2}{3}\frac{E}{V}, \tag{11-35}$$

where E is the average total energy of the gas.

For the Maxwell–Boltzmann gas (or for Fermi and Bose systems in their high-temperature, low-density limits), we can use the result, derived in Section 2-3 and in a Detail at the end of this section, that the average single-particle energy is

$$\langle \varepsilon \rangle = \tfrac{3}{2}k_B T. \tag{11-36}$$

But we know that $\langle \varepsilon \rangle$ satisfies

$$E = N\langle \varepsilon \rangle$$

so

$$PV = Nk_B T, \tag{11-37}$$

the familiar ideal gas law.

We have shown that the quantum distributions reduce to the Maxwell–Boltzmann function if we take e^α very large. The quantum distribution functions can be written as

$$n_i = \frac{1}{e^\alpha e^{\beta\varepsilon_i} \pm 1} = \frac{e^{-\alpha}e^{-\beta\varepsilon_i}}{1 \pm e^{-\alpha}e^{-\beta\varepsilon_i}}, \tag{11-38}$$

where the upper sign refers to fermions and the lower to bosons. The Maxwell distribution results from taking the denominator of the last form in Equation (11-38) equal to unity. This is equivalent to $e^{-\alpha}e^{-\beta\varepsilon_i} \ll 1$. Suppose we go one step further and make a binomial expansion in small x using

$$\frac{1}{1 \pm x} = 1 \mp x + x^2 + \cdots, \tag{11-39}$$

the Taylor expansion in powers of x. The result gives the Maxwell–Boltzmann distribution *and* the first quantum correction:

$$n_i = e^{-\alpha}e^{-\beta\varepsilon_i}\left(1 \mp e^{-\alpha}e^{-\beta\varepsilon_i} + \cdots\right). \tag{11-40}$$

From Equations (11-33) and (11-40) we can show, by a somewhat involved argument that we do not present, how the pressure is changed by the extra term in Equation (11-40). We find that

$$PV = Nk_B T\left(1 \pm \frac{N}{V}\frac{\lambda_{th}^3}{2^{5/2}} + \cdots\right). \tag{11-41}$$

The terms given are the first two in an infinite series, called the *virial expansion*, which is in powers of $(\lambda_{th}^3 N/V)$. This factor is small at low densities and high temperatures. Note the sign of the correction term. The upper sign in Equation (11-41) refers to fermions, the lower to bosons. The pressure for fermions is *higher* than for Maxwell particles, confirming what we expect from the Pauli principle. Because two fermions cannot be in the same spatial location, each fermion moves in a smaller volume than if the particles were distinguishable so that the pressure of the gas on the walls of the container is increased. The result is just what occurs if there is a force of repulsion between Maxwell–Boltzmann particles. Bosons, on the other hand, prefer to be located in the same state. Their statistics gives a result analogous to a Maxwell–Boltzmann gas with an attraction; the particles do not hit the walls as often, so the pressure is reduced slightly over the high-temperature limit.

In Section 11-6 we show the effect of statistics on the Fermi gas pressure in the opposite (low-temperature) limit. In Chapter 13 we examine the low-temperature situation for bosons.

Detail

To compute a sum over single-particle energies of the form $\sum_\varepsilon f(\varepsilon)$, where $f(\varepsilon)$ is an arbitrary function, we need to carry out a procedure that has been discussed in the example at the end of Section 2-3. We briefly repeat portions of that derivation here to establish notation for later use and thereby introduce the *density of states function*. We consider free particles in a large box with sides of length L and volume $V = L^3$. The particles can be fermions, bosons, or classical particles. The energy levels of a particle in a three-dimensional box are given by Equation (11-34) with $\ell_i = 1, 2, 3, \dots$. The sum over ε is just a sum over the three ℓ_i. For a very large box the spacing between energy levels, given by $\hbar^2\pi^2/2mL^2$, is very small so that sums over the ℓ_i can be converted with negligible error to integrals. By this treatment a typical sum becomes

$$\sum_\varepsilon f(\varepsilon) = \sum_{\ell_1 \ell_2 \ell_3} f(\varepsilon) = \tfrac{1}{8} \int_{-\infty}^{+\infty} d\ell_1 \int_{-\infty}^{+\infty} d\ell_2 \int_{-\infty}^{+\infty} d\ell_3 f(\varepsilon), \quad (11\text{-}42)$$

where the factor $\tfrac{1}{8}$ enters because we have extended the ℓ_i integrations over negative as well as positive values. Just as in the case of blackbody radiation, we can consider the ℓ_i as components of a three-dimensional vector ℓ whose length ℓ depends on ε according to Equation (11-34). Since $f(\varepsilon)$ varies only with the magnitude of vector ℓ, we change from the Cartesian integration of Equation (11-42) to spherical coordinates and then change to an integration over ε:

$$\sum_\varepsilon f(\varepsilon) = \frac{4\pi}{8} \int_0^\infty d\ell\, \ell^2 f(\varepsilon) = \frac{\pi}{2} \int_0^\infty d\varepsilon\, \ell^2 \frac{d\ell}{d\varepsilon} f(\varepsilon) \quad (11\text{-}43)$$

with $\ell = \sqrt{\ell_1^2 + \ell_2^2 + \ell_3^2}$. The 4π in Equation (11-43) comes from integrating over all directions of ℓ. The density of states function is defined as the number of levels per unit energy in the range ε to $\varepsilon + d\varepsilon$. This quantity is denoted by $D(\varepsilon)$ in the relation

$$\sum_\varepsilon f(\varepsilon) = \int_0^\infty d\varepsilon\, D(\varepsilon) f(\varepsilon).$$

By Equations (11-34) and (11-43) we have

$$D(\varepsilon) = \frac{\pi}{2}\ell^2 \frac{d\ell}{d\varepsilon} = \frac{\pi}{2}\left(\frac{2mL^2\varepsilon}{\hbar^2\pi^2}\right)\left(\frac{2mL^2}{\hbar^2\pi^2}\right)^{1/2}\frac{1}{2\sqrt{\varepsilon}}$$

or

$$D(\varepsilon) = \frac{m^{3/2}}{\sqrt{2}\,\pi^2\hbar^3}V\varepsilon^{1/2}. \tag{11-44}$$

Note that $D(\varepsilon)$ is proportional to $\varepsilon^{1/2}$, a result that holds for nonrelativistic particles of *any* statistics in three dimensions.

Detail

The single-particle partition function for Maxwell–Boltzmann statistics is

$$Z = \sum_{\varepsilon} e^{-\beta\varepsilon} \text{ with } \beta = \left(k_B T\right)^{-1}. \tag{11-45}$$

This quantity is easily computed by use of the density of states function derived in the previous Detail. We have

$$Z = \int_0^\infty d\varepsilon\, D(\varepsilon)e^{-\beta\varepsilon} = \frac{m^{3/2}V}{\sqrt{2}\,\pi^2\hbar^3}\int_0^\infty d\varepsilon\, \varepsilon^{1/2}e^{-\beta\varepsilon}$$

$$= \frac{m^{3/2}V(k_B T)^{3/2}}{\sqrt{2}\,\pi^2\hbar^3}\int_0^\infty dx\, x^{1/2}e^{-x},$$

taking $x = \beta\varepsilon$. The dimensionless integral has the value $\sqrt{\pi}/2$ so that

$$Z = \frac{V}{\lambda_{\text{th}}^3}, \tag{11-46}$$

where the thermal de Broglie wavelength is defined as

$$\lambda_{\text{th}} = \left(\frac{h^2}{2\pi mk_B T}\right)^{1/2}. \tag{11-47}$$

The probability of finding a particle in state ε_i for Maxwell–Boltzmann statistics is

$$P(\varepsilon_i) = \frac{n_i}{N} = \frac{e^{-\beta\varepsilon_i}}{Z}. \tag{11-48}$$

The probability density for finding a particle in the energy range ε to $\varepsilon + d\varepsilon$ is

$$p(\varepsilon) = D(\varepsilon)\frac{e^{-\beta\varepsilon}}{Z} = \frac{2}{\sqrt{\pi}}\beta^{3/2}\varepsilon^{1/2}e^{-\beta\varepsilon}. \tag{11-49}$$

The average single-particle energy is then obtained as follows:

$$\langle \varepsilon \rangle = \int_0^\infty d\varepsilon \, \varepsilon p(\varepsilon) = \frac{2}{\sqrt{\pi}} \beta^{3/2} \int_0^\infty d\varepsilon \, \varepsilon^{3/2} e^{-\beta \varepsilon} = \frac{2}{\sqrt{\pi}} (k_B T)^{5/2} \beta^{3/2} \int_0^\infty dx \, x^{3/2} e^{-x}$$

$$= \frac{2}{\sqrt{\pi}} (k_B T) \frac{3\sqrt{\pi}}{4} = \frac{3}{2} k_B T. \tag{11-50}$$

In this calculation we again use the substitution $x = \beta \varepsilon$ to make the integral dimensionless. The final result is the same as that found in Section 2-3.

11-4 Photon Statistics

Planck was able to make use of a counting procedure equivalent to that of Maxwell–Boltzmann statistics when he derived his blackbody radiation formula. The oscillators that constituted the distinguishable particles in the various states $nh\nu$ were the radiators in the walls of the cavity. These particles were distinguishable because they were highly localized on the lattice sites in the walls. He calculated the radiation energy density by considering the radiation to be in equilibrium with the oscillators. Einstein pointed out that this argument was inconsistent because the part of the derivation involved in finding the relation between the radiation energy density u_ν and the average energy $\langle \varepsilon \rangle$ of an oscillator, given by

$$u_\nu = \frac{8\pi \nu^2}{c^3} \langle \varepsilon \rangle,$$

was completely classical; that is, it did not take into account the quantization of the oscillators emitting the radiation. It was only in writing down the correct expression for $\langle \varepsilon \rangle$ that the oscillator quantization was invoked.

We have avoided this pitfall by considering the radiation field itself to be quantized. The radiation field energy can be thought of as that of distinguishable oscillators of frequency ν, which can be in states $nh\nu$.

This view is quite valid but can be improved by making greater use of the photon picture involved in the photoelectric effect and in Compton scattering. Instead of thinking of the radiation field as having, for example, one oscillator of frequency ν_1 in an excited state $nh\nu_1$ and another of frequency ν_2 in state $mh\nu_2$, we envision there being n photons each of energy $h\nu_1$ and m of energy $h\nu_2$, and so on. This is the view introduced in Chapter 2, but now we make it explicit in the thermal calculations.

Photons are spin-1 particles and are therefore bosons. The calculation of their average energy must take into account their indistinguishability according to the discussion of Section 11-2. Since photons are massless particles they can easily be created and destroyed while being emitted and absorbed. Thus, the number of photons in a cavity is not constant. Indeed, their number depends on the temperature and frequency. What this means is that the condition imposed by using the Lagrange multiplier α in the derivation of Equation (11-25) is unnecessary here. We can get the result appropriate to photons by taking $\alpha = 0$ in that equation. The average number of photons of frequency ν at temperature T becomes

$$n_\nu = \frac{1}{e^{\beta h\nu} - 1}.$$

The average energy in frequency mode ν is then given as

$$\langle \varepsilon \rangle = n_\nu h\nu = \frac{h\nu}{e^{\beta h\nu} - 1} \tag{11-51}$$

in agreement with Equation (2-38). The rest of the discussion of blackbody radiation follows precisely as in Chapter 2.

11-5   Phonons

It was Einstein (again!) who recognized that the quantization of the oscillators in the blackbody cavity walls had implications concerning the thermal properties of solids themselves as well as the radiation with which they were in equilibrium. Deviations from the classical laws for the heat capacities of solids had already been discovered. The explanation of these deviations was another step toward the verification of the new quantum ideas.

That the thermal properties of radiation and of solids are explicable by the same quantum ideas comes from the very close analogy between photons and the quantized sound waves in a solid.

We first consider a solid by a simplified treatment known as *Einstein's model*. In most crystals each molecule is bound tightly in place by the forces resulting from the surrounding molecules; a molecule oscillates about its lattice site. Einstein's model lets every molecule oscillate *harmonically* with the same frequency ν_0 as if each particle is attached to its lattice site by a single spring with the spring constant the same for every particle. A more accurate picture is one in which all the particles are intercon- nected by springs. We see below how this changes the results, but first let us consider the Einstein picture.

If the particles are highly localized about lattice sites, their wave functions do not overlap much. Hence, particle indistinguishability need not concern us and we can use Maxwell–Boltzmann statistics. Each oscillator has energy levels $nh\nu_0$ and the average energy per oscillator $\langle \varepsilon \rangle$ is calculated in precisely the same way as in Equation (2-42), so that we find

$$\langle \varepsilon \rangle = \frac{3h\nu_0}{e^{\beta h\nu_0} - 1} \text{ with } \beta = \frac{1}{k_B T}. \tag{11-52}$$

The factor 3 arises because each oscillator is three-dimensional and we consider it to have equal frequencies for vibrations in the three spatial directions.

The total energy for N particles in the solid is

$$E = N\langle \varepsilon \rangle.$$

The *heat capacity* C is defined as the total heat required per degree of temperature rise in the entire system. (The *specific heat* is the heat capacity per particle, C/N). Here the heat added is equal to the increase in internal energy E so the heat capacity is

$$C = \frac{dE}{dT} = \frac{3N(h\nu_0)^2}{k_B T^2} \frac{e^{\beta h\nu_0}}{\left(e^{\beta h\nu_0} - 1\right)^2}. \tag{11-53}$$

For large T (small β) we can expand the exponential according to

$$e^{\beta h\nu_0} = 1 + \beta h\nu_0 + \cdots . \tag{11-54}$$

Figure 11-8
Heat capacity of a solid as predicted by the
Einstein model. At high temperatures C
approaches Nk_B, the classical result, but drops
toward zero at low temperatures in general
agreement with experiment.

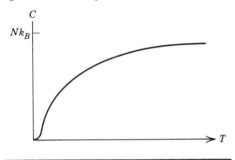

Such an expansion has been used previously in Problem 8 of Chapter 2. Then E
reduces to

$$E = 3Nh\nu_0 \frac{1}{\beta h\nu_0} = 3Nk_BT, \qquad (11\text{-}55)$$

in agreement with the classical law of equipartition, Equation (2-20). The heat
capacity becomes

$$C = 3Nk_B \qquad (k_BT \gg h\nu_0). \qquad (11\text{-}56)$$

This is the law of *Dulong and Petit* (P. L. Dulong and A. T. Petit, 1819); the heat
capacity is independent of temperature at high T. Deviation from this law at low
temperatures was another element along with blackbody radiation in the case against
the classical theory. It was found experimentally that C dropped below Nk_B at low
enough T.

We can see how quantization gives a cure for the inadequacy of Equation (11-56).
Figure 11-8 shows a plot of Equation (11-53). As the temperature drops C begins to
fall for $k_BT \lesssim h\nu_0$; here the thermal energy is of the order of the energy level
separation and the discreteness of the energy levels becomes important. When
$k_BT \ll h\nu_0$, the exponential in Equation (11-53) is large so

$$C \to \frac{3N(h\nu_0)^2}{k_BT^2}e^{-\beta h\nu_0} \qquad (k_BT \ll h\nu_0), \qquad (11\text{-}57)$$

which goes to zero as $T \to 0$. At $T = 0$ all oscillators are in their ground states and if
the system is to absorb energy some oscillators must make the transition to a state $h\nu_0$
higher in energy. Until k_BT approximates $h\nu_0$, this is unlikely so C is exponentially
small in this model.

Measured values of C in real solids do show $C \to Nk_B$ as T becomes large, and C
does indeed drop toward zero at low temperatures in agreement with the Einstein
model. However, at low temperature it is found that C does not obey the exponentially decreasing form given by Equation (11-57).

Figure 11-9

System of springs and masses to be used as a one-dimensional model of a crystal. The equilibrium interparticle spacing is a. Only longitudinal waves are considered.

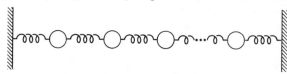

The explanation for this deviation from the Einstein model was given by Debye who assumed that the molecules in a solid were connected together by harmonic forces. The solid was visualized as a set of masses in a kind of three-dimensional bedspring.

One of the characteristics of such a system is that longitudinal and transverse waves can travel through it. These modes are the solid-state version of sound waves. Such waves can have very long wavelengths ($\sim$ length of sample) and very short wavelengths ($\sim$ separation of molecules). For long wavelengths where the lattice discreteness is not important, the modes obey a standard wave equation that shows the frequency and wavelength of the standing waves to be related exactly as they are for blackbody radiation in Equation (2-18). However, here c becomes the velocity of sound in the crystal.

While the theory of sound waves becomes quite analogous to that of the radiation discussed in Chapter 2, there are three differences between sound and electromagnetic modes: (1) The number of modes is not infinite. Very short wavelengths (very high frequencies) are not allowed; the minimum wavelength is equal to twice the distance between particles in the lattice. (2) A relation like Equation (2-18) holds for low frequencies but deviations from it occur for higher frequencies. This phenomenon corresponds to *dispersion* of the sound waves (higher frequencies have smaller velocities) and occurs because the system is not continuous but is made up of discrete masses connected by springs. (3) There are not just two polarizations but three. For some directions of sound waves through a crystal these correspond to two transverse modes and one longitudinal.

A one-dimensional model of a crystal is shown in Figure 11-9 in which a large number of particles are attached together with springs. In this model we consider only longitudinal motion. We can show that in the limit of very large N the frequencies of this system are

$$\nu_n = \nu_0 \left(1 - \cos\frac{\pi n}{N} \right)^{1/2} \tag{11-58}$$

with integer values of $n = 1, 2, 3, \ldots, N$ and $\nu_0 = (1/2\pi)\sqrt{2K/m}$, where K is the spring constant. (See the discussion in Section 12-8.) If we assume n/N to be a continuous variable from 0 to 1, this function can be plotted as in Figure 11-10. If we had just one particle connected by two springs to the wall there would be one frequency in the longitudinal direction having value ν_0. For very many particles the possible waves on the system are spread into a *band* from $\nu = 0$ to $\nu = \sqrt{2}\,\nu_0$.

For small n/N, the cosine in Equation (11-58) can be expanded and only lowest-order terms need be kept. We use

$$\cos x = 1 - \frac{x^2}{2} + \cdots,$$

Figure 11-10
Allowed frequencies of the system shown in
Figure 11-9. The discrete variable n has the
values $1, 2, \ldots, N$; for large N, n/N may be
taken as continuous. The system has a
maximum frequency $\nu_m = \sqrt{2}\,\nu_0$, where $2\pi\nu_0$
$= \sqrt{2K/m}$. The slope of the curve is constant
for small n but decreases at large n, indicating
the presence of dispersion.

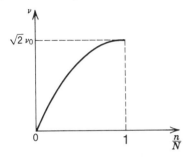

the Taylor series expansion about $x = 0$. Then Equation (11-58) becomes

$$\nu_n = \nu_0\left(1 - 1 + \frac{1}{2}\left(\frac{\pi n}{N}\right)^2 + \cdots\right)^{1/2}$$

$$\approx \frac{\nu_0\pi}{\sqrt{2}\,N}n, \qquad n \ll N. \tag{11-59}$$

This equation describes the initial linear part of the curve in Figure 11-9. If we reduce
the discussion of electromagnetic waves summarized in Equation (2-18) to one
dimension by taking $n_1 = n$, $n_2 = n_3 = 0$ (or refer to Equation (2-13)), we see that
Equation (11-59) has the same form with the velocity of light now replaced by the
velocity of sound given by

$$c_s = \sqrt{2}\,\pi\nu_0 a. \tag{11-60}$$

We have taken the length of the crystal to be Na with a the equilibrium interparticle
spacing. The group velocity of waves along this chain of particles is proportional to
the slope of the ν-versus-n curve, as illustrated by Equations (11-59) and (11-60). The
decrease in the slope of the curve in Figure 11-9 as $n/N \to 1$ corresponds to
dispersion; higher-frequency waves move more slowly.

Let us next consider a three-dimensional model of a solid, with each particle
connected by springs to nearest neighbors in all directions (bedspring model). Then
the frequencies ν may be characterized by three integers n_1, n_2, n_3, and for *small n_i*
the relation turns out to be

$$\nu = \frac{c_s}{2L}\sqrt{n_1^2 + n_2^2 + n_3^2}. \tag{11-61}$$

The total number of frequency modes is still N if we allow only longitudinal waves to move through the crystal. If we assume a crystal with a cubic shape having length L on a side, the number of particles along one edge is

$$N_1 = \frac{L}{a} = N^{1/3},$$

and each n_i then ranges over

$$1 \leq n_i \leq N_1.$$

Small n_i in Equation (11-61) means $n_i \ll N_1$. For large n_i Equation (11-61) is no longer valid and the frequencies begin to show dispersion.

For each set (n_1, n_2, n_3) three polarizations of waves can occur, which in the simplest situation are two transverse and one longitudinal as mentioned above. In most situations the sound velocities of the three polarizations differ. We ignore this, consider c_s to represent some average sound velocity, and simply multiply our results by 3.

The analogy of sound waves to electromagnetic radiation is so strong that the idea of a *phonon* may be introduced. A mode ν (characterized by the three integers n_1, n_2, n_3) may be in an infinity of quantum states having energy

$$\varepsilon_\nu = n_\nu h\nu.$$

We say that such a situation corresponds to the presence of n_ν bosons called phonons, each of energy $h\nu$, that move through the crystal. These massless "particles" are actually *collective modes*, that is, cooperative motions of large groups of particles. These particle-like objects move through a real "aether," the crystal lattice. The phonon–photon analogy carries us quite far. Particle scattering is often used to study the structure of crystals; for example, neutrons shot through a crystal interact with the nuclei at the crystal sites. In the scattering processes we can see analogues of Compton scattering of phonons by neutrons; the neutron can absorb or emit phonons while it passes through the crystal.

At temperature T the average number of phonons is given by the Bose–Einstein distribution function (with the Lagrange multiplier α set equal to zero as for photons)

$$n_\nu = \frac{1}{e^{\beta h\nu} - 1},$$

and the average energy in frequency mode ν is

$$\langle \varepsilon_\nu \rangle = \frac{h\nu}{e^{\beta h\nu} - 1} \tag{11-62}$$

just as for photons. To find the total energy of the solid we sum $\langle \varepsilon_\nu \rangle$ over all integers n_1, n_2, n_3 on which ν depends:

$$E = 3 \sum_{n_1 n_2 n_3} \frac{h\nu}{e^{\beta h\nu} - 1}. \tag{11-63}$$

The factor of 3 is to account for the three polarizations.

We note that there is a maximum frequency ν_m among the values of ν. In the one-dimensional case this is $\sqrt{2}\,\nu_0$ as shown in Figure 11-9. Thus, it is meaningful to consider a high-temperature limit of Equation (11-63). If $k_BT \gg h\nu_m$, the exponential denominator may be expanded in powers of the small quantity $\beta h \nu$ so that

$$e^{\beta h \nu} - 1 = 1 + \beta h \nu + \cdots - 1 \approx \beta h \nu$$

for all ν values. The result for E is

$$E = 3 \sum_{n_1 n_2 n_3} \frac{h\nu}{\beta h \nu} = 3Nk_BT, \qquad (11\text{-}64)$$

where the last equality follows because the total number of integers n_1, n_2, n_3 is N. This is the same classical result that occurs in Einstein's model, Equation (11-55), and leads to the Dulong–Petit law, Equation (11-56), which is experimentally correct at high temperatures.

At lower temperatures, such that $k_BT \lesssim h\nu_m$, we cannot get away with such an easy derivation. However, as we change one of the integers n_i by a single unit while summing over n_1, n_2, n_3, the summand in Equation (11-63) varies very little. This is true because, according to Equation (11-61), ν behaves as n_i/N_1, where N_1 is a very large number. Thus, it is always valid to convert the sum to an integral, so that we have

$$E = 3 \int_0^{N_1} dn_1 \int_0^{N_1} dn_2 \int_0^{N_1} dn_3 \frac{h\nu}{e^{\beta h \nu} - 1}. \qquad (11\text{-}65)$$

Let us now change variables from the n_i to an integral over ν. That is, we want to write

$$3 \int_0^{N_1} dn_1 \int_0^{N_1} dn_2 \int_0^{N_1} dn_3 \cdots = \int_0^{\nu_m} d\nu \, N_\nu \cdots, \qquad (11\text{-}66)$$

where N_ν is another *density of states function* defined as the number of modes at frequency ν per unit frequency interval. To find this function we need detailed knowledge of how ν depends on n_1, n_2, n_3 which for an arbitrary crystal is often known only numerically and is not just a simple generalization of Equation (11-58).

If we are interested only in the *very-low*-temperature behavior of E and C, we can proceed further. For very low T, many of the higher frequencies satisfy $h\nu \gg k_BT$ or $\beta h \nu \gg 1$. For only these frequencies, it is true that

$$n_\nu = \frac{1}{e^{\beta h \nu} - 1} \approx \frac{1}{e^{\beta h \nu}} = e^{-\beta h \nu} \ll 1 \qquad (h\nu \gg k_BT)$$

so that these frequencies contribute negligibly to the sum in Equation (11-63). The conclusion is that only frequencies such that $h\nu \lesssim k_BT$ contribute to E and, if the temperature is low enough, only the low-energy *nondispersive* frequencies given by Equation (11-61) are important. The value of N_ν associated with this relation between ν and the n_i is now essentially the same as in the blackbody case. We find, by examination of Equation (2-19),

$$N_\nu = \frac{12\pi\nu^2}{c_s^3} V \qquad (k_BT \ll h\nu_m).$$

This differs from the N_ν of Equation (2-19) by a factor $\frac{3}{2}$, because of three possible polarizations instead of two, and in the replacement of c by c_s, the sound velocity.

Thus, we find the total energy of the solid to be

$$E = \frac{12\pi}{c_s^3} V \int_0^\infty d\nu \, \nu^2 \frac{h\nu}{e^{\beta h\nu} - 1} \qquad (k_B T \ll h\nu_m). \qquad (11\text{-}67)$$

We are able to replace the upper limit ν_m on the integral by infinity because, at low temperatures, the higher frequencies do not contribute anyway. Let us make the substitution

$$z = \beta h\nu$$

in Equation (11-67) and obtain

$$E = \frac{12\pi V}{c_s^3 h^3} (k_B T)^4 \int_0^\infty dz \frac{z^3}{e^z - 1}.$$

The integral has the value $\pi^4/15$ so the heat capacity dE/dT is given at low temperatures by

$$C = \frac{16}{5} \frac{\pi^5 V}{c_s^3 h^3} k_B (k_B T)^3 \qquad (k_B T \ll h\nu_m). \qquad (11\text{-}68)$$

Solids, at low temperatures, should have heat capacity contributions due to their phonon excitations that are proportional to T^3 (rather than $\exp(-\beta h\nu_0)$ as in Einstein's model) and this is the result that is found experimentally.

Debye was the first to derive the T^3 result. He also defined a convenient characteristic constant for a solid, called the *Debye temperature*, given by

$$\Theta_D = \frac{\hbar c_s}{k_B} \left(6\pi^2 \frac{N}{V} \right)^{1/3}.$$

When expressed in terms of this quantity, Equation (11-68) becomes

$$C = \frac{12}{5} \pi^4 N k_B \left(\frac{T}{\Theta} \right)^3. \qquad (11\text{-}69)$$

The high-temperature limit, Equation (11-56), holds for $T \gg \Theta_D$ and the low-temperature result, Equation (11-69), holds for $T \ll \Theta_D$. Debye temperatures range from about 30 K for solid helium to 1860 K for diamond.

Example

Solid helium can be made only by pressurizing liquid helium at very low temperature. It has a Debye temperature of about 30 K as determined by heat capacity measurements. Its sound velocity is, by the definition of Θ_D,

$$c_s = \frac{k_B \Theta_D}{\hbar} \frac{1}{(6\pi^2 N/V)^{1/3}}.$$

The interparticle separation, $\sim (V/N)^{1/3}$, is about 0.3 nm so that

$$c_s = \frac{(1.38 \times 10^{-23} \text{ J/K})(30 \text{ K})}{1.1 \times 10^{-34} \text{ J} \cdot \text{s}} \frac{3 \times 10^{-10} \text{ m}}{(6\pi^2)^{1/3}} \approx 300 \text{ m/s}.$$

For comparison, the velocity of sound in air is 330 m/s. Most solids have sound velocities at least an order of magnitude larger than this. Solid helium is anomalous in many other ways as well.

Example

The maximum frequency wave in a solid is given approximately, from Equation (11-61), by

$$\nu_m \approx \frac{c_s}{2L} \sqrt{N^2} \approx \frac{c_s}{a}.$$

The velocity of a longitudinal sound wave in copper is about 5000 m/s and the average interparticle spacing is $a \approx 0.2$ nm. Thus, we find

$$\nu_m \approx \frac{5 \times 10^3 \text{ m/s}}{2 \times 10^{-10} \text{ m}} \approx 2 \times 10^{13} \text{ Hz}.$$

This value is considerably above audible frequencies! It corresponds to wavelengths on the order of an interatomic spacing.

11-6 Low-Temperature Fermi – Dirac Systems

Many common assemblies of particles in nature are *degenerate* Fermi systems. This term means that the particles have crowded into their low-lying energy states; this in turn implies that temperature is small on a scale that is explained below. Examples are nuclei, metals, white dwarf and neutron stars, and liquid ^{3}He. Before considering these specific examples we examine the general low-temperature Fermi system.

The distribution function for fermions is, from Equation (11-22),

$$n_i = \frac{1}{e^{\beta\varepsilon_i + \alpha} + 1} \quad \left(\beta = \frac{1}{k_B T}\right). \tag{11-70}$$

The Lagrange multiplier α, determined by Equation (11-23), is temperature and density dependent. At very low temperatures α is found to be negative. To incorporate this feature we write

$$\alpha = -\beta\varepsilon_F(T),$$

where $\varepsilon_F(T)$ is called the *Fermi energy*. Equation (11-70) then becomes

$$n_i = \frac{1}{e^{\beta(\varepsilon_i - \varepsilon_F)} + 1} \tag{11-71}$$

Figure 11-11

Plot of the distribution $n_i = \{\exp[\beta(\varepsilon_i - \varepsilon_F) + 1\}^{-1}$ at three temperatures. For $T = 0$ n_i is a step function with Fermi energy $\varepsilon_F(0)$ being the largest occupied energy level. For $T > 0$, but $T \ll T_F \equiv \varepsilon_F(0)/k_B$, the distribution spreads out to energies of order $k_B T$ above the Fermi energy. The energy $\varepsilon_F(T)$ for which $n_i = \frac{1}{2}$ is a slight bit smaller than $\varepsilon_F(0)$. For large T ($T \gg T_F$) the distribution becomes very spread out and reduces to the Maxwell–Boltzmann value $n_i \sim \exp(-\beta\varepsilon_i)$ while $\varepsilon_F(T)$ becomes negative.

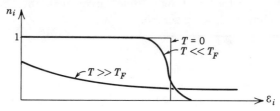

---

when ε_F is used as a parameter.

A plot of Equation (11-71) as a function of ε_i for several temperatures is shown in Figure 11-11. For $T = 0$, n_i is a step function; all particles are in the lowest states consistent with the Pauli principle. The largest occupied energy level at $T = 0$ is $\varepsilon_F(0)$. To see this let us consider a very small positive value of T; note that for $\varepsilon_i < \varepsilon_F$, $\exp[\beta(\varepsilon_i - \varepsilon_F)]$ has a negative argument and goes to zero as $T \to 0$ or $\beta \to \infty$. For those ε_i values n_i is unity. However, for $\varepsilon_i > \varepsilon_F$, $\exp[\beta(\varepsilon_i - \varepsilon_F)]$ has a positive argument and goes to infinity as $\beta \to \infty$. In this region $n(\varepsilon)$ is zero. Raising the temperature from zero removes particles from the levels $\varepsilon_i < \varepsilon_F(0)$ and excites them to higher states.

Let us investigate further the situation at $T = 0$. The average number N of particles in the system is given by Equation (11-23):

$$N = \sum_i n_i = \sum_i \frac{1}{e^{\beta(\varepsilon_i - \varepsilon_F)} + 1},$$

which determines ε_F in terms of N and T. To evaluate the sum we use the density of states function given by Equation (11-44) and obtain

$$N = \sum_i n_i = 2\int_0^\infty d\varepsilon_i\, D(\varepsilon_i)n_i.$$

The factor of 2 in this equation arises from the fact that fermions are spin-$\frac{1}{2}$ particles. In zero magnetic field each energy state is doubly degenerate and can contain two particles, one with spin up and one with spin down just as in atomic orbitals.

Because n_i is a step function that cuts off at $\varepsilon_i = \varepsilon_F(0)$ for $T = 0$, the result is just the integration of Equation (11-44):

$$N = 2\int_0^{\varepsilon_F} D(\varepsilon)\, d\varepsilon = \frac{2\sqrt{2}}{3\pi^2} \frac{m^{3/2}V}{\hbar^3} \varepsilon_F^{3/2}(0). \tag{11-72}$$

This formula serves to define ε_F at $T = 0$. Solving Equation (11-72) for ε_F, we find

$$\varepsilon_F(0) = \frac{\hbar^2}{2m}(3\pi^2\rho)^{2/3}, \tag{11-73}$$

where the particle density is $\rho = N/V$. The Fermi energy is seen to depend on the two-thirds power of the density so that, as more particles are placed into the system, the highest occupied state ε_F increases. As the temperature is raised slightly from $T = 0$, ε_F is no longer the highest occupied state but is the energy for which $n_i = \frac{1}{2}$; ε_F decreases as T increases and for sufficiently large T actually becomes negative (which implies that all states contain on the average less than one-half of a particle).

The total energy of the N-particle system at $T = 0$ is

$$E = \sum_i \varepsilon_i n_i = 2 \int_0^{\varepsilon_F} d\varepsilon\, D(\varepsilon)\varepsilon = \frac{2\sqrt{2}}{5\pi^2} \frac{m^{3/2}}{\hbar^3} V\varepsilon_F^{5/2}(0).$$

By using Equation (11-72) we can write this as

$$E = \tfrac{3}{5} N\varepsilon_F(0) \tag{11-74}$$

so that the average energy per particle E/N is three-fifths of the maximum energy ε_F.

We have seen in Section 11-3 that the Pauli principle causes an increase in the pressure over that of an ideal Maxwell–Boltzmann gas. This effect of particle indistinguishability becomes more predominant as the temperature is lowered. Even at $T = 0$ there is a residual pressure in the system. This *Pauli pressure* can be derived from Equations (11-35) and (11-74) as

$$P = \tfrac{2}{5}\rho\varepsilon_F(0). \tag{11-75}$$

Because of the dependence of ε_F on density ρ, P depends on the $\frac{5}{3}$ power of the density. This result is crucial to the existence of white dwarf stars, as we show below, and has been observed in other Fermi systems.

The calculation of the properties of fermions at low but nonzero temperature is considerably more difficult than at $T = 0$. Rather than carrying out these computations we give a heuristic argument for the temperature dependence of one of the most important of the experimental quantities, the heat capacity. As the temperature is raised particles in states having energies within an interval $k_B T$ below ε_F are raised to states of order $k_B T$ or less above ε_F, as can be inferred from Figure 11-11. If $k_B T \ll \varepsilon_F$ (defining the low-temperature or degenerate regime) then this region of excitation is a very narrow band of width $k_B T$ about ε_F. For an increase of temperature from zero to T a fraction $(k_B T/\varepsilon_F)$ of the N particles each receives an energy increase of order $k_B T$. Thus, the total energy increase is

$$\Delta E \sim N\left(\frac{k_B T}{\varepsilon_F}\right) k_B T \qquad (k_B T \ll \varepsilon_F).$$

The heat capacity is ΔE divided by the temperature increase T or

$$C \sim Nk_B\left(\frac{T}{T_F}\right),$$

where T_F is the *Fermi temperature* defined as $\varepsilon_F(0)/k_B$. A rigorous calculation gives

$$C = \frac{\pi^2}{2} Nk_B\left(\frac{T}{T_F}\right) \qquad (k_B T \ll \varepsilon_F). \tag{11-76}$$

It is easily seen that the Fermi temperature provides a scale to measure the degree of degeneracy. If $T \ll T_F$ the system has strong particle-indistinguishability effects and the low-temperature formula in Equation (11-76) holds; for $T \gg T_F$ the classical-limit formulas become valid.

In the following examples we present some applications of the Fermi formulas to specific systems.

Example

Electrons in a metal: It is often a good approximation to consider the conduction electrons in a metal (these are the outer-shell electrons, which are free to move and which conduct electricity) as if they were noninteracting particles. Because of the Coulomb interactions between electrons and the positive ions and between electrons and other electrons, the justification of this model is nontrivial; however, let us accept it as valid. If we assume that the metal has one conduction electron per atom, the mean separation between electrons is of order $a \approx 0.1$ nm or so. The electron number density is of order $\rho \sim 1/a^3$. The Fermi temperature is found from Equation (11-73) to be

$$T_F = \frac{\varepsilon_F}{k_B} = \frac{\left(3\pi^2\right)^{2/3}}{2} \frac{\hbar^2}{k_B m} \rho^{2/3} \sim \frac{\hbar^2}{k_B m a^2}$$

$$\approx \frac{\left(1 \times 10^{-34} \text{ J} \cdot \text{s}\right)^2}{\left(1.38 \times 10^{-23} \text{ J/K}\right)\left(9.1 \times 10^{-31} \text{ kg}\right)\left(10^{-10} \text{ m}\right)^2} \approx 8 \times 10^5 \text{ K}.$$

$$(11\text{-}77)$$

Thus, we find that electrons in metals are in the low-temperature degenerate regime in almost all experimental situations. The electronic heat capacity is accurately given by Equation (11-76). Of course, the phonons created by atomic thermal vibrations also contribute to the heat capacity. If the experiment is carried out at temperatures that are below both the Debye temperature Θ_D and the Fermi temperature T_F, the total heat capacity of a metal has the form

$$C = N k_B \left[A\left(\frac{T}{T_F}\right) + B\left(\frac{T}{\Theta_D}\right)^3 \right],$$

$$(11\text{-}78)$$

where A and B are numerical constants of order unity.

Example

Protons or neutrons in a nucleus: If we substitute the nucleon mass and $a \sim 10^{-15}$ m into Equation (11-77) in place of the electronic data we find

$$T_F \sim 10^{10} \text{ K}.$$

The nucleons in a nucleus are obviously degenerate under any normal conditions. Only in the early stages of the universe, soon after the Big Bang, could temperatures of this order exist.

Example

Electrons in a white dwarf star: Despite the high temperatures found in stellar interiors, the electrons in a late-stage star known as a white dwarf are quite degenerate because the value of T_F is so large. When a star burns all its hydrogen fuel it is made up mainly of a plasma of helium nuclei and electrons. The gravitational forces between the helium nuclei cause the star to collapse until the Pauli pressure of the electrons given by Equation (11-75) brings a halt to the collapse. The negative (inward) gravitational pressure can be guessed by using a formula analogous to Equation (11-35):

$$P_G \sim \frac{E_G}{V} \sim \frac{GM^2}{R} \frac{1}{R^3},$$

where E_G is the gravitational potential energy, G is the gravitational constant, R is the radius of the star, and $M \sim N m_{\text{He}}$ is the mass of the star made up of N helium atoms each of mass m_{He}. The outward Pauli pressure, from Equations (11-73) and (11-75), is given by

$$P_p \sim \frac{N}{R^3} \frac{\hbar^2}{m_e} \left(\frac{N}{R^3} \right)^{2/3},$$

where m_e is the electron mass and the number of electrons is actually $2N$, twice the number of helium nuclei. Equating P_p and P_G gives the radius of the star:

$$R \sim \frac{\hbar^2}{G m_e m_{\text{He}}^{5/3} M^{1/3}}. \tag{11-79}$$

We have dropped all numerical constants from this expression to get just an order of magnitude estimate. If we put in numbers we get $R = 7 \times 10^2$ km for $M = M_\odot$, the solar mass. This means a white dwarf is a very compact object. However, it is not quite *that* compact. Equation (11-79) is derived on the basis of nonrelativistic theory. We show in Problem 21 at the end of the chapter that an R given by Equation (11-79) results in a *Fermi velocity* ($v_F \equiv \sqrt{2\varepsilon_F/m}$) large enough that relativistic effects must be taken into account. Quite a different radius–mass relation then results. While the basic idea of our calculation remains valid, the relativistic treatment yields the interesting result that, for stellar masses greater than about 1.4 $M_\odot$, the Pauli pressure can no longer stop the gravitational collapse. The star continues its infall, a supernova occurs, and ultimately a neutron star or a black hole is formed. This critical mass is known as the *Chandrasekhar limit*, discovered by S. Chandrasekhar in 1934.

Problems

1. Verify by explicit calculation that the three-particle boson eigenfunction given in the example at the end of Section 11-1 is symmetric under the interchange $2 \leftrightarrow 3$ and the permutation $1, 2, 3 \leftrightarrow 3, 1, 2$. Write down the equivalent Fermi–Dirac function (see Problem 8 in Chapter 9) and test its symmetry in the same two cases. Explain your fermion results.

2. A single spinless particle has potential energy $V(1)$, and has two real eigenstates $\psi_\alpha(1)$ and $\psi_\beta(1)$ with energies ε_α and ε_β, respectively. Two such particles have potential energy $V = V(1) + V(2) + V_{int}(1,2)$, where V_{int} represents an interaction between the two particles. Suppose that $\int\psi_\alpha(\mathbf{r})\psi_\beta(\mathbf{r})\,d\mathbf{r} = 0$, and that the only nonzero integrals involving V_{int} are

    $$\int\psi_\alpha(1)\psi_\beta(2)V_{int}(1,2)\psi_\alpha(1)\psi_\beta(2)\,d\mathbf{r}_1\,d\mathbf{r}_2 = K$$
    $$\text{and } \int\psi_a(1)\psi_\beta(2)V_{int}(1,2)\psi_\beta(1)\psi_\alpha(2)d\mathbf{r}_1\,d\mathbf{r}_2 = J,$$

    where K and J are *positive* constants.

    (a) Evaluate the energy expectation value in the four two-particle states

    $$\psi_\alpha(1)\psi_\alpha(2), \quad \psi_\beta(1)\psi_\beta(2), \quad \text{and} \quad \frac{1}{\sqrt{2}}\left[\psi_\alpha(1)\psi_\beta(2) \pm \psi_\beta(1)\psi_\alpha(2)\right]$$

    in terms of ε_α, ε_β, K, and J.

    (b) Which of these states is allowed if the two particles are identical bosons? fermions?

    (c) Why does the fermion state have the lowest energy?

3. How many distinct throws of a pair of dice can occur? How many ways can a four be thrown? What is the probability of throwing a four? What number has the highest probability of occurring in a throw of a pair of dice?

4. Consider an ensemble made of two systems ($M = 2$) each having two energy levels ε_1 and ε_2. Energy level ε_1 is occupied by a total of two particles ($N_1 = 2$), as is level ε_2 ($N_2 = 2$). Find the total number of ways W of placing the particles in the states if the particles are (a) distinguishable, (b) fermions, or (c) bosons. Use Equations (11-13), (11-21), and (11-24) and also find W by explicit counting.

5. Show, for a Maxwell–Boltzmann system, that $\langle\varepsilon\rangle = -\partial \ln Z/\partial\beta$, where Z is the partition function. Use this formula to derive the result of Equation (11-50).

6. Evaluate $h/\langle p^2\rangle^{1/2}$ and show that the result, up to constant factors, is equal to λ_{th}.

7. (a) Use Equation (5-60) (with position x replaced by energy as a variable) to show that the mean square deviation of the total energy $E = \sum_{i=1}^N \varepsilon^{(i)}$ from its average value is given by

    $$(\Delta E)^2 = N(\langle\varepsilon^2\rangle - \langle\varepsilon\rangle^2),$$

    where $\langle\varepsilon\rangle$ and $\langle\varepsilon^2\rangle$ are single-particle energy averages.

    (b) Show, from the definition of the partition function Z, that the mean square single-particle energy deviation for Maxwell–Boltzmann particles is

    $$(\Delta\varepsilon)^2 = \langle\varepsilon^2\rangle - \langle\varepsilon\rangle^2 = \frac{\partial^2\ln Z}{\partial\beta^2}.$$

    Evaluate this quantity.

(c) Use the results of parts (a) and (b) to show that the fluctuations in energy of a large Maxwell–Boltzmann system diminish with size according to

$$\frac{\Delta E}{\langle E \rangle} = \sqrt{\frac{2}{3N}} .$$

Evaluate this for $N = 100$, 10^9, and 10^{23}.

8. Carry out the details of the maximization procedure for the function W in Equation (11-24) to prove that the Bose–Einstein distribution function is given by Equation (11-25).

9. Consider a system of N distinguishable particles in which each particle has two possible energy levels $\varepsilon_1 = 0$ and $\varepsilon_2 \equiv \Delta$. An example of such a system is a crystal that can form vacancies by exciting particles into interstitial positions.

(a) Find the distribution functions n_1 and n_2 and plot the ratio n_2/n_1 as a function of the parameter $\tau = e^{-\beta\Delta}$, which has range $0 \leq \tau \leq 1$, while the temperature varies from zero to infinity. Explain the behavior of the graph.

(b) Find the average energy $\langle \varepsilon \rangle$ and the specific heat $c = d\langle \varepsilon \rangle/dT$. Plot $\langle \varepsilon \rangle/\Delta$ and c/k_B as functions either of τ or of $k_B T/\Delta$. Explain the behavior of these functions.

10. (a) Consider a two-level system as in Problem 9 but containing spinless "Fermi–Dirac" particles. Find n_1 and n_2. To do this the Lagrange parameter e^α must be eliminated by use of Equation (11-23). Carry the solution out for the only two possible cases $N = 1$ and $N = 2$. Why is $N \geq 3$ not allowed? Plot the ratio n_2/n_1 as a function of the temperature parameter $\tau = e^{-\beta\Delta}$.

(b) Repeat with Bose–Einstein particles for the cases $N = 1$, 2, and 50. The elimination of the Lagrange parameter e^α is not as easy here and may best be achieved by numerical or graphical methods. Plot n_2/n_1 as a function of τ for each of the above values of N. Can you generalize the trend to $N \to \infty$?

(c) Compare the plots of n_2/n_1 from Problem 9 and from parts (a) and (b) of this problem. Explain the physical basis of the behavior. Why do even the $N = 1$ Fermi or Bose distribution functions show quantum behavior?

11. A container filled with H_2 gas has a pressure of 1 atmosphere (1.013×10^5 Pa) at room temperature. To what temperature must the gas be cooled for quantum effects to become important? Estimate this temperature by equating the thermal de Broglie wavelength to the interparticle spacing. Use the ideal gas law to estimate the density. Repeat for the conduction electrons in a metal for which the average interelectronic spacing is fixed at 0.1 nm.

12. Equation (11-49) gives the Maxwell–Boltzmann probability density for finding the single-particle energy in the range ε to $\varepsilon + d\varepsilon$ as $p(\varepsilon) \sim \varepsilon^{1/2}\exp(-\beta\varepsilon)$. Find the most likely energy by maximizing $p(\varepsilon)$. Compare with $\langle \varepsilon \rangle$ of Equation (11-50). Find the half-width of $p(\varepsilon)$ and compare with $\Delta\varepsilon$ evaluated in Problem 7(b).

13. Find the total energy E and heat capacity C of blackbody radiation in a cavity of volume V. Show that C has the same temperature dependence as the low-temperature phonon system. Find the average total number of photons present in the cavity as a function of temperature. Estimate this number for room temperature (300 K) and a volume of 1 cm^3.

14. Atoms adsorbed in a monolayer (a layer one atom thick) on a surface at sufficiently high density sometimes behave like a two-dimensional solid. The low-frequency sound waves obey a relation of the form of Equation (11-61) with $n_3 = 0$. Find the heat capacity of such a solid at high and low temperatures. [Hint: A new density of states N_ν appropriate to standing waves in two dimensions must be derived. Paraphrasing the derivation for

radiation in Section 2-2 or even that for particles in the Detail at the end of Section 11-3 may be helpful.]

15. A scale of temperatures for solids, by which we judge whether T is high or low, is set by the velocity of sound and density, through the Debye constant Θ_D. What is the corresponding characteristic temperature in Einstein's model of a solid?

16. Find the density of states function N_ν, as a function of ν, for *all* frequencies in the one-dimensional phonon system for which ν is given by Equation (11-58). Explain physically why N_ν diverges at $\nu = \nu_m$. Find the heat capacity for low temperatures $k_BT \ll h\nu_m$ and for high temperatures $k_BT \gg h\nu_m$.

17. Aluminum has a Debye temperature of 420 K. Use this to find the velocity of sound in aluminum. What is the specific heat of the phonon system at 300 K? Assume the low-temperature formula in Equation (11-69) is sufficiently accurate.

18. Calculate the Fermi temperature T_F for metallic aluminum. What is the specific heat of the conduction electron system of aluminum at 300 K? Compare this result with the result of Problem 17.

19. Show that the condition $T \ll T_F$ for the degeneracy of a Fermi gas is equivalent to the statement $\lambda_{\text{th}} \gg a$, where a is the mean interparticle spacing.

20. Liquid ^{3}He is a system of fermions that interact quite strongly. Nevertheless, a useful first-stage model is that of a noninteracting gas of helium atoms. Find the Fermi temperature, for a mean interparticle spacing of $a = 0.4$ nm in the liquid. Find the thermal de Broglie wavelength for ^{3}He at $T = 1$ K and compare with the given value of a. Are the results consistent with the conclusions of Problem 19?

21. Show that the electron system of the white dwarf star in the example at the end of Section 11-6 must be treated relativistically. Do this by computing the Fermi velocity for a radius of 700 km and comparing with the speed of light.

22. The rotational motion of diatomic molecules contributes to the specific heat $c = d\langle\varepsilon\rangle/dT$ in a gas of molecules. In this problem you are to calculate this contribution by considering a set of distinguishable rigid rotators each with moment of inertia I and energy levels

$$\varepsilon_j = j(j+1)B, \qquad j = 0, 1, 2, \ldots,$$

where $B = \hbar^2/2I$.

(a) Show that the partition function is

$$Z = \sum_{j=0}^{\infty} (2j+1)e^{-\beta Bj(j+1)}.$$

[Hint: What is the degeneracy of the level ε_j?]

(b) Evaluate Z for high temperature ($\beta B \ll 1$) by taking $\Sigma_j \to \int dj$ and then proving

$$Z = -\frac{1}{\beta B}\int_0^{\infty} dj \frac{d}{dj} e^{-\beta Bj(j+1)}.$$

Discuss the approximation. Find $\langle\varepsilon\rangle$ and C by using the theorem proved in Problem 5.

(c) Evaluate Z, $\langle\varepsilon\rangle$, and C for low temperature ($\beta B \gg 1$) by keeping only the first two terms in the j sum.

T W E L V E

SOLIDS

Solids are materials that are resistant to deforming forces. A crystal is a special kind of solid usually showing facets, that is, surfaces occurring at certain characteristic angles; these give evidence of a regular array or latticework of the atoms making up the material. The term "solid" is more general than "crystal" and can also describe amorphous substances, which may be every bit as hard as crystals but which have no regular lattice structure. Diamond and quartz are crystals, but glass is a solid that is not crystalline. Since the long-range particle periodicity characteristic of a crystal is insufficient to specify what constitutes a solid, we must look further into the distinguishing properties.

The study of solids has been a part of physics for as long as physics has been a science. Before that, alchemists were interested in the nature of so-called "base" metals and whether they could be turned into gold. Navigators found that a lodestone, a natural permanent magnet, could be used as a compass. To satisfy their needs for weapons, utensils, and jewelry, ancient metallurgists learned how to mix copper and tin to form bronze and to mix the right combination of iron and carbon to make steel.

As in these examples, fundamental knowledge has often followed from the need to improve technology. In more recent times the order has often been reversed and technology has entered quickly into areas where new knowledge has been found. In no area of science has this interrelation been more apparent than in solid-state physics. The outstanding examples of this are the developments of the transistor and the integrated circuit, which have brought us into the computer age. These inventions followed directly from the understanding of the nature of semiconductors and could not have occurred without that basic understanding.

As important as solid-state physics is to the development of technology, even more so is its fundamental contribution to our basic understanding of nature. Einstein's explanation of the breakdown of the Dulong–Petit law of heat capacity in solids is only one of several first examples of this. Furthermore, progress in solid-state physics has often stimulated new lines of basic research in other areas. For example, after years of study of the magnetic transition in iron, recent breakthroughs have led us to an understanding of many kinds of phase transitions in a wide range of substances.

Several recent advances in other parts of physics are based on ideas used to describe phase transitions in condensed-matter systems.

The passing of ideas from one area of physics to another is a common theme. We know from our study of diatomic molecules in Chapter 10 that two atoms of hydrogen bind together because the electrons are able to hop from one atom to the other; this is the basis of the covalent bond. The periodic potential energies of electrons in regular crystals also can have sufficiently low barrier heights so that electrons can tunnel from one atom to another. This effect contributes to binding and so the solid behaves somewhat as a gigantic many-atom molecule.

This rapid hopping results in a highly characteristic structure for the energy levels of electrons in solids. The states fall into bands of very closely spaced energy levels separated by forbidden energy regions called band gaps. The basic behavior of crystals depends greatly on how the electrons fit into these states, which are filled according to the Pauli principle. On this basis we are able to understand how substances fall into the categories of metals, insulators, and semiconductors.

To understand the detailed distinctions among the various kinds of solids, experimentalists can probe the materials by measuring such static properties as their pressure–volume curves, heat capacities, and magnetizations. They also can examine their transport properties by setting up nonequilibrium flow conditions. For example, if we establish a voltage across a sample of metal, electricity is conducted through it and we can easily measure its electrical conductivity. When we maintain a temperature difference between the ends of a sample of material, heat flows and we can determine the thermal conductivity coefficient. The dependence on temperature, and on other variables, of these coefficients tells us about the characteristics of the carriers of electricity and heat, and also tells us with what objects these carriers may be colliding as they traverse the solid.

Insulators are able to conduct heat, though not electricity, by means of lattice vibrations. The associated phonons have been introduced previously in Chapter 11. Several properties of phonons have only been quoted there in order to discuss heat capacity as an example of the methods of statistical physics. In this chapter we justify those results by deriving them.

Since the first awards in 1901 the Nobel Prize has gone to many physicists for their work on solids. The achievements have been as significant as any in physics. More research is published each year in solid-state physics than in any other field. Many of the papers are reports of technological applications and many are quite basic. These forms of activity demonstrate to every student of physics the importance of understanding the physics of solids.

12-1   The Structure of Solids

We cannot define a solid simply as a substance that is hard. Suppose that we put water into a closed container with a plunger on one end and then attempt to squeeze it. We would find it to be "hard" as well since under the application of ordinary forces water is essentially incompressible. What differentiates solids from liquids or gases is the ability of the solid to resist *shear* forces.

If we push vertically on a stack of several decks of playing cards, it does not compress much. If we push horizontally near the top of the stack, it may collapse because the cards can slide easily relative to one another. The stack has no shear strength.

The molecules of a gas are basically free particles that only occasionally feel the effects of intermolecular forces as they collide with one another. The molecules of a liquid, such as water, constantly feel forces due to the other molecules. They are in a many-body bound state. A bit of water floating freely in the space-shuttle stays together as a roughly spherical blob. On the other hand, the total kinetic energy shared by the molecules in water is still sufficiently great that each molecule does not find itself bound permanently to a given set of neighbors. Such a particle almost always has an escape route in one direction or another as it rattles about trying to move to a new neighborhood. Even if the molecule momentarily finds it has insufficient energy to escape from a certain small region, collisions are so frequent that it soon has enough, so a new lower energy pathway opens up because of the other particles' random motions. However, if one of the particles tries to leave the surface of the liquid, it usually is unable to do so, because the forces from the other molecules near the surface pull it back. Occasionally, however, some particles do have sufficient kinetic energy to evaporate.

If we try to compress a gas, we have no difficulty because there is so much empty space. But liquids are far denser and compression causes the hard cores of the particles to touch and create resistance. The same is true of solids. A shear force applied to water does not try to push the water into a smaller volume like a compression does. It simply slides the particles relative to one another somewhat like the playing cards mentioned above. This costs very little energy; it is essentially what the individual molecules are already constantly doing on their own.

We are led then to the idea of a solid, with its shear strength, as resulting from the fact that each particle is bound in a potential energy well formed by its neighbors from which it rarely has enough energy to escape. Shear forces are resisted because it takes energy even to slide particles relative to one another. A solid can usually be picked up without a container because of its shear strength.

One way that every particle can be made to sit in a potential energy well formed by its neighbors is to construct a perfect array, that is, a crystal lattice. In the simplest such situation every particle sees precisely the same arrangement of neighboring particles. If one particle is bound then all of them must be.

Consider a situation in two dimensions at absolute zero temperature. Suppose the particles are xenon atoms, each pair of which has a potential energy of interaction described by the Lennard-Jones form, Equation (10-36). If we neglect the small amount of zero-point motion, a pair of these atoms sits at the distance a of the minimum of the potential energy curve and has energy $-\varepsilon$, equal to the depth at the minimum.

The best thing for a third particle to do is to position itself so that the three atoms form an equilateral triangle having sides a. The energy of this trio is -3ε, corresponding to the three bonds formed as shown in Figure 12-1.

A fourth particle can now be added, in our two-dimensional example, at any one of three possible positions so that it similarly has two neighbors each at a distance a. If we neglect the interaction energy between the two particles that are not near neighbors, the total energy of the two triangles is now -5ε. We can continue in this manner until whatever amount of two-dimensional space available is filled with triangles as in Figure 12-2.

Note that every particle in this figure is the center of a hexagon so that it has six neighbors. If we were arranging Ping-Pong balls on a table so that they were touching one another, the most that we could pack about any center ball would be six. Those six would form a hexagon made up of equilateral triangles. We would have packed

Figure 12-1

Three particles with interatomic separation a such that every pair interaction is at minimum energy $-\varepsilon$. The ×'s mark possible positions for a fourth particle.

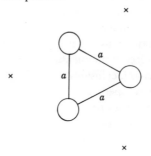

Figure 12-2

Part of a hexagonal or triangular two-dimensional lattice of particles. Particles 1 and 2 are nearest neighbors; particles 1 and 3 are second neighbors.

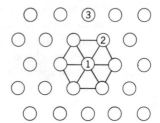

the balls as closely as possible to one another. More balls could be added to fill two-dimensional space as illustrated in Figure 12-3. No denser arrangement of balls in two dimensions is possible. The lattice formed is equivalent to that of Figure 12-2, and is therefore called a hexagonal close-packed lattice, or simply a triangular lattice.

The energy of such a very large xenon lattice is easy to compute. Suppose there are N particles. The energy of interaction of a particle and its six neighbors is -6ε. Each particle is the center of a hexagon so we might think the total energy is $-6\varepsilon N$. However, from looking at any pair of particles in Figure 12-2, it is seen that we have then counted that pair of particles twice, once when one of them.was at the center of a hexagon and once when the other was at the center. We must divide our result by 2 to get $-3\varepsilon N$. Because every bond corresponds to a potential energy minimum, this is the

Figure 12-3

Close packing of Ping-Pong balls. Each ball is at the center of a hexagon and touches all six of its nearest neighbors.

Figure 12-4
Square lattice.

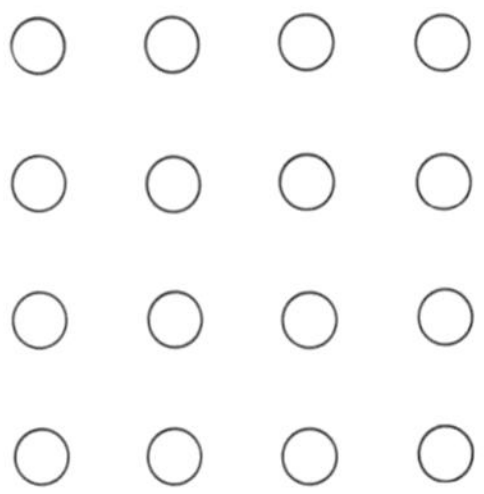

○ ○ ○ ○

○ ○ ○ ○

○ ○ ○ ○

○ ○ ○ ○

Figure 12-5
Three-dimensional arrangement of four particles. Each particle can be a near neighbor of the other three by forming a tetrahedron.

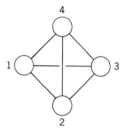

lowest energy possible for any structure in two dimensions for an isotropic interaction like the Lennard-Jones function.

An alternative two-dimensional structure is the square lattice shown in Figure 12-4. It is easy to see that the energy of this structure is $-2\varepsilon N$, which is a third less negative than the energy of the triangular lattice. Also its density is considerably lower.

The above discussion of crystals in two dimensions is far from purely pedagogic. Gaseous xenon and many other elements, when bound to certain carefully chosen planar solid surfaces, behave like two-dimensional gases, liquids, or solids.

We now try to treat three-dimensional systems in the same manner as we have handled two-dimensional solids. One way to start out is by forming an equilateral triangle of particles, as in two-dimensions, and then placing a fourth particle above or below to form a tetrahedron as shown in Figure 12-5. Each of the four particles can have the optimal distance a from the other three particles. Just as two-dimensional space is covered by equilateral triangles, we might hope to cover three-dimensional space by tetrahedrons. Unfortunately, this turns out to be impossible. Suppose we consider the top particle in Figure 12-5 as a central one to be involved in as many tetrahedra as possible, much as any particle in Figure 12-2 is the center of a hexagon made up of equilateral triangles. The next three particles to be added are placed next to the three upper faces of the tetrahedron. If we go on in this way, continuing to make as many tetrahedra as possible centered on one particle, the central particle ends up with 12 neighbors in the structure shown with balls in Figure 12-6. It does not look very symmetric; in fact it has a sort of crack near the top. In 1694, the Scottish astronomer D. Gregory claimed, without proof, that a 13th sphere could be squeezed into such a crack so it would touch the central ball. Newton disputed this, claiming correctly that the maximum number of touching neighbors is 12.

The relevance of this for solids is that a few particles, acting on a local basis, might form a number of tetrahedra. These do not quite fit together with tetrahedra from other nearby systems of particles so that the overall arrangement of particles is filled with cracks. Most particles are bound in potential wells, although those near cracks are not bound so tightly. There is no long-range regularity as in the perfect lattices that we describe below, but the system can form an irregular solid. There exist a variety of types of amorphous, or irregular, solids classified generally as *glasses*, some of which this simple model may actually describe. Even though the glasses have no long-range regular structure, they do have shear strength because each particle is

Figure 12-6

Ping-Pong balls stacked to make as many tetrahedra as possible in an object consisting of a ball and its near neighbors. The geometrical shape is a distorted icosahedron.

bound in a potential well. Generally, there is, for these materials, some lattice structure having lower energy, but the atoms cannot get to that state because they are each trapped in a local potential energy minimum.

Since nature cannot make regular solids by filling three-dimensional space with tetrahedra, as it can fill two-dimensional space with triangles, it usually proceeds to make regular lattice structures in other ways. Rather than maximizing the number of tetrahedra, we look for a way to maximize the density of the entire system such that all nearest neighbors of any atom are at some distance a. It is speculated, but has never been proved, that the densest such arrangement is provided by the face-centered cubic (FCC) lattice. This crystal structure is easily constructed by starting with the two-dimensional triangular lattice of Figure 12-2. The particles in this layer are designated by B's, as indicated in Figure 12-7. Particles are next added in layers above and below this layer. Particles in the next layer above the first go in at sites designated by A. A particle at A forms a tetrahedron with three first-layer particles. All the particles in this layer again form a two-dimensional triangular lattice shifted horizontally a bit from that of the first layer.

A third layer is placed below the first at the positions indicated by the C's. Additional layers are added so that the sequence $\cdots ABCABC \cdots$ is formed.

If we look at the arrangement of a particle and its 12 nearest neighbors in the FCC lattice, as shown in Figure 12-8, we see how this grouping differs from the distorted structure of Figure 12-6. The seven second-layer particles are all on the same level, whereas the levels of these same particles in Figure 12-6 are seen to alternate. Not so many perfect tetrahedra are formed in this argument, but that is more than compensated by the regularity of the packing throughout all space.

We can easily compute the energy of the FCC lattice if we include only nearest-neighbor interactions. Since each particle has 12 neighbors, this energy is $-(\frac{1}{2}) \times 12\varepsilon N = -6\varepsilon N$, where the factor $\frac{1}{2}$ avoids the double counting of the bonds. This energy is the lowest possible for a system with an isotropic interaction.

Figure 12-7

Model of a face-centered cubic lattice. Two-dimensional triangular lattices are stacked one on top of the other in the arrangement $ABCABC \cdots$. The four particles $A1, B1, B2, B3$ form a tetrahedron with all particles separated by the nearest-neighbor distance a.

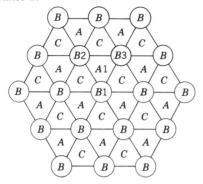

The FCC lattice gets its name from viewing the structure from a different angle. Figure 12-9 shows this alternative view. The demonstration of the equivalence of the two views is left to be shown in Problem 3 at the end of the chapter.

There are many other possible lattices. A simple variation of the FCC case is the three-dimensional hexagonal close-packed lattice. Instead of placing a particle in a C layer as shown in Figure 12-7, it is placed directly over a particle in the A layer resulting in an $\cdots ABAB \cdots$ pattern. With this arrangement each particle again has 12 neighbors. It differs in energy from the FCC only in interactions involving third nearest neighbors. Other possibilities are the simple cubic (SC) lattice shown in Figure 12-10 and the body-centered cubic (BCC) of Figure 12-11.

Figure 12-8

Central particle and its 12 nearest neighbors in an FCC lattice.

Figure 12-9

View of the FCC lattice showing the origin of its name. Each of the faces of a cube contains a particle at its center. Although all sites are equivalent, the particles on the faces are shown in a lighter shade for clarity.

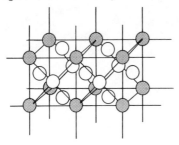

Lattices can be classified according to their symmetry properties of translation, rotation, and reflection. There turn out to be only 14 basic lattice types in three dimensions when thought of in these general terms.

We might wonder why any substance settles into a lattice structure different from the FCC if that is the densest one with the lowest potential energy for an isotropic interaction. There are a variety of answers to this question. As shown in Chapter 10, bonds are often directional, as for example, those of carbon; the structure of the lattice can follow the directionality. Such a directional bias can be changed by applying pressure or otherwise changing the physical state of the substance. Thus, carbon can be found in the form of graphite or diamond, two distinct lattice structures, depending on the preparation of the carbon.

Another possibility is that the substance is made up of more than one element with the relative size of the atoms playing a role in determining the crystal structure. The CsCl crystal has cesium atoms at the corners of cubes with chlorine atoms at the body centers. This might seem to be a BCC arrangement, but it is actually simple cubic. It is necessary to focus on just the cesium atoms or just the chlorine atoms to determine the true structure.

Figure 12-10

Simple cubic lattice.

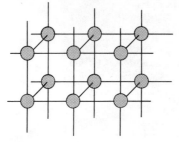

Figure 12-11

Body-centered cubic lattice.

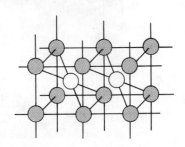

Figure 12-12

Model of the sodium chloride crystal. Sodium atoms are shown in black, chlorine atoms in white. The crystal structure is FCC.

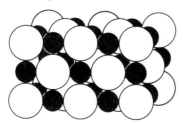

Another answer to the question of a substance's crystal structure is provided by solid helium. While close-packed lattices do occur for this substance, under the appropriate conditions of temperature and pressure a BCC phase also occurs. This happens despite the fact that helium is elemental with a highly isotropic interaction. The cause of this peculiarity is zero-point motion that is so large it destabilizes the closed-packed lattice.

The types of binding that occur in molecules also hold solids together. A solid can often be considered simply to be a large molecule. Covalent bonding occurs in the diamond crystal. The NaCl crystal shown in Figure 12-12 is ionically bound, with Na the positive ion and Cl the negative.

The hydrogen atoms in a water molecule are bound to the oxygen atom covalently. Each hydrogen can also be attracted to an oxygen of *another* water molecule, thus forming a second bond called the *hydrogen bond*. Ice then consists essentially of water molecules connected together by such hydrogen bonds.

Hydrogen itself is another example of a substance that retains its molecular identity in the crystalline state. The H_2 molecules attract each other by the van der Waals force and form what is called a *molecular* solid. Other molecular crystals include, for example, O_2 and the solid states of the noble gases helium, neon, argon, and so on. Under very high pressure, it is expected that the diatomic molecular bonding in solid hydrogen gives up in favor of a homogeneous bonding among all atoms so that the solid undergoes a transition to a metallic phase. This important effect is being looked for in many laboratories around the world.

The binding that occurs in a metal is somewhat analogous to that occurring in the hydrogen molecule, with electrons hopping among all atoms instead of just between two. We investigate this binding and the other properties of metals in the next two sections.

Example

The NaCl crystal is ionically bound. We can try (unsuccessfully) to estimate the contribution of its Coulombic energy to the binding of the crystal by summing over the pair interactions in the lattice. Any given Na^+ atom is surrounded by 6 near-neighbor Cl^- atoms at a distance R_0, 12 second-neighbor Na^+ atoms at $\sqrt{2}\,R_0$, 8 Cl^- atoms at $\sqrt{3}\,R_0$, 6 Na^+ at $\sqrt{4}\,R_0$, and so on. The contribution to the binding from this sequence of interactions is

$$E = \frac{e^2}{R_0}\left(-6 + \frac{12}{\sqrt{2}} - \frac{8}{\sqrt{3}} + \frac{6}{\sqrt{4}} - \cdots\right).$$

The binding energy per particle is N times this result, where N is the number of sodium ions. The sum in parentheses converges very slowly. Its value, upon truncating it after a given numbers of terms, is shown in the following table:

Number of Terms	Sum
1	-6.0
2	2.5
3	-2.1
4	0.9
5	-9.9
6	-0.7

If the complete sum is done, the result is -1.75; special mathematical procedures must be used to evaluate it.

12-2 Bragg Scattering

An experimental means of probing and examining the structure of crystals was provided by Laue early in this century, as we have discussed in Section 2-6. We know that when one irradiates a crystal with x rays they are scattered by the electrons of the individual atoms so that the regular structure of these atoms acts as a three-dimensional diffraction grating. Waves from different atoms interfere with each other constructively in some directions and destructively in others. Scattered waves detected on photographic film show a pattern characteristic of the particular crystal structure being x rayed. Laue invented the idea and the Braggs used it to analyze the structure of many crystals.

Other modes of radiation can be used as probes as well. We have seen in Section 4-2 that electrons can be diffracted from crystals; in 1927 Davisson and Germer, and Thomson proved by this method that electrons behave as waves. Neutrons may also be used. Note, however, that the neutrons interact with the nuclei and not with the atomic electrons. The wavelength of neutrons is normally so short that they can be used to probe smaller distances than can x rays. Because neutrons have spin, they can also reveal the magnetic structure of crystals.

As in Section 2-6 we again consider x rays striking a set of parallel planes of particles occurring in the lattice. Any crystal has an infinite set of such planes found by slicing through the crystal at certain angles. Figure 12-13 uses a two-dimensional representation of a crystal to illustrate some of the many possible planes that could scatter x rays. Obviously, some planes are more dominant than others because the density of particles on them is greater than that on others. The facet that appears on the surface of a crystal represents one of a parallel family of such dominant planes.

From Figure 2-15 we recall that constructive interference occurs when the path difference between the two rays in the figure is equal to an integral number of whole

Figure 12-13

Two-dimensional representation of the many possible scattering planes in a crystal.

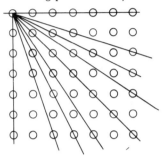

wavelengths. According to Equation (2-49), this happens when the angle θ, between the crystal planes and the incoming or scattered beam, satisfies

$$2d \sin \theta = m\lambda, \qquad m = 0, 1, 2, \ldots, \tag{12-1}$$

where d is the separation between planes.

A narrow beam of x rays hits the crystal. A scattered beam emerges from the crystal for each family of parallel planes satisfying Equation (12-1). Each emerging beam has an intensity depending on the density of atoms in the corresponding set of planes. Photographic film, exposed to the scattered radiation, displays spots, caused by the beams, which can be analyzed to give information about the crystal structure. An example of a Bragg diffraction pattern is shown in Figure 12-14.

Each family of parallel planes can be characterized by a unit vector $\hat{n}$ normal to them. Each possible scattered ray can be characterized by a vector along this direction. To see this, suppose that $\mathbf{k}$ is the wave vector of the incident beam and $\mathbf{k}'$ is the scattered wave vector. The momentum transferred to the lattice is $\hbar \Delta \mathbf{k} \equiv \hbar(\mathbf{k} - \mathbf{k}')$. Because the lattice is so massive, its recoil is negligible and we have

$$|\mathbf{k}| = |\mathbf{k}'| = k = \frac{2\pi}{\lambda}. \tag{12-2}$$

From Figure 12-15, we see that

$$\Delta \mathbf{k} = 2k \sin \theta \, \hat{n}.$$

Substituting from Equation (12-1) for $\sin \theta$, and from Equation (12-2) for λ, gives

$$\Delta \mathbf{k} = \frac{2\pi}{d} m\hat{n} \equiv \mathbf{G}, \qquad m = 0, 1, 2, \ldots . \tag{12-3}$$

Equation (12-3) defines a set of vectors $\mathbf{G}$ whose direction is perpendicular to the family of lattice planes and whose length is inversely proportional to the separation between the planes. There is a set of such vectors for each family of planes in the crystal.

Figure 12-14

Photographic record of the x-ray diffraction pattern due to Bragg scattering from an NaCl crystal.

The infinite, but *discrete*, set of these **G** vectors can be shown to define a lattice, called the *reciprocal lattice*. Each point in this ficticious lattice corresponds to a possible momentum that can be absorbed by the real lattice of particles. There is a one-to-one correspondence between the sites in the reciprocal lattice and the spots in a scattering pattern.

We see below that these vectors show up repeatedly in our analyses of solids. For example, a conduction electron in a metal can be diffracted constructively by the lattice of the atoms, just as an x ray is, if the scattering produces a $\Delta \mathbf{k}$ equal to one of the **G**'s. This phenomenon has important consequences relative to the possible energy states of electrons in crystals.

Figure 12-15

Vectors describing the scattering of x rays from a crystal plane. The momentum transferred to the lattice is $\hbar \, \Delta \mathbf{k}$. The normal to the reflecting plane is $\hat{\mathbf{n}}$.

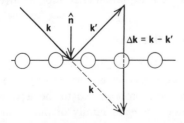

Example

Suppose that the scattering angle in a first-order diffraction ($m = 1$) is $\theta = 42°$ and the x-ray wavelength is 0.2 nm. The separation between the lattice planes can be found from Equation (12-1):

$$d = \frac{m\lambda}{2 \sin \theta} = \frac{1 \times 0.2 \text{ nm}}{2 \sin 42°} = \frac{0.2 \text{ nm}}{2 \times 0.67} = 0.15 \text{ nm}.$$

A photon of this wavelength carries energy

$$E = \frac{hc}{\lambda} = \frac{1.240 \text{ keV} \cdot \text{nm}}{0.2 \text{ nm}} = 6.2 \text{ keV}.$$

Example

If neutrons are used for the Bragg scattering in the above example, each has wavelength 0.2 nm and energy E related to λ by

$$\lambda = \frac{h}{p} = \frac{h}{\sqrt{2ME}} = 0.2 \text{ nm},$$

where p and M are the neutron momentum and mass, respectively. Thus, we find

$$E = \frac{h^2}{2M\lambda^2} = \frac{\left(6.6 \times 10^{-34} \text{ J} \cdot \text{s}\right)^2}{2\left(1.7 \times 10^{-27} \text{ kg}\right)\left(0.2 \times 10^{-9} \text{ m}\right)^2}$$

$$= 3.2 \times 10^{-21} \text{ J} = \frac{3.2 \times 10^{-21} \text{ J}}{1.6 \times 10^{-19} \text{ J/eV}} = 0.02 \text{ eV}.$$

Note how much smaller the neutron energy is than that of the photon of the same wavelength. Neutrons having such low energies are said to be *thermal*, since a $k_B T$ of this value implies a T near room temperature. Nuclear reactors are often the source of neutrons for Bragg scattering. Because most such neutrons originate with energies in the MeV range, it is necessary to slow them down by many inelastic scatterings before using them in a diffraction experiment.

12-3 The Free-Electron Theory of Metals

The most notable properties of metals are their abilities to conduct heat and electricity. The idea that metals contain electrons that can move freely and can carry energy and charge was developed soon after Thomson discovered the electron in 1897. P. Drude first suggested a model that assumed an electron behavior like a classical gas. In Chapter 11 we have examined this assumption and have found that the electrons must be a highly degenerate set of fermions. In 1928 Sommerfeld modified the Drude theory to take degeneracy into account. The resulting model did a surprisingly adequate job of describing many of the properties of metals.

Figure 12-16

Arrangement for measuring the coefficient of thermal conductivity. The temperature gradient is $dT/dz \sim (T_1 - T_2)/L$. The heat current is the flow rate per unit area, $J_T = (1/A)dQ/dT$.

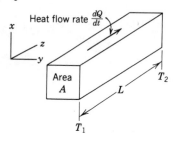

Some of the static properties of a degenerate electron system have been described in Chapter 11. The most important one for metals is that the electron heat capacity has been shown in Equation (11-76) to be linearly proportional to temperature. This characteristic dependence is found experimentally in metals.

Specific heat involves a static or equilibrium measurement. Another class of experiments involves setting up nonequilibrium conditions in order to study how the system returns to equilibrium. The quantities measured are the transport coefficients of the material. For example, if one end of a metal bar is placed at a higher temperature than the other, heat is conducted down the bar in an attempt to reestablish constant temperature. The situation is shown in Figure 12-16. If the gradient of the temperature in the z direction is dT/dz and if the heat current density, that is, the heat flow per unit cross sectional area per second, is J_T, then we have

$$J_T = -\kappa_T \frac{dT}{dz},\qquad(12\text{-}4)$$

where κ_T is called the *coefficient of thermal conductivity*. The minus sign in Equation (12-4) indicates that heat travels from high temperature to low temperature or in the direction of negative dT/dz.

In a similar manner we define the *electrical conductivity* κ_E as the constant of proportionality between an electric field E established in a metal and the charge current density

$$J_E = \kappa_E E,\qquad(12\text{-}5)$$

giving the charge per second per unit area moving in the material. From elementary electrostatics, the field vanishes in a metal at equilibrium. The situation established here is a nonequilibrium one. The charge is flowing in an attempt to set up a distribution that cancels out the field in the metal. If the charge is removed from one end and reinserted in the other continuously, a steady-state nonequilibrium situation can be established as it is in any simple DC resistive circuit.

The transport properties of electrons are usually discussed by considering the states of the particles in momentum space. We need to digress briefly to see what this space is and how it is filled by a set of degenerate free particles. We know from Chapter 11 that the energy states for a Fermi system at $T = 0$ K are filled up to the Fermi energy ε_F and states above that are empty. If we calculate the energies of *traveling-wave* states in a cubical box of volume L^3, we find

$$\varepsilon_{\ell_1\ell_2\ell_3} = \frac{h^2}{2mL^2}\left(\ell_1^2 + \ell_2^2 + \ell_3^2\right).$$

Since $\varepsilon = p^2/2m$, we can write the allowed momenta as

$$p^2 = \frac{h^2}{L^2}\left(\ell_1^2 + \ell_2^2 + \ell_3^2\right).$$

The allowed values of the three Cartesian components of momentum are

$$p_i = \frac{h\ell_i}{L}, \qquad \ell_i = 0, \pm 1, \pm 2, \pm 3, \ldots \tag{12-6}$$

where i is 1, 2, 3 or x, y, z. These results for the energy and momentum are not quite the same as the previous forms of Equations (5-99) because we are now considering traveling rather than standing waves. We consider these differences in more detail in Section 12-4; for now just note that both positive and negative momentum values are allowed corresponding to particles traveling in opposite directions.

Each of these discrete momentum states may be occupied by one electron of each spin type. The three momentum coordinates allow us to construct a space of states similar to the one illustrated in Figure 2-7 for standing waves of electromagnetic radiation. At $T = 0$ K, the states in p space occupied by particles form a sphere, known as the *Fermi sphere*, as shown in Figure 12-17. The radius of the sphere is p_F, the *Fermi momentum*, which is related to ε_F by

$$p_F = \sqrt{2m\varepsilon_F}. \tag{12-7}$$

Figure 12-17
Fermi sphere in p space. The radius of the sphere p_F is related to the Fermi energy ε_F according to $\varepsilon_F = p_F^2/2m$. At $T = 0$ K, states below p_F are all occupied while those above are empty.

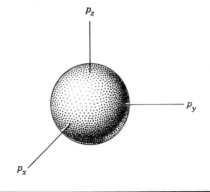

The Fermi velocity is simply

$$v_F = \frac{p_F}{m}.$$ (12-8)

We now consider the electrical conductivity. An electron of charge $-e$ placed in an electric field feels a force $-eE$ and accelerates freely if it does not lose energy by interactions with impurities, ion cores, and other electrons in the metal. These are the resistive interactions of the metal. In a very simplified model we might assume that they provide a frictional type of force whose size may be supposed to depend on the electron's velocity. Newton's law becomes

$$m\frac{d\mathbf{v}}{dt} = -e\mathbf{E} - \alpha\mathbf{v},$$ (12-9)

where α is a constant determined by the nature of the frictional forces. Obviously, the frictional force operates only when the electron is moving and increases in size with the velocity, much as the retarding force of air resistance on a moving automobile. Suppose we consider the situation in which the electron has been given an initial velocity $\mathbf{v}_0$ in the absence of an electric field (and in the absence of any other electrons). Then the electron's equation of motion is

$$\frac{d\mathbf{v}}{dt} = -\frac{\alpha}{m}\mathbf{v},$$

which has solution

$$\mathbf{v} = \mathbf{v}_0\, e^{-(\alpha/m)t}.$$ (12-10)

We see that the velocity returns to zero in a time

$$\tau = \frac{m}{\alpha}.$$ (12-11)

This constant is called the *relaxation time* for obvious reasons. By means of microscopic calculations one can show that τ is not much longer than the *collision time*, the time the electron travels between interactions with impurities, ions, or whatever it is that is interfering with the electron's motion.

In an electric field $\mathbf{E}$, the system relaxes until $d\mathbf{v}/dt$ vanishes, in which case we have a steady-state nonequilibrium condition. From Equation (12-9) the velocity then satisfies

$$\mathbf{v} = -\frac{e\mathbf{E}}{\alpha} = -\frac{e\tau}{m}\mathbf{E}.$$ (12-12)

We have used Equation (12-11) to eliminate α.

If other electrons are present, a given electron is not able to relax to an arbitrary velocity or momentum because of the Pauli principle. In particular, in zero field most electrons certainly cannot relax to zero velocity as implied by Equation (12-10) because that state is probably already occupied. However, the energy loss processes can relax a system of electrons to the many-particle ground state, which involves a filled Fermi sphere centered on zero momentum.

Figure 12-18

Shift of the Fermi sphere by an electric field. The indicated shift is $\delta\mathbf{k} = -e\tau\mathbf{E}/\hbar$ for a field in the $-y$ direction. The shaded region shows the states that are newly occupied; the states in the cross-hatched region are now empty.

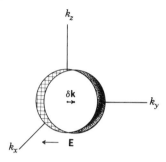

When there is an external field $\mathbf{E}$ every particle cannot have the same small nonzero velocity implied by Equation (12-12), since that violates the Pauli principle. What happens is that every particle in the Fermi sphere is *shifted* in velocity by an identical amount $\delta\mathbf{v}$ equal to the $\mathbf{v}$ of Equation (12-12). So we have

$$\delta\mathbf{v} = -\frac{e\tau}{m}\mathbf{E}. \qquad (12\text{-}13)$$

Thus, the entire Fermi sphere is shifted over by $\delta\mathbf{p} = m\,\delta\mathbf{v}$, as shown in Figure 12-18.

Note that because this shift of origin of the sphere is usually quite small, most of the momentum states that were occupied before the shift are still occupied. However, a thin crescent of states at the left-hand edge of the figure has been emptied and another at the opposite side has been filled.

This kind of effect is characteristic of degenerate Fermi systems; only the particles near the Fermi surface take part in the process. We know from Chapter 11 that the specific heat capacity involves absorption of energy by particles within $k_B T$ of the Fermi surface.

In general, a current density or intensity of a beam is particle density times particle velocity, as we illustrate in Problem 9 at the end of the chapter. Thus, the electrical current density is charge density times particle velocity:

$$J_E = -e\,\delta\rho\,v_F, \qquad (12\text{-}14)$$

where $\delta\rho$ is the density of particles involved in carrying the current; these are the particles in the crescent at the Fermi surface. All these particles move with velocity very nearly equal to the Fermi velocity v_F defined by Equation (12-8).

We can find the density $\delta\rho$ of current carriers by using the relation for the density of states developed in the first Detail at the end of Section 11-3. We have seen that the number of energy levels per unit energy in the range ε to $\varepsilon + \delta\varepsilon$ is

$$D(\varepsilon) = \frac{m^{3/2}}{\sqrt{2}\,\pi^2\hbar^3}\,V\varepsilon^{1/2}.$$

By introducing the Fermi energy ε_F, given by Equation (11-73), we can simplify $D(\varepsilon)$
considerably so that it reads

$$D(\varepsilon) = \frac{3}{4} \frac{\rho V \varepsilon^{1/2}}{\varepsilon_F^{3/2}}.$$

Since, for fermions, there are two particles per energy level, the *density of particles* per
unit energy $R(\varepsilon)$ is obtained by multiplying $D(\varepsilon)$ by $2/V$ to give

$$R(\varepsilon) = \frac{3}{2} \frac{\rho \varepsilon^{1/2}}{\varepsilon_F^{3/2}}. \tag{12-15}$$

The density of current carriers is then the density of particles per unit energy
evaluated at the Fermi energy times the energy width of the crescent:

$$\delta\rho = R(\varepsilon_F)\delta\varepsilon = \frac{3\rho}{2\varepsilon_F} \delta\varepsilon. \tag{12-16}$$

The energy width $\delta\varepsilon$ is related to the velocity shift of Equation (12-13) by

$$\delta\varepsilon = \frac{d\varepsilon}{dv}\bigg|_{v=v_F} \delta v = \frac{1}{2} \frac{d(mv^2)}{dv}\bigg|_{v=v_F} \delta v = mv_F\,\delta v. \tag{12-17}$$

Combining the results of the last few equations, we have for the current

$$J_E \sim ev_F \frac{\rho}{\varepsilon_F} mv_F \frac{e\tau}{m} E \sim \frac{e^2\tau\rho}{m} E, \tag{12-18}$$

where we have dropped constant factors of order unity because our estimate of $\delta\rho$
contains inaccuracies of that order. From Equation (12-5), the electrical conductivity
is

$$\kappa_E = \frac{e^2\tau\rho}{m}. \tag{12-19}$$

Although only the electrons near the Fermi surface contribute to the electrical current,
the density ρ corresponding to *all* electrons, even those away from the Fermi surface,
appears in Equation (12-19). This feature arises because the number of electrons that
are forced by the Pauli principle to be at the Fermi surface depends on the overall
density.

For classical particles (which need not obey the Pauli principle) Equation (12-12) *is*
valid. Furthermore, for such particles, the current is simply $J_E = -e\rho v$ instead of that
given by Equation (12-14). Curiously, the combination of these two equations *also*
leads to Equation (12-19), even though such a simplified derivation cannot be valid
for electrons.

In order to make contact with experiment we need to specify the behavior of τ in
Equation (12-19). The major effect involved in slowing the conduction electrons is
their interaction with vibrating positive ion cores in the solid. We see in Section 12-4
that if the ions form a perfect crystal, the electrons move through the lattice essentially

as an ideal Fermi gas, which has been the basic assumption of this section. It is when the lattice ions vibrate thermally that interactions with the electrons occur. A good picture of the process is obtained by considering an electron as absorbing and reemitting phonons as it moves through the lattice. The relaxation time τ can then be supposed to depend on n_ν, the number of phonons of frequency ν present in the lattice. We know from Section 11-5 that n_ν is given by

$$n_\nu = \frac{1}{e^{\beta h\nu} - 1}. \tag{12-20}$$

The fewer phonons that are present, the longer τ is expected to be, so that we can take

$$\tau \sim \left(\sum_\nu n_\nu \right)^{-1}.$$

For $k_B T/h \gg \nu_m$, a frequency characteristic of lattice vibrations introduced in Section 11-5, or equivalently for $T \gg \Theta_D$, the Debye temperature, also introduced in Section 11-5, we see from Equation (12-20) that

$$n_\nu = \frac{1}{1 + \beta h\nu + \cdots - 1} \approx \frac{k_B T}{h\nu}.$$

Thus, we expect $\tau \sim T^{-1}$ and $\kappa_E \sim T^{-1}$. The *resistivity* $r \equiv 1/\kappa_E$ then behaves like

$$r \sim T, \qquad T \gg \Theta_D.$$

Hence, a metal's electrical resistance decreases with decreasing temperature.

At temperatures low compared to the Debye temperature, one can show by the methods of Section 11-5 that

$$\sum_\nu n_\nu \sim \left(\frac{T}{\Theta_D} \right)^3, \qquad T \ll \Theta_D.$$

We might expect then that $r \sim T^3$; however, there are other details of the angular behavior of the scattering of electrons by phonons that make them even less effective in retarding the current at low temperature so that the actual result is

$$r \sim T^5, \qquad T \ll \Theta_D.$$

At very low temperature, the number of phonons may become so small that impurities in the crystal may account for the frictional forces on the conduction electrons. Such impurities may be foreign atoms, vacancies (absent atoms), and other imperfections in the crystal lattice. For sufficiently small samples the boundaries of the crystal may even be a factor in determining τ. In most of these cases, this contribution to τ is independent of temperature so that r approaches a constant at the lowest temperatures.

A somewhat similar treatment can be given to the thermal conductivity of metals. The heat current is now driven by a temperature difference. One end of the metal sample is hotter than the other. Although we assume that there are equal numbers of

Figure 12-19

Distribution of electrons in a cut through the Fermi sphere along the p_z axis. The distribution is that at a single point in real space. Particles with momentum $+p_z$ have just come from a hotter region; those moving with momentum $-p_z$ have just come from a cooler region. These distinctions show up in the slight differences in the spreads of the distributions around $\pm p_F$.

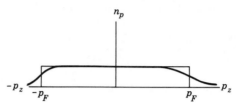

electrons flowing in both directions along the z axis in Figure 12-16, so that a charge does not build up at one end, those flowing from the hot end carry more energy. There is a flow of heat in the absence of a net flow of particles. (Actually there can be a thermoelectric effect in which temperature differences give rise to voltage differences, but we neglect such details here.)

Again it is the electrons near the Fermi surface that are responsible for thermal conduction. Consider the cross-sectional view of the Fermi sphere shown in Figure 12-19. The distribution function n_p in the figure is the number of electrons having momentum p. The electrons deep within the Fermi sphere have no effect on conduction because they come in matched pairs; for every one moving toward $+z$ there is one moving toward $-z$.

The heat current density is the number of heat carriers per unit volume times the velocity of a carrier times the heat carried per particle. The first of these factors is the density of conduction electrons ρ multiplied by the fraction $k_B T/\varepsilon_F$ of the particles that are thermally excited. Only this fraction can carry heat. So we have

$$\text{number of heat carriers per unit volume} = \rho \frac{k_B T}{\varepsilon_F}.$$

The second factor is the Fermi velocity,

$$\text{velocity of a carrier} = v_F.$$

The final factor, the heat carried per particle, is a bit more subtle to compute. The energy carried by a particle moving in the $+p_z$ direction is of order $\varepsilon_F + k_B T_+$ and that in the $-p_z$ direction is of order $\varepsilon_F + k_B T_-$, where T_+ is a bit larger than T_- as explained above. The net energy transported is then $k_B \delta T = k_B(T_+ - T_-)$. To compute δT we note that particles coming from the left are cooled down by collisions with other particles so that they tend toward the local temperature. Particles moving from the right are heated up by collisions. Since the average time between collisions for a particle is assumed to be about τ seconds, the distance between collisions is $\ell = v_F \tau$, a quantity known as the *mean free path*. We can assume then that a temperature change of order δT occurs every distance ℓ and that the gradient in temperature is

$$\frac{dT}{dz} = -\frac{\delta T}{\ell}$$

or

$$\delta T = -\tau v_F \frac{dT}{dz}.$$

Thus, the final factor we need is, in order of magnitude,

$$\text{net energy carried per particle} \sim -k_B \tau v_F \frac{dT}{dz}.$$

Combining these results gives the heat current density

$$J_T \sim -\frac{\rho}{\varepsilon_F} k_B^2 T \tau v_F^2 \frac{dT}{dz}. \tag{12-21}$$

If we set $\varepsilon_F \sim m v_F^2$ and compare with the general form, Equation (12-4), we find the coefficient of thermal conductivity to be

$$\kappa_T = \frac{\rho}{m} k_B^2 T \tau. \tag{12-22}$$

It was noted very early in the history of the study of metals that good electrical conductors are also good conductors of heat. If we take the ratio of κ_T to κ_E from Equations (12-19) and (12-22) we find

$$\kappa_T = \frac{k_B^2 T}{e^2} \kappa_E, \tag{12-23}$$

a rule named the *Wiedemann–Franz law* after G. H. Wiedemann and R. Franz (1853). Over certain temperature ranges Equation (12-23) is found to be obeyed quite well. However, the validity of the relation depends on an implicit assumption, namely, that the relaxation times, the τ's appearing in Equations (12-19) and (12-22), are the same. It turns out that the phonon processes that relax electrons in electrical conductivity are not identical to those involved in thermal conductivity. The τ's then are not quite the same and the Wiedemann–Franz law often breaks down.

In this section we have investigated how we can understand some of the properties of metals on the basis of a free-electron model. The model is found to work pretty well. But surely electrons must interact strongly via the Coulomb force with the ion cores and not just weakly with core vibrations. Interaction with other electrons ought to be present as well. How are we to understand the origins of a free-electron model? The answer to this question is the subject of Section 12-4.

Example

Experimentally, it is found that the resistivity of copper at room temperature is $r = 1/\kappa_E = 1.7 \times 10^{-8} \ \Omega \cdot m$. By using Equation (12-19), we can find a value for the relaxation time τ for electrons in a metal. The average separation of copper atoms is about 0.2 nm and each atom contributes one conduction electron, giving a density of $\rho \approx 1/(2 \times 10^{-10} \text{ m})^3$. We then have

$$\tau = \frac{m}{e^2 \rho r} \approx \frac{9.1 \times 10^{-31} \text{ kg}}{(1.6 \times 10^{-19} \text{ C})^2 \left[1/(2 \times 10^{-10} \text{ m})^3\right](1.7 \times 10^{-8} \, \Omega \cdot \text{m})}$$

$$\approx 2 \times 10^{-14} \text{ s}.$$

In Problem 8 at the end of the chapter the reader is asked to compute the mean free path, the distance an electron travels in this time.

12-4   Energy Bands in Solids

The sodium atom has a single $3s$ electron outside a neon-like closed electronic shell given by $1s^2 2s^2 2p^6$. When a large number of sodium atoms are assembled they form a bound metallic crystal. We attempt to understand this result by examining a simple crystal model. Our primary result is that the outermost atomic states no longer have the energy values found in atomic sodium but are split into bands of energy levels. These states correspond to delocalized electrons that are no longer fixed to individual sodium atoms but are able to move through the entire crystal. The electrons in states of a band behave, in many ways, like free particles as claimed in Section 12-3.

We do find fundamental alterations to the free-particle picture. Between bands of allowed energy states there are forbidden energy regions called band gaps. The existence of these gaps leads to the possibility of understanding the nature of insulators and semiconductors as well as metals.

Just as the quantum tunneling of an electron from one hydrogen atom to the other results in the binding of H_2^+, so also the electrons hopping among the many ions of a metal can lower the total energy of the system.

Equation (10-7) illustrates what happens to the electron energy levels when two hydrogen nuclei are placed close together. The tunneling of the electron through the small barrier midway between the two nuclei results in a splitting of the $1s$ atomic level into two unequal energy levels. The wave functions corresponding to these states no longer describe an electron situated on just one or another atom; they are delocalized. Nevertheless, they are, to a good approximation, combinations of the localized atomic states. The energy banding in metals is basically the same effect; tunneling splits the levels.

Figure 12-20 illustrates the potential energy seen by a single electron in a sodium crystal. One of the crystal edges is shown at the right end of the figure. Note that the potential energy inside the crystal is lower than its value at the edge or at infinity. The

---

Figure 12-20
Schematic diagram of the potential energy curve seen by an electron in metallic sodium. Also shown are the *atomic* $2s$ and $3s$ levels of the sodium atom.

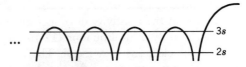

reduced interparticle barrier is what allows the electron to move from nucleus to nucleus. At the edge the electron sees the higher barrier and is unable to escape from the interior.

The core electrons, those in the $1s$, $2s$, or $2p$ states, see hardly any change in the potential energy when the atoms are put together and so their energy levels are affected very little. Only the $3s$ and higher levels are changed.

We now consider the details of a model in which an electron moves along the x axis in just one dimension. The potential energy function centered at one site R_n is the Coulomb potential energy

$$V_n = V(x - R_n) = -\frac{e^2}{4\pi\varepsilon_0|x - R_n|}. \tag{12-24}$$

Note that the charge on the sodium ion is just $+e$. The electron has a total potential energy $W(x)$ given as a sum of all such Coulomb functions as

$$W(x) = V_1 + V_2 + \cdots + V_N \tag{12-25}$$

for N ions.

To find an approximate wave function for an electron having this potential energy we continue to take clues from the analysis of the H_2^+ molecule in Chapter 10. There we have used two wave functions, one of which is the sum of, and the other the difference between, the two atomic functions. Here we assume that the eigenfunction $\psi(x)$ for a single electron is a generalization of this. We try the form

$$\psi(x) = C_1\phi_1(x) + C_2\phi_2(x) + \cdots + C_N\phi_N(x), \tag{12-26}$$

where $\phi_n(x) = \phi_{3s}(x - R_n)$ is the atomic $3s$ function for an electron at the ion core centered at R_n. The constants C_n are to be determined.

If our model crystal has boundaries like a realistic one, the C_n values at and near the right and left edges of the crystal have a fundamentally different character from those near the center. In a very large crystal, say, one with 10^8 sites along an edge, these boundary positions play a relatively unimportant role in determining the energy of a given state; an electron spends only a negligibly small percentage of its time near an edge. To remove the complications of these edges without fundamentally affecting the physics, we use instead what are known as *traveling-wave* or *periodic boundary conditions*. Rather than demanding that the electron wave function be zero to the left of site 1 or to the right of site N, we assume that the wave function begins to repeat itself when it reaches a boundary. Thus, as shown in Figure 12-21, when the electron passes site N by one interatomic distance, denoted by a, it finds itself back at site 1. Similarly, if the electron goes to the left of site 1 it finds itself at site N, and so on. This arrangement can be thought of as the crystal having itself wrapped in a circle and tied head to tail as shown in Figure 12-22. While such a circular arrangement is difficult to visualize in three dimensions, we need not worry about it because it is only a mathematical artifice used to make the solution easier to find. The type of boundary condition used, as noted above, makes negligible difference to the energy levels for a very large system; it does make a large difference in the amount of work necessary to find the energy levels.

Figure 12-21
Periodic boundary conditions. An electron moving to the right of the site N finds itself back at site 1.

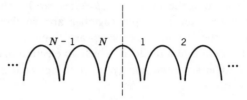

The time-independent Schrödinger equation becomes

$$[T + V_1 + V_2 + \cdots + V_N]\psi(x) = E\psi(x),\qquad(12\text{-}27)$$

where the kinetic energy operator is $T = -(\hbar^2/2m)\ d^2/dx^2$. We can make an immediate simplification if we use the fact that each ϕ_n is a $3s$ function corresponding to the potential energy V_n located at R_n; that is, we have

$$[T + V_n]\phi_n = E_0\phi_n,\qquad(12\text{-}28)$$

where E_0 is the energy of the $3s$ state. Then when the quantity in square brackets in Equation (12-27) acts on ϕ_1 in $\psi(x)$, we get $[E_0 + V_2 + \cdots + V_N]$. When it acts on ϕ_2 we get $[V_1 + E_0 + V_3 + \cdots + V_N]$, and so on.

Thus, Equations (12-26) and (12-27) give us the expression

$$C_1(E_0 + V_2 + \cdots + V_N)\phi_1 + C_2(V_1 + E_0 + V_3 + \cdots + V_N)\phi_2$$
$$+ \cdots + C_N(V_1 + \cdots + V_{N-1} + E_0)\phi_N$$
$$= E(C_1\phi_1 + C_2\phi_2 + \cdots + C_N\phi_N).\qquad(12\text{-}29)$$

What we now want to do is to develop algebraic equations for the C_n's from this result. The way to do this is to multiply Equation (12-29) by ϕ_n and then integrate over all x. Other equations are found by multiplying by other ϕ's. A number of

Figure 12-22
N ion cores set in a ring to give periodic boundary conditions on the wave function of an electron hopping from one ion to another.

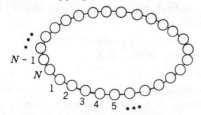

simplifications and approximations are possible in the resulting equations. Since, for example, ϕ_1 is large only in the neighborhood of site 1, and V_2 in the neighborhood of site 2, and so on, we expect any integral involving more than one site to be relatively small. These integrals are analogous to the overlap integrals encountered in Chapter 10. Integrals involving three different sites involve the "leakage" of a function into a region two sites away so that, for example, we have

$$\int \phi_1 V_1 \phi_3 \, dx \approx 0 \quad \text{or} \quad \int \phi_1 V_2 \phi_3 \, dx \approx 0.$$

Integrals involving only two different sites that are nearest neighbors cannot normally be neglected. Thus, the integral

$$J_{12} = \int \phi_1 V_1 \phi_2 \, dx \tag{12-30}$$

is a measure of the probability that an electron hops from one site to its neighbor, and must be kept in our equations. All near-neighbor J's have the same value, which we denote simply as J:

$$J_{12} = J_{21} = J_{n\,n-1} = J. \tag{12-31}$$

There are two other near-neighbor overlap integrals that arise in the operations we are considering. These are of the form

$$Q = \int \phi_2 (V_1 + V_3) \phi_2 \, dx$$

and

$$I = \int \phi_2 \phi_1 \, dx = \int \phi_1 \phi_2 \, dx.$$

Q is the energy shift of a site-2 electron due to its interactions with ion cores 1 and 3. I arises from the terms on the right side of Equation (12-29); it is an interference term that occurs in the normalization integral of the eigenfunction ψ. While Q and I are corrections that ought to be included in any rigorous calculation of the electronic energies, they do not change the basic physics and we simply drop them in this qualitative discussion.

Finally, we assume that each of the atomic functions ϕ_n is normalized as

$$\int \phi_n^2 \, dx = 1.$$

It follows easily from the above discussion of overlap integrals that the result of multiplying Equation (12-29) by ϕ_n and integrating over x is to give the equality

$$C_n E_0 + C_{n-1} J + C_{n+1} J = E C_n$$

or

$$C_n (E_0 - E) + (C_{n-1} + C_{n+1}) J = 0 \tag{12-32}$$

for $n = 1, 2, 3, \ldots, N$.

These results have a very simple structure. If $J = 0$, the solution for the energy is just E_0, the localized atomic energy. However, quantum tunneling of the electron to nearest-neighbor sites is represented by the overlap integral J. Each C_n is coupled to those of the neighboring sites, C_{n-1} and C_{n+1}, by these small hopping terms. Equation (12-32) is easily solved. While we have N homogeneous equations in N unknowns, the usual technique of setting the determinant of the coefficients to zero is an unwieldy way to proceed. A simpler procedure is more useful to us. We anticipate that the electron hopping on a ring of N sites has a solution behaving like a traveling wave. We can guess that such solutions are provided by choosing the form

$$C_n = Ce^{ikR_n}. \tag{12-33}$$

The variable k turns out to be the wave vector for the electronic motion, and C is a normalization constant. By substitution we show that this choice does indeed satisfy Equation (12-32).

The ion sites are taken to be at $0, a, 2a, \ldots, (N-1)a$, or

$$R_n = (n-1)a, \qquad n = 1, 2, \ldots, N, \tag{12-34}$$

where a is the distance between lattice sites. With the *ansatz* of Equation (12-33) plugged into Equation (12-32), we find

$$Ce^{ik(n-1)a}(E_0 - E) + C(e^{ik(n-2)a} + e^{ikna})J = 0$$

or

$$(E_0 - E) + (e^{-ika} + e^{ika})J = 0.$$

The guessed solution, Equation (12-33), has worked and the result for the energy is

$$E \equiv E_k = E_0 + 2J\cos ka. \tag{12-35}$$

The energy levels E_k are still close to E_0, but now there is a tunneling correction dependent on J. We also have a new quantum number k that determines the precise value of the energy level E_k. When the sodium ions are far apart the electron can be localized on any one of the N ions with energy E_0. There are N energy levels all with precisely the same energy. This equality has now been removed by the possibility of tunneling of an electron from one ion to the next, so that we now have N distinct energy levels each characterized by a different value of k. Before we can further understand the nature of these levels we need to determine the values taken on by the quantum numbers k.

The k's are determined by the periodic boundary conditions. When we go to site $N + 1$, we assume that we are back at site 1. From Equation (12-33) this implies

$$C_{N+1} = Ce^{ikNa} = C_1 = C$$

or

$$e^{ikNa} = 1, \tag{12-36}$$

so that

$$k = \frac{2\pi}{Na}\ell, \qquad \ell = 0, \pm 1, \pm 2, \ldots. \tag{12-37}$$

Figure 12-23

Plot of the electronic energy levels E_k for a one-dimensional model of a metal. We assume $J < 0$ and find the width of the band to be $4|J|$. The band is centered on the atomic energy E_0.

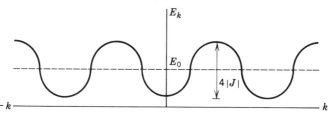

We see that for N large, the allowed k values are very densely spaced just as are the k values of Equation (5-30) or (12-6) for a particle in a large one-dimensional box of size $L = Na$. However, these k values are not precisely equal to those of the particle in a box treated in Chapter 5. The difference is in the factor of 2 in Equation (12-37) and the fact that negative as well as positive ℓ's are allowed. These distinctions arise only because of the different boundary conditions used in the two cases. In Chapter 5 the infinite potential energy well forces the wave function to be zero at the ends of the box, thereby producing standing waves in the box. In the present case we have used traveling-wave boundary conditions. If we use these conditions on a *free* particle in a "box," we indeed get k's identical to those of Equation (12-37) as we find in Problem 13 at the end of the chapter. Positive ℓ's give positive k's, which represent waves traveling to the right; negative k's give waves moving to the left. However, since in the standing-wave case only positive ℓ values occur, we now have twice as many ℓ's. Thus, the factor of 2 is necessary in Equation (12-37) to make the spacing between adjacent k's twice as large. The overall density of k values comes out the same in the two cases. This is enough to ensure that the properties of a large system are independent of which boundary conditions are used.

This relation between quantum numbers of a free particle and those of an electron hopping through a crystal of ion cores is one of the bases for the success of the free-electron theory of metals discussed in Section 12-3. Using k as a continuous variable, we plot the values of E_k given by Equation (12-35) in Figure 12-23. Note that we now have an energy band of allowed energy states.

Since, in the case of highly separated ions, we start out with N identical energy levels corresponding to the electron being localized on any one of the N ions, we still have N distinct states when the ions are close enough for tunneling to take place. However, there are an infinity of k values listed in Equation (12-37). On the other hand, the energy levels of Equation (12-35) are periodic in k. What we intend to show is that certain k values are exactly equivalent to other k values so the set given by Equation (12-37) has multiple redundancy. To see this redundancy we take as our basic set the values

$$k = \frac{2\pi\ell}{Na}, \qquad \ell = \begin{cases} 0, 1, 2, \ldots, N/2 - 1 \\ -1, -2, -3, \ldots, -N/2, \end{cases} \qquad (12\text{-}38)$$

where we have assumed N is even so $N/2$ is an integer. There are exactly N k values in the set given. If k is inside this basic set, the vector $k \pm G$, with G defined by

$$G = \frac{2\pi}{a},\tag{12-39}$$

is always outside the set. This is easily seen by considering several explicit choices of wave vector. For $k = 0$ we see that $k + G = (2\pi/Na)N$ corresponds to $\ell = N$, which is certainly outside the k set given by Equation (12-38). For $k = (2\pi/Na)(-N/2)$ we have

$$k + G = \frac{2\pi}{Na} \cdot \frac{-N}{2} + \frac{2\pi}{Na} \cdot N = \frac{2\pi}{Na}\frac{N}{2},$$

corresponding to $\ell = N/2$, which is just barely outside the set of allowed k values of Equation (12-38). G is the width of the allowed k-vector set. Next, consider the eigenfunctions for the various k values. Upon substituting the C_n values of Equation (12-33), the general eigenfunction $\psi(x)$ of Equation (12-26) is

$$\psi_k(x) = C\left[\phi_1 + e^{ika}\phi_2 + e^{i2ka}\phi_3 + \cdots + e^{i(N-1)ka}\phi_N\right].\tag{12-40}$$

Consider this eigenfunction for a k value outside the basic set:

$$\psi_{k+G}(x) = C\left[\phi_1 + e^{i(k+G)a}\phi_2 + \cdots + e^{i(N-1)(k+G)a}\phi_N\right].$$

Since, by Equation (12-39), a typical factor in this function obeys

$$e^{inGa} = e^{in2\pi} = 1,$$

we see that

$$\psi_{k+G}(x) = \psi_k(x).\tag{12-41}$$

The k values outside the basic set do *not* lead to distinct wave functions but simply reproduce those corresponding to k values in the basic set. Obviously, we also have $E_{k+G} = E_k$. Additional redundant sets of k's can be generated by considering multiples of the G value of Equation (12-39).

The basic set of k values is said to be inside the *first Brillouin zone*, a concept devised by L. Brillouin. Other k values are in second or higher Brillouin zones. The vectors in each zone are perfectly equivalent to those in any other. The unit G and its multiples make up a set of wave vectors known as the *reciprocal lattice* that we have discussed in Section 12-2. The values of these vectors in three dimensions are given in Equation (12-3). The G's that we have been considering can also be defined by the relation

$$e^{iG_m R_n} = 1,\tag{12-42}$$

which is equivalent to the one-dimensional version of Equation (12-3). The fact that the reciprocal lattice vectors, originally associated with Bragg scattering, also arise in the study of the states of electrons in crystals is no coincidence. The relation is explored in more detail below.

We now examine the energy levels for an electron in a crystal a bit more closely. The energies of Equation (12-35) are centered on the atomic energy level E_0, which in the case of sodium metal is the $3s$ energy. They are spread in a *band* about E_0. If we

Figure 12-24

Energy band for k vectors in the first Brillouin zone.

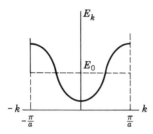

assume that this band has $J < 0$, the energy corresponding to $k = 0$ has the lowest energy, namely, $E_0 - 2|J|$. The largest energy occurs at the two edges of the Brillouin zone where $k = \pm\pi/a$, because there we have

$$-2|J|\cos ka = -2|J|\cos \pi = 2|J|.$$

The energies for k values in the first Brillouin zone are shown in Figure 12-24.

The energy minimum around $k = 0$ is parabolic so that we are able to show the relation of our band energy to free-particle states. To see this, expand the band energies of Equation (12-35) for k around $k = 0$. We use the Taylor series result

$$\cos ka = 1 - \frac{(ka)^2}{2!} + \frac{(ka)^4}{4!} + \cdots .$$

Keeping only the quadratic term leads to

$$E_k = (E_0 - 2|J|) + |J|a^2k^2, \qquad k \sim 0.$$

The last term has the form of a free-particle energy,

$$\varepsilon_k = \frac{\hbar^2 k^2}{2m*}, \tag{12-43}$$

where the electron moves as if it had an *effective mass* identified by the relation

$$|J|a^2 = \frac{\hbar^2}{2m*}$$

or

$$m* = \frac{\hbar^2}{2|J|a^2}. \tag{12-44}$$

Note that this mass has nothing directly to do with the real electron mass; it depends mainly on the tunneling integral J. For small k the particle moves like a free particle but its inertia depends on the overlap of neighboring atomic wave functions and the resulting tunneling rate rather than the real mass. In some circumstances the effective mass can even be less than the true mass m. However, if the tunneling is large because the atomic energy level happens to be very close to the top of the potential energy barrier, the effective mass is nearly equal to the real mass m.

When the tunneling is very large it may be incorrect to start off with an assumed eigenfunction of the form given by Equation (12-26). Such a wave-function *ansatz* is known as the *tight-binding approximation*, in which it is assumed that the atomic energy is a good first estimate of E_k and that the electronic wave function is reasonably close to being an atomic orbital when the electron is in the immediate neighborhood of an ion core. If tunneling is too large this approximation breaks down and one may need to start off with the *nearly-free-electron model*. In this model the basic wave function is that of a free particle and the interaction with the ion cores is considered a perturbing effect on that assumption. Other approaches more nearly exact than either of these are commonly used as well.

When an electron is in one of the higher energy states near the zone boundaries ($k \approx \pm \pi/a$), the energy E_k is again quadratic but with a negative curvature and a *negative* effective mass. Pushing on the electron causes it to slow down! We can understand such an effect much more easily in the nearly-free-electron picture. We note that near the zone boundary the electron's de Broglie wavelength is given by

$$\lambda = \frac{2\pi}{k} \approx \frac{2\pi}{\pi/a} = 2a.$$

The wavelength is now commensurate with the lattice so that strong backscattering is present, and, more importantly, the incident and scattered waves strongly interfere. To see this, consider the condition of strong reflection from two near-neighbor nuclei, as illustrated in Figure 12-25. Two rays of the wave incident from the right are reflected, one from site 1 and one from site 2. There is strong reflection when the wavelengths are such that the two waves' components interfere constructively. This occurs when their path difference, which is $2a$, is equal to one wavelength, or $\lambda = 2a$. Thus, as the electron's k vector approaches the zone boundary, the electron is more and more strongly reflected. The effect is equivalent to an electron possessing a negative effective mass. The electron gains energy but slows down as it does so. Obviously, this discussion is just the one-dimensional version of Bragg scattering where the scattering plane has reduced to a single particle. It is easy to see that the momentum transferred to the lattice in this case satisfies Equation (12-3). Values of k outside the Brillouin zone are not needed because every time an electron reaches a zone boundary it is backscattered to the other end of the zone. Its k value is trapped in the zone.

The above discussion illustrates an important point; the quantity $\hbar k$ of an electron in a crystal often behaves like a momentum but really should not be considered to be one under all circumstances. This situation arises because pushing on the electron may cause the entire crystal, and not just the electron, to gain momentum via interference effects and interaction with the lattice. The vector $\hbar k$ is often called a *quasimomentum* for this reason. Nevertheless, $\hbar k$ behaves sufficiently like a momentum that the free-particle picture often remains reasonable.

Up to now only a single energy band has been considered. We must consider bands corresponding to other atomic orbitals. The energy band under discussion for sodium has been the one spread out around the $3s$ state. Lower energy levels, the $1s$, $2s$, and

Figure 12-25

Electron reflection from the ions of a one-dimensional crystal. The waves are shown at an angle for clarity. The rays reflected from sites 1 and 2 are in phase when the path difference $2a$ is exactly one wavelength.

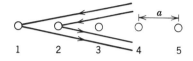

Figure 12-26

Some energy bands in a one-dimensional model of a metal. Note that the value of J_m increases with m so that the higher band is wider. The forbidden region between bands is called a band gap.

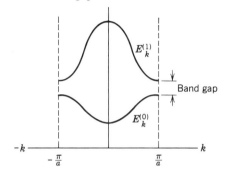

$2p$ states, are so tightly bound that tunneling rarely takes place; the crystal energies are almost unchanged from their atomic values. On the other hand, the higher states, $3p$, $3d$, and so on, are much more loosely bound. Tunneling is very large between different sites. These states form bands that are broader than the $3s$ band. In general, we have several energy bands given by

$$E_k^{(m)} = E_m + 2J_m \cos ka, \qquad (12\text{-}45)$$

where m is the band number and refers to the atomic state, $3s, 3p, \ldots$, on which the band is based.

Note that the overlap integral J_m in Equation (12-45) depends on the band number m. This is in accordance with the idea that the tunneling rate should increase as m increases. In Figure 12-26 we illustrate a typical band structure for two of the bands in our simple sodium model. The higher band is broader than the lower one. We have assumed that the sign of J_m alternates as we proceed from one band to the next. This is not unusual but we do not justify that assumption here.

A further feature of the spectrum is the existence of *band gaps*. These are the forbidden energy regions between bands. Obviously, for a single atom, the forbidden energy regions dominate the allowed regions, which are the discrete levels. The opposite is usually true for the higher bands in a crystal; states where hopping is easily possible have allowed regions that are much wider than the gaps. Band gaps are of fundamental importance to the understanding of metals, insulators, and semiconductors as we see in Section 12-5.

Example

We consider a metal for which $a = 0.2$ nm and m^* is the true electron mass. The tunneling constant J is

$$|J| = \frac{\hbar^2}{2ma} = \frac{\left(1.1 \times 10^{-34} \text{ J} \cdot \text{s}\right)^2}{2\left(9.1 \times 10^{-31} \text{ kg}\right)\left(2 \times 10^{-10} \text{ m}\right)^2 \left(1.6 \times 10^{-19} \text{ J/eV}\right)} = 1.0 \text{ eV}.$$

The width of the band associated with this J is then $4|J| = 4$ eV. The time Δt for an electron to hop from one ion core to the next can be estimated from the uncertainty principle as

$$\Delta t = \frac{\hbar}{|J|} = \frac{1.1 \times 10^{-34} \text{ J} \cdot \text{s}}{(1.0 \text{ eV})(1.6 \times 10^{-19} \text{ J/eV})} = 6.6 \times 10^{-16} \text{ s}.$$

12-5   The Band Theory of Metals, Insulators, and Semiconductors

In Section 12-4 we have constructed a picture of the energy spectrum of an electron in a crystal with its bands and band gaps. When we place many electrons into these states the nature of the resulting substance, that is, whether it is a metal, an insulator, or a semiconductor, depends on the details of the band structure and, even more importantly, on how the electrons fill the allowed states.

In order to investigate how the electrons go about filling the bands we go back to our sodium model of Section 12-4. There is a single 3s electron per sodium atom for a total of N electrons. The 3s band contains N states. Since each state can hold one spin up and one spin down, the band is half-filled as illustrated in Figure 12-27a. If we have two electrons per atom, the band is completely filled, with two electrons for every state, as shown in Figure 12-27b. (While we might think that this is an apt description of magnesium, the element next to sodium in the periodic table, it does not quite work out that way, as is explained shortly.)

If we refer back to Figure 12-18 we see that when electrical conduction takes place the Fermi sphere is shifted slightly in momentum space. On an energy level diagram like the ones we have been using, this shift is shown as illustrated in Figure 12-28. There are slightly more electrons with positive k vectors than with negative k and so there is a general drift of electrons in the positive k direction; a current is set up and we have a *metal*. On the other hand, with a full band as in Figure 12-27b, there is no allowed energy region for the tilt to occur because of the gap. Obviously, materials with full bands like this one are *insulators*.

Figure 12-27
Filling of bands by electrons. A solid with one electron per atom has a half-filled band as indicated by the cross-hatching in (a). This material is a metal. If there are two electrons per atom the band is full, as shown in (b), and the substance is an insulator.

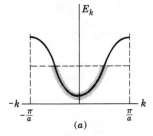 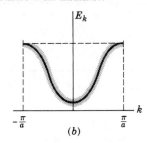

Figure 12-28
Effect of an electric field on the occupation of the energy levels of a band. The application of the field produces a shift so that there is an excess of electrons having positive k values.

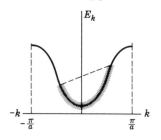

Figure 12-29
Example of overlapping bands in which a substance with two electrons per atom turns out to be a metal instead of an insulator. Generally, k_a and k_b are two distinct directions in three-dimensional k space rather than just $+k$ and $-k$.

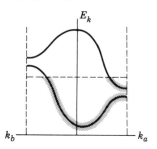

Based on this argument we expect magnesium to be an insulator when it actually turns out to be metal, having relatively low conductivity. The explanation of this phenomenon is found in the peculiar structure of the bands in magnesium. Bands can sometimes overlap in certain regions of k space. An example of such a situation is shown in Figure 12-29. There are now a few electrons in a nonfull band that can provide electrical conduction. Overlapping bands occur in magnesium and several other substances that qualify as metals. On the other hand, diamond has enough electrons to fill a band without any overlap and is an example of an insulator.

All of the above considerations are made at zero temperature. However, if a band gap is narrow, thermal excitations from a full band may be sufficient to provide a few current-carrying electrons. Materials with such thermally activated conductivity are known as *semiconductors*. Examples of these include silicon and germanium. Because of the great importance of semiconductors to technology, we spend the next two sections discussing them in detail.

Example

We list the band gaps of several materials below. Except for diamond all are considered to be semiconductors.

Material	Band Gap (eV)
C (diamond)	5.5
CdS	2.4
Si	1.1
Ge	0.7
Te	0.3
InSb	0.2

Figure 12-30

Valence and conduction bands in an intrinsic semiconductor. A few electrons are thermally excited out of the valence band. Also shown is the distribution function $n(\varepsilon)$. Note that the Fermi energy falls in the band gap.

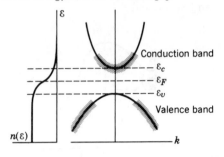

12-6 Semiconductors

Perhaps no technological development has contributed as much to our lives in recent years as the application of semiconductor physics. The entire computer revolution is based on this technology. We discuss these applications in Section 12-7 but here we outline some of the basic ideas of semiconductors.

We have seen in Section 12-5 that semiconductors are characterized by small band gaps so that charge carriers can be thermally activated from the filled band. This filled band contains the electrons from the outermost shells of each atom and so is called the *valence band*. The normally empty band to which charge carriers can be excited is then the *conduction band*.

In an *intrinsic* semiconductor, all the electrons in the conduction band have been thermally excited from the filled valence band. Such a situation occurs in pure crystals. However, semiconductors containing impurity atoms may have localized energy levels within the normally forbidden band gaps. The activation energy of the electron in these states can be considerably less than that of valence electrons. Such a *doped* semiconductor is said to be of the *extrinsic* type. While most technical applications involve extrinsic semiconductors, we first consider the intrinsic type.

The number of charge carriers in the conduction band is given by the usual Fermi function

$$n_k = \frac{1}{e^{\beta(\varepsilon_k - \varepsilon_F)} + 1},\qquad (12\text{-}46)$$

where ε_F is the Fermi energy. Figure 12-30 shows the almost-filled valence band and the partially filled conduction band. The Fermi energy is shown as lying *within* the band gap. This may be a bit of a surprise since, for a gas of free particles at $T = 0$ K, the Fermi level is the energy of the highest filled level, and for $T > 0$ K it is the energy for which n_k has dropped to one-half of its maximum value. But the Fermi energy is more fundamentally the Lagrange multiplier in our statistical treatment of Chapter 11, which determines that the sum over all k of Equation (12-46) comes out to be equal to the number of particles N. We now show that this condition puts ε_F *in* the band gap.

Assume that the top of the valence band is quadratically inverted downward at $k = 0$, indicating negative effective mass. This could just as well be a region of a band out near the Brillouin zone boundary as in Figure 12-27b, rather than at $k = 0$ as in Figure 12-30. The results are the same. The energy ε_c is defined as the bottom of the conduction band, and ε_v is the top of the valence band as shown in the figure. Then the number of particles in the condition band is

$$N_c = \sum_{\substack{k \\ (\varepsilon_k > \varepsilon_c)}} n_k. \qquad (12\text{-}47)$$

If there are N electrons in the valence band at $T = 0$ K, the number of these electrons excited out of the valence band by thermal agitation is N minus the number remaining in the band:

$$N_v = N - \sum_{\substack{k \\ (\varepsilon_k < \varepsilon_v)}} n_k = \sum_{\substack{k \\ (\varepsilon_k < \varepsilon_v)}} (1 - n_k), \qquad (12\text{-}48)$$

where the last equality comes from the fact that the total number of states in the conduction band is equal to the total number of electrons in it at $T = 0$ K when it is full. In an intrinsic semiconductor all the thermally excited electrons must come from the valence band so that we have

$$N_c = N_v. \qquad (12\text{-}49)$$

As we see next, this condition is enough to determine the Fermi energy ε_F.

We can evaluate N_c and N_v easily if we assume that ε_c and ε_v are much more than $k_B T$ away from ε_F. This is often the case for real systems. Since $\varepsilon_k > \varepsilon_c$ in the conduction band, $\beta(\varepsilon_k - \varepsilon_F) \gg 1$ so the exponential dominates the 1 in the denominator of Equation (12-46). We then have

$$n_k \to e^{-\beta(\varepsilon_k - \varepsilon_F)}, \qquad \varepsilon_k > \varepsilon_c. \qquad (12\text{-}50)$$

From Equation (12-47), it follows that

$$N_c = \sum_{\substack{k \\ (\varepsilon_k > \varepsilon_c)}} e^{-\beta(\varepsilon_k - \varepsilon_F)} = e^{\beta(\varepsilon_F - \varepsilon_c)} \sum_{\substack{k \\ (\varepsilon_k - \varepsilon_c > 0)}} e^{-\beta(\varepsilon_k - \varepsilon_c)}. \qquad (12\text{-}51)$$

The last sum on k is easily done by changing to an integral over energy. We know how to do this when energy depends quadratically on wave number k (like a free particle) as it does near the bottom of the conduction band. We have measured energies from ε_c so that the sum in Equation (12-51) is

$$Z = \sum_{\varepsilon > 0} e^{-\beta \varepsilon},$$

which is just the classical partition function of Equation (11-18), evaluated in Equation (11-46). The resulting formula is

$$Z_c = 2V \left(\frac{2\pi m_c k_B T}{h^2} \right)^{3/2}. \qquad (12\text{-}52)$$

In Equation (12-52) we replace the free-electron mass by the effective mass m_c associated with the bottom of the conduction band and we denote the corresponding sum by Z_c. There is an extra factor of 2 in Equation (12-52) to account for the spin degeneracy of each energy level. We end up with the result

$$N_c = e^{\beta(\varepsilon_F - \varepsilon_c)} Z_c. \tag{12-53}$$

To treat the valence band we do a bit of manipulation of Equation (12-48). We write

$$1 - n_k = 1 - \frac{1}{e^{\beta(\varepsilon_k - \varepsilon_F)} + 1} = \frac{e^{\beta(\varepsilon_k - \varepsilon_F)}}{e^{\beta(\varepsilon_k - \varepsilon_F)} + 1} = \frac{1}{e^{-\beta(\varepsilon_k - \varepsilon_F)} + 1}. \tag{12-54}$$

From this expression we see that the *missing* electron population in the valence band is described by a sort of Fermi function, but with a *minus* sign in the exponential instead of a plus sign as in Equation (12-46). However, this minus sign is rather natural since the valence band, as we have drawn it in Figure 12-30, is an inverted quadratic function anyway. Next, note that as ε_k becomes more negative, $\varepsilon_F - \varepsilon_k$ becomes larger positively, and if ε_F is much greater than ε_k for all ε_k below ε_v, then

$$1 - n_k \to e^{-\beta(\varepsilon_F - \varepsilon_k)}$$

and

$$N_v = e^{-\beta(\varepsilon_F - \varepsilon_v)} \sum_{\substack{k \\ (\varepsilon_v - \varepsilon_k > 0)}} e^{-\beta(\varepsilon_v - \varepsilon_k)}.$$

Changing the energy variable from ε_k to $\varepsilon_v - \varepsilon_k$ gives us precisely the same sum as for the conduction band, except that the curvature of the band is determined by an effective mass m_v (a *positive* quantity here). We get

$$N_v = e^{-\beta(\varepsilon_F - \varepsilon_v)} Z_v,$$

where

$$Z_v = 2V \left(\frac{2\pi m_v k_B T}{h^2} \right)^{3/2}.$$

Equating N_c and N_v, according to Equation (12-49), gives an expression for ε_F:

$$e^{\beta(\varepsilon_F - \varepsilon_c)} Z_c = e^{-\beta(\varepsilon_F - \varepsilon_v)} Z_v.$$

Taking the logarithm of both sides gives

$$\beta(\varepsilon_F - \varepsilon_c) + \tfrac{3}{2} \ln m_c = -\beta(\varepsilon_F - \varepsilon_v) + \tfrac{3}{2} \ln m_v$$

or

$$\varepsilon_F = \tfrac{1}{2}(\varepsilon_c + \varepsilon_v) + \tfrac{3}{2} k_B T \ln(m_v/m_c). \tag{12-55}$$

We see that if $T \to 0$ K or if $m_v = m_c$, the Fermi energy ε_F is halfway between the conduction and valence bands. If $m_v \neq m_c$ then ε_F is displaced very slightly from the halfway point at $T > 0$ K. For simplicity in the following we assume that $m_v = m_c$ and that ε_F is halfway between bands.

From Equation (12-53), the number of charge carriers in the conduction band is now seen to be

$$N_c = e^{-\beta E_g/2} Z_c,$$ (12-56)

where

$$E_g = \varepsilon_c - \varepsilon_v$$ (12-57)

is the *band gap*. We can see from Equation (12-56) that if E_g gets too large, the number of charge carriers rapidly drops toward zero.

We might think that Equation (12-56) enumerates all possible charge carriers. However, recall that the valence band is unable to conduct electricity only if it is completely full. When some electrons are thermally excited out of the valence band, it can again contribute to the conductivity. It is a curious feature that it is much more convenient to speak of conduction not by the electrons in the nearly full valence band, but by the *holes* left behind when electrons leave the band. The existence of conduction by these holes is of vital importance to the workings of semiconductor devices.

To begin to understand the concept of holes we again start off with a completely filled valence band. We turn on an electric field. The current density carried by the band is

$$\mathbf{J} = -\frac{e}{V} \sum_k \mathbf{v}_k,$$ (12-58)

where e is a positive number representing the charge magnitude of an electron and where $\mathbf{v}_k$ is the velocity of an electron in state k. We need a bit of discussion about this velocity.

The velocity of an electron having energy ε_k is found from the relation

$$v_{kx} = \frac{1}{\hbar} \frac{d\varepsilon_k}{dk_x}$$ (12-59)

with similar equations for the y and z components. When $\varepsilon_k = \hbar^2 k^2/2m$, as for a free particle, $v_{kx} = \hbar k_x/m$ is just the normal relation between velocity and momentum. When ε_k has a more complicated dependence on k, as can happen in some parts of a band, we justify Equation (12-59) by considering the electron as a wave. The usual group velocity of a wave is derived from the wave's angular frequency ω_k by

$$v_{kx} = \frac{d\omega_k}{dk_x}.$$

From the usual quantum relation $\varepsilon_k = \hbar \omega_k$, Equation (12-59) follows immediately.

Going back to our discussion of the full band current of Equation (12-58) we note that we must have

$$\sum_k \mathbf{v}_k = 0$$

for a full band. When the band is full the total velocity vanishes. For a band energy function ε_k having a complete symmetric shape as a function of a $\mathbf{k}$, this is no surprise

Figure 12-31
Schematic drawing of the silicon crystal indicating its four covalent bonds per atom.

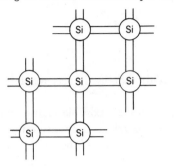

Figure 12-32
Impurity phosphorus atom in a silicon crystal. An extra valence electron is weakly bound to the phosphorus ion.

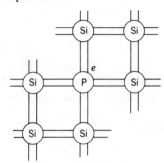

because for every positive $\mathbf{v}_k$ there is a negative $\mathbf{v}_k$. It can be shown that even if the band is unsymmetric this sum vanishes. However, if one electron is missing from momentum state $\mathbf{k}'$, the current density is

$$\mathbf{J} = -\frac{e}{V} \sum_{\substack{k \\ \text{(occupied)}}} \mathbf{v}_k = -\frac{e}{V}\left(\sum_{\substack{k \\ \text{(all)}}} \mathbf{v}_k - \mathbf{v}_{k'} \right) = +\frac{e}{V}\mathbf{v}_{k'}. \qquad (12\text{-}60)$$

The band conducts electricity as if there were a *positive* charge moving with the velocity of the missing electron. A full band of electrons is in essence electrically neutral. Remove one and the equivalent of a positive charge appears. The number of charge carriers in an intrinsic semiconductor includes the holes in the valence bands as well as the electrons in the conduction band. In an intrinsic semiconductor these two numbers are equal.

As it turns out, an extrinsic (or doped) semiconductor is much more useful than the intrinsic type discussed above. A silicon atom has four valence electrons outside a closed shell. The partially filled shell could, if filled, hold eight electrons. So in the solid each atom has four covalent bonds arranged in the same crystal structure as diamond. We show this very schematically in Figure 12-31. Silicon is a semiconductor with a band gap of 1.1 eV. Suppose a small percentage of the silicon atoms are replaced by phosphorus or arsenic, each having five valence electrons. Each of these elements can form the four covalent bonds that silicon has but there is then an extra electron left over. That electron is loosely bound to the impurity nucleus as shown in Figure 12-32. It takes only 0.05 eV to break it away from the phosphorus or arsenic nucleus. This situation introduces new energy states into the band structure. Suppose we plot band energy versus the position in the crystal as in Figure 12-33. The states of the impurity are localized, that is, they correspond to motion of the electron in a restricted region of space. These new states are somewhat like the states of hydrogen because the electron orbits a single ion. Since it does not take much thermal energy to break the extra electron loose, the particle is easily excited into the conduction band and contributes to the number of charge carriers. Because of its ability to contribute to conductivity in this way, this impurity is known as a *donor*. As can be seen from

Figure 12-33

Energy versus position in a semiconductor with donor impurity levels in the band gap. The valence and conduction bands are the same throughout the crystal and the valence band is full. The donor is able to provide an electron to the otherwise empty conduction band.

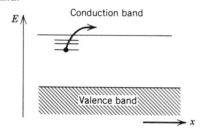

Equation (12-56) the conductivity of an intrinsic semiconductor depends exponentially on the band gap. The donor binding energy is so much smaller than this that donor conduction easily dominates intrinsic conduction in a doped semiconductor.

There is yet another kind of doped semiconductor. Suppose the impurity is aluminum or gallium. These elements have three valence electrons instead of four so that in the silicon crystal structure one of the four covalent bonds is unsatisfied as shown in Figure 12-34. Because of this unsatisfied bond, it is not too energically unfavorable for an electron from someplace else in the crystal to hop into the void and cause the formation of an Al^- or a Ga^- ion. Now a full set of covalent bonds is formed. This electron must have come from the valence band, which now has a hole. The energy levels of the impurity are shown in Figure 12-35 for aluminum and gallium impurities. These energy levels are each about 0.06 eV above the valence band. Because the impurity can locally bind an electron given up by the valence band, it is known as an *acceptor*. Again current due to holes caused by acceptor levels can dominate the intrinsic current.

Figure 12-34

Silicon crystal with an aluminum impurity. One of the covalent bonds is unsatisfied.

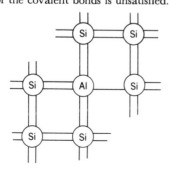

Figure 12-35

Energy versus position with acceptor levels in the band gap. A small amount of thermal energy can cause the excitation of an electron to an acceptor level, leaving a conducting hole behind in the valence band.

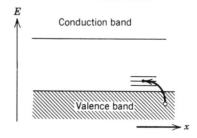

Figure 12-36

Properties of *n*-type and *p*-type semiconductors. The Fermi level in an *n*-type system is closer to the conduction band than to the valence band while the opposite is true in a *p*-type system. The energies of donor and acceptor levels are indicated by ε_d and ε_a, respectively. The relative distributions of electrons and holes are represented by the $n(\varepsilon)$ functions and by the plus and minus signs.

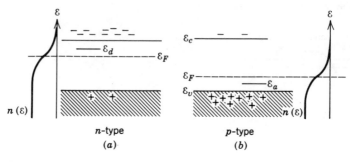

n-type
(*a*)

p-type
(*b*)

Semiconductors doped with donor levels are called *n-type* because the charge carriers are negative electrons in the conduction band. Those doped with acceptor levels are *p-type* because the charge carriers are positively charged holes. Transistors and other semiconductor devices depend fundamentally on the existence of these two types of semiconductor, as we show in Section 12-7.

In order to understand those devices, it is important to know how the presence of impurities modifies the Fermi level. In an *n*-type semiconductor there is clearly an increase in the number of electrons in the conduction band. The tail of the Fermi distribution is enhanced. However, holes in the valence band can be filled, at least partially, by electrons dropping into them from donors. The result of this is that the Fermi level (which still gives the energy at which the distribution function n_k falls to one-half) has moved up toward the conduction band as shown in Figure 12-36*a*. Similarly, in a *p*-type substance the number of holes is enhanced and the number of conduction electrons diminishes so that ε_F moves down toward the valence band as shown in Figure 12-36*b*.

Example

By Equations (12-56) and (12-52) the density of conduction electrons at room temperature ($k_B \times 300$ K $= 1/40$ eV) is

$$\frac{N_c}{V} = 2e^{-40E_g/2}\left[\frac{2\pi(9.1 \times 10^{-31} \text{ kg})(1.38 \times 10^{-23} \text{ J/K})(300 \text{ K})}{(6.6 \times 10^{-34} \text{ J} \cdot \text{s})^2}\right]^{3/2}$$

$$= e^{-40E_g/2}(3 \times 10^{25} \text{ electrons/m}^3) \quad (E_g \text{ given in eV}).$$

For a band gap of 1 eV the exponential factor is 2×10^{-9}, which is a substantial reduction of the number of conduction electrons. However, if the gap is reduced to 0.1 eV, the exponential factor becomes ~ 0.1.

12-7 Semiconductor Devices

Except for the development of nuclear weapons, no other area of technology has had such an impact on society as the transistor and related semiconductor devices. After J. Bardeen, W. H. Brattain, and W. Shockley invented the transistor in 1948, small solid-state devices soon replaced cumbersome vacuum tubes. These devices have been miniaturized to a very small size and can operate very rapidly. Microelectronics, with integrated circuits having hundreds of electronic elements placed on a single chip of silicon, has been developed to an extraordinary degree. The computer, which was slow and monstrous before the transistor, rapidly shrank in physical size, grew in memory, and became enormously faster. Someone has said that this electronics revolution that we are witnessing is much more important than the industrial revolution of the last century because the latter had amplified only human muscle power, whereas the present revolution is expanding the range of the human mind.

Of course, it is not possible to review here more than a few of the general ideas involving these devices. We discuss only simple *p–n* junctions, solar cells, light-emitting diodes, and transistors.

The *diode* or *rectifier* is a system that allows current to flow in only one direction. Such a property is obviously important when using AC to operate objects like calculators or battery chargers that require DC. Rectifiers can also be used as detectors in radio receivers; they aid in the separation of the modulating signal from the carrier wave. They are used in logic circuits in computers as well as in many other electronic systems. The *p–n junction* is a rectifier as we now proceed to describe.

A semiconductor doped with acceptors is placed next to a material doped with donors to form a *p–n* junction. Actually, the two regions may be part of one single crystal, which has had different impurities diffused in from the two ends, but we pretend that it is made up of two separate pieces just placed next to one another. Just before the two pieces touch, the energy level scheme and the electron distribution functions appear as in Figure 12-36. Note especially that the *n* end has a larger concentration of conduction electrons than the *p* end and the *p* end more holes than the *n* end. Placing the two regions together is somewhat like placing a box of oxygen atoms and a box of nitrogen atoms next to one another and opening the connecting wall. The original density gradient in each gas cannot be maintained; the oxygen flows toward the nitrogen and vice versa.

Where this analogy breaks down is in the fact that oxygen and nitrogen are neutral while electrons and holes have charge. The *p* region and the *n* region are both electrically neutral to begin with, but the flow of electrons from the *n* region leaves positively charged donor sites behind; the flow of holes out of the *p* region leaves negative acceptor sites behind. These charged regions set up an electric field that soon halts the flow of electrons and holes. The result is a region on the face of each material which is depleted of local charge carriers, holes on the end of the *p* side and electrons on the end of the *n* side, as shown in Figure 12-37. These two *depletion regions* each have the charge of the fixed impurities of each material. An electric field exists only in the narrow depletion regions. This zone turns out to have a width ranging from 10 to 10^3 nm, as determined by the material and the amount of doping of each region.

We might ask what happened to the electrons that left the *n* region and the holes that escaped the *p* region. Note that the entire system shown in Figure 12-37 remains neutral. The number of bare donors equals the number of bare acceptors in the depletion regions. The electrons and holes from each side have simply recombined and canceled one another.

Figure 12-37

Depletion regions in an n–p junction. Thin charged layers form on the contact surfaces of the junction. The layer on the n side is charged positively because the depletion of electrons bares the positive donor ions. On the p side the loss of holes bares the acceptor ions.

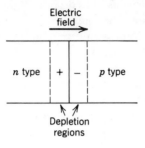

Figure 12-38

Electrical potential $\phi(x)$ seen by a charge in the neighborhood of the depletion regions. The electric field $E = -d\phi/dx$ points to the right and is confined to the charged layers.

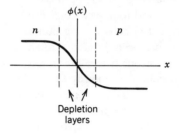

The electrons or holes far from the p–n interface do not see any local gradient in electron density and so have no inclination to flow. Those close to the junction see the gradients in density but do not flow because the electric field opposes any motion on their part to relieve this gradient.

In general, an electric field E_x in the x direction is related to a potential $\phi(x)$ by

$$E_x = -\frac{d\phi(x)}{dx}.$$

Since the electric field is confined to the depletion zone, $\phi(x)$ is constant except in that region, as shown in Figure 12-38. The energy of an electron placed in the potential is changed from ε_k to $\varepsilon_k - e\phi(x)$. Thus, the band edges and the Fermi energies are shifted relative to one another by

$$e\phi_0 = e\left[\phi(-\infty) - \phi(+\infty)\right]$$

as shown in Figure 12-39. Electrons moving from an n region toward a p region now encounter a potential barrier that most cannot surmount. Holes moving from p toward n also find a barrier (seen by viewing Figure 12-39 upside down).

Note in Figure 12-39 that the condition of equilibrium between the two materials is that the Fermi energy ε_F is the same on both sides of the junction. This is a very general thermodynamic principle. Just as heat flows when there is a temperature difference and volume changes when there is a pressure difference, so particles flow when there is a Fermi energy difference. At equilibrium ε_F is the same everywhere. This principle by itself tells us that energy levels readjust as shown.

We can now see how this junction acts as a rectifier. If we apply an external voltage V to the device with the circuit shown in the inset of Figure 12-40, we enhance the internal electric field and the potential energy barrier as shown, preventing electron motion from the n side to the p side and hole motion in the opposite direction. The system is said to be *reverse biased*. Very little current can flow in such circumstances. However, if we provide *forward bias* as illustrated in Figure 12-41, the

Figure 12-39
Energetics of the $p-n$ junction. Because of the charge layer, a potential barrier is set up by the electric field. This raises the band edges by $e\phi_0$ such that the Fermi energy in each region is now the same.

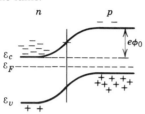

external electric field opposes the internal one, the potential barrier is lowered, and thermally activated electrons and holes can much more easily surmount the barrier. As V increases, the barrier decreases and the current flow increases substantially. Clearly, the system behaves as a rectifier.

It is not difficult to be more quantitative about our analysis of this effect. When $V = 0$, electrons and holes occasionally flow across the junction. There are as many going one way as another and the net current is zero. A few electrons on the n side have sufficient thermal energy to overcome the potential barrier and move from the n side to the p side. This current is called the electron *generation current* I_{eg}. If the few electrons that are on the p side diffuse to the junction, they are swept across by the electric field in the junction. This flow is called the electron *recombination current* I_{er}. Because the net current is zero, we have

$$I_{eg} + I_{er} = 0. \tag{12-61a}$$

Figure 12-40
Reverse-biased $p-n$ junction. The potential barrier increases by eV and very little current flows.

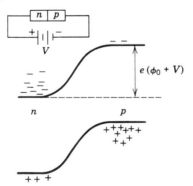

Figure 12-41
Forward-biased $p-n$ junction. The barrier is lowered and thermally activated electrons and holes can easily flow across the junction.

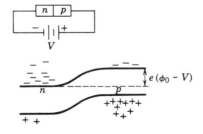

Similar arguments for holes give

$$I_{hg} + I_{hr} = 0. \tag{12-61b}$$

Suppose that a reverse bias voltage V is now applied. The number of electrons in the conduction band on the n side at energy ε is, by Equation (12-50), proportional to the exponential factor $\exp[-\beta(\varepsilon - \varepsilon_F)]$. When the energy is shifted by eV, I_{eg} is therefore reduced by an exponential factor according to

$$I_{eg}(V) = I_{eg}(0)e^{-\beta eV}. \tag{12-62}$$

For any reasonable V this quantity becomes quite small. I_{er}, on the other hand, is not affected by the biasing so that the total electron current becomes just I_{er}. The hole current also is reduced to I_{hr}. In the figures I_{er} is to the left and I_{hr} to the right so that the effects of these two currents add to give a total current

$$I_{reverse} = I_{er} + I_{hr}. \tag{12-63}$$

Under forward bias the two generation currents are each enhanced by the exponential factor so that, for example,

$$I_{eg}(V) = I_{eg}(0)e^{+\beta eV}, \tag{12-64}$$

which dominates I_{er}. We then have a total forward current

$$I_{forward} = \left(I_{eg} + I_{hg} \right)e^{\beta eV}. \tag{12-65}$$

The two situations can be summarized for all V as

$$I(V) = I_0(e^{\beta eV} - 1), \tag{12-66}$$

where the factor

$$I_0 = I_{eg}(0) + I_{hg}(0)$$

is known as the *saturation current*. If V is positive in Equation (12-66), the exponential dominates unity and we have Equation (12-65). If V is negative, the exponential can be dropped; by using Equations (12-61a) and (12-61b) we can see that the resulting current $-I_0$ is correctly equal to that of Equation (12-63). Figure 12-42 graphs the $I–V$ behavior of the $p–n$ junction. Clearly, much more current flows in one direction than the other, as one needs in a rectifier. We also see that the exponential growth of I with V given by Equation (12-66) allows the $p–n$ junction to act as a part of an amplifier.

The $p–n$ junction has several other uses as well. *Solar cells*, known also as *photovoltaic cells*, are appropriately designed $p–n$ junctions. Light shown on the junction region, as in Figure 12-43, produces electron–hole pairs. Electrons on the p side and holes on the n side are swept across the junction by the internal electric field. These excess charges provide an external voltage, the photovoltage, between the ends of the material or between metal conductors attached to the ends. Such cells are used to power calculators, watches, and spacecraft, and to provide commercial electricity. The latter use is still very limited because of the high cost of fabricating solar cells.

Figure 12-42

Current as a function of external voltage in a $p-n$ junction. The current for reverse bias is the small saturation current. Under forward bias the current increases exponentially.

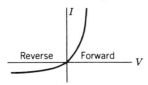

An inversion of the solar cell concept leads us to the *light-emitting diode* (LED). By forward biasing a $p-n$ junction a current of electrons and holes is established across the junction. Some electrons arising from the n side and reaching the p side recombine with the holes there, while holes going in the opposite direction annihilate electrons. Radiation may be released in this process (see Figure 12-44). The doped semiconductor GaAsP emits red light and is often used as a display in calculators and other electronic instruments. It does not burn out like a regular light bulb and requires very little power. It is even possible to build a semiconductor device that acts as a laser and produces coherent light. Such lasers are now commonly used in compact disc players.

While the simple $p-n$ junction diode is obviously an important device, a more complicated semiconductor system, the *transistor*, is even more vital. While transistors come in many forms and perform many functions, we consider only one type and one use—namely, the $p-n-p$ structure used as an amplifier.

An amplifier has many uses; its role is to turn a weak signal into a strong one. In high-fidelity equipment, this might be involved with making the very weak signal from a phonograph cartridge into one sufficiently powerful to drive speakers.

Figure 12-45 shows a double junction, $p-n-p$, and its associated circuit. We show that the AC voltage input V_{in} on the left junction is strongly amplified by the junction

Figure 12-43

Mechanism of a solar cell. When radiation strikes a $p-n$ junction, electrons and holes are produced. The intrinsic electric field sweeps electrons from the p side and holes from the n side. The charges accumulate to establish a photovoltage.

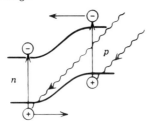

Figure 12-44

Behavior of a light-emitting diode. A forward biased $p-n$ junction has currents of electrons moving to the right and holes to the left. When they recombine with their opposites, radiation is emitted.

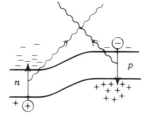

Figure 12-45

p–n–p transistor and circuit to illustrate amplifier operation. The forward-biased p–n junction is known as the emitter; the reverse-biased side is the collector. The connector to the n region is called the base. A small change ΔV_{in} in the input voltage results in a much larger change ΔV_{out} in the output voltage.

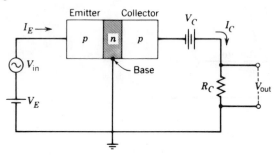

on the right. The left junction, known as the *emitter*, is forward biased so that a large current flows. By Equation (12-66) the current is

$$I_E = I_{E0} e^{\beta e (V_E + V_{in})}, \tag{12-67}$$

where V_E is provided by the battery shown. What the emitter does is to remove electrons from the n region and, more importantly, emit holes into it. These holes drift across the n region and, if the geometry is right, most are picked up at the second junction, the *collector*, before they can reach the *base*. The current at the collector (which is reverse biased by the battery of voltage V_C) is, in the absence of the picked-up emitted part, just the saturation current I_{C0}. Inclusion of that portion of the emitter current picked up by the collector gives for the total collector current

$$I_C = I_{C0} + \gamma I_E \approx \gamma I_E, \tag{12-68}$$

where γ is a geometrical factor that measures the fraction of emitter current picked up by the collector. The final approximate equality assumes that the saturation current is quite small. The fraction γ can be very close to unity.

When the input voltage undergoes a small change ΔV_{in}, it causes a change in the emitter current ΔI_E given by differentiating Equation (12-67):

$$\Delta I_E = I_{E0} e^{\beta e V_E} \beta e \, e^{\beta e V_{in}} \, \Delta V_{in}.$$

If we assume $V_{in} \ll V_E$, we have

$$\Delta I_E \approx I_{E0} \beta e \, e^{\beta e V_E} \, \Delta V_{in}. \tag{12-69}$$

By Equation (12-68), the change in the collector current is just

$$\Delta I_C \approx \gamma \Delta I_E.$$

The change in output voltage becomes

$$\Delta V_{\text{out}} = R_C \, \Delta I_C \approx \gamma R_C \, \Delta I_E \approx \gamma R_C \beta e I_{E0} e^{\beta e V_E} \, \Delta V_{\text{in}}.$$

The resulting amplification ratio is then given as

$$\eta = \frac{\Delta V_{\text{out}}}{\Delta V_{\text{in}}} \approx \frac{\gamma e R_C}{k_B T} I_{E0} e^{\beta e V_E}. \tag{12-70}$$

As we see in an example at the end of this section, this ratio can be more than 100. Note that the right side is independent of V_{in} in this approximation. A sinusoidal V_{in} results in a sinusoidal output voltage V_{out} of larger amplitude; the device is linear. However, if V_{in} gets too large the approximation used in deriving Equation (12-69) breaks down and the output voltage becomes distorted, which is undesirable in, for example, high-fidelity applications.

Transistors have many other applications, including uses as switches and computer memory elements, which we do not discuss here. The reader might take an inventory of the electronic equipment in daily use to see how amazingly dependent we have become on semiconductor devices.

Example

A familiar light-emitting diode gives off red light. From this information we can figure out the band gap E_g in this material, because we know that $E_g = hc/\lambda$. Since the wavelength corresponding to red light is around 650 nm, we have

$$E_g = \frac{1240 \text{ eV} \cdot \text{nm}}{650 \text{ nm}} = 1.9 \text{ eV}.$$

Example

Typical values of the parameters that occur in Equation (12-70) are $R_C = 10^3 \ \Omega$, $I_{E0} = 10 \ \mu\text{A}$, $\gamma \approx 1$, and $k_B T / e = 1/40$ V. If V_E is adjusted so the exponential factor is around 10^3, the voltage amplification ratio is

$$\eta \approx \left(10^3 \ \Omega\right)\left(10^{-5} \ A\right)\left(10^3\right)\left(40 \ V^{-1}\right) = 400.$$

12-8 Phonon Dynamics

Our discussions so far in this chapter have been concerned with systems involving electrons. We now switch emphasis in this section to some properties of solids not dependent on the presence of mobile electrons but dependent rather on the motions of the much more massive ion cores. While these atomic structures do not generally diffuse throughout the crystal they do vibrate about the lattice sites and create

Figure 12-46

Potential energy of atom 2 due to interactions with its neighbors 1 and 3. The result is equal to the sum of the individual potential energies V_{12} and V_{23}.

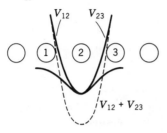

collective wave motions that we have called *phonons*. In Chapter 11 we have seen the influence of phonons on the thermal properties, especially specific heat, of solids. There we use some properties of phonons that are just quoted. In this section we provide further justification of those properties.

The Born–Oppenheimer method mentioned in Section 10-1 in the treatment of molecules is based on solving the electronic Schrödinger equation by assuming fixed nuclei. The electronic energy then acts as a potential energy function for the interaction of the nuclei. This same principle is important for solids. The forces an atom in a crystal feels from neighboring atoms are caused mainly by electronic interactions.

Suppose we consider a one-dimensional line of atoms as a model of a crystal, as we have done above to investigate the electronic structure. Now we want to use the model to study the motion of the entire atom. Figure 12-46 shows the curves that might describe the potential energies of an atom due to its two nearest neighbors in a covalent or van der Waals crystal. The total potential energy is seen to be symmetric and to have a minimum at the equilibrium position of the atom. Such a picture applies to each atom in the crystal (except for the atoms on the surface whose potential energies are not symmetric).

Generally, the motion of an atom about equilibrium is small so that it is reasonable to expand the potential in powers of the distance from the equilibrium position. This argument has been used to discuss diatomic molecules in Section 10-7. There the atoms in a diatomic molecule are considered to be bound harmonically. Here the same result is true; the bottom of the potential energy well of any particle can be fit by a parabola. However, that parabola is determined by interaction with many neighbors, not just one other atom as in the diatomic case.

In this way we justify, to a certain extent, the Einstein model of a crystal treated in Chapter 11. But particles 2 and 3 in Figure 12-46 do not remain stationary; they also move. So the better way to picture a crystal is as a group of particles interconnected by springs—a network of springs—as illustrated in Figure 11-9. One particle's motion causes a neighbor to move, and so on, so that a wave travels along.

To treat waves in our one-dimensional crystal model we again introduce periodic boundary conditions as in Section 12-4, rather than involving ourselves with the complications of free surfaces or container walls. The basic model is shown in Figure

Figure 12-47

One-dimensional model of a crystal with periodic boundary conditions.

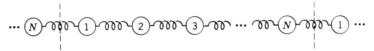

12-47. The particle on the right of the Nth is the first, and that to the left of the first is the Nth. We can consider the entire crystal as strung out in a large circle with the ends tied together by a spring.

We want to solve for the frequencies of waves on our N-body one-dimensional system. We can do this by solving either the classical equations of motion or the Schrödinger equation. The allowed frequencies turn out to be identical in the two approaches. The former approach is simpler and we proceed that way.

The spring constant is K, the mass of an atom m, the position of the nth particle x_n, and the equilibrium separation of the particles a. A particle's displacement from its equilibrium position R_n is

$$u_n = x_n - R_n \tag{12-71}$$

as shown in Figure 12-48 for a particular particle.

The force on particle 2 due to its spring connection with particle 1 is

$$F_{21} = -K(x_2 - x_1 - a). \tag{12-72}$$

We have taken the spring to be at equilibrium ($F_{21} = 0$) when the particles are separated by a lattice distance a, that is, when $x_2 - x_1 = a$. In terms of the u's, we have

$$F_{21} = -K(x_2 - R_2 - x_1 + R_1 + R_2 - R_1 - a) = -K(u_2 - u_1).$$

The classical equation of motion for particle 2 is (denoting d/dt by a dot)

$$m\ddot{x}_2 = -K(u_2 - u_1) + K(u_3 - u_2). \tag{12-73}$$

Figure 12-48

Displacement of a particle in a one-dimensional crystal. In the system shown, all particles are at their equilibrium positions R_n except particle 2, which is displaced a distance $u_2 = x_2 - R_2$. The nearest-neighbor spacing is a.

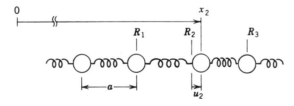

Figure 12-49

Example of a sinusoidal wave at one instant of time. The equilibrium positions are indicated by the vertical lines. The wavelength is 12 lattice spacings.

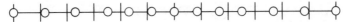

The plus sign in front of the last term means that the force on particle 2 due to particle 3 is pulling it to the right when $u_3 - u_2$ is positive. Since $\ddot{x}_2 = \ddot{u}_2$, Equation (12-73) can be written completely in terms of the u's.

There is nothing special about particle 2 or any particle in the system when periodic boundary conditions are used. Hence the equation for the nth particle is just like Equation (12-73) and can be written as

$$m\ddot{u}_n = -K(u_n - u_{n-1}) + K(u_{n+1} - u_n). \tag{12-74}$$

We expect longitudinal waves to propagate on this chain. An example of a wave at one particular time is shown in Figure 12-49. The particle displacements change sinusoidally as we proceed from one particle to another. The wave pictured might be described at that instant of time as $u_n = s \sin(2\pi n/12)$, where n numbers the particle and s is the amplitude of the motion. The 12 appears because the wavelength of the wave in the figure is $12a$. The time dependence is of the form $\sin \omega t$ or $\cos \omega t$, where ω is the sought-after frequency of the wave. More generally, we can look for traveling-wave solutions having the form

$$u_n = s e^{ikR_n} e^{-i\omega t}, \tag{12-75}$$

where k is the wave number and s is the amplitude. We could use sines and cosines but the exponential form is easier.

Use $R_n = na$ and plug Equation (12-75) into Equation (12-74) to obtain

$$-\omega^2 m s e^{ikna} e^{-i\omega t} = -Ks\left(e^{ikna} - e^{ik(n-1)a}\right)e^{-i\omega t} + Ks\left(e^{ik(n+1)a} - e^{ikna}\right)e^{-i\omega t}.$$

Dividing out $-s e^{ikna} e^{-i\omega t}$ gives

$$\omega^2 m = K\left(1 - e^{-ika}\right) + K\left(1 - e^{ika}\right) = K\left(2 - 2\cos ka\right).$$

Using the trigonometric identity

$$\sin^2 \frac{x}{2} = \frac{1}{2}(1 - \cos x),$$

we have

$$\omega^2 = \frac{4K}{m} \sin^2 \frac{ka}{2}$$

or

$$\omega = 2\sqrt{\frac{K}{m}} \left| \sin \frac{ka}{2} \right|, \tag{12-76}$$

where the absolute value signs occur because a frequency is always a positive quantity by definition.

For small wave numbers (or long wavelengths according to the relation $k = 2\pi/\lambda$) the discrete character of the system, that is, the fact it is made up of many individual masses, is not noticeable to the wave. Then Equation (12-76) can be simplified by using the approximation $\sin x \approx x$ so that

$$\omega \approx 2\sqrt{\frac{K}{m}} \left| \frac{ka}{2} \right|. \tag{12-77}$$

In this long-wavelength limit the phase velocity ω/k and the group velocity $d\omega/dk$ are identical and equal to

$$v_g = \sqrt{\frac{Ka^2}{m}}. \tag{12-78}$$

The velocity does not depend on wave number so there is no dispersion in this limit. For shorter wavelengths the approximation of Equation (12-77) is no longer valid and waves of different frequencies travel at different velocities; the dispersion arises because of the discrete character of this chain of particles.

Just as in the electron case the allowed values of k, which give us the permitted wavelengths and frequencies, are determined by the boundary conditions. With the periodic boundaries we have

$$u_0 = u_N \tag{12-79a}$$

and

$$u_{N+1} = u_1, \tag{12-79b}$$

since to the left of particle 1 is particle N and to the right of particle N is particle 1. From Equations (12-75) and (12-79a) we have

$$se^{-i\omega t} = se^{ikNa}e^{-i\omega t}$$

and

$$e^{ikNa} = 1$$

so that

$$k = \frac{2\pi}{Na}\ell, \qquad \ell = 0, \pm 1, \pm 2, \pm 3, \dots. \tag{12-80}$$

These are precisely the same k values as those allowed by Equation (12-37) for electrons moving through a crystal. And as in that case there are limits on k space. Values of k out of the first Brillouin zone are equivalent to those within it. This is easy to see by adding the vector G of Equation (12-39) to any k value:

$$u_n(k + G) = se^{i(k+G)na}e^{-i\omega t} = \left(se^{ikna}e^{-i\omega t} \right)e^{i2\pi n} = u_n(k).$$

Thus, any region of k vectors having a width G is adequate to specify all possible waves. We choose these vectors to be those in the first Brillouin zone as given by Equations (12-38) and (12-80).

Note again that there are exactly N of these vectors in Equation (12-80); this is as it should be because there are N particles and one expects to find exactly N distinct waves corresponding to the N degrees of freedom of the system.

Figure 12-50

Angular frequency ω versus k for a one-dimensional harmonic chain. Positive and negative k values represent waves traveling in opposite directions along the chain.

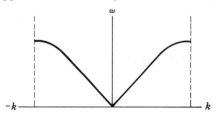

The k values of Equation (12-80) are spaced quite closely together because of the larger number N in the denominator. In plotting ω versus k from Equation (12-76), we can treat k as a continuous variable. Such a plot is shown in Figure 12-50. There are two branches, one corresponding to positive k and one to negative k. Because we have used periodic boundary conditions, we can have traveling waves on our chain; the waves do not meet a wall and undergo reflection resulting in standing waves. Positive k corresponds to waves traveling in the positive direction and negative k to oppositely directed waves.

In Chapter 11, we have derived the contribution of phonons to the heat capacity of crystals. Our discussion there considers a chain of particles connected to walls, rather than having periodic boundary conditions. The use of wall boundary conditions results in standing waves. Negative k has no meaning in such a situation and the spectrum shown in Figure 11-10 has only a positive k branch. However, there are no differences in the thermal properties because, although there is only one branch in the wall-boundary case, the k values are twice as densely distributed. This discussion is identical in spirit to the one given for electrons right after Equation (12-37).

The use of a complex value for the displacement u_n in Equation (12-75) may bother some readers since displacements must actually be real. However, since solutions for plus and minus k correspond to the same frequency, the complex conjugate of Equation (12-75) is also a solution for that frequency. Furthermore, since the sum or difference of two solutions corresponding to the same frequency is also a solution, we have, for example, as another solution

$$u_n = se^{ikR_n}e^{-i\omega t} + se^{-ikR_n}e^{+i\omega t} = 2s\cos(kR_n - \omega t),$$

which is a real displacement representing a traveling wave.

In Chapter 11 we have also discussed another crystal property dependent on phonons, the thermal conductivity. In a metal, the electron contribution to this quantity dominates and thermal conductivity is determined by Equation (12-22). In insulators, however, the phonons carry the heat. It can be shown that the thermal conductivity of a gas of *particles* is

$$\kappa_T = C_V v^2 \tau, \tag{12-81}$$

where C_V is the heat capacity *per unit volume*, v is the velocity of the heat carrier, and τ

is the relaxation time (time between collisions). We find in Problem 11 at the end of the chapter that κ_T of Equation (12-81) reduces to the proper result, Equation (12-22), when applied to a degenerate electron system. While phonons differ from a usual gas of particles (such as electrons) since they can be created and destroyed, Equation (12-81) still applies to them.

From Equation (11-68) we have, for low temperatures (ignoring factors of order 1),

$$C_V \sim \frac{(k_B T)^3 k_B}{(\hbar c_s)^3},$$

where c_s is the velocity of sound in the crystal. The heat carrier velocity in Equation (12-81) for phonons is also c_s.

The mean free path is most commonly determined by phonon–phonon collisions. The computation of it is often quite complicated and beyond the scope of our discussion. One case that is easy to discuss is the situation in which the density of phonons is so low that the phonons hit the walls before they hit each other. Then the relaxation time is the time for a phonon to travel a distance L from one side of the crystal to another. In that case $\tau = L/c_s$ and so

$$\kappa_T \sim \frac{k_B^4 L T^3}{\hbar^3 c_s^2}.$$

A T^3 temperature dependence of κ_T is indeed observed in insulators at the lowest temperatures.

12-9   Magnetism in Solids

Because the atoms that make up a crystal can have electrons with unpaired spins, the solid as a whole can have a net magnetic moment. Because the atoms are in the solid state, the interactions between them can often alter the character of this magnetism considerably. The most notable example of this is ferromagnetism, in which cooperative alignment among the electronic magnetic moments gives some solids, such as iron, a net magnetic moment in the absence of any external magnetic field. Other forms of magnetism occurring in solids include paramagnetism, diamagnetism, and antiferromagnetism.

Suppose we have a solid whose atoms each have a net spin $\frac{1}{2}$ but do not interact with one another magnetically. The only magnetic energy in the system is the direct Zeeman interaction of the spins with an external magnetic field. A solid having such energy levels can have no net magnetic moment in the absence of an external field and is said to be *paramagnetic*. Obviously, there is nothing special here about the spins being in a solid and paramagnetism occurs as well in gases and liquids.

For electrons with spin $\mathbf{S}$, we have seen in Equation (8-20) that the magnetic moment per particle is given by

$$\mathbf{\mu}_S = -g_S \mu_B \mathbf{S}/\hbar,$$

where μ_B is the Bohr magneton and g_S is the electron g-factor. In a magnetic field B the possible energies of the spin are, from Chapter 8,

$$\langle V_M \rangle = -\mu B \sigma, \tag{12-82}$$

where the magnitude of the magnetic moment is

$$\mu = -g_s \mu_B / 2, \tag{12-83}$$

and $\sigma = \pm 1$ is *twice* the magnetic quantum number m_s. By Equation (12-82) and (12-83), spin-down states, that is, those that have $\sigma = -1$ and are antiparallel to B, have a lower energy than those aligned with the field. Thus, the more probable states in a thermal distribution have more particles with spin-down states than with spin-up states. The entire collection of electrons then has a net magnetization, which can be detected.

The magnetization per unit volume of a collection of electrons is defined as

$$M = \mu(\rho_+ - \rho_-), \tag{12-84}$$

where ρ_σ is the density of electrons with magnetic quantum number σ. In equilibrium ρ_- is larger than ρ_+ as explained above, but since μ is negative, M comes out positive. For sufficiently small magnetic fields it turns out that M is proportional to the applied magnetic field or

$$M = \chi B, \tag{12-85}$$

where χ is called the *susceptibility*.

Suppose we have a solid made up of atoms each of which has a tightly bound unpaired electron so that each atom has a net magnetic moment. Because the electrons do not hop from site to site, those on separate atoms really do not see much of one another. Then Fermi statistics is quite irrelevant to their relative spin states. We can consider the situation as one in which the electrons behave classically and obey Boltzmann statistics.

The partition function for a single electron having energy levels given by Equation (12-84) is

$$Z = \sum_{\sigma = \pm 1} e^{\beta \mu B \sigma} = e^{\beta \mu B} + e^{-\beta \mu B} = 2 \cosh \beta \mu B.$$

The magnetization per unit volume M is related to the difference in density between spin-up and spin-down particles, $\rho_+ - \rho_-$, by Equation (12-84). Since σ is $+1$ for spin up and -1 for spin down, we have

$$\rho_+ - \rho_- = \rho \langle \sigma \rangle, \tag{12-86}$$

where ρ is the density of all electrons. The average value of σ in Equation (12-86) is given, according to the general procedures of Chapter 11, by

$$\langle \sigma \rangle = \sum_{\sigma = \pm 1} \frac{\sigma e^{\beta \mu B \sigma}}{Z} = \frac{e^{\beta \mu B} - e^{-\beta \mu B}}{e^{\beta \mu B} + e^{-\beta \mu B}} = \tanh \beta \mu B. \tag{12-87}$$

So the magnetization per unit volume is

$$M = \mu \rho \tanh \beta \mu B. \tag{12-88}$$

For very high temperature, such that $\mu B \ll k_B T$, the argument of the hyperbolic tangent is small. For small x a Taylor expansion gives

$$\tanh x \approx x, \qquad x \ll 1, \tag{12-89}$$

so that we get

$$M \to \mu \rho \beta \mu B = \rho \mu^2 B / k_B T \text{ as } T \to \infty. \tag{12-90}$$

The susceptibility, defined in Equation (12-85), becomes in this limit

$$\chi = \frac{M}{B} = \frac{\rho \mu^2}{k_B T}. \tag{12-91}$$

This expression is known as *Curie's law*, after P. Curie, giving χ for a classical electron system as inversely dependent on the temperature. As T increases, thermal agitation weakens the ability of the external field to align the spins. Many solids exhibit Curie-law behavior.

Because the hyperbolic tangent goes to unity for large values of its argument, Equation (12-88) gives, for low T,

$$M \xrightarrow[T \to 0]{} \mu \rho,$$

which corresponds to all particles aligned with the field. Of course, if Fermi statistics has become important before the low-temperature condition $k_B T \ll \mu B$ has been satisfied, the last limit is not valid.

The magnetism of electrons in a metal cannot be described according to the treatment just given. Those electrons are highly degenerate since the temperature satisfies $T \ll T_F$ (the Fermi temperature is 10^4 or 10^5 K in most metals). We can take the temperature to be essentially 0 K. With no magnetic field, any filled momentum state contains one spin up and one spin down and then the net magnetization M is zero. When a magnetic field is turned on, the Zeeman energy of Equation (12-82) is lowered if some spins align along the magnetic field. But such spins cannot just flip into a different alignment, because then there would be two identical spins in the same momentum state. The spin that is flipping must also be promoted to a higher unoccupied momentum state. As further spins are turned over by increasing the external field they must be given even more additional kinetic energy to avoid violation of the Pauli principle. Even at $T = 0$ K a degenerate Fermi system has a finite susceptibility, unlike the Boltzmann spin system. The behavior of the magnetism of this degenerate fermion system is known as *Pauli paramagnetism*.

Diamagnetism is based not on the Zeeman interaction of Equation (12-82) but on the effect of the Lorentz force on an electron orbit. A free electron in a magnetic field orbits in a circle and establishes a current loop that has a magnetic moment. Examination of Figure 12-51 shows that the magnetic moment is opposed to the external field. Many solids show such a negative magnetization, but we should note that the fact that the electrons in these solids are bound in quantum states in atoms is not taken into account in our simple picture but is crucial to the existence and nature of diamagnetism. The diamagnetic effect is negligible compared to paramagnetic effects if they are present and is seen only in materials consisting of atoms having completely paired electron spins.

Figure 12-51

Lorentz force due to a magnetic field causing an electron to orbit in a circle. The resulting current loop has a magnetic moment μ with a direction opposite to the field.

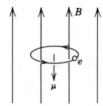

Ferromagnetism is an important subject in human history because of the compass. Ferromagnetic substances have received considerable attention from solid-state physicists because of the magnetic phase transitions that they undergo. At high temperature and in zero magnetic field, a piece of iron is unmagnetized. (With a field on, the iron is paramagnetic at high temperature.) When the temperature is lowered to a certain value, a spontaneous magnetization begins to develop. Below that transition temperature, the system behaves as a permanent magnet. This effect arises because of a magnetic force, having purely quantum mechanical origins, that tends to align all the spins in the same direction. This force is related to the interaction in a covalently bonded molecule. We know from Chapter 10 that, because of the Pauli principle, the singlet spin state is energetically preferred over the triplet. In ferromagnetic substances somewhat analogous interactions give preference to the triplet alignment for pairs and an overall parallel arrangement of the entire spin system.

For some substances the spin–spin force does prefer the singlet or antiparallel pair arrangement like that in the diatomic molecule. At low temperatures we then get a state of a solid in which spins on alternate sites are in opposite directions. Such substances are *antiferromagnets*.

Example

The polarization P of any substance is defined as the fraction of aligned spins. Consider any paramagnetic solid. From Equations (12-84) and (12-90) we have, for high temperatures,

$$P = \frac{\rho_+ - \rho_-}{\rho} = \frac{M}{\mu\rho} = \frac{\mu B}{k_B T}.$$

The electron magnetic moment is -9.3×10^{-24} J/T. At $T = 100$ K and $B = 1$ T the polarization is

$$P = \frac{(9.3 \times 10^{-24} \text{ J/T})(1 \text{ T})}{(1.4 \times 10^{-23} \text{ J/T})(100 \text{ K})} = 6.6 \times 10^{-3}.$$

Problems

1. (a) Compute the number density in terms of the nearest-neighbor distance a for the two-dimensional triangular lattice, the square lattice, and the honeycomb lattice shown in the figure.

    (b) Assuming only nearest-neighbor interactions of strength $-\varepsilon$, find the energy per particle of the three lattices in (a) and of an arbitrary two-dimensional lattice in which each particle has z nearest neighbors.

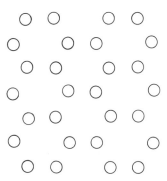

2. (a) In the triangular lattice of Figure 12-2, particles 1 and 2 are nearest neighbors, and particles 1 and 3 are second-nearest neighbors. Each particle has six near neighbors. How many second neighbors does each particle have? How many third neighbors?

    (b) Find the ratio of total second-neighbor interaction energy to total first-neighbor energy if the potential energy falls off like the van der Waals interaction, $-1/r^6$. Is it a good approximation to drop second-neighbor interactions as we have often done?

3. Figures 12-8 and 12-9 show two different views of the FCC lattice. By explicitly numbering the particles in each of your own versions of the two drawings show that the two views are equivalent.

4. (a) Find the volume per particle in terms of the nearest-neighbor distance a for the SC lattice, the FCC lattice, and the BCC lattice.

    (b) What is the energy per particle in each of these lattices if only nearest-neighbor pair interactions, each of energy $-\varepsilon$, are taken into account.

5. Consider a one-dimensional model of a NaCl crystal made up of uniformly spaced alternating positive and negative charges.

    (a) Write an infinite series for the energy as done for the three-dimensional crystal in the example at the end of Section 12-1. Attempt to estimate this energy by considering a large number of terms in the series.

    (b) Evaluate the sum exactly by comparing it with the Taylor series expansion of $\ln(1 + x)$.

6. (a) Show that the vector set

$$\mathbf{R} = a\left(n_1 \hat{\mathbf{x}} + n_2 \hat{\mathbf{y}} + n_3 \hat{\mathbf{z}} \right),$$

    where the n_i are arbitrary integers and $\hat{\mathbf{x}}, \hat{\mathbf{y}}, \hat{\mathbf{z}}$ are Cartesian unit vectors, gives the positions of the particles on a simple cubic lattice. What lattice is represented by

$$\mathbf{R} = (a/2)(n_1\hat{\mathbf{x}} + n_2\hat{\mathbf{y}} + n_3\hat{\mathbf{z}}),$$

with n_1, n_2, n_3 integers whose *sum is even?*

(b)   Find the set of vectors representing all the lattice points of a BCC lattice.

7.   (a)   Show that the vectors

$$\mathbf{R} = a(n_1\hat{\mathbf{u}} + n_2\hat{\mathbf{v}}),$$

where n_1 and n_2 are arbitrary integers and $\hat{\mathbf{u}}$ and $\hat{\mathbf{v}}$ are unit vectors given by $\hat{\mathbf{u}} = \hat{\mathbf{x}}$, $\hat{\mathbf{v}} = (\hat{\mathbf{x}} + \sqrt{3}\,\hat{\mathbf{y}})/2$, give the sites of a two-dimensional triangular lattice.

(b)   The reciprocal lattice vectors for the triangular lattice are given by

$$\mathbf{G} = (2\pi/a)(m_1\hat{\mathbf{u}}^* + m_2\hat{\mathbf{v}}^*),$$

where m_1 and m_2 are integers and $\hat{\mathbf{u}}^*$ and $\hat{\mathbf{v}}^*$ are vectors defined by $\hat{\mathbf{u}}^* \cdot \hat{\mathbf{v}} = \hat{\mathbf{v}}^* \cdot \hat{\mathbf{u}} = 0$, and $\hat{\mathbf{u}}^* \cdot \hat{\mathbf{u}} = \hat{\mathbf{v}}^* \cdot \hat{\mathbf{v}} = 1$. Find $\hat{\mathbf{u}}^*$ and $\hat{\mathbf{v}}^*$. What kind of lattice does the set of $\mathbf{G}$ vectors describe? What is the "nearest-neighbor distance" in the reciprocal lattice?

(c)   Prove that $\exp(i\mathbf{G} \cdot \mathbf{R}) = 1$ in agreement with Equation (12-42).

(d)   A "scattering plane" in a two-dimensional lattice is a *line* of particles. Show that every line of particles has a $\mathbf{G}$ vector normal to it as it should according to the original definition of $\mathbf{G}$ in Equation (12-3).

(e)   Show that every $\mathbf{G}$ has a line of particles perpendicular to it.

8.   (a)   The Fermi energy of copper is about 7 eV. What is the Fermi velocity?

(b)   Given the value of the relaxation time τ in the example at the end of Section 12-3, find the mean free path of a conduction electron. Compare this with the average interatomic spacing.

9.   Show that the current density (intensity) of a particle beam (number per unit area per unit time) is particle density times particle velocity.

10.   The thermal conductivity of copper at room temperature is 400 W/m · K. Use this value to estimate the relaxation time τ. Compare with the value obtained from the electrical conductivity in Section 12-3.

11.   Show that Equation (12-81), when applied to a degenerate Fermi system, reduces properly to Equation (12-22).

12.   (a)   Repeat the derivation of Equation (12-32) for the case $N = 3$. Here the periodic boundary conditions require that particle 3 be a near neighbor of particle 1. Find the energy levels and solve for the C_n's by setting the determinant of the coefficients in the equation to zero.

(b)   What are the k values that make up the first Brillouin zone for $N = 3$? Using these, compare your values for the energies from part (a) to those of the general solution of Equation (12-35).

13.   Consider a free particle in a one-dimensional "box" of length L where the boundary condition is periodic, that is, $\psi(0) = \psi(L)$. Show that the allowed k values satisfy Equation (12-37) with Na replaced by L.

14.   Show that the product of the number of conduction electrons N_c and the number of holes N_v in a semiconductor is independent of the Fermi energy ε_F. The relation for $N_c N_v$ is

called the *law of mass action*. Since the relation is valid for any ε_F, this product is the same independent of degree of doping. If the number of conduction electrons is increased by adding donors, the number of holes must decrease correspondingly. Evaluate $N_c N_v / V^2$ for diamond, silicon, and germanium at room temperature by using the gap data given at the end of Section 12-5. Use effective masses equal to the electron mass.

15. The bulk of solar radiation has a wavelength less than 1 μm. What is the minimum energy gap that a solar cell should possess to take advantage of this? Is silicon appropriate?

16. A small change ΔV_{in} in the input voltage of the transistor circuit of Figure 12-45 produces a change ΔP_E in the emitter circuit power and a corresponding change ΔP_C in the power delivered to the resistor R_C. Show that the power amplification ratio satisfies

$$\frac{\Delta P_C}{\Delta P_E} = \frac{2\gamma^2 \beta e I_E R_C}{1 + \ln(I_E/I_{E0})}.$$

Evaluate this ratio for the conditions given in the example at the end of Section 12-7.

17. A positively charged arsenic donor ion in silicon attracts an electron much like a hydrogen nucleus. However, the intervening silicon behaves like dielectric material having permittivity constant $\varepsilon = 12\varepsilon_0$, where ε_0 is the permittivity in vacuum. The electron also has an effective mass in silicon of 0.2 times the electron mass. The ionization energy of this system is the separation between the lowest donor level and the conduction band as shown in Figure 12-33. Estimate this ionization energy by using the appropriately modified hydrogenic energy levels. Compare your result with the accurate value given in Section 12-6. What is the Bohr radius of this system?

18. Write down the set of equations analogous to Equations (12-74) for a ring of three particles connected by springs. Substitute $u_n(t) = e^{-i\omega t} u_n(0)$, and solve the resulting set of linear equations for the three possible values of ω by the method of setting the determinant of the coefficients to zero. Compare your results with the general solution given by Equation (12-76). One of the solutions is $\omega = 0$. To what values of the $u_n(0)$, and what resulting motion of the particles, does this special solution correspond?

19. Using the general form of Equation (12-81), estimate the temperature dependence of the thermal conductivity of an insulator at very high temperature $(T \gg \Theta_D)$. In your discussion give an argument that shows that the relaxation time is inversely proportional to the number of phonons present in the system.

20. The *nuclear* magnetic moment of the ^{3}He nucleus is -1.1×10^{-26} J/T. What is the polarization of ^{3}He gas in a magnetic field of 1 T at a temperature of 100 K?

21. Find the paramagnetic susceptibility of a spin-1 system at high temperature. The magnetic energy is $-\mu Bm$ with $m = 0, \pm 1$.

SUPERFLUIDS
AND
SUPERCONDUCTORS

Superconductivity and superfluidity are among the most remarkable forms of bulk behavior in matter. Substances having these properties can carry currents of electricity or matter with essentially no resistance. Such capabilities are so out of the realm of ordinary experience that when first observed they were not recognized or were thought to be due to experimental error.

When H. Kamerlingh Onnes succeeded in liquifying helium in 1908 it became possible to observe these effects. Soon after he reached the liquification temperature of 4.2 K, Kamerlingh Onnes was able to cool this very light and transparent material below 2.2 K, the transition temperature to the superfluid state. He must have observed the cessation of boiling, the most apparent signal that this new state of matter has been reached. And yet he made no mention of the effect. It was not until 25 years later that a comment about it finally appeared in the literature. Soon, other strange properties were discovered, some by Kamerlingh Onnes, but most by other workers.

Kamerlingh Onnes used helium as a refrigerant in order to study a variety of substances at very low temperatures. In examining mercury in 1911 he found that its electrical resistance disappeared around 4 K. Although he first assumed that the result was an experimental error, he was able to find the effect in other substances and coined the term *superconductivity* to describe it.

Superconductivity occurs in many different metals and alloys. Transitions to the superconducting state are normally found to occur at temperatures as high as about 20 K and as low as several millidegrees. The recent discovery of a series of alloys with transition temperatures up to and above 100 K may lead to some very important practical applications of this phenomenon. Superfluidity, on the other hand, occurs only in the two isotopes of helium; in ^{4}He, the onset is at 2.2 K, while it does not occur in ^{3}He until 2 mK.

There is a variety of phenomena associated with these two states of matter besides the supercurrents. All these effects are now understood to some degree. For superfluid helium, there is an excellent phenomenological theory, known as the two-fluid model, that successfully correlates many of the properties. Nevertheless, despite a large amount of excellent and fundamental work, there is no comprehensive microscopic theory of superfluid ^{4}He as there is for the superconducting state. While a two-fluid model of a superconductor was used with some success for many years, in 1956 a microscopic theory was finally put forward by Bardeen, L. N. Cooper, and J. R. Schrieffer. This explanation, known as the BCS theory, gave such good agreement with experiment that it was felt that the problem of superconductivity was solved.

Liquid ^{3}He also becomes a superfluid around 2 mK, as discovered by D. D. Osheroff, R. C. Richardson, and D. M. Lee in 1972. Despite the fact that this atom is an isotope of helium carrying no charge, the transition in this case is much more like that in superconductors than in ^{4}He. Because much of the theory existed before the experimental discovery, we now have a rather complete picture of this material.

Superconducting and superfluid systems are properly treated together in the same chapter because of their many similarities. Both involve phase transitions to rather special ordered states that allow friction-free flow. In neither case is the order a spatial one as in a crystal; the substance remains a fluid. (The "fluid" is a gas of electrons in the case of a superconductor.) Both transitions are macroscopic manifestations of the laws of quantum mechanics, and in each case particle statistics, Fermi or Bose, plays a crucial role. Both phenomena reveal themselves at low temperatures because then the thermal de Broglie wavelength becomes large enough to make quantum and statistical effects evident.

13-1 Experimental Characteristics of Superfluids

The properties of liquid ^{4}He that we describe in this section are unique to the substance. The reason for this is not that the fundamental causes are absent in other substances. It is mainly that every other material freezes at a temperature higher than the transition point to the superfluid state. Despite the seemingly wide variety of effects that we describe here, all are shown to be closely related.

In the introduction we have noted one characteristic of liquid ^{4}He below the superfluid transition—namely, the cessation of boiling. In normal liquid helium and in other liquids, local hot spots cause bubbles of vapor to be formed. Below the superfluid transition temperature of 2.17 K, the heat conductivity of the liquid becomes so large that the hot spots necessary for the formation of the bubbles of boiling cannot occur. Under the proper conditions the thermal conductivity of the superfluid can be as large as 2000 times that of copper at room temperature. A drop in temperature of only 1 K in going from the normal liquid to the superfluid state leads to an increase in conductivity by a factor of several million.

As the temperature is lowered in any liquid, we expect to reach the solidification point. This is true in all substances except helium. The condensed phases of both helium isotopes remain liquids all the way to absolute zero. The only way to make solid helium is to pressurize either of the two liquids. At very low temperature, solid ^{4}He finally forms at 25 atm, and solid ^{3}He around 30 atm. A diagram showing the phases of ^{4}He is given in Figure 13-1.

Helium atoms interact with a potential energy approximately described by the Lennard-Jones function, Equation (10-36). The strength of the interaction (as measured by the well-depth value $\varepsilon/k_B \approx 10$ K) is very small compared to other substances. Also the mass of the helium atom is so small that quantum zero-point motion is large. If the separation between atoms is a, the lowest energy of a particle is of order $E_0 \sim \hbar^2/ma^2$, the result of considering the lowest level of a particle in a box.

Figure 13-1

Phase diagram of ^{4}He. Lowering the temperature at atmospheric pressure (along the dotted line) takes the system from gas to normal liquid to superfluid. For temperatures above the critical point, liquid and gas are indistinguishable. The solid does not form, even at $T = 0$ K, until a pressure of 25 atm is applied. The line separating normal fluid from superfluid is known as the λ-line.

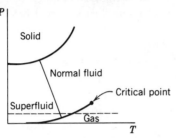

Figure 13-2

Specific heat of liquid ^{4}He. The transition temperature is called the λ-point because of the shape of the curve. At low pressure T_λ occurs at 2.17 K and decreases slightly at higher pressure.

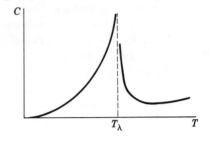

This energy is so large in helium compared to the potential energy that it melts the solid at low pressure. When enough pressure is applied the potential energy grows faster than this kinetic energy until the material is finally able to solidify.

When the normal fluid is cooled through the transition at 2.17 K, the specific heat has a spectacular increase as shown in Figure 13-2. Because of the shape of the curve the transition temperature to the superfluid is known as the λ-*point*. The peak has been studied in great detail and is thought to become infinitely high if the number of atoms in the system is infinite. The transition is of second order since there is no latent heat. At low temperatures the curve shows a T^3 dependence that is characteristic of phonon systems. We know from Chapter 11 that the specific heat of solids depends on temperature in this way. At somewhat higher temperatures, an extra $\exp(-\Delta/k_BT)$ behavior is observed, a form characteristic of an excitation spectrum with a gap Δ, as discussed in Problem 9 in Chapter 11.

As mentioned above, the superfluid has a very high thermal conductivity. There is yet another closely related peculiarity; heat can flow in the form of a wave known as *second sound*. Under normal conditions heat diffuses; it moves away from a hot spot much like molecules spreading away from an open bottle of perfume. However, in superfluid helium, a pulsed heater causes a temperature pulse to travel, largely undistorted, across the container where it can be detected by a thermometer. A heater that is cycled sinusoidally produces a sinusoidal temperature wave that travels across the liquid.

The viscosity of a liquid is related to its resistance to flow. One of the standard ways of measuring this quantity is with a torsion oscillator made up of a plate suspended by a thin rod. As the plate oscillates about the rod axis it drags along any viscous fluid; the rate of damping is easily related to the viscosity. This technique applied to helium shows that the viscosity just below the λ-point is about the same as above, although it then apparently decreases toward zero as the temperature is lowered toward zero.

On the other hand, experimenters using very narrow channels, such as those in very fine filters, have found that above 2.17 K the fluid is unable to flow through, while at any temperature below the λ transition the liquid pours through as if the viscosity were zero. This extraordinary flow property is one of the main reasons for applying the name "superfluid" to this substance. Narrow channels that allow superfluids, but not normal liquids, to pass through are known as *superleaks*. If the flow velocity exceeds a *critical velocity* v_c, viscous drag occurs.

The two kinds of viscosity experiments seem at first sight to be quite irreconcilable, with one showing a viscosity and the other showing none at all. Nevertheless, we see in Section 13-5 how the two-fluid model nicely explains the problem.

When helium is in a container, the vapor above the liquid coats the walls with a thin film that is usually several atomic layers thick. This in itself is not unusual and occurs for any enclosed liquid. However, the superfluid film shows some unique properties. If a breaker holding some of the liquid is raised above the general liquid level, fluid travels via the film flow over the edge of the beaker, accumulates in drops on the bottom, and then drips back into the main body of liquid. This continues until the beaker is empty. This *creeping film* has, in effect, siphoned the bulk liquid out of the small container.

The film can be used for another experiment that shows the amazing flow properties of the superfluid. Suppose the film coats a ring of glass. The film can be made to flow continuously around the ring. Once the flow has started it continues without dissipation indefinitely. Similarly, bulk fluid made to rotate by spinning a bucket continues to flow without friction. Such friction-free flow patterns are known as *persistent currents*.

The *thermomechanical effect* involves two containers of helium connected by a superleak as seen in Figure 13-3. If one side is heated slightly the liquid level on that side rises at the expense of the liquid in the other container. It is possible to set up an arrangement such that the heating causes a thin stream of liquid to spray upward, giving rise to the term *fountain effect*.

We attempt in Section 13-5 to give brief explanations for each of these phenomena in terms of the two-fluid model or by more microscopic theories. Before entering into those explanations, we examine the range of peculiar effects found in superconductors.

Figure 13-3

Thermomechanical or fountain effect. Turning on the heater on the right causes the liquid level on that side to rise while the other side falls.

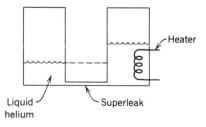

13-2 Experimental Characteristics of Superconductors

Like superfluids, superconductors show a variety of effects. Unlike superfluids, which involve neutral atoms, they are characterized by an array of electromagnetic phenomena because electrons are the basic entities involved. The first effect, as observed by Kamerlingh Onnes, is the sudden drop of the electrical resistivity to zero at the transition point. Electrical currents can then be set up with no internal electric field in the metal. Furthermore, a persistent current can be established in a loop conductor.

Suppose that a piece of superconductor is placed in a magnetic field and the temperature lowered below the transition temperature T_c. The magnetic field is excluded from the interior of the material as illustrated in Figure 13-4. It does this by setting up surface currents that establish an opposing magnetization canceling the interior field. The superconductor is a perfect diamagnet. The exclusion of the field is called the *Meissner effect* after the work of F. W. Meissner and R. Ochsenfeld.

If the external field is increased, superconductivity is destroyed at a *critical field* B_c. Excluding the field has cost the superconductor an energy per unit volume equal to the energy density of the excluded magnetic field $B_c^2/2\mu_0$, where μ_0 is the permeability of the vacuum. When this energy exceeds the difference in energy between normal and superconducting states, the system reverts back to the normal state. An amazing feature is how small this energy turns out to be, only 10^{-8} eV per atom. Considering that the various energies involving electronic interactions are all around 1 eV, including some that are not so accurately calculated, we might think it hopeless to explain such a small effect. Fortunately, it is not.

Like helium, superconductors have a specific heat increase at the transition temperature. However, the peak in superconductors is a finite jump as we see in Figure 13-5. At low T the specific heat falls below the linear dependence of a degenerate Fermi gas and is proportional to $\exp(-\text{constant}/T)$. Such a dependence is, like the situation in the superfluid, characteristic of a spectrum with a gap. If the superconductor is placed in a magnetic field with $B > B_c$, the specific heat reverts to its normal linear behavior.

Another indication of the existence of a gap in the energy spectrum is given by absorption of radiation. A superconductor reflects radiation of frequency less than some critical value lying in the far infrared. Radiation above that frequency is

Figure 13-4

Meissner effect. (a) An external magnetic field penetrates a sphere of superconducting material at $T > T_c$. (b) When the temperature is lowered below T_c the field is expelled from the interior of the sphere.

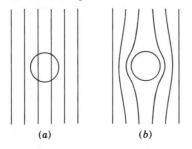

(a)   (b)

Figure 13-5

Specific heat of a superconductor. The curve shows a finite jump at the transition temperature T_c. The exponentially decreasing heat capacity drops below the linear behavior of the normal state at low temperatures.

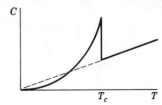

absorbed. If there is a gap in the electronic energy spectrum, an electron cannot accept energy unless the light is able to give it a sufficient amount to overcome the gap.

A seemingly minor property that was crucial in unraveling the mystery of superconductivity is the *isotope effect*. The critical temperature for a set of metals, each of which is made up of one isotope of the same element, depends on the atomic mass M of the isotope according to

$$T_c \sim M^{-\alpha},$$

where α is often about $\frac{1}{2}$. The Debye temperature of a substance is also proportional to $M^{-1/2}$ as can be determined from several of the formulas of Chapter 11. This led theorists to the suggestion that the interaction of electrons with the phonons might have something to do with superconductivity.

Another indication along these same lines is that normal systems such as copper and silver, that are good conductors of heat and electricity, generally become superconductors only at very low temperatures or not at all. Poor normal conductors, such as tin, lead, and alloys, readily become superconductors. We know that the resistivity of a metal depends on the interaction of the electrons with phonons.

The seeming flaw in this argument is that the superconducting transition temperature is generally orders of magnitude lower than the Debye temperature. Nevertheless, we see that an effective force between electrons caused by lattice vibrations acting as intermediary is indeed basic to the occurrence of superconductivity.

One of the most exciting periods in the history of condensed matter physics began in late 1986 and continues to the present: *high-temperature* superconductivity has been discovered in certain oxides. In order to keep up with fast-breaking advances in the subject physicists have had to read current news releases as well as physics journals. At present, the highest critical temperature lies above 100 K and occurs in materials having a layered structure containing planes of copper and oxygen atoms. Such high transition temperatures are technologically remarkable because they can be reached easily and cheaply by using liquid nitrogen, at 77 K, rather than liquid helium at 4 K. This opens the door to a myriad of applications. Room-temperature superconductivity is by no means out of the question.

Example

The critical magnetic field of aluminum is $B_c = 0.01$ T. The magnetic energy density corresponding to this is

$$\frac{B_c^2}{2\mu_0} = \frac{(0.01\text{ T})^2}{2(4\pi \times 10^{-7}\text{N/A}^2)} = 40\text{ J/m}^3.$$

Because the density of aluminum is 6×10^{28} atoms/m^3, this becomes

$$\frac{40\text{ J/m}^3}{(6 \times 10^{28}\text{ atoms/m}^3)(1.6 \times 10^{-19}\text{ J/eV})} = 4 \times 10^{-9}\text{ eV/atom}.$$

Example

We list a table of elements along with their superconducting transition temperatures T_c and their normal-state electrical resistivities. Note that there is a
relation, although not a perfect correlation, between high resistivity and high T_c.

Element	T_c (K)	Resistivity (10^{-4} Ω / m)
Ag	—	1.6
Cu	—	1.7
Au	—	2.4
Mo	0.9	5.7
Ga	1.1	17.4
Al	1.2	2.8
Sn	3.7	11.5
Ta	4.4	15.5
Pb	7.2	22.0
Nb	9.2	12.5

Prior to the discovery of the very-high-temperature superconducting oxide
materials the highest recorded transition temperature had for many years been
23 K in an alloy.

13-3 Superflow and the Energy Gap

We have noted in Sections 13-1 and 13-2 how certain experimental data, including
the specific heats of both superfluid helium and superconductors, indicate the existence of an energy gap in the energy spectra of each of these substances. Such an
energy gap leads to the possibility of friction-free flow and persistent currents.

The explanation is basically quite simple. Suppose we consider a sphere moving
with velocity **v** through a fluid. The sphere feels a drag when immersed in a normal
fluid. This is equivalent, by the principle of Galilean relativity, to the typical
experimental situation in which the fluid flows with friction at velocity $-\mathbf{v}$ past some
stationary object, in this case the sphere. The frictional forces arise from the transfer of
energy from the moving sphere to the liquid. When there is a gap in the spectrum,
then, for sufficiently small velocities, no energy can be transferred—because there are
no fluid states available—and we have a superfluid. For sufficiently large velocities,
enough energy can be transferred to the liquid that the gap can be surmounted. Then
frictional forces are felt by the sphere.

In superfluid flow the walls of the channel through which the helium flows, or other
obstacles, play the role of the sphere. For superconductors phonons and impurity
atoms can cause drag on the electron fluid.

To make the issue more quantitative, we use a discussion similar to one by L. D.
Landau. Denote the energy excitations of the fluid by ε_p as a function of momentum
p. (In a gas of free particles we have, of course, $\varepsilon_p = p^2/2m$; we consider other
functional forms as well.) Let the mass of the sphere be M. When the sphere causes an
excitation, its velocity changes to $\mathbf{v}'$ and the liquid absorbs energy ε_p and momentum

p. Conservation of momentum and energy give

$$M\mathbf{v} = M\mathbf{v}' + \mathbf{p} \tag{13-1a}$$

and

$$\tfrac{1}{2}Mv^2 = \tfrac{1}{2}Mv'^2 + \varepsilon_p, \tag{13-1b}$$

where $\mathbf{v}'$ is the final sphere velocity. Solving Equation (13-1a) for $\mathbf{v}'$ and substituting into Equation (13-1b), we obtain

$$\varepsilon_p + \frac{p^2}{2M} - \mathbf{v} \cdot \mathbf{p} = 0.$$

Since M is a macroscopic mass the second term is negligible and

$$\mathbf{v} \cdot \mathbf{p} = \varepsilon_p \tag{13-2}$$

is the condition for the excitation to take place. The question then becomes whether there is a value of $\mathbf{p}$ and a corresponding value of ε_p such that Equation (13-2) is satisfied for the given value of $\mathbf{v}$. If so, energy and momentum can be transferred to the fluid, which is then viscous. If Equation (13-2) cannot be satisfied, the excitation does not occur and the sphere moves through the fluid without friction.

With μ the cosine of the angle between $\mathbf{v}$ and $\mathbf{p}$, frictional flow requires that

$$v\mu = \frac{\varepsilon_p}{p}. \tag{13-3}$$

The quantity ε_p/p is a characteristic of the fluid and, since it is always positive, takes on values extending from some minimum to infinity. Suppose the minimum value is *larger* than zero. Then, if $v\mu$ is smaller than this minimum, Equation (13-3) cannot be satisfied and no excitation of the fluid can take place. With μ set at its largest value, namely unity, there is some smallest value of v, call it v_c, for which there can be an excitation of the fluid. This *critical velocity* is given by

$$v_c = \text{minimum of } \left(\frac{\varepsilon_p}{p} \right). \tag{13-4}$$

For $\varepsilon_p = p^2/2m$, this gives

$$v_c = \text{minimum of } \left(\frac{p}{2m} \right),$$

which, of course, is zero for $p = 0$. That is, a normal gas of free particles is not a superfluid; excitations can be created at any velocity.

Suppose, however, that there is a gap in the spectrum. For example, suppose

$$\varepsilon_p = \Delta + \frac{p^2}{2m},$$

Figure 13-6
Excitation spectrum with a gap. The system behaves as a superfluid at velocities less than the critical velocity, which is given by the slope of the line through the origin tangent to the curve.

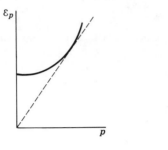

Figure 13-7
Excitation spectrum for superfluid ^{4}He. At small momenta the curve is linear, corresponding to phonon excitations. The sound velocity is denoted by c_s. At larger momenta there is a minimum with gap Δ. The excitations here are called rotons. The dotted line shows the critical velocity construction.

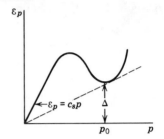

where Δ is a constant. Then the minimum of ε_p/p is easily seen to be

$$v_c = \sqrt{\frac{2\Delta}{m}}.$$

In general, the minimum of ε_p/p satisfies

$$\frac{1}{p}\frac{d\varepsilon_p}{dp} - \frac{\varepsilon_p}{p^2} = 0$$

or

$$\varepsilon_p = \frac{d\varepsilon_p}{dp}p. \qquad (13\text{-}5)$$

Figure 13-6 shows a hypothetical curve of ε_p versus p. Equation (13-5) is the equation of a straight line passing through the origin and yet somewhere tangent to the excitation curve as indicated by the dotted line in the figure. The critical velocity v_c is the slope of this tangent line.

There is no need for the lowest energy of the excitation spectrum ε_p to occur at $p = 0$ as it does in Figure 13-6. Indeed, the actual excitation spectrum of superfluid helium has its gap at finite p as shown in Figure 13-7. The spectrum has a linear region in which $\varepsilon_p = c_s p$, where c_s is a constant, followed by a minimum at an energy Δ. Despite the existence of states between zero and Δ, the tangent construction still gives a nonzero critical velocity. The linear region of the spectrum is typical of a phonon spectrum with c_s the sound velocity, as we have seen in our discussion of solids in Chapter 12. Why there are phonons and no single free-particle modes is fundamental to understanding the nature of superfluidity and is discussed in Section 13-5. The excitations in the region of the minimum of ε_p are called *rotons*.

For superconductors the excitation spectrum has the form

$$\varepsilon_p = \sqrt{\Delta^2 + \left(p^2/2m \right)^2}\,.$$

This form looks much like that shown in Figure 13-6. There is obviously a nonzero critical velocity.

Example

The excitation spectrum of helium shown in Figure 13-7 has the parameter values $\Delta/k_B = 8.65$ K and $p_0/\hbar = 19$ nm^{-1}. The sound velocity is $c_s = 238$ m/s. The critical velocity is then approximately

$$\frac{\Delta}{p_0} = \frac{(8.65 \text{ K})(1.38 \times 10^{-23} \text{ J/K})}{(1.05 \times 10^{-34} \text{ J} \cdot \text{s})(19 \times 10^9 \text{ m}^{-1})} = 60 \text{ m/s}.$$

This critical velocity is rarely observed in actual experiments. Much lower values, on the order of a few centimeters per second, are usually found. The reason is that there is another kind of excitation, the vortex, that can be created at these low velocities. The statistical weight of these vortex excitations is so small that they do not contribute to the specific heat and Landau was not aware of their existence when he presented his criterion for superfluidity.

13-4   The Bose – Einstein Condensation

As mentioned in the introduction, superfluids and superconductors have the common feature that there is some kind of condensation to a new form of order at the transition temperature. Furthermore, this new order is not spatial as in the liquid–solid transition. What other kind of order is there? As we demonstrate in this section, at least one other kind of order can occur—in momentum space. The Bose–Einstein condensation of an ideal gas involves a sudden collapse of a macroscopic number of particles into the lowest single-particle momentum state.

The system to be considered here is a collection of N noninteracting Bose particles. We discuss such a system because it is an easily solvable model showing this peculiar momentum condensation. We do not suggest it as an explicit model either for superfluid helium or for superconductors; unfortunately, things are not so simple in either case. Nevertheless, in the 1930s, London drew attention to this condensation of Bose particles that had been discovered by Einstein and, in a series of papers and in a book, he showed how certain analogies between the ideal Bose gas and the phenomena in liquid helium could be used to understand the latter. Landau, who approached superfluidity from a different point of view, including the use of the ideas presented in Section 13-3, disagreed strongly. London eventually won the argument and we attempt later to show qualitatively how the two views may be reconciled.

Chapter 11, Equation (11-26), tells us that the Lagrange multiplier α and the number of particles in a Bose system are related according to

$$N = \sum_p n_p, \tag{13-6}$$

and

$$n_p = \left(e^{\beta \varepsilon_p + \alpha} - 1\right)^{-1}. \tag{13-7}$$

The expression for n_p, the number of particles having momentum p, differs from the similar one for fermions by only the minus sign in the denominator; but what a difference a sign makes!

Let us first consider the occupation number n_0 corresponding to the lowest single-particle energy ε_0 (which is zero if we use traveling-wave boundary conditions). We have

$$n_0 = \left(e^\alpha - 1\right)^{-1}. \tag{13-8}$$

Obviously, we must have $\alpha > 0$ so that n_0 is finite and positive for bosons. (Recall that for fermions at low temperatures α is negative since it is proportional to minus the Fermi energy.) If α is very small, and in particular if $\alpha = a/N$, where a is a number of order unity, a Taylor series expansion of the exponential shows that

$$n_0 = \left[1 + \left(\frac{a}{N}\right) + \cdots - 1\right]^{-1} \to \frac{N}{a}.$$

That is, the number of particles in the lowest allowed state of energy and momentum can be a substantial fraction of the entire set of particles. Bosons, as we have shown previously in Chapter 11, do not avoid one another like fermions; they actively prefer to be in the same state. What we are going to show is that, while for high temperatures n_0 is no larger than any other occupation number, when we lower the temperature below a value T_c, α becomes of order N^{-1} and the lowest state fills until it contains a substantial fraction of the total set of particles. We then have had a transition to a phase that is partially or, at $T = 0$ K, completely condensed into the zero-momentum state.

It is easy to show that n_p, for any state other than the ground state, is very much smaller than n_0. This is true even for the *first* excited state as is shown in Problem 3 at the end of the chapter. Although any individual n_p is small compared to n_0, the sum of all of them is certainly of order N for any value of α.

Thus, for all values of α we can segregate the sum into two parts:

$$N = n_0 + \sum_p{}' n_p,$$

where the prime on the sum means that we omit the ground state. Since, in the sum, no single state is significantly larger than the rest, it is possible to change from the sum into an integral by using the standard technique of the density of states function. (It is always a good approximation to change a sum into an integral if the summand does not vary much from one term to the next. This is the case once we remove n_0. This process is clarified in Problem 3.)

By using the function $D(\varepsilon)$ of Equation (11-44), we have

$$\sum \cdots \to \int_0^\infty d\varepsilon\, D(\varepsilon) \cdots = AV \int_0^\infty d\varepsilon\, \varepsilon^{1/2} \cdots,$$

where V is the volume of the system and the constant A is defined by

$$A = \frac{m^{3/2}}{\sqrt{2}\,\pi^2 \hbar^3}.$$

We can then write

$$N = n_0 + AV \int_0^\infty d\varepsilon\, \varepsilon^{1/2} \left(e^{\beta\varepsilon + \alpha} - 1 \right)^{-1}.$$

Next, define the new variable

$$z = \beta\varepsilon$$

so that

$$N = n_0 + AV(k_B T)^{3/2} G(\alpha), \tag{13-9}$$

where

$$G(\alpha) = \int_0^\infty dz\, z^{1/2} \left(e^{z+\alpha} - 1 \right)^{-1}. \tag{13-10}$$

For very large α it is easy to show that this integral becomes

$$G(\alpha) \to \frac{\sqrt{\pi}}{2} e^{-\alpha}. \tag{13-11}$$

For $\alpha = 0$ the integral can also be evaluated. It can be shown that

$$G(0) = \frac{\sqrt{\pi}}{2} 2.612. \tag{13-12}$$

When α is of order $1/N$, it is so small that $G(\alpha)$ is essentially equal to $G(0)$ in Equation (13-12). In this case Equation (13-9) gives

$$n_0 = N - AV(k_B T)^{3/2} \frac{\sqrt{\pi}}{2} 2.612 \tag{13-13}$$

or

$$n_0 = N\left[1 - \left(\frac{T}{T_c} \right)^{3/2} \right], \tag{13-14}$$

where

$$T_c \equiv \frac{1}{k_B} \left(\frac{2N}{VA\sqrt{\pi}\,(2.612)} \right)^{2/3} = \frac{2\pi\hbar^2}{k_B m} \left(\frac{\rho}{2.612} \right)^{2/3} \tag{13-15}$$

with ρ the density N/V.

The behavior of n_0/N as a function of temperature is given by Equation (13-14) as shown in Figure 13-8. At $T = 0$ K, $n_0 = N$; all the particles are in the ground state as we might expect. However, the surprising feature is that, as the equation and the figure show, the number in the ground state remains macroscopic (of order N) all the way up to a certain temperature T_c. At that point n_0 becomes small and is no larger than the rest of the terms in the sum. Then the solution, Equation (13-14), which is based on the assumption that α is of order $1/N$, is no longer valid, and we need to

Figure 13-8

Occupation number n_0 for the lowest energy state of the ideal Bose gas. At $T = 0$ K, $n_0 = N$; macroscopic occupation continues up to a critical temperature T_c. Above T_c, n_0 is negligible compared to the total number of particles and is shown as zero in the figure.

Figure 13-9

Plot of the Lagrange multiplier α for an ideal Bose gas. This quantity is of order $1/N$ below T_c but takes on values of order unity above that temperature. The discontinuous nature of its behavior in the limit of an infinite system is characteristic of a phase transition.

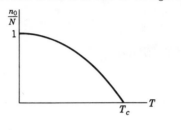

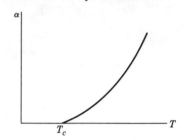

solve for α by using

$$N = AV(k_B T)^{3/2} G(\alpha).$$

Here, we have used the fact that n_0 is very small (of order 1) and have dropped it from Equation (13-9). The reader is asked to show in Problem 5 at the end of the chapter that the result for very large α is just the classical high-temperature, or ideal gas, case (equivalent to that of Equation (11-30)).

Figure 13-9 shows how α behaves over the entire temperature range. The important point is that it has a discontinuity (in first derivative) at T_c. This is a sign that a phase transition occurs at this temperature. As the temperature is lowered through T_c, the transition is one in which particles begin cascading into the zero-momentum state. This is the Bose–Einstein condensation. It is an ordering in momentum space and not in real space.

Other thermodynamic quantities of this ideal Bose gas system have discontinuities at T_c. The specific heat C/N is shown in Figure 13-10. At the transition C has a cusp of finite height. The temperature dependence of this quantity below T_c is $T^{3/2}$.

If we put in the parameters appropriate to helium, as we do in the example below, we find $T_c = 3.1$ K, which is quite close to the superfluid transition temperature of 2.2 K. While the Bose–Einstein condensation may be an appealing model for the behavior of liquid helium, there are a number of reasons why things cannot be so simple. First of all, a lot of details are not right. For example, the specific heat is finite at the transition and is proportional to $T^{3/2}$ below it; the specific heat of helium has an infinite peak at T_λ and is proportional to T^3 at low temperature. The ideal Bose transition is of first order—since it has a latent heat—rather than second order as in helium.

More importantly, the ideal Bose gas does not exhibit superfluidity. The spectrum of excitations depends quadratically on momentum corresponding to free particles. As we have seen in Section 13-3, this gives a zero critical velocity.

The atoms in helium are constantly interacting with one another and certainly cannot be considered an *ideal* Bose gas. It is impossible that n_0 could ever be equal to

Figure 13-10

Specific heat of the ideal Bose gas. C is finite at T_c but has a discontinuous derivative. At low T, C is proportional to $T^{3/2}$. It approaches the classical value $3Nk_B/2$ at very high temperature.

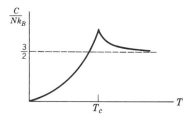

N, even at $T = 0$ K. The wave function corresponding to all particles in the zero-momentum state is a constant as a function of all the particle coordinates. (For one particle, we have $\exp(i\mathbf{p} \cdot \mathbf{r}/\hbar) = 1$ for $\mathbf{p} = 0$.) However, because of the highly repulsive core of the potential energy curve, we know that the wave function for liquid helium cannot be a constant but must vanish whenever any two particles approach too closely. Nevertheless, it is possible to have a *partial* Bose condensation into the zero-momentum state in liquid helium. Theoretical calculations predict that $n_0/N \approx$ 0.1 at absolute zero. Experiments to detect n_0 are very difficult and have often been controversial; however, recent results tend to confirm the theory.

Although the ideal Bose gas is not a good model for liquid helium, the Bose nature of the helium atom and the presence of a condensation are thought to be responsible for the characteristic behavior of the superfluid. We discuss this point more fully in Section 13-5.

Example

For the atomic mass and density of helium, Equation (13-15) gives the Bose condensation temperature as

$$T_c = \frac{2\pi\left(1.05 \times 10^{-34}\ \text{J} \cdot \text{s}\right)^2}{\left(1.38 \times 10^{-23}\ \text{J/K}\right)\left(6.65 \times 10^{-27}\ \text{kg}\right)} \left(\frac{2.2 \times 10^{28}/\text{m}^3}{2.61}\right)^{2/3} = 3.1\ \text{K}.$$

13-5 The Two-Fluid Model of Superfluid Helium

While liquid helium cannot be described as an ideal Bose gas, it is commonly believed that the Bose character of the helium atoms is fundamental to the nature of the superfluid state. Although there is as yet no truly comprehensive microscopic theory of helium, numerous theoretical calculations have indicated how important it is to include Bose statistics in the description. We describe briefly and qualitatively some of the approaches that have been used. Unfortunately, most of the details of these theories are beyond the level of this text; however, there is a less fundamental phenomenological description, the *two-fluid model*, which correlates many features of the data. We try to explain the basis of the model and then show how it makes sense out of the experiments discussed in Section 13-2.

While Landau disputed the view that the Bose condensation was vital to the nature of the superfluid, his countryman N. N. Bogoliubov was able to prove a very fundamental result that supports the view quite strongly. He showed that if a weak interaction is turned on in a previously noninteracting Bose gas, the spectrum of excitations changes from the quadratic free-particle behavior ($\varepsilon_p = p^2/2m$) to linear for small momenta. The linear spectrum, $\varepsilon_p = c_s p$, is just that of sound waves similar to those appearing in the small-momentum region of the plot in Figure 13-7. This behavior is what allows a nonzero critical velocity so that superfluidity can occur according to the Landau argument of Section 13-3. Weak interactions and the existence of a large condensate are essential to the Bogoliubov approach so that this theory, while very suggestive, is not directly applicable to helium either.

It is perhaps not too hard to understand qualitatively why the single-particle excitations, which would destroy superfluidity, are no longer present in the interacting Bose gas. Bosons prefer being in the same state with one another, so that if one atom is pushed on by an external force, all the particles within a de Broglie wavelength λ (which is large at low temperature) want to move in the same way. The collective motion of a sound wave allows this while the single-particle motions are frozen out by this tendency.

R. P. Feynman was able to go even further. He showed on the basis of mathematical as well as physical arguments that helium should have a phonon spectrum and moreover that the roton spectrum, the region of the minimum in Figure 13-7, is just a natural continuation of the phonon part. The excitations in the phonon region are longitudinal compressional waves, but the rotons are pictured as mini-vortices. The wavelength $\lambda = h/p$ corresponding to a roton is small, about 0.3 nm, which is approximately the average interparticle distance. Thus, a roton very likely involves a small number of particles. One model has them moving in a way somewhat analogous to a microscopic smoke ring. Feynman's calculations of the spectrum are in good agreement with experiment and so this aspect of helium is thought to be well understood.

Superfluid helium is condensed into an ordered state in momentum space although we have shown that it cannot be a state with *all* particles in the lowest *single-particle* level even at $T = 0$ K. However, at absolute zero the entire system is certainly in its *many-particle* ground state and is a perfect superfluid. Any experiment to measure viscosity finds it to be zero as long as the critical velocity is not exceeded.

At $T > 0$ K phonons and rotons are excited. Any one of these excitations is not separated by a gap from other excitations of the liquid. It can collide with the walls and deposit energy or take up energy and thereby cause friction. At low temperatures this "gas" of phonons and rotons is not very dense and many of the superfluid properties remain. In fact, it is reasonable to think of helium, when it is below T_λ, as a mixture of two fluids, one a perfect *superfluid* and the other a *normal fluid*. The normal fluid, composed of the gas of excitations, has viscous flow and carries heat. The superfluid is nonviscous and carries no heat. The percentage of superfluid present goes from 100% at $T = 0$ K to 0% at and above T_λ.

The two kinds of viscosity experiments mentioned in Section 13-1 are now easily understood. The torsion oscillator experiment actually measures not simply viscosity but the product of density and viscosity. As the plates oscillate in the fluid they feel no friction from the superfluid component but the normal fluid has density ρ_n, which

Figure 13-11
Superfluid density ρ_s of liquid ^{4}He as a
function of temperature. At $T = 0$ K, ρ_s is
equal to the total density and drops to zero at
T_λ.

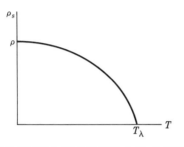

exerts a drag and is carried along as the plates oscillate. This drag diminishes as the
temperature is lowered because the normal density decreases; the result is an apparent
decrease in fluid viscosity. By a slight modification of the torsion oscillator experiment,
the Russian physicist E. L. Andronikashvili was able in 1946 to measure directly the
normal density ρ_n. The superfluid density is given by $\rho_s = \rho - \rho_n$, where ρ is the total
density, and can be plotted as shown in Figure 13-11.

Two misconceptions must be avoided here. First, despite the similarity between the
curve for ρ_s in Figure 13-11 and that for n_0 in Figure 13-8, the two quantities are
quite different. Recall that at $T = 0$ K perhaps only 10% of the particles in real
helium are condensed into the zero-momentum state, and not 100% as in the ideal gas
of Figure 13-8; however, the entire helium system is superfluid then, so that $\rho_s = \rho$.
Second, it is incorrect to think of the superfluid and the normal fluid as made up of
two different sets of atoms. Because a phonon is a collective motion, it involves many
atoms throughout the liquid, and these same atoms can also be involved in the
superfluid component.

The second kind of viscosity experiment discussed in Section 13-1 involves flow
through a small channel known as a superleak. The superleak is so small that the
normal fluid is almost unable to move through the channel; however, the superfluid
flows readily. Since almost everything that gets through is superfluid, this measurement of viscosity gives zero. Normal fluid carries the heat and is largely left behind.
This fluid is now deficient in superfluid and shows an increase in temperature if the
entire system is not maintained at a given value by the external refrigerator. Because
of the removal of the excess heat, some of the normal fluid is quickly converted to
superfluid, which then flows through the superleak. The observer gets the impression
that none of the liquid has viscosity.

The thermomechanical effect also involves having a superleak that allows the
superfluid through but not the normal fluid. When the heater in the chamber on the
right side of Figure 13-3 is turned on, the rise in temperature creates normal fluid at
the expense of the superfluid on that side. There is then a mismatch of superfluid
concentrations on the two sides of the apparatus. This is as though a gas is on the left
at a certain pressure and a partial vacuum of the same gas is on the right. Superfluid
flows through the superleak from left to right. There is also a mismatch in concentrations of normal fluids on the two sides, but since normal fluid cannot readily flow

through the channel, it remains out of equilibrium. The result is an accumulation of the total amount of extra fluid on the right as shown in the figure. We have used heat to pump superfluid from one side of the system to the other!

The creeping film can also be explained as a kind of superleak. The vapor of any fluid coats the inside walls of the container and those of any other vessel inside. The normal fluid is clamped to the wall, but, because the superfluid component moves without friction, the film can act as a siphon, which can, for example, empty a small beaker held above the level of the main liquid reservoir.

The peculiarities of heat conduction can also be analyzed on the basis of the two-fluid model. In this system heat travels mainly by convection. The normal fluid and superfluid can move relative to one another. Normal fluid readily flows away from a hot spot carrying heat with it, while the superfluid flows toward the hot spot so that the overall mass density remains unchanged. The fact that the superfluid backflow carries no heat makes this an efficient technique. Since the normal fluid is made up of the gas of excitations, we can look at this process as a drift or streaming of this gas.

As ordinary sound corresponds to variations in the overall density of a fluid, second sound is a density wave in the gas of the phonon–roton excitations. When the density of excitations is larger than average, there is an excess of normal fluid and the temperature is a bit higher there. Where the density is lower, the temperature is lower. In this way second sound is a temperature wave. The overall density of the fluid is a constant everywhere; this is maintained by having normal and superfluid densities oscillating exactly 180° out of phase.

Since phonons exist in solids as well as in liquid helium, we might expect that second sound could be found there. In most situations second sound in solids is so highly damped out by various mechanisms that it does not occur. However, it has been observed in certain specially prepared samples.

We have now seen, at least qualitatively, how some of the phenomena of super-fluidity can be interpreted. In Section 13-6 we try to do the same thing for superconductivity.

13-6   Cooper Pairs and the BCS Theory

The microscopic theory of superconductivity developed by Bardeen, Cooper, and Schrieffer has been remarkably successful. Its good agreement with experiment has caused the general opinion that superconductivity is a well-understood phenomenon. Since the theory is quite advanced and mathematically complex, our treatment is limited to a simple description of some of the theoretical results.

Cooper made the preliminary discovery that electrons can form bound pairs when there is an attractive interaction between them. It was this result that stimulated the further work that led to the full theory. It may be rather surprising that an attractive interaction can occur between electrons when we know that two isolated electrons always feel a repulsive Coulomb force. The attractive force arises because the electrons are in a crystal lattice.

As the electron passes through the lattice it attracts the neighboring positive ions toward it. Another electron nearby sees the positive grouping and is attracted to it. The resulting attraction can even overcome the bare electron–electron Coulomb repulsion so that the *net* interaction is attractive.

There is an equivalent alternative view of this interaction. When an electron passes by a positive ion core it interacts with it via the Coulomb interaction and can set the

Figure 13-12

Interaction of two electrons caused by the exchange of a phonon through the crystal lattice. If a pair of electrons with zero total momentum exchanges a phonon of momentum **q**, the final momenta are $l = p - q$ and $-l$, respectively.

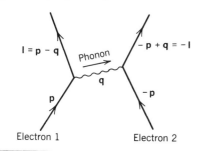

ion vibrating about its site. This in turn sets up lattice waves so that the electron can be said to have *emitted* a phonon. The vibrating ions can affect the motion of a second electron; the second electron *absorbs* the phonon. A simple diagram representing this process is drawn in Figure 13-12. As is shown in Chapter 16, analogous diagrams can be drawn for every force that occurs in nature, especially the fundamental ones like the Coulomb force. The force that results from phonon exchange is attractive and allows the formation of bound pairs of electrons.

In order to understand the various terms that occur in the final results we look at the effective interaction between electrons a bit more closely. Consider the wave function corresponding to a pair of free particles, the first having momentum $\mathbf{p}_1$ and the second momentum $\mathbf{p}_2$. Call it $\phi_{p_1 p_2}(\mathbf{r}_1, \mathbf{r}_2)$, where $\mathbf{r}_1$ and $\mathbf{r}_2$ are the coordinates of the two particles. The total momentum of the system is $\mathbf{P} = \mathbf{p}_1 + \mathbf{p}_2$, which corresponds to the pair moving with a center-of-mass velocity through the crystal while they are orbiting around each other (if indeed we find that they are bound together). For the moment we can just consider the situation in which the center of mass is at rest, that is, $\mathbf{P} = 0$, so that $\mathbf{p}_2 = -\mathbf{p}_1$. The resulting pair wave function can be represented by the simpler notation

$$\phi_p \equiv \phi_{p,\,-p}.$$

(Of course, Cooper pairs can have $\mathbf{P} \neq 0$; indeed, as we see below the pairs are responsible for carrying the supercurrent, so such moving states are essential.)

If the effective attractive interaction between the electrons is written as $V_{12} \equiv V(\mathbf{r}_1, \mathbf{r}_2)$, an important quantity is given by

$$V_{\ell p} = \int\int \phi_\ell^* V_{12} \phi_p \, d\mathbf{r}_1 \, d\mathbf{r}_2. \tag{13-16}$$

It turns out that it is this quantity that is actually represented by the diagram in Figure 13-12. A pair of electrons in momentum state $[\mathbf{p}] \equiv (\mathbf{p}, -\mathbf{p})$ exchanges a phonon that carries momentum **q**. The momentum of the one electron goes from the value **p** to $\boldsymbol{\ell} = \mathbf{p} - \mathbf{q}$, while the other electron that absorbs the phonon goes from $-\mathbf{p}$ to $-\mathbf{p} + \mathbf{q} = -\boldsymbol{\ell}$. The final pair state is then $[\boldsymbol{\ell}] \equiv (\boldsymbol{\ell}, -\boldsymbol{\ell})$ as shown. (Because of $V_{\ell p}$, the simple momentum state ϕ_p is no longer an eigenfunction. The true eigenfunction is a sum of all the different ϕ_p's combined in a manner representing a localized bound state.)

The magnitude of $V_{\ell p}$ depends on a variety of things. For example, it certainly depends on the strength of the interaction an electron has with an ion that must be wiggled to get a phonon excited. Furthermore, we know that the two electrons that are interacting must be near the Fermi surface. Otherwise, their interaction with one another does no good. That is, when one electron interacts with another it feels a force and normally might then change its momentum. However, if it is deep within the Fermi sphere a small kick has no effect because the electron cannot go anywhere in momentum space because of the Pauli principle; all the nearby momentum states are already occupied by other electrons. However, the maximum energy that can be transferred from one electron pair to another by the phonon exchange interaction is on the order of the characteristic phonon energy given by

$$\varepsilon_D \equiv k_B \Theta_D,$$

where Θ_D is the Debye temperature. Phonon energies are quite small compared to the Fermi energy; Θ_D is a few hundred kelvins, while T_F is of order 10^5 K. Thus, only electrons very near the top of the Fermi sphere can interact successfully with one another.

Therefore, we might write

$$V_{\ell p} = \begin{cases} -F, & \varepsilon_\ell \text{ and } \varepsilon_p \text{ in the interval } \left[\varepsilon_F, \varepsilon_F + \varepsilon_D \right] \\ 0, & \text{otherwise,} \end{cases} \tag{13-17}$$

where F is a positive constant that represents, among other things, how strongly an electron is coupled to the ions. The minus sign signifies an attractive potential. This interaction allows electron pairs to interchange energy with one another only in a very small band of energies around the Fermi surface.

Cooper put all these ideas together, solved the Schrödinger equation, and found that there was a bound state for arbitrarily small F. If E is the pair energy then, when there is no interaction, we expect $E = 2\varepsilon_F$, since both electrons are at the Fermi surface. A bound state has an E less than this by a bit. The binding energy can then be defined as

$$E_c = 2\varepsilon_F - E$$

and is found, for small F, to be

$$E_c = 2\varepsilon_D e^{-2/R_0 F}, \tag{13-18}$$

where R_0 is the density of particles per unit energy (Equation (12-15)) evaluated at the Fermi energy $\varepsilon = \varepsilon_F$. This says that the binding energy can be much smaller than the energy ε_D transferred in the exchange of a phonon. The possibility of having a binding energy much smaller than the phonon energy is crucial to the theory of superconductivity since we have already seen that the transition temperature, which must be of order E_c, is usually two orders of magnitude smaller than the Debye temperature. The exponential factor easily accounts for that size reduction.

The BCS theory goes far beyond the consideration of just Cooper's treatment of a pair of electrons. It takes into account the cooperative nature of the pairing. At $T > 0$ K *single* electrons are thermally excited into momentum states above the Fermi surface. Because of the Pauli principle, these states are then not available to be involved in the formation of pair states. As the temperature rises, there is a cooperative blocking of the formation of the bound state causing the binding energy to be temperature dependent. At some temperature T_c, the binding energy E_c goes to zero. This is the transition temperature for the superconducting state.

In the superconducting system the pairs are in a highly coherent state; the formation of a few encourages the formation of others in a cooperative way. This tendency is quite analogous to what happens in a Bose condensation. However, the average radius of the Cooper pair state is quite huge, on the order of 1 μm. This means that the centers of approximately a million other Cooper pairs sit inside the volume encompassed by one of them. While an even number of bound fermions often behaves like a boson, as in the case of the helium atom, the severe entanglement of the Cooper pairs implies that we cannot very well think of the superconducting transition simply as a Bose condensation of Cooper pairs. Such a picture would be valid only if the Cooper pair radius were small compared to the average interpair separation.

The BCS theory predicts that the ground state involves many electron pairs occupying states in the shell of states about the Fermi surface. A single-particle excitation has an energy (measured relative to this ground state) given by

$$E_p = \sqrt{\varepsilon_p^2 + \Delta^2}\,,$$

where $\varepsilon_p = p^2/2m$ and

$$\Delta = 2\varepsilon_D e^{-1/R_0 F},$$

which bears a strong resemblance to Equation (13-18) for the Cooper pair binding energy. When a pair is broken, two electrons are excited so that the smallest energy involved in an excitation is the *energy gap*,

$$2\Delta = 4\varepsilon_D e^{-1/R_0 F}. \tag{13-19}$$

There are no states between the ground state and this energy. However, there is a continuum of states above this energy. In this way superconductors differ from superfluids, which have the phonon modes all the way to zero energy.

For $T > 0$ K, the gap becomes temperature dependent, as we have mentioned above, and diminishes until it finally reaches zero at T_c. This transition temperature is given by

$$k_B T_c = 1.14\varepsilon_D e^{-1/R_0 F}. \tag{13-20}$$

In Section 13-2 we have pointed out that the isotope effect hints at T_c being proportional to the Debye temperature. From Equation (13-20) we see that this is indeed the case through ε_D.

The newly discovered high-temperature (~ 100 K) oxide superconductors discussed in Section 13-2 may operate by mechanisms quite different from those in the BCS theory. Alternative theories for the new materials are being examined at the present time. It can still be said, however, that the BCS theory continues to hold for superconductors with low transition temperatures.

We have now briefly seen a few of the elements of the theory of superconductivity. In Section 13-7 we take a closer look at the theoretical interpretation of the experimental data.

Example

The superconducting transition temperature in aluminum is $T_c = 1.2$ K. The Debye temperature is $\Theta_D = 420$ K. With $\varepsilon_D = k_B\Theta_D$, Equation (13-20) allows an estimate of the dimensionless interaction constant R_0F. We get

$$R_0F = -\frac{1}{\ln(T_c/1.14\Theta_D)} = -\frac{1}{\ln(1.2\text{ K}/1.14 \times 420\text{ K})} = 0.17.$$

A simple form of the BCS theory, the weak-coupling case, gives the results quoted in this section. This case holds when R_0F is less than unity, as occurs for aluminum. The combination of Equations (13-19) and (13-20) gives us a simple relation between two experimentally measurable quantities, T_c and the gap 2Δ. We find

$$\frac{2\Delta}{k_BT_c} = \frac{4}{1.14} = 3.51.$$

Experimentally, this ratio is found to range from 2.8 to 4.6. For aluminum $2\Delta/k_B$ is 4.2 K so that the ratio is 3.4.

13-7 Theoretical Interpretation of Superconducting Experiments

In Section 13-3 we have seen how the existence of an energy gap explains superflow in liquid helium. The condensation of Cooper pairs also provides a spectrum with an energy gap so that we can discuss nonresistive flow in superconductors in a very analogous way. (We point out, however, that this argument is not complete and the question of nonresistive flow and persistent currents is more complicated than we have indicated. However, we are unable to go into the details here.)

The energy gap, equal to 2Δ, also gives rise to the experimentally observed specific heat dependence, which has roughly the form $\exp(-\Delta/k_BT)$ below T_c. The theoretical values of Δ give good agreement with the experimental results from specific heat and radiation absorption experiments.

Because a magnetic field above a critical value B_c can destroy superconductivity, the energy density of the critical field, given by the value $B_c^2/2\mu_0$, must represent the difference W_0 in energy densities between normal and superconducting states. It is easy to calculate the value of this in terms of the energy gap. We make use of methods similar to those used in Chapter 12. Since the pairing involves electrons in a thin shell of states at the Fermi surface, the density of pairs is one-half the number R_0 of particles per unit volume per unit energy interval times the width of the energy shell, given by the gap parameter Δ, or

$$\text{pair density} = \frac{R_0\Delta}{2}.$$

In the superconducting state, each of the pairs is lowered in energy by $-\Delta$, so that the total reduction in energy density is

$$W_0 = -\frac{R_0 \Delta^2}{2} = -2R_0 \varepsilon_D^2 e^{-2/R_0 F}. \tag{13-21}$$

By setting this equal to the magnetic energy density, we can evaluate B_c as we do in an example at the end of this section. The results are quite good. The amazing thing is that W_0 is so small, $\sim 10^{-8}$ eV, as calculated in Section 13-2. There are many energies in normal metal physics that are known to much less accuracy than this. It is only that those effects are the *same* in both normal and superconducting states that allows us to ignore them and focus on this small but obviously vital superconducting effect.

We next look into some of the physics of the Meissner effect. Our discussion involves the use of the differential version of Maxwell's equations in electromagnetic theory.

The Meissner effect is the expulsion of the magnetic field from the interior of the superconductor as shown in Figure 13-4. Surface currents are set up that establish a magnetic field opposing the external field. The field deep inside is canceled out exactly. It turns out that it is not sufficient just to assume that the system is a perfect conductor. A perfect conductor has the property that once a field is established the internal magnetic field is unchanging in time. Thus, it is trapped at the value it had before the material went superconducting; it does not necessarily vanish as the Meissner effect requires.

In Section 12-3 we have treated an equation for the particle velocity of a charge in a resistive medium. In the case of *no resistive forces*, Equation (12-9) combined with the definition of current density, $\mathbf{J} = \rho e \mathbf{v}$, gives

$$\frac{\partial \mathbf{J}}{\partial t} = \frac{\rho e^2}{m} \mathbf{E}.$$

If $\mathbf{J}$ is $\mathbf{J}_s$, the induced current density in the superconducting sample, this equation must be modified because the charge carriers are Cooper pairs; e is replaced by $2e$ and m by $2m$ to give

$$\frac{\partial \mathbf{J}_s}{\partial t} = \frac{2\rho e^2}{m} \mathbf{E}. \tag{13-22}$$

Since we consider an external field to be present, there must also be a current density $\mathbf{J}_e$ that acts as a source for that field. The total field is given in terms of both sets of currents according to the Maxwell equation

$$\nabla \times \mathbf{B} = \mu_0 (\mathbf{J}_s + \mathbf{J}_e). \tag{13-23}$$

We are interested in the value of the field at the position of the superconductor; we assume that the external currents $\mathbf{J}_e$ are zero there and need not appear further in our equations.

The electric field in Equation (13-22) obeys the Faraday law

$$\nabla \times \mathbf{E} = -\frac{\partial \mathbf{B}}{\partial t}.$$

The substitution of $\mathbf{E}$ from Equation (13-22) into the last equation gives the result, valid for a substance with zero resistance:

$$\frac{\partial}{\partial t}\left(\mathbf{B} + \frac{m}{2\rho e^2}\nabla \times \mathbf{J}_s\right) = 0. \tag{13-24}$$

This relation is what the Maxwell equations require. Simply having a perfect conductor obviously requires only that the quantity in parentheses be constant in time. We see in Problem 10 at the end of the chapter that this also causes the field in the interior of the superconductor to be constant in time rather than to vanish as the Meissner effect requires. London proposed that an equation more restrictive than Equation (13-24) must hold for superconductors. This *London equation* is

$$\mathbf{B} + \frac{m}{2\rho e^2}\nabla \times \mathbf{J}_s = 0. \tag{13-25}$$

We accept this equation without proof, and next we see how the Meissner effect follows from it. The BCS theory does indeed give Equation (13-25). The term containing $\mathbf{J}_s$ can be reexpressed in Equation (13-25) by use of Equation (13-23). (Remember that $\mathbf{J}_e$ is zero at the positions we are considering.) Then we use the mathematical identity

$$\nabla \times (\nabla \times \mathbf{B}) = \nabla(\nabla \cdot \mathbf{B}) - \nabla^2\mathbf{B}.$$

By yet another Maxwell equation, we have $\nabla \cdot \mathbf{B} = 0$, so that Equation (13-25) reduces to

$$\mathbf{B} = \lambda^2\nabla^2\mathbf{B}, \tag{13-26}$$

where

$$\lambda^2 = \frac{m}{2\mu_0\rho e^2}. \tag{13-27}$$

We solve Equation (13-26) in the special case in which the superconductor is a semi-infinite block whose face is at $x = 0$ as shown in Figure 13-13. The external field $\mathbf{B} = B\hat{\mathbf{z}}$ is in the z direction, parallel to the face of the block. The variation of B is in the x direction only so that Equation (13-26) reduces to

$$B(x) = \lambda^2\frac{d^2B(x)}{dx^2}.$$

The solution of this equation has $B = \text{constant}$ on the boundary at $x = 0$, while for $x > 0$, inside the material, the solution is

$$B(x) = B(0)e^{-x/\lambda}. \tag{13-28}$$

This result says that the field decays exponentially as one goes into the superconductor

Figure 13-13

Expulsion of the magnetic field in a semi-infinite block of superconducting material. The field, in the z direction, decreases from its exterior value according to $\exp(-x/\lambda)$, where λ is the penetration depth.

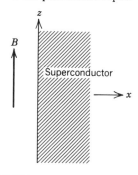

and vanishes deep in the interior. So the Meissner effect is not perfect at the surface; the field actually extends approximately the *penetration depth* λ into the material. Since λ can be as large as 0.1 μm very thin films having a field extending all the way through can be made. Since these systems do not expend much energy expelling the field, they have very high critical fields.

We show via Problem 11 at the end of the chapter that the supercurrent $\mathbf{J}_s$ also resides on the surface within a penetration depth. For later use we note that this result, together with Equation (13-22), implies that the electric field $\mathbf{E}$ always vanishes in the interior of a superconductor.

The unusual properties of superconductors make them useful in many different applications. The recent discovery of oxides that superconduct at high temperature may mean the practical possibility of superconducting electrical transmission lines, magnetically levitated trains, and a host of other uses. In Section 13-8 we see further properties and some of the devices that can be made based on another effect in superconductors.

Example

By setting the ground-state energy density W_0 given by Equation (13-21) equal to the magnetic field energy density, we can find the critical field:

$$\frac{B_c^2}{2\mu_0} = \frac{R_0 \Delta^2}{2},$$

where the density of particles per unit energy at the Fermi surface ($\varepsilon = \varepsilon_F$) is shown in Equation (12-15) to be

$$R_0 = \frac{3\rho}{2\varepsilon_F}.$$

Thus, we obtain

$$B_c = \Delta \sqrt{\frac{3\mu_0 \rho}{2\varepsilon_F}} \ .$$

We recall from the example after Section 13-6 that, for aluminum, $\Delta/k_B = 2.1$ K. Also, since $\rho = 6 \times 10^{28}$ atoms/m^3 and $\varepsilon_F/k_B = 1.4 \times 10^5$ K, we have

$$B_c = (1.38 \times 10^{-23} \text{ J/K})(2.1 \text{ K}) \left[\frac{1.5(4\pi \times 10^{-7} \text{ N/A}^2)(6 \times 10^{28} \text{ m}^{-3})}{(1.38 \times 10^{-23} \text{ J/K})(1.4 \times 10^5 \text{ K})} \right]^{1/2}$$

$$= 7 \times 10^{-3} \text{ T},$$

a result in fairly good agreement with the experimental value of 10×10^{-3} T.

Example

The penetration depth of the magnetic field into a superconductor is given by Equation (13-27). For aluminum we find

$$\lambda = \sqrt{\frac{m}{2\mu_0 \rho e^2}} = \left[\frac{9.1 \times 10^{-31} \text{kg}}{2(4\pi \times 10^{-7} \text{ N/A}^2)(6 \times 10^{28} \text{ m}^{-3})(1.6 \times 10^{-19} \text{ C})^2} \right]^{1/2}$$

$$= 1.1 \times 10^{-8} \text{ m} = 11 \text{ nm}.$$

This is in only qualitative agreement with the experimental value of ~ 50 nm.

13-8   The Josephson Effect

Superconductivity has given rise to the possibility of a variety of devices. Some of them are based on a very interesting discovery due to B. D. Josephson. The *Josephson effect* is worth studying in detail because it illustrates the high coherence of the superconducting state. It is also interesting because it involves a purely quantum mechanical effect, quantum tunneling, in a basic way. *Josephson junctions*, as the elements are called, have been used in a number of applications, such as in the very sensitive measurements of voltage and magnetic field, and as computer memory elements.

In a Josephson junction two pieces of superconducting material are separated by a thin insulating oxide layer as shown in Figure 13-14. Electron pairs in a superconducting state are separated from other pairs in a slightly different state on the other side of the oxide barrier. If the barrier is thin enough, pairs can tunnel, via quantum barrier penetration, from one side of the layer to the other. If the two states differ, or are made to differ by the application of a voltage, a net current arises. The existence of this current not only reveals the existence of quantum tunneling but dramatically illustrates the remarkably coherent character of the superconducting state.

Back in Chapter 5 we have learned that an energy eigenfunction for a single particle corresponding to energy E can be written, for one-dimensional motion, as

$$\Psi(x, t) = \psi(x)e^{-i\omega t},$$

Figure 13-14
Josephson junction. Two pieces of
superconductor are separated by a thin oxide
layer of thickness $2a$.

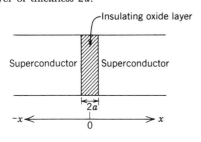

where $\omega = E/\hbar$ and $\psi(x)$ is the time-independent part of the wave function. Because of the highly ordered nature of the many-particle superconducting wave function, we are able to define a similar quantity that is a function of a single variable (in three dimensions it is a function of vector position $\mathbf{r}$). This quantity behaves like a single-particle wave function; but it actually describes the behavior of every particle —actually every Cooper pair—simultaneously. It is called an *order parameter*. Such a function is possible only because the condensed pairs of electrons are so coherent; we cannot push on one without affecting all of them.

In this interpretation $|\psi(x)|^2$, which in a normal wave function is the probability density, here becomes the density of superconducting pairs

$$|\psi(x)|^2 = \rho_s,$$

a real quantity while $\psi(x)$ itself can be complex. Quite generally we can write

$$\psi(x) = \rho_s^{1/2} e^{-i\phi_0},$$

where ϕ_0 is the *phase* of the time-independent part of the order parameter. The full time-dependent order parameter then has the general form

$$\Psi(x, t) = \rho_s^{1/2} e^{-i\phi}, \qquad (13\text{-}29)$$

where the phase ϕ is

$$\phi = \phi_0 + \omega t.$$

In Chapter 5 a probability current is also introduced as

$$j = \frac{\hbar}{2im} \left(\Psi^* \frac{\partial \Psi}{\partial x} - \frac{\partial \Psi^*}{\partial x} \Psi \right). \qquad (13\text{-}30)$$

When Ψ is the superconducting order parameter, this expression multiplied by the electronic charge gives an electric current density. Remember that we are talking about a current carried by Cooper pairs so that we must really multiply by $-2e$ (e is taken as a positive quantity) and replace m by $2m$, where m is the mass of a single electron. The two factors of 2 cancel so that the electric current density is given by

$$J = -ej.$$

Next, we show that the current is dependent on the spatial variation of the *phase* of the order parameter. We then show how that phase can be controlled by applying a voltage in order to generate a tunneling current. Suppose that ρ_s in Equation (13-29) is independent of x. Then combining Equations (13-29) and (13-30) gives

$$J = \frac{e\hbar\rho_s}{m}\frac{\partial\phi}{\partial x}. \tag{13-31}$$

If we put a voltage V across the junction, the resulting change in electronic energy then affects the time-dependent phase. We take the zero of potential at the center of the gap, $x = 0$, in Figure 13-14. Then the left side of the junction is described by

$$\Psi_1(x, t) = \rho_s^{1/2}e^{-i\phi_1}, \tag{13-32}$$

where now

$$\phi_1 = \phi_0 + \left(\omega + \frac{eV}{\hbar}\right)t. \tag{13-33}$$

The potential on the left side is $-V/2$, which is multiplied by $-2e$, the pair charge, to get the associated potential energy given in Equation (13-33). The order parameter on the right side of the gap is described by an equation analogous to Equation (13-32), with 1 replaced by 2 and with the phase on the right side given by

$$\phi_2 = \phi_0 + \left(\omega - \frac{eV}{\hbar}\right)t. \tag{13-34}$$

We assume that the density of Cooper pairs ρ_s and the constant phase ϕ_0 are the same on both sides of the junction. We have also assumed that the zero-potential frequency ω is the same on both sides of the junction. What the potential V has done is to alter the frequency in a different way on the two sides of the gap so that the phase has a gradient; a current is then established according to Equation (13-31). Since the phase difference is time dependent, the current is AC. The only reason that we can maintain such a voltage is because the oxide layer is insulating.

Is any current that might want to flow prevented from doing so by the insulating layer? For thick layers it is, but if the gap is thin enough quantum tunneling through it allows the current to proceed. We assume that the oxide layer represents a constant potential energy barrier. We have treated penetration of such a barrier in Chapter 5 and from that treatment we know that within the oxide layer the left-hand-side order parameter decays exponentially so that

$$\Psi_1(x, t) = \rho_s^{1/2}e^{-i\phi_1}e^{-\alpha(x+a)}, \qquad x > -a. \tag{13-35}$$

The decay constant α is determined by the size of the potential barrier as shown in Chapter 5. Similarly, pairs originating on the right side of the gap are described by

$$\Psi_2(x, t) = \rho_s^{1/2}e^{-i\phi_2}e^{+\alpha(x-a)}, \qquad x < a. \tag{13-36}$$

Figure 13-15

Exponential decay of the order parameters Ψ_1 and Ψ_2 in the potential energy barrier of the oxide layer in a Josephson junction.

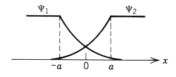

This function decays as we move from $x = a$ to smaller x values. The values of Ψ_1 and Ψ_2 are shown schematically in Figure 13-15.

We expect, from our treatment of the binding of the diatomic molecule in Chapter 10, that an electron pair capable of being in each of two regions has an overall wave function in the intermediate region given by either of the two functions

$$\Psi_{\pm} = (\Psi_1 \pm \Psi_2)/\sqrt{2}$$
$$= \rho_s^{1/2}\left[e^{-i\phi_1}e^{-\alpha(x+a)} \pm e^{-i\phi_2}e^{+\alpha(x-a)}\right]/\sqrt{2}. \qquad (13\text{-}37)$$

We can work with either of these functions; both give the same result for the current. We use Ψ_+.

We find the current in the gap by substituting Ψ_+ into Equation (13-30). After a small amount of algebra we reach the result

$$J = J_0 \sin\theta, \qquad (13\text{-}38)$$

where

$$\theta = \phi_1 - \phi_2 = \frac{2eV}{\hbar}t \qquad (13\text{-}39)$$

and

$$J_0 = -\frac{\alpha\rho_s e\hbar}{m}e^{-2\alpha a}.$$

The details of the calculation are left to Problem 12 at the end of the chapter.

There are several things to note about the Josephson current density of Equation (13-38). If the potential V vanishes, there is no current. (We have assumed that ϕ_0 is the same on both sides of the gap. If it is not—and we have no control over this phase—there is a DC current dependent on the difference in the two ϕ_0 values.) If V is not zero an AC current results. The frequency of this current is

$$\nu_J = \frac{2eV}{h}. \qquad (13\text{-}40)$$

For $V = 1\ \mu V$, this frequency is 484 MHz. Microwave radiation originating from this oscillatory current has been detected. This phenomenon is known as the *AC Josephson effect*.

Figure 13-16

Behavior of a ring of superconducting material when the ring is placed in a magnetic field and the temperature is lowered below T_c. If the field is then turned off, magnetic flux is trapped. The path of integration for Faraday's law is shown in (a). The trapped flux is illustrated in (b).

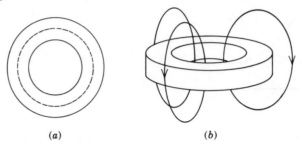

(a)   (b)

Josephson junctions provide the best-known measurement of the fundamental constant ratio e/h; they can also provide very sensitive measurements of voltage. A well-known company spent many millions of dollars developing a computer based on the Josephson effect only to abandon the project in favor of the continued use of silicon. The Japanese are maintaining efforts to develop a Josephson computer.

The AC effect is the result of applying a DC potential difference. If an AC potential is applied then, curiously enough, a DC current results when the applied frequency is ν_J.

A very interesting device can be made from the Josephson junction. Before considering it, we need to look at the phenomenon of *flux quantization*. Suppose we form a ring of superconducting material, place it in a magnetic field, and then reduce the temperature below T_c. While the Meissner effect excludes the field from the superconducting material itself, the field is still able to thread through the hole in the ring. If the external field is turned off, the flux through the hole is unable to decrease. The reason is Faraday's law. The rate of change of flux Φ through an area is the integral of the electric field $\mathbf{E}$ around the perimeter of the area:

$$\oint \mathbf{E} \cdot d\ell = - \frac{\partial \Phi}{\partial t}. \qquad (13\text{-}41)$$

If, as illustrated in Figure 13-16a, we take the circuit of integration around the ring within the material, then, since $\mathbf{E}$ is zero inside any superconductor, we must have $\partial \Phi / \partial t = 0$. That is, the flux is trapped as Figure 13-16b indicates. It is maintained by the flow of surface currents in a thin layer of depth λ in the surface of the superconductor. An important feature here, that we have not derived, is that the amount of flux threading through the ring is *quantized*. It is known that

$$\Phi = \frac{2\pi n\hbar}{q},$$

where n is an integer and q is the charge on the carrier of the supercurrent. Since the charge carriers are Cooper pairs, the value of q is $2e$. The result, that the minimum

Figure 13-17
Superconducting quantum interference device.
The SQUID is a ring of superconductor
interrupted by a Josephson junction.

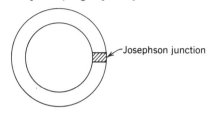

Josephson junction

value of the quantized flux is

$$\Phi_0 = \frac{\pi \hbar}{e} \qquad (13\text{-}42)$$

and not $2\pi\hbar/e$, has been verified experimentally. London originally predicted this
effect before the BCS theory and assumed $q = e$. The result $q = 2e$ is thus a strong
element in the verification of the BCS theory.

Quantized lines of flux can also be trapped within the interiors of certain forms of
superconductors. The Meissner effect is not really violated here because the "hole"
through which the field is threaded is a line of *normal* material. Such superconductors
are designated as *type-II*. The critical fields in these materials are much higher than in
the usual *type-I* form of superconductor, which does not allow such penetration by
magnetic fields without destruction of the superconducting state.

Suppose we now introduce a Josephson junction somewhere in the ring as pictured
in Figure 13-17. The resulting object is the *superconducting quantum interference device* or
SQUID. Analysis shows that the current density in the ring, which necessarily involves
tunneling through the junction, has the form of Equation (13-38). Now, however, the
phase difference θ depends on the magnetic field and is given by

$$\theta = \frac{2\pi\Phi}{\Phi_0},$$

where Φ is the total magnetic flux through the loop. This total flux is made up of the
flux due to any external field as well as that resulting from any surface currents in the
ring. We do not enter into the details of the theory or the use of this device, but only
mention that it has become a very useful tool in condensed-matter physics as well as in
other areas. It can provide an extraordinarily sensitive measurement of magnetic
fields. A flux may be measured to within $10^{-4}\Phi_0$, which is equivalent to a sensitivity
of order of 10^{-13} T!

As we have seen, Josephson junctions are illustrations of fundamental principles in
solid-state physics and quantum mechanics, and at the same time are extremely
practical devices.

Example

The frequency involved in the AC Josephson effect obeys the relation

$$\frac{\nu_J}{V} = \frac{2e}{h} = \frac{2(1.6021 \times 10^{-19} \text{ C})}{6.6256 \times 10^{-34} \text{ J} \cdot \text{s}} = 4.836 \times 10^{14} \text{ Hz/V} = 483.6 \text{ MHz/}\mu\text{V}.$$

The basic unit of quantized flux is just the inverse of this:

$$\Phi_0 = \frac{h}{2e} = \frac{V}{\nu_J} = \frac{1}{4.836 \times 10^{14} \text{ Hz/V}} = 2.068 \times 10^{-15} \text{ Wb}.$$

For a loop of radius ~ 1 mm and area 10^{-6} m^2 this corresponds to a magnetic field of magnitude

$$B = \frac{2.068 \times 10^{-15} \text{ Wb}}{10^{-6} \text{ m}^2} = 2.068 \times 10^{-9} \text{ T}.$$

13-9 Other Superfluid Systems

Other substances, besides ^{4}He, have the potential for superfluidity. As we have already mentioned, most boson systems solidify before they have a chance to become superfluid. The only other commonly available material, which, like ^{4}He, does not solidify even at $T = 0$ K, is ^{3}He. Although ^{3}He is a fermion system it is found to have a superfluid transition, but at a temperature of 2.7 mK, almost three orders of magnitude lower in temperature than the transition in ^{4}He. There is also an "artificial" substance, atomic hydrogen, which is now being studied in hope of finding a Bose condensation. We do not intend to describe either of these substances at length but instead present a brief qualitative picture of each.

In 1972 Osheroff, Richardson, and Lee were running an experiment on solid ^{3}He and found two "glitches" in the melting curve. These were ultimately attributed to the transitions to two different superfluid states in the liquid. After 70 years, superfluidity was no longer restricted to just ^{4}He.

The ^{3}He system is a set of fermions as are the electrons in a metal. The onset of superfluidity in this system involves the pairing of ^{3}He atoms analogous to the superconducting phase change in metals. There is little relation to the λ transition in ^{4}He. However, there are some considerable differences from the superconducting state. Obviously, ^{3}He atoms do not carry any charge and so electrical supercurrents do not exist; instead, the viscosity of the liquid is considerably reduced. In superconductors the electron pairs are in an $\ell = 0$ singlet state. However, in ^{3}He, the pair state has a nonvanishing angular momentum; it is an $\ell = 1$ triplet state. The introduction of this new vector into the problem makes this state quite complex. The ℓ vectors of the ^{3}He atoms can align so that the liquid develops a directional quality known as *texture*.

More than one kind of pairing can take place so that there are two superfluid states, one as described above and the other isotropic. In an external magnetic field a third superfluid state is formed. Because the possibility of $\ell = 1$ superconductivity had been considered long before the discovery of superfluidity in ^{3}He, this system was very quickly understood theoretically. The microscopic description of ^{3}He is now probably more complete than that of ^{4}He.

Hydrogen gas is made up of tightly bound diatomic molecules, so that atomic hydrogen is highly unstable to the formation of H$_2$. The electrons of H$_2$ are in a singlet state; the spins are antiparallel, as we have seen in Chapter 10. Thus, two atoms of hydrogen, approaching each other with antiparallel spins, feel a very strong attraction.

On the other hand, it turns out that two hydrogen atoms approaching one another with parallel spins (the triplet state) feel at most a very weak van der Waals attraction at a distance of several Bohr radii (weaker even than the He–He attraction) and a strong repulsion at short distances. A gas of hydrogen atoms all with parallel electron spins is called *spin-polarized hydrogen* or H↓. The interatomic force is so weakly binding that, like helium, no solid forms, even at $T = 0$ K. Even more astonishing is the prediction that the liquid state does not condense at any temperature; the material is expected to remain a gas all the way to absolute zero at low pressures.

Furthermore, the hydrogen atom, because it is a combination of two fermions, is a boson. In principle then, it is possible to cool H↓ until it undergoes a Bose condensation. Because this system is a dilute gas, its theoretical description is a much simpler problem than that for the dense ^{4}He liquid. Much of this theory already exists.

The experimental problem with atomic hydrogen is stabilizing all the atoms into the parallel-spin state. A strong magnetic field is used to align the electronic spins. (The electron spin has its lowest state in the direction opposite to the magnetic field; this is the reason for the downward pointing arrow in H↓.) Unfortunately, there are several interactions in the gas and on the container walls that can flip spins leading to the destruction of H↓ through the formation of H_2. For this reason the highest density achieved to date is only about 5×10^{24} atoms/m^3, which is two orders of magnitude too low to have the Bose condensation at the convenient temperature of 1 K. For temperatures much lower than this the gas becomes adsorbed onto the container walls where it rapidly recombines. To overcome these problems experimentalists are developing magnetic and laser "traps" that confine the particles to regions away from walls and simultaneously provide unusual cooling procedures. Considerable interest has been shown in this system and there is a continuing effort to reach the conditions necessary to observe the Bose condensation and the presumed accompanying transition to a new superfluid state.

Problems

1. In the neighborhood of the roton minimum the ^{4}He excitation spectrum is given by

$$\varepsilon_p^{\text{roton}} = \Delta + \frac{(p - p_0)^2}{2\mu}$$

   with $\Delta/k_B = 8.65$ K, $p_0/\hbar = 19$ nm^{-1}, and $\mu = 0.16m$ ($m =$ the ^{4}He atomic mass). Use this form to find the numerical correction to the approximation $v_c \approx \Delta/p_0 = 59.8$ m/s for the critical velocity calculated in the example at the end of Section 13-3.

2. Suppose some superfluid system has an excitation spectrum given by

$$\varepsilon_p = A\left[1 - \left(\frac{p}{p_0}\right)^2 + \left(\frac{p}{p_0}\right)^3\right]$$

   in which $A/k_B = 10$ K, $p_0/\hbar = 10$ nm^{-1}, and $p > 0$. Calculate the critical velocity by graphical means. Repeat the calculation analytically.

3. In considering the Bose condensation of an ideal gas, we argue that n_0 is of order N for $T < T_c$, while $n_1, n_2, \ldots$ are all much smaller so that we can separate off n_0 and treat the

sum of the remaining terms as an integral. Using the expression in Equation (12-6) for the momentum and assuming the form $\alpha = a/N$, show that the occupation numbers $n_1, n_2, \ldots$ for the very low excited states are all of order $N^{2/3}$. This is indeed much less than N when N is very large. How many terms having this size are there and of what order is their net contribution to the integral?

4. Compute the total energy and heat capacity for the ideal Bose gas for temperatures less than the Bose condensation temperature T_c.

5. (a) When α is large the exponential of Equation (13-10) is much greater than unity. Show that Equation (13-11) follows. In this case n_0 can be neglected compared to the second term in Equation (13-9).

    (b) Show that the result for α is that of the classical ideal gas of Equation (11-30).

6. An early form for the excitation spectrum of rotons, due to Landau, was

$$\varepsilon_p^{\text{roton}} = \Delta + \frac{p^2}{2\mu}.$$

Compute the heat capacity implied by this spectrum when $k_B T \ll \Delta$. Note that rotons, like phonons, have a Bose distribution $n_p = (e^{\beta \varepsilon_p} - 1)^{-1}$.

7. The radius of a Cooper pair state is about 1 μm. By using the data for aluminum given in the text, calculate how many other Cooper pairs have their centers within the volume occupied by one pair.

8. The superconducting critical temperature of tungsten is 12 mK and the critical field is 10^{-4} T. The mass density of tungsten is 19.3 g/cm^3 and its Debye temperature is 310 K. (a) Evaluate the energy gap 2Δ. (b) Evaluate the energy density of the superconducting state by two different methods: (i) compute $B_c^2/2\mu_0$ and (ii) evaluate W_0 in Equation (13-21). To carry out (ii) you need to estimate ε_F by an appropriate formula from Chapter 11.

9. Evaluate the penetration depth λ of tungsten by using the data given in Problem 8.

10. The London equation of BCS theory, Equation (13-25), leads to the Meissner effect, Equation (13-26). Show that the simpler assumption of perfect conductivity, as embodied in Equation (13-24), leads instead to the result

$$\dot{\mathbf{B}} = \lambda^2 \nabla^2 \dot{\mathbf{B}},$$

where $\dot{\mathbf{B}} \equiv \partial \mathbf{B}/\partial t$. Prove that this implies that the field value becomes fixed at the value it had when the material became a perfect conductor rather than at zero.

11. By using the relations developed in Section 13-7, prove that the current in a superconductor resides within a penetration depth λ of the surface. (You also need the equation of continuity $\nabla \cdot \mathbf{J}_s = \partial \rho_s/\partial t$, the right side of which vanishes in a steady-state condition.)

12. Carry out the steps needed to derive the Josephson relation, Equation (13-38).

13. The flux through a superconducting ring is Φ_0, the basic unit of quantized flux. To what average magnetic field does this correspond, if the ring has a diameter of 2 mm?

14. When a time-dependent potential is placed across a Josephson junction, the phase $eV_0 t/\hbar$ is replaced by $e\int V(t)\, dt/\hbar$. Suppose that $V(t) = V_0 + v \cos \gamma t$, where $v \ll V_0$. Use the Josephson relation, Equation (13-38), and the approximation $\sin(x + \delta x) \approx \sin x + \delta x \cos x$ (for $\delta x \ll x$) to show the DC Josephson effect. That is, show that J has a nonzero time average if the impressed frequency is $\gamma = 2eV_0/\hbar$.

15. Spin-polarized hydrogen has been concentrated to a density of 5×10^{24} m^{-3}. What is the Bose condensation temperature for this density if you assume this system is an ideal gas?

F O U R T E E N

PROPERTIES
AND
MODELS
OF
THE
NUCLEUS

The first investigations of the nucleus were conducted by Rutherford, the founder of nuclear physics. His contributions became influential in the years following the discovery of natural radioactivity. He interpreted this phenomenon as a transformation of matter within the atom, resulting in the emission of two new forms of radiation, and he gave the names *alpha* and *beta* to the two types of emitted particle. His studies of α radiation identified the α particle as a doubly charged helium ion. These emissions were used as beam particles to probe the atom before the coming of the accelerator. The decisive experiments on the scattering of α particles by atoms began in 1909 under Rutherford's direction. The results established the existence of the nucleus and supported the nuclear model of the atom.

Subsequent experiments in Rutherford's laboratory found that protons were produced when α particles collided with the nitrogen atoms in a gas target. These observations of the splitting of the nucleus came in 1919 and gave the first indications of nuclear substructure. Rutherford drew upon this evidence to propose the existence of the *neutron*, a neutral particle supposed to occur along with the proton as a second fundamental constituent of the nucleus. He and his associate J. Chadwick undertook an experimental search for the neutron in order to verify his bold prediction of a new type of nuclear particle. Rutherford's laboratory was dedicated to these explorations of the nucleus at a time when other investigators were focusing their attention in another direction toward the understanding of the quantum theory.

Rutherford's foresight was rewarded when the neutron was finally discovered by Chadwick in 1932. Several other important discoveries also came to light in the same "miraculous year." The chemist H. C. Urey identified the deuterium atom as a heavy form of hydrogen. The deuterium nucleus, the deuteron, provided the simplest

667

proton–neutron system for the study of the force between nuclear particles. Accelerated beams of protons were employed for the first time to cause the disintegration of the nucleus. Accelerators made it possible to probe the structure of the nucleus at controlled energies higher than any obtainable from radioactive sources of α particles. Thus, in a single year, the second constituent of the nucleus was detected, the basic two-body nuclear system was found, and the examination of the nucleus at increasing energy was begun.

Inquiry into the experimental and theoretical problems of nuclear physics quickened after 1932. The pioneering Rutherford passed his inspiration along to the next generation whose leaders included Bohr, Heisenberg, and especially Fermi. These proponents of the quantum theory of the atom advocated the extension of quantum mechanics to the theoretical treatment of the nucleus. It was realized that the nuclear particles were governed by an unknown interaction, and it was believed that the properties of the unknown force could be deduced from experiments on nuclear scattering and nuclear binding. The lack of an underlying theory drove the investigators to adopt a phenomenological approach where *models* could be used to interpret the accumulation of experimental facts. Several of these models have enjoyed a degree of success within their limited domains of validity.

Nuclear models have developed on two separate levels. The basic force between nuclear particles and the application of the force to complex nuclear structure fall into separate areas of speculation. Progress has been made in these areas through the use of different kinds of models. A truly fundamental theory of the nuclear force is only now beginning to germinate by way of an underlying theory of the elementary particles.

The purpose of this first of two chapters on nuclear physics is to examine the main properties of the nucleus and the nuclear force. We take advantage of our experience with atoms and apply our knowledge of quantum mechanics to introduce some of the prevailing models. The phenomenology continues in our treatment of nuclear decays and nuclear reactions in Chapter 15. The specific nature of the fundamental force between the constituents of the nucleus remains submerged until the question can be brought back to the surface in the final chapter on elementary particles.

14-1 Nuclear Particles

A preliminary picture of the nuclear size, charge, and mass was in existence by 1920. The α-particle scattering experiments found the radius of the nucleus to be of order 10^{-14} m, four orders of magnitude smaller than the size of the atom. Moseley's analysis of characteristic x rays established the connection between the nuclear charge and Mendeleev's atomic number Z. The smallness of the electron mass meant that almost all the mass of the atom was concentrated in the nucleus. It was recognized that the atomic mass was close to a whole number of hydrogen mass units, and it was also noted that this *mass number A* was approximately twice as large as the corresponding *charge number Z* for most atoms.

The first tentative nuclear model assumed a bound structure of protons and electrons, the only known particles. It was thought that the nucleus contained A protons to account for the observed mass number, and $A - Z$ electrons to give the total charge of the system the value

$$Ae - (A - Z)e = Ze.$$

The electrons were believed to be present in the nucleus because the β-ray process was known to involve the emission of electrons from the core of the atom.

Figure 14-1

Chadwick's neutron-detection experiment.

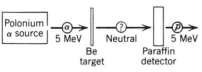

The proton–electron model had several fatal flaws. It was difficult to understand why the magnetic moment of a nucleus containing electrons should have an order of magnitude equal to the nuclear magneton instead of the Bohr magneton. It was also impossible to reconcile certain nuclear spins with the spin-$\frac{1}{2}$ properties of the proton and electron. (For instance, the N atom was known from studies of molecular nitrogen spectra to have an integer-spin bosonic nucleus, while the mass and charge numbers $A = 14$ and $Z = 7$ called for a nuclear constituency of 14 protons and 7 electrons. The odd total number of fermions was not consistent with the existence of a composite boson.) Finally, it was unrealistic to suppose that the electron could be localized in such a small region of space because of the large kinetic energy implied by the uncertainty principle. Rutherford's proposal of a massive uncharged nuclear particle opened the way for a satisfactory picture of the nucleus, and Chadwick's discovery of the neutron removed the electron from consideration as a permanent nuclear particle. These developments made it necessary to regard the emission of β-ray electrons as a separate phenomenon unrelated to the binding of the nucleus.

The discovery of the neutron required a special technique for detecting *neutral* particles. (Neutrons would not leave visible tracks in a material medium because they would not interact electrically with the atoms in the material.) Chadwick's method relied on the observation of the secondary charged particles that appeared when the primary neutral particles reacted with the atoms in the medium of his detector. His experiment used a radioactive polonium source to generate a beam of 5 MeV α particles and employed a beryllium foil to provide a target, as shown in Figure 14-1. Neutral "Be radiation" emerged from the target and passed through sheets of paraffin, ejecting protons from the hydrogenous material with kinetic energies as large as 5 MeV. At first the neutral particles were hypothesized to be photons so that the process in paraffin was presumed to be the Compton scattering of γ rays by protons. The kinematics of the Compton process implied that the incident γ energy must have been as large as 50 MeV to explain the ejection of protons with the observed 5 MeV kinetic energy. Chadwick did not accept the view that a 50 MeV photon could be emitted in the collision of a 5 MeV α particle with a Be nucleus. He also rejected the γ-ray hypothesis because the detection of protons in paraffin greatly exceeded predictions based on the cross section for Compton scattering. He argued instead that the energetic protons were ejected by an unknown form of Be radiation consisting of neutral particles with mass approximately equal to the mass of the proton. This alternative hypothesis enabled him to conclude that the 5 MeV α particles produced 5 MeV neutrons in beryllium and that these neutrons then transferred their energy to protons in the paraffin detector. Similar studies were also performed with nitrogen gas substituted for paraffin. Tracks of recoiling nitrogen nuclei were detected in this version of the experiment, making the γ-ray hypothesis even more untenable and leaving the neutron as the only satisfactory interpretation for the unknown Be radiation. These investigations were convincing enough to establish the existence of the neutron.

Ever since the discovery of the neutron, the accepted view of the nucleus has been to treat the proton and neutron as the basic nuclear particles and to regard the existing species of nuclei as bound configurations of various numbers of these constituents. The positively charged proton and the uncharged neutron are spin-$\frac{1}{2}$ fermions with only slightly different masses:

$$M_p = 938.27231 \text{ MeV}/c^2 \quad \text{and} \quad M_n = 939.56563 \text{ MeV}/c^2.$$

In fact, most of the *nuclear* characteristics of the proton and neutron are *identical*, and so the generic name *nucleon* has been given to the two particles.

Each nucleon has a magnetic moment associated with its spin. These two quantities are known from experiment to have the values

$$\mu_p = 2.792847386 \, \mu_N \quad \text{and} \quad \mu_n = -1.91304275 \, \mu_N$$

in terms of the nuclear magneton defined in Equation (8-56). Let us recall that the general relation between the nuclear magnetic moment μ_I and the corresponding nuclear spin **I** is expressed in Equation (8-57) as

$$\mu_I = g_I \mu_N \frac{\mathbf{I}}{\hbar}.$$

In the case of the proton and the neutron we use the vector **I** to denote a *nucleon spin* with quantum number $i = \frac{1}{2}$, and we quote the results for the magnetic moments as expectation values of the z components of μ_p and μ_n, evaluated in the spin-up state. The nuclear g-factors of the nucleons are therefore given by the values

$$\frac{g_p}{2} = 2.792847386 \quad \text{and} \quad \frac{g_n}{2} = -1.91304275.$$

(The proton g-factor has already come up in our discussion of hyperfine structure in Section 8-12.) These experimental results for the proton and neutron are rather different from the g-factor for electron spin, $g_S/2 = 1.001\ldots$, and are also quite different from the values predicted for charged and neutral spin-$\frac{1}{2}$ particles in Dirac's relativistic quantum theory:

$$\frac{g_p}{2} = 1 \quad \text{and} \quad \frac{g_n}{2} = 0 \quad \text{(Dirac theory)}.$$

The magnetic moments may be taken as evidence that the nucleons are *structured* particles, unlike the electron, where the structure is not simply described by the Dirac equation.

The force that binds nucleons together in the nucleus is very different from the electrostatic force responsible for the structure of atoms. It is clear that the nuclear force must be very complex to explain the observed size, shape, stability, level structure, and reaction behavior of all nuclei. The interactions among the nucleons can be associated with a complicated two-particle potential energy acting between all

Figure 14-2
Qualitative scales of atomic and nuclear size.

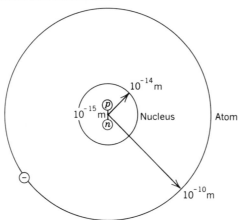

pairs of constituent particles. It is known that the nuclear forces between proton and proton, proton and neutron, and neutron and neutron are essentially *identical*. Evidence for this important simplifying property of the nuclear force has been gathered from proton–proton and neutron–proton scattering experiments and from many other sources. The small size of the nucleus is a qualitative indication that this two-body force has a *very short range*. Figure 14-2 shows a highly schematic (and disproportionate) comparison of the orders of magnitude for the atomic radius, the nuclear radius, and the internucleon range.

Figure 14-3
Potential energy for a system of two nucleons.

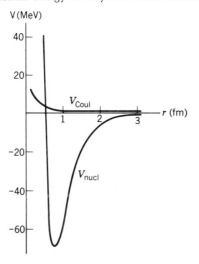

It is obvious that the attractive force between nucleons must be very strong at short range because the nuclear force between protons overwhelms the destabilizing Coulomb repulsion of like charges. The strength of the two-nucleon interaction varies with the separation of the particles and can also depend on such other variables as the momentum and spin of each nucleon. The graph in Figure 14-3 describes a possible model of the nuclear potential energy for a pair of protons in a state with zero total spin. The figure also shows the Coulomb potential energy to indicate the dominance of nuclear attraction for ranges smaller than a few fermi. (The natural scale of length in nuclear physics is defined by the unit 1 fm $= 10^{-15}$ m. This length is written as one femtometer and is read as one *fermi*.)

The units on the vertical axis in Figure 14-3 indicate a nuclear interaction energy in the MeV range. Excitations of the states of nuclei require similar amounts of energy to reach levels much farther apart than those involved in the analogous atomic processes. This upward adjustment in the scale of energy is commensurate with the reduction in the scale of distance suggested by the descent from the atomic to nuclear size in Figure 14-2. The change of the scale of energy has certain physical implications. We witness the excitations of atoms in the optical phenomena of ordinary life, but we observe the excitations of nuclei only under extraordinary circumstances such as high-energy collisions or high-temperature environments. Our initial discussion of the nucleus is restricted accordingly to the properties of the lowest energy state.

Example

Let us calculate the electrostatic repulsion between two protons and compare the two effects shown in Figure 14-3. We express the Coulomb potential energy in terms of the fine structure constant as

$$V_{\text{Coul}}(r) = \frac{e^2}{4\pi\varepsilon_0 r} = \frac{\alpha\hbar c}{r},$$

and we introduce nuclear physics units for the combinations of constants

$$hc = 1240\,\text{MeV}\cdot\text{fm} \quad \text{and} \quad \hbar c = \frac{hc}{2\pi} = 197.3\,\text{MeV}\cdot\text{fm}.$$

A 1 fm proton–proton separation gives a Coulomb energy of magnitude

$$V_{\text{Coul}} = \frac{1}{137}\frac{197\,\text{MeV}\cdot\text{fm}}{1\,\text{fm}} = 1.44\,\text{MeV}.$$

The figure shows that V_{nucl} dominates V_{Coul} at 1 fm by almost two orders of magnitude.

Example

We can use the de Broglie wavelength to demonstrate the relevance of the wave nature of nuclear particles. A proton with mass M_p and kinetic energy K has wavelength

$$\lambda = \frac{h}{p} = \frac{h}{\sqrt{2M_p K}} = \frac{hc}{\sqrt{2KM_p c^2}}.$$

For $K = 5$ MeV and 25 MeV we obtain the results

$$\lambda = \frac{1240 \text{ MeV} \cdot \text{fm}}{\sqrt{2(5 \text{ MeV})(938 \text{ MeV})}} = 12.8 \text{ fm}$$

and

$$\lambda = \sqrt{\frac{5}{25}} \, (12.8 \text{ fm}) = 5.72 \text{ fm}.$$

These values are of the same order as the nuclear radius, so that a proton beam with kinetic energy in the range 5–25 MeV is expected to undergo appreciable diffraction in collisions with nuclei. Note that λ represents a scale of distance for probing the structure of the nucleus, and recall that this scale diminishes with increasing beam energy. The calculation also tells us that protons with kinetic energies of order 5–25 MeV can be bound in nuclei by attractive potential energies of realistic strength, since the wavelengths of the wavefunctions fit within the nuclear diameter.

14-2 Nuclear Systematics

The nuclei of atoms exist in a multitude of different species known as *nuclides*. These systems cannot be charted on the periodic table of the elements because the assignment of atomic number alone does not suffice for a classification of nuclei. Every element corresponds to a specific atom with a particular number of protons, and each may have any one of several possible nuclei with different numbers of constituent neutrons. Nuclides are called *isotopes* if they bear this relation to one another. These varieties of nuclei have different values of the mass number A for the given choice of atomic number Z.

The isotope concept was proposed in 1913 by F. Soddy, another of Rutherford's many colleagues. The name implied that atoms could occupy the "same place" in the periodic table and acknowledged that atoms could be chemically identical and still be physically distinct. Soddy's idea was put forward, almost 20 years before the discovery of the neutron, in an effort to understand why different types of radioactive behavior should be observed for the same element.

The numbers A and Z are used to systematize the properties of the nuclides. We identify the mass number A and the charge number Z for a given nucleus as the number of nucleons and the number of protons, and we define N to be the number of neutrons in the nucleus according to the relation among the three integers

$$A = Z + N. \tag{14-1}$$

The designation of a particular nuclear species is then expressed by the notation

$$^{A}_{Z}\text{X}_{N},$$

where X is the chemical symbol for the atom with atomic number Z. Of course, it is redundant to employ all four symbols since X and Z convey the same information, and since A, Z, and N satisfy Equation (14-1).

The best-known source of nuclear data-at-a-glance is the publication entitled *Chart of the Nuclides*. This compendium displays the salient properties of all existing nuclei in ^AX notation on a plot of proton number Z versus neutron number N. The chart shows the isotopes of all the elements as horizontal entries for each of the various values of Z. A stylized version of this information is constructed in Figure 14-4. We are also interested in subsets of nuclides assembled according to common values of the mass number A. These species with the "same mass" are called *isobars*. Their locations appear on the plot of Z versus N along diagonal lines defined by fixed values of $Z + N$. We show a few of the lower-lying groups of isobars in Figure 14-5.

Stability properties are quoted on the nuclear chart to indicate whether individual nuclides are stable or radioactive. The latter unstable varieties undergo some type of decay, which may be either naturally occurring or artificially induced. A measured *half-life* is identified for every radioactive nucleus, defining the time interval in which a sample of the unstable nuclei decays to half of its original population. A naturally occurring radioactive isotope has a measurable abundance on Earth and must either decay with a very long half-life or exist as part of a disintegration chain originating from some other long-lived decaying nucleus. We include a few half-life data in Figure 14-5 to convey a sense of the enormous range of variation of this quantity.

Let us return to Figure 14-4 and note that the most conspicuous property of the plotted nuclei is the near equality in the numbers of protons and neutrons. We explain this trend later on when we apply the exclusion principle to the nucleus. The figure also shows a secondary tendency for nuclides to fall below the line $Z = N$ toward the region where the neutrons outnumber the protons. This preference for neutrons stems from the Coulomb repulsion between protons and grows with increasing nucleon number. The Coulomb effect causes protons to experience less nuclear binding than neutrons, particularly in the larger nuclei where the nucleons are more likely to be farther apart.

The locations of the stable nuclides are clearly marked in Figure 14-4. (By convention, a nuclide is said to be stable if there is no known decay, or if there is an extremely long half-life expressible only by a very large lower bound.) We see at once that the radioactive entries are in the majority and that the stable species cease to occur at all beyond a particular mass number. The first stable nucleus on the chart is ^1H, the proton, and the last is ^{209}Bi. The figure shows that the stable nuclei occupy central positions in each isotopic or isobaric family, while the radioactive members of the family lie to either side. We can visualize this distribution by drawing an imaginary *valley of stability* through increasing values of the mass number from $A = 1$ to $A = 209$. The stable-nuclide path between ^1H and ^{209}Bi has two noteworthy gaps, at $A = 5$ and $A = 8$. We interpret the nonexistence of stable candidates for these

---

Figure 14-4

Distribution of stable and radioactive isotopes. Data are taken from *Chart of the Nuclides*, 13th edition.

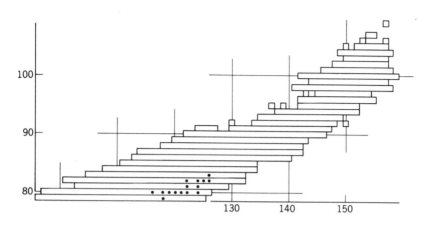

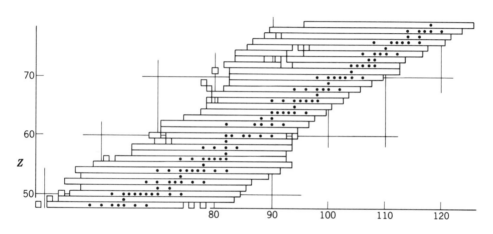

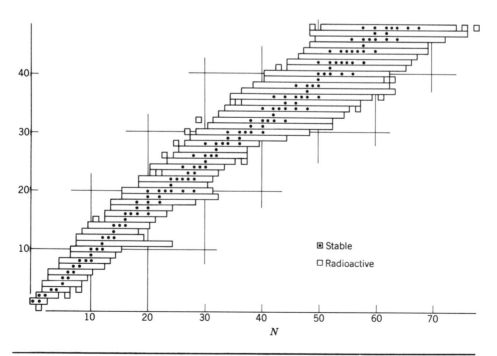

Figure 14-5

Stable and radioactive isobars at low mass number. Isotopic abundances are quoted for the stable nuclides, and radioactive half-lives are given for the unstable isobars.

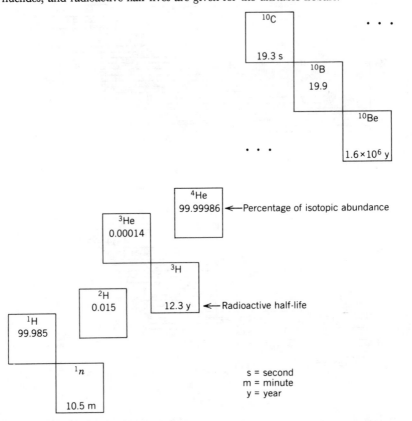

mass numbers as qualitative indications of the unusual stability of the $A = 4$ nuclide ^{4}He. The two gaps tell us that it is not favorable to add one more nucleon to the $A = 4$ system or to bind two such systems together.

We learn several interesting lessons just by counting the nuclides according to the following four possible combinations of proton and neutron numbers. Of the 268 existing stable nuclei (including those that are extremely long-lived), there are

$$
\begin{array}{ll}
\text{159 with even } Z \text{ and even } N & \text{(even–even),} \\
\text{53 with even } Z \text{ and odd } N & \text{(even–odd),} \\
\text{50 with odd } Z \text{ and even } N & \text{(odd–even),} \\
\text{and} \quad \text{6 with odd } Z \text{ and odd } N & \text{(odd–odd).}
\end{array}
$$

odd A even A

The stable odd-A nuclei exist in roughly equal numbers of even–odd and odd–even varieties. This observation is a hint that the nuclear force does not distinguish between protons and neutrons. The distribution of the stable even-A nuclei is more remarkable since almost all these nuclides fall into the even–even category. The odd–odd entries are rare enough to be identified by name:

$$^{2}\text{H}, \quad ^{6}\text{Li}, \quad ^{10}\text{B}, \quad \text{and} \quad ^{14}\text{N} \quad \text{at low } A,$$

together with the extremely long-lived members

$$^{50}\text{V} \quad \text{and} \quad ^{180}\text{Ta} \quad \text{at higher } A.$$

We take the imbalance between even–even and odd–odd nuclei as evidence that the nuclear force has a *pairing* property. The evidence tells us that the force between nucleons in the nucleus has a strong preference for paired-proton and paired-neutron configurations. This feature of the nuclear force must be incorporated in the building of nuclear models.

Figure 14-4 suggests another numerological exercise that also has implications for model building. If we distribute the 268 stable nuclei over approximately 100 elements we find that an average of two or three stable isotopes is expected for each element. In fact, an inspection of the nuclear chart reveals some very marked departures from this expectation. If we follow the valley of stability in the figure we encounter unusually large stable populations of nuclei along the succession of lines $N = 20$, $Z = 20$, $N = 28$, $Z = 28$, $N = 50$, $Z = 50$, and $N = 82$. The most striking occurrence of stability is seen at $Z = 50$ where the element tin is found with ten stable isotopes. These patterns of exceptional stability in nuclei are reminiscent of the shell closures observed in atoms. The numbers 20, 28, 50, and 82 are among the *magic numbers* associated with the shell structure of the nucleus.

We might wonder whether any stable nuclei exist beyond the range of the current nuclear chart. Nuclear stability tends to terminate with increasing numbers of nucleons because of the destabilizing influence of Coulomb repulsion among the growing numbers of protons in the nucleus. However, the shell theory of the magic numbers indicates the possible existence of an "island of stability" off the chart, at coordinates given by the predicted magic numbers $Z = 114$ and $N = 184$. New *superheavy* elements ought to exist in this vicinity if these predictions are correct. No trace of any such element has yet been found in any samples of naturally occurring material. An attempt to synthesize superheavy products in nuclear reactions is also underway using accelerated beams of heavy ions (such as ^{48}Ca or even ^{238}U) incident on heavy targets (such as ^{248}Cm). These studies have not produced any positive evidence in their early stages of investigation.

Example

Let us translate Chadwick's experiment into the language of nuclide notation. We describe the process in which incident α particles produce Be radiation (neutrons) by the nuclear reaction

$$^{4}_{2}\text{He}_{2} + ^{9}_{4}\text{Be}_{5} \rightarrow ^{12}_{6}\text{C}_{6} + ^{1}_{0}n_{1}.$$

Note that conservation laws are imposed on the sum of the charge numbers and the sum of the mass numbers to identify the carbon nucleus in the final state.

The reaction can be expressed in briefer notation as

$$^4\text{He} + {}^9\text{Be} \rightarrow {}^{12}\text{C} + {}^1n,$$

or in even more streamlined fashion as

$$^9\text{Be}(\alpha, n)^{12}\text{C}.$$

We include the superfluous left and right subscripts Z and N in the first version of the reaction to draw attention to the conservation laws for the total charge and the total number of nucleons. Let us also note in passing that ^4He, ^9Be, and ^{12}C are stable nuclides and that the only unstable participant in Chadwick's experiment is the neutron itself. This instability causes the neutron to decay and transform into the proton via the β-radiation process. We see from Figure 14-5 that the transformation is between isobars and that the quoted half-life is 10.5 min.

Example

The observed bosonic property of the nitrogen nucleus ^{14}N is consistent with the assignment of proton and neutron numbers $Z = 7$ and $N = 7$. The total number of fermions is even, and so the collection of nucleons can exist in the form of a composite boson, as determined by experiment.

14-3 Electron Scattering and the Nuclear Radius

The first determinations of nuclear size came from the Rutherford scattering experiments. These estimates of the nuclear radius were obtained by observing deviations from the Rutherford cross section and attributing a deviation to the effect of a nucleus of definite volume.

More refined measurements of nuclear structure became feasible with the development of accelerators. These machines were able to produce beams of particles with de Broglie wavelengths small enough to probe the details of the nucleus at short range. Protons, deuterons, α particles, and other ions could be accelerated in primary beams, and neutrons could be extracted in secondary beams, all for the purpose of bombarding nuclei at variable energies. The accelerators came into use, along with the different kinds of detectors, to provide the "microscopes" for a systematic exploration of the stable nuclei.

All the probes just mentioned are *nuclear* particles whose reactions with nuclei are governed by the nuclear force. Such beam particles are ideally suited for the investigation of this unknown interaction. On the other hand, studies of nuclear size and composition are more clearly interpreted if the *electron* is chosen as the beam particle because the interactions of the electron are dominated by the well-known electromagnetic force. The probing electrons see mainly the protons in the nucleus since the main effect is the electrostatic force between charges. (Electrons also have a weaker magnetic interaction with the neutrons in the nucleus. This effect can be observed in domains where the electric interaction is suppressed.) Unique information about the distribution of the nuclear charge is obtained by performing diffraction

Figure 14-6

Schematic plan of Hofstadter's electron-scattering experiment.

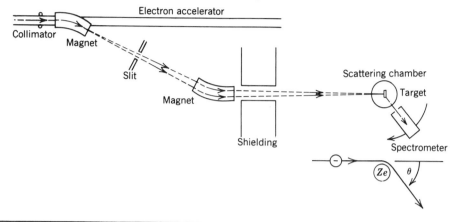

experiments in which electrons are scattered by the nucleus. These data can be used to determine the nuclear radius.

The electron-scattering experiments had to wait for the construction of high-energy electron accelerators. A comprehensive series of investigations of nuclei was finally undertaken by R. Hofstadter and his associates in 1953. Eventually, these studies were extended to include measurements of the internal electromagnetic structure of the proton and the neutron. Thus, the whole range of electron-scattering experiments gave a description of the constituents of the nucleus as well as the nucleus itself.

Figure 14-6 shows a sketch describing Hofstadter's means of observation of the elastic-scattering process

$$e + X \rightarrow e + X.$$

The equipment includes an electron accelerator and deflecting magnets to prepare the high-energy electron beam, a scattering target of species X, and a spectrometer to detect electrons scattered elastically in directions given by the indicated scattering angle θ. This apparatus constitutes an elaborate high-energy device for the study of electron diffraction, since the angular distribution of the scattered electrons has the appearance of a diffraction pattern. We represent these observations by means of the differential cross section $d\sigma/d\Omega$ for elastic electron scattering, an angle-dependent quantity analogous to the Rutherford cross section for the scattering of α particles. Experimental values of $d\sigma/d\Omega$ are plotted in Figure 14-7 for a single beam energy and for several nuclear targets. We see the characteristic features of a diffraction pattern in each of the graphs, as the cross sections fall rapidly from the forward direction at $\theta = 0$ and exhibit small peaks at other angles.

The behavior of $d\sigma/d\Omega$ is similar to the diffraction of light by a spherical obstacle with a dense interior and a diffuse surface. A good characterization of electron scattering can be given in these terms by adopting a spherical model of the nucleus in which the nuclear charge density has the form

$$\rho(r) = \frac{\rho_1}{1 + e^{(r-R)/z_1}}. \qquad (14\text{-}2)$$

Figure 14-7

Differential cross sections for elastic electron scattering at 183 MeV. Data are plotted versus scattering angle θ for calcium, indium, and gold targets.

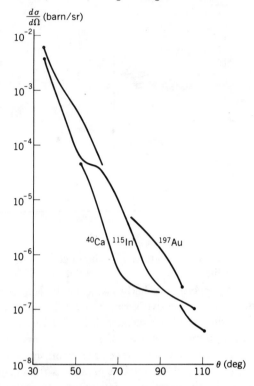

This expression is like a Fermi distribution in which the two parameters R and z_1 control the r dependence. The coefficient ρ_1 is proportional to the central charge density

$$\rho(0) = \frac{\rho_1}{1 + e^{-R/z_1}},$$

so that ρ_1 and $\rho(0)$ are approximately equal for $R \gg z_1$. We can interpret the significance of these features with the aid of Figure 14-8. The illustrated charge density falls through the value $\rho_1/2$ at $r = R$, dropping from 90% to 10% of the maximum density over a small distance given by the indicated surface thickness t. The latter quantity is directly related to the parameter z_1 in Equation (14-2). We leave the derivation of this relation to Problem 3 at the end of the chapter.

The treatment of electron-scattering data for different beam energies and various nuclei leads to the deduction of charge densities like the ones shown in Figure 14-9. We select these particular results to correspond to the cross sections plotted in Figure 14-7. The graphs indicate a decrease of the central charge density $\rho(0)$ and an increase of the radius parameter R as nuclei of increasing nucleon number are considered, while the surface thickness remains essentially unchanged. This analysis determines a nuclear radius R that grows with mass number A according to the

Figure 14-8

Parametrization of the nuclear charge density.

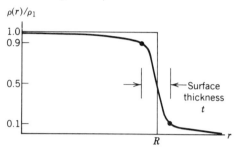

formula

$$R = R_0 A^{1/3} \tag{14-3}$$

and produces constant parameters with the approximate values

$$t = 2.4 \text{ fm} \quad \text{and} \quad R_0 = 1.07 \text{ fm}$$

over the whole survey of nuclei. Other methods of determining the nuclear radius confirm the A dependence of Equation (14-3). In general, these techniques employ the parameter R_0 alone and yield values of R_0 in the range 1.18–1.40 fm.

The decrease of the central charge density $\rho(0)$ is a noteworthy feature of Figure 14-9. This behavior opposes the tendency for neutrons to outnumber protons with

Figure 14-9

Nuclear charge densities deduced from electron scattering. The cases illustrated correspond to the nuclei considered in Figure 14-7.

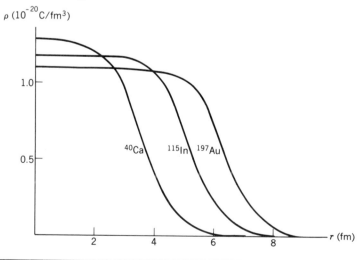

increasing values of A. We can blend these two opposing effects and obtain an effective *nucleon density* with practically the *same* central value for all nuclei. We define this quantity by noting that the density of protons in the nucleus is $\rho(r)/Ze$ and by assuming that protons and neutrons have the same distributions. The result is a density of nucleons given by the expression

$$\frac{A}{Ze}\rho(r).$$

The approximate uniformity of the central value $(A/Ze)\rho(0)$ over all nuclei suggests the approximate uniqueness of the *mass density* for all forms of nuclear matter.

The same conclusion can be drawn from the $A^{1/3}$ behavior of the nuclear radius in Equation (14-3). If we compute the nuclear volume as

$$\tfrac{4}{3}\pi R^3 = \tfrac{4}{3}\pi R_0^3 A$$

and identify the mass density as

$$\frac{M_p A}{\tfrac{4}{3}\pi R^3} = \frac{3}{4\pi}\frac{M_p}{R_0^3},$$

we find that our final result is independent of the number of nucleons.

Example

The following two computations of the nuclear mass density illustrate our conclusions. We take 0.17 nucleons/fm^3 as a reasonable estimate of the central nucleon density, and we multiply by the proton mass to obtain

$$\frac{A}{Ze}\rho(0)M_p = \left(0.17 \text{ fm}^{-3}\right)\left(10^{15} \text{ fm/m}\right)^3\left(1.67 \times 10^{-27} \text{ kg}\right)$$

$$= 2.8 \times 10^{17} \text{ kg/m}^3.$$

Alternatively, we use the result obtained from the $A^{1/3}$ behavior of R to find

$$\frac{3}{4\pi}\frac{M_p}{R_0^3} = \frac{3}{4\pi}\frac{1.67 \times 10^{-27} \text{ kg}}{\left(1.07 \times 10^{-15} \text{ m}\right)^3} = 3.25 \times 10^{17} \text{ kg/m}^3.$$

By either reckoning we estimate a nuclear density of order 3×10^{17} kg/m^3, an enormous value compared to the representative figure 10^3 kg/m^3 for ordinary atomic matter.

14-4 Nuclear Mass and Binding Energy

Thomson's measurements of e/m were able to detect a difference in mass between different positive ions of a given element. These earliest indications of isotope mass splitting showed specifically that neon atoms could have either of the two mass numbers $A = 20$ or $A = 22$. The notion of *atomic weight* had been introduced (as

Figure 14-10

Layout of a typical mass spectrometer.

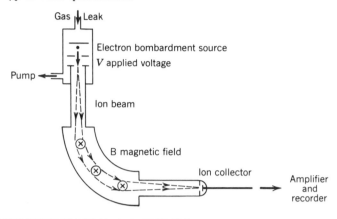

inaccurate terminology) to denote the *average* mass of any naturally occurring element. It was possible to use Thomson's neon masses along with an estimate of the two abundances and obtain an average atomic mass of 20.2, a value consistent with the known atomic weight of neon. Eventually, the isotope concept was put forward and the meaning of atomic weight was made clear. It was found that the atomic weight of any element could be explained by averaging over the abundances of the corresponding stable isotopes.

Thomson's experiments employed the acceleration and deflection of positive ions by electric and magnetic fields. An improved version of his apparatus was developed in 1919 by F. W. Aston in the design of the first *mass spectrograph*. The instrument separated isotopes according to their masses and gave accurate mass determinations for the observed ions. Aston used these *mass spectra* to analyze the isotopic compositions of more than 50 elements. He showed that the isotope masses were nearly equal to integral multiples of the hydrogen mass, and he found that the small deviations from whole-number multiples could also be measured in more refined experiments. Measurements of the atomic mass provided direct information about properties of the nucleus. The measured deviations from whole numbers proved to be especially significant as clues to the binding of the nucleus. Mass spectrometry flourished in the 1930s because of improvements in detector design and vacuum technology. The mass spectrograph demonstrated its practical value as a separator of isotopes in 1935 with the discovery of the rare long-lived uranium nuclide ^{235}U.

Figure 14-10 shows the design of a mass spectrometer for the isotopic analysis of gaseous elements. The apparatus admits a sample of gas at low pressure and bombards the atoms with electrons to convert the sample into positive ions. Electric and magnetic fields guide the charges to an ion detector where the ions are collected separately according to mass. The ions are accelerated to speed v by an applied voltage V and are then deflected in a circular path of radius R by a magnetic field of strength B. Nonrelativistic ions with mass M and charge e acquire a kinetic energy

$$\frac{M}{2}v^2 = eV$$

Figure 14-11

Mass spectrum of xenon indicating the relative abundances of the nine stable isotopes.

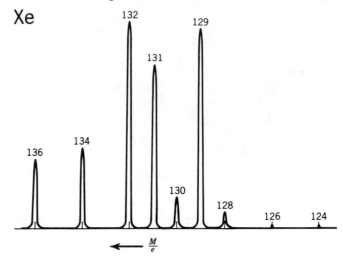

and experience a centripetal force

$$\frac{Mv^2}{R} = Bev.$$

The ion speed can be eliminated between these two equations:

$$v = BR\frac{e}{M} \quad \Rightarrow \quad eV = \frac{M}{2}\left(BR\frac{e}{M}\right)^2.$$

The quantity of interest in the last equality is the mass-to-charge ratio

$$\frac{M}{e} = \frac{(BR)^2}{2V}. \tag{14-4}$$

The mass spectrometer operates at fixed values of B and R and employs a varying voltage V in order to collect ions with various M/e ratios. This technique produces mass spectra like the one sketched in Figure 14-11. Note that the output signal from the ion detector provides a measure of the relative abundance of each isotope.

Masses can be measured more precisely by comparing unknowns with certain carbon-bearing calibration standards. Let us illustrate by considering a sample containing atomic and molecular ions of hydrogen, deuterium, carbon, oxygen, and methane. The resulting mass-spectroscopic lines include the following three M/e doublets:

$$(^1H^1H)^+ - {}^2H^+, \quad (^2H^2H^2H)^+ - {}^{12}C^{++}, \quad \text{and} \quad (^{12}C^1H^1H^1H^1H)^+ - {}^{16}O^+,$$

where the symbols $(\cdots)^+$ refer to singly ionized molecules. We note that the two members of each doublet have the same nominal value of M/e, and we find that a small line splitting appears in the mass spectrum at the location of each pair. We can measure these three splittings and use the results to deduce the masses of ^{1}H, ^{2}H, and ^{16}O relative to the mass of ^{12}C. Problem 5 is included at the end of the chapter to illustrate this procedure.

These investigations of ions in mass spectrometry provide the means of determining the masses of *neutral atoms*, each with its full complement of atomic electrons. A representative listing of atomic masses is provided along with other pertinent nuclear data in Appendix A. By convention, the quoted values have *atomic mass units* (symbol u), defined such that the commonly occurring neutral carbon atom ^{12}C has the ground-state mass value

$$M(^{12}C) = \text{exactly 12 u.}$$

The rest-energy equivalent of this unit of mass turns out to be

$$uc^2 = 931.49432 \text{ MeV.}$$

These numbers are discussed somewhat further in the examples below.

The relativistic concept of rest energy plays a major role in the behavior of nuclei, unlike the situation in the case of atoms. We recall that the rest energies of the nucleus and the electrons are discounted ab initio in the analysis of the energy states of the atom. We cannot argue that these contributions are negligible, for in fact the rest energies of the constituents of the atom are much larger than the excitation energies of the atomic levels. Instead, we note that atomic processes are low-energy phenomena in which the constituents of the atom cannot transform into different species. Since the masses and rest energies maintain their identities, it is consistent to leave these quantities out of the analysis. The same argument does not hold for nuclei because nuclear excitation and nuclear transformation involve comparable energies. We must reinstate the rest energy when we define the nuclear binding energy, and we must take account of changes in mass when we apply the conservation of energy to processes of nuclear transformation.

Binding energy is a property of any aggregate of particles bound together by attractive interactions. The rest energy Mc^2 is identified as the total relativistic energy of such a system in its center of mass frame. This energy consists of the rest energies for the constituent masses M_i, along with the kinetic and potential energies of the system. The total potential energy must be attractive to bind the particles and may be defined to vanish for particles at infinite separation. The energy of separated constituents at rest therefore consists only of the total rest energy $\sum_i M_i c^2$. This quantity must *exceed* Mc^2 if the mass M corresponds to a bound state, and so the difference between these rest energies determines the binding energy of the system:

$$E_b = \sum_i M_i c^2 - Mc^2. \tag{14-5}$$

In practical language, the quantity E_b represents the difference in energy between the disassembled and the assembled collection of particles.

Let us apply the general formula for binding energy to atoms first and then return to nuclei. For a neutral atom with Z electrons we have

$$E_b(\text{atom}) = \left[M(\text{nucleus}) + Zm_e - M(\text{atom}) \right] c^2. \tag{14-6}$$

This atomic binding energy has values like 13.6, 13.6, and 79.0 eV for hydrogen, deuterium, and helium and becomes as large as hundreds of keV for atoms of much larger Z. In all cases we can justifiably ignore such comparatively small quantities whenever we examine the binding properties of the corresponding nuclei. Equation (14-5) tells us that the binding energy of a nuclide with Z protons and N neutrons is given in terms of the masses of these constituents as

$$E_b(\text{nucleus}) = \left[ZM_p + NM_n - M(\text{nucleus}) \right] c^2. \tag{14-7}$$

We can eliminate the nuclear mass between the last two relations and replace the left side of Equation (14-6) by zero in the process to get

$$E_b(\text{nucleus}) = \left[ZM_p + NM_n + Zm_e - M(\text{atom}) \right] c^2.$$

The combination $M_p + m_e$ can be set equal to the mass of the hydrogen atom since it is safe to neglect the small 13.6 eV of atomic binding energy. The final formula for the binding energy of the nuclide ^AX then assumes the following explicit form:

$$E_b(^A\text{X}) = \left[ZM(^1\text{H}) + NM_n - M(^A\text{X}) \right] c^2, \tag{14-8}$$

in which the symbol $M(^A\text{X})$ refers to the mass of the neutral atom. This version of the formula expresses the desired quantity in terms of masses found directly from mass spectrometry. These atomic masses are listed as neutral-atom properties in Appendix A. We can extract the corresponding nuclear masses if we wish by simply subtracting the masses of Z electrons.

Figure 14-12 illustrates an interesting systematic property of the nuclear binding energy. The graph shows the *binding energy per nucleon* E_b/A for the various nuclides, plotted as a function of the number of nucleons. We are struck by the fact that an approximate plateau is reached around the value 8 MeV per nucleon for all nuclei beyond $A = 16$. This behavior indicates a *saturation* phenomenon that reflects the short-range nature of the nuclear force. A long-range force would subject a nucleon to binding interactions with all the other $A - 1$ nucleons in the nucleus and would cause the binding energy of each nucleon to grow with A. Instead, the binding energy per nucleon saturates around a particular energy and tells us that each nucleon experiences binding interactions with only a limited number of nearest neighbors in the nucleus. We also observe a slight decline in the plateau at larger values of A. This secondary feature indicates the growing effect of the Coulomb repulsion between protons to reduce the nuclear binding in larger nuclei. We have already seen such behavior in the tendency for neutrons to outnumber protons at the higher values of A in Figure 14-4.

The formulas for the binding energy can be employed to introduce the useful concept of *neutron separation energy*. This quantity is defined as the binding energy of the last neutron in the assembly of nucleons

$$^{A-1}_{Z}\text{X} + n \leftrightarrow {}^{A}_{Z}\text{X}.$$

We adapt Equation (14-5) to this definition and write the separation energy accordingly:

$$E_n(^A\text{X}) = \left[M(^{A-1}\text{X}) + M_n - M(^A\text{X}) \right] c^2. \tag{14-9}$$

Figure 14-12

Binding energy per nucleon versus nucleon number. A smooth curve is drawn through the plotted points indicating the positions of several of the stable nuclides. Values of E_b are taken from tables compiled by A. H. Wapstra and K. Bos.

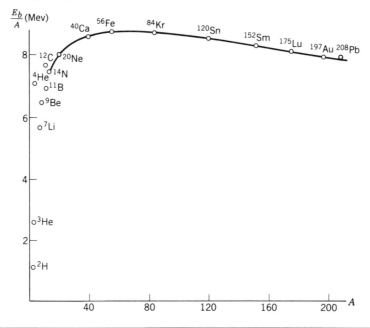

Note that the species X maintains its identity throughout since Z does not change with the removal of a neutron. We can use Equation (14-8) to convert the formula for E_n into the difference of nuclear binding energies for the two isotopes:

$$E_n(^4X) = \{[ZM(^1H) + (A - 1 - Z)M_n]c^2 - E_b(^{A-1}X)\} + M_nc^2$$

$$- \{[ZM(^1H) + (A - Z)M_n]c^2 - E_b(^4X)\}$$

$$= E_b(^4X) - E_b(^{A-1}X). \tag{14-10}$$

Both formulas for E_n are of some interest. The second version is a clear statement of the distinction between the binding energy of the last neutron and the nuclear binding energy. The statement tells us that the latter quantity is the larger of the two energies.

Example

The atomic mass unit u is defined in terms of the carbon standard as $\frac{1}{12}M(^{12}C)$. Let us convert the unit to kilograms with the aid of Avogadro's number:

$$u = \frac{1}{12}\frac{12 \text{ g/mole}}{6.0221 \times 10^{23}/\text{mole}}(10^{-3} \text{ kg/g}) = 1.6606 \times 10^{-27} \text{ kg}.$$

The corresponding rest energy is

$$uc^2 = \frac{(1.6606 \times 10^{-27}\ \text{kg})(2.9979 \times 10^8\ \text{m/s})^2}{1.6022 \times 10^{-13}\ \text{J/MeV}} = 931.50\ \text{MeV}.$$

This conversion factor is used over and over in nuclear physics calculations.

Example

Carbon appears on the nuclear chart with the following data for the two stable isotopes ^{12}C and ^{13}C:

atomic mass (u)	exactly 12	13.00335482
isotopic abundance (%)	98.90	1.10

We can calculate the average atomic mass by combining these numbers as follows:

$$0.9890(12) + 0.0110(13.00335482) = 11.868 + 0.143 = 12.011.$$

This result agrees with the quoted value for the atomic weight of carbon.

Example

Equation (14-8) may be used in conjunction with the atomic masses in Appendix A to compute the binding energy for any of the tabulated nuclei. In the case of ^4He we take $Z = N = 2$ and obtain

$$E_b(^4\text{He}) = [2(1.007825) + 2(1.008665) - 4.002603](931.5\ \text{MeV})$$
$$= 28.30\ \text{MeV}.$$

For ^{12}C we take $Z = N = 6$ and get

$$E_b(^{12}\text{C}) = [6(1.007825) + 6(1.008665) - 12](931.5\ \text{MeV}) = 92.16\ \text{MeV}.$$

Note that large cancellations take place inside the square brackets and that both calculations employ the conversion from uc^2 to MeV. The binding energies per nucleon in the two cases are

$$\frac{E_b}{A}(^4\text{He}) = \frac{28.30\ \text{MeV}}{4} = 7.075\ \text{MeV}$$

and

$$\frac{E_b}{A}(^{12}\text{C}) = \frac{92.16\ \text{MeV}}{12} = 7.680\ \text{MeV}.$$

These results appear to be well on their way toward the approximate universal value 8 MeV. (Actually, the ^4He figure is unusually large, lying well above the rising portion of the data plotted in Figure 14-12.)

Example

Finally, let us illustrate the idea of neutron separation energy by applying the formulas to some of the isotopes of cadmium. Equation (14-9) is particularly suitable because the relevant masses are listed in Appendix A. We consider the separation energies for the two cadmium nuclides ^{113}Cd and ^{114}Cd. For $A = 113$ we find

$$E_n(^{113}\text{Cd}) = \left[M(^{112}\text{Cd}) + M_n - M(^{113}\text{Cd}) \right] c^2$$
$$= (111.902758 + 1.008665 - 112.904400)(931.5 \text{ MeV})$$
$$= 6.542 \text{ MeV},$$

and for $A = 114$ we find

$$E_n(^{114}\text{Cd}) = \left[M(^{113}\text{Cd}) + M_n - M(^{114}\text{Cd}) \right] c^2$$
$$= (112.904400 + 1.008665 - 113.903357)(931.5 \text{ MeV})$$
$$= 9.043 \text{ MeV}.$$

We use these two results again when we continue this illustration at the end of Section 14-5.

14-5   The Semiempirical Mass Formula

The binding energy of a nucleus defines the position of the ground state on a nuclear energy level diagram. This state, and all the higher excited states, should be determinable from a theory of the nuclear force. We recognize, however, that any attempt to solve the quantum problem of the binding of many nucleons would be a most ambitious undertaking. Our approach takes a more deliberate course and turns first to some rather elementary models. We aim these models at the nuclear ground state and direct our attention to the systematic features displayed on the nuclear chart.

Let us begin with an early model, introduced in 1935 by C. F. von Weizsäcker, in which the nucleus is compared to a classical *liquid droplet*. A certain resemblance exists between the two systems since the density and binding energy per nucleon are essentially independent of the number of nucleons in the nucleus, while the density and latent heat of vaporization do not vary with the number of molecules in the liquid. Weizsäcker's model also takes account of the Coulomb forces among the nucleons as well as the quantum effects associated with nucleon spin and fermion antisymmetry. The end result is a parametrization of the atomic mass as a function of A and Z. The predictions pertain to the nucleus as a whole without reference to the individuality of the nuclear constituents.

We construct the Weizsäcker formula by identifying five different contributions to the nuclear binding energy E_b. The first effect is analogous to a heat of vaporization, written as $L_v M_{\text{molecule}} n$, where n denotes the number of molecules in a drop of liquid with latent heat per unit mass L_v. In the case of the nucleus we associate a similar number dependence with the plateau behavior of E_b/A in Figure 14-12, and we write

$$E_{b1} = a_1 A$$

as the leading term in the binding energy. This piece of E_b is called the *volume energy* since the linear dependence on A is related to the volume of a sphere whose radius varies as $A^{1/3}$ in the manner of Equation (14-3).

We have already interpreted the A independence of E_b/A in terms of a fixed number of internucleon bonds for a nucleon in any nucleus. The nucleons on the nuclear surface are not surrounded by as many of these bonds, and so the binding energy is reduced by a *surface correction* of the form

$$E_{b2} = -a_2 A^{2/3}.$$

This dependence on A corresponds to a quadratic behavior in R describing a correction term proportional to the spherical surface area of the nucleus.

These two bulk-liquid contributions to E_b must be supplemented by the *Coulomb energy* due to the repulsion between protons in the nucleus. We use the formula for the potential energy stored in a uniform sphere of charge Ze and radius R,

$$V = +\frac{3}{5}\frac{(Ze)^2}{4\pi\varepsilon_0 R}, \tag{14-11}$$

and we turn again to the relation between R and A in Equation (14-3) to obtain

$$E_{b3} = -a_3 \frac{Z^2}{A^{1/3}}$$

as the electrostatic contribution to E_b. (The derivation of Equation (14-11) is given as Problem 7 at the end of the chapter.) Note the change in sign between the quantities V and E_{b3}.

The classical nature of these first three terms is immediately apparent. It is also clear that the construction of E_b is not yet complete since the three terms alone predict that the most stable nuclei should occur for $Z = 0$ and $A = N$. We must include another effect associated with the fermionic properties of the nucleons. Figure 14-4 has told us that the existing nuclides lie near the line $Z = N$, corresponding to equal numbers of protons and neutrons. We have attributed this systematic feature of the nuclear chart to the influence of the exclusion principle. Let us continue to defer the interpretation of the effect and use empirical grounds to introduce another term called the *symmetry energy*:

$$E_{b4} = -a_4 \frac{(A/2 - Z)^2}{A}.$$

The numerator of this expression behaves as $(N - Z)^2$ and reduces the binding energy as nuclides deviate to *either* side of the line $Z = N$, while the denominator preserves the necessary linear dependence on A. We develop a special model in Section 14-6 to explain this term in context with the meaning of the line $Z = N$.

We have also noted the overwhelming preference for stable even-A nuclides to occur in the even–even, rather than the odd–odd, category. This observation has been taken as evidence for the pairing property of the nuclear force. Let us provide for pairing in phenomenological fashion and include a contribution to E_b called the *pairing energy*, which applies only for even A and favors specifically the even–even nuclei. We introduce such a term by adopting the following expression:

$$E_{b5} = \varepsilon_5 = \begin{cases} \pm \dfrac{a_5}{A^{3/4}} & \begin{matrix} \text{even–even} \\ \text{odd–odd} \end{matrix} \\[2em] 0 & \text{even–odd and odd–even.} \end{cases}$$

It has been found that this formula gives a good empirical representation for the A dependence of the pairing effect.

The nuclear binding energy is assembled from these five terms:

$$E_b(^A\text{X}) = a_1 A - a_2 A^{2/3} - a_3 \frac{Z^2}{A^{1/3}} - a_4 \frac{(A/2 - Z)^2}{A} + \varepsilon_5. \tag{14-12}$$

If we combine this expression with Equation (14-8) we obtain a simplified semiempirical version of Weizsäcker's formula for the atomic mass:

$$M(^A\text{X}) = ZM(^1\text{H}) + (A - Z)M_n$$
$$- \left[a_1 A - a_2 A^{2/3} - a_3 \frac{Z^2}{A^{1/3}} - a_4 \frac{(A/2 - Z)^2}{A} + \varepsilon_5 \right] / c^2. \tag{14-13}$$

(Actually, the fifth term in the binding energy is a more recent addition to the original model.) The formula is called *semiempirical* because the various constants are determined by securing a best fit to the atomic masses and not by invoking any further theoretical arguments. An excellent fit is obtained with the following parameters:

$$a_1 = 15.76 \text{ MeV},$$
$$a_2 = 17.81 \text{ MeV},$$
$$a_3 = 0.7105 \text{ MeV},$$
$$a_4 = 94.80 \text{ MeV},$$
$$a_5 = 39 \text{ MeV}.$$

Remarkably, this five-parameter prescription for $M(^A\text{X})$ is accurate to a few MeV with only occasional exceptions across the whole nuclear chart. The exceptional deviations are significant in their own right as indications of shell structure in nuclei.

We can isolate the Coulomb term E_{b3} from the other contributions to Equation (14-12) by considering the binding energy of the two isobars

$$^A_x\text{X}_y \quad \text{and} \quad ^A_y\text{Y}_x.$$

These species are related by the interchange of proton and neutron numbers Z and N and are known as *mirror nuclides*. The interchange operation $Z \leftrightarrow N$ affects only the a_3 term in the formula for E_b, and so the difference of binding energies turns into the expression

$$E_b(^A\text{X}) - E_b(^A\text{Y}) = a_3 \frac{y^2 - x^2}{A^{1/3}}. \tag{14-14}$$

Hence, the difference in binding energy is the same as the difference in Coulomb energy for any pair of mirror isobars. Furthermore, the nuclei in question have equal radii corresponding to the given value of A. We can take advantage of this observation and determine the common radius by using measurements of E_b in conjunction with the formula for the Coulomb energy in Equation (14-11). We illustrate the procedure in the second example at the end of this section.

The mass formula offers many useful insights into the stability properties of nuclei. We can establish a criterion for stability if we organize the nuclides as *isobars* and examine the variation of the masses with regard to the (A, Z) dependence predicted by the formula. We select a set of isobars by choosing a fixed value of the mass number A, and we note that the chosen nuclides have nearly equal masses. Equation (14-13) describes these species by a *quadratic* function of Z, which *minimizes* as Z varies for constant A. The minimum is determined by the vanishing of the partial derivative

$$\frac{\partial M}{\partial Z}c^2 = \left[M(^1\mathrm{H}) - M_n\right]c^2 + 2a_3\frac{Z}{A^{1/3}} + 2a_4\frac{Z - A/2}{A}.$$

The minimizing charge number has the value

$$Z = Z_A = \frac{A}{2}\frac{a_4 + \left[M_n - M(^1\mathrm{H})\right]c^2}{a_4 + a_3 A^{2/3}}. \tag{14-15}$$

Thus, we predict a stable isobar at Z_A, where the lowest isobar mass occurs for the given mass number A. Of course, this minimizing value of Z should not be expected to coincide with a true integer-valued atomic number. The actual state of affairs is illustrated for three consecutive choices of A in Figure 14-13. The ε_5 term makes no contribution to the formula when A is odd, and so Equation (14-13) represents a single parabola opening upward on the corresponding graph of M versus Z. When A is even the ε_5 term adds a constant shift to the other contributions on the graph, lowering the even–even masses and raising the odd–odd masses relative to the central

Figure 14-13

Isobar masses versus atomic number for three consecutive values of the mass number. The stable isobars are $^{98}_{42}\mathrm{Mo}$ and $^{98}_{44}\mathrm{Ru}$ for $A = 98$, $^{99}_{44}\mathrm{Ru}$ for $A = 99$, and $^{100}_{42}\mathrm{Mo}$ and $^{100}_{44}\mathrm{Ru}$ for $A = 100$.

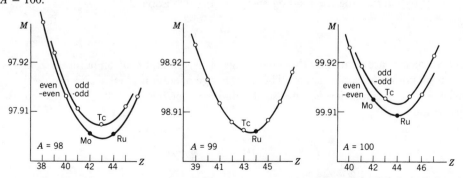

parabolic curve. The figure employs two parabolas in this case to show how the measured masses hop sequentially between the even–even and odd–odd isobars. The minima on the curves lie close to the indicated integer values of Z at which the stable nuclides occur.

Equation (14-13) determines a mass *surface* in the variables A and Z. Figure 14-13 shows three slices at constant A through the surface, with minima corresponding to the quantity Z_A. (Obviously, there are three surfaces: one for odd A, and two more for the even–even and odd–odd varieties with even A.) These graphs furnish a clear picture of the valley of stability to supplement our previous discussion of Figure 14-4.

Let us conclude our observations about stability with the following incidental remarks. We note that nuclides of technetium appear in all three parts of Figure 14-13 and that none of these species is stable. In fact, no stable $^{A}_{43}\mathrm{Tc}$ nuclide exists for any value of A. Technetium at $Z = 43$ and promethium at $Z = 61$ are unique in this respect since these are the only elements with no stable isotopes in the domain of stable elements below bismuth at $Z = 83$.

Example

The constant a_3 is the only parameter in the mass formula to be associated with a specific theoretical prediction. If we compare Equation (14-11) to the expression for E_{b3} and use Equation (14-3), we find

$$\frac{3}{5}\frac{Z^2 e^2}{4\pi\varepsilon_0 R_0 A^{1/3}} = a_3\frac{Z^2}{A^{1/3}} \quad\Rightarrow\quad a_3 = \frac{3}{5}\frac{e^2}{4\pi\varepsilon_0 R_0} = \frac{3}{5}\frac{\alpha\hbar c}{R_0}.$$

(The fine structure constant has been introduced here to simplify the numerical work.) Since a_3 is regarded as known from the empirical fit, the relation yields a determination of the radius parameter:

$$R_0 = \frac{3}{5}\frac{\alpha\hbar c}{a_3} = \frac{3}{5}\frac{197.3\ \mathrm{MeV}\cdot\mathrm{fm}}{(137.0)(0.7105\ \mathrm{MeV})} = 1.216\ \mathrm{fm}.$$

This one-parameter evaluation of the nuclear radius is judged to be in good agreement with the result $R_0 = 1.07$ fm obtained from the two-parameter electron-scattering method of Section 14-3.

Example

The $A = 15$ mirror nuclides $^{15}_{7}\mathrm{N}_8$ and $^{15}_{8}\mathrm{O}_7$ have atomic masses 15.000109 and 15.003065 u, respectively. The difference of binding energies is obtainable from Equation (14-8):

$$\Delta = E_b(^{15}\mathrm{N}) - E_b(^{15}\mathrm{O})$$
$$= \left[7M(^1\mathrm{H}) + 8M_n - M(^{15}\mathrm{N})\right]c^2 - \left[8M(^1\mathrm{H}) + 7M_n - M(^{15}\mathrm{O})\right]c^2$$
$$= \left\{\left[M_n - M(^1\mathrm{H})\right] + \left[M(^{15}\mathrm{O}) - M(^{15}\mathrm{N})\right]\right\}c^2$$
$$= (0.000840 + 0.002956)(931.5\ \mathrm{MeV}) = 3.536\ \mathrm{MeV}.$$

We have identified this quantity with the difference in Coulomb energy. Let us consult Equation (14-11) and take $Z = 7$ and 8 to get

$$V_8 - V_7 = \frac{3}{5}(64 - 49)\frac{e^2}{4\pi\varepsilon_0 R} = 9\frac{\alpha\hbar c}{R}$$

as an alternative expression for the quantity Δ. Note that R has the same value for the two isobars since R depends only on A. We calculate the radius from the result obtained for Δ,

$$R = 9\frac{\alpha\hbar c}{\Delta} = 9\frac{197.3 \text{ MeV} \cdot \text{fm}}{(137.0)(3.536 \text{ MeV})} = 3.666 \text{ fm},$$

and we then predict the radius parameter R_0 from Equation (14-3):

$$R_0 = \frac{R}{A^{1/3}} = \frac{3.666 \text{ fm}}{15^{1/3}} = 1.487 \text{ fm}.$$

Other pairs of mirror nuclides also give values for R_0 in the vicinity of 1.5 fm.

Example

We can get an approximate isolation of the pairing term E_{b5} if we examine the binding energies of three isotopes with consecutive mass numbers $A - 1$, A, and $A + 1$, provided we take the central mass number A to be *odd* and the common atomic number Z to be *even*. Under these conditions $E_b(^{A\pm1}\text{X})$ includes the contribution $+a_5/(A \pm 1)^{3/4}$, while $E_b(^A\text{X})$ contains no ε_5 term. We can then prove to first approximation that the three binding energies obey the relation

$$\frac{1}{2}\left[E_b(^{A+1}\text{X}) + E_b(^{A-1}\text{X})\right] - E_b(^A\text{X}) = \frac{a_5}{A^{3/4}} + \cdots.$$

(The proof is left as Problem 11 at the end of the chapter.) The left side of this relation can be written in terms of two particular neutron separation energies as follows. Equation (14-10) gives the equalities

$$E_n(^{A+1}\text{X}) = E_b(^{A+1}\text{X}) - E_b(^A\text{X}) \quad \text{and} \quad E_n(^A\text{X}) = E_b(^A\text{X}) - E_b(^{A-1}\text{X}),$$

and subtraction of the two quantities gives the result:

$$\tfrac{1}{2}\left[E_n(^{A+1}\text{X}) - E_n(^A\text{X})\right] = \tfrac{1}{2}\left[E_b(^{A+1}\text{X}) + E_b(^{A-1}\text{X})\right] - E_b(^A\text{X}).$$

Thus, our two conclusions lead us to a determination of a_5 using neutron separation data. We refer back to our cadmium calculations at the end of Section 14-4 and find

$$\frac{1}{2}\left[E_n(^{114}\text{Cd}) - E_n(^{113}\text{Cd})\right] = \frac{9.043 - 6.542}{2}\text{MeV} = 1.251 \text{ MeV} = \frac{a_5}{113^{3/4}}.$$

The result, $a_5 = 43.35$ MeV, is a bit larger than the average value given for this constant in the text.

14-6 The Fermi Gas Model

Nucleons behave as fermions and obey the Pauli principle. Hence, the protons and neutrons in the nucleus are influenced by the same constraint of fermion antisymmetry as the electrons in the atom. We know that the Coulomb force and the exclusion principle determine the structure of the atom, and we suppose that a similar construction works just as decisively in the case of the nucleus. Nuclear binding is a very different dynamical problem since the constituents attract each other with a strong short-range force and the system as a whole has no obvious center of attraction. Nevertheless, we are able to adopt an independent-particle central-field approach as one of the avenues in our study of the nucleus. We find that this familiar outlook enables us to draw several important phenomenological conclusions.

Let us begin with a qualitative argument to explain why the nucleus prefers a composition of equal numbers of protons and neutrons. The nuclear force is not needed in any detail since the exclusion principle is the main ingredient in this explanation. We know that the Coulomb repulsion between protons distorts the systematic $Z = N$ property in favor of neutrons for the larger nuclei, so let us consider the smaller nuclei and ignore the Coulomb effect. An individual proton or neutron is then subject to the same average force arising from all the other nucleons. Figure 14-14 shows a short-range central potential energy to represent the influence of this force on a single nucleon. Also shown are the resulting single-particle energy levels, analogous to those found in the central-field model of the atom. (We ignore the angular momenta of the levels since these additional properties do not affect our main conclusion.) The exclusion principle allows us to fill each level with no more than two protons and two neutrons corresponding to spins up and down. The example in the figure describes two occupation schemes for A nucleons with different assignments of Z and N. We see that the binding energy of the last nucleon is greatest for the

---

Figure 14-14

Independent-nucleon potential energy and qualitative energy levels in a central-field model of the nucleus. Coulomb effects are ignored and orbital angular momentum considerations are suppressed. In an $A = 21$ system, the neutron separation energy E_n is greater for the ($Z = 10$, $N = 11$) configuration than for the neutron-laden ($Z = 6$, $N = 15$) case.

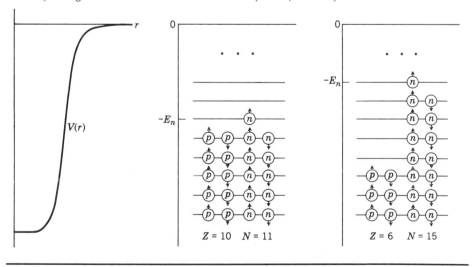

configurations closest to $Z = N$, and we conclude that these circumstances are the most favorable for nuclear stability.

We can employ fermion antisymmetry, and little else by way of input, to justify the expression given for the symmetry energy in Section 14-5. Our procedure retreats temporarily from the independent-particle picture and takes an *average* over the detailed dynamics of the nucleons. The nucleon number A is assumed to be rather large for this purpose so that the nuclear system becomes vast enough to warrant the use of quantum statistics. We treat the nucleus as a large collection of protons and neutrons moving freely in a spherical enclosure defined by the nuclear volume, and we describe the system as a degenerate Fermi gas in which the nucleons occupy their lowest energy states consistent with the exclusion principle. To be specific, we let the zero-temperature Fermi–Dirac distribution function be the sole vehicle for the description of the nuclear particles. The proton and the neutron are distinguishable from each other, and so the exclusion principle and the methods of Fermi–Dirac statistics apply to the two types of nucleon separately. Thus, the resulting nuclear system is regarded as a mixture of proton and neutron Fermi gases.

We can adapt this picture to the independent-nucleon point of view with the aid of Figure 14-15, in which we introduce a *constant* potential energy well for the free motion of each individual bound particle. Since A is large we must incorporate the effect of Coulomb repulsion for the proton and employ wells of different depth for the two types of nucleon. A large number of energy levels is assumed in each case, and all are presumed to be fully occupied by the gases of neutrons and protons up to the indicated Fermi energy levels. Note that the Fermi energy and the number of particles are larger in the case of the neutrons because the Coulomb effect diminishes the nuclear attraction and elevates the well in the case of the protons.

The Fermi gas model of the nucleus follows directly from our zero-temperature results of Section 11-6. We only need to modify the formulas for separate application to neutrons and protons. The nucleus has a spherical volume determined by the mass number A:

$$V = \tfrac{4}{3}\pi R^3 = \tfrac{4}{3}\pi R_0^3 A.$$

We express the nucleon numbers in terms of the respective Fermi energies as in

---

Figure 14-15

Neutron and proton potential energy wells in the Fermi gas model. The proton potential energy is elevated by the effects of Coulomb repulsion.

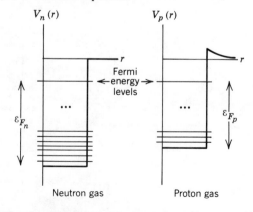

Equation (11-72):

$$Z = \frac{(2M)^{3/2}}{3\pi^2 \hbar^3} V \varepsilon_{F_p}^{3/2} \quad \text{for protons} \tag{14-16}$$

and

$$N = \frac{(2M)^{3/2}}{3\pi^2 \hbar^3} V \varepsilon_{F_n}^{3/2} \quad \text{for neutrons.} \tag{14-17}$$

(The proton and the neutron are assigned the same mass M throughout this discussion.) We then introduce our expression for the nuclear volume and invert these relations to solve for the Fermi energies:

$$
\begin{aligned}
\varepsilon_{F_p} &= \frac{1}{2M} \left(\frac{3\pi^2 \hbar^3}{V} Z \right)^{2/3} = \frac{\hbar^2}{2M} \left(\frac{3\pi^2}{\frac{4}{3}\pi R_0^3 A} Z \right)^{2/3} \\
&= \frac{\hbar^2}{2MR_0^2} \left(\frac{9\pi}{4} \frac{Z}{A} \right)^{2/3}
\end{aligned} \tag{14-18}
$$

and

$$\varepsilon_{F_n} = \frac{\hbar^2}{2MR_0^2} \left(\frac{9\pi}{4} \frac{N}{A} \right)^{2/3}. \tag{14-19}$$

Equation (11-74) gives the total energy of the particles in each gas as a function of the number of particles and the Fermi energy. We use this result to obtain

$$
\begin{aligned}
E_Z &= \frac{3}{5} Z \varepsilon_{F_p} = \frac{3}{10} \frac{\hbar^2}{MR_0^2} Z \left(\frac{9\pi}{4} \frac{Z}{A} \right)^{2/3} \\
&= \frac{3}{10} \left(\frac{9\pi}{4} \right)^{2/3} \frac{\hbar^2}{MR_0^2} A \left(\frac{Z}{A} \right)^{5/3}
\end{aligned}
$$

for the proton gas, and

$$E_N = \frac{3}{10} \left(\frac{9\pi}{4} \right)^{2/3} \frac{\hbar^2}{MR_0^2} A \left(\frac{N}{A} \right)^{5/3}$$

for the neutron gas.

We are concerned with the symmetry energy of the nucleus due to the deviation of the proton and neutron numbers from the equality $Z = N = A/2$. Let us introduce a new variable ζ accordingly by defining

$$N = \frac{A}{2} + \zeta \quad \text{and} \quad Z = \frac{A}{2} - \zeta$$

and evaluate the relevant factors in E_Z and E_N by the following binomial expansions:

$$
\left(\frac{Z}{A}\right)^{5/3} + \left(\frac{N}{A}\right)^{5/3} = \left(\frac{1}{2}\right)^{5/3}\left[\left(1 - \frac{2\zeta}{A}\right)^{5/3} + \left(1 + \frac{2\zeta}{A}\right)^{5/3}\right]
$$

$$
= \left(\frac{1}{2}\right)^{5/3}\left[1 - \frac{5}{3}\frac{2\zeta}{A} + \frac{5}{9}\left(\frac{2\zeta}{A}\right)^2 + \cdots\right.
$$

$$
\left. + 1 + \frac{5}{3}\frac{2\zeta}{A} + \frac{5}{9}\left(\frac{2\zeta}{A}\right)^2 + \cdots\right]
$$

$$
= \left(\frac{1}{2}\right)^{2/3}\left(1 + \frac{20}{9}\frac{\zeta^2}{A^2} + \cdots\right).
$$

The total energy of the system of nucleons then takes the form

$$
E_Z + E_N = \frac{3}{10}\left(\frac{9\pi}{4}\right)^{2/3}\frac{\hbar^2}{MR_0^2}A\left(\frac{1}{2}\right)^{2/3}\left(1 + \frac{20}{9}\frac{\zeta^2}{A^2}\right)
$$

$$
= \frac{3}{40}(9\pi)^{2/3}\frac{\hbar^2}{MR_0^2}\left(A + \frac{20}{9}\frac{\zeta^2}{A}\right) \tag{14-20}
$$

to second order in ζ. This quantity contributes to the rest energy of the nuclide ^AX and must therefore appear in the mass $M(^A\text{X})$ as parametrized in Equation (14-13). We use $\zeta = A/2 - Z$ and rewrite the second-order term in Equation (14-20) in the form

$$
\frac{(9\pi)^{2/3}}{6}\frac{\hbar^2}{MR_0^2}\frac{(A/2 - Z)^2}{A}.
$$

This final result reproduces the contribution of the symmetry energy to the atomic rest energy $M(^A\text{X})c^2$. A numerical exercise is provided in the examples below to complete the comparison.

In closing let us offer some justification for the independent-nucleon theory of the nucleus and indicate especially how the exclusion principle supports this point of view. The interaction between any two nucleons is strong enough at short range to cause a rapid transfer of energy and momentum between the nuclear particles. It would seem that a given nucleon must undergo a prompt dispersal of its energy in nuclear matter and have little opportunity to experience the smoothly varying interaction suggested by Figure 14-14. A competing theory would argue that the behavior of any given particle disappears at once from the memory of formation of a nuclear state, so that the nucleus should be treated as a whole compound instead of a collection of distinguished constituents. In fact, for a low-energy system like the nuclear ground state, the exclusion principle inhibits these energy-transferring rearrangements by preventing the particles from moving into the fully occupied low-lying states. Under such conditions the independent nucleon sees the nucleus as a medium in which the strong nucleon–nucleon interaction is moderated by the effects of fermion antisymmetry. The alternative compound-nucleus point of view becomes more important in nuclear reactions, where the higher energy permits access to the unfilled higher states.

Example

Let us connect the ζ^2 term in Equation (14-20) with the a_4 term in Equation (14-13) by identifying the constant coefficients in the two expressions:

$$a_4 = \frac{(9\pi)^{2/3}}{6} \frac{\hbar^2}{MR_0^2}.$$

If we set $R_0 = 1.2$ fm and take $Mc^2 = 939$ MeV, we find that our statistical model predicts

$$a_4 = \frac{(9\pi)^{2/3}}{6} \frac{(\hbar c)^2}{Mc^2 R_0^2} = \frac{(9\pi)^{2/3}}{6} \frac{(197\text{ MeV} \cdot \text{fm})^2}{(939\text{ MeV})(1.2\text{ fm})^2} = 44\text{ MeV}.$$

Recall from Section 14-5 that the atomic masses are fit with $a_4 = 94.80$ MeV. The agreement is only qualitative, as befits the crudeness of the model.

Example

If we ignore the ζ dependence altogether and set $Z = N = A/2$, we find from Equations (14-18) and (14-19) that the nucleon Fermi energy is independent of A:

$$\varepsilon_F = \frac{\hbar^2}{2MR_0^2}\left(\frac{9\pi}{8}\right)^{2/3}.$$

The numerical value of this quantity is

$$\varepsilon_F = \frac{(9\pi)^{2/3}}{8} \frac{\hbar^2}{MR_0^2} = \frac{3}{4}a_4 = 33\text{ MeV},$$

using the result of the previous example. Figure 14-15 then tells us that the depth of the potential energy well must be of order 40 MeV if a typical nucleon has separation energy around 7 MeV.

14-7 The Nucleon–Nucleon Interaction

The models in the last two sections are concerned with the global properties of all nuclei and not with the details of nuclear structure. Let us set these first observations aside now and take up the structural approach, beginning with the most primitive nuclear system.

The fundamental problem of nuclear physics is the determination of the force between two nucleons. We can probe the basic two-nucleon system to study this unknown interaction by scattering nucleons from a proton target and by examining the properties of the nucleon–nucleon bound state. The *deuteron* is the only existing $A = 2$ nuclide in the latter category. This bound system of proton and neutron is unique since there are no excited *pn* structures and no analogous *pp* or *nn* counterparts. We find that several important features of the nucleon–nucleon interaction can be inferred from the deuteron, while many other details of the force can only be learned from the higher-energy processes of nucleon–nucleon scattering.

Let us compile the following principal characteristics of the deuteron (symbol d) and interpret the various properties afterward:

binding energy	$E_b(^2\mathrm{H}) = 2.225 \text{ MeV}$
nuclear spin and parity	$i^p = 1^+$
magnetic dipole moment	$\mu_d = 0.8574 \, \mu_N$
electric quadrupole moment	$Q_d = 2.82 \times 10^{-3} \, e \cdot \text{barn}$
nuclear radius	$R_d = 2.1 \text{ fm}$

Each of these quantities is known from a number of experimental sources. One way to measure the deuteron binding energy is by observing the energy of the γ rays emitted in the np capture reaction

$$n + p \rightarrow d + \gamma.$$

The deuterons are formed in this process when slow neutrons from a reactor are absorbed by protons in a hydrogenous target. The small value of E_b corresponds to just over 1 MeV per nucleon, the lowest point plotted on the graph in Figure 14-12. The nuclear spin and magnetic moment are determined from measurements of atomic hyperfine structure and from magnetic-resonance experiments. The quadrupole moment is obtained by other applications of the beam-resonance technique. We define the quadrupole moment below and interpret the quantity as an indicator of the nuclear shape. The nuclear radius is measured in electron-scattering experiments as discussed in Section 14-3. Recall that these measurements also provide a picture of the nuclear charge distribution.

We assemble the nuclear spin vector of the deuteron by adding the spins of the proton and neutron to the orbital angular momentum $\mathbf{L}$ of the proton–neutron sysem:

$$\mathbf{I} = \mathbf{L} + \mathbf{S}_p + \mathbf{S}_n. \tag{14-21}$$

A highly schematic picture of this construction is shown in Figure 14-16. Note that $\mathbf{S}_p$ and $\mathbf{S}_n$ are nuclear spin vectors themselves, each with quantum number $i = \frac{1}{2}$ (or, equivalently, $s = \frac{1}{2}$), while the deuteron has nuclear spin quantum number $i = 1$. We can generalize Equation (14-21) immediately and give the following formula for the nuclear spin of any nucleus consisting of Z protons and N neutrons:

$$\mathbf{I} = \sum_{k=1}^{Z} \left(\mathbf{L}_{p_k} + \mathbf{S}_{p_k} \right) + \sum_{k=1}^{N} \left(\mathbf{L}_{n_k} + \mathbf{S}_{n_k} \right). \tag{14-22}$$

We already know that the vector $\mathbf{I}$ and the quantum number i are related by the rules of angular momentum quantization. Thus, the I^2 eigenvalue defines i by the equation

$$I^2 = \hbar^2 i(i + 1),$$

and the I_z eigenvalues define $2i + 1$ quantized orientations of $\mathbf{I}$ by the property

$$I_z = \hbar m_i \quad \text{with } m_i = -i, \ldots, i \text{ in integer steps.}$$

Figure 14-16

Nucleon-spin and orbital contributions to the spin of the deuteron. The quantized angular momentum vectors are represented symbolically, as in the vector model of Figure 8-23.

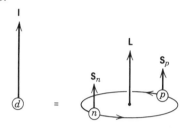

This construction endows a nuclide with half-integral spin if A is odd or integral spin if A is even. We include a listing of nuclear spins in Appendix A, and we observe that the deuteron is an even-A nuclide with three quantized spin orientations for nuclear spin $i = 1$.

We recognize the expressions in Equations (14-21) and (14-22) as exact analogues of the formula for the total angular momentum $\mathbf{J}$ in the theory of complex atoms. We know from Section 9-8 that we can describe an atomic state in the LS-coupling scheme by the notation $^{2s+1}L_j$, provided we can regard ℓ and s as good quantum numbers for the total orbital and spin vectors $\mathbf{L}$ and $\mathbf{S}$. It is valid to adopt the same spectroscopic notation for nuclei, under analogous dynamical conditions, and simply substitute the quantum numbers i and m_i in place of j and m_j.

The nuclear magnetic moment $\boldsymbol{\mu}$ is also composed of contributions from the constituents of the nucleus. In the case of the deuteron we expect a term due to the orbital motion of the proton and another two terms due to the two nucleon spins, as suggested again by Figure 14-16. The general result for a magnetic dipole moment is quoted experimentally as an expectation value of μ_z, evaluated in the state of maximum I_z where $m_i = i$. This measured quantity is a source of information about the state of the bound nucleons, although the interpretation in terms of nuclear structure is likely to be quite complicated for most nuclei. We can interpret the result for the deuteron at once, however, by noting the near equality between μ_d and the sum of the magnetic moments of the proton and neutron:

$$\mu_d = 0.8574\,\mu_N \quad \text{and} \quad \mu_p + \mu_n = 0.8798\,\mu_N.$$

(The data for the second figure are taken from Section 14-1.) It would appear that almost all of the deuteron moment can be explained by assuming a *parallel-spin* $s = 1$ configuration for the proton and neutron in an $\ell = 0$ orbital state. The $\ell = 0$ assignment is expected if the ground state of the two-body system is governed by a central force. The combination of $\ell = 0$ with $s = 1$ is also consistent with the known value $i = 1$ for the nuclear spin. If we can treat ℓ and s as good quantum numbers for two particles subject to the nuclear force, then we can invoke the notation $^{2s+1}L_i$ and refer to the 1^+ deuteron as a $^3S_1$ two-nucleon state. Note that the $\ell = 0$ assumption agrees with the assignment of positive parity.

Two spin-$\frac{1}{2}$ nucleons can form states with total spin quantum numbers $s = 0$ and $s = 1$. The nuclear interaction between the two particles is evidently *spin dependent* since an $\ell = 0$ bound state exists for $s = 1$, but not for $s = 0$. To model this situation we might assume a central interaction for the main part of the nuclear force and add an extra spin-dependent piece to make the nuclear attraction greater for parallel spins. The added feature is analogous to the hyperfine effect discussed in Section 8-12. We have seen how the interaction of spin magnetic moments in Equation (8-58) reduces to a spin–spin coupling, and we might consider the adoption of a similar coupling here. An expression of the form $\mathbf{S}_p \cdot \mathbf{S}_n$ can produce the desired energy splitting for the two total spins $s = 0$ and 1 in the $\ell = 0$ proton–neutron state. Of course, the situation at hand is not exactly like the atomic hyperfine splitting since the underlying dynamics is not of electromagnetic origin.

An $\ell = 0$ orbital state describes a spherically symmetric probability distribution and a spherical shape for the corresponding nucleus. The small deuteron quadrupole moment Q_d implies a small deformation of the basic spherical shape of this nuclide. Let us explore these observations with the aid of the relevant classical and quantum formulas.

We define a classical electric quadrupole moment in terms of a given charge density ρ by the integral

$$Q = \int (3z^2 - r^2)\rho(\mathbf{r})\, d\tau. \qquad (14\text{-}23)$$

It is clear that Q has dimensions of charge times area, and it is convenient to express Q for nuclei in $e \cdot$ barn units. We can interpret Equation (14-23) more readily if we rewrite the polynomial in the integrand as

$$3z^2 - x^2 - y^2 - z^2 = 2\left(z^2 - \frac{x^2 + y^2}{2} \right).$$

This expression samples the shape of the charge by weighting the amount distributed *along* the z axis against the amount distributed *around* that axis. A positive (or negative) value of Q is therefore an indication of an elongated (or flattened) deformation. We add some further classical remarks about the form of Q in the example at the end of the section.

The quantum mechanical electric quadrupole moment is defined by the expectation value of the quadrupole polynomial, taken in the state of maximum I_z. In the case of the deuteron the coordinates in the polynomial are those of the proton:

$$\langle Q \rangle = e \int \Psi^* \left(3z_p^2 - r_p^2 \right) \Psi\, d\tau = \frac{e}{4} \int \Psi^* (3z^2 - r^2) \Psi\, d\tau. \qquad (14\text{-}24)$$

(We relate the two-body coordinate $\mathbf{r}$ and the proton coordinate $\mathbf{r}_p$ as in Figure 14-17, and we then deduce the extra factor of $\frac{1}{4}$ in Equation (14-24) from this relation.) The equation expresses the asphericity of the probability distribution and has the same interpretation as Equation (14-23) with regard to the distribution of charge.

An $\ell = 0$ state produces a *vanishing* result for the quadrupole moment $\langle Q \rangle$. (The proof of this assertion is left as Problem 17 at the end of the chapter.) Hence, the small observed value of Q_d can be explained only if the deuteron wave function is assumed

Figure 14-17

Proton and neutron coordinate vectors and the two-body separation vector. Equal masses are assumed so that the center of mass lies midway between p and n.

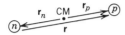

to have a small $\ell \neq 0$ portion supplementing the $^3S_1$ composition of the 1^+ state. We reject the possibility of an $\ell = 1$ contribution because this choice would enter with the wrong parity. We turn next to $\ell = 2$ and conclude that the deuteron state must be a *combination* of $^3S_1$ and $^3D_1$. Note that the spin-triplet property must be maintained so that $s = 1$ and $\ell = 2$ can combine to preserve the $i = 1$ quantum number for the nuclear spin. The assignment of a small probability to the $\ell = 2$ correction results in a fit for Q_d and also accounts for the small discrepancy between the magnetic moments μ_d and $\mu_p + \mu_n$.

We have remarked in Section 14-1 that the interaction between two nucleons is the same for pp, nn, and pn pairs of particles. We have just learned, however, that the deuteron stands alone as a pn bound state. These seemingly inconsistent observations are reconciled by the Pauli principle. We note that the deuteron is in a spin-symmetric state because of the $s = 1$ assignment and is also in a space-symmetric state because of the admixture of the two even values of ℓ. The Pauli principle forbids this combination of symmetries for the identical-fermion pairs pp and nn but has no immediate bearing on the deuteron since the pn system is composed of distinguishable particles.

The admixture of $\ell = 0$ and $\ell = 2$ in the wave function of the deuteron has interesting implications for the nucleon–nucleon interaction. Since the physical state is not characterized by a unique value of ℓ, it follows that ℓ is not exactly a good quantum number so that the fundamental pn interaction must not be due to a purely central force. We conclude that an additional L-nonconserving effect, called the *tensor force*, is also present in the dynamical problem. We have already cited the evidence for a two-nucleon force involving the spins of the two nucleons. The new tensor interaction introduces a further dependence on the angles between the nucleon spins and the two-body coordinate vector **r**. All these complicated aspects of the nucleon–nucleon interaction are apparent in the properties of the primitive $A = 2$ bound system.

We identify other properties of the interaction when we probe the two-nucleon system at higher energy in processes such as pp and np scattering. An experimental finding of some interest is the existence of a large spherically symmetric component in the distribution of protons scattered from protons. This effect is attributed to a strong *repulsive core* at very short range in the nuclear interaction between the two particles. Figure 14-3 shows how such a repulsive interaction dominates the attractive nuclear potential energy of two protons at small values of r. We see this strong short-range phenomenon in spherically symmetric $\ell = 0$ states, and we find that the effect is masked by the repulsive centrifugal potential energy in states with nonzero ℓ.

Signs of a repulsive core are also revealed in electron-scattering studies of the deuteron. A pronounced dip is observed at the center of the deuteron charge distribution, unlike the situation for the larger nuclides in the range considered in

Figure 14-9. This electron-scattering view of the proton–neutron bound state indicates a strong tendency for nucleons to repel and avoid each other at very close range.

It is obvious that a very complex interaction exists between two nucleons. We treat this basic nuclear force by means of models and acknowledge that the force might be too complex to be regarded as fundamental. We might even suppose that a truly elementary interaction is operative and somehow governs the observed nuclear particles at a level beneath that of their revealed behavior.

Example

Let us offer the following construction as background for the classical quadrupole formula in Equation (14-23). First, consider a point charge e at the origin and recall the well-known formula for the electrostatic potential at distance r:

$$\phi(\mathbf{r}) = \frac{e}{4\pi\varepsilon_0 r}.$$

Next, consider an electric dipole configuration consisting of a charge e at $(0, 0, d/2)$ and a charge $-e$ at $(0, 0, -d/2)$. The corresponding potential is the sum of the two point-charge potentials:

$$\phi(\mathbf{r}) = \frac{e}{4\pi\varepsilon_0}\left\{ \frac{1}{\sqrt{x^2 + y^2 + (z - d/2)^2}} - \frac{1}{\sqrt{x^2 + y^2 + (z + d/2)^2}} \right\},$$

where $x^2 + y^2 + z^2 = r^2$. We expand this expression for $d \ll r$ and obtain the following result to first order in d:

$$\phi(\mathbf{r}) = \frac{e}{4\pi\varepsilon_0 r}\left\{ \left(1 - \frac{zd}{r^2} + \frac{d^2}{4r^2}\right)^{-1/2} - \left(1 + \frac{zd}{r^2} + \frac{d^2}{4r^2}\right)^{-1/2} \right\}$$

$$= \frac{e}{4\pi\varepsilon_0 r}\left\{ \left(1 + \frac{zd}{2r^2} + \cdots\right) - \left(1 - \frac{zd}{2r^2} + \cdots\right) \right\} = \frac{ed}{4\pi\varepsilon_0}\frac{z}{r^3} + \cdots.$$

Let us denote the electric dipole moment by the quantity $p = ed$ and express the dipole potential by the familiar formula

$$\phi(\mathbf{r}) = \frac{p}{4\pi\varepsilon_0}\frac{z}{r^3}.$$

Finally, consider two such dipoles with opposite signs and separate locations, taking p to be at $(0, 0, d/2)$ and $-p$ to be at $(0, 0, -d/2)$. The resulting potential is the sum of the two dipole potentials:

$$\phi(\mathbf{r}) = \frac{p}{4\pi\varepsilon_0}\left\{ \frac{z - d/2}{\left[x^2 + y^2 + (z - d/2)^2\right]^{3/2}} - \frac{z + d/2}{\left[x^2 + y^2 + (z + d/2)^2\right]^{3/2}} \right\}.$$

Figure 14-18

Simple distributions of point charges. An arbitrary charge distribution can be assembled in the form of a series whose first three contributions are monopole, dipole, and quadrupole configurations.

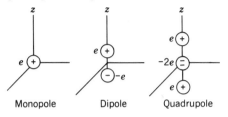

Again, we expand to first order in d:

$$
\begin{aligned}
\phi(\mathbf{r}) &= \frac{p}{4\pi\varepsilon_0 r^3}\left\{\left(z - \frac{d}{2}\right)\left(1 - \frac{zd}{r^2} + \frac{d^2}{4r^2}\right)^{-3/2}\right. \\
&\qquad\qquad\left. - \left(z + \frac{d}{2}\right)\left(1 + \frac{zd}{r^2} + \frac{d^2}{4r^2}\right)^{-3/2}\right\} \\
&= \frac{p}{4\pi\varepsilon_0 r^3}\left\{z\left[\left(1 + \frac{3}{2}\frac{zd}{r^2} + \cdots\right) - \left(1 - \frac{3}{2}\frac{zd}{r^2} + \cdots\right)\right]\right. \\
&\qquad\qquad\left. - \frac{d}{2}\left[(1 + \cdots) + (1 + \cdots)\right]\right\} \\
&= \frac{pd}{4\pi\varepsilon_0 r^3}\left(3\frac{z^2}{r^2} - 1\right) + \cdots = \frac{pd}{4\pi\varepsilon_0}\frac{3z^2 - r^2}{r^5} + \cdots .
\end{aligned}
$$

The final result contains the characteristic quadrupole polynomial $(3z^2 - r^2)$, as in Equations (14-23) and (14-24). We show this succession of three elementary charge distributions in Figure 14-18, and we note that the quadrupole represents a system with zero net charge elongated along the z axis.

14-8   A Simple Model of the Deuteron

The properties of the deuteron can be understood with the aid of the Schrödinger equation. The binding energy is especially interesting because the smallness of this quantity can be explained from a simple description of the proton–neutron interaction.

Let us argue that the spin-dependent aspects of the interaction are only secondary considerations and concentrate on the main effect of a two-body *central* force. We must adopt a model to describe the unknown central potential energy $V(r)$. The selection of the model is not a critical matter since the binding of this nucleus is not very sensitive to the detailed shape of the potential energy function. We find that the deuteron is

Figure 14-19

Square-well model for the binding of the deuteron. The function $rR(r)$ satisfies Equation (14-27), and the $\ell = 0$ stationary-state eigenfunction ψ depends only on r. The dashed curve in the figure shows the limiting form of $rR(r)$ for zero binding energy.

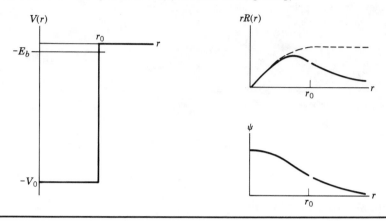

adequately treated with the use of a *square well*, parametrized by a well depth V_0 and a radius r_0. The parametrization of this model is illustrated in Figure 14-19. Note that the energy level of the deuteron is found near the top of the well and that no higher excited states are supposed to exist.

We ignore the neutron–proton mass difference and assign the same mass M to each particle. Consequently, the reduced mass in the central-force problem has the value $\mu = M/2$, and the coordinate vectors in Figure 14-17 satisfy the relation

$$\mathbf{r}_p = -\mathbf{r}_n = \frac{\mathbf{r}}{2}.\qquad (14\text{-}25)$$

We let the deuteron be described as a pure $^3S_1$ state with energy eigenvalue $-E_b$, and we take the $\ell = 0$ spatial eigenfunction to have the form

$$\psi = R(r)Y_{00}(\theta, \phi).\qquad (14\text{-}26)$$

The function $R(r)$ satisfies the general radial differential equation, as formulated in Equation (6-43), for the special situation where $\mu = M/2$, $\ell = 0$, and $E = -E_b$. We can rearrange this equation to read

$$\frac{d^2}{dr^2}(rR) - \frac{M}{\hbar^2}\big[V(r) + E_b\big](rR) = 0.\qquad (14\text{-}27)$$

Since $V(r)$ is a discontinuous step function, the differential equation holds in *piecewise* fashion:

$$\frac{d^2}{dr^2}(rR) + \frac{M}{\hbar^2}(V_0 - E_b)(rR) = 0 \quad \text{for } r < r_0$$

and

$$\frac{d^2}{dr^2}(rR) - \frac{M}{\hbar^2}E_b(rR) = 0 \quad \text{for } r > r_0.$$

Let us define the two wave-number parameters,

$$k = \frac{\sqrt{ME_b}}{\hbar} \quad \text{and} \quad K = \frac{\sqrt{M(V_0 - E_b)}}{\hbar}, \tag{14-28}$$

so that our equations become

$$(rR)'' + K^2(rR) = 0 \quad \text{for } r < r_0$$

and

$$(rR)'' - k^2(rR) = 0 \quad \text{for } r > r_0.$$

We can identify the solutions immediately:

$$rR(r) = \begin{cases} a \sin Kr & r < r_0 \\ be^{-kr} & r > r_0 \end{cases}. \tag{14-29}$$

(The function $\cos Kr$ cannot be used inside the well because an infinite result would follow for $R(0)$. The function e^{+kr} cannot be used outside the well because a divergence would occur as $r \to \infty$.) A sketch of the final solution is included in Figure 14-19. The figure also shows the stationary-state eigenfunction

$$\psi = \frac{R(r)}{\sqrt{4\pi}}$$

as defined in Equation (14-26).

The two pieces of the solution in Equation (14-29) are supposed to match smoothly at the well boundary $r = r_0$. We use this requirement to draw the following implications:

$$\text{continuity of } rR \quad \Rightarrow \quad a \sin Kr_0 = be^{-kr_0}$$

and

$$\text{continuity of } (rR)' \quad \Rightarrow \quad Ka \cos Kr_0 = -kbe^{-kr_0}.$$

These results represent two conditions in the determination of the three unknown constants a, b, and E_b. The normalization of the wave function provides the third constraint needed to complete the solution. We are mainly interested in E_b, so let us take the ratio of our two conditions in order to remove a and b from this part of the problem. The result is a single equation relating E_b to the parameters of the model:

$$K \cot Kr_0 = -k \quad \text{or} \quad \cot\frac{\sqrt{M(V_0 - E_b)}}{\hbar}r_0 = -\sqrt{\frac{E_b}{V_0 - E_b}}. \tag{14-30}$$

The second form of this relation follows from Equations (14-28). The usual methods of quantum mechanics would treat the energy E_b as an unknown quantity to be determined for given parameters V_0 and r_0. Alternatively, we can regard E_b as known and use Equation (14-30) to express a necessary condition on the parametrization of the model. The equation is known as the *range-depth relation* when this point of view is adopted.

We can penetrate Equation (14-30) most effectively if we look first at the limiting situation where $E_b \to 0$. This limit is not far from the physical case, if indeed E_b is as small compared to V_0 as Figure 14-19 suggests. The range-depth relation has the limiting form

$$\cot \frac{\sqrt{MV_0}}{\hbar} r_0 = 0,$$

and so a simplified relation between V_0 and r_0 follows at once:

$$\frac{\sqrt{MV_0}}{\hbar} r_0 = \frac{\pi}{2} \quad \Rightarrow \quad V_0 r_0^2 = \frac{\pi^2 \hbar^2}{4M} \quad (E_b = 0). \qquad (14\text{-}31)$$

We examine this result numerically in the first example at the end of this section and learn that the estimated well depth is at least an order of magnitude larger than the measured binding energy, given a realistic choice for the radius parameter. Equations (14-28) reduce to the limits

$$k \to 0 \quad \text{and} \quad K \to \frac{\sqrt{MV_0}}{\hbar} \to \frac{\pi}{2r_0} \quad \text{as } E_b \to 0.$$

Consequently, the solution in Equation (14-29) becomes

$$rR(r) \to \begin{cases} a \sin \dfrac{\pi r}{2r_0} & r < r_0 \\ b & r > r_0 \end{cases}$$

in the limiting case. The behavior of this solution is indicated by the dashed graph in the figure. Note that when E_b is quite small the solution $rR(r)$ just turns over inside the well. The corresponding broad shape of the eigenfunction ψ implies that this state is not able to probe the details of the two-nucleon force with very much resolution. Hence, the square well is able to give an adequate picture of the binding interaction. The deuteron is said to be *barely bound* because of these properties.

The range-depth relation can be analyzed by graphical means when E_b is allowed to have its actual nonzero value. Let us rewrite Equation (14-30) for this purpose, setting $Kr_0 = u$:

$$\frac{u}{r_0} \cot u = -k \quad \Rightarrow \quad \tan u = -\frac{u}{kr_0}.$$

Figure 14-20 shows graphs of the left and right sides of the final equality, plotted as functions of u. The deuteron solution is determined by the value of u at the first intersection of the two graphs. (Intersections at larger u have no interpretation in this

Figure 14-20

Solution of the deuteron range-depth relation $\tan u = -u/kr_0$. The intersection of the two graphs approaches the asymptote $u = \pi/2$ in the indicated zero-binding limit.

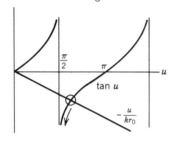

Figure 14-21

Square-well potential energy including an infinitely repulsive core. Realistic values of the parameters in a model of the deuteron are: $V_0 = 70$ MeV, $r_0 = 1.7$ fm, and $r_c = 0.4$ fm.

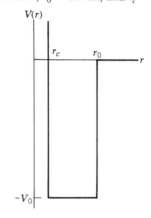

particular model.) The figure indicates the following bounds on the solution:

$$\frac{\pi}{2} < Kr_0 < \pi \quad \text{or} \quad \frac{\pi}{2} < \frac{\sqrt{M(V_0 - E_b)}}{\hbar} r_0 < \pi.$$

Thus, the graphs describe an intersection that moves with E_b and has asymptotic behavior

$$Kr_0 \to \frac{\pi}{2} \quad \text{as } k \to 0$$

in the limit of zero binding energy. This result agrees with the conclusions of the previous paragraph.

Our model for the nucleon–nucleon interaction has not taken account of the short-range repulsive core discussed at the end of Section 14-7. We can make way for this effect if we substitute the modified square well in Figure 14-21 for the original well in Figure 14-19. The problem is altered by the infinite repulsion near $r = 0$ because the modified eigenfunction must vanish everywhere inside the proposed core. This new property gives the probability distribution a vacant region at its center, in qualitative agreement with the electron-scattering results mentioned in Section 14-7. The remainder of the analysis goes through as a straightforward exercise, which we include as Problem 20 at the end of the chapter. The crude model in Figure 14-21 is intended only as a squared-off approximation to a more realistic two-nucleon interaction. Figure 14-3 shows a better model for the case of the $s = 0$ system. The $s = 1$ interaction is expected to have the same general appearance with a somewhat deeper attractive well.

Example

The range-depth relation for zero binding results in the following numerical expression of Equation (14-31):

$$V_0 r_0^2 = \frac{\pi^2}{4} \frac{(\hbar c)^2}{Mc^2} = \frac{\pi^2}{4} \frac{(197 \text{ MeV} \cdot \text{fm})^2}{939 \text{ MeV}} = 102 \text{ MeV} \cdot \text{fm}^2 \quad \text{if } E_b = 0.$$

We can use this calculation to constrain the values of the square-well parameters. Thus, a typical radius $r_0 = 1.6$ fm implies a well depth

$$V_0 = \frac{102 \text{ MeV} \cdot \text{fm}^2}{(1.6 \text{ fm})^2} = 40 \text{ MeV},$$

a figure more than ten times larger than E_b.

Example

Let us return to Equation (14-29) and carry the solution for $rR(r)$ one step further. The constants a and b obey the equality

$$b = a \sin Kr_0 \cdot e^{kr_0},$$

while Equation (14-30) provides the relation

$$\sin^2 Kr_0 = \frac{1}{1 + \cot^2 Kr_0} = \frac{K^2}{K^2 + k^2}.$$

The quantity Kr_0 is known to lie in the interval $(\pi/2, \pi)$, and so the positive root is chosen to give

$$\sin Kr_0 = + \frac{K}{\sqrt{K^2 + k^2}} \quad \text{and} \quad b = a \frac{K}{\sqrt{K^2 + k^2}} e^{kr_0}.$$

The result enables us to rewrite the radial solution in terms of a single multiplicative constant:

$$R(r) = \frac{a}{r} \begin{cases} \sin Kr & r < r_0 \\ \dfrac{K}{\sqrt{K^2 + k^2}} e^{-k(r - r_0)} & r > r_0. \end{cases}$$

This expression is put to use in Problems 18 and 19 at the end of the chapter.

14-9   Magic Numbers

Our treatment of the nucleus takes on a further degree of model dependence as we pass from the two-nucleon system to nuclides with many protons and neutrons.

Figure 14-22

Populations of stable nuclides according to proton number and neutron number.

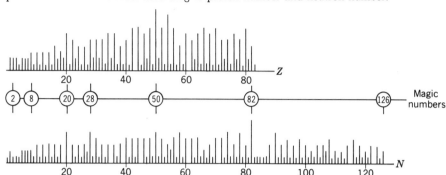

Complex nuclei can be studied only by recourse to some approximate method for handling all the bound particles. We have encountered this same situation in our experience with complex atoms. The independent-particle theory has been a great success in the analysis of atoms and offers a productive strategy for nuclei as well, despite the obvious differences between the two systems. We know that the nucleus has no central core (except in the context of a model) and that the interaction between two nucleons produces strong scattering effects. We might therefore expect the nuclear force to be dominated by the noncentral interactions of pairs of nucleons. Nevertheless, a domain of validity exists for a central-field theory of the nucleus in which an independent nucleon experiences the average force of all the other nuclear particles. We have invoked the exclusion principle in Section 14-6 to argue the merits of this point of view. The central field and the exclusion principle have been brought together successfully in the shell theory of atoms. We find the same methods fruitful again, but to a lesser degree, in the theory of nuclei.

Hints of nuclear shell behavior have already been noted in Figure 14-4. These first indications of the magic numbers have been observed in the larger-than-average populations of stable nuclides along certain rows and columns of the nuclear chart, corresponding to certain numbers of bound protons and neutrons. The magic numbers include these values of Z and N in the following list:

$$2 \quad 8 \quad 20 \quad 28 \quad 50 \quad 82 \quad 126 \ldots$$

We present the same observations again in Figure 14-22 in the form of distributions of the stable nuclides, plotted as functions of Z and N. The figure shows the more obvious accumulations of nuclei at $Z = 20$ (calcium) and 50 (tin), and at $N = 20$ and 82. We note that ^{40}Ca is doubly magic with $Z = N = 20$, and we discover that such other doubly magic examples as ^{4}He $(Z = N = 2)$, ^{16}O $(Z = N = 8)$, and ^{208}Pb $(Z = 82$ and $N = 126)$ are species with unusual stability. The binding energies of magic configurations of nucleons tend to be larger than average, enough to stand out as marked deviations from the smooth graph of E_b/A in Figure 14-12. These fragments of systematic evidence are taken as suggestions of *shell closures* in nuclei.

We find more compelling evidence of closed-shell behavior when we look at nucleon separation energies in the vicinity of the magic numbers. These quantities are

Figure 14-23

Neutron separation energy versus neutron number for several families of isobars. The data are from the tables of Wapstra and Bos, the source used to construct Figure 14-12.

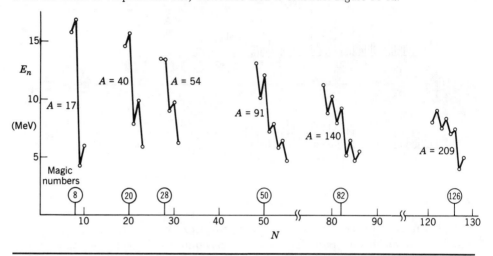

analogous to the ionization energies of atoms. Figure 9-7 has taught us to read the variation of ionization energy with atomic number as a clear indication of atomic shell structure. We present nuclear data of the same sort in Figure 14-23 by plotting the neutron separation energy E_n versus N for several groups of isobars. Only a limited sample of graphs needs to be shown since the results for the selected values of A are typical of many different isobar families. We find in every case that E_n varies with N and drops abruptly when N changes from N^*, a magic number, to $N^* + 1$. The graphs tell us that the binding of the last neutron is relatively large when the nucleus contains N^* neutrons and becomes unusually small with the addition of one more neutron. We conclude that a closed shell of neutrons occurs at the neutron number N^*. A parallel study of the proton separation energy leads to a similar conclusion for protons. We recall again that the ionization energy of atoms demonstrates exactly the same behavior (somewhat more dramatically) whenever the atomic number corresponds to any of the noble gases. We know from Equation (14-10) that E_n is equal to a difference of nuclear binding energies:

$$E_n(^A\text{X}) = E_b(^A\text{X}) - E_b(^{A-1}\text{X}).$$

Hence, a large variation of E_n can be associated with a noticeable departure of E_b from the value predicted in Equation (14-12), the semiempirical formula. We give a numerical illustration of this effect in one of the examples below.

Another way to spot a neutron shell closure is by examining the cross section for the absorption of thermal neutrons. The reaction of interest is the neutron-capture process in which a beam of slow neutrons initiates a nuclear transformation accompanied by the emission of γ rays:

$$n + {}^A\text{X} \rightarrow {}^{A+1}\text{X} + \gamma.$$

This process is also called an (n, γ) reaction, where the transformation of the nucleus is expressed as

$$^A\text{X}(n, \gamma)^{A+1}\text{X}.$$

We define the cross section σ_γ for this type of collision by analogy with the definition given in Section 3-4 for the elastic scattering of charged particles. In the case of neutron capture, we interpret the cross section as the effective area presented by a single target nucleus to a single beam neutron. This quantity can be used to measure the probability for the nucleus to capture the neutron. The (n, γ) cross section becomes quite large when the neutron number of the target nucleus is one unit less than a magic number N^* and becomes quite small when the target neutron number is equal to N^*. These observations constitute good evidence for a closed shell of neutrons at the magic neutron number N^*.

Example

We offer these cross sections for thermal neutron capture in xenon as an illustration:

$$\sigma_\gamma = 2.6 \times 10^6 \text{ barns} \quad \text{for } ^{135}\text{Xe}(n, \gamma)^{136}\text{Xe}$$

and

$$\sigma_\gamma = 0.26 \text{ barn} \quad \text{for } ^{136}\text{Xe}(n, \gamma)^{137}\text{Xe}.$$

Xenon has $Z = 54$, and so the two situations involve target neutron numbers equal to $N^* - 1$ and N^*, with $N^* = 82$. We can appreciate the enormous size of the one cross section relative to the other if we recall the definition

$$1 \text{ barn} = 10^{-28} \text{ m}^2 = (10 \text{ fm})^2.$$

In these units we have

$$2.6 \times 10^6 \text{ barns} = (16000 \text{ fm})^2 \quad \text{and} \quad 0.26 \text{ barn} = (5.1 \text{ fm})^2.$$

The figures in parentheses may be compared with the scale set by the nuclear radius. We recall Equation (14-3) to obtain

$$R = R_0 A^{1/3} = (1.07 \text{ fm})(136)^{1/3} = 5.50 \text{ fm}$$

for $A = 136$, and essentially the same result for $A = 135$. Obviously, ^{135}Xe is very hungry for neutrons while ^{136}Xe is somewhat satiated.

Example

The nuclide $^{88}_{38}\text{Sr}$ has a magic number of neutrons, $N = 50$. We therefore expect the neutron separation energy for this nucleus to be somewhat larger than the value predicted by the semiempirical mass formula. The actual value of E_n can be determined with the aid of Equation (14-10):

$$E_n(^{88}\text{Sr}) = E_b(^{88}\text{Sr}) - E_b(^{87}\text{Sr}) = (768.47 - 757.35) \text{ MeV} = 11.12 \text{ MeV}.$$

(The binding energies are taken from the tables of Wapstra and Bos, the reference used to construct Figures 14-12 and 14-23.) The prediction from the mass formula for an even–even nuclide is given by the expression

$$E_n(^A\mathrm{X}) = a_1 - a_2 A^{2/3}\left(1 - \xi^2\right)$$

$$+ a_3 \frac{Z^2}{A^{1/3}}\left(\frac{1}{\xi} - 1\right) - \frac{a_4}{4}\left[1 - \frac{4Z^2}{A(A-1)}\right] + \frac{a_5}{A^{3/4}},$$

where $\xi = [(A-1)/A]^{1/3}$. (The derivation of this result appears as Problem 10 at the end of the chapter.) For $A = 88$ we have $\xi = 0.9962$, and so we get the following string of terms using the a coefficients quoted in Section 14-5:

$$E_n(^{88}\mathrm{Sr}) = (15.76 - 2.67 + 0.88 - 5.82 + 1.36)\ \mathrm{MeV} = 9.51\ \mathrm{MeV}.$$

The actual value exceeds the predicted value by 1.61 MeV, a substantial 17% deviation.

14-10 The Nuclear Shell Model

The last three sections have prepared us for a theory of nuclei based on the independent-nucleon approach. We follow the example of the theory of atoms, and we employ the central-field approximation and the exclusion principle as the two main pillars of this investigation. The states of a single nucleon are found by adopting a central-field model to describe the interaction between the nucleon and the other $A - 1$ nuclear particles. The exclusion principle governs the protons and neutrons separately and controls the occupation of the single-nucleon energy levels for all A nucleons. These two ingredients of the theory are enough to predict a nuclear shell structure. We find, however, that the nuclear shell model must include other dynamical considerations if the theory is to reproduce the expected magic numbers.

We assume a central potential energy $V(r)$ for each nucleon in the nucleus so that we can apply the properties of angular momentum quantization to the states of each particle. This procedure enables us to label the states by the single-particle quantum numbers $(n\ell m_\ell m_s)$. Hence, the eigenfunction for a nucleon in a stationary state has the familiar form

$$\psi_{n\ell m_\ell m_s} = R_{n\ell}(r)Y_{\ell m_\ell}(\theta, \phi)(\uparrow \text{ or } \downarrow).$$

The radial function $R_{n\ell}(r)$ and the associated energy eigenvalue $E_{n\ell}$ depend on the choice of model for $V(r)$. We have discussed these aspects of the general central-force problem in Section 6-7, and we have formulated the radial differential equation for $R_{n\ell}(r)$ and $E_{n\ell}$ in Equation (6-43). Let us rewrite this equation as

$$\frac{d^2}{dr^2}(rR_{n\ell}) + \frac{2M}{\hbar^2}\left[E_{n\ell} - V_{\mathrm{eff}}(r)\right](rR_{n\ell}) = 0 \tag{14-32}$$

and recall that the effective potential energy includes the ℓ-dependent centrifugal term along with $V(r)$:

$$V_{\mathrm{eff}}(r) = V(r) + \frac{\hbar^2}{2Mr^2}\ell(\ell + 1). \tag{14-33}$$

(Note that the nucleon mass M is substituted for the reduced mass in both formulas.) The index n is introduced in Equation (14-32) as a radial node quantum number according to the interpretation given in Section 6-7. Thus, the choice of ℓ fixes the function $V_{\text{eff}}(r)$, and the index n then enumerates the ascending energy levels and counts the nodes of the corresponding radial solutions. We are reminded of these properties in the example at the end of the section.

Let us emphasize that the *original* definition of the quantum number n is in use in this problem. We should not confuse n with the terminology adopted in the theory of atoms, where the same symbol is used by convention to denote the principal quantum number. It is clear that the notion of a principal quantum number has no logical place in the treatment of nucleons.

The model requires the selection of a function $V(r)$ to describe the nuclear central field. A possible candidate has already been suggested in Figure 14-14. It is possible to parametrize this potential energy by the formula

$$V(r) = -\frac{V_0}{1 + e^{(r-R)/a}}, \qquad (14\text{-}34)$$

where V_0 and a do not vary appreciably from one nucleus to the next, while R varies as $A^{1/3}$. (Observe the similarity between the shape of the potential energy in Figure 14-14 and the shape of the nucleon density in Figure 14-8. We appeal to this analogy, and to Equation (14-2), when we write $V(r)$.) It is more expedient to assume a square-well approximation for $V(r)$ because the corresponding differential equation for $R_{n\ell}(r)$ and $E_{n\ell}$ admits an exact analytical solution. A possible strategy might be to start with a preliminary exact solution for a square-well model of the interaction and then tune in a more realistic numerical solution based on the rounded model of Equation (14-34). The first stage of this procedure generates a collection of energy levels resembling those shown in Figure 14-24. Note that the values of $E_{n\ell}$ are organized according to columns of different ℓ, as in the presentation of the analogous atomic problem in Figure 9-4. The second stage of the procedure shifts these energies to new levels at positions not very far away. Figure 14-25 shows the square-well and rounded-well levels in parallel columns to represent the results of such a method. It is obvious that these results must vary with the choice of mass number A. We describe this implementation of the independent-nucleon model as only one possible strategy, and we understand that any numerical conclusions depend on the parametrization of the model.

Our description of the single-nucleon energy levels applies to the case where the single nucleon is a neutron. We must add the effect of Coulomb repulsion to the model when we consider the interaction of a proton with the remaining $A - 1$ nucleons. This additional feature elevates the potential energy well and introduces a repulsive barrier for the proton, as in the square-well model of Figure 14-15. The resulting single-proton levels are qualitatively similar to those in Figure 14-25 except for a general elevation of all the energies.

The shell-model energy levels emerge from Equation (14-32) with only the two indices n and ℓ. A *degeneracy* exists at every level, as in the atomic problem, because the energy does not depend on the other two single-particle quantum numbers m_ℓ and m_s. Hence, each value of $E_{n\ell}$ corresponds to $2(2\ell + 1)$ degenerate states, distinguished by the up and down orientations of the nucleon spin and by the various assignments of m_ℓ for the given value of ℓ. These degeneracies are indicated next to the two columns of energy levels in Figure 14-25. We are able to deduce a shell structure of the nucleus on the basis of this list of numbers and energies.

Let us consider the process of building up the nuclear ground state by taking Z protons and N neutrons to populate the lowest unoccupied single-nucleon levels. The

Figure 14-24

Single-nucleon energy levels $E_{n\ell}$ in a hypothetical square-well model of the nucleon potential energy.

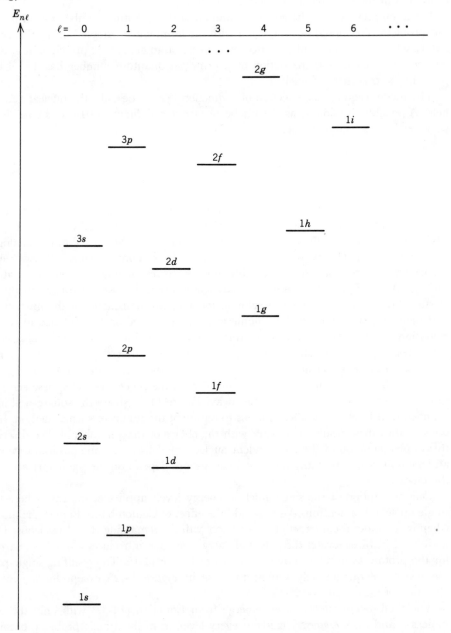

Figure 14-25

Shell-model levels and degeneracies for a square well and a rounded well. The cumulative total of protons or neutrons is supposed to reach a magic number at each of the larger energy gaps. Only the first three encircled predictions agree with the known magic numbers.

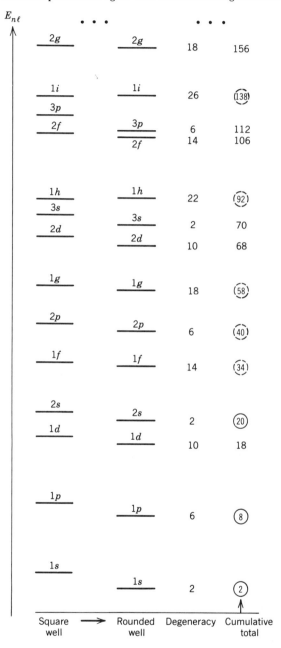

exclusion principle permits no more than $2(2\ell + 1)$ occupants in any shell-model level $E_{n\ell}$ for *both* species of nucleon. This maximum allowed number is given level-by-level, for protons and for neutrons, according to the list of degeneracies in Figure 14-25. Another column on the right side of the figure shows a *running total* of these maximum occupancies, accumulating upward from the lowest level. We predict a closed shell of nucleons of either type wherever we encounter a fully occupied level followed by an appreciable jump to the next higher energy state. The gap between levels implies a reduced binding energy for the next added nucleon. This same effect is observed in the building up of the ground states of atoms, where the analogous gaps are found at the atomic numbers of the noble-gas elements. The nuclear model in the figure predicts a shell closure when either Z or N reaches 2, 8, and 20, in agreement with the first three magic numbers, but fails to continue in the proper sequence thereafter. We are forced to conclude that the nuclear shell model is either incorrect or incomplete in its present form.

The difficulty cannot be resolved by choosing a different central field. A new type of additional ingredient is put forward instead in the form of a *nuclear spin–orbit interaction*. We recall from Sections 8-9, 9-4, and 9-7 that the spin–orbit coupling in atoms causes a splitting of the single-particle energy levels $E_{n\ell}$ for all $\ell \neq 0$ orbital states. The splitting occurs because the atomic interaction contributes a different energy shift for the two allowed values of the total angular momentum quantum number j, as indicated in Figure 8-28. We have expressed this interaction in terms of the central potential energy $V_c(r)$ in Equations (9-36):

$$V_{SL}(\text{atom}) = \frac{1}{2m_e^2 c^2} \frac{1}{r} \frac{dV_c}{dr} \mathbf{S} \cdot \mathbf{L}.$$

We know that the atomic spin–orbit effect has a secure theoretical basis in relativistic quantum mechanics, and we also know that the energy splitting in atoms is rather small especially for small values of the atomic number. In contrast, the nuclear splitting is assumed *a priori* to be *large* and also *inverted*, in the manner shown in Figure 14-26. We ascribe this behavior to a nuclear interaction of the form

$$V_{SL}(\text{nucleus}) = -\frac{a_{SL}^2}{r} \frac{dV}{dr} \mathbf{S} \cdot \mathbf{L}, \tag{14-35}$$

in which the central-field function $V(r)$ appears along with the orbital and spin angular momenta of the nucleon. The minus sign accomplishes the required inversion

---

Figure 14-26
Nuclear spin-orbit splitting of a single-nucleon energy level.

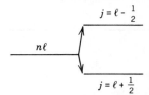

Figure 14-27

Shell-model levels including the effect of nuclear spin–orbit splitting. The degeneracy is given by $2j + 1$ at each level $E_{n\ell j}$. The accumulated population of nucleons corresponds to a magic number at every one of the larger energy gaps.

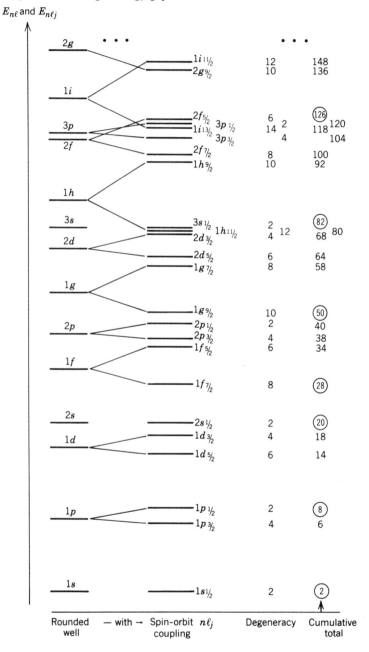

of the split levels, and the phenomenological constant a_{SL}^2 produces the desired amount of energy splitting.

Figure 14-27 shows how the large inverted splitting influences the sequence of shell-model levels. The interaction splits a given single-nucleon energy $E_{n\ell}$ (with $\ell \neq 0$) into a pair of energy levels $E_{n\ell j}$. The figure uses the notation $n\ell_j$ to designate the states with $j = \ell + \frac{1}{2}$ and $j = \ell - \frac{1}{2}$, and assigns the *lower* state to the *larger* value of j. Note that the splitting is large enough to produce *rearrangements* in the final array of energies. An energy level with quantum numbers $(n\ell j)$ is comprised of $2j + 1$ degenerate states corresponding to the various assignments of m_j for the given value of j. The figure lists these degeneracies at all the levels and includes a running total of nucleon populations similar to the scheme employed in Figure 14-25. In this case the accumulated number of nucleons successfully reproduces a known magic number whenever the population process reaches a filled level just before the occurrence of a large energy gap. It is clear that both the large magnitude and the inverted splitting are required of the nuclear spin–orbit effect if the gaps are to fall in place at the magic numbers.

The nuclear spin–orbit interaction was originally proposed on phenomenological grounds after many unsuccessful trials of the independent-nucleon approach. The idea of a large inverted splitting was introduced in 1949 by M. Goeppert-Mayer and J. H. D. Jensen in their contributions to the development of the shell model.

Example

Figure 14-28 describes a square-well model for the nucleon potential energy $V(r)$. Unlike the similar model of the deuteron in Figure 14-19, this illustration admits more than one bound-state energy level. Let us consider only the $l = 0$ case so that $V_{\text{eff}}(r)$ reduces to $V(r)$ in Equation (14-33). Equation (14-32) then becomes essentially the same as Equation (14-27) in the treatment of the deuteron. (The reduced mass and the number of levels are the only differences between the two problems.) Figure 14-28 shows a well large enough to accommodate three bound states, with energy levels E_{1s}, E_{2s}, and E_{3s} (in $E_{n\ell}$ notation). The corresponding solutions of Equation (14-32) can be sketched by following the guidelines established in Section 6-7. An application of the usual arguments about curvatures and nodes results in the three graphs of $rR_{n\ell}(r)$ shown in the figure. We see that $rR_{1s}(r)$ reproduces the shape of our previous graph for the deuteron in Figure 14-19, and we note that each successive function has one additional node. In fact, all three of these $\ell = 0$ radial solutions behave as sine functions inside the well and have parametrizations just like Equation (14-29). We find the two higher energy levels by solving for the values of the variable u at the second and third intersections of the two graphs in Figure 14-20. The qualitative results of this example are extended to the $\ell = 1$ case in Problem 22 at the end of the chapter. A much larger family of energy levels $E_{n\ell}$ can be obtained for each ℓ if the square well is made sufficiently large. We have assumed such a system of levels in Figure 14-24.

14-11 Spins and Moments in the Shell Model

The nuclear shell model makes many other predictions beyond the explanation of the magic numbers. The model is based on a procedure for assigning angular momentum quantum numbers, and so the results are expected to include properties pertaining to

Figure 14-28

Square-well model of the nucleon potential energy with three $\ell = 0$ bound states.

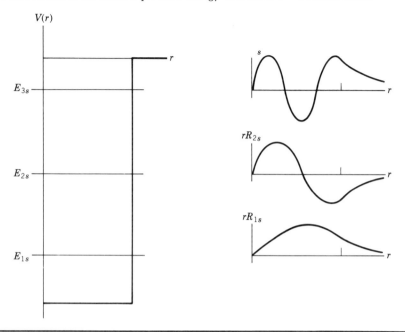

the angular momentum of the assembled nucleons. Predictions of nuclear spins and magnetic moments are particularly interesting because these tests of the model are more stringent than the placement of the magic numbers.

Let us visualize the shell model in terms of the levels shown in Figure 14-27 and use the quantum numbers $(n\ell j m_j)$ to denote the single-nucleon states. The notation implies an identification of the total angular momentum of a nucleon by the familiar formula

$$\mathbf{J} = \mathbf{L} + \mathbf{S}.$$

We extend the same addition of angular momenta over all the A nucleons and express the nuclear spin vector as

$$\mathbf{I} = \sum_{k=1}^{A} \mathbf{J}_k = \sum_{k=1}^{A} (\mathbf{L}_k + \mathbf{S}_k). \qquad (14\text{-}36)$$

Note that the orbital and spin angular momenta are added according to the *jj-coupling scheme*. We recall from our discussion of Figure 9-23 that we are supposed to follow this plan whenever the spin–orbit interaction has a strong effect.

The sum over nucleons in Equation (14-36) behaves like its analogue in atomic physics and reduces immediately to a smaller number of terms. Each level $n\ell_j$ in Figure 14-27 defines a *nuclear subshell* consisting of $2j + 1$ degenerate states. A completely filled subshell contributes zero nuclear spin and positive overall parity, for reasons just like those given in the case of the complex atom. Consequently, the nuclear spin and parity of the ground state are determined solely by the dynamics of the nucleons in the last incomplete subshells.

The pairing effect plays a decisive role in these residual interactions. Two protons or two neutrons with *opposite* values of m_j in a given subshell have a greater probability to be found at small separation where the particles can experience a greater degree of nuclear attraction. The effect produces maximal binding when even numbers of like nucleons pair off with *canceling* angular momenta. These paired protons and paired neutrons make a 0^+ contribution to the total angular momentum and parity of an unfilled subshell. Thus, the sum over nucleons in Equation (14-36) is reduced even further in the state of lowest energy, so that only the *last unpaired proton and neutron* are left as surviving contributors. This consequence of the pairing effect is supported by an abundance of empirical evidence.

We are led at once to the following conclusions about the angular momentum in the nuclear ground state. An even–even nuclide has no unpaired proton or neutron and is predicted to have nuclear spin and parity $i^p = 0^+$ in every case. This prediction is borne out without any known exception. An odd-A nuclide has either Z or N equal to an odd number and therefore has one unpaired proton or neutron in an unfilled subshell with designation $n\ell_j$. These quantum numbers determine i^p, the nuclear spin and parity of the entire nucleus:

$$i = j \text{ (always a half-integer)} \quad \text{and} \quad p = \text{sign of } (-1)^\ell. \tag{14-37}$$

An odd–odd nuclide has an unpaired proton *and* an unpaired neutron. In this situation Equation (14-36) reduces to the addition of two angular momenta:

$$\mathbf{I} = \mathbf{J}_p + \mathbf{J}_n.$$

This coupling problem follows the rules given in Section 9-8 and results in an inequality involving the proton and neutron quantum numbers:

$$|j_p - j_n| \leq i \leq j_p + j_n \tag{14-38}$$

as in Equations (9-44) and (9-45). These upper and lower bounds constitute a prediction for the nuclear spin. An accompanying prediction for the parity is given as $(-1)^{\ell_p}(-1)^{\ell_n}$, the product of the orbital parities for the proton and neutron. All these expectations are illustrated by specific tests of the shell model in the first example at the end of the section.

Our description of the shell model is based on a hypothetical treatment of the central field. We should therefore be prepared for deviations in the actual ordering of the shell-model energies relative to the levels in Figure 14-27. The exact sequence of levels may vary with the mass number A and may also differ between the two species of nucleon. Table 14-1 provides separate lists of shell-model levels for a proton and for a neutron, ordered empirically to agree with observations over a large number of odd-proton and odd-neutron nuclei. Note that the maximum occupancies and running totals are included, to be read upward from the bottom, as in Figure 14-27. These lists exhibit the expected shell closures at the magic numbers. We use some of the entries in this table in the first example below.

The shell model also makes predictions about the magnetic moments of nuclei. The limitations of the model can be grasped by concentrating on the case of an *odd-A* nucleus in the ground state. We visualize this system in terms of $(A - 1)/2$ pairs of nucleons with antiparallel angular momenta, plus a single unpaired proton or neutron. We know from Equations (14-37) that the overall nuclear spin and parity are

Table 14-1   Empirical Level Sequence of Nuclear Subshells

odd-proton			odd-neutron		
			$\cdots$		
			$2g_{7/2}$	8	162
			$1i_{11/2}$	12	154
			$3d_{5/2}$	6	142
$\cdots$			$2g_{9/2}$	10	136
			$1i_{13/2}$	14	126 ◄
			$3p_{1/2}$	2	112
			$2f_{5/2}$	6	110
$3p_{3/2}$	4	104	$3p_{3/2}$	4	104
$2f_{7/2}$	8	100	$1h_{9/2}$	10	100
$1h_{9/2}$	10	92	$2f_{7/2}$	8	90
$3s_{1/2}$	2	82 ◄	$1h_{11/2}$	12	82 ◄
$2d_{3/2}$	4	80	$2d_{3/2}$	4	70
$1h_{11/2}$	12	76	$3s_{1/2}$	2	66
$2d_{5/2}$	6	64	$1g_{7/2}$	8	64
$1g_{7/2}$	8	58	$2d_{5/2}$	6	56
$1g_{9/2}$	10	50 ◄	$1g_{9/2}$	10	50 ◄
$2p_{1/2}$	2	40	$2p_{1/2}$	2	40
$1f_{5/2}$	6	38	$1f_{5/2}$	6	38
$2p_{3/2}$	4	32	$2p_{3/2}$	4	32
$1f_{7/2}$	8	28 ◄	$1f_{7/2}$	8	28 ◄
$1d_{3/2}$	4	20	$1d_{3/2}$	4	20
$2s_{1/2}$	2	16	$2s_{1/2}$	2	16
$1d_{5/2}$	6	14	$1d_{5/2}$	6	14
$1p_{1/2}$	2	8 ◄	$1p_{1/2}$	2	8 ◄
$1p_{3/2}$	4	6	$1p_{3/2}$	4	6
$1s_{1/2}$	2	2 ◄	$1s_{1/2}$	2	2 ◄
$n\ell_j$	maximum occupancy	running total	$n\ell_j$	maximum occupancy	running total

determined by the state of the odd nucleon, and we attribute the magnetic dipole moment of the entire system to this lone unpaired particle. Both orbital and spin parts of the magnetic moment contribute if the particle is a proton, while only the spin magnetic moment contributes if the particle is a neutron. The following analysis treats both possibilities at the same time.

The prediction is immediate when the odd nucleon is in an $\ell = 0$ orbital state. Only the spin contributes in this circumstance and gives

$$\mu = \text{either } \mu_p \text{ or } \mu_n$$

for the two possible cases.

The general formula for the magnetic moment of an odd-A nucleus follows by analogy with the result for the magnetic moment of a one-electron atom in Section 8-11. Let us return to Equation (8-45) and rewrite this defining relation in terms of the orbital and spin magnetic moments of a single bound nucleon:

$$\mu = \mu_L + \mu_S = \frac{\mu_N}{\hbar}(g_L \mathbf{L} + g_S \mathbf{S}). \tag{14-39}$$

The revised expression employs the substitution of magneton units $-\mu_B \rightarrow \mu_N$ and contains the explicit orbital and spin g-factors for the nucleon. In the proton case we take

$$g_L = 1 \quad \text{and} \quad g_S = g_p,$$

and in the neutron case we take

$$g_L = 0 \quad \text{and} \quad g_S = g_n.$$

We then let the measured magnetic dipole moment correspond to the expectation value of μ_z in the nuclear spin state for which $m_i = i$. The evaluation of $\langle \mu_z \rangle$ parallels the treatment in Section 8-11 and leads to the result

$$\langle \mu_z \rangle = \frac{\mu_N}{i+1}\left\{\frac{g_L}{2}\left[i(i+1) + \ell(\ell+1) - \frac{3}{4}\right] + \frac{g_S}{2}\left[i(i+1) - \ell(\ell+1) + \frac{3}{4}\right]\right\} \tag{14-40}$$

for either type of nucleon in a nucleus with quantum numbers i and ℓ. We relegate the derivation of this formula to a few detailed remarks at the end of the section.

Equation (14-40) can assume two different forms since the nuclear spin can satisfy either $i = \ell + \frac{1}{2}$ or $i = \ell - \frac{1}{2}$. Let us eliminate ℓ and examine the resulting dependence on i for the two possibilities. The case $\ell = i - \frac{1}{2}$ gives

$$\begin{aligned}
\langle \mu_z \rangle &= \frac{\mu_N}{i+1}\left\{\frac{g_L}{2}\left[i(i+1) + \left(i - \frac{1}{2}\right)\left(i + \frac{1}{2}\right) - \frac{3}{4}\right]\right. \\
&\quad \left. + \frac{g_S}{2}\left[i(i+1) - \left(i - \frac{1}{2}\right)\left(i + \frac{1}{2}\right) + \frac{3}{4}\right]\right\} \\
&= \frac{\mu_N}{i+1}\left[\frac{g_L}{2}(2i^2 + i - 1) + \frac{g_S}{2}(i+1)\right] \\
&= \mu_N\left[g_L\left(i - \frac{1}{2}\right) + \frac{g_S}{2}\right] \qquad (i = \ell + \tfrac{1}{2}), \tag{14-41}
\end{aligned}$$

and the case $\ell = i + \frac{1}{2}$ gives

$$\begin{aligned}
\langle \mu_z \rangle &= \frac{\mu_N}{i+1}\left\{\frac{g_L}{2}\left[i(i+1) + \left(i + \frac{1}{2}\right)\left(i + \frac{3}{2}\right) - \frac{3}{4}\right]\right. \\
&\quad \left. + \frac{g_S}{2}\left[i(i+1) - \left(i + \frac{1}{2}\right)\left(i + \frac{3}{2}\right) + \frac{3}{4}\right]\right\} \\
&= \frac{\mu_N}{i+1}\left[\frac{g_L}{2}(2i^2 + 3i) + \frac{g_S}{2}(-i)\right] \\
&= \mu_N \frac{i}{i+1}\left[g_L\left(i + \frac{3}{2}\right) - \frac{g_S}{2}\right] \qquad (i = \ell - \tfrac{1}{2}). \tag{14-42}
\end{aligned}$$

Figure 14-29

Magnetic moments versus nuclear spin for odd-proton nuclei. The plotted points fall between the Schmidt lines.

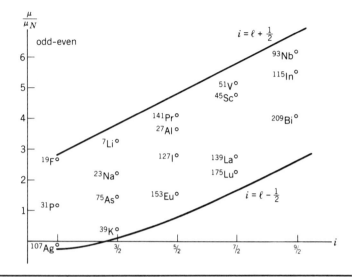

Each of these conclusions can now be specialized for application to an odd-Z or an odd-N nucleus. We set $g_L = 1$ and $g_S = g_p$ to find

$$\langle \mu_z \rangle_p = \begin{cases} \mu_N \left(i - \dfrac{1}{2} + \dfrac{g_p}{2} \right) \\[2ex] \mu_N \dfrac{i}{i+1} \left(i + \dfrac{3}{2} - \dfrac{g_p}{2} \right) \end{cases} \qquad \text{for } i = \ell \pm \tfrac{1}{2}, \qquad (14\text{-}43)$$

and we set $g_L = 0$ and $g_S = g_n$ to find

$$\langle \mu_z \rangle_n = \begin{cases} \mu_N \dfrac{g_n}{2} \\[2ex] -\mu_N \dfrac{i}{i+1} \dfrac{g_n}{2} \end{cases} \qquad \text{for } i = \ell \pm \tfrac{1}{2}, \qquad (14\text{-}44)$$

in the two possible cases where the odd nucleon is either a proton or a neutron.

Our formulas yield four classes of results for the magnetic moments of the odd-A nuclei. We present the predicted dependence on the nuclear spin in Figures 14-29 and 14-30 for each of the four different sets of circumstances. The predictions lie on curves called *Schmidt lines* (after T. Schmidt, an originator of the idea that the moments might be associated with the properties of a single odd nucleon). Experimental values of the moments for several odd-A nuclei are also plotted in the figures. We note a tendency for the experimental points to fall between the Schmidt lines; however, we observe a general failure of the simple theory to give a very good agreement with experiment. Some comparisons with experiment are discussed in the second example below.

Figure 14-30

Magnetic moments versus nuclear spin for odd-neutron nuclei. Almost all plotted points lie between the Schmidt lines.

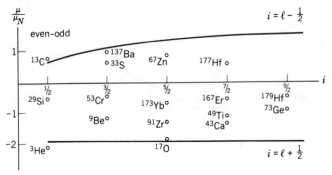

It is clear that the elementary shell model is not entirely satisfactory as a quantum theory of the nucleus. The model adheres to an independent-particle theory in which a fixed spherically symmetric nuclear core interacts with an independent nucleon. An improved self-consistent treatment would allow every nucleon to participate collectively in an interdependent determination of the nuclear state. Some aspects of nucleon interdependence can be included in the independent-particle theory by letting the interaction with the nucleon *deform* the spherically symmetric core. The *collective model* of the nucleus builds on the shell model and incorporates this added feature of a deformed spheroidal nuclear core. The improvement injects some of the philosophy of the liquid-drop model since the spheroidal core has bulk properties like the liquid droplet. The rotational dynamics of the core enriches the theory with new rotational degrees of freedom and thereby opens the way for better agreement with experiment, particularly in the area of magnetic dipole and electric quadrupole moments.

Detail

Equation (8-49) expresses the crucial part of the derivation of $\langle \mu_z \rangle$ in the case of the one-electron atom. Let us transfer the same construction to the magnetic moment of the odd-A nucleus in Equation (14-39). We introduce the nuclear spin vector

$$\mathbf{I} = \mathbf{L} + \mathbf{S}$$

in place of the total angular momentum $\mathbf{J}$ for the atom and write

$$\langle \mu_z I^2 \rangle = \langle \mathbf{\mu} \cdot \mathbf{I} \, I_z \rangle.$$

(This relation is analogous to the expression that follows Equation (8-49) in Section 8-11.) We then continue in parallel with the steps leading to Equation

(8-47):

$$\mathbf{\mu} \cdot \mathbf{I} = \frac{\mu_N}{\hbar} (g_L \mathbf{L} + g_S \mathbf{S}) \cdot (\mathbf{L} + \mathbf{S})$$

$$= \frac{\mu_N}{\hbar} \left[g_L L^2 + (g_L + g_S) \mathbf{S} \cdot \mathbf{L} + g_S S^2 \right]$$

$$= \frac{\mu_N}{\hbar} \left[g_L L^2 + \frac{g_L + g_S}{2} (I^2 - L^2 - S^2) + g_S S^2 \right]$$

$$= \frac{\mu_N}{\hbar} \left[\frac{g_L}{2} (I^2 + L^2 - S^2) + \frac{g_S}{2} (I^2 - L^2 + S^2) \right].$$

Our next move is to insert this result into the expression for $\langle \mu_z I^2 \rangle$ and take the expectation value in the state defined by the eigenvalues

$$I^2 = i(i+1)\hbar^2, \quad L^2 = \ell(\ell+1)\hbar^2, \quad S^2 = \tfrac{3}{4}\hbar^2, \quad \text{and} \quad I_z = i\hbar.$$

These maneuvers produce the following equality:

$$\langle \mu_z \rangle i(i+1)\hbar^2 = \frac{\mu_N}{\hbar} \left\{ \frac{g_L}{2} \left[i(i+1) + \ell(\ell+1) - \frac{3}{4} \right] \hbar^2 \right.$$

$$\left. + \frac{g_S}{2} \left[i(i+1) - \ell(\ell+1) + \frac{3}{4} \right] \hbar^2 \right\} i\hbar.$$

We quote this formula as Equation (14-40) in the text.

Example

Let us compare our shell-model predictions with a very small sample of the known nuclear spins and parities. The doubly magic species ^{16}O and ^{40}Ca have completely filled proton and neutron subshells and are observed to be O$^+$ nuclei as predicted. The odd-A nuclides $^{17}_{8}$O$_9$, $^{39}_{19}$K$_{20}$, and $^{19}_{9}$F$_{10}$ contain an unpaired nucleon and furnish a more interesting test of the shell model. Table 14-1 tells us that the odd neutron in ^{17}O is in a $1d_{5/2}$ subshell and that the odd proton in ^{39}K is in a $1d_{3/2}$ subshell. Equations (14-37) give the corresponding i^p assignments in terms of the quantum numbers j and ℓ of the odd nucleon. Thus, we expect ^{17}O to have $i^p = \tfrac{5}{2}^+$ since $j = \tfrac{5}{2}$ and $\ell = 2$, and we expect ^{39}K to have $i^p = \tfrac{3}{2}^+$ since $j = \tfrac{3}{2}$ and $\ell = 2$. These conclusions agree with experiment, according to the data in Appendix A. On the other hand, the nuclide ^{19}F has an odd proton in a $1d_{5/2}$ subshell, and so the shell-model quantum numbers $j = \tfrac{5}{2}$ and $\ell = 2$ would imply $i^p = \tfrac{5}{2}^+$. A $\tfrac{1}{2}^+$ assignment is observed instead, again as recorded in Appendix A. The odd–odd nuclide $^{14}_{7}$N$_7$ has an odd proton *and* an odd neutron and therefore involves a different test. We consult Table 14-1 and find that both of the odd nucleons are assigned to $1p_{1/2}$ subshells. Since $\ell_p = \ell_n = 1$, the model predicts an even parity from the product of the two odd-parity factors. Since $j_p = j_n = \tfrac{1}{2}$, the model predicts a nuclear spin given as either $i = 0$ or $i = 1$ from Equation (14-38). Appendix A tells us that ^{14}N is a 1^+ nucleus, in agreement with these predictions.

Example

We can easily convert Equations (14-43) and (14-44) into the graphs of Figures 14-29 and 14-30 if we recall the values of the proton and neutron g-factors from Section 14-1. Let us again use the nuclides ^{17}O and ^{39}K to test the predictions of the model. The previous example describes ^{17}O as an odd-neutron nucleus whose quantum numbers $\ell = 2$ and $i = \frac{5}{2}$ satisfy the relation $i = \ell + \frac{1}{2}$. The upper version of Equation (14-44) tells us to expect

$$\frac{\mu}{\mu_N} = \frac{g_n}{2} = -1.913.$$

A table of nuclear data lists the experimental value as -1.893, a result quite close to our prediction. In the case of ^{39}K we have an odd-proton nucleus whose quantum numbers $\ell = 2$ and $i = \frac{3}{2}$ obey the relation $i = \ell - \frac{1}{2}$. The lower version of Equation (14-43) predicts

$$\frac{\mu}{\mu_N} = \frac{i}{i+1}\left(i + \frac{3}{2} - \frac{g_p}{2}\right) = \frac{3}{5}(3 - 2.793) = 0.124.$$

The listed experimental value 0.391 is somewhat further removed from our prediction in this instance. We can see the positions of these nuclides with respect to the Schmidt lines by inspecting the relevant portions of the two figures.

14-12 Charge Independence and Isospin Symmetry

Models are the means of describing the complexities of the nuclear force, in the interaction of two nucleons and in the binding of many nucleons. We know that the force between two nucleons is not purely central and is not independent of spin. However, the two-nucleon problem is not as complex as we might expect, since the nuclear interactions are known from experiment to be the same for pp, pn, and nn pairs of nucleons. Evidently, the nuclear force does not depend on the *charge* of the interacting particles. This property of charge independence represents a new type of *symmetry principle* and *conservation law* in nuclear physics. Our interest in charge independence is focused particularly on the effects of the symmetry in the primitive two-nucleon system. We are also concerned with the manifestations of the symmetry in the structure of complex nuclei. Our main objective is to introduce charge independence as a simplifying ingredient in nuclear physics.

The pp, pn, and nn interactions are said to be identical only with regard to the *nuclear* characteristics of the interacting particles. This assertion presumes that the obviously charge-dependent electromagnetic interactions have already been taken into account wherever these effects may arise. Thus, the proposed nuclear symmetry defines a relationship between the proton and the neutron and allows a substitution of the one for the other as far as the nuclear force is concerned. Of course, the two species of nucleon are not identical in every respect. The particles are distinguished by such electromagnetic properties as charge and magnetic moment so that their electromagnetic interactions are rather different. Hence, the symmetry principle and the conservation law associated with charge independence are supposed to hold in circumstances where electromagnetic effects can be neglected relative to the strong nuclear force. This exact symmetry breaks down and becomes a good approximate symmetry in the presence of electromagnetism. It is reasonable to regard the small difference in mass between the neutron and the proton as a measure of the deviation from exact symmetry. It is then a matter of conjecture that the difference in mass might be due solely to electromagnetic effects.

Charge independence has played a part in the history of nuclear physics since 1932, the year of the neutron. The concept originated with Heisenberg and E. U. Condon, who proposed independently that the proton and the neutron should be treated as differently charged quantum states of a single generic nucleon. They viewed the charge of the nucleon as a mere label with no meaning as a physical parameter in the absence of electromagnetic forces. Their idea was inspired by the near equality of the nucleon masses and was reinforced by the direct observation of charge independence in experiments on pp and np scattering. The notion of a nuclear symmetry was implemented by powerful mathematical techniques in the subsequent work of E. P. Wigner.

Let us begin to appreciate the symmetry between protons and neutrons by considering pairs of *mirror* nuclides related by the interchange of Z and N. We recall from Section 14-5 that the mass formula predicts the same binding energy for these isobars if the distinguishing effects of Coulomb repulsion are disregarded. We therefore expect the interchange $Z \leftrightarrow N$ to have no influence on the level of the nuclear ground state whenever the electromagnetic interactions can be ignored. The mirror nuclides have the remarkable property that the approximate equality of energies persists level-by-level up into the excited states of the two related systems. An excellent example is provided by the pair of nuclides $_3^7\text{Li}_4$ and $_4^7\text{Be}_3$ in Figure 14-31. The two level schemes exhibit parallel patterns of states with the same nuclear spins and parities, and with much the same excitation energies above the respective ground states. We can argue that the Coulomb effect elevates all the levels of the isobar with the larger of the two values of Z. We recognize, however, that the mirror symmetry of energy levels tests only the equality of the pp and nn forces. The comparison gives no information about the relative strength of the pn force because the number of pn bonds is exactly the same in the two isobars. This finding establishes a property of the nuclear force known as *charge symmetry*. Since the equality of forces is tested in only two of the three possible combinations, the symmetry under the interchange $Z \leftrightarrow N$ is less comprehensive than the full symmetry associated with charge independence.

We can compare all three forces if we select a family of nuclides in which the pp, pn, and nn bonds occur in varying numbers. This opportunity is offered by any collection of isobars with an *even* mass number A. We illustrate the arguments in Figure 14-32 by drawing sketches of the internucleon bonds in two proton–neutron systems with odd and even values of A. Figure 14-33 presents physical evidence for the equality of the three interactions in the form of energy levels for the even-A isobars ^{18}O, ^{18}F, and ^{18}Ne. The parallel columns of levels contain states of each isobar for which the i^p quantum numbers are the same and the corresponding excitation energies are very similar. Nuclear states with this property are known as *isobaric analogue states*. Their existence as a family is a clear manifestation of charge independence. We note in passing that there are energy levels in ^{18}F for which there are no isobaric analogues in the level schemes for ^{18}O and ^{18}Ne. This feature is another consequence of the full symmetry of charge independence.

The families of isobar levels in Figures 14-31 and 14-33 have certain structural characteristics in common with the quantum states of a particle in a central field. We can grasp the meaning of the new symmetry principle by developing the close analogy

Figure 14-31

Energy levels and i^p assignments for the mirror nuclides ^{7}Li and ^{7}Be. The excitation energies are given in MeV. The ^{7}Be ground state is unstable and undergoes a β transition to the ground state of ^{7}Li, accompanied by an energy release of 0.86 MeV.

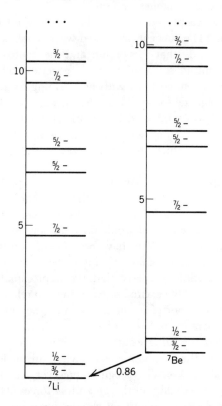

Figure 14-32

Systems of isobars for testing charge symmetry and charge independence. Only the pp and nn forces can be compared in the odd-A nuclides. The pn force can be included in the comparison when A is even because the number of pn bonds does not remain fixed for the different nuclei.

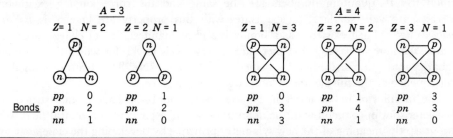

	A = 3			A = 4		
	Z= 1 N = 2	Z = 2 N = 1		Z = 1 N = 3	Z = 2 N = 2	Z = 3 N = 1
Bonds	pp 0	pp 1		pp 0	pp 1	pp 3
	pn 2	pn 2		pn 3	pn 4	pn 3
	nn 1	nn 0		nn 3	nn 1	nn 0

Figure 14-33

Energy levels and i^p assignments for isobars with $A = 18$. The three level schemes contain threefold families of isobaric analogue states. Certain additional levels also occur as indicated in ^{18}F. Energies of excitation are quoted in MeV. Two of the three ground states are unstable so that β transitions take place from ^{18}Ne to ^{18}F to ^{18}O.

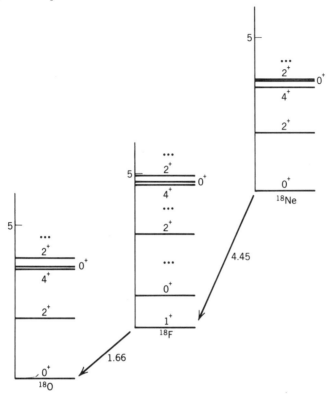

between these two rather different physical systems. To clarify the analogy let us assume for the moment that the isobaric analogue states have *exactly* the same energies. We might suppose that this coincidence is accomplished by the removal of electromagnetic effects. The hypothesis implies sets of *degenerate* states with common i^p assignments in which the nuclei are distinguished by the different combinations of Z and N for the given value of A.

This occurrence of different states with the same energy reminds us of the degeneracy associated with *rotational symmetry* in the case of the particle in a central field. We recall that the energy of a spinless particle may depend on the L^2 quantum number ℓ but cannot vary with the L_z quantum number m_ℓ. In the more general situation where the particle has spin, the energy may vary with j but not with the z-component quantum number m_j. A *multiplet* of $2j + 1$ degenerate states results for each j, corresponding to a set of quantized orientations of the angular momentum vector **J**. We know that the energy cannot depend on these orientations because the rotational symmetry of the central field precludes the existence of a preferred direction for the choice of z axis. The introduction of a constant magnetic field then fixes such an axis in space and causes an energy splitting among the $2j + 1$ states. This lifting of the degeneracy is due to the breaking of rotational symmetry by the application of the external field.

These remarks summarize the ideas of angular momentum quantization and rotational symmetry and recall the meaning of the observed degeneracies in a central field. We now wish to transfer these familiar *spatial* properties of a particle to the *internal* nucleon composition of an isobar family. A new rotational symmetry and a new conservation law are introduced for this purpose. We pretend that electromagnetism can be "turned off" so that isobaric analogue states can be made to coincide at the same energies. The nuclear force is supposed to be responsible for the internal symmetry that underlies this pattern of degeneracies. We then imagine that the actual observed energies depart from the degenerate pattern as we turn the electromagnetic effects back on to break the internal symmetry.

Let us set this proposition aside temporarily and return to our earlier comments about the original interpretation of charge independence. The notion of a generalized two-state nucleon treats the designations *proton* and *neutron* as internal degrees of freedom, like the spin orientations *up* and *down* for a spin-$\frac{1}{2}$ particle. Rotational symmetry in space implies that these spin orientations have the same energy. By analogy, we associate the two quantum states of the nucleon with the up and down orientations of another type of vector quantity. This new vector is fashioned after the familiar spin **S** and is known as the isospin **T**. An abstract three-dimensional *isospin space* is invented so that the orientations of the vector **T** can be defined. We express the charge-independent property of the nuclear force as a rotational *isospin symmetry* in the contrived new space and argue that the equality of masses of the proton and neutron is evidence for the proposed new symmetry. We then claim that the symmetry is a manifestation of a new conservation law for isospin, just as ordinary rotational symmetry is a demonstration of the conservation of angular momentum. The proton and neutron form a degenerate isospin multiplet, called the *nucleon doublet*, in this framework. We can pursue the abstraction a step further and regard the actual neutron–proton mass difference as a lifting of this degeneracy by the breaking of isospin symmetry. Electromagnetism acts as a symmetry-breaking effect since the electromagnetic properties of the nucleon doublet are different for isospin up (the proton) and isospin down (the neutron).

Let us draw the concepts in the last two paragraphs together and formulate isospin in quantum mechanical language. We ensure the desired properties of the isospin vector by imposing quantization rules identical to those enjoyed by the angular momentum **J** (or **L**, or **S**). Thus, the quantum behavior of the isospin **T** is expressed in terms of the quantities T^2 and T_z by the following eigenvalue conditions:

$$T^2 = t(t + 1) \quad \text{where } t \text{ may be } 0, \tfrac{1}{2}, 1, \tfrac{3}{2}, \ldots, \qquad (14\text{-}45)$$

and

$$T_z = -t, -t + 1, \ldots, t - 1, t \quad \text{for a given value of } t. \qquad (14\text{-}46)$$

Note that the choice of t may be integral or half-integral and that T_z has $2t + 1$ quantized values for each choice. (No $\hbar$ factors are used in Equations (14-45) and (14-46) since **T** is an abstract dimensionless quantity with no classical interpretation as an angular momentum.)

Each allowed value of the new *isospin quantum number* t defines a possible *isospin multiplet* consisting of $2t + 1$ isobaric members. The nucleon doublet has two states with isospin up and down, so that the relevant isospin quantum number must be $t = \frac{1}{2}$. The two orientations of **T** correspond to the assignments

$$T_z = -\tfrac{1}{2} \text{ and } +\tfrac{1}{2} \quad \text{for the neutron and proton, respectively.}$$

Exactly the same isospin assignments are adopted level-by-level for the mirror nuclides ^{7}Li and ^{7}Be in Figure 14-31.

We assign a T_z eigenvalue to a particular state of a given nuclide $^A_Z X_N$ according to the general formula

$$T_z = \frac{Z - N}{2}. \tag{14-47}$$

The formalism then implies that the assigned system has an isospin quantum number t and belongs to a multiplet of $2t + 1$ isobar states. The power of isospin symmetry lies in the ability to predict these nuclear systems as isobaric analogue states. An exact symmetry would imply an exact equality of energy levels for the predicted isobars. Figure 14-33 shows an array of $t = 1$ multiplets, or isospin *triplets*, in which the T_z eigenvalues are

$$-1 \text{ in } ^{18}_{8}O_{10}, \quad 0 \text{ in } ^{18}_{9}F_{9}, \quad \text{and} \quad +1 \text{ in } ^{18}_{10}Ne_8.$$

We have already noted that ^{18}F also contains a number of levels without any counterparts in the other two isobars. The figure shows the $i^p = 1^+$ ground state of ^{18}F as one example of this special class. These states occur only for $T_z = 0$ and must therefore be isospin *singlets* with isospin quantum number $t = 0$. The formalism allows for the existence of singlet energy levels when A is even and predicts that such levels should stand alone, unrelated in energy to any other isobars.

We have considered the substitution symmetry of protons and neutrons earlier in this section, and we have associated the interchange of Z and N with the charge symmetry of mirror nuclides. Equation (14-47) tells us that this lesser symmetry operation is just a reversal of the sign of T_z. The rotational aspects of charge independence embody a more powerful use of the isospin degree of freedom. Let us turn to the abstract isospin space for an interpretation of this property and consider an isospin multiplet of nuclear states with isospin quantum number t. The multiplet has $2t + 1$ isobaric analogue states, corresponding to $2t + 1$ orientations of a vector **T** with length $\sqrt{t(t + 1)}$ in the abstract space. Isospin symmetry allows us to *rotate* the vector, and thereby pass from one isobar state to another, without affecting the energy level. This picture has an approximate validity when the symmetry is not exact.

Let us return to the problem of two nucleons and use our knowledge of isospin to make some detailed observations. The deuteron is one such system, but so are the nucleonic configurations pp and nn. Let us also keep in mind that all these systems have eigenfunctions describing the space and spin states of the two particles. If we concentrate on isospin properties first and refer to Equation (14-47), we observe the following possibilities for the value of T_z:

$$pp \text{ has } T_z = 1, \quad pn \text{ has } T_z = 0, \quad \text{and} \quad nn \text{ has } T_z = -1.$$

The total isospin vector for any two-nucleon configuration has the form

$$\mathbf{T} = \mathbf{T}_1 + \mathbf{T}_2,$$

where $\mathbf{T}_1$ and $\mathbf{T}_2$ are isospin-$\frac{1}{2}$ vectors. The analogous addition of spin-$\frac{1}{2}$ vectors results in the quantized vector sums

$$\tfrac{1}{2} + \tfrac{1}{2} = 0 \text{ and } 1.$$

We apply the same rules to isospin and conclude that all two-nucleon states must have either $t = 0$ or $t = 1$. It is obvious that the pp and nn systems belong to $t = 1$ triplets. The pn combination appears in two different isospin states, distinguished by the properties of antisymmetry or symmetry under the exchange of symbols $p \leftrightarrow n$. Let us express all possible isospin states for two nucleons in the notation ξ^A and ξ^S to exhibit these exchange properties. The $t = 0$ singlet state is antisymmetric:

$$\xi^A = \frac{1}{\sqrt{2}} (pn - np). \tag{14-48}$$

The three $t = 1$ triplet states are symmetric:

$$\xi_1^S = pp,$$

$$\xi_0^S = \frac{1}{\sqrt{2}} (pn + np),$$

$$\xi_{-1}^S = nn. \tag{14-49}$$

(The notation $pn/\sqrt{2}$ means there is probability $\frac{1}{2}$ for finding nucleons 1 and 2 to be p and n, respectively.) Note that the triplet states are labeled by their T_z eigenvalues. We recognize these four expressions as exact analogues of the states of total spin $s = 0$ and $s = 1$, obtained for two spin-$\frac{1}{2}$ particles in Section 9-6. Recall that the corresponding antisymmetric and symmetric spin eigenfunctions have been listed as χ^A and $\chi_{m_s}^S$ in Equations (9-25) and (9-26).

The exchange properties of ξ^A and ξ^S can be used to formulate a generalized version of the Pauli principle. Equation (9-22) instructs us to write eigenfunctions for two identical fermions as

$$\psi^S \chi^A \quad \text{or} \quad \psi^A \chi^S.$$

We are reminded that the requirement of fermion antisymmetry is met by multiplying space and spin factors with opposite properties under the interchange of the respective pairs of variables. We can regard all of our systems of nucleons as compositions of identical fermions now that we are adopting the generalized two-state concept for the nucleon. In this language the eigenfunction for a single nucleon becomes a combination of a spatial factor with a spin orientation ($\uparrow$ or $\downarrow$) and an isospin designation (p or n). A state of two nucleons is then described by a triple product of the form

$$\psi^S \chi^S \xi^A \quad \text{or} \quad \psi^S \chi^A \xi^S \quad \text{or} \quad \psi^A \chi^S \xi^S \quad \text{or} \quad \psi^A \chi^A \xi^A,$$

in which the space, spin, and isospin factors are combined to satisfy the required overall antisymmetry under the interchange of the two fermions.

The deuteron provides an interesting application of the generalized Pauli principle. We know from Section 14-7 that this pn system is a superposition of $^3S_1$ and $^3D_1$ states. The orbital assignments $\ell = 0$ and $\ell = 2$ correspond to symmetric distributions with regard to the interchange of the two spatial variables. The triplet spin state is likewise symmetric with respect to the two spin orientations. Hence, we describe the space and spin degrees of freedom by the eigenfunctions ψ^S and χ^S, and we select the singlet ξ^A for the isospin description in order to meet the requirement of generalized antisymmetry. We therefore conclude that the deuteron has isospin quantum number $t = 0$. This result is as expected since the alternative $t = 1$ assignment implies the existence of *impossible* isobaric analogues in the systems pp and nn. (Recall that pp and nn cannot be symmetric in space *and* spin because of the Pauli principle in its original form.)

Isospin gives us a new symmetry to use in the classification of quantum systems. Complexities are simplified and regularities are revealed whenever such an ingredient is introduced in the analysis of complicated phenomena. The theory of nuclear structure spans a class of problems in which isospin symmetry plays an influential role. Isospin principles are also employed extensively, along with other new types of symmetry, in the phenomenology of elementary particles.

Example

Another illustration of a $t = 0$ nuclide is found in the ground state of ^{4}He, the α particle. Equation (14-47) tells us that this $Z = N = 2$ system has $T_z = 0$. In principle, the four isospin-$\frac{1}{2}$ nucleons can have their combined isospin vectors in states with quantum numbers $t = 0$, $t = 1$, or $t = 2$. In fact, the assignments $t = 1$ and $t = 2$ are ruled out because the α particle would then be required to have isobaric analogues. To be specific, there should exist multiplets of $A = 4$ states with $i^p = 0^+$ in which there would be three isobars for $t = 1$ or five isobars for $t = 2$. We know of no such partners of the ^{4}He ground state, and so we take the α particle to be a singlet with isospin quantum number $t = 0$.

Example

The conservation of isospin by the nuclear force is not restricted to collections of protons and neutrons in a single nucleus. We can also apply the principle to two-body nuclear reactions of the form

$$\mathrm{X}_1 + \mathrm{X}_2 \rightarrow \mathrm{X}_3 + \mathrm{X}_4.$$

We let the four nuclides have isospin vectors $\mathbf{T}_1$ to $\mathbf{T}_4$, and we write the conservation of the total isospin as

$$\mathbf{T} = \mathbf{T}_1 + \mathbf{T}_2 = \mathbf{T}_3 + \mathbf{T}_4.$$

This relation among quantized isospin vectors acts as a constraint on the states of the particles in the reaction. Consider, for instance, the (d, α) process

$$^2\mathrm{H} + {}^{12}\mathrm{C} \rightarrow {}^{10}\mathrm{B} + {}^4\mathrm{He}$$

in which the deuteron, the ground-state carbon nucleus, and the α particle participate as $t = 0$ isospin singlets. Isospin symmetry allows the reaction to produce only the $t = 0$ states in ^{10}B. The ground state of ^{10}B is among these allowed possibilities. Excited states with $t = 1$ also exist (as isobaric analogues with ^{10}Be and ^{10}C), but all these ^{10}B states are forbidden in this reaction because of the conservation of isospin.

Problems

1. Suppose that the original γ-ray interpretation holds for the experiment in Figure 14-1 and that the Compton scattering of γ rays causes protons to be ejected from the paraffin detector. Obtain a formula relating the incident γ energy and the kinetic energy K of the observed proton for the case where K is maximum. What γ energy is predicted if K is as large as 5 MeV?

2. Calculate the de Broglie wavelength for an electron with kinetic energy 183 MeV. What electron beam energy corresponds to a 1 fm de Broglie wavelength?

3. Consult Figure 14-8 and prove that the indicated surface thickness t satisfies the relation

$$\frac{t}{z_1} = 4 \ln 3,$$

where z_1 is a parameter in the formula for the nuclear charge density. Use the data in Figure 14-9 to estimate a universal value for the density of nucleons in the interior of the nucleus.

4. Data from the nuclear chart for the three stable neon isotopes ^{20}Ne, ^{21}Ne, and ^{22}Ne are:

atomic mass (u)	19.992436	20.993843	21.991383
isotopic abundance (%)	90.51	0.27	9.22

Compute the atomic weight of neon from these figures and compare the result with the value quoted in the chart.

5. Three doublets are seen in the mass spectrum of a sample containing hydrogen, deuterium, carbon, oxygen, and methane. The splittings observed in the M/e ratios of the ions are:

$$0.001548 \text{ u}/e \quad \text{for} \quad (^1\text{H}^1\text{H})^+ - {}^2\text{H}^+$$
$$0.042306 \text{ u}/e \quad \text{for} \quad (^2\text{H}^2\text{H}^2\text{H})^+ - {}^{12}\text{C}^{++}$$
$$0.036385 \text{ u}/e \quad \text{for} \quad (^{12}\text{C}^1\text{H}^1\text{H}^1\text{H}^1\text{H})^+ - {}^{16}\text{O}^+$$

Use these data to compute the atomic masses of ^{1}H, ^{2}H, and ^{16}O.

6. Use the atomic mass data in Appendix A to compute the binding energy per nucleon for the nuclei ^{40}Ca, ^{56}Fe, ^{120}Sn, and ^{208}Pb. These particular cases are indicated on the graph in Figure 14-12.

7. Derive the formula

$$V = \frac{3}{5} \frac{(Ze)^2}{4\pi\varepsilon_0 R}$$

for the Coulomb energy stored in a uniform solid sphere of charge Ze and radius R. (Determine the work done to construct the sphere by moving infinitesimal spherical shells of charge radially inward from infinity.)

8. Use the semiempirical mass formula to compute the atomic mass of each of the isotopes mentioned in Problem 6. Compare the results with the corresponding entries in Appendix A.

9. Use the atomic mass data for the three pairs of mirror nuclides (^{3}H, ^{3}He), (^{7}Li, ^{7}Be), and (^{11}B, ^{11}C) to make three separate predictions for the constant a_3 in the semiempirical mass formula. Take $M(^{11}$C$)$ to be 11.011433 u and consult Appendix A for the other atomic masses.

10. Show that the semiempirical mass formula leads to the following prediction for the neutron separation energy:

$$E_n(^A\text{X}) = a_1 - a_2 A^{2/3}\left(1 - \xi^2\right) + a_3 \frac{Z^2}{A^{1/3}}\left(\frac{1}{\xi} - 1\right) - \frac{a_4}{4}\left[1 - \frac{4Z^2}{A(A-1)}\right] + \frac{a_5}{A^{3/4}},$$

where $\xi = [(A-1)/A]^{1/3}$. The result takes this form when ^A_ZX is an even–even nuclide.

11. Use the semiempirical mass formula to derive the following leading-order relation involving nuclear binding energies:

$$\frac{1}{2}\left[E_b(^{A+1}\text{X}) + E_b(^{A-1}\text{X})\right] - E_b(^A\text{X}) = \frac{a_5}{A^{3/4}} + \cdots.$$

Take Z even and A odd as conditions in the derivation, and use binomial expansions such as

$$(A-1)^{-1/3} = \frac{1}{A^{1/3}}\left(1 - \frac{1}{A}\right)^{-1/3} = \frac{1}{A^{1/3}}\left(1 + \frac{1}{3A} + \cdots\right)$$

on the way to the final answer.

12. Compute the neutron separation energies for ^{207}Pb and ^{208}Pb taking data from Appendix A. Use the results to make a determination of the constant a_5 in the mass formula.

13. Consider the Fermi gas model of the nucleus and assume that the Coulomb effects are approximated inside the nuclear volume by the constant potential energy $aZ^2/A^{1/3}$. Refer to Figure 14-15 and let the potential energy well for the proton gas be elevated by this amount relative to the depth of the neutron well. Deduce the following relation among the numbers of nucleons,

$$N^{2/3} = Z^{2/3} + bZ^2A^{1/3},$$

and identify the constant b in terms of the given parameters of the model.

14. Consider the np capture reaction $n + p \rightarrow d + \gamma$ in which both of the particles in the initial state are essentially at rest. Obtain a formula for the binding energy of the deuteron in terms of the energy of the γ ray.

15. Points (x, y, z) on the surface of the indicated ellipsoid of revolution satisfy the equation

$$\frac{x^2 + y^2}{a^2} + \frac{z^2}{b^2} = 1.$$

Choose a suitable volume element and show that the volume of the figure is $\frac{4}{3}\pi a^2 b$. Let the ellipsoid contain a uniform distribution of total charge Ze, and derive the formula

$$Q = \frac{2}{5}Ze\left(b^2 - a^2\right)$$

for the classical quadrupole moment.

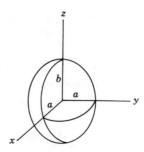

16. The nuclide ^{176}Lu has a comparatively large quadrupole moment, $Q = +8.0$ $e \cdot$ barns. Use this number and the result of Problem 15 to estimate the presumably small axis ratio b/a for the nuclear ellipsoid.

17. Prove that the electric quadrupole moment $\langle Q \rangle$ vanishes in a two-nucleon state described by an $\ell = 0$ wave function.

18. The square-well model of the deuteron produces an eigenfunction of the form

$$\psi = \frac{a}{\sqrt{4\pi}\, r} \begin{cases} \dfrac{\sin Kr}{K} & r < r_0 \\[2mm] \dfrac{}{\sqrt{K^2 + k^2}}\, e^{-k(r-r_0)} & r > r_0 \end{cases}.$$

Use the normalization of the wave function to obtain the constant

$$a = \sqrt{\frac{2k}{1 + kr_0}}\,.$$

The parameters r_0, k, and K are identified in Section 14-8.

19. Derive a formula for the expectation value of r in the deuteron state using the results of the model in Problem 18, and calculate the value of $\langle r \rangle$ for $r_0 = 1.6$ fm.

20. The figure defines the parameters of a square-well model for the deuteron in which the potential energy includes an infinitely repulsive core. Determine the $\ell = 0$ solution of the differential equation for $rR(r)$ with energy $-E_b$ in each of the three intervals $(0, r_c)$, (r_c, r_0), and (r_0, ∞). Use the continuity conditions to derive a relation among V_0, E_b, r_c, and r_0. Obtain a suitably continuous final form for $rR(r)$ containing a single unknown normalizing constant.

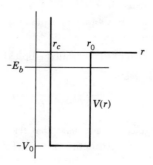

21. Make a prediction of the neutron separation energy for the magic-number nuclide ^{90}Zr based on the semiempirical mass formula. Compare the result with the value determined

from the measured atomic masses. Use $M(^{89}\text{Zr}) = 88.908900$ u and take $M(^{90}\text{Zr})$ from Appendix A.

22. Assume a square-well potential energy as a central-field model in an independent-nucleon theory of the nucleus. Draw a figure to represent the effective potential energy for the $\ell = 1$ case, and let there be three $\ell = 1$ energy levels, designated as E_{1p}, E_{2p}, and E_{3p}. Sketch graphs of the corresponding radial solutions $rR_{1p}(r)$, $rR_{2p}(r)$, and $rR_{3p}(r)$, and compare with the $\ell = 0$ results in the example accompanying Figure 14-28.

23. The next magic number for neutrons after $N = 126$ is supposed to be $N = 184$. Show how the sequence of levels in Figure 14-27 may be continued to generate this prediction.

24. Deduce the predictions of the shell model for the nuclear spins and parities of the odd-A nuclides ^{15}N, ^{23}Na, ^{27}Al, and ^{95}Mo. Are the results in agreement with the listings in Appendix A?

25. What does the shell model predict regarding the nuclear spins and parities of the odd–odd nuclides ^{6}Li, ^{10}B, and ^{50}V? Do the predictions agree with the i^p assignments listed in Appendix A?

26. Calculate the values predicted by the shell model for the magnetic moments of the odd-A nuclides in Problem 24. Compare the results with experiment, using a table of nuclear data.

27. Consult a table of nuclear data to find evidence for isospin symmetry among the nuclides with mass number $A = 10$.

28. Examine the tabulated energy levels of the $A = 17$ system of isobars and find evidence for the existence of isospin multiplets with $t = \frac{1}{2}$ and $t = \frac{3}{2}$ quantum numbers.

NUCLEAR
PROCESSES

*T*he structure of a nucleus is defined by the states of the nuclear quantum system. The nucleus reveals its structure through a variety of interactive processes in which *transformations* of the quantum states occur as observable phenomena. One such process is the emission of γ radiation in the deexcitation of an excited nucleus. This phenomenon is like the radiative deexcitation of an atom and is known as γ *decay*. (Of course, the radiation from a nucleus is much more energetic. The nuclear process results in γ rays characteristic of the emitting species, with typical photon energies in the MeV range.) Radiation from nuclei may also take the form of ejected fragments or particles. These phenomena are called α *decay* or β *decay*, depending on the particular form of emission. The processes of α, β, and γ decay represent the three specific varieties of *radioactivity*. Each type of emission is associated with a certain kind of nuclear instability, and each is accompanied by a particular sort of transition between the energy states of the nuclear system.

Radioactivity involves the *spontaneous disintegration* of an unstable atomic nucleus. The spontaneity of the process implies that the transformation of the system of nucleons takes place in an isolated nucleus, with no stimulus from any external source of energy. Radioactive behavior is *natural* if the emissions are observed for a sample of nuclides occurring in nature and is *artificial* if the radiation is induced in the sample by some sort of external bombardment. Natural radionuclides are generally long-lived, although species with short half-lives are also found in nature as decay products of commonly occurring decay chains.

The history of nuclear physics begins with the discovery of radioactivity. Much of our present understanding of nuclear structure comes from this method of viewing the nucleus. We devote several sections of the chapter to a full discussion of the various aspects of nuclear radiation. Historical remarks and general definitions are considered first, and then the three processes of α, β, and γ decay are taken up in turn.

Nuclear transformations are also observed in the collisions of particles with nuclear targets. The resulting *nuclear reactions* are additional sources of information about the states of the nucleus. We examine the principles behind these investigations later in the chapter. Finally, we consider the *fission* and *fusion* of nuclei as transformations of special interest for the generation of nuclear energy.

These phenomena constitute a modern-day realization of the alchemists' goal. The transmutation of the elements is accomplished as desired, by a change of species within the nucleus of the atom.

15-1 Radioactivity

The first observations of nuclear disintegration were made by accident in the x-ray experiments of A. H. Becquerel in 1896. X rays had just been discovered in the previous year, and Becquerel's laboratory was actively involved in the investigation of their properties. Incident x rays were known to cause fluorescent radiation in atoms, and so a reversal of roles was proposed whereby fluorescent stimulation was suggested as a means of producing x rays. Becquerel selected a fluorescent compound of uranium and indeed found that his sample emitted penetrating radiation. However, he also found that the uranium compound produced the same energetic emissions with *no* external stimulation. It was apparent that this new type of spontaneous radiation was peculiar to the uranium in the sample and was altogether different from the proposed atomic x rays.

The nature of the radiation was studied by Rutherford and Becquerel, and also by Curie and his remarkable colleague M. S. Curie. These investigators established the existence of nuclear radiation in its three forms α, β, and γ. They demonstrated the deflection of α and β rays by a magnetic field and concluded that these emissions had to be charged particles. The α particles turned out to be helium nuclei, and the negative β particles proved to be electrons with properties just like Thomson's cathode rays.

Marie Curie

Figure 15-1

Radioactive decay scheme of ^{80}Br. The nuclear energy levels are connected by β and γ transitions. The amounts of released energy are given in MeV. The designation β^+ actually represents two distinct processes, as described in Section 15-4.

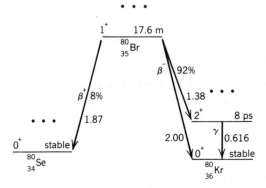

The Curies devoted much of their collaborative work to the identification of the radioactive elements polonium and radium. Their achievements were acknowledged decades later when the element curium was named after them.

The first investigations of radioactivity were conducted with natural sources bearing uranium, thorium, and other heavy elements. Artificially induced radioactivity was demonstrated for the first time in 1934 by J.-F. Joliot and I. Joliot-Curie. They used α particles to bombard the nucleus ^{27}Al and initiated a nuclear reaction producing the radionuclide ^{30}P. The observation of positron emission from this nucleus showed that radioactivity was not restricted just to the heavy elements.

The chronology of interesting events in nuclear physics should be noted. A four-decade period began when natural radioactivity was first observed, prior to the discovery of the nucleus. The neutron came later, followed by artificial radioactivity. Finally, the period of development reached its dramatic climax with the disclosure of nuclear fission.

We have interpreted radioactivity as the disintegration of an unstable nucleus. Hereafter, we cast our description in the language of quantum physics and treat the various phenomena in terms of transitions between nuclear states. The α and β decays bring about definite changes of the nuclear species, while the γ transition causes a change only among the states of a given nucleus. Figure 15-1 shows how a scheme of energy levels can be drawn to illustrate the possible decay processes for a particular radionuclide. We include several pieces of data to indicate the i^p assignments and half-lives of the energy levels and the energy released in the transitions. The figure employs a useful convention whereby the initial and final nuclear levels appear in separate parallel columns and exhibit transitions to the left or right depending on the sign of the charge of the emitted particle. Note that γ transitions stay within their own column in this scheme.

The total charge and the total number of nucleons must be conserved in any radioactive decay. We know that an α particle is a ^{4}He nucleus, and so we express the α decay $X \rightarrow Y + \alpha$ in detail as

$$^A_Z X \rightarrow \,^{A-4}_{Z-2} Y + \,^4_2 He,$$

noting that the charge and mass numbers balance explicitly. The β^- and β^+ decays

Figure 15-2

α and β transitions displayed on a chart of the nuclides.

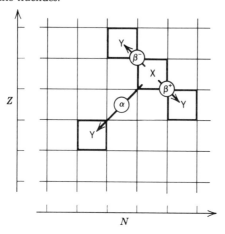

involve electron emission and positron emission, respectively. We learn later on that a new type of neutral particle must also be emitted in each of these processes. It would be premature to divulge this information now, so let us temporarily refer to the two types of β decay as

$$_Z^A X \rightarrow _{Z+1}^A Y + \beta^- \quad \text{and} \quad _Z^A X \rightarrow _{Z-1}^A Y + \beta^+.$$

The emitted β systems do not contain any nucleons; consequently, the mass number A remains unchanged in these transitions. Hence, all β decays are identified as transformations between nuclear isobars, a property illustrated by the $A = 80$ family of nuclides in Figure 15-1.

Each of these nuclear transformations proceeds from a *parent* nucleus X to a *daughter* nucleus Y. Figure 15-2 shows how we can represent such decays with the aid of a chart of the nuclides. If the Z versus N system of axes is employed as in Figure 14-4, then the three different nuclear transitions are described by the following transformations of coordinates on the chart:

$$(Z, N) \xrightarrow{\alpha} (Z - 2, N - 2),$$
$$(Z, N) \xrightarrow{\beta^-} (Z + 1, N - 1),$$
$$(Z, N) \xrightarrow{\beta^+} (Z - 1, N + 1).$$

Parent and daughter are shown executing these transitions in Figure 15-2.

We can use this method of presenting α and β decays to describe the radioactivity of such naturally occurring sources as ^{232}Th or ^{238}U. These heavy radionuclides have the property that their half-lives are of the same order as the age of the Earth. (Appendix A gives values of the half-life in the range 10^{10} years for ^{232}Th and 10^9 years for ^{238}U.) Each unstable species displays its radioactive behavior in a *decay series* where the first nuclear transition in the chain is an α decay with a very long half-life. The series continues through a succession of α and β processes from one unstable nucleus to another until the final stable product of the chain is reached. Figure 15-3

Figure 15-3

Decay series for ^{238}U. Branchings of α and β^- decay modes occur at several locations along the chain between ^{238}U and ^{206}Pb.

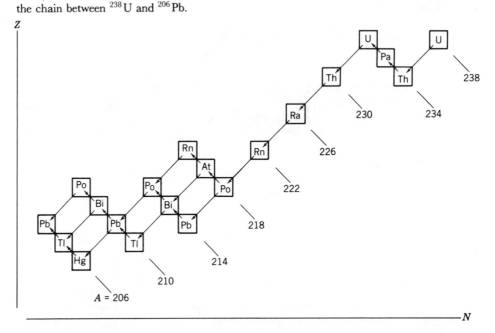

shows how ^{238}U transforms into ^{206}Pb after such a sequence of nuclear decays. We see from the figure that every nuclide in the series has a mass number given by $A = 4n + 2$, where n varies over nine successive integers from 59 down to 51. The ^{238}U decay chain is called the $4n + 2$ series for this reason. Other similarly defined chains are the $4n$ series for ^{232}Th, the $4n + 1$ series for ^{237}Np (half-life $\sim 10^6$ years), and the $4n + 3$ series for ^{235}U (half-life $\sim 10^8$ years). The longer-lived radionuclides ^{232}Th and ^{238}U exist with appreciable abundance in certain natural ores. These sources are primary contributors of background radiation in our environment.

The instability properties of α and β decay are associated with the differences in mass between the parent and daughter nuclides. The two types of process are triggered by two rather different kinds of decay interaction. We see from the half-lives of the various decaying species that radioactive decay is a very slow nuclear phenomenon in either case. Even though α decay is the emission of a nuclear fragment, and is therefore an effect of the strong nuclear force, the rate of decay is substantially inhibited because of the Coulomb repulsion between the fragment and the residual daughter. The β^- and β^+ decays are extremely slow as a general rule because the processes are under the influence of a very weak force.

The decay properties of a known source include its half-life, decay mode, and energy release. These characteristics constitute an observable signature for the corresponding radionuclide. The half-life is especially useful because this property gives the radioactive source a timing mechanism to be used for chronological purposes. A commonly employed application is *radiocarbon dating*, which establishes the age of a carbon-bearing material by comparison with the 5730 year half-life of the β-active isotope ^{14}C. The technique is based on the fact that all living organisms continually exchange CO_2 in the Earth's environment, where a minute fraction of the carbon nuclei are of the ^{14}C variety. These unstable nuclides are formed when neutrons from cosmic-ray interactions are absorbed by the nuclei of nitrogen atoms in the atmosphere. The processes of ^{14}C production and decay are supposed to be at equilibrium in the carbon-dating technique. It is known that the amount of radioactive ^{14}C relative to ordinary ^{12}C in living matter is of order 10^{-12}. It is believed that this value has remained essentially constant for over 30,000 years. Carbon decay in a specimen of plant life must therefore occur at a corresponding constant rate, which turns out to be about one disintegration per second in every 4 g of carbon. The ingestion of carbon ceases when a plant dies, and so the amount of ^{14}C in the specimen decays thereafter at a rate that decreases by half every 5730 years. The time of death (or the age of the specimen) is then determinable from a measurement of the rate of decay for a known mass of carbon. The development of this remarkable technique is attributed to W. F. Libby. A simple numerical illustration of the method is furnished by the following application.

Example

The production of ^{14}C in the atmosphere proceeds according to the (n, p) reaction

$$n + {}^{14}N \rightarrow {}^{14}C + {}^1H.$$

The carbon isotope then reverts to ^{14}N via the decay process

$$^{14}C \rightarrow {}^{14}N + \beta^-.$$

Let us select a 64 g charcoal sample for carbon dating and suppose that β radiation is observed at a rate of 2 disintegrations per second. The decay rate for a living specimen with the same mass of carbon would be

$$(64 \text{ g})\left(\frac{1 \text{ disintegration/s}}{4 \text{ g}} \right) = 16 \text{ disintegrations/s},$$

using the constant rate quoted in the text. (The derivation of this rate is left to Problem 5 at the end of the chapter.) The ratio of the dead rate to the live rate is

$$\frac{2 \text{ disintegrations/s}}{16 \text{ disintegrations/s}} = \frac{1}{8} = \left(\frac{1}{2} \right)^3.$$

The elapsed time since death must therefore be equal to three ^{14}C half-lives:

$$3(5.73 \times 10^3 \text{ y}) = 1.72 \times 10^4 \text{ y}.$$

This result determines the age of the charcoal sample.

15-2 The Exponential Decay Law

The decay of an unstable nucleus is a prime example of a process governed by the probabilistic concepts of quantum mechanics. The instability is a property of the species, and the half-life is a measure of the lifetime of the nuclide. However, the lifetime is not a statement of the exact time of decay for a single unstable nucleus since the decay occurs as a *random* discrete event. Such an act of chance is predictable only in terms of a probability for the occurrence of a certain decay transition within a given interval of time. It follows that the transition probability per unit time determines the radioactive behavior of a sample containing many nuclei. Thus, the instability of the nucleus describes a *statistical* phenomenon, and the half-life of the sample represents an average time for the observation of a random decay.

Let us formulate the quantum principles of radioactive decay by the following arguments. We introduce a fundamental quantity, called the *decay constant* γ, as the transition probability per unit time for the one-step decay $X \rightarrow Y + r$, where r denotes the emitted radiation. This decay rate refers to the radioactive behavior of a single nucleus in the probabilistic sense described above. The decay constant is a fundamental property of the nucleus because, in principle, the quantity is predictable from a quantum theory of the decay transition.

We can determine γ from experiment if we observe the radiation from a source containing many nuclei of the species X. We proceed by defining $N(t)$ as the number of radioactive nuclei in the sample at time t. This number is presumed to be large so that N can be treated as a continuous variable. We then consider the change dN in that number after an infinitesimal time interval dt, and we identify the positive quantity $-dN$ as the corresponding number of decays. The ratio $-dN/N$ represents the probability of decay in terms of the number of decays occurring in the interval dt and the number of decaying nuclei present at time t. We employ γ, the decay probability per unit time, to express the probability in the interval between t and $t + dt$:

$$-\frac{dN}{N} = \gamma \, dt. \qquad (15\text{-}1)$$

This result implies an *activity*, or rate of decay, proportional to the instantaneous number of nuclei in the source:

$$-\frac{dN}{dt} = \gamma N, \qquad (15\text{-}2)$$

a relation deduced first on empirical grounds by Rutherford and Soddy as the original nuclear transformation law. Since γ does not depend on t, it is straightforward to integrate Equation (15-1) and thereby determine $N(t)$:

$$-\int_{N_0}^{N(t)} \frac{dN}{N} = \int_0^t \gamma \, dt \quad \Rightarrow \quad -\ln\frac{N(t)}{N_0} = \gamma t,$$

taking N_0 to be the number of decaying nuclei at $t = 0$. We then solve for the desired quantity to get

$$N(t) = N_0 e^{-\gamma t}. \qquad (15\text{-}3)$$

This exponential decay law for the fundamental one-step transition $X \rightarrow Y$ is described by the *decay curve* in Figure 15-4.

A given nucleus in the sample may undergo the transformation $X \rightarrow Y$ at any time from $t = 0$ onward. We define an average lifetime by computing the mean value of t, averaged with respect to the number of decays in each time interval dt:

$$\langle t \rangle = \frac{-\int_0^\infty t \, dN(t)}{-\int_0^\infty dN(t)} = \frac{\int_0^\infty t e^{-\gamma t} \, dt}{\int_0^\infty e^{-\gamma t} \, dt}.$$

Figure 15-4

Decay curve corresponding to the exponential decay law $N(t) = N_0 e^{-\gamma t}$. The indicated mean life τ and half-life $\tau_{1/2}$ are related to the quantum mechanical decay constant γ.

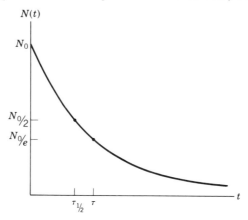

This lifetime is called the *mean life* τ. The integrations in the definition lead to the simple result

$$\tau = \frac{1}{\gamma}, \tag{15-4}$$

so that Equation (15-3) takes the alternative form

$$N(t) = N_0 e^{-t/\tau}. \tag{15-5}$$

The half-life $\tau_{1/2}$ is a different lifetime, already defined as the time required for the sample to decay to half of its initial population. Appendix A includes a list of values of $\tau_{1/2}$ for a large number of radionuclides. We show the times τ and $\tau_{1/2}$ on the graph in Figure 15-4, and we observe that the two parameters satisfy the equations

$$N(\tau) = \frac{N_0}{e} \quad \text{and} \quad N(\tau_{1/2}) = \frac{N_0}{2}.$$

We obtain these results directly from the determination of τ in Equation (15-5) and from the definition of $\tau_{1/2}$. The second expression leads to a simple relation between the two lifetimes and the decay constant:

$$\tau_{1/2} = \tau \ln 2 = \frac{\ln 2}{\gamma}. \tag{15-6}$$

Equations (15-4) and (15-6) are interesting because they connect experimental and theoretical quantities, the lifetime of a sample of nuclei on the one hand and the quantum mechanical decay rate for a single nucleus on the other.

We can measure the activity of a source by counting the rate of emitted charged particles in a detector. Each observed particle, or count, represents a single nuclear decay. A typical counter might be a gas-filled ionization chamber containing an applied electric field. A particle is counted when the gas atoms in the chamber are ionized by the passage of the particle and the ions are collected in the applied field. The rate of observed counts is the same as the rate of decay of the sample $-dN/dt$, except for a possible multiplying correction for the known efficiency of the detector. This detection efficiency may include at least two factors, the fraction of the total solid angle subtended by the detector at the source of radiation and the probability for the particle to produce a signal in the counting circuit.

The decay constant is easily determined by such an experiment. We use the fact that the measured activity depends on the time:

$$-\frac{dN}{dt} = \gamma N = \gamma N_0 e^{-\gamma t} = \left(-\frac{dN}{dt}\right)_0 e^{-\gamma t},$$

where $(-dN/dt)_0$ denotes the activity at $t = 0$. If we take the logarithm on both sides of the equality we get

$$\ln\left(-\frac{dN}{dt}\right) = \ln\left(-\frac{dN}{dt}\right)_0 - \gamma t.$$

Hence, we can deduce the decay constant γ if we plot the counting rate versus t on a semilogarithmic graph and take the slope of the best straight line through the experimental data.

The rate of decay of a radioactive source is usually quoted in curies (symbol Ci), defined by the unit

$$1 \text{ Ci} = 37 \times 10^9 \text{ disintegrations/s.}$$

This large unit of activity is approximately equal to the observed disintegration rate for a ^{226}Ra source with a mass of 1 g.

Many radionuclides display their activities in sequential chains of decays. Figure 15-3 describes one of the longer radioactive series, and Figure 15-5 shows a schematic illustration of a typical short chain. Let us confine our attention to the latter case and formulate the *rate equations* for a general two-step cascade. There are three nuclear states linked by two decay transitions, and so there must be three rates involving two different decay constants. The figure tells us that X_1 decays and X_3 grows, while X_2 experiences both decay and growth. We express the corresponding rates in terms of the decay constants in the figure and find the following relations among the popula-

---

Figure 15-5

Sequential decay processes $X_1 \rightarrow X_2 +$ radiation and $X_2 \rightarrow X_3 +$ radiation. Each transition has its own probability per unit time.

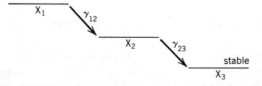

Figure 15-6

Solutions of the rate equations for the populations of the three levels in Figure 15-5. The results apply to the case of vanishing initial populations for X_2 and X_3.

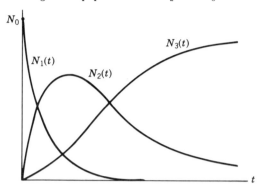

tions of the three nuclear levels:

$$\frac{dN_1}{dt} = -\gamma_{12}N_1,$$

$$\frac{dN_2}{dt} = \gamma_{12}N_1 - \gamma_{23}N_2,$$

$$\frac{dN_3}{dt} = \gamma_{23}N_2. \tag{15-7}$$

The first rate has the form of Equation (15-2) so that the population $N_1(t)$ must behave according to the basic exponential decay law. We note that the rates sum to zero, and we conclude that the solutions of the three coupled differential equations must always obey the constraint

$$N_1 + N_2 + N_3 = N_0, \quad \text{a constant.}$$

Figure 15-6 shows a set of qualitative graphs for the special class of solutions in which the initial populations have the values $N_1(0) = N_0$, $N_2(0) = 0$, and $N_3(0) = 0$. An even more specialized case is treated analytically in one of the exercises below.

Example

The activity of 1 g of ^{226}Ra is supposed to be approximately 1 Ci. To verify this we first convert the 1 g sample into $\frac{1}{226}$ mole, and we then find the number of nuclei from Avogadro's number:

$$N = \left(\tfrac{1}{226} \text{ mole}\right)\left(6.02 \times 10^{23}/\text{mole}\right) = 2.66 \times 10^{21}.$$

Appendix A tells us that ^{226}Ra has a half-life of 1600 years. The rate of decay is then given by combining Equations (15-2) and (15-6):

$$-\frac{dN}{dt} = \gamma N = \ln 2 \frac{N}{\tau_{1/2}}$$

$$= \frac{(\ln 2)(2.66 \times 10^{21})}{(1600 \text{ y})(365 \text{ d/y})(24 \text{ h/d})(3600 \text{ s/h})} = 3.65 \times 10^{10} \text{ s}^{-1},$$

in agreement with the definition of the curie unit of activity.

Example

We can use the exponential decay law in conjunction with the quoted half-lives and abundances of ^{235}U and ^{238}U in order to estimate the age of the Earth. Let us denote these two isotopes as $\hat{X}$ and $\overline{X}$, and express their populations according to Equation (15-5):

$$\hat{N}(t) = \hat{N}_0 e^{-t/\hat{\tau}} \quad \text{and} \quad \overline{N}(t) = \overline{N}_0 e^{-t/\overline{\tau}}.$$

To get an answer we have to make an assumption about the populations of $\hat{X}$ and $\overline{X}$ at $t = 0$, the time of genesis for the Earth and its elements. Let us assume equal initial numbers of the two isotopes, taking $\hat{N}_0 = \overline{N}_0$, and regard our estimate as a consequence of this hypothesis. The ratio of populations then becomes

$$\frac{\overline{N}(t)}{\hat{N}(t)} = e^{t(1/\hat{\tau} - 1/\overline{\tau})}.$$

Thus, we learn that the time variable satisfies the equation

$$t = \frac{\ln\left[\overline{N}(t)/\hat{N}(t)\right]}{1/\hat{\tau} - 1/\overline{\tau}} = \frac{\ln\left[\overline{N}(t)/\hat{N}(t)\right]}{\ln 2\left(1/\hat{\tau}_{1/2} - 1/\overline{\tau}_{1/2}\right)},$$

using Equation (15-6) to get the final result. The value of t refers to the present time when the two isotopic abundances are known to be

$$0.72\% \text{ for } \hat{X} = {}^{235}\text{U} \quad \text{and} \quad 99.27\% \text{ for } \overline{X} = {}^{238}\text{U}.$$

We consult Appendix A to find the two half-lives and use all these data to make our estimate of the age of the Earth:

$$t = \frac{\ln(99.27/0.72)}{(\ln 2)(1/0.704 - 1/4.468)} \times 10^9 \text{ y} = 5.9 \times 10^9 \text{ y}.$$

This conclusion is in reasonable agreement with other geochronological results.

Example

The two-step cascade in Figure 15-5 is somewhat easier to analyze for the behavior of X_2 when X_1 is very long-lived. We assume $\gamma_{12} \ll \gamma_{23}$ and find that the solution for the X_1 population becomes

$$N_1(t) = N_{10} e^{-\gamma_{12}t} \rightarrow N_{10}$$

for times of order $1/\gamma_{23}$. Under these conditions the X_2 population is governed by the rate equation

$$\frac{dN_2}{dt} = R_{12} - \gamma_{23}N_2,$$

where R_{12} denotes the approximately constant value of $\gamma_{12}N_1(t)$:

$$R_{12} = \gamma_{12}N_{10}.$$

We integrate the rate equation as follows:

$$\int_{N_{20}}^{N_2(t)} \frac{dN_2}{R_{12} - \gamma_{23}N_2} = \int_0^t dt \quad \Rightarrow \quad -\frac{1}{\gamma_{23}}\ln\frac{R_{12} - \gamma_{23}N_2(t)}{R_{12} - \gamma_{23}N_{20}} = t,$$

taking N_{20} to be the initial number of X_2 nuclei. The solution for $N_2(t)$ is found from this result in two steps:

$$\frac{R_{12} - \gamma_{23}N_2(t)}{R_{12} - \gamma_{23}N_{20}} = e^{-\gamma_{23}t}$$

and so

$$N_2(t) = \frac{R_{12}}{\gamma_{23}} + \left(N_{20} - \frac{R_{12}}{\gamma_{23}}\right)e^{-\gamma_{23}t} = \frac{\gamma_{12}}{\gamma_{23}}N_{10} + \left(N_{20} - \frac{\gamma_{12}}{\gamma_{23}}N_{10}\right)e^{-\gamma_{23}t}.$$

Our conclusion holds in the interval $0 \leq t \ll 1/\gamma_{12}$. Note that N_2 may either decay or grow during this interval, depending on the relative size of the ratios N_{20}/N_{10} and γ_{12}/γ_{23}.

15-3 α Decay

Some of the earliest observations in nuclear physics involved the participation of the α particle. The emission of α radiation gave evidence of the instability of the nucleus in Becquerel's discovery of radioactivity. The scattering of α particles gave proof of the existence of the nucleus in Rutherford's experiments. These first insights into the nucleus were witnessed with the aid of particles emitted in α decay.

The emission of α particles is generally associated with the heavier unstable nuclei. An inspection of the nuclear chart reveals that α activity is a main mode of instability for the isotopes of elements beyond lead at $Z = 82$ and is only rarely observed in the lighter species. The emitted particle is a nuclear fragment four times as massive as the proton. The emission process is energetically favorable for α particles, compared to

Figure 15-7

Determination of α-particle velocities by means of a 180° magnetic spectrograph. The nonrelativistic momentum is given by $Mv = qBR$ for an emitted particle of known mass M and charge q. A measurement of the magnetic rigidity BR yields a result for the velocity v.

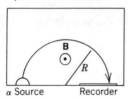

other possible nuclear fragments, because of the large binding energy and unusual stability of the $A = 4$ nuclide. A tendency for α decay to prevail among the heavier nuclides is not hard to understand. The α fragment and the residual daughter are bound together inside the unstable parent nucleus by the attractive short-range nuclear force. At the same time, the Coulomb repulsion between these two charged bodies acts as a disruptive influence and contributes to the probability for nuclear disintegration. Figures 14-4 and 14-12 remind us that these long-range effects of Coulomb repulsion become increasingly significant in nuclei of larger size. The heaviest nuclides may also undergo spontaneous fission, another type of fragmentation process.

Nuclear radiation energy is readily absorbed in matter when the radiation is in the form of α particles. The moving charges lose their kinetic energy in ionizing collisions with the atoms in the material medium. Familiar kinematical arguments tell us that these collisional losses of energy are greater for α particles than for the much lighter β radiation. A *range* in the medium is defined as the distance in which the particle loses all its kinetic energy. An effective absorber for α particles might be an aluminum foil, a sheet of paper, or even a volume of air. The range decreases with the density of the absorbing medium and increases with the energy of the particle. To give an example at a typical energy, a 6 MeV α particle is observed to travel approximately 5 cm in air and 0.05 mm in aluminum.

The velocities of α particles can be measured with accuracy by observing the deflections of the particles in a known magnetic field. An instrument devised for this purpose is the magnetic spectrograph, shown schematically in Figure 15-7. Data from such a device and measurements of range in some absorber can be used in conjunction to obtain a calibration of range versus energy for the particular absorbing medium. The resulting range-energy curve then gives a determination of the energy of an α particle directly from an observation of its range.

The typical α decay $X \rightarrow Y + \alpha$ is initiated in the ground state of the α-active parent X. The observed ranges in air vary between 2 and 9 cm for the different emitters, as the corresponding α-particle energies vary between 4 and 9 MeV. It is also possible in certain nuclei for the transformation to proceed from an excited state X* to the daughter state Y. The resulting α particles are emitted with excessive energy, between 9 and 12 MeV, and are characterized by their distinctive long range.

Figure 15-8

Kinetic energy of emitted α particles versus mass number for several radioactive sources. The data for the selected elements polonium, radium, thorium, uranium, and curium are taken from *Chart of the Nuclides*, 13th edition. All quoted energies are associated with nuclear transitions X → Y from the ground state of X to the ground state of Y. A few of the lighter nuclei emit α particles with energies smaller than the values plotted here; these cases include the very long-lived nuclides ^{144}Nd and ^{148}Sm.

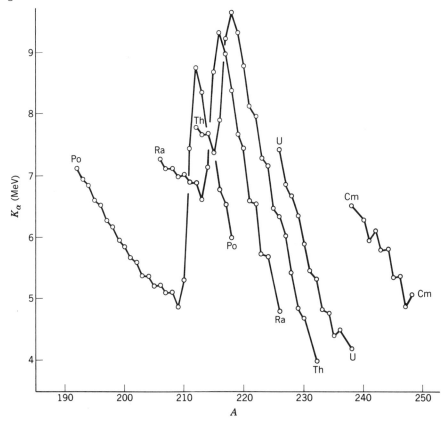

This type of α decay is quite rare because the emission process X* → Y + α must compete favorably with the γ-ray deexcitation X* → X + γ.

Radioactive α emission is said to be *monoenergetic* since every α particle in a given decay X → Y + α is emitted with the *same* kinetic energy K_α. Figure 15-8 shows a plot of these unique values of K_α versus the mass number A, organized by element, for a large number of α-emitting isotopes. One of our two main objectives in this section is to explain the uniqueness of these energies. We see from the figure that almost all the observed values fall between 4 and 9 MeV, a rather narrow energy interval. It is remarkable that the corresponding half-lives have an enormous range of variation, extending over some 24 orders of magnitude. We observe an inverse correlation between half-life and α-particle energy, where the longer-lived nuclides have the smaller values of K_α and vice versa. We can illustrate this behavior by

Figure 15-9

Monoenergetic α decay for a parent nucleus
at rest.

Two-body final state

quoting the following extreme values from the nuclear chart:

$$\tau_{1/2} = 1.40 \times 10^{10} \text{ y} \quad \text{and} \quad K_\alpha = 4.01 \text{ MeV} \quad \text{for } ^{232}\text{Th},$$

while

$$\tau_{1/2} = 0.30 \text{ } \mu\text{s} \quad \text{and} \quad K_\alpha = 8.78 \text{ MeV} \quad \text{for } ^{212}\text{Po}.$$

Our other main goal in this section is to understand the correlation between $\tau_{1/2}$ and K_α, with special concern for the great disparity in orders of magnitude of the one quantity compared to the other.

It is easy to see why the energies of the α particles are unique. We apply the familiar conservation laws of momentum and energy to the process

$$_Z^A\text{X} \rightarrow {}_{Z-2}^{A-4}\text{Y} + {}_2^4\text{He},$$

and we note that the decay produces an α particle in a *two-body final state*. The parent nucleus is taken to be at rest, as in Figure 15-9, so that conservation of momentum requires the emitted α particle and the recoiling daughter nucleus to have equal and opposite momenta. Conservation of the total relativistic energy provides another relation involving the nuclear masses:

$$M_X c^2 = M_Y c^2 + K_Y + M_\alpha c^2 + K_\alpha.$$

We can add Z electron masses to both sides of the equation and convert the expression directly to atomic masses:

$$M(^A\text{X})c^2 = \left[M(^{A-4}\text{Y}) + M(^4\text{He}) \right] c^2 + K_Y + K_\alpha.$$

A criterion for α instability emerges from these relations. The nucleus X is α unstable, so that the decay X $\rightarrow$ Y + α may occur, whenever the mass of X exceeds the sum of the masses of Y and α. We define the α-*disintegration energy*, or *Q-value* for the decay, in terms of the difference in mass:

$$Q = \left[M(^A\text{X}) - M(^{A-4}\text{Y}) - M(^4\text{He}) \right] c^2. \tag{15-8}$$

This quantity represents the total amount of released energy to be shared by the two final particles:

$$Q = K_\alpha + K_Y. \tag{15-9}$$

We may use the nonrelativistic formula for kinetic energy and write

$$
\begin{aligned}
Q &= \frac{p^2}{2M_\alpha} + \frac{p^2}{2M_Y} \\
&= \frac{p^2}{2M_\alpha}\left(1 + \frac{M_\alpha}{M_Y}\right) = K_\alpha\left(1 + \frac{M_\alpha}{M_Y}\right).
\end{aligned} \tag{15-10}
$$

The final solution for K_α is a unique prediction for the α-particle kinetic energy, given in terms of the masses of the particles participating in the decay. We note in passing that relativistic formulas are *required* for the total energy because the rest energies change with the identities of the particles, while nonrelativistic formulas are *allowed* for the kinetic energy because the kinetic energies are much smaller than the rest energies.

Other conservation laws also play a part in the determination of the final state. Let us refer again to Figure 15-9 and introduce the nuclear spins $\mathbf{I}_X$ and $\mathbf{I}_Y$ along with the orbital angular momentum $\mathbf{L}$ for the $Y\alpha$ system. The α particle has spin zero since ^{4}He is an $i = 0$ nuclide. Conservation of angular momentum therefore results in the equality

$$\mathbf{I}_X = \mathbf{I}_Y + \mathbf{L}. \tag{15-11}$$

This constraint among quantized angular momentum vectors implies a condition on the associated quantum numbers. We recognize the vector addition problem from our discussion of Equation (9-44) and immediately establish the following bounds on the orbital quantum number ℓ in terms of the initial and final nuclear spin quantum numbers:

$$|i_X - i_Y| \le \ell \le i_X + i_Y. \tag{15-12}$$

The nuclear states X and Y are also endowed with definite parities. The α particle is known to have even parity and the final orbital parity is given by $(-1)^\ell$. Conservation of parity implies a multiplicative constraint among the various odd and even factors:

$$(\text{parity of } X) = (\text{parity of } Y)(\text{even parity of } \alpha)(\text{orbital parity}).$$

We conclude that ℓ must be even, to give even orbital parity, if X and Y have the same parity, and that ℓ must be odd if X and Y have opposite parity.

The most remarkable feature of α decay is the extraordinarily large variation in the half-life from one nuclide to another, while the corresponding α-particle energies and Q-values vary within only a single order of magnitude. It is apparent that $\tau_{1/2}$ must have a very sensitive dependence on Q. The explanation for this behavior is found among the first applications of quantum mechanics to the nucleus, in the work of G. Gamow and others in 1928. To present the arguments we turn to Figure 15-10 and adopt a potential energy for the $Y\alpha$ system, the final state of the decay $X \rightarrow Y + \alpha$. The two bodies attract each other at short range, where the nuclear force is dominant, and experience like-charge repulsion at longer range. These effects combine to produce a *Coulomb barrier* whose height, for heavy nuclei, reaches 20–30

Figure 15-10

Model of the Coulomb barrier for the α decay $X \rightarrow Y + \alpha$. The potential energy of the $Y\alpha$ system consists of a strongly attractive nuclear contribution at short distance and a repulsive electrostatic contribution for larger separation between Y and α. The α-disintegration energy Q determines the probability for penetration of the barrier.

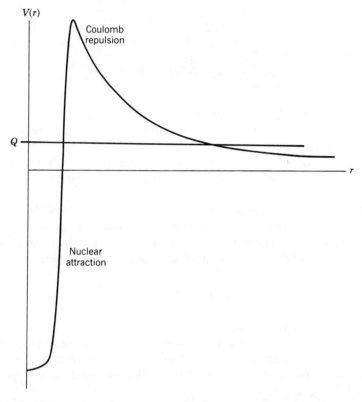

MeV at a separation r of order 10 fm. (We estimate these numbers in the example at the end of the section.) The Q-value is shown in the figure as the energy level of the system, since the total kinetic energy is equal to Q at very large $Y\alpha$ separation, according to Equation (15-9). A typical Q-value is less than 10 MeV, and so a "classical" $Y\alpha$ system cannot leave the region of nuclear binding and reach the distant region of large $Y\alpha$ separation because of the high Coulomb barrier. This trapping of the system at short range corresponds to the existence of the temporarily bound configuration X. The instability of X is attributed to the finite probability for the quantum system to penetrate the Coulomb barrier and enter the decay regime at large r. This line of reasoning suggests a possible theory of α decay based on a solution to the related problem of *quantum tunneling*.

The stationary-state wave function at the energy level Q determines the probability of finding the $Y\alpha$ system in its two configurations on either side of the barrier. The quantity of interest is the transmission probability for penetration from small r to large r. The decay constant for the process $X \rightarrow Y + \alpha$ is directly proportional to this quantity. We have encountered a similar situation in Section 5-9, in our treatment of quantum tunneling for a rectangular barrier in one dimension. We return to our main conclusion in Equation (5-83) and recall that the exponential function

$$\exp\left(-\frac{2a}{\hbar}\sqrt{2m(V_0 - E)}\right)$$

is the primary governing factor in the transmission probability. (This result is only qualitatively germane to α decay, a three-dimensional problem with a sharply peaked Coulomb barrier. The one-dimensional analogue is best suited for transitions X → Y in which both X and Y are even–even nuclides. These species have nuclear spin $i = 0$, and so the final Yα state has *zero* orbital angular momentum, according to Equation (15-12).) Let us apply the one-dimensional analogy as an approximate guide to the behavior of α decay and adapt the exponential factor to the picture in Figure 15-10. We substitute a rectangular step of height V_0 and length a in place of the sharp barrier, and we argue that the decay constant for X → Y + α is proportional to the tunneling factor

$$\exp\left(-\frac{2a}{\hbar}\sqrt{2\mu(V_0 - Q)}\right),$$

where μ is the Yα reduced mass and Q is the energy level. This factor decreases very rapidly with increasing choices of $V_0 - Q$, the height of the barrier above the Q-value. We predict a further reduction in probability if we let the choice of a grow with increasing $V_0 - Q$ to reflect the fact that the quantum tunnel becomes longer with diminishing Q. These features of barrier penetration explain qualitatively why the observed half-lives grow so rapidly with decreasing values of the α-particle energy.

Since α decay is monoenergetic and since the unique energy of the α particles can be measured with high precision, it is feasible to observe the energy and gain precise information about nuclear energy levels. These spectroscopic applications of α decay

Figure 15-11

Decay scheme and α-particle spectrum for ^{232}Th. Nuclear levels are labeled by their i^p quantum numbers and half-lives. The α and γ emissions have energies as shown in MeV.

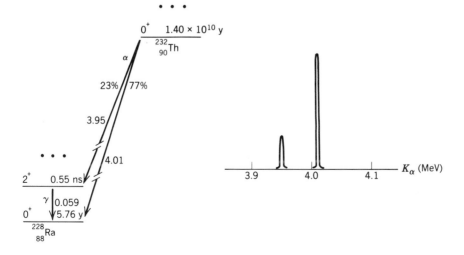

have been employed extensively to study nuclear structure in the heavy elements. A typical radioactive sample may exhibit a *spectrum* of α particles in which several unique energies are observed for the emissions from a single source. A case in point is ^{232}Th decay, where the thorium source emits α particles in two groups, at 4.01 and 3.95 MeV. Figure 15-11 presents these data in a decay scheme along with a schematic display of the α-particle spectrum. The two α groups arise from decays of the type

$$^{232}\text{Th} \rightarrow {}^{228}\text{Ra} + {}^{4}\text{He},$$

where the radium daughter occurs in its ground state and in one of its excited states. The figure shows that the α decay to the excited level is followed promptly by a γ deexcitation to the ground state. Note that the α spectrum maps out the relevant energy levels of ^{228}Ra, while the observed γ-ray energy confirms the spacing between the levels. Other varieties of α-active parent nuclei exist with greater numbers of excited daughter states and produce decay schemes of greater complexity.

Example

Let us use Appendix A again for a computation of Q-values and α-particle kinetic energies. A good illustration is provided by the above-mentioned α decay of ^{232}Th at the higher of the two quoted energies. We look up the atomic masses of ^{232}Th, ^{228}Ra, and ^{4}He, and we then apply Equation (15-8) to obtain the disintegration energy:

$$Q = (232.038054 - 228.031069 - 4.002603)(931.5 \text{ MeV}) = 4.082 \text{ MeV}.$$

The ratio of nuclear masses in the final state is also needed in order to learn how much of Q is carried away by the α particle. We find this quantity from the atomic masses by including the masses of the electrons ($m_e c^2 = 0.5110$ MeV):

$$\frac{M_\alpha}{M_{\text{Ra}}} = \frac{4.002603 - 2(0.5110/931.5)}{228.031069 - 88(0.5110/931.5)} = 0.01755.$$

(We might as well ignore the electron-mass correction since the ratio of atomic masses produces the same answer to five decimal places. In fact, the ratio of mass numbers $4/228$ is just as good to four decimal places.) Equation (15-10) gives the desired kinetic energy

$$K_\alpha = \frac{4.082 \text{ MeV}}{1.01755} = 4.012 \text{ MeV}.$$

This result corresponds to the smallest of the values plotted in Figure 15-8. The Q-value is small compared to the height of the Coulomb barrier in the Ra–He system. Let us estimate the latter quantity by calculating the repulsive Coulomb potential energy at a range equal to the sum of the two nuclear radii (the center-to-center distance in the Yα system of Figure 15-9). We use Equation (14-3) to find the range,

$$r = (228^{1/3} + 4^{1/3})(1.2 \text{ fm}) = 9.2 \text{ fm},$$

and we then compute the Coulomb potential energy for that separation:

$$V(r) = \frac{Z_{He}Z_{Ra}e^2}{4\pi\varepsilon_0 r}$$

$$= \frac{(2)(88)(9\times10^9 \text{ N}\cdot\text{m}^2/\text{C}^2)(1.60\times10^{-19}\text{ C})^2}{(9.2\times10^{-15}\text{ m})(1.60\times10^{-13}\text{ J/MeV})} = 28 \text{ MeV}.$$

This barrier height is seven times as large as the Q-value for the decay. Let us interpret our calculations with the aid of Figure 15-10. We choose a low-lying energy level to represent the small value of Q and observe that a very thick Coulomb barrier is obtained. The resulting quantum tunnel is supposed to be long enough to account for the long half-life of ^{232}Th.

15-4 β Decay

Nuclear β decay has properties and mechanisms quite unlike those associated with the emission of α particles. The radiation from a β source contains nonnuclear charged particles characterized by their small mass and appreciable penetration in matter. The two kinds of β decay involve the nuclear transitions

$$_{Z}^{A}\text{X} \rightarrow {}_{Z+1}^{A}\text{Y} + \beta^- \quad \text{and} \quad {}_{Z}^{A}\text{X} \rightarrow {}_{Z-1}^{A}\text{Y} + \beta^+,$$

in which the proton and neutron numbers change by one unit while the mass number A remains the same, as indicated in Figure 15-2. A more fundamental picture of the phenomena appears to be in effect, where β^- decay transforms a neutron into a proton within the nucleus and β^+ decay reverses this basic transformation. A qualitative definition of the β^- and β^+ instabilities emerges from these observations. The parent nucleus evidently has two much mass, either because of a neutron excess in the case of β^- decay or because of a neutron deficiency in the case of β^+ decay. Each type of β transition alters the system of bound nucleons so that the nuclide $_{Z}^{A}\text{X}_N$ transforms into an isobar of lesser mass, one unit removed in both nucleon numbers Z and N. The transformation proceeds along a line of constant A on the nuclear chart, in a direction toward the valley of stability.

We have mentioned the valley of stability on two occasions in Chapter 14. The Z versus N plot in Figure 14-4 shows the tendency for stable nuclides to lie in a central valley bounded by radioactive isotopes on either side. The semiempirical mass formula in Equation (14-13) determines a mass surface in the variables A and Z and defines a stable valley as the locus of points of minimum mass in each cross section of the surface at constant A. We describe these features by graphs of the atomic mass $M(_{Z}^{A}\text{X})$ as a quadratic function of Z in Figure 14-13, and we note that isobars appear on the graphs for each choice of A. A β-unstable isobar takes its position along the parabolic curve at a value of M above the minimum. A β transition makes its appearance on the graph as a link between the unstable parent and its less-massive daughter one unit away. The link runs from Z to $Z+1$ for β^- decay or from Z to $Z-1$ for β^+ decay. Figure 15-12 shows such a display for the case of the β-active isobars at $A = 80$, the mass number corresponding to the decay scheme in Figure 15-1. Note that the graph of M versus Z has two branches, one for the even–even

Figure 15-12

Atomic mass versus atomic number for isobars at $A = 80$. Even–even and odd–odd nuclides lie on different mass curves separated by the effect of the pairing term in the mass formula. The β^- and β^+ transitions are indicated by right- and left-directed arrows leading in unit steps to the ultimate stable species ^{80}Se and ^{80}Kr.

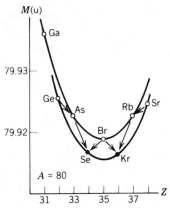

nuclides at lower M and another for the odd–odd nuclides at higher M. We recall from Section 14-5 that the two mass curves are expected for even A because of the pairing effect. We have constructed a similar set of graphs in Figure 14-13 for the isobars with $A = 98$, 99, and 100. The odd-A family at $A = 99$ is plotted again in Figure 15-13 as a graph of M versus Z and also as a partial decay scheme. The various β processes are indicated in both parts of the figure. Observe that the transitions $Z \rightarrow Z - 1$ appear in the decay scheme in two different forms, designated by the notation EC and β^+. We distinguish these processes in due course as we continue through the subsequent discussion.

The conservation laws of angular momentum and energy introduce an immediate new development in the interpretation of β decay. Let us consider angular momentum first and note that a problem arises when we examine the spins of the participating particles and nuclides. The emitted charged particle is either a spin-$\frac{1}{2}$ electron in β^- decay or a spin-$\frac{1}{2}$ positron in β^+ decay. The nuclear transition X $\rightarrow$ Y involves no change in A, and so *both* nuclear spins i_X and i_Y must be given by integers, if A is even, or half-integers, if A is odd. In either case a half-unit of spin is missing when conservation of angular momentum is applied to the decays. The orbital angular momentum also participates in the conservation law; however, this quantity is integer valued and cannot account for the missing half-unit of angular momentum. Conservation of energy presents another glaring problem concerning the energies of the observed electrons or positrons. The emitted β particles are not monoenergetic as in the case of α decay but are distributed in a *continuous spectrum* of energies, Figure 15-14 shows an example of a β^- spectrum in which the kinetic energy of the electrons varies from nearly zero up to a maximum value determined by the energy released in the

Figure 15-13

Sequential β decays in the $A = 99$ system of isobars. The plot of M versus Z reproduces one of the graphs in Figure 14-13. Data in the decay scheme include the values of i^p and $\tau_{1/2}$ along with the released energies in MeV. The β^- decays descend from the left while the electron-capture transitions and β^+ decays descend from the right.

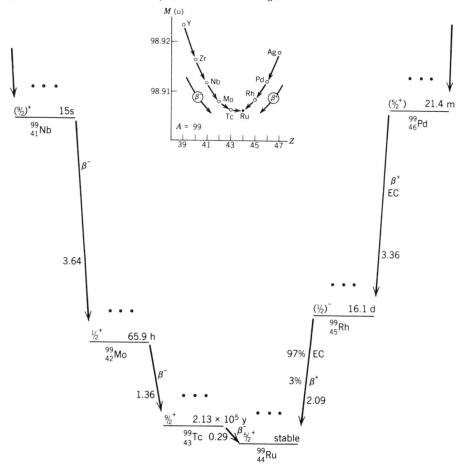

nuclear transition. This display of events reveals that less than half of the total available energy is carried off by the detected charged particles. The missing amounts of angular momentum and energy are explained by arguing that β decay must include another *unobserved* particle in the final state. The argument treats the two processes

$$_Z^A X \rightarrow _{Z+1}^A Y + \beta^- \quad \text{and} \quad _Z^A X \rightarrow _{Z-1}^A Y + \beta^+$$

as *three-body decays* and interprets the symbols β^- and β^+ as two-particle systems. The new third particle in the final state is hypothesized to *share* the total angular momentum and energy and carry away an unobservable portion of these conserved quantities.

The unseen companion of the positron in β^+ decay is known as the *neutrino*. Its antiparticle, the antineutrino, is emitted to accompany the electron in β^- decay. We

Figure 15-14

Electron spectrum in the β decay of ^{210}Bi.

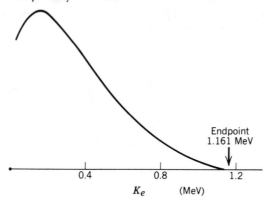

use the designations ν and $\bar{\nu}$, and describe the two nuclear processes as

$$_Z^A X \rightarrow _{Z+1}^A Y + e^- + \bar{\nu} \quad (\beta^- \text{ decay})$$

and

$$_Z^A X \rightarrow _{Z-1}^A Y + e^+ + \nu \quad (\beta^+ \text{ decay}).$$

We invoke a new type of conservation law for nonnuclear particles when we employ this notation. By convention, e^- and ν are called particles, while e^+ and $\bar{\nu}$ are called antiparticles, so that the total number of nonnuclear particles is conserved, along with the total number of nucleons, in each of the decays. Conservation of the total charge requires the neutrino to be a neutral particle. The original problem of the missing angular momentum is resolved by assuming that ν and $\bar{\nu}$ are spin-$\frac{1}{2}$ fermions. Since the electron cannot be a constituent of the nucleus, it follows that the two-body β systems $e^-\bar{\nu}$ and $e^+\nu$ must be *created* in the nuclear transformation and are not regarded as "present in the nucleus" prior to the decay.

The problem of the missing energy in β decay was acknowledged by 1930, before the coming of the neutron. An explanation for the continuous β spectrum was urgently needed since the observations presented a challenge to the fundamental law of conservation of energy. Pauli offered a solution in 1931 by proposing the existence of an unobserved third particle. The neutrino hypothesis was accepted, even though the particle itself could not be observed, because the violation of the sacred conservation laws was not a tolerable alternative. The mass of the neutrino was supposed to be exceedingly small, perhaps to the point of *vanishing*, to explain the observed β spectrum. The interactions of the neutrino were presumed to be practically negligible to explain why the particle should not be detected. Predictions of the interaction cross section were small enough to enable the neutrino to penetrate astronomical distances in matter. Neutrinos were detected for the first time, despite the incredible weakness of their interactions, in an experiment completed in 1956 by F. Reines and C. L. Cowan. These investigators constructed a large scintillating detector in the vicinity of a high-power reactor and observed evidence of the neutrino-absorption process

$$\bar{\nu} + p \rightarrow n + e^+.$$

Antineutrinos were produced with a very large flux from the β-active fission fragments in the reactor, and absorption events were recorded whenever the final-state particles were observed. Positrons and neutrons were detected, through the processes of positron annihilation and neutron capture, by the observation of photons in the scintillating liquid of the detector. This experiment established the existence of the neutrino, and later experiments confirmed the very small prediction for the neutrino interaction cross section.

The neutrino concept certainly does not require the mass of the neutrino to be precisely zero. We must turn to experiment for a determination of the mass, as we should for any particle. The question of a neutrino mass is of great interest in some of the current problems in particle physics. We can safely ignore this question when we study events in the main part of the β spectrum, because the mass of the neutrino is known to be *very* small, much smaller than the mass of the electron.

Let us begin our kinematical investigations by considering β^- decay. We apply conservation of energy to the three-body decay of a nucleus X at rest by writing

$$M_X c^2 = M_Y c^2 + K_Y + m_e c^2 + K_e + E_\nu. \tag{15-13}$$

Note that the relativistic neutrino energy E_ν includes a possible nonzero rest energy $m_\nu c^2$. We use nuclear masses for the nuclides X and Y, and we assume throughout that both parent and daughter are in the ground state. The addition of Z electron masses on each side of the equation results in the equality

$$(M_X + Zm_e)c^2 = \left[M_Y + (Z + 1)m_e\right]c^2 + K_Y + K_e + E_\nu,$$

or

$$M\left(^A_Z X\right)c^2 = M\left(^{\ \ A}_{Z+1}Y\right)c^2 + K_Y + K_e + E_\nu$$

in terms of atomic masses. We therefore define the Q-value, or disintegration energy, for β^- decay by the equations

$$Q = \left[M\left(^A_Z X\right) - M\left(^{\ \ A}_{Z+1}Y\right)\right]c^2$$
$$= K_Y + K_e + E_\nu. \tag{15-14}$$

This result identifies a specific total amount of energy to be shared among the three final particles Y, e^-, and $\bar{\nu}$. In fact, the recoil energy of the massive nucleus Y is usually small enough to be neglected except in situations where K_e and E_ν are also very small. The kinematics of β^+ decay also satisfies Equation (15-13). However, we take a different step when we add Z electron masses to both sides:

$$(M_X + Zm_e)c^2 = \left[M_Y + (Z - 1)m_e\right]c^2 + 2m_e c^2 + K_Y + K_e + E_\nu.$$

Thus, we obtain the equality

$$M\left(^A_Z X\right)c^2 = M\left(^{\ \ A}_{Z-1}Y\right)c^2 + 2m_e c^2 + K_Y + K_e + E_\nu$$

when we rewrite the result in terms of atomic masses. The Q-value for β^+ decay is therefore defined as

$$Q = \left[M\!\left({}_{Z}^{A}\mathrm{X}\right) - M\!\left({}_{Z-1}^{A}\mathrm{Y}\right) - 2m_e \right] c^2$$
$$= K_{\mathrm{Y}} + K_E + E_\nu. \tag{15-15}$$

Equations (15-14) and (15-15) give the total amounts of energy released in the two decays. We note that the nuclides X and Y must give a positive expression for Q if β^- or β^+ decay is to take place. The equations tell us that K_e is strictly less than Q, as needed to solve the problem of the missing energy. The equations also tell us that the maximum value of K_e approaches Q, for zero nuclear recoil, in the limit of vanishing neutrino energy. This maximum kinetic energy defines the *endpoint* in the spectrum for a given β decay, as in the illustration provided in Figure 15-14.

The neutrino-mass question comes to the surface when the β spectrum is examined close to its endpoint. This region of maximum K_e and *minimum* E_ν is sensitive to the rest energy of the neutrino because the decay probability is a function of the neutrino momentum

$$p_\nu = \frac{\sqrt{E_\nu^2 - m_\nu^2 c^4}}{c}.$$

The fact that E_ν and $m_\nu c^2$ appear together in this expression implies that a test for a nonzero m_ν is conceivable if events can be detected at sufficiently small values of E_ν. (We note in passing that p_ν can only *approach* zero for $m_\nu = 0$, because a zero-mass neutrino cannot be at rest in any Lorentz frame. The β spectrum distinguishes the $m_\nu = 0$ case since the approach to the endpoint varies gradually with K_e only when $p_\nu = E_\nu/c$.) This idea is under investigation in a current experiment to study triton β decay,

$$ {}^{3}\mathrm{H} \rightarrow {}^{3}\mathrm{He} + e^- + \bar{\nu},$$

near the endpoint of the electron spectrum. Indications suggest a *lower* bound for $m_\nu c^2$ around 30 eV, a rather small neutrino rest energy. This question remains unsettled, however, since there are strong arguments for a much smaller neutrino mass from other investigations.

A nucleus may undergo a second kind of β^+ transformation in the process of *electron capture*. This phenomenon is illustrated in Figure 15-15 by the reaction

$$ e^- + {}_{Z}^{A}\mathrm{X} \rightarrow {}_{Z-1}^{A}\mathrm{Y} + \nu $$

in which the electron is absorbed from an atomic orbital state to initiate the nuclear transformation. The process is most favored when capture occurs for a K electron, because the $\ell = 0$ orbital assignment maximizes the probability of finding the electron in the volume of space occupied by the nucleus. An electron capture is detected when an atomic x ray is observed; the x ray results from the transition of an outer electron to fill the inner hole left by the captured electron. We neglect the atomic binding energy in the initial state and apply conservation of energy to the process by writing

$$M_{\mathrm{X}}c^2 + m_e c^2 = M_{\mathrm{Y}}c^2 + K_{\mathrm{Y}} + E_\nu. \tag{15-16}$$

The addition of $Z - 1$ electron masses produces the equation

$$\left(M_{\mathrm{X}} + Zm_e \right)c^2 = \left[M_{\mathrm{Y}} + (Z-1)m_e \right]c^2 + K_{\mathrm{Y}} + E_\nu,$$

Figure 15-15

Initial and final states of the electron-capture process.

Atomic orbit

or

$$M\left(_Z^A\text{X}\right)c^2 = M\left(_{Z-1}^A\text{Y}\right)c^2 + K_\text{Y} + E_\nu.$$

The Q-value for electron capture is therefore given in terms of atomic masses by the formulas

$$Q = \left[M\left(_Z^A\text{X}\right) - M\left(_{Z-1}^A\text{Y}\right) \right]c^2$$
$$= K_\text{Y} + E_\nu. \tag{15-17}$$

This amount of energy released to the final $\text{Y}\nu$ state must be a positive quantity, so that the atomic mass of X must exceed the atomic mass of Y, if electron capture is to occur.

Equations (15-15) and (15-17) refer to β^+ transitions between the *same* parent and daughter. The difference in Q-values implies that electron capture is energetically allowed whenever β^+ decay takes place. The two processes compete under these circumstances, as suggested in some of the transitions shown in Figure 15-13. Since β^+ decay gives the smaller Q-value, it is possible to have masses of X and Y for which Equations (15-15) and (15-17) have opposite signs. In this situation the unstable nucleus undergoes electron capture as its *only* mode of β^+ transition. We illustrate some of the possibilities in the examples below.

Our discussion of β decay presumes that both X and Y are in their ground states. A β transition often leaves the daughter nucleus in an excited state Y*, which then deexcites by the γ-emission process $\text{Y*} \rightarrow \text{Y} + \gamma$. We observe such a transition by detecting a γ ray in delayed coincidence with the observed β particle. The Q-value formulas determine the *total* released energy, including γ-ray energy, in this situation.

Example

The isobars $_{12}^{27}\text{Mg}$, $_{13}^{27}\text{Al}$, and $_{14}^{27}\text{Si}$ engage in β processes similar to those of the $A = 99$ system in Figure 15-13. A glance at the tabulated atomic masses tells us that ^{27}Al is the stable isobar:

26.984342 u for ^{27}Mg,   26.981539 u for ^{27}Al,   and   26.986704 u for ^{27}Si.

The ^{27}Mg nuclide undergoes β^- decay,

$$^{27}_{12}\text{Mg} \rightarrow {}^{27}_{13}\text{Al} + e^- + \bar{\nu},$$

with released energy given by Equation (15-14):

$$Q = (26.984342 - 26.981539)(931.5 \text{ MeV}) = 2.611 \text{ MeV}.$$

One mode of instability for ^{27}Si is the electron-capture (EC) reaction

$$e^- + {}^{27}_{14}\text{Si} \rightarrow {}^{27}_{13}\text{Al} + \nu.$$

The disintegration energy is found from Equation (15-17):

$$Q = (26.986704 - 26.981539)(931.5 \text{ MeV}) = 4.811 \text{ MeV}.$$

We obtain a result larger than twice the rest energy of the electron, and so we also expect β^+ decay

$$^{27}_{14}\text{Si} \rightarrow {}^{27}_{13}\text{Al} + e^+ + \nu.$$

Equation (15-15) tells us that the Q-value is found by computing

$$Q(\beta^+) = Q(\text{EC}) - 2m_e c^2 = 4.811 \text{ MeV} - 2(0.5110 \text{ MeV}) = 3.789 \text{ MeV}.$$

Each of these released energies corresponds to a difference of nuclear ground-state energy levels. Actually, some of the observed β decays lead to excited daughter states in ^{27}Al and continue on to the ground state by γ emission. The same state of affairs prevails in Figure 15-13.

Example

Figure 15-1 includes all three β processes in the decays of the radionuclide ^{80}Br. Let us consult the table of atomic masses and calculate the relevant Q-values. For β^- decay, $^{80}\text{Br} \rightarrow {}^{80}\text{Kr} + e^- + \bar{\nu}$, we find

$$Q(\beta^-) = \left[M\left({}^{80}_{35}\text{Br}\right) - M\left({}^{80}_{36}\text{Kr}\right) \right] c^2$$
$$= (79.918528 - 79.916376)(931.5 \text{ MeV}) = 2.005 \text{ MeV}.$$

For electron capture, $e^- + {}^{80}\text{Br} \rightarrow {}^{80}\text{Se} + \nu$, we find

$$Q(\text{EC}) = \left[M\left({}^{80}_{35}\text{Br}\right) - M\left({}^{80}_{34}\text{Se}\right) \right] c^2$$
$$= (79.918528 - 79.916521)(931.5 \text{ MeV}) = 1.870 \text{ MeV}.$$

For β^+ decay, $^{80}\text{Br} \rightarrow {}^{80}\text{Se} + e^+ + \nu$, we find

$$Q(\beta^+) = Q(\text{EC}) - 2m_e c^2 = 1.870 \text{ MeV} - 2(0.5110 \text{ MeV}) = 0.848 \text{ MeV}.$$

The first two of these released energies are entered on the decay scheme in the figure.

15-5 The Weak Nuclear Interaction

The β instabilities of nuclei are caused by a fundamental type of force known as the *weak interaction*. We call the force fundamental because its influence applies to the most basic particles in nature. The weak interaction affects the behavior of nuclear particles and exhibits properties of strength and range on a scale very different from the strong nuclear force. Despite its feeble strength, the weak nuclear force is able to reveal many of its characterizing features through the commonplace phenomenon of nuclear β decay. In return, much is learned about the structure of the nucleus through the use of the β-decay interaction as a nuclear probe.

The weak force shows its strength in the properties of the neutrino, whose processes are only of the weak variety. We can appreciate the weakness of the force by inspecting the extraordinarily small cross sections for the interactions of neutrinos in matter. We realize some of the consequences when we also learn that neutrinos are produced in certain nuclear reactions that occur typically in the interiors of stars. The stars are almost transparent to the passage of neutrinos, and so the weakly interacting particles easily escape the stellar interiors. The result is a continuing inundation of the universe with a debris of almost-massless neutral particles.

The primitive weak phenomena in all nuclear β decays are the neutron and proton β processes

$$n \rightarrow p + e^- + \bar{\nu} \quad \text{and} \quad p \rightarrow n + e^+ + \nu \quad \text{inside the nucleus.}$$

The β decay of the neutron also occurs in the *free* state and accounts for the observed instability of this basic nuclear particle. Both nuclear and nonnuclear particles are engaged in these manifestations of the weak interaction. The proton and neutron have already been classified by the name nucleon. This subclass of particles is part of a larger family whose members interact strongly with each other. The electron and neutrino do not participate in the strong nuclear interaction and are called *leptons* to distinguish their behavior from that of the nucleons. We define a lepton number L according to the particle and antiparticle assignments

$$L = +1 \text{ for } e^- \text{ and } \nu \quad \text{and } L = -1 \text{ for } e^+ \text{ and } \bar{\nu}.$$

The nuclear β decays are then supposed to obey a conservation law for the total number of leptons along with the familiar law conserving the total number of nucleons. We have learned to identify p and n as two different charged isospin states of a single generic nucleon in order to implement the symmetry properties of the strong nuclear force. It also proves fruitful in the context of the weak interaction to let an analogous generic lepton be comprised of the two differently charged quantum states e^- and ν. We enlarge upon this notion later when we encounter other types of leptons.

The construction of a weak interaction theory has had a long history. Theoretical developments began in nuclear physics in 1934 when Fermi proposed his theory of β-decay transitions in nuclei. His picture of the nuclear β interaction had a number of very serious shortcomings. Nevertheless, the tentative theory was phenomenologically instructive as a guide to the β spectra and selection rules for a large body of nuclear decays. Several crucial discoveries were made in experiment and in theory during the 1950s. These ingredients became additional aspects of the picture as attention then turned toward the understanding of the weak decays of the elementary particles.

The Fermi theory draws a close analogy between the weak interaction in nuclear β decay and the electromagnetic interaction in atomic γ emission. Let us represent the

Enrico Fermi

deexcitation of the atom $A^* \to A + \gamma$ as a radiative transition of a bound electron with specific subshell quantum numbers:

$$e(n_i \ell_i) \to e(n_f \ell_f) + \gamma.$$

In similar fashion let us express the nuclear β decay $X \to Y + e^- + \bar{\nu}$ in terms of the bound neutron-to-proton transformation

$$n(X) \to p(Y) + e^- + \bar{\nu}.$$

We illustrate these two fundamental processes in Figure 15-16 by *space–time diagrams*, which show the passage with time from the respective initial states to the corresponding final states. In each case the emitted radiation is created by virtue of the relevant interaction and is not regarded as "present in the emitting source" prior to the transition. We know from Section 7-5 that the oscillation of the electric dipole moment $\langle -e\mathbf{r} \rangle$ in a transition state describes the atomic photon-emission process and that the dipole transition amplitude determines the intensity of the radiation. Let us express the formula for the amplitude in terms of the electron states in the figure by writing

$$\int \psi_{\alpha_f}^* \mathbf{r} \psi_{\alpha_i} \, d\tau,$$

where α_i and α_f represent initial and final sets of subshell quantum numbers $(n\ell m_\ell m_s)$ according to Equation (9-9). In the Fermi theory there is an analogous

Figure 15-16

Atomic γ emission and nuclear β decay as bound electron and nucleon single-particle transitions. The electron states α_i and α_f denote initial and final subshell quantum numbers, while the nucleon states X and Y refer to parent and daughter nuclei.

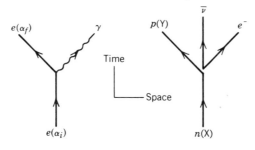

amplitude of the form

$$\int \psi_Y^* \psi_X \, d\tau$$

to describe the transition of the nucleon in the nuclear transformation X → Y. We can use this integral expression to deduce selection rules for nuclear β decay. The resulting constraints on the quantum numbers are analogous to the electric dipole conditions for atomic radiation.

The transforming nucleon in Figure 15-16 is a bound particle with definite initial and final orbital quantum numbers. These ℓ values label the angle dependence of the nucleon eigenfunctions in the Fermi transition amplitude. The amplitude vanishes if the ℓ values are not *equal*, because of the orthogonality of the spherical harmonics in the integral $\int \psi_Y^* \psi_X \, d\tau$. This observation means that the nucleon does not change its orbital state, so that the overall parity of the nucleus remains unaltered in the transformation. An *allowed transition* is further defined as one in which the pair of leptons is emitted with orbital angular momentum $\ell = 0$. This type of decay is favored because there is no centrifugal potential energy barrier to inhibit the emission of the created particles. The nuclear β process is said to be *forbidden* whenever the $\ell = 0$ condition on the leptons is not met. (In this context forbiddenness implies a high degree of suppression rather than an outright prohibition of the decay. Forbidden transitions actually occur for many unstable species on the nuclear chart. The degree of suppression grows with increasing ℓ, and so the half-lives tend to be longer for these nuclides.)

Let us consider only *allowed* transitions and look for constraints on the change in the nuclear spin as X transforms into Y. The emitted β system consists of two spin-$\frac{1}{2}$ particles; therefore, the pair of leptons must appear in states with two possible values for the total spin quantum number. We recall our analysis of the two electrons in the helium atom and conclude that the electron–neutrino system can emerge with spins in the $s = 0$ singlet and the $s = 1$ triplet states. We isolate the $s = 0$ case first and use conservation of the total angular momentum to argue that the nuclear spin cannot change in the transition X → Y, since the leptons are emitted with $\ell = 0$. The conclusion, that $i_X = i_Y$ or $\Delta i = 0$, is known as a *Fermi selection rule*. We then turn to the other possibility in which the $\ell = 0$ lepton pair is created with total spin $s = 1$. In this case we express the conservation of angular momentum for the process $X \to Y + \beta$ by writing

$$i_X = i_Y + 1.$$

The addition of vectors constrains the quantized angular momenta such that the nuclear spin quantum numbers must obey either $i_X = i_Y$ or $i_X = i_Y \pm 1$. We also note that the conservation law does not permit transitions from $i_X = 0$ to $i_Y = 0$. The result, that $\Delta i = 0$ or ± 1, is a *second* type of selection rule due originally to Gamow and E. Teller. We distinguish the two cases $s = 0$ and $s = 1$ by calling the one a Fermi transition and the other a Gamow–Teller transition.

We observe that the Fermi transition amplitude $\int \psi_Y^* \psi_X \, d\tau$ contains no operation to change the nuclear spin. Another amplitude must therefore be introduced to accommodate the $\Delta i = \pm 1$ transitions included in the Gamow–Teller selection rule. The required spin modification does not affect the nuclear parity, and so the parity of the nuclear state remains unchanged in *both* Gamow–Teller and Fermi transitions.

Let us summarize our conclusions about the allowed decays and the emission of β radiation in an $\ell = 0$ state. We separate the two lepton spin-state possibilities by the following rules:

$$s = 0, \quad \Delta i = 0, \quad \text{no} \tag{15-18}$$

for Fermi transitions, and

$$s = 1, \quad \Delta i = 0 \text{ or } \pm 1 \text{ (but } 0 \nrightarrow 0), \quad \text{no} \tag{15-19}$$

for Gamow–Teller transitions. Both selection rules are designated "no" to indicate no change in the parity of the nuclear state. We recall that the total spin quantum numbers $s = 0$ and $s = 1$ refer to antiparallel and parallel spins. Let us apply this familiar picture to the emission of the pair of leptons and represent the selection rules schematically with the aid of Figure 15-17. Specific examples of allowed Fermi and

Figure 15-17

Selection rules for Fermi and Gamow–Teller transitions in the decay $X \to Y + e^- + \bar{\nu}$. Neutron β decay involves both of these contributions.

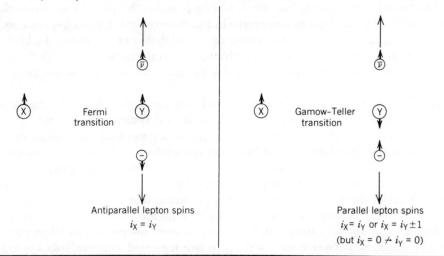

Gamow–Teller β decays are given below. We have already remarked that forbidden decays also occur. These processes can be associated with changes in the nuclear parity and with changes by more than one unit in the nuclear spin.

We have learned in Section 7-5 that the overall parity of an interacting system obeys a multiplicative conservation law in the radiative transitions of atoms. We also know that these processes are governed by the electromagnetic interaction. It is natural to ask whether the overall parity is similarly respected by the weak interaction in the β decays of nuclei. We discuss the experimental and theoretical responses to this important question in Chapter 16, when we turn to the more recent discoveries in weak interaction physics.

Example

The β^+ decay of ^{42}Sc is described as

$$^{42}_{21}\text{Sc} \rightarrow {}^{42}_{20}\text{Ca} + e^+ + \nu,$$

in which both initial and final nuclides have 0^+ for their i^p quantum numbers. This allowed $0 \rightarrow 0$ transition is a Fermi process, obeying the selection rules given in Equations (15-18). The Gamow–Teller selection rules in Equations (15-19) are illustrated by the β^- decay of ^{32}P,

$$^{32}_{15}\text{P} \rightarrow {}^{32}_{16}\text{S} + e^- + \bar{\nu}.$$

The i^p values are 1^+ for ^{32}P and 0^+ for ^{32}S, implying $\Delta i = -1$ with no change in parity. Most allowed β decays proceed via a combination of Fermi and Gamow–Teller transitions. The decay of the neutron is a case in point since the emitted leptons e^- and $\bar{\nu}$ may have parallel or antiparallel spins, and the nucleon spin may or may not flip in the $n \rightarrow p$ transformation. Forbidden transitions are also much in evidence across the nuclear chart. An interesting example is furnished by the β decays of ^{36}Cl,

$$^{36}_{17}\text{Cl} \rightarrow {}^{36}_{18}\text{Ar} + e^- + \bar{\nu} \quad \text{and} \quad {}^{36}_{17}\text{Cl} \rightarrow {}^{36}_{16}\text{S} + e^+ + \nu,$$

in which the i^p quantum numbers transform from 2^+ to 0^+. This two-unit change in nuclear spin cannot be balanced by the spins of the leptons alone, and so a nonzero value of ℓ is needed in the orbital state of the leptons. A suppression of ^{36}Cl decay results, accounting qualitatively for the long half-life of the nuclide, $\tau_{1/2} = 3 \times 10^5$ y.

Example

A very crude calculation can be made along the following lines to estimate the strength of neutrino interactions in matter. Let us examine this question in terms of the antineutrino absorption reaction $\bar{\nu} + p \rightarrow n + e^+$, as studied in the Reines–Cowan experiment. We construct the absorption cross section σ_ν as an effective area presented to an incoming antineutrino by a target proton, incorporating the probability for an interaction to occur. The primitive weak nucleon transformation is represented by the β decay of the neutron $n \rightarrow p + e^- + \bar{\nu}$.

The average time of decay is found from the 10.5 min half-life:

$$\tau = \frac{\tau_{1/2}}{\ln 2} = \frac{(10.5 \text{ min})(60 \text{ s/min})}{\ln 2} = 909 \text{ s}.$$

Let us use the reciprocal of this time interval to set a rough measure of the probability per unit time for the $p \to n$ transition in the $\bar{\nu}$ absorption process. We neglect the mass of the antineutrino and choose a representative $\bar{\nu}$ energy of 3 MeV. The antineutrino then has momentum $p_\nu = E_\nu/c$ and de Broglie wavelength

$$\lambda = \frac{h}{p_\nu} = \frac{hc}{E_\nu} = \frac{1240 \text{ MeV} \cdot \text{fm}}{3 \text{ MeV}} = 413 \text{ fm}.$$

We take the antineutrino to be localized within a distance of order λ while the particle moves with speed approaching c. The ratio of these parameters determines a "transit time" for the antineutrino:

$$t = \frac{\lambda}{c} = \frac{4.13 \times 10^{-13} \text{ m}}{3 \times 10^8 \text{ m/s}} = 1.38 \times 10^{-21} \text{ s}.$$

We multiply this quantity by the interaction probability per unit time to estimate the probability for an antineutrino interaction:

$$P = t \cdot \frac{1}{\tau} = \frac{1.38 \times 10^{-21} \text{ s}}{909 \text{ s}} = 1.52 \times 10^{-24}.$$

The antineutrino collides with a target proton if the radius of the target area A is within the range of localization of the antineutrino. The collision then causes the absorption of the antineutrino with a probability given by P. We use these arguments to estimate the target area

$$A = \pi \lambda^2 = \pi \left(4.13 \times 10^{-13} \text{ m}\right)^2 = 5.36 \times 10^{-25} \text{ m}^2$$

and the absorption cross section

$$\sigma_\nu = AP = \left(5.36 \times 10^{-25} \text{ m}^2\right)\left(1.52 \times 10^{-24}\right) = 8.15 \times 10^{-49} \text{ m}^2.$$

Let us pursue the arguments a bit farther to estimate the mean free path ℓ_ν for neutrino interactions in matter. This quantity is deduced from the cross section by identifying $\sigma_\nu \ell_\nu$ as the volume swept out by antineutrinos between absorptive interactions with protons in the given medium. It follows that the mean free path is given by the formula

$$\ell_\nu = \frac{1}{\sigma_\nu n},$$

where n denotes the number of protons per unit volume. We take the absorbing medium to be liquid hydrogen, and we find n from the density and the proton mass:

$$n = \frac{70 \text{ kg/m}^3}{1.67 \times 10^{-27} \text{ kg}} = 4.2 \times 10^{28} \text{ m}^{-3}.$$

Our estimation then gives

$$\ell_\nu = \frac{1}{\left(8.15 \times 10^{-49} \text{ m}^2\right)\left(4.2 \times 10^{28} \text{ m}^{-3}\right)} = 2.9 \times 10^{19} \text{ m}.$$

Thus, the extremely small $\bar{\nu}$ cross section translates into an incredibly long mean free path. This average distance between interactions is large enough to call for the use of light-year units:

$$\ell_\nu = \frac{2.9 \times 10^{19} \text{ m}}{9.46 \times 10^{15} \text{ m/lt-y}} = 3.1 \times 10^3 \text{ lt-y}.$$

Of course, our estimates are based on an extremely gross calculation. The actual cross section is somewhat larger, but only by a single order of magnitude. For 3 MeV antineutrinos, σ_ν is approximately 10^{-47} m^2 and ℓ_ν is of order 100 light-years. We offer these figures to demonstrate the extraordinary weakness of the weak interaction. The observation of antineutrinos begins to become feasible, as in the Reines–Cowan experiment, when the $\bar{\nu}$ flux upon the target is sufficiently great to achieve an appreciable enhancement of the detection probability.

15-6 γ Decay

The electromagnetic interaction affects the nucleus through the process of γ deexcitation. Nuclear γ radiation is emitted whenever a transition occurs from an excited level to a state of the *same* nuclide at lower energy. The prior state of excitation of the nucleus is often the result of some nuclear reaction or disintegration, as in the example shown in Figure 15-18. If the excited level is not so high as to allow the prompt ejection of a nuclear particle, then the radiative process offers the main available mode of decay. Photons produced in these electromagnetic transitions have energies of MeV order, a typical scale of energy for nuclear excitation. A particular nucleus yields a spectrum of γ rays characteristic of a certain system of excited states. The resulting decay scheme is unique and serves as a signature for the radiating species.

Measurements of γ-ray wavelengths may be made using crystal spectrometers like those employed for atomic x rays, provided the energies of the emitted photons are not far beyond the usual x-ray regime. These Bragg techniques are used for γ rays with energies up to 1 MeV. Radiation of moderate energy may also be detected by introducing a thin-foil material (called a radiator) in the path of the photons and by observing the ejection of electrons produced by the photoelectric effect or the Compton effect. If the γ energy exceeds $2m_e c^2$, the detection of photons can be performed by letting the γ rays produce electron–positron pairs in a thin-foil absorber.

Figure 15-18

Sequential β and γ transitions in the decay of
^{60}Co. The β process leads to an excited state
of ^{60}Ni, which then decays to the ground state
by the emission of γ rays. The released
energies are quoted in MeV for each decay.

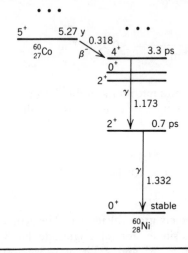

We express the γ decay of an excited nucleus as

$$_Z^A X^* \to {_Z^A} X + \gamma,$$

and we note that the final level need not be the ground state. This process is not fundamentally different from the radiative deexcitation of an atom, although the energy scale is appreciably larger for nuclear radiation. Other numerical distinctions come into play when we consider the enormous range of lifetimes found in γ decay. The vastly different probabilities for the various radiative transitions depend on the γ-ray energies *and* the i^p quantum numbers of the initial and final nuclear states. It is instructive to organize all these decays with the aid of selection rules similar to those found for the emission of electromagnetic radiation in atoms.

We have taken the electric dipole transition to be the only important mechanism in the deexcitation of atoms. Sources of radiation other than the oscillations of the electric dipole moment must also be considered in nuclear γ decay. Figure 14-18 shows the electric dipole as one of a sequence of configurations that make up a complex distribution of charge. Nuclei may also have quadrupole, octupole, or higher electric moments as nonnegligible contributors to the radiation field. We classify these *multipole* structures as 2^k-poles, taking $k = 1, 2, 3, \ldots$ for the sequence of dipole, quadrupole, octupole, ... configurations. Magnetic multipoles also exist as distinct physical systems with the same 2^k designations. We note that no $k = 0$ radiation exists because such a monopole field would be spherically symmetric and not transverse, as required for the propagation of electromagnetic waves. Classical examples of electric

Figure 15-19

Oscillating electric and magnetic dipole moments.

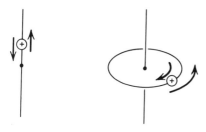

and magnetic dipole sources of radiation are shown in Figure 15-19. We associate electric dipole radiation with the linear oscillation of an electric charge and magnetic dipole radiation with an oscillating current loop. The analogous $k = 1$ quantum systems are described by the expectation values of the electric and magnetic dipole moments, oscillating in a transition state. Similar descriptions apply to the multipoles beyond $k = 1$.

The multipole index k is equal to the angular momentum (in $\hbar$ units) carried away by each photon in a 2^k-pole radiation field. This interpretation of k as an angular momentum quantum number generalizes our remarks in Section 7-5, where we consider the special case $k = 1$ and find that the electric dipole photon radiates one $\hbar$ unit of angular momentum. The probability for radiative nuclear decay depends sensitively on the quantum number k, as large assignments of k imply highly suppressed rates of decay and long radiative lifetimes. We can illustrate this property with the aid of the following qualitative list of ranges of the mean life τ, organized by multipole for given extremes of the photon energy ε:

electric dipole $E1$	$\tau = 10^{-16}$ to 10^{-14} s	
magnetic dipole $M1$	10^{-14} to 10^{-12} s	
electric quadrupole $E2$	10^{-11} to 10^{-8} s	as ε ranges from
magnetic quadrupole $M2$	10^{-9} to 10^{-6} s	1 MeV down
electric octupole $E3$	10^{-6} to 10^{-2} s	to 200 keV.
magnetic octupole $M3$	10^{-4} to 10 s	

$\cdots$

We use the conventional notation Ek and Mk to symbolize the various electric and magnetic 2^k-poles. Note that τ grows through many orders of magnitude as k increases, and observe that the electric multipoles have the faster decays for each value of k.

Electric and magnetic multipoles carry definite parity as well as definite angular momentum. We know that electric dipole radiation has *odd* parity since the behavior of the $E1$ transition amplitude is linked to the space-inversion property of the odd-parity dipole vector $e\mathbf{r}$. Magnetic dipole radiation can be attributed to the oscillations of a spin magnetic moment in a transition state. We therefore assign *even* parity to the $M1$ mode, since space inversion has no effect on spin. These observations can be generalized systematically for all electric and magnetic multipoles, with the following conclusions:

Electric 2^k-poles have even/odd parity for even/odd k

and

magnetic 2^k-poles have even/odd parity for odd/even k.

Thus, we associate *opposite* parities with the Ek and Mk modes for each value of k.

Selection rules in nuclear γ decay follow from the conservation of overall angular momentum in the nucleus-plus-radiation system. We let the participants in the process $X^* \rightarrow X + \gamma$ have nuclear spin and multipole quantum numbers i^*, i, and k, and we then use these assignments to denote quantized angular momentum vectors in the conservation law

$$\mathbf{i^*} = \mathbf{i} + \mathbf{k}.$$

This familiar vector constraint implies that the multipole index k must obey the condition

$$|i^* - i| \le k \le i^* + i. \tag{15-20}$$

We have already learned that no radiation multipole exists for $k = 0$. It follows that the quantum numbers $i^* = i = 0$ cannot satisfy the vector equality for any k, so that no $0 \rightarrow 0$ radiative decay is possible.

Radiation may be emitted with or without a change of parity in the nuclear state. Conservation of the overall parity calls for the emission of an odd-parity multipole when X^* and X have opposite parity, or an even-parity multipole when X^* and X have equal parity. We can be more specific about the multipoles if we invoke the space-inversion properties of the various electric and magnetic modes. Thus, we find that parity conservation permits

$E1, M2, E3, \ldots$ radiation with a change in nuclear parity,

and

$M1, E2, M3, \ldots$ radiation with no change in nuclear parity.

Our list of lifetimes correlates each of these multipoles with a range of values of τ. The enormous variations indicate vividly the role of the selection rules in nuclear γ decay.

The list of τ values suggests the existence of nuclear states with large radiative lifetimes. These metastable systems may live long enough to allow the direct measurement of such state properties as the mean life, the nuclear spin, and the magnetic moment. An excited configuration of the nucleus is called an *isomeric state*, or a nuclear isomer, if its mean life is long enough to be measurable. The radiative decay is then said to occur in an *isomeric transition*, and the phenomenon is known as *nuclear isomerism*. We expect isomeric transitions to be accompanied by radiation in the higher multipoles, where the probabilities for decay are greatly suppressed, and where the changes in the nuclear spin are given by several units.

The deexcitation of a nuclear state may also occur by a *radiationless* mechanism known as *internal conversion*. In this alternative process, the nucleus releases energy in a transition to a lower state, and the energy is absorbed at once in the electronic configuration of the associated atom. The exchange of energy results in the ejection of atomic electrons instead of the emission of a nuclear γ ray. An analogous nonradiative type of transition takes place in atoms through the Auger effect, as discussed in Section 3-7.

Detail

Transition probabilities for nuclear and atomic radiation vary in powers of a dimensionless *radiation scale parameter* ξ. In the nuclear case the parameter is defined by the ratio of the nuclear radius and the γ-ray wavelength:

$$\xi = \frac{R}{\lambda/2\pi}.$$

We interpret ξ as a means of comparing scales of length for the only two lengths available in the radiation problem. We then recall Equation (2-51) and (14-3) so that we can express ξ in terms of the photon energy and nuclear mass number:

$$\xi = 2\pi R \frac{\varepsilon}{hc} = \frac{R_0 A^{1/3}\varepsilon}{\hbar c}.$$

Values of this nuclear parameter are less than $\frac{1}{10}$ for most nuclei, while values of the analogous atomic ratio are two orders of magnitude *smaller* for all atoms. (We illustrate this claim in the first example below.) Since ξ is raised to a certain power to determine the probability for a certain type of transition, it follows that that type of transition is suppressed much less in nuclei than in atoms. This comparison of scales tells us why more radiation multipoles are considered in the case of nuclear γ decay.

Example

A typical nuclear γ energy might be as large as 2 MeV. The corresponding wavelength is two orders of magnitude larger than a nuclear diameter:

$$\lambda = \frac{hc}{\varepsilon} = \frac{1240 \text{ MeV} \cdot \text{fm}}{2 \text{ MeV}} = 620 \text{ fm}.$$

This comparison of lengths is the idea behind the radiation scale parameter ξ. Let us look at the upper end of the range of ξ by choosing a large mass number.

We take $A = 6^3$ and again set $\varepsilon = 2$ MeV to get

$$\xi = \frac{R_0 A^{1/3} \varepsilon}{\hbar c} = \frac{(1.2 \text{ fm})(6)(2 \text{ MeV})}{197 \text{ MeV} \cdot \text{fm}} = 0.073.$$

The probability for γ decay depends on ξ raised to a power, where the exponent increases with the multipole index k. The smallness of ξ implies small decay rates and long lifetimes for radiation in the high multipoles. We find an even smaller scale parameter when we consider the radiative deexcitation of atoms. Let us estimate ξ for atoms by computing a ratio of typical lengths:

$$\xi = \frac{2\pi R}{\lambda} = 2\pi \frac{10^{-10} \text{ m}}{6 \times 10^{-7} \text{ m}} = 1 \times 10^{-3}.$$

Electric dipole transitions take the smallest power of ξ and give the only appreciable modes of decay for such a small scale parameter. This argument justifies the neglect of all but the $E1$ mode in the radiation from atoms.

Example

The β^- decay of ^{137}Cs leads predominantly to an isomeric state of ^{137}Ba, as indicated in Figure 15-20. This $\frac{11}{2}^-$ level has a measurable half-life of 2.55 min. Its isomeric transition to the $\frac{3}{2}^+$ ground state is labeled as such in the figure. The transition occurs with a change in nuclear parity, while the nuclear spin decreases by four units. Equation (15-20) tells us that the multipole quantum number is in the range $4 \leq k \leq 7$, and parity conservation then implies that the allowed multipoles are $M4$, $E5$, $M6$, and $E7$. The $M4$ multipole has the smallest value of k and is expected to give the dominant decay mode. Larger values of Δi are not uncommon in other isomeric transitions on the nuclear chart. The correspondingly longer half-lives are listed in hours, days, and even years.

Figure 15-20
Isomeric transition in ^{137}Ba following the β decay of ^{137}Cs. Energies for the β and γ transitions are given in MeV.

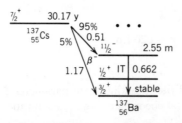

15-7   Resonance Radiation

The γ-active states of nuclei have radiative lifetimes ranging from attoseconds to years. The variations span 25 orders of magnitude, and so the methods for measuring such disparate intervals of time must vary accordingly. We can determine mean lives longer than 1 μs directly from observations of γ-ray counting rates. We can measure lifetimes greater than 10 ps, for excited nuclei resulting from β decay, by detecting delayed coincidences between the emitted β and γ rays. Very short lifetimes can be determined by an indirect measurement based on the concept of *natural width*. This property of an unstable energy level is an important general consideration in the study of nuclear excitations.

An unstable state has an average lifetime given by the mean life τ. The uncertain duration of the system implies an uncertainty in the energy of the state, in keeping with the energy–time uncertainty principle. We identify the natural width of the energy level with the energy uncertainty Γ and connect the width to the lifetime by the inverse relation

$$\Gamma = \frac{\hbar}{\tau}. \tag{15-21}$$

Hence, a measurement of Γ amounts to a determination of τ, and vice versa. Widths are of order 10^{-7} eV in atoms, as typical values of τ are around 10^{-8} s. Much larger widths are found in nuclear states since some of the lifetimes are extremely short. To give an example, an $E1$ mean life of 10^{-16} s implies a natural width around 10 eV.

The natural width of an unstable level can be obtained, in principle, from the following *idealized* experiment. We consider a source particle and a target particle of the *same* species and assume that both particles are constrained from moving. The emission of a photon by the source may then result in an observable *resonant* absorption of the photon by the target. Both the distribution of emitted radiation and the cross section for γ-ray absorption have widths that include the natural width of the excited state. Thus, we can measure the natural width if we eliminate thermal line broadening from such an idealized absorption experiment. The main idealization is, of course, the assumption of no motion for the particles in their interactions with the photon. A free emitter and a free absorber must undergo *recoil* and exchange the photon *off* resonance.

Let us examine this phenomenon of resonance radiation (or resonance fluorescence) more closely. Figure 15-21 shows how emission and absorption can operate in tandem between corresponding energy levels in two similar quantum systems. The interplay of processes couples the systems, emitter and absorber, by an exchange of photon energy. We achieve a resonance when the emitter deexcites and produces a photon with just the energy needed to excite the absorber. The energy-level difference ΔE is exactly the same for both systems, as required for the resonant transfer of the photon. However, the recoil of each system causes the emitted and absorbed photon energies ε_e and ε_a to be different from ΔE *and* from each other. This mismatch influences the resonance mechanism to an extent determined by the natural width of the excited state.

The figure describes the kinematics involved in the conservation of energy and momentum. Let us look first at the photon emission process $X^* \rightarrow X + \gamma$ and observe that the energies obey the relation

$$E^* = E + \varepsilon_e + \frac{p^2}{2M}, \tag{15-22}$$

Figure 15-21

Resonance radiation and the effects of recoil. The emitted and absorbed photons have unequal energies ε_e and ε_a, both different from the transition energy ΔE.

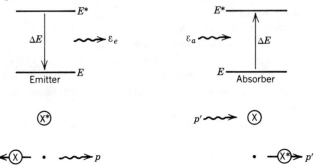

while the photon energy determines the recoil momentum of X:

$$p = \frac{\varepsilon_e}{c}.$$

This problem has been studied in Section 3-6, and the solution for the energy of the emitted photon has been given in Equation (3-61). We rewrite our result as

$$\varepsilon_e = \Delta E - \frac{(\Delta E)^2}{2Mc^2}, \tag{15-23}$$

and we note that the photon energy is *less* than the transition energy by an amount equal to the kinetic energy of the recoiling system X. We then turn to the absorption process $\gamma + X \rightarrow X^*$ and find that the recoil effect requires the incident photon energy to be *greater* than the transition energy by the same amount:

$$\varepsilon_a = \Delta E + \frac{(\Delta E)^2}{2Mc^2}. \tag{15-24}$$

(The proof is left to Problem 18 at the end of the chapter.) Figure 15-22 summarizes our conclusions in terms of a schematic spectrum of energies for photons emitted and absorbed at the transition energy ΔE. Note that the energies occur in broadened distributions instead of sharp peaks, due in part to the natural width of the excited energy level E^*. We see that resonance radiation may occur only when the two peaks in the figure have sufficient overlap. The criterion is realized if the indicated width Γ is large compared to the recoil kinetic energy

$$K = \frac{(\Delta E)^2}{2Mc^2}. \tag{15-25}$$

We meet this condition for resonance when we consider the visible radiation from atoms, and indeed we observe resonance fluorescence as a common atomic phenome-

Figure 15-22

Energy distribution of photons emitted and absorbed as in Figure 15-21. The peaks are spread by the natural width Γ and are shifted to either side of ΔE by the effects of recoil.

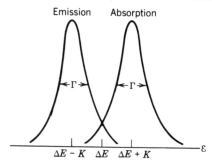

non. The analogous nuclear problem finds K much larger than Γ, however. It would therefore seem that resonance radiation is not an observable phenomenon in nuclei.

The mechanism for resonance can be recovered by letting the emitter approach the absorber, so that the energy of the emitted photon is Doppler shifted to the larger value required for absorption. We can determine the desired shift by noting the connection between Equations (15-23) and (15-24):

$$\varepsilon_a = \varepsilon_e \frac{1 + \Delta E / 2Mc^2}{1 - \Delta E / 2Mc^2} = \varepsilon_e \left(1 + \frac{\Delta E}{Mc^2} \right)$$

to first order in the small ratio $\Delta E / Mc^2$. We then recall from Equation (1-9) that the Doppler effect produces a blue-shifted light frequency, with transformation factor $\sqrt{(c + u)/(c - u)}$, when the emitter moves at speed u toward the absorber. This relation between frequencies can be cast immediately in terms of the photon energies ε_e and ε_a to obtain

$$\varepsilon_a = \varepsilon_e \sqrt{\frac{c + u}{c - u}} = \varepsilon_e \frac{c + u}{\sqrt{c^2 - u^2}} = \varepsilon_e \left(1 + \frac{u}{c} \right),$$

to first order in u/c. The two expressions for ε_a tell us that the Doppler effect offsets the recoil effect if the emitter approaches the absorber at velocity

$$u = \frac{\Delta E}{Mc}. \tag{15-26}$$

Resonance fluorescence has been observed in nuclei by experimental methods based on this idea.

The resonance condition may also be achieved through the Mössbauer effect, a technique discovered in 1958 in the experiments of R. L. Mössbauer. The effect is based on the *recoil-free* property of the emission and absorption of γ rays for nuclei bound in *solids*. Embedding the emitter and absorber in separate crystals forces the crystal lattice to take up the recoil of the nucleus in both parts of the resonance mechanism. An equivalent kinematical result is obtained by substituting the very large mass of the crystal in place of the nuclear mass in the formula for K. The mismatch between ε_e and ε_a disappears under these conditions, as in the idealized case of the emitter and absorber kept at rest. The recoilless behavior of the Mössbauer system makes it possible to achieve a resonance for an extremely sharp γ transition and enables us to measure the natural width of the resulting absorption line.

The Mössbauer effect exercises the criterion for nuclear resonance radiation with great sensitivity. The technique can be used to detect and measure minute shifts in the energies of photons caused, for instance, by an altered nuclear environment or a small relative motion within the system. These perturbations on the photon energy are easily seen in an absorption experiment, since the slightest mismatch between emitter and absorber causes the Mössbauer device to respond off resonance.

Mössbauer's original observations were made in a low-temperature experiment using the 129 keV γ ray in the spectrum of ^{191}Ir. The effect has since been observed in numerous γ transitions of many different nuclides. The capability to detect small γ-ray energy shifts was put to immediate use in several types of precision experiments. A test of the theory of relativity was among the first applications of the effect. Use of the technique has also spread into biology and metallurgy. To cite one example, the 14.4 keV γ ray in ^{57}Fe has been used to probe the environments of iron nuclei in samples of organic and metallic matter.

Example

Resonance radiation is best discussed with the aid of numerical illustrations. Let us start with the relation between mean life and natural width:

$$\Gamma\tau = \hbar = 1.055 \times 10^{-34} \text{ J} \cdot \text{s} = 0.6582 \times 10^{-15} \text{ eV} \cdot \text{s}.$$

In atoms we expect to find $\tau \sim 10^{-8}$ s and $\Gamma \sim 10^{-7}$ eV, as noted above. Visible light corresponds to a transition energy around 2 eV, while atomic rest energies are at least as large as 10^9 eV. If we use these numbers in Equation (15-25) to assess the recoil effect, we get

$$K = \frac{(\Delta E)^2}{2Mc^2} = \frac{(2 \text{ eV})^2}{2(10^9 \text{ eV})} = 2 \times 10^{-9} \text{ eV}.$$

Our estimates find K much less than Γ and imply that recoil has no adverse consequences for resonance fluorescence in atoms. The rather different situation in nuclei is nicely illustrated by Mössbauer's nucleus ^{191}Ir. The 129 keV γ ray comes from a state with half-life $\tau_{1/2} = 0.13$ ns. The corresponding natural width is quite small:

$$\Gamma = \frac{\hbar}{\tau} = \frac{0.658 \times 10^{-15} \text{ eV} \cdot \text{s}}{0.13 \times 10^{-9} \text{ s}/\ln 2} = 3.5 \times 10^{-6} \text{ eV}.$$

Hence, the energy distribution of emitted and absorbed photons forms two narrow peaks, unlike the broad ones shown in Figure 15-22. The recoil effect separates these peaks by the comparatively large energy $2K$, where

$$K = \frac{(0.129 \text{ MeV})^2 (10^6 \text{ eV/MeV})}{2(191)(931.5 \text{ MeV})} = 4.68 \times 10^{-2} \text{ eV}$$

for a free Ir nucleus. (The substitution of the γ-ray energy for the transition energy is a very good approximation here.) Resonance fluorescence can be observed via the Doppler effect if emission and absorption occur at a relative velocity given by

$$u = \frac{\Delta E}{Mc^2} c = \frac{0.129 \text{ MeV}}{(191)(931.5 \text{ MeV})} c = 7.25 \times 10^{-7} c = 218 \text{ m/s},$$

according to Equation (15-26).

15-8 Introduction to Nuclear Reactions

Nuclear structure is studied by observing the excitations and transformations of the nucleus. The goals of the investigation are to determine the states of the nucleus and to deduce the properties of the nuclear interaction. We know that radioactive decays can furnish this kind of information, but only for the species that actually participate in the various decay processes. We gain the capability to examine the states of any nucleus at any energies when we turn to nuclear reactions as the means of exploration.

A nuclear reaction may be any of the possible consequences of a collision at short range between a nuclear projectile x and a nuclear target X. We describe such a process as

$$x + \text{X} \rightarrow \text{Y} + y \quad \text{or} \quad \text{X}(x, y)\text{Y},$$

for situations where the reaction yields a two-body final state composed of a residual nucleus Y and an outgoing nuclear particle y. The product Y may be either stable or radioactive. In either event we call the process a two-body *reaction* whenever X and Y are different nuclides. Nuclear *scattering* is then classified as a special case in which X and Y are of the same species. We divide scattering into its categories, elastic and inelastic, and we express the two possibilities as

$$\text{X}(x, x)\text{X} \quad \text{and} \quad \text{X}(x, x')\text{X}^*.$$

In the inelastic case, X* denotes an excited state and x' refers to a scattered particle whose energy is reduced by the excitation of the nuclear target.

A typical two-body reaction experiment might entail the detection of y as a function of the outgoing angle, for a number of choices of the incident energy. Typical beam particles might be protons, neutrons, deuterons, α particles, or other ions. A charged projectile must have enough kinetic energy to penetrate the region of Coulomb repulsion around the nucleus in order to produce a nuclear interaction with the target. For insufficient energy the outcome of the charged-particle collision is expected to agree with the predictions of Rutherford scattering. If we consider protons in collision with light nuclei as an example, and recall our calculation of short-range Coulomb repulsion at the end of Section 14-1, we find that a beam energy of several

MeV is needed to gain access to the nuclear regime. A suitable proton accelerator must be supplied to meet this requirement. We can probe the nucleus at lower energy if we use neutrons instead, since neutrons are not affected by electrostatic repulsion. A nuclear reactor serves as a suitable source of neutrons for these low-energy experiments.

Nuclear reactions were first observed in 1919 in Rutherford's laboratory, when energetic particles were found to be produced by α particles passing through air. These occasional events were interpreted as α-particle collisions with nitrogen nuclei, yielding protons in the final state:

$$\,^4_2\text{He} + \,^{14}_7\text{N} \rightarrow \,^{17}_8\text{O} + \,^1_1\text{H} \quad \text{or} \quad \,^{14}\text{N}(\alpha, p)\,^{17}\text{O}.$$

Radioactive α emitters were the only available sources of energetic charged particles until the coming of the accelerator. One of the earliest machines was the high-voltage generator, developed for the acceleration of protons by J. D. Cockcroft and E. T. S. Walton. Particles from this device were used to initiate nuclear disintegrations for the first time in 1932. The accelerated protons bombarded a lithium target and produced pairs of α particles via the reaction

$$\,^1_1\text{H} + \,^7_3\text{Li} \rightarrow \,^4_2\text{He} + \,^4_2\text{He}.$$

These observations were of special interest since the reaction gave one of the earliest quantitative tests of Einstein's mass–energy relation.

A nuclear reaction may proceed from the given initial state xX to any final state, as long as all the relevant conservation laws are obeyed. The total charge and total number of nucleons must be conserved, to balance the values of Z and A as in the two processes cited in the previous paragraph. Energy and momentum must satisfy the familiar conservation laws in all collisions. Every process other than elastic scattering has different total kinetic energies in the initial and final states. Such reactions are called *exoergic* if

$$\sum K_{\text{final}} > \sum K_{\text{initial}},$$

or *endoergic* if

$$\sum K_{\text{final}} < \sum K_{\text{initial}}.$$

A certain minimum, or threshold, energy is needed to initiate the reaction in the latter circumstance. Conservation of the total angular momentum constrains the orbital and nuclear spin quantum numbers of the participating nuclides, and conservation of the overall parity adds further conditions to these constraints. Finally, since charge independence is a valid symmetry of the strong nuclear interaction, a further conservation law is also applied to the total isospin.

The conservation laws lead directly to a number of important kinematical relations. Let us apply these considerations to the general two-body reaction $X(x, y)Y$, taking the target nucleus X to be at rest in the lab frame as in Figure 15-23a. The determining ingredients in the kinematics are the beam kinetic energy K_x and the energy released in the reaction. This latter quantity is also known as the *Q-value*. We express the conservation of the total relativistic energy according to the notation in the figure by writing

$$K_x + M_x c^2 + M_\text{X} c^2 = K_y + M_y c^2 + K_\text{Y} + M_\text{Y} c^2.$$

Figure 15-23

Kinematics of the two-body reaction $x + X \rightarrow Y + y$ in the laboratory frame and in the center of mass frame.

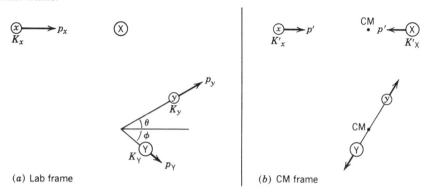

(a) Lab frame | (b) CM frame

We then define the Q-value by the equations

$$Q = (M_x + M_X - M_y - M_Y)c^2$$
$$= K_y + K_Y - K_x. \tag{15-27}$$

Note that the nuclear masses may be replaced by their atomic-mass counterparts in the formula for Q, since the requisite electron masses make no net contribution. Note also that the Q-value for a reaction may be either positive or negative, unlike the strictly positive difference of rest energies required for a decay. According to our remarks in the previous paragraph, we have $Q > 0$ in an exoergic reaction and $Q < 0$ in an endoergic reaction. The remaining $Q = 0$ possibility applies in the special case of elastic scattering.

Equations (15-27) suggest a way to determine an unknown nuclear mass, provided the other three masses in the reaction are known and the final kinetic energies are measured. We concede that the small recoil of the residual nucleus Y is difficult to detect, so that K_Y is usually not measurable. Fortunately, we are able to remove K_Y from consideration by applying conservation of momentum. This constraint on the variables is expressed by the two equations

$$p_x - p_y\cos\theta = p_Y\cos\phi \quad \text{and} \quad p_y\sin\theta = p_Y\sin\phi,$$

using the quantities defined in Figure 15-23a. We assume that the kinetic energies are small compared to the rest energies so that we can employ the nonrelativistic formulas

$$K_x = \frac{p_x^2}{2M_x}, \quad K_y = \frac{p_y^2}{2M_y}, \quad \text{and} \quad K_Y = \frac{p_Y^2}{2M_Y}.$$

Note that relativistic energy conservation must be applied to obtain Equations (15-27) and that approximations to the kinetic energies may then be adopted when the momenta are nonrelativistic. It is possible to eliminate all the kinematical variables of the nucleus Y by straightforward use of the foregoing equations. We quote the following result and leave the derivation to Problem 20 at the end of the chapter:

$$Q = K_y\left(1 + \frac{M_y}{M_Y}\right) - K_x\left(1 - \frac{M_x}{M_Y}\right) - 2\frac{\sqrt{M_x M_y}}{M_Y}\sqrt{K_x K_y}\cos\theta. \quad (15\text{-}28)$$

In application, the mass ratios M_x/M_Y and M_y/M_Y are often replaced, without loss of accuracy, by the corresponding ratios of mass numbers. We can determine Q from experiment by measuring the final energy K_y and angle θ, for a given beam energy K_x. We then consult Equations (15-27) and use the three known mass values to find the unknown fourth mass.

It is instructive to analyze the two-body reaction in the CM coordinate system, where the total momentum is zero and the center of mass is at rest. Figure 15-23b describes the process in terms of the transformed kinematical variables, denoted as primed quantities in the new system. We continue to assume nonrelativistic momenta so that we can pass between the CM and lab frames by means of the nonrelativistic rule for velocity addition:

velocity of x in lab = velocity of x in CM + velocity of CM in lab.

The target X is at rest in the lab; therefore, the center of mass must be moving through the lab with velocity p'/M_X since this quantity must have the same magnitude as the velocity of X in the CM system. Hence, the velocity-addition rule takes the form

$$\frac{p_x}{M_x} = \frac{p'}{M_x} + \frac{p'}{M_X} = p'\frac{M_X + M_x}{M_x M_X}$$

and gives an immediate formula for the momentum of x in the two frames:

$$p_x = p'\frac{M_X + M_x}{M_X}. \quad (15\text{-}29)$$

The colliding particles in the CM system have total kinetic energy

$$K' = K'_x + K'_X = \frac{p'^2}{2M_x} + \frac{p'^2}{2M_X} = \frac{p'^2}{2M_x}\frac{M_X + M_x}{M_X}.$$

We use Equation (15-29) to connect this energy with K_x, the beam kinetic energy in the lab:

$$K_x = \frac{p_x^2}{2M_x} = \frac{p'^2}{2M_x}\left(\frac{M_X + M_x}{M_X}\right)^2 = K'\frac{M_X + M_x}{M_X}. \quad (15\text{-}30)$$

The result shows that K_x exceeds K' and reminds us that a portion of the beam energy is permanently allocated to the motion of the center of mass.

An endoergic reaction has a negative Q-value associated with the mass inequality $M_y + M_Y > M_x + M_X$. The threshold for an endoergic process is defined such that y

and Y are produced *at rest* in the CM frame. This configuration of the final state in Figure 15-23*b* exists when K' assumes its minimum value K'_{th}, as determined from the conserved relativistic energy:

$$K'_{th} + (M_x + M_X)c^2 = (M_y + M_Y)c^2.$$

When we compare with Equations (15-27), we find that the threshold energy in the CM frame is given directly by the excess mass:

$$K'_{th} = -Q \quad (Q < 0). \tag{15-31}$$

We then use Equation (15-30) to obtain the corresponding beam energy:

$$K_{th} = K'_{th} \frac{M_X + M_x}{M_X} = -Q \frac{M_X + M_x}{M_X}. \tag{15-32}$$

This formula for the threshold of the endoergic reaction in the lab satisfies the inequality $K_{th} > K'_{th}$ and again reminds us that part of the beam energy is not available for conversion into rest energy. Thus, it is possible to convert *all* the kinetic energy of the colliding particles into excess mass in the final state *only* in the CM frame.

Finally, let us return to Equation (15-28) and show how the expression can be put to spectroscopic use in reactions of the general type $X(x, y)Y$. We may take the residual nucleus Y to be in its ground state, or any of its excited states. Each higher level has a well-defined excitation energy above the level of the ground state, and so each excited state has a definite mass value M_Y. We let the incident beam have a fixed energy K_x. We then arrange the spectrometer to detect the outgoing particle y at a fixed angle θ, so that we obtain a unique determination of K_y for each possible state of the nuclide Y. Hence, the energy spectrum for the detected particle serves as an image of the energy levels of Y. An example of this form of nuclear spectroscopy is provided by the (d, p) reaction

$$^2_1\text{H} + ^A_Z\text{X} \rightarrow ^{A+1}_Z\text{X} + ^1_1\text{H},$$

in which the excited states of the nucleus ^{A+1}X are mapped out by the distribution of outgoing proton energies. We demonstrate some of the analysis in the following illustration.

Example

Figure 15-24 shows a portion of the energy spectrum of protons produced in the reaction $^{27}\text{Al}(d, p)\,^{28}\text{Al}$. The experiment employs a beam of 2.10 MeV deuterons along with a proton detector set at 90°. The figure also includes a portion of the energy level diagram for the nucleus ^{28}Al. We refer to Equations (15-27) and (15-28), and we take

$$K_x = K_d = 2.10 \text{ MeV}, \quad K_y = K_p, \quad \text{and} \quad \theta = 90°$$

to obtain

$$Q = \left[M(^2\text{H}) + M(^{27}\text{Al}) - M(^1\text{H}) - M(\text{Y}) \right]c^2$$

$$= K_p \left(1 + \frac{M_p}{M_Y} \right) - K_d \left(1 - \frac{M_d}{M_Y} \right),$$

Figure 15-24

Schematic energy spectrum of protons from the reaction $^2\text{H} + {}^{27}\text{Al} \rightarrow {}^{28}\text{Al} + {}^1\text{H}$. The energy levels of ^{28}Al are also shown. Every level is matched to a proton "line" whose intensity is proportional to the excitation probability for the corresponding state in ^{28}Al.

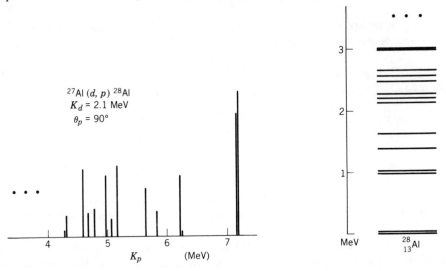

noting that Y may be any ^{28}Al state. Let us solve for K_p so that we can predict values for the detected proton energy:

$$K_p = \frac{Q + K_d(1 - M_d/M_Y)}{1 + M_p/M_Y}.$$

If we choose Y to be the ^{28}Al ground state, we find

$$Q = (2.014102 + 26.981539 - 1.007825 - 27.981913)(931.5 \text{ MeV})$$
$$= 5.499 \text{ MeV},$$

using masses from Appendix A. We then predict the energy for protons detected at 90° to be

$$K_p = \frac{5.50 + 2.10(1 - 2/28)}{1 + 1/28} \text{ MeV} = 7.19 \text{ MeV}.$$

(It is adequate to compute the ratios M_p/M_Y and M_d/M_Y from the mass numbers since K_d is not given very accurately.) Let us also consider an excited state of ^{28}Al at 1.0140 MeV on the energy level diagram. The excitation of this state results in a correspondingly larger value for the rest energy $M(\text{Y})c^2$, and so the Q-value is reduced accordingly:

$$Q = 5.499 \text{ MeV} - 1.014 \text{ MeV} = 4.485 \text{ MeV}.$$

Protons are therefore predicted at 90° with energy

$$K_p = \frac{4.48 + 2.10(1 - 2/28)}{1 + 1/28} \text{ MeV} = 6.21 \text{ MeV}.$$

The two calculated values of K_p appear in the figure as the first and fourth lines at the high end of the proton spectrum.

15-9 Nuclear Reactions and Nuclear Structure

A reaction of the general form $X(x, y)Y$ may yield any two-body system allowed by the conservation laws. Hence, the collision of a beam particle x with a target nucleus X produces a particular Yy final state as a *random* event, subject to the probabilistic principles of quantum mechanics. The probability for the reaction is expressed by means of a *reaction cross section*. This quantity is measurable in a suitably designed experiment and is predictable in a theory of the nuclear reaction. Any theoretical treatment of the nuclear process necessarily includes a model of the nucleus itself. The resulting phenomenology brings experiment and theory together for the interpretation of nuclear structure.

Let us continue to concentrate on two-body final states as we formulate these ideas. The events in a given reaction $x + X \rightarrow Y + y$ are described in principle by a quantum mechanical amplitude $\chi(Yy, xX)$. (We follow the practice adopted in Section 8-7 and use right-to-left notation to designate the amplitude.) The construction of the complex function χ is such that $|\chi|^2$ determines the reaction probability. We cannot be specific about the form of the function without a detailed theory, except to say that χ depends on the beam energy K_x and the angle θ of the detected particle y. We suppose that the detector subtends an infinitesimal solid angle $d\Omega$, as in our discussion of Rutherford scattering in Section 3-4. The reaction cross section $d\sigma$ is then defined by complete analogy with the Rutherford cross section:

$$d\sigma = \frac{\text{\# } y \text{ particles detected in } d\Omega/\text{time}}{\text{\# } x \text{ particles incident on X/area} \cdot \text{time}}.$$

We make an appropriate definition of the amplitude χ so that we can write the differential reaction cross section as

$$\frac{d\sigma}{d\Omega}(Yy, xX) = |\chi(Yy, xX)|^2. \tag{15-33}$$

This quantity is a measurable and calculable function of the variables K_x and θ. The total reaction cross section σ_r is obtained by integrating over all directions of the detected particle:

$$\sigma_r = \sigma(Yy, xX) = \int \frac{d\sigma}{d\Omega}(Yy, xX) \, d\Omega. \tag{15-34}$$

The usual interpretation of the cross section for scattering carries over for reactions as well. The reaction cross section represents the effective target area presented to a beam particle x by a target nucleus X for a specific reaction leading to the final state Yy.

Figure 15-25

Formation and disintegration of the compound nucleus in the reaction $x + X \rightarrow [W] \rightarrow Y + y$.

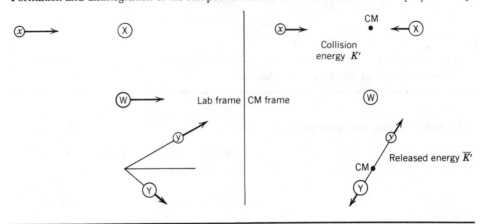

These equations adapt at once to the special case of elastic scattering, with total cross section

$$\sigma_e = \sigma(Xx, xX) = \int \frac{d\sigma}{d\Omega}(Xx, xX)\, d\Omega, \tag{15-35}$$

by taking Y and y to be the same as X and x.

Quantum mechanics makes predictions about the process of interest through the associated amplitude. In principle, we might argue that $\chi(Yy, xX)$ should be obtained by solving the Schrödinger equation to find the wave functions for the initial and final states. In practice, we must acknowledge the difficulty of such a multinucleon problem and turn instead to the use of approximate models. A prospective model is viable in this phenomenological approach if the ideas have intuitive appeal and if the results are subject to direct experimental test.

A useful picture of the interacting nuclear system is provided by the *compound model*, a nuclear reaction mechanism proposed by Bohr in 1936. The basic entity in the model is a temporary composite structure called the *compound nucleus*. This object is supposed to form in the collision of x with X and then decay into the final system consisting of Y and y. We describe the two-stage process of formation and decay by the notation

$$x + X \rightarrow [W] \rightarrow Y + y.$$

We visualize the two stages in the lab and CM frames by the sequence of events shown in Figure 15-25. The model makes three assumptions: the total energy of the colliding system is rapidly distributed throughout the whole compound structure, the compound is rather long-lived relative to the time required for x to traverse the region of nuclear interaction, and the disintegration of the compound is completely independent of the circumstances of formation. These premises are understood to have limits of validity that depend on the energy and mass number of the system.

The model refers to a compound state of the nucleus without specification of the states of the individual nucleons. These constituents are presumed to execute complex

Figure 15-26

Total cross section and absorption cross section of ^{27}Al for incident neutrons. Elastic scattering is the dominant contribution to the total cross section at these energies.

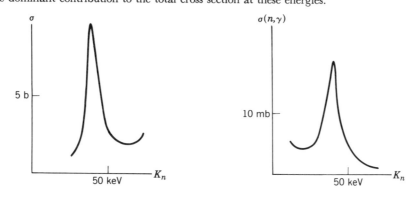

random motions upon collision and share their energies quickly by virtue of their strong short-range interactions. This mode of distribution of the total energy permits the rapid formation of the compound nuclear state and brings about its subsequent disintegration into any one of several possible final channels. The random selection of a decay channel is supposed to occur without memory of the formation state and is assumed to result from the random exchanges of energy among the nucleons. This collective view of interacting constituents in the compound model is fundamentally different from the independent-nucleon theory invoked in Section 14-10 as the basis for the nuclear shell model.

The best evidence for temporary compound nuclear states is the existence of resonances. We observe these phenomena in a given reaction as well-defined enhancements of the cross section at specific values of the energy. A resonant nuclear state represents a system of nucleons whose configuration is especially favorable for the distribution of a definite amount of total energy. We interpret a resonance as a quantum state of the compound nucleus, and we assign the state a nuclear spin and parity according to the experimental evidence. The temporary nature of the state implies a lifetime τ and width Γ satisfying the uncertainty relation

$$\Gamma \tau = \hbar.$$

The lifetime may be smaller than an $E1$ decay time of 10^{-16} s but must be larger than a nuclear traversal time of 10^{-22} s. A resonance is seen as a peak in the cross section, and so the short lifetime of the compound state can be found by measuring the resonance width. Figure 15-26 shows an illustration of a resonance for low-energy neutrons incident on ^{27}Al nuclei. The compound nucleus [^{28}Al] is clearly in evidence, as a resonant state occurs at the same value of the neutron energy in both the scattering and the capture processes ^{27}Al(n, n) ^{27}Al and ^{27}Al(n, γ) ^{28}Al. We discuss the numerical aspects of these graphs in the first example at the end of the section.

A formalism for nuclear reactions has been developed by Wigner and G. Breit. The theory includes an expression, known as the Breit–Wigner formula, for the cross section in the neighborhood of a resonance. Let us examine the construction of this formula from the point of view of the CM frame in Figure 15-25. The initial collision energy and the final released energy are identified in the figure as K' and $\overline{K}'$, respectively. At resonance we relate the two energies to ΔE, the excitation energy of the resonance, by using conservation of energy:

$$K' + M_x c^2 + M_X c^2 = M_W c^2 + \Delta E = \overline{K}' + M_y c^2 + M_Y c^2, \qquad (15\text{-}36)$$

where $M_W c^2$ denotes the rest energy of the compound nucleus in its ground state. Note that the Q-value in Equations (15-27) can be equated to the energy difference $\overline{K}' - K'$ at resonance. We then define the new energy

$$W = K' + (M_x + M_X - M_W)c^2 = \overline{K}' + (M_y + M_Y - M_W)c^2, \qquad (15\text{-}37)$$

and we observe that W is equal to ΔE when the collision is on resonance so that W serves as a *variable* for the description of the cross section *off* resonance. The momentum of the colliding particles determines the de Broglie wavelength

$$\lambda = \frac{h}{p'}.$$

This kinematical quantity varies with K' and therefore depends on W. We use λ to introduce an energy-dependent area

$$\sigma_0(W) = \pi \left(\frac{\lambda}{2\pi} \right)^2, \qquad (15\text{-}38)$$

anticipating the appearance of the area as one of the factors in the cross section. The Breit–Wigner reaction cross section is given in terms of σ_0 by the formula

$$\sigma_r = \sigma(Yy, xX) = \sigma_0(W) \frac{\Gamma_y \Gamma_x}{(W - \Delta E)^2 + (\Gamma/2)^2}. \qquad (15\text{-}39)$$

We note that the second factor achieves a maximum value when the variable passes through resonance at $W = \Delta E$. Thus, the formula describes the W dependence of a resonance peak whose sharpness is controlled by the value of the parameter Γ. A similar result follows for elastic scattering by letting the final state Yy be replaced by the elastic state Xx:

$$\sigma_e = \sigma(Xx, xX) = \sigma_0(W) \frac{\Gamma_x^2}{(W - \Delta E)^2 + (\Gamma/2)^2}. \qquad (15\text{-}40)$$

We have identified the parameter Γ as the width of the resonance. The other quantities Γ_x and Γ_y are called *partial widths* for the two states Xx and Yy. The sum of all such parameters over all possible final states is defined to be equal to Γ, the *total width*. Hence, the ratios Γ_x/Γ and Γ_y/Γ express the probabilities for decay of the resonance into the channels Xx and Yy. These properties of Equations (15-39) and (15-40) give an excellent description of the cross sections near resonance to support the notion of the compound nucleus as an idealized representation of the interacting system.

Another useful picture of the system is provided by the *direct model*, in which the collision process is described as a *direct reaction*. This view of the process holds in energy regimes away from resonances, where the collision energy cannot be so effectively distributed over the whole system. We assume instead that the incident projectile x interacts rapidly within a small region of the target nucleus and detaches the ejected particle y, leaving the product nucleus behind. Examples of direct processes are the (d, p) stripping reaction

$$^2\text{H} + {}^A\text{X} \rightarrow {}^{A+1}\text{X} + {}^1\text{H}$$

Figure 15-27

Energy levels of ^{15}N. The excited state at 12.49 MeV is observed as a resonance for the energies indicated in the reactions $^{14}\text{C}(p, n)^{14}\text{N}$, $^{14}\text{N}(n, n)^{14}\text{N}$, and $^{11}\text{B}(\alpha, n)^{14}\text{N}$.

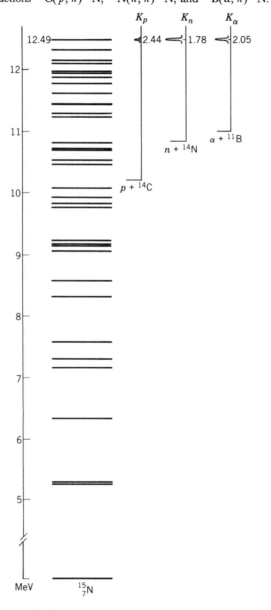

and the (p, d) pickup reaction

$$^1\text{H} + {}^A\text{X} \rightarrow {}^{A-1}\text{X} + {}^2\text{H}.$$

Both of these collisions involve the transfer of a single nucleon between the projectile and the target. Such processes can be analyzed as direct reactions to test the shell model or some other independent-nucleon theory of the nucleus.

It is apparent that the compound and direct models provide alternative contexts for the analysis of nuclear structure in a nuclear reaction. Let us take a last look at resonances and nucleon transfers as the basic alternative mechanisms. We wish to consider a few of the processes pertaining to the ^{15}N system for illustration, so let us display some of the energy levels of this nucleus in Figure 15-27 for reference. We find abundant evidence of excited states in the compound nucleus $[^{15}\text{N}]$ by observing many resonances in the following reactions:

$$\alpha + {}^{11}\text{B} \rightarrow {}^{14}\text{N} + n \quad \text{or} \quad {}^{14}\text{C} + p$$
$$d + {}^{13}\text{C} \rightarrow {}^{14}\text{N} + n \quad \text{or} \quad {}^{14}\text{C} + p \quad \text{or} \quad {}^{11}\text{B} + \alpha$$
$$p + {}^{14}\text{C} \rightarrow {}^{14}\text{N} + n$$
$$n + {}^{14}\text{N} \rightarrow {}^{14}\text{N} + n \quad \text{or} \quad {}^{14}\text{C} + p \quad \text{or} \quad {}^{11}\text{B} + \alpha.$$

The thresholds for the states $p + {}^{14}\text{C}$, $n + {}^{14}\text{N}$, and $\alpha + {}^{11}\text{B}$ are located at the energies shown in the figure, and the resonant states of ^{15}N are found at excitation energies above these levels. We examine some of the numerical details of one of these resonances in the second example below. We also find the transfer of nucleons at work when we use spectrometers to detect protons and neutrons in such ^{15}N-producing reactions as

$$\alpha + {}^{12}\text{C} \rightarrow {}^{15}\text{N} + p,$$
$$d + {}^{14}\text{C} \rightarrow {}^{15}\text{N} + n,$$
$$d + {}^{14}\text{N} \rightarrow {}^{15}\text{N} + p.$$

The low-lying excited levels of the residual ^{15}N nucleus can be determined from the proton and neutron spectra by techniques described at the end of Section 15-8.

Example

The resonance in Figure 15-26 appears in $n + {}^{27}\text{Al}$ collisions at neutron bombarding energy $K_n = 35$ keV and locates a unique excited level of the compound nucleus $[^{28}\text{Al}]$. To inspect the quantitative details, we transform the process to the CM frame as in the right half of Figure 15-25 and identify the nuclei X and W to be ^{27}Al and ^{28}Al. Equation (15-30) gives the transformed collision energy:

$$K' = \frac{M_\text{X}}{M_\text{X} + M_n} K_n = \frac{27}{28}(0.035 \text{ MeV}) = 0.034 \text{ MeV}.$$

The desired quantity is the excitation energy ΔE of the ^{28}Al resonant state. We find ΔE from K' by using conservation of energy in the CM system, as in

Equation (15-36):

$$K' + M_n c^2 + M_X c^2 = M_W c^2 + \Delta E.$$

The resulting excitation energy is found with the aid of the mass table in Appendix A:

$$
\begin{aligned}
\Delta E &= (M_n + M_X - M_W)c^2 + K' \\
&= (1.008665 + 26.981539 - 27.981913)(931.5 \text{ MeV}) + 0.034 \text{ MeV} \\
&= 7.723 \text{ MeV} + 0.034 \text{ MeV} = 7.757 \text{ MeV}.
\end{aligned}
$$

This excited state occurs well up among the energy levels of ^{28}Al. Let us refer to the diagram in Figure 15-24 and note that the $n + {}^{27}$Al threshold is located at the energy $(M_n + M_X - M_W)c^2 = 7.723$ MeV. The resonance at $K_n = 35$ keV appears just beyond this energy, above the range of levels shown in the figure.

Example

An elastic-scattering resonance occurs prominently in the cross section for ^{14}N(n,n) ^{14}N at $K_n = 1.779$ MeV. We wish to find the corresponding excitation energy in ^{15}N, and so we first compute the CM collision energy from Equation (15-30), taking X to be ^{14}N:

$$K' = \frac{M_X}{M_X + M_n} K_n = \frac{14}{15}(1.779 \text{ MeV}) = 1.660 \text{ MeV}.$$

We then turn to Equation (15-36), taking W to be ^{15}N, and again use Appendix A:

$$
\begin{aligned}
\Delta E &= (M_n + M_X - M_W)c^2 + K' \\
&= (1.008665 + 14.003074 - 15.000109)(931.5 \text{ MeV}) + 1.660 \text{ MeV} \\
&= 10.83 \text{ MeV} + 1.66 \text{ MeV} = 12.49 \text{ MeV}.
\end{aligned}
$$

Note that the $n + {}^{14}$N threshold lies at a level 10.83 MeV above the ^{15}N ground state. The same resonance is also observed in the reactions ^{11}B(α, n) ^{14}N and ^{14}C(p, n) ^{14}N. We can determine the beam energies in these processes by reversing the above procedure. The CM collision energies are found by taking X to be ^{11}B in the first case,

$$
\begin{aligned}
K' &= \Delta E - (M_\alpha + M_X - M_W)c^2 \\
&= 12.49 \text{ MeV} - (4.002603 + 11.009305 - 15.000109)(931.5 \text{ MeV}) \\
&= 12.49 \text{ MeV} - 10.99 \text{ MeV} = 1.50 \text{ MeV},
\end{aligned}
$$

and by taking X to be ^{14}C in the second case,

$$
\begin{aligned}
K' &= \Delta E - (M_p + M_X - M_W)c^2 \\
&= 12.49 \text{ MeV} - (1.007825 + 14.003242 - 15.000109)(931.5 \text{ MeV}) \\
&= 12.49 \text{ MeV} - 10.21 \text{ MeV} = 2.28 \text{ MeV}.
\end{aligned}
$$

Note that the $\alpha + {}^{11}B$ and $p + {}^{14}C$ thresholds have excitation levels in the ${}^{15}N$ system at 10.99 and 10.21 MeV, respectively. The beam energies at resonance in the two collisions are

$$K_\alpha = \frac{M_X + M_\alpha}{M_X} K' = \frac{15}{11}(1.50 \text{ MeV}) = 2.05 \text{ MeV}$$

and

$$K_p = \frac{M_X + M_p}{M_X} K' = \frac{15}{14}(2.28 \text{ MeV}) = 2.44 \text{ MeV}.$$

We indicate the resonance schematically for these values of K_n, K_α, and K_p alongside the ${}^{15}N$ level at 12.49 MeV in Figure 15-27.

15-10 Nuclear Fission

The *splitting* of the nucleus was discovered in the 1930s in experiments on the reactions of neutrons with heavy nuclei. Neutrons had come into use as nuclear projectiles, with the capability to probe the structure of the nucleus at low energy. It was found that neutrons could be slowed down to *thermal* velocities by letting the particles undergo elastic collisions with atoms in a moderating medium. It was also learned that thermal neutrons could initiate nuclear reactions with substantial cross sections. Fermi pioneered these studies and was the first to see indications of the fission process. His discovery came in 1934 in the course of investigations of heavy nuclei, especially uranium, under irradiation with slow neutrons. Fermi's original intention was to synthesize the nuclei of unknown elements heavier than uranium. He observed that the radioactive products of these reactions displayed a complex variety of β activities that could not be explained simply in terms of the expected β decays of the heavy nuclides. O. Hahn and F. Strassmann repeated Fermi's experiment in 1938 and proved by radiochemical analysis that the reaction products had the properties of elements in the *middle* of the periodic table. In 1939, L. Meitner and O. R. Frisch interpreted the basic process as a *binary* splitting of the compound nucleus, resulting in a pair of nuclear fragments of intermediate mass. They predicted a large release of energy and offered an explanation for the mechanism based on the liquid-drop model of the nucleus. In the same year their ideas were adopted by Bohr and J. A. Wheeler in the first theory of nuclear fission.

The fission process was found to yield neutrons along with the final nuclear products. This circumstance suggested the possibility of a chain reaction, where the neutrons produced in the fission of one nucleus could induce subsequent fissions in other nuclei in a sample of fissile material. The prospects for the generation of energy in very large quantity were appreciated by many as early as 1939. It was recognized that a chain reaction might be put to practical use as a controlled mechanism in a nuclear reactor, or as a catastrophic device in a nuclear bomb. The first example of a controlled chain reaction was demonstrated under Fermi's direction in 1942, and the first test of a nuclear weapon was performed in 1945.

Neutrons may initiate the fission of a heavy nucleus at incident energies as small as 0.025 eV, corresponding to thermal motion at 0°C. The amount of energy released in a single thermal-neutron fission reaction is around 200 MeV. To get a rough

Figure 15-28

Schematic evolution of the fission process including the ejection of prompt photons and neutrons. The residual nuclei are radioactive and emit secondary β and γ radiation.

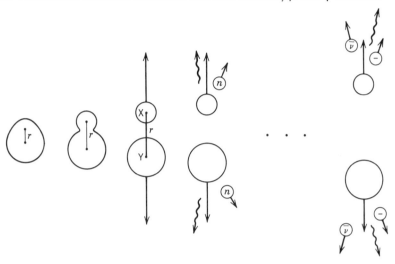

understanding of this result, let us consult Figure 14-12 and compare the binding energies of nuclei at intermediate and large mass numbers. The figure shows that a nucleus has about 1 MeV *more* binding energy per nucleon at medium A than at high A. Consequently, in a system of approximately 200 nucleons, we find the rest energy to be approximately 200 MeV *greater* when the nucleons are found in a single heavy nucleus than when the system is divided into two medium-mass fragments. This excess rest energy determines the total energy released for distribution among all the final products of the fission reaction.

The fissioning heavy nucleus has a neutron-rich composition so that the fission fragments contain many more neutrons than protons. The fragments are highly unstable and deexcite by the prompt emission of γ rays along with the all-important neutrons. A typical fission sequence passes through these early stages according to the scenario shown in Figure 15-28. The instabilities continue after the neutron-ejection stage because the residual nuclear products are still rich in neutrons and are therefore radioactive. The fission products decay to their final stable states by the emission of β and γ radiation, often in chains of sequential transitions. These last phases of deexcitation are also indicated in the figure.

The fission of uranium by thermal neutrons is energetically possible in the naturally occurring isotope ^{235}U. The more common species ^{238}U fissions under neutron bombardment, provided the incident neutrons exceed a threshold energy around 1 MeV. Many of the heavy nuclides have appreciable fission cross sections for thermal neutrons. However, only ^{233}U, ^{235}U, and ^{239}Pu have the combined properties of large cross section and long half-life. These nuclei are the main fissile materials available, either naturally or synthetically, for the generation of energy on a large scale.

The absorption of thermal neutrons by ^{235}U results in the (n, γ) radiative-capture reaction as well as the (n, f) fission process. A single $n + {}^{235}$U collision may lead to

Figure 15-29

Fission yield in percentage versus mass number for the thermal-neutron fission of ^{235}U.

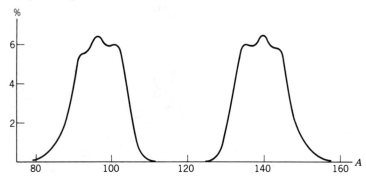

either of these channels in random quantum fashion, with probabilities favoring fission over radiation by an approximate $5:1$ ratio. We can picture this competition of processes in two stages: the formation of the compound nucleus $[^{236}$U$]$ and the immediate disintegration of the intermediate state into either the fission or the radiation channel.

The fission process does not yield a unique pair of fragment nuclides. Instead, we observe a *distribution* of mass-number pairs, comprising a "light group" and a "heavy group," as indicated by the fission-yield graph in Figure 15-29. In the fission of ^{235}U, there is a probability in excess of 1% for the occurrence of any light fragment with A between 85 and 107, together with a corresponding heavy fragment with A between 129 and 151. We note that a symmetric yield into equal-mass fragments is extremely improbable and that the largest probabilities occur for asymmetric pairs in the vicinity of the mass numbers 96 and 140.

A typical fission event might display the following chain of representative nuclear products:

$$n + {}^{235}\text{U} \rightarrow [{}^{236}\text{U}] \rightarrow {}^{95}\text{Sr} + {}^{141}\text{Xe} \rightarrow {}^{94}\text{Sr} + {}^{140}\text{Xe} + 2n.$$

This example illustrates the essential fact that the excess of neutrons in the original heavy nucleus is passed along as a large unstable surplus of neutrons in each nuclear fragment. The instability is partially relieved by the ejection of prompt neutrons, and the remaining neutron excess is eventually stabilized by subsequent β^- decays. Our representative example might exhibit the following series of β transitions:

$$^{94}\text{Sr} \rightarrow {}^{94}\text{Y} \rightarrow {}^{94}\text{Zr(stable)} \quad \text{and} \quad {}^{140}\text{Xe} \rightarrow {}^{140}\text{Cs} \rightarrow {}^{140}\text{Ba} \rightarrow {}^{140}\text{La} \rightarrow {}^{140}\text{Ce(stable)}.$$

Radiative transitions also participate in this cascade of deexcitation processes. The bookkeeping for the total release of energy can be drawn up roughly as follows:

kinetic energy of nuclear products	165 MeV
kinetic energy of fission neutrons	5
energy of prompt γ rays	7
β-decay energy including neutrinos	17
secondary γ-decay energy	6
total fission energy	200 MeV

We note at once that the largest share of the energy is apportioned to the residual nuclei.

A single fission event may yield two or three ejected neutrons. The many different fission channels may contain different numbers of neutrons, and so an average number must be used to identify the neutron yield of a particular fissile nucleus. An average number of neutrons equal to 2.4 is quoted as the yield of ^{235}U.

The binary fission of a compound nucleus is a quantum process in which the fragmentation of the compound and the selection of the pair of fragments proceed at random. The probabilities for the many different pairs are determined by the complex dynamics of the nuclear interactions. Models must be employed to describe these phenomena since no completely satisfactory theory of nuclear fission exists. Our schematic picture in Figure 15-28 suggests a useful model to show how fission might evolve. Let us choose a particular pair of nuclear fragments X and Y and take r to be the variable distance between the two centers. The description begins at $r = 0$ where the fragments coincide in a single spherical compound nucleus. The variable then increases during the evolution of fission in the manner shown in the left half of the figure. Scission occurs over a small interval around the separation point $r = R_X + R_Y$, a distance determined by the radii of the two nuclear fragments. Beyond this interval, the attractive nuclear force is not operative and the nuclei are under the influence of Coulomb repulsion alone.

Let us introduce an r-dependent potential energy of the form shown in Figure 15-30 to represent the regions of attraction and repulsion. The configuration of the compound nucleus is supposed to vary from deformation to fragmentation and beyond as r varies across these regions. The potential energy $V(r)$ can be inserted in the Schrödinger equation, so that a stationary-state eigenfunction can be obtained, to describe the fission of a given compound nucleus into a specific pair of fragments. Note that $V(r)$ has a *fission barrier* much like the Coulomb barrier in α decay. The nucleus disintegrates with a certain decay constant, through the process known as *spontaneous* fission, if the stationary state has an energy level below the top of the barrier. Many of the heavy nuclides, including ^{236}U, exhibit this behavior in their ground state as a less probable alternative to α decay. Thermal neutrons may also cause *induced* fission in the same compound nuclear system, as in the process of main interest

$$n_{\text{thermal}} + {}^{235}\text{U} \rightarrow \left[{}^{236}\text{U}\right] \rightarrow \text{X} + \text{Y}.$$

We identify this state of the system with an excitation energy above the ground state, and we note that the energy level lies *higher* than the fission barrier. Thus, there is a distinction between spontaneous and induced fission associated with the two energy levels shown in the figure.

Figure 15-30

Fission-barrier model describing the deformation and fragmentation of the compound nucleus as a function of fragment separation. The potential energy $V(r)$ pertains to a specific pair of fragment nuclei X and Y.

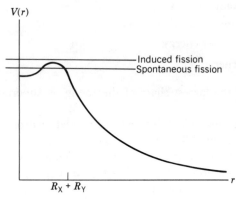

The old fission theory of Bohr and Wheeler is still useful as a qualitative guide to the problem of deformation and fragmentation. Their treatment begins with the liquid-drop model of a spherical nucleus and focuses on the surface energy and the Coulomb energy as the only radius-dependent considerations. The corresponding forces of surface tension and electrostatic repulsion are expected to compete with each other as the excitation of the droplet deforms the original spherical shape. We have expressed the surface and Coulomb contributions to the semiempirical mass formula through the terms

$$a_2 A^{2/3} \quad \text{and} \quad a_3 \frac{Z^2}{A^{1/3}}$$

in Equation (14-13). Let us recall the formula for the nuclear radius,

$$R = R_0 A^{1/3},$$

so that we can identify the surface and electrostatic energies by the formulas

$$V_s = 4\pi R^2 s \quad \text{and} \quad V_e = \frac{3}{5} \frac{Z^2 e^2}{4\pi \varepsilon_0 R}. \tag{15-41}$$

In V_s we introduce a surface energy s per unit area, and in V_e we use a result quoted in Equation (14-11). A small deformation of the spherical droplet implies a small growth in the radius R. We refer to Figures 15-28 and 15-30 and observe that this small deformation corresponds to a small departure of the separation variable r from an equilibrium position at $r = 0$. It is clear from Equations (15-41) that the sum of V_s and V_e may either increase or decrease with a growth in R, since the one term increases as the other decreases. Bohr and Wheeler argue that V_s has the larger role, so that the nucleus *resists* deformation, if the ratio Z^2/A is *less* than a certain critical value. (We compute this number in the first example at the end of this section and obtain a value approximately equal to 50.) This criterion establishes the existence of the stable-equilibrium well shown in the neighborhood of $r = 0$ in Figure 15-30. A fission barrier must then develop as r increases, because the repulsive Coulomb potential energy of the separated fragments eventually becomes the only effect for large r.

This picture accommodates both types of fission, spontaneous and induced, depending on the energy level of the compound nucleus. We have already associated spontaneous fission with the problem of barrier penetration for a level below the top of the barrier. The large masses severely diminish the penetration probability so that spontaneous fission usually competes very unfavorably with α decay. We have also associated induced fission with levels above the barrier. We find that thermal-neutron fission occurs for ^{235}U, but not for ^{238}U, because the thermal-neutron energy level exceeds the barrier for the compound nucleus [^{236}U], but not for the compound nucleus [^{239}U]. Thresholds exist in all fission reactions of the type ^{238}U(n, f) because of this property.

The Bohr–Wheeler theory does not agree with experiment in several important respects. The most glaring defect is its failure to explain the observed asymmetry in the distribution of fragment masses. This question remains to be solved in a satisfactory theory of nuclear fission.

Let us turn from theory to practice, and discuss the fission process in the context of nuclear reactors. The design of a reactor takes advantage of the fact that most of the total fission energy is carried away as kinetic energy by the final nuclear products. These energetic massive nuclei quickly dissipate their energies in collisions with atoms in the fuel element of the reactor. The exchanged energy can be drawn off as heat for the conversion of water into steam, and the steam can be used to turn turbines and produce electricity.

The essential feature of the fission process in the development of a chain reaction is the production of neutrons. These secondary particles are able to sustain the sequence of reactions if each fission, on the average, causes another subsequent fission. The *critical mass* defines the minimum amount of fissile material required for a chain reaction in a given configuration of the material.

The fission sequence may produce an occasional *delayed* neutron, with about 1% probability, instead of the usual prompt variety. The delay occurs because the neutron does not come from a primary fission fragment, as in Figure 15-28, but comes instead from some other nuclear product in the sequence after one of the intervening β decays. The intervening time may amount to a delay of many seconds in the emission of the neutron. This effect provides a means by which the chain reaction in a reactor can be controlled. In practice, the shape of a fissile material is designed to allow room for neutron absorbers (such as boron or cadmium), whose location can be varied to control the determination of the critical mass. It is clear, however, that no such mechanical device is fast enough to regulate the prompt neutrons. The design of the reactor may instead incorporate a configuration of material in which the mass stays subcritical for prompt neutrons alone and achieves criticality when delayed neutrons participate in the process.

Fission reactions induced by thermal neutrons tend to have much larger cross sections than the same reactions initiated by fast neutrons. The reduced cross section for fast neutrons poses a problem for reactor design because the typical fission neutron has an energy in the MeV range. A moderator may be introduced in the fission fuel to solve this problem. The moderating medium contains light atoms (such as deuterium or carbon), which undergo elastic collisions with the neutrons and thereby degrade the neutron energies to thermal levels.

Enrichment of the fissile material presents another problem if naturally occurring uranium is used as the reactor fuel. The fissioning uranium species for thermal neutrons is the rarer nuclide ^{235}U. This isotope can be separated from the more abundant ^{238}U by gaseous diffusion or mass spectrography, since these techniques are sensitive to the small difference in mass between the two nuclei. Another solution is to manufacture the fissile material from an abundant source. The breeder reactor operates on this principle and makes fissionable ^{239}Pu from nonfissionable ^{238}U in the following series of steps:

$$n + {}^{238}\text{U} \rightarrow {}^{239}\text{U} + \gamma$$
$$^{239}\text{U} \rightarrow {}^{239}\text{Np} + e^- + \bar{\nu}$$
$$^{239}\text{Np} \rightarrow {}^{239}\text{Pu} + e^- + \bar{\nu}.$$

The ^{239}Pu nucleus fissions under bombardment by slow *or* fast neutrons, and so the use of a moderator is not essential if plutonium is the reactor fuel.

Figure 15-28 reminds us that the secondary nuclei in the fission sequence are radioactive. Some of these nuclides are long-lived hazardous sources of radiation. The disposition of this waste material is a very serious consideration for the nuclear reactor program.

Example

It is not difficult to show that $[{}^{236}\text{U}]$ should have a fission barrier. We consider the general situation and return to Equations (15-41). These formulas apply to a spherical droplet of radius R, with surface and electrostatic forces in equilibrium. Let us perturb this structure and deduce the critical condition for stability. We consider a small increase in R and find the perturbing forces from the expressions

$$F_s = -\frac{dV_s}{dR} = -8\pi R s \quad \text{and} \quad F_e = -\frac{dV_e}{dR} = \frac{3}{5}\frac{Z^2 e^2}{4\pi\varepsilon_0 R^2}.$$

The directions of these two forces are, respectively, inward and outward, as anticipated. Hence, the equilibrium of the droplet is restorable if the expressions obey the inequality

$$8\pi R s > \frac{3}{5}\frac{Z^2 e^2}{4\pi\varepsilon_0 R^2},$$

or

$$\frac{Z^2}{R^3} < \frac{8\pi s}{\frac{3}{5}e^2/4\pi\varepsilon_0}.$$

Let us use the formula $R = R_0 A^{1/3}$ and rearrange this criterion as follows:

$$\frac{Z^2}{R_0^3 A} < \frac{8\pi s}{\frac{3}{5}e^2/4\pi\varepsilon_0} \quad \Rightarrow \quad \frac{Z^2}{A} < 2\frac{4\pi R_0^2 s}{\frac{3}{5}e^2/4\pi\varepsilon_0 R_0}.$$

We can obtain a value for the combination of constants by recalling the connection between Equations (15-41) and the corresponding terms in the mass formula:

$$4\pi R^2 s = 4\pi R_0^2 s A^{2/3} = a_2 A^{2/3}$$

and

$$\frac{3}{5}\frac{Z^2 e^2}{4\pi\varepsilon_0 R} = \frac{3}{5}\frac{e^2}{4\pi\varepsilon_0 R_0}\frac{Z^2}{A^{1/3}} = a_3\frac{Z^2}{A^{1/3}}.$$

We express the constants as

$$4\pi R_0^2 s = a_2 \quad \text{and} \quad \frac{3}{5}\frac{e^2}{4\pi\varepsilon_0 R_0} = a_3,$$

and we thereby deduce the critical maximum value for Z^2/A:

$$\left(\frac{Z^2}{A}\right)_c = 2\frac{a_2}{a_3} = 2\frac{17.81 \text{ MeV}}{0.7105 \text{ MeV}} = 50.13,$$

using the values quoted for a_2 and a_3 in Section 14-5. The Bohr–Wheeler argument tells us that the compound nucleus [$_Z^A$W] has a stable equilibrium configuration and, hence, a fission barrier if Z^2/A is less than the critical value. The choice of numbers $Z = 92$ and $A = 236$ gives

$$\frac{Z^2}{A} = \frac{(92)^2}{236} = 35.9,$$

so that [$_{92}^{236}$U] meets the desired criterion.

Example

Let us associate the two energy levels in Figure 15-30 with the ground state of ^{236}U and the excitation state of the system $n_{\text{thermal}} + {}^{235}$U. We can compute the energies of interest with the aid of a table of masses. (We turn again to Appendix A and appeal to the original sources of Appendix A as well.) First, consider the neutron separation energy for the ground state of ^{236}U:

$$\left[M_n + M({}^{235}\text{U}) - M({}^{236}\text{U})\right]c^2$$

$$= \left(1.008665 + 235.043928 - 236.045566\right)(931.5 \text{ MeV})$$

$$= 6.546 \text{ MeV}.$$

We take this result to represent the difference in energy between the levels in the figure, and we thus neglect the very small kinetic energy of the thermal neutron. (Note that the curve in the figure plays no role in this exercise. Remember, the potential energy $V(r)$ refers to the pair of interacting fragments X and Y.) Next, let us focus on the particular fission channel

$$n + {}^{235}\text{U} \rightarrow {}^{94}\text{Sr} + {}^{140}\text{Xe} + 2n$$

and calculate the Q-value for a reaction leaving the residual nuclei in their ground states:

$$
\begin{aligned}
Q &= \left[M_n + M({}^{235}\text{U}) - M({}^{94}\text{Sr}) - M({}^{140}\text{Xe}) - 2M_n \right] c^2 \\
&= (235.043928 - 93.91523 - 139.92144 - 1.008665)(931.5 \text{ MeV}) \\
&= 185.0 \text{ MeV}.
\end{aligned}
$$

This calculation determines the total amount of energy released in the form of energetic residual nuclei, fast neutrons, and prompt γ rays. The ground states of ${}^{94}\text{Sr}$ and ${}^{140}\text{Xe}$ are β unstable, and so the subsequent decay chains are further sources of released energy. A final quantity of interest is the following ratio of released energy and fissile rest energy:

$$\frac{Q}{M({}^{235}\text{U})c^2} = \frac{185 \text{ MeV}}{(235.0)(931.5 \text{ MeV})} = 8.45 \times 10^{-4}.$$

The result tells us that the prompt release of energy in this particular channel amounts to a mere 0.08% conversion of mass to energy in the fission of a single ${}^{235}\text{U}$ nucleus.

15-11 Fusion and Thermonuclear Energy

Nuclear fusion occurs when light nuclei combine in a reaction to form a more massive nucleus in the final state. Nuclear binding energies are such that the fusion of a typical pair of nuclei at low mass number results in the release of a large quantity of energy. These reactions are of great interest because of their capability to generate useful energy on a very large scale. The process has fundamental significance in astrophysics since fusion is the main contributor to the energy produced in stars. The process also has practical importance as a potential source of controlled nuclear energy and as a proven mechanism for a nuclear weapon.

We have employed Figure 14-12 in Section 15-10 to explain how fission causes the release of energy from nuclei at high mass number. It is apparent that a similar argument can be applied at the other end of the figure to explain why energy is also liberated by the fusion of low-mass nuclei. In either case the reaction products lie toward the intermediate range of mass numbers where the nucleons have greater binding energy. Thus, both fission and fusion take advantage of an excess in rest energy to produce a large amount of released energy.

An idealized example of a fusion reaction is provided by the (very improbable) collision of two protons and two neutrons to yield an α particle and a prompt γ ray. Figure 14-12 tells us that each nucleon in ${}^4\text{He}$ has about 7 MeV of binding energy, so that the hypothetical process releases four of these units, approximately 28 MeV, to the final $\alpha\gamma$ system. We note that an exceptionally large amount of energy is generated when an α particle is produced because of the especially tight binding of the ${}^4\text{He}$ nucleus.

The more realistic fusion reactions occur in binary collisions. A good example is deuteron–triton fusion

$$d + t \rightarrow \alpha + n,$$

in which the formation of the α particle again results in a large release of energy. This fusion reaction proceeds rapidly, via the strong nuclear interaction, and generates an energy of 17.6 MeV. Let us describe the nuclear process by writing

$$^2\mathrm{H} + {}^3\mathrm{H} \rightarrow {}^4\mathrm{He} + n + 17.6\,\mathrm{MeV},$$

and let us also consider the other deuteron-induced processes

$$^2\mathrm{H} + {}^2\mathrm{H} \rightarrow \begin{cases} ^3\mathrm{He} + n + 3.3\,\mathrm{MeV} \\ ^3\mathrm{H} + {}^1\mathrm{H} + 4.0\,\mathrm{MeV} \end{cases}$$

and

$$^2\mathrm{H} + {}^3\mathrm{He} \rightarrow {}^4\mathrm{He} + {}^1\mathrm{H} + 18.3\,\mathrm{MeV}.$$

This collection of four reactions can conceivably operate together as a *combined reaction*, in which the nuclei $^3\mathrm{H}$ and $^3\mathrm{He}$ serve as catalysts for the basic process of ternary deuteron fusion

$$3d \rightarrow \alpha + n + p.$$

A net Q-value of 21.6 MeV is obtained in the overall release of energy. To confirm this result, let us sum the four reactions, cancel the catalysts $^3\mathrm{H}$ and $^3\mathrm{He}$, and divide by 2 as follows:

$$6\left(^2\mathrm{H}\right) + {}^3\mathrm{H} + {}^3\mathrm{He} \rightarrow 2\left(^4\mathrm{He} + n + {}^1\mathrm{H}\right) + {}^3\mathrm{He} + {}^3\mathrm{H} + 43.2\,\mathrm{MeV}$$

$$\Rightarrow 3\left(^2\mathrm{H}\right) \rightarrow {}^4\mathrm{He} + n + {}^1\mathrm{H} + 21.6\,\mathrm{MeV}.$$

The practical implementation of these combined processes is among the goals of the current research program in controlled fusion energy. The fusion reaction generates less energy than a typical fission process; however, the yield of energy per unit mass of nuclear fuel is greater in the fusion reaction by more than a factor of 3. Further advantages in favor of fusion are the nonoccurrence of direct radioactive by-products and the relatively low cost of fusible fuel.

Every fusion reaction involves a collision of particles with positive charge. The collision energies must therefore be large enough to overcome the effects of Coulomb repulsion. Let us illustrate this fact with the aid of Figure 15-31 by considering dt fusion, $d + t \rightarrow \alpha + n$. Since the reaction is exoergic, an appreciable nuclear interaction might be expected even for a dt collision at rest. The figure tells us instead that the reaction cross section is severely affected by the Coulomb barrier for low deuteron bombarding energies and that the maximum cross section is finally reached around 100 keV. We can undertake a fusion experiment in a regime of energy favorable for the reaction, simply by accelerating a beam of particles onto a target. Unfortunately, such a method is not suitable for the generation of useful energy because the energy released in the reaction is largely dissipated in the ionization of the target. In principle, this problem can be circumvented by the use of an ionized target. In

Figure 15-31

Cross section for the reaction $^2\text{H} + ^3\text{H} \rightarrow \,^4\text{He} + n$ versus deuteron beam energy.

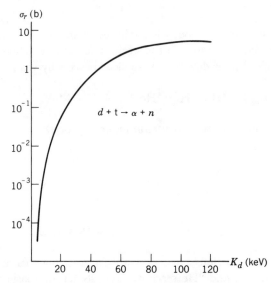

practice, fusion is rather improbable under these conditions, as elastic *dt* scattering is the dominant process at these energies.

We can achieve fusion on a large scale if we can somehow confine the interacting nuclei to a small region in which many collisions can occur with sufficient energy. These conditions of confinement are satisfied in a *thermal* medium, where an average particle energy around the desired value $k_B T = 100$ keV corresponds to a temperature around 10^9 K. (Fusion may also occur in the medium if the temperature is below this range, provided there are enough interacting nuclei available in the high-energy tail of the thermal distribution.) The process is said to be *thermonuclear* because of this association between fusion and thermal motion at high temperature.

The confined system of particles fits the description of a fully ionized gaseous medium known as a *plasma*. Matter in the plasma state consists of coexisting gases of nuclei and electrons, the debris of atoms ionized by collisions at high temperature. Neutral matter on Earth can be maintained in this phase only under extraordinary laboratory conditions. The plasma state exists much more commonly in the high-temperature interiors of stars, because of the containment provided by the force of gravitational attraction.

The origin of the energy emitted by stars was one of the first great mysteries of astrophysics. The problem was originally put in terms of the Sun, a modest star of well-known size, mass, and temperature. In time, other information about the Sun, such as composition, gas density, interior temperature, age, and radiation rate, also became rather well known. It was apparent that a chemical reaction could not account for all these properties. It was especially obvious that a new mechanism was needed to explain the generation of a very large quantity of energy. H. A. Bethe found a solution to the problem, for all the normal stars, in 1938. He proposed the existence of *cycles* of thermonuclear processes in which the basic effect was the fusion of four protons to make an α particle. The resulting *proton chain* was supposed to cause

Figure 15-32

Reactions in the proton cycle. The three processes execute the basic proton fusion chain
$4p \rightarrow \alpha + 2e^+ + 2\nu + 2\gamma$.

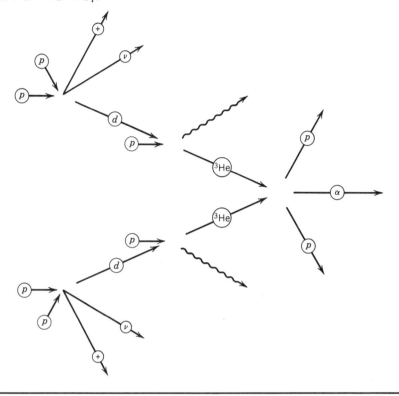

the "burning" of hydrogen in the formation of helium, an example of the *synthesis* of
nuclei by the processes of thermonuclear fusion in stars. The evaluation of the various
reaction rates predicted two specific cycles whose relative influence was expected to
depend on the interior temperature of a given star.

Hydrogen burning proceeds via the *proton cycle* in stars of the main sequence, for
masses up to one solar mass and for temperatures in the range $(0.8-1.5) \times 10^7$ K.
The proton cycle consists of the following reactions:

$$^1\text{H} + {}^1\text{H} \rightarrow {}^2\text{H} + e^+ + \nu,$$
$$^1\text{H} + {}^2\text{H} \rightarrow {}^3\text{He} + \gamma,$$
$$^3\text{He} + {}^3\text{He} \rightarrow {}^4\text{He} + 2({}^1\text{H}).$$

Figure 15-32 shows how this sequence of processes takes four protons through three
successive reaction stages to yield one α particle along with liberated energy in the
form of positrons, neutrinos, and γ rays. We note that two copies of the first two
processes are needed to carry out the cycle, and we find that an energy of 26.7 MeV is
released as the net result. This amount includes the γ rays produced when the
positrons annihilate with electrons in the stellar plasma. A 2% share of the released
energy is carried away by the emitted neutrinos and is not observed in the luminosity

of the star. We note that the cycle starts at a very slow rate since the first process is governed by the weak interaction. In effect, the first reaction is equivalent to the β^+ transition $p \to n$ in the presence of a spectator proton. Nevertheless, the generation of energy proceeds apace in stars like the Sun because of the ample abundance of hydrogen.

Protons are able to penetrate the Coulomb barrier and interact with larger nuclei when the temperature exceeds 2×10^7 K, a value beyond that found in the interior of the Sun. The *carbon cycle* is the main mechanism for stellar energy if the interior temperature of the star is in this regime. The cycle begins with protons incident on carbon and proceeds through the following steps:

$$^1\text{H} + {}^{12}\text{C} \to {}^{13}\text{N} + \gamma$$
$$^{13}\text{N} \to {}^{13}\text{C} + e^+ + \nu$$
$$^1\text{H} + {}^{13}\text{C} \to {}^{14}\text{N} + \gamma$$
$$^1\text{H} + {}^{14}\text{N} \to {}^{15}\text{O} + \gamma$$
$$^{15}\text{O} \to {}^{15}\text{N} + e^+ + \nu$$
$$^1\text{H} + {}^{15}\text{N} \to {}^{12}\text{C} + {}^4\text{He}.$$

We note that once again four protons are fused in stages to make one α particle. The other essential property of this cycle is the remarkable fact that ^{12}C is regenerated, and not consumed, during the reaction sequence. Thus, carbon participates as a catalyst, so that no large abundance of carbon is needed to sustain the hydrogen-burning chain. The overall reaction rate is much greater, while the loss of released energy to neutrinos is only somewhat larger, in the carbon cycle than in the proton cycle. Both of these cycles contribute, although the latter one dominates, in the production of solar energy.

A star may exhaust its supply of hydrogen and contract, to reach temperatures in excess of 10^8 K. Helium burning begins in this range of greater temperature, as the helium nuclei become able to penetrate the higher Coulomb barrier and fuse together in the formation of carbon. We find that the net result of the cycle of reactions is the ternary fusion process

$$3({}^4\text{He}) \to {}^{12}\text{C},$$

and we note that nucleosynthesis is well underway at this stage. Newly formed carbon acts as a catalyst in the carbon cycle for these temperatures, and carbon fusion also sets in as the ^{12}C composition of the star increases. Thus, heavier nuclei appear in a succession of fusions performed by the (α, γ) reactions

$$^4\text{He} + {}^{12}\text{C} \to {}^{16}\text{O} + \gamma,$$
$$^4\text{He} + {}^{16}\text{O} \to {}^{20}\text{Ne} + \gamma,$$
$$^4\text{He} + {}^{20}\text{Ne} \to {}^{24}\text{Mg} + \gamma.$$

At 10^9 K and beyond, carbon burning and oxygen burning commence in the reactions

$$^{12}\text{C} + {}^{12}\text{C} \to {}^{20}\text{Ne} + {}^4\text{He},$$
$$^{16}\text{O} + {}^{16}\text{O} \to {}^{28}\text{Si} + {}^4\text{He}.$$

The synthesis of nuclei continues via the fusion process until the top of the curve in Figure 14-12 is reached with the formation of ^{56}Fe.

The building up of nuclei beyond iron takes place through a series of neutron-capture reactions and occasional β decays. These stages of nucleosynthesis depend critically on the availability of neutrons in the stellar plasma. Neutron captures proceed slowly, allowing time for the decay of unstable nuclei, if the number of accessible neutrons is small. These slowly progressing syntheses come to an end at the last stable nucleus ^{209}Bi. Very large numbers of neutrons may also become available in rarer situations. A rapid succession of neutron captures can then occur, with no intervening β decays, to carry the formation of heavier nuclei to the highest possible mass numbers. The existence in nature of heavy elements like thorium and uranium is supposed to be due to such a mechanism.

We have noted that the proton cycle generates energy in stars quite slowly since the weak interaction controls the first step in the cycle. Laboratory fusion reactions must rely instead on the strong interactions of the nuclear particles. Candidate processes include the dt fusion reaction $d + t \rightarrow \alpha + n$, and the deuteron fusion chain $3d \rightarrow \alpha + n + p$. It is clear that the extraordinary conditions of the plasma phase must be established in the laboratory if fusion is to be useful as a source of controlled thermonuclear energy. Containment of the plasma for indefinite times is accomplished in stars by the force of gravity. Terrestrial facilities do not possess this capability and must depend on magnetic and inertial schemes of confinement in order to maintain the interacting medium at sufficiently high temperature and density. In magnetic confinement, the hot ionized medium is insulated from the colder walls of the gas chamber by specially designed configurations of magnetic fields. The main technological problems in this scheme are to achieve a combination of fusion conditions and containment times and to preserve a balance between the energy released for use and the energy stored for maintenance of the plasma. Inertial confinement is based on the employment of laser beams to induce fusion. The laser light is arranged to bombard a small pellet of deuterium and tritium from all directions at once, so that the resulting implosion can ionize and compress the vaporized material to the temperatures and densities needed for fusion. These techniques are currently under investigation, and the prospects for developing controlled thermonuclear fusion are hopeful.

Example

We can appreciate a few of the advantages and problems of nuclear fusion by calculating some numbers. Let us consider the dt fusion reaction, $^2\mathrm{H} + {}^3\mathrm{H} \rightarrow {}^4\mathrm{He} + n$ and begin with the Q-value:

$$Q = (2.014102 + 3.016049 - 4.002603 - 1.008665)(931.5 \text{ MeV})$$
$$= 17.59 \text{ MeV}.$$

The Coulomb barrier may be estimated by using the formula $R = R_0 A^{1/3}$ to find the center-to-center distance for $^2\mathrm{H}$ and $^3\mathrm{H}$ in contact,

$$r = (1.2 \text{ fm})(2^{1/3} + 3^{1/3}) = 3.2 \text{ fm},$$

and by computing the Coulomb potential energy at the distance r:

$$V(r) = Z_2 Z_3 \frac{e^2}{4\pi\varepsilon_0 r} = (1)(1)\frac{1.44 \text{ MeV} \cdot \text{fm}}{3.2 \text{ fm}} = 0.45 \text{ MeV}.$$

Actually, the collision energy does not need to be as large as this to give an appreciable fusion cross section, as Figure 15-31 indicates. Let us ignore the collision energy by taking the initial dt system to be fused at rest, and let us construct a variant of Equation (15-10), so that we can predict unique energies for the final α particle and neutron:

$$Q = K_\alpha + K_n = K_n\left(1 + \frac{M_n}{M_\alpha}\right)$$

$$\Rightarrow K_n = \frac{17.6 \text{ MeV}}{1 + 1/4} = 14.1 \text{ MeV} \quad \text{and} \quad K_\alpha = 3.5 \text{ MeV}.$$

Such energetic neutrons would be able to escape from the plasma. A shroud of lithium could be provided to recover the energy and regenerate ^{3}H through the exoergic capture reaction

$$n + {}^6\text{Li} \rightarrow {}^4\text{He} + {}^3\text{H}.$$

Escaping neutrons would also create other problems, since their interactions outside the plasma would induce radioactivity and cause radiation damage in the surrounding structures.

Example

The reactions of the proton cycle generate the proton fusion chain

$$4p \rightarrow \alpha + 2e^+ + 2\nu + 2\gamma$$

in the manner illustrated in Figure 15-32. The released energy is given by the difference of rest energies $(4M_p - M_\alpha - 2m_e)c^2$. Let us reexpress the basic process by introducing four electrons on both sides:

$$4(p + e^-) \rightarrow (\alpha + 2e^-) + 2(e^- + e^+) + 2\nu + 2\gamma.$$

This maneuver enables us to account for the subsequent annihilation of the emitted positrons via the familiar reaction

$$e^- + e^+ \rightarrow 2\gamma.$$

We can now identify a total release of energy Q in the form of neutrinos and γ rays (including those from $e^- e^+$ annihilation), and also allow for α-particle recoil, by introducing atomic masses and writing the simple expression

$$4M(^1\text{H})c^2 = M(^4\text{He})c^2 + Q.$$

The calculation of this quantity gives

$$Q = \left[4M(^1\text{H}) - M(^4\text{He})\right]c^2 = [4(1.007825) - 4.002603](931.5 \text{ MeV})$$
$$= 26.73 \text{ MeV}$$

or

$$\frac{26.7 \text{ MeV}}{4} = 6.68 \text{ MeV per proton in a single fusion.}$$

The proton chain supplies the energy radiated by the Sun, with a rate of radiation known to be 4.0×10^{26} W. Let us use this number to compute the rate of proton consumption by fusion in the Sun:

$$\frac{4.0 \times 10^{26} \text{J}/s}{(6.68 \text{ MeV}/\text{proton})(1.60 \times 10^{-13}\text{J}/\text{MeV})} = 3.7 \times 10^{38} \text{ protons}/s.$$

We can estimate the number of available protons in the Sun by supposing that protons constitute the entire 2.0×10^{30} kg solar mass:

$$\frac{2.0 \times 10^{30} \text{ kg}}{1.67 \times 10^{-27} \text{ kg}/\text{proton}} = 1.2 \times 10^{57} \text{ protons.}$$

The Sun can then be expected to exhaust its supply of protons in a time estimated as

$$\frac{1.2 \times 10^{57}}{3.7 \times 10^{38} \text{ s}^{-1}} = 3.2 \times 10^{18} \text{ s.}$$

It is comforting to know that the prediction exceeds 10^{11} years.

Problems

1. Positron-emitting radionuclides are produced artificially in (α, n) reactions with ^{10}B and ^{27}Al targets. Identify the radioactive nucleus and the corresponding nuclear decay product for each of these reaction-and-decay processes.

2. The $4n$ series begins with the natural radionuclide ^{232}Th and ends with the stable nuclide ^{208}Pb. The chain of decays proceeds through the following steps:

$$\alpha, \beta^-, \beta^-, \alpha, \alpha, \alpha, \alpha, \beta^-, \alpha \text{ and } \beta^-, \beta^- \text{ and } \alpha,$$

where a branching of α and β decay modes occurs toward the end of the series. Draw up a partial chart of the nuclides and make a plot of this decay chain, identifying each of the nuclides in the series.

3. Derive the formulas

$$\tau = \frac{1}{\gamma} \quad \text{and} \quad \tau_{1/2} = \ln 2 \cdot \tau$$

relating the mean life τ, the decay constant γ, and the half-life $\tau_{1/2}$ for the exponential decay of a radioactive nucleus.

4. Calculate the mass of a 100 Ci ^{90}Sr source.

5. The production and decay of radioactive ^{14}C are in equilibrium in the Earth's environment. Let the equilibrium mass ratio for ^{14}C relative to ordinary ^{12}C be 1.5×10^{-12} in

organic matter, and show that the rate of ^{14}C decay in a live sample is 1 disintegration/s for every 4 g of carbon.

6. Let the sequential decay scheme in Figure 15-5 have decay constants such that $\gamma_{12} \gg \gamma_{23}$, and assume vanishing initial populations for X_2 and X_3. Simplify the rate equations in this situation, and determine the populations $N_1(t)$, $N_2(t)$, and $N_3(t)$ for $0 \leq t \ll 1/\gamma_{23}$. Sketch a graph for each of these solutions.

7. Let the initial populations for the two-step cascade in Figure 15-5 be such that $N_2 = N_3$ $= 0$ at $t = 0$. Obtain the corresponding solutions to the rate equations, valid for any choice of the parameters γ_{12} and γ_{23}. Determine the form of the solutions in the special case $\gamma_{12} \gg \gamma_{23}$, for times much less than $1/\gamma_{23}$. These limiting results should agree with those from Problem 6.

8. The energies of α particles are determined by means of a magnetic spectrograph, shown in Figure 15-7, in which an emitted α particle moves in a semicircular orbit of radius R because of the application of a magnetic field B. The product BR (the so-called magnetic rigidity) is found to be 0.3316 T · m and 0.3149 T · m for the α sources ^{210}Po and ^{226}Ra, respectively. Use these data to compute the corresponding α-particle energies. Take the mass of the α particle to be 6.645×10^{-27} kg, and observe that a nonrelativistic calculation is valid.

9. Use data from Appendix A to compute the kinetic energies of α particles emitted in the decays of ^{210}Po and ^{235}U. The results should agree with information plotted in Figure 15-8.

10. ^{223}Ra is a known α emitter. Recently, this nucleus has been observed to undergo another more exotic mode of radioactive decay in which ^{14}C fragments are emitted instead of α particles. The two decay modes are ^{223}Ra $\rightarrow$ ^{219}Rn $+$ ^{4}He and ^{223}Ra $\rightarrow$ ^{209}Pb $+$ ^{14}C, where the α decay dominates by a factor of order 10^9. Calculate the kinetic energy of the emitted fragment in each case, using $M(^{209}$Pb$) = 208.981080$ u and $M(^{219}$Rn$) =$ 219.009485 u along with other mass data obtainable from Appendix A. Estimate the height of the Coulomb barrier for each of the decays.

11. The nuclide ^{32}P is β^- active. Identify the decay process and use masses given in Appendix A to compute the Q-value.

12. The atomic mass of ^{41}Ca is 40.962278 u. Show that ^{41}Ca is unstable with respect to electron capture and compute the corresponding disintegration energy. Is β^+ decay possible for this nuclide?

13. The isobars $^{36}_{16}$S, $^{36}_{17}$Cl and $^{36}_{18}$Ar have atomic masses 35.967079, 35.968307 and 35.967546 u. Identify all the β instabilities in this system and calculate the corresponding Q-values.

14. Isobar masses at $A = 64$ are:

$^{64}_{27}$Co	$^{64}_{28}$Ni	$^{64}_{29}$Cu	$^{64}_{30}$Zn	$^{64}_{31}$Ga
63.935812 u	63.927968 u	63.929766 u	63.929146 u	63.936837 u

Identify every possible β transition in this system and compute the Q-value for each process.

15. The nuclide ^{40}K is β unstable. Identify all possible β processes and compute the corresponding disintegration energies. The relevant atomic masses are listed in Appendix A.

16. Refer to Figure 15-18 and classify the three indicated transitions according to the selection rules for β and γ decay.

17. Predict the radiation multipole for each of the γ-ray transitions shown in the figure.

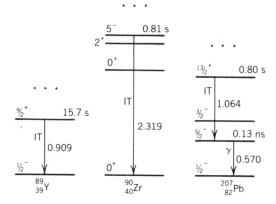

18. Refer to the photon absorption process in Figure 15-21, and show that the incident photon energy ε_a must exceed the transition energy ΔE by the amount $(\Delta E)^2 / 2Mc^2$.

19. A 14.4 keV γ ray is emitted in a transition to the ground state in ^{57}Fe. The half-life of the excited state is 98 ns. Calculate the natural width of the excited state and the recoil kinetic energy of the nucleus. Determine the relative velocity required between emitter and absorber to meet the criterion for resonance radiation.

20. The Q-value for the two-body reaction in Figure 15-23 can be expressed in terms of measurable kinematical quantities by the equation

$$Q = K_y \left(1 + \frac{M_y}{M_Y} \right) - K_x \left(1 - \frac{M_x}{M_Y} \right) - 2 \frac{\sqrt{M_x M_y}}{M_Y} \sqrt{K_x K_y} \cos \theta,$$

in which mass numbers may be substituted for the various nuclear masses. Derive this result using nonrelativistic kinetic-energy formulas.

21. Consider a present-day version of Rutherford's (α, p) reaction:

$$^4\mathrm{He} + {}^{14}\mathrm{N} \rightarrow {}^{17}\mathrm{O} + {}^1\mathrm{H}.$$

Take the incident α particles (from ^{214}Po) to have energy 7.69 MeV, and let the outgoing protons be detected at $\theta = 90°$. Calculate the Q-value using mass data from Appendix A, and predict the kinetic energy of the observed protons. Determine the threshold energy required for accelerator-produced α particles in the same endoergic reaction.

22. The energy levels in ^{46}Sc can be deduced from the energy spectrum of outgoing protons, detected at fixed angle, in the reaction ^{45}Sc(d, p) ^{46}Sc. Let 6.974 MeV deuterons be incident on a ^{45}Sc target, and consider proton detection at 37.5°. Predict the energy of the protons for the case where ^{46}Sc is left in the ground state and for the case where ^{46}Sc is left in an excited state at excitation energy 0.4441 MeV. The atomic mass of ^{46}Sc is 45.955174 u, and the other relevant masses are listed in Appendix A.

23. Identify the compound nucleus in each of the following reactions: ^{10}B $(\alpha, p)^{13}$C, ^{23}Na $(\alpha, p)^{26}$Mg, ^{32}S $(\alpha, p)^{35}$Cl, ^{65}Cu $(p, n)^{65}$Zn, ^{27}Al $(d, \alpha)^{25}$Mg, and ^{31}P $(d, p)^{32}$P. Which reactions are exoergic, and which are endoergic?

24. The Breit–Wigner cross section near resonance is

$$\sigma_r = \sigma_0(W) \frac{\Gamma_y \Gamma_x}{(W - \Delta E)^2 + (\Gamma/2)^2}$$

for the reaction $x + X \rightarrow Y + y$. Take σ_0 to be essentially constant across the resonance, and show that Γ gives the full width of the resonance peak at half-maximum, as indicated in the drawing. Refer to the graphs in Figure 15-26 and estimate the full width Γ and the partial widths Γ_n and Γ_γ, using the fact that, at the energies given, $^{27}\mathrm{Al} + n$ and $^{28}\mathrm{Al} + \gamma$ are the only channels open for reactions initiated by $n + ^{27}\mathrm{Al}$ collisions.

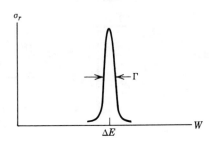

25. Resonances are observed in the reaction $^{27}\mathrm{Al}(\alpha, p)^{30}\mathrm{Si}$ at α-particle bombarding energies 3.95, 4.84, and 6.57 MeV. Compute the excitation energy for each of the corresponding levels of the compound nucleus.

26. A beam of α particles is incident on a $^9\mathrm{Be}$ target, and a resonance is seen at 1.732 MeV beam energy. Calculate the excitation energy of the corresponding state of the compound nucleus. The same resonant state occurs in neutron collisions with a $^{12}\mathrm{C}$ target. Compute the value of the neutron beam energy at the position of the resonance.

27. Use the Maxwell–Boltzmann distribution of particle kinetic energies to deduce a velocity distribution function for thermal neutrons. (The second example in Section 2-3 is a useful place to start.) Determine the temperature dependence of the most probable velocity, and show that the corresponding neutron kinetic energy is equal to $k_B T$. Calculate the value of the energy and the velocity for neutrons at $0°\mathrm{C}$.

28. Assume that the energy released in the fission of a single $^{235}\mathrm{U}$ nucleus is 200 MeV, and compute the total amount of energy obtained from a 1 g sample of $^{235}\mathrm{U}$. Convert the answer to an energy in megawatt-hours.

29. The fission of $^{239}\mathrm{Pu}$ by thermal neutrons yields

$$^{104}\mathrm{Mo} + ^{133}\mathrm{Te} + 3n \quad \text{and} \quad ^{94}\mathrm{Sr} + ^{143}\mathrm{Ba} + 3n,$$

as well as many other possible sets of nuclear products. Calculate the prompt energy release in the two instances, given the masses (in u) 93.91523, 103.91358, 132.91097, and 142.92055 for the Sr, Mo, Te, and Ba nuclides.

30. Consider the symmetric fission of the compound nucleus $\left[^A_Z\mathrm{W}\right]$ into a pair of like nuclei. Deduce an expression for the height of the Coulomb barrier in the form $CZ^2/A^{1/3}$, and calculate a value for the constant C.

31. Calculate the energies released in the fusion reactions $^2\mathrm{H} + ^6\mathrm{Li} \rightarrow 2(^4\mathrm{He})$ and $^1\mathrm{H} + ^{11}\mathrm{B} \rightarrow 3(^4\mathrm{He})$.

32. Compute the energy released in each of the four stages of the cycle that make up the ternary deuteron fusion process $3d \rightarrow \alpha + n + p$, and confirm the result quoted in the text for the net Q-value.

33. Determine the energy released at each stage of the carbon cycle. Relevant masses not given in Appendix A are $M(^{13}\mathrm{N}) = 13.005739$ u and $M(^{15}\mathrm{O}) = 15.003065$ u.

SIXTEEN

ELEMENTARY
PARTICLES

*T*he search for the basic components of matter has been the most enduring of all scientific endeavors. The investigation of this idea has always turned on questions of *scale*, since the operative measure of smallness has usually been set in advance by the limitations of current experiment. Thus, in Dalton's time, it was realistic to picture matter as a composition of atoms and to regard these smallest units as identical constituents for a given element and as distinct constituents for different elements. The atom was known by 1900 to be divisible and to have a structure containing electrons. Distinguishing properties and unifying features could be drawn among the atoms on the basis of these different structures. Soon, Rutherford's experiments confirmed the existence of the atomic nucleus and reduced the examination to a smaller scale where the fundamental constituents of the nucleus could also be identified.

The family of elementary particles had a very modest size in the early 1930s, when all matter was supposed to be made of electrons, protons, and neutrons. Photons were included in the picture as quanta to account for the radiation emitted by atoms, and neutrinos were admitted by virtue of their conjectured role in β decay. The new quantum theory was also in place and provided a theoretical framework for the whole system of known particles. This simple organizational scheme became obsolete, however, as many new particles were soon discovered. The new developments called for revisions of theory in particular, since the nature of the *fundamental interactions* of all the particles proved to be the central issue.

Our study of the nucleus has already introduced us to the reciprocal domains of high energy and small scale. We proceed now to higher regimes of energy as we turn to the examination of matter and the interactions of particles at shorter range. Relativity must also play an essential part in any investigation involving collisions at high energy. We find that, as the collision energy increases, a succession of thresholds occurs where the energies are sufficient to create the masses of new particles. Our small family of elementary particles thus becomes a rather large array of species, whose ranks still continue to grow in number. Nature has furnished some of the evidence for these particles in cosmic-ray processes, while the great body of experimental information about the known particles has come primarily from accelerators and detectors, the sophisticated machinery of high-energy physics.

Elementary particle theory has its own elaborate apparatus for treating phenomena at high energy. The theory is concerned with the principles pertaining to the four main forces of nature—the strong, electromagnetic, weak, and gravitational interactions of particles. We have already been introduced to the first three forces in such problems as the nuclear binding of protons and neutrons, the emission of photons by atomic and nuclear systems, and the β decay of radioactive nuclei. Our goal is to understand the distinguishing features and, especially, the *unifying* characteristics of these very different interactions.

Particle physics is currently enjoying a period of unprecedented fundamental progress, following decades of previous developments in phenomenology. The theoretical discoveries include a general formalism for a theory of the interactions of particles, a basic principle to underlie all theories, and a demonstrable realization of such a theory in successful agreement with experiment. Achievements in the laboratory have been just as momentous. The experiments have given timely confirmation of the most crucial theoretical predictions and continue to point the way toward possible new physics.

We examine these more recent discoveries in the latter part of the chapter, after we present the necessary phenomenological background. The phenomenology is important in its own right because the observations describe the properties of the particles, and because the interpretations indicate the organization of the large and growing particle family. A classification scheme emerges from this investigation, bearing new conservation laws and quantum numbers. These attributes of the particles represent new *symmetry* properties of the fundamental interactions. We learn that it is possible to implement the concepts of symmetry by introducing a *substructure* for the particles themselves. Thus, a collection of uniquely designed constituents takes its place in the theory on a scale smaller than that of the observed particles. This substructure reveals its existence in certain kinds of experiments. The notions of symmetry and substructure finally culminate in a theory of the fundamental interactions, where the basic principles operate among the new particle subentities at the more primitive level.

16-1 Introduction to High-Energy Physics

Particle phenomena were studied in cosmic-ray processes before the coming of the high-energy accelerator. Several of the original elementary particles were discovered in these investigations. The existence of cosmic radiation was proposed in 1911 by V. F. Hess in order to explain the presence of ionization in the atmosphere. It was suggested that radioactive elements on Earth might be responsible; however, this hypothesis was ruled out when it was found that the ionization increased with altitude. Hess argued on the basis of his experiments that the radiation must be coming from outer space.

Cosmic rays are classified as primary and secondary forms of radiation. The primary component refers to the extremely energetic radiation that impinges from space on the upper atmosphere of the Earth. The secondary component is produced copiously as the result of collisions of primary radiation with particles in the atmosphere. The primaries are known to be particles with positive charge because of their characteristic deflections in the Earth's magnetic field. Most of these particles are protons, although surprising abundances of other nuclei are also observed. The flux of the radiation does not vary with time or direction in space as the particles enter the magnetic field of the Earth. The most striking aspect of the primary radiation is its enormous range of energy, spanning 12 orders of magnitude from 10^8 to 10^{20} eV. Solar, galactic, and even extragalactic sources of the radiation are possible. The Sun is a likely contributor, but only in the lower regimes of energy, and then only during periods of solar activity. It is believed that the particles are produced with large amounts of nuclear energy in astrophysical events and are then accelerated to even greater energies by configurations of varying magnetic fields in space.

Secondary radiation is produced in showers as the primary particles penetrate the Earth's atmosphere. The secondaries usually include pions and muons, new members soon to be added to the growing family of elementary particles. The particle-production events may occur as reactions or decays according to the following typical scenario. A primary proton interacts with the nucleus of an atom in the atmosphere, and a portion of the large collision energy is transformed to create several energetic unstable pions. The charged pions decay into the lighter muons, and the neutral pions decay into photons. Muons are also unstable and decay into electrons or positrons as they descend to the Earth. Many muons actually survive the descent without decay, since time dilation allows the faster ones to live longer and travel farther in the Earth's frame of reference. Photons from neutral pion decay can make electron–positron pairs in the vicinity of atmospheric nuclei. The positrons may then annihilate with other electrons to generate more photons. These processes cascade through many stages and result in an extensive shower of electrons and muons at the surface of the Earth.

Cosmic rays remain interesting even in the age of the accelerator, chiefly because the radiation is still the only source of certain kinds of information about particles. The observations offer a possible view of super-energetic phenomena and ultramassive particles, beyond the range of existing accelerators. Special interest is attached to the primaries since these particles provide unique samples of galactic and extragalactic matter for inspection on Earth. Finally, the origin of cosmic radiation continues to be one of the intriguing unsolved puzzles in astrophysics.

Experiments in elementary particle physics have come to rely on accelerators as controlled sources of particles of increasing energy. The following historical sketch should give some orientation for these developments. The probing of the nucleus by accelerated particles began in 1932 with the use of protons from the Cockcroft–Walton generator, a fact already noted in Chapter 15. The cyclotron was designed for similar purposes by E. O. Lawrence in 1929, and subsequent versions of the design evolved in size to meet the demands for higher energy. In effect, the development of the cyclotron set in motion the transition of laboratories from nuclear to particle physics. This type of machine reached its limit for the acceleration of protons at energies around tens of MeV. The cyclotron was superseded in 1945 by the invention of the synchrocyclotron and its immediate successor the proton synchrotron. The innovative features of these accelerators were developed independently by V. I. Veksler and E. M. McMillan. Their contributions to machine design made it possible to accelerate beams of protons to hundreds of MeV and, eventually, to energies far beyond.

Detectors also had to be devised so that the subnuclear particles could be seen in the experiments. Two particular types of apparatus, the cloud chamber and the bubble chamber, should be mentioned because of their place in history and because of their instrumental relation to each other. The cloud chamber was constructed in 1911 by C. T. R. Wilson and was put to immediate use in nuclear and cosmic-ray

Ernest Lawrence

experiments. Electronic counters were added to the instrument so that the operation of the chamber could be triggered by incoming particles. This modification was built into the design of the cloud chamber by P. M. S. Blackett in 1931. Most of the early pictures of elementary particle processes were taken with the aid of these detectors. The bubble chamber was invented in 1952 by D. A. Glaser and proved to be a more sensitive device for observing the tracks of particles. High-energy physics enjoyed a time of remarkable productivity as a result of the introduction of the bubble chamber in the accelerator laboratory.

Each of the high-energy devices operates on the basis of classical principles. The Cockcroft–Walton generator accelerates protons through a large potential difference, built up by a high-voltage transformer and voltage-multiplying circuits. Machines with the same basic design are still employed today as low-energy injectors for the higher-energy accelerators.

The cyclotron is a circular accelerator, shown schematically in Figure 16-1, in which protons are guided in semicircular orbits by magnetic fields and are boosted to successively higher velocities by intervening electric fields. The figure shows how the E field in the gaps of the cyclotron alternates in direction synchronously with each semicircular pass of the proton, and how each accelerating boost enlarges the radius of the orbit. Equation (1-45) governs the relativistic motion of the proton and reduces to the following nonrelativistic expression for the angular velocity:

$$ p = eBR \quad \Rightarrow \quad \frac{v}{R} = \frac{eB}{M}. $$

This quantity remains constant as long as the nonrelativistic approximation holds. Hence, the time for each semicircular orbit does not change with the orbit radius, as

Figure 16-1

Schematic design of the cyclotron. Protons from the ion source S follow semicircular paths in the D-shaped regions of magnetic field B. Acceleration is produced by an electric field E that reverses direction in synchronism with the orbiting time of the particles.

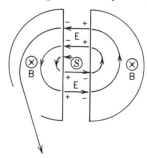

required to ensure synchronism with the reversal of the electric field. Relativity eventually takes effect with the replacement of the proton mass M by the relativistic form $\gamma_v M$, an increasing function of the velocity v. The orbiting time must therefore begin to grow with v, so that the faster protons arrive at the gaps behind schedule and the synchronous operation of the cyclotron breaks down.

Synchronism can be recovered by varying the frequency of the applied E field. The synchrocyclotron incorporates this mechanism to achieve the acceleration of protons at much larger energies. The strength of the B field may also be varied to gain the same effect. The relation $p = eBR$ holds for a stable circular orbit and suggests that, if R is kept fixed, a small increase in B can cause a small increase in the momentum p. The proton synchrotron makes use of this notion, along with a slowly changing E-field frequency to achieve synchronization, and along with certain focusing techniques to stabilize an equilibrium orbit of fixed radius. Accelerators of this type are not limited in their design energy by any theoretical considerations.

Both the cloud chamber and the bubble chamber are optical detectors in which the paths of charged particles are seen as visible tracks. The tracks are made in the cloud chamber from droplets of liquid suspended in a gas. In operation, a volume of air and vapor in the chamber is allowed to expand and undergo a reduction in temperature. The vapor becomes saturated so that a condensation of droplets can occur along the ionized path left by a charged particle passing through the chamber. The supplemental use of electronic counters enables the expansion of the chamber and the photograph of the track to be initiated by the arrival of the incoming particle. The tracks are made in the bubble chamber from bubbles of gas suspended in a liquid. In operation, the liquid is kept below the boiling point and is then allowed to expand. The reduction in pressure lowers the boiling point so that localized boiling can occur to form bubbles along the path of a charged particle moving through the chamber. The visualization of tracks makes it possible to analyze the kinematics of the observed particles in both devices. These procedures have been used extensively to reconstruct events in high-energy collisions.

It should not be assumed that every accelerator is designed for the bombardment of a fixed target. The colliding-beam accelerator is another kind of machine, in which two beams circulate in opposite directions in separate storage rings and undergo collisions where the rings intersect. This design has certain advantages from the standpoint of usable energy. Since the center of mass of the colliding particles is taken to be at rest, all the collision energy is available in the CM system for the initiation of reactions and, especially, for the creation of new particles. A most impressive example of such an accelerator is the proton–antiproton collider at the CERN laboratory in Switzerland. The colliding particles in this machine can have up to 270 GeV kinetic energy in each of the two beams. Of course, many more collisions can occur for a single beam incident on a dense target. The energy advantage is overriding, however, if high energy is the main goal of the design. A comparison of energies can be drawn up with the aid of a previous example, discussed at the end of Section 1-10. The formula of interest is the relation between the total kinetic energies K and K' in the two frames where the target particle is at rest and where the center of mass is at rest. The earlier example gives this relation in terms of the two total relativistic energies. A simple conversion of energies leads us to the desired result:

$$K = 2K'\left(1 + \frac{K'}{4Mc^2}\right). \tag{16-1}$$

Note that the formula applies to collisions of equal-mass particles and that K appears as a *quadratic* function of K'. Some of the striking consequences of this result are demonstrated in the following numerical illustration.

Example

Equation (16-1) can be used to determine how large the single-beam energy K must be in a fixed-target accelerator to produce the equivalent CM energy of a given colliding-beam accelerator with collision energy K'. Let the colliding particles be protons in each instance and, for convenience, let the proton mass be approximated as 1 GeV/c^2. The formula then becomes

$$K = 2K'\left(1 + \frac{K'}{4}\right),$$

where both energies are expressed in GeV. The numerical results are listed for comparison in the following table:

Colliding-Beam Energies	K' (GeV)	K (GeV)
10 MeV on 10 MeV	0.02	0.0402 (40.2 MeV)
100 MeV on 100 MeV	0.2	0.42 (420 MeV)
1 GeV on 1 GeV	2	6
10 GeV on 10 GeV	20	240
100 GeV on 100 GeV	200	20,400 (20.4 TeV)
270 GeV on 270 GeV	540	147,000 (147 TeV)

The quadratic dependence of K on K' begins to be noticeable when K' is comparable to Mc^2, and becomes an enormous effect at the higher values of K'.

The calculation of the last entry deserves closer inspection:

$$K = 2(540)\left(1 + \frac{540}{4}\right) = 2(540)(136) = 146,880.$$

The energies of the colliding beams in this case are the same as those for the protons and antiprotons in the giant CERN collider.

16-2 Particles and Fields

The main goal of particle physics is to understand the interactions of the fundamental particles. We know that the strong nuclear interaction accounts for the binding of protons and neutrons at short range, with energies of the order of several MeV per nucleon. We have already speculated that the proton and neutron are not fundamental and that the force between the particles is a demonstration of a more basic interaction among more elementary entities. The electromagnetic interaction is supposed to be a fundamental mechanism for the emission and absorption of photons and the behavior of charged particles. The Coulomb force yields the structure of the atom as one of its manifestations. We know that this force has infinite range and produces atomic binding energies of the order of several eV. Nuclear β decay is attributable to the weak interaction. We have assessed its extraordinary weakness in Section 15-5 by estimating the cross section for the absorption of neutrinos in matter. The weak force is known to have a very short range by the fact that leptons are created in the immediate vicinity of a nucleon in the $n \rightarrow p$ transition. We are able to judge the relative strengths of the strong, electromagnetic, and weak interactions when we compare the cross sections in typical reactions or the lifetimes in typical decays. It is obvious that the interactions differ very substantially as to range, strength, and energy scale. Consequently, the prospects for unifying the forces of nature according to some common principle may appear to be quite remote. In fact, the most notable development in the theory of elementary particles has been the achievement of a fundamental unification among some of the forces.

The theoretical apparatus of particle physics is provided by the relativistic quantum theory of fields. Our understanding of particle theory depends on this kind of machinery as much as our progress in experiment relies on the accelerator. The formalism of field theory is too advanced for the purposes of this text. Fortunately, the theory produces certain graphic techniques that furnish an intuitive picture of the essential ideas. We wish to adopt these visual techniques so that we can gain a qualitative understanding of particle behavior. We find that the Schrödinger equation must be set aside because the Schrödinger theory is limited to the nonrelativistic treatment of a fixed number of particles. We turn instead to quantum field theory in order to incorporate relativity with the principles of quantum mechanics and provide for the creation and destruction of particles.

Relativistic quantum mechanics has been discussed briefly in Section 8-10 in the context of Dirac's theory for spin-$\frac{1}{2}$ particles. We recall that the prediction of antiparticles is among the conclusions of this remarkable theory. The Dirac equation for a free particle has solutions for any relativistic energy above the value mc^2, or below the value $-mc^2$. The antiparticle concept stems from the latter class of solutions. We can illustrate Dirac's interpretation of the negative energy states by means of the energy level diagram in Figure 16-2. The nonexistence of a lowest state

Figure 16-2

Energy levels for a free Dirac particle. The vacuum is interpreted as a sea of fully occupied negative energy states, and a hole in the sea is interpreted as an antiparticle with positive energy.

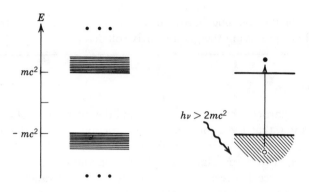

seems to imply that a particle can make an unending series of radiative transitions to lower energy and emit photons indefinitely. To prevent this instability, Dirac's hypothesis defines the no-particle state (the vacuum) as a system in which every negative level is fully occupied by a spin-up and a spin-down fermion. A particle at positive energy is then unable to fall below the level $E = mc^2$ because the exclusion principle prohibits transitions to the filled levels at negative energy. The argument goes on to consider the effect on the vacuum state of an incident photon with energy greater than $2mc^2$. The figure shows that the absorption of the photon excites a particle to a positive energy level and leaves a *hole* in the negative energy sea. The absence of a negative-energy particle is interpreted as the presence of a positive-energy *antiparticle*. We can express this absorption process in terms of electrons and positrons by writing

$$\gamma + (\text{Dirac sea}) \to e + (\text{Dirac sea with a hole})$$

or

$$\gamma + \text{vac} \to e + \bar{e}.$$

Thus, the phenomenon shown in the figure has the same behavior as the process of electron-positron pair production. Dirac's argument relies only on the Pauli principle, and so the prediction of antiparticles applies to fermions of all sorts.

Confirmation of Dirac's prediction came in 1932 with the discovery of the positron by C. D. Anderson. The new particle was found in a cosmic-ray experiment using a cloud chamber with an applied magnetic field. The events of interest showed tracks originating in a thin lead plate and curving in opposite directions in the applied field, as indicated in Figure 16-3. Since no entry track was visible, Anderson concluded that the process was initiated by a cosmic-ray photon and that the oppositely charged tracks were due to electron–positron pair production.

The first principles of quantum field theory were introduced before 1930 by Dirac, Heisenberg, Pauli, and others. Their work was beset by profound difficulties that delayed the formulation of a consistent theory for many years. The most urgent

Figure 16-3

Anderson's discovery of the positron. The
pair-production process $\gamma \rightarrow e^- + e^+$ takes
place in the vicinity of a lead nucleus.

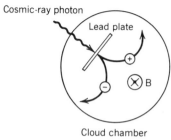

problem was to ensure that the observable properties of an interacting system could be
expressed as finite quantities at any stage of approximation. A theory was said to be
renormalizable if the mathematical formalism guaranteed this behavior. The first
fundamental particle theory to be so constructed was *quantum electrodynamics*, the name
given to the theory of the electromagnetic interaction. This feat was achieved during
the 1940s through the work of Feynman, J. Schwinger, S.-I. Tomonaga, and F. J.
Dyson. Very precise calculations were performed with the theory, and the results were
compared with measurements of similar precision. The credibility of quantum electrodynamics drew support from the fact that extraordinary agreement was obtained in
every case. Two of the more stringent tests of theory and experiment were the
determinations of the Lamb shift and the g-factor for electron spin, quantities already
mentioned in Section 8-10. Quantum field theory has also been applied to the other
interactions. It would be premature to discuss the consequences until more has been
said about these other forces later in the chapter.

Let us turn our qualitative comments about field theory to better advantage now,
and extract some tangible benefits for later use. The theory assigns a field to each
fundamental particle and casts the interactions of particles in terms of the corresponding interactions of fields. The field varies in space and time and has the quantum
mechanical properties of an operator that creates and destroys the associated particle
in states of definite momentum and energy. Thus, as an example, the electromagnetic
field (actually, the electromagnetic vector potential) acts as a quantized field and
creates and destroys quanta identified as photons. Interactions can be represented
according to a scheme, introduced by Feynman, in which the procedures of the theory
are summarized by space–time diagrams. These *Feynman diagrams* are much the same
in appearance as the ones drawn in Figure 15-16. The intuitive content of the
diagrams is immediately apparent and compelling. In fact, the appeal to intuition is
our sole motive in referring to fields at all. We intend to use the diagrams as "pictures
of the theory," and ignore the fact that we know nothing at all about the formalism
behind them. This naive approach serves our interests throughout the chapter.

Figure 16-4 illustrates how these visual aids are applied to two particular processes
in quantum electrodynamics. We interpret the lines in such diagrams as world-lines in
space–time, where time runs in the vertical direction. The bremsstrahlung diagram
shows the emission of a photon as an electron passes through the electric field of a

Richard Feynman

nucleus. The accelerating electron is a source of electromagnetic radiation, and the emitted photon is a quantum of electromagnetic energy and momentum belonging to this radiation field. The scattering diagram shows the world-lines of two electrons in the act of mutual electromagnetic interaction. Each particle undergoes a change of energy and momentum, and so a *transfer* of these quantities must proceed from one particle to the other in order to satisfy the conservation laws in the overall collision. The transfer of energy and momentum is evidently accomplished by the exchange of a photon, because the electromagnetic interaction is the cause of the scattering and because the photon is the quantum of the electromagnetic field. We may suppose that the photon originates in the radiation field of the one accelerating electron and is absorbed into the field of the other. The bremsstrahlung photon and the exchanged photon are said to be *real* and *virtual*, respectively, since the one can be observed in a detector while the other cannot.

Figure 16-5 summarizes the main points to assume in our heuristic introduction to quantum field theory. The basic diagrams of quantum electrodynamics are drawn as *vertices* describing photon emission and photon absorption by a particle of charge e. Each vertex contains the world-line of a moving charge, and so the corresponding element of the theory has the form of a *current*, as indicated in the figure. The coupling of the photon to the current represents the interaction between the electromagnetic field and the charged particle. This interaction is proportional to the charge e, since e is an explicit factor in the current itself. A *coupling constant*, given by e, is therefore included in the figure beside each of the vertices. Powers of e can then be associated with diagrams in which several vertices are connected together. For instance, there are two vertices in the scattering diagram in Figure 16-4, and so the corresponding amplitude for the elastic scattering of electrons must be proportional to e^2. It is

Figure 16-4

Feynman diagrams for bremsstrahlung emission and electron–electron scattering.

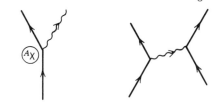

Figure 16-5

Vertices describing the basic photon–current interaction in quantum electrodynamics. The strength of the interaction is proportional to the charge e.

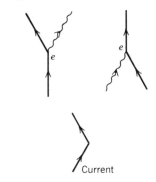

conventional to introduce the dimensionless fine structure constant

$$\alpha = \frac{e^2}{4\pi\varepsilon_0\hbar c}$$

and rewrite the dependence on the charge so that a power of α is substituted for every two powers of e. Graphical techniques based on currents, quanta, and coupling constants are similarly employed for the other fundamental particle interactions.

Example

Let us recall that the interaction of an electron with an applied magnetic field is expressed in terms of the magnetic moment of the particle. We can represent the interaction with the aid of Figure 16-6 by showing the behavior of the electron current under the influence of the applied field. The indicated diagrams describe

Figure 16-6

Feynman diagrams defining the magnetic moment of the electron. The indicated contributions to g_S are of order $\alpha^0, \alpha^1, \alpha^2, \ldots$. The number of distinct diagrams increases with each ascending power of α.

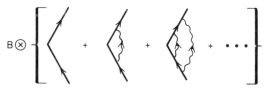

the contributions to the magnetic moment in terms of an increasing number of virtual photons. The leading term in this series corresponds to the Dirac prediction $g_S = 2$, since the associated diagram contains no exchanged photons and, hence, no corrections due to quantum electrodynamics. Succeeding diagrams must contain an increasing *even* number of vertices as in the sample shown in the figure. Consequently, the set of all such contributions to the magnetic moment gives the deviation from $g_S = 2$ in the form of a power series in e^2 or, equivalently, a power series in α. These terms are expected to diminish in size as the powers grow because of the smallness of the fine structure constant.

16-3   Mesons and the Nuclear Force

Many of the early ideas about particles were motivated by observations taken from nuclear physics. The theory of the nuclear force was originally regarded as fertile ground for particle concepts, since the interactions of nucleons were presumed to be explainable in a fundamental theory of the strong interaction. Quantum field theory had been used to express the theory of the electromagnetic interaction, and so the same methods were expected to be applicable to the force between nucleons. The two conspicuous aspects of this force, which any proposed nuclear field theory would have to accommodate, were the properties of short range and charge independence.

A new kind of field, called the *meson field*, was introduced in 1935 by H. Yukawa to act as a mediator for the interactions of nuclear particles. The field was assumed to be composed of quanta, called *mesons*. The exchange of a meson was supposed to furnish the basic mechanism for the strong force between a pair of nucleons, as in the exchange diagram shown in Figure 16-7. These notions were patterned closely after the properties of photons, the quanta of the electromagnetic field. Yukawa made allowance for the short range of the force by endowing his hypothesized mesons with mass. Charge independence could also be incorporated in the theory by letting the mesons exist with different charges. The analogy with quantum electrodynamics went further to include a mesonic version of the bremsstrahlung process. The theory predicted meson emission, also shown in Figure 16-7, in which a quantum of the field was allowed to materialize as a real meson whenever a nucleon passed within range of another nuclear particle, in circumstances where the collision energy was enough to create the meson mass. In 1947 the predicted meson was discovered as a cosmic-ray particle in the execution of this particular process.

Figure 16-7

Exchange of a virtual meson between nucleons and meson emission by a nucleon in the field of a nucleus.

The mass of the meson and the range of the nuclear force can be related to each other. Let us use the Feynman diagram for virtual meson exchange in Figure 16-7 to deduce the relation. The meson is emitted by the one nucleon and absorbed by the other in the indicated time interval Δt. Consequently, there is a temporary increase in the total energy of the two-nucleon system by an amount at least as large as the meson rest energy mc^2. This momentary violation of energy conservation cannot be detected since the exchange of the virtual meson cannot be observed. Thus, the energy of the system is uncertain for a time Δt by an amount $\Delta E \geq mc^2$, and so the uncertainty principle can be invoked to determine the time interval:

$$\Delta t = \frac{\hbar}{\Delta E} \leq \frac{\hbar}{mc^2}.$$

The range of the meson cannot exceed the distance

$$c \, \Delta t \leq \frac{\hbar}{mc},$$

because the velocity of the exchanged particle must be less than c. This argument serves to define the spatial extent of the meson field, and hence the range of the nuclear force, as

$$r_0 = \frac{\hbar}{mc}. \tag{16-2}$$

The result is such that $2\pi r_0$ is equal to the meson Compton wavelength h/mc. We note that r_0 and m are inversely related. Hence, if the general formula is also applied in the case of the electromagnetic field, the range of the field and the mass of the quantum (the photon) are linked with parameters $r_0 = \infty$ and $m = 0$, as expected.

We reach the same conclusion if we employ a relativistic wave equation to describe the meson field. Let us recall the relativistic relation between energy and momentum, given in Equation (1-35), and make the usual operator assignments

$$p^2 \to -\hbar^2 \nabla^2 \quad \text{and} \quad E \to i\hbar \frac{\partial}{\partial t}.$$

In this case the differential operations are assumed to act on a meson wave function $\Phi(\mathbf{r}, t)$. We obtain the partial differential equation for Φ by the familiar substitution:

$$-c^2 p^2 + E^2 = m^2 c^4 \quad \Rightarrow \quad \nabla^2 \Phi - \frac{1}{c^2} \frac{\partial^2}{\partial t^2} \Phi = \frac{m^2 c^2}{\hbar^2} \Phi. \tag{16-3}$$

This candidate for a relativistic wave equation is attributed to O. Klein and W. Gordon and predates the Dirac equation. We wish to interpret the parameter given by Equation (16-2) as a range from the meson field. Let us insert the expression for r_0 into Equation (16-3) and make the interpretation in terms of a time-independent wave function for a *static* meson state. The equation becomes

$$\nabla^2 \Phi = \frac{1}{r_0^2} \Phi \tag{16-4}$$

when the time-derivative term is deleted. A spherically symmetric solution to the equation is given by the function

$$\Phi(r) = \phi_0 \frac{e^{-r/r_0}}{r}. \tag{16-5}$$

We leave the proof of this result to Problem 7 at the end of the chapter. Equation (16-5) describes a static meson field whose exponential fall-off with r is controlled by the value of r_0. This scale parameter therefore plays the role of a range, determined by the mass of the meson in accord with Equation (16-2).

The meson theory of the nuclear force called for *three* differently charged species of meson, with charges $+e$, 0, and $-e$, in order to allow for the property of charge independence. Evidence for this feature of the nucleon–nucleon interaction came originally from a cyclotron experiment undertaken in 1949 to investigate neutron-proton elastic scattering. The collision energy was intended to be large enough to reveal the details of the nuclear force on a scale comparable to the range r_0. The cyclotron was used to accelerate deuterons, and these particles were stripped of their protons to produce a beam of energetic neutrons. Hydrogenous material was placed in the beam to furnish a dense target of protons, and detectors were supplied to determine the angle dependence of the scattered neutrons. A regime of energy was chosen so that neutron diffraction would result, and indeed a large cross section was observed for scattering in the forward direction. It was especially interesting to discover that the cross section also showed equally large scattering in the *backward* direction.

We show the results of the experiment and interpret the results in the two parts of Figure 16-8. These illustrations employ the center of mass system as the ideal frame for picturing the forward–backward symmetry of np scattering. Large backward scattering is shown in the upper part of the figure as a growth of the differential cross section toward the 180° direction. The symmetry of this angular distribution is explained in the lower part of the figure in terms of two kinds of contribution to the neutron–proton interaction. The expected diffraction of neutrons at forward angles is seen as *direct* scattering, associated with the effect of a short-range meson field and the exchange of an *uncharged* meson. The observation of scattered neutrons at backward angles is interpreted as *charge-exchange* scattering, in which the nuclear force *transforms* neutrons into protons inside the region of interaction. This feature of the force between nucleons is evidence for a *charge-changing* property of the meson field, which necessitates the exchange of *charge-bearing quanta* between the interacting particles. The symmetry of the scattering for forward and backward angles implies an equality of contributions from the exchange of the corresponding neutral and charged mesons. These observations enable us to express the charge independence of the nucleon–nucleon interaction as a characteristic ingredient of the meson theory. Evidently, charge independence is a symmetry property of the interactions of mesons with nucleons, which constrains the nucleons to emit the differently charged mesons with equal probabilities. It is interesting that the quanta of the meson field can carry *charge* as well as momentum and energy. This notion is a bold extension of the ideas of quantum field theory. Quantum electrodynamics has no such behavior, since the emission of photons does not cause a change in the charge of the emitting particle. The notion has real substance because charged mesons do exist among the observed particles and because their properties indeed conform to the symmetrical predictions implied by charge independence.

Figure 16-8

Differential cross section for *np* elastic scattering. The data are plotted versus neutron-scattering angle in the CM system, and the results are interpreted in terms of direct and charge-exchange scattering. A symmetrical distribution of scattered neutrons implies a symmetry of the nuclear force resulting from the exchange of neutral and charged mesons.

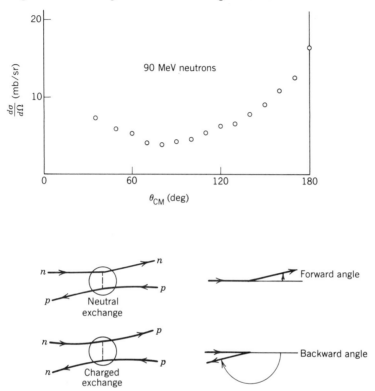

Meson theory has not been a great success, primarily because the strong interactions of nucleons are far more complicated than the simple exchange of single mesons suggests. In fact, meson theory is not fundamental at all, as quantum electrodynamics is, since the particles of the theory are not fundamental and, hence, not qualified to serve as basic elements in a field theory. The connection between the range of a field and the mass of an exchanged quantum is still a very useful idea. The property of charge independence is also very important as a symmetry to simplify the phenomenology of nucleons and mesons. We give more attention to this property in other sections of the chapter.

Example

Let us examine the range-mass relation and extract a numerical estimate for the meson mass. Equation (16-2) can be rearranged to give the rest energy:

$$mc^2 = \frac{\hbar c}{r_0} = \frac{197.3 \text{ MeV} \cdot \text{fm}}{r_0}.$$

Then, if we choose a typical range for the nuclear force to be 1.50 fm, we obtain

$$mc^2 = \frac{197 \text{ MeV} \cdot \text{fm}}{1.50 \text{ fm}} = 131 \text{ MeV}$$

as the predicted value for the rest energy of the exchanged meson.

16-4   Muons and Pions

New particles began to make their existence known in cosmic-ray events in the 1930s, starting with Anderson's discovery of the positron in 1932. The muon was the next to be seen in 1937, in experiments performed by Anderson and S. H. Neddermeyer. These studies of the secondary cosmic rays found evidence of penetrating charged particles whose radiative behavior in matter was not understood in terms of properties of the known particles. The anomaly was at first thought to be due to a failure of quantum electrodynamics, the theory of the radiative process. The experiments indicated a more constructive alternative, however, in which the observed penetrability of the particles in matter was explained by proposing a new species of particle, with charge $\pm e$ and with mass two orders of magnitude larger than the electron mass. The new particle was temporarily given the name mesotron, or μ meson, because it was originally believed to be the predicted quantum of Yukawa's field theory. This belief collapsed when it was demonstrated that the particles penetrated matter with little interaction, so that the candidate could not qualify for identification as a strongly interacting meson. Instead, the observations established the existence of the two muons μ^- and μ^+ as new particles unrelated to the nuclear force. The detailed properties of the muons were eventually determined in the course of accelerator experiments during the 1950s.

The oppositely charged muons are spin-$\frac{1}{2}$ fermions with mass 106 MeV/c^2. We can rule out the possibility that these are Yukawa's mesons because the quanta exchanged in Figures 16-7 and 16-8 must be integer-spin bosons in uncharged as well as charged varieties. The muons are unstable and decay via the β process

$$\mu^\mp \to e^\mp + 2 \text{ neutrinos},$$

with a lifetime of 2.2×10^{-6} s. These decays tell us that the muon should be classified as a member of the *lepton* family along with the electron and neutrino. Lepton conservation implies that μ^- and μ^+ are to be designated as lepton and antilepton, in keeping with our previous assignments for e^- and e^+. We can be more specific about these designations in Section 16-5 when we learn about some of the further properties of neutrinos. The similarity between the leptons μ and e may tempt us to conclude that the muon is simply an excited state of the less massive electron. It would then follow that the radiative transition $\mu \to e + \gamma$ should occur as the dominant mode of decay. In fact, the radiative mode is not observed, as though the particles are under the influence of some internal attribute that distinguishes μ from e and requires muon decay to proceed via the weak interaction.

The charged forms of Yukawa's meson were discovered in cosmic rays in 1947. Their tracks were detected in photographic exposures taken at high altitudes in experiments conducted by C. F. Powell. The mesons were found with charge $\pm e$ and were called π mesons, or pions, π^+ and π^-. It was evident from the tracks that the

Successive decays $\pi \rightarrow \mu \rightarrow e$ in a cosmic-ray
exposure

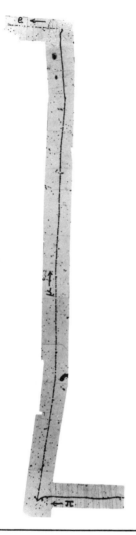

particles were produced in meson-emission reactions, like the one in Figure 16-7, in
which high-energy nucleons emitted pions in collisions with atomic nuclei in the upper
atmosphere. The photographic plates also showed the decays of the unstable charged
pions into muons. The processes were interpreted as

$$\pi^+ \rightarrow \mu^+ + \nu \quad \text{and} \quad \pi^- \rightarrow \mu^- + \bar{\nu},$$

representing new forms of lepton-conserving β decay governed by the fundamental
weak interaction. It soon became possible to make pions in the laboratory in
accelerator experiments. The neutral meson π^0 was seen for the first time in these
pion-production reactions. The observable signature for this unstable particle was its

distinctive and rapid decay into γ rays

$$\pi^0 \to \gamma + \gamma,$$

an electromagnetic decay process. Pion production was studied with the aid of accelerated beams of protons and neutrons in such reactions as

$$p + p \to \begin{cases} p + n + \pi^+ \\ p + p + \pi^0 \end{cases} \quad \text{and} \quad n + p \to \begin{cases} n + n + \pi^+ \\ n + p + \pi^0 \\ p + p + \pi^- \end{cases}.$$

These interactions of nucleons were observed to obey the principle of nucleon-number conservation and to occur with substantial cross sections. Thus, it was established that the new mesons were indeed strongly interacting particles of the kind predicted by Yukawa's nuclear field theory.

The pion masses are 140 MeV/c^2 for $\pi^\pm$ and 135 MeV/c^2 for π^0. These values are quite consistent with expectations based on the range of the nuclear force. The decays of the charged and uncharged pions are very different, however. This fact is emphasized by the large disparity between the two lifetimes, 2.6×10^{-8} s for $\pi^\pm$ and 8.3×10^{-17} s for π^0. We know that the weak and electromagnetic interactions are responsible for the two modes of decay, and we can argue that the observed lifetimes are indicative of the distinctions between the two interactions. The pion is the first in a succession of new unstable elementary particles whose production is controlled by the strong interaction and whose decay is governed by an interaction of lesser strength.

The charged pions have been used extensively as beam particles to probe the strong interactions between mesons and nucleons. Scattering processes such as

$$\pi^+ + p \to \pi^+ + p \quad \text{and} \quad \pi^- + p \to \begin{cases} \pi^- + p \\ \pi^0 + n \end{cases}$$

have given indications of several resonant states in the pion–nucleon system. The cross sections have a scale typical of the strong interactions, and the resonances stand out as unique excitations of the interacting particles. Pion–nucleon collisions also lead to other kinds of final states, revealing the existence of new types of particles. We return to this development in another section.

Pions have spin and parity quantum numbers that represent intrinsic qualities of the particles. We must turn to experiment to learn such properties, in the same spirit as for the nuclear spin and parity i^p in the case of nuclei. Experiments sensitive to these characteristics tell us that the pion is a spin-0 meson with odd parity. Such particles are designated by 0^- quantum numbers and are called *pseudoscalar* mesons. The arguments for this assignment are discussed in two of the following exercises.

Example

We can get a very rough estimate of the scale expected for a strong interaction cross section by computing the area of a disk of radius r_0, the range of the nuclear force. A typical choice for r_0 gives the result

$$\pi r_0^2 = \pi \left(1.5 \times 10^{-15} \text{ m}\right)^2 = 7.1 \times 10^{-30} \text{ m}^2 = 71 \text{mb}.$$

In fact, the cross sections for the elastic scattering of protons by protons and pions by protons vary with the collision energy and have values of order tens of millibarns.

Example

The spin of the pion is known from a comparison of the cross sections for the two inversely related processes

$$p + p \rightarrow \pi^+ + d \quad \text{and} \quad \pi^+ + d \rightarrow p + p.$$

The rate of each reaction is determined by the probability for producing the respective final particles and is proportional to the number of final states available at the given energy. In particular, the cross section for the first of the two reactions contains a spin-multiplicity factor for the deuteron, given by

$$2i_d + 1 = 3 \quad \text{for } i_d = 1,$$

along with a similar factor $(2s_\pi + 1)$ pertaining to the unknown pion spin s_π. The cross sections are also proportional to the squared moduli of the complex reaction amplitudes

$$|\chi(\pi^+ d, pp)|^2 \quad \text{and} \quad |\chi(pp, \pi^+ d)|^2.$$

These quantities are identical if the processes are compared at the same values of the CM energy. The *ratio* of cross sections therefore contains the factor $(2s_\pi + 1)$ as the only unknown ingredient. This sort of analysis can be applied to the two reactions to give the result $s_\pi = 0$.

Example

The parity of the pion is known from an observation of the reaction

$$\pi^- + d \rightarrow n + n$$

and from the fact that the overall parity cannot change in a process governed by the strong interaction. The reaction of interest proceeds from an initial *bound* state corresponding to a π-mesic atom, in which the π^- meson replaces the electron in its atomic orbit around the deuteron. A strong $\pi^- d$ interaction occurs in the initial $\ell = 0$ orbital state, so that all the initial angular momentum quantum numbers are known quantities:

$$s_\pi = 0, \quad i_d = 1, \quad \text{and} \quad \ell = 0.$$

The total angular momentum quantum number must therefore have the value $j = 1$ in the initial state, and conservation of angular momentum implies $j = 1$ in the final state as well. The spins of the two final neutrons are constrained by the Pauli principle, so that the only allowed spin state is the *triplet* with total spin quantum number $s = 1$. (The $s = 0$ alternative is ruled out because $s = 0$

and $j = 1$ would imply $\ell = 1$ for the nn orbital state. The combination $s = 0$ and $\ell = 1$ would then give a spin-antisymmetric and space-antisymmetric wave function for the two identical fermions.) These arguments result in a unique set of final angular momentum quantum numbers,

$$s = 1, \quad \ell = 1, \quad \text{and} \quad j = 1,$$

since no other value of ℓ satisfies the constraints. The corresponding parity is odd, and so the overall parity of the $\ell = 0$ $\pi^- d$ system must also be odd because of parity conservation. We assemble the initial parity multiplicatively from the three separate factors

(unknown pion parity)(even deuteron parity)(even orbital parity).

Thus, we find the overall parity to be odd if we assign the pion an odd intrinsic parity.

16-5 Neutrinos

The weak decays of the muon and pion call attention again to the neutrino and the idea of lepton conservation. We have identified the neutrino ν to be the neutral, almost massless, spin-$\frac{1}{2}$ fermion that occurs undetected along with the emitted positron in nuclear β^+ decay. We recall that the particle ν is designated as a lepton. The antilepton $\bar{\nu}$ is then distinguished from ν according to the Dirac interpretation of an antiparticle. We must now refine these bookkeeping procedures to make allowance for other kinds of leptons and, especially, other kinds of neutrinos.

The lepton concept must be enlarged in order to accommodate the results of the two-neutrino experiment of 1962. Let us divulge the conclusion before we describe the experiment itself. We consider neutrino processes, which may be either reactions or decays, and we observe the neutrino that appears in association with a muon to be *intrinsically different* from the neutrino that accompanies an electron. The two varieties are discriminated accordingly by the names muon-neutrino (symbol ν_μ) in the one case, and electron-neutrino (symbol ν_e) in the other. This distinction means that neutrinos from the muonic decays of charged pions are *not* the same as neutrinos from the β decays of nuclei. Hence, $\pi^\pm$ decay must be written as

$$\pi^+ \to \mu^+ + \nu_\mu \quad \text{and} \quad \pi^- \to \mu^- + \bar{\nu}_\mu,$$

while neutron β decay takes the form

$$n \to p + e^- + \bar{\nu}_e.$$

Neutrinos have served as beam particles through the use of reactors since the 1950s and through the use of accelerators since the 1960s. The neutrino-induced production of leptons may yield either muons or electrons in accordance with the particular type of incident neutrino. Thus, there are reactions that lead either to μ^- and μ^+,

$$\nu_\mu + n \to p + \mu^- \quad \text{and} \quad \bar{\nu}_\mu + p \to n + \mu^+,$$

or to e^- and e^+,

$$\nu_e + n \to p + e^- \quad \text{and} \quad \bar{\nu}_e + p \to n + e^+,$$

depending on the choice of neutrinos and antineutrinos in the incident beams.

The two-neutrino experiment was among the first to use an accelerator as a source of high-energy neutrinos. A proton beam from the accelerator was directed onto a nuclear target to produce fast forward pions, and the decays of the charged pions provided the desired neutrinos. Meters of steel were placed in the beamline to absorb muons and other unwanted particles. The resulting neutrino beam was then allowed to pass through an array of aluminum plates so that observations could be made of the occasional interactions between the beam particles and the nuclear material in the plates. These neutrino collisions were found to yield *muons* in every case. Equal numbers of electrons and muons would have been seen if there had been no difference between ν_μ and ν_e. The observation of muons alone showed instead that the incident neutrinos, originating in π decay, were identifiable as distinct muon-neutrinos.

The interpretation of this experiment suggests a classification of leptons into subgroups, or *leptonic generations*, in which a distinction is introduced on the basis of an enlarged set of conserved lepton quantum numbers. These new lepton numbers are regarded as internal attributes that express the electron-lepton and muon-lepton content of the various particles. Thus, the conserved lepton number L, introduced in Section 15-5, turns into the pair of *separately conserved* lepton numbers L_e and L_μ. These quantum numbers are assigned to the different generations as follows:

$$\begin{bmatrix} e^- \\ \nu_e \end{bmatrix} \text{ and } \begin{bmatrix} e^+ \\ \bar{\nu}_e \end{bmatrix} \text{ have } \left(L_e = +1 \text{ and } -1, L_\mu = 0 \right),$$

while

$$\begin{bmatrix} \mu^- \\ \nu_\mu \end{bmatrix} \text{ and } \begin{bmatrix} \mu^+ \\ \bar{\nu}_\mu \end{bmatrix} \text{ have } \left(L_e = 0, L_\mu = +1 \text{ and } -1 \right).$$

We see that the separate conservation laws for L_e and L_μ are obeyed in the β decays of the neutron and pion, as described above. The charged pions also exhibit the rare electronic decay modes

$$\pi^+ \to e^+ + \nu_e \quad \text{and} \quad \pi^- \to e^- + \bar{\nu}_e$$

as a further illustration. We note that the new scheme of conserved quantum numbers immediately accounts for the nonoccurrence of the radiative decay $\mu^\mp \to e^\mp + \gamma$. The observed β instability of the muon involves the emission of two neutrinos, whose identities must be specified to conform to these ideas. If we write μ^- decay as

$$\mu^- \to e^- + \bar{\nu}_e + \nu_\mu,$$

we see that lepton conservation is satisfied by the assignment ($L_e = 0, L_\mu = +1$) on both sides of the process.

A third leptonic generation has recently been identified, following the discovery in 1975 of the new massive charged leptons τ^- and τ^+. If we grant these particles their own distinct neutrinos ν_τ and $\bar{\nu}_\tau$, and if we introduce a third conserved lepton quantum number L_τ, we obtain

$$\begin{bmatrix} \tau^- \\ \nu_\tau \end{bmatrix} \text{ and } \begin{bmatrix} \tau^+ \\ \bar{\nu}_\tau \end{bmatrix} \text{ with } \left(L_e = 0, L_\mu = 0, L_\tau = +1 \text{ and } -1 \right)$$

as the third generation of leptons. The extended principles of lepton conservation are supposed to hold in the τ-decay processes

$$\tau^- \to \begin{cases} e^- + \bar{\nu}_e + \nu_\tau \\ \mu^- + \bar{\nu}_\mu + \nu_\tau \\ \pi^- + \nu_\tau, \end{cases}$$

to cite only a few examples. All these remarks presume the existence of a distinct tauon-neutrino. This conjecture has not been confirmed directly by experiment.

Several unanswered questions about neutrinos still stand out. The topics of greatest concern are the mass of the neutrino, solar neutrinos, and neutrino oscillations. These problems have important bearing on the theory of weak phenomena and on the theory of astrophysical processes.

Until recent times the mass of the neutrino has not been accessible to decisive measurement, so that values of the mass have been found only in terms of upper bounds. We have mentioned a possible approach to this question in Section 15-4 in the context of a current study of the endpoint in the β spectrum of triton decay. The quoted *lower* bound of 30 eV refers to the rest energy $m_{\nu_e} c^2$ for the electron-antineutrino emitted in the process

$$^3\text{H} \to {}^3\text{He} + e^- + \bar{\nu}_e.$$

This bound is disputed to be much too large, on grounds related to an entirely different experiment discussed below. The neutrino-mass issue has great cosmological significance, even if the mass happens to be extremely small. Neutrinos inundate the universe and can make an enormous contribution to the force of gravity if the mass is nonvanishing. Masses can be assigned to ν_e, ν_μ, and ν_τ, and gravitational effects can be estimated, in order to evaluate the critical conditions required to make the universe *close*. Closure would imply a slowing down and, eventually, a *reversal* in the expansion of the universe, so that the evolution following the Big Bang would terminate inexorably in a Big Crunch at some time in the distant future.

Neutrinos from stars like the Sun have their origin in the various reactions that take place deep within the stellar interiors. Consequently, the detection of these neutrinos on Earth offers a unique view of the processes of stellar energy generation. The flux of solar neutrinos is currently being studied for this purpose with the aid of large underground detectors. The mechanism used for the detection of neutrinos is the ν_e-capture reaction

$$\nu_e + {}^{37}\text{Cl} \to {}^{37}\text{Ar} + e^-.$$

Neutrinos are produced in the Sun by several different processes, including certain branches of the proton fusion chain, and a small number are captured by the chlorine nuclei in a large sample of the liquid C_2Cl_4, a dry-cleaning fluid. (Unfortunately, the capture reaction has a high neutrino threshold energy, so that neutrinos from the main proton cycle fall below the level of sensitivity of the detector.) The rate of neutrinos captured per nucleus is found from the quantity of radioactive ^{37}Ar present after an exposure of the sample. A useful measure of this quantity is the solar neutrino unit (symbol SNU), defined as

$$1 \text{ SNU} = 10^{-36} \text{ neutrino captures per second per } {}^{37}\text{Cl nucleus.}$$

Theory and experiment are in disagreement over the value of the solar neutrino rate. Models of solar energy generation predict a rate of approximately 6 SNU, while the measured result is closer to 2 SNU. This factor-of-3 shortfall in the number of detected neutrinos is an indication of something unusual in the behavior of the Sun, or in the behavior of the neutrinos.

Oscillation in a beam of neutrinos is a phenomenon in which the lepton-number composition of the particles varies with the length of the beam. This effect may occur when a source emits electron-neutrinos, and some ν_e's transform into ν_μ's as the beam propagates along. The resulting neutrino beam may then be observed to have a variable ν_e content along its path. A similar kind of behavior definitely takes place in certain other neutral systems to be discussed later on. Neutrino oscillation provides a plausible explanation for the solar neutrino problem. It is possible that the Sun produces ν_e's at the predicted rate and that an oscillation from ν_e to ν_μ occurs on the way to the detector on Earth. (This transformation is most likely to occur inside the volume of the Sun.) A shortage of detected neutrinos is then observed because the arriving ν_μ's cannot participate in the ^{37}Cl capture reaction. A recent analysis of the problem of neutrino conversion inside the Sun indicates values of the mass for ν_e and ν_μ far below the lower bound for ν_e found in the triton β-decay experiment.

Neutrinos provide many means of access to information about the weak interaction. Phenomena such as β decays and reactor processes are able to probe the interaction at low energy, and neutrino beams from accelerators are able to explore the properties of the interaction at high energy. Experiments in the latter category have contributed to many recent developments in the theory of the fundamental particles.

16-6 Nonconservation of Parity

Conservation laws are related to the basic dynamical principles that govern a physical system. We may identify a conserved quantity from principle if the dynamical theory is known, or we may recognize the conserved nature of such a quantity from empirical evidence if the theory is not fully established. In either case, the conservation law supplies a scheme of organization, known as a *symmetry*, that reduces the complexity of the given system. We use conserved quantities for this purpose in classical mechanics when we let constants of the motion act as parameters to designate the orbits of particles. We also use such quantities in quantum mechanics when we let quantum numbers serve as labels to specify the states of the system. Many of our more familiar conserved quantities are associated with spatial behavior. Conservation of angular momentum is a case in point. The conservation law stems from rotational symmetry, and the angular momentum quantum numbers identify the various quantum states. Parity is another independent spatial property of a quantum system. Our aim is to examine its meaning as a symmetry so that we can understand the circumstances in which the associated conservation law is *violated*.

Let us first recall the problem of particle motion in one dimension with potential energy $V(x)$, where V behaves as an even function satisfying $V(-x) = V(x)$. This property implies the existence of energy eigenfunctions having symmetric and anti-symmetric behavior under the reflection $x \rightarrow -x$. Thus, states of definite energy exist with definite even and odd parity as a result of the reflection symmetry of V.

We then recall that the parity property in three dimensions pertains to *space inversion* in all three coordinates $(x, y, z) \rightarrow (-x, -y, -z)$. This operation differs from *mirror reflection*, where only one coordinate changes sign. Figure 16-9 illustrates the distinction and shows how space inversion is represented as a mirror reflection

Figure 16-9

Transformation of the coordinate axes by space inversion and by mirror reflection. A reflection in one axis, followed by a 180° rotation about the same axis, is equivalent to the space-inversion operation in which all three axes are reflected.

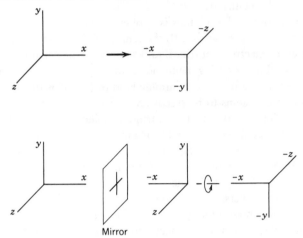

accompanied by a 180° rotation about the reflected axis. The central-force problem is governed by a potential energy $V(r)$, where V remains unchanged under the inversion $\mathbf{r} \to -\mathbf{r}$. Consequently, stationary states exist in the form of angular momentum eigenfunctions with orbital quantum number ℓ *and* with definite parity given by $(-1)^{\ell}$. Again, the parity of the states originates in the inversion symmetry of V. Since inversion is the same as reflection accompanied by rotation, as in Figure 16-9, and since rotational symmetry is already a property of V, it follows that the parity of the states is associated directly with the mirror symmetry of the interaction expressed by V.

We have introduced parity as a mathematical aspect of wave functions, and we have learned that parity is an intrinsic observable of particles as in the case of the nucleus and, more recently, the pion. Parity can also be used to classify the inversion properties of certain observables. The position vector $\mathbf{r}$ is the basic odd-parity quantity inasmuch as inversion is defined by the operation $\mathbf{r} \to -\mathbf{r}$. The momentum vector $\mathbf{p}$ also has odd parity because of its connection to $\mathbf{r}$ through the operator assignment $\mathbf{p} \to (\hbar/i)\nabla$. Both $\mathbf{r}$ and $\mathbf{p}$ belong to a class of *polar vector* quantities whose members change sign under space inversion. The angular momentum vector $\mathbf{L} = \mathbf{r} \times \mathbf{p}$ is a product of two odd factors and therefore has even parity. The spin vector $\mathbf{S}$ is also even since spin has the quantum mechanical properties of angular momentum. Both $\mathbf{L}$ and $\mathbf{S}$ belong to a class of *axial vector* quantities whose members do not change sign under space inversion. We can use these considerations of parity to advantage whenever inversion-symmetric interactions determine the states of the system.

Our investigation of parity is concerned with the issue of symmetry under the operation $\mathbf{r} \to -\mathbf{r}$. We know from our remarks on Figure 16-9 that an examination of mirror behavior is sufficient if rotational symmetry is already known to hold. It is advantageous to recognize this property since mirror reflection is easier to visualize. The question of interest then asks whether a given physical process has a mirror image that also occurs in nature as an equally operative physical process. To illustrate, let us imagine taking a device made of several moving parts and using the image in a mirror to assemble a reflected version of the original machinery. We could then compare the two machines and argue that the original and the duplicate should function in mirror-symmetric states of motion. In a similar spirit, let us consider two optically active materials whose respective molecules rotate the plane of polarization of light in opposite directions. It would be surprising if the two species of molecules could exist as exact mirror images of each other and yet could rotate the light in opposite directions by different amounts. These mirror-related phenomena have no distinguishable physical effects and therefore offer no operational way to tell right- and left-handed systems apart. We note that any existing asymmetric physical behavior could be used as a means of differentiating between right- and left-handedness. These models of familiar macroscopic symmetry do not prepare us for the mirror asymmetry that exists among subnuclear particles.

The nucleon and the pion exist, and possess internal structure, by virtue of the strong interaction. Both the strong and electromagnetic interactions are known from experiment to offer no evidence for any violation of reflection symmetry. Since these two strongest interactions respect the parity property, it makes sense to assign an intrinsic parity to the particles and treat this quantum number as a conserved quantity. We have already applied the concept of parity conservation in the context of these particular interactions to deduce the parity of the pion and to establish the parity selection rule for radiative decay. The weak interaction has yet to be examined for its behavior under the parity operation. It is in this domain that we find decisive evidence for a breakdown of reflection symmetry.

A puzzle developed early in the 1950s out of experiments on the properties of certain unstable particles, known at the time as tau and theta mesons. The two species were distinguished by their different modes of weak decay. It was observed that tau decayed into three pions while theta decayed into two pions, and it was shown that these final states had odd and even overall parities, respectively. This observation suggested that odd and even parities should be assigned to the parent tau and theta particles themselves. The puzzling aspect of the situation was the fact that the two unstable mesons were identical in every other respect. In particular, the masses of tau and theta were equal, and the circumstances of their production in strong interaction processes were the same. This so-called tau-theta puzzle was resolved in 1956 by T. D. Lee and C. N. Yang. They argued that tau and theta were one and the same decaying particle (now known as the K meson, or kaon) and that the unique-parity particle was able to decay into pion states with different parities because the weak decay processes did not conserve parity. This hypothesis of Lee and Yang expressed the proposition of *parity nonconservation by the weak interaction*. Specific experiments were recommended to look for other violations of parity. One of the proposed tests was to investigate the nuclear β decays of unstable nuclei whose magnetic moments could be aligned in an applied magnetic field.

Confirmation of the Lee–Yang idea came promptly in an experiment directed by C. S. Wu on the β decay of ^{60}Co,

$$^{60}\text{Co} \rightarrow {}^{60}\text{Ni} + e^- + \bar{\nu}_e.$$

The cobalt source was placed in a magnetic field to orient the nuclear moments, and the sample was kept at low temperature to inhibit the thermal disordering of the aligned spins. The nuclide ^{60}Co was an excellent choice because of its large nuclear

Figure 16-10

^{60}Co decay and its mirror image. Electrons are emitted to the south in correlation with a west-to-east spinning motion of the sample of nuclear spins. The image reverses the spinning motion, but not the electron direction, to give an unrealized decay configuration.

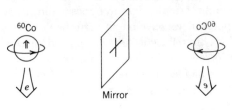

spin ($i^p = 5^+$) and long half-life ($\tau_{1/2} = 5.3$ y). Particular advantage was taken of the ^{60}Co decay sequence (shown in Figure 15-18), in which β decay occurred from the ^{60}Co ground state to an excited state of ^{60}Ni, and γ decay followed to the ^{60}Ni ground state. The β and γ radiation could be detected together in the experiment, and the observed distribution of γ rays from the oriented nuclei could be used to monitor the polarization of the sample. It was found that when the spins were aligned predominantly in one direction, the electrons in the β-decay process were emitted preferentially in the opposite direction. This observation was interpreted as a violation of reflection symmetry. Conservation of parity would have called for equal numbers of electrons emitted parallel and antiparallel to the spin orientation (as was observed for the γ rays emitted by the same aligned sample).

We can understand the asymmetric emission of electrons by the polarized source as a violation of mirror symmetry by examining the argument sketched in Figure 16-10. The sample is shown with its spins pointing north, representing a spinning motion from west to east, while the electrons are shown radiating to the south. In the mirror, the image of the experiment shows the electrons emitted again to the south, but in correlation with a spinning motion from east to west. The mirror image does not correspond to an observed process of electron emission. Since the image process does not occur in nature, it follows that the decay of the nucleus is not reflection symmetric.

Parity violation may be used to define right- and left-handedness, since the symmetry between the two is broken in β decay. Let us suppose that we are in communication with an experimenter in another galaxy, and we wish to tell the person how to make a left-hand screw. The instructions would be to align a ^{60}Co sample and observe the direction of the emitted electrons. The threads should then be cut so that the screw advances in the electron direction as it turns in the sense of the spinning nuclei. The violation of reflection symmetry is needed to convey this distinction between right and left. It is also essential that the distant experimenter uses ^{60}Co and not anti-^{60}Co in order that parity nonconservation alone serves to distinguish right from left.

Mirror-symmetry violation is a built-in property of the neutrino. We see the evidence for this parity-nonconserving feature of the particle when we examine β decay and find that the neutrino possesses a unique *handedness*. Figure 16-11 demon-

Figure 16-11

Experiment to determine the neutrino spin orientation. Electron capture occurs in an $\ell = 0$ initial state, and so the z component of the total angular momentum can have either of the quantum numbers $m = +\frac{1}{2}$ or $m = -\frac{1}{2}$. The γ ray in the final state is observed to be left-handed, with $m_\gamma = -1$. Therefore, the collinear neutrino must have $m_\nu = +\frac{1}{2}$, and the initial-state quantum number must have been $m = -\frac{1}{2}$. The bookkeeping depends on the fact that the initial and final nuclei have $i = 0$ and on the fact that there are no orbital contributions to the angular momentum. Since the neutrino is in a spin-up state with respect to a z axis opposite to its momentum, it follows that the neutrino is a left-handed particle.

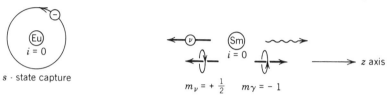

strates this property in a schematic rendition of a classic experiment, performed by M. Goldhaber and others in 1958. A two-step process is shown all at once, beginning with an electron-capture transition from a metastable state of ^{152}Eu to an excited state of ^{152}Sm, and ending with a γ-ray transition to the ground state of ^{152}Sm. The nuclear spins in the two interaction stages are given as follows:

$$e^- + {}^{152}\text{Eu}(i = 0) \rightarrow {}^{152}\text{Sm}^*(i = 1) + \nu_e$$
$$\hookrightarrow {}^{152}\text{Sm}(i = 0) + \gamma.$$

Hence, the net effect is the indicated final state consisting of ν_e and γ along with $\text{Sm}(i = 0)$. A detection scheme for γ rays is contrived to guarantee that this final system has collinear momenta, as shown in the figure. The γ rays are observed to have unique left-handed polarization, and it is inferred that the neutrinos can exist in only one of the two possible spin states normally allowed for a spin-$\frac{1}{2}$ particle. The experiment draws the conclusion that the neutrino ν_e is a *left-handed* particle, having its spin orientation opposite to its momentum. Such a configuration of spin and momentum is intrinsically asymmetric under mirror reflection.

This experiment is construed as an observation of the left-handed nature of the neutrino and, by inference, the right-handed nature of the antineutrino. We note in passing that the properties of unique handedness and absolute masslessness are intimately connected. To see the connection, we consider a massive particle whose spin and momentum are parallel in some Lorentz frame. It is possible to transform to a frame moving faster than the particle and to find that the momentum is reversed but the spin is not. The original right-handed configuration of spin and momentum is thus observed as left-handedness in the new frame. A zero-mass particle always has speed c and cannot be transformed in this manner. Therefore, its spin and momentum directions must maintain their parallel or antiparallel relationship in all Lorentz frames. These remarks are not meant to imply that the neutrino must be massless. We

Figure 16-12

Spin orientation of the muon-neutrino in the decay $\pi^+ \to \mu^+ + \nu_\mu$.

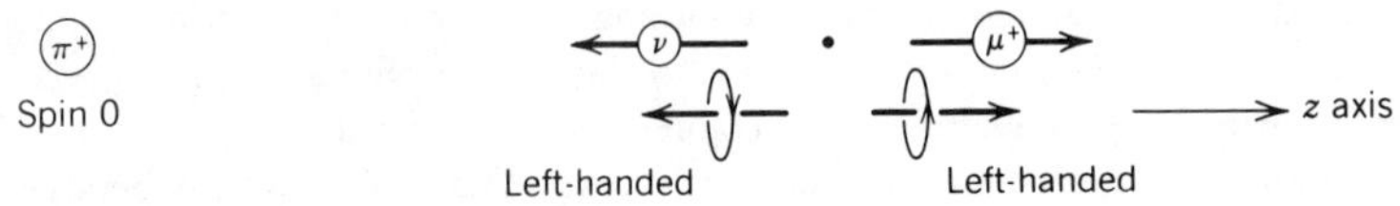

argue instead that the neutrino mass may be very small and that a very fast frame is then required to outrun the neutrino and reverse its observed handedness.

Parity nonconservation is a basic principle of weak interaction theory. The violation of mirror symmetry occurs in distinct patterns, as evidenced by the handedness properties of the neutrinos. These patterns of asymmetry are built into the current theory of the fundamental particles. Thus, the asymmetry has the same respected status in particle physics as any of the equally well-established symmetries and conservation laws.

Example

A simpler observation of neutrino spin orientation is described in Figure 16-12. In this case we consider the muon-neutrino emitted in pion decay, $\pi^+ \to \mu^+ + \nu_\mu$. The experiment in the figure shows that the muons are observed with their z component of angular momentum given by $m_\mu = -\frac{1}{2}$. Angular momentum conservation then implies $m_\nu = +\frac{1}{2}$ for the neutrinos, since the pion is a spin-0 meson. This observation tells us that ν_μ is an intrinsically left-handed particle.

16-7 The Weak Interaction

Let us recall the beginnings of weak interaction theory so that we can start to assemble a picture of the fundamental weak interaction. The weak theory is rooted in an analogy with quantum electrodynamics. Since the electromagnetic interaction can be formulated in terms of the emission and absorption of photons, it is natural to suppose that a similar idea also holds for the β-decay interaction.

The original β-decay theory for elementary particles was an outgrowth of the formalism for nuclear β decay. The first ideas were put forward in the Fermi theory of 1934, as discussed in Section 15-5. Fermi's picture drew a parallel between β decay and radiative decay by taking the emission of a pair of leptons in a weak nuclear transition to be similar to the emission of a photon in an electromagnetic transition. Parity nonconservation was a subsequent development that had to be built into this primitive theory.

We can rely on Feynman diagrams to visualize the weak theory, as we have done in the case of quantum electrodynamics in Section 16-2. In fact, such a representation of nuclear β decay has already been offered in Figure 15-16. This diagram from the previous chapter is reproduced for our present purposes in Figure 16-13. We see that

Figure 16-13

Neutron-decay diagram composed of weak nucleon and lepton currents. The weak current–current interaction occurs at very short range, and the strength of the coupling is given by the Fermi constant G_F.

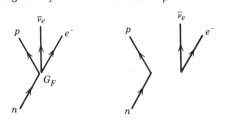

Figure 16-14

Muon decay given by the coupling of weak μ-lepton and e-lepton currents. The strength of the coupling is determined by the same G_F as in neutron decay.

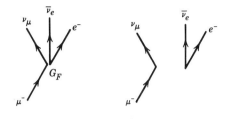

the interaction involves the coupling at very short range between a nucleon world-line and an emitted lepton pair. These coupled elements of the theory are identified as *weak currents*, by analogy with the electromagnetic current shown in Figure 16-5. The neutron-to-proton world-line constitutes a *weak nucleon current* in which the interacting particle undergoes a change of identity as the weak transition takes place. This feature of the weak theory is in contrast with the identity-preserving and charge-preserving properties of the electromagnetic current. The pair of emitted leptons $\bar{\nu}_e e$ is represented in similar fashion by a *weak lepton current*, whose orientation is contrived to fit the given diagram. We display these weak currents in Figure 16-13 as separate elements of the interacting system.

The weak nucleon current of the original Fermi theory is a polar-vector quantity. Its structure is the same as the electromagnetic current except for the operation that changes the identity of the nucleon. An analogous weak nucleon current of axial-vector form is also needed to allow for parity nonconservation. To interpret the interaction in Figure 16-13, it is understood that both polar-vector and axial-vector contributions of opposite parity appear in the nucleon current and that the lepton current incorporates the unique handedness of the neutrino. The resulting construction is known as the *universal Fermi interaction*. This version of the weak theory is due to Feynman and M. Gell-Mann and also to R. E. Marshak and E. C. G. Sudarshan. The current–current form of the weak interaction has been in use since 1958 and has been found to give a satisfactory description of all low-energy β-decay processes.

The *universal* character of the Fermi theory refers to the adaptability of the interaction to the substitution of other weak currents. Figure 16-14 shows an example involving the coupling of two weak *lepton* currents, one for μ-leptons and one for e-leptons. We note that the nucleon current of Figure 16-13 is replaced by the μ-lepton current so that the universal interaction describes the β decay of the muon. In all applications of weak current–current coupling, the strength of the interaction is expressed in terms of a universal coefficient called the Fermi constant G_F. Its value is given by

$$G_F = 8.95 \times 10^{-50} \text{ MeV} \cdot \text{m}^3$$

Figure 16-15

Exchange of weak quanta in the β decay of the neutron and in the neutrino reaction $\bar{\nu}_\mu + p \rightarrow \mu^+ + n$.

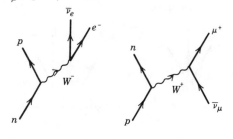

and is often written in terms of the proton mass as

$$G_F = 1.03 \times 10^{-5} \left(\frac{\hbar}{M_p c} \right)^2 \hbar c.$$

We introduce this quantity as a coupling constant at the vertices of the Feynman diagrams in Figures 16-13 and 16-14.

We have represented the weak theory by means of an interaction between currents, as in the case of quantum electrodynamics. We now wish to indicate how the interaction may be formalized in a quantum field theory, where a mediating field is introduced and virtual field quanta are exchanged. The theory is supposed to employ a weak vector field, by analogy with the electromagnetic field, for use as a mediator between any pair of weak currents. The quanta exchanged in this field are then supposed to be analogous to the photon, except that the charge-changing property of the weak currents requires the new quanta to carry units of charge $\pm e$ as well as momentum and energy. Figure 16-15 shows the exchange of these weak quanta, denoted by $W^\pm$, between certain weak currents of the universal Fermi theory.

The weak interaction is distinguished by features other than the charged nature of the quanta and the corresponding charge-changing behavior of the interacting particles. The weak and electromagnetic theories also differ markedly as to the range of the respective fields and the mass of the exchanged quanta. We recall that these properties of a quantum field theory are connected by the range-mass relation in Equation (16-2). We know that the electromagnetic interaction has infinite range and that the photon has zero mass. We also know that the weak interaction has very short range, and so we expect the weak field to have very massive quanta. These considerations imply that weak decays like the one in Figure 16-15 occur at energies too low to be useful for probing the very short-range details of the weak field. A weakly interacting probe of very high energy is required if the weak interaction is to be sampled at distances short enough to reveal the mass of the weak quantum. Reactions induced by high-energy neutrinos are suitable for this purpose. An example of such a neutrino reaction is included in Figure 16-15 to illustrate the exchange of a weak quantum between a scattered pair of weak currents.

The interaction of a weak current with the massive quanta of a weak vector field can be formulated in a legitimate quantum field theory. This accomplishment entails the introduction of other principles beyond the ideas suggested here. A renormalizable theory results, in which the principles are artfully devised to generate the desired interaction. The achievement of this goal is one of the recent triumphs in the theory of the fundamental interactions.

Example

The scale of the weak interaction is set by the Fermi constant G_F, a very small quantity. We wish to make a comparison with the fine structure constant α of quantum electrodynamics, but we cannot do so directly since α is dimensionless while G_F is not. Figure 16-15 suggests an approach based on the exchange of weak quanta. Let us assign a coupling constant g to each vertex where a weak quantum is coupled to a weak current. If we compare the diagrams for neutron decay between Figures 16-13 and 16-15, we see that G_F must be proportional to g^2. The dimensions of G_F can then be assembled from the mass of the exchanged quantum and from fundamental constants. A reasonable expression with these properties may be written in the form

$$G_F = \frac{g^2}{\hbar c}\left(\frac{\hbar}{m_W c}\right)^2 \hbar c,$$

where $g^2/\hbar c$ is treated as a dimensionless factor and where the range $\hbar/m_W c$ is included to secure the proper dimensions. The mass of the weak quantum $W^\pm$ is now known to be around 81 GeV$/c^2$. Let us use this result to compute the range of the weak field:

$$r_W = \frac{\hbar c}{m_W c^2} = \frac{197\ \text{MeV} \cdot \text{fm}}{81 \times 10^3\ \text{MeV}} = 2.4 \times 10^{-3}\text{fm}.$$

We can obtain a numerical evaluation of $g^2/\hbar c$ from the quoted value of G_F:

$$\frac{g^2}{\hbar c} = \frac{G_F}{r_W^2 \hbar c} = \frac{8.95 \times 10^{-50}\ \text{MeV} \cdot \text{m}^3}{(2.4 \times 10^{-18}m)^2(197 \times 10^{-15}\ \text{MeV} \cdot \text{m})} = \frac{1}{13}.$$

We note that r_W is very small, and we see that $g^2/\hbar c$ is analogous to the fine structure constant

$$\alpha = \frac{e^2}{4\pi\varepsilon_0 \hbar c} = \frac{1}{137}.$$

Thus, the two comparable quantities in the weak and electromagnetic theories are found to differ by only a factor of 10. We must therefore explain the enormous difference in strength between the two interactions by turning to some other distinguishing property of the two theories. We conclude that the very weak strength of the weak interaction must be attributed to the extremely short range of the weak field.

16-8 Strangeness

We are about to describe a proliferation of particles, and we wish to prepare for this outburst by introducing a preliminary classification scheme. The elementary particles are conveniently organized into two mutually exclusive families called the *leptons* and the *hadrons*. Leptons have already been defined in Section 15-5 as particles that do not engage in any strong interaction processes. The first part of Table 16-1 contains a complete list of the known leptons, including the new members recently encountered in this chapter. The strongly interacting particles comprise the much more numerous family of hadrons. A partial list of some of the existing hadrons is given in the other two parts of the table. These particles have properties that are more complicated than the leptons since they participate in *all* the fundamental interactions. The table indicates some of their characteristics in the form of several *new quantum numbers*. Our goal is to interpret these quantities by uncovering the corresponding symmetries and conservation laws.

We make a further subdivision of the hadrons according to fermion and boson properties. The strongly interacting particles with half-integral and integral spins are called *baryons* and *mesons*, respectively. Table 16-1 lists one particular group of hadrons in each of these two categories. A conserved *baryon number B* is defined accordingly:

$$B = 0 \text{ for mesons}, \quad B = +1 \text{ for baryons}, \quad B = -1 \text{ for antibaryons}.$$

We note that the baryon is a generalization of the nucleon. This interpretation is borne out by the presence of the proton and neutron in the list of spin-$\frac{1}{2}$ baryons. A conservation law for the total baryon number is adopted by extension of the law from nuclear physics in which the total nucleon number, or mass number, is conserved. Baryon number conservation is ultimately an assertion of the stability of the lightest baryon, the proton, against any known form of decay. We return to this interesting notion in another context later on.

Our first new quantum number is the strangeness S appearing in the hadronic parts of Table 16-1. We introduce S coincidentally with the arrival of the new baryons $\Lambda, \Sigma^+, \ldots$ (called hyperons) and the new mesons $K^+, K^0, \ldots$ (called kaons). Let us suppose that we have a beam of π^- mesons of increasing energy incident on protons in a liquid-hydrogen bubble chamber. Figure 16-16 shows the thresholds for particle production as the collision energy grows through values large enough to create the masses of the various new particles. The plotted quantity is the total relativistic energy of the colliding system in the CM frame, and so the thresholds are given by the total rest energies for the indicated production channels. The figure includes the familiar $\pi^- p$ processes

$$\pi^- + p \to \begin{cases} \pi^0 + n \\ \\ \pi^- + p \end{cases} \quad \text{and} \quad \pi^- + p \to \begin{cases} \pi^0 + \pi^0 + n \\ \pi^0 + \pi^- + p \\ \pi^+ + \pi^- + n \end{cases}$$

Table 16-1 Properties of Leptons and Hadrons

Leptons	Mass (MeV/c^2)	Lifetime (s)
e	0.5110	Stable
ν_e	$< 4.6 \times 10^{-5}$	Stable
μ	105.7	2.20×10^{-6}
ν_μ	< 0.50	Stable
τ	1784	3.4×10^{-13}
ν_τ	< 164	

Hadrons are classified as baryons and mesons. The tabulated quantum numbers refer to charge Q/e, strangeness S, isospin t and T_z, and hypercharge Y.

$\frac{1}{2}^+$ Baryons	Q/e	S	t	T_z	Y	Mass (MeV/c^2)	Lifetime (s)
p	1	0	$\frac{1}{2}$	$\frac{1}{2}$	1	938.3	Stable
n	0	0	$\frac{1}{2}$	$-\frac{1}{2}$	1	939.6	898
Λ	0	-1	0	0	0	1116	2.63×10^{-10}
Σ^+	1	-1	1	1	0	1189	0.80×10^{-10}
Σ^0	0	-1	1	0	0	1192	5.8×10^{-20}
Σ^-	-1	-1	1	-1	0	1197	1.48×10^{-10}
Ξ^0	0	-2	$\frac{1}{2}$	$\frac{1}{2}$	-1	1315	2.90×10^{-10}
Ξ^-	-1	-2	$\frac{1}{2}$	$-\frac{1}{2}$	-1	1321	1.64×10^{-10}

0^- Mesons	Q/e	S	t	T_z	Y	Mass (MeV/c^2)	Lifetime (s)
$\pi^\pm$	± 1	0	1	± 1	0	139.6	2.60×10^{-8}
π^0	0	0	1	0	0	135.0	8.3×10^{-17}
$K^\pm$	± 1	± 1	$\frac{1}{2}$	$\pm\frac{1}{2}$	± 1	493.7	1.24×10^{-8}
$K^0/\overline{K}^0$	0	± 1	$\frac{1}{2}$	$\mp\frac{1}{2}$	± 1	497.7	$(0.89 \times 10^{-10}$ and $5.18 \times 10^{-8})^a$
η	0	0	0	0	0	548.8	$(0.88 \text{ keV})^b$

[a] The neutral kaon lifetimes pertain to K_S and K_L.
[b] The η lifetime is not measured directly, so the width is quoted instead.

Figure 16-16

Thresholds for particle production in reactions initiated by $\pi^- p$ and $K^- p$ collisions. The production channels for the two interacting hadron systems have strangeness $S = 0$ and $S = -1$, respectively.

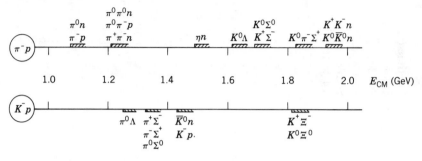

at the lower energies. The new particles make their first appearance at the thresholds for the *associated production* reactions

$$\pi^- + p \to K^0 + \Lambda \quad \text{and} \quad \pi^- + p \to \begin{cases} K^0 + \Sigma^0 \\ \\ K^+ + \Sigma^- \end{cases}.$$

Our attention is drawn especially to the fact that the observed hyperons and kaons are *not* seen singly in final states such as $K^0 + n$, or $\pi^0 + \Lambda$, or $\pi^+ + \Sigma^-$, even though these systems have thresholds at lower energies. Apparently, the new *strange particles* are produced with some intrinsic property that prevents their occurrence until there is enough energy to create these particles in association with each other. A new kind of conservation law is in evidence here. The conserved quantity is the strangeness quantum number S, and the assignment of a value of S to each hadron is such that the total strangeness is conserved in every hadronic reaction. Thus, if we take $S = 0$ for pions and nucleons, $S = +1$ for K^+ and K^0, and $S = -1$ for Λ, Σ^0, and Σ^-, we see that conservation of strangeness holds in the observed $\pi^- p$ reactions. We also note that the conservation law would be violated by the production of unassociated hyperons or kaons. Other strange particles observed at the higher $\pi^- p$ thresholds in Figure 16-16 are the hyperon Σ^+ and the kaons K^- and $\overline{K}^0$. It is clear that we must assign $S = -1$ to each of these particles if strangeness is to be conserved in the indicated final states.

Our search for more strange particles can be continued along lines suggested in the second portion of Figure 16-16. If we let a beam of K^- mesons be incident on protons in our liquid-hydrogen bubble chamber, we find thresholds as shown for the various final states $\pi^0 + \Lambda, \pi^0 + \Sigma^0, \ldots$. All these systems have total strangeness $S = -1$. The new baryons Ξ^0 and Ξ^- also appear, with strangeness $S = -2$, in the reactions

$$K^- + p \to \begin{cases} K^0 + \Xi^0 \\ \\ K^+ + \Xi^- \end{cases}$$

at the higher energy thresholds in the figure. It is noteworthy that the new strange baryons are $\frac{1}{2}^+$ particles like the nucleons and that the new kaons are 0^- mesons like the pions. The organization of the hadrons in Table 16-1 incorporates this observation. We find that the strangeness-conserving $\pi^- p$ and $K^- p$ reactions, indicated by the thresholds in the figure, have cross sections in the millibarn range. We therefore regard the reactions as strong interaction processes.

All strange particles are unstable. Their various modes of decay contain further information about the validity of strangeness as a conserved quantity. Let us list the dominant modes for the strange baryons and charged kaons as follows:

$$\Lambda \rightarrow \begin{cases} p + \pi^- \\ n + \pi^0 \end{cases}$$

$$\Sigma^+ \rightarrow \begin{cases} p + \pi^0 \\ n + \pi^+ \end{cases} \qquad \Sigma^0 \rightarrow \Lambda + \gamma \qquad \Sigma^- \rightarrow n + \pi^-$$

$$\Xi^0 \rightarrow \Lambda + \pi^0 \qquad \Xi^- \rightarrow \Lambda + \pi^-$$

and

$$K^+ \rightarrow \begin{cases} \mu^+ + \nu_\mu \\ \pi^+ + \pi^0 \\ \pi^+ + \pi^+ + \pi^- \\ \pi^+ + \pi^0 + \pi^0 \\ \pi^0 + e^+ + \nu_e \\ \pi^0 + \mu^+ + \nu_\mu \end{cases} \qquad K^- \rightarrow \begin{cases} \mu^- + \bar{\nu}_\mu \\ \pi^- + \pi^0 \\ \pi^- + \pi^- + \pi^+ \\ \pi^- + \pi^0 + \pi^0 \\ \pi^0 + e^- + \bar{\nu}_e \\ \pi^0 + \mu^- + \bar{\nu}_\mu \end{cases}$$

(Neutral kaons decay in a special way to be discussed in Section 16-9.) The lifetimes for all but the Σ^0 decay are in the range 10^{-10} to 10^{-8}s, long enough for the various charged particles to form measurable tracks in a bubble chamber. Figure 16-17 shows an example of particle production and decay, as recorded by a photograph taken in such a detector. The orders of magnitude of the lifetimes are characteristic of decays that proceed via the *weak* interaction. The main observation to associate with this remark is the fact that strangeness is *not* conserved in these decays. If we examine each of the listed strange-particle decays (except the one for Σ^0), we find that strangeness conservation is violated by exactly one unit in every case. The exceptional Σ^0 decay occurs with a much shorter lifetime and yields a photon in the final state. These circumstances, taken together with the strangeness-conserving nature of the process, tell us that Σ^0 decay is governed by the electromagnetic interaction.

Strangeness is assigned as an internal attribute of every hadron, so that the assignment is completely independent of the other kinematical properties of the particle. We handle strangeness as an additive quantum number in the same way that we treat charge and baryon number. It should be emphasized, however, that charge is a special type of particle property. Every particle has a charge assignment, where the charge is directly measurable in terms of the response of an electromagnetic current to the application of an electromagnetic field. On the other hand, strangeness and baryon number are defined as useful concepts only for hadrons. A more fundamental interpretation of these quantum numbers remains for us to uncover.

Conservation of strangeness is a new property of the strong interaction and also of the electromagnetic interaction. The weak interaction acknowledges the property of

Figure 16-17

Associated production and decay of strange particles in a liquid-hydrogen bubble chamber. The production-decay sequence consists of the processes

$$\pi^- + p \rightarrow K^0 \ + \Lambda$$
$$\Big\downarrow\rightarrow p + \pi^-$$
$$\Big\downarrow\rightarrow \pi^+ + \pi^- \ .$$

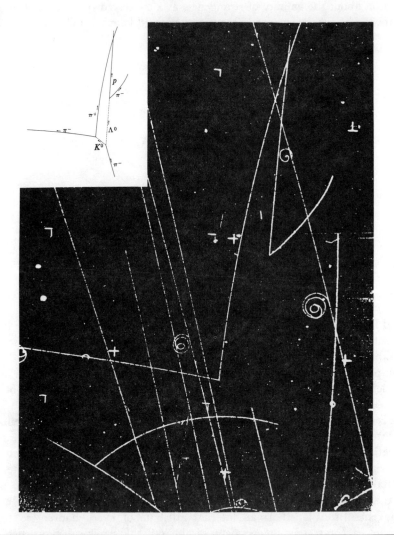

strangeness but does not observe the conservation law. Instead, the strangeness quantum number changes in a definite way, by a single unit, in every instance of strange-particle weak decay. We have already seen the breakdown of a conservation law due to the weak interaction in our discussion of parity. Here, another conserved quantity of an internal nature is found to be violated in the same class of physical processes. It is interesting to note that the strong interaction determines the existence of the hadrons and obeys the conservation law, while the weak interaction causes their decay and exhibits a breaking of the law.

Strange particles were first seen in cosmic-ray events by G. D. Rochester and C. C. Butler in 1947, the year of the pion. Hyperons and kaons began to be produced in accelerator reactions in the next decade, and associated production was interpreted during this period. The idea that strange particles were produced in pairs was originally proposed by A. Pais. Subsequently, the concept of strangeness was fully developed in terms of a symmetry principle by Gell-Mann. The identification of this new internal quantum number was the first step in the organization of a body of symmetries governing all the hadrons.

Example

We can apply the methods of relativistic kinematics from Section 1-11 to study the associated production of strange particles. Let us consider π^- mesons incident on protons at rest and determine the threshold beam energy for the $K^0 + \Lambda$ final state. In the reaction $\pi^- + p \rightarrow K^0 + \Lambda$, we have the two momentum four-vectors

$$\mathscr{P} = \begin{bmatrix} \mathbf{P}_\pi \\ i\dfrac{E_\pi + M_p c^2}{c} \end{bmatrix} \quad \text{and} \quad \mathscr{P}' = \begin{bmatrix} 0 \\ i(m_K + M_\Lambda)c \end{bmatrix}$$

for the initial $\pi^- p$ system in the lab frame and the final $K^0 \Lambda$ system *at rest* in the CM frame. Lorentz invariance relates $\mathscr{P}$ and $\mathscr{P}'$ by the equality

$$p_\pi^2 - \frac{\left(E_\pi + M_p c^2\right)^2}{c^2} = -(m_K + M_\Lambda)^2 c^2,$$

where p_π and E_π satisfy the relation

$$E_\pi^2 = c^2 p_\pi^2 + m_\pi^2 c^4.$$

We are led to the threshold result for E_π in a few algebraic steps:

$$\left(m_K + M_\Lambda\right)^2 c^4 = E_\pi^2 + 2 E_\pi M_p c^2 + M_p^2 c^4 - c^2 p_\pi^2$$

$$= 2 E_\pi M_p c^2 + \left(m_\pi^2 + M_p^2\right) c^4$$

$$\Rightarrow \quad E_\pi = \frac{\left(m_K + M_\Lambda\right)^2 - \left(m_\pi^2 + M_p^2\right)}{2 M_p} c^2.$$

The beam kinetic energy at threshold is given by the formula

$$K_\pi = E_\pi - m_\pi c^2 = \frac{\left(m_K + M_\Lambda\right)^2 - \left(m_\pi + M_p\right)^2}{2M_p} c^2$$

$$= \frac{m_K + M_\Lambda + m_\pi + M_p}{2M_p} \left(m_K + M_\Lambda - m_\pi - M_p\right)c^2.$$

If we consult Table 16-1 to find the various particle masses, we obtain

$$K_\pi = \frac{498 + 1116 + 140 + 938}{2(938)}(498 + 1116 - 140 - 938)\text{MeV} = 769 \text{ MeV}.$$

Other illustrations of this four-vector technique are offered as problems at the end of the chapter.

16-9 Neutral K Decay and CP Symmetry

The mesons K^0 and $\overline{K}^0$ are produced as distinct particles in strangeness-conserving strong reactions. Their characteristics are distinguished by the assignment of different strangeness quantum numbers $S = +1$ and $S = -1$. We have postponed our discussion of the decays of the neutral kaons because we have not yet described the symmetry principles that pertain to these remarkable phenomena. The analysis of this problem is an elegant application of basic quantum mechanics, which leads us directly to a major experimental discovery.

We are concerned with the properties of the states of particles under the three symmetry operations of charge conjugation, space inversion, and time reversal (symbols C, P, and T). Charge conjugation C changes particle to antiparticle by the transformation $X \to \overline{X}$. Space inversion P reflects the spatial coordinates according to the familiar parity operation $\mathbf{r} \to -\mathbf{r}$. Time reversal T runs the time variable in the reversed sense through the substitution $t \to -t$. Our problem is to ascertain the behavior of the fundamental interactions in a given system as each operation is allowed to act on the wave function and produce a transformed state. Thus, a particular physical process is turned into three other processes under the action of the different symmetry operations, and each of the three is compared in turn to the one given. The comparison asks specifically whether the transformed copy process is physical, and whether the transition probability is the same as for the original process. We recall that we have already pursued this line of inquiry with regard to the P operation in Section 16-6. If the copy and the original are found to have equal probability, then the interaction governing the process is said to be C-, P-, or T-invariant, depending on the transformation in question. We can also subject the interacting system to any combination of the three operations. In this way we may learn, for instance, that CP-invariance holds where C and P symmetries are violated, and we may then argue that the two violations exactly compensate each other.

An important theorem applies to these considerations. It can be proved from very general properties of relativistic quantum field theory that every interaction is invariant under the composite transformation CPT. This invariance means that the combined symmetry with respect to the three operations together is always respected, even in situations where any of the three factors exhibit violations. One of the main consequences of $CP\!T$-invariance is the requirement that particle and antiparticle have the same mass and the same lifetime. It is clear that any observed breakdown of this symmetry would be cause for alarm, calling for a reevaluation of the principles of quantum field theory.

Our principal concern is the behavior of the weak interaction with regard to C, P, and T. We already know that P-invariance is violated in β decay: the mirror image of a certain decaying sample of aligned nuclei does not exist as an observed physical system. We can see immediately that C-invariance is also violated by the inherent left-handedness of the neutrino: the application of C to the neutrino yields a left-handed antineutrino, a nonexistent particle. Note, however, that when CP is applied to the left-handed neutrino, a legitimate right-handed antineutrino results, so that C and P appear to have compensating violations in the weak interaction. No direct and decisive experimental evidence for time-reversal asymmetry is known, although tests of this property continue to be explored. We may therefore expect CP-invariance to hold as a consequence of T-invariance and the $CP\!T$ theorem. These notions furnish the background for our analysis of the neutral kaon system.

Neutral kaons decay predominantly into $\pi^+\pi^-$ and $\pi^0\pi^0$ final states. The mesons K^0 and $\bar{K}^0$ have the same mass, charge, and baryon number, and so it is natural to ask what distinguishes the two particles. We know that the strangeness quantum numbers are different, since K^0 has $S = +1$ and $\bar{K}^0$ has $S = -1$. We also know, however, that strangeness is not conserved in the decays of strange particles. Consequently, the strangeness distinction between K^0 and $\bar{K}^0$ becomes inoperative when the weak decay interaction takes effect. Let us apply quantum mechanics directly to this problem and consider the *superposition* of the mass-degenerate K^0 and $\bar{K}^0$ states. We are especially interested in the *mixed-strangeness* combinations

$$K_1 = \frac{1}{\sqrt{2}}(K^0 - \bar{K}^0) \quad \text{and} \quad K_2 = \frac{1}{\sqrt{2}}(K^0 + \bar{K}^0), \qquad (16\text{-}6)$$

because these states have definite properties under the CP operation. If we accept this assertion for the moment, we can perceive a subtle distinction between K^0 and $\bar{K}^0$. Let us solve Equation (16-6) for the states of definite strangeness:

$$K^0 = \frac{1}{\sqrt{2}}(K_1 + K_2) \quad \text{and} \quad \bar{K}^0 = -\frac{1}{\sqrt{2}}(K_1 - K_2). \qquad (16\text{-}7)$$

These wave functions correspond to different superpositions of definite-CP states. This interplay between strangeness and CP properties has very interesting experimental implications.

Let us examine the behavior of the states under C and P before we turn to the experimental situation. Charge conjugation changes particle to antiparticle according to the transformations

$$C K^0 = \bar{K}^0 \quad \text{and} \quad C \bar{K}^0 = K^0. \qquad (16\text{-}8)$$

Space inversion then introduces a minus sign for the 0^- particles:

$$CP K^0 = -\bar{K}^0 \quad \text{and} \quad CP \bar{K}^0 = -K^0. \qquad (16\text{-}9)$$

The *CP* properties of the states in Equations (16-6) follow from these operations:

$$CP\,K_1 = K_1 \quad \text{and} \quad CP\,K_2 = -K_2. \tag{16-10}$$

Thus, we find that the K_1 superposition is *CP*-even, while the K_2 superposition is *CP*-odd.

Next, we consider the $\pi^+\pi^-$ final state that results from neutral kaon decay. This system must have angular momentum zero and orbital quantum number $\ell = 0$. The overall *even* parity of the state is determined from the orbital parity, which is even, and the intrinsic parities of the two pions, which are odd. Thus, when we apply C and P to the $\pi^+\pi^-$ state, we find that C interchanges π^+ and π^-, and then P interchanges them back, such that the sign of the wave function remains unaltered. Consequently, the $\pi^+\pi^-$ final state in neutral K decay is *CP*-even, and the same is true for the $\pi^0\pi^0$ final state. If *CP*-invariance is a valid symmetry of the weak interaction, then the K_1 portion of Equations (16-7) must be responsible for the observed 2π decays since the K_2 state is forbidden to have the *CP*-even 2π mode. Decay into *three* pions is allowed for K_2; however, the Q-value is much smaller for 3π than for 2π, so that the probability should be correspondingly smaller for K_2 decay than for K_1 decay. A substantially longer lifetime should therefore be observed for K_2 relative to K_1. These arguments lead us to the conclusion that, although K^0 and $\overline{K}^0$ are the particles produced with definite strangeness in strong reactions, they are *not* the particles observed in weak decays. Instead, K^0 and $\overline{K}^0$ *mix* to form two distinct decay states K_1 and K_2, whose lifetimes are quite different because of *CP* invariance.

The principles of $K^0 - \overline{K}^0$ mixing were developed in 1955 by Gell-Mann and Pais. (Their analysis actually predated the overthrow of parity and assumed *C*-invariance instead of *CP*-invariance to reach the same general conclusions.) The mixing theory claimed that the familiar neutral K decays into two pions were due to the *CP*-even component K_1, and that other decay modes should also exist corresponding to a longer-lived state with the *CP*-odd property of the component K_2. These predicted decays were detected in 1956, in apparent fulfillment of *CP*-invariance. The two definite-*CP* states were found to have the following principal modes of decay:

$$K_1 \to \begin{cases} \pi^+ + \pi^- \\ \pi^0 + \pi^0 \end{cases} \quad \text{and} \quad K_2 \to \begin{cases} \pi^+ + e^- + \bar{\nu}_e \\ \pi^- + e^+ + \nu_e \\ \pi^+ + \mu^- + \bar{\nu}_\mu \\ \pi^- + \mu^+ + \nu_\mu \\ \pi^0 + \pi^0 + \pi^0 \\ \pi^+ + \pi^- + \pi^0 \end{cases},$$

with lifetimes around 10^{-10} and 10^{-8} s, respectively. This interpretation of the neutral kaon system remained intact until 1964.

The logic of $K^0 - \overline{K}^0$ mixing follows from the quantum mechanical principle of superposition. We can pursue the arguments further and show how neutral kaons participate in the process of *regeneration*, an extraordinary phenomenon in which a member of a system is severed away and later grows back. The following discussion of the process is illustrated in Figure 16-18. We suppose that π^- mesons are incident on a slab of material and that the collisions in matter produce K^0's via the familiar

Figure 16-18

Regeneration of short-lived neutral kaons. Reactions in slab A produce K^0's whose K_1 component decays away, leaving a pure K_2 beam incident on slab B. The K^0 and $\overline{K}^0$ parts of the K_2 state have different strong reactions in the slab. The $\overline{K}^0$ portions are removed in these processes so that a beam of K^0's emerges. A regenerated K_1 component is contained in the emergent beam.

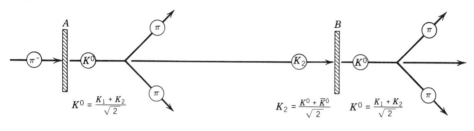

associated-production reaction

$$\pi^- + p \rightarrow K^0 + \Lambda.$$

Equations (16-7) tell us that *half* of the emerging K^0's are in the K_1 state, which decays rapidly into two pions. The surviving half consists of kaons in the longer-lived K_2 state. These particles are directed at a second slab of material, located along the beam-line at a distance such that the transit time between slabs is greater than the K_1 lifetime and less than the K_2 lifetime. We can therefore assume that we have a pure K_2 beam incident on the second slab. These kaons interact strongly with matter through their K^0 and $\overline{K}^0$ components, so that strangeness conservation becomes operative at this stage of the regeneration mechanism. The $\overline{K}^0$ portion of the K_2 state can make strange baryons in the reactions

$$\overline{K}^0 + p \rightarrow \begin{cases} \pi^+ + \Lambda \\ \pi^0 + \Sigma^+ \\ \pi^+ + \Sigma^0 \end{cases},$$

while the K^0 portion engages primarily in the scattering process

$$K^0 + p \rightarrow K^0 + p.$$

Thus, the strong reactions in the second slab *remove* most of the $\overline{K}^0$ half from the incident K_2 state and leave mostly K^0's in the transmitted beam. Equations (16-7) again tell us that this K^0 beam is an equal-parts mixture of K_1 and K_2. The remarkable conclusion is summarized by the observation that the K_1 state dies out at one end of the beam-line and then grows back at the other through the intervention of the strong interaction. The rebirth of K_1 is seen by detecting again the rapid decay into two pions, as shown in the figure. This process of regeneration of short-lived neutral kaons has been observed in the laboratory. The steps in the analysis are reminiscent of those followed in the Stern–Gerlach thought experiments of Section 8-7. Recall that we have applied superposition to spin states in the case of the

Stern–Gerlach problem. Here, we superpose states of definite strangeness to describe K_1 regeneration. Fundamental quantum principles are at work in both of these situations, as the particles exhibit wave behavior similar to the polarization states of light.

The mixing theory of Gell-Mann and Pais predicted that the CP-odd K_2 state could not decay into two pions. An experiment to place improved bounds on this forbidden mode was undertaken by J. W. Cronin, V. L. Fitch, and others in 1964. They discovered that a small fraction of all the observed decays of the long-lived neutral kaon led to the unexpected $\pi^+\pi^-$ final state. A main task of the experiment was to demonstrate that the detected 2π mode was not the result of regeneration. The similarly forbidden $\pi^0\pi^0$ state was also seen in another investigation in 1967. These findings constituted proof of the *violation* of CP-invariance in the weak interaction. The results, in conjunction with the CPT theorem, implied a breakdown of time-reversal symmetry in the neutral kaon system.

The experiments associate definite lifetimes with the two distinct physical decay states. These decaying systems do not correspond exactly to the definite-CP states K_1 and K_2. The decaying particles are therefore renamed K_S and K_L (K-short and K-long) and are called by those names in Table 16-1. The experimental results may be interpreted by attributing the violation of CP-invariance to the fact that the physical decay states K_S and K_L are not CP-pure. The departure of K_S and K_L from K_1 and K_2 is expressed in this view by altering the superposition of states to read

$$K_S = \frac{K_1 + \varepsilon K_2}{\sqrt{1 + |\varepsilon|^2}} \quad \text{and} \quad K_L = \frac{K_2 + \varepsilon K_1}{\sqrt{1 + |\varepsilon|^2}}. \tag{16-11}$$

We learn from experiment that the quantity $|\varepsilon|$ is of order 10^{-3} if we adopt this interpretation. The equations describe the short-lived state as mostly CP-even and the long-lived state as mostly CP-odd. The observed CP-violation in the decays

$$K_L \rightarrow \begin{cases} \pi^+ + \pi^- \\ \pi^0 + \pi^0 \end{cases}$$

is then presumed to be due to the small CP-even K_1 contribution in the expression for K_L. Equations (16-6) can be employed to write K_S and K_L in terms of the mass-degenerate states of definite strangeness K^0 and $\overline{K}^0$. This remark is developed more fully in the example below.

CP-invariance implies equal probabilities for the two long-lived modes

$$K_L \rightarrow \begin{cases} \pi^- + e^+ + \nu_e \\ \pi^+ + e^- + \bar{\nu}_e \end{cases}.$$

CP-violation is seen in these decays, as positrons are observed in slightly greater number than electrons. The asymmetry between e^+ and e^- is interesting because the asymmetric behavior provides an absolute means of distinguishing particles and antiparticles. We could use this effect to communicate the difference between left and right to an experimenter in a distant galaxy, a thought experiment begun in Section 16-6. The first instruction would be to establish the distinction between e^+ and e^- by observing the charge asymmetry in K_L decay. This step would furnish the standard for differentiating matter from antimatter. The β decay of aligned ^{60}Co (or anti-^{60}Co) could then be used to distinguish left from right.

Example

Our discussion of neutral kaon mixing and *CP*-violation can be neatly cast in the language of the Schrödinger equation. First, let the weak interaction be turned off and consider the simple time-dependent wave functions for stable kaons at rest:

$$K^0 = Ae^{-im_0c^2t/\hbar} \quad \text{and} \quad \overline{K}^0 = \overline{A}e^{-im_0c^2t/\hbar}.$$

Note that each state is parametrized by the *same* mass m_0. The t dependence satisfies the equations

$$\frac{i\hbar}{c^2}\frac{d}{dt}K^0 = m_0K^0 \quad \text{and} \quad \frac{i\hbar}{c^2}\frac{d}{dt}\overline{K}^0 = m_0\overline{K}^0.$$

These equations can be consolidated in matrix form as

$$\frac{i\hbar}{c^2}\frac{d}{dt}\begin{bmatrix} K^0 \\ \overline{K}^0 \end{bmatrix} = \begin{bmatrix} m_0 & 0 \\ 0 & m_0 \end{bmatrix}\begin{bmatrix} K^0 \\ \overline{K}^0 \end{bmatrix}.$$

Next, let the weak interaction be turned on, and introduce the mixing mechanism by altering the form of the "mass matrix:"

$$\begin{bmatrix} m_0 & 0 \\ 0 & m_0 \end{bmatrix} \rightarrow \begin{bmatrix} m & u \\ v & m \end{bmatrix}.$$

The diagonal elements remain equal because of *CPT*-invariance, but *CP*-violation requires the off-diagonal elements to be different. This step produces a matrix equation for K^0 and $\overline{K}^0$ that couples and mixes the t dependences of the two-states:

$$\frac{i\hbar}{c^2}\frac{d}{dt}\begin{bmatrix} K^0 \\ \overline{K}^0 \end{bmatrix} = \begin{bmatrix} m & u \\ v & m \end{bmatrix}\begin{bmatrix} K^0 \\ \overline{K}^0 \end{bmatrix}.$$

The physical decay states K_S and K_L have unique and distinct lifetimes, by definition. Consequently, their t dependences must be *unmixed*, and so the Schrödinger equation for these states must have the form

$$\frac{i\hbar}{c^2}\frac{d}{dt}\begin{bmatrix} K_S \\ K_L \end{bmatrix}\begin{bmatrix} m_S & 0 \\ 0 & m_L \end{bmatrix}\begin{bmatrix} K_S \\ K_L \end{bmatrix}.$$

We can relate the mass-degenerate states K^0 and $\overline{K}^0$ to the physical decay states K_S and K_L with the aid of equations in the text. First, we note that Equations (16-6) and (16-11) are expressible in terms of matrices:

$$\begin{bmatrix} K_1 \\ K_2 \end{bmatrix} = \frac{1}{\sqrt{2}}\begin{bmatrix} 1 & -1 \\ 1 & 1 \end{bmatrix}\begin{bmatrix} K^0 \\ \overline{K}^0 \end{bmatrix}$$

and

$$\begin{bmatrix} K_S \\ K_L \end{bmatrix} = \frac{1}{\sqrt{1+|\varepsilon|^2}}\begin{bmatrix} 1 & \varepsilon \\ \varepsilon & 1 \end{bmatrix}\begin{bmatrix} K_1 \\ K_2 \end{bmatrix}.$$

Matrix multiplication then produces the desired result:

$$\begin{bmatrix} K_S \\ K_L \end{bmatrix} = \frac{1}{\sqrt{1 + |\varepsilon|^2}} \begin{bmatrix} 1 & \varepsilon \\ \varepsilon & 1 \end{bmatrix} \frac{1}{\sqrt{2}} \begin{bmatrix} 1 & -1 \\ 1 & 1 \end{bmatrix} \begin{bmatrix} K^0 \\ \overline{K}^0 \end{bmatrix}$$

$$= \frac{1}{\sqrt{2(1 + |\varepsilon|^2)}} \begin{bmatrix} 1 + \varepsilon & -(1 - \varepsilon) \\ 1 + \varepsilon & 1 - \varepsilon \end{bmatrix} \begin{bmatrix} K^0 \\ \overline{K}^0 \end{bmatrix}.$$

The observed *CP*-violation parameter ε and the observed mass parameters m_S and m_L are ultimately connected to the quantities m, u, and v given in the mass matrix. We leave the derivation of these relations to Problem 17 at the end of the chapter. The given quantities are complex-valued, and so the derived parameters must have the same property. We interpret the meaning of the complex masses m_S and m_L by defining the following real and imaginary parts:

$$m_S c^2 = m_{S0} c^2 - i \frac{\hbar}{2\tau_S} \quad \text{and} \quad m_L c^2 = m_{L0} c^2 - i \frac{\hbar}{2\tau_L}.$$

The solutions for K_S and K_L are then written as

$$K_S = A_S e^{-i m_S c^2 t / \hbar} = A_S e^{-i m_{S0} c^2 t / \hbar} e^{-t/2\tau_S}$$

and

$$K_L = A_L e^{-i m_L c^2 t / \hbar} = A_L e^{-i m_{L0} c^2 t / \hbar} e^{-t/2\tau_L}.$$

These states have decaying probabilities:

$$|K_S|^2 = |A_S|^2 e^{-t/\tau_S} \quad \text{and} \quad |K_L|^2 = |A_L|^2 e^{-t/\tau_L}.$$

Thus, we see that m_{S0} and m_{L0} are the masses of the decaying particles K_S and K_L, and τ_S and τ_L are the corresponding lifetimes.

16-10 Isospin

The list of baryons and mesons in Table 16-1 includes the assignment of several internal quantum numbers. These properties are associated with the symmetries and conservation laws that govern the behavior of every hadron. We have just examined how the strangeness S is treated as one such conserved quantity. Another internal attribute with its own unique symmetry and conservation law is the isospin, denoted for each hadron in the table by the pair of quantum numbers t and T_z. We have already introduced the isospin properties of the strongly interacting particles in Section 14-12 in our discussion of the charge independence of the nuclear force. The binding of protons and neutrons in the nucleus is only one of many manifestations of the strong interaction. Our purpose now is to extend the domain of isospin symmetry beyond the nucleons so that the concept embraces the whole family of hadrons.

Let us begin by organizing the tabulated baryons and mesons according to two charts, as in Figure 16-19. The members of each group have spatial properties in common, since the baryons are $\frac{1}{2}^+$ particles and the mesons are 0^- particles. (These

Figure 16-19

Mass levels of the eight $\frac{1}{2}^{+}$ baryons and the eight 0^{-} mesons listed in Table 16-1.

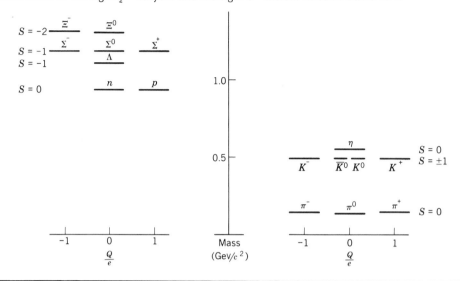

spin and parity assignments are deduced by observing the spatial distributions of outgoing particles in the reactions and decays of the various species.) The figure is a mass-level diagram, in which the eight $\frac{1}{2}^{+}$ baryons and eight 0^{-} mesons exhibit their masses at four different levels and three different levels, respectively. The striking feature of these displays is the fact that approximate *mass degeneracies* exist among certain subsets of the particles. This pattern of *degenerate multiplets* subdivides the two hadron systems into singlets, doublets, and triplets, each with a strangeness assignment and a value for the mass. The charge and the slight deviation in mass appear to be the only distinguishing characteristics among the members of each of the multiplets.

We have seen multiplets of states of different charge and nearly equal energy before in the isobaric analogue levels of nuclei. Recall that these degenerate energy states refer to nuclear isobars with the same i^{p} assignment and with different values of Z and N for a given value of the mass number A. The existence of such related nuclear states reflects the charge independence of the nuclear force and, more generally, the charge independence of the strong interaction. We know that this property of the nuclear force is expressed in terms of a symmetry with respect to rotations in isospin space. Recall that the conserved isospin vector $\mathbf{T}$ behaves as a quantized angular momentum, whereby T^{2} has eigenvalues given by $t(t+1)$, and T_{z} has $2t+1$ discrete values for each integral or half-integral choice of the isospin quantum number t. Rotational isospin symmetry then implies that the $2t+1$ quantized orientations of the vector $\mathbf{T}$ correspond to different states of equal energy. We establish a correspondence between these degenerate isospin states and the nearly equal analogue levels of isobaric nuclei. In the process, we argue that nuclear isospin multiplets result from charge independence and rotational symmetry in isospin space, just as degeneracies of angular momentum states arise in the central-force problem from rotational symmetry in ordinary space. This analogy is implemented by transferring the properties of angular momentum quantization directly to the isospin vector

T. We may also argue that slight deviations from degeneracy within an isospin multiplet are attributable to the influence of the electromagnetic interaction. These effects break isospin symmetry, just as the imposition of a magnetic field acts to separate degenerate states in the central-force problem. Thus, the strong interaction is indifferent to the isospin direction, while the electromagnetic interaction selects the z direction in isospin space and breaks the symmetry by distinguishing states of different T_z and different charge.

We assign isospin quantum numbers to nuclei in accordance with the basic $t = \frac{1}{2}$ doublet identification of the nucleon, where the proton has $T_z = +\frac{1}{2}$ (isospin up) and the neutron has $T_z = -\frac{1}{2}$ (isospin down). The z component of isospin for a nuclear state is then determined by the proton and neutron numbers as in Equation (14-47):

$$T_z = \frac{Z - N}{2}.$$

Let us reconstruct this relation into the form

$$T_z = Z - \frac{A}{2} = \frac{Q}{e} - \frac{A}{2},$$

or

$$\frac{Q}{e} = T_z + \frac{A}{2}, \tag{16-12}$$

by introducing the mass number $A = Z + N$ and by using

$$Z = \frac{Q}{e}$$

to define the charge in units of e.

We now proceed to assign the isospin quantum numbers of hadrons simply by observing the charge independence of the masses in multiplets of the sort shown in Figure 16-19. To illustrate, we note the near degeneracy of the pions π^-, π^0, and π^+ and take $t = 1$ to denote a triplet of states with $T_z = -1$, 0, and $+1$. All the tabulated $\frac{1}{2}^+$ baryons and 0^- mesons are similarly organized according to the following assignments of internal quantum numbers:

			$T_z = -1$	$T_z = -\frac{1}{2}$	$T_z = 0$	$T_z = +\frac{1}{2}$	$T_z = +1$
$B = 1$	$S = 0$	N doublet $t = \frac{1}{2}$		n		p	
	$S = -1$	Λ singlet $t = 0$			Λ		
	$S = -1$	Σ triplet $t = 1$	Σ^-		Σ^0		Σ^+
	$S = -2$	Ξ doublet $t = \frac{1}{2}$		Ξ^-		Ξ^0	

			$T_z = -1$	$T_z = -\frac{1}{2}$	$T_z = 0$	$T_z = +\frac{1}{2}$	$T_z = +1$
$B = 0$	$S = 0$	π triplet $t = 1$	π^-		π^0		π^+
	$S = +1$	K doublet $t = \frac{1}{2}$		K^0		K^+	
	$S = -1$	$\overline{K}$ doublet $t = \frac{1}{2}$		K^-		$\overline{K}^0$	
	$S = 0$	η singlet $t = 0$			η		

These properties are entered in Table 16-1.

The values of the charge Q/e, the baryon number B, the strangeness S, and the z component of isospin T_z obey the relation

$$\frac{Q}{e} = T_z + \frac{B + S}{2} \tag{16-13}$$

for each of the tabulated baryons and mesons. This relation among quantum numbers is due to Gell-Mann and K. Nishijima and is meant to be a general formula for all the hadrons. Since B and S appear together in the equation, it is convenient to define a new quantity called the *hypercharge*,

$$Y = B + S, \tag{16-14}$$

and write the Gell-Mann–Nishijima relation in the form

$$\frac{Q}{e} = T_z + \frac{Y}{2}. \tag{16-15}$$

A listing of hypercharge quantum numbers is also included in Table 16-1. We note that the assigned values of Y range symmetrically between -1 and $+1$ for all listed baryons and mesons. We also note the close similarity between Equation (16-15) for hadrons and Equation (16-12) for nuclei.

It is instructive to display the hadrons by plotting their quantum-number assignments on a graph of hypercharge versus z component of isospin. Figure 16-20 shows such a presentation for the $\frac{1}{2}^{+}$ baryons and 0^{-} mesons. The two particle patterns are known as *octets*, since there are eight members in each group. Note that the designs are symmetrical in the variables Y and T_z and are identical in form. The use of the hypercharge Y, instead of the strangeness S, accounts for the similar appearance of the baryon and meson octets. These graphs give a complete picture of the quantum-number content of the various hadrons. We observe that the isospin t is not directly indicated. The value of this quantum number is evident nevertheless, since each isospin multiplet appears on the graph as a horizontal subset of particles with

Figure 16-20

Octets of $\frac{1}{2}^{+}$ baryons and 0^{-} mesons.

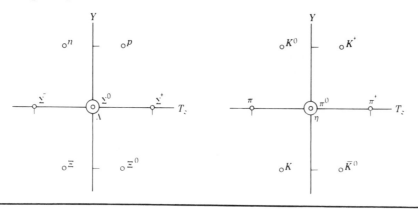

multiplicity $2t + 1$. The symmetry of these patterns suggests an organization of the hadron families on a higher level beyond the scheme put forward so far.

Let us turn now to the use of isospin as a conserved quantity in elementary particle phenomenology. We can test the conservation law, and the isospin assignment of the pion, by comparing the two reactions

$$n + p \rightarrow \pi^0 + d \quad \text{and} \quad p + p \rightarrow \pi^+ + d.$$

The deuteron is a $t = 0$ nucleus, and the pion is taken to be a $t = 1$ triplet of mesons. Both final states should then have $t = 1$, and the same should be true of the initial states if isospin is conserved. We therefore expect the cross sections to have the ratio

$$\frac{\sigma(np \rightarrow \pi^0 d)}{\sigma(pp \rightarrow \pi^+ d)} = \frac{1}{2},$$

because we see from Equations (14-48) and (14-49) that the np state has $t = 1$ in half of its composition, while the pp state has $t = 1$ exclusively. This prediction for the cross sections in the two reactions is confirmed by experiment.

Charge independence can be put to good use in the study of the interactions of pions with nucleons. The scattering processes

$$\pi^+ + p \rightarrow \pi^+ + p \quad \text{and} \quad \pi^- + p \rightarrow \begin{cases} \pi^- + p \\ \pi^0 + n \end{cases}$$

are especially useful as sources of information about the πN system. It is also possible to examine many other combinations of initial and final states of pions and nucleons. We learn from charge independence that only *two* independent complex amplitudes are required to express *all* possible combinations. We can understand this property of pions and nucleons by adding isospins and invoking isospin symmetry. The πN system has isospins $\mathbf{T}_\pi$ and $\mathbf{T}_N$, and so the total isospin is expressed as the vector sum

$$\mathbf{T} = \mathbf{T}_\pi + \mathbf{T}_N. \tag{16-16}$$

This prescription instructs us to add quantized vectors corresponding to isospin-1 and isospin-$\frac{1}{2}$. The vector addition produces two values for the total isospin quantum number according to the rule

$$1 + \frac{1}{2} = \frac{1}{2} \text{ and } \frac{3}{2}$$

(just like the addition of angular momenta $\mathbf{J} = \mathbf{L} + \mathbf{S}$, where $\ell = 1$ and $s = \frac{1}{2}$ combine to form $j = \frac{1}{2}$ and $j = \frac{3}{2}$). The main thrust of charge independence can now be recognized. Rotational isospin symmetry tells us that the strong interaction is indifferent about the orientation of the vector $\mathbf{T}$ in isospin space. Only the magnitude of $\mathbf{T}$ matters, so that a separate independent amplitude need not be introduced for every possible value of T_z. We have just found $\mathbf{T}$ to have two allowed values of the magnitude $\sqrt{t(t + 1)}$, one for $t = \frac{1}{2}$ and one for $t = \frac{3}{2}$. Consequently, only two independent quantum mechanical amplitudes are needed to describe all $\pi N \rightarrow \pi N$ processes. If we designate these quantities as $\chi_{1/2}$ and $\chi_{3/2}$, we find that the amplitudes for the elastic and charge-exchange scattering processes cited above are given by the formulas

$$\chi(\pi^+ p, \pi^+ p) = \chi_{3/2},$$

$$\chi(\pi^- p, \pi^- p) = (\chi_{3/2} + 2\chi_{1/2})/3,$$

$$\chi(\pi^0 n, \pi^- p) = \sqrt{2}\,(\chi_{3/2} - \chi_{1/2})/3. \qquad (16\text{-}17)$$

It is clear that isospin symmetry is a rather powerful idea since the number of different reactions of interest is reduced to a fundamental set labeled only by the isospin quantum number t.

We do not need a derivation of the numerical coefficients in Equations (16-17) to comprehend the basic meaning of the formulas. The first equation says that the $\pi^+ p$ system can only be in a $t = \frac{3}{2}$ state, an obvious fact since the z components of isospin add up to give

$$T_z = 1 + \tfrac{1}{2} = \tfrac{3}{2}$$

for π^+ and p. (The same argument applies to the $\pi^- n$ system where

$$T_z = -1 - \tfrac{1}{2} = -\tfrac{3}{2}.$$

Charge symmetry tells us that the sign of T_z is immaterial and that the elastic scattering amplitudes for $\pi^+ p$ and $\pi^- n$ satisfy the identity

$$\chi(\pi^+ p, \pi^+ p) = \chi(\pi^- n, \pi^- n) = \chi_{3/2}.$$

This prediction is verifiable and agrees with experiment.) We cannot draw such an immediate conclusion about the $\pi^- p$ and $\pi^0 n$ states. Both of these systems have $T_z = -\frac{1}{2}$:

$$T_z = -1 + \tfrac{1}{2} \text{ for } \pi^- p \quad \text{and} \quad T_z = 0 - \tfrac{1}{2} \text{ for } \pi^0 n.$$

Since $T_z = -\frac{1}{2}$ occurs in one of the four substates for $t = \frac{3}{2}$ and also in one of the two substates for $t = \frac{1}{2}$, it is apparent that both $\pi^- p$ and $\pi^0 n$ correspond to *superpositions* of the $t = \frac{3}{2}$ and $t = \frac{1}{2}$ states. The actual mixtures needed to form the two πN systems are well-defined and lead directly to the results quoted in Equations (16-17). We pursue these remarks somewhat further in Section 16-11.

Example

Our discussions of the internal quantum numbers have been aimed primarily at the conservation laws pertaining to the strong interaction. Let us consider some weak decays now so that we can demonstrate a few violations of the laws. In Λ decay, $\Lambda \rightarrow p + \pi^-$, we have a transition of overall quantum numbers of the form

$$S = -1 \text{ and } T_z = 0 \rightarrow S = 0 \text{ and } T_z = -\tfrac{1}{2}.$$

The changes of strangeness and isospin are $\Delta S = 1$ and $\Delta T_z = -\frac{1}{2}$. In charged K decay, $K^+ \rightarrow \pi^+ + \pi^0$, we have another transition of the form

$$S = 1 \text{ and } T_z = \tfrac{1}{2} \rightarrow S = 0 \text{ and } T_z = 1,$$

and so the changes are $\Delta S = -1$ and $\Delta T_z = \frac{1}{2}$. An examination of the weak decays of all the strange particles reveals a pattern of violations:

$$\Delta S = \pm 1 \text{ and } \Delta T_z = \mp \tfrac{1}{2}, \quad \text{or} \quad \Delta S = -2 \Delta T_z.$$

This selection rule for strangeness-changing decays is consistent with Equation (16-13), the Gell-Mann–Nishijima formula, by virtue of the conservation of charge and baryon number.

Example

The conservation law for isospin is a vector constraint, as is the conservation of angular momentum. We implement the law in strong-interaction processes as follows, with comments appropriate in each instance:

$\pi^- + p \rightarrow K^0 + \Lambda$, a $t = \frac{1}{2}$ reaction, even though $\pi^- p$ also has a $t = \frac{3}{2}$ part.

$K^- + p \rightarrow \eta + \Lambda$, a $t = 0$ reaction, even though $K^- p$ also has a $t = 1$ part.

$\Lambda + \overline{\Lambda} \nrightarrow \pi^0 + \eta$, a violation of isospin conservation.

$K^+ + p \nrightarrow \pi^+ + \Sigma^+$, a violation of conservation of isospin and strangeness.

$\pi^+ + p \rightarrow K^+ + \Sigma^+$, a pure $t = \frac{3}{2}$ allowed reaction.

Other systems are considered in Problems 18 and 19 at the end of the chapter.

16-11 Baryon and Meson Resonances

High-energy spectroscopy is concerned with the mass levels and quantum states of baryons and mesons. These particle properties can be ascertained by investigating the reactions of colliding hadrons. The products of a given reaction may exhibit new hadronic states as resonances in the reaction cross section. We have seen resonances in nuclear reactions, and we have interpreted the phenomena as excited nuclear states representing unique short-lived configurations of the interacting nucleons. Hadron resonances occur in a higher-energy range, as expected for the excitation of matter on a smaller scale. The widths of resonant hadronic states are considerably larger than their nuclear counterparts, and the lifetimes are correspondingly shorter, as appropriate for the smaller size of the resonating systems. Typical widths and lifetimes are of order 100 MeV and 10^{-23} s, respectively. Thus, the hadronic excitations are in the domain of the strong interaction, where the instabilities of the resonances are many orders of magnitude stronger than the weak instabilities of the listed hadrons in Table 16-1. The resonances are produced via the strong interaction and decay by the same route. These systems exist with specific mass and quantum-number assignments and are accorded hadron status on a par with the more familiar particles, despite their extremely transitory nature.

Excited states of the nucleon are seen in the interactions of pions with protons. Prominent resonances of the πN system appear as distinct bumps in the *total* cross sections for the $\pi^+ p$ and $\pi^- p$ reactions

$$\pi^\pm + p \rightarrow \text{all possible final states.}$$

These processes are subject to the isospin conservation laws discussed in Section 16-10. The reactions can therefore be reduced to a description in terms of *two* basic cross sections, corresponding to the two values of the total isospin quantum number $t = \frac{1}{2}$ and $t = \frac{3}{2}$. If we denote these quantities as $\sigma_{1/2}$ and $\sigma_{3/2}$, we find that the observed $\pi^{\pm}p$ total cross sections are given by the formulas

$$\sigma(\pi^+p) = \sigma_{3/2} \quad \text{and} \quad \sigma(\pi^-p) = \tfrac{1}{3}\left(\sigma_{3/2} + 2\sigma_{1/2}\right). \qquad (16\text{-}18)$$

(Note that the formulas are just like the ones for the scattering amplitudes in Equations (16-17). The results follow from a theorem that relates the *total* cross section *directly* to the imaginary part of the amplitude for elastic scattering. These total cross sections are also expressible in more familiar terms as sums of the squared moduli of all the contributing amplitudes.) We can easily invert the relations to obtain $\sigma_{1/2}$ and $\sigma_{3/2}$:

$$\sigma_{1/2} = \tfrac{3}{2}\sigma(\pi^-p) - \tfrac{1}{2}\sigma(\pi^+p) \quad \text{and} \quad \sigma_{3/2} = \sigma(\pi^+p). \qquad (16\text{-}19)$$

In these formulas, the fundamental isospin cross section σ_t refers to a hypothetical process in which the colliding πN system is prepared in a state of definite isospin t. Recall that π^+p is already such a state and that π^-p is a superposition of isospins $\frac{1}{2}$ and $\frac{3}{2}$.

The total cross sections for π^+p and π^-p are sketched in Figure 16-21, along with the derived isospin cross sections $\sigma_{1/2}$ and $\sigma_{3/2}$. Both sets of quantities depend only on the πN collision energy. We show the $\pi^{\pm}p$ information as a function of the pion beam energy K_π, and we use the total CM energy E_{CM} in the graphs of $\sigma_{1/2}$ and $\sigma_{3/2}$. The latter variable is the same as the one used to display the various reaction thresholds in Figure 16-16. The graphs on the left side of Figure 16-21 are obtained from experiment, and those on the right are then constructed with the aid of Equations (16-19). We are particularly interested in the second set of cross sections. These graphs demonstrate a clear separation of the resonance bumps, and thus indicate a unique isospin assignment for each resonant state.

The most striking feature of the πN system is the resonance peak in the cross section at $K_\pi = 190$ MeV (or $E_{\text{CM}} = 1.23$ GeV). We see from Figure 16-21 that the cross sections at this energy have the approximate values

$$\sigma(\pi^+p) = 200 \text{ mb} \quad \text{and} \quad \sigma(\pi^-p) = 67 \text{ mb}.$$

The 3 : 1 ratio of these results is readily understood from isospin considerations. A $t = \frac{3}{2}$ assignment for the resonance is obvious, first of all, since $\sigma_{3/2}$ is much larger than $\sigma_{1/2}$. If we neglect $\sigma_{1/2}$ at this energy, we obtain

$$\sigma(\pi^+p) = 3\sigma(\pi^-p)$$

directly from Equations (16-19). The elastic and charge-exchange scattering processes can also be examined from the same viewpoint. If we neglect the $t = \frac{1}{2}$ amplitude $\chi_{1/2}$ in Equations (16-17), we get the following ratios for the elastic and charge-

Figure 16-21

Pion–nucleon total cross sections. The graphs on the left pertain to the processes $\pi^{\pm} + p \rightarrow$ all possible final states, and the graphs on the right refer to πN interactions in the separated isospin states $t = \frac{1}{2}$ and $t = \frac{3}{2}$.

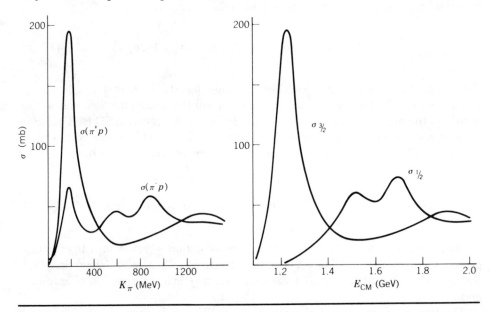

exchange cross sections:

$$\left|\chi(\pi^{+}p, \pi^{+}p)\right|^{2} : \left|\chi(\pi^{-}p, \pi^{-}p)\right|^{2} : \left|\chi(\pi^{0}n, \pi^{-}p)\right|^{2} = 1 : \tfrac{1}{9} : \tfrac{2}{9}.$$

The experimental results confirm this prediction and verify decisively the $t = \frac{3}{2}$ hypothesis for the resonant state.

Three other resonance bumps also appear with clearly established isospin quantum numbers in Figure 16-21. The resonant states are found to have unique spins and parities by analyzing the spatial distributions of particles scattered at the different resonance energies. Furthermore, the bumps occur at specific values of the CM energy, so that the resonances correspond to definite assignments of mass. Thus, a given resonant state is defined by a mass level and a complete set of hadron quantum numbers, including the usual spatial properties of spin and parity, and also the internal attributes of isospin, baryon number, and strangeness. We recognize these characteristics to be exactly the same as those possessed by the more familiar (and more nearly stable) hadrons, and we therefore regard the resonances as full-fledged hadronic particles. The four prominent resonances at energies below $E_{\text{CM}} = 2$ GeV are designated by the following symbols and specifications. The $t = \frac{3}{2}$ states are called

$$\Delta\!\left(1232, \tfrac{3}{2}^{+}\right) \quad \text{and} \quad \Delta\!\left(1950, \tfrac{7}{2}^{+}\right),$$

and the $t = \frac{1}{2}$ states are called

$$N\!\left(1520, \tfrac{3}{2}^{-}\right) \quad \text{and} \quad N\!\left(1680, \tfrac{5}{2}^{+}\right).$$

Figure 16-22

Meson–nucleon reaction yielding two or more mesons in the final state. The detection of two mesons provides information about the meson–meson interaction.

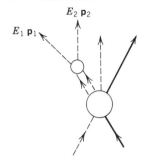

The parenthetical information gives the mass in MeV/c^2 along with the spin and parity. Each of these hadrons has baryon number $B = 1$ and strangeness $S = 0$, because each occurs as an excitation in the pion–nucleon system. The isospin assignments tell us that there should be four degenerate particles for $t = \frac{3}{2}$, with $T_z = -\frac{3}{2}, -\frac{1}{2}, \frac{1}{2}$, and $\frac{3}{2}$,

$$\Delta^-, \Delta^0, \Delta^+, \text{ and } \Delta^{++},$$

and two degenerate particles for $t = \frac{1}{2}$, with $T_z = -\frac{1}{2}$ and $\frac{1}{2}$,

$$N^0 \text{ and } N^+.$$

The indicated charges of these baryons are in accord with the Gell-Mann–Nishijima formula, Equation (16-13). Numerous other baryon resonances with strangeness quantum numbers $S = 0$ and $S \neq 0$ are also known.

Meson resonances also occur, in states whose existence could be detected if meson–meson collisions could be observed. Unfortunately, such experiments cannot be performed directly because stable meson targets do not exist. It is possible to see these resonances indirectly by investigating those meson–nucleon reactions in which at least two mesons appear in the final state. Figure 16-22 illustrates this sort of process and suggests how a particular pair of final mesons may be analyzed in order to gain access to the meson–meson interaction. We can undertake such a study by measuring events in bubble-chamber photographs, for example. Measurements of momentum and energy for the selected pair of mesons can be used to determine a value from each event for the *invariant mass* of the two-meson system. This quantity is defined in terms of the kinematic observables shown in the figure by recalling the energy–momentum relation for a single particle, Equation (1-35), and by introducing the analogous formula

$$\left(m_{12}c^2\right)^2 = \left(E_1 + E_2\right)^2 - c^2(\mathbf{p}_1 + \mathbf{p}_2)^2 \qquad (16\text{-}20)$$

for the case of two particles. We note that the expression $m_{12}c^2$ is Lorentz-invariant and is equal to the total energy of the two mesons in their own CM frame. The mass m_{12} is a useful variable to employ in a plot of the distribution of the measured events.

Figure 16-23

Mass distributions of pion pairs showing the ρ meson resonance at 770 MeV/c^2.

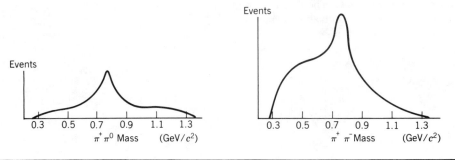

A resonance in the meson–meson system appears in such a plot as a peak in the distribution, indicating an exceptional population of events for a specific value of the invariant mass.

Evidence for the resonance known as the ρ meson is obtainable by an invariant-mass analysis of the reactions

$$\pi^+ + p \to \begin{cases} \pi^+ + \pi^0 + p \\ \pi^+ + \pi^+ + \pi^- + p. \end{cases}$$

Figure 16-23 shows two typical distributions in which a prominent peak appears at the mass value 770 MeV/c^2 in the $\pi^+\pi^0$ system of the first reaction, and in both $\pi^+\pi^-$ systems of the second reaction. No such phenomenon is seen, however, for the $\pi^+\pi^+$ combination in the second process. These results are consistent with a $t = 1$ isospin assignment for the resonance, where the states with $T_z = -1$, 0, and 1 correspond to the charged mesons ρ^-, ρ^0, and ρ^+. The ρ^+ state is observed in $\pi^+\pi^0$, and the ρ^0 state is observed in $\pi^+\pi^-$. The absence of an effect in $\pi^+\pi^+$ rules out the possibility of a $t = 2$ assignment, and so, of the total isospins possible for two isospin-1 pions, $t = 0$, 1, and 2, only the $t = 1$ hypothesis agrees with experiment. Further analysis shows the spin and parity of the resonance to be 1^-, the quantum numbers of a *vector meson*. The $t = 1$ ρ meson has strangeness $S = 0$ and, of course, baryon number $B = 0$. Other meson resonances are found in many different varieties with a profusion of spatial and internal quantum numbers.

Vector meson resonances can also be seen in processes initiated by the collision of beams of electrons and positrons. In these reactions electron–positron pair annihilation takes place, producing electromagnetic energy, and systems of hadrons then materialize out of this energy. The total cross section $\sigma(e^-e^+)$ includes the processes

$$e^- + e^+ \to \text{all possible hadrons}$$

and exhibits a rich spectrum of resonance peaks corresponding to uncharged hadronic states with internal quantum numbers $B = 0$ and $S = 0$. These meson resonances are produced via the electromagnetic interaction and are *vector* particles since they bear the same 1^- spin and parity properties as the electromagnetic field.

Example

Let us examine the hadrons $\Delta(1232, \frac{3}{2}^+)$ and $\rho(770, 1^-)$ and show that both occur as $\ell = 1$ resonant states in their respective interacting systems. The baryon Δ^{++} is a $\frac{3}{2}^+$ resonance of the $\pi^+ p$ system. Since π^+ and p are 0^- and $\frac{1}{2}^+$ particles, the parity of the $\pi^+ p$ orbital state must be odd in order to conserve parity. It follows that the orbital quantum number ℓ must be odd, and only $\ell = 1$ is allowed by conservation of angular momentum. The meson ρ^+ is a 1^- resonance of the $\pi^+ \pi^0$ system. In this case, $\ell = 1$ is required immediately so that the two 0^- particles conserve both angular momentum and parity. The meson ρ^0 appears in $\pi^+ \pi^-$ but is forbidden to occur in $\pi^0 \pi^0$. This conclusion follows from Bose symmetry, which demands an exchange-symmetric wave function for the two identical pions. An $\ell = 1$ pair of π^0 mesons is not allowed because the state would have odd spatial symmetry and would therefore be exchange antisymmetric.

Example

Two different kinematic variables are used to display the baryon resonances in Figure 16-21. The beam energy (or beam momentum) is the natural choice for the presentation of cross-section data, but the total CM energy is more convenient for interpreting the resonances as particles of definite mass. We can readily pass from one variable to the other with the aid of techniques discussed in Section 1-11. Consider $\pi^\pm p$ collisions where the beam pion (mass m) has momentum $\mathbf{p}$ and energy E, and where the target proton (mass M) is at rest. The total momentum four-vector in the lab frame is

$$\mathscr{P} = \begin{bmatrix} \mathbf{p} \\ i(E + Mc^2)/c \end{bmatrix}.$$

Let E_{CM} be the total CM energy so that $\mathscr{P}$ transforms into the four-vector

$$\mathscr{P}' = \begin{bmatrix} \mathbf{0} \\ iE_{\mathrm{CM}}/c \end{bmatrix}$$

in the CM frame. We invoke Lorentz invariance and obtain

$$-\frac{E_{\mathrm{CM}}^2}{c^2} = p^2 - \frac{E^2 + 2EMc^2 + M^2c^4}{c^2}$$

$$= -\frac{2EMc^2 + (m^2 + M^2)c^4}{c^2},$$

using the relativistic relation between p and E. The result takes the form

$$E_{CM}^2 = 2Mc^2(K + mc^2) + (m^2 + M^2)c^4 = 2Mc^2K + (m + M)^2c^4$$

when we introduce the pion beam kinetic energy $K = E - mc^2$. This formula enables us to relate the energy axes in the left and right halves of Figure 16-21.

16-12   Quarks

An abundance of hadrons and an organizational scheme have begun to take shape in our survey of the elementary particles. We have identified a few of the existing baryon and meson species, and we have drawn special attention to the internal quantum numbers that characterize the particles. More than a hundred different hadronic states actually occur. All these particles can be grouped into families of common spin and parity like the eight $\frac{1}{2}^+$ baryons and the eight 0^- mesons in Table 16-1. The groupings may be represented by plotting the isospin and hypercharge quantum numbers T_z and Y, as in the octet patterns shown in Figure 16-20. We have noted that this classification suggests the influence of a *higher symmetry*, encompassing the independent symmetries associated with the separate conservation laws of isospin and strangeness.

It should be acknowledged at once that the hundred or more existing hadrons are far too numerous to be regarded as fundamental particles. This remark obviously applies to the baryon and meson resonances, since these states occur as short-lived composite systems of interacting hadrons. We must concede the same point regarding the proton and neutron, because we know that even these more primitive particles are structured entities. The evidence for nucleon structure begins with the fact that the magnetic moments of p and n are not equal to the values expected for elementary Dirac particles. The proton and neutron are also known to have an extended size and, by inference, an internal *substructure*. We learn about these properties in electron-scattering experiments with nucleon targets, similar to the studies discussed in Section 14-3 in the case of nuclei. Thus, it is clear that the nucleons are no more fundamental than the various resonances and that all hadrons should be treated on an equal footing. We have already espoused this democratic point of view in Section 16-11.

The two notions of higher symmetry and hadron substructure were drawn together into a single concept by Gell-Mann. In 1961, he organized the hadrons according to a generalization of the concept of isospin by invoking a rotational symmetry in an internal vector space of *eight* dimensions. This extension of the familiar three-dimensional isospin space was designed to incorporate hypercharge among the eight necessary internal degrees of freedom. Gell-Mann called his mathematical framework the Eightfold Way. The proposal played the role of a periodic table for the elementary particles, as the scheme was used to classify known varieties and predict new hadrons. In 1964, Gell-Mann showed that the mathematical aspects of his higher-symmetry scheme could be encoded in the quantum numbers of a hypothesized set of three fundamental particles, called quarks. He demonstrated that the existing baryons could be constructed from simple combinations of three quarks, and that the existing mesons could be assembled by combining quarks and antiquarks. The concept was originally developed as a means of synthesizing the quantum numbers of the observed hadrons in terms of more basic elements. In this sense the quarks were contrived to implement a model of the higher internal symmetry. Soon, however, it

Murray Gell-Mann

became clear from experiment that quarks and antiquarks had real identities as fundamental constituents in the substructure of hadrons.

Quarks and antiquarks are distinguished from conventional hadrons by the fact that they are assumed to carry *fractional* charge and baryon number. The rules of the quark model tell us from the outset that the postulated particles have values of Q/e and B in *one-third units*. Three quarks, called u, d, and s, are introduced with the following assignments of quantum numbers:

u	$B = \frac{1}{3}$	$t = \frac{1}{2}$	$T_z = \frac{1}{2}$	$S = 0$	$Y = \frac{1}{3}$	$Q/e = \frac{2}{3}$
d	$\frac{1}{3}$	$\frac{1}{2}$	$-\frac{1}{2}$	0	$\frac{1}{3}$	$-\frac{1}{3}$
s	$\frac{1}{3}$	0	0	-1	$-\frac{2}{3}$	$-\frac{1}{3}$

We note that the entries in the last column are consistent with the Gell-Mann–Nishijima formula, $Q/e = T_z + Y/2$. The corresponding antiquarks $\bar{u}$, $\bar{d}$, and $\bar{s}$ have suitably conjugated quantum numbers as follows:

$\bar{u}$	$B = -\frac{1}{3}$	$t = \frac{1}{2}$	$T_z = -\frac{1}{2}$	$S = 0$	$Y = -\frac{1}{3}$	$Q/e = -\frac{2}{3}$
$\bar{d}$	$-\frac{1}{3}$	$\frac{1}{2}$	$\frac{1}{2}$	0	$-\frac{1}{3}$	$\frac{1}{3}$
$\bar{s}$	$-\frac{1}{3}$	0	0	1	$\frac{2}{3}$	$\frac{1}{3}$

We observe from the listings for T_z and S that the role of each quark is to supply a *single* nonzero quantum number in the synthesis of any chosen hadron. This use of quarks as ingredients in a composition is called an assignment of *flavors*. Thus, the u quark is u-flavored with isospin up ($T_z = \frac{1}{2}$), the d quark is d-flavored with isospin down ($T_z = -\frac{1}{2}$), and the s quark is s-flavored with strangeness ($S = -1$). These

Figure 16-24

Flavor assignments of quarks and antiquarks.

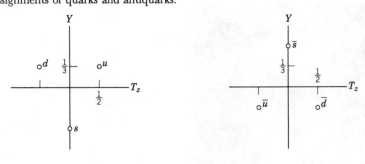

flavor designations are conveniently plotted on graphs of the hypercharge versus the z component of isospin, as in Figure 16-24. The resulting *triplet* and *antitriplet* patterns represent the basic structures of Gell-Mann's higher symmetry.

Let us consider the construction of hadron states in terms of our newly defined building blocks. Quarks and antiquarks are assumed to be spin-$\frac{1}{2}$ fermions, so that odd and even numbers of them are needed to make baryons and mesons. We use three quarks (qqq) to assemble the quantum numbers of baryons, because of the $B = \frac{1}{3}$ definition for the baryon number of quarks, and we use quark–antiquark ($q\bar{q}$) combinations to construct the quantum numbers of mesons. Other compositions, such as $qqqq\bar{q}$ with $B = 1$ and $qq\bar{q}\bar{q}$ with $B = 0$, may also be considered. These so-called exotic states are not essential for the purposes of this discussion.

The internal quantum numbers of the 0^- mesons are obtained by the blending of flavors in the following $q\bar{q}$ systems:

$$
\begin{aligned}
\pi^+ &= u\bar{d} & Y = S = 0 &\qquad T_z = 1 \\
\pi^0 &= \frac{1}{\sqrt{2}}(d\bar{d} - u\bar{u}) & 0 &\qquad 0 \\
\pi^- &= -d\bar{u} & 0 &\qquad -1 \\
K^+ &= u\bar{s} & 1 &\qquad \tfrac{1}{2} \\
K^0 &= d\bar{s} & 1 &\qquad -\tfrac{1}{2} \\
\overline{K}^0 &= s\bar{d} & -1 &\qquad \tfrac{1}{2} \\
K^- &= s\bar{u} & -1 &\qquad -\tfrac{1}{2} \\
\eta &= \frac{1}{\sqrt{6}}(u\bar{u} + d\bar{d} - 2s\bar{s}) & 0 &\qquad 0
\end{aligned}
\tag{16-21}
$$

The set of all possible combinations of three quarks and three antiquarks also includes a ninth independent structure:

$$
\eta' = \frac{1}{\sqrt{3}}(u\bar{u} + d\bar{d} + s\bar{s}) \qquad Y = S = 0 \qquad T_z = 0 \tag{16-22}
$$

(This *singlet* state is separate from the octet of pseudoscalar mesons listed in Table 16-1. Note that the π^0, η, and η' systems are formed by superposing $u\bar{u}$, $d\bar{d}$, and $s\bar{s}$

states.) It is implicit in each of these expressions that the spatial states of the constituent particles must have total spin and orbital quantum numbers $s = 0$ and $\ell = 0$ so that a vanishing total angular momentum is obtained, as required for a spin-0 composite system. The odd parity of the 0^- mesons is then due to the intrinsic even and odd parities of quarks and antiquarks. This property of opposite parities holds true for all spin-$\frac{1}{2}$ Dirac particles and antiparticles.

The three-quark baryon structures are somewhat more difficult to assemble. A convenient starting place is the $t = \frac{3}{2}$ Δ multiplet:

$$\Delta^{++} = uuu,$$

$$\Delta^+ = \frac{1}{\sqrt{3}}(duu + udu + uud),$$

$$\Delta^0 = \frac{1}{\sqrt{3}}(ddu + dud + udd),$$

$$\Delta^- = ddd. \tag{16-23}$$

Note that the uuu system has $T_z = \frac{3}{2}$ as required, and note that the successors to this state follow by applying the systematic isospin-lowering operation $u \to d$ to each quark. There are exactly ten different combinations of the three quarks u, d, and s:

$$
\begin{array}{llll}
ddd \quad ddu \quad duu \quad uuu & \text{with } Y = 1 \\
\quad\quad dds \quad dus \quad uus & \text{with } Y = 0 \\
\quad\quad\quad\quad dss \quad uss & \text{with } Y = -1 \\
\quad\quad\quad\quad\quad\quad sss & \text{with } Y = -2
\end{array}
$$

Superposed permutations of these states can be put in one-to-one correspondence with the ten members of an existing family of $\frac{3}{2}^+$ resonances, whose members include the Δ multiplet. The unique $Y = -2$ system, identified as

$$\Omega^- = sss,$$

is especially interesting because the proven existence of this hadron validates the underlying higher-symmetry scheme. These three-quark expressions must be supplemented by spin and orbital configurations appropriate for composite states with spin and parity $\frac{3}{2}^+$. Positive parity is obtained if each quark is in an s state, and spin-$\frac{3}{2}$ is the result if the quarks are assembled with parallel spins. The procedure is illustrated by the composition

$$\Delta^{++} = uuu \uparrow \uparrow \uparrow,$$

in which three s-state quarks have parallel spins so that the z component of total angular momentum corresponds uniquely to a $\frac{3}{2}^+$ state.

Combinations of three quarks can also be deduced for the $\frac{1}{2}^+$ baryon octet of Table 16-1 and Figure 16-20. We see at once that the flavor mixtures uud and udd reproduce the internal quantum numbers of the proton and neutron. Since these same forms appear, along with their permutations, in Δ^+ and Δ^0, it is necessary to take specific combinations of permutations if p and n are to have their own independent states. If we take

$$p = \frac{1}{\sqrt{2}}(uud - udu) = \frac{1}{\sqrt{2}}u(ud - du) \qquad (16\text{-}24a)$$

and

$$n = \frac{1}{\sqrt{2}}(dud - ddu) = \frac{1}{\sqrt{2}}d(ud - du), \qquad (16\text{-}24b)$$

we find that p and n are orthogonal to Δ^+ and Δ^0, ensuring the independence of the states. Strange baryons in the octet may then be constructed by the systematic substitution of an s-flavored quark. Thus, we derive Σ^+ from p by substituting s for d to get

$$\Sigma^+ = \frac{1}{\sqrt{2}}u(us - su).$$

The $\frac{1}{2}^+$ assignment of these baryons is achieved by assembling s-state quarks in states with two spins parallel and one spin antiparallel.

So far we have treated quarks only as bearers of flavor in a model of the states of hadrons. We may wonder whether the constituents are just fictitious artifacts of the model, or whether they really exist as observable particles. The latter question has to be decided by experiment. If a free quark could be separated from others of its kind, and if the quark could be bound in an atom, the distinctive one-third unit of charge would be a readily detectable signature. Recent experiments conducted by W. M. Fairbank have succeeded in isolating a fractional charge on a metal sphere; however, no other similar studies have confirmed this observation. Millikan himself is supposed to have seen, and discarded, a fractional charge on an oil drop as long ago as 1910. Quarks bound in atoms could also manifest themselves in atomic emission spectra and in mass spectrograms, but no such indications have ever been recorded. To date, searches of all kinds have failed to offer convincing evidence, not only for the occurrence of free quarks in bulk matter but also for the production of free quarks in cosmic-ray events and in accelerator experiments.

None of these negative results is in conflict with the accepted belief that quarks really do exist as bound constituents inside hadrons. Evidence for the validity of this

Figure 16-25

Inelastic electron scattering as a probe of baryon substructure.

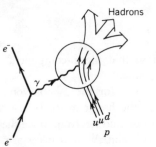

Figure 16-26

Construction of $q\bar{q}$ states by adding vectors in the (Y, T_z) plane.

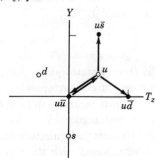

picture has been gathered since 1968 in reactions of the form

$$e^- + p \rightarrow e^- + \text{hadrons},$$

in which the inelastic scattering of electrons is studied at very high energy. Figure 16-25 shows schematically how such a process employs very-high-energy virtual photons to probe the small-scale structure of the proton. The experiments indicate that the nucleon contains point-like particles whose properties fit the description of u and d quarks. The conclusion to draw from these findings is most remarkable. The quarks are evidently *confined* in hadrons as permanently bound constituents and are not found as liberated particles. This point of view is a fundamental feature of the modern theory of the strong interaction.

Example

Quark-combining and flavor-bookkeeping may be visualized with the aid of graphs of Y versus T_z. Let us refer to Figure 16-26 for illustration and construct the $q\bar{q}$ composite states by the addition of vectors in the (Y, T_z) plane. The figure shows the quark triplet centered at the origin and the antiquark anti-triplet centered at the location of the u quark. The indicated vector additions generate the flavor-blended systems $u\bar{u}$, $u\bar{d}$, and $u\bar{s}$. If we relocate the center of the antitriplet at the site of the d quark, and then at the site of the s quark, we obtain the combinations $d\bar{u}$, $d\bar{d}$, and $d\bar{s}$, and then $s\bar{u}$, $s\bar{d}$, and $s\bar{s}$. The nine resulting composite systems have the internal quantum numbers of the octet states in Equations (16-21) and Figure 16-20, and the remaining singlet state described by Equation (16-22).

16-13   The Electromagnetic and Weak Interactions of Quarks

If hadrons are composed of bound quarks it follows that the interactions of hadrons are fundamentally expressed in terms of the interactions of quarks. All electromagnetic hadron processes are thereby attributed to radiative quark transitions accompanied by the emission of photons. Likewise, all weak hadronic processes are due to weak quark transitions in concert with the emission of weak quanta. The analogy between the electromagnetic and weak interactions has been introduced with the aid of Feynman diagrams in Section 16-7. This parallel can now be developed further using diagrams based on quarks. In fact, the overall success of weak interaction theory in the context of the quark model constitutes a large share of the evidence that supports the reality of quarks.

Figure 16-27 shows two examples of the electromagnetic interaction of quarks. The diagram for electron–proton elastic scattering shows the exchange of a virtual photon, where the indicated coupling of the photon is taken to each of the proton's u and d quarks in turn. The hyperon decay $\Sigma^0 \rightarrow \Lambda + \gamma$ is represented as a radiative quark transition from one uds system to another, where the emission of a real photon can take u into u, d into d, or s into s, each with its own amplitude. These transition amplitudes are determined by the coupling between the photon and the electromagnetic current of each quark.

Weak processes have already made their appearance in the Feynman diagrams of Figure 16-15. Recall that the exchanged weak quanta carry charge and that the corresponding hadronic transitions are charge-changing phenomena in all the cases

Figure 16-27

Quark–photon couplings in ep elastic
scattering and Σ^0 decay.

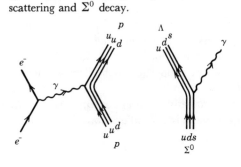

considered so far. The quark model offers a description in which the weak interaction
causes changes in flavor among the constituents of the interacting hadrons. We can
display these processes as weak transitions from one quark flavor to another, using
diagrams like the ones in Figure 16-28. Note that the hadron transitions $n \to p$ and
$p \to n$ proceed via the coupling of W^- and W^+ to the weak charge-changing quark
currents $d \to u$ and $u \to d$, while the remaining u and d quarks act as spectators. The
β decays of strange particles take place in similar fashion, as illustrated in Figure
16-29. In these instances, the quark flavor transition is associated with the
strangeness-changing charge-changing weak current $s \to u$. Strange particles also
decay to final states consisting only of hadrons. These nonleptonic modes have their
own quark Feynman diagrams. Again, Figure 16-29 describes such decays of the
strange hadrons in terms of the flavor transition $s \to u$.

Strangeness-changing decays are observed to be universally smaller in amplitude
than comparable strangeness-conserving decays. Processes suitable for comparison are

$$\Lambda \to p + e^- + \bar{\nu}_e \qquad\qquad n \to p + e^- + \bar{\nu}_e$$
$$\text{and} \qquad\qquad \text{versus} \qquad\qquad \text{and}$$
$$K^- \to \pi^0 + e^- + \bar{\nu}_e \qquad\qquad \pi^- \to \pi^0 + e^- + \bar{\nu}_e$$
$$\text{where } s \to u \qquad\qquad\qquad \text{where } d \to u.$$

Figure 16-28

Weak quark transitions in the β decay of the neutron and in the neutrino reaction $\bar{\nu}_\mu + p$
$\to \mu^+ + n$.

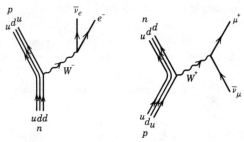

Figure 16-29

Quark Feynman diagrams representing the decays of strange particles. The upper three diagrams show the β decays $\Lambda \to p + e^- + \bar{\nu}_e$ and $K_L \to \pi^- + e^+ + \nu_e$ or $\pi^+ + e^- + \bar{\nu}_e$. The lower three diagrams show the nonleptonic decays $\Lambda \to p + \pi^-$ and $K_S \to \pi^+ + \pi^-$.

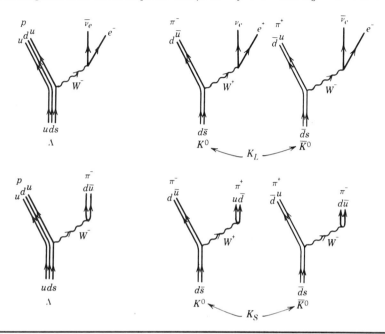

N. Cabibbo has shown that these phenomena conform to a unified picture based on Gell-Mann's higher-symmetry scheme. The ideas can be translated into the language of quarks so that a direct comparison can be drawn between the weak quark currents $s \to u$ and $d \to u$. Note that the strangeness change is $\Delta S = 1$ for the one current and $\Delta S = 0$ for the other. The quarks s and d have the same charge, $Q/e = -\frac{1}{3}$, and so a superposition of the two flavors is allowed. This mixing of strangeness states fixes a connection between the violation and conservation of strangeness in the weak transitions. According to the Cabibbo argument, the flavor change from s or d to u is such that the accompanying emission of W^- always selects the specific *rotated* combination of s and d given by the expression

$$d \cos\theta + s \sin\theta.$$

The coupling of the weak quantum to the corresponding *Cabibbo current* is shown schematically in the left half of Figure 16-30. The flavor structure of the current can be expressed as

$$\begin{bmatrix} u \\ d \cos\theta + s \sin\theta \end{bmatrix} \quad \text{with} \quad \frac{Q}{e} = \begin{cases} \frac{2}{3} \\ -\frac{1}{3} \end{cases}. \tag{16-25}$$

This prescription stipulates that *all* $s \to u$ transitions are reduced in amplitude relative to $d \to u$ transitions by the ratio of the respective coefficients $\sin\theta/\cos\theta$. The

Figure 16-30

Couplings of generalized weak quark currents. The first diagram contains the Cabibbo current, and the second introduces the concept of charm.

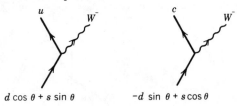

observed suppression of strangeness-changing decays is thus parameterized by a single constant given by the Cabibbo angle θ. Good agreement with experiment is obtained for a value of θ around $13°$.

A problem arises when we consider the possibility of the two apparently similar decays

$$K^- \to \pi^0 + e^- + \bar{\nu}_e \quad \text{and} \quad K^- \to \pi^- + \nu_e + \bar{\nu}_e.$$

The first of these processes is observed, as noted above; however, the second decay is not known to occur. We see that the transitions in the respective quark–antiquark systems are

$$s\bar{u} \to u\bar{u} \quad \text{and} \quad s\bar{u} \to d\bar{u},$$

so that the relevant quark flavor transitions must be

$$s \to u \quad \text{and} \quad s \to d.$$

Both are strangeness-changing effects; however, the one is charge-changing while the other is charge-conserving. The observed suppression in the second case can be explained by invoking the following flavor-transition mechanism, put forward by S. L. Glashow, J. Iliopoulos, and L. Maiani. Their proposal calls for the cancellation of a pair of conspiring contributions to the strangeness-changing charge-conserving current and also introduces a new quark flavor called *charm*. The mechanism employs *two* rotated combinations of s and d:

$$d \cos\theta + s \sin\theta \quad \text{and} \quad -d \sin\theta + s \cos\theta.$$

We note that the second expression is orthogonal to the first and that the first expression is part of the Cabibbo current. A cancellation of contributions to the $s \to d$ current takes place via the schematic procedure shown in Figure 16-31. The replacement of s and d by the two rotated forms produces the indicated substitutions, in which the $s \to d$ transition appears with coefficient $\sin\theta \cos\theta$ in the first contribution and $-\cos\theta \sin\theta$ in the second. The argument goes on to hypothesize a new quark flavor with $Q/e = \frac{2}{3}$, denoted by c. This new quark is conceived to be a partner for the second of the two rotated combinations of s and d in a new charge-changing

Figure 16-31

Cancellation mechanism for strangeness-changing charge-conserving quark currents.

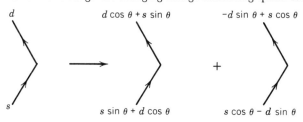

current:

$$\left[-d \sin \theta \overset{c}{+} s \cos \theta \right] \quad \text{with} \quad \frac{Q}{e} = \begin{cases} \frac{2}{3} \\ -\frac{1}{3} \end{cases}. \tag{16-26}$$

Thus, two related weak quark currents are defined by introducing the rotations of the two flavors. We show these flavor transitions together in Figure 16-30.

A charm quantum number C is also defined along with the introduction of the c-flavored quark. Table 16-2 lists the quantum-number assignments for the quarks u, d, s, and c and includes a new column for the values of C. Like isospin and strangeness, charm refers to another quantity conserved by the strong interaction. The entries in the table tell us that the Gell-Mann–Nishijima formula for the charge must now be modified to include C:

$$\frac{Q}{e} = T_z + \frac{Y}{2} + \frac{C}{2}. \tag{16-27}$$

Table 16-2   Quantum Numbers of Quarks and Antiquarks[a]

	B	t	T_z	S	Y	C	Q/e
u	$\frac{1}{3}$	$\frac{1}{2}$	$\frac{1}{2}$	0	$\frac{1}{3}$	0	$\frac{2}{3}$
d	$\frac{1}{3}$	$\frac{1}{2}$	$-\frac{1}{2}$	0	$\frac{1}{3}$	0	$-\frac{1}{3}$
s	$\frac{1}{3}$	0	0	-1	$-\frac{2}{3}$	0	$-\frac{1}{3}$
c	$\frac{1}{3}$	0	0	0	$\frac{1}{3}$	1	$\frac{2}{3}$
$\bar{u}$	$-\frac{1}{3}$	$\frac{1}{2}$	$-\frac{1}{2}$	0	$-\frac{1}{3}$	0	$-\frac{2}{3}$
$\bar{d}$	$-\frac{1}{3}$	$\frac{1}{2}$	$\frac{1}{2}$	0	$-\frac{1}{3}$	0	$\frac{1}{3}$
$\bar{s}$	$-\frac{1}{3}$	0	0	1	$\frac{2}{3}$	0	$\frac{1}{3}$
$\bar{c}$	$-\frac{1}{3}$	0	0	0	$-\frac{1}{3}$	-1	$-\frac{2}{3}$

[a]Flavors are listed according to baryon number, isospin and z component, strangeness, hypercharge, charm, and charge. Evidence exists for a b-flavored quark with $Q/e = -\frac{1}{3}$, and speculation abounds for a t-flavored quark with $Q/e = \frac{2}{3}$.

The charm hypothesis implies that the c quark should manifest itself in the existence of hadrons containing the new flavors c and $\bar{c}$. These quantum numbers may occur in c-flavored baryon and meson systems, such as udc and $c\bar{u}$, or in hidden-charm $c\bar{c}$ states.

The first definitive evidence for charm came with the discovery of the J/ψ resonance in 1974. This unusual vector meson was seen in independent experiments performed by the research groups of S. C. C. Ting and B. Richter (hence the split personality implied in the name of the particle). The resonance was detected in the collisions of high-energy particles as a very sharp yield of e^-e^+ pairs at an energy $E_{\text{CM}} = 3.1$ GeV in the two-lepton CM frame. Hadronic decays of the resonance were also in evidence, but not to the degree expected for such a massive unstable hadron. This suppression of hadronic modes suggested that the observed meson must be composed of new quark flavors unlike those found in the available hadrons. The resonance was identified as a bound $c\bar{c}$ system, with an energy below the threshold for decay into any as-yet-unknown particles of nonzero charm. Thus, the J/ψ meson was interpreted to have charm $C = 0$ like the other known hadrons. The $c\bar{c}$ system was given the generic name *charmonium*. The bound charm degree of freedom was expected to become observable if enough energy could be added to dissociate a $c\bar{c}$ state into separate c- and $\bar{c}$-flavored hadrons. An intensive search for charm in other forms immediately yielded other charmonium states and eventually led to new c-flavored nonstrange mesons with compositions $c\bar{u}$ and $c\bar{d}$, as well as c-flavored strange mesons of the $c\bar{s}$ variety. Charm spectroscopy flourished from the start and has continued to yield a flood of information about systems containing the c quark.

New flavors beyond charm have also come to light since 1977. Another vector meson known as the Υ resonance was discovered in high-energy collisions as a narrow peak in the yield of $\mu^-\mu^+$ pairs at $E_{\text{CM}} = 9.5$ GeV. The circumstances of this discovery were just like those associated with the J/ψ meson. The Υ meson was therefore interpreted as a $b\bar{b}$ bound system, in which a new quark b was introduced, with $Q/e = -\frac{1}{3}$, to fit the analysis of the resonance and to suit the analogy with charm. A whole family of similarly defined Υ states has subsequently been observed. One of these recent discoveries was found at an energy above the threshold for decay into pairs of b- and $\bar{b}$-flavored hadrons. Yet another quark t, with $Q/e = \frac{2}{3}$, has also been postulated to act as a partner for b. Direct experimental evidence for the existence of this new flavor is anticipated at higher energy.

High-energy electron–positron colliding beams are ideal for the production of such uncharged vector mesons as J/ψ and Υ. These resonances make their appearance in processes of the form

$$e^- + e^+ \to \text{hadrons}$$

and are seen as conspicuous features of the total cross section $\sigma(e^-e^+)$. Electron–positron storage rings have been constructed at several laboratories around the world to serve as factories for c- and b-flavored particles. Observations of $c\bar{c}$ and $b\bar{b}$ systems are presented convincingly in the ratio of cross sections

$$R = \frac{\sigma(e^-e^+)}{\sigma(e^-e^+ \to \mu^-\mu^+)}, \qquad (16\text{-}28)$$

plotted as a function of E_{CM}, the e^-e^+ CM energy. Figure 16-32 shows a graph of R, indicating the 1^- charmonium states and the Υ family at their various resonance energies. The $t\bar{t}$ system is expected to occur on such a graph at higher values of the energy.

Figure 16-32

Ratio of cross sections $\sigma(e^- e^+)/\sigma(e^- e^+ \to \mu^- \mu^+)$ versus $e^- e^+$ CM energy. The two charmonium states and the first three Υ states are off scale. The fourth Υ decays into b- and $\bar{b}$-flavored hadrons.

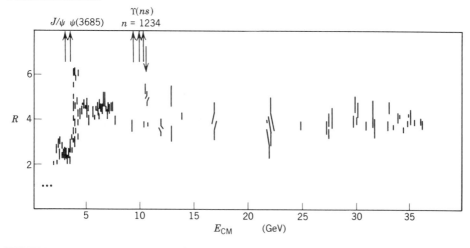

The t and b quarks are designed to contribute t and b flavors in the composition of whole collections of new hadrons. Quantum numbers and conserved quantities have to be defined accordingly, extending the lists given in Table 16-2. The new quantum numbers associated with t and b have been given the names *truth* and *beauty*. These two new species of flavor terminate a succession of three pairwise generations of quarks,

$$(u, d), \quad (c, s), \quad \text{and} \quad (t, b),$$

and complete the parallel with the already-introduced three generations of leptons

$$(e^-, \nu_e), \quad (\mu^-, \nu_\mu), \quad \text{and} \quad (\tau^-, \nu_\tau).$$

We note that the pairs of Q/e values are $(\frac{2}{3}, -\frac{1}{3})$ for the quarks and $(-1, 0)$ for the leptons. The symmetry between these two groups of particles is a striking coincidence to be explored later on.

16-14 The Electroweak Interaction

Several areas of common ground are shared by the electromagnetic and weak interactions. We have learned that both types of phenomena can be treated theoretically in terms of the fundamental couplings of currents and quanta. The hadrons participate in these processes through the electromagnetic and weak currents that represent the flavor transitions of quarks. A similar picture based on currents has also been presented earlier regarding the behavior of the leptons. The likenesses between the treatments of the two interactions encourage the view that the two theories may have a common origin in a single body of principles. The electromagnetic-to-weak analogy has been entertained since the time of Fermi, and the possibility of a single theory has evolved during the intervening decades.

We have also noted that the electromagnetic and weak interactions are very different as to their strength and range. Electrodynamics describes a force of infinite range mediated by the exchange of zero-mass photons, while the weak force is known to have an extremely short range so that the exchanged weak quanta are presumably very massive. It would seem that the analogy between the two interactions should break down over these rather substantive distinctions. Despite the enormous disparity in strength and range, it has been shown that the two forces actually do represent different manifestations of a single *unified interaction*. The *electroweak theory* that draws together the electromagnetic and weak interactions is in agreement with all experimental tests and is currently regarded as an established theory. This unification of forces may be compared to the outcome of Maxwell's theory, in which the principles of electricity and magnetism are united in a single formalism. New physical laws are at work in the unified theory. These ideas have had a revolutionary influence on all recent developments in particle physics.

Our presentation of the electroweak theory is divided into two parts. First, in this section, we describe the structure of the theory and examine the observable consequences in the interactions of quarks, leptons, and quanta. We continue to rely on Feynman diagrams to illustrate our descriptive approach. Later, in Section 16-16, we turn to the deeper questions of principle that underlie the whole successful formulation.

The unified theory of the electroweak interaction is a renormalizable quantum field theory in which the interactions of quarks and leptons are mediated by a unified electroweak field having *four* charge-specific degrees of freedom. Four different charge-bearing quanta are associated with this generalized field. The quanta

$$W^+, \quad W^0, \quad W^-, \quad \text{and} \quad B^0$$

are required to be *massless* particles by the guiding principle of the theory, just as the photon is required to be massless in the theory of the electromagnetic interaction. Two of these quanta become the expected W^+ and W^- of the weak interaction. The remaining two are neutral and can therefore be mixed in various combined states. The couplings to the electromagnetic currents of quarks and leptons select a particular combination of B^0 and W^0 for identification as the photon:

$$\gamma = B^0 \cos \theta_w + W^0 \sin \theta_w. \tag{16-29}$$

This relation has the form of a rotation defined by the *weak mixing angle* θ_w, a parameter of the theory to be determined by experiment. Another rotated combination of B^0 and W^0, orthogonal to γ, identifies another neutral entity:

$$Z = -B^0 \sin \theta_w + W^0 \cos \theta_w. \tag{16-30}$$

This expression defines an independent neutral weak quantum and leads to effects not anticipated by any previous discoveries in weak interaction phenomenology.

One of the two vital ingredients of the electroweak theory is the principle by which the quanta are introduced originally as massless particles. The other is a mechanism by which the weak quanta W^+, W^-, and Z *acquire mass* while the electromagnetic quantum γ retains its requisite zero-mass property. The guiding principle is known as *gauge symmetry*, and the mass-acquisition mechanism is called *spontaneous symmetry breaking*. The renormalizability of the theory is achieved through the one ingredient, and the short range of the weak force is obtained through the other.

The predictions of this remarkable theory include a relation between the masses of the charged and neutral weak quanta,

$$\frac{m_W}{m_Z} = \cos\theta_w, \tag{16-31}$$

and a relation between the fine structure constant $\alpha = e^2/4\pi\varepsilon_0\hbar c$ and the Fermi constant G_F of Section 16-7:

$$\alpha = \frac{\sqrt{2}}{\pi}\frac{G_F}{(\hbar c)^3}\left(m_W c^2 \sin\theta_w\right)^2. \tag{16-32}$$

These formulas involve the weak mixing angle, a quantity determined by the analysis of a variety of weak processes. Equation (16-32) can be used to predict the mass of $W^{\pm}$ in terms of the known quantities α and G_F, and Equation (16-31) can then be applied to predict a result for the mass of Z. Both of these masses are expected to be quite large, of order 100 GeV/c^2, in keeping with the very short range of the weak force. Equation (16-32) is especially interesting in this regard because the relation connects the strengths of the two unified interactions. The dependence on the large mass m_W demonstrates clearly how the extremely small size of G_F is associated with the extremely short range of the weak interaction.

The electroweak theory includes electrodynamics. Hence, all the predicted couplings between the photon and the electromagnetic currents of the fundamental charged particles have their usual form. The theory also describes all the usual charge-changing weak transitions in terms of the anticipated couplings of $W^{\pm}$ to the various charge-changing currents. These *charged currents* are shown in Figure 16-33. Note that the quark flavor transitions involve the modified quarks d', s', and b'. These

Figure 16-33
Coupling of W^- to the weak charged currents of quarks and leptons.

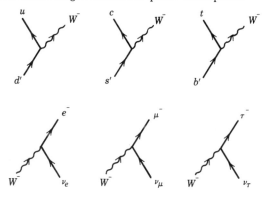

Figure 16-34
Weak neutral currents of quarks and leptons coupled to Z.

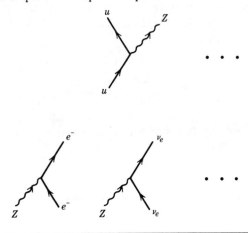

flavors with $Q/e = -\frac{1}{3}$ are combinations of d, s, and b, obtained by making rotations like the ones introduced in Section 16-13. (An extended Cabibbo-rotation scheme is presently in use to accommodate the three generations of quarks. This new flavor-mixing procedure replaces the single Cabibbo angle θ of Section 16-13 by four different angles. One of the four plays a direct role in the parametrization of CP-violation.)

Figure 16-35
Neutrino scattering processes involving contributions due to weak neutral currents. A charged-current term also occurs in $\nu_e e^- \to \nu_e e^-$, but not in $\nu_\mu e^- \to \nu_\mu e^-$.

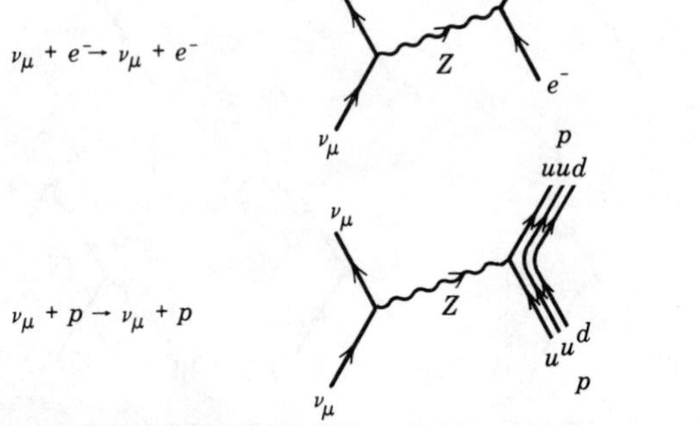

$$\nu_e + e^- \to \nu_e + e^-$$

$$\nu_\mu + e^- \to \nu_\mu + e^-$$

$$\nu_\mu + p \to \nu_\mu + p$$

The theory also predicts a new weak phenomenon brought about by the presence of the neutral quantum Z. This fourth quantum in the theory couples to *charge-conserving* weak currents, also called weak *neutral currents*, as illustrated in Figure 16-34. An entire body of new weak processes is introduced by this piece of the unified interaction, with probabilities determined by the unambiguous predictions of the unified theory. The couplings to the weak neutral currents are distinct from electromagnetic couplings, which also involve charge-conserving currents, because of the parity-violating effects in the weak interaction. Examples of such neutral-current phenomena are shown in Figure 16-35.

Several different ideas had to be brought together by several people to build this intricate theory. Yang conceived the guiding principle, and Schwinger proposed the unification concept, both in the 1950s. The fourfold structure of the unified electroweak field was devised by Glashow in 1961. The mechanism by which the weak quanta acquired their mass was incorporated in the theory by S. Weinberg in 1967, and a very similar model was put forward a year later by A. Salam. Little attention was given to these contributions at first, because the crucial renormalizability of the theory was regarded as doubtful. Skepticism gave way to enthusiasm in 1971, however, when a proof of this property was carried out by G. 't Hooft. The theory received support from experiment in 1973, when neutral currents were discovered in neutrino reactions of the type

$$\nu_\mu + p \rightarrow \nu_\mu + \text{hadrons}.$$

Further confirmation of a piece of the theory came in 1974 with the discovery of charm in the J/ψ resonance.

The keystones of the electroweak theory, the new quanta $W^\pm$ and Z, were not discovered until 1983. An accelerator with especially high energy had to be constructed to create these massive particles, and the CERN proton–antiproton collider was dedicated to that purpose. The essential innovation in the design of the collider was a technique developed by S. van der Meer for accelerating and storing antiprotons at very high energy. Collisions were achieved between 270 GeV protons and 270 GeV antiprotons, and evidence of the weak quanta was found, in experiments conducted by C. Rubbia and others. The basic production mechanism was interpreted to be quark–antiquark annihilation, as illustrated in Figure 16-36. The detection of

Figure 16-36
Production of weak quanta in proton–antiproton collisions.

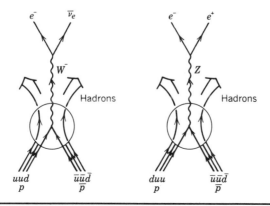

the particles, first W and then Z, was performed through the materialization of the massive quanta into pairs of leptons. Quoted values for the two masses have been given as

$$m_W = 80.8 \text{ GeV}/c^2 \quad \text{and} \quad m_Z = 92.9 \text{ GeV}/c^2,$$

in striking agreement with the predictions of the theory.

Example

The experimental value of $\sin^2\theta_w$ is around 0.23, and so θ_w is about $29°$. We can use this result to make predictions for m_W and m_Z. Let us rewrite Equation (16-32) as

$$m_W c^2 \sin\theta_w = \sqrt{\frac{\pi\alpha}{\sqrt{2}\, G_F/(\hbar c)^3}} \, ,$$

and let us recall from Section 16-7 the expression for G_F in terms of the proton mass:

$$\frac{G_F}{(\hbar c)^3} = \frac{1.03 \times 10^{-5}}{\left(M_p c^2\right)^2}.$$

The combination of formulas results in the following prediction:

$$m_W c^2 \sin\theta_w = \sqrt{\frac{\pi/137}{\sqrt{2}\left(1.03 \times 10^{-5}\right)}} \, M_p c^2$$

$$= (39.7)(0.938 \text{ GeV}) = 37.2 \text{ GeV}.$$

The masses of the weak quanta are then found to be

$$m_W = \frac{37.2 \text{ GeV}/c^2}{\sin 29°} = 78 \text{ GeV}/c^2$$

and

$$m_Z = \frac{78 \text{ GeV}/c^2}{\cos 29°} = 90 \text{ GeV}/c^2,$$

where the second prediction is based on Equation (16-31). Both results are in agreement with the findings from CERN.

16-15 Color and the Strong Interaction

Two immediate questions confront us when we adopt a theory of hadrons based on the quark hypothesis. The quarks are taken to be real constituents, permanently bound in hadronic states and not observed as liberated particles. If we accept this view, we must ask how the mechanism of *permanent confinement* works. The quarks are also known to have spin-$\frac{1}{2}$ properties. If we accept the fact that they behave as fermions, we must ask how *fermion antisymmetry* is obeyed in the formation of baryons. These questions lead us to the concept of color and to the theory of the strong interaction where color plays the main role. (In this context color is a fanciful name for a new quantum number. The concept is unrelated to any attribute affecting our visual senses.) The fundamental treatment of the strong interaction involves the dynamics of color in the binding of quarks. Accordingly, the theory is called *quantum chromodynamics*. Strong processes occur among the observable hadrons as outward manifestations of this basic interaction of confined constituents.

Let us begin with the issue of fermion antisymmetry and illustrate the problem by referring to the hadron Δ^{++}. Recall from Section 16-12 that one of the states of this spin-$\frac{3}{2}$ particle is described as

$$\Delta^{++} = uuu \uparrow \uparrow \uparrow,$$

where each u quark is in an s state with spin up. The description in its present form violates the Pauli principle, because both the space and spin factors in the wave function are symmetric under the exchange of quark variables. Color is introduced at this point to contribute an additional quark degree of freedom. If we assign a different color to each quark, we see that the violation of the exclusion principle disappears since no two of the quarks in Δ^{++} are in the same flavor–space–spin–color state. We need *three* different values for the color quantum number to accomplish this trick. (In fact, there are three colors of quarks in all, for a variety of reasons. One of the pieces of pertinent information from experiment is discussed below.) The quark colors are conventionally designated by means of the indices R, Y, and B, using notation inspired by the three primary colors red, yellow, and blue. We assume that this same threefold multiplicity is attached to every quark flavor, so that we have *color triplets* of quarks (u_R, u_Y, u_B), (d_R, d_Y, d_B), (s_R, s_Y, s_B), and so on.

We can now return to the Δ^{++} problem and construct an eigenfunction for three u quarks. The desired expression must be exchange symmetric in space and spin, and exchange antisymmetric in color. The Slater determinant of Section 9-5 can be adapted to these properties by writing

$$\Delta^{++} = \frac{1}{\sqrt{3!}} \begin{vmatrix} u_R(1) & u_Y(1) & u_B(1) \\ u_R(2) & u_Y(2) & u_B(2) \\ u_R(3) & u_Y(3) & u_B(3) \end{vmatrix}. \tag{16-33}$$

Six terms are written in this eigenfunction. Each is a product of three quark factors, whose flavor and color are explicit, and whose s-state and spin-up specifications are implicit in the quark labels (1), (2), and (3). We note that the expression remains unchanged when the color indices are rotated through the cycle

because of the cyclic symmetry of the determinant. The color portion of the eigenfunction is called a *singlet* since its unique form is maintained under such rotations. In this respect color acts like charge and combines in a *color-neutral* system to make a hadron. Color-singlet states are assumed for *all* hadrons. This stipulation prevents the added color degree of freedom from causing an undesired proliferation in the number of observable hadronic states. The color-singlet property is ensured for baryons if the three flavors are taken with colors R, Y, and B in equal parts, as in Equation (16-33). Color-singlet mesons also call for equal-parts admixtures of color. For instance, the color composition of the π^+ state $u\bar{d}$ is expressed as

$$\pi^+ = \frac{1}{\sqrt{3}}\left(u_R\bar{d}_R + u_Y\bar{d}_Y + u_B\bar{d}_B\right), \tag{16-34}$$

where the flavors u and $\bar{d}$ refer to s-state constituents and where the usual $s = 0$ spin eigenfunction $(1/\sqrt{2})(\uparrow\downarrow - \downarrow\uparrow)$ is implied. Note that the baryon state in Equation (16-33) is *antisymmetric* under exchange of any pair of color indices and that the meson state in Equation (16-34) is *symmetric* under the same operation. We obtain color-singlet expressions for all known baryons and mesons by adhering throughout to color-antisymmetric and color-symmetric constructions in the two situations.

Experimental evidence for three colors of quarks is found in electron–positron annihilation into hadrons. We have already introduced the cross-section ratio R in Equation (16-28) and Figure 16-32 to express the energy dependence of this process relative to the energy dependence of $e^-e^+ \rightarrow \mu^-\mu^+$. The quark model takes the reaction $e^-e^+ \rightarrow$ hadrons through the intermediate step $e^-e^+ \rightarrow q\bar{q}$ and lets the $q\bar{q}$ pairs materialize into the various hadronic final states. Figure 16-37 shows this hadronization mechanism in a schematic derivation of R. The different quark–anti-quark pairs $u\bar{u}, d\bar{d}, s\bar{s}, \ldots$ have successive *thresholds* in the variable E_{CM}. Each pair contributes to the numerator of R with a factor of 3, owing to the multiplicity of colors for each quark. We observe that, if the e^-e^+ collision energy is far enough above the threshold for a given $q\bar{q}$, then the comparison of the processes

$$e^- + e^+ \rightarrow q + \bar{q} \quad \text{and} \quad e^- + e^+ \rightarrow \mu^- + \mu^+$$

is rather insensitive to the difference in mass between q and μ. The energy dependence for $e^-e^+ \rightarrow q\bar{q}$ therefore becomes the same as the energy dependence for $e^-e^+ \rightarrow \mu^-\mu^+$, and so the particular $q\bar{q}$ contribution to R becomes independent of E_{CM} in this regime. If the quarks are point-like, as the muons are, then the $q\bar{q}$ and $\mu^-\mu^+$ processes differ only through the quark charge Q_q. This charge affects the coupling of the quark current to the virtual photon in the manner indicated in the figure. We can summarize these arguments in the derivation of R by writing

$$R = \frac{\sigma(e^-e^+)}{\sigma(e^-e^+ \rightarrow \mu^-\mu^+)} \rightarrow \frac{3\sum_q |\chi(q\bar{q}, e^-e^+)|^2}{|\chi(\mu^-\mu^+, e^-e^+)|^2} \rightarrow 3\sum_q \left(\frac{Q_q}{e}\right)^2. \tag{16-35}$$

The amplitudes in the intermediate step correspond to the diagrammatic model shown in the figure. This result predicts a succession of rising plateaus as each $q\bar{q}$ threshold is exceeded. Our expectations are illustrated in the lower part of Figure 16-37. The actual behavior of R in Figure 16-32 shows good agreement with the stepwise shape of this prediction. We note particularly that the factor of 3 arising from color is needed to make the agreement possible.

If we compare Figures 16-32 and 16-37 more closely, we see that R exhibits sharp deviations from Equation (16-35) wherever the successive $q\bar{q}$ contributions have their thresholds. Thus, as E_{CM} decreases, the Υ family of $b\bar{b}$ states appears at the onset of

Figure 16-37

Quark model for the ratio R in Figure 16-32. Quarks and muons couple to the virtual photon with charges Q_q and e. Quarks contribute to the sum over q as indicated in the graph of R.

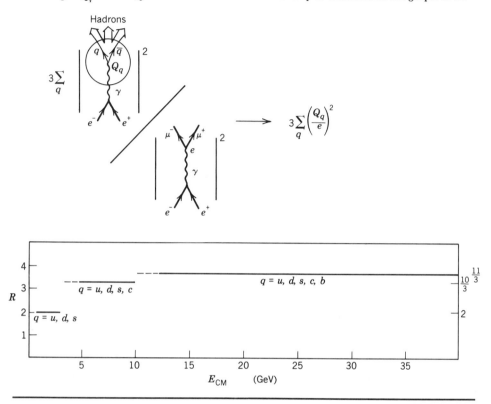

the $b\bar{b}$ term in R, and then the $c\bar{c}$ charmonium states occur near the lower $c\bar{c}$ threshold. We can estimate the masses of the constituent b and c quarks on these grounds to be around 5 and 1.5 GeV/c^2, respectively. It is obvious that the succeeding quarks s, d, and u are much lighter than b and c. It should also be noted that the whole complex issue of quark masses is subject to interpretation, since the quarks are not observed as free particles.

The original problem of fermion antisymmetry was resolved in the 1960s with the introduction of the color concept by O. W. Greenberg, and by M.-Y. Han. and Y. Nambu. Color dynamics was promoted a decade later as the basis for the theory of the strong interaction by Weinberg and Gell-Mann, among others.

The strong interaction accounts for the binding of quarks and antiquarks to make mesons and the binding of three quarks to make baryons. These bound systems of color-bearing constituents are formed dynamically in color-singlet states, as noted above. Quantum chromodynamics is the renormalizable quantum field theory that promises to explain such phenomena. We devote the rest of the section to a purely descriptive account of this complex and compelling theory. The underlying principle is again furnished by gauge symmetry, as in the theories of electromagnetism and electroweak unification. The resulting formalism describes the strong interaction of quarks in terms of transitions and couplings in the *color* degree of freedom of each

Figure 16-38

Quark and antiquark color transitions with the corresponding gluon couplings. The gluons execute changes of color represented by pairs of color indices.

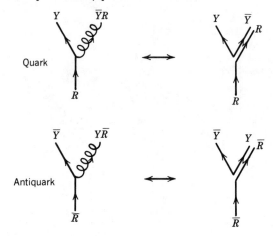

quark. These color-changing effects propagate from quark to quark by the transmission of the *color-changing quanta* that make up the mediating strong field. The quanta are called *gluons* since they are devised to account for the binding of quarks. Figure 16-38 shows how the color quantum numbers behave when gluons are emitted in the color transitions of quarks and antiquarks. Like the photon in quantum electrodynamics, the gluons must occur in the theory as *massless* quanta because gauge symmetry demands this property. Unlike the photon, however, the quanta of the gluon field are not emitted from hadrons as observable particles because the gluons carry color, and color is contrived to be permanently confined inside the hadrons.

Quantum chromodynamics and quantum electrodynamics, QCD and QED, are analogous theories to the extent that their principles share gauge symmetry as a basic concept. The gauge symmetries of the two theories are different, however, enabling the gluons to have a certain very important feature not shared by the photon. Fundamental charged particles emit and absorb photons, and so the photon is said to be coupled to the charge degree of freedom in QED. Since the photon carries no charge, there exists no coupling of photons to photons. By analogy, gluons are coupled to the color degree of freedom in QCD. The gluons must carry color since their couplings to quarks are designed to cause quark color transitions. Consequently, gluons can interact with gluons through their own intrinsic color quantum numbers. QCD departs fundamentally from QED in this respect and leads to a remarkable property called *asymptotic freedom* as a result. By virtue of the gluon–gluon interaction, the coupling of color in QCD approaches *zero* strength at arbitrarily small distances or, equivalently, at arbitrarily large momenta. This property has been established as a rigorous consequence of the theory in investigations by D. J. Gross and F. Wilczek and by H. D. Politzer.

The fact that the coupling of colors vanishes asymptotically at very short range encourages the view that the same coupling may grow without bound for very long range. This belief holds the key to the notion of *color confinement*, whereby free quarks

Figure 16-39

Meson photoproduction in a model of confined quarks.

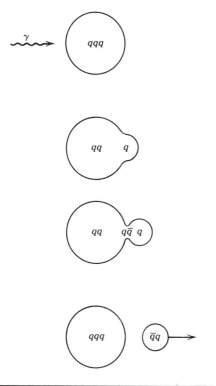

are never seen, and observable hadrons are always found with quarks bound in color-singlet (or zero-color) states. Quantum chromodynamics is a very complicated relativistic quantum field theory. A few rigorous conclusions have been extracted from the theory so far, while confinement remains the central problem to be solved.

Approximations to QCD and models of quark confinement have been rather successful. These studies support the idea that the configurational energy of a quark system increases as the quarks separate. We illustrate the situation with the aid of Figure 16-39 by showing the effect of an incident γ ray on a bound qqq baryon. The added energy excites the quark system, producing a tendency toward quark separation, but no amount of additional energy can suffice to achieve quark liberation. Instead, the added energy produces $q\bar{q}$ pairs in the strong gluon field so that the original baryon fragments into a meson–baryon final state.

The ideas of QCD are readily translated for application to the phenomenology of *heavy-quark* systems. The large masses of the constituent quarks allow the use of the nonrelativistic Schrödinger equation in models based on a suitable central potential energy. Expressions of the form

$$V(r) = -\frac{a_1}{r} + a_2 r \qquad (16\text{-}36)$$

Figure 16-40.

Potential-energy model for heavy quark–antiquark binding and mass levels of charmonium. The ^3P states of the $c\bar{c}$ system exhibit spin–orbit splitting.

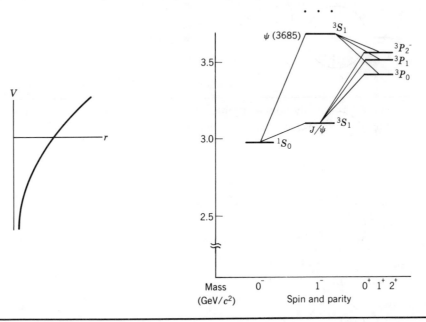

can be applied to analyze the states of the $c\bar{c}$ charmonium system and the $b\bar{b}$ Υ system. This type of potential energy function is sketched in Figure 16-40. The first term in V mimics Coulomb attraction, in keeping with the analogy between QCD and QED, and the second term supplies a linear barrier to simulate the effect of confinement. The familiar angular momentum techniques of atomic physics hold for $c\bar{c}$ and $b\bar{b}$. Consequently, the resulting states are identifiable by the familiar spectroscopic notation $^{2s+1}L_j$ for total spin s, orbital angular momentum ℓ, and total angular momentum j. Thus, the 1^- vector meson J/ψ is a $^3S_1$ $c\bar{c}$ state having $s = 1$, $\ell = 0$, and $j = 1$. The excited state $\psi(3685)$ occurs with the same set of vector-meson quantum numbers. These two charmonium states are distinguished by their radial quantum numbers $n = 1$ and $n = 2$. A spectrum showing most of the known $c\bar{c}$ states and radiative transitions is included in Figure 16-40. A similar approach can be taken with the $b\bar{b}$ system to analyze the family of Υ states.

The strong interaction, as just described, bears little resemblance to the old theory of the nuclear force, in which protons and neutrons are bound by the exchange of mesons. This original problem in hadron physics is evidently not a fundamental strong-interaction problem, simply because the interacting hadrons are not fundamental particles. Each hadron owes its existence to the chromodynamics of permanently confined quarks. The hadrons themselves are color-neutral, and so the exchange of gluons does not occur from one hadron to another unless there is some overlap of the hadronic quark systems. Hence, the fundamental strong interaction can become operative in the force of attraction between nucleons at short range, but the complex mechanism involves the participation of many quarks. This situation may be likened to the relationship between atoms and molecules. The fundamental electromagnetic interaction accounts for the existence of atoms, and the van der Waals force then binds the neutral atoms together at short range to form molecules. It is speculated that a similarly successful theory of the nucleon–nucleon force can be derived from quantum chromodynamics. It is also conceivable that the structure of nuclei may be usefully cast in terms of quarks. Both of these objectives are currently under active investigation.

Example

Equation (16-35) is easily applied to make predictions for the cross-section ratio R. If we let the sum over q include $u\bar{u}$, $d\bar{d}$, and $s\bar{s}$ terms, we obtain

$$3\sum_q \left(\frac{Q_q}{e}\right)^2 = 3\left[\left(\tfrac{2}{3}\right)^2 + \left(-\tfrac{1}{3}\right)^2 + \left(-\tfrac{1}{3}\right)^2\right] = 2.$$

When the sum is extended to take in $c\bar{c}$, and then $b\bar{b}$, the results are

$$3\sum_q \left(\frac{Q_q}{e}\right)^2 = 2 + 3\left(\frac{2}{3}\right)^2 = \frac{10}{3}$$

and

$$3\sum_q \left(\frac{Q_q}{e}\right)^2 = \frac{10}{3} + 3\left(-\frac{1}{3}\right)^2 = \frac{11}{3}.$$

These values of R are indicated on the right side of the graph in Figure 16-37. We observe that each of the numbers would be reduced by one-third in the absence of color.

16-16 Gauge Symmetries

Conservation laws and symmetry principles have appeared throughout the chapter with the introduction of each new internal quantum number. Every case considered so far has been an application of a *global symmetry*, where the relevant quantity obeys its conservation law in a uniform manner over all space and for all time. To cite an example, the conservation of isospin in strong reactions is attributed to a rotational symmetry in which the rotations in the three-dimensional isospin space are independent of locations in space and time. The conservation of charge in all reactions has also been treated (so far) in terms of a similarly global symmetry principle. The interactions of particles must be described in a way that accommodates all these conserved quantities.

A more powerful concept emerges when a conserved quantity can be associated with a local type of conservation law. Charge is a prime example of such a quantity. A corresponding *local symmetry* then becomes operative and implements the conservation law in a manner that may vary from one space–time point to another. This variability forces the system to include a mediating field whose response to the symmetry operation is such as to *compensate* for the variation of the symmetry from point to point. Thus, the required field enables the local symmetry to propagate through the system and provide a mechanism for interaction. Gauge symmetry is another name used to describe this behavior. The mediating field is referred to as a *gauge field*, and the mediators of the field are called *gauge quanta*. There can be no limit to the space–time extent over which the local symmetry may vary. Therefore, the compensating property of the gauge field has to have infinite range, so that the gauge quanta must be *massless*. A theory based on gauge symmetry is said to possess *gauge invariance*, since observable properties of the system are not altered by the local symmetry operation.

A gauge-invariant theory *defines* an interacting system. The interaction medium is furnished by the gauge field, and the structure of the interaction is dictated by the nature of the underlying gauge symmetry. The resulting relativistic quantum field theory ultimately proves to be renormalizable by virtue of the constraints imposed by the local symmetry. Theories of the fundamental interactions of particles are con-structed from gauge principles in order to take advantage of this crucial property. The electromagnetic, electroweak, and strong interactions are cases in point. We have already described the highlights of these theories, and now we wish to devote an entire section to an explanation of the general principles.

Let us examine electrodynamics as the prototype of a gauge theory. Electromag-netism is chosen because the gauge behavior of the theory is particularly straightfor-ward to understand. We begin with the classical fields, and then we confine our attention to a nonrelativistic particle when we extend the problem to quantum mechanics.

Classical electromagnetic theory is completely contained in Maxwell's equations for the coupled $\mathbf{E}$ and $\mathbf{B}$ fields:

$$\nabla \cdot \mathbf{E} = \frac{\rho}{\varepsilon_0} \quad \text{(Coulomb's law)} \tag{16-37a}$$

$$\nabla \times \mathbf{B} - \frac{1}{c^2}\frac{\partial \mathbf{E}}{\partial t} = \mu_0 \mathbf{J} \quad \text{(Ampere's law and displacement current)} \tag{16-37b}$$

$$\nabla \times \mathbf{E} + \frac{\partial \mathbf{B}}{\partial t} = 0 \quad \text{(Faraday's law)} \tag{16-37c}$$

$$\nabla \cdot \mathbf{B} = 0 \quad \text{(absence of magnetic monopoles)} \tag{16-37d}$$

We consider fields in vacuum in the presence of charge and current densities ρ and $\mathbf{J}$. These quantities satisfy a *local* charge–current conservation law, obtained by differen-tiating Equations (16-37a) and (16-37b) as follows:

$$\frac{\partial}{\partial t}\nabla \cdot \mathbf{E} = \frac{1}{\varepsilon_0}\frac{\partial \rho}{\partial t}$$

and

$$\nabla \cdot (\nabla \times \mathbf{B}) - \frac{1}{c^2}\nabla \cdot \frac{\partial \mathbf{E}}{\partial t} = \mu_0 \nabla \cdot \mathbf{J}.$$

When we use

$$\nabla \cdot (\nabla \times \mathbf{B}) = 0 \quad \text{and} \quad \frac{1}{c^2} = \mu_0 \varepsilon_0,$$

we obtain the desired local relation between charge and current:

$$\frac{\partial \rho}{\partial t} + \nabla \cdot \mathbf{J} = 0. \tag{16-38}$$

Note that the result hinges on the presence of Maxwell's displacement current, given by the $\partial \mathbf{E}/\partial t$ term added to Ampere's law. We recall that the prediction of electromagnetic waves also follows because of this contribution.

The last two of Maxwell's equations involve no source terms and may therefore be construed as kinematical relations among the various field components. We can use *vector* and *scalar potentials* to convey the effects of these formulas and thereby simplify the whole system of coupled equations. A vector potential $\mathbf{A}$ is introduced by setting

$$\mathbf{B} = \nabla \times \mathbf{A}, \tag{16-39}$$

so that Equation (16-37d) is automatically satisfied. A scalar potential ϕ is then employed to define

$$\mathbf{E} = -\nabla \phi - \frac{\partial \mathbf{A}}{\partial t}, \tag{16-40}$$

so that Equation (16-37c) is also secured. This second result follows with the aid of the identity

$$\nabla \times (\nabla \phi) = 0.$$

We obtain a reduced, but still coupled, set of differential equations for $\mathbf{A}$ and ϕ when we return to the first two of Maxwell's equations and insert Equations (16-39) and (16-40).

The potentials are not uniquely defined by this procedure. In fact, this is the place where gauge symmetry makes its first appearance. We observe that $\mathbf{B}$ is not changed in Equation (16-39) if we replace $\mathbf{A}$ by

$$\mathbf{A}' = \mathbf{A} + \nabla \Lambda, \tag{16-41}$$

and that $\mathbf{E}$ is not changed in Equation (16-40) if we also replace ϕ by

$$\phi' = \phi - \frac{\partial \Lambda}{\partial t}. \tag{16-42}$$

The new quantity $\Lambda(\mathbf{r}, t)$ is introduced as an *arbitrary* scalar function of space and time in the two expressions. These assertions hold because

$$\nabla \times \mathbf{A}' = \nabla \times \mathbf{A}$$

and

$$-\nabla \phi' - \frac{\partial \mathbf{A}'}{\partial t} = -\nabla \phi + \nabla \frac{\partial \Lambda}{\partial t} - \frac{\partial \mathbf{A}}{\partial t} - \frac{\partial}{\partial t} \nabla \Lambda = -\nabla \phi - \frac{\partial \mathbf{A}}{\partial t}.$$

The substitutions $\mathbf{A} \rightarrow \mathbf{A}'$ and $\phi \rightarrow \phi'$ in Equations (16-41) and (16-42) specify the *gauge-transformation* properties of the electromagnetic potentials. We see that these operations have no effect on the observable fields $\mathbf{E}$ and $\mathbf{B}$, and we conclude that classical electromagnetic theory is a gauge-invariant formalism. It is possible to exploit this freedom in the definition of the potentials and generate uncoupled (or otherwise simplified) differential equations for the determination of $\mathbf{A}$ and ϕ.

These formal characteristics of the electromagnetic potentials take on greater significance when the quantum behavior of a charged particle is examined in a gauge-invariant theory. Let us focus on the nonrelativistic problem and start with the Schrödinger equation for a free particle:

$$\frac{1}{2m}\left(\frac{\hbar}{i}\nabla\right)^2 \Psi = i\hbar\frac{\partial}{\partial t}\Psi. \tag{16-43}$$

The phase of the complex-valued wave function $\Psi(\mathbf{r}, t)$ is not a measurable quantity. We may therefore argue that the description of the particle cannot be affected if Ψ is replaced by a phase-altered wave function of the form

$$\Psi'(\mathbf{r}, t) = e^{i\alpha}\Psi(\mathbf{r}, t).$$

It is easy to see that Ψ' also obeys Equation (16-43), *provided* the phase α is a constant. Let us suppose, however, that the phase is allowed to *vary* with $\mathbf{r}$ and t and that the observable predictions of the theory are required to remain unchanged. In this case the differential operators act on $\alpha(\mathbf{r}, t)$ as well as Ψ, so that Ψ' no longer satisfies the given free-particle equation. We express such a variable-phase alteration of Ψ as

$$\Psi'(\mathbf{r}, t) = e^{iQ\Lambda(\mathbf{r}, t)/\hbar}\Psi(\mathbf{r}, t), \tag{16-44}$$

where Λ is an arbitrary function and Q is a parameter to be identified in due course. The prescription defines a *gauge transformation* of the wave function and creates *differences* in the phase of Ψ between different locations in space and time. We can prevent these arbitrary effects from becoming observable by introducing the electromagnetic potentials to serve as *gauge fields*. If Λ in Equation (16-44) is taken to be the same arbitrary function as in Equations (16-41) and (16-42), the gauge transformation of $\mathbf{A}$ and ϕ exactly compensates for the arbitrary phase variation of Ψ. The resulting formalism for Ψ, $\mathbf{A}$, and ϕ therefore constitutes a gauge-invariant theory.

We have to modify the free-particle equation to accomplish these ends. If the replacements

$$\frac{\hbar}{i}\nabla \rightarrow \frac{\hbar}{i}\nabla - Q\mathbf{A} \quad \text{and} \quad i\hbar\frac{\partial}{\partial t} \rightarrow i\hbar\frac{\partial}{\partial t} - Q\phi \tag{16-45}$$

are made in Equation (16-43), the Schrödinger equation becomes

$$\frac{1}{2m}\left(\frac{\hbar}{i}\nabla - Q\mathbf{A}\right)^2\Psi = \left(i\hbar\frac{\partial}{\partial t} - Q\phi\right)\Psi. \tag{16-46}$$

We now wish to demonstrate how this approach gives a gauge-invariant procedure for the determination of Ψ. The gauge-transformed version of the left side of Equation

(16-46) is analyzed in detail as follows:

$$\left(\frac{\hbar}{i}\nabla - Q\mathbf{A}'\right)^2 \Psi' = \left[\frac{\hbar}{i}\nabla - Q(\mathbf{A} + \nabla\Lambda)\right]\cdot\left[\frac{\hbar}{i}\nabla - Q(\mathbf{A} + \nabla\Lambda)\right]e^{iQ\Lambda/\hbar}\Psi$$

$$= \left[\frac{\hbar}{i}\nabla - Q(\mathbf{A} + \nabla\Lambda)\right]$$

$$\cdot\left\{e^{iQ\Lambda/\hbar}(Q\nabla\Lambda)\Psi + e^{iQ\Lambda/\hbar}\left[\frac{\hbar}{i}\nabla - Q(\mathbf{A} + \nabla\Lambda)\right]\Psi\right\}$$

$$= \left[\frac{\hbar}{i}\nabla - Q(\mathbf{A} + \nabla\Lambda)\right]\cdot e^{iQ\Lambda/\hbar}\left(\frac{\hbar}{i}\nabla - Q\mathbf{A}\right)\Psi$$

$$= e^{iQ\Lambda/\hbar}(Q\nabla\Lambda)\cdot\left(\frac{\hbar}{i}\nabla - Q\mathbf{A}\right)\Psi$$

$$+ e^{iQ\Lambda/\hbar}\left[\frac{\hbar}{i}\nabla - Q(\mathbf{A} + \nabla\Lambda)\right]\cdot\left(\frac{\hbar}{i}\nabla - Q\mathbf{A}\right)\Psi$$

$$= e^{iQ\Lambda/\hbar}\left(\frac{\hbar}{i}\nabla - Q\mathbf{A}\right)^2 \Psi.$$

The analysis of the right side of the equation proceeds along similar lines:

$$\left(i\hbar\frac{\partial}{\partial t} - Q\phi'\right)\Psi' = \left[i\hbar\frac{\partial}{\partial t} - Q\left(\phi - \frac{\partial\Lambda}{\partial t}\right)\right]e^{iQ\Lambda/\hbar}\Psi$$

$$= -e^{iQ\Lambda/\hbar}\left(Q\frac{\partial\Lambda}{\partial t}\right)\Psi + e^{iQ\Lambda/\hbar}\left[i\hbar\frac{\partial}{\partial t} - Q\left(\phi - \frac{\partial\Lambda}{\partial t}\right)\right]\Psi$$

$$= e^{iQ\Lambda/\hbar}\left(i\hbar\frac{\partial}{\partial t} - Q\phi\right)\Psi.$$

Thus, the structure of Equation (16-46) is such that an arbitrary phase variation of Ψ is reconciled by the corresponding gauge behavior of $\mathbf{A}$ and ϕ. The phase-altering factor therefore passes through all the operations in the equation, so that Ψ' satisfies the gauge-transformed version of the original equation for Ψ.

The interpretation of the parameter Q becomes apparent when Equation (16-46) is rewritten:

$$\frac{1}{2m}\left(\frac{\hbar}{i}\nabla - Q\mathbf{A}\right)^2 \Psi + Q\phi\Psi = i\hbar\frac{\partial}{\partial t}\Psi. \tag{16-47}$$

The familiar features of the Schrödinger equation for an interacting particle appear in this result, as the second term on the left evidently contains a potential energy of the form

$$V = Q\phi.$$

If we let the scalar potential ϕ be an electrostatic potential, we recognize Q to be the *charge* of the particle and V to be the usual Coulomb interaction. Note that the term $Q\phi\Psi$ in the equation represents a coupling in which the particle and the gauge field interact through the coupling parameter Q. The remarkable conclusion to draw from this demonstration is that the requirement of gauge invariance dictates both the existence and the form of the interaction. Gauge symmetry operates in exactly the same way in conjunction with the Dirac equation to generate the electromagnetic interaction for a relativistic spin-$\frac{1}{2}$ charged particle. Quantum electrodynamics is the resulting gauge-invariant theory.

Quantum chromodynamics is obtained as the theory of the strong interaction through a different application of gauge invariance. We again consider local phase variations in the spirit of Equation (16-44), and we also let the gauge transformations act on the colors of quarks to produce changes in those degrees of freedom. There are eight distinct color alterations possible among the three quark colors. Consequently, eight gauge fields are needed, with gauge behavior different from Equations (16-41) and (16-42), in order to compensate for the arbitrary phase adjustments of the quarks. The color-changing aspects of the quark transformations are fundamental to the gauge symmetry of QCD. These features set this theory apart from the simpler gauge symmetry of QED. The result is a vastly more complex structure for the description of the strong interaction. Eight color-changing zero-mass gluon quanta are associated with the eight gauge fields. These gluons couple to the colors of quarks, and also to the colors carried by other gluons, so that the gauge-invariant theory describes gluon–quark and *also* gluon–gluon interactions. The latter situation has no analogue in QED. The gluon–gluon interaction is the source of the property of asymptotic freedom for gluon couplings at short range and is a crucial ingredient in the mysterious property of color confinement. The gluon–gluon interaction comes about because the various transformations of the internal degrees of freedom are not commutative operations. Gauge theories based on *noncommuting* local variations of phase are named generically after Yang and R. L. Mills, the first investigators to make such an extension of gauge symmetry.

While one type of Yang–Mills theory implements a particular gauge symmetry among the *colors* of quarks, another type invokes a different set of gauge principles for the behavior of quark *flavors*. We are led in the first case to the theory of the strong interaction, as just discussed, and in the second case to the theory of electroweak unification. This second application of noncommutative gauge symmetry pertains to local phase variations in which the noncommuting transformations occur within the doublets of quark flavors

$$\begin{bmatrix} u \\ d' \end{bmatrix} \qquad \begin{bmatrix} c \\ s' \end{bmatrix} \qquad \begin{bmatrix} t \\ b' \end{bmatrix},$$

and also within the doublets of leptons

$$\begin{bmatrix} e^- \\ \nu_e \end{bmatrix} \qquad \begin{bmatrix} \mu^- \\ \nu_\mu \end{bmatrix} \qquad \begin{bmatrix} \tau^- \\ \nu_\tau \end{bmatrix}.$$

(Recall that the quark entries d', s', and b' represent rotated combinations of d, s, and b in the generalized Cabibbo-rotation scheme for three generations of quarks. We have already referred to this scheme in our discussion of Figure 16-33.) It is easy to see that four different flavor transformations are called for and that the four required gauge fields correspond to the previously introduced electroweak quanta. We illustrate this remark by itemizing the four types of quark and lepton transitions as follows:

$$u \rightarrow u + \gamma/Z \qquad\qquad e^- \rightarrow e^- + \gamma/Z$$
$$u \rightarrow d' + W^+ \qquad\qquad e^- \rightarrow \nu_e + W^-$$
$$\qquad\qquad \cdots \qquad\qquad\qquad\qquad\qquad \cdots$$
$$d' \rightarrow u + W^- \qquad\qquad \nu_e \rightarrow e^- + W^+$$
$$d' \rightarrow d' + \gamma/Z \qquad\qquad \nu_e \rightarrow \nu_e + Z$$

(We indicate the alternative emission of γ or Z by the notation γ/Z.)

The historical obstacle to a successful weak interaction theory looms up at this juncture. Gauge symmetry demands *zero-mass* quanta and produces a renormalizable formalism, while phenomenology requires *massive* quanta, in apparent conflict with renormalizability. The impasse is resolved ingeniously by adopting gauge symmetry as the underlying dynamical principle and by assuming that the ground state of the theory does not possess the adopted symmetry. The procedure is called the *spontaneous breakdown* of gauge invariance. Remarkably, this breaking of the local symmetry leaves the renormalizability of the gauge-invariant theory intact. The miracle continues, as the spontaneous breakdown of gauge symmetry causes the appropriate gauge quanta to *acquire mass*. This central ingredient of the theory comes into play through the so-called Higgs mechanism, a phenomenon discovered by several people including (and named after) P. W. Higgs. The credits for assembling these pieces of the puzzle have already been given in Section 16-14 to Glashow, Weinberg, Salam, and 't Hooft.

The application of Yang–Mills theory to electroweak unification is too involved to describe any further in this text. Let us be content to illustrate spontaneous symmetry breaking in the context of a simpler gauge symmetry instead. We reconsider local phase transformations of the sort defined in Equation (16-44) for this purpose.

Let us suppose that the system includes a relativistic charged scalar particle whose wave function is $\Phi(\mathbf{r}, t)$. In the case of a free particle, Φ obeys the Klein–Gordon equation

$$\left(\frac{\hbar}{i} \nabla \right)^2 \Phi - \frac{1}{c^2} \left(i\hbar \frac{\partial}{\partial t} \right)^2 \Phi = -m^2 c^2 \Phi, \qquad (16\text{-}48)$$

as in Equation (16-3). We want Φ to be subject to local phase variations of the form

$$\Phi(\mathbf{r}, t) \rightarrow \Phi'(\mathbf{r}, t) = e^{iQ\Lambda(\mathbf{r},\, t)/\hbar} \Phi(\mathbf{r}, t).$$

Gauge invariance requires that we again incorporate the gauge fields $\mathbf{A}$ and ϕ and assume the gauge behavior given in Equations (16-41) and (16-42). The interaction of the particle with $\mathbf{A}$ and ϕ is then described by the modified Klein–Gordon equation

$$\left(\frac{\hbar}{i} \nabla - Q\mathbf{A} \right)^2 \Phi - \frac{1}{c^2} \left(i\hbar \frac{\partial}{\partial t} - Q\phi \right)^2 \Phi = -m^2 c^2 \Phi \qquad (16\text{-}49)$$

when the replacements in Equations (16-45) are made. If we define the real and imaginary parts of Φ by writing

$$\Phi = \Phi_1 + i\Phi_2,$$

we see that Φ_1 and Φ_2 satisfy Equation (16-49) independently.

Figure 16-41

Mass-squared paraboloids $m^2(\Phi_1, \Phi_2) = \mu^2 + \lambda(\Phi_1^2 + \Phi_2^2)$ for μ^2 positive and negative. The parameter λ is taken to be positive.

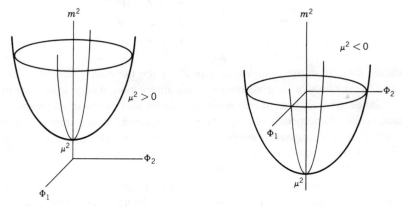

The mechanism for breaking gauge invariance can be introduced formally by letting the mass parameter m contain a contribution proportional to the probability density $\Phi^*\Phi$:

$$m^2 = \mu^2 + \lambda\Phi^*\Phi$$

or

$$m^2(\Phi_1, \Phi_2) = \mu^2 + \lambda\left(\Phi_1^2 + \Phi_2^2\right). \qquad (16\text{-}50)$$

This hypothesis injects an ingredient of nonlinear behavior that couples the determination of Φ_1 and Φ_2 through Equation (16-49). We observe that the expression $\Phi^*\Phi$ is independent of the phase of Φ and is therefore left unchanged by local phase variations. Consequently, $m^2(\Phi_1, \Phi_2)$ is a gauge-invariant quantity so that the incorporation of m^2 in Equation (16-49) does not conflict with the governing principle of gauge symmetry. Figure 16-41 shows two versions of the paraboloidal surface given by m^2 as a function of Φ_1 and Φ_2. We are interested in the lowest energy state of the system, and we suppose that this state occurs for the smallest possible value of the mass m. If we take the parameter μ^2 to be positive, we find

$$m_{\min} = \mu \quad \text{when } \Phi_1 = \Phi_2 = 0.$$

We note, however, that smaller values of m are obtained if μ^2 is allowed to be *negative*. The figure tells us that the least nonnegative result for m^2 is given by $m = 0$, when Φ_1 and Φ_2 lie anywhere on the circle

$$\Phi_1^2 + \Phi_2^2 = -\frac{\mu^2}{\lambda} \quad \text{with } \mu^2 < 0. \qquad (16\text{-}51)$$

This conclusion determines only the *modulus* of Φ, while the phase of Φ is left unspecified in accordance with gauge symmetry. If we now select a particular Φ for the ground state of the system, such as

$$\Phi_1 = \sqrt{-\frac{\mu^2}{\lambda}} \qquad \Phi_2 = 0, \tag{16-52}$$

we see that the choice immediately *breaks* the symmetry by singling out a *specific phase* in the solution. This spontaneous breakdown of gauge invariance affects the range of the gauge fields. The broken symmetry manifests itself through *finite-range* potentials A and ϕ whose quanta are no longer massless. The Higgs mechanism operates in similar fashion to give mass to the weak quanta $W^{\pm}$ and Z in the electroweak theory. The mechanism is more complicated since the electroweak gauge transformations are more intricate than the simple phase variation considered here.

Spontaneous symmetry breaking is not an uncommon phenomenon in nature. The example most often cited is that of a ferromagnetic medium of infinite extent. Such a system has an underlying rotational symmetry in three-dimensional space and presents a ground state in which all spins are aligned in the same direction. The choice of direction is arbitrary since the rotational symmetry allows a continuous range of possible alignments. This symmetry is broken spontaneously when a particular alignment of spins is chosen. The following example is taken from nonrelativistic classical mechanics and gives an even simpler illustration.

Example

Spontaneous breakdown does not always have to be associated with gauge invariance, or with other forms of continuous symmetry. Let us consider as an alternative a classical mass-on-a-spring in one-dimensional motion, with potential energy

$$V(x) = \frac{k}{2}x^2 + \frac{\lambda}{4}x^4.$$

This expression includes the usual harmonic term with spring constant k and an additional anharmonic contribution with positive parameter λ. The system is in stable equilibrium at $x = 0$, provided k is positive. In general, we find a vanishing force wherever the potential energy has zero slope:

$$0 = F(x) = -\frac{dV}{dx} = -(kx + \lambda x^3) = -x(k + \lambda x^2).$$

The $x = 0$ solution does not correspond to stable equilibrium if we allow k to be *negative*. In this case, stability occurs at the points

$$x = \pm x_0, \quad \text{where} \quad x_0 = \sqrt{-\frac{k}{\lambda}}.$$

Figure 16-42 shows the shape of $V(x)$ in the two cases. We expect to find *symmetric* stable minima on either side of $x = 0$ for $k < 0$, because $V(x)$ is symmetric under the parity operation $x \rightarrow -x$. We break this symmetry spontaneously when we choose, say, $x = +x_0$ to be the position of equilibrium for

---

Figure 16-42

Anharmonic potential energy $V(x) = (k/2)x^2 + (\lambda/4)x^4$ for k positive and negative. The case of negative k leads to a choice of stable minima at $x = \pm x_0$, where $x_0 = \sqrt{-k/\lambda}$.

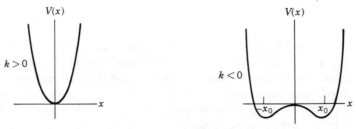

---

small oscillations of the particle. Note that the broken symmetry is discrete, and not continuous, in this illustration.

16-17 Grand Unification

Gauge theory has revolutionized our understanding of the elementary particles. The general notion of local symmetry inspires a procedure that applies in different ways to each of the fundamental forces. Quantum chromodynamics is generally accepted as the theory of the strong interaction, and electroweak unification is established as the proper framework for the electromagnetic and weak interactions. These two gauge theories, taken together, comprise the so-called Standard Model for the behavior of quarks and leptons. No experimental contradictions and no mathematical inconsistencies are known to be in conflict with this combined theory.

Despite its resounding success the Standard Model is not regarded as the ultimate fundamental theory, for a variety of reasons. The electroweak portion of the model does not meet all the qualifications of a truly unified theory, since the precise unification of electromagnetic and weak interactions is not completely specified. We can see this immediately by noting that the weak mixing angle θ_w is left unpredicted. In fact, the arbitrary separate treatment of the strong and electroweak theories involves an excessive number of such free parameters. The presumption of a 1 : 3 ratio between the charges of quarks and leptons is another arbitrary and, hence, unsatisfying feature of the overall theory. We should be able to eliminate these elements of arbitrariness, and gain more predictive power, by turning to a higher level of unification based on a greater degree of local symmetry. The higher symmetry would make the unified theory simpler and would introduce constraints reducing the number of free parameters. *Grand unified theories* are designed to accomplish these objectives. The theories are so named because they embed the Standard Model in a *single* gauge theory and thus serve to unify the strong, electromagnetic, and weak interactions. This speculative notion has an obvious appeal as the next step to take toward an ultimate theory.

The proposed schemes for grand unification differ as to their specific details. It is not yet possible to say that any of these proposals has emerged as the obvious choice. One of the leading candidates is the model developed in 1974 by Glashow and H. M. Georgi. Their theory may not be correct in all its predictions; however, there is reason

Figure 16-43

Variation of coupling strength with energy scale. Strong, electromagnetic, and weak interactions approach equal strength at $m_X c^2$, the unification scale.

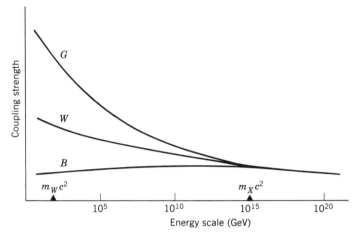

to believe that the ideas contained in this simplest of models are at least pointing in the right direction.

Scales of distance and scales of energy are decisive considerations in grand unified theories. The unification concept presumes the existence of a regime of the variables where the forces of interest obey the assumed symmetry *exactly*. In this domain the forces are expected to exhibit *no* differences in strength. Such a situation can be realized only if the strengths of the interactions are able to *vary* from one regime to another. Renormalizability is the all-important property of a theory that enables us to investigate this question. In the case of the electromagnetic and weak interactions, we suppose that the two forces possess their unifying symmetry at very high energies, so far beyond the range of practical experiment that the rest energies of the weak quanta are negligible by comparison. These conditions are such that the effects of spontaneous breakdown become relatively unimportant and the underlying electroweak gauge symmetry becomes exact. Hypothetical experiments at the very large scales of energy would be able to probe matter electromagnetically and weakly over distance scales very much smaller than $\hbar/m_W c$, the range of the weak force. In these domains the electromagnetic and weak interactions would appear to have the same (essentially infinite) range, and the interacting particles would seem to be very much alike. We then become aware of the large differences between the two forces via the breaking of gauge symmetry when we descend in energy toward values comparable to $m_W c^2$.

We illustrate this state of affairs in Figure 16-43 by showing the behavior of the various coupling strengths as functions of the energy scale. The curves labeled W and B refer to the couplings of the gauge-field quanta W^+, W^0, W^-, and B^0. (Recall that the electromagnetic and neutral-weak interactions are associated with mixings of W^0 and B^0 to form the observed quanta γ and Z.) Note that the W curve falls with increasing energy scale, so that the W couplings diminish in strength at short range. This property of asymptotic freedom is expected for these couplings because of the noncommutative nature of the W gauge transformations. It is known that the W and

B curves intersect at a very high energy, of order 10^{15} GeV, indicated on the graph by the quantity $m_X c^2$. The behavior of the strong gluon interaction is shown, with the label G, in the same figure. This coupling also enjoys the property of asymptotic freedom and must also fall with increasing energy scale. Remarkably, the G curve merges with the W and B curves in the vicinity of the same high energy determined by $m_X c^2$. This coalescence of the three independent couplings is ideally suited for a scheme of unification of the strong, electromagnetic, and weak interactions. The parameter m_X is called the *unification mass* in such a framework. The grand unified theory would be seen as an exact symmetry for scales of energy beyond $m_X c^2$, where forces of equal strength act between seemingly similar particles. A single curve describes the strength of this unified interaction, as indicated by the high-energy portion of the graph sketched in the figure.

Grand unified theories treat quarks and leptons as fundamentally similar particles in the regime where the higher gauge symmetry is exact. The well-known distinctions between these particles are then supposed to evolve at the lower energy scales, where spontaneous breakdown of the symmetry takes effect. The unifying interaction embraces the familiar gluon, photon, and weak gauge couplings to quarks and leptons, corresponding to the usual lower gauge symmetries with their associated strong, electromagnetic, and weak gauge fields. New symmetries are also present in the unified theory, because the higher level of unification includes operations that transform *quarks into leptons*. Each of these transformations involves a certain change in color *and* flavor, and each of the possibilities calls for the introduction of a new kind of color-changing flavor-changing gauge quantum. The additional quanta give rise to unanticipated interaction phenomena with couplings prescribed by the particular higher gauge symmetry chosen for grand unification.

The model of Georgi and Glashow defines gauge transformations within the following five- and ten-fold collections of colors, flavors, and leptonic quantum numbers:

$$\begin{bmatrix} (\bar{d})_{R,Y,B} \\ e^- \\ \nu_e \end{bmatrix} \quad \text{and} \quad \begin{bmatrix} (\bar{u})_{R,Y,B} \\ (u)_{R,Y,B} \\ (d)_{R,Y,B} \\ e^+ \end{bmatrix}. \tag{16-53}$$

These assignments group the quarks u and d together with the leptons e^- and ν_e. The other two generations of quarks and leptons participate in the model in similar fashion. When we analyze the unified interaction for all possible quark-to-quark and lepton-to-lepton transformations, we reproduce the usual strong, electromagnetic, and weak gauge couplings. Some of these familiar couplings are shown in Figures 16-33 and 16-38. In addition, we also generate new classes of *quark-to-lepton* and *quark-to-antiquark* transformations. Some of these so-called leptoquark and diquark transitions are illustrated in Figure 16-44. Note that the transformations occur with the emission or absorption of color- and flavor-bearing quanta, represented by the symbols X and Y. These quanta have Q/e values equal to $\frac{4}{3}$ and $\frac{1}{3}$, respectively. Note also that the indicated transitions *violate* the conservation laws for baryon number and lepton number. The new gauge quanta acquire very large masses via the spontaneous breakdown of gauge symmetry. Thus, the mass of X is selected in Figure 16-43 to specify the energy scale $m_X c^2$, above which the higher gauge symmetry becomes exact and below which the effects of spontaneous breakdown become important. It is

Figure 16-44

Couplings of X and Y gauge quanta in quark–lepton and quark–antiquark transitions. Changes of color are included in the diquark couplings.

Leptoquark transitions   Diquark transitions

noteworthy, to say the least, that the value of m_X should lie so many orders of magnitude beyond m_W, the mass of the weak gauge quantum.

The embedding of the three familiar forces in a single unified interaction imposes restrictions on several of the arbitrary features of the Standard Model. In particular, the adoption of a single coupling strength for a single gauge symmetry fixes the relative strengths for the couplings of the G, W, and B quanta. This determination of couplings is made initially at energy scales beyond $m_X c^2$. The three curves in Figure 16-43 can then be followed below the unification scale to ascertain the strengths of the couplings in the longer-range regime of current experiment. The same procedure can be used to predict a value for the experimentally measurable weak mixing angle. In the Georgi–Glashow model, the result is

$$\sin^2\theta_w = \tfrac{3}{8} \tag{16-54}$$

at the unification scale. The value then evolves with the coupling strengths at longer range to yield a prediction much closer to the experimental figure, $\sin^2\theta_w = 0.23$.

The model of Georgi and Glashow is an example of a grand unified theory in which the charges of quarks and leptons are related. This relation provides an argument for the quantization of electric charge. We note that the five- and ten-fold assignments of quark and lepton quantum numbers have sums of charges given by

$$3Q_{\bar{d}} + Q_{e^-} = 0$$

and

$$3Q_{\bar{u}} + 3Q_u + 3Q_d + Q_{e^+} = 0,$$

where the factors of 3 are due to the three colors. Both of these equalities imply the relation

$$Q_d = \tfrac{1}{3}Q_{e^-}. \tag{16-55}$$

In fact, the model goes on to explain why such sums of charges *must* equal zero for *any* assignment of quark and lepton quantum numbers. We can use this property of the model to argue that the charge relation is a necessary consequence of the higher level of symmetry. One more element of arbitrariness in the Standard Model is thus removed by the adoption of a unified interaction.

Figure 16-45

Diagrams that contribute to proton decay.

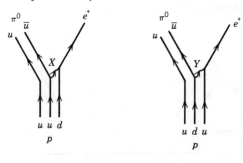

Nonconservation of baryon and lepton numbers is built into the Georgi–Glashow theory. We can see this feature by noting that u, d, and $\bar{u}$ are assigned along with e^+ to the *same* collection of interconnected particle states. Similar constructions are found in other models of grand unification. This provision leads us directly to the most striking conclusion of the grand unified theories, the prediction of baryon- and lepton-number violating *nucleon decay*. Figure 16-45 shows two of the possible mechanisms for the proton decay mode

$$p \rightarrow \pi^0 + e^+.$$

We note that the diagrams contain the vertices from Figure 16-44, connected by the exchange of the superheavy X and Y gauge quanta. A prediction can be made for the proton lifetime τ_p, based on the specific choice of unification model. (Actually, the prediction is somewhat blurred by uncertainties stemming from the parametrization of the strong interaction and the treatment of the hadronic bound states.) The resulting formula for τ_p depends sensitively on the unification mass m_X. It is remarkable that values for τ_p in excess of 10^{30} years are obtainable for input selections of m_X larger than 10^{14} GeV/c^2. A mean life in this range is amenable to experiment, and a unification mass in the corresponding regime is in line with the scenario described in Figure 16-43. This consistent set of circumstances is favorable for a decisive experimental test of the ideas of grand unification.

The possible violation of baryon number is an intriguing development, particularly in view of the comparable situation regarding the conservation of electric charge. We have learned that charge conservation is associated with an exact *gauge* symmetry, whose validity can be tested by setting very low experimental bounds on measurements of the photon mass. Baryon number does not appear to have its origin in any analogous unbroken gauge invariance, and so an exact conservation law would have to correspond to an exact *global* symmetry. We are inclined to view this sort of invariance as somewhat implausible, now that we have found a scheme, operative on an extraordinarily small scale of distance, in which quarks and leptons look very much alike. Baryon and lepton numbers cannot be regarded as sacred conserved quantities in such a framework. We note in passing that the difference in quantum numbers $B - L$ continues to be conserved, at least in the context of the Georgi–Glashow model.

Of course, the possible instability of the proton must ultimately be decided in the laboratory. Several experiments have been mounted around the world to search for evidence of proton decay. Since the lifetime of an unstable proton must be very long, an enormous sample of matter is needed to collect detectable numbers of decay events in practical intervals of time. To illustrate, let us assume 10^{30} nucleons in a ton of material, and let us take the lifetime to be $\tau_p = 10^{30}$ y. Under these conditions we may find one observable proton decay in the sample per year. One of the ongoing experiments is located in an abandoned salt mine near Lake Erie. A volume of pure water containing 8000 metric tons is monitored by phototubes to detect the light radiated by the expected products of proton decay. The apparatus is buried deep underground to reduce the detection of events due to cosmic-ray muons entering the sample. Unfortunately, neutrinos cannot be eliminated, and their interactions can simulate the desired $p \rightarrow \pi^0 e^+$ events. In fact, the only decay candidates seen so far are attributable to this source of background. The experiment to date has set a lower limit for the partial mean life in the $\pi^0 e^+$ mode greater than 10^{32} years. This result is at least a factor of 10 larger than any of the predictions based on the simplest Georgi–Glashow model.

Grand unified models belong to a general class of spontaneously broken gauge theories that predict the existence of *magnetic monopoles*. These objects have been found mathematically among the solutions of the classical field equations of the theory. The prediction gives an enormous estimate for the mass, of order 10^{16} GeV/c^2, and provides a unique relation between the magnetic pole strength and the electric charge of the particle. Such objects seem to occur very seldom, if at all, in the real world. Only a single candidate has been seen in the laboratory, in an experiment performed by B. Cabrera in 1982. Since their incidence is so rare, it is incumbent on the theory that some reason be offered for their suppression.

Proton decays also happen very infrequently (if indeed they happen at all), and so the new baryon-nonconserving interaction contained in the grand unified theories must be extremely weak. Such B-violating phenomena would have been more prevalent during the hotter and denser early stages of the universe just after the Big Bang. Hence, it is possible that cosmological relics of grand unification have been left behind and are in evidence at the present time. A most notable case in point is the observed excess of baryons over antibaryons, matter over antimatter, in the composition of the universe. A collection of effects, including B-violation, may have generated this *baryon asymmetry* dynamically, starting from an initial configuration of material with *zero* net baryon number.

If we take B to be exactly conserved, then we can have unequal numbers of baryons and antibaryons only if we postulate the known excess as an initial condition on the Big Bang. Instead, let us adopt a grand unified theory in which B is violated and insist on equal numbers at the start. Grand unification then takes its course, beginning with very brief initial time intervals and very large thermal energies beyond the unification scale at $m_X c^2$. Matter in this epoch is supposed to consist of quarks and leptons in equilibrium with all the quanta of the higher gauge symmetry. Spontaneous symmetry breaking eventually sets in as time evolves and the universe cools. Violations of C and CP are also introduced as essential ingredients in this scenario. Let us illustrate their role by considering the decays of the superheavy gauge quanta, even though these processes are not the only (or the most important) contributors to the asymmetry. The diquark decay modes of X and $\bar{X}$ can differ because of C- and CP-violations, so that the rates of production from

$$X \rightarrow q + q \quad \text{and} \quad \bar{X} \rightarrow \bar{q} + \bar{q}$$

are not the same. Consequently, we can have equal amounts of X and $\bar{X}$ in the equilibrium period and still generate unequal numbers of quarks and antiquarks thereafter. A small net excess of the one over the other would survive subsequent processes of matter–antimatter annihilation and perhaps explain why our part of the universe contains essentially no antibaryons.

Grand unification has many satisfying and promising aspects. Nevertheless, it is conceivable that the strong, electromagnetic, and weak interactions cannot be unified properly until the unification of forces takes account of the force of gravity. By the same token, it is possible that a consistent quantum theory of gravity cannot exist in isolation from the other forces of nature. These areas of speculation belong to the next more comprehensive stage in the problem of unification. The *final* solution of this problem is called the Theory of Everything.

Example

The energy scale of grand unification is extraordinarily large, and the associated scale of distance is extremely small:

$$\frac{\hbar}{m_X c} = \frac{\hbar c}{m_X c^2} = \frac{0.2 \text{ GeV} \cdot \text{fm}}{10^{15} \text{ GeV}} = 2 \times 10^{-16} \text{ fm}.$$

We may wonder whether the effects of gravity are negligible in a relativistic quantum theory at such short range. Let us argue that the gravitational interaction becomes important when the potential energy of two equal masses is comparable with the rest energy of either mass, and when the separation is comparable to the corresponding Compton wavelength. We express these comparisons by the two conditions

$$G\frac{M^2}{R} = Mc^2 \quad \text{and} \quad R = \frac{\hbar}{Mc}.$$

The combination of conditions provides a criterion for the scale of the mass M in terms of the gravitational constant G:

$$GM^2\frac{Mc}{\hbar} = Mc^2 \quad \Rightarrow \quad Mc^2 = \sqrt{\frac{\hbar c^5}{G}}.$$

This argument determines the rest energy of the so-called Planck mass:

$$Mc^2 = \sqrt{\frac{(1.05 \times 10^{-34} \text{ J} \cdot \text{s})(3.00 \times 10^8 \text{ m/s})^5}{6.67 \times 10^{-11} \text{ N} \cdot \text{m}^2/\text{kg}^2}}$$

$$= \frac{1.96 \times 10^9 \text{J}}{1.60 \times 10^{-10} \text{ J/GeV}} = 1.22 \times 10^{19} \text{ GeV}.$$

The unification scale $m_X c^2$ falls several decades short of this value. It is noteworthy, however, that the unification mass m_X lies closer to the Planck mass than to the weak mass m_W.

Problems

1. Positively charged cosmic-ray particles approach the Earth from various directions and are deflected in the Earth's magnetic field. Give a qualitative argument to explain why more particles penetrate the field in the polar regions and why the particles tend to arrive on Earth preferentially from the west.

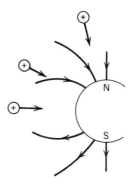

2. Make a quantitative comparison of the forces of gravitational attraction and electrostatic repulsion between a pair of electrons, and between a pair of protons. Note that the separation of the particles drops out of the comparison.

3. Consider the collision of protons with protons in a fixed-target accelerator and in a colliding-beam accelerator. Derive the relation

$$K = 2K'\left(1 + \frac{K'}{4Mc^2}\right),$$

where K is the single-beam kinetic energy in the first machine, and K' is the total kinetic energy in the second machine. Take the nonrelativistic limit of the formula, and compare with the result found in Section 15-8.

4. Generalize the derivation in Problem 3 to the case of unequal colliding masses. Let K be the beam kinetic energy for a particle of mass m incident on a target particle of mass M, and let K' be the total kinetic energy in the CM frame.

5. Assign momenta and energies for the emission of a photon by a free electron according to the diagram, and show that the conservation laws of momentum and energy cannot be satisfied. The conclusion implies that the photon must be virtual in this case. How is it possible that the photon is real for the emission processes in Figures 15-16 and 16-4?

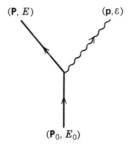

6. Represent the g-factor for electron spin as in Figure 16-6, and draw all Feynman diagrams of order α and α^2.

7. Show that the spherically symmetric meson wave function

$$\Phi(r) = \phi_0 \frac{e^{-r/r_0}}{r}$$

is a solution to the static version of the Klein–Gordon equation.

8. Calculate the minimum energy of photons incident on a fixed proton target for the photoproduction of neutral pions in the reaction $\gamma + p \to \pi^0 + p$.

9. Let neutrons be incident on a fixed proton target and calculate the threshold kinetic energies for each of the pion-production reactions

$$n + p \to \begin{cases} n + n + \pi^+, \\ n + p + \pi^0, \\ p + p + \pi^-. \end{cases}$$

Table 16-1 gives sufficiently accurate values for the masses of the particles.

10. Calculate the momenta of the final particles in the charged pion decays

$$\pi^+ \to \mu^+ + \nu_\mu \quad \text{and} \quad \pi^+ \to e^+ + \nu_e.$$

Assume that the pion is at rest and neglect the neutrino masses.

11. Consider the decay of π^0 mesons in flight and obtain expressions for the maximum and minimum energies of the emitted γ rays in terms of the π^0 velocity.

12. The reaction used to detect solar neutrinos is the endoergic capture process

$$\nu_e + {}^{37}\text{Cl} \to {}^{37}\text{Ar} + e^-.$$

Compute the threshold energy for incident neutrinos using atomic mass data from Appendix A. One source of solar neutrinos is the proton fusion reaction

$${}^1\text{H} + {}^1\text{H} \to {}^2\text{H} + e^+ + \nu_e.$$

Calculate the maximum energy of neutrinos produced for protons at rest, and determine whether the chlorine detector is sensitive to these neutrinos.

13. Neutrinos and antineutrinos are intrinsically left- and right-handed, respectively, if the particles have zero mass. Use this fact to prove that the π^0 decay mode $\pi^0 \to \nu\bar{\nu}$ cannot possibly occur for massless neutrinos.

14. Let π^- mesons be incident on protons at rest and derive a formula for the beam energy at the threshold for the production of a final state of total mass M. Consider the reactions

$$\pi^- + p \to \pi^+ + \pi^- + n \quad \text{and} \quad \pi^- + p \to K^+ + \Sigma^-$$

and compute the threshold energies in each case.

15. Consider a π^- beam, and a K^- beam, incident on a proton target in separate experiments to produce Ξ^0 baryons. Identify the threshold reactions and calculate the threshold beam energies for each experiment.

16. Draw a bubble-chamber picture for the reaction $K^- + p \to \pi^- + \Sigma^+$, in which Σ^+ decays by its proton mode, and the γ rays from π^0 decay convert to $e^- e^+$ pairs inside the chamber.

17. Assume that the neutral kaon states $\begin{bmatrix} K^0 \\ \overline{K}^0 \end{bmatrix}$ and $\begin{bmatrix} K_S \\ K_L \end{bmatrix}$ obey the matrix Schrödinger equations

$$\frac{i\hbar}{c^2} \frac{d}{dt} \begin{bmatrix} K^0 \\ \overline{K}^0 \end{bmatrix} = \begin{bmatrix} m & u \\ v & m \end{bmatrix} \begin{bmatrix} K^0 \\ \overline{K}^0 \end{bmatrix}$$

and

$$\frac{i\hbar}{c^2} \frac{d}{dt} \begin{bmatrix} K_S \\ K_L \end{bmatrix} = \begin{bmatrix} m_S & 0 \\ 0 & m_L \end{bmatrix} \begin{bmatrix} K_S \\ K_L \end{bmatrix}.$$

Use the transformation between these sets of states to derive the relations

$$\left(\frac{1-\varepsilon}{1+\varepsilon} \right)^2 = \frac{u}{v}, \quad m_S = m - \sqrt{uv}, \quad \text{and} \quad m_L = m + \sqrt{uv},$$

where ε is the *CP*-violation parameter.

18. Why should the reaction $d + d \rightarrow \alpha + \pi^0$ not be observed? The experimental upper bound on the cross section for this process is of the same order as the observed cross section for $d + d \rightarrow \alpha + \gamma$. Explain qualitatively why these two reactions are expected to proceed at comparable rates.

19. The $\Sigma\pi$ system has nine different charge combinations $\Sigma^+ \pi^+$, $\Sigma^+ \pi^0$, $\Sigma^+ \pi^-$,.... How many independent amplitudes for elastic and charge-exchange scattering are there in this system, and how are the amplitudes characterized? How many independent amplitudes are needed to describe the production of $\Sigma\pi$ states in $K^- p$ collisions?

20. Determine the $\pi^\pm p$ cross section ratios, expressed as

$$|\chi(\pi^+ p, \pi^+ p)|^2 : |\chi(\pi^- p, \pi^- p)|^2 : |\chi(\pi^0 n, \pi^- p)|^2,$$

under the assumption that the isospin amplitudes $\chi_{1/2}$ and $\chi_{3/2}$ are equal, and under the assumption that $\chi_{3/2}$ vanishes.

21. Calculate the values of the pion beam kinetic energy in $\pi^\pm p$ reactions for the observation of the πN resonances $\Delta(1232, \frac{3}{2}^+)$, $N(1520, \frac{3}{2}^-)$, $N(1680, \frac{5}{2}^+)$, and $\Delta(1950, \frac{7}{2}^+)$.

22. Show that, for a system of two pions, the isospin t and the angular momentum ℓ are correlated so that t and ℓ must be either both odd or both even.

23. Refer to the three-quark formulas for the states Δ^{++}, Δ^+, Δ^0, and Δ^-, and use the flavor substitution $u \rightarrow s$ to deduce expressions for all possible strange baryon states. Identify the quantum numbers T_z and Y for the resulting ten hadrons, and plot their locations on a graph of Y versus T_z.

24. Construct all nine $q\bar{q}$ systems vectorially, using the addition of vectors in the (Y, T_z) plane. Show that the result contains states with quantum numbers in the antitriplet pattern, and note that a sextet (six-fold) pattern of quantum numbers remains when the antitriplet is removed.

25. Continue the vectorial construction begun in Problem 24, and generate all qqq systems by applying the method of vector addition to the composition $q(qq)$. Specifically, construct the $q(qq)$ states by adding the triplet of q vectors in the (Y, T_z) plane to the antitriplet and sextet results obtained from the qq system. Show that the final 9 triplet–antitriplet states form octet and singlet patterns and that the final 18 triplet–sextet states form octet and decuplet (ten-fold) patterns.

26. What combination of quarks is needed to construct the antibaryon $\overline{\Xi^-}$? Consider the photoproduction of $\overline{\Xi^-}$ via the reaction

$$\gamma + p \rightarrow \overline{\Xi^-} + X,$$

and deduce the identity of the least-massive particles comprising the unspecified system X. Calculate the corresponding threshold energy for photons incident on target protons at rest.

27. Draw quark Feynman diagrams to represent the decays $\pi^+ \rightarrow \pi^0 + e^+ + \nu_e$, $K^+ \rightarrow \pi^0 + e^+ + \nu_e$, $K^+ \rightarrow \mu^+ + \nu_\mu$, $K^+ \rightarrow \pi^+ + \pi^0$, $\Sigma^- \rightarrow n + e^- + \bar{\nu}_e$, and $\Sigma^- \rightarrow n + \pi^-$.

28. Identify all quantum numbers for the c-flavored hadron systems udc, $c\bar{u}$, $c\bar{d}$, $\bar{c}u$, $\bar{c}d$, $c\bar{s}$, and $\bar{c}s$.

29. Draw Feynman diagrams to show the effects of weak neutral currents in the processes $e^- + e^+ \rightarrow \mu^- + \mu^+$, $e^- + p \rightarrow e^- + p$, and $\nu_\mu + p \rightarrow \nu_\mu + $ hadrons.

30. Assume a value of 40 GeV/c^2 for the t-quark mass, and show how this undiscovered sixth quark may contribute in the cross-section ratio R for electron–positron annihilation.

31. Gluons change the colors of quarks and antiquarks in the manner of Figure 16-38. Show that eight different gluons are needed to describe all possible color transitions for three colors of quarks and antiquarks.

32. The known 1^- bound states of the $b\bar{b}$ system are designated as $\Upsilon(nS)$, where $n = 1$–4. Explain the notation, assuming a central potential energy like the one shown in Figure 16-40. The explanation should include sketches of the radial functions corresponding to the four different Υ states.

33. Use Maxwell's equations to deduce the equations satisfied by the electromagnetic potentials $\mathbf{A}$ and ϕ in terms of the source densities $\mathbf{J}$ and ρ. Assume that $\mathbf{A}$ and ϕ obey the subsidiary (Lorentz) condition

$$\nabla \cdot \mathbf{A} + \frac{1}{c^2}\frac{\partial \phi}{\partial t} = 0,$$

and show that decoupled equations result relating $\mathbf{A}$ to $\mathbf{J}$ and ϕ to ρ.

34. Consider expectation values taken in the states of a nonrelativistic particle of charge Q, and show that $\langle \mathbf{p} - Q\mathbf{A} \rangle$ and $\langle E - Q\phi \rangle$ are invariant under gauge transformations.

35. Let the classical one-dimensional motion of a particle of mass m be governed by the potential energy $V(x) = (k/2)x^2 + (\lambda/4)x^4$, and consider the case $k < 0$. Deduce the frequency of small oscillations about the position of stable equilibrium, chosen to be at the point $x = +x_0$ as described in Figure 16-42.

APPENDIX A

TABLE
OF
NUCLEAR
PROPERTIES

*I*nformation is tabulated for selected isotopes according to atomic number Z, chemical symbol, and mass number A. The atomic mass M refers to the mass of the neutral atom, and values of M are listed in terms of the atomic mass unit u. Other data quoted are the nuclear spin and parity i^p, the isotopic abundance (%), and the radioactive half-life $\tau_{1/2}$. The last two items share the last column in the table, depending on the stable or unstable character of the particular nuclide. Unstable species are indicated by an asterisk next to the value of A. Their half-lives are given in seconds (s), minutes (m), hours (h), days (d), and years (y). Sources of information are *Chart of the Nuclides*, 13th edition (1984), and *Atomic Mass Evaluation*, by A. H. Wapstra and K. Bos (1977).

Z Atom	A	M (u)	i^p	$\% / \tau_{1/2}$
0 n	1*	1.008665	$\frac{1}{2}+$	10.5 m
1 H	1	1.007825	$\frac{1}{2}+$	99.985
	2	2.014102	$1+$	0.015
	3*	3.016049	$\frac{1}{2}+$	12.3 y
2 He	3	3.016029	$\frac{1}{2}+$	0.00014
	4	4.002603	$0+$	99.99986
3 Li	6	6.015121	$1+$	7.5
	7	7.016003	$\frac{3}{2}-$	92.5
4 Be	7*	7.016930	$\frac{3}{2}-$	53.28 d
	9	9.012182	$\frac{3}{2}-$	100
	10*	10.013535	$0+$	1.6×10^6 y
5 B	10	10.012936	$3+$	19.9
	11	11.009305	$\frac{3}{2}-$	80.1
	12*	12.014353	$1+$	20.2 ms
6 C	12	12.000000	$0+$	98.90
	13	13.003355	$\frac{1}{2}-$	1.10
	14*	14.003242	$0+$	5730 y
7 N	14	14.003074	$1+$	99.63
	15	15.000109	$\frac{1}{2}-$	0.37
	16*	16.006099	$2-$	7.13 s

Z Atom	A	M (u)	i^p	$\% / \tau_{1/2}$
8 O	16	15.994915	$0+$	99.762
	17	16.999131	$\frac{5}{2}+$	0.038
	18	17.999160	$0+$	0.200
9 F	19	18.998403	$\frac{1}{2}+$	100
	20*	19.999982	$2+$	11.0 s
10 Ne	20	19.992436	$0+$	90.51
	21	20.993843	$\frac{3}{2}+$	0.27
	22	21.991383	$0+$	9.22
11 Na	22*	21.994435	$3+$	2.605 y
	23	22.989768	$\frac{3}{2}+$	100
12 Mg	24	23.985042	$0+$	78.99
	25	24.985838	$\frac{5}{2}+$	10.00
	26	25.982594	$0+$	11.01
13 Al	27	26.981539	$\frac{5}{2}+$	100
	28*	27.981913	$3+$	2.25 m
14 Si	28	27.976927	$0+$	92.23
	29	28.976495	$\frac{1}{2}+$	4.67
	30	29.973770	$0+$	3.10
15 P	31	30.973762	$\frac{1}{2}+$	100
	32*	31.973908	$1+$	14.28 d

Z Atom	A	M (u)	i^p	$\% / \tau_{1/2}$
16 S	32	31.972071	0^+	95.02
	34	33.967867	0^+	4.21
	35*	34.969032	$\frac{3}{2}^+$	87.2 d
17 Cl	35	34.968853	$\frac{3}{2}^+$	75.77
	36*	35.968307	2^+	3.01×10^5 y
	37	36.965903	$\frac{3}{2}^+$	24.23
18 Ar	37*	36.966776	$\frac{3}{2}^+$	34.8 d
	39*	38.964315	$\frac{7}{2}^-$	269 y
	40	39.962383	0^+	99.600
19 K	39	38.963707	$\frac{3}{2}^+$	93.2581
	40*	39.963999	4^-	1.25×10^9 y
	41	40.961825	$\frac{3}{2}^+$	6.7302
20 Ca	40	39.962590	0^+	96.941
	44	43.955481	0^+	2.086
	45*	44.956189	$\frac{7}{2}^-$	165 d
21 Sc	45	44.955910	$\frac{7}{2}^-$	100
	47*	46.952410	$\frac{7}{2}^-$	3.34 d
22 Ti	44*	43.959693	0^+	47 y
	46	45.952630	0^+	8.0
	48	47.947948	0^+	73.8
23 V	50*	49.947161	6^+	$> 3.9 \times 10^{17}$ y
	51	50.943962	$\frac{7}{2}^-$	99.750
24 Cr	51*	50.944769	$\frac{7}{2}^-$	27.70 d
	52	51.940510	0^+	83.79
	53	52.940652	$\frac{3}{2}^-$	9.50
25 Mn	54*	53.940360	3^+	312.2 d
	55	54.938047	$\frac{5}{2}^-$	100
26 Fe	54	53.939613	0^+	5.8
	55*	54.938295	$\frac{3}{2}^-$	2.68 y
	56	55.934940	0^+	91.72
27 Co	59	58.933198	$\frac{7}{2}^-$	100
	60*	59.933820	5^+	5.272 y
28 Ni	58	57.935347	0^+	68.27
	60	59.930789	0^+	26.10
	63*	62.929670	$\frac{1}{2}^-$	100 y
29 Cu	63	62.929599	$\frac{3}{2}^-$	69.17
	64*	63.929766	1^+	12.70 h
	65	64.927793	$\frac{3}{2}^-$	30.83
30 Zn	64	63.929146	0^+	48.6
	65*	64.929244	$\frac{5}{2}^-$	243.8 d
	66	65.926035	0^+	27.9
31 Ga	69	68.925580	$\frac{3}{2}^-$	60.1
	70*	69.926028	1^+	21.1 m
	71	70.924701	$\frac{3}{2}^-$	39.9

Z Atom	A	M (u)	i^p	$\% / \tau_{1/2}$
32 Ge	70	69.924250	0^+	20.5
	72	71.922080	0^+	27.4
	74	73.921177	0^+	36.5
33 As	74*	73.923930	2^-	17.78 d
	75	74.921593	$\frac{3}{2}^-$	100
34 Se	75*	74.922524	$\frac{5}{2}^+$	119.78 d
	78	77.917306	0^+	23.5
	80	79.916521	0^+	49.6
35 Br	79	78.918336	$\frac{3}{2}^-$	50.69
	81	80.916290	$\frac{3}{2}^-$	49.31
36 Kr	82	81.913483	0^+	11.6
	84	83.911508	0^+	57.0
	86	85.910615	0^+	17.3
37 Rb	85	84.911793	$\frac{5}{2}^-$	72.17
	87*	86.909188	$\frac{3}{2}^-$	4.89×10^{10} y
38 Sr	86	85.909267	0^+	9.86
	88	87.905619	0^+	82.58
	90*	89.907746	0^+	29 y
39 Y	88*	87.909503	4^-	106.61 d
	89	88.905850	$\frac{1}{2}^-$	100
40 Zr	90	89.904703	0^+	51.45
	92	91.905037	0^+	17.17
	94	93.906314	0^+	17.33

Z Atom	A	M (u)	i^p	$\% / \tau_{1/2}$
41 Nb	93	92.906376	$\frac{9}{2}^+$	100
	94*	93.907282	6^+	2.0×10^4 y
42 Mo	95	94.905840	$\frac{5}{2}^+$	15.92
	96	95.904678	0^+	16.68
	98	97.905406	0^+	24.13
43 Tc	97*	96.906362	$\frac{9}{2}^+$	2.6×10^6 y
	99*	98.906252	$\frac{9}{2}^+$	2.13×10^5 y
44 Ru	101	100.905581	$\frac{5}{2}^+$	17.0
	102	101.904348	0^+	31.6
	104	103.905422	0^+	18.7
45 Rh	103	102.905499	$\frac{1}{2}^-$	100
	105*	104.905684	$\frac{7}{2}^+$	35.4 h
46 Pd	105	104.905075	$\frac{5}{2}^+$	22.33
	106	105.903475	0^+	27.33
	108	107.903896	0^+	26.46
47 Ag	107	106.905095	$\frac{1}{2}^-$	51.84
	108*	107.905956	1^+	2.42 m
	109	108.904757	$\frac{1}{2}^-$	48.16
48 Cd	112	111.902758	0^+	24.13
	113*	112.904400	$\frac{1}{2}^+$	9×10^{15} y
	114	113.903357	0^+	28.73
49 In	113	112.904061	$\frac{9}{2}^+$	4.3
	115*	114.903880	$\frac{9}{2}^+$	4.4×10^{14} y

Z Atom	A	M (u)	i^p	$\% / \tau_{1/2}$
50 Sn	116	115.901747	0^+	14.7
	118	117.901609	0^+	24.3
	120	119.902200	0^+	32.4
51 Sb	121	120.903823	$\frac{5}{2}^+$	57.3
	123	122.904220	$\frac{7}{2}^+$	42.7
	125*	124.905259	$\frac{7}{2}^+$	2.76 y
52 Te	126	125.903314	0^+	18.95
	128*	127.904467	0^+	$> 5.5 \times 10^{24}$ y
	130*	129.906232	0^+	2.4×10^{21} y
53 I	127	126.904478	$\frac{5}{2}^+$	100
	129*	128.904986	$\frac{7}{2}^+$	1.6×10^7 y
54 Xe	129	128.904780	$\frac{1}{2}^+$	26.4
	131	130.905075	$\frac{3}{2}^+$	21.2
	132	131.904147	0^+	26.9
55 Cs	133	132.905433	$\frac{7}{2}^+$	100
	137*	136.907075	$\frac{7}{2}^+$	30.17 y
56 Ba	136	135.904556	0^+	7.854
	137	136.905816	$\frac{3}{2}^+$	11.23
	138	137.905236	0^+	71.70
57 La	138*	137.907114	5^+	1.06×10^{11} y
	139	138.906346	$\frac{7}{2}^+$	99.91

Z Atom	A	M (u)	i^p	$\% / \tau_{1/2}$
58 Ce	140	139.905433	0^+	88.48
	142*	141.909241	0^+	$> 5 \times 10^{16}$ y
59 Pr	140*	139.909079	1^+	3.39 m
	141	140.907657	$\frac{5}{2}^+$	100
60 Nd	142	141.907731	0^+	27.13
	144*	143.910084	0^+	2.1×10^{15} y
	146	145.913114	0^+	17.19
61 Pm	145*	144.912754	$\frac{5}{2}^+$	17.7 y
	146*	145.914717	3^-	5.53 y
62 Sm	147*	146.914895	$\frac{7}{2}^-$	1.06×10^{11} y
	152	151.919729	0^+	26.7
	154	153.922206	0^+	22.7
63 Eu	151	150.919847	$\frac{5}{2}^+$	47.8
	153	152.921226	$\frac{5}{2}^+$	52.2
64 Gd	156	155.922119	0^+	20.47
	158	157.924100	0^+	24.84
	160	159.927051	0^+	21.86
65 Tb	159	158.925341	$\frac{3}{2}^+$	100
	160*	159.927171	3^-	72.4 d
66 Dy	162	161.926795	0^+	25.5
	163	162.928726	$\frac{5}{2}^-$	24.9
	164	163.929172	0^+	28.2

Z Atom	A	M (u)	i^p	% / $\tau_{1/2}$
67 Ho	165	164.930319	$\frac{7}{2}-$	100
	166*	165.932296	$0-$	26.80 h
68 Er	166	165.930292	$0+$	33.6
	167	166.932047	$\frac{7}{2}+$	22.95
	168	167.932369	$0+$	26.8
69 Tm	169	168.934212	$\frac{1}{2}+$	100
	171*	170.936442	$\frac{1}{2}+$	1.92 y
70 Yb	172	171.936379	$0+$	21.9
	173	172.938208	$\frac{5}{2}-$	16.12
	174	173.938860	$0+$	31.8
71 Lu	175	174.940771	$\frac{7}{2}+$	97.40
	176*	175.942680	$7-$	3.7×10^{10} y
72 Hf	177	176.943219	$\frac{7}{2}-$	18.6
	178	177.943697	$0+$	27.1
	180	179.946547	$0+$	35.2
73 Ta	181	180.947995	$\frac{7}{2}+$	99.988
	182*	181.950170	$3-$	114.5 d
74 W	182	181.948205	$0+$	26.3
	184	183.950932	$0+$	30.67
	186	185.954361	$0+$	28.6
75 Re	185	184.952955	$\frac{5}{2}+$	37.40
	187*	186.955749	$\frac{5}{2}+$	4.5×10^{10} y

Z Atom	A	M (u)	i^p	% / $\tau_{1/2}$
76 Os	189	188.958142	$\frac{3}{2}-$	16.1
	190	189.958442	$0+$	26.4
	192	191.961477	$0+$	41.0
77 Ir	191	190.960594	$\frac{3}{2}+$	37.3
	193	192.962944	$\frac{3}{2}+$	62.7
78 Pt	194	193.962685	$0+$	32.9
	195	194.964796	$\frac{1}{2}-$	33.8
	196	195.964956	$0+$	25.3
79 Au	197	196.966573	$\frac{3}{2}+$	100
	199*	198.968756	$\frac{3}{2}+$	3.14 d
80 Hg	199	198.968285	$\frac{1}{2}-$	17.0
	200	199.968330	$0+$	23.1
	202	201.970647	$0+$	29.65
81 Tl	203	202.972348	$\frac{1}{2}+$	29.524
	204*	203.973856	$2-$	3.78 y
	205	204.974427	$\frac{1}{2}+$	70.476
82 Pb	206	205.974469	$0+$	24.1
	207	206.975900	$\frac{1}{2}-$	22.1
	208	207.976655	$0+$	52.4
83 Bi	209*	208.980403	$\frac{9}{2}-$	$> 10^{19}$ y
	210*	209.984125	$1-$	5.01 d

Z Atom	A	M (u)	i^p	$\% / \tau_{1/2}$
84 Po	209*	208.982432	$\frac{1}{2}-$	102 y
	210*	209.982877	$0+$	138.38 d
	218*	218.008971	$0+$	3.11 m
85 At	211*	210.987500	$\frac{9}{2}-$	7.21 h
	213*	212.992926	$\frac{9}{2}-$	0.11 μs
86 Rn	220*	220.011378	$0+$	55.6 s
	222*	222.017576	$0+$	3.8235 d
87 Fr	221*	221.014241	$\frac{5}{2}-$	4.8 m
	223*	223.019734	$\frac{3}{2}+$	22 m
88 Ra	223*	223.018507	$\frac{1}{2}+$	11.434 d
	226*	226.025408	$0+$	1600 y
	228*	228.031069	$0+$	5.76 y
89 Ac	227*	227.027751	$\frac{3}{2}-$	21.773 y
	228*	228.031020	$3+$	6.13 h
90 Th	229*	229.031758	$\frac{5}{2}+$	7300 y
	231*	231.036299	$\frac{5}{2}+$	25.52 h
	232*	232.038054	$0+$	1.40×10^{10} y
91 Pa	231*	231.035885	$\frac{3}{2}-$	3.28×10^4 y
	233*	233.040244	$\frac{3}{2}-$	27.0 d

Z Atom	A	M (u)	i^p	$\% / \tau_{1/2}$
92 U	233*	233.039632	$\frac{5}{2}+$	1.592×10^5 y
	235*	235.043928	$\frac{7}{2}-$	7.04×10^8 y
	238*	238.050788	$0+$	4.468×10^9 y
93 Np	237*	237.048171	$\frac{5}{2}+$	2.14×10^6 y
	239*	239.052932	$\frac{5}{2}+$	2.350 d
94 Pu	239*	239.052162	$\frac{1}{2}+$	2.411×10^4 y
	240*	240.053812	$0+$	6560 y
	244*	244.064200	$0+$	8.2×10^7 y
95 Am	241*	241.056827	$\frac{5}{2}-$	432 y
	243*	243.061378	$\frac{5}{2}-$	7370 y
96 Cm	244*	244.062751	$0+$	18.11 y
	248*	248.072345	$0+$	3.40×10^5 y
97 Bk	247*	247.070300	$\frac{3}{2}-$	1400 y
	249*	249.074984	$\frac{7}{2}+$	320 d
98 Cf	249*	249.074848	$\frac{9}{2}-$	351 y
	252*	252.081622	$0+$	2.64 y
99 Es	253*	253.084822	$\frac{7}{2}+$	20.47 d
	254*	254.088021	$7+$	276 d
100 Fm	253*	253.085181	$\frac{1}{2}+$	3.0 d
	257*	257.095103	$\frac{9}{2}+$	100.5 d

BIBLIOGRAPHY

Some of the following references contain material at a higher level. A suitable adjustment of level is made whenever any such material is used in this book.

Parallel Resources

Eisberg, Robert, and Robert Resnick. *Quantum Physics of Atoms, Molecules, Solids, Nuclei, and Particles*. New York: Wiley, 1985.

Feynman, Richard P., Robert B. Leighton, and Matthew Sands. *The Feynman Lectures on Physics*, Vols. 1, 2, and 3. Reading, MA: Addison-Wesley, 1963, 1964, and 1965.

Ford, Kenneth W. *Classical and Modern Physics*, Vol. 3. New York: Wiley, 1974.

Gasiorowicz, Stephen. *The Structure of Matter*. Reading, MA: Addison-Wesley, 1979.

Krane, Kenneth S. *Modern Physics*. New York: Wiley, 1983.

Leighton, Robert B. *Principles of Modern Physics*. New York: McGraw-Hill, 1959.

McGervey, John D. *Introduction to Modern Physics*. New York: Academic Press, 1983.

Richtmyer, F. K., E. H. Kennard, and John N. Cooper. *Introduction to Modern Physics*. New York: McGraw-Hill, 1969.

Tipler, Paul A. *Modern Physics*. New York: Worth, 1978.

Weidner, Richard T., and Robert L. Sells. *Elementary Modern Physics*. Boston: Allyn and Bacon, 1973.

Young, Hugh D. *Fundamentals of Waves, Optics, and Modern Physics*. New York: McGraw-Hill, 1976.

Mathematical Sources

Boas, Mary L. *Mathematical Methods in the Physical Sciences*. New York: Wiley, 1983.

Dwight, Herbert Bristol. *Tables of Integrals and Other Mathematical Data*. New York: Macmillan, 1961.

Sokolnikoff, Ivan S. *Advanced Calculus*. New York: McGraw-Hill, 1939.

Background Readings

Bernstein, Jeremy. *Einstein*. New York: Viking Press, 1973.

Childs, Herbert. *An American Genius: The Life of Ernest Orlando Lawrence*. New York: Dutton, 1968.

Curie, Eve. *Madame Curie: A Biography*. Translated by Vincent Sheean. Garden City: Double-day, 1937.

de Broglie, Louis. *Matter and Light*. New York: Dover, 1946.

Fermi, Laura. *Atoms in the Family: My Life with Enrico Fermi*. Chicago: University of Chicago Press, 1954.

Heisenberg, Werner. *The Physical Principles of the Quantum Theory*. Chicago: University of Chicago Press, 1930.

Hermann, Armin. *The Genesis of Quantum Theory (1899–1913)*. Cambridge, MA: MIT Press, 1971.

Jammer, Max. *The Conceptual Development of Quantum Mechanics*. New York: McGraw-Hill, 1966.

Pagels, Heinz R. *The Cosmic Code*. New York: Simon and Schuster, 1982.

Pagels, Heinz R. *Perfect Symmetry: The Search for the Beginning of Time*. New York: Simon and Schuster, 1985.

Pais, Abraham. '*Subtle is the Lord . . .*': *The Science and the Life of Albert Einstein*. New York: Oxford University Press, 1982.

Schwartz, Joseph, and Michael McGuinness. *Einstein for Beginners*. New York: Pantheon Books, 1979.

Segre, Emilio. *From X Rays to Quarks*. San Francisco: W. H. Freeman, 1980.

Shankland, R. S. "The Michelson–Morley Experiment." *Scientific American*, pp. 107–114 (November 1964).

Swenson, Loyd S. *The Ethereal Aether*. Austin: University of Texas Press, 1972.

Thomson, George. *The Electron*. Oak Ridge: United States Atomic Energy Commission, 1972.

Classical Sources

Griffiths, David J. *Introduction to Electrodynamics*. Englewood Cliffs, NJ: Prentice-Hall, 1981.

Resnick, Robert, and David Halliday. *Physics*. New York: Wiley, 1977.

Sears, Francis Weston. *An Introduction to Thermodynamics, the Kinetic Theory of Gases, and Statistical Mechanics*. Cambridge, MA: Addison-Wesley, 1955.

References on Relativity

Kacser, Claude. *Introduction to the Special Theory of Relativity*. Englewood Cliffs, NJ: Prentice-Hall, 1967.

Mermin, N. David. *Space and Time in Special Relativity*. New York: McGraw-Hill, 1968.

Taylor, Edwin F., and John Archibald Wheeler. *Spacetime Physics*. San Francisco: W. H. Freeman, 1966.

References on Quantum Mechanics

Gasiorowicz, Stephen. *Quantum Physics*. New York: Wiley, 1974.

Merzbacher, Eugen. *Quantum Mechanics*. New York: Wiley, 1970.

Schiff, Leonard I. *Quantum Mechanics*. New York: McGraw-Hill, 1968.

References on Statistical Mechanics

ter Haar, D. *Elements of Statistical Mechanics*. New York: Rinehart, 1954.

Huang, Kerson. *Statistical Mechanics*. New York: Wiley, 1963.

Landau, L. D., and E. M. Lifshitz. *Statistical Physics*. Translated by J. B. Sykes and M. J. Kearsley. Reading, MA: Addison-Wesley, 1969.

Reif, F. *Fundamentals of Statistical and Thermal Physics*. New York: McGraw-Hill, 1965.

References on Atoms and Molecules

Bransden, B. H., and C. J. Joachain. *Physics of Atoms and Molecules*. New York: Longman, 1983.

Fano, U., and L. Fano. *Physics of Atoms and Molecules: An Introduction to the Structure of Matter*. Chicago: University of Chicago Press, 1972.

Herzberg, Gerhard. *Molecular Spectra and Molecular Structure: I. Spectra of Diatomic Molecules*. New York: Van Nostrand Reinhold, 1950.

Karplus, Martin, and Richard N. Porter. *Atoms and Molecules: An Introduction for Students of Physical Chemistry*. Reading, MA: W. A. Benjamin, 1970.

Lengyel, Bela A. *Lasers*. New York: Wiley-Interscience, 1971.

Levine, Ira N. *Quantum Chemistry*. Boston: Allyn and Bacon, 1974.

Pauling, Linus, and E. Bright Wilson. *Introduction to Quantum Mechanics*. New York: McGraw-Hill, 1935.

Slater, John C. *Quantum Theory of Matter*. New York: McGraw-Hill, 1968.

Svelto, Orazio. *Principles of Lasers*. New York: Plenum Press, 1976.

White, Harvey Elliott, *Introduction to Atomic Spectra*. New York: McGraw-Hill, 1934.

References on Condensed Matter

Amoros, Jose Luis, Martin J. Buerger, and Marisa Canut de Amoros. *The Laue Method*. New York: Academic Press, 1975.

Ashcroft, Neil W., and N. David Mermin. *Solid State Physics*. New York: Holt Rinehart and Winston, 1976.

Blakemore, J. S. *Solid State Physics*. Philadelphia: W. B. Saunders, 1969.

Dobbs, E. R., and G. O. Jones. "Theory and Properties of Solid Argon." *Reports on Progress in Physics* **20**:516–564 (1957).

Feynman, R. P. "Application of Quantum Mechanics to Liquid Helium." In *Progress in Low Temperature Physics*, edited by C. J. Gorter, Vol. 1. Amsterdam: North-Holland, 1955.

Kittel, Charles. *Introduction to Solid State Physics*. New York: Wiley, 1971.

London, Fritz. *Superfluids*, Vols. 1 and 2. New York: Dover, 1961, and Wiley, 1954.

Lynton, E. A. *Superconductivity*. New York: Wiley, 1962.

McKelvey, John P. *Solid-State and Semiconductor Physics*. New York: Harper & Row, 1966.

Schrieffer, J. R. *Theory of Superconductivity*. New York: W. A. Benjamin, 1964.

Silvera, Isaac F., and Jook Walraven. "The Stabilization of Atomic Hydrogen." *Scientific American*, pp. 66–74 (January 1982).

Wilks, J. *An Introduction to Liquid Helium*. Oxford: Clarendon Press, 1970.

Ziman, J. M. *Principles of the Theory of Solids*. Cambridge: The University Press, 1965.

References on Nuclei

Bethe, Hans A., and Philip Morrison. *Elementary Nuclear Theory*. New York: Wiley, 1956.

Blatt, John M., and Victor F. Weisskopf. *Theoretical Nuclear Physics*. New York: Wiley, 1952.

Cohen, Bernard L. *Concepts of Nuclear Physics*. New York: McGraw-Hill, 1971.

Enge, Harald A. *Introduction to Nuclear Physics*. Reading, MA: Addison-Wesley, 1966.

Evans, Robley D. *The Atomic Nucleus*. New York: McGraw-Hill, 1955.

Kaplan, Irving. *Nuclear Physics*. Reading, MA: Addison-Wesley, 1963.

Krane, Kenneth S. *Introductory Nuclear Physics*. New York: Wiley, 1987.

Mayer, Maria Goeppert, and J. Hans D. Jensen. *Elementary Theory of Nuclear Shell Structure*. New York: Wiley, 1955.

References on Elementary Particles

Griffiths, David. *Introduction to Elementary Particles*. New York: Harper & Row, 1987.

Halzen, Francis, and Alan D. Martin. *Quarks and Leptons*. New York: Wiley, 1984.

Perkins, Donald H. *Introduction to High Energy Physics*. Reading, MA: Addison-Wesley, 1982.

Powell, C. F., P. H. Fowler, and D. H. Perkins. *The Study of Elementary Particles by the Photographic Method*. New York: Pergamon Press, 1959.

Ryder, Lewis. *Elementary Particles and Symmetries*. New York: Gordon and Breach, 1975.

Segre, Emilio. *Nuclei and Particles*. Reading, MA: W. A. Benjamin, 1977.

Sources of Tabulated Data

Cohen, E. Richard, and Barry N. Taylor. *The 1986 Adjustment of the Fundamental Physical Constants: A Report of the CODATA Task Group on Fundamental Constants*. Elmsford, NY: Pergamon Press, 1987.

Herman, Frank, and Sherwood Skillman. *Atomic Structure Calculations*. Englewood Cliffs, NJ: Prentice-Hall, 1963.

Lederer, C. Michael, and Virginia S. Shirley, eds. *Table of Isotopes*. New York: Wiley, 1978.

Moore, Charlotte E. *Atomic Energy Levels*, Vols. 1, 2, and 3. Washington: National Bureau of Standards, 1949, 1952, and 1958.

Particle Data Group. "Review of Particle Properties." *Physics Letters* **170B**:1–350 (1986).

Thekaekara, M. P., R. Kruger, and C. H. Duncan. "Solar Irradiance Measurements from a Research Aircraft." *Applied Optics* **8**:1713–1732 (1969).

Walker, F. William, Dudley G. Miller, and Frank Feiner. *Chart of the Nuclides*. San Jose: General Electric Co., 1984.

Wapstra, A. H., and K. Bos. "The 1977 Atomic Mass Evaluation in Four Parts: Part I. Atomic Mass Table." *Atomic Data and Nuclear Data Tables* **19**:177–214 (1977).

General Reference Books

Besancon, Robert M., ed. *The Encyclopedia of Physics*. New York: Van Nostrand Reinhold, 1985.

Gray, H. J., and Alan Isaacs, eds. *A New Dictionary of Physics*. London: Longman, 1975.

Lerner, Rita G., and George L. Trigg, eds. *Encyclopedia of Physics*. Reading, MA: Addison-Wesley, 1981.

ANSWERS

Chapter 1

1. $\dfrac{d_1 + d_2}{\lambda} \left(\dfrac{u}{c}\right)^2$

3. $\frac{7}{25}$ m, $\frac{3}{5}$ m, $\tan^{-1}(\frac{5}{7}\sqrt{11}\,)$

5. $\sqrt{6} \times 10^{-5}$ m

9. $c/3$

11. $c/3$, $3\sqrt{8}$ m

13. $(20$ m, 20 m$/c)$
    and $(10$ m, 10 m$/c)$

15. $\frac{5}{4}$ m$/c$, $\frac{35}{37}$ c,
    $\frac{37}{16}$ m$/c$, $\frac{3}{4}$ m$/c$

16. $\frac{63}{65}c$, $\frac{65}{63}$ m$/c$, $\frac{20}{63}$ m$/c$

20. $\dfrac{v'}{\sqrt{2}}$, $\dfrac{v'/\sqrt{2}}{\sqrt{1 - v'^2/2c^2}}$,

    $\dfrac{\sqrt{2}\,v'}{1 + v'^2/2c^2}$, $\dfrac{v'}{\sqrt{2}}\dfrac{\sqrt{1 - v'^2/2c^2}}{1 + v'^2/2c^2}$

26. 1.13×10^{-20}

29. $\dfrac{\sqrt{M^2 - 4m^2}}{M}\,$c

33. $c/5$, $\frac{5}{7}c$, $c/5$

35. $\sqrt{\left(m_1 + m_2\right)^2 + 2K_1 m_2/c^2}$

38. $(2 + m/2M)mc^2$, 279.7 MeV

Chapter 2

2. 1.72×10^{17} W, 3.82×10^{26} W,
   5770 K, 502 nm

3. $\dfrac{c}{2L}\sqrt{n_1^2 + n_2^2}$

10. 355 nm, 1053 nm,

11. $0.5682c$

12. 5, 1.288×10^{-5} J/K$^5 \cdot$ m$^3 \cdot$ s

15. 5.93×10^{-9} eV, 5×10^{30} s^{-1}

18. 5×10^{19} s^{-1}, 1×10^{18} m$^{-2} \cdot$ s^{-1},
    1.32 eV, 337 nm

19. 5.80×10^{-6} m, 5.79×10^{-6} m

21. 140 keV, 95°

25. $2(1 + m/M)mc^2$

26. $\varepsilon\dfrac{E + \sqrt{E^2 - m_e^2 c^4}}{E - \sqrt{E^2 - m_e^2 c^4} + 2\varepsilon}$,
    7.0 GeV

Chapter 3

1. 6.6×10^{23} moles^{-1}

5. 2.80×10^{-6} m, 8.46×10^{-14} kg, 25

6. 42 nm

10. 1.25×10^{-14} m, 6.02×10^{28} m^{-3}

11. 0.74 s^{-1}

12. 1.90×10^{-14} m, 5.90×10^{28} m^{-3},
    0.0950 sr, 2.02×10^{-7}

16. 7.25×10^{-27} kg · m/s, 4.34 m/s

17. $E_4 \rightarrow E_3$,
    $E_n \rightarrow E_4$ $(n \geq 6)$,
    $E_n \rightarrow E_5$ $(n \geq 12)$

18. 0.0442 nm

20. $E_n \rightarrow E_1$ $(n \geq 2)$,
    $E_n \rightarrow E_2$ $(n \geq 5)$

21. 2.847×10^{-13} m, -2529 eV

22. Al, Mn, Co

24. 656.1 nm, 102.6 nm, 121.6 nm

Chapter 4

1. 7.8 nm, 0.18 nm

3. $134°, 78°$

4. $110°$, no

5. 3.97×10^{-15} W/m^2,
   1.73×10^{-6} N/C

7. 5 MeV

8. 9.4 MeV

9. $-4E_0, a_0/4$

11. $\sqrt{\dfrac{2\hbar}{m}} \sqrt{\dfrac{2H}{g}}$, 3.67×10^{-16} m

14. $4|A|^2 \cos^2(\dfrac{kd}{2}\sin\theta)$

17. $8\ln 2$

19. $A_0\dfrac{\pi}{a}$ for $|x| < a$

Chapter 5

3. $\dfrac{\hbar k}{m}|A|^2$, 0

6. $\dfrac{1}{4} + \dfrac{1}{2\pi}$

7. $\pm n\dfrac{\hbar\pi}{2ma}$

9. $|c|^2 + |c'|^2 = 1$

10. $\dfrac{1}{\sqrt{a}}(\sin\dfrac{2\pi x}{a}e^{-4iE_1 t/\hbar}$
    $+ \cos\dfrac{3\pi x}{a}e^{-9iE_1 t/\hbar})$

11. $\frac{1}{6}$

13. $\frac{1}{2}\mathrm{erf}(\frac{1}{2})$

14. 0.01 nm

18. $-\dfrac{V_0}{4} + \dfrac{V_0}{16}(\dfrac{x - 2x_0}{x_0})^2$

22. 0.056 nm

27. $|E_n - E_{n'}| \cdot |cc'|$

28. $0, \dfrac{2\pi^2 - 3}{24}(\dfrac{a}{\pi})^2, 0, (\dfrac{2\hbar\pi}{a})^2$

29. $-\dfrac{48}{25\pi^2}a\cos\dfrac{5E_1}{\hbar}t,$
    $\dfrac{24\hbar}{5a}\sin\dfrac{5E_1}{\hbar}t$

Chapter 6

1. $\dfrac{b}{2E}(1 + \sqrt{1 + 2EL^2/\mu b^2})$

2. $-\dfrac{b}{2E}(1 \pm \sqrt{1 + 2EL^2/\mu b^2})$
   for $E < 0$,
   $\dfrac{b}{2E}(-1 + \sqrt{1 + 2EL^2/\mu b^2})$
   for $E > 0$

4. $\dfrac{\hbar}{2i\mu R^2}(\Psi^*\dfrac{\partial\Psi}{\partial\phi} - \dfrac{\partial\Psi^*}{\partial\phi}\Psi)$

5. $\dfrac{\hbar m}{\mu R^2}|A|^2$

6. $\dfrac{\hbar^2}{2\mu R^2}(\dfrac{m}{2})^2, \sin\dfrac{m}{2}\phi$, no

10. 0.015 eV, 28 μm

11. $0, \hbar$

12. $1/4\pi$

15. $0°, 180°;$
    $0°, 90°, 180°;$
    $54.7°, 125.3°$

18. $\hbar^2, \hbar^2$

19. $\hbar^2[\ell_1(\ell_1 + 1)|a_1|^2 + \ell_2(\ell_2 + 1)|a_2|^2],$
    $\hbar(m_1|a_1|^2 + m_2|a_2|^2),$
    $\hbar^2|\ell_1(\ell_1 + 1) - \ell_2(\ell_2 + 1)| \cdot |a_1 a_2|,$
    $\hbar|m_1 - m_2| \cdot |a_1 a_2|$

Chapter 7

1. $-\dfrac{\mu}{m_e}Z^2\dfrac{E_0}{4}$

2. $(2a^3)^{-1/2}$

4. $A(1 - \rho/2)e^{-\rho/2}$

7. $4a, 5a, -\dfrac{\mu}{m_e}Z^2\dfrac{E_0}{2}$

9. $5/e^2$

10. $(Z/n)\alpha c$

11. $-e\mathbf{r}\dfrac{M + Zm_e}{M + m_e}$

14. $\frac{16}{27}\left(\frac{6}{5}\right)^6 a$

16. $(0, 0, 3ea)$

Chapter 8

1. $\dfrac{\mu_0}{2}\dfrac{Ir_0^2}{\left(z^2 + r_0^2\right)^{3/2}}$,

$\dfrac{1}{2\pi\varepsilon_0}\dfrac{q\,dz}{\left(z^2 - d^2/4\right)^2}$

4. $2, 1$

6. 1.16×10^{-4} eV, ± 0.33 nm
for $\Delta m = \pm 1$

7. $(5/4\alpha)c$

8. 2.24 mm

9. 5.59 GHz

11. 0.25, 0.067

14. $1/405a^3$

15. 1.34×10^{-5} eV, 4.47×10^{-6} eV

19. 4.39×10^{-6} eV, 4.53×10^{-5} eV

20. $\pm 3, \pm 2, \pm 1, 0$ times $\mu_B B$

22. $\mu_B B m_j$

23. 9.43×10^{-7} eV;
$\pm 11.58, \pm 8.27, \pm 4.96, \pm 1.65$
times 10^{-8} eV for $j = \frac{7}{2}$;
$\pm 6.20, \pm 3.72, \pm 1.24$
times 10^{-8} eV for $j = \frac{5}{2}$

Chapter 9

2. 79.0 eV below

3. -77.5 eV

4. $6s, 4f, 5d, 6p$

5. 1.26, 1.84, 2.26

7. 0.103 nm, 0.108 nm, 0.122 nm;
12.09 keV, 11.52 keV, 10.19 keV

11. $-\dfrac{3\zeta}{4}\hbar^2, \dfrac{\zeta}{4}\hbar^2$

12. 1.85 eV, 3.38 eV, 3.84 eV, 3.88 eV,
4.34 eV, 4.53 eV, 4.54 eV, 4.54 eV

13. 6.3 μeV, 2132.4 μeV

16. 4, 3 (three ways),
2 (four ways), 1 (three ways), 0

17. $^3P_{0,1,2}$, $^1P_1$

19. $^3F_2$

20. $^2S_{1/2}, {}^1S_0, {}^2D_{3/2}, {}^3F_2$

24. ± 1.74 pm for $\Delta m_j = \pm 1$

Chapter 10

2. 21 ps, 5.6 meV

4. $(1 + \gamma^2)^{-1/2}$, 3.8×10^{-2}

5. (b) 1, (c) $\frac{1}{2}$

6. (a) $> e/7$

8. 12.8 eV, 0.41

9. (c) 0.77, 0.64

10. 3.91 eV, 0.02 eV

14. 120 K

15. (b) U(meV) $=$
$1.3\left(\dfrac{0.3\text{ nm}}{h}\right)^9 - 9.6\left(\dfrac{0.3\text{ nm}}{h}\right)^3$;
(c) 0.26 nm, -0.01 eV;
(d) 1.2×10^2 K

17. 2.4×10^{-13} s   and 7.6×10^{-13} s
compared with 1.5×10^{-16} s

19. 1.6 meV, 2.0 meV

20. 8.3×10^{14} rad/s, 4.5 eV

21. $A = 9.73$ eV, $a = 24.9$ nm^{-1},
$R_0 = 0.113$ nm

22. 0.128 nm

24. 1.0, 1.3

Chapter 11

3. $36, 3, \frac{1}{12}, 7$

4. 96, 1, 9

11. 0.1 K, 6×10^5 K

13. 6×10^8

17. 3.6 km/s, 1.2×10^{-21} J/K

18. 6.2×10^4 K, 3.3×10^{-25} J/K

20. 4.8 K, 10^3 nm

21. 2.1×10^9 m/s

Chapter 12

1. (a) $\dfrac{1}{a^2}, \dfrac{2}{\sqrt{3}\,a^2}, \dfrac{4}{3\sqrt{3}\,a^2}$;
(b) $-3\varepsilon, -2\varepsilon, -3\varepsilon/2, -z\varepsilon/2$

2. (a) 6, 6; (b) $\frac{1}{27}$

4. (a) $a^3, a^3/\sqrt{2}, 4a^3/3\sqrt{3}$;
(b) $-3\varepsilon, -6\varepsilon, -4\varepsilon$

8. 1.6×10^6 m/s, 31 nm

10. 5×10^{-14} s

14. $3 \times 10^{-42}, 2 \times 10^{32}, 1 \times 10^{39}$

15. 1.24 eV

16. 100
17. 0.02 eV, 3.2 nm
19. proportional to T^{-1}
20. 8×10^{-6}

Chapter 13

1. -0.9 m/s
2. 131 m/s
7. 3×10^6
8. (a) 3.6 μeV;
   (b) 4.0×10^{-3} J/m^3, 4.3×10^{-3} J/m^3
9. 15 nm
13. 2.6×10^{-9} T
15. 47 mK

Chapter 14

1. 51 MeV
2. 6.76 fm, 1240 MeV
4. 20.18
5. 1.007825 u, 2.014102 u,
   15.994915 u
6. 8.551 MeV, 8.790 MeV,
   8.505 MeV, 7.868 MeV,
8. 39.9632 u, 55.9359 u,
   119.903 u, 207.987 u
9. 0.367 MeV, 0.4498 MeV,
   0.5590 MeV
12. 6.738 MeV, 7.368 MeV, 17.2 MeV
13. $\dfrac{2MR_0^2}{\hbar^2}(\dfrac{4}{9\pi})^{2/3}a$
16. 1.31 (taking $R_0 = 1.2$ fm)
19. $(1 + kr_0)/2k$, 2.96 fm
21. 10.62 MeV, 11.98 MeV
24. $\frac{1}{2}^-, \frac{5}{2}^+, \frac{5}{2}^+, \frac{5}{2}^+$

26. -0.264 μ_N, 4.793 μ_N,
    4.793 μ_N, -1.913 μ_N

Chapter 15

4. 0.730 g
8. 5.302 MeV, 4.781 MeV
9. 5.304 MeV, 4.602 MeV
11. 1.711 MeV
13. 0.709 MeV, 1.144 MeV, 0.122 MeV
15. 1.312 MeV, 1.505 MeV, 0.483 MeV
19. 4.67 neV, 1.95 meV, 81.4 m/s
21. -1.191 MeV, 4.43 MeV, 1.531 MeV
25. 13.11 MeV, 13.89 MeV, 15.39 MeV
26. 11.85 MeV, 7.48 MeV
27. 0.0235 eV, 2120 m/s
28. 22.8 MW $\cdot$ h
29. 195.9 MeV, 185.4 MeV
30. 0.19 MeV
31. 22.37 MeV, 8.683 MeV
33. 1.943 MeV, 1.199 MeV, 7.551 MeV,
    7.297 MeV, 1.732 MeV, 4.966 MeV

Chapter 16

8. 144.7 MeV
9. 292.6 MeV, 279.9 MeV, 287.0 MeV
10. 29.8 MeV/c, 69.80 MeV/c
12. 0.813 MeV, < 0.420 MeV
14. 172.4 MeV, 904 MeV
19. three, two
20. $1:1:0$ and $0:2:1$
21. 190 MeV, 612 MeV,
    885 MeV, 1407 MeV
26. $\bar{d}\bar{s}\bar{s}$, $\Lambda + \Lambda$, 6258 MeV
30. $\frac{15}{3}$ beyond 80 GeV
35. $(1/2\pi)\sqrt{-2k/m}$

PHOTO
CREDITS

Einstein, page 3. Courtesy of Philip Rosen.

Space-travel cartoon, page 7. From *Einstein for Beginners*, by Joseph Schwartz, illustrated by Michael McGuinness. Illustration copyright © 1979 by Michael McGuinness. Reprinted by permission of Pantheon Books, a Division of Random House, Inc.

Planck, page 74. Permission of AIP Niels Bohr Library.

Rutherford, page 123. Permission of AIP Niels Bohr Library.

Bohr, page 123. Permission of AIP Niels Bohr Library.

Electron diffraction pattern, Figure 4-4, page 187. From *The Electron*, by George Thomson, published by U.S. Atomic Energy Commission, 1972. Permission of U.S. Department of Energy. Courtesy of Oak Ridge National Laboratory.

Heisenberg, page 196. Permission of AIP Niels Bohr Library.

Schrödinger, page 220. Photography by Francis Simon. Permission of AIP Niels Bohr Library.

Born, page 226. Permission of AIP Niels Bohr Library.

Probability distributions, page 365. From *Introduction to Atomic Spectra*, by Harvey Elliott White, published by McGraw-Hill, 1934, and from "Pictorial Representations of the Electron Cloud for Hydrogen-like Atoms" by H. E. White, in *Physical Review* 37, published by American Institute of Physics, 1931. Reproduced with permission of McGraw-Hill and American Institute of Physics.

Zeeman lines, page 376. From *Introduction to Atomic Spectra*, by Harvey Elliott White, published by McGraw-Hill, 1934. Reproduced with permission of McGraw-Hill.

Dirac, page 421. Permission of AIP Niels Bohr Library.

Pauli, page 441. Courtesy of Philip Rosen.

Rotational spectrum, Figure 10-19, page 531. From *Physics of Atoms and Molecules* by B. H. Bransden and C. J. Joachain, published by Longman, 1983. Reproduced with permission of Longman Group Ltd.

Vibrational–rotational spectrum, Figure 10-20, page 531. From *Atoms and Molecules: An Introduction for Students of Physical Chemistry* by Martin Karplus and Richard N. Porter, published by W. A. Benjamin, 1970. Reproduced with permission of Benjamin/Cummings.

Molecular electronic spectrum, Figure 10-21, page 532. Photograph by R. Colin. From *Physics of Atoms and Molecules* by B. H. Bransden and C. J. Joachain, published by Longman, 1983. Reproduced with permission of Longman Group Ltd.

Two-dimensional hexagonal lattice, Figure 12-3, page 578. Photograph by W. J. Mullin.

Icosahedral structure, Figure 12-6, page 580. Photograph by W. J. Mullin.

Face-centered-cubic lattice, Figure 12-8, page 581. Photograph by W. J. Mullin.

X-ray diffraction pattern, Figure 12-14, page 586. From *The Laue Method* by Jose Luis Amoros, Martin J. Buerger and Marisa Canut de Amoros, published by Academic Press, 1975. Reproduced with permission of Academic Press.

Mme. Curie, page 741. Permission of AIP Niels Bohr Library.

Fermi, page 768. Permission of AIP Niels Bohr Library.

Lawrence, page 818. Permission of AIP Niels Bohr Library.

Feynman, page 824. Permission of AIP Niels Bohr Library.

Decays $\pi \rightarrow \mu \rightarrow e$, page 831. From *The Study of Elementary Particles by the Photographic Method*, by C. F. Powell, P. H. Fowler, and D. H. Perkins, published by Pergamon Press, 1959, and from "Mesons" by C. F. Powell, in *Reports on Progress in Physics* **13**, published by The Institute of Physics, 1950. Reproduced with permission of Pergamon Press and The Institute of Physics. Courtesy of P. H. Fowler.

Strange-particle processes, Figure 16-17, page 850. From *Nuclei and Particles*, by Emilio Segre, published by W. A. Benjamin, 1977. Reproduced with permission of Benjamin/Cummings and Lawrence Berkeley Laboratory, University of California.

Gell-Mann, page 871. Permission of AIP Niels Bohr Library.

NAME INDEX

SUBJECT INDEX

1 hydrogen	30 zinc	59 praseodymium	88 radium
2 helium	31 gallium	60 neodymium	89 actinium
3 lithium	32 germanium	61 promethium	90 thorium
4 beryllium	33 arsenic	62 samarium	91 protactinium
5 boron	34 selenium	63 europium	92 uranium
6 carbon	35 bromine	64 gadolinium	93 neptunium
7 nitrogen	36 krypton	65 terbium	94 plutonium
8 oxygen	37 rubidium	66 dysprosium	95 americium
9 fluorine	38 strontium	67 holmium	96 curium
10 neon	39 yttrium	68 erbium	97 berkelium
11 sodium	40 zirconium	69 thulium	98 californium
12 magnesium	41 niobium	70 ytterbium	99 einsteinium
13 aluminum	42 molybdenum	71 lutetium	100 fermium
14 silicon	43 technetium	72 hafnium	101 mendelevium
15 phosphorus	44 ruthenium	73 tantalum	102 nobelium
16 sulfur	45 rhodium	74 tungsten	103 lawrencium
17 chlorine	46 palladium	75 rhenium	104 rutherfordium*
18 argon	47 silver	76 osmium	105 hahnium*
19 potassium	48 cadmium	77 iridium	106 **
20 calcium	49 indium	78 platinum	107 **
21 scandium	50 tin	79 gold	
22 titanium	51 antimony	80 mercury	109 **
23 vanadium	52 tellurium	81 thallium	
24 chromium	53 iodine	82 lead	
25 manganese	54 xenon	83 bismuth	
26 iron	55 cesium	84 polonium	
27 cobalt	56 barium	85 astatine	
28 nickel	57 lanthanum	86 radon	*Elements with disputed names
29 copper	58 cerium	87 francium	**Elements without names